BIOLOGICAL
NITROGEN
FIXATION

BIOLOGICAL NITROGEN FIXATION

Edited by
Gary Stacey, Robert H. Burris
and Harold J. Evans

Chapman & Hall

New York

London

First published in 1992 by
Chapman and Hall,
an imprint of
Routledge, Chapman and Hall, Inc.
29 West 35 Street
New York, NY 10001

Published in Great Britain by
Chapman and Hall
2-6 Boundary Row
London SE1 8HN

Library of Congress Cataloging in Publication Data

Biological nitrogen fixation/edited by Gary Stacey, Harold J. Evans,
    and Robert H. Burris.
        p.   cm.
    Includes bibliographical references and index.
    ISBN 0-412-02421-7
    1. Nitrogen—Fixation. I. Stacey, G. S. (Gary S.) II. Evans, H.
J. (Harold J.) III. Burris, Robert H. (Robert Harza), 1914–   .
    QR89.7.B53   1991 1992
    589.9'0135—dc20   90-22839
                                                                       CIP

British Library cataloguing data also available.

Contents

Contributors

Daniel J. Arp
Laboratory for Nitrogen Fixation
  Research
Department of Botany and Plant
  Pathology
Oregon State University
Corvallis, OR 97331, U.S.A.

Dwight D. Baker
School of Forestry and Environmental
  Studies
Yale University
New Haven, CT 06511, U.S.A.

W. Mark Barbour
Center for Legume Research
Department of Microbiology
University of Tennessee
Knoxville, TN 37996-0845, U.S.A.

J. H. Becking
Research Institute ITAL
P.O. Box 48, 6700 AA Wageningen
The Netherlands

Paul E. Bishop
Department of Microbiology
North Carolina State University
Raleigh, NC 27695-7615, U.S.A.

Ton Bisseling
Department of Molecular Biology
Agricultural University
Dreijenlaan 3, 6703 HA Wageningen
The Netherlands

Peter J. Bottomley
Department of Microbiology
Oregon State University
Corvallis, OR 97331-3804, U.S.A.

William J. Buikema
Department of Molecular Genetics and
  Cell Biology
University of Chicago
920 East 58th St.
Chicago, IL 60637, U.S.A.

Robert H. Burris
Department of Biochemistry
University of Wisconsin
Madison, WI 53706, U.S.A.

Dennis R. Dean
Department of Anaerobic Microbiology
Virginia Polytechnic Institute and State
  University
Blacksburg, VA 24060, U.S.A.

Claudine Elmerich
Unité de Physiologie Cellulaire,
CNRS URA 1300
Département des Biotechnologies
Institut Pasteur
25 rue du Dr. Roux
75724 Paris Cedex 15, France

Harold J. Evans
Laboratory for Nitrogen Fixation
    Research
Oregon State University
Corvallis, OR 97331, U.S.A.

Henk J. Franssen
Department of Molecular Biology
Agricultural University
Dreigenlaan 3, 6703 HA Wageningen
The Netherlands

Robert Haselkorn
Department of Molecular Genetics and
    Cell Biology
University of Chicago
920 East 58th St.
Chicago, IL 60637, U.S.A.

Susan Hill
Agriculture Food Research Council
Division of Nitrogen Fixation
University of Sussex
Brighton, East Sussex BN1 9RQ
United Kingdom

Marty R. Jacobson
Worcester Foundation for Experimental
    Biology
222 Maple Avenue
Shrewsbury, MA 01545

Jan W. Kijne
Department of Plant Molecular Biology
Leiden University
Nonnensteeg 3, 2311 VJ Leiden
The Netherlands

A. L. Lobo
Department of Microbiology
Stocking Hall, Cornell University
Ithaca, NY 14853, U.S.A.

Sharon Long
Department of Biological Sciences
Herrin Laboratory, Stanford University
Stanford, CA 94305-5020, U.S.A.

Paul W. Ludden
Department of Biochemistry
University of Wisconsin
Madison, WI 53706, U.S.A.

M. J. Merrick
Agriculture Food Research Council
Division of Nitrogen Fixation
University of Sussex
Brighton, East Sussex BN1 9RQ
United Kingdom

Beth C. Mullin
Center for Legume Research
Department of Botony
University of Tennessee
Knoxville, TN 37996, U.S.A.

Jan-Peter Nap
Department of Molecular Biology
Agricultural University
Dreijenlaan 3, 6703 HA Wageningen
The Netherlands

William E. Newton
Department of Biochemistry
Virginia Polytechnic Institute and State
    University
Blacksburg, VA 24061, U.S.A.

Donald A. Phillips
Department of Agronomy and Range
    Science
University of California
Davis, CA 95616, U.S.A.

R. Premakumar
Department of Microbiology
North Carolina State University
Raleigh, NC 27695-7615 U.S.A.

John A. Raven
Department of Biological Sciences
University of Dundee
Dundee DD1 4HN
United Kingdom

Gary P. Roberts
Department of Bacteriology
University of Wisconsin
Madison, WI 53706, U.S.A.

Janet I. Sprent
Department of Biological Sciences
University of Dundee
Dundee DD1 4HN, United Kingdom

Gary Stacey
Center for Legume Research
Department of Microbiology
University of Tennessee
Knoxville, TN 37996-0845 U.S.A.

Larry R. Teuber
Department of Agronomy and Range
    Science
University of California
Davis, CA 95616, U.S.A.

Claire Vieille
Unité de Physiologie Cellulaire
CNRS URA 1300, Département des
    Biotechnologies
Institut Pasteur
25 rue du Dr. Roux
75724 Paris Cedex 15
France

Shui-Ping Wang
Center for Legume Research
Department of Microbiology
University of Tennessee
Knoxville, TN 37996-0845, U.S.A.

Dietrich Werner
Fachbereich Biologie-Botanik
Phillipis-Universität
Marburg, Federal Republic of Germany

M. G. Yates
Agriculture Food Research Council
Division of Nitrogen Fixation
University of Sussex
Brighton, East Sussex BN1 9RQ
United Kingdom

J. P. W. Young
John Innes Institute
AFRC Institute of Plant Science
    Research
Colney Lane, Norfolk NR4 7UH
United Kingdom

Wolfgang Zimmer
Unité de Physiologie Cullulaire
CNRS URA 1300
Départment des Biotechnologies
Institut Pasteur
25 rue du Dr. Roux
75724 Paris Cedex 15
France

Stephen H. Zinder
Department of Microbiology
Stocking Hall, Cornell University
Ithaca, NY 14853, U.S.A.

Preface

Biological N_2 fixation involves the enzymatic reduction of N_2 to ammonia. The ammonia produced then can be incorporated by enzymatic means for the growth and maintenance of the cell. N_2 fixation is unique to bacteria; animals and plants that fix N_2 must do so in association with bacteria. An understanding of biological N_2 fixation is essential to elucidate the dynamics of the global nitrogen cycle. With the exception of water, nitrogen is generally considered the most limiting nutrient for growth of plants in their natural environment; therefore, input of nitrogen via biological N_2 fixation often has profound ecological effects. Capability of fixing N_2 is recognized as a process of great agronomic importance, and a variety of leguminous plants and some nonleguminous plants can obtain their nitrogen from the air by symbiotic association with microorganisms. In addition to these rather practical applications of research on biological N_2 fixation, there has been a considerable development of new technology and information from research on N_2 fixation that is benefiting other areas of biological research. Few other areas of biological research, from metallochemistry to plant breeding, require the same breadth of investigation.

This volume presents current summaries (literature reviews completed early 1990) on various fields of biological N_2-fixation research written by experts in their field. The Contents gives some impression of the breadth of this rapidly developing area of research. A student reading these chapters will be exposed to many modern methods of biological analysis, because unraveling the complexities of biological N_2 fixation has required application of most extant methodologies and the development of new methods. Methods from traditionally different areas of research have been brought together to uncover new information about N_2 fixation (e.g., genetics and protein chemistry). We hope that each chapter will interest both the novice and expert. The field of biological N_2 fixation is at an exciting stage. In the last few years, tremendous strides have been made in our basic understanding of the processes underlying this critically important reaction that converts N_2 to ammonia. This volume emphasizes recent advances and focuses on subjects such as the structure of the nitrogenase proteins, the enzymatic role of the associated cofactor (FeMoco), the demonstration that a third kingdom of microorganisms (i.e., Archaebacteria) can fix N_2 and that there are alternative enzyme systems for N_2 fixation, and the story of how bacteria and plants interact to establish a N_2-fixing symbiosis.

The editors hope that this book will provide an authoritative and current summary of the field of biological N_2 fixation research, that it will point to the areas most in need of further research focus, and that it thus will aid in enchancing our understanding of N_2 fixation. Furthermore, we hope that this book will serve an important role in exciting a new generation of researchers into pursuing the multitude of important problems in N_2 fixation still awaiting resolution.

A project of this scope requires the cooperation of many participants. The editors want to thank each of the authors for contributing a chapter in the face of demanding schedules. We would like to give special recognition to Sharon Long, who first proposed the need for a current summary of N_2-fixation research. Special gratitude also goes to Greg Payne of Chapman and Hall, who helped guide us through the process of preparing a book of this size. Our thanks also to Lorraine Coffey, whose secretarial skills were of tremendous help. Lastly, we want to thank the readers of this volume, whose interest and enthusiasm for N_2 fixation are so critical to the continuing growth and development of this important area of biological research.

Gary Stacey
Harold J. Evans
Robert H. Burris

Introduction

Highlights in Biological Nitrogen Fixation during the Last 50 Years

Harold J. Evans and *Robert H. Burris*

I. Introduction

The classical book *Root Nodule Bacteria and Leguminous Plants* by Fred, Baldwin, and McCoy,[85] and Wilson's[212] monograph *The Biochemistry of Symbiotic Nitrogen Fixation* present interesting historical accounts of the use of leguminous plants in agriculture in ancient Greece and Rome and in medieval periods. The most spectacular highlight in the field before the twentieth century was the discovery by Hellriegel and Wilfarth[95] that symbiotic N_2 fixation by legumes was dependent upon "ferments" in the soil, which were responsible for nodulation of roots. The nodulating bacteria from *Pisum* were isolated by Beijerinck.[16] The one hundredth anniversary of Hellriegel and Wilfarth's monumental discovery was celebrated by a symposium sponsored by the Royal Society[22] and by the 7th International Congress on N_2 Fixation.[36] Undoubtedly it was the discovery of the fundamental aspects of N_2 fixation by free-living bacteria and nodulated legumes that provided the basis and motivation for the marked increase in research activities that led to the books by Fred, Baldwin, and McCoy[85] and Wilson.[212] These authors discussed such subjects as basic microbiological methods for culture and examination of rhizobia, cross-inoculation groups of rhizobia, histology and cytology of nodules, inoculation methods, the biochemistry of rhizobia, the interaction of host legumes and rhizobia, the chemical mechanism of N_2 fixation, and some physical and chemical characteristics of the nitrogenase system that were deduced from *in vivo* experiments. Considering the methodology that was available prior to 1940, it is impressive to consider the quantity and quality of information the early workers accumulated.

A series of treatises and monographs on biological N_2 fixation have been published since 1940. Many of these will be referred to in this volume. These, and the large number of journal articles on the subject since World War II, provide a vast literature for the field. In addition to the general treatments, books on methodology by Vincent[207] and Bergersen[19] have been especially valuable to basic and applied researchers over the world. Since 1974 a series of international symposia and conferences encompassing broad aspects of N_2 fixation have been convened and proceedings published. Before the formal symposia were organized, small groups of some of the now more senior researchers in the field met in obscure places, including Butternut Lake in northern Wisconsin; Sanibel Island, Florida; and Camp Sagehen in northern California. At these meetings, initial reports of some of the fundamental discoveries concerning the biochemistry of N_2 fixation were presented.

It is the purpose of this chapter to point out briefly what we consider to be representative and significant developments in the field during the past 50 years. More detailed information on the present status of the subject is considered in later chapters.

II. Symbiotic N_2 Fixation Systems

A. Legume–Rhizobium Symbiosis

i. Infection and Nodule Formation

Although major advances have been made in our understanding of the various facets of the infection process, much useful and still valid information was known 50 years ago.[212] With the exception of peanut (*Arachis hypogea* L.), one is left with the conclusion from the older literature that invasion of root hairs is the primary method of entry of rhizobia into legume roots, but it now is known that the infection process in several legumes occurs through penetration of the mucigel and primary layers of cell walls, and through breaks in the epidermis or at surface fractures where roots emerge.[184]

A definite highlight in our understanding of the recognition process is the continuing development of the concept that recognition involves interaction of lectins with sugar moieties in *Rhizobium* cell surface polysaccharides. Hamblin and Kent[89] initially demonstrated that a lectin from *Phaseolus vulgaris* L. exhibited a capacity to bind *Rhizobium* as well as red blood cells. With a fluorescent marker attached to soybean lectin, Bohlool and Schmidt[33] showed that 22 of 25 strains of nodulating *B. japonicum* were bound to the labeled lectin, whereas none of the 23 strains that lacked the capacity to nodulate soybeans were capable of lectin binding. Dazzo and Brill[61] utilized labeled lectin antibody to estimate the quantity of lectin present on root surfaces, and they demonstrated that concentrations of nitrate that inhibited nodulation also inhibited *Rhizobium* attachment to root surfaces. The extent of attachment was related to the amount of lectin present. This and other evidence has led to an intensive study of the binding of *R. leguminosarium* biovar *trifolii* (hereafter referred to as *R. trifolii*) to clover (*Trifolium repens*

L.) roots.[62] Trifoliin A, a root lectin, accumulates on root hairs and functions as the site of *Rhizobium* cell recognition.[62] Dazzo et al.[62] argue that the bacterial receptors for trifoliin A are polysaccharides. They have shown that wild-type *R. trifolii* cells are bound to trifoliin A transiently but a mutant of *R. trifolii* lacking hair-curling ability (Hac⁻) bound 95% less lectin than the wild-type bacteria. The introduction of host-specific genes (*hsn*) from *R. trifolii* into a strain of *R. leguminosarum* incapable of nodulating white clover conferred the capability to infect white clover roots and a capacity to synthesize acidic polysaccharides with characteristics comparable to those present in *R. trifolii*.[157]

Recently, Diaz et al.[63] have introduced the pea (*Pisum sativum* L.) lectin gene into white clover, thus creating transgenic plants capable of synthesis of pea lectin in white clover roots. This change in the clover resulted in the acquisition of nodulation susceptibility to *R. leguminosarum*, whereas control clover plants lacking the acquired pea lectin gene were nodulated only by *R. trifolii*. Impressive evidence continues to accumulate strongly supporting the general aspects of the lectin recognition mechanism.

ii. Bacteroids at the Site of N_2 Fixation

Before it was established that N_2 fixation takes place in nodule bacteroids, some proposed that peribacteroid membranes were the site of N_2 fixation. Convincing evidence that N_2 fixation occurs in bacteroids was provided from several sources.[20] After nodules of *Ornithopsis sativus* L. were exposed to $^{15}N_2$ and their bacteroids isolated in an isotonic medium, the greatest isotope concentration was found in the bacteroids and not in the membranes.[106] During the same period, a special press was designed in Bergersen's laboratory that made possible the preparation of nodule breis under anaerobic conditions. Using these preparations, Bergersen and Turner[23,24] separated bacteroid cells from the breis, washed them, and showed they retained the capability for $^{15}N_2$ fixation provided that O_2 was supplied during the reaction period. The addition of succinate, fumarate or pyruvate to bacteroids stimulated both N_2 fixation and H_2 evolution.

Using the acetylene reduction method to detect nitrogenase activity, Koch et al.[114] developed a procedure in which anaerobic conditions, use of insoluble polyvinylpolypyrrolidone[127] and a buffered ascorbate medium were utilized to prepare active nodule breis and washed bacteroids.[114] This procedure removed phenols from crushed nodules, thus preventing the inactivation of bacteroid proteins by formation of phenol–peptide bond complexes. The nitrogenase activity of bacteroids was markedly enhanced by succinate and β-hydroxybutyrate, but leghemoglobin was not essential under *in vitro* assay conditions. From the active bacteroids, cell-free extracts were prepared with N_2-fixing and C_2H_2-reducing activities that were increased strikingly by the addition of an ATP-generating system and $Na_2S_2O_4$.[114,115] These investigations firmly established that N_2 fixation in legume nodules takes place in the bacteroid cells.

iii. Leghemoglobin

In 1939, Kubo[119] discovered that the long observed red pigment in legume nodules had the spectral characteristics of a hemoglobin. Evans and Russell[84] have summarized evidence correlating the concentrations of this pigment in legume nodules with N_2-fixing effectiveness. Since 1960, a series of investigations have elucidated the physiological role of leghemoglobins. The chemical composition, amino acid sequence, kinetics of binding and release of O_2, genetic origin, and many other properties have been studied in detail in Canberra by Appleby.[9,10] The role of leghemoglobin in nodule metabolism has been reviewed by Bergersen,[20] another researcher from Canberra who has pioneered important investigations in this field. Appleby[8] reported that leghemoglobin exhibited an unusually high affinity for O_2 and suggested that this protein probably functioned in the facilitated diffusion of O_2 within nodules. There is no doubt that leghemoglobin is present in the cytoplasm of nodule cortex cells, but whether it exists inside the peribacteroid membranes that enclose bacteroids continues to be a subject of argument. According to Appleby,[10] leghemoglobin is present both inside and outside the peribacteroid membranes of soybean nodules.

Evidence supporting a role of leghemoglobin in the facilitated diffusion of O_2 was summarized by Appleby,[9,10] whose calculations indicate that the free dissolved O_2 concentration in typical soybean nodule plant cytoplasm is 11 nM. In contrast, the concentration of oxyleghemoglobin in the same tissue, is 600 μM. Therefore, the molar ratio of oxyleghemoglobin to free dissolved O_2 is near 55,000. The respiring bacteroids within the peribacteroid membranes must obtain their O_2 from an environment containing dissolved O_2 near 11 nM. The 55,000-fold higher concentration of oxyleghemoglobin serves as a reservoir for replenishing the free O_2 supply by dissociation of O_2 from this pigment. The enormous difference in concentration of leghemoglobin-bound O_2 and free O_2 at the bacteroid surface and the known association and dissociation kinetics of reactions of O_2 with leghemoglobin provide the basis for the conclusion that this protein plays its physiological role in the facilitated diffusion of O_2 within the nodule tissue. This is supported by several experiments; for example, the addition of leghemoglobin to *Phaseolus* nodule bacteroids maintained at low levels of dissolved O_2 greatly increased nitrogenase activity.[200] The O_2 delivery system, in which leghemoglobin participates, allows a high-affinity bacteroid oxidase to function efficiently at a continuously replenished low free O_2 concentration that does not harm the O_2-sensitive bacteroid nitrogenase.

iv. Oxygen Exclusion Barrier

Tjepkema and Yocum[197] reported that layers of cortex cells near the surface of soybean (*Glycine max* (L.) Merr.) nodules serve as a barrier to free O_2 diffusion. An O_2 microelectrode inserted into a whole nodule revealed a sharp decline in O_2 partial pressure when the electrode reached a depth of less than 0.5 mm from the surface. These results demonstrated the existence of a barrier to free O_2 diffusion. Although the mechanism of regulation and the precise nature of the barrier has not been clarified, strong evidence is available showing that a variety of stress conditions result in striking increases in the diffusion resistance of nodules to O_2.[139,206] Restriction of the phloem supply, addition of high concentrations of nitrate salts, exposure to 10% acetylene, lowering of the root temperature, and exposure to water stress all result in marked decreases in nitrogenase activity.[103] These effects are at least partially reversed by an increase in the external O_2 partial pressure, indicating that nitrogenase activity is limited by the internal O_2 concentration.

The general picture that emerges is that N_2 fixation by legume nodule bacteroids takes place in an environment in which O_2 is continuously limited by the high respiratory O_2 demands of the bacteroids and the limited rate of O_2 diffusion into the internodule tissue, which is regulated by a diffusion barrier as described by Tjepkema and Yocum.[197] The nodule responds to a variety of stress conditions by changes in the O_2 diffusion barrier, which further restricts the O_2 supply, thus decreasing the bacteroid respiratory rate and N_2-fixing activity. Although elucidation of the signal mechanism for regulation of the barrier awaits further research, the revelation of interrelationships among bacteroid O_2 demands, leghemoglobin role, diffusion-barrier-regulated O_2 supply, and a functional nitrogenase system represents a highlight in our progress toward understanding the N_2-fixing system[139,175] (D. Layzell, personal communication, 1989).

v. Vascular Export System

In our opinion, information obtained from the detailed light and electron microscopic examination of the nodule vascular tissue of white clover and the elucidation of the structural and functional relationships to the bacteria-infected tissues of nodules represents a highlight in this area of research.[153] In this species, and several others, the phloem and xylem of the nodular vascular tissue are surrounded by endodermal and specialized pericycle cells, the latter with walls consisting of a "labyrinth of branched protuberances." The authors propose that these cells function as glands which concentrate amides that are synthesized utilizing fixed N derived from N_2 fixation in the bacteroid-containing tissue. From the pericycle cells, amides are released into the vascular xylem for export. The glandlike function of the pericycle cells is supported by data showing that the concentration of asparagine and glutamine in the nodule bleeding sap from severed vascular elements of broad bean (*Vicia faba* L.) nodules is 10-fold or more higher than that in either root bleeding sap or nodule bacterial tissue. Furthermore, the development of branched protuberances from the walls of the glandlike pericycle cells is initiated about the same time that leghemoglobin synthesis begins, suggesting that N_2 fixation and gland function are related. This research helps explain the relationship of nodule structure and function in those legumes that export the major proportion of their fixed N in the form of amides. This information may also apply to those species that export most of their fixed N as ureides.

vi. Isotope Discrimination

Kohl and Shearer[116] initially observed that the ^{15}N abundance in red clover (*Trifolium pratense* L.) and soybean grown hydroponically in a nutrient medium free of fixed N was greater than that of the atmosphere. In soybeans, the isotope effect gave a delta value of $+0.98$ whereas in clover the effect was $+1.88$. As summarized recently by Rasmussen et al.,[163] similar findings have been published by several researchers. The basis for this unexpected accumulation of the isotope with the higher mass remains obscure. An extension of these studies[163] has revealed substantial ^{15}N enrichment in nodules of common bean (*Phaseolus vulgaris* L.), birdsfoot trefoil (*Lotus corniculatus* L.), mung bean (*Vigna radiata* L.), and crown vetch (*Coronilla varia* L.). Variation in the extent of enrichment has sometimes been correlated in a positive way with differences in effectiveness of inocula. Obviously the extent of isotope discrimination needs to be considered when low rates of N_2 fixation are being measured with ^{15}N as a tracer.

vii. Stem-Nodulated Legumes

Although the stem-nodulated legumes, which includes some species of *Sesbania, Aeschynomene* and *Neptunia,* have been used for many years as green manure crops, these symbioses recently have received renewed attention.[5] Legumes with nodulated stems are capable of substantial rates of N_2 fixation in waterlogged soils where the O_2 supply for root nodules may be limiting. Furthermore, the effect of soil nitrate concentration on nodulation and N_2 fixation of stem-nodulated legumes is much less than this effect on legumes that limit nodulation to their roots. It has been proposed[5] that genetic engineering methods might be used to transfer stem-nodulating capability to additional agriculturally important legumes with the objective of improving their N_2-fixing potential.

viii. Diversity Among Rhizobia

During the past several years it has been revealed that there is enormous nutritional and metabolic diversity in both *Rhizobium* and *Bradyrhizobium*.[152] For this and other reasons, the revision in the taxonomy of the traditional *Rhizobium* genus has been necessary (see Chapter 1). Dilworth and Glenn[65] pointed out that substrates utilized for growth include alcohols, sugars, polysaccharides, and a variety of organic and amino acids. They also report that *R. leguminosarum* is capable of utilizing for growth a series of phenolic compounds, including *p*-hydroxybenzoate, 3,4-dihydroxybenzoate, 3,5-dihydroxybenzoate, and ferulate. This and other information indicates the broad range in the capabilities of *Rhizobium* and *Bradyrhizobium* for utilizing carbon and nitrogenous compounds that may be derived from degradation of soil organic materials.

During the period between 1888, when rhizobia were first isolated, and 1979, all rhizobia were considered as heterotrophs dependent upon fixed carbon for growth. This concept ended when the Nitrogen Fixation Laboratory group in Corvallis, Oregon, discovered that those isolates of *B. japonicum* strains that efficiently recycle H_2 as nodule occupants exhibit a capability to grow as chemolithotrophs in a mineral salts medium utilizing H_2 as the only source of energy and CO_2 as carbon source.[90,122] Furthermore, Hup^+ (H_2 uptake positive) strains of *B. japonicum* synthesize ribulose 1,5-bisphosphate carboxylase and utilize this enzyme in the fixation of CO_2. The genes necessary for H_2 oxidation and chemolithotrophic growth in *B. japonicum* were cloned, and these were expressed when transferred into Hup^- isolates of *B. japonicum* and *R. meliloti*.[51,121]

Additional versatility of the Hup^+ *B. japonicum* has been established by Neal et al.,[146,147] who showed that NO_3^-, NO_2^- and N_2O served as electron acceptors for H_2 oxidation. NO_3^- also functioned in place of O_2 for H_2-dependent chemolithotrophic growth of a Hup^+ strain of *B. japonicum*. Furthermore, the capacity of some isolates of *Bradyrhizobium* to function as denitrifiers has been reported.[128] It is clear that there is great metabolic diversity among *Rhizobium* and *Bradyrhizobium*, and this helps to explain the widespread distribution and persistence of these important soil bacteria in a variety of environmental conditions over the world.

ix. Rhizobia as the Source of Genes for Nitrogenase

Before 1975, insufficient evidence was available to establish with certainty the source of genetic information for synthesis of a functional symbiotic nitrogenase system. Generally, it was reported that neither the leguminous host nor rhizobia in pure culture were capable of N_2 fixation. As discussed previously, the nitrogenase system had been shown to be present in nodule bacteroids, but the question of whether the genetic information for nitrogenase synthesis required input from the plant had not been answered. This problem was approached by Bishop and Evans,[27] who prepared antibodies to the purified soybean nodule dinitrogenase component of nitrogenase and found that they reacted with extracts of *B. japonicum* cells cultured anaerobically with nitrate to produce a definite precipitin band. Precipitin bands from interaction of extracts of aerobic cells with the dinitrogenase antibodies were less convincing. During the same year, laboratories in Australia, Canada, and the United States published convincing evidence that pure cultures of certain isolates of *Bradyrhizobium* exhibited very definite N_2 fixation. This subject has been reviewed by Child.[55] The special conditions necessary for expression of activity include very low partial pressures of O_2 over the cultures, limited amounts of fixed N in the culture medium, and use of selected rhizobial strains. Generally, the rates of N_2 fixation are very low and insufficient to provide adequate fixed N to support growth of the bacteria. The conditions for optimal expression of nitrogenase activity in laboratory cultures of *Azorhizobium sesbaniae* have been examined by Stouthamer et al.[188] Although research since 1975 has provided proof that the genes necessary for expression of nitrogenase reside in rhizobial cells, it is obvious that fixation, at rates needed to support growth of a legume, demands a complex O_2 regulatory system of the type that is present in the nodules of legumes.[9,10]

B.   Other Nitrogen-Fixing Symbioses

i.   Parasponia–Rhizobium

For many years it was thought that symbioses between plants and *Rhizobium* or *Bradyrhizobium* were limited to the legumes. In 1973, Trinick[201] discovered that a member of the family Ulmaceae (later correctly identified as *Parasponia*[4]) was nodulated by a bacterium now placed in the *Bradyrhizobium* "cowpea miscellany" group. From nodules of *Parasponia andersonii*, Appleby et al.[11] isolated and characterized a hemoglobin with properties similar to those of the leghemoglobin that had been described from legume nodules. They concluded that the physiological role of the *Parasponia* hemoglobin is probably the same as the role in facilitated O_2 diffusion proposed for leghemoglobins in nodules of legumes. The *Parasponia–Bradyrhizobium* symbiosis provides a unique experimental model for use in genetic engineering research that we hope will lead to extension of effective symbiotic nitrogen fixation to important nonleguminous plants useful in agriculture and forestry.

ii. Actinorhizal Systems

That many dicotyledonous plants are nodulated by microorganisms other than rhizobia and fix N_2 has been known for many years. Fred, Baldwin, and McCoy[85] discussed this subject and presented illustrations of nodulation of several nonleguminous plants. Research on the symbioses between the filamentous bacteria of the genus *Frankia* and actinorhizal hosts increased dramatically after Callaham et al.[50] developed the methodology necessary for culture of the nodulating bacteria in laboratory media. In a review of this area, Torrey[198] pointed out that plants nodulated by *Frankia* include more than 200 species distributed in 23 genera of eight families. In both laboratory cultures and in nodules, nitrogenase activity is associated with vesicles. In laboratory cultures, nitrogenase activity is expressed over a wide span of O_2 concentrations ranging up to 20% O_2.[196] However, no activity is observed under anaerobic conditions. It was concluded that walls of the vesicles play an important role in regulating the O_2 supply to the nitrogenase proteins located inside the vesicles.[199]

Also, it has been established that several, but not all, actinorhizal nodules contain hemoglobin that is postulated to function in the facilitated diffusion of O_2.[195] In addition to advances in understanding the physiology of actinorhizal symbioses, progress is being made in the elucidation of taxonomic relationships and molecular genetics of *Frankia*.[120] The striking increase in activities in several areas of research on actinorhizal symbioses has been fueled by the realization of the present and potential importance of these symbiotic associations in the provision of fixed N_2 to forests and woodlands over the world.

iii. Cyanobacteria–Plant Associations

Nitrogen-fixing associations between cyanobacteria and plants have been known for a long time, but investigations into the breadth of distribution, physiology, and economic importance of these symbioses have increased tremendously during the past 30 years. Undoubtedly, the availability of (1) the acetylene reduction assay for detecting nitrogenase activity, (2) the $^{15}N_2$ method for confirming N_2 fixation, (3) improved electron microscopic methods for study of the morphology of the symbiotic associations, and (4) greatly improved analytical methods for physiological and biochemical studies have contributed to interest and progress.

The host plants for these associations include angiosperms, gymnosperms, pteridophytes, bryophytes, fungi, and algae.[186] The cyanobacterial constituents of the associations, for the most part, are species of *Nostoc* and *Anabaena*. However, other genera of cyanobacteria participate in associations with fungi to form N_2-fixing lichens.[180,186]

The magnitude of fixed N provided to the environment by these associations varies greatly. *Gunnera,* the only genus among angiosperms that is known to form an association with cyanobacteria, has been estimated[179] to fix up to 72 kg N ha^{-1} yr^{-1}. Some cycads, which are classified among the gymnosperms, are nodulated by either *Nostoc* or *Anabaena* sp. and have been estimated[86] to fix up to 18 kg N ha^{-1} yr^{-1}. Several species of *Azolla,* a representative of pteridophytes, form associations with *Anabaena azollae.* These have been used in China and other parts of southeast Asia for many years as a "green manure" to supply nitrogen to rice fields. The *Azolla* symbiosis has been reported[186] to contribute from 50 to 80 kg N ha^{-1} yr^{-1}. The propagation and distribution of *Azolla* requires much hand labor, and therefore the utility of this association for supplying nitrogen to rice is most practical in parts of the world where hand labor is relatively inexpensive. The physiology and morphological aspects of the *Azolla–Anabaena* symbiosis have been investigated intensively for several years by Peters and associates.[154,155]

Among the bryophytes, representative of the genera *Anthosceros, Blasia, Cavicularia,* and *Sphagnum* form associations that fix N_2 at appreciable rates.[186] In addition, Dalton and Chatfield[58] discovered a *Nostoc–Porella* N_2-fixing association that is quite common in forests and woodlands of the northwestern portion of the United States. Meeks et al.[138] have utilized ^{13}N-labeled N_2 to investigate the pathway of fixed N assimilation in the *Anthoceros–Nostoc* symbiosis and have concluded that NH_4^+ is the initial product of fixation and is assimilated by the glutamate synthase pathway.

The magnitude of fixation by the bryophyte–cyanobacteria associations obviously depends upon the extent of cover within habitats where they exist. Assuming a 100% cover of *Sphagnum* moss, this association has been estimated[15] to fix up to 94 kg N ha^{-1} yr^{-1}. The importance of N_2-fixing lichens is related to their capacities to exist in extreme environments such as those on barren, rock surfaces, in deserts, and in arctic environments. Estimated low fixation rates of 0.5–1.5 kg N ha^{-1} yr^{-1} may be very important under severe environmental conditions.[6]

III. Associative Nitrogen Fixation

A. *Bacteria–Plant Root Associations*

Döbereiner et al.[70] reported that *Paspalum notatum* roots often were associated with populations of N_2-fixing *Azotobacter paspali.* These bacteria are located in mucilaginous sheaths on outer root surfaces and are incorporated into outer layers of the root cortex. In the same year Raju et al.[162] identified N_2-fixing *Enterobacter cloacae* on roots of unusually dark green maize (*Zea mays* L.) growing in a nitrogen-deficient soil near Corvallis, Oregon. This organism fixed N_2 in laboratory cultures with O_2 partial pressures ranging from 0.005 to 0.015 atm. Under anaerobic conditions, rates were sufficient to support vigorous bacterial growth. Döbereiner and Day[69] reported that root systems of several tropical grasses include populations of bacteria, many of which are classified as *Azospirillum* spp., that fix N_2 at sufficient rates (estimated from acetylene reduction assays) to meet the needs for plant growth. In addition, von Bülow and Döbereiner[209] concluded that N_2-fixing *Azospirillum* spp. on roots of maize grown in Brazil reduced acetylene at rates that were converted to estimates of 2 kg N ha^{-1} d^{-1}. These and other reports of associative N_2 fixation by soil bacteria on roots of grasses prompted questions about the methodology employed in such investigations. Often the high rates of acetylene reduction had been observed after researchers had preincubated root samples under limited O_2 prior to measuring C_2H_2 reduction. This procedure, according to Barber et al.,[14] allowed an average of 30-fold proliferation of the population of N_2-fixing bacteria on roots, and thus distorted estimates of the acetylene-reducing capability of roots in their natural habitats. The methodology and potential for associative N_2 fixation by plants was discussed by Evans and Barber[77] and by van Berkum and Bohlool.[203] More recently, van Berkum and Sloger[204] pointed out that severe nitrogen deficiency symptoms persisted in grasses inoculated with *Azospirillum* spp. and tested under their conditions. They questioned a unique role of these bacteria in associative N_2 fixation. An extensive literature has accumulated on *Azospirillum* spp. and other associative N_2-fixing organisms, and one is referred to the series of proceedings edited by Klingmüller.[107–110] Okon et al.[150] in a discussion of inoculation with *Azospirillum* spp., state that "Statistically significant increases in crop yield have been demonstrated in field experiments in many countries by various research groups." In the case of maize and sorghum (*Sorghum vulgare* Pers.) inoculated with *Azospirillum* spp., they pointed out that results of immediate acetylene reduction assays and $^{15}N_2$ incorporation measurements (in pots) have not indicated sufficient fixation of N_2 to account for the total observed growth responses. The effects of inoculation on root morphology and development, uptake of minerals, and hormone supply to plants have been suggested as factors contributing to growth responses. The recent description of *Acetobacter diazotrophicus,* a new species that colonizes the roots and stems of sugar cane (*Saccharum officinarum* L.) and that grows at low pH, documents the continuing development in this field of investigation.[87] Although the controversy regarding the magnitude of fixation by associative systems persists, there is no doubt that Döbereiner and associates have made major contributions by discovery and characterization of N_2-fixing bacteria that colonize the root systems of nonlegumes, especially tropical grasses.

B. Bacterial–Animal Associates

In addition to associations between N_2-fixing bacteria and plants, a variety of N_2-fixing associations are known between microorganisms and both terrestrial and marine animals. Nitrogenase activity has been detected in excreta of humans and guinea pigs.[21] In general, it is concluded from acetylene reduction assays that rates of N_2 fixation are low. However, more quantitative studies of the magnitude of N_2 fixataion need to be carried out. Breznak et al.[38] demonstrated low, but convincing rates of acetylene reduction in guts of termites. The significance of the preceding associations requires more study.

IV. Free-Living Nitrogen-Fixing Bacteria

During the past two decades, there has been a spectacular increase in the number of confirmed N_2-fixing free-living microorganisms. This undoubtedly is related to the use of improved methods for detecting and confirming N_2-fixing capability. A list that includes free-living diazotrophs and designates the type of habitat in which N_2 fixation takes place was compiled by Postgate.[159] With the exception of the N_2-fixing photosynthetic bacteria and cyanobacteria, most of the free-living diazotrophic bacteria are heterotrophs requiring habitats capable of providing a source of utilizable carbon substrates necessary to meet the energy demands for N_2 fixation. Aerobic and microaerobic heterotrophic N_2-fixing bacteria—of which *Azotobacter* spp. and free-living *Azosporillum* spp., respectively, are typical—are common in many soils. Diazotrophic capability now has been extended to some chemoautrophic bacteria, including those that oxidize methane[60] and sulfur[132] and those that utilize H_2 as a source of energy for autotrophic growth.[26] Faculative anaerobic N_2-fixing bacteria of the genera *Klebsiella* and *Enterobacter* require nitrogenous compounds for growth under aerobic conditions but fix N_2 at rates sufficient to support vigorous growth under anaerobic conditions. The conditions for fixation of N_2 by *Klebsiella pneumoniae* were first described in Wilson's laboratory.[134] An example of the occurrence of these types of bacteria was revealed in research by Aho et al.,[3] who discovered that 52% of the facultative anaerobic bacteria they cultured from decaying white fir [*Abies concolor* (Lindl. and Gord.)] trees from forests in Oregon exhibited a capacity for N_2 fixation under anaerobic conditions. The most prevalent isolate from this O_2-limited habitat was identified as *Enterobacter agglomerans*.

More information on the occurrence and economic potential of N_2-fixing representatives among genera of heterotrophic, chemoautotrophic, and photosynthetic bacteria and cyanobacteria are given in reviews by Stewart.[185,187] In general, it is concluded that N_2 fixation by heterotrophic organisms in most natural habitats is limited by the availability of carbon substrates. This helps explain why azotobacters and clostridia have been estimated[140] to contribute less than 0.5 kg N ha^{-1} yr^{-1}. On the other hand, the photosynthetic process in free-living cyanobacteria provides sufficient energy in the form of photosynthates to support N_2 fixation at rates[185] up to 25 or more kg N $ha^{-1} yr^{-1}$.

In addition to economic considerations, representatives of free-living N_2-fixing bacteria have been utilized widely as model organisms for laboratory investigations. *Clostridium pasteurianum, Azotobacter vinelandii,* and *A. chroococcum* were used extensively in the isolation and characterization of nitrogenase. Also, *A. chroococcum* was the primary organism employed in investigations of respiratory and conformation protection of nitrogenase.[159] *Klebsiella pneumoniae, A. vinelandii, A. chroococcum,* and *Anabaena* species have served as primary organisms for genetic research.[93,159]

V. Role of Trace Elements in Nitrogen-Fixing Organisms

The fixation of N_2 by nodulated legumes involves symbiosis between the host plant and an effective strain of *Rhizobium* or *Bradyrhizobium*. Therefore, it seems logical to ask whether the trace element requirements for symbiosis are different from those of the host plant per se. With the exception of some trace elements, the mineral requirements of nodulated legumes are not appreciably different from those of plants in general.[183]

A. Molybdenum

In 1930, Bortels[34] discovered that Mo is essential for growth of *Azotobacter chroococcum,* and seven years later[35] demonstrated that three species of legumes cultured symbiotically also required Mo. Anderson's discovery[7] that additions of small amounts of sodium molybdate to nodulated *Trifolium subterraneum* L. growing in an acid soil in Australia greatly increased growth and N_2 fixation led to reports of field responses from addition of molybdate to legumes in several countries. Evans and Purvis[82] applied small amounts of molybdate to alfalfa *Medicago sativa* L. growing in an acid Norton soil from northern New Jersey and observed alleviation of what appeared to be nitrogen deficiency symptoms as well as marked increases in weight and nitrogen contents of plants. For the last three decades, application of a few ounces of molybdate salts to acid soils used for culture of legumes has been common in various parts of the world. It is now well established that Mo is required by legumes for N_2 fixation and also utilization of nitrates, but the magnitude of the requirement for symbiotic growth appears to be greater than that for plants provided with nitrogenous compounds.[83] The specific role of Mo in legume nitrogenase was clarified when the enzyme from *B. japonicum* bacteroids was purified and separated into a dinitrogenase and dinitrogenase reductase.[25,113] A highly purified dinitrogenase from soybean nodules contained 1.3 and 28.8 g atoms of Mo and Fe, respectively, for each mole of enzyme.[104] Further consideration of properties of nitrogenase components from different sources is included in Chapter 9.

B. Cobalt

There is conclusive evidence that Co is essential for symbiotically cultured legumes,[1,2,164] laboratory cultures of several rhizobial isolates,[130] and cyanobacteria.[100] But there is no convincing evidence of the essentiality of this element for higher plants provided with fixed nitrogen. Further details on this subject are included in reviews by Evans and Kliewer[81] and Evans and Russell.[84] Cobalt is needed by rhizobia for the synthesis of vitamin B_{12} coenzyme, which functions as an essential cofactor for methylmalonyl mutase and ribonucleotide reductase.[84] Cobalt also is required for the synthesis of a cobalamin cofactor necessary for methionine synthesis in *Rhizobium meliloti* and presumably other rhizobia.[178] As a consequence of insufficient Co, *R. meliloti* cells elongate abnormally, a symptom postulated to be caused by a lesion in DNA synthesis and therefore cell division.[57] The authors are aware of no evidence that green parts of higher plants synthesize vitamin B_{12} coenzymes and also have found no evidence for the occurrence of enzymes in green parts of plants requiring a vitamain B_{12} derivative as a cofactor.

Several field responses of nodulated legumes to Co fertilization have been reported in various parts of the world. In Australia, responses of *Trifolium subterraneum*[160] and *Lupinus angustifolius* L.[168] were described. As far as we are aware, no responses from the addition of Co to field-grown legumes has been reported in the United States.

C. Nickel

An important biological role of Ni in organisms was suggested by the discovery in 1975 that this element is a constituent of urease.[66] As summarized in a review,[59] Ni now has been demonstrated to be an important element in a variety of organisms, including legumes, nonlegumes, fungi, and bacteria. Ni plays a primary role in the metabolism of arginine released during protein degradation and turnover. In germinating seed, for example, the degradation of protein yields arginine, ornithine, and urea, the last of which is converted to CO_2 and NH_3 when sufficient Ni is present for normal urease function. Striking effects of Ni deficiency on urease activity have been observed in soybean.[75,112,158] Similar observations have been made with several other organisms.[59] A common consequence of Ni deficiency is an accumulation in tissues of urea at concentrations that are toxic.[75]

Ni also is a constituent of several hydrogenases, particularly those that function primarily in H_2 uptake reactions.[59] Many N_2-fixing organisms contain uptake hydrogenases that participate in oxidation and energy conservation of the H_2 that is produced as an obligatory by-product of the N_2-fixation reaction.[80] A role of Ni in the efficient utilization of energy by *B. japonicum* was demonstrated by Klucas et al.[112] When this bacterium was grown in a highly purified medium and provided with H_2 as the only source of energy, there was a definite requirement for Ni for growth and expression of hydrogenase activity.[112] Furthermore, nodulated soybean plants cultured with purified nutrients required added Ni for optimal expression of H_2 uptake activity in nodule bacteroids. Ni is now known to be an essential constituent of the uptake hydrogenase from *B. japonicum* and other N_2-fixing bacteria.[80] Some soils from the United States may contain insufficient Ni for maximum activities of hydrogenase in bacteroids from nodules and urease in leaves from soybean plants.[59]

D. Other Elements

Recently, selenium has been shown to be a constituent of the hydrogenase from *B. japonicum*.[37] The addition of selenite to a highly purified medium markedly increased the activity of the uptake hydrogenase, but further research is required to establish the physiological importance of this element in the H_2-recycling process in *Bradyrhizobium*.

As indicated in a recent discussion by Thomas et al.,[190] sodium is known to be an essential element for growth of cyanobacteria. Recent research[190] indicates a role for Na in N_2 fixation in these microorganisms, but its specific role in N_2 fixation remains to be determined.

VI. The Iron Molybdenum Cofactor (FeMo-co)

All Mo-requiring nitrogenases from sources investigated may be separated into dinitrogenase (MoFe protein) and dinitrogenase reductase (Fe protein) components (see Chapter 10). A major advance toward understanding the composition and properties of dinitrogenase was made by Shah and Brill,[174] who successfully extracted a novel metal cluster from dinitrogenase. Since its discovery, this cofactor has been referred to as FeMo-co. The biological activity of FeMo-co was established by demonstrating that addition of an anaerobic cofactor preparation to an extract of a *nif⁻* strain of *A. vinelandii* (UW45) restored nitrogenase activity. FeMo-co can be extracted from the dinitrogenases isolated from several different N_2-fixing bacteria, and the electron spin resonance signal of the isolated FeMo-co is essentially similar to that of the native dinitrogenase from which FeMo-co was obtained. Hoover et al.[101] reported that FeMo-co is located at the site of substrate reduction. The extreme lability of FeMo-co has retarded determination of its composition by standard chemical methods. However, use of a series of *A.*

vinelandii mutants lacking capacities for synthesis of functional FeMo-co and the development of an assay that detects the synthesis of FeMo-co have supported definite progress toward an understanding of FeMo-co biosynthesis and composition. In addition to Mo, the gene products of *nif A, nif B, nif V, nif N* and *E,* and *nif H* are all required for FeMo-co biosynthesis.[101] Although the precise role of all these gene products in FeMo-co biosynthesis has not been clarified, it has been established that the gene product of *nif V* is homocitrate, a molecule that is an endogenous component of FeMo-co. Although many questions about the chemical composition and mechanism of interaction of FeMo-co with the apodinitrogenase remain, progress during the last two years is substantial.

VII. Alternative Nitrogenases

Prior to 1980, researchers in the field generally believed that *A. vinelandii* and other N_2 fixers contained a single nitrogenase consisting of dinitrogenase (MoFe protein) and dinitrogenase reductase (Fe protein). As a consequence of the pioneering research of Bishop and associates[28] and the papers that followed the discovery, it now is established that *A. vinelandii* harbors three genetically different nitrogenases,[29,30,56] referred to as nitrogenase 1, nitrogenase 2, and nitrogenase 3. The first paper by Bishop et al.[28] reported that limitation of the supply of Mo to certain Nif⁻ mutants of *A. vinelandii* resulted in the appeaance of four new cell proteins, the expression of which was repressed by ammonium compounds. From these and other results, they concluded that Mo deprivation suppressed the synthesis of conventional nitrogenase 1 and allowed expression of an alternative nitrogenase 2. Furthermore, it was suggested[29] that V (vanadium) may be involved in catalysis by the new nitrogenase 2. Several of the major laboratory groups involved in nitrogenase research over the world were skeptical and some were severely critical of the interpretations that had been made. As a consequence of this, a sabbatical leave beginning in 1984 was arranged that provided an opportunity for Dr. Bishop to personally become involved in research at the AFRC Laboratory at Brighton, UK, that hopefully would help to resolve the controversy. In collaborative research with the AFRC Laboratory group,[105] and in a paper by Bishop et al.,[31] it was reported that *A. vinelandii* Nif⁻ mutants lacking *nif H, D,* and *K* expressed nitrogenase activity and fixed N_2 when deprived of Mo. This constituted convincing evidence of the validity of Bishop's[28,29] earlier results and interpretations. After this, the alternative nitrogenase was purified from *A. chroococcum* and shown to contain V as a constituent of a VFe cofactor in the larger component of the nitrogenase 2 complex.[71,169] The VFe cofactor also is present in *A. vinelandii.*[30] More recently, Chisnell et al.[56] have shown that *A. vinelandii* contains a nitrogenase 3 in which neither Mo nor V seem to play an important role.

A more complete treatment of these exciting developments during the past decade is presented in Chapter 10.

VIII. Hydrogen Evolution and Consumption during Nitrogen Fixation

The evolution of H_2 from nodules of soybean plants initially was reported by Hoch et al.[97] Later, they concluded that the production of this gas was catalyzed by nitrogenase or a closely related enzyme. Insight into the metabolism of H_2 in nodules from peas (*Pisum sativum* L.) was provided by Dixon,[67,68] who demonstrated that nodules formed by a strain of *R. leguminosarum* failed to evolve H_2, because the H_2 produced during N_2 fixation was consumed by an O_2-dependent H_2 oxidation system. Dixon proposed several advantages of the H_2 recycling process, including (1) H_2-dependent ATP synthesis, (2) protection of nitrogenase from O_2 damage, (3) prevention of H_2 inhibition of the nitrogenase reaction, and (4) conservation of carbon substrates in nodules capable of using H_2 as a source of energy. The magnitude and extent of energy loss from legume nodules was pointed out by Schubert and Evans,[173] who surveyed nodule samples from a large number of different species and reported that the great majority of them lost 30–50% of their nitrogenase electron flux as H_2. Those nodules that evolved H_2 during N_2 fixation lacked an effective H_2 oxidation system. Since 1976 much research has been devoted to (1) purification and characterization of hydrogenase, (2) the pathway of electron transport from H_2 to O_2, (3) the distribution of H_2 oxidation capability among rhizobial strains, (4) the effect of hosts on expression of H_2-recycling capacity, (5) the benefits that may be realized from use of inoculation strains capable of synthesis of the H_2 oxidation system and establishing the conditions necessary for expression of H_2 oxidation capability in laboratory cultures of *Rhizobium* and *Bradyrhizobium*. These subjects have been reviewed by Eisbrenner and Evans,[72] O'Brian and Maier,[149] and Evans et al.,[80] and in Chapter 11.

In addition, several other noteworthy developments have occurred in this area of research. Simpson et al.[182] reported that H_2 increased CO_2 fixation in a Hup$^+$ strain of *B. japonicum*, and that the time courses of derepression of H_2 uptake and ribulose 1,5-bisphosphate carboxylase activities were positively correlated. During the same period, Hanus et al.[90] and Lepo et al.[122] demonstrated that Hup$^+$ strains of *B. japonicum* grew as chemolithoautotrophs in a medium containing mineral salts, ammonium nitrogen, and CO_2 and H_2 as sources of carbon and energy, respectively. This research and that by Purohit et al.[161] demonstrated that CO_2 was fixed in the chemolithotrophic cultures by the ribulose-1,5-bisphosphate carboxylase/oxygenase system. The properties of highly purified hydrogenase from *B. japonicum* have been described by Harker et al.,[92] Stults et al.,[189] and Arp.[12]

The genes required for H_2 oxidation in *B. japonicum* were cloned and characterized by Cantrell et al.[51] and Haugland et al.[94] Lambert et al.[121] demonstrated intra- and interspecies transfer and expression of H_2 uptake and chemolithoautotrophic capability. The structural genes for the 34.5- and 65.9-kDa subunits of hydrogenase have been sequenced by Sayavedra-Soto et al.[170] in Corvallis. It remains to construct stable Hup^+ strains of *R. meliloti* and *R. trifolii*.

In 1985, strains of Hup^+ and Hup^- *B. japonicum*, isogenic with the exception of hydrogenase, were constructed by the Corvallis group. These were used to inoculate soybeans planted in large concrete tiles; the experiments were designed to determine the benefits of H_2 oxidation capability on growth and N_2 fixation by symbiotically cultured soybeans. In these controlled trials, each treatment was replicated 10 times. Use of inoculum with H_2 recycling capability increased the N contents and dry weight of the soybeans 9–11%.[79] Further discussion of H_2 recycling in N_2 fixation organisms is presented in Chapter 11.

IX. Genetics of Biological Nitrogen Fixation

A rapid expansion of research in the genetics of N_2-fixing organisms has occurred since the initial transfer in 1971 of functional *nif* genes from *Klebsiella pneumoniae* to *Escherichia coli*.[159] Knowledge of the genetics of N_2-fixing microorganisms in 1975 is illustrated by the following statement regarding genetics research in the proceedings of a workshop sponsored by the National Science Foundation.[76] "Thorough basic (genetic) studies are sadly lacking in other (i.e., other than *K. pneumoniae*) N_2-fixing microorganisms including azotobacters, blue-green algae, enterobacters, clostridia, bacilli, spirilla, photosynthetic bacteria, and rhizobia." The proportion of overall research effort now being devoted to genetics of N_2-fixing microorganisms is illustrated by the contents of the proceedings of 1985 and 1988 international symposia on N_2 fixation, which show that 42% of the invited papers concerned some aspect of molecular genetics.[36,78] It is estimated that input into genetic investigations in the field of N_2 fixation has increased at least fourfold during the period 1976–1988.

Some areas where impressive progress has been made include (1) organization and regulation of *nif* genes in *Klebsiella pneumoniae* and *Azotobacter*[159] (see Chapters 19–21); (2) characterization of the genes in *Rhizobium* and *Bradyrhizobium* that are related to nodulation, metabolism, and expression of cell surface components[125,126] (see Chapters 14,17); (3) the organization of *nif* and related genes in cyanobacteria[93] (see Chapter 4); (4) the finding that certain flavones and isoflavones produced by host plants are necessary for the activation of nodulation genes in *Rhizobium* and *Bradyrhizobium*[125,156] (see Chapters 14,17); (5) characterization of the alternative nitrogenases (see Chapter 19), (6) discovery of nodule proteins, referred to as nodulins, that are uniquely involved in nodule development and function[126] (see Chapter 15).

In this regard, it is of great interest that van de Wiel et al.[205] recently have shown that ENOD2, an early nodulin in pea and soybean nodules, is located in the cell walls of nodule parenchyma. This nodulin contains two groups of repeating pentapeptides and is postulated to be involved in the O_2 diffusion barrier that participates in the regulation of the O_2 supply to the central nodule tissue where the bacteroids are located.

It seems obvious that future research progress in biological N_2 fixation will be increasingly dependent upon the integrated utilization of physiological, biochemical, and molecular genetical research approaches.

X. The Biochemistry of N_2 Fixation

A. *Response to Nitrogen, Oxygen, and Hydrogen Pressure*

The earliest studies of biological nitrogen fixation were concerned primarily with investigations of the culture and physiology of the microorganism involved, the process of infection of leguminous plants, the specificity of infection, nodule development and symbiotic interactions, and the practical aspects of legume inoculation. Burk[41-43] conducted some of the first studies on the influence of the pN_2 and pO_2 on N_2 fixation. He employed *Azotobacter chroococcum* primarily as his test organism and reported that its fixation increased almost linearly as the pN_2 was increased to 0.50 atm and the pO_2 was held at 0.05 atm. (Burk[43] reported a Michaelis constant of 0.21 atm N_2 for the azotobacter.) Burk[42] found that the response of the organism to changes in the pO_2 was not specific for N_2 fixation and gave no indication of the chemical mechanism of the process. Burk and Lineweaver[44] observed that O_2 at 60% in the atmosphere was inhibitory to the azotobacter.

Rather similar, but more exhaustive, experiments were performed by P. W. Wilson and colleagues[210,214,215] with red clover (*Trifolium pratense* L.) plants. Such experiments were much more laborious than those with the azotobacter but revealed additional information. The pN_2 supporting a half-maximal fixation rate by red clover was near 0.05 atm. Thus, under an atmosphere of air, the nitrogenase enzyme in clover operates near its potential maximum rate.

Red clover seeds were surface sterilized, germinated, and planted in a sterilized sand substrate in 9-L serum bottles. The gas phase in the bottles was renewed weekly, and in the interim CO_2 had to be added batchwise to support photosynthesis. A deficiency of CO_2 was indicated when cresol red in 0.001 M $NaHCO_3$ turned red in a vial suspended in the bottle. When the pN_2 was decreased, argon or helium normally was added to bring the pressure to 1 atm. H_2 was added in lieu of A or He in one experiment on the assumption that it was inert relative to the nitrogenase reaction. Plants exposed to H_2 grew less and fixed less N_2 than expected. A careful examination of the response with a variety of sources of H_2 indicated that the H_2 per se was an inhibitor of N_2 fixation. The inhibition was specific for N_2 and did not influence growth on ammonium or nitrate ions; the inhibition was competitive.[211]

Although high levels of O_2 inhibited the growth of red clover,[214] the inhibition was nonspecific, and O_2 inhibited growth of the plants on both combined N and N_2. The work of Wilson and colleagues was convincing, and subsequent workers have chosen not to repeat the demanding and time-consuming work with higher plants, but rather to use free-living N_2-fixing bacteria for their studies. One can speculate that if such organisms had been used exclusively, H_2 inhibition might not have been discovered.

B. Carbohydrate Nitrogen Relationship

The Wilson group also explored the carbohydrate–nitrogen relationship in symbiotic N_2 fixation. This had been a continuing research problem for over 40 years, but issues remained controversial. Wilson has a chapter in his 1940 monograph[212] on the carbohydrate–nitrogen relationship; there he has plotted in three dimensions how nitrogen versus carbohydrate influences nodulation and the amount of N_2 fixed. Wilson and coworkers examined the "nitrogen hunger period" in soybeans and verified that it resulted from an excess of carbohydrate in the plants before N_2 fixation was well established. A balance between photosynthetic production of carbohydrate and nitrogen available to the plants could be reestablished by shading the plants to reduce carbohydrate production or by feeding a low concentration of nitrate to boost the plants' level of fixed nitrogen.

C. Ureides

Wilson's group examined the change in the balance of nitrogenous constituents of soybean plants with alteration of growth conditions. With the rather crude methods available in the 1930s, they measured van Slyke distributions of nitrogen to include nonbasic amino N, basic nonamino N, basic amino N, amide + NH_3−N, and other N. From the data, Umbreit concluded that a ureide was an important intermediate in the metabolism of the N_2-fixing soybean. As Wilson stated,[212] "If the basic nonamino compound proves to be a ureide, as present studies indicate, this agrees with the views of numerous investigators whose experiments have led them to the conclusion that ureides, purines, and related compounds play a significant role in the nitrogen metabolism of many plants, especially the legumes." The journal editor to whom Umbreit submitted his manuscript was unconvinced and rejected the paper. It took almost 30 years before the concept was accepted that ureides are a major form of fixed N transported in N_2-fixing soybeans.[135]

D. Intermediates in Biological N_2 Fixation

There was early speculation about the probable intermediates in biological N_2 fixation, and special importance was attached to the "key intermediate" that marked the end product of the fixation process and the product that combined with a carbon chain to yield organically bound nitrogen. There were proponents of oxidized products, such as nitrate, and reduced products, such as ammonia. The experimental data characteristically were sketchy and nonspecific for the hypotheses featuring asparagine, aspartic acid, and other amino acids. The ammonia hypothesis was pushed most vigorously, especially by Winogradsky.[216] He found pea nodules dried at 50°C and then ground would evolve NH_3, whereas roots treated this way did not evolve NH_3. No evidence was presented that the nodules were fixing N_2 at the time of the tests. Kostytschew and colleagues[117,118] reported that *Azotobacter agile* released NH_4^+ into the culture medium and concluded that the NH_4^+ was an intermediate in N_2 fixation and probably the first product rather than a decomposition product of the cells. Horner and Burk[102] reported that young cultures of the azotobacter excreted NH_4^+ and organic nitrogen compounds whether they were supplied N_2 or nitrate and concluded that the excretion of NH_4^+ was not specific for the ammonia hypothesis.

The most vigorously championed hypothesis of the 1930s and 1940s was the hydroxylamine hypothesis. It was first seriously suggested by Blom,[32] but then was supported by numerous contributions from Virtanen's laboratory.[208] He suggested that N_2 was converted to NH_2OH and that this combined with oxalacetic acid to form an oxime that was reduced further to aspartic acid. Support rested heavily on the demonstration of oxalacetic acid in the leguminous plants and aspartic acid as an excretion product from roots of pea plants fixing N_2. Isolation of an intermediate oxime also was reported. The major difficulty with the hypothesis was that other investigators could not repeat the experimental observations of Virtanen and coworkers. Excretion observed in other laboratories was vanishingly small, and as a result the isolation of aspartic acid and the oxime could not be verified. Although the lines of evidence cited in support were not specific for the hydroxylamine hypothesis, the hypothesis was rather widely accepted.

When enriched ^{15}N, a stable isotope of nitrogen, became available, it supported use of a different and highly specific technique for probing the mechanism of N_2 fixation. Burris and Miller[49] showed that there was no detectable exchange between ^{15}N and ^{14}N in N_2-fixing cultures of *A. vinelandii;* hence, ^{15}N could be used as a valid tracer in studying the fixation of N_2. ^{15}N-enriched N_2 was supplied for a short time to an N_2-fixing culture of *A. vinelandii,* and the nitrogenous compounds of the cells were fractionated. Among the amino acids, glutamic acid carried the highest enrichment in ^{15}N. This was compatible with the ammonia hypothesis, not with Virtanen's hydroxylamine hypothesis, which demanded that aspartic acid be the primary amino acid labeled. This same high ^{15}N labeling of glutamic acid after short exposure to $^{15}N_2$ was observed with *Rhodospirillum rubrum, Clostridium pasteurianum, Chromatium* sp., *Chlorobium* sp., *Nostoc muscorum,* and nodules excised from soybeans.[213]

Additional support for the ammonia hypothesis was derived from the facts that (1) ammonia was used in preference to other compounds by N_2-fixing cells, (2) ammonia excreted in cultures of *C. pasteurianum* had a much higher concentration of ^{15}N than any other compound isolated, (3) kinetic studies strongly supported the role of ammonia, and (4) only ammonia accumulated in N_2-fixing cell-free preparations of *C. pasteurianum* and other organisms.[48,213]

If a culture of *A. vinelandii* was offered $^{15}NH_4NO_3$ so that the origin of N in the cells could be ascertained, it was apparent that virtually all the ammonia was consumed before any nitrate was utilized. Ammonium ions were used immediately when added to N_2-fixing cells as if ammonium already had been the product being assimilated from N_2 fixation. In contrast, the N_2-fixing cells had to adapt to the use of nitrate. When NH_4^+ was present, it was used to the virtual exclusion of other nitrogenous compounds present. The data indicated that nitrate was converted to NH_4^+ before being assimilated. Even cultures of *A. vinelandii* that had been grown on NO_3^- preferentially assimilated NH_4^+ when $^{15}NH_4NO_3$ was supplied.

Most N_2-fixing organisms excrete very little NH_4^+ during active growth. However, the strictly anaerobic *C. pasteurianum* may be deficient in its production of α-ketoglutaric acid, a compound needed for assimilation of NH_4^+ via the glutamic dehydrogenase or GOGAT (glutamine oxoglutarate aminotransferase) system. When α-ketoglutaric acid is not available, the organism excretes NH_4^+ and this furnishes an opportunity to measure the ^{15}N concentration of NH_4^+ directly rather than only the ^{15}N of the products of NH_4^+ assimilation. When Zelitch et al.[219] determined the ^{15}N concentration in NH_4^+ from *C. pasteurianum* cultures, it proved to be far higher than that of other products. This furnished direct support for the ammonia hypothesis.

When the kinetics of ^{15}N assimilation from $^{15}N_2$ were followed for 5 min in *A. vinelandii,* and the percentage of the total ^{15}N fixed in various compounds was plotted against time, only the "amide fraction" (free NH_4^+ + NH_4^+ released by hydrolysis of amides) showed a negative slope.[48] At an infinitely short time, 100% of the ^{15}N fixed should be in the initial product, and this percentage should decrease with time as the first product is used in forming other products. All other products should exhibit positive slopes in such plots. As only the "amide fraction" and the free NH_4^+ fraction exhibited negative slopes, the ammonia hypothesis was supported.

When active cell-free preparations of nitrogenase became available, it was possible to remove the enzymes involved in ammonia assimilation. Under these conditions, ammonia accumulated as the first demonstrable product of N_2 fixation.[48]

At the time most of these observations were made, the GOGAT system had not been described, and it was assumed that nitrogen was assimilated by the action of glutamic dehydrogenase to form glutamate. Subsequently, it became apparent that under most situations, NH_4^+ is in such low concentrations that the GOGAT pathway is favored over the glutamic dehydrogenase pathway.

The hydroxylamine hypothesis has faded from the picture. Although one can hardly argue convincingly that N_2 is reduced in one step to $2NH_4^+$, it may well be that intermediates with reduction levels between N_2 and $2NH_4^+$ remain bound to the nitrogenase enzymes rather than being released as free intermediates. Thorneley et al.[191] have presented evidence that a bound dinitrogen hydride intermediate yielding a compound at the hydrazine reduction level is formed by N_2-fixing preparations from *Klebsiella pneumoniae*. This is compatible with the theory that NH_4^+ is the end product of biological N_2 fixation that is assimilated.

E. Cell-Free N_2 Fixation

Bach and colleagues[13] recorded the first serious report of cell-free N_2 fixation. They claimed in 1934 that ground *Azotobacter chroococcum* extracts fixed N_2. Burk spent some time in Bach's laboratory, but neither he nor Bach and his colleagues were successful in repeating the observations that cell-free preparations increased in total nitrogen. Likewise, Roberg's[167] attempts, and those of others, to repeat the experiments were unsuccessful.

Interest in obtaining cell-free N_2 fixation was revived by the availability of ^{15}N-enriched N_2, as this provided a far more sensitive tool than increase in total N for measuring N_2 fixation. The attempts to obtain cell-free fixation have been summarized.[46] The method with $^{15}N_2$ did not reveal any fixation under conditions reported by Bach et al.[13] Magee and Burris[133] recorded data from a number of experiments that showed fixation at the rather conservative level of 0.015 at. % ^{15}N excess or greater; fixation at this level occurred in about 27% of their experiments. Hoch and Westlake[99] had encouraging cell-free fixation with *Clostridium pasteurianum* in 1958, and they found levels of fixation of 0.585, 0.644, 0.812, 0.823, 0.929, and 1.251 at. % ^{15}N excess after exposures of sonicated and centrifuged preparations for 1 hr at 30°C to 95 at. % $^{15}N_2$ at a pN_2 of 0.05 atm. These levels of fixation were impressive.

Carnahan et al.[52,53] succeeded in obtaining consistent fixation with extracts from *C. pasteurianum*. Although cell-free fixation had been obtained earlier, the key point is that the preparations of Carnahan et al. were consistent and reproducible in other laboratories, and hence they were highly useful in studying the nature of nitrogenase and the mechanism of its action. Schneider et al.[171] promptly reproduced the results of Carnahan et al. with *C. pasteurianum* and also reported their success with preparations from blue-green algae and *Rhodospirillum rubrum*. The clostridial preparations were the most consistently active, and for a few years such preparations dominated studies of cell-free N_2 fixation. Important considerations in achieving reproducible fixation with the preparation of Carnahan et al. included the following: Dry the cells in a rotary evaporator at relatively high temperature (cold can inactivate dinitrogenase reductase); keep manipulations strictly anaerobic; and supply a high level of pyruvate as an energy source.

F. Separation of Components of Nitrogenase and Their Purification

With modification of techniques, it became possible to prepare cell-free nitrogenase from a number of organisms. With this came attempts to purify nitrogenase. Mortenson[143] made the interesting and fundamental discovery that nitrogenase consists of two proteins rather than one protein. He stated in 1965, "Nitrogen fixation in cell-free extracts of *C. pasteurianum* is catalyzed by at least two protein components. . . . Neither nitrogen fixation nor the electron-requiring ATP utilization occurs unless both components are present." The race then was on to purify the components and to establish their nature. Contributions came from several laboratories, so it is difficult to attribute priority of discovery of purification methods. In 1972, Tso et al.[202] reported purification of the Fe and MoFe proteins from *C. pasteurianum* essentially to homogeneity and found specific activities, respectively, of about 3100 and 2500 nmoles C_2H_4 formed/(min $\times$ mg protein) at 30°C. This was the highest activity reported for a preparation at that time, but highly active preparations had been recovered by others with different techniques. Characteristically, each nitrogenase required some modification of purification methods. In 1976, Winter and Burris[217] published a table of activities that had been reported in the literature.

Preparation of soluble nitrogenase from nodules required a more extensive modification of the methods. The occurrence of high levels of phenolic compounds complicated the recovery of active cell-free nitrogenase from nodules. Loomis and Battaile[127] had described a method for using polyvinylpolypyrrolidone to complex the phenolic compounds. This method was adopted by Koch et al.[114,115] and Klucas et al.[111] in their successful isolation of nitrogenase from legume nodules and by Benson et al.[18] in their isolation from alder nodules of nitrogenase and hydrogenase. With care to maintain anaerobic conditions, preparations of many different nitrogenases can be recovered at room temperature. There is a tendency for groups to work on nitrogenase preparations from a single source, but this precludes observation of differences among organisms that are interesting from the standpoint of comparative biochemistry.

G. Homology Among Nitrogenases

Interesting observations on the comparative biochemistry of nitrogenases were made by Emerich[74] when he purified nitrogenase components from eight organisms and then examined which nonhomologous combinations of the Fe protein and MoFe protein from these organisms produced active nitrogenases. About 85% of the combinations showed nitrogenase activity, and this supported the concept that there is a marked homology among the nitrogenases, indicative of a common evolutionary pathway. This concept subsequently has been substantiated by genetic studies of a variety of nitrogenases. The combinations between the strictly anaerobic *C. pasteurianum* and component proteins from the aerobic *A. vinelandii* were uniformly devoid of nitrogenase activity. Further examination of this response[73] revealed that the inactivity was not based on the inability of the Fe protein and MoFe protein to bind together. Rather, it was based upon their inability to dissociate. In each cycle of N_2 fixation the dinitrogenase and dinitrogenase reductase bind together, MgATP is hydrolyzed and an electron is transferred, and the protein components dissociate. The tight binding between the Fe protein of *C. pasteurianum* and the MoFe protein of *A. vinelandii* precludes dissociation of the complex (data from Thorneley and Lowe[192] indicate that this is the rate-limiting step for nitrogenase), and thus blocks the activity of nitrogenase.

H. Nitrogenase and Its Operation

Purification of nitrogenase has evolved as new methods have been developed. The first requirement is to disrupt the cells to release nitrogenase. As indicated, cells of *C. pasteurianum,* dried anaerobically in an evaporator rotating in a bath at about 45°C, can be extracted with buffer to yield an active preparation of nitrogenase. Cells can also be disrupted by sonic oscillation, by osmotic shock, by grinding with abrasives, by freezing and thawing, or by extrusion through a French press. After disruption, unbroken cells and debris are sedimented by centrifugation. The crude supernatant is useful for some measurements, but usually purification of the dinitrogenase and dinitrogenase reductase is necessary. Purification commonly is achieved with polyethylene glycol, and separation on columns of Sephadex or DEAE cellulose, by crystallization of A. *vinelandii* MoFe protein, and by preparative gel electrophoresis.

After purification had been achieved, it was possible to establish the physicochemical properties of the MoFe and Fe components of nitrogenase. The MoFe proteins carry 2 Mo and generally about 32 Fe and a comparable number of acid-labile S atoms per molecule. Their molecular weights are about 215,000–240,000, and their specific activities 1500–2500 nmoles C_2H_4 formed/(min × mg protein). The MoFe center generally is accepted as the site of N_2 binding and reduction.

Dinitrogenase reductases (the Fe proteins) have a 4 Fe 4 acid-labile S center that is shared between two equivalent protein subunits, each of about 30,000 molecular weight. Dinitrogenase reductases are considerably more labile to O_2 than are the dinitrogenase proteins to which they transfer electrons.

I. MgATP Is Required

The initial detailed reports of consistent cell-free fixation of N_2 indicated that ATP inhibited N_2 fixation in preparations from C. *pasteurianum*. Hence, there was considerable skepticism when it was reported[137] that ATP ("high-energy" phosphate) was required for fixation by cell-free preparations of nitrogenase from C. *pasteurianum*. Tests with firefly luciferase showed that the extracts produced ATP with added pyruvate.

J. Electron Donors and Lability to Oxygen

It was not clear in the initial papers on consistent cell-free N_2 fixation what compounds were serving as electron donors to effect the reduction of N_2. However, Mortenson et al.[144] and Mortenson[142] reported that an iron-containing compound they termed ferredoxin supported reduction of protons to H_2 in C. *pasteurianum*. Eventually it was found that ferredoxin would support N_2 reduction and that it was of the same family of iron compounds as the methemoglobin-reducing factor of Hill and the photosynthetic pyridine nucleotide reductase (PPNR). Carter et al.[54] isolated a ferredoxin from the bacteroids of soybean nodules that could function in both the reduction of C_2H_2 to C_2H_4 and the reduction of protons to H_2. No measurable quantity of flavodoxin was detected in bacteroid extracts. Flavodoxin[17,96,177] can serve as an electron donor for reduction of N_2 in the azotobacter, and Burns and Bulen[45] found that $Na_2S_2O_4$ functioned effectively as reductant in nitrogenase systems reconstructed from their components.

The isolated nitrogenase components are labile to O_2, so they must be prepared and handled under anaerobic conditions. $Na_2S_2O_4$ at a concentration of about 0.001 M can aid in keeping solutions anaerobic. Some protection against inactivation is afforded by the Shethna[176] protein, isolated from *A. vinelandii*.

K. Sequence of Electron Transfer

When the components of nitrogenase had been separated and purified, it became feasible to study the sequence of electron transfer in the system. Early work on the system emphasized preparations from *C. pasteurianum*, and hence the role of ferredoxin as an electron donor was prominent. Several research groups contributed to establishing the sequence of electron transfer. A particularly useful tool was low-temperature electron paramagnetic resonance (EPR), as both the Fe protein and the MoFe protein have characteristic EPR signals, and the signals differ for the oxidized and reduced forms of the proteins. Nicholas et al.[148] demonstrated EPR signals originating in particles from *A. vinelandii* at $g = 1.94$, 2.00, and 4.3. The $g = 1.94$ signal also was apparent in a nitrogenase preparation from *C. pasteurianum*. Hardy et al.[91] observed the EPR signals of the MoFe protein; they reported resonances at g values of 2.01, 3.67, and 4.30 at 4 °K and noted that $Na_2S_2O_4$ enhanced the 3.67 and 4.3 resonances and introduced a resonance at $g = 1.94$. They observed no EPR signal for the clostridial Fe protein.

With further purification of the nitrogenase components, it was possible to apply EPR measurements in a more defined way. Orme-Johnson et al.[151] used preparations both from *C. pasteurianum* and *A. vinelandii* and showed that the EPR signal for the Fe protein of nitrogenase was altered by the addition of $Mg \cdot ATP$. In the absence of $Mg \cdot ATP$, but with $Na_2S_2O_4$[151] (Fig. 2), the MoFe protein and the Fe protein showed characteristic EPR spectra; when the MoFe and Fe proteins were mixed, the spectrum of the MoFe protein remained unchanged. In the presence of $Mg \cdot ATP$ and $Na_2S_2O_4$[151] (Fig. 3), when the MoFe and Fe proteins were mixed, the MoFe protein was reduced and its 4.29 and 3.77 peaks virtually disappeared. When the $Na_2S_2O_4$ was exhausted, these peaks reappeared. It was concluded that,

(i) $Mg \cdot ATP$ binds to the reduced Fe protein. The EPR spectrum is substantially altered during this binding. Significant $Mg \cdot ATP$ hydrolysis does *not* occur as a result of this. (ii) The reduced Fe protein will not reduce the MoFe protein, but the reduced Fe protein–$Mg \cdot ATP$ complex does reduce the MoFe protein. (iii) The reduced MoFe protein and reduced Fe protein, in the presence of $Mg \cdot ATP$, are reoxidized by substrates; this is apaparent in the EPR spectra when the reductant is exhausted.

These observations established the electron transfer pathway in the nitrogenase system, although it remained for later observations to clarify that electrons were passed one at a time from the Fe protein to the MoFe protein, and that the $Mg \cdot ATP$ was hydrolyzed and the protein complex dissociated when the transfer occurred.

L. Inhibitors

As mentioned earlier, H_2 is a specific and competitive inhibitor of N_2 fixation, and it has been particularly useful in studying the mechanism of fixation. N_2O is the only other inhibitor of N_2 fixation that has been demonstrated to be competitive. Not only is it a competitive inhibitor, but it also is a substrate for nitrogenase. Molnar et al.[141] reported that N_2O is a specific inhibitor, and Repaske and Wilson[165] established that the inhibiton is competitive. The inhibitor constant, K_i, is about 0.108 atm.[166] Other substrates and inhibitors of nitrogenase were examined,[166] including CO, C_2H_2, CN^-, and N_3^-. As all substrates for nitrogenase compete for electrons, they are mutually interactive, but only H_2 and N_2O are competitive inhibitors of N_2. Even at an infinite concentration, N_2 does not block H_2 production completely,[181] and extrapolation to infinite concentration indicates that in this limiting case 25% of the electrons are used to produce H_2 and 75% to reduce N_2. N_2O is competitive with N_2, and N_2 is competitive with N_2O. We are aware of only two nonreciprocal responses among substrates and inhibitors for nitrogenase. C_2H_2 is noncompetitive with N_2, but N_2 is competitive when C_2H_2 is serving as the substrate.[166] N_2O is competitive with C_2H_2, whereas C_2H_2 is noncompetitive with N_2O. NO is a potent noncompetitive inhibitor of the reduction of both N_2 and C_2H_2.

M. Formation of HD

The nitrogenase system supplied D_2 in the presence of N_2 forms HD.[98] This initially was referred to as an exchange reaction, but Bulen[39] showed that an electron was required for each HD formed, so the reaction clearly was not an exchange. He also suggested that H_2 inhibition and HD formation were manifestations of the same reaction. Burgess et al.[40] concluded that N_2 was not required for HD formation. Apparently their diluent argon was contaminated with N_2, because when their experiments were repeated with care to exclude N_2, a clear need for N_2 to support HD formation was demonstrated again.[123]

N. Substrates Other than N_2

N_2O was the first compound other than N_2 to be demonstrated to serve as a substrate for nitrogenase.[145] Subsequently, H^+,[45] C_2H_2,[64,172] azide,[172] CN^-,[91] CH_3NC,[91] and cyclopropene[136] were demonstrated to serve as substrates. It is apparent that nitrogenase is a versatile enzyme system. As these substrates all accept electrons from reduced dinitrogenase, they compete among themselves for the electrons.

The reduction of C_2H_2 has served as a very useful assay for nitrogenase activity, because the product, C_2H_4, can be analyzed very easily and in low concentration with a gas chromatographic unit equipped with a H_2 flame ionization detector. Reduction of C_2H_2 was discovered independently in the United States and Australia in 1965. The history of the independent discovery has been reviewed.[47]

O. Mechanism of Nitrogenase

There is consensus on a general model for nitrogenase action that has evolved over a number of years. It is recognized that the Fe protein accepts electrons from a donor such as ferredoxin or flavodoxin, and that it binds $2 \, Mg \cdot ATP$. It transfers electrons, one at a time, to the MoFe protein. The MoFe protein and the Fe protein form a complex, the electron is transferred, and $2 \, Mg \cdot ATP$ are hydrolyzed to $2 \, Mg \cdot ATP + 2 \, Pi$. The MoFe protein and Fe protein dissociate and the process is repeated. When the MoFe protein has accumulated an adequate number of electrons, it binds an N_2 molecule, reduces it, and releases NH_4^+. The MoFe protein then accepts additional electrons from the Fe protein to repeat the cycle.

A study of HD formation[88] has thrown considerable light on the general reactions of nitrogenase and specifically on the nature of H_2 inhibition. Lowe and Thorneley[128,129,193,194] have made extensive kinetic measurements to elaborate their detailed model of nitrogenase action. Their data indicate that the dissociation of the MoFe protein–Fe protein complex is the rate-limiting step of the process. The mechanism for nitrogenase action has been stated here only in briefest outline, and it is elaborated in Chapter 18.

P. Control of Nitrogenase

The nitrogenase reaction has a high energy demand; a minimum of 16 $Mg \cdot ATP$ are hydrolyzed under ideal conditions in the reaction $N_2 + 8H^+ + 8 \, e^- \rightarrow 2NH_3 + H_2$, and the actual energy used is considerably higher under natural conditions. Hence, it is important that the nitrogenase system be under the control of the N_2-fixing cells. When NH_4^+ is added to a N_2-fixing system, characteristically the nitrogenase reaction is turned off. One of the most interesting and best-defined control systems was described first in *R. rubrum*.[131] Control is centered on dinitrogenase reductase, which is turned off by ADP-ribosylation of arginine 100 or 101 in one unit of the dimer of dinitrogenase reductase and is reactivated by removal of the inactivating group. This phenomenon and genetic control of nitrogenase are developed in Chapter 3.

Q. Nomenclature

The literature is replete with terms to describe the two nitrogenase protein components, and we have used a number of them interchangeably in this chapter. Use of some of the terms has faded with time. Fraction 1 and 2 or protein 1 and 2 merely refer to the sequence of elution of the protein from a DEAE cellulose column. MoFe protein and Fe protein are more informative terms. Rr1, Rr2, and so on, furnish a useful shorthand for describing the MoFe and Fe proteins, respectively, in this case from *Rhodospirillum rubrum*. The most informative nomenclature, based on the enzymatic reaction, designates the overall complex as nitrogenase and the MoFe protein as dinitrogenase (the "type species" substrate is dinitrogen, and enzymes coventionally are named relative to their substrate). The Fe protein is designated dinitrogenase reductase, as the general consensus is that its function is the reduction of dinitrogenase.

References

1. Ahmed, S., and Evans, H. J., "Effect of cobalt on the growth of soybeans in the absence of supplied nitrogen," *Biochem. Biophys. Res. Commun. 1*, 271–275 (1959).

2. Ahmed, S., and Evans, H. J., "The essentiality of cobalt for soybean plants grown under symbiotic conditions," *Proc. Natl. Acad. Sci. USA 47*, 24–36 (1961).

3. Aho, P. E., Seidler, R. J., Nelson, A. D., and Evans, H. J., "Association of nitrogen-fixing bacteria with decay in white fir," *Proceedings of the 1st International Symposium on Nitrogen Fixation*, Vol. 2, W. E. Newton and C. J. Nyman, eds. Pullman: Washington State University Press, 1976, pp. 629–640.

4. Akkermans, A. D. L., Abdulkadir, S., and Trinick, M. J., "Nitrogen-fixing root nodules in *Ulmaceae*," *Nature 274*, 190 (1978).

5. Alazard, D., Ndoye, I., and Dreyfus, B., "*Sesbania rostrata* and other nodulated legumes," *Nitrogen Fixation: Hundred Years After*, H. Bothe, F. J. DeBruijn, and W. E. Newton, eds. Stuttgart: Gustav Fischer, 1988, pp. 765–769.

6. Alexander, V., "Nitrogen fixing lichens in tundra and taiga ecosystems," *Current Perspectives in Nitrogen Fixation*, A. H. Gibson and W. E. Newton, eds. Canberra: Australian Academy of Science, 1981, p. 257.

7. Anderson, A. J., "Molybdenum deficiency on a South Australian ironstone soil," *J. Aust. Inst. Agr. Sci. 8*, 73–75 (1942).

8. Appleby, C. A., "The oxygen equilibrium of leghaemoglobin," *Biochim. Biophys. Acta 60*, 226–235 (1962).

9. Appleby, C. A., "Leghaemoglobin and *Rhizobium* respiration," *Ann. Rev. Plant Physiol. 35*, 443–478 (1984).

10. Appleby, C. A., "Plant hemoglobin properties, function and genetic origin," *Nitrogen Fixation and CO₂ Metabolism*, P. W. Ludden and J. E. Burris, eds. New York: Elsevier, 1985, pp. 41–51.

11. Appleby, C. A., Tjepkema, J. D., and Trinick, M. J., "Hemoglobin in a nonleguminous plant *Parasponia:* Possible genetic origin and function in nitrogen fixation," *Science 220*, 951–953 (1983).

12. Arp, D. J., "*Rhizobium japonicum* hydrogenase: Purification to homogeneity from soybean nodules, and molecular characterization," *Arch. Biochim. Biophys. 237*, 504–512 (1985).

13. Bach, A. N., Jermolieva, Z. V., and Stepanian, M. P., "Fixation de l'azote atmosphérique par l'intermédiaire d'enzymes extraités de cultures d'*Azotobacter chroococcum*," *Compt. Rend. Acad. Sci. (USSR) 1*, 22–24 (1934).

14. Barber, L. E., Tjepkema, J. D., Russell, S. A., and Evans, H. J., "Acetylene reduction (nitrogen fixation) associated with corn inoculated with *Spirillum*," *Appl. Environ. Microbiol. 32*, 108–113 (1976).

15. Basilier, K., and Granhall, U., "Nitrogen fixation in wet minerotrophic moss communities of a subartic mire," *Oikos 31*, 236–246 (1978).

16. Beijerinck, M. W., "Die Bakterien der papilionaceen Knollchen," *Bot. Zeit. 46*, 725–735 (1888).

17. Benemann, J. R., Yoch, D. C., Valentine, R. C., and Arnon, D. I., "The electron transport system in nitrogen fixation by Azotobacter. I. Azotoflavin as an electron carrier," *Proc. Natl. Acad. Sci. USA 64*, 1079–1086 (1969).

18. Benson, D. R., Arp, D. J., and Burris, R. H., "Cell-free nitrogenase and hydrogenase from actinorhizal root nodules," *Science 205*, 688–689 (1979).

19. Bergersen, F. J., *Methods for Evaluating Biological Nitrogen Fixation*, New York: Wiley, 1980.

20. Bergersen, F. J., *Root Nodules of Legumes: Structure and Function*, Chichester: Research Studies Press, 1982, p. 164.

21. Bergersen, F. J., and Hipsley, E. H., "The presence of N₂-fixing bacteria in the intestines of man and animals," *J. Gen. Microbiol. 60*, 61–65 (1970).

22. Bergersen, F. J., and Postgate, J. R., *A Century of Nitrogen Fixation Research: Present Status and Future Prospects*. London: The Royal Society, 1987, p. 231.

23. Bergersen, F. J., and Turner, G. L., "Nitrogen fixation by the bacteroid fraction of breis of soybean root nodules," *Biochim. Biophys. Acta 141*, 507–515 (1967).

24. Bergersen, F. J., and Turner, G. L., "Comparative studies of nitrogen fixation by soybean root nodules, bacteroid suspensions and cell-free extracts," *J. Gen. Microbiol. 53*, 205–220 (1968).

25. Bergersen, F. J., and Turner, G. L., "Gel filtration of nitrogenase from soybean root nodule bacteroids," *Biochim. Biophys. Acta 214*, 28–36 (1970).

26. Berndt, H., Ostwal, K. P., Schumann, C., Mayer, F., and Schlegel, H. G., "Identification and physiological characterization of the nitrogen-fixing bacteria *Corynebacterium autotrophicum* GZ 29," *Arch. Microbiol. 108*, 17–26 (1976).

27. Bishop, P. E., and Evans, H. J., "Immunological evidence for the capability of free-living *Rhizobium japonicum* to synthesize a portion of a nitrogenase component," *Biochim. Biophys. Acta 381*, 248–256 (1975).

28. Bishop, P. E., Jarlenski, D. M. L., and Hetherington, D. R., "Evidence for an alternative nitrogen fixation system in *Azotobacter vinelandii*," *Proc. Natl. Acad. Sci. USA* *77*, 7342–7346 (1980).

29. Bishop, P. E., Jarlenski, D. M. L., and Hetherington, D. R., "Expression of an alternative nitrogen fixation system in *Azotobacter vinelandii*," *J. Bacteriol. 150*, 1244–1251 (1982).

30. Bishop, P. E., and Joerger, R. D., "Genetics and molecular biology of alternative nitrogen fixation systems," *Ann. Rev. Plant Physiol. Plant Mol. Biol.*, 41:109–125 (1990).

31. Bishop, P. E., Premakumar, R., Dean, D. R., Jacobson, M. R., Chisnell, J. R., Rizzo, T. M., and Kopczynski, J., "Nitrogen fixation by *Azotobacter vinelandii* strains having deletions in structural genes for nitrogenase," *Science 232*, 92–94 (1986).

32. Blom, J., "Ein Versuch, die chemischen Vorgange bei der Assimilation des molekularen Stikstoffs durch Mikroorganism zu erklaren," *Zent. Bakt., 2 Abt., 84*, 60–86 (1931).

33. Bohlool, B. B., and Schmidt, E. L., "Lectins: A possible basis for specificity in the *Rhizobium*–legume nodule symbiosis," *Science 185*, 269–271 (1974).

34. Bortels, H., "Molybdenum as a catalyzer in biological nitrogen fixation," *(translated)* *Arch. Mikrobiol. 1*, 333–342 (1930).

35. Bortels, H., "The effect of molybdenum and vanadium compounds on leguminosae" *(trans.)*, *Arch. Mikrobiol. 8*, 13–26 (1937).

36. Bothe, H., DeBruijn, F. J., and Newton, W. E., eds. *Nitrogen Fixation: Hundred Years After*. Stuttgart: Gustav Fischer, 1988, p. 878.

37. Boursier, P., Hanus, F. J., Papen, H., Becker, M. M., Russell, S. A., and Evans, H. J., "Selenium increases hydrogenase expression in autotrophically cultured *Bradyrhizobium japonicum* and is a constituent of the purified enzyme," *J. Bacteriol. 170*, 5594–5600 (1988).

38. Breznak, J. A., Brill, W. J., Mertins, J. W., and Coppel, H. C., "Nitrogen fixation in termites," *Nature* (Lond.) *244*, 577–580 (1973).

39. Bulen, W. A., "Nitrogenase from *Azotobacter vinelandii* and reactions affecting mechanistic interpretations," *Proc. 1st Int. Symp. Nitrogen Fixation*. Pullman, Washington: Washington State University Press, 1976, pp. 177–186.

40. Burgess, B. K., Wherland, S., Newton, W. E., and Stiefel, E. I., "Nitrogenase reactivity: Insight into the nitrogen-fixing process through hydrogen-inhibition and HD-forming reactions," *Biochemistry 20*, 5140–5146 (1981).

41. Burk, D., "The influence of nitrogenase upon the organic catalysis of nitrogen fixation by Azotobacter," *J. Phys. Chem. 34*, 1174–1194 (1930).

42. Burk, D., "The influence of oxygen gas upon the organic catalysis of nitrogen fixation by Azotobacter," *J. Phys. Chem. 34*, 1195–1209 (1930).

43. Burk, D., "Azotase and nitrogenase in *Azotobacter*," *Ergeb. Enzymforsch. 3*, 23–56 (1934).

44. Burk, D., and Lineweaver, H., "The influence of fixed nitrogen on Azotobacter," *J. Bacteriol. 19*, 389–414 (1930).

45. Burns, R. C., and Bulen, W. A., "ATP-dependent hydrogen evolution by cell-free preparations of *Azotobacter vinelandii*," *Biochim. Biophys. Acta 105*, 437–445 (1965).

46. Burris, R. H., "Biological nitrogen fixation," *Ann. Rev. Plant Physiol., 17,* 155–184 (1966).

47. Burris, R. H., "The acetylene-reduction technique," *Nitrogen Fixation by Free-living Micro-organisms,* Vol. 6, W. D. P. Stewart, ed. Cambridge: Cambridge University Press, pp. 249–257 (1975).

48. Burris, R. H., "Nitrogen fixation," *Plant Biochemistry,* 3rd ed., J. Bonner and J. E. Varner, eds. New York: Academic Press, 1976, pp. 887–908.

49. Burris, R. H., and Miller, C. E., "Application of N^{15} to the study of biological nitrogen fixation," *Science 93,* 114–115 (1941).

50. Callaham, D., del Tredici, P., and Torrey, J. G., "Isolation and cultivation *in vitro* of the actinomycete causing root nodulation in *Comptonia,*" *Science 199,* 899–902 (1978).

51. Cantrell, M. A., Haugland, R. A., and Evans, H. J., "Construction of a *Rhizobium japonicum* gene bank and use in the isolation of a hydrogen uptake gene," *Proc. Natl. Acad. Sci. USA 80,* 181–185 (1983).

52. Carnahan, J. E., Mortenson, L. E., Mower, H. F., and Castle, J. E., "Nitrogen fixation in cell-free extracts of *Clostridium pasteurianum,*" *Biochim. Biophys. Acta 38,* 188–189 (1960).

53. Carnahan, J. E., Mortenson, L. E., Mower, H. F., and Castle, J. E., "Nitrogen fixation in cell-free extracts of *Clostridium pasteurianum,*" *Biochim. Biophys. Acta 44,* 520–535 (1960).

54. Carter, K. R., Rawlings, J., Orme-Johnson, W. H., Becker, R. R., and Evans, H. J., "Purification and characterization of a ferredoxin from *Rhizobium japonicum* bacteroids," *J. Biol. Chem. 255,* 4213–4223 (1980).

55. Child, J. J., "Nitrogen fixation by free-living *Rhizobium* and its implications," *Recent Advances in Biological Nitrogen Fixation,* N. S. Subba Rao, ed. London: Arnold, (1980), pp. 325–343.

56. Chisnell, J. R., Premakumar, R., and Bishop, P. E., "Purification of a second alternative nitrogenase from a *nifHDK* deletion strain of *Azotobacter vinelandii,*" *J. Bacteriol. 170,* 27–33 (1988).

57. Cowles, J. R., Evans, H. J., and Russell, S. A., "B_{12} coenzyme-dependent ribonucleotide reductase in *Rhizobium* species and the effects of cobalt deficiency on the activity of the enzyme," *J. Bacteriol. 97,* 1460–1465 (1968).

58. Dalton, D. A., and Chatfield, J. M., "A new nitrogen-fixing cyanophyte–hepatic association: *Nostoc* and *Porella,*" *Am. J. Bot. 72,* 781–784 (1985).

59. Dalton, D. A., Russell, S. A., and Evans, H. J., "Nickel as a micronutrient element for plants," *BioFactors 1,* 11–16 (1988).

60. Dalton, H., "Chemoautotrophic nitrogen fixation," *Nitrogen Fixation. Proc. Phytochem. Soc. Eur., Symp. No. 18,* W. D. P. Stewart and J. R. Gallon, eds. London: Academic Press, 1980, pp. 177–195.

61. Dazzo, F. B., and Brill, W. J., "Regulation by fixed nitrogen of host–symbiont recognition in the *Rhizobium*–clover symbiosis," *Plant Physiol. 62,* 18–21 (1978).

62. Dazzo, F. B., Hollingsworth, R. I., Sherwood, J. E., Abe, M., Hrabak, E. M., Gardiol, A. E., Pankratz, H. S., Smith, K. B., and Yang, H., "Recognition and infection of clover root hairs by *Rhizobium trifolii,*" *Nitrogen Fixation Research Progess,* H. J. Evans, P. J. Bottomley, and W. E. Newton, eds. Dordrecht: Nijhoff, 1985, pp. 239–246.

63. Diaz, C. L., Melchers, L. S., Hooykaass, P. J. J., Lugtenberg, G. J. J., and Kijne, J. W., "Root lectin as a determinant of host–plant specificity in the *Rhizobium*–legume symbiosis," *Nature 338,* 579–581 (1989).

64. Dilworth, M. J., "Acetylene reduction by nitrogen-fixing preparations from *Clostridium pasteurianum,*" *Biochim. Biophys. Acta 127,* 285–294 (1966).

65. Dilworth, M. J., and Glenn, A. R., "Control of carbon substrate utilization by rhizobia," *Current Perspectives in Nitrogen Fixation,* A. H. Gibson and W. E. Newton, eds. Canberra: Australian Academy of Science, 1981, pp. 244–251.

66. Dixon, N. E., Gazzola, C., Blakeley, R. L., and Zerner, B., "A metalloenzyme. A simple biological role for nickel," *J. Am. Chem. Soc. 97,* 4131–4133 (1975).

67. Dixon, R. O. D., "Hydrogen uptake and exchange by pea root nodules," *Ann. Bot. 31,* 179–188 (1967).

68. Dixon, R. O. D., "Hydrogenase in pea root nodule bacteroids," *Arch. Mikrobiol. 62,* 272–283 (1968).

69. Döbereiner, J., and Day, J. M., "Associative symbioses in tropical grasses: Characterization of microorganisms and dinitrogen-fixing sites," *Proceedings of the 1st International Symposium on Nitrogen Fixation,* Vol. 2, W. E. Newton and C. J. Nyman, eds. Pullman: Washington State University Press, 1967, pp. 518–538.

70. Döbereiner, J., Day, J. M., and Dart, P. J., "Nitrogenase activity and oxygen sensitivity of the *Paspalum notatum–Azotobacter paspali* association," *J. Gen. Microbiol. 71,* 103–116 (1972).

71. Eady, R. R., Robson, R. L., Richardson, T. H., Miller, R. W., and Hawkins, M., "The vanadium nitrogenase of *Azotobacter chroococcum.* Purification and properties of the VFe protein," *Biochem. J. 244,* 197–207 (1987).

72. Eisbrenner, G., and Evans, H. J., "Aspects of hydrogen metabolism in nitrogen-fixing legumes and other plant-microbe associations," *Annual Reviews of Plant Physiology,* Vol. 34, W. R. Briggs, R. L. Jones, and V. Walbot, eds. Palo Alto, CA: Annual Reviews, 1983, pp. 105–136.

73. Emerich, D. W., and Burris, R. H., "Interactions of heterologous nitrogenase components that generate catalytically inactive complexes," *Proc. Natl. Acad. Sci. USA 73,* 4369–4373 (1976).

74. Emerich, D. W., and Burris, R. H., "Complementary functioning of the component proteins of nitrogenase from several bacteria," *J. Bacteriol. 134,* 936–943 (1978).

75. Eskew, D. L., Welch, R. M., and Cary, E. E., "Nickel: An essential micronutrient for legumes and possibly all higher plants," *Science 222,* 621–623 (1983).

76. Evans, H. J., ed., *Enhancing Biological Nitrogen Fixation.* Washington, D.C.: Division of Biological Medical Sciences, National Science Foundation, 1975, p. 52.

77. Evans, H. J., and Barber, L. E., "Biological nitrogen fixation for food and fiber production: Some immediately feasible possibilities," *Science 197,* 332–339 (1977).

78. Evans, H. J., Bottomley, P. J., and Newton, W. E., *Nitrogen Fixation Research Progess.* Dordrecht: Martinus Nijhoff, 1985, p. 731.

79. Evans, H. J., Hanus, F. J., Haugland, R. A., Cantrell, M. A., Xu, L. S., Russell, S. A., Lambert, G. R., and Harker, A. R., "Hydrogen recycling in nodules affects nitrogen fixation and growth of soybeans," *Proceedings of the World Soybean Conference III,* R. Shibles, ed. Boulder, CO: Westview Press, 1985, pp. 935–942.

80. Evans, H. J., Harker, A. R., Papen, H., Russell, S. A., Hanus, F. J., and Zuber, M., "Physiology, biochemistry and genetics of the uptake hydrogenase in *Rhizobium*," *Ann. Rev. Microbiol. 41*, 335–361 (1987).

81. Evans, H. J., and Kliewer, M., "Vitamin B_{12} compounds in relation to cobalt requirements of higher plants and nitrogen-fixing organisms. Symposium of B_{12} co-enzymes," *Ann. N.Y. Acad. Sci. 112* (Part 2), 735–755 (1964).

82. Evans, H. J., and Purvis, E. R., "Molybdenum status of some New Jersey soils with respect to alfalfa production," *Argon J. 53*, 70–71 (1951).

83. Evans, H. J., Purvis, E. R., and Bear, F. E., "Molybdenum nutrition of alfalfa," *Plant Physiol. 25*, 555–566 (1950).

84. Evans, H. J., and Russell, S. A., "Physiological chemistry of nitrogen fixation by legumes," *Chemistry and Biochemistry of Nitrogen Fixation*, J. Postgate, ed. London: Plenum Press, 1971, pp. 191–244.

85. Fred, E. B., Baldwin, I. L., and McCoy, E., *Root Nodule Bacteria and Leguminous Plants*. Madison: The University of Wisconsin Press, 1932, p. 343.

86. Gallon, J. R., and Chaplain, A. E., *An Introduction to Nitrogen Fixation*. London: Cassell Educational, 1987, p. 56.

87. Gillis, M., Kersters, K., Hoste, B., Janssens, D., Kroppenstedt, R. M., Stephan, M. P., Teixeira, K. R. S., Do¨bereiner, J., and De Ley, J., "*Acetobacter diazotrohicus* sp. nov., a nitrogen-fixing acetic acid bacterium associated with sugarcane," *Int. J. Syst. Bacteriol. 39*, 361–364 (1989).

88. Guth, J. H., and Burris, R. H., "Inhibition of nitrogenase-catalyzed NH_3 formation by H_2," *Biochemistry 22*, 5111–5122 (1983).

89. Hamblin, J., and Kent, S. P., "Possible role of phytohaemagglutinin in *Phaseolus vulgaris*," *Nat. New Biol. 245*, 28–30 (1973).

90. Hanus, F. J., Maier, R. J., and Evans, H. J., "Autotrophic growth of H_2-uptake positive strains of *R. japonicum* in an atmosphere supplied with hydrogen gas," *Proc. Natl. Acad. Sci. USA 76*, 1788–1792 (1979).

91. Hardy, R. W. F., Burns, R. C., and Parshall, G. W., "The biochemistry of N_2 fixation," *Bioinorganic Chemistry*, R. F. Gould, ed. Washington, D.C.: Amer. Chem. Soc., 1971, pp. 219–247.

92. Harker, A. R., Xu, L. S., Hanus, F. J., and Evans, H. J., "Some properties of the nickel-containing hydrogenase of chemolithotrophically grown *Rhizobium japonicum*," *J. Bacteriol. 159*, 850–856 (1984).

93. Haselkorn, R., Golden, J. W., Lammers, P. J., and Mulligan, M. E., "Organization of the genes for nitrogen fixation in the cyanobacterium *Anabaena*," *Nitrogen Fixation Research Progress*, H. J. Evans, P. J. Bottomley, and W. E. Newton, eds. Dordrecht: Nijhoff, 1985, pp. 485–490.

94. Haugland, R. A., Cantrell, M. A., Beaty, J. S., Hanus, F. J., Russell, S. A., and Evans, H. J., "Characterization of hydrogen uptake genes of *Rhizobium japonicum*," *J. Bacteriol. 159*, 1006–1012 (1984).

95. Hellriegel, H., and Wilfarth, H., *Untersuchungen uber die Stickstoff-Nahrung der Gramineen und Leguminosen*. Zeitschrift fur der verschiedige Rubenzücker des Deutsches Reichs (Beilageheft). 1888.

96. Hinkson, J. W., and Bulen, W. A., "A free-radical flavoprotein from Azotobacter. Isolation, crystallization, and properties," *J. Biol. Chem. 242*, 3345–3351 (1967).

97. Hoch, G. E., Little, H. N., and Burris, R. H., "Hydrogen evolution from soybean root nodules," *Nature 179,* 430–431 (1957).

98. Hoch, G. E., Schneider, K. C., and Burris, R. H., "Hydrogen evolution and exchange and conversion of N_2O to N_2 by soybean root nodules," *Biochim. Biophys. Acta 37,* 273–279 (1960).

99. Hoch, G. E., and Westlake, D. W. S., "Fixation of N_2 by extracts from *Clostridium pasteurianum,*" *Fed. Proc. 17,* 243 (1958).

100. Holm-Hansen, O., Gerloff, G. C., and Skoog, F., "Cobalt as an essential element for blue-green algae," *Plant Physiol. 7,* 665–667 (1954).

101. Hoover, T. R., Imperial, J., Ludden, P. W., and Shah, V. K., "Mini Review: Biosynthesis of the iron-molybdenum cofactor of nitrogenase," *BioFactors 1,* 199–205 (1988).

102. Horner, C. K., and Burk, D., "The nature and amount of extracellular nitrogen in *Azotobacter* cultures," *Third Comm. Int. Soc. Soil Sci. Trans. A.* 168–174 (1939).

103. Hunt, S., King, B. J., and Layzell, D. B., "Effects of gradual increases in O_2 concentration on nodule activity in soybean," *Plant Physiol. 91,* 315–321 (1989).

104. Israel, D. W., Howard, R. L., Evans, H. J., and Russell, S. A., "Purification and characterization of the Mo-Fe protein component of nitrogenase from soybean nodule bacteroids," *J. Biol. Chem. 249,* 500–508 (1974).

105. Kennedy, C., Robson, R., Jones, R., Woodley, P., Evans, D., Bishop, P., Eady, R., Gamal, R., Humphrey, R., Ramos, J., Dean, D., Brigle, K., Toukdarian, A., and Postgate, J., "Genetical and physical characterization of *nif* and *ntr* in *Azotobacter chroococcum* and *A. vinelandii,*" *Nitrogen Fixation Research Progress,* H. J. Evans, P. J. Bottomley, and W. E. Newton, eds. Dordrecht: Nijhoff, 1985, pp. 469–476.

106. Kennedy, I. R., *Primary Products of Symbiotic Nitrogen Fixation. A Study of the Rate of ^{15}N Distribution, and of Some Transformation Mechanisms,* Ph.D. dissertation. Perth: University of Western Australia, 1965.

107. Klingmüller, W., ed., *Azospirillum I. Proceedings of First Bayreuth Azospirillum Workshop.* Basel: Birkhauser Verlag, 1982.

108. Klingmüller, W., ed., *Azospirillum II. Proceedings of the Second Bayreuth Azospirillum Workshop.* Basel: Birkhauser Verlag, 1983.

109. Klingmüller, W., ed., *Azospirillum III. Proceedings of the Third Bayreuth Azospirillum Workshop.* Berlin: Springer-Verlag, 1985.

110. Klingmüller, W., ed., *Azospirillum IV. Proceedings of the Fourth Bayreuth Azospirillum Workshop.* Berlin: Springer-Verlag, 1988.

111. Klucas, R. V., and Evans, H. J., "An electron donor system for nitrogenase-dependent acetylene reduction by extracts of soybean nodules," *Plant Physiol. 42,* 1458–1460 (1968).

112. Klucas, R. V., Hanus, F. J., Russell, S. A., and Evans, H. J., "Nickel: a micronutrient element for hydrogen-dependent growth of *Rhizobium japonicum* and for expression of urease activity in soybean leaves," *Proc. Natl. Acad. Sci. USA 80,* 2253–2257 (1983).

113. Klucas, R. V., Koch, B., Russell, S. A., and Evans, H. J., "Purification and some properties of the nitrogenase from soybean (*Glycine max*) nodules," *Plant Physiol. 43,* 1906–1912 (1968).

114. Koch, B. L., Evans, H. J., and Russell, S. A., "Reduction of acetylene and nitrogen gas by breis and cell-free extracts of soybean root nodules," *Plant Physiol. 42,* 466–468 (1967).

115. Koch, B., Evans, H. J., and Russell, S. A., "Properties of the nitrogenase system in cell-free extracts of bacteroids from soybean root nodules," *Proc. Natl. Acad. Sci. USA* *58*, 1343–1350 (1967).

116. Kohl, D. H., and Shearer, G., "Isotopic fractionation associated with symbiotic N₂ fixation and uptake of NO₃⁻ by plants," *Plant Physiol.* *66*, 51–56 (1980).

117. Kostytschew, S., and Ryskaltschuk, A., "Les produits de la fixation de l'azote atmosphérique par l'*Azotobacter agile*," *Compt. Rend. Acad. Sci. (Paris) 180*, 2070–2072 (1925).

118. Kostytschew, S., Ryskaltschuk, A., and Schwezowa, O., "Biochemische Untersuchungen über *Azotobacter agile*," *Z. Physiol. Chem.* *154*, 1–17 (1926).

119. Kubo, H., "Über Hämaprotein as den Wurzelknollchen von Leguminosen," *Acta Phytochim. (Jap.) 11*, 195–200 (1939).

120. Lalonde, M., Simon, L., Bousquet, J., and Séguin, A., "Advances in the taxonomy of *Frankia*: Recognition of species *alni* and *elaegni* and novel subspecies of *pommerii* and *vandijkii*," *Nitrogen Fixation: Hundred Years After*, H. Bothe, F. J. de Bruijn, and W. E. Newton, eds. Stuttgart: Gustav Fischer, 1988, pp. 671–680.

121. Lambert, G. R., Cantrell, M. A., Hanus, F. J., Russell, S. A., Haddad, K. R., and Evans, H. J., "Intra- and interspecies transfer and expression of *Rhizobium japonicum* hydrogen uptake genes and autotrophic growth capability," *Proc. Natl. Acad. Sci. USA* *82*, 3232–3236 (1985).

122. Lepo, J. E., Hanus, F. J., and Evans, H. J., "Further studies on the chemoautotrophic growth of hydrogen uptake positive strains of *R. japonicum*," *J. Bacteriol. 141*, 664–670 (1980).

123. Li, J.-L., and Burris, R. H., "Influence of pN₂ on HD formation by various nitrogenases," *Biochemistry 22*, 4472–4480 (1983).

124. Liang, J., and Burris, R. H., "Interactions among N₂, N₂O, and C₂H₂ as substrates and inhibitors of nitrogenase from *Azotobacter vinelandii*," *Biochemistry 27*, 6726–6732 (1988).

125. Long, S. R., "*Rhizobium*–legume nodulation: Life together in the underground," *Cell 56*, 203–214 (1989).

126. Long, S. R., "*Rhizobium* genetics," *Annu. Rev. Genet. 23*, 483–506 (1989).

127. Loomis, W. D., and Battaile, J., "Plant phenolic compounds and the isolation of plant enzymes," *Phytochemistry 5*, 423–438 (1966).

128. Lowe, D. J., and Thorneley, R. N. F., "The mechanism of *Klebsiella pneumoniae* nitrogenase action. Pre-steady-state kinetics of H₂ formation," *Biochem. J. 224*, 877–886 (1984).

129. Lowe, D. J., and Thorneley, R. N. F., "The mechanism of *Klebsiella pneumoniae* nitrogenase action. The determination of rate constants required for the simulation of the kinetics of N₂ reduction and H₂ evolution," *Biochem. J. 224*, 895–901 (1984).

130. Lowe, R. H., and Evans, H. J., "Cobalt requirements for growth of rhizobia," *J. Bacteriol. 85*, 210–211 (1962).

131. Ludden, P. W., and Burris, R. H., "Activating factor for the iron protein of nitrogenase from *Rhodospirillum rubrum*," *Science 194*, 424–426 (1976).

132. Mackintosh, M. E., "Nitrogen fixation by *Thiobacillus ferrooxidans*," *J. Gen. Microbiol. 105*, 215–218 (1978).

133. Magee, W. E., and Burris, R. H., "Oxidative activity and nitrogen fixation in cell-free preparations from *Azotobacter vinelandii*," *J. Bacteriol. 71*, 635–643 (1956).

134. Mahl, M. C., Wilson, P. W., Fife, M. A., and Ewing, W. H., "Nitrogen fixation by members of the tribe Klebsielleae," *J. Bacteriol. 89*, 1482–1487 (1965).

135. Matsumoto, T., Yatazawa, M., and Yamamoto, Y., "Incorporation of nitrogen-15 into allantoin in nodulated soybean plants supplied with molecular nitrogen-15," *Plant Cell Physiol. 18*, 459–462 (1977).

136. McKenna, C. E., McKenna, M. C., and Higa, M. T., "Chemical probes of nitrogenase. 1. Cyclopropene nitrogenase-catalyzed reduction to propene and cyclopropane," *J. Am. Chem. Soc. 98*, 4657–4659 (1976).

137. McNary, J. E., and Burris, R. H., "Energy requirements for nitrogen fixation by cell-free preparations from *Clostridium pasteurianum*," *J. Bacteriol. 84*, 598–599 (1962).

138. Meeks, J. C., Enderlin, C. S., Joseph, C. M., Steinberg, N., and Weeden, Y. M., "Use of ^{13}N to study N_2-fixation and assimilation by cyanobacterial–lower plant associations," *Nitrogen Fixation Research Progress*, H. J. Evans, P. J. Bottomley, and W. E. Newton, eds. Dordrecht: Nijhoff, 1985, pp. 301–307.

139. Minchin, F. R., Sheehy, J. E., and Witty, J. F., Factors limiting N_2 fixation by the legume–rhizobium symbiosis," *Nitrogen Fixation Research Progress*, H. J. Evans, P. J. Bottomley, and W. E. Newton, eds. Dordrecht: Nijhoff, 1985, pp. 285–291.

140. Mishustin, E. N., and Shil'nikova, V. K., *Biological Nitrogen Fixation*. London: Macmillan (English ed.), 1971.

141. Molnar, D. M., Burris, R. H., and Wilson, P. W., "The effect of various gases on nitrogen fixation by *Azotobacter*," *J. Am. Chem. Soc. 70*, 1713–1716 (1948).

142. Mortenson, L. E., "Nitrogen fixation: Role of ferredoxin in anaerobic metabolism" *Ann. Rev. Microbiol. 17*, 115–138 (1963).

143. Mortenson, L. E., "Nitrogen fixation in extracts of *Clostridium pasteurianum*," *Nonheme Iron Proteins: Role in Energy Conversion*, A. San Pietro, ed. Yellow Springs, Ohio: Antioch Press, 1965, pp. 243–269.

144. Mortenson, L. E., Valentine, R. C., and Carnahan, J. E., "An electron transport factor from *Clostridium pasteurianum*," *Biochem. Biophys. Res. Commun. 7*, 448–452 (1962).

145. Mozen, M. M., and Burris, R. H., "The incorporation of ^{15}N-labelled nitrous oxide by nitrogen fixing agents," *Biochim. Biophys. Acta 14*, 577–578 (1954).

146. Neal, J. L., Allen, G. C., Morse, R. D., and Wolf, D. D., "Nitrate, nitrite, nitrous oxide and oxygen-dependent hydrogen uptake by *Rhizobium*," *FEMS Microbiol. Lett. 17*, 335–338 (1983).

147. Neal, J. L., Allen, G. C., Morse, R. D., and Wolf, D. D., "Anaerobic nitrate-dependent chemolithotrophic growth by *Rhizobium japonicum*," *Can. J. Microbiol. 29*, 316–320 (1983).

148. Nicholas, D. J. D., Wilson, P. W., Heinen, W., Palmer, G., and Beinert, H., "Use of electron paramagnetic resonance spectroscopy in investigations of functional metal components in micro-organisms," *Nature 196*, 433–436 (1962).

149. O'Brian, M. R., and Maier, R. J., "Hydrogen metabolism in *Rhizobium*: Energetics, regulation, enzymology and genetics," *Adv. Microbiol. Physiol. 29*, 1–52 (1988).

150. Okon, Y., Fallik, E., Sarig, S., Yahalom, E., and Tal, S., "Plant growth promoting effects of *Azospirillum*," *Nitrogen Fixation: Hundred Years After*, H. Bothe, F. J. deBruijn, and W. E. Newton, eds. Stuttgart: Gustav Fischer, 1988, pp. 741–746.

151. Orme-Johnson, W. H., Hamilton, W. D., Ljones, T., Tso, M.-Y. W., Burris, R. H., Shah, V. K., and Brill, W. J., "Electron paramagnetic resonance of nitrogenase and nitrogenase components from *Clostridium pasteurianum* W5 and *Azotobacter vinelandii* OP," *Proc. Natl. Acad. Sci. USA, 69,* 3142–3145 (1972).

152. Parke, D., and Ornston, L. M., "Nutritional diversity of *Rhizobiaceae* revealed by auxanography," *J. Gen. Microbiol. 130,* 1743–1750 (1984).

153. Pate, J. S., Gunning, B. E. S., and Briarty, L. G., "Ultrastructure and functioning of the transport system of the leguminous root nodule," *Planta 85,* 11–34 (1969).

154. Peters, G. A., "The *Azolla–Anabaena* symbiosis," *Genetics Engineering for Nitrogen Fixation,* A. Hollaender, ed. New York: Plenum Press, 1977, pp. 259–281.

155. Peters, G. A., Kaplan, D., Meeks, J. C., Buzby, K. M., Marsh, B. H., and Corbin, J. L., "Aspects of nitrogen and carbon interchange in the *Azolla–Anabaena* symbioses," *Nitrogen Fixation and CO₂ Metabolism,* P. W. Ludden and J. E. Burris, eds. Amsterdam: Elsevier, 1985, pp. 213–222.

156. Peters, N. K., Frost, J. W., and Long, S. R., "A plant flavone, luteolin, induces expression of *Rhizobium meliloti* nodulation genes," *Science 233,* 977–980 (1986).

157. Philips-Hollingsworth, S., Hollingsworth, R. I., Dazzo, F. B., Djordjevic, M. A., and Rolfe, B. G., "The effect of interspecies transfer of *Rhizobium* host-specific nodulation genes on acidic polysaccharide structure and *in situ* binding by host lectin," *J. Biol. Chem. 264,* 5710–5714 (1989).

158. Polacco, J. C., "Nitrogen metabolism in soybean tissue culture. II. Urea utilization and urease synthesis require Ni^{+2}," *Plant Physiol. 59,* 827–830 (1977).

159. Postgate, J. R., *The Fundamentals of Nitrogen Fixation.* Cambridge: Cambridge University Press, 1983, p. 252.

160. Powrie, J. K., "A field response by subterranean clover to cobalt fertilizer," *Aust. J. Sci. 23,* 198–199 (1960).

161. Purohit, K., Becker, R. R., and Evans, H. J., "D-ribulose-1,5-bisphosphate carboxylase/oxygenase from chemolithotrophically-grown *Rhizobium japonicum* and inhibition by D-4-phosphoerythronate," *Biochim. Biophys. Acta 715,* 230–239 (1982).

162. Raju, P. N., Evans, H. J., and Seidler, R. J., "An asymbiotic nitrogen-fixing bacterium from the root environment of corn," *Proc. Natl. Acad. Sci. USA 69,* 3474–3478 (1972).

163. Rasmussen, L. J., Peters, G. K., and Burris, R. H., "Isotope fractionation in biological nitrogen fixation," *Phykos 28,* 64–79 (1989).

164. Reisenauer, H. M., "Cobalt in nitrogen fixation by a legume," *Nature 186,* 375–376 (1960).

165. Repaske, R., and Wilson, P. W., "Nitrous oxide inhibition of nitrogen fixation by *Azotobacter,*" *J. Am. Chem. Soc. 74,* 3101–3103 (1952).

166. Rivera-Ortiz, J. M., and Burris, R. H., "Interactions among substrates and inhibitors of nitrogenase," *J. Bacteriol. 123,* 537–545 (1975).

167. Roberg, M., "Beiträge zur Biologie von *Azotobacter:* III. Zur Frage eines ausserhalb der Zelle den Stickstoffbindenden Enzyms," *Jahrb. Wiss. Bot. 83,* 567–592 (1936).

168. Robson, A. D., "Cobalt and nitrogen fixation in *Lupinus angustifolius* L. I. Growth, nitrogen concentrations and cobalt distribution," *The New Phytologist 83,* 53–62 (1979).

169. Robson, R. L., Eady, R. R., Richardson, T. H., Miller, R. W., Hawkins, M., and Postgate, J. R. "The alternative nitrogen fixation system of *Azotobacter chroococcum* is a vanadium enzyme," *Nature* (Lond.) *322,* 388–390 (1986).

170. Sayavedra-Sota, L. A., Powell, G. K., Evans, H. J., and Morris, R. O., "Nucleotide sequence of the genetic loci encoding subunits of *Bradyrhizobium japonicum* uptake hydrogenase," *Proc. Natl. Acad. Sci. USA 85*, 8395–8399 (1988).

171. Schneider, K. C., Bradbeer, C., Singh, R. N., Wang, L.-C., Wilson, P. W., and Burris, R. H., "Nitrogen fixation by cell-free preparations from microorganisms," *Proc. Natl. Acad. Sci. USA 46*, 726–733 (1960).

172. Schöllhorn, R., and Burris, R. H., "Study of intermediates in nitrogen fixation," *Fed. Proc. 25*, 710 (1966).

173. Schubert, K. R., and Evans, H. J., "Hydrogen evolution: A major factor affecting the efficiency of nitrogen fixation in nodulated symbionts," *Proc. Natl. Acad. Sci. USA 73*, 1207–1211 (1976).

174. Shah, V. K., and Brill, W. J., "Isolation of an iron-molybdenum cofactor from nitrogenase," *Proc. Natl. Acad. Sci. USA 74*, 3249–3253 (1977).

175. Sheehy, J. E., Minchin, F. R., and Witty, J. F., "Control of nitrogen fixation in a legume nodule: An analysis of the role of oxygen diffusion in relation to nodule structure," *Ann. Bot. 55*, 549–562 (1985).

176. Shethna, Y. I., "Non-heme iron (iron–sulfur) proteins of *Azotobacter vinelandii*," *Biochim. Biophys. Acta 205*, 58–62 (1970).

177. Shethna, Y. I., Wilson, P. W., and Beinert, H., "Purification of a nonheme iron protein and other electron transport components from azotobacter extracts," *Biochim. Biophys. Acta 113*, 225–234 (1966).

178. Shinji, I., Sato, K., and Shimizu, S., "The relationship of cobalt requirement to vitamin B_{12}-dependent methionine synthesis in *Rhizobium meliloti*," *Agric. Biol. Chem. 41*, 2229–2234 (1977).

179. Silvester, W. B., "Endophyte adaptation in *Gunnera–Nostoc* symbiosis," *Symbiotic Nitrogen Fixation in Plants*, P. S. Nutman, ed. Cambridge: Cambridge University Press, 1967, pp. 521–538.

180. Silvester, W. B., "Dinitrogen fixation in plant associations excluding legumes," *A Treatise on Dinitrogen Fixation*, Vol. IV, R. W. F. Hardy and A. H. Gibson, eds. New York: Wiley, 1977, pp. 141–190.

181. Simpson, F. B., and Burris, R. H., "A nitrogen pressure of 50 atmospheres does not prevent evolution of hydrogen by nitrogenase," *Science 224*, 1095–1097 (1984).

182. Simpson, F. B., Maier, R. J., and Evans, H. J., "Hydrogen-stimulated CO_2 fixation and coordinate induction of hydrogenase and ribulosebisphosphate carboxylase in a H_2-uptake positive strain of *R. japonicum*," *Arch. Microbiol. 123*, 1–8 (1979).

183. Smith, F. W., "Mineral nutrition of legumes," *Nitrogen Fixation in Legumes*, J. M. Vincent, ed. Sydney: Academic Press, 1982, pp. 155–172.

184. Sprent, J. I., and de Faria, S. M., "Mechanisms of infection of plants by nitrogen fixing organisms," *Plant and Soil 110*, 157–165 (1988).

185. Stewart, W. D. P., "Present-day nitrogen-fixing plants," *Ambio 6*, 166–178 (1977).

186. Stewart, W. D. P., "Symbiotic nitrogen fixing cyanobacteria," *Nitrogen Fixation*, W. D. P. Stewart and J. R. Gallon, eds. London: Academic Press, 1980, pp. 239–277.

187. Stewart, W. D. P., "Nitrogen fixation—its current relevance and future potential," *Isr. J. Bot. 31*, 5–44 (1982).

188. Stouthamer, A. H., Stam, H., deVries, W., and van Vlerken, M., "Some aspects of nitrogen fixation in free-living cultures of *Rhizobium*," *Nitrogen Fixation: Hundred*

Years After, H. Bothe, F. J. deBruijn, and W. E. Newton, eds. Stuttgart: Gustav Fischer, 1988, pp. 257–261.

189. Stults, L. W., O'Hara, E. B., and Maier, R. J., "Nickel is a component of hydrogenase in *Rhizobium japonicum,*" *J. Bacteriol. 159,* 153–157 (1984).

190. Thomas, J., Apte, S. K., and Reddy, B. R., "Sodium metabolism in cyanobacterial nitrogen fixation and salt tolerance," *Nitrogen Fixation: Hundred Years After,* H. Bothe, F. J. deBruijn, and W. E. Newton, eds. Stuttgart: Gustav Fischer, 1988, pp. 195–201.

191. Thorneley, R. N. F., Eady, R. R., and Lowe, D. J., "Biological nitrogen fixation by way of an enzyme-bound dinitrogen–hydride intermediate," *Nature 272,* 557–558 (1978).

192. Thorneley, R. N. F., and Lowe, D. J., "Nitrogenase of *Klebsiella pneumoniae.* Kinetics of the dissociation of oxidized iron protein from molybdenum–iron protein: Identification of the rate-limiting step for substrate reduction," *Biochem. J. 215,* 393–403 (1983).

193. Thorneley, R. N. F., and Lowe, D. J., "The mechanism of *Klebsiella pneumoniae* nitrogenase action. Pre-steady-state kinetics of an enzyme-bound intermediate in N_2 reduction and NH_3 formation," *Biochem. J. 224,* 887–894 (1984).

194. Thorneley, R. N. F., and Lowe, D. J., "The mechanism of *Klebsiella pneumoniae* nitrogenase action. Simulation of the dependences of H_2-evolution rate on component-protein concentration and ratio and sodium dithionite concentration," *Biochem. J. 224,* 903–909 (1984).

195. Tjepkema, J. D., "Hemoglobins in the nitrogen-fixing root nodules of actinorhizal plants," *Can. J. Bot. 61,* 2924–2929 (1983).

196. Tjepkema, J. D., Ormerod, W., and Torrey, J. G., "Vesicle formation and acetylene reduction activity in *Frankia* sp. CP11 cultured in defined nutrient media," *Nature* (Lond.) *287,* 633–635 (1980).

197. Tjepkema, J. D., and Yocum, C. S., "Measurement of oxygen partial pressure within soybean nodules by oxygen microelectrode," *Planta 119,* 351–360 (1974).

198. Torrey, J. G., "The site of nitrogenase in *Frankia* in free-living culture and in symbiosis," *Nitrogen Fixation Research Progress,* H. J. Evans, P. J. Bottomley, and W. E. Newton, eds. Dordrecht: Nijhoff, 1985, p. 731.

199. Torrey, J. G., and Callaham, D., "Structural features of the vesicle of *Frankia* sp Cp11 in culture," *Can. J. Microbiol. 28,* 749–757 (1982).

200. Trinchant, J. C., Birot, A. M., and Rigaud, J., "Oxygen supply and energy-yielding substrates for nitrogen fixation (acetylene reduction) by bacteroid preparations," *J. Gen. Microbiol. 125,* 159–165 (1981).

201. Trinick, M. J., "Symbiosis between *Rhizobium* and the non-legume *Trema aspera,*" *Nature 244,* 459–460 (1973).

202. Tso, M.-Y., Ljones, T., and Burris, R. H., "Purification of the nitrogenase protein from *Clostridium pasteurianum,*" *Biochim. Biophys. Acta 267,* 600–604 (1972).

203. van Berkum, P., and Bohlool, B. B., "Evaluation of nitrogen fixation by bacteria in association with roots of tropical grasses," *Microbiol. Rev. 44,* 491–517 (1980).

204. van Berkum, P., and Sloger, C., "A critical evaluation of the characteristics of associative nitrogen fixation in grasses," *Nitrogen and the Environment,* K. A. Malik, S. H. Mujtaba Naqvi, and A. I. H. Aleem, eds. Faisalabad, Pakistan: Nuclear Institute for Agriculture and Biology, 1985, pp. 139–158.

205. van de Wiel, C., Scheres, B., Franssen, H., van Lierop, M.-J., van Lammeren, A., van Kammen, A., and Bisseling, T., "The early nodulin transcript ENOD2 is located

in the nodule parenchyma (inner cortex) of pea and soybean root nodules," *EMBO J.* *9*, 1–7 (1990).

206. Vessey, J. K., Walsh, K. B., and Layzell, D. B., "Oxygen limitation of N_2 fixation in stem-girdled and nitrate-treated soybean," *Physiol. Plant. 73*, 113–121 (1988).

207. Vincent, J. M., *A Manual for the Practical Study of Root-Nodule Bacteria*. Oxford: Blackwell, 1970.

208. Virtanen, A. I., *Cattle Fodder and Human Nutrition*. Cambridge, Cambridge University Press, 1938.

209. von Bülow, F. W., and Döbereiner, J., "Potential for nitrogen fixation in maize genotypes in Brazil," *Proc. Natl. Acad. Sci. USA 72*, 2389–2393 (1975).

210. Wilson, P. W., "Mechanism of symbiotic nitrogen fixation. I. The influence of pN_2," *J. Am. Chem. Soc. 58*, 1256–1261 (1936).

211. Wilson, P. W., "Mechanism of symbiotic nitrogen fixation," *Ergeb. Enzymforsch. 8*, 13–54 (1939).

212. Wilson, P. W., *The Biochemistry of Symbiotic Nitrogen Fixation*. Madison: University of Wisconsin Press, 1940.

213. Wilson, P. W., and Burris, R. H., "Biological nitrogen fixation—A reappraisal," *Ann. Rev. Microbiol. 7*, 415–432 (1953).

214. Wilson, P. W., and Fred, E. B., "Mechanism of symbiotic nitrogen fixation. II. The pO_2 function," *Proc. Natl. Acad. Sci. USA 23*, 503–508 (1937).

215. Wilson, P. W., and Umbreit, W. W., "Mechanism of symbiotic nitrogen fixation. III. Hydrogen as a specific inhibitor," *Arch. Mikrobiol. 8*, 440–457 (1937).

216. Winogradsky, S., "Sur l'origine de l'ammoniac degagee par les fixateurs d'azote," *Zent. Bakt, 2 Abt. 97*, 399–413 (1938).

217. Winter, H. C., and Burris, R. H., "Nitrogenase," *Annu. Rev. Biochem. 45*, 409–426 (1976).

218. Zablotowicz, R. M., Eskew, D. L., and Focht, D. D., "Denitrification in *Rhizobium*," *Can. J. Microbiol. 24*, 752–759 (1978).

219. Zelitch, I., Rosenblum, E. D., Burris, R. H., and Wilson, P. W., "Isolation of the key intermediate in biological nitrogen fixation by clostridium" *J. Biol. Chem. 191*, 295–298 (1951).

1
Phylogenetic Classification of Nitrogen-Fixing Organisms

J. P. W. Young

I. Introduction

A. Why Classify Bacteria by Their Evolutionary Relationships?

N_2 fixation has not been found in eukaryotes, but it is very widely distributed among eubacteria and archaeobacteria. Consequently, any discussion of the classification of N_2-fixing organisms must cover the classification of the bacteria in general. Until recently, it has been customary to subdivide bacteria into orders, families, and so on, on the basis of conspicuous morphological and physiological properties,[67] and previous discussions of the diversity of diazotrophs have by and large adopted this approach. Postgate's reviews in 1981[94] and 1982[95] are the most comprehensive listings that I am aware of, though there have been more recent but less complete treatments.[33,41] In those reviews, symbiotic bacteria were discussed separately from phototrophic (photosynthetic) N_2 fixers. Although it has been evident for a long time that this classification does not reflect the true evolutionary relationships among bacteria, only in the last few years have sufficient molecular data accumulated to allow a comprehensive restructuring of our concepts.

The key to this process has been the 16S ribosomal RNA (rRNA) molecule, which is universal and conserved in function. Divergence of the nucleotide sequence of this molecule, or of the genes encoding it, has been assessed by hybridization, by oligonucleotide cataloging, and increasingly by sequencing. The result is that we now have the outlines of an evolutionary tree for bacteria, which confirms certain of the old physiological categories but dismantles many of them and reassembles the components into new groupings. This tree, its advantages, and its problems are discussed in Woese's notable review of bacterial evolution.[139]

This approach can only be useful, though, if bacterial genomes really do have a single consistent evolutionary tree and if we can reconstruct this tree with a high degree of confidence. There are two main ways in which molecular phylogeny might fail us: Sequence similarity might reflect factors other than common ancestry, and the phylogeny of one gene might not be representative of the history of the genome as a whole.

Convergent evolution is much less of a problem at the genotypic level than it is at the phenotypic level, since the mapping from genotype to phenotype is highly redundant. Thus, the action of similar selective forces may well cause the same metabolic capability to evolve independently in different organisms, but it is virtually certain that the gene sequences responsible will retain characteristic differences. The rate of sequence evolution is constrained by selection, of course, and varies between one gene and another, between different parts of the same gene, and between one line of descent and another. This means that some sequences change fast enough to show significant variation at the intraspecific level (pseudogenes, for example, or silent sites in codons), whereas sequences that are crucial for function are highly constrained and remain recognizable in all organisms (such as some parts of the rRNA molecules). This variation in rate allows sequence comparisons, if well chosen, to be useful at all levels of taxonomic resolution, but it has also meant that increasingly sophisticated analyses have had to be developed to extract the maximum phylogenetic information.

If the evolutionary history of a set of sequences can be reconstructed reliably, the next important question is the extent to which the evolution of one molecule, such as the 16S rRNA, truly reflects the relationships of bacterial genomes as a whole. If it does, then evolutionary trees derived from other genes or phenotypic properties should be broadly consistent. Two potential causes of discrepancy are gene duplication and gene transfer, and these will be discussed later in this chapter with specific reference to the dinitrogenase reductase gene *nifH*. The phylogeny of this gene is broadly consistent with that of 16S rRNA, but some aspects require special pleading of one kind or another.

The issue of gene transfer deserves further comment, as it is crucial to our whole philosophy of bacterial classification, but since I have discussed it elsewhere[144] I will give only the briefest outline here. In sexually reproducing eukaryotes, the species is a natural biological unit defined by intraspecific gene exchange and interspecific barriers, but it is not clear whether bacterial species are, or could be, natural units in the same sense. The broad host range of some plasmids has been cited as evidence against such a concept, on the grounds that it provides a potential for extensive long-range gene flow. Another counterargument stresses the opposite problem: The apparent clonality within bacterial species suggests that there may be insufficient exchange to hold the species together as a coherent evolutionary unit. On the other hand, the case for the reality of bacterial species is strengthened by the recent observations that there has apparently been extensive recombination within *Escherichia coli* populations on a fine genetic scale,[35] and that DNA repair genes provide an "adjustable" barrier to gene flow between *E. coli* and the related species *Salmonella typhimurium*.[102] The order of genes on the chromosome has also been proposed as a potential barrier to interspecific recombination.[66] More direct evidence for the reality of species comes from DNA–DNA hybridization: There is often a more or less clear gap between intraspecific comparisons, which show more than 70% relatedness, and interspecific comparisons, which fall considerably lower. Indeed, this has been used as the basis for the "official" definition of a bacterial species.[132] It is probably not wise to insist on a rigid numerical standard, however, because some natural species may be genetically broader than others: Biological evolution and sequence evolution do not always proceed at the same relative rate.

Though the concept of a species may correspond to a real entity, the biological meaning of the higher taxa (genus, family, class, and so on) must be more dubious. Molecular comparisons now give us common yardsticks that allow us to ask whether there really are hierarchical strata of relatedness. If there are not, we shall just have to settle for arbitrary cutoffs: we need the names even if the bacteria do not. To be most useful, the groups should be based on sets of phenotypic properties, but it is important that the names refer to discrete and coherent parts of the phylogenetic tree. Since many of the traditional names violate this principle (*Rhizobiaceae, Rhodospirillaceae, Azotobacteraceae,* for example) it is not appropriate to use them in this chapter. Instead, I shall refer to phylogenetic groups, using informal names because in many cases the appropriate taxonomic level remains to be decided.

A classification based on evolutionary relationships has many advantages over one based on superficial similarities. If it is constructed on sufficient data, then it should be stable. It is not dependent on the arbitrary choice of "significant" properties. (Why should phototrophy be given precedence over diazotrophy?) It has predictive value: related organisms will tend to share many attributes. It allows the evolution of genes, of biochemical pathways, and of adaptations to be traced and explained. It draws attention to anomalous gene distributions that might reflect horizontal transfer. One claim that is sometimes made for traditional classifications is that they are "practical": They allow a new isolate to be given a name with little experimental effort. This is true, but it is only helpful if a taxon defined by a small number of characteristics is also relatively homogeneous in other properties, and this is only likely to be the case if the taxon is phylogenetically coherent. In any case, a conscientious attempt at diagnosis using classical methods is not necessarily easy. For example, "the second most abundant bacterium on the root surface of young sugar beet plants was identified as a *Phyllobacterium* sp. (*Rhizobiaceae*) based on a comparison of the results of 39 conventional identification tests, 167 API tests, 30 antibiotic susceptibility tests, and sodium dodecyl sulfate-polyacrylamide gel electrophoretic fingerprints of total cellular proteins."[70] It would have been at least as quick, easy, cheap, and convincing (i.e., "practical") to have used the polymerase chain reaction and DNA sequencing to determine part of the 16S rRNA gene sequence of a few representative strains. Of course there is no substitute for a good basic phenotypic description, but sequencing will provide a practical substitute for the multiplicity of arcane chemicals that bacteriologists have pressed into service to generate "fingerprints." Bacterial taxonomists now emphasize the need to adopt a "polyphasic" approach in the delineation of taxa (meaning that one should use a range of different techniques), but they agree that phylogenetic consistency is of fundamental importance.[82,132]

B. The Use of Ribosomal RNA Data for Reconstructing Phylogeny

Three approaches have been used in the study of bacterial relationships based on rRNA similarity: hybridization,[26] cataloging,[139] and sequencing.[139] In the first, labeled rRNA of one strain is hybridized with total DNA from each of a number of others, and the melting temperature of the hybrid molecule allows an estimate of similarity for each pairwise combination. In the second method, purified 16S rRNA is cleaved enzymatically, and the sequence of each of the resulting short oligonucleotides is determined; the "catalog" of such oligonucleotides is characteristic for each organism and can be compared with all previously known catalogs. The third approach is complete or partial sequencing of the rRNA, or of the DNA that encodes it, followed by sequence alignment and comparison. Catalogs have been determined for the widest range of organisms, and the classification adopted in this chapter is based primarily on this method. Hybridization data are available for a much larger number of isolates and species, however, and provide a more detailed picture, especially of the proteobacteria. Four rRNA superfamilies have been defined by hybridization; these correspond to groupings of the proteobacteria ("purple bacteria") based on catalogs: Superfamily I is the enteric/vibrio branch of the gamma subdivision; II is the pseudomonad branch of the gamma; III is the beta; and IV is the alpha subdivision.

One disadvantage of hybridization is that it depends on pairwise comparisons: Every new strain has to be compared experimentally against each of a large (and increasing) number of others before it can be reliably located. Sequence data, by contrast, are determined just once for each strain; all comparisons can then be made by computer against a cumulative database of sequences. Sequencing rRNA or the corresponding genes was formerly very time-consuming, but recent advances have made it very much easier, and it is now the method of choice for most purposes. Clearly, sequence data will provide more detailed and reliable information than the other two approaches, but although published sequences are accumulating very rapidly, it will be some time before they supplant the existing, very extensive, hybridization and catalog data, and all three are used in the following account. Besides the 16S rRNA, comparative sequencing of the 5S rRNA molecule has also been used to construct phylogenies. Since this molecule is only about 120 nucleotides long, these phylogenies are not always reliable, but in general they correspond reasonably well with those based on 16S rRNA, and they can sometimes provide useful extra insights.[129]

The computational techniques used to align sequences and to reconstruct phylogenies from them are themselves undergoing continual development, and there is no consensus on a single correct method. The studies cited here have used a number of different approaches, and the details of some trees may depend on the methods chosen, and will certainly depend on the sample of sequences chosen for analysis. There is general agreement, though, on the broad outlines of the classification presented here.

II. Nitrogen-Fixing Organisms and Their Evolutionary Relationships

A major consequence of the study of molecular evolution has been the recognition of the Archaeobacteria as a separate kingdom, and the subdivision of the remaining bacteria (the Eubacteria) into a number of broad lineages. Although some details will doubtless be revised in the light of further evidence, these groups represent the current state of our understanding of the natural relationships among bacteria, and will form the basis for the arrangement presented in this chapter. Woese calls these eubacterial groups "phyla,"[139] but they, and their subdivisions, do not for the most part have any formal standing in bacterial taxonomy. There are formal names for major groups of bacteria, of course, but since these do not always correspond to discrete branches of the rRNA tree, I will not use them here. Instead, I will follow Woese's nomenclature, except that I will use informal vernacular versions of the formal names for the "purple bacteria" (proteobacteria[118]) and Gram positives (firmibacteria and thallobacteria[81]).

The relationships among the major groups of bacteria are indicated in Figure 1-1. It is evident from this figure that N_2 fixation is a very widely distributed property, although all the groups also include many nonfixing genera. In this chapter I aim to list all known N_2-fixing species, arranging them in their evolutionary groups. I will not attempt to chart the evolution of N_2 fixation, or the reasons for it; these issues are broached in Chapter 12 and elsewhere.[97] The full list of species is in Table 1-1, but this should be read in conjunction with the following text sections, in which many of the relationships are discussed in more detail. Names that have not been validly published (in the *Approved Lists of Bacterial Names*[113] or subsequently in the *International Journal of Systematic Bacteriology*) are shown in quotes

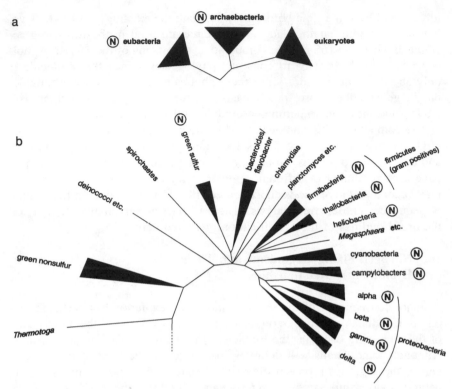

Figure 1-1. (a) The tree of all known organisms, showing the relationships of the three primary kingdoms deduced from small subunit (16S-like) rRNA sequences.[139] The root is thought, from other evidence,[54] to lie between the eubacteria and the other kingdoms. (b) Major phylogenetic groups of the eubacteria, based on 16S rRNA sequences[139] (not strictly to scale). Campylobacters are placed (tentatively) by their apparent affinity with proteobacteria.[63] Groups that include nitrogen-fixing organisms are marked (N).

("   "), whereas brackets ([   ]) indicate that the organism is not related to the type species of the genus to which it has been assigned.

The four volumes of *Bergey's Manual of Systematic Bacteriology*[67,114,120,138] form an authoritative source for information on the nomenclature, relationships, and properties of all bacteria. I shall frequently cite the appropriate volume of the *Manual* rather than the original publication, to keep the bibliography within bounds and to avoid the misleading implication that I have read all the primary literature.

There is now phylogenetic information of some sort on virtually all known diazotrophs, and the only organism I have been totally unable to place is *Propionispira arboris*.[108] It was isolated from wetwood disease of living cottonwood trees (*Populus deltoides*) and is a Gram-negative diazotrophic anaerobe with nonsporing, curved rod-shaped cells and a very low genomic

Table 1.1   The Known Nitrogen-Fixing Organisms Arranged in Phylogenetic Groups

		Literature References[a]		
Genus	Species	Name	Phylogeny	Fixation
Green sulfur bacteria				
Chlorobium	*limicola* (also *limicola* f. sp. *thiosufatophilum*)	120	139	120
	phaeobacteroides	120	129	94
	vibrioforme	120	139.	120
Chloroherpeton	*thalassium*	120	139	120
Pelodictyon	*luteolum*	120	120	50,95
Prosthecochloris	*aestuarii*	120	43	122
Firmibacteria				
Bacillus	*azotofixans*	111	111	111
	macerans	114	114	114
	polymyxa	114	114	114
Clostridium	*aceticum*	114	139	107
	acetobutylicum	114		107
	arcticum	114		114
	beijerinckii	114		107
	butylicum	114		107
	butyricum	114	139	114
	felsineum	114		107
	formiaceticum	114		114
	kluyveri	114		107
	"*lactoacetophilum*"[b]			107
	"*madisoni*"			107
	pasteurianum	114	139	114
	"*pectinovorum*"			107
	"*saccharobutyricum*"			95
	"*tetanomorphum*"	114		107
	tyrobutyricum	114		95
Desulfotomaculum	*orientis*	114	28	93
	nigrificans	114	28	84
	ruminis	114	28	93
Thallobacteria				
Arthrobacter	sp. "*fluorescens*"		117	18,19
Frankia	spp.	138	48	138
Streptomyces	spp.		117	32,63
Propionibacterium	*freudenreichii*	114	48	8
	jensenii	114		8
	(*peterssonii* is now *jensenii*)			
	(*shermanii* is now *freudenreichii*)			
Heliobacteria				
"*Heliobacillus*	*mobilis*"	11	11	11
Heliobacterium	*chlorum*	120	139	120
"*Heliospirillum*	*gestii*"	11		50
Cyanobacteria		120	139,46	120,104[c]
Cyanothece group	(Section I)			
Gloeocapsa group	(Section I)			
Gloeothece	(Section I)			

Table 1.1   Continued

Genus	Species	Literature References[a]		
		Name	Phylogeny	Fixation
Synechococcus group	(Section I)			
Synechocystis group	(Section I)			
Chroococcidiopsis	(Section II)			
Dermocarpa	(Section II)			
Myxosarcina	(Section II)			
Pleurocapsa group	(Section II)			
Xenococcus	(Section II)			
Lyngbya	(Section III)			
Oscillatoria	(Section III)			
Pseudanabaena	(Section III)			
Spirulina	(Section III)			
(Trichodesmium included in *Oscillatoria)*				
Anabaena	(Section IV)			
Aphanizomenon	(Section IV)			
Calothrix	(Section IV)			
Cylindrospermum	(Section IV)			
Nodularia	(Section IV)			
Nostoc	(Section IV)			
Scytonema	(Section IV)			
Chlorogloeopsis	(Section V)			
Fischerella	(Section V)			
Geitleria	(Section V)			
Stigonema	(Section V)			
Prochloron	*didemni*	120	127	120
Campylobacter				
Campylobacter	*nitrofigilis*	79	124	79
Proteobacteria: alpha subdivision				
Acetobacter	*diazotrophicus*	45	45	45
Agrobacterium	*tumefaciens*	67,27	134,140	59,60
Ancylobacter	*aquaticus*	101	17	77
[Aquaspirillum][c]	*fasciculus[d]*	67		67
	itersonii	67	139	67
	magnetotacticum[d]	120		120
	peregrinum[d]	67		67
Azorhizobium	*caulinodans*	34	34	34
Azospirillum	*amazonense*	38	38	38
	brasilense	67	140	67
	halopraeferens	103	103	103
	lipoferum	67	103	67
Beijerinckia	*derxii*	67		67
	fluminensis	67		67
	indica	67	34	67
	mobilis	67		67
Bradyrhizobium	*japonicum*	58	51,55	67
	sp. (other hosts)	67	51,55	67
"[Chromobacterium]	*folium"*	44	44	14
"Methylocystis	*echinoides"[d]*	120		120
	"parvus"		126	106,83

Table 1.1   Continued

Genus	Species	Literature References[a]		
		Name	Phylogeny	Fixation
"Methylosinus	*sporium"*			83
	"trichosporium"	67	126	83
(Microcyclus aquaticus				
and *M. eburneus* are				
now *Ancylobacter aquaticus)*				
Mycoplana	*bullata*	114	27	90a
	dimorpha	114	27	90a
"Photorhizobium	*thompsonum"*	35a,37	144a	35a,37
"[Pseudomonas]	*azotocolligans"*	30	30	30
[Pseudomonas]	*diazotrophicus*	131	136	131
[Pseudomonas]	*paucimobilis*	30	30	9
"Renobacter	*vacuolatum"*	72	17	77
Rhizobium	*fredii*	109	55,133	109
	galegae	74	55	74
	leguminosarum	67	140	67
	loti	56	55	56
	meliloti	67	51	67
Rhodobacter	*adriaticus*	53,120		120
	capsulatus	53,120	140	76
	sphaeroides	53,120	140	76
	sulfidophilus	53,120		76
	veldkampii	120		120
Rhodomicrobium	*vannielii*	120	140	76
Rhodopila	*globiformis*	53	140	76
Rhodopseudomonas	*acidophila*	120	140	76
	blastica	120	120	76
	marina	120,78		78
	palustris	120	140	76
	rutila	120	120	120
	sulfoviridis	120	120	76
	viridis	120	140	76
Rhodospirillum	*fulvum*	53	126	76
	molischianum	53	140	76
	photometricum	53	140	76
	rubrum	53	140	76
(Sinorhizobium		23		
see *Rhizobium)*				
Xanthobacter	*agilis*	57,5		57
	autotrophicus	67		67
	flavus	67	34	67
Proteobacteria:				
beta subdivision				
Alcaligenes	*faecalis*	67	67	21
	latus	67	67	16
	paradoxus	67	67	65
"Azoarcus"	spp.	103a	103a	103a
Derxia	*gummosa*	67	29	67
Herbaspirillum	*seropedicae*	7	103	7
"Lignobacter"	sp.		27	25
[Pseudomonas]	*saccharophila*	67	67,30	10

Table 1.1   Continued

Genus	Species	Literature References[a]		
		Name	Phylogeny	Fixation
Rhodocyclus	*gelatinosus*	53,120	142	76
	tenuis	53,120	142	76
Thiobacillus	*ferrooxidans*	120	71	75,100
Proteobacteria:				
gamma subdivision				
Amoebobacter	*roseus*	120		95
Azomonas	*agilis*	67	26	67
	insignis	67	26	67
	macrocytogenes	67	26	67
Azotobacter	*armenaicus*	67	26	67
	beijerinckii	67	26	67
	chroococcum	67	26	67
	nigricans	67	26	67
	paspali	67	26	67
	vinelandii	67	26	67
Beggiatoa	*alba*	120	119	120
Chromatium	*gracile*	120		120
	minus	120		120
	minutissimum	120		120
	vinosum	120	141	120
	violascens	120		120
	warmingii	120	141	120
	weissei	120	141	120
Citrobacter	*freundii*	67	2	67
Ectothiorhodospira	*shaposhnikovii*	120	141	120
	vacuolata	120		120
Enterobacter	*aerogenes*	67	67	67
	agglomerans (Pantoea?)	67,42	67,42	99
	cloacae	67	67	67
Erwinia	*herbicola (Pantoea?)*	67,42	67,42	89
(Escherichia intermedia is				
now *Citrobacter freundii*)				
Klebsiella	*pneumoniae*	67	67	
	oxytoca	67	67	62,67
	planticola	67	67	
	terrigena	67	67	
Lamprobacter	*modestohalophilus*	120		120
"*Methylobacter*	*capsulatus*" Y^d	67	17	86[e]
Methylococcus	*capsulatus*	67	126	106,83
	luteus	67		106
	thermophilus	67		106
	ucrainicus	67		106
Methylomonas	*methanica*	67	126	106
	"*rubra*"			106
(Pantoea see *Erwinia*				
and *Enterobacter*)		42		
Pseudomonas	*stutzeri*	67	67	9
Thiocapsa	*pfennigii*	120	141	120
	roseopersicina	120	141	120

Table 1.1   Continued

Genus	Species	Literature References[a]		
		Name	Phylogeny	Fixation
Thiocystis	*violacea*	120	139	120
Vibrio	*cincinnatiensis*	4		128
	diazotrophicus	67		67
	natriegens	67	67	135
	pelagius	67		128
Proteobacteria:				
delta subdivision				
Desulfobacter	*curvatus*	137,6	28	137
	hydrogenophilus	137,6	28	137
	latus	137,6	28	137
Desulfovibrio	*africanus*	67		84,98
	baculatus	67		84,98
	desulfuricans	67	28	84,98
	gigas	67	28	98
	salexigens	67	28	98
	thermophilus	67		84,98
	vulgaris	67	28	98
Archaeobacteria				
Halobacterium	*halobium*	120	139	92[e]
Methanobacterium	*formicicum*	120		92[e]
	ivanovii	5		92,115
	thermautotrophicum	120		92[e]
Methanococcus	*aeolicus*	120		120
	maripaludis	120	139,120	120
	thermolithotrophicus	120	139,120	13,120,115
	vannielii	120		92[e]
	voltae	120	139,120	116
Methanolobus	*tindarius*	120		64
Methanoplanus	sp.	120		92[e]
Methanosarcina	*barkeri*	120	139,120	80,120
Methanothermus	"*facilis*"			92[e]
	fervidus	120		92[e]
Unknown phylogenetic				
position				
Propionispira	*arboris*	108,3		108

[a]Three types of references are given: for identification and naming, for phylogenetic position (preferably based on 16S rRNA), and for evidence that the organism fixes nitrogen.

[b]" " indicates that the name has not been validly published; [ ] indicates that the organism is not related to the type species of the genus to which it is assigned.

[c]cyanobacteria in Sections IV and V have heterocysts and are therefore assumed to fix nitrogen, but this has not always been examined experimentally.

[d]The phylogenetic position of this organism is uncertain; it is listed, for convenience, with supposed relatives that belong in this group.

[e]DNA from this species hybridizes with *nif* gene probes, but there is as yet no direct evidence for nitrogen fixation.

G + C content (36.7%). Although it has been described in considerable detail,[108] I am unaware of any data that allow it to be assigned to a particular phylogenetic branch. All other known diazotrophs can be placed, at least tentatively, in one of the following categories.

A. Green Sulfur Bacteria

The green sulfur phototrophic bacteria form a distinct and coherent evolutionary branch,[139] and N_2 fixation has been demonstrated in several genera[120] (see Table 1-1). Some strains of both *Chlorobium limicola* and *C. vibrioforme* have the ability to utilize thiosulfate; these are known as forma specialis *thiosuphatophilum*.[120] These species are genetically heterogeneous (by DNA–DNA hybridization[120]) and probably merit subdivision.

Incidentally, the green bacterium *Chloroflexus aurantiacus* is not related to this group,[43,139] nor does it fix N_2.[49] It represents a group commonly known as "green nonsulfur" (see Fig. 1-1) but more accurately described as "multicellular filamentous green bacteria."[120]

B. Firmibacteria

The classical division of bacteria into Gram positive and Gram negative reflects a half-truth, and the Gram positives are the half that is true. They really do form a coherent phylogenetic group, and this is divided into two subgroups with widely different genomic G + C (guanine plus cytosine) contents. Murray calls these subgroups the classes *Firmibacteria* and *Thallobacteria,* forming together the division *Firmicutes*.[81] These are convenient names and correspond to real evolutionary branches, so I will adopt them here. However, I will use them as vernacular terms rather than to indicate a specific formal rank.

The firmibacteria, or low-G + C Gram-positive bacteria, are dominated by *Bacillus* and *Clostridium*. Each of these genera has a large number of named species, a minority of which have been shown to fix N_2. At present, the internal phylogenies of these genera are not clearly established, so it is not possible to say whether fixation is widespread or restricted in distribution. Extensive surveys of 16S rRNA sequences are under way in various laboratories but are not yet published.

The known N_2-fixing species of *Bacillus* are *B. polymyxa, B. macerans,* and *B. azotofixans,* but there are also some unnamed strains that show little DNA homology with these species.[110] N_2-fixing clostridia have been placed in quite a few different species (see Table 1-1).

Desulfotomaculum also belongs here; it is not related to the other sulfate-reducing anaerobes, which are in the delta subdivision of the proteobacteria. Like them, though, it includes some N_2-fixing strains.

C. Thallobacteria

The thallobacteria, or high-G + C Gram-positive bacteria, are the actino-mycetes: streptomycetes, mycobacteria, nocardias, and corynebacteria, among others. N_2 fixation has been demonstrated in *Streptomyces*[32,63] and in *Ar-throbacter,*[18,19] although there is no information on the genetic relationships between these N_2-fixing strains and type strains of the genera. Diazotrophy has also been reported in some species of *Propionibacterium,*[8] though tests on other strains proved negative.[94] Better known is the N_2-fixing capability of *Frankia* in symbiotic root nodules formed on plants in various dicot fam-ilies. Sequences of 16S rRNA indicate that *Streptomyces, Arthrobacter, Propionibacterium,* and *Frankia* are not closely related,[48,117] so it is plausible that N_2 fixation may eventually be demonstrated in other actinomycetes.

All actinorhizal symbiotic bacteria are at present placed in the single ge-nus *Frankia,* which is closely related (by 16S rRNA sequence) to *Geoder-matophilus* and *"Blastococcus".*[48] No species names are recognized in *Fran-kia* except the type species *F. alni,* and even this is not clearly delimited.[138] Earlier, some species names were assigned on the basis of host range, but this criterion is seen to be inadequate now that pure cultures are available.[138] Isolates from three broad host compatibility groups (*Alnus,* Elaeagnaceae, and Casuarinaceae) are indeed unrelated by DNA hybridization, but isolates within each group are not necessarily related either, and nine "genomic spe-cies" have been defined, with the likelihood of many more as other strains are examined.[40] Lalonde et al.[69] have summarized a great deal of information relevant to the taxonomy of *Frankia,* but their proposals to retain *F. elaeagni* and to name two subspecies of *F. alni* are probably premature in view of our rapidly growing knowledge of this genus. The biology of *Frankia* is discussed in Chapter 7.

D. Heliobacteria

Woese called this group the photosynthetic subdivision of the Gram-positive bacteria,[139] but since this is ungainly, since we are urged these days to use *phototrophic* rather than *photosynthetic,* and since the organisms in question do not have Gram-positive cell walls anyway, I propose to call them simply *heliobacteria.* They are indeed related to the Gram positives, but they are phototrophic, and their chlorophyll pigments are distinctly different from those of their more distant cousins the cyanobacteria. The three named spe-cies—*Heliobacterium chlorum, "Heliobacillus mobilis",* and *"Heliospiril-lum gestii"*—were isolated from soil under anaerobic conditions that se-lected for photoheterotrophs able to fix N_2. Although they have been placed in separate genera on the basis of morphological and physiological differ-ences, their 16S rRNA sequences are very similar.[11]

E. Cyanobacteria

The cyanobacteria, the *Prochlorales,* and the chloroplasts of eukaryotes are all oxygen-evolving phototrophs, and it is clear that they form a distinct and coherent phylogenetic group.[127,139] Unfortunately, that is about the only thing that is clear about the classification of cyanobacteria.

A major hindrance to the development of a workable taxonomy for the cyanobacteria is that most of the names were assigned when the cyanobacteria (blue-green algae) were regarded as plants. Cohn suggested more than a century ago that they would be better grouped with bacteria, but even now, with overwhelming evidence that they are true bacteria, the transfer of their nomenclature is not complete. The fundamental problem is that the Botanical Code requires type specimens to be dead, whereas the Bacteriological Code requires type cultures to be alive. Since none of the old botanical names correspond to living cultures, it is not possible to use them directly as valid bacterial names, nor, more important, to use modern bacteriological and molecular techniques to ascertain the biological relationships of the species they represent. Actually, modern molecular biology does provide a potential approach to this problem, in that the polymerase chain reaction has been used to amplify and sequence nucleic acids from dead specimens,[88] but it would be an unrewarding task to attempt to resurrect all the blue-green algal species by this means.

While the taxonomists attempt to resolve their impasse,[125] experimentalists need a workable nomenclature to describe the cultures they are using. This was provided by Rippka et al.,[104] who classified 178 strains into 22 genera, distinguished by morphological and developmental criteria. They did not attempt to divide these genera into species. This classification was intended to be provisional, but although it has been updated,[105,120] it is still the basis for practical purposes today. Now that cyanobacteria have finally been accepted as true bacteria, as attested by their inclusion in *Bergey's Manual of Systematic Bacteriology,*[120] they would be a worthy test case for the powerful methods of modern phylogenetic bacterial systematics.

The genera of Rippka et al.[104] are based on a small number of morphological and developmental characters, and these might not always reflect evolutionary relatedness. The molecular phylogeny that can be constructed from 16S rRNA sequences is now extensive but still far from comprehensive (Fig. 1-2[46]). It has many deep branches, and the relative order of these branches must be regarded as uncertain, since the addition of new sequences to essentially the same data set leads to a certain amount of rearrangement.[127] Sections II, IV, and V of Rippka et al. maintain their identity in this phylogeny, but I and III are clearly heterogeneous (see Figure 1-2). Sections IV and V are defined by the presence of heterocysts, and in these sections N_2 fixation is, almost by definition, universal and aerobic. N_2 fixation can

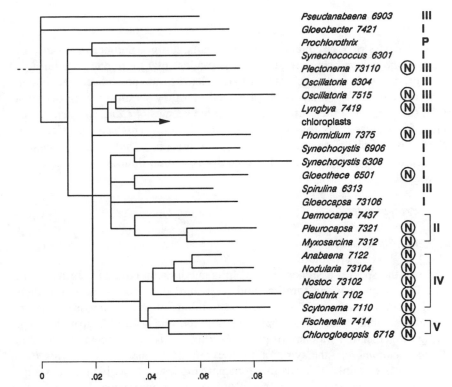

Figure 1-2. Phylogeny of some cyanobacteria[46] and *Prochlorothrix hollandica*[127] based on 16S rRNA sequences. Pasteur Culture Collection (PCC) strain numbers, nitrogen-fixing ability (N), and classification into Sections I to V are taken from Rippka et al.[104] Scale units are fixed point mutations per sequence position. Some short, ambiguous branch segments have been amalgamated.

also be demonstrated in most of the other genera, though not in all strains and usually only under microaerobic conditions.[104,120] Its phylogenetic distribution appears haphazard on the basis of our current knowledge (see Figure 1-2).

Prochloron didemni is the name given to phototrophic obligate symbionts of certain marine invertebrates (sea squirts), and *"Prochlorothrix hollandica"* is a free-living freshwater filamentous bacterium.[120] Like chloroplasts, they contain chlorophyll *b* as well as chlorophyll *a,* and for this reason they have been separated from the (other?) cyanobacteria into the order Prochlorales.[120] Howwever, their relationship to each other, and to chloroplasts, and their precise phylogenetic position within or alongside the cyanobacteria, are all uncertain at present.[127] There is a report of C_2H_2 reduction, and hence presumably N_2 fixation, in a *Prochloron* symbiotic system.[120]

F. Campylobacters

Now that 16S rRNA sequence data are available for a number of species of *Campylobacter* and some relatives, it is evident that these bacteria define a new deep branch on the eubacterial tree, related to the proteobacteria, though rather distantly.[90,124] It has been proposed that they should be regarded as a fifth subdivision of the proteobacteria.[82] One species of *Campylobacter*, isolated from grass roots, has been shown to fix N_2.[79] This species, *C. nitrofigilis*, is definitely on the *Campylobacter* branch,[124] but it is not closely related to the type species, and eventually it will no doubt be renamed as a new genus.

G. Proteobacteria

i. Alpha Subdivision

Most of the species assigned to the alpha subdivision have the ability to fix N_2. They include many familiar and well-studied N_2 fixers, although their phylogenetic arrangement may be less familiar, as it cuts across the traditional assemblages of phototrophs, symbionts, and so on. A grouping that brings together the methanotrophic *"Methylocystis"* with the phototrophic *Rhodopseudomonas,* the symbiotic *Bradyrhizobium,* and the free-living *Beijerinckia* is certainly a departure from past modes of thought, but it is supported by comparisons of the 16S rRNA of these organisms (Figure 1-3).

Most of the literature on these organisms is still organized along traditional lines, so in the following paragraphs I shall consider first the phototrophic bacteria and then the legume symbionts, the methylotrophs, and other bacteria. However, it is important to keep in mind the relationships shown in Figure 1-3, and to realize that in many aspects of their genetics, biochemistry, and physiology, we might expect *Rhizobium* to resemble *Agrobacterium* more than it does *Bradyrhizobium, Rhodopseudomonas* to resemble *Bradyrhizobium* rather than *Rhodobacter;* and so on. These relationships should be evident in the N_2-fixation genes too, unless these have been subject to extensive lateral transfer.

Nearly all the phototrophic purple nonsulfur bacteria fix N_2,[76] and most are in the alpha subdivision of the proteobacteria (though a few are found in the beta subdivision). They are not a particularly tight group, in terms of genetic relatedness, and all of them have nonphototrophic relatives (see Figure 1-3). The current nomenclature is due to Imhoff, Trüper, and Pfennig,[53] who improved the previous classification considerably by creating the new genera *Rhodobacter* and *Rhodopila* for some species formerly called *Rhodopseudomonas,* and moving *Rhodopseudomonas gelatinosa* and *Rhodospirillum tenue* into *Rhodocyclus* (in the beta subdivision). However, the gen-

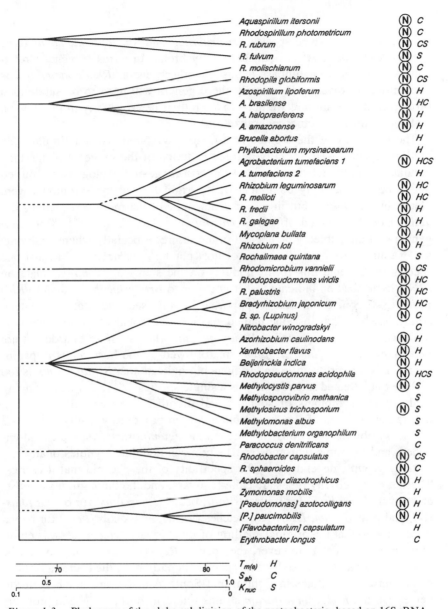

Figure 1-3. Phylogeny of the alpha subdivision of the proteobacteria, based on 16S rRNA similarity. The outline of the tree is based on catalog data[51,139,140] (C), with additional data from hybridization[24,27,30,34,38,40,55,103] (H) and sequencing[126,134] (S). Dotted lines indicate uncertainty of placement or inconsistency between methods. Some adjustments have been made to reconcile the data, but the scales indicate approximate DNA-rRNA hybrid melting temperature $T_{m(e)}$ (in °C), oligonucleotide catalog similarity S_{ab}, and fixed point mutations per sequence position on each branch K_{nuc}. The tree includes all those nitrogen-fixing species (N) whose relative position can be ascertained, but some non-fixing species are omitted.

era *Rhodospirillum* (type species *Rsp. rubrum*) and *Rhodopseudomonas* (type species *Rps. palustris*) are still extremely broad: In terms of genetic relatedness, they each encompass a host of other genera. *Rhodobacter* too is uncomfortably diverse. Eventually these genera will need to be subdivided further; in the meantime, we need to keep their heterogeneity in mind when making comparisons and generalizations.

The taxonomy of the legume root-nodule symbionts is in a similar state of flux. Species were initially defined in terms of the range of plants they nodulated, but the relationship between host range and phylogeny has turned out to be unexpectedly complicated.[145] Originally, all the symbionts were placed in *Rhizobium*, but in recent years some of the more conspicuous deviants have been hived off (*Bradyrhizobium*,[58] *Azorhizobium*[34]). More new genera will undoubtedly be needed in the future, especially when symbionts are examined from a wider range of host plants. *Azorhizobium* has just one species so far, and this forms nodules on the stems of *Sesbania*. Data on rRNA hybridization indicate that it is closer to *Bradyrhizobium* than to *Rhizobium*,[34] a conclusion that is supported by *nifH* sequence data (see later). *Bradyrhizobium* has just one named species too: Isolates that nodulate soybean (*Glycine max*) are called *B. japonicum*, whereas those that do not are known simply as *Bradyrhizobium* sp. followed by the host genus in parentheses. DNA hybridization and sequencing indicate that *B. japonicum* is too diverse to be treated satisfactorily as a single species,[52,121] but a wide-ranging study of the symbionts of soybean and of other legumes will be needed before new species of *Bradyrhizobium* can be proposed usefully.

The symbiotic functions of *Rhizobium leguminosarum* are plasmid-encoded, and the same (chromosomal) strain can nodulate legumes in any one of three groups, depending on the specificity of the plasmid that it carries. These host-specificity types are called biovars, and the three groups of hosts are clovers (*Trifolium*) for biovar *trifolii*, beans (*Phaseolus*) for biovar *phaseoli*, and peas (*Pisum, Lathyrus*), vetches (*Vicia*), and lentils (*Lens*) for biovar *viciae*. (Note that the original spelling of this biovar, *"viceae"*,[67] was etymologically incorrect.) However, the species *R. leguminosarum* may not be the only symbiont of those legume species; in particular, the bacteria isolated from *Phaseolus* nodules in Central and South America are much more diverse than the *R. leguminosarum* strains found in Europe[91,143] and probably should be regarded as a number of different species.

Alfalfa (*Medicago sativa*) and related legumes are nodulated by *R. meliloti*. This species, like *R. leguminosarum* biovar *phaseoli*, is more diverse at the center of diversity of its host plants (in this case the eastern Mediterranean) than elsewhere, but virtually all isolates fall into one or the other of two distinct genetic groups.[36] The familiar laboratory strains are all in one group, together with almost all the other strains isolated from alfalfa.

The other named species of *Rhizobium* are *R. loti,*[56] *R. fredii,*[109] and *R. galegae.*[74] Recently, a proposal has been made to move *R. fredii* (the fast-growing symbionts of soybeans) into a new genus, *Sinorhizobium.*[23] This proposal was based on a numerical taxonomic study that showed that fast-growing soybean isolates were less similar to *R. leguminosarum* than were *R. meliloti* or *Agrobacterium tumefaciens.* However, the evidence from DNA–DNA[133] and DNA–rRNA hybridization[55] is that *R. fredii* is quite closely related to *R. meliloti* and *R. leguminosarum,* and is certainly closer than either *R. loti* or *R. galegae* (see Figure 1-3). On that basis, these latter species are better candidates for new genera, but it is probably best to retain them all in *Rhizobium* for the present, acknowledging that *Agrobacterium* falls within the range of variation represented by this genus.

We have seen that symbiotic and phototrophic organisms are intermingled in the alpha subdivision (see Figure 1-3), so it would be exciting, but not too astounding, to find an organism that combined both properties. Such organisms are now known to exist: some N_2-fixing stem-nodule symbionts of the legume *Aeschynomene* have recently been shown to be phototrophic.[35a,37,67a,112] One of them, strain BTAi1, has been tentatively named *"Photorhizobium thompsonum"*[37] or *"P. thompsonianum,"*[35a] but DNA sequence data for part of the 16S rRNA gene show that it is very closely related to some strains of both *Bradyrhizobium japonicum* and *Rhodopseudomonas palustris,* and could appropriately be named *Bradyrhizobium* sp. *Aeschynomene.*[144a] It seems that we have here a cluster of organisms that includes symbionts, phototrophs, and phototrophic symbionts: a striking illustration of the need to update our traditional concept of these bacterial families.

In *Agrobacterium,* as in *Rhizobium,* the nature of the plant interaction is determined by plasmid-borne genes. Strains carrying Ti plasmids are called *A. tumefaciens* and cause crown galls, whereas those carrying Ri plasmids are called *A. rhizogenes* and cause hairy roots. Strains having neither type of effect are called *A. radiobacter.* Most, though not all, *Agrobacterium* strains fall into one of two clusters (or biotypes) on the basis of DNA and rRNA relatedness.[27,67] Ti plasmids are found in both of these clusters, but Ri plasmids appear to be confined to cluster 2.[67] Two cluster 1 strains, C58 and B6, have been reported to fix N_2.[59,60] In itself this is not surprising, in view of the close relationship with *Rhizobium,* but these strains show no hybridization with a *Klebsiella nifHDK* probe, which has usually picked up the homologous genes even in much more distant organisms. A similar circumstance has been recorded in *"Methylosinus"*[86] (see later). Further investigation is clearly needed, and may be facilitated by the reported cloning of DNA from *A. tumefaciens* C58 that confers apparent diazotrophy on *E. coli.*[59] Two further *Agrobacterium* strains, NCTC10590 and 18-1, also fixed N_2.[60] Both were 3-ketolactose positive, which is characteristic of agrobacteria in cluster 1, and these strains did hybridize with *nif* probes in the expected fashion.

Mycoplana is a soil bacterium with a branching morphology that led some workers to group it with actinomycetes. However, it is gram negative, and 16S rRNA hybridization indicates that it is a close relative of *Rhizobium*.[24,27] N_2 fixation has been demonstrated in both the known species.[90a]

Some plants in the families Rubiaceae and Myrsinaceae have leaf nodules that contain bacteria. Many authors have suggested that these might represent a N_2-fixing symbiosis, but there is no clear evidence for N_2 fixation *in planta*.[73] Bettelheim et al.[14] isolated bacteria that definitely did fix N_2 in culture and showed by immunofluorescence that these definitely corresponded to the bacteria found in leaf nodules. They identified them as *Chromobacterium lividum*, but this is definitely wrong. *C. lividum* (now *Janthinobacterium*) is the beta subdivision, whereas the organisms of Bettelheim et al. (also known as "*[Chromobacterium] folium*") are in the *Zymomonas* branch of the alpha subdivision.[44] They have not been compared with other relatives of *Zymomonas*, such as *[Pseudomonas] azotocolligans*. Knösel also isolated bacteria from leaf nodules, and called his isolates *Phyllobacterium*.[67] It is not known whether these are capable of N_2 fixation,[48] but they are certainly not the same as the Bettelheim organisms, because they are more closely related to *Agrobacterium*[27] and are specifically close to *Brucella*.[24] *Phyllobacterium* has recently been shown to be widespread and abundant in the rhizosphere of sugar beet; *nifHDK* probes did not hybridize detectably with DNA from these rhizosphere isolates.[70]

Methylotrophs are microbes that can obtain energy and carbon from reduced one-carbon compounds. Those bacteria that use methane (methanotrophs) have been divided into three groups (I, II, and X) on the basis of ultrastructure, metabolism, and base composition,[83] whereas those that use other C_1 compounds are much more diverse and include some species, such as *Rhodopseudomonas acidophila*, that are more commonly classified on the basis of other conspicuous properties. Methylotrophic bacteria have important uses in biotechnology, and their physiology has been extensively studied, although their phylogeny has remained obscure. New 16S rRNA sequence data[126] show that many methylotrophs belong in the alpha subdivision. Others fall into the beta subdivision (though none of these have been shown to fix N_2) and the gamma subdivision (see later). Nonpigmented type II obligate methanotrophs in the genera "*Methylosporovibrio*," "*Methylosinus*," and "*Methylocystis*"—together with the type I strain *Methylomonas albus* BG8—form a coherent group on the basis of 16S rRNA sequence data, and they are specifically related to *Rhodopseudomonas acidophila* (see Fig. 1-3). *Methylobacterium* species (pink-pigmented facultative type II methylotrophs) form a separate cluster that is also related to the *Rhodopseudomonas* branch, as was indicated by earlier DNA–rRNA hybridization data.[34] N_2 fixation[106] and *nifH*-related DNA hybridization[86] have been reported in *"Methylosinus"* and *"Methylocystis"*; the other genera have not been examined. Interestingly, one strain (*"Methylosinus trichosporium"* PG) readily fixed N_2 but showed no detectable hybridization with the *Klebsiella nifH* DNA probe,[86] a phenomenon also reported in certain *Agrobacterium* strains (see earlier). *Ancylobacter aquaticus* (which includes *A. eburneus*[72]) and *"Renobacter vacuolatum"* are facultative methylotrophs from fresh water and soil, characterized by a curved shape and by gas vacuoles in some strains. For 55 years *Ancylobacter* was called *Microcyclus,* until it was realized that this name had been preempted by a fungal pathogen of rubber trees, named in 1904.[101] Both *Ancylobacter* and *Renobacter* grow on nitrogen-free medium, though attempts to demonstrate nitrogenase activity were unsuccessful.[77] There is significant DNA homology between the two genera,[72] and their 5S rRNA sequences[17] place them on the *Nitrobacter winogradskyi* branch of alpha proteobacteria, which also includes *Bradyrhizobium* and *Rhodopseudomonas*.

The root-associated genus *Azospirillum* can be placed by 16S rRNA cataloging as a relative of *Rhodospirillum*.[140] This confirms earlier studies based on DNA-rRNA hybridization.[26] More recent DNA–rRNA hybridization studies have demonstrated that *Azospirillum brasilense* and *A. lipoferum* are closely related, whereas *A. amazonense* and *A. halopraeferens* are more distant.[103] *A. irakense* has been assigned to the genus on the basis of phenotype,[61] but this has not yet been confirmed at the genotypic level. Strains of *Conglomeromonas largomobilis* subsp. *largomobilis* were shown to fix N_2, but it was established in the same paper[39] that this species should be included in *Azospirillum lipoferum* (to the great relief, no doubt, of those who have to talk about it). *Herbaspirillum seropedicae*[7] was originally described as a species of *Azospirillum,* but it is in fact in the beta subdivision of the proteobacteria.[103]

Aquaspirillum is a genus of aerobic freshwater helical rods, but not all the species are closely related. *A. serpens,* the type species of its genus, is in the beta subdivision;[48] it does not fix N_2. [*Aquaspirillum*] *itersonii,* however, which does include N_2-fixing strains, is related to *Azospirillum* in the alpha subdivision.[139] The phylogenetic positions of the other N_2-fixing aquaspirilla (see Table 1-1) are unknown.[67]

DNA–rRNA hybridization studies show that *Xanthobacter* is a close relative of *Azorhizobium,*[34] *Beijerinckia* is a more distant member of the same cluster,[34] and *Acetobacter* is on a separate branch in the alpha subdivision.[44] All these bacteria are found in soil, and they all fix N_2.

The genus *Pseudomonas* has for many years been a dumping ground for bacteria of uncertain affinity. As a consequence, many of the bacteria bearing this name are totally unrelated to the true genus *Pseudomonas,* that is, to *P. fluorescens* and its immediate relatives in the gamma subdivision.[67]

By heroic efforts, the true affiliation of many of these bacteria has been determined,[29,30] but at the same time other authors persist in dumping new, poorly characterized isolates into the genus. Sometimes there are sufficient data to indicate the true nature of these strains. "[*Pseudomonas*] *azotocolligans*" and [*P.*] *paucimobilis*, which fix N_2, are on a branch of the alpha subdivision which also includes [*Flavobacterium*] *capsulatum* (itself misnamed!), and *Zymomonas*[30] (Figure 1-3). [*Pseudomonas*] *diazotrophicus*,[131] which was isolated from the roots of wetland rice, is shown by DNA–rRNA hybridization to fall within *Rhizobium;*[136] it also has an insertion sequence that is characteristic of *Rhizobium meliloti*.[136] [*Pseudomonas*] *saccharophila*[10] is in the beta subdivision.[29] The phylogenetic position of other root-associated diazotrophic "pseudomonads"[20,47] is less clear.

ii. Beta Subdivision

Only a minority of species in the beta subdivision has been demonstrated to fix N_2. The three species of *Rhodocyclus* are in this subdivision, but the type species, *R. purpureus*, is almost the only purple phototroph that has not been shown to fix N_2.[76] The other two species, *R. tenuis* (formerly *Rhodospirillum*[53]) and *R. gelatinosus* (formerly *Rhodopseudomonas*[53]) do fix N_2[76] but are not closely related to each other[142] (Figure 1-4).

Other N_2 fixers in this subdivision are *Derxia* (gummy aerobes from tropical soils) and some species or strains of *Alcaligenes* and [*Pseudomonas*] *saccharophila*. This latter species is related to [*P.*] *acidivorans* and [*P.*] *testosteroni* but not to the true *Pseudomonas* species, which are in the gamma subdivision.[30] Sundman[123] isolated some lignan-degrading bacteria from soils rich in decaying wood. She called them *Agrobacterium*, but in fact they are not related to this genus.[27] One of these, now known as "*Lignobacter*" strain K17, is related to *Alcaligenes*[27] and has a plasmid-encoded *nif* gene cluster.[25]

Herbaspirillum seropedicae is a root-associated azospirillumlike organism but is said to belong to rRNA superfamily III, which is equivalent to the beta subdivision, though more detailed information has not been published.[103] "*Azoarcus*" is another root-associated diazotroph; a number of related but distinct strains were isolated from the interior of kallar grass (*Leptochloa fusca*) roots in Pakistan.[103a] DNA–rRNA hybridization[103a] and 16S rRNA sequence data (B. Reinhold-Hurek, personal communication) show that *Azoarcus* is in the beta subdivision but is not closely related to *Herbaspirillum* or to any other characterized genus.

Thiobacillus is the name given to bacteria that derive energy from the oxidation of sulfur compounds. If they also use ferrous compounds as electron donors, they are called *T. ferrooxidans*. A number of strains of *T. ferrooxidans* are known to fix N_2[75] or to have *nif* genes,[100] and one of these (ATCC19859) appears from its 5S rRNA sequence to be a peripheral member of the beta subdivision.[71] This does not mean that the other potential N_2

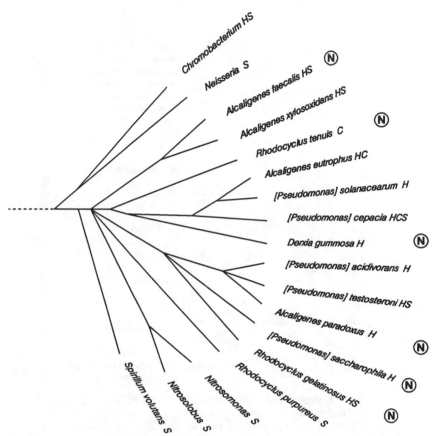

Figure 1-4. Phylogeny of the beta subdivision of the proteobacteria, based on 16S rRNA sequences[31] (S), with additional data from hybridization[29,30,67] (H) or catalogs[142] (C). The basal branch point corresponds approximately to a $T_{m(e)}$ of 68°C, an S_{ab} of 0.48, or a K_{nuc} of 0.06–0.08 (see legend to Figure 1–3). The tree includes most nitrogen-fixing species (N), but many non-fixers are omitted.

fixers are also in this subdivision: the name *Thiobacillus* has been applied to a very diverse range of bacteria that are scattered throughout the alpha, beta, and gamma subdivisions.[71,119] Even within the "species" *T. ferrooxidans,* some strains have no detectable DNA homology with each other.[120] It is not too surprising, then, that another strain of this "species" (ATCC33020) has a *nifH* gene sequence that most closely resembles that of *Bradyrhizobium,* which is in the alpha subdivision (*nifH* evolution is discussed later in this chapter).

iii. Gamma Subdivision

The gamma subdivision includes many well-studied bacteria,[141] and N_2 fixation has been reported in a number of widely separated groups: the enterics/

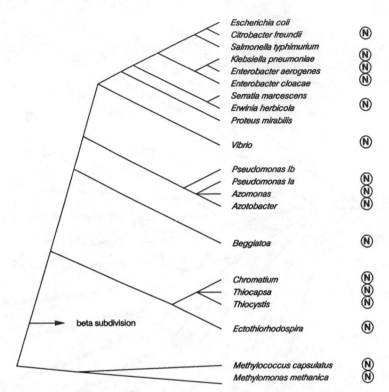

Figure 1-5. Phylogeny of the nitrogen-fixing organisms (N) in the gamma sub-division of the proteobacteria (not to scale). The internal branching of the enteric bacteria is deduced from biosynthetic enzyme envolution;[1,2] the kinship of pseudomonads, azomonads, and azotobacters is from DNA–rRNA hybridization[26] and enzyme evolution;[15] the relationship between enterics, vibrios, pseudomonads, beggiatoas, and purple sulfur phototrophs is from 16S rRNA catalogs;[141] and the outlying position of the methylotrophs is from 16S rRNA sequences.[126]

vibrios, the *Azotobacter/Azomonas/Pseudomonas* group, the purple sulfur phototrophs, *Beggiatoa,* and *Methylococcus/Methylomonas* (Figure 1-5).

The best-studied N_2-fixation genes are, of course, those of *Klebsiella pneumoniae* M5a1. This name is so well known that it seems churlish, and is certainly futile, to point out that M5a1, being indole positive, is not really a strain of *K. pneumoniae* at all, but rather of *K. oxytoca*.[67] Some *nif* gene studies have used an independently isolated strain, identified as *K. oxytoca* NG13, that is closely related to M5a1.[62]

Diazotrophy is found only in certain *Klebsiella* strains, however, and in some *Citrobacter;* in the related enteric genera *Escherichia* and *Salmonella* it has not been reported in any natural isolates, although diazotrophic strains have been created by gene transfer in the laboratory.[96] (N_2 fixation was reported in *Escherichia intermedia,*[95] but this is now regarded as a synonym of *Citrobacter freundii*).[67] Some natural strains of *Enterobacter aerogenes* and *E. cloacae* fix N_2.[67] These species have an unusual bifunctional enzyme in the tryptophan pathway that suggests they are specifically related to the *Escherichia/Salmonella/Klebsiella/Citrobacter* group.[1,2] Aerogenic strains of *Enterobacter agglomerans* also have this enzyme, but anaerogenic strains, including the type strain, do not. This type strain is very similar to the type strain of *Erwinia herbicola,*[12] and it has recently been proposed that these strains, and many others that have been assigned arbitrarily to one or the other of these virtually synonymous species, should be placed in the new genus *Pantoea,* in which two species have so far been defined.[42] N_2-fixing *Enterobacter agglomerans* have been found in the guts of termites,[99] and many isolates of *Erwinia herbicola* are able to fix N_2;[89] it is probable that all these, unless aerogenic, belong in *Pantoea.*

The genus *Vibrio* is related to the enteric bacteria. Representatives of 27 of the 31 named species have been examined for N_2 fixation, and it was detected in four species and some other, unclassified isolates, all of marine or saline water origin.[128,135]

The purple sulfur phototrophic bacteria form two related groups on the basis of 16S rRNA catalogs: *Chromatium* and its relatives (including *Thiocapsa, Thiocystis,* and presumably, *Lamprobacter*), and the species of *Ectothiorhodospira.* N_2 fixation has been demonstrated in the genera just mentioned and in *Amoebobacter,* and it is likely that it occurs in the others too.

The ability to fix N_2 is part of the definition of *Azotobacter,* which occurs in soil and forms cysts, and *Azomonas,* which is found in soil and fresh water and does not form cysts. These genera are closely related to each other and to *Pseudomonas* (the genuine genus this time!). They appear to be closest to the subgroup Ia (nonfluorescent) pseudomonads,[15] and some diazotrophic strains of one of these, *P. stutzeri,* have been reported.[9] N_2 fixation has not been demonstrated in any of the subgroup Ib (fluorescent) *Pseudomonas* species. Other "pseudomonads" that fix N_2 are, in fact, unrelated bacteria that belong in the alpha or beta subdivisions.

Beggiatoas look like colorless oscillatorias, and it was suggested at one time that they might be apochlorotic cyanobacteria (see ref. 119), but two different strains have now been shown by 16S rRNA cataloging[141] or 5S rRNA sequence[119] to belong among the gamma proteobacteria. One of these was the type strain of the only official species, *Beggiatoa alba.* This strain fixes N_2, as do various other strains, some of which probably represent other species as yet unnamed.[120]

Nitrogenase activity has been demonstrated in several species of *Methylococcus.*[106] *M. capsulatus* (Bath) is a type X obligate methanotroph with well-documented N_2-fixation ability;[83] its 16S rRNA sequence suggests that it is a very peripheral member of the gamma subdivision[126] and is specifically, but rather distantly, related to *Methylomonas methanica*, which is a type I methanotroph.[126] Some other strains of *M. methanica* have DNA homologous to *nifH*,[86] and nitrogenase has been demonstrated in both *M. methanica* and *"M. rubra"*.[106] Analysis of 5S rRNA sequences[17] corroborates the relationship between *Methylococcus capsulatus* and *Methylomonas methanica*, and indicates that *M. methanica* is a member of a fairly tight cluster of obligate methanotrophs that include *"M. rubra"*, *"Methylocystis parvus,"* *"Methylosinus trichosporium,"* and *"Methylobacter capsulatus"* strain Y. The dilemma that this poses is that 16S rRNA sequences place *"Methylocystis parvus"* and *"Methylosinus trichosporium"* unambiguously in the alpha subdivision.[126] They have, for example, the small loop between nucleotides 200 and 220 (*E. coli* numbering) that is characteristic of alpha sequences,[139] whereas *Methylomonas methanica* has a larger stem loop of the beta/gamma type. Since the analyses involve different strains, different genes, and different phylogenetic reconstruction methods, it is not possible to resolve this conflict without further study.

iv. Delta Subdivision

Most of the bacteria so far assigned to the delta subdivision are the obligately anaerobic sulfate reducers (excluding *Desulfotomaculum*, which is a firmicute). N_2 fixation has been reported in a number of species of *Desulfovibrio*[98] and *Desulfobacter*[137] (see Table 1-1). A phylogeny of the sulfate reducers has been published[28] and is shown in simplified form in Figure 1-6, with the addition of *Bdellovibrio*, which also falls in the delta subdivision,[139] as does *Myxococcus* and some related organisms. The short basal segments in the original figure are collapsed to a single branch point in Figure 1-6 because the branching order is not certain, as indicated by the inconsistent placement of *Myxococcus* within, or outside, the sulfur-reducing group.[28,139]

H. Archaeobacteria

The discovery of N_2 fixation in the archaeobacteria[13,80] is surprising. Indeed, it opens the possibility that the common ancestor of all current life forms was a N_2-fixing organism, if one accepts, first, that the presence of nitrogenase in both eubacteria and archaeobacteria represents common ancestry rather than lateral transfer or independent evolution and, second, that eubacteria split off from the archaeobacteria before the eukaryotes did[54] (see Figure 1-1). On an alternative view[68] of the branching pattern in early evolution, in which the archaeobacteria are regarded as two independent groups, one might surmise that nitrogenase arose slightly later: in the common ancestor of the methanogens/eubacteria branch after this separated from the eukaryotes/eocytes (sulfur-dependent archaeobacteria).[92]

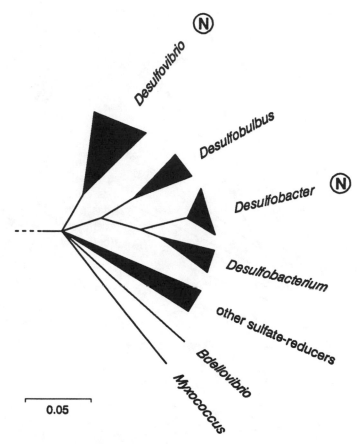

Figure 1-6. Phylogeny of the delta subdivision of the proteobacteria, based on 16S rRNA sequences. Simplified from reference 28, with uncertain branch points amalgamated and *Bdellovibrio* added.[139] Scale units are fixed point mutations per sequence position on each branch (K_{nuc}). Genera marked (N) include nitrogen-fixing species.

N_2 fixation has been detected in several disparate methanogens, and a number of others have DNA that hybridizes with *nif* gene probes,[92] so fixation is probably a general property of the methanogen group. By 16S rRNA analysis, the extreme halophiles are a group that originated within the methanogen tree (Figure 1-7), but N_2 fixation has not yet been reported in the halophiles, although some strains of *Halobacterium* do hybridize with *nif* gene probes.[92] No *nif* gene hybridization has been detected in the other main branch of the archaeobacteria, the sulfur-dependent thermophiles.[92] (This kingdom was originally called *archaebacteria,* and most authors still use that spelling. Strictly speaking, however, this is etymologically incorrect and the logically consistent spelling is *archaeobacteria.*[81])

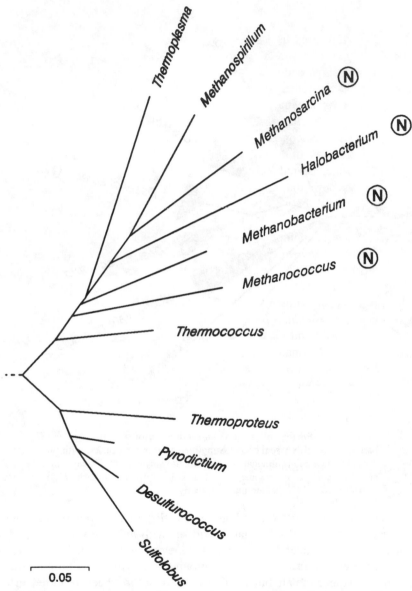

Figure 1-7. Phylogeny of the archaeobacteria, based on 16S rRNA sequences.[139] Scale units are fixed point mutations per sequence position on each branch (K_{nuc}). Nitrogen fixation or DNA hybridization with *nif* genes has been demonstrated in the genera marked (N), though not necessarily in the strains used to construct the tree.

III. The Phylogeny of *nifH* Genes

The classification used in this chapter has leaned very heavily on the ribosomal RNA genes. It would be good to be able to confirm that the evolutionary relationships deduced in this way were really applicable to the genome as a whole rather than to these genes alone. For many specific comparisons, the correctness of this phylogeny is borne out by other genetic, physiological, or biochemical data, but the only other genes for which extensive comparative sequence data are available across the whole gamut of diazotrophs are those of the N_2-fixing system itself, in particular *nifH*, which encodes dinitrogenase reductase. Does the phylogeny of *nifH* reflect the ancestry of the bacteria that now carry it, or must we postulate that it has been transferred laterally between widely diverged branches of the evolutionary tree?

The first serious attempt to answer this question was made by Hennecke et al.[51] Relatively few *nif* gene sequences were available at that time, but these authors showed that the available data supported the 16S rRNA tree. The *nifH* genes of *Rhizobium, Bradyrhizobium, Azotobacter,* and *Klebsiella* closely mirrored the known relationships among these genera, and those of *Anabaena* and *Clostridium* fell, as expected, well outside this cluster. Limited data from *nifD* and *nifK* were also consistent.

More recently, Normand and Bousquet[85] have analyzed a much more extensive data set consisting of 22 published *nifH* sequences from 18 different species. They concluded that *Frankia* and *Anabaena* probably acquired their *nifH* genes relatively recently by lateral transfer from a proteobacterium. They based this conclusion on the finding that the *nifH* of *Frankia* was most closely related to that of *Anabaena,* and the nearest relatives of this pair were found among the proteobacteria. They felt that, on a purely phylogenetic hypothesis, *Frankia nifH* ought instead to have resembled the *nifH* genes of *Clostridium,* another firmicute (Gram-positive bacterium). Unfortunately, the data analysis presented by Normand and Bousquet does not provide clear support for their hypothesis and is in fact equally consistent with explanations that do not postulate lateral transfer.

Before tackling this question, it should be pointed out that many features of the *nifH* tree are entirely consistent with the 16S rRNA phylogeny. All the branches within the alpha and gamma subdivisions of the proteobacteria, as well as the relationship between these subdivisions, are concordant. The only apparent anomaly in the proteobacteria is *Thiobacillus ferrooxidans*. The published *nifH* sequence[100] is unambiguously related to that of *Bradyrhizobium,* in the alpha subdivision, whereas 5S rRNA sequence places *T. ferrooxidans* on the edge of the beta subdivision.[71] The two reports refer to different strains, however, and the name *T. ferrooxidans* has been applied to such a diverse range of bacteria[120] that we should not assume that they really belong to the same species!

The first problem with Normand and Bousquet's evidence for lateral transfer of *nifH* is that the order of some of the critical branches is not defined unambiguously by the *nifH* data. This is illustrated by the standard error bars that the authors show on their cluster analysis (their Figure 5A[85]). Branch points with overlapping error bars are certainly not significantly different, and Figure 1-8 shows this dendrogram redrawn with these branch points amalgamated. This makes it clear that *Frankia* and *Anabaena* are not necessarily internal to the proteobacterial cluster. This point is also made by Normand and Bousquet's own analyses, which treat the same distance matrix by different methods: while the Fitch-Margoliash and Neighbor-Joining methods group *Frankia* with the alpha proteobacteria to the exclusion of the gammas, UPGMA groups it with the gammas, and the Kitch program groups the alphas and gammas without *Frankia*!

The evidence that the *nifH* genes of *Frankia* and *Anabaena* are specifically related is marginal, but if it is true it supports the 16S rRNA phylogeny, which suggests a specific but distant relationship between cyanobacteria and firmicutes. Normand and Bousquet assume that the clostridial sequences represent the "true" firmicute *nifH* and that the *Frankia* gene is therefore aberrant, but it is just as likely that the reverse is true. Indeed, if one extrapolates from the relative lengths of branches within the proteobacteria, one would expect a firmicute *nifH* gene to resemble the proteobacterial genes much more than the clostridial ones do. The clostridial *nifH*1 and *nifH*2 are highly diverged from other eubacterial *nifH* genes, suggesting either that they have been subject to unusually rapid evolution or that the clostridial genes branched off before the major eubacterial lineages were established.

This brings us to the second major difficulty that arises when we attempt to construct phylogenies of genes such as *nifH*: gene duplication. We know that duplication has occurred during the evolution of *nifH* because some organisms have more than one *nifH* gene. For phylogenetic reconstruction this raises the problem of distinguishing between orthologous comparisons, in which the genes are descended from the same gene in the last common ancestor, and paralogous comparisons, where they belong to different gene families that have arisen through gene duplication. Apparent discrepancies between the phylogeny of the gene and of the organism can arise from paralogy. For example, if the only "*nifH*" known from *Azotobacter vinelandii* was the gene from its third nitrogenase cluster (*anfH*), we would conclude that its phylogeny was grossly aberrant. In fact, this gene clusters with *Clostridium pasteurianum nifH*3 and *Methanococcus lithotrophicus nifH*2, and it is plausible that, as Normand and Bousquet[85] suggest, these genes represent a separate branch that arose by duplication in the common ancestor of eubacteria and archaeobacteria. These genes are so different from the

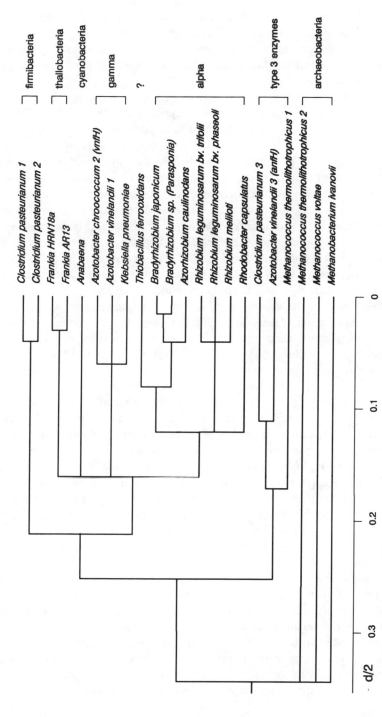

Figure 1-8. Phylogeny of NifH protein sequences calculated from the estimated pairwise amino acid substitutions per site (d) by the UPGMA clustering method. (Redrawn from Normand and Bousquet[85] after amalgamating branch points that are not significantly different.)

majority of known *nifH* genes that their paralogous status is obvious, but more subtle cases could easily be overlooked. Unfortunately, confirming orthology by other criteria, such as conserved chromosomal map position, will often not be possible when dealing with a system as phylogenetically widespread as nitrogenase.

A third factor that makes genes less than perfect as phylogenetic tools is the wide variation in G + C content of bacterial genomes. Clostridia have extremely low G + C contents: The third base positions of the *C. pasteurianum nifH*1 and *nifH*2 genes are only 19% G + C. *Frankia nifH,* by contrast, is at the opposite extreme: 94% G + C in the third position. Clearly, it is impossible for the *nifH* genes of these species to be closely related, and difficult even for amino acid sequences of the corresponding proteins. We do not really know why some organisms have evolved extreme G + C contents, or how fast such evolution can occur. Nevertheless, it is noteworthy, as Hennecke et al.[51] point out, that the G + C of the *nifH* genes is typical of the genomes in which they are found. This is consistent with the idea that these genes are an established component of the genome, and certainly does not suggest that they are recent acquisitions from foreign sources.

Threading together the arguments of the last few paragraphs, we can put forward an explanation of the current distribution of *nifH* genes that does not involve any lateral gene transfer. The common ancestor of eubacteria and archaeobacteria had a gene that gave rise both to the "type 3" *nifH* genes and, by one or more gene duplication events, to the other dinitrogenase reductase genes in both kingdoms. Early in the eubacterial branch, another duplication separated what we now think of as the "normal" eubacterial *nifH* genes from a lineage that is represented only (as far as our present knowledge goes) in *Clostridium pasteurianum,* where a series of subsequent duplications has given rise to a family of closely related genes: *nifH*1, *nifH*2, *nifH*4, *nifH*5, and *nifH*6.[130] (The functional dinitrogenase reductase gene in *C. pasteurianum* appears to be *nifH*1, and there is no indication that this species has anything more closely resembling the *nifH* genes of other eubacteria, but the extremely low G + C content makes hybridization studies difficult).[22] A recent duplication of *nifH* in the alpha proteobacteria has produced the *vnfH* gene associated with the vanadium dinitrogenase of *Azotobacter chroococcum.*

This hypothesis involves a number of arbitrary assumptions and is unlikely to be correct in every respect. That is not the point. The point is that gene duplication provides an alternative to the (equally arbitrary) assumption of lateral gene transfer in instances where the phylogeny of a gene does not seem to correspond to the phylogeny of the genome. The existence of duplicate genes proves that at least some gene duplication has been involved in the evolution of *nifH* genes. Although many experiments have demonstrated artificial transfer of *nif* genes between species (usually fairly closely related species or genera),[96] it is much harder to prove that long-range lateral transfer has played a role in the present distribution of *nif* genes. A convincing example would require the gene to be conspicuously "foreign" in its current environment (differing in G + C content or codon usage, for example), perhaps embedded in the remains of a transposable element, and identifiably derived from a particular group of putative donors. Only relatively recent events are likely to satisfy these criteria; ancient transfers are likely to remain speculative. Evidence for transfer cannot be claimed simply because genes from two organisms are, in the author's opinion, surprisingly similar. Both transfer and duplication are plausible events in a gene's ancestry, and in many cases the evidence needed to decide between them is not available.

One final point: The evolution of *nifH* is not synonymous with the evolution of the *nif* genes as a whole. This is apparent from the sequence comparisons of the nitrogenase genes and predicted gene products of *Azotobacter vinelandii*, given in Table 19-5 of this book. Although the dinitrogenase reductase (VnfH) associated with the vanadium dinitrogenase is closely similar to that of the molybdenum enzyme (NifH, 90% amino acid identity), the dinitrogenase enzymes themselves are grossly different (about 30% identity between VnfD and NifD, and between VnfK and NifK); indeed, the vanadium enzyme is much more like the alternative (type 3) dinitrogenase (AnfD and AnfK), and both of these have an extra subunit, VnfG or AnfG, that has no counterpart in the molybdenum enzyme. It appears that the vanadium gene cluster has recently "adopted" a duplicate (*vnfH*) of the molybdenum dinitrogenase reductase gene (*nifH*). This kind of trade between duplicate systems within a genome may complicate the story still further.

IV. Conclusion

Implicit in the title of this chapter is a view of the world that is manifestly false. N_2 fixation is found in the eubacteria and the methanogen/halophile part of the archaeobacteria; that is, in half of all life (see Figure 1-1). Within these groups, N_2 fixation has been reported or inferred in almost 100 genera distributed over most of the major phylogenetic divisions. There are good reasons for supposing that the genes involved have a very ancient origin, perhaps even in the ancestor of all today's organisms. In these circumstances, it is surely presumptuous to refer to "the classification of N_2-fixing organisms"; this would be not far short of a classification of all bacteria.

Of course, most bacteria have not been shown to fix N_2. On the other hand, very few have been shown definitively to be incapable of it. Most bacteria have simply not been examined: N_2 fixation is not one of the myriad of standard tests routinely applied when a new species is described. I suspect that if it had been, this chapter might have been impossibly long. There is a problem, of course, in that many diazotrophs do not show their ability except under special growth conditions: if *Rhizobium* had been isolated from soil no one would have suspected its talents. However, *nif* gene probes could be much more widely used as a preliminary screen; they have been successful, even over wide taxonomic ranges, in picking up putative homologies that have subsequently proved real.

If N_2 fixation is so widespread, why do we persist in thinking of it as a specialization? I think this stems from our "colicentric" view of bacteria: *Escherichia coli* K-12 does not fix N_2, therefore anything that does must be a specialist. Hence the feeling that one can treat the classification of N_2-fixing organisms as a self-contained discussion of a few specialized groups. No doubt there are many bacteria like *E. coli* K-12 that lead lives in which N_2 fixation is unnecessary or impossible, and so the loss of *nif* genes has probably been a frequent evolutionary event. (Postgate and Eady[97] discuss the apparently haphazard distribution of diazotrophy within genera.) The assumption in the past has been that most bacteria are incapable of N_2 fixation, and so it is not worth looking for it where it is not brazenly evident. In fact, the available evidence provides no grounds that justify this negative attitude over the opposite, positive one. If instead we begin to look on N_2 fixation as a normal attribute of bacteria, albeit often lost, new findings may fill in some of the blanks on the bacterial map, and the distribution of N_2 fixation may prove to be less patchy than it appears at present.

I hope that this chapter will be a useful source of information and references on diazotrophs, but more than that, I hope it will encourage bacterial physiologists, biochemists, and molecular geneticists to take a different perspective, to compare *Rhodocyclus* with *Derxia* as well as with *Rhodobacter,* and *Azorhizobium* with *Xanthobacter* as well as with *Bradyrhizobium.* It is only through an appreciation of the evolutionary context that we can hope to understand why organisms do things the way they do.

Acknowledgments

The fact that a phylogenetic approach to bacterial classification is now possible is largely due to Dr. De Ley, Dr. Woese, and the many colleagues who have shared their vision of 16S rRNA as a master key to unlock the secrets of history. I would also like to acknowledge all those authors whose research or contributions to *Bergey's Manual* I have cited only indirectly, and to apologize to them for the consequent damage to their citation records. I certainly could not claim to be an expert on all the bacteria I have discussed and must thank numerous colleagues whose generous comments have helped to fill some of the gaps: Eunice Allen, Dwight Baker, Paul Bishop, Bert Eardly, Claudine Elmerich, Dick Hanson, Hauke Hennecke, Susan Hill, Brion Jarvis, Michael Madigan, Colin Murrell, John Postgate, G. R. K. Sastry,

Janet Sprent, Hans Trüper, Liz Wellington, the editors, and an anonymous but very knowledgeable reviewer.

References

1. Ahmad, S., and Jensen, R. A. (1989). Utility of a bifunctional tryptophan pathway enzyme for the classification of the herbicola-agglomerans complex of bacteria. *Int. J. Syst. Bacteriol. 39,* 100–104.

2. Ahmad, S., Weisburg, W. G., and Jensen, R. A. (1990). Evolution of aromatic amino acid biosynthesis and application to the fine-tuned phylogenetic positioning of enteric bacteria. *J. Bacteriol. 192,* 1051–1061.

3. Anonymous (1983). Validation of the publication of new names and new combinations previously effectively published outside the IJSB. List No. 11. *Int. J. Syst. Bacteriol. 33,* 672–674.

4. Anonymous (1986). Validation of the publication of new names and new combinations previously effectively published outside the IJSB. List No. 20. *Int. J. Syst. Bacteriol. 36,* 354–356.

5. Anonymous (1988). Validation of the publication of new names and new combinations previously effectively published outside the IJSB. List No. 24. *Int. J. Syst. Bacteriol. 38,* 136–137.

6. Anonymous (1988). Validation of the publication of new names and new combinations previously effectively published outside the IJSB. List No. 26. *Int. J. Syst. Bacteriol. 38,* 328–329.

7. Baldani, J. I., Baldani, V. L. D., Seldin, L., and Döbereiner, J. (1986). Characterization of *Herbaspirillum seropedicae* gen. nov., sp. nov., a root-associated nitrogen-fixing bacterium. *Int. J. Syst. Bacteriol. 36,* 86–93.

8. Baranova, N. A., and Gogotov, I. N. (1974). Fixation of molecular nitrogen by propionic bacteria (in Russian). *Mikrobiologiya 43,* 791–794.

9. Barraquio, W. L., Dumont, A., and Knowles, R. (1988). A method for counting free-living aerobic N_2-fixing H_2-oxidizing bacteria. In: *Nitrogen Fixation: Hundred Years After* (H. Bothe, F. J. de Bruijn, and W. E. Newton, eds.). Stuttgart: Gustav Fischer, p. 790.

10. Barraquio, W. L., Padre, B. C., Watanabe, I., and Knowles, R. (1986). Nitrogen fixation by *Pseudomonas saccharophila* Doudoroff ATCC 15946. *J. Gen. Microbiol. 132,* 237–241.

11. Beer-Romero, P., and Gest, H. (1987). *Heliobacillus mobilis,* a peritrichously flagellated anoxyphototroph containing bacteriochlorophyll *g. FEMS Microbiol. Lett. 41,* 109–114.

12. Beji, A., Mergaert, J., Gavini, F., Izard, D., Kersters, K., Leclerk, H., and De Ley, J. (1988). Subjective synonymy of *Erwinia herbicola, Erwinia milletiae* and *Enterobacter agglomerans* and redefinition of the taxon by genotypic and phenotypic data. *Int. J. Syst. Bacteriol. 38,* 77–88.

13. Belay, N., Sparling, R., and Daniels, L. (1984). Dinitrogen fixation by a thermophilic methanogenic bacterium. *Nature 312,* 286–288.

14. Bettelheim, K. A., Gordon, J. F., and Taylor, J. (1968). The detection of a strain of *Chromobacterium lividum* in the tissues of certain leaf-nodulated plants by the immunofluorescence technique. *J. Gen. Microbiol. 54,* 177–184.

15. Byng, G. S., Berry, A., and Jensen, R. A. (1986). Evolution of aromatic biosynthesis and fine-tuned phylogenetic positioning of *Azomonas, Azotobacter* and rRNA group I pseudomonads. *Arch. Microbiol. 144*, 222–227.

16. Bowien, B., and Schlegel, H. G. (1981). Physiology and biochemistry of aerobic hydrogen-oxidizing bacteria. *Ann. Rev. Microbiol. 35*, 405–452.

17. Bulygina, E. S., Galchenko, V. F., Govorukhina, N. I., Netrusov, A. I., Nikitin, D. I., Trotsenko, Y. A., and Chumakov, K. M. (1990). Taxonomic studies on methylotrophic bacteria by 5S ribosomal RNA sequencing *J. Gen. Microbiol. 136*, 441–446.

18. Cacciari, I., Giovannozzi-Sermanni, G., Grapelli, A. and Lippi, D. (1971). Nitrogen fixation by *Arthrobacter* species. Part I. Taxonomic study and evidence of nitrogenase activity of two new strains. *Ann. Microbiol. Enzimol.* 21, 97–105.

19. Cacciari, I., Lippi, D., Pietrosanti, T. and Pietrosanti, W. (1986). Ammonium assimilation in an *Arthrobacter* sp. "fluorescens." *Arch. Microbiol.* 145, 113–115.

20. Chan, Y.-K., Wheatcroft, R. and Watson, R. J. (1986). Physiological and genetic characterization of a diazotrophic pseudomonad. *J. Gen. Microbiol.* 132, 2277–2285.

21. Chen, J.-M. and Ye, Z.-H. (1983). Transfer and expression of *Klebsiella nif* genes in *Alcaligenes faecalis,* a nitrogen-fixing bacterium associated with rice root. *Plasmid* 10, 290–292.

22. Chen, K. C.-K., Chen, J. S. and Johnson, J. L. (1986). Structural features of multiple *nifH*-like sequences and very biased codon usage in nitrogenase genes of *Clostridium pasteurianum. J. Bacteriol.* 166, 162–172.

23. Chen, W. X., Yan, G. H. and Li, J. L. (1988). Numerical taxonomic study of fast-growing soybean rhizobia and a proposal that *Rhizobium fredii* be assigned to *Sinorhizobium* gen. nov. *Int. J. Syst. Bacteriol.* 38, 392–397.

24. De Ley, J., Manneheim, W., Segers, P., Lievens, A., Denijn, M., Vanhoucke, M. and Gillis, M. (1987). Ribosomal ribonucleic acid cistron similarities and taxonomic neighborhood of *Brucella* and CDC group Vd. *Int. J. Syst. Bacteriol.* 37, 35–42.

25. Derylo, M., Glowacka, M., Skorupska, A. and Lorkiewicz, Z. (1981). Nif plasmid from *Lignobacter. Arch. Microbiol.* 130, 322–324.

26. De Smedt, J., Bauwens, M., Tytgat, R. and De Ley, J. (1980). Intra- and intergeneric similarities of ribosomal ribonucleic acid cistrons of free-living, nitrogen-fixing bacteria. *Int. J. Syst. Bacteriol.* 30, 106–122.

27. De Smedt, J. and De Ley, J. (1977). Intra- and intergeneric similarities of *Agrobacterium* ribosomal ribonucleic acid cistrons. *Int. J. Syst. Bacteriol.* 27, 222–240.

28. Devereux, R., Delaney, M., Widdel, F. and Stahl, D. A. (1989) Natural relationships among sulfate-reducing eubacteria. *J. Bacteriol. 171*, 6689–6695.

29. De Vos, P., and De Ley, J. (1983). Intra- and intergeneric similarities of *Pseudomonas* and *Xanthomonas* ribosomal ribonucleic acid cistrons. *Int. J. Syst. Bacteriol. 33*, 487–509.

30. De Vos, P., Van Landschoot, A., Segers, P., Tytgat, R., Gillis, M., Bauwens, M., Rossau, R., Goor, M., Pot, B., Kersters, K., Lizzaraga, P., and De Ley, J. (1989). Genotypic relationships and taxonomic localization of unclassified *Pseudomonas* and *Pseudomonas*-like strains by deoxyribonucleic acid:ribosomal ribonucleic acid hybridizations. *Int. J. Syst. Bacteriol. 39*, 35–49.

31. Dewhirst, F. E., Paster, B. J., and Bright, P. L. (1989). *Chromobacterium, Eikenella, Kingella, Neisseria, Simonsiella,* and *Vitreoscilla* species comprise a major branch of

the beta group *Proteobacteria* by 16S ribosomal ribonucleic acid sequence comparison: transfer of *Eikenella* and *Simonsiella* to the family *Neisseriaceae* (emend.). *Int. J. Syst. Bacteriol. 39*, 258–266.

32. Ding, J., Sun, H., Su, F., Xu, Q., Huang, Y., and Lin, P. (1981). Studies on nitrogen fixation by actinomycetes. *Acta Microbiol. Sin. 21*, 424–427.

33. Dixon, R. O. D., and Wheeler, C. T. (1986). *Nitrogen Fixation in Plants*. Blackie, Glasgow.

34. Dreyfus, B., Garcia, J. L., and Gillis, M. (1988). Characterization of *Azorhizobium caulinodans* gen. nov., sp. nov., a stem-nodulating nitrogen-fixing bacterium isolated from *Sesbania rostrata*. *Int. J. Syst. Bacteriol. 38*, 89–98.

35. DuBose, R. F., Dykhuizen, D. E., and Hartl. D. L. (1988). Genetic exchange among natural isolates of bacteria: recombination within the *phoA* gene of *Escherichia coli*. *Proc. Natl. Acad. Sci. USA 85*, 7036–7040.

35a. Eaglesham, A. R. J., Ellis, J. M., Evans, W. R., Fleischman, D. E., Hungria, M., and Hardy, R. W. F. (1990). The first photosynthetic N_2-fixing rhizobium: characteristics. In *Nitrogen Fixation: Achievements and Objectives* (P. M. Gresshoff, L. E. Roth, G. Stacey, and W. L. Newton, eds.). New York: Chapman and Hall, pp. 805–811.

36. Eardly, B. D., Materon, L. A., Smith, N. H., Johnson, D. A., Rumbaugh, M. D., and Selander, R. K. (1990). Genetic structure of natural populations of the nitrogen-fixing bacterium *Rhizobium meliloti*. *Appl. Envir. Microbiol. 56*, 187–194.

37. Ellis, J. M., Eardly, B. D., Hungria, M., Hardy, R. W. F., Rizzo, N. W., and Eaglesham, A. R. J. (1989). Photosynthetic N_2-fixing *Rhizobium*. *12th North American Symbiotic Nitrogen Fixation Conference Proceedings*, p. 101. Ames, Iowa.

38. Falk, E. C., Döbereiner, J., Johnson, J. L., and Krieg, N. R. (1985). Deoxyribonucleic acid homology of *Azospirillum amazonense* and emendation of the description of the genus *Azospirillum*. *Int. J. Syst. Bacteriol. 35*, 117–118.

39. Falk, E. C., Johnson, J. L., Baldani, V. L. D., Döbereiner, J., and Kreig, N. R. (1986). Deoxyribonucleic and ribonucleic acid homology studies of the genera *Azospirillum* and *Conglomeromonas*. *Int. J. Syst. Bacteriol. 36*, 80–85.

40. Fernandez, M. P., Meugnier, H., Grimont, P. A. D., and Bardin, R. (1989). Deoxyribonucleic acid relatedness among members of the genus *Frankia*. *Int. J. Syst. Bacteriol. 39*, 424–429.

41. Gallon, J. R., and Chaplin, A. E. (1987). *An Introduction to Nitrogen Fixation*. London: Cassell.

42. Gavini, F., Mergaert, J., Beji, A., Mielcarek, C., Izard, D., Kersters, K., and De Ley, J. (1989). Transfer of *Enterobacter agglomerans* (Beijerinck 1888) Ewing and Fife 1972 to *Pantoea* gen. nov. as *Pantoea agglomerans* comb. nov. and description of *Pantoea dispersa* sp. nov. *Int. J. Syst. Bacteriol. 39*, 337–345.

43. Gibson, J., Ludwig, W., Stackebrandt, E., and Woese, C. R. (1985). The phylogeny of the green photosynthetic bacteria: absence of a close relationship between *Chlorobium* and *Chloroflexus*. *Syst. Appl. Microbiol. 6*, 152–156.

44. Gillis, M., and De Ley, J. (1980). Intra- and intergeneric similarities of the ribosomal ribonucleic acid cistrons of *Acetobacter* and *Gluconobacter*. *Int. J. Syst. Bacteriol. 30*, 7–27.

45. Gillis, M., Kersters, K., Hoste, B., Janssens, D., Kroppenstedt, R. M., Stephan, M. P., Teixeira, K. R. S., Döbereiner, J., and De Ley, J. (1989). *Acetobacter diazotro-*

phicus sp. nov., a nitrogen-fixing acetic acid bacterium associated with sugarcane. *Int. J. Syst. Bacteriol.* 39, 361–364.

46. Giovannoni, S. J., Turner, S., Olsen, G. J., Barns, S., Lane, D. J. and Pace, N. R. (1988). Evolutionary relationships among cyanobacteria and green chloroplasts. *J. Bacteriol.* 170, 3584–3592.

47. Haahtela, K., Helander, I., Nurmiaho-Lassila, E.-L. and Sundman, V. (1983). Morphological and physiological characteristics and lipopolysaccharide composition of N_2-fixing (C_2H_2-reducing) root-associated *Pseudomonas* sp. *Can. J. Microbiol.* 29, 874–880.

48. Hahn, D, Lechevalier, M. P., Fischer, A. and Stackebrandt, E. (1989). Evidence for a close phylogenetic relationship between members of the genera *Frankia, Geodermatophilus* and *Blastococcus* and emendation of the family *Frankiaceae*. *System. Appl. Microbiol.* 11, 236–242.

49. Heda, G. D. and Madigan, M. T. (1986). Utilization of amino acids and lack of diazotrophy in the thermophilic anoxygenic phototroph *Cloroflexus aurantiacus*. *J. Gen. Microbiol.* 132, 2469–2673.

50. Heda, G. D. and Madigan, M. T. (1987). Nitrogen metabolism and N_2 fixation in phototrophic green bacteria. In *Green Photosynthetic Bacteria* (J. M. Olson, J. G. Ormerod, J. Amesz, E. Stackebrandt and H. G. Trüper, eds.) pp. 175–187. Plenum Press.

51. Hennecke, H., Kaluza, K., Thöny, B., Fuhrmann, M., Ludwig, W. and Stackebrandt, E. (1985). Concurrent evolution of nitrogenase genes and 16S rRNA in *Rhizobium* species and other nitrogen fixing bacteria. *Arch. Microbiol.* 142, 342–348.

52. Hollis, A. B., Kloos, W. E. and Elkan, G. H. (1981). DNA:DNA hybridization studies of *Rhizobium japonicum* and related *Rhizobiaceae*. *J. Gen. Microbiol.* 123, 215–222.

53. Imhoff, J. F., Trüper, H. G. and Pfennig, N. (1984). Rearrangement of the species and genera of the phototrophic "purple nonsulfur bacteria." *Int. J. Syst. Bacteriol. 34*, 340–343.

54. Iwabe, N., Kuma, K.-I., Hasegawa, M., Osawa, S., and Miyata, T. (1989). Evolutionary relationship of archaebacteria, eubacteria, and eukaryotes inferred from phylogenetic trees of duplicated genes. *Proc. Natl. Acad. Sci. USA 86*, 9355–9359.

55. Jarvis, B. D. W., Gillis, M., and De Ley, J. (1986). Intra- and intergeneric similarities between the ribosomal ribonucleic acid cistrons of *Rhizobium* and *Bradyrhizobium* species and some related bacteria. *Int. J. Syst. Bacteriol. 36*, 129–138.

56. Jarvis, B. D. W., Pankhurst, C. E., and Patel, J. J. (1982). *Rhizobium loti,* a new species of legume root nodule bacteria. *Int. J. Syst. Bacteriol. 32*, 378–380.

57. Jenni, B., and Aragno, M. (1987). *Xanthobacter agilis* sp. nov., a motile, dinitrogen-fixing, hydrogen-oxidizing bacterium. *Syst. Appl. Microbiol. 9*, 254–257.

58. Jordan, D. C. (1982). Transfer of *Rhizobium japonicum* Buchanan 1980 to *Bradyrhizobium* gen. nov., a genus of slow-growing, root nodule bacteria from leguminous plants. *Int. J. Syst. Bacteriol. 32*, 136–139.

59. Kanvinde, L., and Sastry, G. R. K. (1990). *Agrobacterium tumefaciens* is a diazotrophic bacterium. *Appl. Envir. Microbiol. 56*, 2087–2092.

60. Kanvinde, L., Soliman, M. H., Wardhan, H., Nowell, L., Fox, D., and Sastry, G. R. K. (1987). Studies on the diazotrophic nature of *Agrobacterium*. In *Molecular Genetics of Plant-Microbe Interactions* (D. P. S. Verma and N. Brisson, eds.). Dordrecht: Martinus Nijhoff, pp. 309–312.

61. Khammas, K. M., Ageron, E., Grimont, P. A. D., and Kaiser, P. (1989). *Azospirillum irakense* sp. nov., a nitrogen-fixing bacterium associated with rice roots and rhizosphere soil. *Res. Microbiol. 140*, 679–693.

62. Kim, Y.-M., Ahn, K.-J., Beppo, T., and Uozumi, T. (1986). Nucleotide sequence of the *nifLA* operon of *Klebsiella oxytoca* NG13 and characterization of the gene products. *Mol. Gen. Genet. 205*, 253–259.

63. Knapp, R., and Jurtshuk, P. (1988). Characterization of free-living nitrogen-fixing *Streptomyces* species and factors which affect their rates of acetylene reduction. *Abstr. Annu. Meet. Am. Soc. Microbiol. 88*, 219.

64. König, H., Nusser, E., and Stetter, K. O. (1985). Glycogen in *Methanolobus* and *Methanococcus. FEMS Microbiol. Lett. 28*, 265–269.

65. Kravchenko, I. K., and Kalininskaya, T. A. (1988). Nitrogen-fixing bacteria from strongly saline Takyr soil. (in Russian). *Mikrobiologiya 57*, 279–283.

66. Krawiec, S. (1985). Concept of a bacterial species. *Int. J. Syst. Bacteriol. 35*, 217–220.

67. Krieg, N. R., ed. (1984). *Bergey's Manual of Systematic Bacteriology*, vol. 1. Baltimore: Williams & Wilkins.

67a. Ladha, J. K., Pareek, R. P., So, R., and Becker, M. (1990). Stem nodule symbiosis and its unusual properties. In *Nitrogen Fixation: Achievements and Objectives* (P. M. Gresshoff, L. E. Roth, G. Stacey, and W. L. Newton, eds.). New York: Chapman and Hall, pp. 633–640.

68. Lake, J. A. (1988). Origin of the eukaryotic nucleus determined by rate-invariant analysis of rRNA sequences. *Nature 331*, 184–186.

69. Lalonde, M., Simon, L., Bousquet, J., and Séguin, A. (1988). Advances in the taxonomy of *Frankia:* recognition of species *alni* and *elaeagni* and novel subspecies *pommerii* and *vandijkii*. In *Nitrogen Fixation: Hundred Years After* (H. Bothe, F. J. de Bruijn, and W. E. Newton, eds.). Stuttgart: Gustav Fischer, pp. 671–680.

70. Lambert, B., Joos, H., Dierickx, S., Vantomme, R., Swings, J., Kersters, K., and Van Montagu, M. (1990). Identification and plant interaction of a *Phyllobacterium* sp., a predominant rhizobacterium of young sugar beet plants. *Appl. Envir. Microbiol. 56*, 1093–1102.

71. Lane, D. J., Stahl, D. A., Olsen, G. J., Heller, D. J., and Pace, N. R. (1985). Phylogenetic analysis of the genera *Thiobacillus* and *Thiomicrospira* by 5S rRNA sequences. *J. Bacteriol. 163*, 75–81.

72. Larkin, J. M., and Borrall, R. (1984). Deoxyribonucleic acid base compositions and homology of "*Microcyclus,*" *Spirosoma*, and similar organisms. *Int. J. Syst. Bacteriol. 34*, 211–215.

73. Lersten, N. R., and Horner, H. T. (1976). Bacterial leaf nodule symbiosis in angiosperms with emphasis on Rubiaceae and Myrsinaceae. *Bot. Rev. 42*, 145–214.

74. Lindström, K. (1989). *Rhizobium galegae*, a new species of legume root nodule bacteria. *Int. J. Syst. Bacteriol. 39*, 365–367.

75. Mackintosh, M. E. (1978). Nitrogen fixation by *Thiobacillus ferrooxidans. J. Gen. Microbiol. 105*, 215–218.

76. Madigan, M., Cox, S. S., and Stegeman, R. A. (1984). Nitrogen fixation and nitrogenase activities in members of the family *Rhodospirillaceae. J. Bacteriol. 157*, 73–78.

77. Malik, K. A., and Schlegel, H. G. (1981). Chemolithoautotrophic growth of bacteria able to grow under N_2-fixing conditions. *FEMS Microbiol. Lett. 11*, 63–67.

78. Mangels, L. A., Favinger, J. L., Madigan, M. T., and Gest, H. (1986). Isolation and characterization of the nitrogen-fixing marine photosynthetic bacterium *Rhodopseudomonas marina* var. *agilis*. *FEMS Microbiol. Lett. 36*, 99–104.

79. McClung, C. R., Patriquin, D. G., and Davis, R. E. (1983). *Campylobacter nitrofigilis* sp. nov., a nitrogen-fixing bacterium associated with roots of *Spartina alterniflora* Loisel. *Int. J. Syst. Bacteriol. 33*, 605–612.

80. Murray, P. A., and Zinder, S. H. (1984). Nitrogen fixation by a methanogenic archaebacterium. *Nature 312*, 284–286.

81. Murray, R. G. E. (1984). The higher taxa, or, a place for everything . . . ? In *Bergey's Manual of Systematic Bacteriology* (N. R. Kreig, ed.), Vol. 1. Baltimore: Williams & Wilkins, pp. 31–34.

82. Murray, R. G. E., Brenner, D. J., Colwell, R. R., De Vos, P., Goodfellow, M., Grimont, P. A. D., Pfennig, N., Stackebrandt, E., and Zavarzin, G. A. (1990). Report of the ad hoc committee on approaches to taxonomy within the proteobacteria. *Int. J. Syst. Bacteriol. 40*, 213–215.

83. Murrell, J. C., and Dalton, H. (1983). Nitrogen fixation in obligate methanotrophs. *J. Gen. Microbiol. 129*, 3481–3486.

84. Nazina, T. N., Rozanova, E. P., and Kalininskaya, T. A. (1979). Fixation of molecular nitrogen by sulfate-reducing bacteria from oil strata. (in Russian). *Mikrobiologiya 48*, 133–136.

85. Normand, P., and Bousquet, J. (1989). Phylogeny of nitrogenase sequences in *Frankia* and other nitrogen-fixing microorganisms. *J. Mol. Evol. 29*, 436–447.

86. Oakley, C. J., and Murrell, J. C. (1988). *nifH* genes in the obligate methane oxidizing bacteria. *FEMS Microbiol. Lett. 49*, 53–57.

87. Olsen, G. J. and Woese, C. R. (1989). A brief note concerning archaebacterial phylogeny. *Can. J. Microbiol. 35*, 119–123.

88. Pääbo, S. (1989). Ancient DNA: extraction, characterization, molecular cloning, and enzymatic amplification. *Proc. Natl. Acad. Sci. 86*, 1939–1943.

89. Papen, H., and Werner, D. (1979). N_2 fixation in *Erwinia herbicola*. *Arch. Microbiol. 120*, 25–30.

90. Paster, B. J., and Dewhirst, F. E. (1988). Phylogeny of campylobacters, wolinellas, *Bacteroides gracilis*, and *Bacteroides ureolyticus* by 16S ribosomal ribonucleic acid sequencing. *Int. J. Syst. Bacteriol. 38*, 56–62.

90a. Pearson, H. W., Howsley, R., and Williams, S. T. (1982). A study of nitrogenase activity in *Mycoplasma* species and free-living actinomycetes. *J. Gen. Microbiol. 128*, 2073–2080.

91. Piñero, D., Martinez, E., and Selander, R. K. (1988). Genetic diversity and relationships among isolates of *Rhizobium leguminosarum* biovar *phaseoli*. *Appl. Envir. Microbiol. 54*, 2825–2832.

92. Possot, O., Henry, M., and Sibold, L. (1986). Distribution of DNA sequences homologous to *nifH* among archaebacteria. *FEMS Microbiol. Lett. 34*, 173–177.

93. Postgate, J. R. (1970). Nitrogen fixation by sporulating sulphate-reducing bacteria including rumen strains. *J. Gen Microbiol. 63*, 137–139.

94. Postgate, J. R. (1981). Microbiology of the free-living nitrogen-fixing bacteria, excluding cyanobacteria. In *Current Perspectives in Nitrogen Fixation* (A. H. Gibson and W. E. Newton, eds.) pp. 217–228. Australian Academy of Science, Canberra.

95. Postgate, J. R. (1982). *The Fundamentals of Nitrogen Fixation.* Cambridge University Press.

96. Postgate, J. R., Dixon, R. A., Hill, S. and Kent, H. (1987). *nif* genes in alien backgrounds. *Phil. Trans. R. Soc. Land.* B 317, 227–243.

97. Postgate, J. R. and Eady, R. R. (1988). The evolution of biological nitrogen fixation. In *Nitrogen Fixation: Hundred Years After* (H. Bothe, F. J. de Bruijn and W. E. Newton, eds.), pp. 31–40. Gustav Fischer, Stuttgart.

98. Postgate, J. R. and Kent, H. M. (1985) Diazotrophy within *Desulfovibrio. J. Gen Microbiol.* 131, 2119–2122.

99. Potrikus, C. J. and Breznak, J. A. (1977). Nitrogen-fixing *Enterobacter agglomerans* isolated from guts of wood-eating termites. *Appl. Envir. Microbiol.* 33, 392–399.

100. Pretorius, I.-M., Rawlings, D. E. and Woods, D. R. (1986). Identification and cloning of *Thiobacillus ferrooxidans* structural *nif* genes in *Escherichia coli. Gene* 45, 59–65.

101. Raj, H. D. (1983). Proposal of *Ancylobacter* gen. nov. as a substitute for the bacterial genus *Microcyclus* Ørskow 1928. *Int. J. Syst. Bacteriol.* 33, 397–398.

102. Rayssiguier, C., Thaler, D. S. and Radman, M. (1989). The barrier to recombination between *Escherichia coli* and *Salmonella typhimurium* is disrupted in mismatch-repair mutants. *Nature* 342, 396–401.

103. Reinhold, B., Hurrek, T., Fendrik, I., Pot, B., Gillis, M., Kersters, K., Thielemans, S., and De Ley, J. (1987). *Azospirillum halopraeferens* sp. nov., a nitrogen-fixing organism associated with roots of kallar grass [*Leptochloa fusca* (L.) Kunth]. *Int. J. Syst. Bacteriol. 37*, 43–51.

103a. Reinhold-Hurek, B., Hurek, T., Gillis, M., Hoste, B., Kersters, K., and De Ley, J. (1990). Diazotrophs repeatedly isolated from roots of Kallar grass form a new genus, *Azoarcus*. In *Nitrogen Fixation: Achievements and Objectives* (P. M. Gresshoff, L. E. Roth, G. Stacey, and W. L. Newton, eds.). New York: Chapman and Hall, p. 432.

104. Rippka, R., Deruelles, J., Waterbury, J. B., Herdman, M., and Stanier, R. Y. (1979). Generic assignments, strain histories and properties of pure cultures of cyanobacteria. *J. Gen. Microbiol.* 111, 1–61.

105. Rippka, R. (1988). Recognition and identification of cyanobacteria. In *Methods in Enzymology,* vol. 167: *Cyanobacteria* (L. Packer and A. N. Glazer, eds.). San Diego: Academic Press, pp. 28–67.

106. Romanovskaya, V. A., Lyudvichenko, E. S., Sokolov, I. G., and Malashenko, Yu. R. (1980). Fixation of molecular nitrogen by methane oxidizing bacteria. (in Russian). *Mikrobiol. Zh. (Kiev) 42*, 683–688.

107. Rosenblum, E. D., and Wilson, P. W. (1948). Fixation of isotopic nitrogen by *Clostridium. J. Bacteriol. 57*, 413–414.

108. Schink, B., Thompson, T. E., and Zeikus, J. G. (1982). Characterization of *Propionispira arboris* gen. nov. sp. nov., a nitrogen-fixing anaerobe common to wetwoods of living trees. *J. Gen. Microbiol. 128*, 2771–2779.

109. Scholla, M. H., and Elkan, G. H. (1984). *Rhizobium fredii* sp. nov., a fast-growing species that effectively nodulates soybeans. *Int. J. Syst. Bacteriol. 34*, 484–486.

110. Seldin, L., and Dubnau, D. (1985). Deoxyribonucleic acid homology among *Bacillus polymyxa*, *Bacillus macerans*, *Bacillus azotofixans*, and other nitrogen-fixing *Bacillus* strains. *Int. J. Syst. Bacteriol. 35*, 151–154.

111. Seldin, L., Van Elsas, J. D., and Penido, E. G. C. (1984). *Bacillus azotofixans* sp. nov., a nitrogen-fixing species from Brazilian soils and grass roots. *Int. J. Syst. Bacteriol. 34*, 451–456.

112. Shanmugasundaram, S., Suguna, S., and Fleischman, D. (1989). Genetic evidence for the presence of bacterial photosynthetic gene(s) in *Rhizobium* sp. isolated from *Aeschynomene indica*. *12th North American Symbiotic Nitrogen Fixation Conference Proceedings*, p. 56. Ames, Iowa.

113. Skerman, V. B. D., McGowan, V., and Sneath, P. H. A., eds. (1980). Approved lists of bacterial names. *Int. J. Syst. Bacteriol. 30*, 225–420.

114. Sneath, P. H. A., ed. (1986). *Bergey's Manual of Systematic Bacteriology*, Vol. 2. Williams & Wilkins, Baltimore.

115. Souillard, N., Magot, M., Possot, O., and Sibold, L. (1988). Nucleotide sequence of regions homologous to *nifH* (nitrogenase Fe protein) from the nitrogen-fixing archaebacteria *Methanococcus thermolithotrophicus* and *Methanobacterium ivanovii*: evolutionary implications. *J. Mol. Evol. 27*, 65–76.

116. Souillard, N., and Sibold, L. (1986). Primary structure and expression of a gene homologous to *nifH* (nitrogenase Fe protein) from the archaebacterium *Methanococcus voltae*. *Mol. Gen. Genet. 203*, 21–28.

117. Stackebrandt, E., and Charfreitag, O. (1990). Partial 16S rRNA primary structure of five *Actinomyces* species: phylogenetic implications and development of an *Actinomyces israelii*-specific oligonucleotide probe. *J. Gen. Microbiol. 136*, 37–43.

118. Stackebrandt, E., Murray, R. G. E., and Trüper, H. G. 1988). *Proteobacteria* classis nov., a name for the phylogenetic taxon that includes the "purple bacteria and their relatives." *Int. J. Syst. Bacteriol. 38*, 321–325.

119. Stahl, D. A., Lane, D. J., Olsen, G. J., Heller, D. J., Schmidt, T. M., and Pace, N. R. (1987). Phylogenetic analysis of certain sulfide-oxidizing and related morphologically conspicuous bacteria by 5S ribosomal ribonucleic acid sequences. *Int. J. Syst. Bacteriol. 37*, 116–122.

120. Staley, J. T., ed. (1989). *Bergey's Manual of Systematic Bacteriology*, Vol. 3. Baltimore: Williams & Wilkins.

121. Stanley, J., Brown, G. G., and Verma, D. P. S. (1985). Slow-growing *Rhizobium japonicum* comprises two highly divergent symbiotic types. *J. Bacteriol. 163*, 148–154.

122. Stewart, W. D. P., Rowell, P., and Rai, A. N. (1980). Symbiotic nitrogen-fixing cyanobacteria. In *Nitrogen Fixation* (W. D. P. Stewart and J. R. Gallon, eds.). San Diego: Academic Press, pp. 239–277.

123. Sundman, V. (1964). A description of some lignolytic soil bacteria and their ability to oxidize simple phenolic compounds. *J. Gen. Microbiol. 36*, 171–183.

124. Thompson, L. T., Smibert, R. M., Johnson, J. L., and Kreig, N. R. (1988). Phylogenetic study of the genus *Campylobacter*. *Int. J. Syst. Bacteriol. 38*, 190–200.

125. Trüper, H. G. 1986). International Committee on Systematic Bacteriology Subcommittee on the Taxonomy of Phototrophic Bacteria. *Int. J. Syst. Bacteriol. 36*, 114–115.

126. Tsuji, K., Tsien, H. C., Hanson, R. S., DePalma, S. R., Scholtz, R., and LaRoche, S. (1990). 16S ribosomal RNA sequence analysis for determination of phylogenetic relationship among methylotrophs. *J. Gen. Microbiol. 136*, 1–10.

127. Turner, S., Burger-Wiersma, T., Giovannoni, S. J., Mur, L. R., and Pace, N. R. (1989). The relationship of a prochlorophyte *Prochlorothrix hollandica* to green chloroplasts. *Nature 337*, 380–382.

128. Urdaci, M. C., Stal, L. J., and Marchand, M. (1988). Occurrence of nitrogen fixation among *Vibrio* spp. *Arch. Microbiol. 150*, 224–229.

129. Van den Eynde, H., Van de Peer, Y., Perry, J., and De Wachter, R. (1990). 5S rRNA sequences of representatives of the genera *Chlorobium, Prosthecochloris, Thermomicrobium, Cytophaga, Flavobacterium, Flexibacter* and *Saprospira* and a discussion of the evolution of eubacteria in general. *J. Gen. Microbiol. 136*, 11–18.

130. Wang, S-Z., Chen, J.-S., and Johnson, J. L. (1988). The presence of five *nifH*-like sequences in *Clostridium pasteurianum:* sequence divergence and transcription properties. *Nucleic Acids Res. 16*, 439–453.

131. Watanabe, I., So, R., Ladha, J. K., Katayama-Fijimura, Y., and Kuraishi, H. (1987). A new nitrogen-fixing species of pseudomonad: *Pseudomonas diazotrophicus* sp. nov. isolated from the root of wetland rice. *Can. J. Microbiol. 33*, 670–678.

132. Wayne, L. G., Brenner, D. J., Colwell, R. R., Grimont, P. A. D., Kandler, O., Krichevsky, M. I., Moore, L. H., Moore, W. E. C., Murray, R. G. E., Stackebrandt, E., Starr, M. P. and Trüper, H. G. (1987). Report of the ad hoc committee on reconciliation of approaches to bacterial systematics. *Int. J. Syst. Bacteriol. 37*, 463–464.

133. Wedlock, D. N., and Jarvis, B. D. W. (1986). DNA homologies between *Rhizobium fredii*, rhizobia that nodulate *Galega* sp., and other *Rhizobium* and *Bradyrhizobium* species. *Int. J. Syst. Bacteriol. 36*, 550–558.

134. Weisburg, W. G., Woese, C. R., Dobson, M. E. and Weiss, E. (1985). A common origin of rickettsiae and certain plant pathogens. *Science 230*, 556–558.

135. West, P. A., Brayton, P. R., Twilley, R. R., Bryant, T. N., and Colwell, R. R. (1985). Numerical taxonomy of nitrogen-fixing "decarboxylase-negative" *Vibrio* species isolated from aquatic environments. *Int. J. Syst. Bacteriol. 35*, 198–205.

136. Wheatcroft, R., and Watson, R. J. (1988). Distribution of insertion sequence IS*Rm1* in *Rhizobium meliloti* and other gram-negative bacteria. *J. Gen. Microbiol. 134*, 113–121.

137. Widdel, F. (1987) New types of acetate-oxidizing sulfate-reducing *Desulfobacter* species: *Desulfobacter hydrogenophilus* sp. nov., *Desulfobacter latus* sp. nov., and *Desulfobacter curvatus* sp. nov. *Arch. Microbiol. 148*, 286–291.

138. Williams, S. T., ed. (1989). *Bergey's Manual of Systematic Bacteriology*, Vol. 4. Baltimore: Williams & Wilkins.

139. Woese, C. R. (1987) Bacterial evolution. *Microbiol. Rev. 51*, 221–271.

140. Woese, C. R., Stackebrandt, E., Weisburg, W. G., Paster, B. J., Madigan, M. T., Fowler, V. J., Hahn, C. M., Blanz, P., Gupta, R., Nealson, K. H., and Fox, G. E. (1984). The phylogeny of purple bacteria: the alpha subdivision. *Syst. Appl. Microbiol. 5*, 315–326.

141. Woese, C. R., Weisburg, W. G., Hahn, C. M., Paster, B. J., Zablen, L. B., Lewis, B. J., Macke, T. J., Ludwig, W. and Stackebrandt, E. (1985). The phylogeny of purple bacteria: the gamma subdivision. *System. Appl. Microbiol. 6*, 25–33.

142. Woese, C. R., Weisburg, W. G., Paster, B. J., Hahn, C. M., Tanner, R. S., Krieg, N. R., Koops, H.-P., Harms, H. and Stackebrandt, E. (1984). The phylogeny of purple bacteria: the beta subdivision. *Syst. Appl. Microbiol. 5*, 327–336.

143. Young, J. P. W. (1985). *Rhizobium* population genetics: enzyme polymorphism in isolates from peas, clover, beans and lucerne grown at the same site. *J. Gen. Microbiol. 131*, 2399–2408.

144. Young, J. P. W. (1989). The population genetics of bacteria. In *Genetics of Bacterial Diversity* (D. A. Hopwood and K. F. Chater, eds.), pp. 417–438. Academic Press, London.

144a. Young, J. P. W., Downer, H. L. and Eardly, B. D. (1991). Phylogeny of the phototrophic rhizobium strain BTAi1 by polymerase chain reaction-based sequencing of a 16S rRNA gene segment. *J. Bacteriol. 173*, 2271–2277.

145. Young, J. P. W. and Johnston, A. W. B. (1989). The evolution of specificity in the legume-rhizobium symbiosis. *Trends Ecol. Evol. 4*, 341–349.

2

Physiology of Nitrogen Fixation in Free-Living Heterotrophs

Susan Hill

I. Introduction

Heterotrophic diazotrophs fix N_2 by utilizing the energy generated from the oxidation of organic compounds. The dividing line between free-living and associative diazotrophs is not clearcut. Many associative or symbiotic types do fix N_2 in the free-living state. Conversely, N_2 fixation and growth by free-living heterotrophs in natural habitats may depend upon the activity of other living organisms. However, the aim of this chapter is to review aspects of the physiology of N_2 fixation in heterotrophs that are growing in the free-living state.

The physiologies of diazotrophs are markedly affected by the properties and the requirements of the nitrogenase complex. These are the exceptional oxygen sensitivity and the metal content of the component proteins (Fe, Mo, or V) and the need for adequate supplies of reducing power and MgATP for nitrogenase activity. Physiological investigations specifically relating to the recently discovered alternative nitrogenases are just beginning. Therefore, the information in this chapter relates to free-living heterotrophs expressing the Mo-dependent nitrogenase. The occurrence of the alternative nitrogenases, particularly with relevance to metal supply, is covered elsewhere (see Chapter 19).

Physiological studies attempt to elucidate the reactions of organisms to changes in their environment by investigating their genetic and biochemical makeup. The importance of the environment in such studies is illustrated by the use of chemostats, where the operator can define the physical parameters and select the nutritional limitation.[17] However, owing to the interactions between apparent responses to O_2, N, and energy status during diazotrophy, the factors limiting growth may be less clear. Aspects of the underlying physiology relating to these interactions are described here. The study of ecology (see Knowles[104] and Postgate[147]) introduces additional complexities into the system and these are not covered here, except incidentally.

Genera containing free-living N_2 fixers are grouped in Table 2-1 with respect to the mode of their catabolic activity in relation to O_2 during N_2 fixation. This grouping differs, in some places markedly, from the phylogenetic arrangement based on ribosomal RNA gene sequences described in Chapter 1 (see Chapter 1 for discussion). In Table 2-1 the affiliation of each genus to the phylogenetic grouping of Table 1-1 (Chapter 1) is indicated by a capital letter. The names of the species that fix N_2 are shown in Table 1-1. The number of prokaryotes found to fix N_2 has greatly expanded over recent years (see Postgate[147] and Chapter 1). The recognition of the importance of microaerophilic conditions for N_2 fixation by many aerobic organisms and of the rhizosphere as a habitat harboring N_2 fixers have contributed to this expansion. It is generally considered that the contribution to global N_2 fixation by free-living heterotrophs is small, since there is a scarcity of suitable carbon and energy sources (see Postgate[147]). However, the recent estimate of the global production of CH_4 (approximately 1×10^{15} g yr^{-1}) and the projected increases (1% per year)[65] suggests that methanotrophs might make a significant contribution to global biological N_2 fixation (approximately 1.2×10^{14} g N yr^{-1} [147]). Little is known about the methanotrophs' present contribution.[116] If the majority of this CH_4 was harnessed, it could increase the global N_2 fixation rate by 5% (assuming an efficiency of 10 mg N per g CH_4 consumed).

Energy to support N_2 fixation in free-living heterotrophic diazotrophs is principally generated by the oxidation of organic compounds. The oxidation of H_2 via the uptake hydrogenase may provide an additional source. Some of the organisms listed in Table 2-1 also have chemolithotrophic capabilities. The first group in Table 2-1 comprises the obligate anaerobes, which use either fermentation or anaerobic respiration for N_2 fixation and growth. The second group contains the facultative anaerobes. These use fermentation and in some cases microaerobic respiration for diazotrophy. The third group embraces the obligate aerobes. They are divided into two subgroups, which relate to the organisms' tolerance toward O_2 when fixing N_2. The larger subgroup contains the microaerophilic diazotrophs; however, these are usually aerotolerant when a fixed N source is provided. The smaller subgroup comprises those that are aerotolerant. This grouping is discussed in the next section.

II. Interactions of O_2 with N_2 Fixation

N_2 fixation and growth in obligate anaerobes occurs only in the absence of O_2, although in some an occasional short exposure to O_2 can be tolerated (see following discussion). In natural systems the provision of a suitable carbon an energy source for anaerobic diazotrophy may depend upon the aerobic activity of neighbors.[120] In such habitats the ambient O_2 concentra-

Table 2–1   Free-Living Heterotrophic N_2 Fixers and Their Distribution Arranged into
Groups Reflecting Modes of Catabolism and Response to O_2 During Diazotrophy

Genus	Habitat	Ref.
Obligate Anaerobes Using Fermentation		
Clostridium (B)[f]	Ubiquitous in soil, incuding marine sediments and decaying vegetation	13,115
Obligate Anaerobes Using Fermentations and Fumarate Respiration		
Propionispira	Wet wood	161
Obligate Anaerobes Using SO_4^{2-} Respiration		
Desulfotomaculum (B)	Widespread in soil, mud, and	149,150,181
Desulfobacter (J)	water, including brackish	
Desulfovibrio (J)	and marine, especially when high in organic matter	
Facultative Anaerobes Using Fermentation and, in some, Aerobic Respiration		
Propionibacterium (C)	Dairy products, silage	7
Bacillus (B)	Soil and rhizoshere	162
Citrobacter (I)	Rhizosphere and digests of	60,130
Enterobacter (I)	paper, food, and forest	
Erwinia (I)	residues	
Klebsiella (I)		
Vibrio (I)	Marine water and mud, associated with plant roots and animals	178
Obligate Aerobes Using Aerobic Respiration		
Microaerophilic N_2 fixers		
Acetobacter (G)	Roots and stems of sugar cane	55
Alcaligenes (H)[a]	Rhizoshere, especially of rice	191
Aquaspirillum (G)	Fresh water, rhizosphere	46
Arthrobacter (C)	Soil	18
Azorhizobium (G)[b,e]	Soil, fresh water	Ch. 8
Azospirillum (G)[a,d]	Rhizosphere and roots of cereals	46
Beggiatoa (I)	Marine and fresh water, rice rhizosphere	140
Bradyrhizobium (G)[a,e]	Soil	Ch. 8
Campylobacter (F)	Rhizosphere, salt marsh	121
Derxia (H)[a,b]	Mildly acid subtropical soil, rhizosphere of forage grasses	172
Herbaspirillum (H)	Rhizosphere and roots of cereals	46
Methylobacter (I)[c]	Methanotrophs, widespread,	116,134
Methylococcus (I)	soil, marine and fresh	
Methylocystis (G)	water, and muds	
Methylomonas (I)[c]		
Methylosinus (G)[b]		
Pseudomonas (G,H,I)[a]	Soil, associated with roots	

Table 2–1   Continued

Genus	Habitat	Ref.
Rhizobium (G)[e]	Soil	Ch. 8
Xanthobacter (G)[a]	Soil, mud, water	182,116
Lignobacter (H)	Decaying wood and pulp mill effluents	39
Aerotolerant N₂ fixers		
Beijerinckia (G)	Acid tropical and subtropical soils, phyllosphere of tropical plants	172
Azomonas (I)	Soil, fresh water	172
Azotobacter (I)	Soil, fresh water, activated sludge (A. paspali—rhizosphere of grass Paspalum)	172
Frankia (C)[e]	Soil	Ch. 7

[a]Some species have chemolithotrophic capabilities.

[b]Moderate microaerophiles.

[c]nifH genes identified by hydridization but N₂ fixation not demonstrated.

[d]some species can fix N₂ anaerobically using N₂O respiration.

[e]Major contribution to N₂ fixation as symbionts.

[f]Capital letters refer to the divisions of classification in Table 1-1 of Chapter 1: Firmibacteria (B), Thallobacteria (C), Campylobacter (F), Proteobacteria: subdivisions alpha (G), beta (H), gamma (I), and delta (J).

tion will be influenced by the apparent K_s for O_2 of the O_2-consuming processes, as well as the rate of O_2 transfer through the matrix. Thus, if the affinity for O_2 is very high, the difference between an anaerobic and an O_2-limited environment may be physically indistinguishable. On the other hand, the physiologies within such environments are distinct.

In facultative anaerobes, diazotrophy occurs under anaerobiosis, but in some of the enteric N_2 fixers, limiting O_2 can be tolerated and may be beneficial. Under conditions where a fermentable carbon and energy source is present in excess, O_2 does not markedly stimulate nitrogenase activity.[59,71,130] On the other hand, where the concentration of the C source limits anaerobic growth, microaerobic catabolism can be coupled to N_2 fixation and growth.[60,71] O_2-dependent diazotrophy has also been found in dark-grown facultative anaerobic photobacteria[164] (see Chapter 3). The extent of the benefits of microaerobic catabolism for diazotrophy in facultative anaerobes may be widespread.[130]

The grouping of aerobic diazotrophs into aerotolerant and microaerophilic types (see Table 2-1) is based crudely upon an organism's ability to grow on an N-free medium in air. Those that can do so are aerotolerant, whereas those that require a subatmospheric O_2 concentration for consistent diazotrophy are termed microaerophilic. Rarely (e.g., *Beggitoa, Campylobacter*) is this obligate microaerophilic habit extended to growth on fixed N sources, although in some organisms (e.g., *Azospirillum*[135]) it is preferred. The dividing line between aerotolerant and microaerophilic types is not clearcut; for example, some strains (Na-dependent) of *Azotobacter chroococcum* require microaerobic conditions for the initiation of diazotrophy[136] and the relative tolerance toward O_2 varies among microaerophilic N_2 fixers (see note *b* of Table 2-1).

The optimum condition for aerobic diazotrophy is where the supply of O_2 just meets the respiratory demand. At lower rates of O_2 supply nitrogenase activity is energy limited. At higher rates of aeration nitrogenase activity is inhibited. Thus, a bell-shaped curve is obtained when nitrogenase activity is plotted against O_2 concentration. The inhibition of nitrogenase activity by O_2 is reversible in many organisms (see later discussion), but at elevated rates of O_2 supply the synthesis of nitrogenase is prevented (see later discussion).

Two factors need to be considered when the optimum O_2 supply for a diazotroph is being assessed. The first is the supply of O_2 to the organism. This is influenced by the atmospheric O_2 concentration, the rate of O_2 diffusion through the liquid, the agitation rate of the liquid, and the solubility of O_2 within the liquid. The second is the demand for O_2 by the organism, which can be influenced by its physiological state, the population density, and the genotype of the strain. For experimental purposes, equipment that controls the desired dissolved O_2 concentration (DOC) can overcome difficulties arising from these variables.[75] With regard to some microaerophiles the conventional polarographic or galvanic O_2 electrodes may not be sufficiently sensitive for measuring DOC; therefore, alternative means, such as the spectral properties of leghaemoglobin[11,75] and the O_2-dependent light emission by photobacteria have been employed.[94]

The value of the optimum DOC for nitrogenase activity in various diazotrophs ranges widely, from air saturation (e.g., in heterocystous cyanobacteria) to four orders of magnitude below (e.g., in *K. pneumoniae*).[75] Clumping or clustering of respiring organisms causes O_2 gradients to develop, so that protection from O_2 is gained within the cluster. Examples of this phenomenon are colony dimorphism by *Derxia gummosa* on solid media[69] and pellicle formation by *Azospirillum* in semisolid media.[46] The latter involves aerotactic behavior.[155]

A. *Flow of Electrons to O_2 (Respiratory Protection)*

Respiratory activity as a device to prevent O_2 inhibition of nitrogenase activity was formulated by Dalton and Postgate[31,32] to explain the particular behavior of *Azotobacter* when fixing N_2.[188] *Azotobacter* has the ability to adjust, by adaptation, its respiratory rate to match a wide range of rates of O_2 supply. It can exhibit one of the highest known rates of respiration, particularly when fixing N_2, but such activity occurs at the expense of a high rate of carbon and energy source consumption (discussed later). Although the respiratory activity of microaerophilic diazotrophs neither reaches such elevated rates nor shows obvious adaptation to increases in O_2 supply, it nevertheless serves to prevent O_2 inhibition of nitrogenase activity.

In this section aspects of the route of carbon catabolism and of the composition of the respiratory chain in the facultative anaerobe *K. pneumoniae*, in the microaerophile *Azospirillum*, and in the aerotolerant aerobe *Azotobacter* are described. The effects of changes in O_2 supply are considered in particular, because the basis of aerotolerance probably rests in the flexibility of synthesis and activity of the various components. Although much more is known about *Azotobacter* than about the other two organisms, the order in which they are considered reflects their ascending tolerance toward O_2 when fixing N_2. The optimum DOC for nitrogenase activity in *K. pneumoniae* is 0.03 μM O_2,[73] whereas in *Azospirillum brasilense* it is between 6 and 9 μM O_2.[131] In *Azotobacter* for efficiency of N_2 fixation is maximal near 10 μM O_2,[32] but steady-state populations can be grown in an air-saturated medium (225 μM O_2).[143]

i. Carbon Catabolism

The anaerobic fermentation of glucose in *K. pneumoniae* strain M5a1 proceeds via the Embden–Meyerhof–Parnas pathway, followed by the action of pyruvate–formate–lyase and formate hydrogenlyase.[72] Some of the pyruvate is oxidized by a *nif*-specific pyruvate flavodoxin oxidoreductase (the *nifJ* product) to yield reducing power for N_2 fixation (discussed later). The main products of glucose fermentation are acetate, ethanol, H_2, and CO_2. The effect of providing limiting O_2 on this catabolism during diazotrophy has not been examined in detail, except that acetate production was not altered[72] and formate plus lactate can support microaerobic nitrogenase activity.[165] Attempts to establish diazotrophy during the obligate aerobic catabolism of a nonfermentable carbon soure, such as succinate, have so far been unsuccessful and may reside in a limitation of reducing power for N_2 fixation (Hill, unpublished). On the other hand, in a *K. pneumoniae* rhizosphere isolate, microaerobic diazotrophic growth and C_2H_2 reduction were supported by malate.[60] Thus different isolates may vary in their aerobic capabilities for N_2 fixation.

Organic acids are the preferred carbon and energy source of *Azospirillum*. The species of this genus differ markedly in the pattern of sugar and carbohydrate utilization (see Döbereiner and Pedrosa[46] for a review). Only *A. amazonense* can utilize disaccharides, whereas *A. brasilense*, unlike *A. ama-*

zonense and *A. lipoferum,* cannot take up glucose. Sugar catabolism proceeds through either the Embden–Meyerhof–Parnas or the Entner–Doudoroff pathways, and these organisms have an operative TCA cycle and also contain key enzymes of the glyoxylic acid cycle. Poly-β-hydroxybutyrate (PHB) synthesis in *A. brasilense* is in *Azotobacter* (discussed later) occurs under an O_2 limitation and is independent of the N source.[46,170] Metabolism of PHB in *A. brasilense* supports N_2 fixation, growth, and resistance to environmental stress.[170]

Azotobacter can utilize a wide range of sugars, alcohols, and organic acids.[172] Catabolism occurs by the Entner–Doudoroff and pentose phosphate pathways[35,76,127] and by the TCA[83] and glyoxylate cycles.[42] *A. vinelandii* also possess a *meta*-cleavage pathway for the degradation of benzoate.[95] Although the precise effects of DOC on the levels and activities of catabolic enzymes are unclear, possibly because various cultural conditions have been used,[20,35,61,83,127] moderate increases in oxygenation appear to enhance the level of enzymes associated with the TCA cycle (isocitrate dehydrogenase and 2-oxoglutarate dehydrogenase).[83] This is consistent with the finding that mutants of *A. chroococcum,* which can no longer fix N_2 on sugars in air (Fos⁻), have defects in enzymes that ensure an adequate replenishment of TCA cycle intermediates[152,153,154] (discussed later).

Limiting the supply of O_2 to *Azotobacter* leads to the accumulation of PHB. This phenomenon is not restricted to diazotrophy.[36] The work of Dawes and colleagues[35,163] concerning the regulation of glucose catabolism in *A. beijerinckii* has demonstrated that PHB synthesis serves as an electron sink permitting growth when O_2 is limiting. The synthesis and breakdown of PHB occurs by a precisely regulated cycle that, by oxidizing NADH, permits glucose metabolism and the activity of the TCA cycle to continue during O_2 limitation. Unlike the catabolic activity of fermentative organisms where partially oxidized carbon compounds are excreted[66] significant excretion by *Azotobacter*[20,83] occurs only when the oxygenation is exceptionally high.[78,108]

ii. *Respiratory Activity*

The respiratory chain of *K. pneumoniae*[166] is very similar to that of the better studied one in *Escherichia coli.*[4,41] It is branched and terminates with an *o*-type and a *d*-type cytochrome oxidase.[166] There are probably two sites of proton translocation.[86] Kinetic studies of O_2 uptake by intact *K. pneumoniae* using leghaemoglobin or myoglobin as the O_2 sensor indicate that a high-affinity oxidase (apparent K_s of 0.11 μ*M* O_2) occurs in N_2-fixing populations.[12] An additional oxidase of lower affinity (apparent K_s of 25 μ*M* O_2) appears in populations where O_2 partially inhibits nitrogenase activity.[12] The only oxidase, detected spectrophotometrically, during anaerobic or microaerobic N_2 fixation is of the *d*-type.[166] The purified form of this oxidase complex from *K. pneumoniae* has an exceptionally high affinity for O_2 (apparent K_m of 20 nM)[166] which is compatible with a critical role in supporting microaerobic N_2 fixation.

This role has been confirmed by exploiting the ability of *E. coli* to fix N_2 when carrying the *K. pneumoniae nif* genes. Two functions for the *d*-type oxidase were revealed by using a well-characterized *E. coli* mutant that lacked cytochrome *d* (*cyd*⁻). They are to support energy-requiring processes and to protect anaerobic processes from O_2 inhibition.[77] Preliminary investigations suggest that lactate and formate dehydrogenases may provide electrons for this microaerobic respiratory activity.[77,165]

Three presumptive terminal cytochrome oxidases of the *o*-, *d*-, and aa_3-types as well as soluble *c*-type cytochrome have been identified in NH_4^+-grown *A. brasilense*. The aa_3-type cytochrome oxidase is seen mainly during growth with high DOC,[17,135,155] whereas the *o*-type cytochrome oxidase may have a role in O_2 recognition for aerotactic behavior.[135] Membranes from NH_4^+-assimilating populations grown at high DOC (near 240 μM) contain greater levels of succinate and NADH oxidase activities than those grown at low DOC (near 10 μM).[133,135] Kinetics of O_2 uptake by membranes from high-DOC-grown cells indicated the presence of two terminal oxidases (apparent K_m of 1.5 and 28 μM O_2).[135] The O_2 electrode was not sensitive enough to resolve the kinetics of O_2 uptake by membranes from cells grown at low O_2. However, a study in which leghaemoglobin or myoglobin was used as the O_2 sensor revealed a single terminal oxidase (apparent K_s of 0.005 μM) in intact *A. brasilense* grown on N_2 at DOTs between 0.44 and 11 μM.[12] During growth at the highest DOT a second terminal oxidase was identified (apparent K_s of 0.5 μM O_2).[12] Proton translocation experiments on NO_2^--grown *A. brasilense* suggest that energy conservation with O_2 as the terminal electron acceptor occurs at three sites.[33] Clearly more information regarding the composition of the respiratory chain during diazotrophic growth is needed to characterize the components that are associated with microaerobic N_2 fixation.

The respiratory chain of *Azotobacter* has been extensively investigated (for reviews see refs. 63,88,160,187,188). Therefore, only a summary is presented here with references only to more recent work. The components of this branched chain include NADH, NADPH, succinate, and malate dehydrogenases, all of which, except the last, are closely associated with flavin,[138] an uptake hydrogenase, a pool of ubiquinone-8, and at least seven cytochromes of the *a*-, *c*-, and *d*-types. The chain is branched after the ubiquinone cytochrome b_1 segment and ends in two terminal oxidases, which apparently differ in their affinities for O_2 (discussed later). The $c_4 + c_3$ branch terminates in an *o*-type cytochrome oxidase, which has been purified. It contains protohaem, but apparently no Cu.[185] Its K_m for O_2 has not been measured.[185] The other branch terminates in a *d*-type cytochrome oxidase, which is considered to be the dominant oxidase under condition of high aeration. Energy conservation is associated with NADH dehydrogenase (site I), the uniquinone segment (site II), the $c_4 + c_5$, a_1 o branch (site III), and hydrogenase (site IV).

Increases in the level of oxygenation to cultures lead to a loss of respiratory control (the ability of ADP plus phosphate to stimulate O_2 uptake) and to losses of energy conservation at sites I and III, as well as to increases in the level and activity of NADH and NADPH hydrogenases. Under these conditions the rate-limiting step in oxidase activity appears to be associated with these dehydrogenases. Besides this potential for uncoupling of respiration, additional, as yet unidentified means, such as futile cycles, may participate in dissipating unwanted electrical membrane potential or ATP.

Measurements of cytochrome content have been made by several methods on various types of culture. They show that, in general, levels of cytochrome d increase upon an increase in oxygenation,[1,47,61,85,117,132] which is consistent with the role of this oxidase in supporting the high rates of O_2 uptake in *Azotobacter* (discussed later). However, in one case, cytochrome d levels rose when O_2 became limiting at the end of batch growth.[1] Such a rise is similar to the behavior of members of the Enterobactericae when subjected to an O_2 limitation[66,166] and suggests that the synthesis of cytochrome d in *Azotobacter* is under dual control.[188]

Estimates of the apparent K_s for O_2 uptake in whole organisms and membrane preparation are shown in Table 2-2. The accuracy of such estimates depends upon the sensitivity of the O_2 sensor; thus, comparisons of values from different experimenters may be misleading. Nevertheless, separate studies suggest that, in intact organisms, the apparent K_s for O_2 changes in parallel with the ambient DOC provided for growth (see Table 2-2) and can be decreased by supplementing the growth medium with carboxylic acids (see Table 2-2).

A period of adaptation in intact organisms is required for the apparent K_s for O_2[139] and the V_{max} for respiratory activity[47] to respond to a change in O_2 supply. Such adjustments probably provide a means for the organism to cope with increased oxygenation and to maximize respiratory activity at low aeration rates. In addition, the potential for respiration to respond immediately to moderate changes in O_2 supply may exist because the DOC during growth appears to be poised between the apparent K_s and twice its value.[139]

However, the ability of intact *A. chroococcum* to display a lowering in the apparent K_s for O_2 uptake when provided with one of a range of carboxylic acids was shown by Ramos and Robson[152-154] to be an important feature of aerotolerant diazotrophy. Ramos and Robson[152] isolated three classes of mutants (Fos⁻) that were unable to fix N_2 when grown on sugars in air. This phenotype was restricted to diazotrophic growth and was corrected by

Table 2–2 Changes in the Apparent Affinity for O_2 in Intact Organisms and Membrane Preparations of Azotobacter

Organism	Growth		O_2 uptake measurements					Ref.
	Type of culture[a]	DOC (μM)	Sensor[b]	Reductant for intact organisms	K_s (μM)	Reductant for membrane preparations	K_m (μM)	
Effect of DOC during growth								
A. vinelandii 0	C	0.04–1.0	L,M	Glucose	0.48	–[c]	–	12
		0.2–1.0	M	Glucose	35	–	–	
A. vinelandii ATCC 9046	C	2.2	E	Glucose	17	–	–	132
		12.5		Glucose	83	–	–	
		78.1		Glucose	00[g]	–	–	
A. vinelandii OP	B	6	E	Parahydr oxybenzoate	10	–	–	139
		140			93	–	–	
		6		Mannitol	10	–	–	
		140		Mannitol	59	–	–	
		7		Succinate	10	–	–	
		106		Succinate	55	–	–	
Effect of reductant								
A. chroococcum MD1 or Fos⁻ class I	B	NR[d]	E	Sucrose	20	NADH, succinate, or malate	<0.5	152
				Sucrose plus carboxylic acid (e.g., acetate or pyruvate)	5–10			
Fos⁻ class II(250)	B	NR		Sucrose plus acetate	27			

Table 2–2 Continued

	Growth			O_2 uptake measurements				
Organism	Type of culture[a]	DOC (μM)	Sensor[b]	Reductant for intact organisms	K_s (μM)	Reductant for membrane preparations	K_m (μM)	Ref.
Fos⁻ class III(189)	B	NR		Sucrose plus succinate	5			
				Sucrose plus pyruvate	46			
				Sucrose plus acetate	12			
A. vinelandii OP or 11[e]	B	NR	E	–	–	NADH or malate	18	80
A. vinelandii OP	B	NR	E	–	–	Ascorbate plus TMPD[f]	3.1	

[a]Batch (B) or chemostat (C) growth.

[b]Sensors for O_2 uptake measurements were either O_2 electrodes (E) or leghaemoglobin (L) or myoglobin (M).

[c]Not determined.

[d]Not recorded.

[e]Strain 11 is an ascorbate–TMPD oxidase-negative mutant in which the branch of the respiratory chain terminating in the o-type cytochrome oxidase is inoperative.

[f]N,N,N^1,N^1-tetramethylene-p-phenylenediamine.

[g]Infinity.

97

either lowering the pO_2 or adding a carboxylic acid that increased the apparent affinity for O_2.

Class I mutants were corrected by the full range of carboxylic acids or by the introduction of a plasmid, which constitutively expressed the NifA of *K. pneumoniae*. This behavior suggested that the mutation somehow rendered nitrogenase synthesis hypersensitive to O_2 repression. A representative of class II (Fos252) did not respond to acetate or pyruvate and was found to possess a defective citrate synthase,[153] whereas a representative of class III (Fos189) did not respond to pyruvate and was found to have a defective phosphoenol pyruvate carboxylase.[154] In contrast to whole cells, respiratory kinetics in membrane preparations from mutants and the wild type were similar, indicating that the altered respiratory properties of the mutants were not directly attributable to membrane components.[152] This elegant work indicates that the replenishment of TCA cycle intermediates and the maintenance of a high respiratory activity at low DOC are important features of aerotolerance in *Azotobacter*. The underlying mechanism remains to be elucidated. Ramos and Robson[152] speculated that the influence of carboxylic acids may rest upon the control of electron flux. They[152] also drew attention to the role of Ca^{2+} in aerotolerance. Earlier work (reviewed in Ref. 152) showed that diazotrophy in *Azotobacter* requires a raised Ca^{2+} level, which could be replaced by carboxylic acids. Ramos and Robson[152] found that the addition of extra Ca^{2+} to the growth medium lowered the apparent K_s for O_2 uptake in intact *A. chroococcum* and restored aerotolerant diazotrophy to Fos⁻ mutants, except to those that were not corrected by pyruvate. Whether the $2H^+/Ca^{2+}$ antiporter and the electrogenic Ca^{2+} porter of *Azotobacter*[194] have roles in the uncoupling of respiration, as has been suggested,[152] remains to be established.

Mutants that are defective in terminal oxidase function have provided a means of assigning roles for the *o*- and *d*-type cytochrome oxidases with regard to energy conservation and respiratory protection. Mutants of *A. vinelandii* (from *N*-methyl-*N'*-nitro-*N*-nitrosoguanidine mutagenesis) with defects in the branch terminating in the *o*-type cytochrome oxidase still fix N_2 in air.[80,81] By comparing the respiratory kinetics in membranes prepared from the wild type and from one of these mutants (strain II), the apparent K_m for the *o*-type cytochrome oxidase was found to be lower than that of the *d*-type cytochrome oxidase (see Table 2-2). Consistent with the loss of function of the limb of the respiratory chain terminating in the *o*-type oxidase, the mutant displayed a higher steady-state DOC and a lower biomass than the wild type during N_2-fixing O_2-limited growth.[122]

Recently, mutants in *A. vinelandii* have been constructed in which Tn5-B20 is inserted into or near to the region of the chromosome that hybridizes to the *cydAB* genes of *E. coli*.[97] Insertion in the region hybridizing to the *E. coli cydB* gene led to loss of the spectrum attributed to cytochromes *d* and $b_{595}(a_1)$. These mutants were unble to grow on solid N-free medium in air. Interestingly, poor growth in air also occurred when the medium was supplemented with NH_4^+. Good growth was restored on both media when the pO_2 was lowered. Since this microaerophilic requirement was not specific for N_2 fixation, respiratory activity involving the *d*-type oxidase appears to protect other enzymes, in addition to nitrogenase, from inhibition by O_2.

Mutants carrying Tn5-B20 inserted near the putative *cydA* of *A. vinelandii* made more cytochrome *d* than the wild type.[97] A similar phenotype was found in a class of mutants isolated after chemical mutagenesis.[123] These mutants grow normally in air.[97,123] However, the transposon insertion mutants failed to grow under the microaerobic condition.[97] Clearly, further examination of these insertion mutants should be very rewarding.

B. Behavior of Azotobacter *in C-Limited Chemostats*

Dalton and Postgate's observations[32] on the behavior of *A. chroococcum* in C-, P-, and N_2-limited chemostats rationalized the concept of respiratory protection. N_2-limited[32,47] and O_2-limited N_2-fixing[78,114] populations adapted to an O_2 stress by increasing their rate of respiration, but at the expense of inefficient N_2 fixation.[32] On the other hand, C- or P-limited N_2-fixing populations, unlike equivalent NH_4^+-assimilating ones, lost viability when subjected to an O_2 stress, unless either a C source (for the C-limited)[32] or a fixed N source (for the P-limited)[114] was provided. Under neither limitation did the respiratory activity increase upon the imposition of an O_2 stress.[110,114] Although the fatality in the P-limited population has been attributed to inadequate respiratory control,[32,114] such a proposal is not consistent with the low ADP/ATP ratio found in P-limited batch cultures.[38]

Carbon-limited N_2-fixing populations of *A. vinelandii* can be maintained at a DOC of air saturation (225 μM).[143] In such cultures the O_2 gradient between the medium and the site of N_2 fixation must be very steep. The detailed characterization by Oelze and colleagues[108–110,143,144,158] of the steady-state growth of *A. vinelandii*, when maintained over a range of DOC, strongly suggests that mechanisms in addition to respiratory activity are involved in achieving such a steep O_2 gradient. Up to a DOC of 68 μM the respiratory activity[110,143] and the rate of sucrose consumption increased,[109] and the kinetics of NADH, NADPH, and malate oxidase activities in membrane preparations[144] changed proportionally in response to increases in the DOC provided for growth. Above a DOC of 68 μM increases or changes in these parameters were much less marked. Thus, it was argued that respiration alone cannot account for the protection of nitrogenase from O_2 inhibition.[110]

The specific activity of nitrogenase, measured in situ, was independent of the DOC[45] and of the type of C source supporting growth.[110] On the other hand, the specific activity was directly proportional to the dilution rate (D),[45,110] although the levels of nitrogenase polypeptides were not greatly affected by D or DOC.[45] Thus, Kuhla and Oelze[110] argued that the rate-limiting step in N_2 fixation is the flux of suitable reducing power for nitrogenase activity. When they applied a defined O_2 stress, the proportion of nitrogenase remaining active, was dependent only upon the initial nitrogenase activity and not on the respiratory activity, which did not apparently alter.[110] Therefore, Kuhla and Oelze[110] concluded that the greater the flux of electrons to nitrogenase the greater the stability of the enzyme toward O_2 (also discussed later). The mechanism regulating this electron flux is unknown.

An extensive intracytoplasmic membrane can develop in *Azotobacter*. Biochemically, this membrane does not apparently differ from the cytoplasmic membrane.[144] Whether and how extensively the development of the intracytoplasmic membrane is influenced by the N and O_2 status has proved controversial,[188] possibly because the DOC was not always controlled. However, Post et al.,[142] who used N_2-fixing and NH_4^+-assimilating C-limited populations maintained over a range of DOC (2–225 μM), have shown that in both types of population the cell size and the extent of the intracytoplasmic membrane increased with DOC. Only in the N_2-fixing population did the ratio of intracytoplasmic to cytoplasmic membrane increase over the whole range of DOCs used. Such changes might result in steeper O_2 gradients developing within the bacterial cell so that at high DOCs the spatial separation of O_2 and nitrogenase (which is apparently located on the inside of cytoplasmic membranes[82]) could be achieved within a single cell. It seems likely that a combination of temporal, spatial, and morphological factors contribute to aerotolerance of diazotrophy in *Azotobacter*.

C. Inhibition of Nitrogenase Activity by O_2 (Conformational Protection and Autoprotection)

The inhibition of nitrogenase activity *in vivo* can range from slight to complete depending upon the severity of the O_2 stress. Upon the removal of the O_2 stress and in the presence of a protein synthesis inhibitor the return of activity can be complete, incomplete, or absent. The reversibility of O_2 inhibition of nitrogenase activity, which was first characterized in *Azotobacter*,[47,186] indicates that the enzyme can exist in an apparently inactive form that is protected from O_2 damage. Further work (see Robson and Postgate[160] and Yates[188]) revealed that this protection in *Azotobacter* probably involves protein–protein interactions; it is sometimes referred to as conformational protection. Reversible inhibition by O_2 occurs in other free-living heterotrophic diazotrophs (Table 2-3), including obligate anaerobes. So far protection by conformational protection involving redox proteins has been found only in *Axotobacter,* and there is no evidene for a modification of the Fe

Table 2–3 *Free-Living Heterotrophic Diazotrophs Showing Reversible Inhibition of Nitrogenase Activity by Either O_2 or $NH_4{}^+$*

	References for inhibition by	
Diazotroph	O_2	$NH_4{}^+$
Azotobacter chroococcum	186,47[a]	22
Azotobacter vinelandii	43	112, 57, 101
Derxia gummosa	69	ND[d]
Azorhizobium caulinodans	ND	110a
Xanthobacter flavus	75	ND
Azospirillum brasilense	68	67, 52, 118[b]
Azospirillum lipoferum	68	67, 52[b]
Azospirillum amazonense	ND	168, 67
Herbaspirillum seropedicae	ND	51
Methylococcus capsulatus	ND	126
Methylocystis parvus	ND	126
Methylosinus sporium	ND	126
Methylosinus trichosporium	ND	126, 190
Beggiatoa alba	ND	140
Klebsiella pneumoniae	71,[a] 56	175,[c] 57[c]
Vibrio spp	178	ND
Desulfovibrio gigas	148[a]	148
Clostridium pasteurianum	ND	57[c]

[a]The reversibility occurs in the presence of a protein synthesis inhibitor.

[b]The Fe protein nitrogenase in these organisms undergoes covalent modification.

[c]The addition of $NH_4{}^+$ does not inhibit nitrogenase activity.

[d]Not determined.

protein of nitrogenase, as occurs during $NH_4{}^+$ inhibition in some diazotrophs[68] (see Table 2-3).

Nitrogenase in crude extracts of *Azotobacter* sediments as an air-tolerant complex that contains a third redox protein (called the Shethna, FeS II, or protective protein). The biochemical characterization of this complex has recently been reviewed.[188] Therefore, only a summary of pertinent points relating to the physiology is given here. The formation of the air-tolerant complex of *Azotobacter* nitrogenase requires the FeS II protein and Mg^{2+} but the precise ratio of the components remains unresolved. Complex formation appears to depend on controlled oxidation of the component proteins. The Fe protein of nitrogenase can reduce the FeS II protein; therefore, changes in electron flux to nitrogenase *in vivo* may regulate complex formation. Protection from O_2 damage *in vivo* occurs rapidly, and the efficacy is independent of the magnitude and duration of the O_2 stress, but diminishes during growth at high DOC.[43] The FeS II protein may have other functions *in vivo*, since its synthesis is regulated neither by $NH_4{}^+$ nor by O_2.[159]

In crude extracts of *A. chroococcum,* the V nitrogenase, unlike the Mo nitrogenase, does not sediment as a complex and is more sensitive to exposure to oxygen.[49] Thus the V nitrogenase may not form an O_2-stable complex with the FeS II protein. This might account for the somewhat greater O_2 sensitivity of V-dependant compared to Mo-dependent diazotrophic growth in *A. chroococcum*.[49]

In addition to conformational protection of the Mo nitrogenase by the FeS II protein in *Azotobacter* the reversibility of O_2 inhibition probably involves, in both this genus and other organisms, the diversion of electrons away from nitrogenase to oxygen. The proportion of electron flux to nitrogenase compared to that supplying respiration is small in *Azotobacter* (approximately 10% in C-sufficient cultures[180] or less than 1% under C-limited conditions[110]), but somewhat larger in *K. pneumoniae* (approximately 20% in C-sufficient cultures[56,71,72]). In *K. pneumoniae* a diversion of electrons was considered to account for the effect of cyanide on respiration, which is partially inhibited, and on nitrogenase, which apparently becomes more O_2 tolerant.[56] On the other hand, the earlier experiments of Drozd and Postgate[47] with *A. chroococcum* and the more recent ones[77] with Cyd$^+$ and Cyd$^-$ *E. coli* strains carrying *K. pneumoniae nif* genes indicate that the effect of lowering the respiratory rate is to decrease the O_2 tolerance of C_2H_2 reduction. The difference in these results probably arises because O_2 uptake and C_2H_2 reduction were not measured under the same conditions of DOC. At low DOC the relatively cyanide-insensitve *d*-type cytochrome oxidase, compared to the cyanide-sensitive *o*-type oxidase, is particularly active, in *K. pneumoniae* or in *E. coli* carrying *nif* genes, because of its high O_2 affinity. Thus, in the presence of cyanide an enhancement of respiratory activity at low DOC could lead to a greater O_2 gradient and thus an apparently greater O_2 tolerance of C_2H_2 reduction.

A further possibility to explain reversible inhibition by O_2 of nitrogenase activity *in vivo* involves the reduction of O_2 by the Fe protein of nitrogenase ("autoprotection").[173] O_2 is reduced to H_2O_2 and more slowly to H_2O by a greater than fourfold molar excess of the reduced Fe protein of nitrogenase without loss of activity. This reduction occurs with purified proteins from either *K. pneumoniae* or *A. chroococcum* and in the presence of either MgADP or MgATP.[173] Such activity could contribute significantly to O_2 uptake by intact organisms[173] and might account for the response of nitrogenase activity to a defined and controlled O_2 stress in C-limited populations of *A. vinelandii*.[110] In such populations the electron flux to nitrogenase apparently influences the degree of switch-off by O_2.[110] This switch-off may represent electron flux for O_2 reduction by the Fe protein of nitrogenase.

The mechanism and products of O_2 inactivation of nitrogenase are not fully characterized.[160,173] The oxidation of the reduced Fe protein of nitrogenase (from *K. pneumoniae* or *A. chroococcum*) by excess O_2, which inactivates the protein, generates H_2O_2 probably via O_2^-.[173] The damage has not been fully characterized, but initially does not involve the 4Fe4S cluster.[173] Thus, the nitrogenase complex may generate, and be damaged by, the potentially toxic activated O_2 species associated with aerobic metabolism.[192] Increased levels of superoxide dismutase and catalase are in some cases correlated with nitrogenase levels or DOC.[44,136,151] However, mutants with elevated levels of catalase do not exhibit a greater tolerance of diazotrophy to O_2 than the wild type.[46,79]

D. Regulation of Nitrogenase Synthesis by O_2

Elevated levels of oxygenation that inhibit nitrogenase activity also prevent *de novo* synthesis of nitrogenase in the heterotrophs so far examined (*K. pneumoniae, Azotobacter,* and Rhizobiaceae). In the Rhizobiaceae regulation by O_2 status is more rigorous than that by fixed N status. This probably reflects their symbiotic habit. In contrast, in the free-living heterotroph *K. pneumoniae*, regulation by O_2 and N status apparently share common features. The present understanding of the mechanism of *nif* transcriptional control by O_2 in *K. pneumoniae* and Rhizobiaceae is reviewed in Chapter 21. In addition, O_2 regulation regulation of nitrogenase synthesis was the subject of a recent review.[75] Therefore, in this section the mechanism of O_2 regulation of *nif* transcription in *K. pneumoniae* is briefly outlined, and additional features of O_2 regulation of nitrogenase synthesis that are not covered in Chapter 21 are described, namely, the stability of *nif* mRNA and the involvement of nitrogenase activity.

In *K. pneumoniae* the mechanism of *nif* transcriptional regulation by O_2 status comprises two tiers. Transcription from p*nifL* is less sensitive to O_2 than that at the other *nif* promoters. The former transcriptional regulation probably involves DNA topology, whereas the latter apparently shares common features with that mediated by fixed N status. Upon an elevation in either O_2 or N status NifL prevents the transcriptional activtion by NifA of other *nif* promoters. The mechanisms whereby NifL apparently senses changes in the environmental O_2 and N status are unknown. In addition, the *nifX* gene product has been shown recently to have a role in negative regulation of *nif* transcription or *nif* mRNA stability in response to O_2 or N status.[58] This introduces another component into the regulatory mechanism.

So far, a *nifL*-like gene has not been found in any other diazotroph, and the mechanism of O_2 regulation in other free-living heterotrophs (besides Rhizobaceae; see Chapter 21) has yet to be explored. However, part of the ORF (open-reading-frame) preceding *nifA* in *A. vinelandii* shows some homology with the *K. pneumoniae nifL* gene.[10] In *K. pneumoniae* the rates of repression by O_2 and NH_4^+ of ongoing synthesis of either *nifHDKY* mRNA or *nifHDK* polypeptides are similar.[19,27,91] On the other hand, in *A. chroococcum* the rate of repression of ongoing synthesis of *nifHDK* polypeptides by O_2 is considerably faster than that by NH_4^+.[146,159] Thus, the mechanisms regulating *nifHDK* expression by O_2 and N status in *A. chroococcum* may differ from those in *K. pneumoniae*.

The *nif* mRNAs of *K. pneumoniae,* apart from *nifLA* mRNA, are exceptionally stable under anaerobic N-limited conditions. This stability apparently involves NifL or NifA and is rapidly lost upon a rise in O_2 or N status or a moderate rise in temperature.[28,91] In strains that lack a functional NifL, *nif* mRNA stability is maintained upon a rise in O_2 or N status, but not upon a rise in temperature.[28] Since NifA is temperature sensitive,[14,193] the stability of *nif* mRNA may involve NifA rather than NifL.[28] In *A. chroococcum* during nitrogenase derepression *nif* mRNAs are also unusually stable.[146] Following an O_2 stress the subsequent recovery of *nifHDK* polypeptide synthesis occurs in the presence of rifampicin, which inhibits the synthesis of RNA and the synthesis of other polypeptides.[146] Therefore, in this organism the stability of *nif* mRNA is apparently retained upon O_2 treatment.

The mechanism whereby NifL prevents NifA activation at the various *K. pneumoniae nif* promoters, is probably as O_2-sensitive as nitrogenase activity. A DOC of 100 nM O_2, which is near the apparent K_s for O_2 uptake in intact *K. pneumoniae,* inhibits both p*nifH* expression and nitrogenase activity by approximately 50%.[73] Furthermore, significant nitrogenase activity is detected during derepression conducted at this DOC. Therefore, none of the synthetic processes that are required for nitrogenase activity are markedly more O_2-sensitive than expression form p*nifH*.[73] Similarly, in *A. chroococcum* there appears to be a correlation between the levels of O_2 required to inhibit nitrogenase activity and nitrogenase synthesis. Nitrogenase activity is inhibited by 95% at the minimum DOC required to repress its synthesis.[146,159]

When *K. pneumoniae* is subjected to N starvation under anaerobic conditions, *nif* expression and nitrogenase poplypeptides synthesis are curtailed (discussed later). So, elevations in O_2 status that inhibit nitrogenase activity could initiate N starvation and hence prevent nitrogenase synthesis. Thus, in addition to NifL-mediated regulation, O_2 control of nitrogenase synthesis may involve its activity. This has been demonstrated in a NifL$^-$ *K. pneumoniae* strain.[76] The involvement of N nutrition in the regulation of nitrogenase synthesis by O_2 status could account in part for the rapid recovery of *nif* mRNA and polypeptide synthesis that occurs upon the removal of an O_2 stress in *K. pneumoniae*.[19] A similar rapid recovery of nitrogenase polypeptide synthesis occurred in *A. chroococcum*.[159] In both cases the recovery was accompanied by a rapid return of nitrogenase activity and hence N_2 fixation.

III. Supplies of Electrons for Nitrogenase Activity

The topic of electron transport to N_2 has recently been reviewed[62] and is also covered elsewhere in this volume (see Introduction and Chapters 18 and

20). However, in the context of the physiology of free-living heterotrophs, the following points are relevant:

1. For activity the nitrogenase complexes require a supply of electrons at redox potentials between about -400 and -500 mV.[37]

2. Reduced ferrodoxins or the hydroquinone form of flavodoxins can fulfill this role. These electron carriers are found in obligate anaerobic, facultative anaerobic, and obligate aerobic diazotrophs.

3. During anaerobic catabolism there is a need to remove reducing power so that fermentation processes can proceed in order to yield ATP for biosynthesis. The reductions of both ferrodoxins and flavodoxins are associated with these catabolic processes. In those diazotrophs so far studied, the electrons for nitrogenase are generated from the oxidation of pyruvate, which yields reduced ferrodoxin in *C. pasteurianum* and reduced flavodoxin in *K. pneumoniae*. In the latter organism this electron transport pathway is *nif* specific. The extremely O_2-labile pyruvate oxidoreductase (*nifJ* product) and the flavodoxin (*nifF* product) have been purified and support nitrogenase activity *in vivo* (also see Wahl and Orme-Johnson[179]).

4. During aerobic catabolism reducing power is required for the conservation of energy through respiratory activity. Thus, in aerobic diazotrophy electron donation to nitrogenase represents a diversion of electron flow away from respiration (this is only of the order of 1–10%[110,180]). In no aerobe has the route of electron transfer from the level of a carbon intermediate to nitrogen been established. It has been proposed that the reduction of ferrodoxin or flavodoxin could occur by reversed electron transport, which may involve respiratory activity. Even the recently identified ATP-dependent reverse respiratory electron flow to NAD^+ has been included in a scheme.[64] However, the reconstruction of such pathways *in vio* to yield respectable nitrogenase activity has generally failed. The reasons for this inactivity are discussed elsewhere.[62] Recently, electron transfer from formate to nitrogenase in *Methylosinus trichosporium* has been demonstrated with partially purified components.[25]

IV. Energy Requirement for N_2 Fixation *In Vivo*

The reaction between N_2 and H_2 to form NH_3 is exergonic and therefore should proceed spontaneously. However, the reaction is restricted kinetically owing to the associated high energy of activation, which is related to the stability of the N_2 molecule. Industrially, the reaction is speeded up by a metal catalyst and high temperature and pressure. Biologically, the catalytic reduction of N_2 by nitrogenase depends upon a source of low-potential electrons. These are supplied by the reduced form of the Fe protein of nitrogenase, involving a process requiring the concomitant hydrolysis of ATP.

The reduction of N_2 by nitrogenase is accompanied by the reduction of H^+. The proportion of total electron flux yielding H_2 can vary, but 25% for this purpose is apparently mandatory (see Chapter 18). The principal factors that can increase the electron flux for H_2 evolution are as follows: (1) a decrease in the rate of total electron flow to nitrogenase, which can occur by lowering the ratio of effective Fe protein to MoFe protein of nitrogenase and by diminishing the concentrations of ATP or reductant;[188] (2) the presence of inhibitory levels of H_2, which specifically inhibit N_2 reduction;[188] and (3) the expression of the alternative nitrogenases in place of the Mo nitrogenase (see Chapter 19). *In vitro* the most efficient stoichiometry in terms of electrons and ATP is as follows:

$$N_2 + 8H^+ + 8e^- + 16ATP = 2NH_3 + H_2 + 16ADP + 16Pi$$

There may be an additional energy source *in vivo* for providing a suitable supply of electrons to reduce the Fe protein of nitrogenase (discussed earlier). In this section, methodologies, results, and discussions of the estimates of the apparent ATP requirement for N_2 fixation *in vivo* are summarized.

A. The Carbon and Energy Source Expenditure for N_2 Reduction

Estimate of the energy required for N_2 fixation *in vivo* are based upon comparing the efficiency of carbon and energy source utilization in N_2-fixing and NH_4^+-assimilating populations (Table 2-4, column 3). Differences in the consumption of the carbon and energy source for processes other than growth are minimized in carbon- and energy-limited chemostats. Strictly speaking, such chemostat populations are either carbon or energy limited. For example, during anaerobic fermentative growth they are generally energy limited, whereas during aerobic growth the type of limitation depends on the degree of reduction of the carbon and energy source (see de Vries et al.[40] for discussion).

Table 2-4 (column 3) lists examples of these efficiencies in heterotrophs where catabolism proceeds by either fermentation or respiration. The efficiencies are generally higher during growth supported by respiratory activity because of the greater energy conservation associated with this process. The data in Table 2-4, column 4, show the apparent carbon and energy source expenditure for N_2 fixation, and are computed from the data in column 3. This expenditure is lowest in the microaerophiles *A. brasilense* and *A. caulinodans* and in *Azotobacter* during growth at a low DOC. The higher values for *Azotobacter* reflect the expenditure of carbon and energy source for respiratory protection.

Table 2–4 Carbon and Energy Source Requirement for N_2 Fixation in Heterotrophs Grown in Chemostats Limited by the Carbon and Energy source

(1) Organism	(2) Carbon and energy source	(3) Efficiency of N incorporation during growth (mg N per g carbon and energy source used) with		(4) Carbon and energy expenditure for N_2 fixation (g carbon and energy used for 1 g N_2 reduced)	(5) Ref.
		N_2	NH_4^+		
Anaerobic Growth supported by Fermentation					
Clostridium pasteurianum	Sucrose	11	22	46	29
Klebsiella pneumoniae	Glucose	8	19	72	72
Anaerobic Growth Supported by N_2O Respiration					
Azospirillum brasilense	Malate	26[a]	52[a]	19	33
Aerobic Growth					
Klebsiella pneumoniae (O_2 limited)	Glucose	15	35	38	71,72
Azorhizobium caulinodans (at 2 μM O_2)	Succinate	23	40	19	167
Azospirillum brasilense (at 9 μM O_2)	Malate	26[a,b]	48[a]	19	100,33
Azotobacter chroococcum (at 5–10 μM O_2)	Mannitol	33	40	5.3	32
Azotobacter vinelandii (at 2–10 μM O_2)	Sucrose	16[b]	63	47	143,108
Azotobacter vinelandii (at 180 μM O_2)	Sucrose	7[b]	38	117	143,108
Azotobacter vinelandii (at 180 μM O_2)	Sucrose	29[a]	–	–	109

[a]Calculated from the theoretical maximum molar growth yields (i.e., the value at infinite dilution rate) and either the reported or assumed N content of the dry weight (13%).

[b]Includes excreted fixed N during growth.

Table 2–5   Ratios Associated with the Energy Expenditure for N_2 Fixation In Vitro and In Vivo

N_2-Fixing System	ATP:N_2	H_2:N_2	ATP:e^-	% ATP Utilized for H_2 Evolution	Ref.
In vitro under optimum conditions	16	1	2	25	Ch. 18
In vivo during:					
Anaerobic glucose-limited growth of *K. pneumoniae*	29	2.5	2.7	45	72
Anaerobic glucose-supported NH_3 excretion by *K. pneumoniae*	22	1.3	2.6	31	3
Aerobic succinate-limited growth by *A. caulinodans*	40	1.5	4.4	33	41

With *A. brasilense* neither the efficiencies of N incorporation (Table 2-4, column 3) nor the carbon and energy source expenditure for N_2 fixation (Table 2-4, column 4) are lowered by changing the terminal electron acceptor from O_2 to N_2O. Danneberg et al.[33] rationalized these results thermodynamically, since the redox potential of the N_2/N_2O couple ($E_0' = +1.36$ V) is higher than that of the $H_2O/^1/_2O_2$ couple ($E_0' = +0.81$ V). However, these results were not consistent with the lower levels of proton translocation that were obtained with N_2O instead of O_2 as the terminal electron acceptor.[33]

B. The Apparent ATP/e^- Ratio for N_2 Fixation in Obligate and Facultative Anaerobes

When the coupling of catabolism to energy conservation is known, as is the case for substrate phosphorylation during fermentation, then the apparent ATP requirement for N_2 fixation (the ATP/N_2 ratio) can be calculated from the estimate of the carbon and energy expenditure for N_2 fixation. The apparent ATP/N_2 ratio during growth was 20 in *C. pasteurianum*[29] and 29 in *K. pneumoniae*.[72] These values are significantly higher than the minimum of 16 obtained *in vitro* (Table 2-5). In *K. pneumoniae* the high ATP/N_2 ratio [which excludes the ATP requirement for NH_3 assimilation by glutamine synthetase (GS)] was attributed to an elevated level of H_2 evolution by nitrogenase, because the organisms were energy limited. Indirect evidence suggested that about 45% of the ATP expenditure was associated with H^+ reduction.[72] More recently, Kleiner[99] has identified a further likely ATP requirement associated with N_2 fixation *in vivo*. His analysis[99] of the energy cost for cyclic NH_3/NH_4^+ retention sugesets that at least 4 ATP per N_2 molecule fixed (equivalent to 0.67 ATP/e^-) may be used to retain the NH_3 before it is assimilated. It is interesting to note that when N_2-fixing cultures of *A. vinelandii, K. pneumoniae,* and *C. pateurianum* were treated with excess NH_4^+ under $^{15}N_2$, approximately 95% of the fixed ^{15}N appeared in the medium as $^{15}NH_4^+$ (the form of fixed ^{15}N was checked only for *A. vinelandii*).[57] This excretion was presumably a result of the relaxation of the energy-driven NH_4^+ uptake[99] when excess NH_4^+ was provided.

A mutant of *K. pneumoniae* that was unable to assimilate NH_3 allowed Andersen and Shanmugam[3] to estimate the apparent ATP/N_2 ratio in non-growing cells supplied with excess glucose. They also measured the proportion of H_2 evolved to N_2 reduced during N_2 fixation by using a Hyd^- derivative of the NH_3-excreting strain that failed to produce H_2 from fermentation. The values of their ATP/N_2 and H_2/N_2 ratios are shown in Table 2-5, where they are compared with those found during growth of wild-type *K. pneumoniae*. Interestingly, the derived ATP/e^- ratios are similar, and when provision is made for the ATP requirement for NH_3 retention they are close to the value of 2 found *in vitro*.

C. Maintenance Energy in Obligate Aerobes

The efficiency of N_2 incorporation in *Azotobacter* is greatly influenced by the DOC (Table 2-4) because of the carbon and energy source consumption associated with respiratory protection. The magnitude of this consumption is reflected in estimates of the maintenance energy coefficient (ME). Values of ME for gowth limited by the supply of the carbon and energy source for *Azotobacter* and *A. brasilense* are shown in Table 2-6. The data illustrate four points.

1. The ME in N_2-fixing populations of *Azotobacter* increases as the DOC is increased. However, as indicated in an earlier section, the increases at high values of DOC are not adequate to account for the protection of nitrogenase from O_2 inhibition by respiratory activity alone.

2. When acetate is used instead of sucrose, the ME in *Azotobacter* is significantly lower, which may have some relevance to the involvement of carboxylic acids in aerotolerance (see earlier discussion).

3. The disparate values of ME in *Azotobacter* when glucose is the carbon and energy source probably reflect the different ways in which the cultures were supplied with O_2. The higher value was found when a constant rate of O_2 was supplied, but unlike the other culture, the DOC was not controlled. Thus, when the dilution rate was increased there was a proportional decrease in the DOC. Such changes could increase the ME and probably account for the negative instead of positive slope that Nagai and Aiba[128] found when they plotted the specific rate of glucose consumption versus growth rate to obtain their value of ME.

Table 2–6 Maintenance Energy for Carbon and Energy-Limited Growth in Chemostats Supplied with Either N_2 or NH_4^+ as N Source

| Organism | Carbon source | DOC[a] (μM) with N source as | | Maintenance energy (ME) (m mol C. g dry wt^{-1} h^{-1}) with N source as | | Ref. |
		N_2	NH_4^+	N_2	NH_4^+	
Azotobacter chroococcum	Mannitol	5–10	40	34.9	1.3	32
Azotobacter vinelandii	Sucrose	12	–	11.8	–	109
A. vinelandii	Sucrose	108	–	71.4	–	109
A. vinelandii	Sucrose	192	–	105	–	109
A. vinelandii	Acetate	108	–	65.2	–	109
A. vinelandii	Glucose	108	–	73.5	–	109
A. vinelandii	Glucose	>[b]	–	120	–	128
Azospirillum brasilense	Malate	9	NR[d]	1.32	1.2	100,33
A. brasilense	Malate	N_2O^c	N_2O^c	1.32	0.96	33

[a]Dissolved O_2 concentration (DOC).

[b]DOC decreased as D was decreased.

[c]N_2O as the terminal electron acceptor.

[d]Not recorded.

4. During diazotrophy in *A. brasilense* the ME is much lower than that in *Azotobacter* and is near the value found in a population assimilating NH_4^+. This result is consistent with the microaerophilic habit of *A. brasilense*.

D. The Apparent ATP/e⁻ Ratio for N_2 Fixation in Obligate Aerobes

During growth of aerobic diazotrophs and, in particular that of *Azotobacter*, the coupling of catabolism to energy conservation cannot be measured reliably and probably differs in N_2-fixing and NH_4^+-assimilating populations. In addition, in those organisms possessing an uptake hydrogenase, the proportion of H_2 produced to N_2 reduced by nitrogenase and the energy saving during H_2 recycling may vary. Hence, calculation of the ATP/N_2 ratio from estimates of the carbon and energy source expenditure for N_2 fixation involves a number of parameters that are not easily measured.

De Vries et al.[41] measured the H_2/N_2 ratio using a mutant of *Azorhizobium caulinodans* that lacked the uptake hydrogenase. The value of this ratio (1.2–1.9) was similar during C-, O_2-, or Mg^{2+}- limited growth. The molar growth yields during succinate-limited growth of the wild type[167] and the Hup⁻ strain[41] were similar, when corrections were made for the accumulation of PHB that occurred in the latter. Stam et al.[167] calculated a value for the ATP/N_2 ratio of 40 for *A. caulinodans*, by determining the number of reducing equivalents that were available for respiratory activity and by assuming constant energy coupling at two sites of oxidative phosphorylation (which fitted their estimates of H^+ translocation and spectroscopic studies). Using these values for the ATP/N_2 and H_2/N_2 ratios, de Vries et al.[41] obtained an ATP/e^- ratio near 4. Because this value is twice the one found *in vitro* (see Table 2-6), de Vries et al.[41] suggested that in part the difference reflects the energy requirement for providing a suitable electron donor to reduce the Fe protein of nitrogenase in this aerobe.

In addition to the values for the H_2/N_2 ratio quoted earlier for Hup$^-$ *A. caulinodans* and Hyd$^-$ *K. pneumoniae*, a value near 1 was found during sucrose-limited growth of a Hup$^-$ strain of *A. chroococcum*.[2] In *A. chroococcum* only extremes of O_2 limitation and O_2 excess increased the ratio.[180,188] Thus, in contrast to earlier belief,[72,167] nitrogenase operates *in vivo* at near maximum efficiency over a wide range of conditions. Whether any of the Hup$^-$ strains examined so far have been grown under energy-limited conditions is unclear. De Vries et al.[41] concluded that growth in apparently succinate-limited chemostats of Hup$^-$ mutants of *A. caulinodans* was probably N-limited because, unlike the wild type, these strains accumulated PHB. With sucrose-limited *A. chroococcum* the nutritional status during growth is likely to be influenced by the O_2 supply (discussed later) owing to the uncoupling of respiration associated with O_2 removal. A comparison of the H_2/N_2 ratio during anaerobic growth of a Hyd$^-$ *K. pneumoniae* strain under glucose and intrinsic N_2 limitations should indicate whether the energy status can influence the H_2/N_2 ratio *in vivo*.

E. The Benefits of Uptake Hydrogenase for N_2 Fixation

The evolution of hydrogen by nitrogenase *in vivo* was unequivocally established in 1976 (see Yates[188]). Since that time much work has focused on proving or disproving Dixon's hypotheses relating to the benefits of the uptake hydrogense for N_2 fixation. Present knowledge regarding symbiotic N_2-fixing organisms is reviewed in Chapter 11 and in Yates,[188] who also covers the topic for free-living heterotrophs. Thus, only a summary of the work relating to the latter is presented here.

There is no evidence to support Dixon's hypothesis that hydrogenase activity is beneficial by removing the H_2 inhibition of N_2 reduction. However, there is evidence to support his proposals that hydrogenase activity could recoup lost reducing power for nitrogenase activity or for respiration to remove O_2 and to produce ATP. Exogenous H_2 can support C_2H_2 reduction in *Azotobacter*,[188] in *A. brasilense*,[137] and in *Methylosinus trichosporium*.[24] It can also improve the tolerance toward O_2 of nitrogense activity in *Azotobacter*[188] but it has this effect in neither *A. brasilense*[137] nor *Pseudomonas saccharophila*.[8] Interestingly, in *P. saccharophila* the activity of hydrogenase is much more O_2 tolerant *in vivo* than that of nitrogenase, and a similar pattern is seen in *Azospirillum amazonense*.[50]

In *A. vinelandii,* H_2 and NADH are equally efficient at supporting ATP synthesis in membrane vesicles, but the former is more O_2 sensitive than the latter.[111] The pathway of electron transfer to O_2 during H_2 oxidation apparently involves cytochromes *b*, *c*, and *d*, and terminates specifically by way of the *d*-type oxidase.[184] The pathway in *A. brasilense* may proceed independently of C-dependent respiration, yield less energy, and terminate in a specific high-affinity oxidase.[137] In *A. caulinodans* H_2 oxidation is also associated with lower levels of energy conservation than C-dependent respiration.[167]

The regulation of hydrogenase synthesis appears to be influenced by the N, C, O_2, and H_2 status to varying degrees in different organisms. Exogenous H_2 usually enhances hydrogenase activity, as does a C limitation, whereas high O_2 and fixed N levels are repressive (for *Azotobacter* see ref. 188; for *Azospirillum* sp. see refs. 174 and 50; for *P. saccharophila* see ref. 8; for *A. caulinodans* see ref. 40; for *M. trichosporium* see ref. 24). However, Yates[188] concludes that the underlying regulatory mechanism probably involves the redox status of the cell.

The H_2/N_2 ratio in aerobic N_2 fixers is near 1 over a wide range of conditions (discussed earlier). Thus, under optimum conditions of O_2 supply, the proportion of total electron flux through respiration could be increased by $<3\%$ if hydrogenase recycled all the reducing power lost by nitrogenase as H_2.[188] The potential benefit from H_2 recycling during N_2 fixation therefore appears to be small. The only significant evidence, in free-living heterotrophs, for a benefit has come from studies on the survival of isogenic Hup⁻ and Hup⁺ strains of *A. chroococcum* in mixed populations during growth in chemostats.[189] The Hup⁺ strain became dominant in C- and P-limited populations when growth was dependent upon N_2 fixation. In similar populations supplied with NH_4^+ there was little change in the composition of the mixture. On the other hand, the Hup⁻ strain became dominant during diazotrophic growth in O_2-, S-, or Fe-limited populations. The results with the C and P limitations are consistent with the proposed roles of hydrogenase in augmenting ATP synthesis and removing O_2. The result with the O_2 limitation suggests that in *Azotobacter* H_2 recycling is less well coupled to energy transduction than C-dependent respiration.[189] This result constrasts with the measurement of ATP synthesis *in vitro* (see earlier).

V. The N Status and Nitrogenase Function

The mechanism of regulation of *nif* expression by fixed N has been most extensively studied in *K. pneumoniae*. In this organism it involves a two-tier model composed of the Ntr regulatory cascade and the *nif* specific regulatory gene products NifL and NifA (see Chapter 21). The mechanism whereby the N status is sensed differs within the two tiers. The Ntr regulatory cascade responds to the intracellular α-ketoglutarate and glutamine pools, whereas the environmental stimulus and the mechanism regulating the sensing of "fixed N" by NifL is unknown. Similarities and modifications to this model have been found in other free-living diazotrophs (see Chapter 21). During growth the NH_3 produced by N_2 fixation is assimilated in many free-living heterotrophs by the glutamine synthetase (GS) –glutamate synthase (GOGAT) pathway (see refs. 92, 147, and 124). GS requires ATP for activity ($ATP/NH_3 = 1$) but has a higher affinity for NH_3 than glutamate dehydrogenase (GDH). The latter enzyme, if present, is the main route for the assimilation of plentiful NH_3. However, in *Bacillus macerans* and *Bacillus polymyxa* GDH plays a substantial role in NH_4^+ assimilation during N_2 fixation.[92]

The aim of this section is to categorize the environmental N status so as to highlight spects of N metabolism, particularly in relation to the regulation of N_2 fixation. From a simplistic point of view there appear to be three potential sources of fixed N to satisfy synthetic necessities in diazotrophs: first, exogenous sources, which can be of the inorganic or organic types; second NH_3 form N_2 fixation; and third, endogenous supplies coming from recycling. In a population where nitrogenase is expressed, the level of its synthesis or activity can vary, depending upon environmental or physiological factors that influence the N status. The N status can be roughly divided into five subtypes: (1) NH_4^+ in excess, (2) N supplied from fixed N and N_2 fixation, (3) intrinsic N_2 limitation, (4) operational N_2 limitation, and (5) N starvation. Although in some cases the site of regulation is clear, in others it remains obscure, and therefore it could be at the level of transcription, translation, protein maturation, protein conformational change, protein interactions, or enzyme activity. During nitrogenase function the energy and the N status are probably closely interlinked; therefore, evidence that invokes a role for energy status in *nif* regulation is also considered.

A.   Excess NH_4^+

Nitrogenase activity in cultures of several free-living heterotrophs can be inhibited by the addition of NH_4^+ (see Table 2-3). This phenomenon, termed *NH_4^+ switch-off*, is reversible. It can involve a metabolized form of NH_4^+, but it may occur by more than one mechanism.[51] The better-understood mechanism, elucidated in the phototroph *Rhodospirillum rubrum* by Ludden and coworkers (see Chapter 3), comprises the reversible ADP-ribosylation of a conserved arginyl residue of the Fe protein of nitrogenase. This occurs by the reciprocal action of the *draT* and *draG* gene products (see Chapter 3). Physiological and biochemical studies and gene hybridization experiments have indicated that this mechanism of modification occurs in *A. brasilense* and *A. lipoferum*, but apparently not in *Azotobacter, M. trichosporium, A. amazonense, H. seropedice,* and *A. caulinodans* (see Table 2-3).

In *Azotobacter* the addition of NH_4^+ depresses the membrane potential, which could account for the inhibition of nitrogenase activity.[112] The efficacy of NH_4^+ in switching off nitrogense activity in *Azotobacter* depends upon the DOC, the pH, the age of the culture, the nature of the provided C source,[101] and the apparent C/N ratio within the cells.[22] N-starved cultures of *A. chroococcum,* like those of *R. rubrum,* show little switch-off of nitrogenase activity by NH_4^+ addition.

Although nitrogenase activity in cultures of *C. pasteurianum* and *K. pneumoniae* is not inhibited by added NH_4^+ the Fe proteins of the nitrogenases of these organisms and that of *A. vinelandii* can undergo modification by the *draT* product of *R. rubrum.*[119] Both *A. vinelandii* and *K. pneumoniae* show cross-reactivity toward antiserum prepared against the product of *draG.*[119] Clearly, the screening of diazotrophs for *draT* and *draG* genes should resolve some uncertainties.

B. N Supplied from Fixed N and N_2 Fixation

Batch cultures of free-living heterotrophic diazotrophs when supplied with an excess of carbon and energy source but a limiting supply of fixed N, such as NH_4^+, show a typical diauxic growth pattern. After growth on the fixed N there is a lag before nitrogenase activity is detected and growth is resumed. The length of this lag is considerably longer than that associated with the derepression or induction of carbon catabolic enzymes (e.g., the induction of β-galactosidase in *E. coli* occurs within 1–2 min[54]). The time taken for nitrogenase to be synthesized can be influenced by the environmental N status (see later). It can vary in different organisms; for example, in *Azotobacter* it is generally shorter (20–60 min)[103,159] than in *K. pneumoniae* (1–6 h),[19,89,91,129,175] although this difference may have an environmental rather than biological origin.

A severe N limitation in *K. pneumoniae* can elicit a stringent response, which is characterized by a rise in ppGpp concentration,[129,156,157] an increase in the stability of bulk mRNA[90] and a decrease in the rate of total RNA synthesis.[19] About one hour after the onset of N deprivation the synthesis of the exceptionally stable *nif* mRNAs can be detected.[19,90,91] Transcription of the *nifLA* regulon precedes transcription of the *nif* structural operon.[19,27] The latter transcript can vary in length, although the significance of this variation is unknown.[23,107] Although the transcription and translation of *nif* genes are closely linked,[19] there is a further delay before nitrogenase activity is detected.[19,90,159] This delay is probably associated with the time required for protein maturation (see Chapter 20) before nitrogenase activity can commence.

Three observations implicated ppGpp in the mechanism of *nif* regulation. First, levels of ppGpp are higher in N_2-grown compared to NH_4^+-grown *A. vinelandii, K. pneumoniae,* and *C. pasteurianum.*[98] Second, during derepression of *nif* in *K. pneumoniae* the level of ppGpp rises.[129,156,157] Third, derepression of *nif* is poorer in relaxed mutants of *K. pneumoniae* (*relA⁻*), which are defective in ppGpp synthesis.[156,157] However, there appears to be no obligate involvement of ppGpp in *nif* derepression, because the addition of sufficient glutamine to prevent the expansion of the ppGpp pool enhanced, rather than depressed, *nif* expression in *K. pneumoniae.*[129]

The potential for an added fixed N source to bring about either an enhancement or the repression of nitrogenase synthesis is dependent upon whether the subsequent limitation is due to N or another factor. For example, in batch cultures of *K. pneumoniae* the derepression of *nif*, as indicated earlier, can be markedly speeded up by providing a moderate concentration of an amino acid such as glutamine or aspartate.[129,175] In this case, the added fixed N is probably used for synthetic purposes, but the population is still N-limited. On the other hand, a similar concentration of aspartate when added to the medium supply of either a SO_4^{2-}-limited or C-limited population caused the complete repression of nitrogenase.[175] In these populations the concentration of fixed N was more than sufficient to support the SO_4^{2-}-limited or C-limited growth. In contrast, when the population was limited by the rate of N_2 fixation (intrinsic N_2 limitation, see later), then specific nitrogenase activity was unaffected, although biomass was increased.[175] Here the subsequent limitation was still N_2.

The poising of nitrogenase activity at different levels can be achieved in chemostat populations supplied with excess N_2 by varying the concentration of NH_4^+ in the medium supply. With SO_4^{2-}-limited populations of *A. chroococcum*[48] and of *K. pneumoniase*[175] nitrogenase, activity was inversely proportional to the supplied NH_4^+ concentration. Similarly, with sucrose-limited populations of *A. vinelandii,* nitrogenase activity increased in response to an elevation in the C/N ratio (sucrose/NH_4^+) in the inflowing medium.[16] However, at a particular ratio, nitrogense activity varied inversely with the DOC.[16] Such a response to DOC is consistent with the reciprocal changes in respiratory rate.[15] In *Azotobacter* these differences in nitrogenase activity could be due to the ability of NH_4^+ to regulate activity (see earlier). However, in sucrose-limited *A. vinelandii,* immunological data suggested that the levels of nitrogenase polypeptides were correlated with differences in activity.[16] These observations, at least in *K. pneumoniae,* are consistent with the model of *nif* regulation that invokes an equilibrium, controlled by N status, of the interaction between NifL with NifA (see Chapter 21). The relevance of this model to the effect of N status on *nif* expression in *Azotobacter* is as yet unknown (see Chapter 21).

C. Intrinsic N_2 Limitation

The maximum growth rate of free-living heterotrophs is generally slower during N_2 fixation than when an easily assimilated fixed N source is provided. The rate-limiting step has not been identified in any diazotroph. The early studies on *A. chroococcum*, growing diazotrophically in chemastats, indicated that when all nutrients are provided in excess the growth rate is controlled by some intrinsic process associated with nitrogenase activity.[32,70] Simiar intrinsic N_2-limited populations have been established with the anaerobe *C. pasteurianum*[29] and during anaerobic growth of *K. pneumoniae*.[175] As indicated earlier, such a population of *K. pneumoniae* is apparently N limited because the addition of aspartate to the growth medium increases the steady-state biomass without affecting the specific activity of nitrogenase.[175] Since the rate of carbon and energy source consumption during N_2 fixation is greater than that during NH_4^+ assimilation (see earlier), the rate of ATP production by catablism was considered to be the rate-limiting step in *C. pasteurianum*.[29] In *K. pneumoniae* the rate of glucose consumption is not maximal under an intrinsic N_2 limitation, as it can be markedly increased by lowering the concentration of the supplied N_2 to yield an operational N limitation (see later) (A. T. Smith and S. Hill, unpublished).

A characteristic of intrinsic N_2-limited growth in an chemostat is that the biomass decreases when the dilution rate (D) is increased. Thus, if a well-defined operational limitation (e.g., a C limitation) is established at a low value of D, then when D is increased the subsequent steady state may be intrinsically N_2 limited, even though the composition of the medium is unchanged. This occurred during anaerobic glucose-limited growth of *K. pneumoniae* and was accompanied by a large rise in the residual glucose concentration. A glucose limitation was regained by halving the supplied glucose concentration.[72] During sucrose-limited growth of *A. vinelandii* the significant rise in residual sucrose concentration when D was increased was interpreted by Kuhla and Oelze[109] as a change in the K_s for sucrose. They suggested that such a change could result from an alteration in the sucrose transport activity of the culture.[109] The influence of halving the supplied sucrose concentration on the residual sucrose concentration at the higher D should either confirm that the K_s had changed or reveal that the culture was intrinsically N_2 limited. Even under conditions where the carbon and energy source apparently limits growth, the probability that populations of aerobic diazotrophs are actually N_2 limited was initially proposed for *A. chroococcum*[32] and more recently considered for *A. vinelandii*[110] and *A. caulinodans*.[41] Whether this limitation occurs because oxygen inhibits nitrogenase activity

Table 2–7   Increases in Specific Nitrogenase Activity (sp. act.) in Response to an Operational N Limitation

Organism	N Supply	Excess N_2 Growth Limitations	Sp.act.	Operational N Limitations Atmospheric pN_2 (kPa)[a]	Sp.act.	Ratio of Sp.act.[e]	Ref.
Chemostat growth							
K. pneumoniae	N_2	intrinsic	32	1.2	104	3.3	176
K. pneumoniae	N_2	intrinsic	51	3.0	113	2.0	Up[d]
K. pneumoniae	N_2	glucose	64	35[b]	170	5.0	72
K. pneumoniae	$NH_4^+ + N_2$	sucrose	32	0	283	8.8	177
C. pasteurianum	$NH_4^+ + N_2$	sucrose	100	0	459	4.6	177
C. pasteurianum	$NH_4^+ + N_2$	intrinsic	180	0	300	1.7	30
A. chroococcum	$NH_4^+ + N_2$	intrinsic	5	0	13	2.6	32
Batch growth							
R. palustris[c]	yeast extract + N_2	intrinsic	9.7	0	29.3	5.0	6
K. pneumoniae nif in S. typhimurium	N_2	intrinsic	75	0	154	2.1	145

[a]Balance to 101 kPa with Ar.

[b]Balance to 101 kPa with H_2.

[c]Similar results were found in three other nonsulfur purple bacteria.

[d]A. T. Smith and S. Hill, unpublished.

[e]Ratio of specific activities, under an operational N limitation: under excess N_2.

or there is a competition for electrons between respiration and nitrogen fixation has yet to be resolved.

D.   Operational N Limitation

An operational N limitation occurs when the supplies of fixed N plus N_2 limit growth. It is recognized by a lowering in the biomass in proportion to decreases in the concentration of the supplied nitrogen. The lowering in biomass in diazotrophs is accompanied by a rise in nitrogenase-specific activity. Table 2-7 lists examples of organisms and culture conditions under which a rise in the level of nitrogenase specific activity is attributable to an operational N limitation. In addition, the inhibition of N_2 fixation by either introducing exogenous H_2 to a chemostat (see Table 2-7) or prolonging the exposure of batch cultures to C_2H_2[34] has led to increases in the level of nitrogenase specific activity. The latter observation has important practical implications when prolonged C_2H_2-reducing assays are used to assess the nitrogenase activity of natural systems.[34]

The elevation of the specific activity of an enzyme within a population in response to a limitation in the supply of a substrate for that enzyme (as shown in Table 2-7) has been referred to as hyperderepression or hyperinduction.[147] It is not a phenomenon unique associated with nitrogenase. The increased level of the cytochrome *d*-type oxidase observed under an O_2 limitation in Enterobacteriacae, first observed by Moss[125] is a classic example, and evidence now suggests that the increase in synthesis in *E. coli* occurs at the level of transcription.[53]

Only in *Rhodopseudomonas palustris* has this apparent hyperderepression of nitrogenase been investigated to determine whether the greater activity is due to enhanced synthesis. Qualitative estimates by EPR and immunological techniques indicated that the level of both nitrogense proteins were elevated.[6] Furthermore, the higher nitrogenase-specific activity was correlated with a greater expansion in the ppGpp pool.[195]

That strains are selected for high activity under the harsh conditions of N limitation seems unlikely, because in *A. chroococcum*[32] and in *C. pasteurianum*[30] nitrogenase activity decreased when an excess of N_2 was provided. Theoretically, at least in *K. pneumoniae,* a gradual adjustment of the antagonism of NifA by NifL in response to N status (see Chapter 21) could account for the apparent transition from derepression to hyperderepression of *nif* expression. However, the regulation of posttranscriptional events leading to active nitrogense have received relatively little attention and could also be involved. Interestingly, in *K. pneumoniae* the functional stability if *nifHDKY* mRNA is greater during the deprivation of fixed N accompanying *nif* derepression than during diazotrophic growth.[90]

E. N Starvation

Derepression of nitrogenase synthesis in *K. pneumoniae* can be speeded up by providing a low concentration of a nonrepressing amino acid (see earlier). This suggests that the rate of synthesis is limited by the availability of fixed N. If, on the other hand, the derepression is conducted in the absence of added fixed N and under argon in place of N_2, then the derepression is curtailed.[76] The time at which nitrogenase activity is first detected is not affected, but the subsequent rise in nitrogenase synthesis requires N_2 fixation. The mechanism by which nitrogenase function appears to regulate *nif* expression does not involve NifL and appears to be at the level of transcription,[76] although posttranscriptional events have not been ruled out.

Thus, in contrast to the hyperderepression that occurs under an operational N limitation, severe N starvation can apparently halt *nif* expression. Under conditions of severe N limitation the stringent response, characterized by the presence of ppGpp, appears to maintain intracellular levels of ATP and may suppress translation defects during *nif* derepression in *K. pneumon-*

iae.[157] In addition, although as yet unproven, the stringent response could play an important role in mobilizing cellular fixed N by the action of proteases[21,113] so as to provide a source of fixed N for nitrogenase synthesis during derepression. Under conditions of N starvation the deprivation of fixed N may limit the synthesis of these proteases, as well as that of nitrogenase.

F. The Energy Status and Nitrogenase Synthesis

Ammonia production by nitrogenase should be maximized by providing an adequate electron flux to sustain activity. Such a flux depends on the ATP pool and the ATP/ADP ratio (see ref. 117). Consequently, when nitrogenase is active, changes in the energy status could indirectly influence nitrogenase synthesis through an effect on N status. Alternatively, the effect of energy status on nitrogenase synthesis might be more direct.

The evidence indicating that nitrogenase synthesis is affected by energy status is summarized in Table 2-8. Generally, perturbations leading to an elevation in energy status are correlated with a rise in nitrogenase activity. This rise, in several instances, is associated with an increase in an aspect of nitrogenase synthesis. In one case the effect was posttranscriptional.[169]

Comparisons of ATP pool size and the ATP/ADP ratio during N_2-fixing and during NH_4^+-assimiating growth have been made. During chemostat growth N_2 fixation appeared to decrease the ATP pool and markedly lower the ATP/ADP ratio (see Table 2-8). In batch growth there was no significant change with N source (see Table 2-8). Whether these differences reflect the organisms, the method of growth, or the techniques of sampling for adenylates is unknown.

VI. Conclusions

Aspects of the reactions of free-living heterotrophic N_2 fixers to the change in O_2, energy, or N status are described in this chapter. The dividing lines between these reactions are becoming more blurred, which in one sense introduces more complexity, but in another may reveal opportunities for new approaches.

Although reactions to O_2 status of heterotrophic diazotrophs have been characterized, the underlying mechanisms remain unclear. For instance, the distribution of the ability among facultative anaerobes to harness respiratory-driven ATP production for N_2 fixation is as yet unknown. Because of the high energy requirement for N_2 fixation, this facility may be widespread. Perhaps some can couple an obligate aerobic catabolism to N_2 fixation for growth.

Table 2–8 Correlations of Changes in the ATP Pool with Nitrogenase Level following a Perturbation in Energy Status

| Organism | Environmental Conditions for | | Effect on | | Observed Site of Regulation of Nitrogenase synthesis | Ref. |
	Control	Perturbation	ATP Pool	Nitrogenase Activity		
Derepression						
K. pneumoniae wt. or relA⁻	Anaerobic	Plus adenine and aspartate	Inc.[a]	Inc.[a]	Polypeptide synthesis[c]	157
K. pneumoniae wt.	Anaerobic	Plus glutamine	Inc.	Inc.	pnifH expressions[c]	129
K. pneumoniae hisA⁻	Anaerobic	Plus extra histidine	Inc.	Inc.	Post pnifL, H or N expressions[c]	84,169
Growth						
K. pneumoniae wt.	Anaerobic	Microaerobic	Inc.	Inc.	pnif L or H expression	74
Rhizobium sp	O₂ limited	Increased aeration	Inc.	Inc.	ND[a]	26
A. chroococcum	O₂ limited	Increased aeration	ND	Inc.	Polypeptide synthesis[c]	146
A. chroococcum Hup⁻	O₂ limited	C-limited	Inc.	Inc.	ND	PC[d]
A. vinelandii	O₂ limited	O₂ shock	ND	Inc.	Polypeptide levels	102
R. capsulatus	N limited	Increased light	ND	Inc.	Polypeptide levels	87
C. pasteurianum	NH₄⁺ assimilating	N₂ fixing				
K. pneumoniae						
A. vinelandii			Dec[b]	Inc.	ND	177
B. polymyxa	NH₄⁺ assimilating	N₂ fixing				
B. azotofixans			None	ND	ND	93

[a] Inc. = increase; Dec. = decrease; ND = not determined.

[b] Correlated with a marked decrease in ATP/ADP ratio.

[c] Synthetic events specific for nitrogenase.

[d] O. M. Aguilar and M. G. Yates, personal communication.

The characterization of obligate aerobic diazotrophy is far from complete. One fundamental issue that requires verification is the mechanism whereby reducing power is generated for N_2 fixation. Moreover, it is not clear whether or how this process competes for electrons otherwise destined for the respiratory chain. Another intriguing aspect is the O_2 affinities of the terminal electron acceptors of the respiratory chain. In both *K. pneumoniae* and *A. vinelandii* the alternative terminal cytochrome oxidases are of the *d*-and the *o*-types. The *o*-types are probably very different. In *A. vinelandii* the *o*-type oxidase is closely associated with cytochrome *c* and is coupled to energy conservation, whereas in *K. pneumoniae* the *o*-type is associated with neither a *c*-type cytochrome nor a coupling site. The *K. pneumoniae* *o*-type oxidase shows cross-reactivity to antibodies raised against the *E. coli* *o*-type oxidase, whereas the *A. vinelandii* protein does not.[106] On the other hand, the *d*-type oxidases of *K. pneumoniae* and *A. vinelandii* do share some structural similarity as they both cross-react with antibodies prepared against the *E. coli* *d*-type oxidase.[106] The *d*-type oxidase of *K. pneumoniae* has been purified and shown to have an exceptionally low apparent Km for O_2. Although the *A. vinelandii* *d*-type oxidase has not been characterized, the evidence from membrane preparations suggests that it has a relatively low affinity for O_2. In both *A. vinelandii* and *K. pneumoniae* the *d*-type oxidase appears to be essential for aerobic diazotrophy. However, only in *A. vinelandii* is it needed for aerotolerant growth on fixed N. How do these *d*-type cytochrome oxidases differ in structure and function?

Much has still to be investigated regarding the mechanisms in *Azotobacter* whereby the DOC and the C source can influence the apparent K_s for O_2 uptake and the uncoupling of respiration. Can the "uncoupling of respiration" be explained in terms of the proposed proton or redox slip processes associated with the respiratory chain (see ref. 96), or are specific futile cycles involved? What aspects of the flexibility of respiratory activity in *Azotobacter* are specific to growth on N_2?

The respiratory chains in microaerophilic heterotrophic N_2 fixers apart from Rhizobiaceae are not as well characterized as those in *K. pneumoniae* and *A. vinelandii*. The respiratory chain of *A. brasilense* apparently differs by having a cytochrome aa_3-type terminal oxidase and thus may share common features with that of the Rhizobiaceae (see ref. 5). Since species of *Bacillus* also have an aa_3-type oxidase,[141] this raises the interesting possibility that if microaerobic diazotrophy occurs in this genus, it may differ in character from that found in *K. pneumoniae*.

There appear to be three potential sources of fixed N to satisfy synthetic necessities in diazotrophs. Although exogenous sources, and NH_3 from N_2 fixation, can supply the demands for growth, and under some conditions both sources are used apparently simultaneously, it is not clear how the organism allocates the available fixed N for synthetic purposes associated with N_2 fixation and growth. The importance of endogenous supplies of fixed N coming from recycling with respect to N_2 fixation has been indicated in cyanobacteria during heterocyst differentiation[183] but has received little attention in free-living heterotrophs, other than during encystment in *A. vinelandii*.[105] In *Anabaena* protein degradation associated with heterocyst differentiation is apparently independent of the stringent response.[183] On the other hand, this may not be so in free-living heterotrophs. In *E. coli* some protease synthesis is associated with the stringent response or N starvation,[113] so in the Enterobacteriaceae possibilities exist for N recycling to support *nif* derepression.

This review reflects the classical approach to physiology where parameters relating to the organism's biochemical activity and growth are correlated with changes in the environment. The rapidly developing field of molecular genetics is providing fruitful techniques for investigating the physiology of whole organisms, provided that the number of variables within an experimental system are manageable. Correlations of the type mentioned here when directed toward a comparison between a specific mutant and the wild type may reveal novel routes of investigation for more fundamental research.

Acknowledgments

I thank M. J. Merrick, B. E. Smith, and especially M. G. Yates for useful discussion and for reading the manuscript; P. Young and R. R. Eady for access to their manuscripts before publication, and R. J. Foote for typing. Finally, a chapter on the physiology of N_2 fixation in free-living heterotrophs remains incomplete without a particular acknowledgment to Professor J. R. Postgate for his contribution to this subject as the assistant director and subsequently the director of the AFRC Unit of Nitrogen Fixation at the University of Sussex, U.K.

References

1. Ackrell, B. A. C., and Jones, C. W. "The respiratory system of *Azotobacter vinelandii* 2 oxygen effects," *Eur. J. Biochem. 20*:29–35 (1971).

2. Aguilar, O. M., Yates, M. G., and Postgate, J. R., "A comparison of H_2-uptake hydrogenase negative (Hup⁻) mutants and wild type (Hup⁺) *Azptobacter chroococcum* under different growth conditions," in C. Veeger and W. E. Newton (eds.), *Advances in Nitrogen Fixation Research*. Nijhoff/Junk, The Hague, 1984, p. 213.

3. Andersen, K., and Shanmugam, K. T., "Energetics of biological nitrogen fixation: Determination of the ratio of formation of H_2 to NH_4^+ catalysed by nitrogenase of *Klebsiella pneumoniae in vivo. J. Gen. Microbiol. 103*:107–122 (1977).

4. Anraku, Y., and Gennis, R. B., "The aerobic respiratory chain of *Escherichia coli,*" *Trends Biochem. Sci. 12*:262–266 (1987).

5. Appleby, C. A., "Leghaemoglobin and rhizobium respiration," *Ann. Rev. Plant. Physiol. 35*:443–478 (1984).

6. Arp, D. J., and Zumft, W. G., "Overproduction of nitrogenase by nitrogen-limited cultures of *Rhodopseudomonas palustris*," *J. Bacteriol. 153*:1322–1330 (1983).

7. Baranova, N. A., and Gorotov, I. N. "Fixation of molecular nitrogen by propionic bacteria," Microbioilogy *43*:675–678 (1974).

8. Barraquio, W. L., and Knowles, R., "Effect of hydrogen, sucrose and oxygen on uptake hydrogenase in nitrogen-fixing and ammonium-grown *Pseudomonas saccharophila*," *J. Gen. Microbiol. 134*:893–901 (1988).

9. Barraquio, W. L., Dumont, A., and Knowles, R., "Enumeration of free-living aerobic N_2-fixing H_2-oxidizing bacteria by using a heterotrophic semisolid medium and most probably number technique," *Appl. Environ. Microbiol. 54*:1313–1317 (1988).

10. Bennett, L. T., Cannon, F., and Dean, D. R., "Nucleotide sequence and mutagenesis of the *nifA* gene from *Azotobacter vinelandii*," *Mol. Microbiol. 2*:315–321 (1988).

11. Bergersen, F. J., and Turner, G. L., "Systems utilizing oxygenated leghaemoglobin and myoglobin as sources of free dissolved O_2 at low concentrations for experiments with bacteria," *Anal. Biochem. 96*:165–174 (1979).

12. Bergersen, F. J., and Turner, G. L., "Properties of terminal oxidase systems of bacteroids from root nodules of soybean and cowpea and of N_2-fixing bacteria grown in continuous culture," *J. Gen. Microbiol. 118*:232–252 (1980).

13. Bogdahn, M., and Kleiner, D., "Inorganic nitrogen metabolism in two cellulose-degrading clostridia," *Arch. Microbiol. 145*:159–161 (1986).

14. Buchanan-Wollaston, V., Cannon, M. C., Beynon, J. J., and Cannon, F. C., "Role of the *nifA* gene product in the regulation of *nif* expression in *Klebsiella pneumoniae*," *Nature (Lond.) 294*:776–788 (1981).

15. Bühler, T., Monter, U., Sann, R., Kuhla, J., Dingler, D., and Oelze, J., "Control of respiration and growth yield in ammonium-assimilating cultures of *Azotobacter vinelandii*," *Arch. Microbiol. 148*:242–246 (1987).

16. Bühler, T., Sann, R., Monter, U., Dingler, C., Kuhla, J., and Oelze, J., "Control of dinitrogen fixation in ammonium-assimilating cultures of *Azotobacter vinelandii*," *Arch. Microbiol. 148*:247–251 (1987).

17. Burris, R. H., Albrecht, S. L., and Okon, Y., "Physiology and biochemistry of *Spirillum lipoferum*," in J. Döbereiner, R. H. Burris, A. Hollaender, A. A. Franco, C. A. Neyra, and D. B. Scott (eds.), *Limitations and Potentials for Biological Nitrogen Fixation in the Tropics*, Plenum, New York, 1978, pp. 306–316.

18. Cacciari, I., Lippi, D., and Bordeleav, L. M., "Effect of oxygen on batch and continuous culture of a nitrogen-fixing *Arthrobacter* sp.," *Can. J. Microbiol. 25*:746–751 (1979).

19. Cannon, M., Hill, S., Kavanagh, E., and Cannon, F., "A molecular genetic study of *nif* expression in *Klebsiella pneumoniae* at the level of transcription, translation and nitrogenase activity," *Mol. Gen. Genet. 198*:198–206 (1985).

20. Carter, I. S., and Dawes, E. A., "Effect of oxygen concentration and growth rate on glucose metabolism, poly-β-hydroxybutyrate biosynthesis and respiration of *Azotobacter beijerinckii*," *J. Gen. Microbiol. 110*:393–400 (1979).

21. Cashel, M., and Rudd, K. E., "The stringent response," in F. C. Neidhardt (ed.), *Escherichia coli and Salmonella typhimurium Cellular and Molecular Biology*, Americn Society for Microbiology, Washington, D.C., 1987, pp. 1410–1438.

22. Cejudo, F. J., and Paneque, A., "Effect of nitrogen starvation on ammonium-inhibition of nitrogenase activity in *Azotobacter chroococcum*," *Arch. Microbiol. 149*:481–484 (1988).

23. Chang, C. L., Davis, L. C., Rider, M., and Takemoto, D. J., "Characterization of *nifH* mutations of *Klebsiella pneumoniae*," *J. Bacteriol. 170*:4015–4022 (1988).

24. Chen, Y.-P. and Yoch, D. C., "Regulation of two nickel-requiring (inducible and constitutive) hydrogenases and their coupling to nitrogenase in *Methylosinus trichosporium* OB3b," *J. Bacteriol. 169*:4778–4783 (1987).

25. Chen, Y.-P., and Yoch, D. C., "Reconstitution of the electron transparent system that couples formate oxidation to nitrogenase in *Methylosinus trichosporium* OB3b," *J. Gen. Microbiol. 134*:3123–3128 (1988).

26. Ching, T. M., Bergersen, F. J., and Turner, G. L., "Energy status, growth and nitrogenase activity in continuous cultures of *Rhizobium* sp. strain CB756 supplied with NH_4^+ and various rates of aeration," *Biochim. Biophys. Acta 636*:82–90 (1981).

27. Collins, J. J., and Brill, W. J., "Control of *Klebsiella pneumoniae nif* mRNA synthesis," *J. Bacteriol. 162*:1186–1190 (1985).

28. Collins, J. J., Roberts, G. P., and Brill, W. J., "Post transcriptional control of *Klebsiella pneumoniae nif* mRNA stability by the *nifL* product." *J. Bacteriol. 168*:173–178 (1986).

29. Daesch, G., and Mortenson, L. E., "Sucrose catabolism in *Clostridium pasteurianum* and its relation to N_2 fixation," *J. Bacteriol. 96*:346–351 (1967).

30. Daesch, G., and Mortenson, L. E., "Effect of ammonia on the synthesis and function of the N_2-fixing enzyme system in *Clostridium pasteurianum*," *J. Bacteriol. 110*:103–109 (1972).

31. Dalton, H., and Postgate, J. R., "Effect of oxygen on growth of *Azotobacter chroococcum* in batch and continuous cultures," *J. Gen. Microbiol. 54*:463–473 (1969).

32. Dalton, H., and Postgate, J. R., "Growth and physiology of *Azotobacter chroococcum* in continuous culture," *J. Gen. Microbiol. 56*:307–319 (1969).

33. Danneberg, G., Zimmer, W., and Bothe, H., "Energy transduction efficiencies in nitrogenous oxide respiration of *Azospirillum brasilense* sp. 7," *Arch. Microbiol. 151*:445–453 (1989).

34. David, K. A. V., and Fay, P., "Effects of long-term treatment with acetylene on nitrogen-fixing microorganisms," *Appl. Environ. Microbiol. 34*: 640–646 (1977).

35. Dawes, E. A., "Carbon metabolism," in P. H. Calcott (ed.) *Continuous Culture of Cells,* Vol. 2, CRC Press, Boca Raton, Florida, 1981, pp. 1–38.

36. Dawes, E. A., "The effect of environmental oxygen concentration on the carbon metabolism of some aerobic bacteria," in I. S. Kulaev, E. A. Dawes, and D. W. Tempest (eds.), *Environmental Regulation of Microbial Metabolism,* Academic Press, New York, 1985, pp. 121–126.

37. Deistung, J., and Thorneley, R. N. F., "Electron transfer to nitrogenase: characterization of flavodoxin from *Azotobacter chroococcum* and comparison of its redox potentials with those of flavodoxins from *Azotobacter vinelandii* and *Klebsiella pneumoniae*," *Biochem. J. 239*:69–75 (1986).

38. de la Rubia, T., Gonzalez-Lopez, J., Moreno, J., Martinez-Toledo, M. V., and Ramos-Cormenzana, A., "Adenine nucleotide contents and energy charge of *Azotobacter vinelandii* grown at low phosphate concentrations," *Arch. Microbiol. 147*:354–357 (1987).

39. Derylo, M., Glowacka, M., Skorupska, A., and Lorkiewicz, Z., "Nif plasmid from *Lignobacter*," *Arch. Microbiol. 130*:322–324 (1981).

40. de Vries, W., Stam, H., Stouthamer, A. H., "Hydrogen oxidation and nitrogen fixation in rhizobia, with special attention focused on strain ORS571. *Antonie van Leenwenhoek 50*:505–524 (1984).

41. de Vries, W., Ras, J., Stam, H., van Vlerken, M. M. A., Hilgert, U., de Bruijn, F. J., and Stouthamer, A. H., "Isolation and characterization of hydrogenase-negative mutants of *Azorhizobium caulinodans* ORS571," *Arch. Microbioil. 150*:595–599 (1988).

42. Dilworth, M. J., and Kennedy, I. R., "Oxygen inhibition in *Azotobacter vinelandii* some enzymes concerned in acetate metabolism," *Biochim. Biophys. Acta 67*:240–253 (1963).

43. Dingler, C., and Oelze, J., "Reversible and irreversible inactivation of cellular nitrogenase upon oxygen stress in *Azotobacter vinelandii* growing in oxygen controlled continuous culture," *Arch. Microbiol. 141*:80–84 (1985).

44. Dingler, C., and Oelze, J., "Super oxide dismutase and catalase in *Azotobacter vinelandii* grown in continuous culture at different dissolved oxygen concentrations," *Arch. Microbiol. 147*:291–294 (1987).

45. Dingler, C., Kuhla, J., Wassink, H., and Oelze, J., "Levels and activities of nitrogenase proteins in *Azotobacter vinelandii* grown at different dissolved oxygen concentrations," *J. Bacteriol. 170*:2148–2152 (1988).

46. Döbereiner, J., and Pedrosa, F. O., "Nitrogen-fixing bacteria in non-leguminous crop plants," Springer-Verlag, New York, 1987.

47. Drozd, J., and Postgate, J. R., "Effect of oxygen on acetylene reduction, cytochrome content and respiratory activity of *Azotobacter chroococcum*," *J. Gen. Microbiol. 63*:63–73 (1970).

48. Drozd, J. W., Tubb, R. S., and Postgate, J. R., "A chemostat study of the effect of fixed nitrogen sources on nitrogen fixation membranes and free amino acids in *Azotobacter chroococcum*," *J. Gen. Microbiol. 73*:221–232 (1972).

49. Eady, R. R., Robson, R. L., Richardson, T. H., Miller, R. W., and Hawkins, M., "The vanadium nitrogenase of *Azotobacter chroococcum*," *Biochem. J. 244*:197–207 (1987).

50. Fu, C., and Knowles, R., "Oxygen tolereance of uptake hydrogenase in *Azospirillum* spp." *Can. J. Microbiol. 32*:897–900. 1986.

51. Fu, H., and Burris, R. H., "Ammonium inhibition of nitrogense activity in *Herbaspirillum seropedicae*," *J. Bacteriol. 171*:3168–3175 (1989).

52. Fu, H., Hartmann, A., Lowery, R. G., Fitzmaurice, W. P., Roberts, G. P., and Burris, R. H., "Posttranslational regulatory system for nitrogenase activity in *Azospirillum* spp.," *J. Bacteriol. 171*:4679–4685 (1989).

53. Georgiou, C. D., Dueweke, T. J., and Gennis, R. B., "Regulation of expression of the cytochrome *d* terminal oxidase in *Escherichia coli* is transcriptional," *J. Bacteriol. 170*:961–966 (1988).

54. Glass, R. E., "Gene function: *E. coli* and its heritable elements," Croom Helm, London, 1982.

55. Gillis, M., Kersters, K., Hoste, B., Janssens, D., Kroppenstedt, R. M., Stephan, M. P., Teixeira, K. R. S., Döbereiner, J., and De Ley, J., "*Acetobacter diazotrophicus* sp. nov., a nitrogen-fixing acetic acid bacterium associated with sugarcane," *Int. J. Syst. Bact. 39*:361–364 (1989).

56. Goldberg, I., Nadler, V., and Hochman, A., "Mechanism of nitrogenase switch-off by oxygen," *J. Bacteriol. 169*:874–879 (1987).

57. Gordon, J. K., Shah, V. K., and Brill, W. J., "Feedback inhibition of nitrogenase," *J. Bacteriol. 148*:884–888 (1981).

58. Gosink, M. M., Franklin, N. M., and Roberts, G. P., "The product of the *Klebsiella pneumoniae nifX* gene is a negative regulator of the nitrogen fixation (nif) regulon," *J. Bacteriol. 172*:1441–1447 (1990).

59. Guerinot, M. L., and Patriquin, D. G., "N₂-fixing vibrios isolated from the gastrointestinal tract of sea urchins," *Can. J. Microbiol. 27*:311–317 (1981).

60. Haahtela, K., Kari, K., and Sundman, V., "Nitrogenase activity (acetylene reduction) of root-associated cold-climate *Azospirillum, Enterobacter, Klebsiella* and *Pseudomonas* species during growth on various carbon sources, and at various partial pressures of oxygen," *Appl. Environ. Microbiol. 45*:563–570 (1983).

61. Haaker, H., and Veeger, C., "Regulation of respiration and nitrogen fixation in different types of *Azotobacter vinelandii*," *Eur. J. Biochem. 63*:499–507 (1976).

62. Haaker, H., and Klugkist, J., "The bioenergetics of electron transport to nitrogense," *FEMS Microbiol. Rev. 46*:57–71 (1987).

63. Haddock, B. A., and Jones, C. W., "Bacterial respiration," *Bacteriol.* Rev. *41*:47–99 (1977).

64. Häger, K.-P., and Bothe, H., "Reduction of NAD^+ by the reversed respiratory electron flow in *Azotobacter vinelandii*," *Biochim. Biophys. Acta 892*:213–223 (1987).

65. Hanson, R. S., "Distribution in nature of reduced one carbon components and microbes that utilize them," in J. C. Murrell and H. Dalton (eds.), *The Methane and Methanol Utilizers,* Plenum, New York, 1991.

66. Harrison, D. E. F., "Physiological effects of dissolved oxygen tension and redox potential on growing populations of microorganisms," *J. Appl. Chem. Biotechnol. 22*:417–440 (1972).

67. Hartmann, A., Fu, H. A., and Burris, R. H., "Regulation of nitrogenase activity by ammonium chloride in *Azospirillum* spp.," *J. Bacteriol. 165*:864–870 (1986).

68. Hartmann, A., and Burris, R. H., "Regulation of nitrogenase activity by O_2 in *Azospirillum brasilense* and *Azospirillum lipoferum*," *J. Bacteriol. 169*:944–948 (1987).

69. Hill, S., "Influence of oxygen concentration on the colony type of *Derxia gummosa* grown on nitrogen-free media," *J. Gen. Microbiol. 67*:77–83 (1971).

70. Hill, S., Drozd, J. W., and Postgate J. R., "Environmental effects on the growth of nitrogen-fixing bacteria," *J. Appl. Chem. Biotechnol. 22*:541–558 (1972).

71. Hill, S., "Influence of atmospheric oxygen concentration on acetylene reduction and efficiency of nitrogen fixation in intact *Klebsiella pneumoniae*," *J. Gen. Microbiol. 93*:335–345 (1976).

72. Hill, S., "The apparent ATP requirement for nitrogen fixation in growing *Klebsiella pneumoniae*," *J. Gen. Microbiol. 95*:297–312 (1976).

73. Hill, S., Turner, G. L., and Bergersen, F. J., "Synthesis and activity of nitrogenase in *Klebsiella pneumoniae* exposed to low concentrations of oxygen," *J. Gen. Microbiol. 130*:1061–1067 (1984).

74. Hill, S., Turner, G. L., and Bergersen, F. J., "Synthesis and activity of nitrogenase in *Klebsiella pneumoniae* exposed to low concentration of oxygen," in C. Veeger and

W. E. Newton (eds.) *Advances in Nitrogen Fixation Research*, Nijhoff/Junk, The Hague, 1984, p. 260.

75. Hill, S., "How is nitrogense regulated by oxygen?" *FEMS Microbiol. Rev. 54*:111–130 (1988).

76. Hill, S., and Kavanagh, E., "Oxygen control and autoregulation of the *nif* structural operon in *Klebsiella pneumoniae* involves nitrogenase function," in H. Bothe, F. J. de Bruijn, and W. E. Newton (eds.), *Nitrogen Fixation: One Hundred Years After*, Gustave Fischer, Stuttgart, 1988, p. 309.

77. Hill, S., Viollet, S., Smith, A. T., and Anthony, C., "Roles for enteric *d*-type cytochrome oxidase in N$_2$ fixation and microerobiosis," *J. Bacteriol. 172*:2071–2078 (1990).

78. Hine, P. W., and Lees, H., "The growth of nitrogen-fixing *Azotobacter chroococcum* in continuous culture under intense aeration," *Can. J. Microbiol. 22*:611–618 (1976).

79. Hochman, A., Reich, I., and Nadler, V., "Effect of oxygen on nitrogenase of *Rhodopseudomonas capsulata* and *Klebsiella pneumoniae*," in H. J. Evans, P. J. Bottomley, and W. E. Newton (eds.), *Nitrogen Fixation Research Progress*, Martinus Nijhoff, Dordrecht, 1985, p. 442.

80. Hoffman, P. S., Morgan, T. V., and Dervartanian, D. V., "Respiratory-chain characteristics of mutants of *Azotobacter vinelandii* negative to tetramethyl-*p*-phenylenediamine oxidase," *Eur. J. Biochem. 100*:19–27 (1979).

81. Hoffman, P. S., Morgan, T. V., and Dervartanian, D. V., "Respiratory properties of cytochrome-*c*-deficient mutants of *Azotobacter vinelandii*," *Eur. J. Biochem. 110*:349–354 (1980).

82. Howard, K. S., Hales, B. J., and Socolofsky, M. D., "*In vivo* interactions between nitrogenase molybdenum-iron protein and membrane in *Azotobacter vinelandii* and *Rhodospirillum rubrum*," *Biochim. Biophys. Acta 812*:575–585 (1985).

83. Jackson, F. A., and Dawes, E. A., "Regulation of the tricarboxylic acid cycle and poly-β-hydroxybutyrate metabolism in *Azotobacter beijerinckii* grown under nitrogen or oxygen limitation," *J. Gen. Microbiol. 97*:303–312 (1976).

84. Jensen, J. S., and Kennedy, C., "Pleiotropic effect of *his* gene mutations on nitrogen fixation in *Klebsiella pneumoniae*," *EMBO J. 1*:197–204 (1982).

85. Jones, C. W., Brice, J. M., Wright, V., and Ackrell, B. A. C., "Respiratory protection of nitrogenase in *Azotobacter vinelandii*," *FEBS Lett. 19*:77–81 (1973).

86. Jones, C. W., Brice, J. M., and Edwards, C., "Bacterial cytochrome oxidases and respiratory chain energy conservation," in H. Degn, D. Lloyd, and G. C. Hill (eds.), *Function of Alternative Terminal Oxidases*, Pergamon Press, Oxford, 1978, pp. 89–97.

87. Jovanneau, Y., Wong, B., and Vignais, P. M., "Stimulation by light of nitrogenase synthesis in cells of *Rhodopseudomonas capsulata* growing in N-limited continuous culture," *Biochim. Biophys. Acta 808*:149–155 (1985).

88. Jurtshuk, P., Jr., and Yang, T.-Y., "Oxygen reactive hemoprotein components in bacterial respiratory systems," in C. J. Knowles (ed.), *Diversity of Bacterial Respiratory Systems*, CRC Press, Boca Raton, Florida, 1980, pp. 137–159.

89. Kahn, D., Hawkins, M., and Eady, R. R., "Nitrogen fixation in *Klebsiella pneumoniae*: Nitrogenase levels and the effects of added molybdate on nitrogenase derepressed under molybdenum deprivation," *J. Gen. Microbiol. 128*:779–787 (1982).

90. Kahn, D., Hawkins, M., and Eady, R. R., "Metabolic control of *Klebsiella pneumoniae* mRNA degradation by the availability of fixed nitrogen," *J. Gen. Microbiol.* 128:3011–3018 (1982).

91. Kaluza, K., and Hennecke, H., "Regulation of nitrogenase messenger RNA synthesis and stability in *Klebsiella pneumoniae*," *Arch. Microbiol.* 130:38–43 (1981).

92. Kanamori, K., Weiss, R., and Roberts, J. D., "Role of glutamate dehydrogense in ammonia assimilaton in nitrogen-fixing *Bacillus macerans*," *J. Bacteriol.* 169:4692–4695 (1987).

93. Kanamori, K., Weiss, R., and Roberts, J. D., "Efficiency factors and ATP/ADP ratios in nitrogen-fixing *Bacillus polymyxa* and *Bacillus azotofixans*," *J. Bacterioil.* 172:1962–1968 (1990).

94. Kavanage, E., and Hill, S., "The automatic maintenance of low dissolved oxygen using a photobacterium oxygen sensor for the study of microaerobiosis," *J. Appl. Bacteriol.*, 69:539–549 (1990).

95. Keil, H., "Molecular cloning and expression of a novel catechol 2,3-dioxygense gene from the benzoate meta-cleavage pathway in *Azotobacter vinelandii*," *J. Gen. Microbiol.* 136:607–613 (1990).

96. Kell, D. B. "Proton motive energy-transducing systems: Some physical principles and experimental approaches," in C. Anthony (ed.), *Bacterial Energy Transduction*, Academic Press, New York, 1988, pp. 429–490.

97. Kelly, M. J. S., Poole, R. K., Yates, M. G., and Kennedy, C., "Cloning and mutagenesis of genes encoding the cytochrome *bd* complex in *Azotobacter vinelandii*, mutants deficient in cytochrome *d* are defective in aerobic nitrogen fixation," *J. Bacteriol.*, 172:6010–6019.

98. Kleiner, D., and Phillips, S., "Relative levels of guanosine 5'-diphosphate 3'-diphosphate (ppGpp) in some N$_2$-fixing bacteria during derepression and repression of nitrogense," *Arch. Microbiol.* 128:341–342 (1981).

99. Kleiner, D., "Bacterial ammonium transport," *FEMS Microbiol. Rev.* 32:87–100 (1985).

100. Kloss, M., Iwannek, K. H., and Fendrik, I., "Physiological properties of *Azospirillum brasilense* sp7 in a malate-limited chemostat," *J. Appl. Microbiol.* 29:447–457 (1983).

101. Klugkist, J., and Haaker, H., "Inhibition of nitrogenase activity by ammonium chloride in *Azotobacter vinelandii*," *J. Bacteriol.* 157:148–151 (1984).

102. Klugkist, J., Haaker, H., Wassink, H., and Veeger, C., "The catalytic activity of nitrogenase in intact *Azotobacter vinelandii* cells," *Eur. J. Biochem.* 146:509–519 (1985).

103. Klugkist, J., Haaker, H., and Veeger, C., "Studies on the mechanism of electron transport to nitrogenase in *Azotobacter vinelandii*," *Eur. J. Biochem.* 155:41–46 (1986).

104. Knowles, R. "The significance of asymbiotic dinitrogen fixation by bacteria," in R. W. F. Hardy and A. H. Gibson (eds.), *A Treatise on Dinitrogen Fixation*, (section IV: Agronomy and Ecology), Wiley, New York, 1977, pp. 33–83.

105. Kossler, W., and Kleiner, D., "Degradation of the nitrogenase protein during encystment of *Azotobacter vinelandii*," *Arch. Microbiol.* 145:287–289 (1986).

106. Kranz, R. G., and Gennis, R. B., "Immunological investigation of the distribution of cytochromes related to the two terminal oxidases of *Escherichia coli* in other gram-negative bacteria." *J. Bacteriol.* 161:709–711, (1985).

107. Krol, A. J. M., Hontelez, J. G. T., Roozendaal, B., and van Kammen, A., "On the operon structure of the nitrogense genes of *Rhizobium leguminosarum* and *Azotobacter vinelandii*," *Nucl. Acids Res.* 10:4147–4157 (1982).

108. Kuhla, J., Dingler, C., and Oelze, J., "Production of extracellular nitrogen-containing components by *Azotobacter vinelandii* fixing dinitrogen in oxygen-controlled continous culture," *Arch. Microbiol. 141*:297–302 (1985).

109. Kuhla, J., and Oleze, J., "Dependency of growth yield, maintenance and K_s-values on the dissolved oxygen concentration in continuous cultures of *Azotobacter vinelandii*," *Arch. Microbiol. 149*:509–514 (1988).

110. Kuhla, J., and Oelse, J., "Dependence of nitrogense switch-off upon oxygen stress on the nitrogenase activity in *Azotobacter vinelandii*," *J. Bacteriol. 170*:5325–5329 (1988).

110a. Kush, A., Elmerich, C., and Aubert, J.-P., "Nitrogenase of *Sesbania rhizobium* strain ORS571: Purification, properties and "switch off" by ammonia," *J. Gen. Microbiol. 131*:1765–1777 (1985).

111. Laane, C., Haaker, H., and Veeger, C., "On the efficiency of oxidative phosphorylation in membrane vesicles of *Azotobacter vinelandii* and of *Rhizobium leguminosarum* bacteroids," *Eur. J. Biochem. 97*:369–377 (1979).

112. Laane, C., Krone, W., Konings, W., Haaker, H., and Veeger, C., "Short-term effect of ammonium chloride on nitrogen fixation by *Azotobacter vinelandii* and of bacteroids of *Rhizobium leguminosarum*," *Eur. J. Biochem. 103*:39–46 (1980).

113. Lazdunski, A. M., "Peptidases and proteases of *Escherichia coli* and *Salmonella typhimurium*," *FEMS Microbiol. Rev. 63*:265–276 (1989).

114. Lees, H., and Postgate, J. R., "The behaviour of *Azotobacter chroococcum* in oxygen- and phosphate-limited chemostat culture," *J. Gen. Microbiol. 75*:161–166 (1973).

115. Leschine, S. B., Holwell, K., and Canale-Parola, E., "Nitrogen fixation by aerobic cellulolytic bacteria," *Science 242*:1157–1159 (1988).

116. Lidstrom, M. E., Murrell, J. C., and Toukdarian, A. E., "Dinitrogen fixation in methylotrophic bacteria," in P. W. Ludden and J. E. Burris (eds.), *Nitrogen Fixation and CO$_2$ Metabolism*, Elsevier, New York, 1985, pp. 245–251.

117. Lesinkova, L. L., and Khmel, I. A. "Effect of cultural conditions on cytochrome contact of *Azotobacter vinelandii* cells," *Microbiology 36*:762–767 (1967).

118. Ljungström, E., Yates, M. G., and Nordlund, S., "Purification of the activating factor for the Fe-protein of nitrogenase from *Azospirillum brasilense*," *Biochim. Biophys. Acta 994*:210–214 (1989).

119. Ludden, P. W., Roberts, G. P., Lowery, R. G., Fitzmaurice, W. P., Saari, L. L., Lehman, L., Lies, D., Woehle, D., Wirt, H., Murrell, S. A., Pope, M. R., and Kanemoto, R. H., "Regulation of nitrogenase activity by reversible ADP-ribosylation of dinitrogenase reductase," in H. Bothe, F. J. de Bruijn, and W. E. Newton (eds.), *Nitrogen Fixation: Hundred Years After*, Gustav Fischer, Stuttgart, New York, 1988, pp. 157–162.

120. Lynch, J. M., and Harper, S. H. T., "Straw as a substrate for cooperative nitrogen fixation," *J. Gen. Microbiol. 129*:251–253 (1983).

121. McClung, C. R., Patriquin, D. G., and Davis, R. E., "*Campylobacter nitrofigilis* sp. nov. a nitrogen-fixing bacterium associated with roots of *Spartina alterniflora* Loisel," *Int. J. syst. Bacteriol. 33*:605–612 (1983).

122. McInerney, M. J., Holmes, K. S., andd Dervartanian, D. V., "Effect of O$_2$ limitation on growth and respiration of the wild type and an ascorbate-tetramethyl-*p*-phenylene diamine-oxidase-negative mutant strain of *Azotobacter vinelandii*," *J. Bioenerg. Biomembr. 14*:451–456 (1982).

123. McInerney, M. J., Holmes, K. S., Hoffman, P., and Dervartanian, D. V., "Respiratory mutants of *Azotobacter vinelandii* with elevated levels of cytochrome *d*," *Eur. J. Biochem. 141*:447–452 (1984).

124. Merrick, M. J., "Regulation of nitrogen assimilation by bacteria," in J. A. Cole and S. J. Ferguson (eds.), *SGM symposium, Volume 42: The Nitrogen and sulphur cycles,* Cambridge University Press, Cambridge, 1988, pp. 331–361.

125. Moss, F., "Adaptation of the cytochromes of *Aerobacter aerogenes* in response to environmental oxygen tension," *Austral. J. Exp. Biol. 34*:395–406 (1956).

126. Murrell, J. C., "The rapid switch-off of nitrogenase activity in obligate methane-oxidizing bacteria," *Arch. Microbiol. 150*:489–495 (1988).

127. Nagai, S., Nishizawa Y., Onodera, M., and Aibas, S., "Effect of dissolved oxygen on growth yield and aldolase activity in chemostat cultures of *Azotobacter vinelandii,*" *J. Gen. Microbiol. 66*:197–203 (1971).

128. Nagai, S., and Aiba, S., "Reassessment of maintenance and energy uncoupling in the growth of *Azotobacter vinelandii,*" *J. Gen. Microbiol. 73*:531–538 (1972).

129. Nair, M., and Eady, R. R., "Nitrogenase synthesis in *Klebsiella pneumoniae:* Enhanced *nif* expression without accumulation of guanosine 5′-diphosphate 3′-diphosphate," *J. Gen. Microbiol. 130*:3063–3069 (1984).

130. Neilson, A. H., and Allard, A.-S., "Acetylene reduction (N_2-fixation) by Enterobacteriacease isolated from industrial waste waters and biologicl treatment systems," *Appl. Microbiol. Biotechnol. 23*:67–74 (1985).

131. Nelson, L. M., and Knowles, R., "Effect of oxygen and nitrate on nitrogen fixation and denitrification by *Azospirillum brasilense* grown in continuous culture," *Can. J. Microbiol. 24*: 1395–1403 (1978).

132. Nishizawa, Y., Nagi, S., and Aiba, S., "Effect of dissolved oxygen on electron transport system of *Azotobacter vinelandii* glucose-limited and oxygen-limited chemostat cultures," *J. Gen. Appl. Microbiol. 17*:131–140 (1971).

133. Nur, I., Okon, Y., and Henis, Y., "Effect of dissolved oxygen tension on production of carotenoids, poly-β-hydroxybutyrate, succinate oxidase and superoxide dismutase by *Azospirillum brasilense* Cd grown in continuous culture," *J. Gen. Microbiol. 128*:2937–2943 (1982).

134. Oakley, C., and Murrell, J. C., "*nifH* genes in the obligate methane oxidizing bacterium," *FEMS Microbiol. Lett. 49*:53–57 (1988).

135. Okon, Y., Nur, I., and Henis, Y., "Effect of oxygen concentration on electron transport components and microaerobic properties of *Azospirillum brasilense,*" in W. Klingmüller (ed.), *Azospirillum II: Genetics, Physiology, Ecology,* Birkhäuser, Basel. 1983, pp. 115–125.

136. Page, W., Jackson, L., and Shivprasad, S., "Sodium-dependent *Azotobacter chroococcum* strains are aeroadaptive, microaerophilic, nitrogen-fixing bacteria," *Appl. Environ. Microbiol. 54*:2123–2128 (1988).

137. Pedrosa, F. O., Stephan, M., Döbereiner, J., and Yates, M. G., "Hydrogen-uptake hydrogenase activity in nitrogen-fixing *Azospirillum brasilense,*" *J. Gen. Microbiol. 128*:161–166 (1982).

138. Peterson, J. B., "On the role of flavin in malate oxidation by *Azotobacter vinelandii* respiratory membrane," *Biochim. Biophys. Acta 933*:107–113 (1988).

139. Peterson, J. B., "Respiratory differences associated with aeration in *Azotobacter vinelandii,*" *Can. J. Microbiol. 35*:918–924 (1989).

140. Polman, J. K., and Larkin, J. M., "Properties of *in vivo* nitrogense activity in *Beggiatoa alba*," *Arch. Microbiol. 150*:126–130 (1988).

141. Poole, R. K., "Bacterial cytochrome oxidases," in C. Anthony (ed.), *Bacterial Energy Transduction*, Academic Press, New York, 1988, pp. 231–291.

142. Post, E., Golecki, J. R., and Oelze, J., "Morphological and ultrastructural variations in *Azotobater vinelandii* growing in oxygen-controlled continuous culture," *Arch. Microbiol. 133*:75–82 (1982).

143. Post, E., Kleiner, D., and Oelze, J., "Whole cell respiration and nitrogenase activities in *Azotobacter vinelandii* growing in oxygen controlled continuous culture," *Arch. Microbiol. 134*:68–72 (1983).

144. Post, E., Vakalopoulou, E., and Oelse, J., "On the relationship of intracytoplasmic membranes in nitrogen-fixing *Azotobacter vinelandii*," *Arch. Microbiol. 134*:265–269 (1983).

145. Postgate, J. R., and Krishnapillai, V., "Expression of Klebsiella *nif* and *his* genes in *Salmonella typhimurium*," *J. Gen. Microbiol. 98*:379–385 (1977).

146. Postgate, J. R., Eady, R. R., Dixon, R. A., Hill, S., Kahn, D., Kennedy, C., Partridge, P., Robson, R., and Yates, M. G., "Some aspects of the physiology of dinitrogen fixation," in H. Bothe and A. Trebst (eds.), *Biology of Inorganic Nitrogen and Sulfur*, Springer-Verlag, Berlin, 1981, pp. 103–115.

147. Postgate, J. R., *The Fundamentals of Nitrogen Fixation*, Cambridge University Press, Cambridge, UK, 1982.

148. Postgate, J. R., and Kent, H. M., "Derepression of nitrogen fixation in *Desulfovibrio gigas* and its stability to ammonia or oxygen stress *in vivo*," *J. Gen. Microbiol. 130*:2825–2831 (1984).

149. Postgate, J. R., and Kent, H. M., "Diazotrophy within *Desulfovibrio*," *J. Gen. Microbiol. 134*:2119–2122 (1985).

150. Postgate, J. R., Kent, H. M., Hill, S., and Blackburn, H., "Nitrogen fixation by *Desulfovibrio gigas* and other species of *Desulfovibrio*," in P. W. Ludden and J. E. Burris (eds.), *Nitrogen Fixation and CO₂ Metabolism*, Elsevier, New York, 1985, pp. 225–234.

151. Puppo, A., and Rigaud, J., "Hypothesis, superoxide dismutase an essential role in the protection of the nitrogen fixation process?" *FEBS Lett. 201*:181–189 (1986).

152. Ramos, J. L., and Robson, R. L., "Isolation and properties of mutants of *Azotobacter chroococcum* defective in aerobic nitrogen fixation," *J. Gen. Microbiol. 131*:1449–1458 (1985).

153. Ramos, J. L., and Robson, R. L., "Lesions in citrate synthase that affect aerobic nitrogen fixation by *Azotobacter chroococcum*," *J. Bacterial. 162*:746–751 (1985).

154. Ramos, J., and Robson, R. L., "Cloning of the gene for phosphoenol-pyruvate carboxylase from *Azotobacter chroococcum*, an enzyme important in aerobic nitrogen fixation," *Mol. Gen. Genet. 208*:481–484 (1987).

155. Reiner, O., and Okon, Y., "Oxygen recognition in aerotatic behaviour of *Azospirillum brasilense* Cd." *Can. J. Microbiol. 32*:829–834 (1986).

156. Riesenberg, G., and Kari, C., "Isolation and characterisation of prototrophic relaxed mutants of *Klebsiella pneumoniae*," *Mol. Gen. Genet. 181*:476–483 (1981).

157. Riesenberg, D., Erdei, S., Kondorosi, E., and Kari, C., "Positive involvement of ppGpp in derepression of the *nif* operon in *Klebsiella pneumoniae*," *Mol. Gen. Genet. 185*:198–204 (1982).

158. Röckel, D., Hernando, E., Vakalopoulou, E., Post, E., and Oelze, J., "Localization and activities of nitrogenase, glutamine synthetase and glutamate synthase in *Azotobacter vinelandii* grown in oxygen-controlled continuous culture," *Arch. Microbiol.* *136*:74–78 (1983).

159. Robson, R. L., "O.$_2$-repression of nitrogenase synthesis in *Azotobater chrococcum*," *FEMS Microbiol. Lett.* *5*:259–262 (1979).

160. Robson, R. L., and Postgate, J. R., "Oxygen and hydrogen in biological nitrogen fixation," *Ann Rev. Microbiol.* *34*:183–207 (1980).

161. Schink, B., Thompson, T. E., and Zeikus, J. G., "Characterizaton of *Propionispira arboris* gen. nov. sp. nov., a nitrogen-fixing anaerobe common to wet woods of living trees," *J. Gen. Microbiol.* *128*:2771–2779 (1982).

162. Seldin, L., and Penido, E. G., "Identification of *Bacillus azotofixans* using AP1 tests," *Antonie van Leeuwenhoek 52*:403–409 (1986).

163. Senior, P. J., and Dawes, E. A., "The regulation of poly-β-hydroxybutyrate metabolism in *Azotobacter beijerinckii*," *Biohem. J.* *134*:225–238 (1973).

164. Siefert, E., and Pfenning, N., "Diazotrophic growth of *Rhodopseudomonas acidophila* and *Rhodopseudomonas capsulata* under microaerobic conditions," *Arch. Microbiol.* *125*:73–77 (1980).

165. Smith, A. T., "Oxygen, respiration and nitrogen fixation in the facultative anaerobe *Klebsiella pneumoniae*," Ph.D. thesis, University of Southampton, 1989.

166. Smith, A., Hill, S., and Anthony, C., "The purification, characterization and role of the *d*-type cytochrome oxidase of *Klebsiella pneumoniae* during nitrogen fixation," *J. Gen. Microbiol.* *136*:171–180 (1990).

167. Stam, H., van Verseveld, H. W., de Vries, W., and Stouthamer, A. H., "Hydrogen oxidation and efficiency of nitrogen fixation in succinate-limited chemostat culture of *Rhizohium* ORS571," *Arch. Microbiol.* *139*:56–60 (1984).

168. Song, S-D., Hartman, A., and Burris, R. H., "Purification and properties of the nitrogenase of *Azospirillum amazonense*," *J. Bateriol.* *164*:1271–1277 (1985).

169. Stougaard, J., and Kennedy, C., "Regulation of nitrogenase synthesis in histidine auxotrophs of *Klebsiella pneumoniae* with altered levels of adenylate nucleotides," *J. Bacteriol.* *170*:250–257 (1988).

170. Tal, S., and Okon, Y., "Production of the reverse material poly-β-hydroxybutyrate and its function in *Azospirillum brasilense* Cd.," *Can. J. Microbiol.* *31*:608–613 (1985).

171. Tempest, D. W., "The continuous cultivation of micro-organisms. 1. Theory of the chemostat," in J. R. Norris and D. W. Ribbons (eds.), *Methods in Microbiology*, Vol. 2, Academic Press, New York, 1970, pp. 259–276.

172. Thompson, J. P., and Skerman, V. B. D., "Azotobacteraceae: The taxonomy and ecology of the aerobic nitrogen-fixing bacteria," Academic Press, New York, 1979.

173. Thorneley, R. N. F., and Ashby, G. A., "Oxidation of nitrogenase iron protein by dioxygen without inactivation could contribute to high respiration rates of Azotobacter species and facilitate nitrogen fixation in other aerobic environments," *Biochem. J.* *261*:181–187 (1989).

174. Tibelius, K. H., and Knowles, R., "Effect of hydrogen and oxygen on uptake-hydrogenase activity in nitrogen-fixing and ammonium-grown *Azospirillum brasilense*," *Can. J. Microbiol.* *29*:1119–1125 (1983).

175. Tubb, R. S., and Postgate, J. R., "Control of nitrogenase synthesis in *Klebsiella pneumoniae*," *J. Gen. Microbiol.* *79*:103–117 (1973).

176. Tubb, R. S., "Ammonia regulation of nitrogenase synthesis in *Klebsiella,*" Ph.D. thesis, University of Sussex, UK, 1974.

177. Upchurch, R. G., and Mortenson, L. E., "*In vivo* energetics and control of nitrogen fixation: changes in the adenylate energy charge and adenosine 5'-diphosphate/adenosine 5'-triphosphate ratio of cells during growth on dinitrogen versus growth on ammonia," *J. Bacteriol. 143*:274–284 (1980).

178. Urdaci, M. C., Stal, L. J., and Marchand, M., "Occurrence of nitrogen fixation among *Vibrio* spp." *Arch. Microbiol. 150*:224–229 1988.

179. Wahl, R. C., and Orme-Johnson, W. H., "Clostridial pyruvate oxidoreductase and the pyruvate-oxidizing enzyme specific to nitrogen fixation in *Klebsiella pneumoniae* are similar enzymes," *J. Biol. Chem. 282*:10489–10496 (1987).

180. Walker, C. C., Partridge, C. D. P., and Yates, M. G., "The effect of nutrient limitation on hydrogen production by nitrogenase in continuous culture of *Azotobacter chroococcum,*" *J. Gen. Microbiol. 124*:317–327 (1981).

181. Widdel, F., "New types of acetate-oxidizing sulfate-reducing *Desulfobacter* species *D. hydrogenophilus* sp. nov. *D. latus* sp. nov. and *D. curvatus* sp. nov.," *Arch. Microbiol. 148*:286–291 (1987).

182. Wiegel, J., Wilke, D., Baumgarten, J., Opitz, R., and Schlegel, H. G., "Transfer of the nitrogen-fixing hydrogen bacterium *Corynebacterium autotrophicum* Baumngarten *et al.* to *Xanthobacter* gen. nov.," *Int. J. Syst. Bacteriol. 28*:513–581 (1978).

183. Wood, N. B., and Haselkorn, R., "Control of phycobiliprotein proteolysis and heterocyst differentiation in Anabaena," *J. Bacteriol. 141*:1375–1385 (1980).

184. Wong, T.-Y., and Maier, R. J., "Hydrogen-oxidizing electron transport components in nitrogen-fixing *Azotobacter vinelandii,*" J. Bacteriol. 159:348–352 (1984).

185. Yang, T. "Biochemical and biophysical properties of cytochrome *o* of *Azotobacter vinelandii,*" *Biochim. Biophys. Acta 848*:342–351 (1986).

186. Yates, M. G., "Control of respiration and nitrogen fixation by oxygen and adenine nucleotides in N_2-grown *Azotobacter chroococcum,*" *J. Gen. Microbiol. 60*:393–401 (1970).

187. Yates, M. G., and Jones, C. W., "Respiration and nitrogen fixation in *Azotobacter,*" *Adv. Microbiol. Physiol. 11*:97–135 (1974).

188. Yates, M. G., "The role of oxygen and hydrogen in nitrogen fixation," in J. A. Cole and S. Ferguson (eds.), *SGM Symposium Vol. 42, The Nitrogen and Sulphur Cycles,* Cambridge University Press, Cambridge, 1988, pp. 383–416.

189. Yates, M. G., and Campbell, F. O., "The effect of nutrient limitation on the competition between an H_2-uptake hydrogenase positive (Hup$^+$) recombinant strain of *Azotobacter chroococcum* and the Hup$^-$ mutant present in mixed populations," *J. Gen. Microbiol. 135*:221–226 (1989).

190. Yoch, D. C., Li, J. D., Hu, C. Z., and Scholin, C., "Ammonia switch-off of nitrogenase from *Rhodopseudomonas sphaeroides* and *Methylosinus trichosporium:* no evidence for Fe protein modification," *Arch. Microbiol. 150*:1–5 (1988).

191. You, C-B., Li, X., Wang, Y.-W., Qui, Y.-S., Mo, Z.-Z., and Zhang, Y.-L., "Associative dinitrogen fixation of *Alcaligenes faecalis* with rice plants," *Biological N_2 Fixation Newsletter,* Sydney, Australia, 11:92–103 (1983).

192. Youngeman, R. J., "Oxygen inactivation: Is the hydroxyl radical always biological relevant?" *Trends Biochem. Sci. 9*:280–283 (1984).

193. Zhu, J., and Brill, W. J., "Temperature sensitivity of the regulation of nitrogenase synthesis by *Klebsiella pneumoniae*," *J. Bacteriol. 145*:1116–1118 (1981).

194. Zimniak, P., and Barnes, E. M., "Characterization of a calcium proton antiporter and an electrogenic calcium transporter in membrane vesicle from *Azotobacter vinelandii*," *J. Biol. Chem. 255*:10140–10143 (1980).

195. Zumft, W. G., and Neumann, S., "Changes in guanosine tetraphosphate level during nitrogense expression in the phototrophic bacterium *Rhodopseudomonas palustris*," *FEBS Lett. 154*:121–126 (1983).

3

Nitrogen Fixation by Photosynthetic Bacteria

Gary P. Roberts and *Paul W. Ludden*

I. Introduction

The ability of *photosynthetic bacteria* to fix N_2 was first noted by Gest and Kamen as a result of their studies on H_2 metabolism by *Rhodospirillum rubrum*.[31,32] Their results were confirmed and extended by Lindstrom et al.,[65] who surveyed the abilities of photosynthetic bacteria, cyanobacteria, and other bacteria to fix N_2. N_2 fixation is widespread among all three of the major groups of photosynthetic bacteria and is the rule rather than the exception among these organisms. Madigan and coworkers[79] quantitated the rates of N_2 fixation among the purple, nonsulfur photosynthetic bacteria and found that most strains are vigorous fixers (Figure 3.1). However, several species of nonsulfur photosynthetic bacteria that are unable to use N_2 as a nitrogen source have been reported and characterized.[79]

The photosynthetic bacteria consist of three families in the order Rhodospirillales, all of which produce bacteriochlorophyll and carry out anoxygenic photosynthesis.[47,103] The most extensively studied family has been the Rhodospirillaceae (the purple nonsulfur bacteria), a group unable to use elemental sulfur as the electron donor for photosynthesis and including *Rhodospirillum rubrum* and *Rhodobacter capsulatus*. The Chromatiaceae (or purple sulfur bacteria), a group capable of using elemental and nonelemental sulfur as electron donors for photosynthetic growth, includes species of the genus *Chromatium* among others. The Chlorobiaceae, or green sulfur bacteria, is the least studied with respect to N_2-fixing ability and includes *Chlorobium thiosulfatifilum*. Not discussed in this review is N_2 fixation by the cyanobacteria. These bacteria are phylogenetically distinct from the photosynthetic bacteria and will be covered in Chapter 4.

More recent approaches to phylogeny of bacteria have relied on similarities of 16S ribosomal RNA sequences and have resulted in the recognition of photosynthetic bacteria as the progenitors of many of the families of Gram-negative bacteria.[28,33] These studies have also resulted in the grouping of the

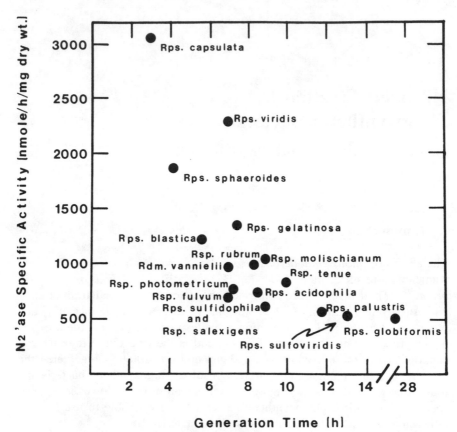

Figure 3-1. Whole cell nitrogenase activities by photosynthetic bacteria. (From Madigan et al.[79])

photosynthetic bacteria with other classes of bacteria. For this review, the close relationship of the fast-growing rhizobia and the nonsulfer photosynthetic bacteria is perhaps the most germane. In fact, a species of bacteriochlorophyll-producing *Rhizobium* has been isolated by Eaglesham and coworkers.[24]

Whereas all the preceding groups of photosynthetic bacteria are Gram-negative, Gest and coworkers[8] have recently isolated and characterized a Gram-positive N_2-fixing bacterium possessing an unusual variant of bacteriochlorophyll. These organisms, for which the genus *Heliobacterium* has been proposed, share cell wall characteristics with the Gram-positive organisms and have 16S ribosomal RNA sequences that would indicate their particular similarity to *Bacillus subtilis*.

The photosynthetic bacteria, especially the nonsulfur phototrophs, are characteristically diverse in their metabolism. They are able to use a wide variety of carbon sources, including C-1 compounds, for growth and as energy sources. The purple nonsulfur bacteria are also diverse in their ability to adapt to growth conditions and are able to grow photosynthetically with CO_2 as the carbon source and H_2 as the reductant. They will also grow phototrophically with light as the energy source and a fixed carbon source such as succinate or malate as the carbon and reductant source. These organisms will grow aerobically via respiration and will ferment a limited range of carbon sources anaerobically. Conditions for N_2 fixation by photosynthetic bacteria can be met when cells are growing under any of these conditions.[39,53,79,80,84,90]

Although the analysis of photosynthetic bacteria has not been central to the development of our understanding of the enzymology or genetics of N_2 fixation, their diverse metabolism raises interesting questions regarding the coordination of photosynthetic electron transport and N_2 fixation. Much of the attention to N_2 fixation by photosynthetic bacteria has dealt with the inhibition of nitrogenase activity by fixed nitrogen sources that was first observed by Gest and Kamen.[31,54] This posttranslational regulation, now known to be caused by the reversible ADP-ribosylation of the dinitrogenase reductase,[78] is not limited to the photosynthetic bacteria and has established ADP-ribosylation as a mechanism for enzyme regulation.

II. The Nitrogenase Enzyme System of Photosynthetic Bacteria

Although the details of the enzymology of nitrogenase are discussed elsewhere in this volume (see Chapter 18), the properties of nitrogenase are described briefly here to review which aspects of the general model for the enzyme have been established for the nitrogenase of photosynthetic bacteria. This discussion deals entirely with the molybdenum nitrogenase, as no reports of the vanadium or iron nitrogenase systems in photosynthetic bacteria have been published at the time of this review, although Paschinger[101] reported growth of a tungsten-resistant *R. rubrum* strain on N_2.

Among the photosynthetic bacteria, the complete nitrogenase has been purified to homogeneity from only two organisms, *R. rubrum* and *R. capsulatus*, both of which are nonsulfur purple bacteria.[37,72,96] Like the enzymes from the more studied examples, *Azotobacter vinelandii*, *Clostridium pasteurianum*, and *Klebsiella pneumoniae*, the nitrogenases from *R. rubrum* and *R. capsulatus* consist of a dinitrogenase (also called molybdenum–iron protein, MoFe protein, or component I) and a dinitrogenase reductase (also called the iron protein, the Fe protein, or component II). Dinitrogenase from photosynthetic bacteria, like that from other organisms, as an $\alpha_2\beta_2$ tetramer of the *nifKD* gene products. The first dinitrogenase protein purified was that of the purple sulfur bacterium, *Chromatium*, but its dinitrogenase reductase has yet to be purified to homogeneity.[2] This protein was investigated for its spectral and redox properties and found to exhibit what has become known as the characteristic EPR signal of dinitrogenase at $g = 3.65$. The *Chromatium* protein was also among the first to be be analyzed for its oxidation states by treatment with oxidizing dyes.[2] The proteins from *R. rubrum* and *R. capsulatus* were purified subsequently but analyzed less thoroughly than the *Chromatium* protein. To the extent that they have been studied, the dinitrogenases of photosynthetic bacteria are functionally identical to those of the more studied cases. Emerich and Burris[25] showed that dinitrogenases from *R. rubrum* and *Chromatium* would function with dinitrogenase reductases from other organisms, and Shah and Brill[120] showed that the FeMo-co isolated from *R. rubrum* protein would activate the apo-dinitrogenase from *A. vinelandii*. The metal content reported for the dinitrogenase proteins of photosynthetic bacteria is consistent with the presence of FeMo-co plus additional redox centers, presumably the four iron–sulfur centers referred to as the "P" clusters. The dinitrogenase proteins from photosynthetic bacteria are O_2-labile, as are all dinitrogenases studied to date, but otherwise they present no unusual problems during purification. The proteins are abundant, comprising about 1% of the total cell protein. Various combinations of ion-exchange chromatography, gel filtration, precipitation, and electrophoresis have been used in the purification of these dinitrogenases.[37,72,96] No dinitrogenase from a photosynthetic bacterium has yet been crystallized.

Dinitrogenase reductases are α_2 dimers of the *nifH* gene product with a native molecular weight of approximately 65,000.[37,72,96] They contain a single 4 iron–4 sulfur center that can be oxidized and reduced by a single electron.[66] The reduced protein exhibits an EPR signal in the $g = 2$ region. These proteins bind MgATP and MgADP and undergo a conformational shift upon MgATP binding that can be observed by Walker and Mortenson's assay[133] in which the iron–sulfur center becomes accessible to iron chelators in the presence of the nucleotide. The nucleotide specificity of the *R. rubrum* protein[69] has been studied in some detail with respect to its regulation and, except for minor differences, it behaves like the dinitrogenase reductase from *A. vinelandii* in its abilities to bind, hydrolyze, or be inhibited by various analogs of ATP and ADP.[138] Dinitrogenase reductase from *R. rubrum* was shown to function with dinitrogenases from other organisms[25] and antibody raised against the *R. rubrum* protein is highly cross-reactive with dinitrogenase reductases from all other sources. Likewise, the sequence of the *nifH* genes of *R. capsulatus* and *R. rubrum* have been determined and found to be highly similar to the *nifH* genes of other organisms. Like other dinitrogenase reductases, the enzymes from photosynthetic organisms are extremely O_2-labile and their purification demands rigorous exclusion of O_2. These proteins also comprise approximately 1% of the total cell protein and are purified by combinations of ion exchange and gel filtration chromatography and by electrophoresis.

The utilization of substrates by nitrogenase from *R. rubrum* has been examined and found to correspond with that from other organisms. N_2, C_2H_2, N_3^-, CN^-, and protons are all reduced by the enzyme and all reductions, except that of protons, are inhibited by CO.[90]

The nitrogenase system from *Chromatium* has also been prepared by Winter and Ober[140,141] in a particulate form that contains both dinitrogenase and dinitrogenase reductase activities. This particulate form may be similar to the O_2-stable particulate form of nitrogenase that can be isolated from extracts of *A. vinelandii* prepared by disruption in a French pressure cell. This particulate form may suggest that nitrogenase is found in a supramolecular structure *in vivo* in *Chromatium*.

III. Regulation of Nitrogenase Activity by Reversible ADP-Ribosylation

A. Observation of Posttranslational Regulation of Nitrogenase Activity in R. rubrum.

When Gest and Kamen carried out manometric studies of H_2 metabolism by *R. rubrum,* they noted that there was an especially vigorous H_2 evolution by the cultures when the cells were grown on glutamate.[31] They also noted that N_2 did not behave as an inert gas in the presence of glutamate-grown cells and that the cells appeared to consume N_2. This led to the hypothesis, soon confirmed, that *R. rubrum* might be able to grow with N_2 as the nitrogen source. The N_2-fixing ability of *R. rubrum* was confirmed by demonstrating the incorporation of $^{15}N_2$ into cellular material. These initial experiments by Gest and Kamen were then repeated and extended to other species of photosynthetic bacteria by Lindstrom et al.[65]

Gest and Kamen also noted that fixed-nitrogen sources such as ammonia inhibited both the uptake of N_2 and the evolution of H_2 by these glutamate-grown cells. It was not until much later that Gest hypothesized that the H_2 evolution of these cells might be mediated by nitrogenase itself,[100] a fact confirmed when Burns and Bulen demonstrated the ability of partially purified nitrogenase to evolve H_2.[11] Thus, H_2 was not only an inhibitor of nitrogenase, a fact known for many years from the work of Wilson (see Burris[12]), but also a product. The demonstration that the vigorous H_2 evolution by N_2-fixing cultures of photosynthetic bacteria was mediated *in vivo* by the nitrogen-fixing enzyme itself was demonstrated in *R. capsulatus* by Wall and Gest,[136] who showed that point mutations in the *nif* genes resulted in loss of both activities and that revertants regained both activities.

The ability of ammonia to inhibit *in vivo* N_2 uptake was curious, as this property was not exhibited by other organisms, such as *C. pasteurianum*. Ormerod et al.[99] studied the effect in more detail and showed that the inhibition did not require protein synthesis, as N_2 uptake by cultures treated with protein synthesis inhibitors could still be inhibited by ammonia. Schick[115] carried out extensive studies of the N_2 and H_2 metabolisms of *R. rubrum* and showed that the inhibitory effect of ammonia was reversible and that the length of the inhibitory period was correlated with the concentration of ammonia added. The undemonstrated presumption was that once ammonia was assimilated by the culture, nitrogenase activity resumed. The recovery of activity was fairly rapid and did not appear to be due to synthesis of new nitrogenase. Once the acetylene reduction assay was available, Neilson and Nordlund[93] investigated the inhibition of nitrogenase by ammonia and were able to establish more precisely the time course of inhibition and recovery of enzyme activity *in vivo*. They also extended their studies to show that other fixed-nitrogen sources such as glutamine and asparagine would inhibit nitrogenase activity as well as ammonia.

As these experiments with whole-cell nitrogenase activity were proceeding, methods for extracting nitrogenase activity from whole cells were developed. Schneider et al.[116] first reported positive results for nitrogenase activity in extracts of *R. rubrum*. Burns and Bulen[11] then developed methods for preparing consistently active extracts once it was understood that ATP was required for nitrogenase activity and that dithionite would serve as an electron donor to the enzyme *in vitro*. Burns and Bulen observed that, unlike nitrogenases from *C. pasteurianum* and *A. vinelandii*, nitrogenase activity in *R. rubrum* was most active when Mg^{2+} was present at five times the concentration of ATP in the assay mixture. They also noted that the membrane fragments of extracts were highly inhibitory to nitrogenase activity in crude extracts. Munson and Burris[90] further investigated the properties of crude extracts of *R. rubrum* and found that the nitrogenase activity was nonlinear, with the activity increasing with time. These workers also observed that the most consistently active preparations of nitrogenase from *R. rubrum* were obtained from cells that had been grown on limiting ammonia and thus were starved for nitrogen prior to collection.

It was then found that Mn^{2+} stimulated the activity of crude extracts over and above the stimulation by excess Mg^{2+}, and this led to the discovery of an "activating factor" for nitrogenase from *R. rubrum*.[71] The activation was found to apply specifically to the dinitrogenase reductase protein, as the dinitrogenase protein was active as isolated.[71,95] Analysis of the activation process led to the finding that inactive dinitrogenase reductase from *R. rubrum* was modified by a nucleotide that was eventually identified as adenosinediphosphoribose (ADP-ribose) by Pope et al.[104,105]

B. Model for the Regulation of Dinitrogenase Reductase by ADP-ribosylation

The model for the regulation of dinitrogenase reductase is shown in Figure 3-2. Dinitrogenase reductase is depicted as identical subunits and the un-

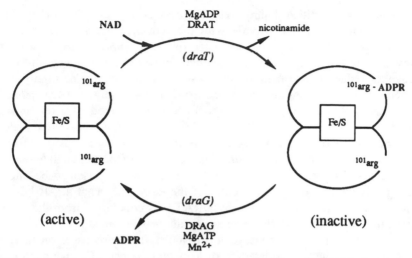

Figure 3-2. Model for regulation of nitrogenase activity by reversible ADP-ribosylation.

modified form on the left is the active form of the enzyme. *D*initrogenase *R*eductase *A*DP-ribosyl *T*ransferase (DRAT) carries out the transfer of the ADP-ribose moiety of NAD to arginine-101 of dinitrogenase reductase in an ADP-dependent reaction. The modification of one of the two available arg-101 residues of the dimer results in complete inactivation of the enzyme. Activation of the enzyme occurs when *D*initrogenase *R*eductase *A*ctivating *G*lycohydrolase (DRAG) removes the ADP-ribose from the enzyme in an MgATP- and M^{2+}-dependent reaction. *In vivo*, ammonia, darkness, and the oxidizing dye phenazine methosulfate all lead to ADP-ribosylation of dinitrogenase reductase.[55] Evidence for this model and the regulation of DRAT and DRAG is presented in following sections.

C. Evidence that ADP-ribose Is the Modifier of Dinitrogenase Reductase at Arg-101

The modifier of dinitrogenase reductase was isolated by heating the modified protein at alkaline pH and it was purified by boronate affinity chromatography and HPLC; it was identified as ADP-ribose using mass spectral and NMR analysis.[105] The linkage to the protein was established as an N-glycosidic bond from the 1″ carbon of ribose to the guanidinium group of arginine by mass spectral analysis of the gly-arg(ADP-ribose)-gly tripeptide obtained after proteolysis of the modified protein.[104] These results were confirmed by comparison of the isolated compound with authentic arg-ADP-ribose prepared by treating arginine and NAD with cholera toxin. The isolated material was found to be an approximately 40:60 mixture of $\alpha:\beta$ anomers of the *N*-glycosidic bond; the anomeric specificity of the DRAG and DRAT enzymes are discussed in the following sections.

Treatment of modified dinitrogenase reductase with subtilisin alone yielded a single ADP-ribosylated peptide with the sequence gly-arg(ADP-ribose)-gly-val-ile-thr.[104] Although the sequence of the *R. rubrum* protein was not known at the time this result was obtained, inspection of sequences of dinitrogenase reductases from other organisms revealed a single, highly conserved region containing this sequence from gly-99 to thr-104 (*A. vinelandii* sequence).[43] When the *R. rubrum* sequence was obtained, this region was found to correspond to gly-100 to thr-105. Thus the functionally equivalent arginine residue occurs at position 101 in *R. rubrum* and at position 100 in *K. pneumoniae* and *A. vinelandii*. This region is of particular interest because cys-97 of the *A. vinelandii* protein had been proposed to be one of the ligands for the iron–sulfur center of the enzyme.[42] That arg-101 is the unique site for ADP-ribosylation by DRAT was concluded from the observation that no other ADP-ribosylated peptide was obtained after proteolysis of the modified protein. It was then proved by the demonstration that the dinitrogenase reductase isolated from a mutant stain of *K. pneumoniae* in which arg-100 of dinitrogenase reductase was converted to a histidine residue was unable to be ADP-ribosylated by DRAT.[67] Other ADP-ribosyl transferases are known to modify dinitrogenase reductase at a number of sites *in vitro*.

D. Properties of DRAG

DRAG was initially detected as an "activating factor" that would stimulate the activity of crude extracts of *R. rubrum*.[71,95] DRAG is found associated with the membrane fraction of the crude extract, suggesting that the enzyme is bound to the chromatophores.[71,95] The enzyme can be solublized from the particulate fraction by washing with buffer containing 0.5 M NaCl. Significant purification of DRAG is achieved by elution from buffer-washed pellets and the enzyme is then further purified to homogeneity by ion exchange, hydroxylapatite, affinity, and gel filtration chromatography.[112,152] Upon purification and characterization, DRAG was shown to be a monomeric enzyme with a molecular weight of 32,000. Its extreme lability to O_2 (it has a 90-sec half-life in air) suggests the presence of an oxidizable prosthetic group; although dithiolthreitol is required for stability of DRAG, it does not protect from O_2 or renature O_2-treated enzyme and thus the O_2 lability cannot be explained by a reactive thiol.[111] When *K. pneumoniae* is transformed with a plasmid encoding DRAG, a functional gene product is produced.[29] Thus, if DRAG does contain a prosthetic group, it is one that is produced by *K. pneumoniae*, an organism that does not normally contain a gene coding for DRAG. The sensitivity of DRAG activity to redox suggests a role for the redox state of the enzyme in the regulation of its activity *in vivo*.

The removal of ADP-ribose from reduced, native dinitrogenase reductase requires MgATP and free divalent metal.[111] The role of MgATP appears to be binding to dinitrogenase reductase and causing a conformational change that allows DRAG to recognize the ADP-ribosylated site. This conclusion is based on the finding that the removal of ADP-ribose from oxidized or O_2-denatured dinitrogenase reductase and the hydrolysis of N-glycosidic bonds between ADP-ribose and arginine analogs does not require MgATP.[106] All activities of DRAG require free divalent metal with Mn^{2+} (K_m 0.5 mM) being the most effective, followed by Fe^{2+} and Mg^{2+};[98,106,111] Ca^{2+} inhibits DRAG activity. It is not clear which metal ion functions *in vivo*.

DRAG will carry out the hydrolysis of a range of α-N-glycosidic bonds between ADP-ribose and arginine-containing groups.[106,107] DRAG removes ADP-ribose from both oxidized and reduced dinitrogenase reductase both *in vivo* and *in vitro* and the K_m for the protein as substrate is estimated to be about 35–40 μM.[111] The ADP-ribosyl peptides isolated from protease digests of dinitrogenase reductase are excellent substrates for the enzyme with K_ms of 13 μM. N^ω-ADP-ribosyl-arginine is a poor substrate, but when the α-amino group is blocked by dansylation, it becomes an excellent substrate.[106,107] Carboxymethylation of the arginyl residue does not affect the K_m or V_{max} of the substrate. Neither N^ω-ribosyl dansyl arginine nor N^ω-(ribose-5-phosphate)-dansylarginine are substrates for the enzyme.[107]

When ADP-ribosyl dansylargininemethylester is used as the substrate, the products of the N-glycohydrolysis reaction have been established as ADP-ribose and dansylarginine methyl ester; thus the guanidinium group is regenerated during activation of dinitrogenase reductase and is available for future cycles of regulation.[106,111]

DRAG is inhibited by ADP-ribose, α-NAD (but not β-NAD), O_2, and mercurials.[111] With the physiological substrate as the target, ADP is a potent inhibitor of the enzyme. DRAG is not inhibited by ammonium ion or glutamine *in vivo*.[111]

E.   Purification and Properties of DRAT

Once ADP-ribose was established as the covalently attached nucleotide on dinitrogenase reductase, a search for a NAD-dependent ADP-ribosyl transferase analogous to the ADP-ribosylating toxins such as cholera toxin was initiated and resulted in isolation of DRAT.[70] DRAT was shown to be electrophoretically and immunologically distinct from DRAG[68] and the uniqueness of each protein has been confirmed by the isolation of genes encoding each one.[27] DRAT is an O_2-stable, soluble, monomeric enzyme with a molecular weight of 30,000. The enzyme has been purified to homogeneity by ion exchange, affinity, gel filtation, and high-pressure liquid chromatography with an overall purification of 20,000-fold.[68] Thus, like DRAG, DRAT is a protein of very low abundance.

DRAT is far more specific than DRAG in its substrate specificity and will transfer ADP-ribose only to dinitrogenase reductases; neither small molecules nor the hexapeptide modification site of dinitrogenase reductase (gly-arg-gly-val-ile-thr) will serve as acceptors.[68] In contrast to the ADP-ribosylating toxins, DRAT has no detectable NAD glycohydrolase activity. All purified dinitrogenase reductases that have been tested, except that from *Azospirillum amazonense*, are capable of being ADP-ribosylated by *R. rubrum* DRAT (Lowery and Ludden, unpublished results) and the sequence cys-gly-arg-gly-val/ile-ile-thr is found in all dinitrogenase reductases that have been sequenced (e.g., Pretorius et al.[108]), including the dinitrogenase reductases of the alternate system.[48,110] The sequence of the *A. amazonense* enzyme has not been determined.

That arg-100 (*A. vinelandii* numbering) is the unique site of modification was demonstrated when it was shown that the mutant dinitrogenase reductase (Kp2) from *K. pneumoniae* strain UN1041 was unable to be ADP-ribosylated by DRAT as the result of an arg → his substitution at position 100.[67] Interestingly, the dinitrogenase reductase from *K. pneumoniae* strain UN1041 is inactive with respect to electron transfer to dinitrogenase, but catalyzes unproductive dinitrogenase-dependent ATP hydrolysis.[17,67] Kp2(UN1041) is normal with respect to its migration on gel filtration columns and exhibits an EPR signal identical to that of the wild-type enzyme. Although it is able to bind MgATP, the rate of ATP-dependent Fe^{2+} release in Walker and Mortenson's assay is distinctly slower than that of the wild-type enzyme. The results with Kp2(UN1041) suggest that arginine-100 not only is a target for disruption of dinitrogenase reductase activity, but that it plays an essential role in dinitrogenase reductase function.

DRAT will use β-NAD, nicotinamide guanine dinucleotide, nicotinamide hypoxanthine dinucleotide, and etheno-NAD as ADP-ribose donors, but will not use NADP or reduced forms of either NAD or NADP.[68]

The role of MgADP in DRAT activity is thought to be analogous to the role of MgATP in DRAG activity. Although ADP-ribosylation of *R. rubrum* dinitrogenase reductase is dependent on ADP, the modification of the *A. vinelandii* and *K. pneumoniae* dinitrogenase reductases can be observed in the absence of ADP.[69] In both of these cases, the effect of ADP is to lower the K_m for NAD in the reaction. Further support for the hypothesis that ADP acts on dinitrogenase reductase rather than on DRAT comes from the observation that the nucleotide specificity for ADP-ribosylation of dinitrogenase reductase mimics the specificity of ADP analogs in inhibiting nitrogenase activity.[69] ATP effectively inhibits DRAT activity[69] and the converse activation and inhibition effects of ATP and ADP on DRAT and DRAG suggest a role for these nucleotides in the regulation of the ADP-ribosylation cycle *in vivo*. In opposition to this hypothesis, however, is the observation that changes in relative levels of ATP and ADP *in vivo* have not been found to be necessary or sufficient to trigger modification *in vivo*.[102]

F. Effect of ADP-Ribosylation on Dinitrogenase Reductase

The effect of ADP-ribosylation on dinitrogenase function appears to be the prevention of complex formation with dinitrogenase.[91] ADP-ribosylated dinitrogenase reductase coelectrophoreses with the unmodified protein on nondenaturing gels and comigrates with the unmodified protein on gel filtration columns. Thus ADP-ribosylation does not lead to aggregation or subunit separation. The ADP-ribosylated protein is as O_2-labile as the unmodified protein and copurifies with it.[72,77] The modified protein exhibits a normal EPR spectrum and can be oxidized and reduced.[73] It binds ATP and releases its iron–sulfur center to chelators in Walker and Mortenson's assay.[35]

The ability of ADP-ribosylated dinitrogenase reductase to form a complex with dinitrogenase was tested by examining the effect of ADP-ribosylation on the tight complex that is formed between dinitrogenase from *A. vinelandii* (Av1) and dinitrogenase reductase from *C. pasteurianum*.[25] ADP-ribosylated *C. pasteurianum* dinitrogenase reductase was found not to inhibit the *A. vinelandii* nitrogenase system, whereas the unmodified protein was a potent inhibitor.[91] Similarly, it was not possible to ADP-ribosylate dinitrogenase reductase from *C. pasteurianum* when it was complexed to the dinitrogenase of *A. vinelandii*. The inability of ADP-ribosylated *C. pasteurianium* dinitrogenase reductase to inhibit *A. vinelandii* nitrogenase is taken as evidence that ADP-ribosylation inhibits the binding of the two proteins and, by analogy, it suggests that homologous pairs of nitrogenase proteins are prevented from forming productive complexes when the dinitrogenase reductase component is ADP-ribosylated.

Although the ADP-ribosylated protein is incompetent to transfer electrons to dinitrogenase, modified dinitrogenase reductase functions in the *in vitro* synthesis of the iron–molybdenum cofactor. Likewise, the dinitrogenase reductase from *K. pneumoniae* strain UN1041 is incompetent in electron transfer to dinitrogenase, but functions in FeMo-co synthesis *in vivo* and *in vitro* (Shah, Imperial, and Ludden, unpublished results). Apparently, arg-100 is not essential for the role that dinitrogenase reductase plays in iron–molybdenum cofactor biosynthesis.

Only one of the two identical arginyl-101 residues in the dinitrogenase reductase dimer can be modified at one time and the modification of one arginine per dimer results in complete loss of activity. The inability to modify both arginines on the same protein dimer raises a question regarding the proximity of the two arginine-101 residues on the protein. Since there is no evidence to suggest that ADP-ribosylation results in a conformational change of the native protein that would hide the second residue, an attractive alternative would be that the two residues are close to each other so that the ADP-ribosylation of one would sterically block the modification of the other. The proximity of the modification site to the site of iron–sulfur center ligation (cys-98) demands that arg-101 be near the interface of the subunits because the iron–sulfur center bridges the two subunits. The answer to this question awaits the determination of the structure of the protein by X-ray crystallography.

G. Evidence for ADP-Ribosylation of Nitrogenase in Other Organisms

Although ADP-ribosylation of nitrogenase has been studied primarily in *R. rubrum*, there is evidence for this system in other organisms as well. The system has been thoroughly documented in *R. capsulatus*.[38,39,52,147] DRAG has been isolated from this organism and ADP-ribose has been identified as the bound nucleotide at arg-100 of the dinitrogenase reductase isolated from this organism. Likewise, the ADP-ribosylation of dinitrogenase reductase has been established in *Azospirillum lipoferum* and *A. brazilense* and shown not to operate in *A. amazonense*.[30,40,76] Strong evidence for the system has also been presented for *Chromatium*,[34] *R. palustris*,[150,151] *Rhodopseudomonas viridis*,[46] and *Ectothiorhodospira*.[10] In *R. spheroides*,[36,50] *Azorhizobium sesbaniae*,[62] and *A. chroococcum*[16] physiological responses consistent with a covalent modification of dinitrogenase reductase have been observed, but it has not been possible to obtain biochemical evidence for the system in extracts. Ironically, three of the organisms that have provided most of our understanding of the enzymology of nitrogenase, *C. pasteurianum*, *A. vinelandii*, and *K. pneumoniae*, appear to lack this system of regulation at least in strains used for nitrogenase research. A more extensive discussion of regulation of nitrogenase by ADP-ribosylation in nonphototrophs is given in a recent review.[78]

IV. Physiology of Photosynthetic Bacteria Relevant to Nitrogen Fixation

Photosynthetic bacteria are able to employ a range of modes of growth including light-dependent photosynthetic growth, fermentative growth on pyruvate or fructose,[117,130] and aerobic respiration. Some are able to grow by respiring nitrate or other alternate electron acceptors and some will grow anaerobically with trimethylamine oxide or dimethylsulfoxide as electron acceptors from fermentative metabolism. N_2 fixation can be demonstrated under all growth conditions in one or more N_2-fixing bacteria. The mechanisms by which the *in vivo* needs for nitrogenase activity (reductant, ATP, and an anaerobic environment) are met differ dramatically for these modes of growth. The effect of each of these growth modes on the expression of nitrogenase and its regulation by the ADP-ribosylation cycle must also be considered.

Nitrogenase synthesis is repressed by high levels of O_2 (depending on the respiratory ability of the organisms) and utilizable sources of fixed nitrogen in the photosynthetic bacteria. The mechanisms of regulation of *nif* expression is discussed in Section VIII of this chapter and is not considered in detail here. One practical consideration regarding the analysis of nitrogenase in the purple bacteria is that when glutamate is supplied as the nitrogen source, it is metabolized in a way that allows expression of nitrogenase. Thus, expression of nitrogenase can be tested under conditions that do not demand nitrogenase activity for growth.

A. Physiological Effects of the ADP-Ribosylation Cycle

The ADP-ribosylation cycle described in Section III was first observed as the inhibition of nitrogenase activity (H_2 evolution) by fixed-nitrogen sources.[31,32] Subsequent studies have established that other sources of fixed nitrogen such as glutamine and asparagine also lead to the inhibition of *in vivo* nitrogenase activity.[93] This *in vivo* loss of activity has been termed "switch-off" by Zumft[151] and in this review, switch-off will refer specifically to the *in vivo* loss of nitrogenase activity due to the ADP-ribosylation of dinitrogenase reductase. Other signals that lead to switch-off are darkness (for light-grown cells) and nonphysiological agents such as the uncoupler CCCP and the oxidizing dye phenazine methosulfate. Switch-off by darkness is a particularly useful feature of the system because light leaves no residue in the medium when it is removed and there is no barrier to uptake by the cell when it is added.

A key observation with respect to the ADP-ribosylation system is the finding that cells grown with N_2 or glutamate as the nitrogen source respond to the various stimuli by inhibiting nitrogenase activity, whereas cells that are completely starved for nitrogen do not,[15] even though both DRAT and DRAG activities can be measured in extracts of both types of cells.[68,70,128]

That the ADP-ribosylation system is responsible for inhibition *in vivo* has been established both biochemically and genetically. It was shown that the extent of modification (ADP-ribosylation) of dinitrogenase reductase correlated inversely with whole-cell nitrogenase activity of glutamate-grown cells in response to darkness or ammonia.[55] Furthermore, it was shown that the ADP-ribosylation cycle could be carried out for repeated cycles in response to light and darkness; thus, the ADP-ribosylation of dinitrogenase reductase is not a one-time event.[55] Mutants in which the expression of the *draTG* region has been eliminated by a polar mutation in the *draT* gene do not exhibit the switch-off response upon treatment with darkness or ammonia, thus proving the requirement of the *draTG* gene products for the switch-off *in vivo* and the ADP-ribosylation cycle (Liang and Roberts, unpublished). It should be noted that Yakunin and Gogotov[143,144] have maintained that ADP-ribosylated dinitrogenase reductase from *R. capsulatus* will function normally in the transfer of electrons to dinitrogenase if it is reduced by ferredoxin rather than dithionite *in vitro*. However, Yoch and coworkers have been unable to repeat these experiments.[148]

When *R. rubrum* is grown in the dark anaerobically with pyruvate as the fermentable carbon source, the cells must adapt the ADP-ribosylation system if nitrogenase is to function. Schultz et al.[118] have demonstrated that *R. rubrum* grown in the dark will respond to ammonia by switching off the nitrogenase activity that is present. The whole-cell nitrogenase activity of dark-grown cells is much lower than that of photosynthetically grown cells, and it is not known if the ammonia-dependent inhibition that is observed for dark-grown cells represents the ADP-ribosylation of all of the dinitrogenase reductase present under these conditions or if 90% of the dinitrogenase reductase is already ADP-ribosylated because of the dark conditions and the ammonia treatment results in the ADP-ribosylation of the remaining 10%.

Evidence that DRAG is regulated *in vivo* comes from experiments in which dinitrogenase was labeled with ^{32}P-ADP-ribose *in vivo* and the label then chased by adding an excess of ^{31}P phosphate to the cell medium.[55] As long as inhibitory conditions were maintained, dinitrogenase reductase remained labeled even though the total cell phosphate pools chased out very rapidly. Thus, DRAG was not catalyzing the removal of ADP-ribose during switch-off conditions. Evidence that DRAT is regulated *in vivo* comes from experiments with mutants lacking DRAG (Nielsen, Liang, Ludden, Roberts, unpublished results). Mutants that lack DRAG accumulate dinitrogenase reductase in the active, unmodified form, under conditions that favor activation of nitrogenase, indicating that DRAT is regulated *in vivo*.

The mechanism by which signals such as darkness and ammonia are transduced to regulate DRAT and DRAG activity has not been established. Although the ratio of ATP to ADP is a good candidate for involvement in the regulation, changes in the ATP/ADP ratio that correlate with inhibition have not been observed.[102] Likewise, the redox state of dinitrogenase reductase might be suspected, but the ADP-ribosylation cycle has been shown to occur *in vivo* in a NifD⁻ strain of *R. rubrum,* in which it must be assumed that dinitrogenase reductase remains in the reduced state.[75] Kanemoto[56] investigated the possible role of glutamine and other amino acids in regulation of ADP-ribosylation and did not observe a consistent correlation of glutamine concentration with ADP-ribosylation, although it has been suggested that glutamine may be the effector of nitrogenase regulation under some conditions.[64] Surprisingly, when the *draTG* genes of *R. rubrum* are introduced into *K. pneumoniae* on an expression vector, they reversibly regulate this heterologous nitrogenase in response to added fixed nitrogen.[29] The response is remarkably similar to that seen in *R. rubrum* and involves ADP-ribosylation of dinitrogenase reductase. This result suggests that the fixed nitrogen signal transduction system of *K. pneumoniae* is capable of interacting with DRAG and DRAT, resulting in the regulation of their activity. The ability of DRAT and DRAG to function in *K. pneumoniae* provides a well-characterized genetic background in which to examine the function of these gene products.

V. Electron Donors to Nitrogenase in Photosynthetic Bacteria

Only in *K. pneumoniae* has the path of electrons to nitrogenase been established definitively by a combination of biochemical and genetic studies.[94,121] In this organism the *nifJ* gene codes for a pyruvate:flavodoxin oxidoreductase and the *nifF* gene encodes a *nif*-specific flavodoxin. In *C. pasteurianum*, biochemical evidence that strongly supports the function of a similar pyruvate-dependent system has been presented but it is not known whether the pyruvate oxidizing system is *nif*-specific.[88,89] In *A. vinelandii*, a *nifF*-like gene has been identified, but its gene product is not essential for growth of the organism on N_2.[9]

In the photosynthetic bacteria, relatively little work has been done to establish the path of electrons to nitrogenase. Because there is light-driven electron transport in photosynthetic bacteria, an attractive hypothesis has been that the electron donor to nitrogenase is generated by light reactions. This seems unlikely since the midpoint potential of the primary electron acceptor of photosynthetic electron transport in the purple bacteria is only -150 mV. Furthermore, no one has been able to develop a system in which light-generated reductant can be supplied to nitrogenase using chromatophores isolated from *R. rubrum* or any other photosynthetic bacterium. In the green sulfur bacteria, however, the light-generated reductant has a much lower redox potential, perhaps sufficient to drive nitrogenase.[109]

Pyruvate will replace dithionite as the electron donor to nitrogenase in crude extracts of *R. rubrum* and this activity is stimulated by coenzyme A and by ferredoxins isolated from *R. rubrum*.[74] It has not been demonstrated that any component of this system is *nif*-specific and *nifF*- or *nifJ*-like genes have not been identified in any of the phototrophs studied to date.

In *R. sphaeroides* it has been suggested that electrons for nitrogenase are generated by reverse electron flow that is mediated by the light-generated membrane potential of the photosynthetic chromatophores.[36] This model is based on *in vivo* experiments and a reconstitution of this system *in vitro* should be possible, but has not been reported.

Several ferredoxins that might serve as the direct *in vivo* electron donor to dinitrogenase reductase have been isolated and characterized in *R. ru-*

brum, R. capsulatus, and other organisms. *R. rubrum* has several ferredoxins and two soluble ferredoxins have been shown to function efficiently as electron donors to nitrogenase *in vitro.*[148] A ferredoxin has been isolated from *R. capsulatus* and characterized in detail.[113] *R. rubrum* has been reported to produce a flavodoxin under iron-starvation conditions,[21] but it is not known whether this flavodoxin is produced under nitrogen-fixing conditions or whether it functions as an electron donor to nitrogenase.

The elucidation of the path of electrons to nitrogenase in photosynthetic bacteria awaits a combined genetic and biochemical investigation.

VI. Ammonia Assimilation in the Photosynthetic Bacteria

N_2 fixation and ammonia assimilation are obviously related reactions and the regulation of their expression and activity are tightly coordinated in the enteric bacteria, in which they have been most extensively studied. The photosynthetic bacteria that have been studied assimilate ammonia via the glutamine synthetase–glutamate synthase pathway.[92] Glutamine synthetases have been purified and characterized from several representatives of the purple bacteria,[3,4,13,14,26,49,124,125,149] and all fit into the paradigm established for the enteric glutamine synthetase.[126,129] The enzymes are dodecamers of a single subunit and available evidence suggests that they are regulated by the reversible adenylylation of the enzyme. The existence or role of uridyltransferase (the *glnD* product) in regulation of glutamine synthetase expression or activity has not been established, but the *glnB* gene has tentatively been identified.[61] Conclusive demonstration of its functional homology to the enteric version awaits biochemical demonstration. Evidence for the *ntr* system in photosynthetic bacteria is discussed in Section VIII.

Although the *R. rubrum* glutamine synthetase is regulated by adenylyation, it differs from the *E. coli* enzyme in that adenylylation of the enzyme affects both the biosynthetic activity and the γ-glutamyl transferase activity of the enzyme, and it is not possible to determine the adenylylation state of the enzyme by assaying the Mg^{2+}- and Mn^{2+}-dependent transferase activities of an enzyme preparation.[97]

Glutamate synthase has been purified from *R. rubrum* and shown to be similar to the *E. coli* enzyme in its activity and molecular properties.[146]

VII. Organization of *nif* Genes

Because the significant majority of analyses on the physiology and genetic organization of the *nif* genes in photosynthetic bacteria has been done with *R. capsulatus,* this section will focus on this organism and tersely allude to other organisms at the end. The analysis of *R. capsulatus* has been the product of a wide variety of classical and molecular genetic techniques,[41] where

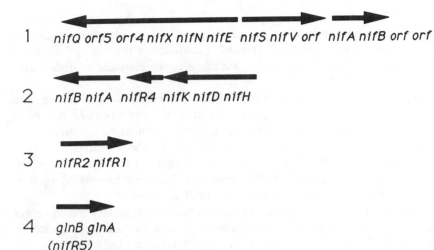

1 nifQ orf5 orf4 nifX nifN nifE nifS nifV orf nifA nifB orf orf

2 nifB nifA nifR4 nifK nifD nifH

3 nifR2 nifR1

4 glnB glnA
   (nifR5)

Figure 3-3. Characterized clusters of the *nif* and *gln* genes in *R. capsulatus*. The top three clusters shown are those containing known *nif* genes as described in the text. The bottom cluster contains the *glnB* gene (previously designated *nifR4*), whose product is involved in the transcriptional regulation of *nif* expression. The arrows indicate the suggested transcriptional units. The term *orf* is used to denote open reading frames, where neither the gene product nor its function have been examined.

the initial work involved mutant isolation and mapping by Wall and coworkers[134,136,137] with some clustering of mutations noted using the gene transfer agent (GTA) indigenous to *R. capsulatus*. The small size of this vector inhibited its use for large-scale mapping, and more global organization of these genes was identified by Haselkorn and coworkers using R'-plasmids generated from *R. capsulatus* DNA.[6] Finally, a variety of labs have been involved in the cloning, sequencing, and functional characterization of these regions.[1,7,51,86,87] The general conclusion has been that the *nif* genes exist in at least three clusters and that some of the genes are duplicated, as determined by either mutational analysis or sequence comparison. This is in contrast to the case in *K. pneumoniae*, where all the known *nif* genes are found at a single locus and there is no apparent duplication of functionally homologous genes.[22,83] It remains a shortcoming of these analyses that biochemical characterization of the products of few of these genes have been described.

The first linkage group, containing copies of *nifQ, X, N, E, S, V, A*, and *B*, with four other open reading frames (see Figure 3-3), was detected by both Wall[134,137] and Haselkorn,[6,7] who demonstrated hybridization of this region to a *K. pneumoniae nif* probe. The sequence analysis of this region by Klipp, Puhler, and coworkers has been particularly instructive and has been the basis for most of the gene assignments noted here[57,86,87] and in

Figure 3-3. Within this group, only the *nifA* gene has been verified as necessary for the Nif$^+$ phenotype and even in this case another functional *nifA* gene has been demonstrated.[81] It is of some interest that one of the open reading frames, ORF5, has a sequence motif appropriate for an iron-sulfur-containing protein.[86]

The second cluster includes a *nifHDK* region, a second *nifAB* region, and a regulatory gene with sequence similarity to *ntrC* (*nifR4*, discussed later).[1,7,57,139] These *nifHDK* genes have been shown to be functional, because insertions in these genes cause a Nif$^-$ phenotype. Surprisingly, these insertions revert to a Nif$^+$ phenotype at high frequency and this observation, coupled with the detection of other regions hybridizing to *nifHDK* probes, has suggested the presence of a cryptic *nifHDK* region.[119]

The other two linkage groups include a pair of regulatory genes (*nifR1* and *2*) identified by both mutational characterization and sequence analysis[6,58,136] and another potential linkage group identified by a single mutation.[6,58] The former is considered in the next section but the latter will require further analyses before its biological function can be interpreted.

The outstanding genetic questions yet to be addressed for the *nif* system of *R. capsulatus* are the existence and locations of the other *nif* genes and the proper interpretation of the genetic results, which have indicated the presence of duplicate copies of the *nifAB* and *nifHDK* regions.

Much less is known about the *nif* genes in the other photosynthetic bacteria. In *R. sphaeroides*, many loci have been identified genetically and placed on a circular map, but functional and sequential characterization of these mutations is lacking.[122,123] In *R. rubrum*, only the *nifHDK* genes have been cloned and sequenced, but mutational and hybridization analysis suggests that this is the only activatable copy of these genes in this organism,[63,75] in apparent contrast to the case in *R. capsulatus*.

VIII. Transcriptional Regulation of *nif* Genes

In the photosynthetic bacteria there is good evidence for transcriptional regulation in response to the amount of fixed nitrogen, the degree of anaerobicity, and the availability of photosynthetic energy (reviewed by Haselkorn[41]). Because the majority of these analyses have been performed with *R. capsulatus*, this section will focus on work with that organism.

The paradigm for nitrogen regulation, in gram-negative bacteria at least, is the model developed for the enterics (as reviewed in Dixon et al.[22] and Merrick[83]; also see Chapter 21 in this volume). In this scheme, there is a two-component regulatory system (the *ntrBC* gene products) that becomes a transcriptional activation system for certain operons under nitrogen-limiting conditions. The actual decision-making system has not been completely described, but it is clear that at least the *glnB* product plays a role in the generation of the activator, a phosphorylated *ntrC* product. This activator, in concert with a minor sigma factor, the *rpoN* (*ntrA*) gene product, acts to stimulate transcription of the nitrogen-regulated promoters in general and of the *nifLA* region specifically.

As described later, a similar system seems to be functioning in *R. capsulatus*, based on similarity of mutant phenotypes and nucleic acid sequences. However, some caution in interpretation of the similarities is warranted by the fact that (1) the enteric system is itself by no means fully understood and it is likely that more factors in the "sensing system" remain to be recognized and characterized; (2) even our current understanding of the *R. capsulatus* system indicates that there are significant differences between the two systems, as noted later; and (3) the physiological background of the enterics differs substantially from that of the photosynthetic bacteria.

Initial analyses of *nif* genetics in photosynthetic bacteria has utilized genetic screens seeking both Nif⁻ and Nif^c (constitutive) mutants, which allowed the identification of regions involved in global nitrogen regulation.[60,134,136] The phenotypes of these mutations, together with the most recent sequence and transcriptional analyses of these regions, has allowed the following general picture to be drawn. The *nifR1* and 2 genes appear to be sequentially[51] similar to the *ntrBC* genes of the enterics. Their functional similarity is evidenced by the fact that mutants in these genes are unable to express *nifH*[58] or transport methyl ammonium ions.[136] It is clear, however, that these product functions are not completely identical in the different organisms, since affected mutants in these genes do have the ability to utilize amino acids as nitrogen sources,[134] in rather striking distinction to the case in the enterics.

The sequence of the *nifR4* gene is similar to that of the *rpoN* gene of the enterics.[51] On the other hand, there is the surprising difference that the *nifR4* gene is itself transcriptionally regulated in response to fixed nitrogen and O_2,[58] unlike the *rpoN* gene, which is constitutively expressed.[82] Mutational analysis also suggests that the *nifR4* is functionally distinct from *rpoN* in that the former is apparently specific for *nif* expression,[51] whereas the latter is required for expression of a wide range of other systems (see Chapter 21, this volume).

Finally, the *nifR5* gene, identified by the isolation of ammonia-constitutive mutants, behaves as an analog of *glnB* in that its product seems to be involved in sensing nitrogen status of the cell and then influencing the activity of the *nifA* and *nifR4* products.[60,61] It has also been demonstrated that *nifR5* is sequentially similar to *glnB* of *K. pneumoniae* and that it is transcriptionally upstream of *glnA*.[61]

This is by no means a complete list of candidate regulatory genes in *R. capsulatus*, as other interesting nitrogen regulatory loci have been noted (for example, see Allibert et al.[5]), but their interpretation is as yet unclear. It should also be noted that although *nif* promoters have not been rigorously identified in these bacteria, there appear to be sequences similar to those implicated in binding of the *nifA* gene product in enteric *nif* operons.[51,81]

In *K. pneumoniae*, there is clearly a *nif*-specific response to fixed nitrogen or aerobicity, at the levels of both transcription and posttranscription, that depends on the *nifL* product and its effect on *nifA* activation of the other *nif* transcripts.[18,19,23,44] There is currently no evidence for either a *nifL* gene or a *nif*-specific regulation in photosynthetic bacteria in response to O_2. Indeed, both the absence of a *nifL*-like sequence near either of the functional *nifA* genes[81] and the "less than global" effects of some of the *nifR* functions noted earlier suggest these latter genes may replace *nifL* for at least some of its functions in these organisms.

The regulation of anaerobic gene expression has only recently been the subject of intense investigation, even in the enterics, so that much less is known than for nitrogen regulation. Although the enteric *fnr* system seems to be a "global" anaerobic regulatory system,[127] it does not seem to have a direct role in the regulation of *nif*.[45] In *K. pneumoniae* it is therefore unclear which factors are responsible for the regulation of *nifLA* transcription, nor is it clear if the *nifL*-mediated response to O_2 requires other factors. In none of the photosynthetic bacteria have any genes or proteins been identified as being required for these regulatory responses.

It has been argued, however, that these bacteria perform at least part of their regulation of anaerobic gene expression by the modulation of DNA supercoiling,[59] following a model for the enteric bacteria.[145] Although such a model remains attractive, it has recently been experimentally challenged by a novel assay for local supercoiling *in vivo*,[20] and the authors conclude that the supercoiling results reflect the properties of heavily expressed, not anaerobically regulated, genes. In any event, the presence of both general (*fnr*) and *nif*-specific (*nifL*) O_2-responsive regulators in *K. pneumoniae* suggests that supercoiling regulation must at least be supplemented by specific gene products, as yet unidentified in the photosynthetics.

There is little doubt that there will also be transcriptional regulation of *nif* functions in response to the illumination of the culture.[53] To date, however, no mutants affected in this regulation have been isolated, nor have any loci been correlated with these effects. Finally, there is a report for *R. capsulatus* that molybdenum is necessary for the accumulation of at least the products of the *nifHDK* operon, although the mechanisms of such a regulatory system are unclear.[39]

IX. Genetics and Regulation of Auxiliary Functions

This section briefly covers the genetic characterization in the photosynthetic bacteria of four categories of genes of relevance to N_2 fixation in vivo: *glnA* (encoding glutamine synthetase), *draTG* (encoding the system responsible for the posttranslational regulation of nitrogenase activity), *frd* (encoding ferredoxins), and *hup* (encoding hydrogenase).

The gene encoding glutamine synthetase (*glnA*) of *R. capsulatus* has been cloned and mutationally characterized[119,135,139] and it is known to be downstream of the characterized *glnB* regulatory locus.[61] As is the case in enterics, mutations in *glnA* cause the constitutive synthesis of nitrogenase in the presence of ammonia. Unlike the enterics, however, this is apparently a physiological result of reduced glutamine pools and not a consequence of enhanced transcription of linked regulatory genes. The *glnA* locus has also been identified in *R. sphaeroides,* both genetically and physically.[123]

The genetic characterization of the *R. rubrum* genes encoding the reversible ADP-ribosylation system of dinitrogenase reductase has recently been reported.[27] The genes (*draT*, encoding DRAT, and *draG*, encoding DRAG) are located very close to the *nifHDK* operon in this organism, but are divergently transcribed. Preliminary evidence suggests that there are several differentially regulated transcripts through the *draTG* region and that the two genes may not always be coordinately regulated (S. Montgomery, and G. Roberts, unpublished results). Contrary to the initial report,[27] mutations in *draT* are viable. As noted earlier, the analysis of constructed strains lacking DRAG shows that DRAT product is itself subject to posttranslational regulation. When the *draTG* genes are cloned onto a P_{tac}-based expression vector in *K. pneumoniae,* the surprising result is obtained that the nitrogenase activity in this organism shows reversible posttranslational regulation by ammonia that is superficially very similar to that seen in *R. rubrum.*[29] This result confirms the *draTG* gene products as the basis for this regulation. It might also suggest that there are proteins in *K. pneumoniae* capable of transmitting information on the nitrogen status of the cell to the heterologous DRAG–DRAT system.

There have been several reports concerning the efficacy of ferredoxins isolated from *R. capsulatus*[148] or *R. rubrum*[37] in the *in vitro* reduction of dinitrogenase reductase, and the accumulation of at least one ferredoxin from *R. capsulatus* shows a correlation with nitrogenase activity.[114] Based on sequence analysis, it has further been hypothesized that some of the open-reading frames scattered among the *nif* genes might encode ferredoxins.[86] However, given the capacity of dinitrogenase reductase to accept electrons from a range of low-potential electron donors *in vitro,* as well as the difficulties in predicting gene product function from nucleic acid sequences, positive identification of the physiological reductant of dinitrogenase reductase must await biochemical characterization of appropriate mutant strains. The isolation of genes encoding ferredoxin I from *R. capsulatus*[113,114] and *Rhodopseudomonas palustris*[85] will make such analyses possible.

Uptake hydrogenases have been isolated from a number of photosynthetic bacteria (see Vignais et al.[131] for a review), but only in *R. capsulatus* have the structural genes been isolated.[142] The region containing the structural genes is apparently highly linked to other operons encoding *hup*-related functions (cited in Vignais et al.[132]).

X. Summary

Thus, although the biochemistry of N_2 fixation in the photosynthetic bacteria is apparently rather similar to that of other diazotrophs, the regulatory systems controlling these functions are novel. The analysis of the interaction of these varied levels of transcriptional and posttranslational regulatory system promises to be rewarding.

Acknowledgments

Work from the authors' laboratories has been generously supported by The USDA Competitive Grants Program and the National Science Foundation. The authors wish to express their appreciation to their colleagues and co-workers who have contributed to this area of research in their laboratories, including Mark Pope, Theresa Paul, Roy Kanemoto, Eric Triplett, Leonard Saari, Stefan Nordlund, Michael Kahn, Scott Murrell, Robert Lowery, Wayne Fitzmaurice, Lori Lehman, Doug Lies, Haian Fu, Diana Woehle, Heidi Wirt, Liang Jihong, Gary Nielsen, and Robert Burris.

References

1. Ahombo, G., Willison, J. C., and Vignais, P. M., "The *nifHDK* genes are contiguous with a *nifA*-like regulatory gene in *Rhodobacter capsulatus*," *Mol. Gen. Genet. 205*, 442–445 (1986).

2. Albrecht, S. L., and Evans, M. C. W., "Measurement of the oxidation reduction potential of the EPR detectable active centre of the molybdenum iron protein of *Chromatium* nitrogenase," *Biochem Biophys. Res. Commun. 55*, 1009–1014 (1973).

3. Alef, K., Burkardt, H.-J., Horstmann, H.-J., and Zumft, W. G., "Molecular characterization of glutamine synthetase from the nitrogen-fixing phototrophic bacterium *Rhodopseudomonas palustris*," *Z. Naturforsch. 36C*, 246–253 (1981).

4. Alef, K., and Zumft, W. G., "Regulatory properties of glutamine synthetase from the nitrogen-fixing phototrophic bacterium *Rhodopseudomonas palustris*," *Z. Naturforsch. 36c*, 784–789 (1981).

5. Allibert, P., Willison, J. C., and Vignais, P. M., "Complementation of nitrogen-regulatory (*ntr*-like) mutations in *Rhodobacter capsulatus* by an *Escherichia coli* gene: Cloning and sequencing of the gene and characterization of the gene product," *J. Bacteriol. 169*, 260–271 (1987).

6. Avtges, P., Kranz, R. G., and Haselkorn, R., "Isolation and organization of genes for nitrogen fixation in *Rhodopseudomonas capsulata,*" *Mol. Gen. Genet. 201,* 363–369 (1985).

7. Avtges, P., Scolnik, P. A., and Haselkorn, R., "Genetic and physical map of the structural genes (*nifHDK*) coding for the nitrogenase complex of *Rhodopseudomonas capsulata,*" *J. Bacteriol. 156,* 251–256 (1983).

8. Beer-Romero, P., Favinger, J., and Gest, H., "Distinctive properties of bacilliform photosynthetic heliobacteria," *FEMS Microbiol. Lett. 49,* 451–454 (1988).

9. Bennett, L. T., Jacobson, M. R., and Dean, D. R., "Isolation, sequencing, and mutagenesis of the *nifF* gene encoding flavodoxin from *Azotobacter vinelandii,*" *J. Biol. Chem. 263,* 1364–1369 (1988).

10. Bognar, A., Desrosiers, L., Ligman, M., and Newman, E. B., "Control of nitrogenase in the photosynthetic autotrophic bacterium, *Ectothiorhodospira* sp.," *J. Bacteriol. 152,* 706–713 (1982).

11. Burns, R. C., and Bulen, W. A., "A procedure for the preparation of extracts from *Rhodospirillum rubrum* catalyzing N_2 reduction and ATP-dependent H_2 evolution," *Arch. Biochem. Biophys. 113,* 461–463 (1966).

12. Burris, R. H., "H_2 as an inhibitor of N_2 fixation," *Physiol. Veg. 23,* 843–848 (1985).

13. Caballero, F. J., Dejudo, F. J., Florencio, F. J., Cardenas, J., and Castillo, F., "Molecular and regulatory properties of glutamine synthetase from the phototrophic bacterium *Rhodopseudomonas capsulata* E1F1," *J. Bacteriol. 162,* 804–809 (1985).

14. Caballero, F. J., Igeño, Cárdenas, J., and Castillo, F., "Regulation of reduced nitrogen assimilation in *Rhodobacter capsulatus* E1F1," *Arch. Microbiol. 152,* 508–511 (1989).

15. Carithers, R. P., Yoch, D. C., and Arnon, D. I., "Two forms of nitrogenase from the photosynthetic bacterium *Rhodospirillum rubrum,*" *J. Bacteriol. 137,* 779–789 (1979).

16. Cejudo, F. J., de la Torre, A., and Paneque, A., "Short-term ammonium inhibition of nitrogen fixation in *Azotobacter,*" *Biochem. Biophys. Res. Commun. 123,* 431–437 (1984).

17. Chang, C. L., Davis, L. C., Rider, M., and Takemoto, D. J., "Characterization of *nifH* mutations of *Klebsiella pneumoniae,*" *J. Bacteriol. 170,* 4015–4022 (1988).

18. Collins, J. J., and Brill, W. J., "Control of *Klebsiella pneumoniae nif* RNA synthesis," *J. Bacteriol. 162,* 1186–1190 (1986).

19. Collins, J. J., Roberts, G. P., and Brill, W. J., "Posttranscriptional control of *Klebsiella pneumoniae nif* mRNA stability by the *nifL* product," *J. Bacteriol. 168,* 173–178 (1986).

20. Cook, D. N., Armstrong, G. A., and Hearst, J. E., "Induction of anaerobic gene expression in *Rhodobacter capsulatus* is not accompanied by a local change in chromosomal supercoiling as measured by a novel assay," *J. Bacteriol. 171,* 4836–4843 (1989).

21. Cusanovich, M. A., and Edmondson, D. E., "The isolation and characterization of *Rhodospirillum rubrum* flavodoxin," *Biochem. Biophys. Res. Commun. 45,* 327 (1971).

22. Dixon, R. A., Austin, S., Buck, M., Drummond, M., Hill, S., Holtel, A., MacFarlane, S., Merrick, M., and Minchin, S., "Genetics and regulation of *nif* and related genes in *Klebsiella pneumoniae,*" *Phil. Trans. R. Soc. Lond. 317,* 147–158 (1987).

23. Drummond, M. H., and Wootton, J. C., "Sequence of *nifL* from *Klebsiella pneumoniae:* Mode of action and relationship to two families of regulatory proteins," *Mol. Microbiol. 1,* 37–44 (1987).

24. Ellis, J., Hungria, M., Rizzo, N., Eaglesham, A., and Hardy, R. W. R., "First photosynthetic nitrogen fixing rhizobium," *Plant Physiol. 89*, 180 (1989).

25. Emerich, D. W., and Burris, R. H., "Interactions of heterologous nitrogenase components that generate catalytically inactive complexes," *Proc. Natl. Acad. Sci. USA 73*, 4369–4373 (1976).

26. Englehardt, H., and Klemme, J.-H., "Purification and structural properties of adenylylated and deadenylylated glutamine synthetase from *Rhodopseudomonas sphaeroides*," *Arch. Microbiol. 133*, 202–205 (1982).

27. Fitzmaurice, W. P., Saari, L. L., Lowery, R. G., Ludden, P. W., and Roberts, G. P., "Genes coding for the reversible ADP-ribosylation system of dinitrogenase reductase from *Rhodospirillum rubrum*," *Mol. Gen. Genet. 218*, 340–347 (1989).

28. Fox, G. E., Stackebrandt, E., Hespell, R. B., Gibson, J., Maniloff, J., Dyer, T. A., Wolfe, R. S., Balch, W. E., Tanner, R. S., Magrum, L. J., Zablen, L. B., Blakemore, R., et al., "The phylogeny of prokaryotes," *Science 209*, 457–463 (1980).

29. Fu, H.-A., Burris, R. H., and Roberts, G. P., "Reversible ADP-ribosylation is demonstrated to be a regulatory mechanism in prokaryotes by heterologous expression," *Proc. Natl. Acad. Sci. USA 87*, 1720–1724 (1990).

30. Fu., H.-A., Fitzmaurice, W. P., Lehman, L. J., Roberts, G. P., and Burris, R. H., "Regulation of nitrogenase activity in azospirilla, herbaspirilla and acetobacter: and cloning of *draG*- and *draT*-homologous genes of *A. lipoferum* SpBr 17 (Abs)," *Nitrogen Fixation: Hundred Years After (Proceedings of the 7th International Congress on Nitrogen Fixation)*, H. Bothe, F. J. de Bruijn and W. E. Newton, eds. Stuttgart: Gustav Fischer, 1988, p. 336.

31. Gest, H., and Kamen, M. D., "Photoproduction of molecular hydrogen by *Rhodospirillum rubrum*," *Science 109*, 558–559 (1949).

32. Gest, H., Kamen, M. D., and Bregoff, H. M., "Studies on the metabolism of photosynthetic bacteria. V. Photoproduction of hydrogen and nitrogen fixation by *Rhodospirillum rubrum*," *J. Biol. Chem. 182*, 153–170 (1950).

33. Gibson, J., Stackebrandt, E., Zablen, L. B., Gupta, R., and Woese, C. R., "A phylogenetic analysis of the purple photosynthetic bacteria," *Curr. Microbiol. 3*, 59–64 (1979).

34. Gotto, J. W., and Yoch, D. C., "Regulation of nitrogenase activity by covalent modification in *Chromatium vinosum*," *Arch. Microbiol. 141*, 40–43. (1985).

35. Guth, J. H., and Burris, R. H., "Comparative study of the active and inactive forms of dinitrogenase reductase from *Rhodospirillum rubrum*," *Biochim. Biophys. Acta 749*, 91–100 (1983).

36. Haaker, H., Laane, C., Hellingwerf, K., Houwer, B., Konings, W. N., and Veeger, C., "Short-term regulation of the nitrogenase activity in *Rhodopseudomonas sphaeroides*," *Eur. J. Biochem. 127*, 639–645 (1982).

37. Hallenbeck, P. C., Jouanneau, Y., and Vignais, P. M., "Purification and molecular properties of a soluble ferredoxin from *Rhodopseudomonas capsulata*," *Biochim. Biophys. Acta 681*, 168–176 (1982).

38. Hallenbeck, P. C., Meyer, C. M., and Vignais, P. M., "Nitrogenase from the photosynthetic bacterium *Rhodopseudomonas capsulata*: Purification and molecular properties," *J. Bacteriol. 149*, 708–717 (1982).

39. Hallenbeck, P. C., Meyer, P. C., and Vignais, P. M., "Regulation of nitrogenase in the photosynthetic bacterium *Rhodopseudomonas capsulata* as studied by two-dimensional gel electrophoresis," *J. Bacteriol. 151*, 1612–1616 (1982).

40. Hartmann, A., Fu, H., and Burris, R. H., "Regulation of nitrogenase activity by ammonium chloride in *Azospirillum* sp.," *J. Bacteriol. 165*, 864–870 (1986).

41. Haselkorn, R., "Organization of the genes for nitrogen fixation in photosynthetic bacteria and cyanobacteria," *Ann. Rev. Microbiol. 40*, 525–547 (1986).

42. Hausinger, R. P., and Howard, J., "Thiol reactivity of the nitrogenase Fe-protein from *Azotobacter vinelandii*," *J. Biol. Chem. 258*, 13486–13492 (1983).

43. Hausinger, R. P., and Howard, J. B., "The amino acid sequence of the nitrogenase iron protein from *Azotobacter vinelandii*," *J. Biol. Chem. 257*, 2483–2490 (1982).

44. Henderson, N., Austin, S., and Dixon, R. A., "Role of metal ions in negative regulation of nitrogen fixation by the *nifL* gene product from *Klebsiella pneumoniae*," *Mol. Gen. Genet. 216*, 484–491 (1989).

45. Hill, S., "Redox regulation of enteric *nif* expression is independent of the *fnr* gene product," *FEMS Microbiol. Lett. 25*, 5–9 (1985).

46. Howard, K. S., Hales, B. J., and Socolofsky, M. D., "Nitrogen fixation and ammonia switch-off in the photosynthetic bacterium *Rhodopseudomonas viridis*," *J. Bacteriol. 155*, 107–112 (1983).

47. Imhoff, J. F., Truper, H. G., and Pfennig, N., "Rearrangement of the species and genera of the phototrophic purple nonsulfur bacteria", *Int. J. Syst. Bacteriol. 34*, 340–343 (1984).

48. Joerger, R. D., Jacobson, M. R., Premakumar, R., Wolfinger, E. D., and Bishop, P. E., "Nucleotide sequence and mutational analysis of the structural genes (*anf*HDGK) for the second alternative nitrogenase from *Azotobacter vinelandii*," *J. Bacteriol. 171*, 1075–1086 (1989).

49. Johansson, B. C., and Gest, H., "Adenylylation/deadenylylation control of the glutamine synthetase of *Rhodopseudomonas capsulata*," *Eur. J. Biochem. 81*, 365–371 (1977).

50. Jones, B. L., and Monty, K. J., "Glutamine as a feedback inhibitor of the *Rhodopseudomonas sphaeroides* nitrogenase system," *J. Bacteriol. 139*, 1007–1013 (1979).

51. Jones, R., and Haselkorn, R., "The DNA sequences of the *Rhodobacter capsulatus ntrA, ntrB* and *ntrC* gene analogues required for nitrogen fixation," *Mol. Gen. Genet. 215*, 507–516 (1989).

52. Jouanneau, Y., Roby, C., Meyer, C. M., and Vignais, P. M., "ADP-ribosylation of dinitrogenase reductase in *Rhodobacter capsulatus*," *Biochemistry 28*, 6524–6530 (1989).

53. Jouanneau, Y., Wong, B., and Vignais, P. M., "Stimulation by light of nitrogenase synthesis in cells of *Rhodopseudomonas capsulata* growing in N-limited continuous cultures," *Biochim. Biophys. Acta 808*, 149–155 (1985).

54. Kamen, M. D., and Gest, H., "Evidence for a nitrogenase system in the photosynthetic bacterium *Rhodospirillum rubrum*," *Science 109*, 560 (1949).

55. Kanemoto, R. H., and Ludden, P. W., "Effect of ammonia, darkness, and phenazine methosulfate on whole-cell nitrogenase activity and Fe protein modification in *Rhodospirillum rubrum*," *J. Bacteriol. 158*, 713–720 (1984).

56. Kanemoto, R. H., and Ludden, P. W., "Amino acid concentrations in *Rhodospirillum rubrum* during expression and switch-off of nitrogenase activity," *J. Bacteriol. 169*, 3035–3043 (1987).

57. Klipp, W., Masepohl, B., and Puhler, A., "Identification and mapping of nitrogen fixation genes of *Rhodobacter capsulatus:* Duplication of a *nifA-nifB* region," *J. Bacteriol. 170*, 693–699 (1988).

58. Kranz, R. G., and Haselkorn, R., "Characterization of the *nif* regulatory genes in *Rhodopseudomonas capsulata* using *lac* gene fusions," *Gene 40*, 203–215 (1985).

59. Kranz, R. G., and Haselkorn, R., "Anaerobic regulation of nitrogen fixation genes in *Rhodopseudomonas capsulata*," *Proc. Natl. Acad. Sci. USA 83*, 6805–6809 (1986).

60. Kranz, R. G., and Haselkorn, R., "Ammonia-constitutive nitrogen fixation mutants of *Rhodobacter capsulatus*," *Gene 71*, 65–74 (1988).

61. Kranz, R. G., Pace, V. M., and Caldicott, I. M., "Inactivation, sequence, and *lacZ* fusion analysis of a regulatory locus required for repression of nitrogen fixation genes in *Rhodobacter capsulatus*," *J. Bacteriol. 172*, 55–62 (1990).

62. Kush, A., Elmerich, C., and Aubert, J. P., "Nitrogenase of *Sesbania Rhizobium* strain ORS571: Purification, properties and 'switch-off' by ammonia," *J. Gen. Microbiol. 131*, 1765–1777 (1985).

63. Lehman, L. J., Fitzmaurice, W. P., and Roberts, G. P., "The cloning and functional characterization of the *nifH* gene of *Rhodospirillum rubrum*," *Gene 95*,143–147 (1990).

64. Li, J., Hu, C.-Z., and Yoch, D. C., "Changes in amino acid and nucleotide pools of *Rhodospirillum rubrum* during switch-off of nitrogenase activity initated by NH^+_4 or darkness," *J. Bacteriol. 169*, 231–237 (1987).

65. Lindstrom, E. S., Tove, S. R., and Wilson, P. W., "Nitrogen fixation by the green and purple sulfur bacteria," *Science 112*, 197–198 (1950).

66. Ljones, T., and Burris, R. H., "ATP hydrolysis and electron transfer in the nitrogenase reaction with different combinations of the iron protein and the molybdenum–iron protein," *Biochim. Biophys. Acta 275*, 93–101 (1972).

67. Lowery, R. G., Chang, C. L., Davis, L. C., McKenna, M. C., Stephens, P. J., and Ludden, P. W., "Substitution of Histidine for Arginine-101 of Dinitrogenase Reductase Disrupts Electron Transfer to Dinitrogenase," *Biochemistry 28*, 1206–1212 (1989).

68. Lowery, R. G., and Ludden, P. W., "Purification and properties of the dinitrogenase reductase inactivating ADP-ribosyltransferase from *Rhodospirillum rubrum*," *J. Biol. Chem. 263*, 16714–16719 (1988).

69. Lowery, R. G., and Ludden, P. W., "Effect of nucleotides on the activity of dinitrogenase reductase ADP-ribosyltransferase from *Rhodospirillum rubrum*," *Biochemistry 28*, 4956–4961 (1989).

70. Lowery, R. G., Saari, L. L., and Ludden, P. W., "Reversible regulation of the iron protein of nitrogenase from *Rhodospirillum rubrum* by ADP-ribosylation *in vitro*," *J. Bacteriol. 166*, 513–518 (1986).

71. Ludden, P. W., and Burris, R. H., "Activating factor for the iron protein of nitrogenase from *Rhodospirillum rubrum*," *Science 194*, 424–426 (1976).

72. Ludden, P. W., and Burris, R. H., "Purification and properties of nitrogenase from *Rhodospirillum rubrum* and evidence for phosphate, ribose and an adenine-like unit covalently bound to the iron protein," *Biochem. J., 175*, 251–259 (1978).

73. Ludden, P. W., and Burris, R. H., "Removal of an adenine-like molecule during activation of dinitrogenase reductase from *Rhodospirillum rubrum*," *Proc. Natl. Acad. Sci. USA 76*, 6201–6205 (1979).

74. Ludden, P. W., and Burris, R. H., "ATP and electron donors to nitrogenase from *Rhodospirillum rubrum in vivo* and *in vitro*," *Arch. Microbiol. 130*, 155–158 (1981).

75. Ludden, P. W., Lehman, L., and Roberts, G. P., "Reversible ADP-ribosylation of dinitrogenase reductase in a NifD⁻ mutant of *Rhodospirillum rubrum*," *J. Bacteriol. 171*, 5210–5211 (1989).

76. Ludden, P. W., Okon, Y., and Burris, R. H., "The nitrogenase system of *Spirillum lipoferum*," *Biochem. J. 173*, 1001–1003 (1983).

77. Ludden, P. W., Preston, G. G., and Dowling, T. E., "Comparison of active and inactive forms of iron protein from *Rhodospirillum rubrum*," *Biochem. J. 203*, 663–668 (1982).

78. Ludden, P. W., and Roberts, G. P., "Regulation of nitrogenase activity by reversible ADP-ribosylation," *Current Topics in Cellular Regulation*, Vol. 30, B. Horecker, E. Stadtman, P. B. Chock, and A. Levitzki, eds. Orlando, FL: Academic Press, 1989, pp. 23–55.

79. Madigan, M., Cox, S. S., and Stegeman, R. A., "Nitrogen fixation and nitrogenase activities in members of the family *Rhosodpirillaceae*," *J. Bacteriol. 157*, 73–78 (1984).

80. Madigan, M. T., Wall, J. D., and Gest, H., "Dark anaerobic dinitrogen fixation by a photosynthetic microorganism," *Science 204*, 1430 (1979).

81. Masepohl, B., Klipp, W., and Puhler, A., "Genetic characterization and sequence analysis of the duplicated *nifA/nifB* gene region of *Rhodobacter capsulatus*," *Mol. Gen. Genet. 212*, 27–37 (1988).

82. Merrick, M., and Stewart, W. D. P., "Studies on the regulation of the *Klebsiella pneumoniae ntrA* gene," *Gene 35*, 297–303 (1985).

83. Merrick, M. J., "Organization and regulation of nitrogen fixation genes in *Klebsiella* and *Azotobacter*," *Nitrogen Fixation: Hundred Years After*, H. Bothe, F. J. de Bruijn and W. E. Newton, eds. Stuttgart: Gustav Fischer, 1988, pp. 293–302.

84. Meyer, J., Kelley, B. C., and Vignais, P. M., "Aerobic nitrogen fixation by *Rhodopseudomonas capsulata*," *FEBS Lett. 85*, 224–228 (1987).

85. Minami, Y., Wakabayashi, S., Yamada, F., Wada, K., Zumft, W. G., and Matsubara, II., "Ferredoxins from the photosynthetic purple nonsulfur bacterium *Rhodopseudomonas palustris*. Isolation and amino acid sequence of ferredoxin I," *J. Biochem. 96*, 585–592 (1984).

86. Moreno-Vivian, C., Hennecke, S., Puhler, A., and Klipp, W., "Open reading frame 5 (ORF5), encoding the ferredoxinlike protein and *nifQ* are cotranscribed with *nifE*, *nifN*, *nifX*, and ORF4 in *Rhodobacter capsulatus*," *J. Bacteriol. 171*, 2591–2598 (1989).

87. Moreno-Vivian, C., Masepohl, B., Schmehl, M., Klipp, W., and Puhler, A., "Nucleotide sequence of *Rhodobacter capsulatus nif* gene regions carrying homologous genes to *Klebsiella pneumoniae*, *nifE*, *nifN*, *nifX*, *nifQ*, *nifS* and *nifV*," *Nitrogen Fixation: Hundred Years After*, H. Bothe, F. J. de Bruijn and W. E. Newton, eds. Stuttgart: Gustav Fischer, 1988, p. 177.

88. Mortenson, L. E., "Ferredoxin and ATP, requirements for nitrogen fixation in cell-free extracts of *Clostridium pasteurianum*," *Proc. Natl. Acad. Sci. USA 52*, 272–279 (1964).

89. Mortenson, L. E., Valentine, R. C., and Carnahan, J. E., "An electron transport factor from *Clostridium pasteurianum*," *Biochem. Biophys. Res. Commun. 7*, 448–452 (1962).

90. Munson, T. O., and Burris, R. H., "Nitrogen fixation by *Rhodospirillum rubrum* grown in nitrogen limited continuous culture," *J. Bacteriol. 97*, 1093–1098 (1969).

91. Murrell, S. A., Lowery, R. G., and Ludden, P. W., "ADP-ribosylation of dinitrogenase reductase from *Clostridium pasteurianum* prevents its inhibition of nitrogenase from *Azotobacter vinelandii*," *Biochem. J. 251*, 609–612 (1988).

92. Nagatani, H., Shimizu, M., and Valentine, R. C., "The mechanism of ammonia assimilation in nitrogen fixing bacteria," *Arch. Mikrobiol. 79*, 164–175 (1971).

93. Neilson, A. H., and Nordlund, S., "Regulation of nitrogenase synthesis in intact cells of *Rhodospirillum rubrum:* Inactivation of nitrogen fixation by ammonia, L-glutamine and L-asparagine.," *J. Gen. Microbiol. 91*, 53–62 (1975).

94. Nieva-Gomez, E., Roberts, G. P., Klevickis, S., and Brill, W. J., "Electron transport to nitrogenase in *Klebsiella pneumoniae*," *Proc. Natl. Acad. Sci. USA 77*, 2555–2558 (1980).

95. Nordlund, S., Eriksson, U., and Baltscheffsky, H., "Necessity of a membrane component for nitrogenase activity in *Rhodospirillum rubrum*," *Biochem. Biophys. Acta 462*, 187–195 (1977).

96. Nordlund, S., Eriksson, U., and Baltscheffsky, H., "Properties of the nitrogenase system from a photosynthetic bacterium, *Rhodospirillum rubrum*," *Biochim. Biophys. Acta 504*, 248–254 (1978).

97. Nordlund, S., Kanemoto, R. H., Murrell, S. A., and Ludden. P. W., "Properties and regulation of glutamine synthetase from *Rhodospirillum rubrum*," *J. Bacteriol. 161*, 13–17 (1985).

98. Nordlund, S., and Noren, A., "Dependence on divalent cations of the activation for inactive Fe protein of nitrogenase from *Rhodospirillum rubrum*," *Biochim. Biophys. Acta 791*, 21–27 (1984).

99. Ormerod, J. G., Ormerod, K. S., and Gest, H., "Light-dependent utilization of organic compounds and photoproduction of molecular hydrogen by photosynthetic bacteria; relationships with nitrogen metabolism," *Arch. Biochem. Biophys. 94*, 449–463 (1961).

100. Ormerod, J. G., and Gest, H., "Symposium on metabolism of inorganic compounds: Hydrogen photosynthesis and alternative metabolic pathways in photosynthetic bacteria," *Bacteriol. Rev. 26*, 51–66 (1962).

101. Paschinger, H., "A changed nitrogenase activity in *Rhodospirillum rubrum* after substitution of tungsten for molybdenum," *Arch. Microbiol. 101*, 379–389 (1974).

102. Paul, T. D., and Ludden, P. W., "Adenine nucleotide levels in *Rhodospirillum rubrum* during switch-off of whole-cell nitrogenase activity," *Biochem., J. 224*, 961–969 (1984).

103. Pfennig, N., and Truper H. G., "The phototrophic bacteria," *Bergey's Manual of Determinative Bacteriology*, 8th ed. R. E. Buchanan and N. E. Gibbson, eds. Baltimore: The Williams & Wilkins Co., 1974, pp. 24–64.

104. Pope, M. R., Murrell, S. A., and Ludden, P. W., "Covalent modification of the iron protein of nitrogenase from *Rhodospirillum rubrum* by adenosine diphosphoribosylation of a specific arginyl residue," *Proc. Natl. Acad. Sci. USA 82*, 3173–3177 (1985).

105. Pope, M. R., Murrell, S. A., and Ludden, P. W., "Purification and properties of the heat released nucleotide modifying group from the inactive Fe protein of nitrogenase from *Rhodospirillum rubrum*," *Biochemistry 24*, 2374–2380 (1985).

106. Pope, M. R., Saari, L. L., and Ludden, P. W., "*N*-glycohydrolysis of adenosine diphosphoribosyl arginine linkages by dinitrogenase reductase activating glycohydrolase (activating enzyme) from *Rhodospirillum rubrum*," *J. Biol. Chem. 261*, 10104–10111 (1986).

107. Pope, M. R., Saari, L. L., and Ludden, P. W., "Fluorometric assay for ADP-ribo sylarginine cleavage enzymes," *Anal. Biochem. 160*, 68–77 (1987).

108. Pretorius, I.-M., Rawlings, D. E., O'Neill, E. G., Jones, W. A., Kirby, R., and Woods, D. R., "Nucleotide sequence of the gene encoding the nitrogenase iron protein of *Thiobacillus ferrooxidans*," *J. Bacteriol. 169*, 367–370 (1987).

109. Prince, R. C., and Olson, J. M., "Some thermodynamic and kinetic properties of the primary photochemical reactants in a complex from a green photosynthetic bacterium," *Biochim. Biophys. Acta 423,* 357–362 (1976).

110. Robson, R., Woodley, P., and Jones, R., "Second gene (*nifH*) coding for a nitrogenase iron protein in *Azotobacter chroococcum* is adjacent to a gene coding for a ferredoxin-like protein," *EMBO 5,* 1159–1163 (1986).

111. Saari, L. L., Pope, M. R., Murrell, S. A., and Ludden, P. W., "Studies on the activating enzyme for iron protein of nitrogenase from *Rhodospirillum rubrum,*" *J. Biol. Chem. 261,* 4973–4977 (1986).

112. Saari, L. L., Triplett, E. W., and Ludden, P. W., "Purification and properties of the activating enzyme for iron protein of nitrogenase from the photosynthetic bacterium *Rhodospirillum rubrum,*" *J. Biol. Chem. 259,* 15502–15508 (1984).

113. Schatt, E., Jouanneau, Y., and Vignais, P. M., "Isolation of the ferredoxin I gene from *Rhodobacter capsulatus,*" *Nitrogen Fixation: Hundred Years After,* H. Bothe, F. J. de Bruijn, and W. E. Newton, eds. Stuttgart: Gustav Fischer, 1988, p. 178.

114. Schatt, E., Jouanneau, Y., and Vignais, P. M., "Molecular cloning and sequence analysis of the structural gene of ferredoxin I from the photosynthetic bacterium *Rhodobacter capsulatus,*" *J. Bacteriol. 171,* 6218–6226 (1989).

115. Schick, H.-J., "Substrate and light dependent fixation of molecular nitrogen in *Rhodospirillum rubrum,*" *Arch. Mikrobiol. 75,* 89–101 (1971).

116. Schneider, K. C., Bradbeer, C., Singh, R. N., Wang, L. C., Wilson, P. W., and Burris, R. H., "Nitrogen fixation by cell-free preparations from microorganisms," *Proc. Natl. Acad. Sci. USA 46,* 726–733, (1960).

117. Schon, G., and Voelskow, H., "Pyruvate fermentation in *Rhosospirillum rubrum* after transfer from aerobic to anaerobic conditions in the dark," *Arch. Microbiol. 107,* 87–92 (1976).

118. Schultz, J. E., Gotto, J. W., Weaver, P. F., and Yoch, D. C., "Regulation of nitrogen fixation in *Rhodospirillum rubrum* grown under dark, fermentative conditions," *J. Bacteriol. 162,* 1322–1324 (1985).

119. Scolnik, P. A., and Haselkorn, R., "Activation of the extra copies of genes coding for nitrogenase in *Rhodopseudomonas capsulata,*" *Nature 307,* 289–292 (1984).

120. Shah, V. K., and Brill, W. J., "Isolation of an iron molybdenum cofactor (FeMo-co) from nitrogenase," *Proc. Natl. Acad. Sci. USA 74,* 3249–3253 (1977).

121. Shah, V. K., Stacey, G., and Brill, W. J., "Electron transport to nitrogenase—Purification and characterization of pyruvate–flavodoxin oxidoreductase, the *nifJ* gene product," *J. Biol. Chem. 258,* 12064–12068 (1983).

122. Shestakov, S., Frolova, V., Mitronova, T., and Belavina, N., "*Rhodobacter sphaeroides* mutants with derepressed nitrogenase," *Microbiology (Microbiologiya) 57,* 67–71 (1988).

123. Shestakov, S., Zinchenko, V., Babykin, M., Kopteva, A., Kameneva, S., Frolova, V., Shestopalov, V., and Bondarenko, O., "Genetic studies on the regulation of nitrogen fixation in *Rhodobacter sphaeroides,*" *Nitrogen Fixation: Hundred Years After,* H. Bothe, F. J. de Bruijn, and W. E. Newton, eds. Stuttgart: Gustav Fischer, 1988, pp. 163–169.

124. Soliman, A. E.-H., Nordlund, S., Johansson, B. C., and Baltscheffsky, H., "Purification of glutamine synthetase from *Rhodospirillum rubrum* by affinity chromatography, *Acta Chem. Scand. 35,* 63–64 (1981).

125. Soliman, A. E.-H., Noren, A., and Nordlund, S., "Metabolic regulation of nitrogen fixation in *Rhodospirillum rubrum:* Studies on Ca^{2+}-efflux during 'switch-off' (Abs)," *Nitrogen Fixation: Hundred Years After,* H. Bothe, F. J. de Bruijn, and W. E. Newton, eds. Stuttgart: Gustav Fischer, 1988. p. 179.

126. Stadtman, E. R., and Chock, P. B., "Interconvertible enzyme cascades in metabolic regulation," *Curr. Top. Cell. Regul. 13,* 53–95 (1978).

127. Stewart, V., "Nitrate respiration in relation to facultative metabolism in Enterobacteria," *Microbiol. Rev. 52,* 190–232 (1988).

128. Triplett, E. W., Wall, J. D., and Ludden, P. W., "Expression of the activating enzyme and Fe protein of nitrogenase from *Rhodospirillum rubrum,*" *J. Bacteriol. 152,* 786–791 (1982).

129. Tyler, B., "Regulation of the assimilation of nitrogen compounds," *Annu. Rev. Biochem. 47,* 1127–1162 (1978).

130. Uffen, R. L., and Wolfe, R. S., "Anaerobic growth of purple nonsulfur bacteria under dark conditions," *J. Bacteriol. 104,* 462–472 (1970).

131. Vignais, P. M., Colbeau, A., Willison, J. C., and Jouanneau, Y., "Hydrogenase, nitrogenase and hydrogen metabolism in the photosynthetic bacteria," *Adv. Microbiol. Physiol. 26,* 155–234 (1984).

132. Vignais, P. M., Willison, J. C., and Jouanneau, Y., "Nitrogen fixation, nitrogenase and hydrogenase in the photosynthetic bacteria," *Nitrogen Fixation: Hundred Years After,* H. Bothe, F. J. de Bruijn, and W. E. Newton, eds. Stuttgart: Gustav Fischer, 1988, pp. 147–155.

133. Walker, G. A., and Mortenson, L. E., "An effect of magnesium adenosine 5'-triphosphate on the structure of azoferredoxin from *Clostridium pasteurianum,*" *Biochem. Biophys. Res. Commun. 53,* 904–909 (1973).

134. Wall, J. D., and Braddock, K., "Mapping of *Rhodopseudomonas capsulata nif* genes," *J. Bacteriol. 158,* 404–410 (1984).

135. Wall, J. D., and Gest, H., "Derepression of nitrogenase activity in glutamine auxotrophs of *Rhodopseudomonas capsulata,*" *J. Bacteriol. 137,* 1459–1463 (1979).

136. Wall. J. D., Goldenberg, A., Figueredo, A., Rapp, B. J., and Landrum, D. C., "Genetics of nitrogen fixation in photosynthetic bacteria," *Nitrogen Fixation Research Progress,* H. J. Evans, P. J. Bottomley, and W. E. Newton, eds. Dordrecht: Nijhoff, 1985, pp. 497–503.

137. Wall, J. D., Love, J., and Quinn, S. P., "Spontaneous Nif⁻ mutants of *Rhodopseudomonas capsulata,*" *J. Bacteriol. 159,* 652–657 (1984).

138. Weston, M. F., Kotake, S., and Davis, L. C., "Interaction of nitrogenase with nucleotide analogs of ATP and ADP and the effect of metal ions on ADP inhibition," *Arch. Biochem. Biophys. 225,* 809–817 (1983).

139. Willison, J. C., Ahombo, G., Chabert, J., Magnin, J.-P., and Vignais, P. M., "Genetic mapping of the *Rhodopseudomonas capsulata* chromosome shows non-clustering of genes involved in nitrogen fixation," *J. Gen. Microbiol. 131,* 3001–3015 (1985).

140. Winter, H. C., and Arnon, D. I., "The nitrogen fixation system of photosynthetic bacteria. I. Preparation and properties of a cell-free extract from *Chromatium,*" *Biochim. Biophys. Acta 197,* 170–179 (1970).

141. Winter, H. C., and Ober, J. A., "Isolation of particulate nitrogenase from *Chromatium* strain D," *Plant Cell Physiol. 14,* 769–773 (1973).

142. Xu, H.-W., Love, J., Borghese, R., and Wall, J. D., "Identification and isolation of genes essential for H_2 oxidation in *Rhodobacter capsulatus*," *J. Bacteriol. 171*, 714–721 (1989).

143. Yakunin, A. F., Tsygankov, A. A., and Gogotov, I. N., "Regulation of two forms of nitrogenase in a continuous culture," *Microbiology 56*, 719–724 (1987).

144. Yakunin, A. G., and Gogotov, I. N., "Formation of two forms of nitrogenase and effects of its switch-off by ammonia in purple bacteria," *Microbiol. Sci. 5*, 78–81 (1988).

145. Yamamoto, N., and Droffner, M. L., "Mechanisms determining aerobic or anaerobic growth in the facultative anaerobe *Salmonella typhimurium*," *Proc. Natl. Acad. Sci. USA 82*, 2077–2081 (1985).

146. Yelton, M. M., and Yoch, D. C., "Nitrogen metabolism in *Rhodospirillum rubrum*: Characterization of glutamate synthase," *J. Gen. Microbiol. 123*, 335–342 (1981).

147. Yoch, D. C., "Regulation of nitrogenase A and R concentrations in *Rhodopseudomonas capsulata* by glutamine synthetase," *Biochem. J. 187*, 273–276 (1980).

148. Yoch, D. C., and Arnon, D. I., "Comparison of two ferredoxins from *Rhodospirillum rubrum* as electron carriers for the native nitrogenase," *J. Bacteriol. 121*, 743–745 (1975).

149. Yoch, D. C., Cantu, M., and Zhang, A. M., "Evidence for a glutamine synthetase–chromatophore association in the phototroph *Rhodospirillum rubrum*: Purification, properties, and regulation of the enzyme," *J. Bacteriol. 154*, 632–639 (1983).

150. Zumft, W. G., "Regulation of nitrogenase activity in the anoxygenic phototrophic bacteria," *Nitrogen Fixation Research Progress*, H. J. Evans, P. J. Bottomley, and W. E. Newton. Dordrecht: Martinus Nijhoff, 1985, pp. 551–557.

151. Zumft, W. G., and Castillo, F., "Regulatory properties of the nitrogenase from *Rhodopseudomonas palustris*," *Arch. Microbiol. 117*, 53–60 (1978).

152. Zumft, W. G., and Nordlund, S., "Stabilization and partial characterization of the activating enzyme for dinitrogenase reductase (Fe-protein) from *Rhodospirillum rubrum*," *FEBS Lett. 127*, 79–82 (1981).

4

Nitrogen Fixation in Cyanobacteria

Robert Haselkorn and *William J. Buikema*

I. Introduction

Cyanobacteria have inhabited much of the surface of the earth for billions of years. Today they are responsible for a significant proportion of the biological fixation of nitrogen. Hence, the study of N_2 fixation in cyanobacteria has been pursued from both fundamental and applied perspectives. All diazotrophs must protect nitrogenase from inactivation by O_2. Cyanobacteria, which produce O_2 as a by-product of photosynthesis, have evolved diverse mechanisms for this purpose, ranging from cellular differentiation in filamentous strains to temporal regulation in unicellular strains. The filamentous heterocystous cyanobacteria also provide a rare opportunity to study the development of a simple pattern at the cellular level. Unicellular forms appear to contribute significantly to the nitrogen balance of the oceans. Useful applications of cyanobacteria include their promulgation in rice paddies as a source of combined nitrogen, either in free-living form or in symbiotic associations with plants such as *Azolla*.

In general, both biochemical and genetic studies of N_2 fixation in cyanobacteria have lagged behind comparable work on *Klebsiella pneumoniae,* *Azotobacter* spp., and *Rhizobium* spp., principally because of the absence of systems of genetic analysis for cyanobacteria. That situation changed in 1984 when a system for transferring DNA from *Escherichia coli* to *Anabaena* by conjugation was introduced.[87] Recent developments of this system include isolation of wild-type genes encoding functions essential for N_2 fixation by complementation of mutants,[30] transposon tagging of *nif* genes,[9] interposon inactivation of genes,[26] and the introduction of gene fusions capable of revealing the cellular pattern of gene expression.[16,30] Electroporation has recently been shown to be effective in introducing DNA into *Anabaena* species.[74] The DNA so introduced is subject to restriction, so it must be methylated by passage through *E. coli* carrying the appropriate methylases (see later).[15] It is too early to tell whether electroporation will make the kinds of experiments now done by conjugation more efficient.

Most of the information about cyanobacterial *nif* genes is based on studies of one strain, *Anabaena* PCC 7120. The cloned *nif* genes from that strain have been used to determine the organization of the structural genes for nitrogenase in other strains. Extended nucleotide sequence information is available only for *Anabaena* 7120, where counterparts of most of the known *nif* genes of *K. pneumoniae* have been identified.[45,51,62] In addition, the *nif* gene region of *Anabaena* 7120 contains several open reading frames similar to ones found in the *nif* gene region of *Azotobacter* spp. (D. Norris, unpublished), as well as a unique gene encoding a heterocyst-specific ferredoxin,[5] and two DNA elements that interrupt the *nifHDK* and the *nifBSU* operons.[23,24,25] Finally, a complete physical map of the *Anabaena* 7120 chromosome has been constructed.[2] In addition to the 6.4 Mb chromosome, *Anabaena* 7120 contains three megaplasmids (400-, 200-, and 100-kb) and several small ones.[66]

In the sections that follow, we shall review the biochemical requirements for N_2 fixation in *Anabaena spp.*, compare the physiology of N_2 fixation in unicellular and filamentous cyanobacteria, and describe the organization of the *nif* genes in *Anabaena spp.* and their rearrangement during heterocyst differentiation. Recent work on the genetics of heterocyst differentiation and the use of *nif* gene probes to study the taxonomy of cyanobacteria in N_2-fixing symbiotic associations will also be described. This review is not comprehensive in a historical sense; it emphasizes recent work on the molecular genetics of *Anabaena*. For physiological work we lean on the publications of H. Böhme.[4,7] Other recent accounts of electron flow to nitrogenase in heterocysts have been provided by Houchins[31] and by Bothe and Neuer.[10]

II. Requirements for Nitrogen Fixation

Cyanobacteria have the same general requirements for N_2 fixation common to all diazotrophs: a nitrogenase complex including iron and molybdenum coenzymes, ATP, a source of low potential electrons, and an anaerobic atmosphere. Characterization of cyanobacterial nitrogenase has a long but incomplete history. Fleming identified polypeptides, newly synthesized in *Anabaena* heterocysts, that had the same molecular weight as dinitrogenase and dinitrogenase reductase subunits purified from *Azotobacter*.[19] Subsequently, Tsai and Mortenson purified the *Anabaena* nitrogenase complex and showed that it contained two separable components analogous to the *Klebsiella* and *Azotobacter* enzymes.[76] Peterson and Wolk used an improved procedure for the isolation of *Anabaena* heterocysts that preserved nitrogenase activity and showed, using native gel electrophoresis, that the complex contained an iron component and a molybdenum–iron component.[60]

These studies were followed by the cloning and sequencing of the *nifHDK* genes of *Anabaena*.[25,38,44,45,48,62] Sequence comparisons show that the nitrogenase proteins of *Anabaena* are very similar to those of *Klebsiella, Azotobacter, Clostridium,* and *Rhizobium* (see Chapter 20).

The early work on *nif* gene cloning from *Anabaena* revealed a second, unlinked *nifH* gene that was termed *nifH**,[62] whose sequence is very similar to that of *nifH* (S. J. Robinson, personal communication). To date, the function of *nifH** in *Anabaena* is unknown, but one possibility is that this gene's product is part of an alternative nitrogenase. Long ago an attempt was made to starve *Anabaena* for molybdenum, in order to accumulate a Mo-free precursor of nitrogenase. The attempt failed, in the sense that Mo-starved cells continued to reduce C_2H_2 and to grow on N_2 as a sole nitrogen source. However, Mo starvation was successful with the nonheterocystous cyanobacterium *Plectonema;* therefore, *Anabaena* was not analyzed further.[55] In retrospect, it seems possible that *Anabaena* has an alternative, Mo-independent nitrogenase and that *nifH** is part of the alternative system.

Two complementary lines of evidence indicate that cyanobacteria make the same Mo–Fe cofactor found in *Klebsiella* and *Azotobacter*. First, acid extracts of Mo-grown *Plectonema* restore C_2H_2 reduction activity to crude extracts of *Azotobacter* deficient in Mo–Fe cofactor. Second, purified *Klebsiella* Mo–Fe protein restores C_2H_2 reduction activity to extracts of Mo-starved *Plectonema*.[55] The latter result shows as well that the recognition between dinitrogenase and dinitrogenase reductase is highly conserved. Additionally, *Anabaena* contains counterparts of the *nifB*[50] and *nifN* genes (D. Norris, unpublished), whose functions in *Klebsiella* are required for Mo–Fe cofactor synthesis.

ATP production for N_2 fixation in cyanobacteria is not fully understood. *Anabaena* vegetative cells carry out oxygenic photosynthesis. That is, they have two photosystems and an electron transport chain that transfers electrons from photosystem II (PSII) through cytochromes and plastocyanin to photosystem I (PSI), resulting in a proton gradient that produces ATP. PSII obtains electrons for re-reduction of oxidized chlorophyll from water, generating O_2 in the process. Since O_2 is inimical to N_2 fixation, PSII must be inactivated in N_2-fixing cells. The precise mechanism by which this is accomplished in heterocysts is still not known. Destruction of phycobiliproteins, which collect and transfer energy to the PSII reaction center, is part of the story.[88] PSI remains active in heterocysts, however, and it is believed to produce ATP by cyclic photophosphorylation.[73] ATP can also be produced by oxidative phosphorylation, a process that consumes O_2. Heterocysts have all the components of an electron transport chain from reduced pyridine nucleotide to O_2.[31,73] The reduced pyridine nucleotide might be NADPH from the oxidative pentose pathway or NADH (see later).

Loss of PSII usually means loss of the ability to reduce NADP$^+$ to NADPH photochemically. That indeed occurs in heterocysts. However, an alternate source of NADPH is made available by importing carabohydrate from neighboring vegetative cells.[33,81] This material, still unidentified but possibly a disaccharide such as sucrose or maltose, is converted to glucose-6-phosphate, 6-phosphogluconate and ribose-5-phosphate, producing NADPH in the process.[1,7,79] It has been suggested that this NADPH provides electrons for N$_2$ reduction, reducing a special ferredoxin by way of ferredoxin:nucleotide oxidoreductase. For this to be so, the ratio of NADPH to NADP$^+$ must be high,[1] as is the case within heterocysts.[4] Some of the NADPH is likely to be used in the synthesis of the glycolipids that comprise one layer of the heterocyst envelope. Additionally, NADPH may be used to reduce NAD$^+$ via a transhydrogenase,[65] with the resulting NADH providing the reducing end of the electron transport chain that consumes O$_2$.

Much of the biochemical work on the flow of electrons to nitrogenase in heterocysts is summarized in Figure 4-1, most of the details of which were kindly provided by Herbert Böhme. The principal findings on which the drawing is based are the following[1,4,7,31]: High C$_2$H$_2$ reduction activity could be observed in a cell-free system prepared from heterocysts when supplemented with glucose, fructose, fructose-6-phosphate, fructose bis-phosphate, glucose-6-phosphate, 6-phosphogluconate, ribose-5-phosphate or dihydroxyacetone phosphate. All these compounds support formation of NADPH as shown on the left side of Figure 4-1. There is no net synthesis of nucleic acid in heterocysts, so it is likely that the ribose-5-phosphate is cycled, via transaldolase and transketolase reactions, back to glucose-6-phosphate to yield more NADPH. In the dark, ATP is produced by oxidative phosphorylation. In the light, the relative contributions of oxidative phosphorylation and cyclic photophosphorylation to ATP production have yet to be determined for *Anabaena*, although recent experiments favor oxidative phosphorylation in the unicellular *Gloeothece*.[42]

Pyruvate alone was inactive in supporting C$_2$H$_2$ reduction. This result is in contrast to the system in *Klebsiella*, in which the *nifJ* gene product is a pyruvate:flavodoxin oxidoreductase, the flavodoxin subsequently reducing the nitrogenase complex. However, when pyruvate was supplemented with stoichiometric amounts of oxaloacetate, up to 50% of the maximum rate of C$_2$H$_2$ reduction was observed.[7] The right side of Figure 4-1 interprets this result to mean that pyruvate plus oxaloacetate yield both acetylCoA (for glycolipid synthesis?) and isocitrate, which then is converted to oxoglutarate and NADPH. The oxoglutarate is critical for nitrogen assimilation; together with glutamine it produces glutamate in the reaction catalyzed by glutamate synthase (GOGAT). The ammonia produced by nitrogenase is assimilated by reaction with glutamate to yield glutamine (using glutamine synthetase). The now classic ^{13}N experiments of Wolk's group showed that the major

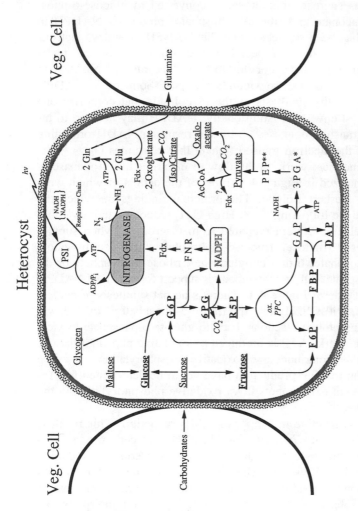

Figure 4-1. Heterocyst metabolism and electron donation to nitrogenase in *Anabaena*. See text for details of the experiments leading to the schemes outlined. Substrates that support high nitrogenase activity in a cell-free system prepared from heterocysts are printed in boldface and underlined. Metabolites that are underlined but not boldface yielded less than 50% of the activity observed with G6P. Other metabolites were inactive or inhibited (*) the G6P-dependent activity. Abbreviations: G6P: glucose 6-phosphate, 6PG: 6-phosphogluconate, R5P: ribose 5-phosphate, F6P: fructose 6-phosphate, FBP: fructose 1,6-bisphosphate, GAP: glyceraldehyde 3-phosphate, DAP: dihydroxyacetone phosphate, 3PGA: 3-phosphoglycerate, PEP: phosphoenol pyruvate, AcCoA: acetyl coenzymeA, Glu: glutamate, Gln: glutamine, Fdx: ferredoxin, FNR: ferredoxin:NADP oxidoreductase, ox. PPC: oxidative pentose phosphate cycle, PSI: photosystem I.

route of N assimilation in heterocysts is through glutamine, the form in which fixed N is exported to vegetative cells.[47,75,86]

Some further comments on the regulation of electron flow are needed. The activity of phosphofructokinase is very low. Therefore, the activity of DAP (and to a lesser extent of GAP) can best be understood if they form fructose bisphosphate and the latter is catabolized (as in pseudomonads) via glucose-6-phosphate. Two ferredoxins are shown. One is needed for the GOGAT reaction that yields glutamate.[10] Another serves as the immediate donor to nitrogenase and is the product of a heterocyst-specific *nif* gene called *fdxH*.[5] The drawing shows a ferredoxin:NADP oxidoreductase (FNR) linking the pool of NADPH to the *fdxH* gene product.[1] As noted later, there is another ferredoxinlike protein encoded among the *Anabaena nif* genes (*fdxN*, located within the *nifBSU* operon) that could be inserted between NADPH and *fdxH* or that, alternatively, could link *fdxH* to the PSI reaction center.[50]

There remains a paradox concerning the regulation of the dehydrogenases. Both G6PD and 6PGD are inhibited *in vitro* by NADPH.[1] G6PD is also inhibited by reduced thioredoxin and, for that reason, by light.[4,7] Yet we know that heterocysts are full of reductant, going back to the classical studies of their ability to reduce tetrazolium dyes[70] and silver salts in photographic emulsions.[18] How do the dehydrogenases stay active?

Anaerobiosis is a universal requirement for N_2 fixation. O_2 rapidly inactivates dinitrogenase and dinitrogenase reductase. Although recent results suggest that partially oxidized nitrogenase can be re-reduced and restored to activity, oxidation eventually leads to irreversible inactivation (and precipitation) of the enzyme (see Chapter 18). In the next section we shall consider how different cyanobacteria cope with this problem. Here we simply point out that O_2 also blocks *nif* gene expression at the transcriptional level in both *Klebsiella* and *Rhodobacter*.[35,36] Whether the same is true for *Anabaena* has yet to be determined, although DNA supercoiling has been invoked in the control of *nif* gene expression in *Gloeothece*.[22]

An elaborate protocol for the detection of nitrogenase activity in cyanobacteria under anaerobic conditions was introduced by Rippka and Waterbury.[63] Their purpose was to determine which of a large number of strains had the capacity to fix N_2, even if they did not differentiate heterocysts. Their method involves the accumulation of carbohydrate reserves followed by the addition of DCMU to inhibit O_2 evolution from PSII and sparging with argon to remove traces of atmospheric O_2. When this protocol was used with *Anabaena,* morphological differentiation of heterocysts was incomplete, very high levels of nitrogenase activity were observed, and the nitrogenase activity was fully sensitive to O_2.[28] The nitrogenase polypeptides were rapidly destroyed by protease following the admission of air to such anaerobically induced cultures. The very high levels of nitrogenase activity and the sensitivity to O_2 were interpreted to mean that nitrogenase was probably induced in undifferentiated vegetative cells under these anaerobic conditions. Later, however, following the discovery of rearrangement of the *nifHDK* operon, it was found that only a small percentage of the cells (3–5%) undergoes the necessary DNA rearrangement under anaerobic conditions (M. E. Mulligan, G. J. Schneider, and R. Haselkorn, unpublished observations). Only these cells can synthesize active nitrogenase (see later). Therefore, even in the absence of complete morphological differentiation of heterocysts, there is some degree of functional differentiation of *Anabaena* under anaerobic conditions. This differentiation does not, however, provide protection against air.

Inactivation of nitrogenase by O_2 can be observed even in fully differentiated heterocysts. When the marine cyanobacterium *Anabaena* CA was transferred to an atmosphere of 1% CO_2–99% O_2, nitrogenase activity was reduced to less than 10% of the original level. If incubation in hyperbaric O_2 was continued, the activity of nitrogenase recovered to 65–80% of its original level.[54,61,68] Recovery of activity requires new protein synthesis but not new synthesis of nitrogenase.[61] *Anabaena* CA has the unusual property of being repressed with respect to differentiation and nitrogenase synthesis by nitrate but not ammonia.[11] However, recovery from O_2 inactivation of nitrogenase is blocked by ammonia,[68] further distinguishing the recovery process from new synthesis of nitrogenase.

The molecular basis of O_2 inactivation and recovery is not known, but several interesting observations have been made. Hyperbaric O_2 treatment results in a reduction in mobility of the denatured dinitrogenase reductase subunit on polyacrylamide gels. The apparent molecular weight is increased from 36 to 38 kD. Upon recovery, the mobility increases to its previous value.[67,71] However, these mobility changes are not affected by the addition of chloramphenicol or ammonia, both of which prevent recovery.[71] It can be suggested that inactivation is accompanied by modification of the dinitrogenase reductase subunit and that reversal of that modification is not sufficient for reactivation.

What might the modification be? The obvious suggestion, by analogy with the photosynthetic bacteria, is that the protein is ADP-ribosylated.[41] Recent results on the inactivation of nitrogenase by ammonia (at high pH) in *Anabaena variabilis* have been interpreted this way.[17] But neither group has shown that *Anabaena* dinitrogenase reductase incorporates ^{32}P or labeled adenosine. Moreover, in both examples,[17,71] all the dinitrogenase reductase protein changes mobility, whereas in *Rhodospirillum rubrum* only one of the two subunits is modified.[41]

III. Modes of Nitrogen Fixation in Cyanobacteria

All the major cellular forms of cyanobacteria include species capable of fixing N_2. As noted before, many filamentous nonheterocystous forms, such as *Plectonema boryanum,* can fix N_2 under anaerobic or microaerobic conditions.[63,69] Although this ability may be significant in nature (e.g., in particular layers of complex mats or during the night), it has not been studied extensively in the laboratory in recent years.

Several unicellular species have attracted more attention. It was reported that marine *Synechococcus* could carry out photosynthesis and N_2 fixation in the same cell.[49] Earlier work on *Gloeocapsa*[89] and *Gloeothece,* which grow extremely slowly, made the same point.[21] Subsequently, it was shown that when the cells are synchronized with a light/dark cycle, photosynthesis is confined to the daytime as expected, but N_2 fixation occurs extensively at night.[49,52,53] The latter result was unanticipated because extensive studies of *Anabaena* showed that light (and accompanying CO_2 fixation and ATP production) was essential for N_2 fixation. The unicellular cyanobacteria, in contrast, accumulate enough carbohydrate during the day to fuel N_2 fixation most of the night. As dawn approaches, the carbohydrate reserve is exhausted and N_2 fixation stops. Cessation of N_2 fixation during the day is not due to inactivation of nitrogenase by photosynthetically generated O_2, however. In *Gloeothece,* O_2 is *required* for N_2 fixation both in the dark and in the light.[42] It appears that oxidative phosphorylation is the major source of both ATP and reductant for nitrogenase in that organism.

The circadian appearance of nitrogenase has also been observed in a freshwater *Synechococcus* isolated from a rice paddy. In that organism it has been shown further that the diurnal appearance of nitrogenase activity is accompanied by the synthesis of *nifHDK* messenger RNA, so it is likely that nitrogenase is destroyed during the (oxygenic) daytime and must be synthesized anew each night.[32]

In the unicellular cyanobacteria examined, the *nifHDK* genes have the contiguous organization observed in *Klebsiella.*[34] These organisms lack the 11-kb element interrupting the *nifD* gene in most of the *Anabaena* and *Nostoc* strains examined to date. Nothing has been reported on the organization of other genes for N_2 fixation in the unicellular cyanobacteria.

Several new isolates of unicellular strains have been reported (J. B. Haskell and L. A. Sherman, personal communication). One of these, BH63, has a doubling time of 8–10 h on nitrate or 20–26 h on N_2 as nitrogen source. It forms nice colonies on agar plates, tolerates light intensity above 100 $\mu E/m^{-2}/sec^{-1}$ and can grow heterotrophically on glycerol. It may be an extremely useful organism for studies of N_2 fixation in unicellular cyanobacteria.

IV. *nif* Gene Organization in *Anabaena*

The organization of the genes whose products are required for N_2 fixation in *Klebsiella, Azotobacter* and other diazotrophs is described in detail in

Chapter 20. Here we present the features of *nif* gene organization in *Anabaena* 7120 that differ from those of other diazotrophs. As will be seen, much remains to be discovered.

The *nif* genes of *Klebsiella, Azotobacter, Rhizobium,* and *Rhodobacter* species were isolated by various combinations of classical and molecular genetic techniques. Generally speaking, one started with Nif⁻ mutants, isolated wild-type *nif* gene-containing DNA fragments by complementation and then walked along those fragments by transposon mutagenesis and/or DNA sequencing. Until recently, it was not possible to use this procedure with *Anabaena* because there was no system for introducing DNA to test complementation. The conjugation system of Wolk and Elhai has now made it feasible both to complement mutants and to use transposons for insertional mutagenesis and gene tagging (see later). Formerly, it was only possible to isolate DNA fragments from *Anabaena* libraries on the basis of sequence similarity to *Klebsiella* probes. In this way, DNA similar to *nifH, D, K,* and *S* was cloned and mapped.[45,62] Subsequent sequencing revealed other *nif* genes in the same neighborhood and their organization could then be related to that of *Klebsiella*.

Our current picture of the major *nif* region in the *Anabaena* 7120 vegetative cell chromosome is shown in Figure 4-2. The orientation of the *nifHDK* operon is shown right to left as it is in all the early maps of the *Klebsiella nif* gene region. The *nifD* gene is interrupted by an 11-kb element that is excised during heterocyst differentiation,[25] and the *fdxN* gene is interrupted by a 55-kb element that is similarly excised.[23,26] (See later for a discussion of the distribution of these elements in cyanobacterial strains from different habitats.) The final arrangement of the *nif* genes in *Anabaena* 7120 heterocysts is shown in the lower part of Figure 4-2.

There may be additional *nif* genes beyond the region shown in Figure 4-2. DNA sequencing has not gone further than the *nifB* gene on the right or *fdxH* on the left. Roughly 10 kb from the *nifB* gene is the *rbcLS* operon, encoding the large and small subunits of ribulose bisphosphate carboxylase (Rubisco). This operon is transcribed in vegetative cells but not under conditions of heterocyst induction.[29]

The first operon on the right includes four open reading frames: *nifB, fdxN, nifS,* and *nifU*.[51] In *Klebsiella, nifB* is required for Fe–Mo cofactor synthesis and *nifS* and *nifU* have unknown functions required for synthesis of active nitrogenase. The *fdxN* gene has no counterpart in *Klebsiella,* but there is a similar gene in *Rhizobium*.[50] Since it is transcribed in *Anabaena* only under N⁻ conditions,[24] we expect its ferredoxinlike product to play a role in electron transfer to nitrogenase. Cyanobacterial ferredoxins are plantlike in that they contain 2Fe–2S centers. By contrast, the *fdxN* gene would encode a typical bacterial ferredoxin with two 4Fe–4S centers. In fact, the protein whose amino acid sequence is most similar to the product of *fdxN*

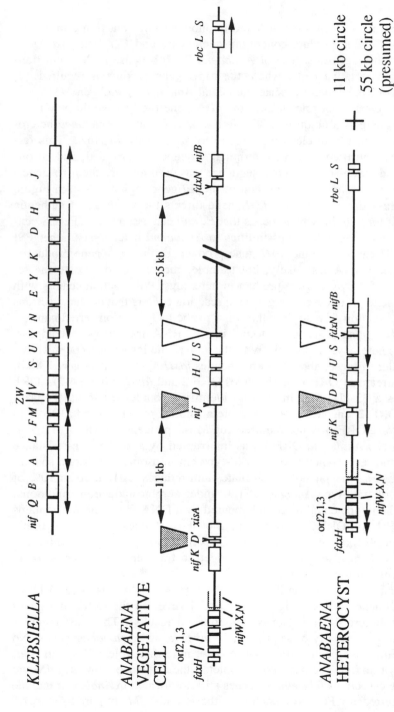

Figure 4-2. Organization of the *nif* gene region of *Klebsiella*, *Anabaena* vegetative cells, and *Anabaena* heterocysts after excision of the 11-kb and 55-kb elements. Shaded triangles indicate the location of short, direct repeats within which the recombination events leading to excision occur. Horizontal arrows below each map indicate transcription units. Sizes of the transcripts produced by genes between *nifW* and *nifK* have not been determined. Sequencing of the *nifN–nifK* region has not been completed. (See Chapter 11 for a description of the *Nif* gene products.)

175

is the small iron–sulfur protein found in photosystem I reaction centers (the 9 kDa *psaC* gene product containing centers F_A and F_B). The *fdxN* gene cannot be the real *psaC* gene of *Anabaena* 7120 because *fdxN* is not transcribed in vegetative cells, where the *psaC* gene product is required.[24,51] Further work is needed to locate the actual *Anabaena psaC* gene.

It is not clear what role, if any, the *fdxN* gene product would play in N_2 fixation. There is a unique 2Fe–2S ferredoxin, found in *Anabaena* heterocysts, that can donate electrons directly to nitrogenase *in vitro*.[7] This ferredoxin was purified, its amino-terminal sequence determined, and an oligonucleotide probe based on that sequence was used to clone the *fdxH* gene.[5] That gene was found at the left end of the *nif* gene region shown in Figure 4-2. Expressed in *E. coli*, the *fdxH* gene directs the synthesis of authentic heterocyst ferredoxin, which means that *E. coli* can assemble a 2Fe–2S center.[6] We can imagine two possibilities in *Anabaena* heterocysts: either PSI or NADPH can reduce the *fdxH* gene product, or the *fdxN* gene product is an intermediary. A less likely, but testable, model would involve the replacement of the *psaC* gene product in heterocyst PSI reaction centers with the *fdxN* gene product, creating special reaction centers that donate electrons to the N_2 fixation ferredoxin rather than to the CO_2 fixation ferredoxin.

The *fdxH* gene sequence is found in two RNA transcripts that appear, following heterocyst induction, with the same kinetics as *nifHDK* mRNA.[5] The smaller of these, about 0.6 kb, encodes *fdxH* alone. The larger, 1.8 kb, starts upstream of ORF 1 and includes ORF 2 and *fdxH* (Figure 4-2).[8] ORF 1 encodes a 33-kDa protein with no identified counterpart in other diazotrophs. ORF 2 encodes a 9-kDa protein similar to one called ORF 6 in *Azotobacter*. ORF 6 is not required for Mo-dependent growth of *A. vinelandii* on N_2. *Anabaena* ORF 1 was interrupted by a kanamycin-resistance cassette and recombined into the wild-type chromosome. The strain so constructed has a Nif⁻ phenotype on plates and reduces C_2H_2 at about 10% of the level of wild-type *Anabaena* 7120 under aerobic induction conditions. The *nif* phenotype could be complemented by a DNA fragment that contains only ORF 1 and ORF 2, so one or both of those "genes" encodes a protein needed for N_2 fixation.[8] Based on the C_2H_2 reduction phenotype, we believe that the ORF product participates in electron transfer to nitrogenase or in the O_2 protection mechanism.

Between ORF 1 and the 3′ end of the *nifK* gene there is about 4 kb of DNA. Sequencing to the right from ORF 1 revealed the ORFs and genes shown in Figure 4-2 (D. Norris, unpublished results). These assignments are based on similarity of sequence to *Klebsiella* or *Azotobacter* genes. No experiments have been carried out to determine their role, if any, in *Anabaena* N_2 fixation. We have not completed the sequence between *nifN* and *nifK*. The essential *Klebsiella nif* genes still not located in *Anabaena* include *nifV*, required for Fe–Mo cofactor synthesis and *nifM*, required for maturation of dinitrogenase reductase, as well as the regulatory genes *nifA* and *nifL*.

One or more genes required for aerobic N_2 fixation in *Anabaena* may have been located by transposon tagging.[9] Tn5 was introduced into *Anabaena* 7120 by conjugation and was found to hop to numerous chromosomal locations. One such Tn5 insert created a Nif⁻ phenotype. The restriction fragments in which the Tn5 was found in that mutant did not correspond to any of the known fragments in Figure 4-2.[9] Further work is needed to identify the gene interrupted in that mutant and to determine its map location. Other Nif⁻ mutants have also been collected following penicillin selection (see the following section on Heterocyst Differentiation), but none of these has been characterized yet.

The subject of *nif* gene rearrangement during *Anabaena* heterocyst differentiation has been reviewed recently.[27] Here we summarize the salient features shown in Figure 4-2. The vegetative cell DNA of *Anabaena* 7120 contains two elements interrupting the *nif* gene region. One of 11 kb interrupts the *nifD* gene; the other of 55 kb interrupts the *fdxN* gene. Both are excised as a result of site-specific recombination between short, directly repeated sequences at the end of each element.[23,25] The sequence within which recombination occurs differs in the two elements.[24] The 11-kb element has been detected in many strains of *Anabaena* and *Nostoc* while the 55-kb element appears to have a more restricted distribution in nature[46] (J. C. Meeks, personal communication; C. D. Carrasco, J. S. Brusca, and J. W. Golden, personal communication).

Among the better-known laboratory strains of *Nostoc* and *Anabaena*, *Nostoc* MAC and *Anabaena cylindrica* have both elements and undergo both rearrangements during heterocyst differentiation (C. D. Carrasco and J. W. Golden, personal communication). More recently isolated *Nostoc* strains also contain at least part of the 11-kb element. *Anabaena* 29413 contains the 11-kb element but not the 55-kb element. Excision of the 11-kb element requires the product of a gene within the element, called *xisA*.[37] In a mutant of *Anabaena* 7120 containing an inactivated *xisA* gene, heterocyst differentiation and the 55-kb excision are unaffected, but the 11-kb element fails to excise and the heterocysts are unable to synthesize nitrogenase.[26] *xisA*-dependent excision of the 11-kb element from its *Anabaena* DNA surroundings can be observed in *E. coli*.[37] This excision in *E. coli* requires only *xisA* among the possible ORFs encoded by the element itself. The cloned *xisA* gene from *Anabaena* 29413 can produce rearrangement of the *Anabaena* 7120 substrate in *E. coli* (J. S. Brusca and J. W. Golden, personal communication). The 11-kb element also contains a smaller fragment that can invert in E. coli, but this subelement appears not to be required for excision in *E. coli* and does not invert in *Anabaena* (A. J. Ryncarz and P. J. Lammers, personal communication). Efforts to observe excision of the 11-kb element from its surrounding sequences *in vitro* have thus far been unsuccessful. A gene equivalent to *xisA,* responsible for excision of the 55-kb element from the *fdxN* gene, has recently been identified (T. S. Ramasubramanian and J. W. Golden, personal communication). A reduced 55-kb element substrate was constructed, containing only 147 bp surrounding the left border and 322 bp surrounding the right border, flanking an antibiotic resistance cassette. Introduced by conjugation into *Anabaena* 7120, the cassette excised perfectly during heterocyst differentiation (T. S. Ramasubramanian and J. W. Golden, personal communication).

Expression of *xisA* from a plasmid introduced into *Anabaena* 7120 resulted in a "cured" strain lacking the 11-kb element. Such a strain grows as well as the parent strain on both N^+ and N^- media.[12a] Similar observations on growth rates were made by J. C. Meeks (personal communication) using two Nif$^+$ revertants of *Nostoc* MAC that he isolated, one of which retained the 11-kb element and one of which was "cured."

A 2-kb region near the middle of the 11-kb element in *Anabaena* 7120 is transcribed very weakly during heterocyst induction. This region gives rise to mRNA of approximately 1000, 900, and 300 bases.[38a] As mentioned earlier, these transcripts are not required for the *xisA*-promoted recombination of the 11-kb element borders in *E. coli,* but they may be required in *Anabaena.* The 2-kb region (An 207.3) encodes three ORFs, one of which has features shared by the lambda protein called integrase. This ORF, however, is not present unless the invertible subelement referred to earlier is inverted. Another of the ORFs is similar to genes encoding cytochrome P450.[38a] The nucleotide sequence of An207.3 is reasonably well conserved in *Nostoc* MAC and *Anabaena* 29413. It is present but less similar in *Calothrix* 7601.[38a]

One final comment about the *nif* gene organization in *Anabaena* heterocysts: In general, the *nif* genes of *Klebsiella* and other diazotrophs are clustered according to biochemical function, except for *nifB,* which is found in a variety of contexts. *Anabaena* is the only organism (compare Figure 20-8) in which *nifB, nifS,* and *nifU* are transcribed together in a single operon. Does this mean that *nifS* and *nifU,* like *nifB,* actually participate in Fe–Mo cofactor synthesis?

In the course of determining the nucleotide sequences of *nif* and other genes of *Anabaena* 7120, a number of instances of short, tandemly repeated sequences were discovered. Many of these seem to be related to each other; most are heptamer repeats. Following the *nifB* gene and the *nifS* gene, Mulligan found six or seven repeats of the sequence CCCCAAT.[51] There are many instances of this heptamer repeat in the *Anabaena* 7120 chromosome, based on probing a large cosmid library with a heptamer repeat fragment. One cosmid was sequenced in part and found to have 25 tandem repeats of CCCCAAT (M. E. Mulligan, unpublished). Repeats of the element ACCCATT were found on the noncoding strand in the region between ORF

1 and *nifK* (E. Söderbäck, unpublished). Closer to *nifK*, repeats of an octamer, CCCCTA$\overset{T}{\underset{C}{}}$A were found (C. Bauer, unpublished). The related sequences CCCCAGT and ACC$\overset{G}{\underset{T}{}}A\overset{G}{\underset{A}{}}$T were found, each repeated six or seven times, downstream of the *psbB* gene of *Anabaena* 7120.[39] The *hetA* gene described by Holland et al.,[30] required for envelope polysaccharide synthesis, is followed by four copies of the sequence TTCAAAA and 11 copies of the sequence TTCCCCA, the latter recognized as a variant of Mulligan's element CCCCAAT. There is no evidence yet regarding the function of these tandemly repeated sequences. In some circumstances, their location would be consistent with a role in the termination of transcription. Another possibility is that the repeats might serve as targets for proteins involved in modifying DNA topology, such as DNA gyrase. Because the *psbB* gene is transcribed by itself, analysis of the possible function of the tandem repeats probably will be attempted first with that gene. Three families of short, tandemly repeated sequences were recently noted in *Calothrix* 7601.[43] Two of these sequences are related to the *Anabaena* elements discussed earlier. Tandeau de Marsac et al. point out that the tandemly repeated elements are found in a wide range of filamentous cyanobacteria in characteristic locations, making them useful for taxonomic studies (RFLPs).[43]

V. Heterocyst Differentiation

The conversion of a photosynthesizing vegetative cell to a N_2-fixing heterocyst requires the coordinated regulation of many sets of genes. The heterocyst is surrounded by a double-layered envelope that includes novel polysaccharide and glycolipid components. In addition, it turns off photosystem II, stops fixing CO_2, breaks down phycobiliproteins, turns on the systems for generation of ATP and reductant for nitrogenase, and induces the synthesis of nitrogenase polypeptides and cofactors.

The changes in gene expression implied by this extensive program must involve modification of the transcription apparatus. Thus far, however, the nature of that modification remains unknown. The major vegetative cell RNA polymerase has been purified and the genes encoding most of its components have been cloned.[3,12,64] The *rpoD* gene encoding the promoter-recognizing σ factor is very similar in sequence to the *rpoD* genes of *E. coli* and *B. subtilis*, particularly in a highly conserved region called the *rpoD* box.[12] This region was used as a probe to clone a second σ-factor gene that contained a similar sequence. The latter gene was inactivated in wild-type *Anabaena* 7120 by insertion of a drug-resistance cassette, but the strain so created was capable of heterocyst differentiation and N_2 fixation (B. Brahamsha, unpublished observations). Other *Anabaena* DNA fragments contain sequences similar to those of the vegetative σ-factor gene, so it is possible that a differentiation-specific σ factor will still be found.

An alternative approach to the study of genes required for heterocyst differentiation is based on the conjugation system mentioned briefly in the introduction. Now that an efficient conjugation system has been developed, it is possible to complement mutants with a pooled library of wild-type DNA fragments, using the ability to complement a particular mutant as the means of selecting the corresponding wild-type gene. The system developed by Wolk and Elhai has three components:[87] a shuttle vector built from a cryptic plasmid capable of replication in *Anabaena* fused to part of pBR322, a broad host range plasmid to make the apparatus for conjugal transfer of DNA, and a third plasmid that mobilizes the shuttle vector for conjugal transfer. The shuttle vector contains cloning sites for inserted fragments of *Anabaena* DNA. A recent improvement includes genes encoding methylase activities, capable of protecting DNA against the *Ava* restriction enzymes, in the *E. coli* strain in which the shuttle vector is maintained.[15] Additionally, a low copy number shuttle vector with a *cos* site and a multiple cloning site flanked by termination sequences that block transcription into the cloned insert has been constructed.[13] These modifications make the cosmid library of cloned *Anabaena* DNA fragments more representative because they minimize the expression of *Anabaena* genes that are toxic to *E. coli*. Methylation increases the efficiency of conjugation sufficiently that a pooled library of cosmids can be transferred *en masse* and the complementing cosmid isolated from the colonies capable of growth under selective conditions.[13,84]

These principles have been applied in Wolk's laboratory to study the timing and the spatial distribution of transcripts of several genes expressed only in heterocysts. One set of experiments concerns a gene called *hetA,* which is required for envelope polysaccharide synthesis.[30] This gene was isolated by complementation of a Nif⁻ mutant, whose defect was traced to the envelope polysaccharide. The complementing wild-type DNA fragment carries an open reading frame encoding a 588 amino acid polypeptide. Dot blots of total RNA showed that the *hetA* gene is expressed maximally seven hours after induction of differentiation (i.e., very early in the process). Fusions of *hetA* to the luciferase genes *luxAB* were used to monitor expression of *hetA* in individual cells, employing an image processor attached to a microscope. Light emission could be detected from individual cells that later were seen to become proheterocysts; that is, the spatial pattern of *hetA* expression could be observed prior to the characteristic morphological differentiation.[16]

It is known that *nif* gene expression is confined to heterocysts in filaments that differentiate under aerobic conditions. Fusions of the *nifHDK* promoter to either ß-galactosidase (*lac*) or luciferase (*lux*) genes, introduced by conjugation, can be used to answer the question of whether *nif* gene expression is environmentally or developmentally regulated.[16] The heterocyst provides an environment that is anaerobic and, prior to the onset of N_2 fixation, low in combined nitrogen. What happens when *Anabaena* 7120 is starved for combined nitrogen under anaerobic conditions? There is no obvious morphological differentiation of heterocysts, but there is abundant synthesis of nitrogenase. All the nitrogenase activity is rapidly destroyed by admission of air to such a culture.[28,63] Nevertheless, *nifH*::*luxAB* fusion expression is confined to morphologically distinguishable cells (i.e., to "virtual" proheterocysts[16]). Moreover, the excision of the 11-kb element from within the *nifD* gene occurs in only 3–5% of the cells in a culture induced anaerobically; only these cells are capable of synthesizing nitrogenase (G. J. Schneider, M. E. Mulligan, and R. Haselkorn, unpublished observations). The conclusion from these lines of evidence is that *nif* gene expression is developmentally regulated in *Anabaena*.

The *hetA* gene required for envelope polysaccharide synthesis was the first to be isolated by complementation of a heterocyst-defective mutant of *Anabaena* 7120.[30,84] The modifications introduced by Buikema make it possible to scale up the complementation step of the gene isolation procedure.[13] Chemical mutagenesis and several rounds of penicillin selection were used to isolate aerobic Nif⁻ mutants of *Anabaena* 7120. Among these mutants were a number that displayed abnormal heterocyst morphology: thickened envelopes, weakened cell attachments, odd shape, or, in one case, no signs of differentiation at all. Each of these mutants was complemented *en masse* with the library of wild-type DNA fragments, selecting for ability to fix N_2 and resistance to an antibiotic carried by the cosmid vector. From each complemented colony, plasmid DNA could be reisolated and used to transform *E. coli*. Amplification of the complementing plasmid in *E. coli* provided the starting material for further analysis (i.e., trimming down the insert and repeating the complementation test) to define the minimum length of DNA needed for complementation. In this way, individual genes required for heterocyst differentiation have been cloned.[13]

One mutant was analyzed in some detail.[13a] This mutant fails to initiate heterocyst differentiation when transferred to nitrogen-free medium. The wild-type *Anabaena* gene (*hetR*) that complements the mutation in this strain is expressed at a low level in vegetative cells, although it is not required for vegetative growth. Its transcript increases in abundance within the first few hours of heterocyst induction. The transcript appears to be monocistronic. The gene's translation product is not similar in sequence to any known protein, including σ factors and DNA-binding transcriptional activator proteins. The most interesting result obtained with this cloned gene to date is the following: when the wild-type gene is introduced on a multicopy plasmid into either wild-type *Anabaena* or the original mutant, on nitrogen-free medium the filaments produce clusters of heterocysts, two to five cells in a row, instead of the usual single heterocysts at approximately 10-cell intervals. Heterocysts are believed to suppress differentiation of vegetative cells by producing an inhibitor that diffuses along the filament.[80] When the concentration of inhibitor falls below some threshold in a particular cell, that cell can differentiate.[77,78,85] Normally, only the cell midway between two heterocysts meets that criterion. Since the time required for heterocyst differentiation is approximately equal to the time required for a vegetative cell to divide, the spacing pattern is maintained at one heterocyst every 10 cells. In an alternate model for the formation of the heterocyst pattern,[82] pumps are envisioned that are activated under conditions of nitrogen deprivation and would be responsible for draining nitrogen from adjacent cells. The last cells to activate their pumps would then become heterocysts. Wolk has recently reviewed the cellular interactions that regulate heterocyst differentiation and has proposed an experimental test of the two models that exploits the *lux* gene fusions described earlier.[83] In the strains with multiple copies of the *hetR* gene, we suggest that the elevated level of the *hetR* product raises the threshold level of the putative inhibitor needed to suppress differentiation. Two, three, or more cells would then fall below the threshold and all would differentiate. Needless to say, the nature of the *hetR* gene product and its role in differentiation are under intense scrutiny.

The efficient conjugation system makes other types of genetic experiments possible as well.[9] The transposon Tn5 was conjugated into *Anabaena* 7120 on a nonreplicating plasmid (pBR322) and the exconjugants were selected for resistance to neomycin, carried by Tn5. Most of the Neo^r exconjugants had Tn5 inserted at unique chromosomal locations; one appeared to have integrated the Tn5 into a plasmid. The inserted Tn5 might be mutagenic; screening the population of exconjugants revealed one Nif$^-$ mutant whose insert was mapped to a DNA fragment not linked to the known *Anabaena nif* gene region. The Tn5 insert made it possible to clone the mutated fragment easily and then to use the flanking region of *Anabaena* DNA to clone the corresponding wild-type fragment.

The conjugation system can also be used for specific gene inactivation. In one particularly interesting example, Golden and Wiest inserted an antibiotic-resistant cassette into the cloned *xisA* gene and introduced that construction by conjugation, selecting for resistance to the antibiotic. They found that most of the exconjugants had integrated the entire shuttle vector by a single recombination event at the *xisA* gene, resulting in tandem duplication of *xisA*, only one copy of which was interrupted. By patiently waiting for the second recombination event, between tandemly duplicated regions that resulted in excision of the vector sequences, they eventually isolated a strain containing a single copy of the (interrupted) *xisA* gene. This strain failed to excise the 11-kb element and had a Nif$^-$ phenotype, as expected.[26]

Cai and Wolk have introduced a method to accelerate the second step of this gene inactivation procedure.[14] They added the conditionally lethal *sacB* gene from *B. subtilis* to the vector. This gene encodes levan sucrase, which kills its host only in the presence of sucrose. Thus, following conjugation and isolation of antibiotic-resistant single recombinants, the relatively rare double recombinants, in which the vector sequences have been deleted, are selected by growth on 5% sucrose and antibiotic. The sucrose- and antibiotic-resistant survivors are usually the desired double recombinants containing the single inactivated target gene.

VI. Nitrogen-Fixing Cyanobacteria in Symbiotic Associations

Molecular probes have added a new dimension to the study of N_2-fixing associations involving cyanobacteria. Three examples will illustrate the power of the newer analysis: the *Anabaena–Azolla* association and the *Nostoc–Anthoceros* and *Nostoc–cycad* associations.

The aquatic fern *Azolla* is found all over the temperate and tropical world, frequently used as a "green manure" fertilizer for rice. It is a simple plant consisting of bilobed leaves attached to a floating stem.[59] Between the dorsal and ventral epidermal layers of each leaf there is a single cavity filled with cyanobacteria. Communication between host and symbiont is probably accomplished through specialized hairs that grow from the host tissue and penetrate the cyanobacterial clumps. Transferred materials include carbohydrate from host to symbiont and ammonia from symbiont to host. The symbiont has a high level of nitrogenase and *nif* gene expression but very little glutamine synthetase.[58] The level of *glnA* (glutamine synthetase) gene transcripts in the symbiont is very low.[57] Thus, the initial product of symbiont N_2 fixation, ammonia, is made directly available to the host.

Several reports of the isolation of the *Azolla* symbiont in pure culture have appeared.[56,72] Others have described unsuccessful attempts to culture the symbiont.[20] *Azolla* can be freed of the symbiont by growth on antibiotic containing medium. The resulting plants require nitrate to survive. Many attempts have been made to reconstitute the symbiotic association by adding free-living cyanobacteria, either historic strains or the alleged *Azolla* symbiont, back to the cured plants. These attempts are generally recognized to have been unsuccessful.[59]

Studies of *nif* gene organization in the symbiont of *Azolla* show why these attempts failed. These studies employed restriction fragment length polymorphisms (RFLPs), which allow inferences regarding identity or distinction between pairs of organisms. Thus, for example, the distribution of restriction sites in and around the *nifHDK* genes of *Anabaena* said to be isolated in pure culture from *Azolla* was compared with the distribution of these sites in the freshly isolated symbiont. These distributions were clearly different, leading to the conclusion that the *Anabaena* grown out from the *Azolla* association was not the principal symbiont.[20,46,57] In fact, the symbiont has never been cultured.

On the other hand, RFLP analysis of freshly isolated symbionts from *Azolla* collected in different parts of the world indicate that all the symbionts are related. They can be assigned to one of two superfamilies, regardless of location.[20] This result suggests that the establishment of the *Azolla* association occurred long ago and that the association is stable. Indeed, it is known that the cyanobacterial partner is transmitted from generation to generation through the megaspore.[59] By contrast, thalli of the hornwort *Anthoceros* form associations with free-living *Nostoc* that can be dissected and reconstituted. In these cases it can be shown by RFLP analysis that the freshly isolated cyanobacterium, the cultured organism, and the cyanobacteria from the reconstituted association contain the same *nif* genes and therefore are likely to be the same organism.[46]

In another symbiotic association, the roots of cycads are colonized by *Nostoc* species. In contrast to the *Azolla* symbiosis, new plants do not retain the parental strains of *Nostoc* and, as might be expected for an association that is re-created frequently in nature, independent isolates of the cyanobacterial partner from the same soil region reveal different RFLPs.[40] RFLP analysis has thus proven to be an effective method for investigating the ecological relationships in these types of symbioses.

Acknowledgments

We are grateful to Herbert Böhme for most of the details of Figure 4-1 and its explanation, to Shree Apte and Jack Meeks for further advice, and to James Golden, Peter Lammers, Conrad Halling, Kristin Bergsland, and Bianca Brahamsha for constructive criticism of this manuscript. The previously unpublished results from our laboratory mentioned in this article were made possible by grants from the National Institutes of Health (GM 21823 and GM40685).

References

1. Apte, S. K., Rowell, P. and Stewart, W. D. P., "Electron donation to ferredoxin in heterocysts of the N_2-fixing alga *Anabaena cylindrica*," *Proc. R. Soc. Lond. B.*, *200*, 1–25 (1978).

2. Bancroft, I., Wolk, C. P., and Oren, E. V., "Physical and genetic maps of the genome of the heterocyst-forming cyanobacterium *Anabaena* sp. strain PCC 7120," *J. Bacteriol.*, *171*, 5940–5948 (1989).

3. Bergsland, K. J., and Haselkorn, R., "Evolutionary relationships among cyanobacteria, chloroplasts and eubacteria: evidence from the *rpoCl* gene of *Anabaena* sp. strain PCC 7120," *J. Bacteriol., 173*, in press (1991).

4. Böhme, H., "Regulation of electron flow to nitrogenase in a cell-free system from heterocysts of *Anabaena variabilis*," *Biochim. Biophys. Acta, 891*, 121–128 (1987).

5. Böhme, H., and Haselkorn, R., "Molecular cloning and nucleotide sequence analysis of the gene coding for heterocyst ferredoxin from the cyanobactrium *Anabaena* sp. strain PCC 7120," *Mol. Gen. Genet., 214*, 278–285 (1988).

6. Böhme, H., and Haselkorn, R., "Expression of *Anabaena* ferredoxin genes in *Escherichia coli*," *Plant Mol. Biol., 12*, 667–672 (1989).

7. Böhme, H., and Schrautemeier, B., "Electron donation to nitrogenase in a cell-free system from heterocysts of *Anabaena variabilis*," *Biochim. Biophys. Acta, 891*, 121–128 (1987).

8. Borthakur, D., Basche, M., Buikema, W. J., Borthakur, P. B., and Haselkorn, R., "Expression, nucleotide sequence and mutational analysis of two open reading frames in the *nif* gene region of *Anabaena* sp. strain PCC 7120," *Mol. Gen. Genet., 221*, 227–234 (1990).

9. Borthakur, D., and Haselkorn, R., "*Tn5* mutagenesis of *Anabaena* sp. strain PCC 7120: Isolation of a new mutant unable to grow without combined nitrogen," *J. Bacteriol, 171*, 5759–5761 (1989).

10. Bothe, H., and Neuer, G., "Electron donation to nitrogenase in heterocysts," *Meth. Enzymol., 167*, 496–501 (1988).

11. Bottomley, P. J., Grillo, J. F., Van Baalen, C. and Tabita, F. R., "Synthesis of nitrogenase and heterocysts by *Anabaena* sp. CA in the presence of high levels of ammonia," *J. Bacteriol., 140*, 938–943 (1979).

12. Brahamsha, B., and Haselkorn, R., "Isolation and characterization of the gene encoding the principal sigma factor of the vegetative cell RNA polymerase from the cyanobacterium *Anabaena* sp. strain PCC 7120," *J. Bacteriol., 173*, 2442–2450 (1991).

12a. Brusca, J. S., Chastain, C. J. and Golden, J. W., "Expression of the *Anabaena* sp. strain 7120 *xisA* gene from a heterologous promoter results in excision of the *nifD* element," *J. Bacteriol., 172*, 3925–3931 (1990).

13. Buikema, W. J. and Haselkorn, R., "Isolation and complementation of nitrogen fixation mutants of the cyanobacterium *Anabaena* sp. strain PCC 7120," *J. Bacteriol., 173*, 1879–1885.

13a. Buikema, W. J. and Haselkorn, R., "Characterization of a gene controlling heterocyst development in the cyanobacterium *Anabaena* 7120," *Genes & Development, 5*, 321–330 (1991).

14. Cai, Y., and Wolk, P., "Use of a conditionally lethal gene in *Anabaena* sp. strain PCC 7120 to select for double recombinants and to entrap insertion sequences," *J. Bacteriol., 172*, 3138–3145 (1990).

15. Elhai, J. and Wolk, C. P., "Conjugal transfer of DNA to cyanobacteria," *Methods Enzymol., 167*, 747–754 (1988).

16. Elahi, J. and Wolk, C. P., "Developmental regulation and spatial pattern of expression of the structural genes for nitrogenase cyanobacterium *Anabaena*," *EMBO J., 9*, 3379–3388 (1990).

17. Ernst, A., Reich, S., and Böger, P., "Modification of dinitrogenase reductase in the cyanobacterium *Anabaena variabilis* due to C starvation and ammonia," *J. Bacteriol.*, *172*, 748–755 (1990).

18. Fay, P., Stewart, W. D. P., Walsby, A. E., and Fogg, G. E., "Is the heterocyst the site of nitrogen fixation in blue-green algae?" *Nature (London)*, *220*, 810–812 (1968).

19. Fleming, H., and Haselkorn, R., "Differentiation in Nostoc muscorum: Nitrogenase is synthesized in heterocysts," *Proc. Natl. Acad. Sci. U.S.A.*, *70*, 2727–2731 (1973).

20. Franche, C., and Cohen-Bazire, G., "Evolutionary divergence in the *nif*H, D, K gene region among nine symbiotic *Anabaena azollae* and between *Anabaena azollae* and some free-living heterocystous cyanobacteria," *Symbiosis, 3,* 159–178 (1987).

21. Gallon, J. R., "Nitrogen fixation by photoautotrophs," in *Nitrogen Fixation* (W. D. P. Stewart and J. R. Gallon, eds.). New York: Academic Press, 1980, pp. 197–238.

22. Gallon, J. R., and Chaplin, A. E., "Recent studies on N_2 fixation by non-heterocystous cyanobacteria," in *Nitrogen Fixation: Hundred Years After,* (H. Bothe, F. J. de Bruijn, and W. E. Newton, eds.). Stuttgart: Gustav Fischer, 1988 pp. 183–188.

23. Golden, J. W., Carrasco, C. D., Mulligan, M. E., Schneider, G. J., and Haselkorn, R., "Deletion of a 55-kilobase-pair DNA element from the chromosome during heterocyst differentiation of *Anabaena* sp. strain PCC 7120," *J. Bacteriol.*, *170*, 5034–5041 (1988).

24. Golden, J. W., Mulligan, M. E., and Haselkorn, R., "Different recombination site specificity of two developmentally regulated genome rearrangements," *Nature, 327,* 526–529 (1987).

25. Golden, J. W., Robinson, S. J., and Haselkorn, R., "Rearrangement of nitrogen fixation genes during heterocyst differentiation in the cyanobacterium *Anabaena,*" *Nature, 314,* 419–423 (1985).

26. Golden, J. W., and Wiest, D. R., "Genome rearrangement and nitrogen fixation in *Anabaena* blocked by inactivation of *xisA* gene," *Science, 242,* 1421–1423 (1988).

27. Haselkorn, R., "Excision of elements interrupting nitrogen fixation operons in cyanobacteria," in *Mobile DNA* (D. E. Berg and M. M. Howe, eds.). Washington, DC: American Society for Microbiology, 1989, pp. 735–742.

28. Haselkorn, R., Mazur, B., Orr, J., Rice, D., Wood, N., and Rippka, R., "Heterocyst differentiation and nitrogen fixation in cyanobacteria (blue-green algae)," in *Nitrogen Fixation* (W. E. Newton and W. H. Orme-Johnson, eds.). Baltimore: University Park Press, 1980, pp. 259–278.

29. Haselkorn, R., Rice, D., Curtis, S. E., and Robinson, S. J., "Organization and transcription of genes important in *Anabaena* heterocyst differentiation," *Ann. Microbiol. Paris, 134B,* 181–193 (1983).

30. Holland, D., and Wolk, C. P., "Identification and characterization of *hetA*, a gene that acts early in the process of morphological differentiation of heterocysts," *J. Bacteriol.*, *172*, 3131–3137 (1990).

31. Houchins, J. P., "Electron transfer chains of cyanobacterial heterocysts," *Nitrogen Fixation and CO$_2$ Metabolism,* Ludden, P. W. and J. E. Burris (eds). Elsevier Sci. Publ., New York. 261–268 (1985).

32. Huang, T.-C., Chow, T.-J., and Hwang, I.-S., "The cyclic synthesis of the nitrogenase of *Synechococcus* RF-1 and its control at the transcription level," *FEMS Microbiol. Lett., 50,* 127–130 (1988).

33. Juttner, F., "^{14}C-labeled metabolites in heterocysts and vegetative cells of *Anabaena cylindrica* filaments and their presumptive function as transport vehicles of organic carbon and nitrogen," *J. Bacteriol.*, *155*, 628–633 (1983).

34. Kallas, T., Rebiere, M. C., Rippka, R., and Tandeau de Marsac, N., "The structural nif genes of the cyanobacteria *Gloeothece* sp. and *Calothrix* sp. share homology with those of *Anabaena* sp., but the *Gloeothece* genes have a different arrangement," *J. Bacteriol.*, *155*, 427–431 (1983).

35. Kong, Q. T., Wu, Q. L., Ma, Z. F. and Shen, S. C., "Oxygen sensitivity of the *nifLA* promoter of *K. pneumoniae*," *J. Bacteriol.*, *166*, 353–356 (1986).

36. Kranz, R. G., and Haselkorn, R., "Anaerobic regulation of nitrogen-fixation genes in *Rhodopseudomonas capsulata*," *Proc. Natl. Acad. Sci. U.S.A.*, *83*, 6805–6809 (1986).

37. Lammers, P. J., Golden, J. W., and Haselkorn, R., "Identification and sequence of a gene required for a developmentally regulated DNA excision in *Anabaena*," *Cell*, *44*, 905–911 (1986).

38. Lammers, P. J. and Haselkorn, R., "Sequence of the *nifD* gene coding for the α subunit of dinitrogenase from the cyanobacterium *Anabaena*," *Proc. Natl. Acad. Sci. U.S.A.*, *80*, 4723–4727 (1983).

38a. Lammers, P. J., Mclaughlin, S., Papin, S., Trujillo-Provencio, C. and Ryncarz, A. J., "Developmental rearrangement of cyanobacterial nif genes—nucleotide sequence, open reading frames, and cytochrome P-450 homology of the *Anabaena* sp. strain PCC 7120 *nifD* element," *J. Bacteriol.*, *172*, 6981–6990 (1990).

39. Lang, J. D., and Haselkorn, R., "Isolation, sequence and transcription of the gene encoding the photosystem II chlorophyll-binding protein, CP-47, in the cyanobacterium *Anabaena* 7120," *Plant Mol. Biol.*, *13*, 441–456 (1989).

40. Lindblad, P., Haselkorn, R., Bergman, B., and Nierzwicki-Bauer, S. A., "Comparison of DNA restriction fragment length polymorphisms of *Nostoc* strains in and from cycads," *Arch. Microbiol.*, *152*, 20–24 (1989).

41. Lowery, R. G., and Ludden, P. W., "Purification and properties of dinitrogenase reductase ADP-ribosyl transferase from the photosynthetic bacterium *Rhodospirillum rubrum*," *J. Biol. Chem.*, *263*, 16714–16719 (1988).

42. Maryan, P. S., Eady, R. R., Chaplin, A. E., and Gallon, J. R., "Nitrogen fixation by *Gloeothece* sp. PCC 6909: respiration and not photosynthesis supports nitrogenase activity in the light," *J. Gen. Microbiol.*, *132*, 789–796 (1986).

43. Mazel, D., Houmard, J., Castets, A. M., and Tandeau de Marsac, N., "Highly repetitive DNA sequences in cyanobacterial genomes," *J. Bacteriol.*, *172*, 2755–2761 (1990).

44. Mazur, B. J., and Chui, C.-F., "Sequence of the gene coding for the β-subunit of dinitrogenase from the blue-green alga *Anabaena*," *Proc. Natl. Acad. Sci. U.S.A.*, *79*, 6782–6786 (1982).

45. Mazur, B. J., Rice, D., and Haselkorn, R., "Identification of blue-green algal nitrogen fixation genes by using heterologous DNA hybridization probes," *Proc. Natl. Acad. Sci. U.S.A.*, *77*, 186–190 (1980).

46. Meeks, J. C., Joseph, C. M., and Haselkorn, R., "Organization of the nif genes in cyanobactria in symbiotic association with *Azolla* and *Anthoceros*," *Arch. Microbiol.*, *150*, 61–71 (1988).

47. Meeks, J. C., Wolk, C. P., Lockau, W., Schilling, N., Shaffer, P. W., and Chien, W.-S., "Pathways of assimilation of [^{13}N]N_2 and ^{13}NH$_4^+$ by cyanobacteria with and without heterocysts," *J. Bacteriol.*, *134*, 125–130 (1978).

48. Mevarech, M., Rice, D. and Haselkorn, R., "Nucleotide sequence of a cyanobacterial *nifH* gene coding for nitrogenase reductase," *Proc. Natl. Acad. Sci. U.S.A., 77*, 6476–6480 (1980).

49. Mitsui, A., Kumazawa, S., Takahashi, A., Ikemoto, H., Cao, S., and Arai, T., "Strategy by which nitrogen-fixing unicellular cyanobacteria grow photoautotrophically," *Nature, 323*, 720–722 (1986).

50. Mulligan, M. E., Buikema, W. J., and Haselkorn, R., "Bacterial-type ferredoxin genes in the nitrogen fixation regions of the cyanobacterium *Anabaena* sp. strain PCC 7120 and *Rhizobium meliloti,*" *J. Bacteriol., 170*, 4406–4410 (1988).

51. Mulligan, M. E., and Haselkorn, R., "Nitrogen-fixation (*nif*) genes of the cyanobacterium *Anabaena* sp. strain PCC 7120: the *nifB-fdxN-nifS-nifU* operon," *J. Biol. Chem., 264*, 19200–19207 (1989).

52. Mullineaux, P. M., Chaplin, A. E., and Gallon, J. R., "Effects of a light to dark transition on carbon reserves, nitrogen fixation and ATP concentrations in cultures of *Gloeocapsa (Gloeothece)* sp. 1430/3," *J. Gen. Microbiol., 120*, 227–232 (1980).

53. Mullineaux, P. M., Gallon, J. R., and Chaplin, A. E., "Acetylene reduction (nitrogen fixation) by cyanobacteria grown under alternating light-dark cycles," *FEMS Microbiol. Lett., 10*, 245–247 (1981).

54. Murry, M. A., Hallenbeck, P. C., Esteva, D., and Benemann, J. R., "Nitrogenase inactivation by oxygen and enzyme turnover in *Anabaena cylindrica,*" *Can. J. Microbiol., 29*, 1286–1294 (1983).

55. Nagatani, H. H., and Haselkorn, R., "Molybdenum independence of nitrogenase component synthesis in the non-heterocystous cyanobacterium *Plectonema,*" *J. Bacteriol., 134*, 597–605 (1978).

56. Newton, J. W., and Herman, A. I., "Isolation of cyanobacteria from the aquatic fern, *Azolla,*" *Arch. Microbiol., 120*, (1979).

57. Nierzwicki-Bauer, S. A., and Haselkorn, R., "Differences in mRNA levels in *Anabaena* living freely or in symbiotic association with *Azolla,*" *Embo J., 5*, 29–35 (1986).

58. Orr, J. and Haselkorn, R., "Regulation of glutamine synthetase activity and synthesis in free-living *Anabaena* and in symbiotic association," *J. Bacteriol., 152*, 626–635 (1982).

59. Peters, G. A., and Calvert, H. E., "The *Azolla-Anabaena* symbiosis," in *Algal Symbiosis* (L. J. Goff, eds.). Cambridge: Cambridge University Press, 1983, pp. 109–145.

60. Peterson, R. B., and Wolk, C. P., "High recovery of nitrogenase activity and [55]Fe-labeled nitrogenase in heterocysts isolated from *Anabaena variabilis,*" *Proc. Natl. Acad. Sci. U.S.A., 75*, 6271–6275 (1978).

61. Pienkos, P. T., Bodmer, S., and Tabita, F. R., "Oxygen inactivation and recovery of nitrogenase activity in cyanobacteria," *J. Bacteriol., 153*, 182–190 (1983).

62. Rice, D., Mazur, B. J., and Haselkorn, R., "Isolation and physical mapping of nitrogen fixation genes from the cyanobacterium *Anabaena* 7120," *J Biol Chem, 257*, 13157–13163 (1982).

63. Rippka, R., and Waterbury, J. B., "Anaerobic nitrogenase synthesis in non-heterocystous cyanobacteria," *FEMS Microbiol. Lett., 2*, 83–86 (1977).

64. Schneider, G. J., Tumer, N. E., Richaud, C., Borbely, G., and Haselkorn, R., "Purification and characterization of RNA polymerase from the cyanobacterium *Anabaena* 7120," *J. Biol. Chem., 262*, 14633–14639 (1987).

65. Schrautemeier, B., "How is heterocyst metabolism adapted to nitrogen fixation?" in *Nitrogen Fixation: Hundred Years After* (H. Bothe, F. J. de Bruijn, and W. E. Newton, eds.). Stuttgart: Gustav Fischer, 1988, p. 216.

66. Simon, R. D., "Survey of extrachromosomal DNA found in the filamentous cyanobacteria," *J. Bacteriol., 136,* 414–418 (1978).

67. Smith, R. L., Van Baalen, C., and Tabita, F. R., "Alteration of the Fe protein of nitrogenase by oxygen in the cyanobacterium *Anabaena* sp. strain CA," *J. Bacteriol., 169,* 2537–2542 (1987).

68. Smith, R. L., Van Baalen, C., and Tabita, F. R., "Control of nitrogenase recovery from oxygen incativation by ammonia in the cyanobacterium *Anabaena* sp. strain CA (ATCC 33047)," *J. Bacteriol., 172,* 2788–2790 (1990).

69. Stewart, W. D. P., "Some aspects of structure and function in N₂-fixing cyanobacteria," *Annu. Rev. Microbiol., 34,* 497–536 (1980).

70. Stewart, W. D. P., Haystead, A., and Pearson, H. W., "Nitrogenase activity in heterocysts of blue-green algae," *Nature, 224,* 226–228 (1969).

71. Tabita, F. R., Van Baalen, C., Smith, R. L., and Kumar, D., "In vivo protection and recovery of nitrogenase from oxygen in heterocystous cyanobacteria," in *Nitrogen Fixation: Hundred Years After* (H. Bothe, F. J. de Bruijn, and W. E. Newton, eds.). Stuttgart: Gustav Fischer, 1988, pp. 189–194.

72. Tel-Or, E., Sandovsky, T., Kobilar, D., Arad, H. and Weinberg, R., "The unique properties of the symbiotic *Anabaena* in the waterfern *Azolla*," in *Photosynthetic Prokaryotes: Cell Differentiation and Function* (G. C. Papageorgiou and L. Packer, eds.). New York: Elsevier Scientific Publishing Co., 1983, pp. 303–314.

73. Tel-Or, E., and Stewart, W. D. P., "Photosynthetic components and activities of nitrogen-fixing isolated heterocysts of *Anabaena cylindrica*," *Proc. R. Soc. Lond. Ser. B, 198,* 61–86 (1977).

74. Thiel, T. and Poo, H., "Transformation of a filamentous cyanobacterium by electroporation," *J. Bacteriol., 171,* 5743–5746 (1989).

75. Thomas, J., Meeks, J. C., Wolk, C. P., Shaffer, P. W., Austin, S. M., and Chen, W. S., "Formation of glutamine from [¹³N]ammonia, [¹³N]dinitrogen, and [¹⁴C]glutamate by heterocysts isolated from *Anabaena cylindrica*," *J. Bacteriol., 129,* 1545–1555 (1977).

76. Tsai, L. B. and Mortenson, L. E., "Interaction of the nitrogenase components of *Anabaena cylindrica* with those of *Clostridium pasteurianum*," *Biochem. Biophys. Res. Commun., 81,* 280–287 (1978).

77. Wilcox, M., Mitchison, G. J., and Smith, R. J., "Pattern formation in the blue-green alga, *Anabaena*. I. Basic mechanisms," *J. Cell Sci., 12,* 707–723 (1973).

78. Wilcox, M., Mitchison, G. J., and Smith, R. J., "Pattern formation in the blue-green alga, *Anabaena*. II. Controlled proheterocyst regression," *J. Cell Sci., 13,* 637–649 (1973).

79. Winkenbach, F., and Wolk, C. P., "Activities of enzymes of the oxidative and reductive pentose phosphate pathways in heterocysts of a blue-green alga," *Plant Physiol., 52,* 480–483 (1973).

80. Wolk, C. P., "Physiological basis of the pattern of vegetative growth of a blue-green alga," *Proc. Natl. Acad. Sci. U.S.A., 57,* 1246–1251 (1967).

81. Wolk, C. P., "Movement of carbon from vegetative cells to heterocysts in *Anabaena cylindrica*," *J. Bacteriol., 96,* 2138–2143 (1968).

82. Wolk, C. P., "Intercellular interactions and pattern formation in filamentous cyanobacteria," in *Determinants of spatial organization, 37th Symp. Soc. Developmental Biology* (S. Subtelny and I. R. Konigsberg, eds.). New York: Academic Press, 1979, pp. 247–266.

83. Wolk, C. P., "Alternative models for the development of the pattern of spaced heterocysts in *Anabaena (Cyanophyta)*," *Pl. Syst. Evol., 164,* 27–31 (1989).

84. Wolk, C. P., Cai, Y., Cardemil, L., Flores, E., Hohn, B., Murry, M., Schmetterer, G., Schrautemeier, B., and Wilson, R., "Isolation and complementation of mutants of *Anabaena* sp. strain PC 7120 unable to grow aerobically on dinitrogen," *J. Bacteriol., 170,* 1239–1244 (1988).

85. Wolk, C. P. and Quine, M. P., "Formation of one-dimensional patterns by stochastic processes and by filamentous blue-green algae," *Dev. Biol., 46,* 370–382 (1975).

86. Wolk, C. P., Thomas, J., and Shaffer, P. W., "Pathway of nitrogen metabolism after fixation of ^{13}N-labeled nitrogen gas by the cyanobacterium, *Anabaena cylindrica*," *J. Biol. Chem., 251,* 5027–5034 (1976).

87. Wolk, C. P., Vonshak, A., Kehoe, P., and Elhai, J., "Construction of shuttle vectors capable of conjugative transfer from *Escherichia coli* to nitrogen-fixing filamentous cyanobacteria," *Proc. Natl. Acad. Sci. U.S.A., 81,* 1561–1565 (1984).

88. Wood, N. B. and Haselkorn, R., "Proteinase activity during heterocyst differentiation in nitrogen-fixing cyanobacteria," in *Limited Proteolysis in Microorganisms: Biological Function, Use in Protein Structural and Functional Studies* (H. Holzer and G. N. Cohen, eds.) Bethesda, MD: Fogarty International Center Publication, 1978, pp. 159–166.

89. Wyatt, J. T., and Silvey, J. K. G., "Nitrogen fixation by *Gloeocapsa*," *Science, 165* (1969).

5

Nitrogen Fixation by Methanogenic Bacteria

A. L. Lobo and *S. H. Zinder*

I. Introduction

A. Methanogenic Bacteria and Archaebacteria

Bacteria capable of producing methane (methanogenic bacteria or methanogens) were first described at the beginning of the twentieth century. The taxonomic and phylogenetic position of methanogens was unclear until 1977, when Balch et al.[2] made the revolutionary proposal that methanogens are related to neither eubacteria nor eucaryotes, but rather belong to a third urkingdom (primaray kingdom) termed *archaebacteria* by C. Woese.[55] The basis for this contention was the comparison of "catalogs" of oligonucleotide sequences derived from ribonuclease T_1-digested 16S ribosomal RNA from methanogens with analogous oligonucleotide catalogs derived from other organisms.[2] More recent studies, based on comparison of nearaly complete sequences of 16S rRNA, confirm a three-kingdom phylogeny.[55] This scheme is part of a general overhaul of bacterial taxonomy by Woese[55] and others (described in Chapter 1, this volume) which is now nearly universally accepted.

The archaebacteria also include certain obligately halophilic bacteria (the halobacteria), which cluster fairly closely with methanogens.[55] The other major branch of the archaebacteria includes several extreme thermophiles, many of which use elemental sulfur as an electron donor or acceptor. The archaebacteria are clearly prokaryotic in terms of their cell size and architecture (no nucleus, submicroscopic flagella, etc.). However, the molecular machinery for gene expression in archaebacteria bears some resemblance to eukaryotes. Other properties of the archaebacteria are unique, such as ether-linked isoprenoid lipids and cell walls composed of protein, glycoprotein, or pseudomurein. Woese[55] has hypothesized that the ancestors of each of the three urkingdoms diverged from a common ancestor at an early stage in cellular evolution, so that there is no specific relataionship between any two

Table 5-1. Major Reactions Carried Out by Methanogenic Bacteria[a]

Reactants	Products	$\Delta G^{o\prime}$ kj/CH4	Organisms
$4H_2 + HCO_3^- + H^+$	$CH_4 + 3H_2O$	-135	Almost all
$4HCO_2^- + H^+ + H_2O$	$CH_4 + 3HCO_3^-$	-145	Many
$4CH_3OH$	$3CH_4 + HCO_3^+ + H_2O + H^+$	-105	Methanosarcina and relatives
$4CH_3NH_3^+ + 3H_2O$	$3CH_4 + HCO_3^- + 4NH_4^+ + H^+$	-75	Methanosarcina and relatives
$CH_3COO^- + H_2O$	$CH_4 + HCO_3^-$	-31	Methanosarcina and Methanothrix

[a]Other substrates shown to be used by methanogens include ethanol, isopropanol, and other short-chain alcohols, methylated sulfides, and carbon monoxide.

of the urkingdoms. See recent reviews[19,53,55] for more information about properties and phylogeny of archaebacteria.

On the basis of 16S rRNA sequence comparisons, methanogens fall into three orders: the Methanomicrobiales, the Methanococcales, and the Methanobacteriales.[19,53] The S_{AB} (similarity) coefficients for comparison of 16S rRNA oligonucleotide sequence catalog between any two methanogen orders are relatively low, ranging from 0.22 to 0.28,[1] as compared to 0.27 between enteric bacteria and *Bacillus,* indicating that the methanogens are phylogenetically diverse. This phylogenetic diversity in methanogens is also reflected in diversity of their cell wall structures, which is comparable to that in eubacteria.[1]

Methanogens are a physiologically homogeneous group in that all are strict anaerobes that produce methane from a limited number of simple substrates (Table 5-1). Nearly all methanogens can use H_2–CO_2, and most of those can also use formate as an electron donor for CO_2 reduction. Recently, it was found that some methanogens can also use some short-chain primary or secondary alcohols as electron donors for CO_2 reduction.[54] Only a limited number of methanogens can use methanol or methylamines for methanogenesis, and only two genera can use acetate. The production of methane is a very specialized mode of metabolism and involves a variety of unique cofactors, electron carriers, and enzymological steps.[19,41] Iron–sulfur proteins are very abundant in methanogens, and crude extracts are often dark brown because of their presence.

Methanogens carry out important terminal reactions in many anoxic habitats, including muds, sediments, flooded soils, animal gastrointestinal tracts, anaerobic bioreactors, and geothermal springs. Because of their limited catabolic repertoire, methanogens often depend on other microbial groups to break down complex organic matter to substrates that they can utilize, mainly

H_2–CO_2 or acetate. In anaerobic sediments and anaerobic bioreactors, approximately two thirds of the methane is derived from acetate; one third is from H_2CO_2.[59] Methylamines, which are break-down products of choline and the osmoprotectant betaine, may be quantitatively important methanogenic substrates in marine environments.[16] The removal of H_2 by methanogens allows the reactions of many of H_2-producing anaerobes to be thermodynamically favorable, a process termed *interspecies hydrogen transfer*.[56] Conversion of acetic acid to methane and CO_2 in anaerobic habitats can be important to maintaining neutral pH, and there is growing appreciation that acetate consumption can also improve the thermodynamics of acetate-producing anaerobic reactions.[3]

B. Discovery of Diazotrophy in Methanogenic Archaebacteria

Nitrogenase has been found to occur in diverse eubacteria, while diazotrophy in a eucaryote has never been demonstrated convincingly.[36] Since methanogens and halobacteria were studied long before they were recognized as archaebacteria, the possibility has existed for several decades that diazotrophy could be discovered in them. There have been no reports of diazotrophy in halobacteria, many of which have complex nutrition. In 1954, Pine and Barker[33] showed that an anaerobic culture that converted ethanol to methane, called *Methanobacterium omelianskii,* could incorporate $^{15}N_2$. However, the subsequent discovery that this culture was a syntrophic association of two bacteria[9] left in doubt the identity of the N_2-fixing member(s) of the coculture.

Following that report, 30 years elapsed until there was another description of N_2 fixation in methanogens. J. Postgate[37] attributed this hiatus to "the sheer awkwardness of the methanogens and perhaps the relative dullness of adding another diazotroph to the list." During the 1970s, culturing methanogens became less awkward due to advances in anaerobic glove-box design and improved culture vessels and stoppers.[1] Moreover, the nutritional requirements of the methanogens became better characterized, especially an unanticipated requirement for nickel,[43,52] so that it was possible to grow many methanogens to high density in chemically defined growth media without resorting to the addition of complex nutrients containing combined N, such as yeast extract or rumen fluid. Finally, the recognition that methanogens belonged to a separate urkingdom from eubacteria and the growing appreciation for their unique biochemistry made the determination of their basic properties less "dull," and caused many more laboratories to enter the field of methanogenesis. These factors all converged in the early 1980s, so that the time was obviously very ripe for the independent discovery of diazotrophy in methanogens by four laboratories reported in 1984–1985.

In the case of the discovery in our laboratory,[30] we were actually examining the formation and utilization of a glycogen-like polysaccharide by *Methanosarcina thermophila*.[31] This finding was of interest because methanogens were not considered to be capable of utilizing carbohydrates. We found that adding limiting amounts of ammonium to the culture medium led to enhanced formation of this polysaccharide, a typical response of many organisms to N starvation and C sufficiency. However, the amounts of the polysaccharide formed by this thermophilic methanogen were always less than 1% of the total dry weight, making physiological studies and chemical charactrization difficult. Because another strain of *Methanosarcina* might accumulate more of this polysaccharide, we tested a mesophilic strain: *Methanosarcina barkeri* strain 227.[28]

When P. Murray, the graduate student working on this project, inoculated strain 227 into medium containing a limiting amount of ammonium, the culture continued to grow rather than showing the limitation we had seen in strain TM-1. Subsequent experiments confirmed growth in medium lacking added ammonium. We next showed that growth of cultures was severely diminished by replacing N_2 with Ar unless ammonium was also added, in which case the replacement had no effect. N_2 fixation is usually demonstrated by the acetylene (C_2H_2) reduction technique,[10] but we were aware that C_2H_2 is a potent inhibitor of methanogenesis[32,49] and were concerned that methanogens might reduce acetylene nonspecifically using one of their many other iron–sulfur redox proteins. We therefore demonstrated diazotrophy directly by $^{15}N_2$ incorporation. $^{15}N_2$ incorporation was not detected in cells that were also provided with ^{14}N-ammonium. Finally, since Pine and Barker's description of diazotrophy in a methanogen[33] had been tarnished by its being a mixed culture, and since there could be a diazotrophic eubacterium cross-feeding N to the methanogen,[36] we carefully checked the culture for purity using microscopical and cultural methods, and we showed that vancomycin, an antibiotic that inhibits cell wall synthesis in eubacteria but not in archaebacteria, did not inhibit diazotrophic growth.

At the same time, Belay, Sparling, and Daniels[4] at the University of Iowa were examining the nutrition of the newly described *Methanococcus thermolithotrophicus*,[17] including utilization of N_2 sources. "Control" cultures, to which no combined N_2 source was added, were found to grow. In this case, diazotrophy was confirmed by the acetylene reduction assay, but by using C_2H_2 at a much lower partial pressure (0.003 atm) than the standard partial pressure used (0.1 atm), so that methanogenesis was not significantly inhibited (see later). C_2H_2 was not reduced by ammonium-grown cells. The two papers on N_2 fixation by *M. barkeri* and in *M. thermolithotrophicus* were published together in the same issue of *Nature*.[4,30] *M. thermolithotrophicus* is presently the most thermophilic organism in which diazotrophy has been demonstrated.

Soon afterward, M. Bomar, K. Knoll, and F. Widdel[8] at the University of Konstanz, Germany, described N_2 fixation in *Methanosarcina barkeri* strain Fusaro. They had been enriching for diazotrophic acetate-oxidizing sulfate reducers, and one enrichment made methane instead of sulfide and was dominated by a *Methanosarcina*-like organism. This prompted them to test the already isolated *Methanosarcina barkeri* strain Fusaro for diazotrophy. They demonstrated diazotrophy by inhibition of growth by replacing N_2 with Ar, by $^{15}N_2$ incorporation, and by C_2H_2 reduction.

L. Sibold and his colleagues at the Pasteur Institute[44] used a different approach. They used probes for the structural genes of nitrogenase (*nifH*, *nifD*, and *nifK*) from *Klebsiella pneumoniae* and a *nifH* probe from *Anabaena*, and hybridized these probes against Southern blots of restriction enzyme-digested methanogen DNA. They found fairly strong hybridization of *nifH* probes with DNA from *Methanococcus voltae* and *Methanobacterium ivanovii* and weaker hybridization against DNA from *Methanosarcina barkeri* strain MS and *Methanobacterium thermoautotrophicum* strain ΔH. They found much weaker hybridization under less stringent hybridization conditions with *nifD* and *nifK* probes and DNA from *M. voltae* and *M. ivanovii* and no hybridization of these probes with DNA from the other two methanogens. Curiously, they were initially unable to demonstrate diazotrophic growth of the two species that gave the best hybridization. Eventually, *M. ivanovii* was shown to be capable of diazotrophic growth,[27] whereas *M. voltae* has never been shown to fix N_2.

These discoveries were the first confirmed examples of N_2 fixation among members of the archaebacteria. The identification of DNA sequences homologous to eubacterial *nif* genes complemented the physiological findings of the other three groups. The discovery of N_2 fixation in methanogenic archaebacteria raises questions about the origins and evolution of the N_2-fixing system. Did diazotrophy originate before or after the divergence of the archaebacteria and eubacteria, or was it transferred from one urkingdom to the other? In light of the strong structural and functional homology among eubacterial Mo nitrogenases, would archaebacterial nitrogenases be similar, or would they have subtle or major differences?

C. *Methanogens for Which N_2 Fixation Has Been Reported*

Since the initial descriptions of confirmed diazotrophy among methanogens, the list of N_2-fixing methanogens continues to grow. Table 5-2 lists the methanogens for which diazotrophy has been described, as well as some tested by the authors. The poorest criterion for the demonstration of diazotrophic growth is growth in medium presumably free of fixed N, since it is easy to be fooled by efficient N scavengers.[36] All three orders of the methanogens have members that can fix N_2. It is interesting that many of

Table 5-2. Methanogens in Which Diazotrophic Growth Has Been Reported

Methanogen	Criteria for Diazotrophy[a]	Reference
Methanosarcina barkeri 227	G, N, A	24,30
Methanosarcina barkeri Fusaro	G, N, A	8
Methanosarcina barkeri strain MS	G	Personal observation[b]
Methanosarcina acetivorans	G	Personal observation
Methanococcus		
thermolithotrophicus	G, N, A	4,5
Methanolobus tindarius	G	22
Methanobacterium ivanovii	G, A	27
Methanobacterium		
thermoautotrophicum		
strain Marburg	G	13
Methanobacterium bryantii strain		
M.o.H.	G, N, A	5
Methanospirillum hungatei strain		
GP1	G, N, A	5
Methanospirillum hungatei MS1	G	Personal observation

[a]Abbreviations: G, growth under nitrogen-fixing conditions; N, incorporation of $^{15}N_2$ into cellular material under nitrogen-fixing conditions; A, acetylene reduction by cells or extracts under nitrogen-fixing conditions.

[b]Species for which diazotrophy was not detected by the authors: *Methanococcus jannaschii*, *Methanococcus vannielii*, *Methanobacterium formicicum* JR, *Methanosarcina* sp. TM-1 and *Methanothrix* sp. CALS-1

the diazotrophic methanogens grow on hydrogen, which can be an inhibitor of nitrogenases.[36]

Of the cultures listed, the report of diazotrophy by *Methanobacterium thermoautotrophicum* strain Marburg is troublesome. Fardeau et al.[13] reported that the culture continued to grow in a chemostat when the growth medium feeding the culture was replaced by medium lacking ammonia. One of us (S. Zinder) was unable to demonstrate diazotrophic growth of batch cultures of strain Marburg obtained from two different laboratories, including the laboratory of Fardeau et al. It is possible that strain Marburg only grows diazotrophically in chemostat culture. Another possible explanation is that some ammonia was contaminating the gas (a $H_2/CO_2/N_2$ mixture) that was continuously passing through the chemostat culture. This may explain why growth yields of N_2-grown *M. thermolithotrophicus* or *M. thermoautotrophicum* strain Marburg were nearly equal to those for ammonium-grown cells, in contrast to results obtained for batch cultures of methanogens (see later) or eubactrial diazotrophs.[36] The finding of hybridization between eubacterial *nifH* probes and DNA from strain Marburg[35] (see later) indicates that the culture is or was once capable of diazotrophy. Diazotrophic growth

Table 5-3. Effect of Nitrogen Source on Growth Yields (YCH_4 = g Dry Weight per Mole CH$_4$) and Doubling Time (t_d) in Methanogens

Organism	Methanogenic Substrate	NH$_{4+}$ as N-source		$_2$ as N-source		Reference
		t_d	YCH_4	t_d	YCH_4	
Methanosarcina barkeri strain 227	Methanol	14	3.7	22	2.3	24
Methanosarcina barkeri strain 227	Acetate	30	2.8	72	1.2	24
Methanosarcina barkeri strain Fusaro	Methanol	26	4.6	37	1.7	8
Methanococcus thermolithotrophicus	H$_2$–CO$_2$	0.6	0.94	4.5	0.36	5
Methanococcus thermolithotrophicus	Formate	0.6	1.8	4.5	0.8	5
Methanobacterium bryantii	H$_2$–CO$_2$	10	1.0	23	0.23	5
Methanospirillum hungatei	H$_2$–CO$_2$	8	1.1	27	0.26	5

in strain Marburg would be desirable since this rapid-growing methanogen is very well characterized in terms of its biochemistry and genetics.[29,57]

Because many laboratory strains of methanogens have been cultured with combined nitrogen sources for several years, *nif* genes may have been rendered nonfunctional by mutations in those methanogens that have *nif*-like sequences but that cannot grow diazotrophically. This may be the case for *M. voltae* and *M. thermoautotrophicum* strain Marburg. In this regard, *M. bryantii* strain M.o.H., the methanogen in the *Methanobacterium omelianskii* coculture, has been cultured for nearly 50 years on media containing fixed N, first together with its syntrophic partner and then alone, yet was found to be capable of diazotrophic growth,[5] corroborating Pine and Barker's[33] original result. Because of potential genetic instability, independent subcultures of the same strain may differ in ability to fix N.

II. Physiology

A. Growth Parameters

N$_2$ fixation is an energetically costly process in eubacteria, usually requiring 16 or more moles of ATP per mole N$_2$ fixed,[36] and is reflected in reduced growth rates and yields when N$_2$ is used as a N source. Table 5-3 shows that diazotrophy exacts a heavy toll on both growth rate and yield in batch cultures of methanogens. The effects were most pronounced for *M. barkeri* growing on acetate (t_d = 72 h), which provides very little energy and is a poor reductant (Table 5-1). It is difficult to calculate an ATP cost of diazotrophy in methanogenic cells based on these decreases in cell yield in batch cultures. Ideally, the growth yields of N_2 or ammonium-grown cultures should be compared for cultures grown at identical rates in a chemostat so that they have identical maintenance energies. More important, there are no values presently known for moles ATP synthesized per mole CH_4 produced by methanogens. If it is assumed that *M. barkeri* strain 227 growing on methanol conserves 1 mole of ATP per mole CH_4, a value close to 50 ATP per N_2 fixed would be calculated from the data in Table 5-3 using the method of Hill.[15]

B. Trace Metal Requirements for Diazotrophic Growth in Methanogens

The involvement of molybdenum in nitrogenase has been recognized for several decades.[36] Recent evidence has confirmed early studies that there also exist alternative nitrogenases[6,7] that use vanadium as a cofactor,[40] and recently, a third alternative nitrogenase was described in which the only metal detected was iron[7,11] (see Chapter 19). Therefore, it is of interest to determine the effects of Mo, V, and other trace metals on diazotrophic growth in methanogens.

The first study on trace metal effects on diazotrophy in methanogens was that of Bomar et al.,[8] who detected only a slight effect (a growth reduction of 10–20%) of Mo deficiency on diazotrophic growth in *M. barkeri* strain Fusaro, even after nine transfers of the culture in Mo-deficient medium. Also, 0.5 μM tungstate, an antagonist of Mo function, did not inhibit diazotrophic growth. In our studies[24] we found that we needed to wash all glassware with sulfuric acid and to use ultrapure water and Mo-starved cells of *M. barkeri* strain 227 before we saw any effect of Mo deprivation. As little as 10nM molybdate stimulated diazotrophic growth, but growth in medium without added molybdate was nearly half of that with 1 μM added molybdate. Mo deficiency had no detectable effect on growth with ammonium. Tungstate at 100 μM severely inhibited diazotrophic growth in Mo-free medium, but not growth of ammonium-grown cells. These results are best explained by a Mo requirement for diazotrophic growth and by sufficient Mo contaminating the growth medium to allow some diazotrophic growth, as well as to meet Mo requirements of ammonium-grown cells.[43] Mo also stimulated diazotrophic growth of *Methanosarcina barkeri* strain Fusaro[42] and *Methanococcus thermolithotrophicus*.[27]

Although we could never find reproducible stimulation of diazotrophic growth in *M. barkeri* strain 227 by vanadate or vanadyl sulfate, Scherer[42] showed a clear stimulation of growth and methanogenesis by either Mo or V in both *M. barkeri* strains Fusaro and 227. The optimum concentration of V was near 2 μM with little stimulation by concentrations of less than 0.4 μM. Growth was somewhat slower with V and growth yields were also lower, suggesting that the putative V nitrogenase in methanogens is less efficient than the Mo nitrogenase. Other elements tested that did not stimulate diazotrophic growth of *M. barkeri* included W,[24,42] Re[42,] and Cr.[24]

C. Incorporataion of $^{15}N_2$ into Whole Cells of Methanogens

Belay et al.[5] used $^{15}N-NMR$ to follow the fate of $^{15}NH_4^+$ and $^{15}N_2$ incorporated into *Methanococcus thermolithotrophicus*. They determined an apparent K_m value for growth on N_2 of 0.11 ± 0.05 atm, a value typical for eubacterial nitrogenases.[39] In cells incubated with $^{15}NH_4^+$ for half a generation time, they found that the amino group of glutamate was the major labeled species in ethanol extracts from cells. After incubation for three generations, when the cells reached stationary phase, another peak of equal intensity appeared that did not overlap with any of those of known amino acids. Similar products were seen in cells incubated with 0.14 atm $^{15}N_2$. However, they consistently found that cells incorporated much less $^{15}N_2$ than expected in early points in the incubation, with much more being incorporated later, and hypothesized that a large pool of a ^{14}N-compound existed that cells used in preference to $^{15}N_2$ at early time points. This phenomenon may be related to the low rates of acetylene reduction seen in cells and extracts of methanogens (see later). Similar patterns of incorporation were obtained with *Methanobacterium bryantii* and *Methanospirillum hungatei*. From these results, it is difficult to assign a pathway for N incorporation, although it is likely that glutamate is involved. These results are consistent with earlier studies on ammonium assimilation in methanogens,[21] in which glutamate was found to be the most abundant amino acid in soluble pools from *M. barkeri* and *Methanobacterium thermoautotrophicum*. Both organisms contained significant activities of glutamine synthetase, glutamate synthase, and alanine dehydrogenase. Glutamate dehydrogenase was not detected in either organism.[21]

III. Enzymology

A. Rates of Reduction of C_2H_2 and N_2

C_2H_2 is one of several triple-bonded molecules reduced by nitrogenase, and measurement of ethylene produced by reduction of C_2H_2 is accepted as a valid index of nitrogenase activity.[10] Mo nitrogenases reduce C_2H_2 to ethylene at approximately the same rate per electron as N_2 to ammonium if one considers N_2 reduction to be an eight-electron process because of the production of at least 1 mole of hydrogen per N_2 reduced.[36] Alternative nitro-

Table 5-4. C_2H_2 Reduction Rates for Whole Cells and Crude Cell-free Extracts of Methanogens and Other Bacteria.

Sample	Rate[a]	Reference
Methanococcus thermolithotrophicus cells	15.	5
Methanosarcina barkeri strain 227 cells	1.4	24
Methanosarcina barkeri strain Fusaro cells	0.7[b]	8
Rhodopseudomonas palustris cells	500	26
Methanococcus thermolithotrophicus extracts	3–30	27
Methanosarcina barkeri strain 227 extracts	5	24
Methanosarcina barkeri strain 227 extracts	20–135	Personal observation[c]
Methanobacterium ivanovii extracts	3.0	27
Clostridium pasteurianum extracts	3000–5000	12

[a]Nmoles ethylene formed min^{-1} [mg protein]$^{-1}$.

[b]Rate calculated by authors assuming that 50% of cell dry weight is protein.

[c]Freshly harvested cells and maintaining a protein concentration −20 mg/ml.

genases are characterized by somewhat lower rates of C_2H_2 reduction relative to N_2 and higher rates of hydrogen production.[11,40]

The C_2H_2 reduction assay is clearly specific for nitrogenase in methanogens, since ammonium-grown cells have not been found to show any acetylene-reducing activity.[4,8,24,27] However, 0.1 atm acetylene, the partial pressure typically used for acetylene reduction assays,[10] completely inhibits methanogenesis.[32,49] Therefore, C_2H_2 must be used at lower partial pressures, since methanogenesis provides ATP needed for N_2 fixation. All researchers[4,8,24,27] have found 0.003 atm C_2H_2 to be an optimal partial pressure, a balance point at which there is only slight inhibition of methanogenesis while providing as high a concentration as possible of C_2H_2 for nitrogenase. Even considering these factors, the rates of C_2H_2 reduction by whole cells are inordinately low (Table 5-4), ranging from 1 to 15 nmol ethylene formed h^{-1} [mg protein]$^{-1}$, with the highest value being for *M. thermolithotrophicus,* which has a doubling time of 4.5 h when fixing N_2. For comparison, a *Rhodopseudomonas palustris* culture with a doubling time near 22h,[26] similar to that of diazotrophic *M. barkeri* on methanol,[24] showed *in vivo* rates nearly 400 times greater.

The problem of C_2H_2 inhibition of methanogenesis should be obviated in an *in vitro* assay of nitrogenase in cell extracts, since the enzyme would then be provided with reducing power from dithionite and ATP from a regenerating system. However, rates detected in studies of C_2H_2 reduction by crude extracts (Table 5-4) were similar to those found for whole cells. Lobo and Zinder[24] found rates near 5 nmol ethylene formed h^{-1} [mg protein]$^-$1 for *M. barkeri* extracts, and *K. pneumoniae* extracts prepared and assayed using identical procedures showed rates nearly 100-fold higher. Various changes in the composition of the reaction mixture did not increase activity, and addition of methanogen extracts to active *K. pneumoniae* extracts did not inhibit them.[24] C_2H_2 reduction in extracts from *M. barkeri* strain 227 required ATP and was severely inhibited by added ADP, similar to eubacterial nitrogenases.[36] Magot et al.,[27] in the case of *M. thermolithotrophicus,* attributed low rates to switch-off of the nitrogenase (see later). We have recently obtained higher rates of C_2H_2 reduction in *M. barkeri* extracts when we have maintained protein concentrations in excess of 20 mg/ml and used freshly harvested cells rather than extracts from frozen cells. These higher activities are still not high enough to be equivalent to those needed to provide enough N for growth (500–1000 nmol ethylene formed h^{-1} [mg protein]$^{-1}$).

Another potential explanation for low C_2H_2 reducing activity is that C_2H_2 is a poor substrate relative to N_2 for methanogen nitrogenases. We have been intrigued by that possibility, especially since alternative nitrogenases show somewhat lower activity toward C_2H_2.[11,40] Initial experiments in crude extracts of *M. barkeri* strain 227, using a sensitive assay for ammonium, were hampered by high background concentration of ammonium in extracts and low enzyme activity.[24] More recent experiments (Lobo and Zinder, in preparation), using partially purified nitrogenase components, indicate that N_2 is a better substrate than C_2H_2 (NH_4^+/ethylene formed ≈ 1.1), but not to the extent that would account for the very low activity found. At this point, no single satisfactory explanation exists for the low C_2H_2 reducing activity of the methanogen nitrogenases. This phenomenon may also be associated with degradation of the enzyme during harvesting and preparation of extracts, the potential unsuitability of dithionite as an electron donor, or inadequacy of the reaction mixture usually used. We have found that simply diluting crude extracts to concentrations below 10 mg protein ml^{-1} can cause loss of activity in *M. barkeri,* even after reconcentrating to the original protein concentration.

B. Nitrogenase Proteins

The eubacterial Mo nitrogenase component 1 is an $\alpha_2\beta_2$ tetramer with subunit molecular weights near 55,000 and 60,000 kDa, whereas component 2 is a homodimer with a subunit molecular weight of 30–35,000 kDa.[36] Therefore, if methanogens have nitrogenases that are similar in structure to eubacterial ones and if expression of those nitrogenases is repressed by ammonia, one would expect to see three additional bands representing these nitrogenase structural proteins in sodium dodecyl sulfate (SDS) polyacrylamide gel electrophoretograms of extracts from diazotrophic cells.

Magot et al.[27] compared Coomassie Blue–stained SDS gels from ammonium-grown and N_2-grown cells of both *Methanococcus thermolithotrophicus* and *Methanobacterium ivanovii*. In the case of *M. thermolithotrophicus,* a strongly staining band at 29 kDa was seen as well as a more weakly staining band at 55 kDa. In the case of *M. ivanovii,* bands were seen at 38, 54, and 56 kDa. Immunoblots were prepared using antibody to *Azotobacter vinelandii* Mo nitrogenase components. Antibody to Av2 faintly stained bands at 29 and 32 kDa from diazotrophic *M. thermolithotrophicus,* and a band at 38 kDa from diazotrophic *M. ivanovii,* and no reaction was observed in lanes from ammonia-grown cells. No specific reaction was obtained with antibodies to Av1. If the band at 55 kDa from *M. thermolithotrophicus* represents a component-1 analog, the absence of a second band may indicate that there is only a single subunit type, that the two subunit types are of equal molecular weight, or that one of the subunits is obscured by another heavily staining band. Magot et al.[27] noted that rates of C_2H_2 reduction by *M. thermolithotrophicus* extracts were lowest when the second band near 30 kDa reacted the most strongly with antibodies and proposed that low C_2H_2 reducing activity in extracts was due to switch-off by covalent modification of a single subunit of that component[25,34] (see later). No physiological evidence for switch-off in *M. thermolithotrophicus* was presented.

Lobo and Zinder[24] compared SDS gels from diazotrophic and ammonium-grown cells of *M. barkeri* strain 227 and found a heavily staining band near 26 kDa and a single band near 55 kDa specific to diazotrophic cells. A faintly staining band near 31 kDa was also usually detected in diazotrophic cells. Subsequent analysis (Lobo and Zinder, in preparation) has shown that the actual component-2 analog, in strain 227 is represented by the band near 31 kDa. This is supported by "Western" immunoblotting using antibodies to *Rhodospirillum rubrum* component 2 (Rr2, kindly provided by P. Ludden), which only stained the band at 31 kDa in lanes from N_2-grown cells while no reaction was obtained in lanes from ammonium-grown cells. This band also co-purified with nitrogenase activity (see later). The identity of the heavily staining band at 26 kDa is not known. It may represent a proteolytic cleavage fragment of the 31 kDa protein, which may also explain the low C_2H_2 reducing activity seen in extracts. If so, it has lost the ability to react with antibodies to Rr2.

We have recently obtained a 20–30-fold purification of nitrogenase from *M. barkeri* strain 227 using protamine precipitation and FPLC chromatography inside an anaerobic glove-box (Lobo and Zinder, in preparation). We have found that the enzyme is binary and shows subunit properties similar to those of eubacterial nitrogenases. One component has a molecular weight near 240 kDa, as determined by Superose 12 gel permeation chromatography, and shows two heavily staining bands near 57 and 62 kDa in SDS gels, making it similar to eubacterial component 1. The other protein has a subunit molecular weight near 31 kDa but shows an anomalously high molecular weight of 120 kDa from Superose 12 chromatography, suggesting that it is a tetramer rather than a dimer. Molecular weight determinations using other methods should be made before a firm conclusion can be drawn. The highest specific activity we have obtained in a purified enzyme preparation is approximately 1000 nmol ethylene formed h^{-1} [mg protein]$^{-1}$. This value should be compared to the values of 60,000–150,000 nmol ethylene formed h^{-1} [mg protein]$^{-1}$ typically found for purified eubacterial nitrogenases.[12,36] We have obtained no evidence for production of ethane from C_2H_2, a reaction carried out by both the vanadium alternative nitrogenase and the non-Mo, non-V alternative nitrogenase.

C. Ammonium Switch-Off

Since diazotrophy is an energetically costly process, many eubacterial diazotrophs regulate nitrogenase activity in response to ammonium and other forms of fixed N. This regulation is usually evidenced by a rapid switch-off of C_2H_2 reducing activity in response to added ammonia. This response is reversible and is found in many eubacterial diazotrophs, especially many purple photosynthetic bacteria[20,60] (see Chapter 3). In the best studied example of switch-off, in *Rhodospirillum rubrum,* switch-off is mediated by the addition of ADP-ribose to an arginine residue on one subunit of the dinitrogenase reductase.[34] This covalent modification leads to a splitting of the band for the dinitrogenase reductase into two bands on SDS polyacrylamide gels.

Since diazotrophy is costly to methanogens, it is possible that they regulate activity of their nitrogenases in response to fixed N. Lobo and Zinder[24] found that addition of 5 mM ammonium caused a rapid switch-off of C_2H_2 reduction by cell suspensions of *M. barkeri* strain 227. More recent studies (Lobo and Zinder, in preparation) have shown that ammonium concentrations as low as 10 μM can cause a transient switch-off of C_2H_2 reduction. We reported switch-off in response to 5 mM glutamine[24] but have since concluded that this was due to small amounts ammonium contaminating the glutamine preparation used. No effect on switch-off was detected after addition of 1 mM methionine sulfoxamine, an inhibitor of glutamine synthetase known to block switch-off in eubacteria,[20] perhaps due to an inability of strain 227 to take it up.

We examined Western blots for possible covalent modification of component 2 (dinitrogenase reductase) analog in strain 227 and found no splitting of the band in lanes from switched-off cells (Lobo and Zinder, in preparation). Also, C_2H_2 reducing activity in extracts from switched-off cells was identical with those from extracts prepared before ammonium addition. These results are consistent with a mechanism of switch-off not involving covalent modification of the component 2 analog, as has been proposed for some eubacterial species,[58] but it is possible that any modification may have been reversed during harvesting and preparation of extracts from switched-off cells. Diazotrophy is clearly a highly regulated process in *M. barkeri* strain 227, showing both apparent repression of nitrogenase and switch-off by ammonium, and represents one of the most highly regulated metabolic activities detected in methanogenic archaebacteria.

IV. Molecular Biology

A. Genes Homologous to Eubacterial nif Genes in Methanogens and Other Archaebacteria

L. Sibold and colleagues have greatly extended their initial finding of DNA fragments in four methanogens homologous to eubacterial *nif* genes. They used an *Anabaena nifH* probe to examine DNA from a wider variety of archaebacteria[35] and detected hybridization of varying intensity to DNA from each one of 14 methanogens tested. These included some species that have been shown to be diazotrophic, such as *Methanococcus thermolithotrophicus, Methanosarcina barkeri* strains MS and 227, and *Methanolobus tindarius*. Also included were *Methanobacterium thermoautotrophicum* strain Marburg, which may not grow diazotrophically in batch culture (see earlier), and *Methanothermus fervidus*, which grows at temperatures up to 97°C[19] but which, to our knowledge, has not been tested for diazotrophy. They also found weak hybridization to DNA from two species of *Halobacterium*, and no hybridization to any of the sulfur-dependent thermophilic archaebacteria tested.

B. Sequence and Transcription of nif-Homologous Genes

Making use of the homology to *nif* probes, Souillard and Sibold[46] cloned and sequenced an open reading frame (ORF) from *M. voltae* homologous to eubacterial *nifH*. The nucleotide and deduced amino acid sequence corresponded to a 30-kDa polypeptide with strongly conserved regions characteristic of eubacterial *nifH* genes, yet S_{AB} values between the *M. voltae* *nifH* amino acid sequence and those of eubacteria were lower (0.5) than the lowest between any two eubacteria (0.6). Despite the presence of this ORF clearly resembling *nifH*, *M. voltae* has not been shown to fix N_2, and its complex nutritional requirements[52] make diazotrophy unlikely, and perhaps these requirements indicate genetic instability in this strain.

The Sibold group later cloned and sequenced two different *nifH*-like ORFs from the N_2-fixing methanogen *M. thermolithotrophicus*, only one of which (called *nifH1*) was expressed in diazotrophic cells.[47,48] They also cloned DNA fragments containing a *nifH*-like ORF from *M. ivanovii*,[47] and two different *nifH*-like regions from *M. barkeri* 227.[45] As in the case of *M. voltae*, the newly derived sequences were distantly related to eubacterial sequences but

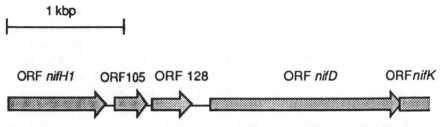

Figure 5-1. Map of open reading frames (ORFs) homologous to eubacterial *nif* genes and associated ORFs in *Methanococcus thermolithotrophicus*. The direction of the arrows indicates the direction of transcription of the ORFs. Adapted from Ref. 47.

contained some highly conserved regions, especially those near cysteines involved in iron–sulfur centers.

Sequencing downstream from *nifH1* in M. thermolithotrophicus revealed two ORFs 105 and 128 codons long, respectively (Figure 5-1), with no homology detected to known *nif* genes. Downstream from the two ORFs was a *nifD*-like ORF followed by a *nifK*-like ORF that presumably continued past the cloned fragment. There was an 8-bp overlap between the *nifD* and *nifK*-like ORF. Overlap is often seen in cotranscribed genes that have products that are synthesized in stoichiometrically equal amounts. Sequence similarity of the *nifD*-like ORF to corresponding eubacterial *nifD* genes was lower than that for *nifH1* ($S_{AB} = 0.37$–0.42).

Using the cloned gene for *nifH1* from *M. thermolithotrophicus* as a probe, Souillard and Sibold[48] detected a hybridizing 1.8-kb messenger RNA fragment in "northern" blots of RNA. This mRNA was only present in N_2-grown cells, indicative of repression by ammonium of nitrogenase synthesis at the level of transcription. The mRNA contained *nifH1* and the two smaller ORFs. The *nifD* and *nifK*-like ORFs were transcribed separately from *nifH1*, in contrast to eubacteria, in which the three structural genes for nitrogenase are usually cotranscribed.[36] Transcription starts were found, using primer extension and S1 mapping, to be 87–88 bp before the *nifH1*, and 85 bp before the *nifD* ORF start codons. Twenty-one bp before the transcription start of the *nifH1* ORF was the sequence 5′-TTTATATA-3′, a putative archaebacterial promoter.[50] No promoterlike elements were found upstream from the *nifD* ORF transcription start.

Two other observations made by this group are worth noting. First, while the amino acid sequence of *nifH1* gene product from *Methanococcus thermolithotrophicus* is distantly related to *nif* gene amino acid sequences from eubacteria ($S_{AB} = 0.56$–0.60), it is even more distantly related to *nifH* gene products from *Methanococcus voltae* ($S_{AB} = 0.52$) or *Methanobacterium ivanovii* ($S_{AB} = 0.51$), and the *nifH2* gene product from the same organism ($S_{AB} = 0.47$). This suggests a more profound divergence of *nifH1* sequences than that observed in eubacteria. Second, an exceptionally high degree of sequence similarity (S_{AB} = 0.70–0.71) was found between *nifH1* from *M. thermolithotrophicus* and those of two eubacterial genes: *nifH3* (*anfH*) from *A. vinelandii*, which codes for the component 2 for the non-Mo non-V alternative nitrogenase,[18] and *nifH3* of *Clostridium pasteurianum*.[51] The amino acid sequence similarities between methanogen nitrogenases and eubacterial ones are represented graphically in Figure 1-8, this volume.

V. Summary, Conclusions, and Perspective

In the five years between the first reports of diazotrophy in methanogens and the preparation of this chapter, the picture emerging of methanogen nitrogenases is one in which similarities to eubacterial nitrogenases are significant, whereas differences are subtle. Diazotrophic growth is of a similar energetic cost to methanogens as eubacteria and is stimulated by Mo or V. Both genetic and biochemical evidence indicate a binary enzyme with overall structure similar to eubacterial nitrogenases. Evidence has accrued for regulation by ammonium of nitrogenase synthesis at the level of transcription and regulation of enzyme activity by switch-off, similar to many free-living eubacterial diazotrophs. Differences in substrate specificity and subunit structure apparently occur but need to be verified by more intensive study.

Particularly intriguing is the possibility that there is some specific relataionship between methanogen nitrogenases and the non-Mo, non-V alternative nitrogenases. The amino acid sequence of the *nifH1* from *Methanococcus thermolithotrophicus* showed the highest similarity with the *anfH* that codes for the non-Mo, non-V alternative nitrogenase from *Azotobacter vinelandii*. Chisnell et al.[11] purified this enzyme and found that it reduced N_2 at a higher rate than C_2H_2 (NH_4^+/ethylene $\approx$ 1.3), that it was unstable in general and labile to freezing, and that even when highly purified, it showed much lower specific activity than either the Mo or V nitrogenases from the same organism. Similarly, methanogen cells and extracts show low C_2H_2 reducing activity. Our recent results with *Methanosarcina barkeri* strain 227 (Lobo and Zinder, in preparation) also indicate an enzyme that shows low specific activity, is labile to freezing, and shows a preference for N_2 over C_2H_2. However, it should be recalled that a Mo requirement for diazotrophic growth has been demonstrated in methanogens,[24,27,42] suggesting that methanogen nitrogenases contain Mo. Clearly, the metal content of methanogen nitrogenases needs to be examined.

The antiquity of N_2 fixation is the subject of some debate.[36] The strong homology of Mo nitrogenases in diverse eubacteria could be due to nitrogenase having an ancient origin and being highly conserved, or it could be due to a recent origin of nitrogenase followed by horizontal transfer to various species. Hennecke et al.[14] constructed a phylogenetic tree for eubacteria based on *nifH* sequence S_{AB} values and showed that it closely resembled a tree made from 16S rRNA S_{AB} values, supporting an ancient origin. From the work of Sibold and colleagues,[44-48] and as shown in Figure 1-8 this volume, the methanogen nitrogenase sequences represent the most divergent ones yet determined, indicating that nitrogenase existed in eubacteria and archaebacteria before the groups underwent significant radiation and suggesting that the ancestor of the eubacteria and archaebacteria was capable of diazotrophy, perhaps using a nitrogenase with more similarity to "alternative" ones. The one exception to this divergence is the *nifH1* ORF in *M. thermolithotrophicus*, which bears a strong resemblance to *anfH* from *A. vinelandii* and *nifH3* of *C. pasteruainum*. This similarity brings up the possibility of lateral transfer of *nif* genes between methanogens and eubacteria,[47] perhaps clostridia that share similar habitats. It is also of interest that N_2 fixation is widespread in methanogens, whereas neither genetic nor physiological evidence for N_2 fixation in sulfur-dependent archaebacteria has yet been found.

The widespread ability to fix N_2 in methanogens indicates that they can play an important role in the N_2 cycle in the many anoxic habitats in which they already play important roles in the carbon cycle, including flooded soils and rice paddies important to agriculture. In this regard, it is interesting that three out of four cultures of methanogens isolated from a rice paddy were found to be capable of diazotrophic growth.[38] The contribution of methanogens to N_2 fixation in these anoxic environments would certainly be underestimated when using the standard C_2H_2 reduction technique, since 0.1 atm C_2H_2 would inhibit both methanogenesis and C_2H_2 reduction by methanogens. The inhibition of methanogenesis by C_2H_2 might also inhibit the rate of C_2H_2 reduction by many other anaerobes not directly affected by C_2H_2 but metabolically coupled to methanogens through interspecies hydrogen or acetate transfer. This may lead to significant underestimation of rates of N_2 fixation in anoxic methanogenic habitats.[32]

There is considerable interest in conversion to methane of lignocellulosic wastes, which are often low in combined N. The recent isolation of a diazotrophic celluloytic *Clostridium*[23] raises hopes of devising consortia of diazotrophic anaerobes to convert cellulosic wastes to methane and leave behind a residue higher in N content than it originally had. It should be pointed out that this *Clostridium* grows very slowly, as does *M. barkeri* strain 227 on acetate ($t_d = 72$ h), the precursor of roughly two thirds of the methane produced by anaerobic bioreactors.[59] Therefore, such a system may only be useful when high rates of conversion of substrates to methane are not desired.

While it is somewhat disappointing that archaebacterial nitrogenases are so similar to eubacterial nitrogenases, they still represent the most divergent ones found. Comparing the nitrogenase enzymes and their genes in methanogenic archaebacteria with those of eubacteria will contribute to our understanding of the evolution and function of biological N_2 fixation.

Acknowledgments

The authors' research in this area was supported by USDA/Hatch funds and by grant N00014-86-K-0876 from the Office of Naval Research. The typing assistance of Shirley Cramer is appreciated.

References

1. Balch, W. E., Fox, G. E., Magrum, L. J., Woese, C. R., and Wolfe, R. S., "Methanogens: reevaluation of a unique biological group," *Microbiol. Rev. 43*, 260–293 (1979).

2. Balch, W. E., Magrum, L. J., Fox, G. E., Wolfe, R. S., and Woese, C. R., "An ancient divergence among the bacteria," *J. Mol. Evol. 9*, 305–311 (1977).

3. Beaty, P. S., and McInerney, M. J., "Effects of organic acid anions on the growth and metabolism of *Syntrophomonas wolfei* in pure culture and in defined consortia," *Appl. Environ. Microbiol. 55*, 977–983 (1989).

4. Belay, N., Sparling, R., and Daniels, L., "Dinitrogen fixation by a thermophilic methanogenic bacterium," *Nature (Lond.) 312*, 286–288 (1984).

5. Belay, N., Sparling, R., Choi, B. -S., Roberts, M., Roberts, J. E., and Daniels, L., "Physiological and ^{15}N-NMR analysis of molecular nitrogen fixation by *Methanococcus thermolithotrophicus, Methanobacterium bryantii* and *Methanospirillum hungatei.*," *Biochim. Biophys. Acta 971*, 233–245 (1988).

6. Bishop, P. E., Jarlenski, D. M. L., and Hetherington, D. R., "Evidence for an alternative nitrogen fixation system in *Azotobacter vinelandii.*," *Proc. Natl. Acad. Sci. USA 77*, 7342–7346 (1980).

7. Bishop, P. E., Premakumar, R., Dean, D. R., Jacobson, M. R., Dalton, D. A., Chisnell, J. R., and Wolfinger, E. D., "Alternative nitrogen fixation systems in *Azotobacter vinelandii.* In *Nitrogen Fixation: Hundred Years After,* Boethe, H., de Bruijn, F. J., and Newton, W. E., (eds). Stuttgart: Gustav Fischer, 1988, pp. 71–79.

8. Bomar, M., Knoll, K., and Widdel, F., "Fixation of molecular nitrogen by *Methanosarcina barkeri.*," *FEMS Microbiol. Ecol. Lett. 31*, 47–55 (1985).

9. Bryant, M. P., Wolin, E. A., Wolin, M. J., and Wolfe, R. S., "*Methanobacillus omelianskii,* a symbiotic association of two species of bacteria," *Arch. Mikrobiol. 59*, 20–31 (1967).

10. Burris, R. H. "Nitrogen fixation-assay methods and techniques," *Methods Enzymol. 24B*, 415–431 (1972).

11. Chisnell, J. R., Premakumar, R., and Bishop, P. E., "Purification of a second alternative nitrogenase from a *nifHDK* deletion strain of *Azotobacter vinelandii,*" *J. Bacteriol. 170*, 27–33 (1988).

12. Eady, R. R., "Methods for studying nitrogenase." In *Methods for Evaluating Biological Nitrogen Fixation,* Bergersen, F. J., (ed.) New York: John Wiley & Sons, 1980, pp. 216–264.

13. Fardeau, M. -L., Peillex, J. -P., and Belaich, J. -P., "Energetics of the growth of *Methanobacterium thermoautotrophicum* and *Methanococcus thermolithotrophicus* on ammonium chloride and dinitrogen," *Arch. Microbiol. 148*, 128–131 (1987).

14. Hennecke, H., Kaluza, K., Thöny, B., Fuhrmann, M., Ludwig, W., and Stackebrandt, E., "Concurrent evolution of nitrogenase genes and 16S rRNA in *Rhizobium* species and other nitrogen fixing bacteria," *Arch. Microbiol. 142*, 342–348 (1985).

15. Hill, S. "The apparent ATP requirement for nitrogen fixation in growing *Klebsiella pneumoniae.*" *J. Gen. Microbiol. 95*, 297–312 (1976).

16. Hippe, H., Caspari, D., Fiebig, K., and Gottschalk, G., "Utilization of trimethylamine and other N-methyl compounds for growth and methane formation by *Methanosarcina barkeri*," *Proc. Natl. Acad. Sci. USA 76*, 494–498 (1979).

17. Huber, H., Thomm, M., König, H., Thies, G., and Stetter, K. O., "*Methanococcus thermolithotrophicus*, a novel thermophilic lithotrophic methanogen," *Arch. Microbiol. 132*, 47–50 (1982).

18. Joerger, R. D., Jacobson, M. R., Premakumar, R., Wolfinger, E. D., and Bishop, P. E., "Nucleotide sequence and mutational analysis of the structural genes (*angHDGK*) for the second alternative nitrogenase from *Azotobacter vinelandii*," *J. Bacteriol. 171*, 1075–1086 (1989).

19. Jones, W. J., Nagle, D. P., Jr., and Whitman, W. B., "Methanogens and the diversity of archaebacteria," *Microbiol. Rev. 51*, 135–177 (1987.)

20. Kanemoto, R. H., and Ludden, P. W., "Effect of ammonia, darkness, and phenazine methosulfate on whole-cell nitrogenase activity and Fe protein modification in *Rhodospirillum rubrum*, *J. Bacteriol. 158*, 713–720 (1984).

21. Kenealy, W. R., Thompson, T. E., Schubert, K. R., and Zeikus, J. G., "Ammonia assimilation and synthesis of alanine, aspartate, and glutamate in *Methanosarcina barkeri* and *Methanobacterium thermoautotrophicum*," *J. Bacteriol. 150*, 1357–1365 (1982).

22. Konig, H., Nusser, E., and Stetter, K. O., "Glycogen in *Methanolobus* and *Methanococcus*," *FEMS Microbiol. Lett. 28*, 265–269 (1985).

23. Leschine, S. B., Holwell, K., and Canale-Parola, E., "Nitrogen fixation by anaerobic cellulolytic bacteria," *Science 242*, 1157–1160, (1988).

24. Lobo, A. L., and Zinder, S. H., "Diazotrophy and nitrogenase activity in the archaebacterium *Methanosarcina barkeri* 227," *Appl. Environ. Microbiol. 54*, 1656–1661 (1988).

25. Ludden, P. W., and Burris, R. H., "Purification and properties of nitrogenase from *Rhodospirillum rubrum* and evidence for phosphate, ribose and an adenine-like unit covalently bound to the iron protein," *Biochem. J. 175*, 251–259 (1978).

26. Madigan, M., Cox, S. S., and Stegeman, R. A., "Nitrogen fixation and nitrogenase activities in members of the family *Rhodospirillacea*," *J. Bacteriol. 157*, 73–78 (1984).

27. Magot, M., Possot, O., Souillard, N., Henriquet, M., and Sibold, L., "Structure and expression of *nif* (nitrogen fixation) genes in methanogens." In *Biology of anaerobic bacteria*, Dubourgier, H. C., et al., eds. Amsterdam: Elsevier Science Publishers, 1986, pp. 193–199.

28. Mah, R. A., Smith, M. R., and Baresi, L., "Studies on an acetate-fermenting strain of *Methanosarcina*," *Appl. Environ. Microbiol 35*, 1174–1184 (1978).

29. Meile, L., Jenal, U., Studer, D., Jordan, M., and Leisinger, T., "Characterization of YM1, a virulent phage of *Methanobacterium thermoautotrophicum* Marburg," *Arch. Microbiol. 152*, 105–110 (1989).

30. Murray, P. A., and Zinder, S. H., "Nitrogen fixation by a methanogenic archaebacterium," *Nature (Lond.) 312*, 284–286, (1984).

31. Murray, P. A., and Zinder, S. H., "Polysaccharide reserve material in the acetotrophic methanogen *Methanosarcina thermophila* strain TM-1: accumulation and mobilization," *Arch. Microbiol 147*, 109–116 (1987).

32. Oremland, R. S., and Taylor, B. F., "Inhibition of methanogenesis in marine sediments by acetylene and ethylene: validity of the acetylene reduction assay for anaerobic microcosms," *Appl. Microbiol. 30*, 707–709 (1975).

33. Pine, M. T., and Barker, H. A., "Studies on the methane bacteria. XI. Fixation of atmospheric nitrogen by *Methanobaterium omelianskii*," *J. Bacteriol. 68*, 589–591 (1954).

34. Pope, M. P., Murrell, S. A., and Ludden, P. W., "Covalent modification of the iron protein of nitrogenase from *Rhodospirillum rubrum* by adenosine diphosphoribosylation of a specific arginine residue," *Proc. Natl. Acad. Sci. USA 82*, 3173–3177 (1985).

35. Possot, O., Henry, M., and Sibold, L., "Distribution of DNA sequences homologous to *nifH* among archaebacteria," *FEMS Microbiol. Lett. 34*, 173–177 (1986).

36. Postgate, J. R., *The Fundamentals of Nitrogen Fixation*. Cambridge: Cambridge University Press, 1982.

37. Postgate, J. R., "New kingdom for nitrogen fixation," *Nature 312*, 194 (1984).

38. Ragjagopal, B. S., Belay, N., and Daniels, L., "Isolation and characterization of methanogenic bacteria from rice paddies," *FEMS Microbiol. Ecol. 53*, 153–158 (1988).

39. Rivera-Ortiz, J. M., and Burris, R. H., "Interactions among substrates and inhibitors of nitrogenase," *J. Bacteriol. 123*, 537–545 (1975).

40. Robson, R. L., Eady, R. R., Richardson, T. H., Miller, R. W., Hawkins, M., and Postgate, J. R., "The alternative nitrogen fixation system of *Azotobacter chroococcum* is a vanadium enzyme," *Nature (Lond.) 322*, 388–390 (1986).

41. Rouvière, P. E., and Wolfe, R. S., "Novel biochemistry of methanogenesis," *J. Biol. Chem. 263*, 7913–7916 (1988).

42. Scherer, P., "Vanadium and molybdenum requirement for the fixation of molecular nitrogen by two *Methanosarcina* strains," *Arch. Microbiol. 151*, 44–48 (1989).

43. Schönheit, P., Moll, J., and Thauer, R. K., "Nickel, cobalt, and molybdenum requirement for growth of *Methanobacterium thermoautotrophicum*," *Arch. Microbiol. 123*, 105–107 (1979).

44. Sibold, L., Pariot, D., Bhatnagar, L., Henriquet, M., and Aubert, J. P., "Hybridization of DNA from methanogenic bacteria with nitrogenase structural genes (*nifHDK*)," *Mol. Gen. Genet. 200*, 40–46 (1985).

45. Sibold, L., and Souillard, N., "Genetic analysis of nitrogen fixation in methanogenic archaebacteria." In *Nitrogen Fixation: Hundred Years After*, Boethe, H., de Bruijn, F. J., and Newton, W. E., eds., Stuttgart: Gustav Fischer, 1988, pp. 705–710.

46. Souillard, N., and Sibold, L., "Primary structure and expression of a gene homologous to *nifH* (nitrogenase Fe protein) from the archaebacterium *Methanococcus voltae*," *Mol. Gen. Genet. 203*, 21–28 (1986).

47. Souillard, N., Magot, M., Possot, O., and Sibold, L., "Nucleotide sequence of regions homologous to *nifH* (nitrogenase Fe protein) from the nitrogen fixing archaebacteria *Methanococcus thermolithotrophicus* and *Methanobacterium ivanovi:* evolutionary implications," *J. Mol. Evol. 27*, 65–76 (1988).

48. Souillard, N., and Sibold, L., "Primary structure, functional organization and expression of nitrogenase structural genes of the thermophilic archaebacterium *Methanococcus thermolithotrophicus,*" *Molec. Microbiol. 3,* 541–551 (1989).

49. Sprott, G. D., Jarrell, K. F., Shaw, J. M., and Knowles, R., "Acetylene as an inhibitor of methanogenic bacteria," *J. Gen. Microbiol. 128,* 2453–2462 (1982).

50. Thomm, M., and Wich, G., "An archaebacterial promoter element for stable RNA genes with homology to the TATA box of higher eukaryotes," *Nucl. Acids Res. 16,* 151–163 (1988).

51. Wang, S. Z., Chen, J. S., and Johnson, J. L., "The presence of five *nifH*-like sequences in *Clostridium pasteurianum:* sequence divergence and transcription properties," *Nucl. Acids Res. 16,* 439–454 (1988).

52. Whitman, W. B., Ankwanda, E., and Wolfe, R. S., "Nutrition and carbon metabolism of *Methanococcus voltae,*" *J. Bacteriol. 149,* 852–863 (1982).

53. Whitman, W. B., "Methanogenic bacteria." *The Bacteria, vol. VIII.* New York: Academic Press, 1985, pp. 4–65.

54. Widdel, F., "Growth of methanogenetic bacteria in pure culture with 2-propanol and other alcohols as hydrogen donors," *Appl Environ. Microbiol. 51,* 1056–1062 (1986).

55. Woese, C. R., "Bacterial evolution," *Microbiol. Rev. 51,* 221–271 (1987).

56. Wolin, M. J., and Miller, T. L., "Interspecies hydrogen transfer: 15 years later," *ASM News 48,* 561–565 (1982).

57. Worrell, V. E., Nagle, D. P., Jr., McCarthy, D., and Eisenbraun, A., "Genetic transformation system in the archaebacterium *Methanobacterium thermoautotrophicum* Marburg," *J. Bacteriol. 170,* 653–656 (1988).

58. Yoch, D. C., Li, J.-D., Hu, C. -Z., and Scholin, C., "Ammonia switch-off of nitrogenase from *Rhodobacter sphaeroides* and *Methylosinus trichosporium:* no evidence for Fe protein modification," *Arch. Microbiol. 150,* 1–5 (1988).

59. Zinder, S. H., "Microbiology of anaerobic conversion of organic wastes to methane: recent developments," *ASM News 50,* 294–298 (1984).

60. Zumft, W. G., and Castillo, F., "Regulatory properties of the nitrogenase from *Rhodopseudomonas palustris,*" *Arch. Microbiol. 117,* 53–60 (1978).

Note Added in Proof: The results from our laboratory on nitrogenase in *Methanosarcina barkeri* described as "Lobo and Zinder, in preparation" have been published (*J. Bacteriol. 172,* 6789–6796, 1990). The unidentified peak described in NMR studies on ^{15}N incorporation by *Methanococcus thermolithotrophicus*[5] was identified as β-glutamate (*Biochmim. Biophys. Acta 992,* 320–326, 1989), which is accumulated as an osmolite. Finally, since the preparation of this manuscript, we learned of the untimely death of Lionel Sibold. We would like to dedicate this chapter to him in recognition of his pioneering studies on the evolution of nitrogenase genes.

6

Associative Nitrogen-Fixing Bacteria

Claudine Elmerich, Wolfgang Zimmer, and Claire Vieille

I. Introduction

It has been established that substantial amounts of nitrogen are fixed in tropical agriculture, in particular in the farming of lowland rice, sugar cane, and pasture grass, where little or no fertilizer is used.[25,54,260] Since the initial discovery of *Azotobacter paspali-Paspalum notatum* association,[55] N_2-fixing bacteria associated with forage grasses and cereal crops, in both natural and cultivated ecosystems, have been extensively studied. This led to the identification of new bacterial genera and species, including *Azospirillum*, *Klebsiella*, *Bacillus*, *Acetobacter*, and *Pseudomonas*-like bacteria. The association of plants with these N_2-fixing bacteria, without formation of differentiated structures, is designated *associative symbiosis*. The bacteria contribute to some extent to plant growth and can be considered as plant growth-promoting rhizobacteria (PGPR). *Spirillum*-like bacteria, first discovered in association with tropical grasses in Brazil[53,270] and later classified in the genus *Azospirillum*, have been the most extensively studied. This review will deal mostly with this genus.

II. The Ecology and Physiology of N_2-Fixing Organisms Associated with Grasses

A. The Distribution of Nitrogen-Fixing Bacteria in the Rhizosphere of Grasses

The acetylene reduction technique has been used *in situ* to examine various ecosystems for N_2-fixation activity.[102] Despite early controversial reports,[30] it appears that certain environmental conditions favor proliferation of N_2-fixing heterotrophs in the rhizosphere.[53,58,109,256,284] It is difficult to quantify the contribution of each specific N_2-fixing bacterium in associative symbiosis. Because of the absence of differentiated structures on the root system and of the proliferation of bacteria in the rhizosphere, it is often impossible to determine which organism(s) is (are) actually responsible for plant growth promotion and/or for N_2 fixation. It is not clear whether these N_2-fixing bacteria arc on the root surface (the rhizoplane) or elsewhere in the rhizosphere[5,57] (the zone around the roots). However, it is out of the scope of this review to discuss the techniques used and the difficulties encountered for enumerating soil bacteria, for determining the relative number of N_2-fixers, and for accurate identification. The use of selective media enabled isolation of specific bacterial strains and comparative studies of their relative distribution.[8,56,206,207,215,280] The proper taxonomy of the isolated strains was not always easily established (this volume, Chapter 1). Numerous new species have been isolated, and refined selective conditions should lead to the characterization of as yet unknown genera and species. The distribution and frequency of N_2-fixing bacteria differ greatly with the geographical regions. Among the most common isolates are *Azospirillum*, *Azotobacter*, *Bacillus*, *Beijerinckia*, and *Enterobacteriaceae*.[5,206,207] Representatives from other genera include *Acetobacter*, *Alcaligenes*, *Campylobacter*, *Herbaspirillum*, and *Pseudomonas*-like bacteria.[5,52,57] *Azospirillum*, *Klebsiella* sp., *Azotobacter paspali*, *Alcaligenes*, and *Campylobacter* (see Part V) have been described as having a tight association with the root system. Other *Azotobacter* spp. and *Beijerinckia* are present in soil but are considered to be free-living.[57,208] Thus, it is unclear whether all N_2-fixing bacteria associated with soil core should be included in this report on associative symbiosis.

B. Ecology, Taxonomy, and Basic Physiology

i. Azospirillum

The ecological distribution of *Azospirillum* is extremely wide.[56,57] Strains have been found in association with monocotyledons—including maize, rice, sugar cane, sorghum, such forage grasses as *Digitaria* and Kallar grass,[56,99,205,206,278]—and with dicotyledons.[200] The genus was defined by Tarrand and colleagues in 1978.[252] It now comprises five species, characterized on the basis of phenotypic properties and DNA relatedness: *A. brasilense*,[252] *A. lipoferum*,[252] *A. amazonense*,[72,143] *A. halopraeferens*,[204] and *A. irakense*.[119] A detailed review of the physiological properties of *A. brasilense*, *A. lipoferum*, and *A. amazonense* was recently published.[57] These bacteria are aerobic nonfermentative chemoorganotrophs. They are Gram negative, though some *A. brasilense* strains are Gram variable.[252] The cells are vibrioid to S-shaped, from 2 to 4 μm in length, and motile by a polar flagellum in liquid media. On solid media, *A. brasilense*, *A. lipoferum*, and *A. irakense* possess several lateral flagellae. They contain granules of poly-beta-hydroxybutyrate (PHB). The G + C content of DNA varies between

66 and 70 mol %. Their optimal growth temperature is between 33° and 41°C. *A. lipoferum* and *A. halopraeferens* require biotin; other species are prototrophs. *A. amazonense* prefers low pH. As indicated by its species name, *A. halopraeferens* is halotolerant, which is also a property of *A. irakense*. *A. brasilense* and *A. lipoferum* accumulate carotenoids, which are responsible for a pink coloration in old cultures,[70] but other species are not pigmented. In general, members of the *Azospirillum* genus use organic acids (malate, lactate, pyruvate, succinate) as carbon sources. They can be differentiated by their sugar utilization spectrum.[148,149,276] *A. brasilense* cannot utilize glucose, though some strains have been reported to use it. The sugar utilization spectrums of *A. irakense* and *A. amazonense* are very similar, and the two species can use disaccharides such as sucrose, lactose, maltose, and trehalose. *A. irakense* grows with pectin as the sole carbon source.[119] Autotrophic growth under aerobic conditions, with H_2 as the energy source has been demonstrated for *A. lipoferum* but not for *A. brasilense*.[57,145] Nitrogen fixation occurs under conditions of microaerobiosis. Utilization of N_2, ammonia, and nitrate is discussed in Part IV; other features of these bacteria related to their association with plants are discussed in Parts V and VI.

ii. Enterobacteria: Klebsiella, Enterobacter, Erwinia

Gram-negative, facultative anaerobic enterobacteria are commonly isolated from cereal crops and forage grasses. Though it is extremely difficult to draw general rules, they are found more frequently in association with wheat, in particular in temperate climates.[123,188,206] A survey of forage grass in Texas showed the majority of N_2-fixers to be *Enterobacter cloacae* and *Klebsiella pneumoniae*.[280] The same species were isolated from grasses in Finland, together with *Azospirillum* and *Pseudomonas*-like bacteria.[97,99] *Klebsiella* and *Enterobacter* have also been isolated from rice.[130] Three N_2-fixing species of *Klebsiella* have now been described, *K. pneumoniae*, *K. oxytoca*, and *K. planticola*. However, the taxonomic status of *Klebsiella* is unclear, and *K. oxytoca* and *K. planticola* are often confused with *K. pneumoniae*.[4,212] In contrast to *K. pneumoniae*, *K. oxytoca* is indole positive and *K. planticola* can utilize hydroxyproline. In addition, some strains are psychotroph. The famous *K. pneumoniae* M5a1 strain, employed by geneticists, should be referred to as *K. oxytoca*.[62,195] The possibility that most *K. pneumoniae* isolates from soil belong to *K. planticola* or *K. oxytoca* cannot be ruled out. *K. oxytoca* strains have been isolated from rice in Japan,[109,120] and *K. planticola* strains have been isolated from rice in the Philippines.[130] The main N_2-fixing *Enterobacter* species is *E. cloacae*, but some strains of *E. agglomerans*, closely related to N_2-fixing *Erwinia herbicola*, also have been isolated.[183,188,208,238] These bacteria have the general physiological properties of *Enterobacteriaceae* and can fix N_2 under anaerobic conditions, although it is possible that traces of oxygen stimulate N_2 fixation (this volume, Chapter 2).

iii. Other Bacteria

Besides *Azospirillum* and *Enterobacteriaceae*, there are a number of free-living N_2-fixing bacteria isolated from the rhizosphere. Some of them belong to well-known families or genera (*Bacillus*, *Azotobacter*, *Pseudomonas*) whose biology in terms of general physiology will not be discussed here. For others, either identified as new genera (*Herbaspirillum*) or as new N_2-fixing species of a known genus (*Acetobacter diazotrophicus*), knowledge is still limited.

Bacillus. Facultative anaerobic, N_2-fixing, spore-forming bacteria in soil are often detected, in particular in association with wheat in temperate climates.[206] Until recently, two species have been described, *B. polymyxa* and *B. macerans*. These species displayed nitrogenase activity but could not grow with N_2 as the sole nitrogen source.[233] In contrast, a new species, *B. azotofixans*, able to use molecular nitrogen was isolated from wheat in Brazil and from sugar cane in Hawaii.[234] Nitrogen fixation occurs under anaerobic conditions.

Aerobic Nitrogen Fixers. The ecology of aerobic fixers has been reviewed previously.[57] *Azotobacter* and *Beijerinckia* have been found in sugar cane, corn, oat, and soybean rhizospheres, but the bacteria were not found to be associated with the root system.[125,208] The only very specific association is *A. paspali-Paspalum notatum*.[55,187]

Pseudomonas-like Bacteria. Several strains, classified as *Pseudomonas* and capable of fixing nitrogen under microaerobiosis, have been isolated from rice, sorghum, and other grasses.[13,97,127,273] Among them, new species have been identified; their taxonomic status is discussed in Chapter 1. A strain showing up to 93% DNA relatedness with *Pseudomonas stutzeri* has been isolated.[127] Several isolates from wetland rice[13] have been classified as a new species, designated *Pseudomonas diazotrophicus*.[273]

Herbaspirillum. The bacteria now classified in the genus *Herbaspirillum* were first reported as a new species of *Azospirillum*, with whom they shared common features: vibrioid morphology, G + C content of 66 to 67 mol %, preferential growth on organic acids, and formation of a pellicle under the surface in N-free semisolid medium.[57] However, RNA/RNA hybridization showed that these bacteria were in fact a new genus. The species name given was *H. seropedicae*.[6]

Other microaerobic N_2-fixers include a novel species of *Acetobacter*, *A. diazotrophicus*, which was isolated from sugar cane in Brazil.[52] The bacteria produced an orange brown pigment and reduced acetylene at low oxygen tension.[93] Plant association with two unusual bacterial species has also been reported. A N_2-fixing strain, *Alcaligenes faecalis*, which associates with rice,[283] has been described in China, and in Canada a strain of *Campylobacter*, associated with a seaweed *Spartina alterniflora*,[151,152] has been described.

III. Genetics and Molecular Biology

As illustrated in Part II, the bacterial species involved in associative symbiotic ecosystems have been isolated recently, and for most of them little information on their biology is available, as yet. It is difficult to predict which organism has the most potential in the future for applied purposes of associative symbiosis. It is therefore important to develop genetics and molecular biology in parallel with physiological and biochemical studies. In contrast to ecologists, geneticists should limit their investigation to a small number of strains. Powerful genetic tools have long been available for *Enterobacteriaceae*, *Bacilli*, and *Pseudomonas*, and they will not be described here. Indeed, the model system for genetics of N_2 fixation was established with *K. pneumoniae* strain M5a1 (this volume, Chapters 20 and 21), a strain closely related to *Klebsiella* strains isolated from plants. Similarly, the molecular biology of *Azotobacter* is well developed (this volume, Chapters 19 and 20). The molecular genetics of *Azospirillum* is the most extensively developed of the bacteria isolated from root systems.

A. Plasmids and Bacteriophages

i. DNA Content

The DNA content of *A. vinelandii*[197,220] and *A. chroococcum*,[214] which are large cells, has long been believed to be much higher than the *Escherichia coli* genome. This was recently confirmed. These bacteria may contain about 40 identical copies of the chromosome per cell. However, the amount of DNA per cell has not been measured in *Azospirillum*. On the basis of electrophoretic migration of plasmids, it has also been suggested that the *Azospirillum* genome is composed of nonidentical minichromosomes or megaplasmids.[279] Further analysis of this hypothesis is awaited.

ii. Plasmids in Azospirillum

Franche and Elmerich[83] first reported the presence of plasmids in *A. brasilense* and *A. lipoferum*, and this topic has been extensively reviewed.[57,63,65,66] In general, the strains examined contain from one to six plasmids, with M_r ranging from a few MDa to more than 300 MDa. The presence of a 90-MDa plasmid was noted frequently in *A. brasilense* strains, whereas *A. lipoferum* carries a 150-MDa plasmid.[7,83,193,269] Plasmids are also present in the three other *Azospirillum* species. Using the technique of Kado and Liu,[116] we have detected three plasmids of 100, 90, and 19 MDa in *A. amazonense* Y1, two plasmids of 14 and 12 MDa in *A. halopraeferens* Au4, and two plasmids of 115 and 3 MDa in *A. irakense* KBC1 and KA3 (Onyeocha and Zimmer, unpublished).

iii. Molecular Genetics of Azospirillum plasmids

Until recently, little information on *Azospirillum* plasmids was available in terms of physical and genetic characterization. They appeared to be cryptic and nontransferable. Spontaneous and acridine orange curing of some plasmids has been reported.[57] In particular the 115-MDa plasmid of strain *A. brasilense* Sp7 is easily lost.[83] Random transposon mutagenesis has been used to tag plasmids in *A. brasilense* strain Sp7 and *A. lipoferum* strain 4B.[7,265] This led to the successful transfer of the 150-MDa plasmid of *A. lipoferum* 4B into *Agrobacterium tumefaciens*.[7] Small plasmids have been cloned, in particular a plasmid of 5.9 kb from *A. brasilense* R07[157] and a plasmid (pAL1) of 5.4 kb from *A. lipoferum* RG20.[243] This did not lead to further utilization or characterization.

iv. The 90-MDa Plasmid of Azospirillum brasilense Sp7

The medium-sized plasmids and megaplasmids in *Azospirillum* were assumed to carry important information for the interaction with the host plant, by analogy with other soil-bacteria–harboring plasmids, in particular *Agrobacterium* and *Rhizobium*.[196] This was indeed the case of a plasmid of 90 MDa (148 kb) from *A. brasilense* strain Sp7, designated p90, whose physical map was recently established[175] (Figure 6-1). This plasmid was studied because it was found to carry a DNA region that hybridizes with the *R. meliloti* symbiotic plasmid (pSym) *nodPQ* genes.[268,269] As described in Part VI, p90 also carries other functions related to exopolysaccharide synthesis.[153,155] Moreover, it harbors two different ampicillin resistance genes.[175] Plasmid p90 was tagged with transposon Tn5-Mob, and its transfer to *A. tumefaciens* was promoted by a mobilizing plasmid, pJB3JI. Plasmid p90 was recovered from *Agrobacterium* as a cointegrate with the helper plasmid, suggesting that its origin of replication cannot function in *Agrobacterium*, in contrast to the 150-MDa plasmid of strain 4B. Hybridization experiments with plasmid clones of p90 used as probes showed that each fragment hybridized with a plasmid of similar size in other *Azospirillum* strains, suggesting that all the plasmids of 90 MDa in this genus contain conserved regions, which presumably code for common functions.[175] In addition, attempts to obtain strains cured of this particular plasmid were unsuccessful.

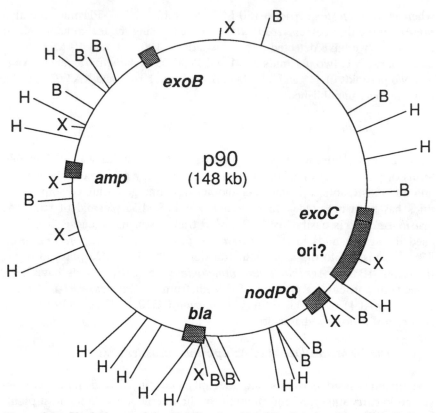

Figure 6-1. Physical map of the 90-MDa plasmid (p90) of *A. brasilense* strain Sp7 and localization of a few markers according to Onyeocha et al.[175] Restriction sites: B: *Bam*HI, H: *Hind*III, X: *Xho*I. *nodPQ* is the locus that hybridizes with the *nodPQ* genes of *R. meliloti*,[175,268] *bla* and *amp* are nonhomologous loci involved in ampicillin resistance,[175] *exoB* and *exoC* are loci that functionally complement mutations in the *exoB* and *exoC* genes of *R. meliloti*,[153,155,175] ori indicates a DNA region necessary for the maintenance of the plasmid as an independent replicon.[175]

This led to speculation that p90 may carry essential functions, which remain to be determined.

v. Localization of nif Genes on Plasmids in Soil Bacteria

As N$_2$-fixation genes are localized on symbiotic plasmids in fast-growing rhizobia, it is often generalized that all N$_2$-fixing bacteria carrying plasmids harbor the *nif* genes on these plasmids. However, this is apparently not the case in *Azotobacter* or in *Azospirillum*, where it could not be demonstrated that *nif* genes were plasmid borne.[193,214,269] In contrast, *nif* genes are carried on plasmids in N$_2$-fixing *E. agglomerans* strains isolated from the rhizosphere of wheat.[238] The physical map of a 111-kb *nif*-containing plasmid, pAE3, originating from *E. agglomerans* 333 has been established.[239,241] In *P. stutzeri* strain CMT.9 isolated from *Sorghum*, curing of a 30-MDa plasmid led to the concomitant loss of N_2-fixing activity.[127] This strongly suggested that at least some of the *nif* genes in this particular strain are plasmid borne and raised the possibility of plasmid-mediated genetic exchange in the rhizosphere, which could lead to a natural selection of N_2-fixing hybrids.

vi. Bacteriophages

Azotobacter and *Azospirillum* bacteriophages have already been reviewed.[57,65,66] Numerous bacteriophages of *Azotobacter* have been studied.[65,253] Azotophages have the characteristic of being maintained in a pseudolysogenic state.[253] The isolation of phages from soil samples is often straightforward,[92] though in the case of *Azospirillum* the distribution of phages seems to be limited to particular Brazilian soils.[57,91,92] Phage A11, titrating on a limited range of host strains of either *A. lipoferum* or *A. brasilense*, was isolated in Brazil[57] and has been characterized. It is a temperate, nontransducing bacteriophage, maintained as a plasmid in the prophage state.[68] Unfortunately, A11 does not titrate on *A. brasilense* Sp7, where the genetics is the most developed. *A. brasilense*, including Sp7, and *A. lipoferum* strains tested were found to be lysogenic for defective prophages, and no mature phage particle was detected after induction by mitomycin C.[83] A different phage, Ab1, was obtained after spontaneous induction of *A. brasilense* strain Sp7 (ATCC29145),[91] though this phage is not produced by the strain Sp7 of our collection. Interestingly, the phage particles titrate on the host lysogen.[91] This could be a novel example of pseudolysogeny.[91] A third phage, Ab2, also titrating on Sp7, was isolated from Brazilian soil.[92] No information is available yet on phages specific for other bacterial strains isolated from associative symbiotic ecosystems.

B. Genetic Tools in Azospirillum

i. Conjugation, Transformation

Endogenous plasmids and phages cannot be used for genetic analysis in *Azospirillum*.[66] Broad host range IncP plasmids have been introduced into *Azospirillum* by conjugation,[18,82,157,158,194,237] whereas broad host range IncW and IncQ plasmids are not stable in this background.[157] The IncP plasmids, RP4 and its derivative R68-45, which displays enhanced chromosomal mobilization, have been used to promote genetic transfer between *A. brasilense* auxotrophs and antibiotic-resistant mutants.[18,82,158] The marker transfer appeared to be nonpolar, which suggested chromosome mobilization from several origins, as reported for *P. aeruginosa*.[100] Linkage between a few markers has been established. However, the low frequency of recombination, between 10^{-6} and 10^{-7}, prevented this technique from being further exploited. DNA-mediated transformation has not been reported for *A. lipoferum* but has been successful for *A. brasilense* with linear[159] or plasmid[74] DNA. A protocol using electroporation of plasmid DNA was recently developed for *A. brasilense*.[263] The authors were unable to obtain transformants with *A. lipoferum*.

ii. Mutagenesis

The isolation of various types of mutants (mainly auxotrophs, Nif⁻, antimetabolite and antibiotic resistant—see Parts V and VI) of *Azospirillum*, after classical chemical or UV irradiation mutagenesis has been reviewed.[57,63,66] Studies using a large spectrum of mutagens led to suggestions that *Azospirillum* lacked a SOS repair system.[45] Collections of *A. lipoferum* and *A. brasilense* mutants isolated after random transposon mutagenesis have been obtained[1,89,240,261,264,265] using methods developed in *Rhizobium*.[235,236] The authors have constructed Tn5-containing plasmids (e.g., pSUP2021[236] and pGS9[235]) that have a broad host-range but that cannot replicate outside enteric bacteria and thus behave as suicide vectors. Transposition events in *Azospirillum* are selected by the acquisition of kanamycin resistance after conjugation with *E. coli* strains containing the plasmid. It was shown by hybridization that the resistance phenotype resulted from the physical integration of the transposon into the host genome. Different types of mutants were isolated, in particular, auxotrophs, Nif⁻, and mutants impaired in indole acetic acid production and in flocculation and calcofluor binding activity.[156,240,265] Using Tn5–*lacZ*, several mutants with impaired motility and chemotaxis were obtained.[261] Endogenous plasmids were tagged with suicide vectors containing a Tn5-Mob transposon.[7,265,269] Random transposon mutagenesis is a very effective technique for isolating mutants, and large collections can be obtained. The frequency of the tranposition event in *Azospirillum* is in the range of 10^{-7}, thus much lower than in *R. meliloti*. Higher frequencies of transposition could be obtained with some mutant strains such as strain 7030, a derivative of *A. brasilense* Sp7.[261] The protocol for Tn5 site-directed mutagenesis, also developed in *Rhizobium*,[236] has been applied to *Azospirillum*. It was particularly useful for genetic analyses of *nif* regions in *A. brasilense* Sp7.[89,192] In this case, mutagenesis was performed in *E. coli* containing a vector carrying an *A. brasilense* DNA fragment.[236] The mutant DNA fragments were subsequently introduced in the host genome by homologous recombination.

iii. Gene Bank Construction and Gene Cloning

Several gene libraries of *Azospirillum* DNA have been constructed in phages[86,199,226,275] or in broad-host-range plasmids or cosmid vectors.[71,76,80,155] A plasmid library has also been obtained.[175] This led to the cloning of nitrogen assimilation and fixation genes (see Part IV), genes for amino acid biosynthesis,[17,75] an ampicillin resistance gene,[267] and genes possibly involved in the association with the plant (see Parts V and VI). Bazzicalupo and colleagues[17] identified, by genetic complementation, plasmid clones complementing histidine, pyrimidine, and cysteine mutants of *A. brasilense* Sp6. One plasmid complemented *E. coli hisA, hisB, and hisF* mutations. The analysis of the nucleotide sequence of the *A. brasilense hisB* gene and flanking regions revealed the presence of two other open reading frames homologous to *hisD* and *hisH* of *E. coli*.[75] Thus, at least five genes of the histidine biosynthetic pathway of *A. brasilense* are clustered in the order *hisDBHFA* and are probably organized in a polycistronic operon.[75] The fact that amino acid biosynthetic genes of *A. brasilense* are expressed in *E. coli* indicates that these genes may be transcribed from their own promoter in this bacterium. However, this remains to be proved. A broad-host-range cloning vehicle, pAF300, was designed to identify putative promoter regions of *A. brasilense*, by fusing random DNA fragments to a promoterless chloramphenicol resistance gene.[77] This led to the cloning of several fragments with promoter activity in *E. coli* and *Azospirillum*, but no promoter sequence has been determined.[77]

IV. Nitrogen Metabolism

A. Pathways of Ammonia Assimilation

i. Enzymes Involved

In Gram-negative bacteria three enzymes are involved in ammonia assimilation: glutamine synthetase (GS), NADPH-dependent glutamate synthase (GOGAT), and NAD(P)H-dependent glutamate dehydrogenase (GDH), which catalyze the following reactions:

$$GS:\ glutamate + NH_3 + ATP \rightarrow glutamine + ADP + Pi$$
$$GOGAT:\ glutamine + 2\text{-oxo-glutarate} + NADPH + H^+ \rightarrow 2\ glutamate + NADP^+$$
$$GDH:\ 2\text{-oxo-glutarate} + NH_3 + NAD(P)H + H^+ \leftrightarrow glutamate + NAD(P)^+$$

Ammonia assimilation requires GS activity and GS is a key enzyme in the regulation of the nitrogen flux in the cell. Its functioning coupled with GOGAT activity ensures synthesis of glutamine and glutamate, which are precursors of all the nitrogenous compounds of the cell. GDH is not always essential. The activities of the three enzymes in *Azospirillum* have been measured in crude extracts of cells grown under different physiological conditions.[9,90,169,277] GS activity in *Azospirillum*, as in most Gram-negative bacteria, is controlled by adenylylation. The maximum adenylylation was observed in cultures grown in the presence of ammonia, whereas minimal adenylylation was observed under conditions of N_2 fixation.[90,169,277] GS biosynthesis was repressed two- to eight fold by high ammonia concentrations. In contrast, the NADH-dependent GDH activity increased in cells grown on ammonia.[201] These data, the phenotypes of GS mutants and $^{13}NH_4^+$ assimilation kinetics data, strongly suggested that GS–GOGAT is the major route of ammonia assimilation under conditions of N_2 fixation.[90,277] *A. brasilense* Sp6 mutants deficient in GOGAT activity have been isolated.[9] These mutants were also impaired in N_2 fixation, and growth on nitrate or several other organic nitrogen sources, including histidine, alanine, adenine, and xanthine.[9] However, the structural genes for GOGAT were probably not mutated. GOGAT was purified to homogeneity from *A. brasilense* Sp6.[201] The native enzyme is a flavoprotein with a M_r of 740,000. It is composed of two subunits of M_r 50,000 and 135,000 and has an alpha$_4$–beta$_4$ quaternary structure, similar to that of the *E. coli* enzyme.[201] A GDH was purifed to homogeneity from *A. brasilense* RG.[150] The enzyme was cold-labile and displayed activity with NADPH or NADH. It might be different from the GDH characterized in other *Azospirillum* strains. *A. brasilense* and *A. lipoferum* contain a specific ammonia carrier, which is energy dependent and is derepressed when ammonia is limiting.[105]

ii. Cloning of the glnB-glnA Structural Genes

The structural gene for GS in enteric bacteria, *glnA*, belongs to a complex regulon, which contains the *ntrBC* genes. *NtrBC* gene products, together with sigma-54, are responsible for the transcriptional activation of a number of operons, including N_2-fixation genes (this volume, Chapter 21). *A. brasilense* Sp7 glutamine auxotrophs impaired both in GS activity and nitrogen fixation have been isolated.[90] A plasmid that restores a wild-type phenotype to one of these mutants has been isolated from a gene bank of strain Sp7. This plasmid contained the GS structural gene *glnA*[29,67] and the protein P_{II} structural gene *glnB*.[49] The P_{II} protein plays a role in the adenylylation cascade and the synthesis of GS in enteric bacteria (this volume, Chapter 21). As this plasmid did not contain *ntrBC* genes, it appears that the organization of *gln* and *ntr* genes in *Azospirillum* is similar to that observed in *Rhizobiaceae*, where *glnB* and *glnA* are contiguous [40,147] (Figure 6-2[1]). The nucleotide sequence of *A. brasilense* glnA[28] and glnB[49] was established. Transcription analysis and promoter mapping showed that *glnA* can be transcribed

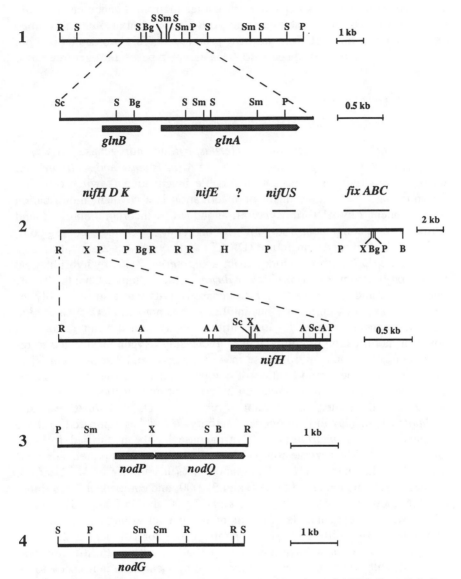

Figure 6-2. Physical map of the *gln*, *nif*, *fix*, *nod* gene regions of *A. brasilense* Sp7. Restriction sites: A: *Ava*II, B: *Bam*HI, Bg: *Bgl*II, P: *Pst*I, R: *Eco*RI, S: *Sal*I, Sc: *Sac*I, Sm: *Sma*I, X: *Xho*I. (1) Physical map of the *glnB-glnA* region.[28,29,49,80] (2) Physical map of the major *nif* cluster.[48,73,89,199,226] (4) Physical map of *nodG* (C. Vieille and Elmerich, in preparation). (3) Physical of *nodPQ*.[268]

with *glnB* from an NtrA-dependent promoter or from a canonical *E. coli*–type promoter, located upstream from *glnB*.[49] GlnA can also be transcribed as a monocistronic unit from its own promoter.[28] This promoter does not function in *E. coli* and appears to be a novel type of nitrogen-regulated promoter.[49]

B. Biochemistry and Genetics of N_2 Fixation

i. Nitrogenase and ADP Ribosylation

In all *Azospirillum* species and in *Herbaspirillum*, nitrogenase activity is derepressed at low oxygen tension.[57,66,85] In *A. brasilense* and *A. lipoferum*, nitrogenase activity is subject to reversible inactivation ("switch-off–on") by ammonia.[104] The mechanism of inactivation is a covalent modification of component 2 by ADP-ribosylation, as judged by incorporation of ^{32}P and formation of a component 2 subunit with higher M_r.[104] Genes encoding component-2 ADP-ribosyl transferase (DRAT) and component-2-activating glycohydrolase (DRAG) were cloned from *A. lipoferum* Br17, by hybridization with a probe from *Rhodospirillum rubrum*[86] (this volume, Chapter 3). *A. amazonense* and *H. seropedicae* nitrogenase activity was also reversibly inactivated *in vivo* by ammonia, but inhibition was not complete even at high ammonia concentrations.[85,104] In addition, inactivation did not result in a covalent modification of component 2. Increasing oxygen tension led to reversible inactivation of nitrogenase in *A. brasilense* and *A. lipoferum*.[103] In this case no covalent modification of component 2 was observed.[103] In contrast, anaerobiosis led to inactivation by covalent modification.[103] As a consequence of the switch-off mechanism, crude extracts of *A. brasilense* and *A. lipoferum* display low nitrogenase activity,[103,142,189] as compared to those obtained from *A. amazonense*.[245] Nitrogenase has been purified from *A. amazonense*.[245] The enzyme complex contains two different proteins in common with other nitrogenases. Component 1 is a tetramer of M_r 210,000, formed of two monomers of 50,000 and 55,000, and component 2 is a dimer of M_r 67,000, formed of two subunits of 31,000 and 35,000.[245] The origin of the two different subunits of component 2 is unclear and *A. amazonense* may possess two different structural *nifH* genes for component 2.[245] The nitrogenase of *A. brasilense* has been partially purified. Purification was more difficult due to the low activity of component 2, which needs to be reactivated by Mn^{2+} ions or by an activating factor.[142,189] The nitrogenase polypeptides were characterized by immunoprecipitation and immunoblots.[104,163]

ii. H_2-Dependent Nitrogen Fixation

Nitrogenase catalyzes the reduction of protons to H_2 concomitantly with N_2 reduction and can be considered as a hydrogenase with an ATP-dependent evolution activity. Uptake hydrogenase can play a role in conserving energy by recycling H_2 released by nitrogenase. This has been shown for uptake hydrogenases found in *A. brasilense* and *A. amazonense*.[20,38,87,190] Moreover uptake hydrogenase enables some organisms, including *A. lipoferum*, *Derxia gummosa*, and *P. diazotrophicus*, to grow chemolithoautotrophically even under N_2-fixing conditions.[96,145] Expression of the uptake hydrogenase is enhanced by electron donor limitation.[38,87] Like nitrogenase, hydrogenase activity is sensitive to oxygen. However, when using O_2 as the respiratory electron acceptor, a higher concentration of O_2 was necessary to support maximal activity of hydrogenase as compared to nitrogenase.[190] Therefore, hydrogenase-dependent oxygen reduction is presumably unable to remove enough oxygen to protect nitrogenase, a protective mechanism previously suggested.[51]

iii. Nitrogen Fixation Genes

The molecular organizations of the *nif* genes of two *Enterobacteriaceae*, *E. agglomerans* 333, and *K. oxytoca* NG3 have been studied. In both organisms, *nif* genes are clustered and their organization is similar to that of *K. pneumoniae* M5a1.[120,241] In *E. agglomerans*, however, the *nif* genes are located on a plasmid of 111 kb, pAE3, and their order is similar to that of M5a1, except for *nifJ*, which is contiguous with *nifQ*.[241] Hybridization studies with *K. pneumoniae* nitrogenase structural genes (*nifHDK* probe) were performed with several strains of *A. brasilense* and *A. lipoferum*.[71,199,226] This led to the cloning of the corresponding genes. The *nif* genes of *A. brasilense* Sp7 (ATCC 29145) and of strain Cd (ATCC 29710) have been further characterized. Strain Cd is extremely similar if not identical to Sp7 and could simply be a pigmentation mutant of Sp7.[70] Nif⁻ mutants have been isolated after chemical or transposon mutagenesis,[112,189,240,264] but most of the mutations did not map in known regions and they were not identified. In strain Sp7, a 30-kb DNA region containing *nifHDK*, *nifE*, and *nifUS* has been characterized by hybridization with heterologous probes and Tn5 site-directed mutagenesis of cloned fragments[67,89,192,199] (Figure 6-2 [2]). The order of *nif* genes is similar to that of *K. pneumoniae* and *Azotobacter*. Interestingly, a DNA region containing genes hybridizing with *Rhizobium fix ABC* probes was found in *Azospirillum*[81,89] (Figure 6-2 [2]). This region contains functional genes in *Azospirillum*, since inactivation by Tn5 led to a Nif⁻ phenotype.[89] The *nifH* gene of strain Sp7 was sequenced and displayed the classical feature of other *nifH* genes.[48,73] The *nifHDK* cluster was shown to be organized as a single transcription unit by genetic complementation of Tn5 mutants and transcription analysis.[48,192] Under conditions of N_2 fixation, transcription proceeded from an NtrA-like dependent promoter.[48,73] This promoter is preceded by two upstream elements (UAS), typical of genes controlled by *nifA* (this volume, Chapter 21).

iv. Nif Gene Regulation

Regulation of N_2 fixation in enteric bacteria such as *K. oxytoca* is likely to be similar to that in *K. pneumoniae*.[120] The *nifA* gene of *K. oxytoca* NG3 was sequenced and the translation product differed from M5a1 NifA by four amino acids.[120] In *Azospirillum*, the regulatory mechanisms are not totally elucidated. Pedrosa and Yates[189] isolated, after chemical mutagenesis, *A. brasilense* Sp7 Nif⁻ mutants with NtrC-like or NifA-like phenotypes. In NtrC-like mutants, N_2 fixation was restored after introduction of plasmids, which contained *K. pneumoniae ntrBC* genes (pGE10) or a constitutive *nifA* gene (pCK3), whereas in NifA-like mutants N_2 fixation was complemented only by *K. pneumoniae nifA*.[189] The NtrC-like mutants could not use nitrate as nitrogen source, and thus resembled the GOGAT-defective mutants isolated by Bani and colleagues.[9] Mutants with a Nif constitutive phenotype (i.e., fixing nitrogen in the presence of ammonia), also displaying impaired histidine transport, have been isolated as ethylenediamine-resistant mutants.[79] The isolation of another NifA-like mutant, obtained after random Tn5 mutagenesis of *A. brasilense* strain ATCC 29710, has been reported.[242] This mutant was Nif⁻, did not produce nitrogenase component 2, and a Nif⁺ phenotype was restored after introduction of *K. pneumoniae nifA* (plasmid pCK3).[242] However, identity with *K. pneumoniae nifA* was not proven by hybridization. A DNA region containing *nifA-nifB* was cloned from *A. brasilense* Sp7 by hybridization with *Bradyrhizobium japonicum* probes (Liang and Elmerich, unpublished). Strong similarity of *nifA* was confirmed by nucleotide sequencing. Inactivation of *nifA* and *nifB* by site-directed Tn5 mutagenesis led to mutants with a Nif⁻ phenotype. The physical map of the *nifBA* clone from Sp7 was different from that of the NifA-like clone from strain ATCC 29710. Moreover, Sp7 *nifA* hybridized with DNA fragments of identical size in the two strains. This strongly suggests that in addition to *nifA*, *Azospirillum* contains another regulatory gene leading to a NifA-like phenotype, a situation also encountered in *Azotobacter*.[221] Genes of *A. brasilense* Sp7, sharing homology with *B. japonicum ntrBC*, were recently cloned (Liang and Elmerich, unpublished), but their involvement in N_2 fixation has not yet been established. Thus, in addition to an unknown *nifA*-like regulatory gene,[242] *Azospirillum* contains genes homologous to *nifA* and *ntrBC*. Nif genes are probably transcribed from NtrA-dependent promoters, as has been established for *nifH*,[48,73] which suggests the presence of a gene encoding an NtrA-like product in *Azospirillum*. The role of *glnBA* gene products, whose transcription is also under the control of an NtrA-dependent promoter, in N_2 fixation also remains to be elucidated.

C. *Nitrate Assimilation and Denitrification*

In addition to N_2 fixation, most soil and rhizosphere bacteria have the ability to reduce nitrate. These organisms can use nitrate as a nitrogen source and/ or as an electron acceptor of the respiratory chain, when oxygen is limiting or absent.[248] Some fermentative bacteria, like the *Enterobacteriaceae*, remove excess reducing equivalents by converting nitrate via nitrite to ammonia (reviewed in refs. 39 and 247). Although the conversion of nitrate to nitrite does not affect the total amount of nitrogen available for plants, the reduction of nitrate by the denitrification process—via nitrite, nitric oxide (NO), nitrous oxide (N_2O), and N_2—results in a loss of combined nitrogen in the soil. Most strains of *A. brasilense* and *A. lipoferum* are denitrifiers. This property is also widespread in *Pseudomonas*-like bacteria. In contrast, *A. amazonense* and *A. irakense* are reported to be unable to denitrify.[119,143] However, in the case of *A. amazonense* strains QR242 and Y1, denitrification activity was found, but lower than in *A. lipoferum* and *A. brasilense*.[165] Denitrification by *Azospirillum* in association with wheat is favored by high pH values, high temperature, and low oxygen pressure.[165]

i. *Nitrate Reductases*

The first step in the assimilation and dissimilation of nitrate is the reduction of nitrate to nitrite catalyzed by nitrate reductase. Two genetically distinct enzymes are synthesized, depending on growth conditions in *P. aeruginosa*, *K. pneumoniae*, and *R. meliloti*.[33,113,229] Ammonia but not oxygen represses synthesis of the assimilatory nitrate reductase in *Klebsiella*, whereas synthesis of the dissimilatory enzyme is not affected by ammonia but is prevented by oxygen.[262] In most organisms, the dissimilatory nitrate reductase is a Mo-containing enzyme (reviewed in ref. 110). *Azospirillum* can grow with nitrate as the sole nitrogen source under aerobic and anaerobic conditions.[164,166,228] It is not clear whether *Azospirillum* possesses an assimilatory nitrate reductase in addition to the dissimilatory enzyme. In addition, there are no data available about the biochemistry of the assimilatory nitrite reductase in *Azospirillum*. It was found that as soon as nitrate reductase is synthesized, nitrogenase biosynthesis is repressed.[26]

ii. *Nitrate Respiration*

In the absence of oxygen, most strains of *Azospirillum*[259] as well as several *Pseudomonas*-like bacteria, such as *P. diazotrophicus* isolated from rice roots,[37] can use nitrate, nitrite, and nitrous oxide as electron acceptors in the respiratory chain. In the case of *A. brasilense*, it was established that all the steps of denitrification, from nitrate to dinitrogen, were coupled to ATP formation under anaerobic conditions.[164,246,288] Measurements of molar growth yields in malate-limited continuous cultures and of proton translocation across the bacterial membrane with these different electron acceptors showed that the energy transduction efficiencies are almost the same for O_2 and for N_2O.[44] The respecting values for the respiratory reduction of nitrate to nitrite were two thirds lower and for the reduction of nitrite to nitrous oxide, one third lower. From these data it was concluded that conversion of nitrate to nitrite and of nitrous oxide to dinitrogen by *Azospirillum* proceeds at the internal face of the cytoplasmic membrane, whereas reduction of nitrite to nitrous oxide takes place at the periplasmic face of the cytoplasmic membrane.[44]

iii. Dissimilatory Nitrite and Nitrous Oxide Reductases

The dissimilatory nitrite and nitrous oxide reductases of *Azospirillum* are expressed only under anaerobic conditions and their activity is maximal at a pH between 7.3 and 8.5.[288] The dissimilatory nitrite reductase of *Azospirillum* catalyzes the reduction of nitrite to nitric oxide after solubilization from the membrane.[286] This enzyme contains cytochrome c/d.[43,131] The nitrous oxide reductase is a Cu-containing enzyme, as is that of *Pseudomonas perfectomarina*.[42]

V. Bacterial Factors Involved In Root Colonization and Plant Growth

Proliferation of bacteria in the rhizosphere depends on many parameters, including the availability of organic materials to support their growth and the physiological properties of the bacteria to ensure that they can establish themselves and survive in the rhizosphere. Cyst formation, siderophore, bacteriocin, and phytohormone production, and motility and chemotactic response to root exudate might be important parameters to ensure the competitive establishment of a particular bacterial species in the rhizosphere. Other physiological properties might also be important in the colonization process of the root system, such as the attachment to the root cortex and penetration of the root tissues.[57,64,154,168,187] The soil and rhizosphere environments are susceptible to change with seasons and crops. This change influences most, if not all, the parameters of the medium, including water status, temperature, pH, availability of O_2, carbon and nitrogen sources, and minerals. Consequently, well-adapted bacteria must be able to respond to nutrient and pH gradients to colonize their niche competitively and to survive under unfavorable conditions.

A. Bacterial Survival in the Soil

i. Cyst Formation

Cell differentiation into dormant or resistant forms has been extensively described for the Gram-positive spore-forming *Bacilli* and for the Gram-negative *A. vinelandii*, which undergo encystment under unfavorable conditions. In the case of *Azospirillum* Sp7, Eskew and colleagues[70] described the formation of cystlike structures in old cultures, often called C-forms, which might be functionally analogous to *Azotobacter*, *Derxia*, or *Azomonas* cysts.[210,219] Like *Azotobacter* cysts, *Azospirillum* C-forms are ovoid and larger than vegetative cells, devoid of flagella, not motile, and contain PHB granules.[19] PHB can represent up to 57% of the dry weight.[23] Acetylene reduction activity was shown to decline concomitantly with the appearance of C-forms.[70,184] However, Berg and colleagues[19] reported that C-forms of *A. brasilense* associated with plant callus cultures were able to fix N_2. C-forms often include several cells in the same structure. The capsule surrounding the C-forms is composed mainly of polysaccharides.[19] In liquid cultures, encapsulated cells aggregate into macroscopic clumps and flocculate. Culture conditions to achieve maximum flocculation have been defined.[23,217] When flocculated, cysts are embedded in a fibrillar matrix probably containing cellulose and other capsular polysaccharides (CPS).[217] The greater resistance of encapsulated *Azospirillum* to desiccation and heat[132,217,218] suggested that CPS played a role in survival of these bacteria under unfavorable conditions. In addition, encystment was associated with the production of a melaninlike pigment, which is generally found in structures resistant to deleterious environmental conditions.[218] Noncapsulating mutants of *A. brasilense* and *A. lipoferum* were isolated by plating on a medium containing Congo red.[16] Wild-type colonies stain red on this medium, and white colonies contained mutants, probably impaired in exopolysaccharide synthesis and unable to differentiate into C-forms. These mutants were not affected in root infectious activity, since they induced the same pattern of adsorption onto roots and deformation of root hairs as the wild type.[16] Thus, cyst formation in *Azospirillum* may be related not to root colonization, but to survival in the soil.

ii. Siderophore Production

Although iron is an abundant element in soil, it is generally present in highly insoluble forms. Several bacterial genera, including PGPR and phytopathogens, synthesize siderophores, which are mainly hydroxamate or phenolate compounds, with high affinity for FeIII.[135] Phenolate-type siderophores have been reported as present in *Azotobacter* and *Azospirillum*. In most cases the siderophores identified were derivatives of dihydroxybenzoic acid (DHBA), and include azotochelin and aminochelin in *A. vinelandii*,[181] spirillobactin in *A. brasilense* RG,[3] and 2,3-DHBA and 3,5-DHBA lysine and leucine conjugates in *A. lipoferum* D2.[222] Production of salicylic acid was also reported in *A. lipoferum* D2.[222] The synthesis of a hydroxamate-type siderophore, azotobactin, related to *Pseudomonas* pseudobactin was reported in *A. vinelandii*.[47] Synthesis of siderophores in *Azotobacter* and *Azospirillum* is coupled with the synthesis of several membrane proteins, some of which are probably involved in iron transport.[3,179,222] In addition, siderophores are involved in Mo transport. In N_2 fixers, molybdenum, like iron, is directly involved in the structure and function of nitrogenase component 1. Molybdenum is also associated with nitrate reductase. These are indications that azotochelin plays a role in Mo transport in *A. vinelandii*.[180] Similarly, the synthesis of the 3,5-DHBA in *A. lipoferum* D2 and of a catechol-type siderophore in cowpea *Rhizobia* is enhanced under molybdenum starvation conditions.[185,223] It thus appears that, in these bacteria, uptake of Mo is mediated by the Fe transport system. The role of *Azospirillum* siderophore production as a competitive agent against soil bacteria has not been examined. In contrast, *A. vinelandii* siderophores charged with iron, enhanced growth of the phytopathogenic bacteria *A. tumefaciens* and *Erwinia carotovora*.[178] Thus, the behavior of *A. vinelandii*, which has a synergistic effect on the growth of plant pathogens, is different from that of the PGPR *P. putida*, which inhibits pathogens.[135]

iii. Bacteriocin Production

Production of bacteriocins by N_2 fixers has been examined in *Rhizobium* and in *Azospirillum*.[174] Although many *A. brasilense* and *A. lipoferum* strains are able to produce bacteriocins, it has not been demonstrated that production occurs in the soil. In addition, the activity of crude bacteriocin introduced into sterilized soils seemed to disappear rapidly.[174]

iv. Adsorption to Soil Particles

When soil is flooded, adsorption of free-living bacteria to soil particles prevents them from being washed down to deeper soil areas unfavorable for survival. *A. brasilense* Cd adsorbed strongly to soil and peat particles and could not be desorbed by washing.[14] However, in the presence of a bacterial attractant (e.g., a chemical gradient caused by exudates of plant roots), adsorption was decreased.[14]

B. Root Colonization Process

i. Motility and Chemotaxis

Azospirillum spp., *Pseudomonas* spp., *Herbaspirillum*, and *Campylobacter* have polar flagella, whereas *Azotobacter* spp. have peritrichous flagella.[57,138,151,219] These flagella confer motility, which is a prerequisite for chemotactic and/or aerotactic responses. Some *A. brasilense* mutants impaired in motility, obtained after Tn5–*lac* mutagenesis, were devoid of polar flagella (Fla⁻) and others had impaired chemotactic responses (Che⁻).[261] *Azospirillum* is attracted by low oxygen tension and many chemical substances. Barak and colleagues[10] established experimental conditions to distinguish between aerotactic and chemotactic responses. In the aerotactic response to oxygen gradient, *Azospirillum* tends to migrate to zones with low dissolved oxygen,[171] as observed in semisolid nitrogen-free medium, where the bacteria form a pellicle below the surface.[167] Chemoattractants for *A. brasilense* Cd include aspartic acid, glutamic acid, glycine, arabinose, galactose, succinate, and malate.[10] The main attractant for *A. lipoferum* is sucrose, which is not utilized by this bacterium.[108] Sucrose is produced in wheat root exudates, which also contain an invertase that hydrolyzes sucrose into fructose and glucose. These two sugars are growth substrates for *A. lipoferum*, and together they were found to be better chemoattractants than sucrose.[108] Experiments with various *Azospirillum* growth substrates, including organic acids and sugars, showed that several are not chemoattractant for the bacterium.[10,203] Thus, the chemotactic effect of a substrate does not necessarily reflect the ability of a bacterium to metabolize this substrate. Reinhold and colleagues[203] suggested that the chemotactic response was strain-specific. In particular, they observed that *A. brasilense* and *A. lipoferum* strains isolated from C4 plants were attracted by substances present at a high level in C4 plant root exudates. An analogous observation has been reported for *Azospirillum* strains isolated from C3 plants.[203]

ii. Attachment to the Root Cortex

Inoculation of pearl millet roots with different soil bacteria revealed that root hairs adsorbed *Azospirillum* more effectively than species of other genera.[258] These observations imply the presence of a specific recognition mechanism between *Azospirillum* and the host plant. Indeed, it was demonstrated that different *Azospirillum brasilense* and *A. lipoferum* strains bind to wheat germ agglutinin, the best studied lectin from gramineae. In contrast, soybean seed lectin was not recognized by *Azospirillum*.[46] Flow cytometry has been used to quantify the binding of fluorescein-labeled lectin to *A. brasilense*. The affinity of *A. brasilense* to agglutinins of different plants decreases in the following order: *Griffonia simplicifolia* II agglutinin, *Griffonia simplicifolia* I agglutinin, *Triticum vulgaris* agglutinin, *Glycine max* agglutinin, *Canavalia ensiformis* agglutinin, *Limax flavus* agglutinin, and *Lotus tetragonolobus* agglutinin.[281] These observations suggest that lectin recognition involves surface polysaccharides, as in the *Rhizobium*-legume symbiosis. The finding that genes of *A. brasilense* can complement the absence of alfalfa nodulating activity in exopolysaccharide-defective mutants of *R. meliloti*[155] is in agreement with this hypothesis.

Fimbriae. In the nitrogen-fixing enterobacteria *Klebsiella* and *E. agglomerans*, two types of fimbriae (type 1 and type 3) were reported to be involved in the attachment to roots of grasses (for references see ref. 98). Type 1 fimbriae are known to bind mannosides, like those found in the cell walls of plants,[59] but the antigenic receptor of type 3 fimbriae on plant roots has not been identified.

Fibrillar Material. A correlation between the percentage of fibrillated *R. leguminosarum* cells and their attachment to root hair tips has been reported,[244] suggesting that the production of fibrils is of major importance for adhesion of *Rhizobium* to the root surface. In electron micrographs of *Azospirillum* colonized roots, the bacteria appear to be anchored to the roots by fibrillar material.[88,258]

iii. Colonization of Root Tissue

Colonization of root tissue by diazotrophic bacteria is beneficial for both partners of the association. It lowers competition for substrates, since only a few bacterial species are able to penetrate the root. This results in a higher, specialized bacterial population and facilitates metabolite transfer between partners. Colonization of maize root tissue by *Azospirillum* has been demonstrated.[170] Using fluorescent antibody staining techniques, it has been established that *Azospirillum* colonizes the intercellular spaces between the epidermis and the cortex.[224] The same technique was used to demonstrate the colonization of monoaxenically grown rice by *Beijerinckia*.[50] The aerenchyma of the roots of field-grown *Spartina alterniflora* living in association with *Campylobacter nitrofigilis* is colonized by bacteria.[151] *Alcaligenes faecalis* strain A15, isolated from rice roots, has even been found in the cytoplasm of root cells.[283] The apparently intact structure of the aerenchymatic tissue colonized by *Azospirillum* raises the question of how bacteria penetrate the roots. It has been proposed that *Azospirillum* enters the roots at points of emergence of lateral roots.[186] This hypothesis is strongly supported by recent studies with colonized kallar grass roots.[202] Electron micrographs of root tissue of monoaxenically grown *Panicum maximum* revealed that *Azospirillum* seems to penetrate the middle lamella of cortical cells.[257] This is presumably an active process and might be dependent on pectinolytic enzymes, which *Azospirillum* is known to produce.[119,254,257]

C. Production of Phytohormones and Other Plant Growth-Promoting Substances

The production of phytohormones by plant-associated bacteria (reviewed in ref. 57) and phytopathogenic microorganisms[162] is well documented. In 1979, production of auxins, cytokininlike and gibberellinlike substances, was proposed for *A. brasilense*, since the increased number of root hairs and of lateral roots observed after inoculation with this bacterium could be mimicked by application of a mixture of indole-3-acetic acid (IAA), kinetin and giberillic acid GA_3.[255] Whereas IAA production by bacteria could be established by chromatographic techniques,[211,255] release of cytokininlike or gibberellinlike substances remained uncertain, since they were detected only by bioassays in very old cultures.[255] Experiments using the same techniques suggested that *A. vinelandii* releases cytokininlike and gibberellinlike substances in addition to IAA.[94]

i. Auxins

Auxins like IAA and indole-3-lactic acid can be easily detected by reversed-phase HPLC and colorimetric reactions.[251] Enzyme immunoassays[274] and specific bioassays are also available, such as auxin-stimulated ethylene formation by pea epicotyl sections.[137] Release of auxins by several soil bacteria into the culture medium has been studied intensively (e.g., *A brasilense*,[255] *A. chroococcum*,[61] *A. vinelandii*,[133] *A. paspali*,[12,31] and *Rhizobium* spp.[118,272]).

ii. Auxin Metabolism Mutants

To investigate the role of IAA production in plant-microbe interaction, IAA over- and low-producing mutants of *A. brasilense* were isolated by selecting for resistance or sensitivity to 5-fluoro-tryptophan.[106,146] IAA overproducing mutants had no altered effect on root growth of the host plant as compared to the wild-type strain.[117] However, low IAA producers had lost the ability to increase the number and the length of lateral roots.[11] Mutants of *A. lipoferum*, obtained by Tn5 random mutagenesis,[1] have not been tested for their effect on the morphology of wheat roots.

iii. Auxin Biosynthesis

Although biosynthesis of IAA in bacteria and in plants always uses tryptophan (Trp) as a precursor, there are several pathways for conversion of

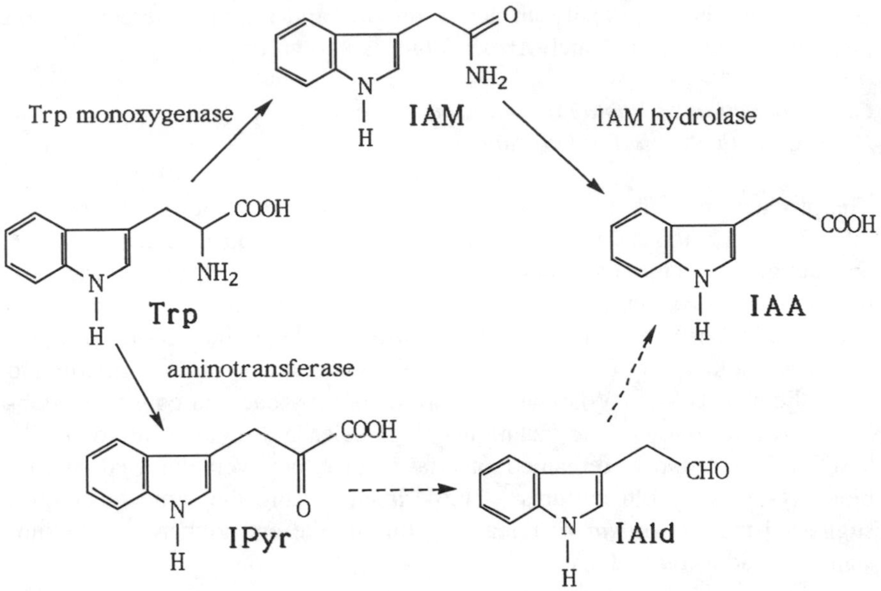

Figure 6-3. Pathways of indole-3-acetic acid (IAA) biosynthesis. See text for detail. Trp: tryptophan, IAM: indole-3-acetamide, IPyr: indole-3-pyruvate, IAld: indole-3-acetaldehyde.

Trp into IAA. In the best-investigated pathway, first described in the phytopathogen *P. syringae*.[144] Trp is converted into indole-3-acetamide (IAM) by a Trp monooxygenase (Figure 6-3). The second step is the formation of IAA, catalyzed by an IAM hydrolase. In *P. syringae*, the genes coding for Trp monooxygenase (*iaaM*) and for IAM hydrolase (*iaaH*) have been sequenced.[282] They are cotranscribed and localized on a 52-kb plasmid.[182] Genes encoding the same enzymes are located on the tumor-inducing plasmid (pTi) of *A. tumefaciens*.[122] Recently, the *B. japonicum* gene for IAM hydrolase, *bam*, was also isolated and sequenced.[231,232] It was shown that IAM can be converted to IAA by many different *Bradyrhizobium* spp. but not by *Rhizobium* spp.[230] No hybridization signal was detected between *A. brasilense* total DNA and intragenic probes from the IAA synthesis genes, *iaaM* and *iaaH* from *P. syringae* and *bam* from *Bradyrhizobium* (Zimmer and Elmerich, in preparation). As *Azospirillum* was additionally found to be unable to utilize IAM, it is likely that the pathway of IAA synthesis in *Azospirillum* is different from that of *Bradyrhizobium* and *P. syringae*. Another pathway of IAA synthesis, involving the intermediates indole-3-pyruvate (IPyr) and indole-3-acetaldehyde (IAld), is widespread in plants[160,225] and is also likely to exist in several soil bacteria (see Figure 6-3). The initial reaction by converting Trp into IPyr is catalyzed by nonspecific aromatic aminotransferases in bacteria. Four of these enzymes have been identified in *A. lipoferum*[216] and in *R. meliloti*[121] The rest of this pathway has not yet been detected in *Azospirillum* or *Rhizobium*, since the intermediates IPyr and IAld are unstable. However, the release of the corresponding reduced compounds, indole-3-lactate and indole-3-ethanol, has been reported for both bacteria.[41,69]

iv. Cytokinins

As in *Rhizobium* (reviewed in ref. 162), it is not clear to what extent *Azospirillum* produces cytokinins. Using the cytokinin-specific production of beta-cyanin by *Amaranthus* seedlings,[22] it was shown that the release of cytokinin into the medium of one- or three-day-old cultures of *A. brasilense* and *A. lipoferum* was lower than 0.5 ng/ml.[285] Attempts to identify cytokinins in cultures of *Azospirillum* by radioimmunoassays revealed only traces of these compounds.[124] Three cytokinins have been isolated recently from *Azotobacter* and tentatively identified as trans-zeatin, isopentenyladenosine, and isopentenyladenine by thin-layer chromatography, tobacco callus bioassay, and enzyme immunoassay.[250] The genes and enzymes involved in biosynthesis of bacterial cytokinins are known only in phytopathogens, such as *A. tumefaciens* and *P. syringae* (reviewed in ref. 162). One of the main reactions of the biosynthetic pathway is the transfer of an isoprenoid side chain to an adenine derivative, which is catalyzed by an isopentenyl transferase. Diversity in cytokinins produced by these organisms results from additional modifications, such as hydroxylation of the isopentenyl side chain and glucosylations.[136,162]

v. Gibberellins

As compared to data for gibberellin metabolism in plants,[107] there are only a few reports on gibberellin production in bacteria. Production of gibberellinlike substances by *A. brasilense* and *A. vinelandii* after prolonged cultivation (10 days) has been suggested,[94,255] since extracts of culture supernatants caused elongation of lettuce hypocotyls. Though this test has been described to be gibberellin specific,[84] it is worth noting the elongation of hypocotyls is often stimulated by auxins, which are also produced by these bacteria. Using the gibberellin-specific release of alpha-amylase by barley endosperm as the bioassay,[115] it was not possible to detect gibberellins in the culture medium of one- and three-day-old cultures of *A. brasilense* or *A. lipoferum*. The minimum sensitivity of this method was estimated to detect 25 pg GA_3 per ml of culture.[285] Recently, concentrations of 20–40 pg per ml culture of the gibberellins A_1, A_3, and iso-A_3 were detected in *A. lipoferum* cultures, by combined gas chromatography and mass spectrometry.[27]

vi. Other Plant Growth–Affecting Substances

In addition to IAA, abscisic acid has been detected by radioimmunoassay in supernatants of old *Azospirillum* cultures, but at concentrations as low as 0.4 ng/ml.[124] In addition to classical phytohormones, it has been shown that nitrite at a concentration range of 0.1–10 mM [e.g., that generated by the dissimilatory nitrate reductase of *Azospirillum* (Part IV.C)] mimics the effect of IAA in several plant tests for auxins.[285,287] Since the effect of nitrite could be enhanced by ascorbate, it was suggested that nitrite interacted with ascorbate in the plant cells and that the reaction product was responsible for the observed auxinlike response in bioassays.

D. Changes in Root Physiology and Morphology

Several soil bacteria cause changes in the root physiology and morphology of their host plant. The response of roots to colonization by associated bacteria has been studied with *Azospirillum*.[57,168,187]

i. Changes in Root Physiology

Wheat roots respond with an increased proton efflux after inoculation with *Azospirillum*.[15] As acidification of the rhizosphere was proposed to be a major mechanism for mobilization of minerals by plants,[213] the changed proton efflux of roots colonized by *Azospirillum* could contribute to the stimulated mineral uptake by the host plant described by several authors.[140,161] An additional change in root physiology observed with tomato seedling roots, inoculated with *Azospirillum*, was an enhanced specific activity of the enzymes involved in the tricarboxylic acid cycle.[101] In contrast, phythopathogenic organisms cause a reduction in the activity of these enzymes in infected plant tissues.[95]

ii. Changes in Root Morphology

The major morphological changes in roots of plants inoculated with *Azospirillum* are the increased number of root hairs and lateral roots,[255] a reduced space between the root tip and the root hair zone,[161] and an increase of branched root hairs.[111] *Sorghum* displayed these responses to an inoculum of 10^7–10^8 bacteria per plant. Higher concentrations, such as 10^9 bacteria per plant, cause inhibition of root development.[161] Most of the stimulation of root growth is assumed to result from the release of phytohormones by *Azospirillum* (Part V.C). By measuring the extent of the root network under field conditions, it was observed that wheat roots colonized by *Azospirillum* are concentrated in a smaller soil area than the roots of uninoculated wheat or plants inoculated with *Bacillus* sp.[128] However, though the roots of plants colonized by *Azospirillum* are not spread as far as the roots of uninoculated plants, the increased surface and the changes in root physiology confer to the plant a more efficient mineral uptake, which contributes to crop improvement (Part VII).

VI.   Genes Common to *Azospirillum* and *Rhizobiaceae*

In addition to *nif* and *fix* genes identified in *Azospirillum* (Part IV), other loci that may be involved in *Azospirillum*–grass interactions have been identified. In particular, loci homologous or functionally equivalent to *Rhizobium nod* and *exo* genes have been cloned. Interestingly, some of the common regions are plasmid-borne in *Azospirillum*.

A.   *Isolation and Phenotype of Exopolysaccharide Mutants*

i.   *Exopolysaccharide Involvement in Virulence and Nodulation in Rhizobium and Agrobacterium*

The attachment of bacteria to the root surface is a critical step in *Agrobacterium* virulence and *Rhizobium* nodulation ability, since mutants impaired in attachment either are noninfectious or produce ineffective nodules.[34,134,198] *Rhizobiaceae* produce two major exopolysaccharides (EPS), succinoglycan and beta-D-glucan. Several genes involved in EPS production have been identified in *Agrobacterium* and *Rhizobium* and are functionally equivalent in the two species.[34] The *exo* genes are essentially involved in succinoglycan production.[34,134] In *A. tumefaciens*, only EPS mutants affected in beta-D-glucan synthesis or secretion (*chvA*, *chvB*, and *exoC* mutants) are noninfectious.[34,35,198] These mutants are impaired in root surface attachment. Equivalent genes in *R. meliloti* (*ndvA*, *ndvB*, and *exoC*) appear to have an analogous function during nodulation, since their mutants show an altered attachment and induce the formation of nonfixing empty nodules.[60,134] In *R. meliloti*, the other major EPS, succinoglycan, also appears to be involved in effective nodulation, since *exo* mutants induce the formation of ineffective nodules on alfalfa.[134]

ii.   *Role of Exopolysaccharides in Azospirillum*

Azospirillum spp. synthesize exopolysaccharides and capsular polysaccharides.[46] Several *A. brasilense* and *A. lipoferum* mutants, which do not fluoresce under UV light in the presence of calcofluor (Cal⁻), have been isolated.[16,156] Most of these mutants were affected in EPS synthesis. Some of them synthesized a more acidic EPS that prevented flocculation in liquid medium.[156] However, they did not show any difference in root attachment or motility as compared to wild-type *Azospirillum* strains.[16,156] Michiels and colleagues[155] identified two *A. brasilense* Sp7 loci able to restore the Cal⁺ phenotype to *R. meliloti exoB* and *exoC* mutants. The two loci were called *exoB*- and *exoC*-correcting loci, respectively. In addition, the *exoB*-correcting locus could complement the *R. meliloti exoB* mutant defective for nodulation of alfalfa. These two *A. brasilense* loci appear not to be involved in calcofluor fluorescence since mutants of these loci still fluoresce under UV.[155] This suggests that these loci are not related to the Cal⁻ mutants described earlier. Surprisingly, the *exoB*- and *exoC*-correcting loci did not hybridize with *R. meliloti exoB* and *exoC* probes. Moreover, no sequence similar to *R. meliloti exoB* seems to be present in the *Azospirillum* genome.[175] *Azospirillum exoB*- and *exoC*-correcting loci mutants were not affected in root hair attachment.[155] *A. brasilense* Sp7 genomic DNA fragments hybridized with *A. tumefaciens chvA* and *chvB* and *R. meliloti exoC* probes.[175,271]

B. *Azospirillum nod* Genes

In *R. meliloti*, the genes necessary for nodulation include *nodDABCIJ*, referred to as the common nodulation (*nod*) genes,[126,141] and *nodQPGEFH*, referred to as host-specific nodulation (*hsn*) genes.[36,126,227] *A. brasilense* Sp7 genomic DNA fragments were found to hybridize with *R. meliloti* DNA probes containing either *nodDABC* or *nodPGEFH*.[81] Two *Azospirillum* fragments hybridizing with the *nodPGEFH* probe were cloned to yield plasmids pAB502 and pAB503.[67] As established by DNA sequencing, pAB502 contained genes homologous to *R. meliloti nodPQ* [Figure 6-2(3)]. These genes are functional in *Azospirillum*.[268] No particular phenotype can yet be attributed to *Azospirillum nodPQ*. Plasmid pAB503 contains a functional gene homologous to *R. meliloti nodG* (unpublished data) [Figure 6.2-4]. There is no obvious explanation for the presence of such genes in *Azospirillum*, since this bacterium does not have the same host specificity as *R. meliloti*. These *hsn* genes could have a more general function, and not only be involved in the nodulation process.

C. Role of Plasmids in the Bacteria-Plant Interaction

The *nod* and *exo* genes of *R. meliloti* are located on two different megaplasmids, except for *exoC* and *exoD*, which are chromosomal.[78] The *A. brasilense* Sp7 genes homologous to *R. meliloti nodPQ* as well as the *exoB*- and *exoC*-correcting loci are plasmid borne. Elmerich and colleagues[67] showed, by hybridization, that the *nodPQ* genes of *A. brasilense* Sp7 were carried by the 90-MDa resident plasmid (p90) (see Figure 6.1). Moreover, a probe carrying these genes hybridized with a plasmid of similar size in all *A. brasilense* strains tested, except strain RO7, where hybridization with the 115-MDa plasmid occurred.[269] This suggests that genes homologous to *Rhizobium nodPQ* are widespread in *Azospirillum* spp. The same conclusion can be drawn for *exoB*- and *exoC*-correcting loci, which also have been mapped on *A. brasilense* Sp7 p90[153,175] (see Figure 6.1). *R. meliloti nod* genes were shown to be involved in specific chemotaxis toward luteolin, the main inducer of *R. meliloti nod* genes,[32] whereas the *R. leguminosarum* symbiotic plasmid was involved in chemotaxis toward apigenin and naringenin.[2] *A. brasilense* Sp7 Tn5 mutants impaired in general chemotaxis were recently isolated.[261] Two of the mutants had a Tn5 insertion localized on the resident p90.[261] Moreover, nonmotile Tn5-mutant of strain Sp7 was isolated, in which the Tn5 insertion was carried by one of the two megaplasmids of more than 300 Mda.[261] Other functions involved in survival in the rhizosphere or in its colonization have been mapped to plasmids. Johnston and colleagues[114] described the existence of bacteriocinogenic plasmids in *R. leguminosarum* spp. In *R. phaseoli*, the ability to synthesize a brown pigment, probably melaninlike, is strongly associated with the nodulation ability and with the presence of one of the resident plasmids.[21] *Azospirillum* mutants unable to differentiate into C-forms can be obtained at high frequency.[16] However, this frequency could be increased by plasmid curing. All these observations suggested that several genes involved in bacteria-plant interaction and survival of the bacteria in the soil are plasmid borne in *Azospirillum*.

VII. Crop Improvement

Early field inoculation experiments performed with *A. chroococcum* showed improvements of 8–12% in maximum crop yield (reviewed in ref. 30). With time, it became clear that the slight improvement in plant growth observed in some experiments carried out in the presence of mineral fertilizers could be due not only to biological N_2 fixation but also to production phytohormones (Part V.C) and/or inhibition of phytopathogens.

A. Contribution of Biological Nitrogen Fixation to Crop Productivity

The ^{15}N isotope dilution method has been used to show that in the association between *A. paspali* and *Paspalum notatum*, 10% of the total nitrogen accumulated in the plant was contributed by biological N_2 fixation.[24] The ^{15}N isotope dilution technique was also applied to rice plants, which can be grown with little of no N fertilizer. It was shown that 20–30% of total nitrogen in rice plants were contributed by biological N_2 fixation.[266] Indeed, high numbers of several nitrogen-fixing bacteria could be isolated from the rhizosphere and the rhizoplane of rice. They are *A. irakense*,[119], *E. clo-*

acae,[130] *K. planticola*,[130] and *P. diazotrophicus*.[13,273] The recently identified *Acetobacter diazotrophicus*[93] isolated from roots and stems of sugar cane is assumed to be the major contributor to the 17% nitrogen supplied by biological fixation.[139] A comparative study, carried out with the ^{15}N dilution technique, showed that in temperate climates the diazotrophic *Bacillus* C-11-25 supplied more nitrogen to wheat than *A. brasilense*.[209]

B. Effect of Azospirillum on Crop Yield

Depending on the conditions chosen (e.g., size of the bacterial inoculum, time of inoculation, amount and concentration of organic matter and of minerals in the soil, combination of bacterial strains and the host plant), increases of 10–30% were reported in grain and forage yields with wheat, corn, and sorghum, after inoculation with *Azospirillum*.[172,173] The effect on the growth of cereals could be reinforced when a combined inoculation with *Azospirillum* and a vesicular-arbuscular mycorrhizal fungus (VAM) was used.[249] This was thought to be due to an enhanced N and P uptake by the plant, facilitated by the associative organisms. Indeed, crop yield of plants inoculated with *A. brasilense* and the VAM-fungus *Glomus fasciculatum* and grown without fertilizer were comparable to that of noninoculated plants supplemented with N and P fertilizers.[176] In addition, the enhanced level of Fe and Zn in the inoculated plants indicated that the associative organisms also favored the uptake by the host plant of other soil minerals. It was also stated that the establishment and persistence of *A. brasilense* in the endorhizosphere of sorghum was enhanced when roots were colonized by *Glomus*.[177] Field inoculation of sorghum separately performed with *Herbaspirillum seropedicae* and *Azospirillum* spp. both originally isolated from sorghum roots, resulted in an increased plant dry weight only in the case of *Azospirillum* spp.[191] Moreover, 60 days after inoculation, only the *Azospirillum* spp. were found to be established among the natural root-associated bacteria.

With the exception of some highly specific associations, it thus appears that, for obtaining a significant increase in crop yield by inoculation with nitrogen-fixing bacteria, it is necessary to select a suitable bacterial strain for a particular environment and a particular plant. Otherwise, inoculation can be ineffective or may even result in decreased yields, as has been observed in several comparative studies.[128,129,191,209]

Acknowledgments

The authors wish to thank Prof. J.-P. Aubert and Dr. I. Onyeocha for helpful discussions, and Mrs. M. Ferrand for typing the manuscript. Unpublished experiments from our laboratory were performed in collaboration with Drs. M. de Zamaroczy and I. Onyeocha, Messrs. F. Delorme and Y. Y. Liang.

WZ is the recipient of a postdoctoral fellowship from the Deutsche Forschungs Gemeinschaft and CV is appointed by the French Ministry of Agriculture.

References

1. Abdel-Salam, M.S., and Klingmüller, W., "Transposon Tn5 mutagenesis in *Azospirillum lipoferum*: isolation of indole acetic acid mutants," *Mol. Gen. Genet.*, *210*, 165–170 (1987).

2. Armitage, J. P., Gallagher, A., and Johnston, A. W. B., "Comparison of the chemotactic behaviour of *Rhizobium leguminosarum* with and without the nodulation plasmid," *Molecular Microbiology 2*, 743–748 (1988).

3. Bachhawat, A. K., and Ghosh, S., "Iron transport in *Azospirillum brasilense*: role of the siderophore spirilobactin," *J. Gen. Microbiol.*, *133*, 1759–1765 (1987).

4. Bagley, S. T., Seidler, R. J., and Brenner, D. J., "*Klebsiella planticola* sp. nov.: a new species of *Enterobacteriaceae* found primarily in non-clinical environments," *Curr. Microbiol.*, *6*, 105–109 (1981).

5. Balandreau, J., "Microbiology of the association," *Can. J. Microbiol.*, *29*, 851–859 (1983).

6. Baldani, J. L., Baldani, V. L. D., Seldin, L., and Döbereiner, J., "Characterization of *Herbaspirillum seropedicae* gen. nov., sp. nov., a root associated nitrogen fixing bacterium," *Int. J. Syst. Bacteriol.*, *36*, 86–93 (1986).

7. Bally, R., and Givaudan, A., "Mobilisation and transfer of *Azospirillum lipoferum* plasmid by the Tn5-Mob transposon into a plasmid-free *Agrobacterium tumefaciens* strain," *Can. J. Microbiol.*, *34*, 1354–1357 (1988).

8. Bally, R., Thomas-Bauzon, D., Heulin, T., Balandreau, J., Richard, C. and De Ley, J., "Determination of the most frequent N_2-fixing bacteria in the rice rhizosphere," *Can. J. Microbiol.*, *29*, 881–887 (1983).

9. Bani, D., Barberio, C., Bazzicalupo, M., Favilli, F., Gallori, E., and Polsinelli, M., "Isolation and characterization of glutamate synthase mutants of *Azospirillum brasilense*," *J. Gen. Microbiol.*, *119*, 239–244 (1980).

10. Barak, R., Nur, I., and Okon, Y., "Detection of chemotaxis in *Azospirillum brasilense*," *J. Appl. Microbiol.*, *53*, 399–403 (1983).

11. Barbieri, P., Zanelli, T., Galli, E., and Zanetti, G., "Wheat inoculation with *Azospirillum brasilense* Sp6 and some mutants altered in nitrogen fixation and indole-3-acetic acid production," *FEMS Microbiol. Lett.*, *36*, 87–90 (1986).

12. Barea, J. M., and Brown, M. E., "Effects on plant growth produced by *Azotobacter paspali* related to synthesis of plant regulating substances," *J. Appl. Bacteriol.*, *37*, 583–593 (1974).

13. Barraquio, W. L., Ladha, J. K., and Watanabe, I., "Isolation and identification of N_2-fixing *Pseudomonas* associated with wetland rice," *Can. J. Microbiol.*, *29*, 867–873 (1983).

14. Bashan, Y., and Levanony, H., "Adsorption of the rhizosphere bacterium *Azospirillum brasilense* Cd to soil, sand and peat particles," *J. Gen. Microbiol.*, *134*, 1811–1820 (1988).

15. Bashan, Y., Levanony, H., and Mitiku, G., "Changes in proton efflux of intact wheat roots induced by *Azospirillum brasilense* Cd," *Can. J. Microbiol.*, *35*, 691–697 (1989).

16. Bastarrachea, F., Zamudio, M., and Rivas, R., "Non-encapsulated mutants of *Azospirillum brasilense* and *Azospirillum lipoferum*," *Can. J. Microbiol.*, *34*, 24–29 (1988).

17. Bazzicalupo, M., Fani, R., Gallori, E., Turbanti, L., and Polsinelli, M., "Cloning the histidine, pyrimidine and cysteine genes of *Azospirillum brasilense*: Expression of pyrimidine and three clustered histidine genes in *Escherichia coli*," *Mol. Gen. Genet.*, *206*, 76–80 (1987).

18. Bazzicalupo, M., and Galori, E., "Genetic analysis in *Azospirillum*," *Azospirillum II: Genetics, Physiology, Ecology*, W. Klingmüller, ed., Basel: Birkhäuser Verlag, 1983, EXS 48, pp. 24–28.

19. Berg, R. H., Tyler, M. E., Novick, N. J., Vasil, V., and Vasil, I. K., "Biology of *Azospirillum*-sugarcane association: enhancement of nitrogenase activity," *Appl. Environ. Microbiol.*, *39*, 642–649 (1980).

20. Berlier, Y. M., and Lespinat, P. A., "Mass-spectrometric kinetic studies of the nitrogenase and hydrogenase activities in *in vivo* cultures of *Azospirillum brasilense* Sp7. *Arch. Microbiol.*, *125*, 67–72 (1980).

21. Beynon, J. L., Beringer, J. E., and Johnston, A. W. B., "Plasmids and host-range in *Rhizobium leguminosarum* and *Rhizobium phaseoli*," *J. Gen. Microbiol.*, *120*, 421–429 (1980).

22. Biddington, N. L., and Thomas, T. H., "A modified *Amaranthus betacyanin* bioassay for rapid determination of cytokinins in plant extracts," *Planta 111*, 183–186 (1973).

23. Bleakley, B. H., Gaskins, M. H., Hubbell, D. H., and Zam, S. G., "Floc formation by *Azospirillum lipoferum* grown on poly-beta-hydroxybutyrate," *Appl. Environ. Microbiol.*, *54*, 2986–2995 (1988).

24. Boddey, R. M., Chalk, P. M., Victoria, R. L., Matsui, E., and Döbereiner, J., "The use of the ^{15}N isotope dilution technique to estimate the contribution of associated nitrogen fixation nutrition of *Paspalum notatum* cv. batatais," *Can. J. Microbiol.*, *29*, 1036–1045 (1983).

25. Boddey, R. M., and Döbereiner, J., "Association of *Azospirillum* and other diazotrophs with tropical gramineae," *Nonsymbiotic Nitrogen Fixation and Organic Matter in the Tropics*, New Dehli, Indian Society of Soil Science, 1982, pp. 28–47.

26. Bothe, H., Klein, B., Stephan, M. P., and Döbereiner, J., "Transformations of Inorganic Nitrogen by *Azospirillum* ssp," *Arch. Microbiol.*, *130*, 96–100 (1981).

27. Bottini, R., Fulchieri, M., Pearce, D., Pharis, R. P., "Identification of gibberellins A1, A3, and iso-A3 in cultures of *Azospirillum lipoferum*," *Plant Physiol.*, *90*, 45–47 (1989).

28. Bozouklian, H., and Elmerich, C., "Nucleotide sequence of the *Azospirillum brasilense* Sp7 glutamine synthetase structural gene," *Biochimie 68*, 1181–1187 (1986).

29. Bozouklian, H., Fogher, C., and Elmerich, C., "Cloning and characterization of the *glnA* gene of *Azospirillum brasilense* Sp7," *Ann. Inst. Pasteur/Microbiol.*, *137*, 3–18 (1986).

30. Brown, M. E., "Seed and root bacterization," *Ann. Rev. Phytopath.*, *12*, 181–197 (1974).

31. Brown, M. E., "Role of *Azotobacter paspali* in association with *Paspalum notatum*," *J. Appl. Bacteriol.*, *40*, 341–348 (1976).

32. Caetano-Anollés, G., Crist-Estes, D. K., and Bauer, W., "Chemotaxis of *Rhizobium meliloti* to the plant flavone luteolin requires functional nodulation genes," *J. Bacteriol.*, *170*, 3164–3169 (1988).

33. Cali, B. M., Micca, J. L., and Stewart, V., "Genetic regulation of nitrate assimilation in *Klebsiella pneumoniae* M5al," *J. Bacteriol.*, *171*, 2666–2672 (1989).

34. Cangelosi, G. A., Hung, L., Puvanesarajah, V., Stacey, G., Ozga, D. A., Leigh, J. A., and Nester, E. W., "Common loci for *Agrobacterium tumefaciens* and *Rhizobium meliloti* exopolysaccharide synthesis and their roles in plant interactions," *J. Bacteriol.*, *169*, 2086–2091 (1987).

35. Cangelosi, G. A., Martinetti, G., Leigh, J. A., Lee, C., Theines, C., and Nester, E. W., "Role of *Agrobacterium tumefaciens* ChvA protein in export of beta-1,2-glucan," *J. Bacteriol.*, *171*, 1609–1615 (1989).

36. Cervantes, E., Sharma, S. B., Maillet, F., Vasse, J., Truchet, G., and Rosenberg, C., "The *Rhizobium meliloti* host-range *nodQ* gene encodes a protein which shares homology with translational elongation and initiation factors," *Mol. Microbiol. 3*, 745–755 (1989).

37. Chan, Y.-K., "Denitrification by a diazotrophic *Pseudomonas* species," *Can. J. Microbiol.*, *31*, 1136–1141 (1985).

38. Chan, Y. K., Nelson, L. M., and Knowles, R., "Hydrogen metabolism of *Azospirillum brasilense* in nitrogen-free medium," *Can. J. Microbiol.*, *26*, 1126–1131 (1980).

39. Cole, J. A., "Assimilatory and dissimilatory reduction of nitrate to ammonia," *Symp. Soc. Gen. Microbiol.*, *42*, 281–329 (1987).

40. Colonna-Romano, S., Riccio, A., Guida, M., Defez, R., Lamberti, A., Iaccarino, M., Arnold, W., Priefer, U., and Pühler, A., "Tight linkage of *glnA* and a putative regulatory gene in *Rhizobium leguminosarum*," *Nucl. Acids Res.*, *15*, 1951–1964 (1987).

41. Crozier, A., Arruda, P., Jasmin, J. M., Montero, A. M., and Sandberg, G., "Analysis of indole-3-acetic acid and related indoles in culture medium from *Azospirillum lipoferum* and *Azospirillum brasilense*," *Appl. Environ. Microbiol.*, *54*, 2833–2837 (1988).

42. Danneberg, G., "Untersuchungen zur Denitrification an *Azospirillum brasilense* Sp7," Thesis, Cologne, FRG, 1987, p. 128.

43. Danneberg, G., Kronenberg, A., Neuer, G., and Bothe, H., "Aspects of nitrogen fixation and denitrification by *Azospirillum*," *Plant Soil 90*, 193–202 (1986).

44. Danneberg, G., Zimmer, W., and Bothe, H., "Energy transduction efficiencies in nitrogenous oxide respiration of *Azospirillum brasilense* Sp7," *Arch. Microbiol.*, *151*, 445–453 (1989).

45. Del Gallo, M., Gratani, L., and Morpurgo, G., "Mutation in *Azospirillum brasilense*," *Azospirillum III: Genetics, Physiology, Ecology*, Klingmüller, W., ed. Berlin: Springer-Verlag, 1985, pp. 85–97.

46. Del Gallo, M., Negi, M., and Neyra, C. A., "Calcofluor- and lectin-binding exocellular polysaccharides of *Azospirillum brasilense* and *Azospirillum lipoferum*," *J. Bacteriol.*, *171*, 3504–3510 (1989).

47. Demange, P., Bateman, A., Dell, A., and Abdallah, M., "Structure of azotobactin D, a siderophore of *Azotobacter vinelandii* strain D (CCM 289)," *Biochemistry 27*, 2745–2752 (1988).

48. de Zamaroczy, M., Delorme, F., and Elmerich, C., "Regulation of transcription and promoter mapping of the structural genes of nitrogenase (*nifHDK*) of *Azospirillum brasilense* Sp7," *Mol. Gen. Genet.*, *220*, 88–94 (1989).

49. de Zamaroczy, M., Delorme F., and Elmerich, C., "Characterization of three different nitrogen regulated promoters regions for the expression of *glnB* and *glnA* in *Azospirillum brasilense*," *Mol. Gen. Genet.*, *224*, 421–430 (1990).

50. Diem, H. G., Schmidt, E. L., and Dommergues, Y. R., "The use of fluorescent-antibody technique to study the behaviour of *Beijerinckia* isolate in the rhizosphere and spermosphere of rice," *Ecol. Bull.*, Stockholm, *26*, 312–318 (1978).

51. Dixon, R. O. D., "Hydrogenase in legume root nodule bacteroîds: occurrence and properties," *Arch. Mikrobiol.*, *85*, 193–201 (1972).

52. Döbereiner, J., "Isolation and identification of root associated diazotrophs," *Nitrogen fixation with non-legumes*, F. A. Skinner, R. M. Boddey, and I. Fendrik, eds., Dordrecht: Kluwer Academic Press, 1989, pp. 103–108.

53. Döbereiner, J., and Day, J. M., "Associative symbioses in tropical grasses: Characterization of microorganisms and dinitrogen fixing sites," *Proceedings of the 1st International Symposium on Nitrogen Fixation*, Newton, W. E., and Nyman, C. J., eds. Pullman: Washington State University Press, 1976, pp. 518–538.

54. Döbereiner, J., Day, J. M., and Dart, P. J., "Nitrogenase activity in the rhizosphere of sugar cane and other tropical grasses," *Plant Soil*, *37*, 191–196 (1972).

55. Döbereiner, J., Day, J. M., and Dart, P. J., "Rhizosphere associations between grasses and nitrogen-fixing bacteria: effect of O_2 on nitrogenase activity in the rhizosphere of *Paspalum notatum*," *Soil Biol. Biochem.*, *5*, 157–159 (1973).

56. Döbereiner, J., Marriel, I. E., and Nery, M., "Ecological distribution of *Spirillum lipoferum* Beijerinck," *Can. J. Microbiol.*, *22*, 1464–1476 (1976).

57. Döbereiner, J., and Pedrosa, F. O., "Nitrogen-fixing bacteria in non-leguminous crop plants," *Brock/Springer Series in Contemporary Bioscience*. Madison, WI: Science Tech. Publishers; Berlin: Springer Verlag, 1987.

58. Dommergues, Y., Balandreau, J., Rinaudo, G., and Weinhard, P., "Non-symbiotic nitrogen fixation in the Rhizospheres of rice, maize and different tropical grasses," *Soil Biol. Biochem.*, *5*, 83–89 (1973).

59. Duguid, J. P., and Old, D. C., "Adhesive properties of enterobacteriaceae," *Bacterial adherence, receptors and recognition*, Beachey, E. H. ed. London: Chapman & Hall, series B, vol. 6, 1980, pp. 185–217.

60. Dylan, T., Ielpi, L., Stanfield, S., Kashyap, L., Douglas, C., Yanofsky, M., Nester, E., Helinski, D. R., and Ditta, G., "*Rhizobium meliloti* genes required for nodule development are related to chromosomal virulence genes in *Agrobacterium tumefaciens*," *Proc. Natl. Acad. Sci. USA*, *83*, 4403–4407 (1986).

61. El-Essawy, A. A., El-Sayed, M. A., Mohamed Y. A. H., and El-Shanshoury, A., "Effect of combined nitrogen in the production of plant growth regulators by *Azotobacter chroococcum*," *Zbl. Microbiol.*, *139*, 327–333 (1984).

62. Elmerich, C., "Génétique et régulation de la fixation de l'azote," *Phys. Vég.*, *17*, 883–906 (1979).

63. Elmerich, C., "*Azospirillum* genetics," *Molecular Genetics of the Bacteria-Plant Interaction*, A. Pühler, ed. Berlin: Springer Verlag, 1983, pp. 367–372.

64. Elmerich, C., "Molecular biology and ecology of diazotrophs associated with nonleguminous plants," *Biotechnology*, *2*, 967–978 (1984).

65. Elmerich, C., "*Azotobacter* and *Azospirillum* genetics and molecular biology," *Current Developments in Biological Nitrogen Fixation*, N.S. Subba Rao, ed. New Delhi: Oxford & IBH Publishers, 1984, pp. 315–346.

66. Elmerich, C., *"Azospirillum," Nitrogen Fixation-Molecular Biology*, W. J. Broughton and A. Pühler, eds. Oxford: Clarendon Press, 1986, pp. 106–126.

67. Elmerich, C., Bozouklian, H., Vieille, C., Fogher, C., Perroud, B., Perrin, A., and Vanderleyden, J., *"Azospirillum*: genetics of nitrogen fixation and interaction with plants," *Phil. Trans. R. Soc.*, *317*, 183–192 (1987).

68. Elmerich, C., Quiviger, B., Rosenberg, C., Franche, C., Laurent, P., and Döbereiner, J., "Characterization of a temperate bacteriophage for *Azospirillum*," *Virology*, *122*, 29–37 (1982).

69. Ernstsen, A., Sandberg, G., Crozier, A., and Wheeler, C. T., "Endogenous indoles and the biosynthesis and metabolism of indole-3-acetic acid in cultures of *Rhizobium phaseoli*," *Planta 171*, 422–428 (1987).

70. Eskew, D. L., Focht, D. D., and Ting, I. P., "Nitrogen fixation, denitrification, and pleomorphic growth in a highly pigmented *Spirillum lipoferum*," *Appl. Environ. Microbiol.*, *34*, 582–585 (1977).

71. Fahsold, R., Singh, M., and Klingmüller, W., "Cosmid cloning of nitrogenase structural genes of *Azospirillum lipoferum*," *Azospirillum III: genetics, physiology, ecology*, W. Klingmüller, ed. Berlin: Springer Verlag, 1985, pp. 30–40.

72. Falk, E. C., Döbereiner, J., Johnson, J. L., and Krieg, N. R., "Deoxyribonucleic acid homology of *Azospirillum amazonense* Magalhaes *et al.* 1984 and emendation of the description of the genus *Azospirillum*," *Int. J. Syst. Bacteriol.*, *35*, 117–118 (1985).

73. Fani, R., Allotta, G., Bazzicalupo, M., Ricci, F., Schipani, C., and Polsinelli, M., "Nucleotide sequence of the gene encoding the nitrogenase iron protein (*nifH*) of *Azospirillum brasilense* and identification of a region controlling *nifH* transcription," *Mol. Gen. Genet.*, *220*, 81–87 (1989).

74. Fani, R., Bazzicalupo, M., Coianiz, P., and Polsinelli, M., "Plasmid transformation of *Azospirillum brasilense*," *FEMS Microbiol. Lett.*, *35*, 23–27 (1986).

75. Fani, R., Bazzicalupo, M., Damiani, G., Bianchi, A., Schipani, C., Sgaramella, V., and Polsinelli, M., "Cloning of the histidine genes of *Azospirillum brasilense*: Organization of the *ABFH* gene cluster and nucleotide sequence of the *hisB* gene," *Mol. Gen. Genet.*, *216*, 224–229 (1989).

76. Fani, R., Bazzicalupo, M., Gallori, E., Turbanti, L., and Polsinelli, M., "Construction of a gene bank of *Azospirillum brasilense*," *Azospirillum III: Genetics, Physiology, Ecology*, Klingmüller, W., ed. Berlin: Springer Verlag, 1985, pp. 1–9.

77. Fani, R., Bazzicalupo, M., Ricci, F., Schipani, C., and Polsinelli, M., "A plasmid vector for the selection and study of transcription promoters in *Azospirillum brasilense*," *FEMS Microbiol. Lett.*, *50*, 271–276 (1988).

78. Finan, T. M., Kunkel, B., De Vos, G. F., and Signer, E. R., "Second symbiotic megaplasmid in *Rhizobium meliloti* carrying exopolysaccharide and thiamine synthesis genes," *J. Bacteriol.*, *167*, 66–72 (1986).

79. Fischer, M., Levy, E., and Geller, T., "Regulatory mutation that controls *nif* expression and histidine transport in *Azospirillum brasilense*," *J. Bacteriol.*, *167*, 423–426 (1986).

80. Fogher, C., Bozouklian, H., Bandhari, S. K., and Elmerich, C., "Construction of a genomic library of *Azospirillum brasilense* and cloning of the glutamine synthetase gene." In *Azospirillum III: genetics, physiology, ecology*, Klingmüller, W., ed. Berlin: Springer Verlag, 1985, pp. 41–57.

81. Fogher, C., Dusha, I., Barbot, P., and Elmerich, C., "Heterologous hybridization of *Azospirillum* DNA to *Rhizobium nod* and *fix* genes," *FEMS Microbiol. Lett.*, *30*, 245–249 (1985).

82. Franche, C., Canelo, E., Gauthier, D., and Elmerich, C., "Mobilization of the chromosome of *Azospirillum brasilense* by plasmid R68-45," *FEMS Microbiol. Lett.*, *10*, 199–202 (1981).

83. Franche, C., and Elmerich, C., "Physiological properties and plasmid content of several strains of *Azospirillum brasilense* and *Azospirillum lipoferum*. Ann. Microbiol. (Inst. Pasteur), *132A*, 3–18 (1981).

84. Frankland, B., and Warening, P. F., "Effect of gibberellic acid on hypocotyl growth of lettuce seedlings," *Nature 185*, 225–226 (1960).

85. Fu, H., and Burris, R., "Ammonium inhibition of nitrogenase activity in *Herbaspirillum seropedicae*," *J. Bacteriol.*, *171*, 3168–3175 (1989).

86. Fu, H. A., Fitzmaurice, W. P., Roberts, G. P., and Burris, R. H., "Cloning and expression of *draTG* genes from *Azospirillum lipoferum*," *Gene 86*, 95–98 (1990).

87. Fu, C., and Knowles, R., "Oxygen tolerance of uptake hydrogenase in *Azospirillum* spp.," *Can. J. Microbiol.*, *32*, 897–900 (1986).

88. Gafny, R., Okon, Y., and Kapulnik, Y., "Absorption of *Azospirillum brasilense* to corn roots," *Soil Biol. Biochem.*, *18*, 69–75 (1986).

89. Galimand, M., Perroud, B., Delorme, F., Paquelin, A., Vieille, C., Bozouklian, H., and Elmerich, C., "Identification of DNA regions homologous to nitrogen fixation genes *nifE*, *nifUS* and *fixABC* in *Azospirillum brasilense* Sp7," *J. Gen. Microbiol.*, *135*, 1047–1059 (1989).

90. Gauthier, D., and Elmerich, C., "Relationship between glutamine synthetase and nitrogenase in *Spirillum lipoferum*," *FEMS Microbiol. Lett.*, *2*, 101–104 (1977).

91. Germida, J. J., "Spontaneous induction of bacteriophage during growth of *Azospirillum brasilense* in complex media," *Can. J. Microbiol.*, *30*, 805–808 (1984).

92. Germida, J. J., "Population dynamics of *Azospirillum brasilense* and its bacteriophage in soil," *Plant Soil*, *90*, 117–128 (1986).

93. Gillis, M., Kersters, K., Hoste, B., Janssens, D., Kroppenstedt, R. M., Stephan, M. P., Teixeira, K. R. S., Döbereiner, J. and de Ley J., "*Acetobacter diazotrophicus* sp. nov., a nitrogen-fixing acetic acid bacterium associated with sugarcane," *Int. J. Syst. Bacteriol.*, *39*, 361–364 (1989).

94. Gonzalez-Lopez, J., Salmeron, V., Martinez-Toledo, M. V., Ballesteros, F., and Ramos-Cormenzana, A., "Production of auxins, gibberellins and cytokinins by *Azotobacter vinelandii* ATCC 12837 in chemically-defined media and dialysed soil media," *Soil Biol. Biochem.*, *18*, 119–120 (1986).

95. Goodman, R. N., Kiraly, Z., and Wood, K. R., *The Biochemistry and Physiology of Plant Disease*, Columbia: University of Missouri Press, 1986.

96. Gowda, T. K. S. and Watanabe, I., "Hydrogen-supported N_2 fixation of *Pseudomonas* sp. and *Azospirillum lipoferum* under free-living conditions and in association with rice seedlings," *Can. J. Microbiol.*, *31*, 317–321 (1985).

97. Haahtela, K., Helander, I., Nurmiaho-Lassila, E.-L., and Sundman, V., "Morphological and physiological characteristics and lipopolysaccharide composition of N_2-fixing (C_2H_2-reducing) root-associated *Pseudomonas* sp.," *Can J. Microbiol.*, *29*, 874–880 (1983).

98. Haahtela, K., and Korhonen, T. K., "*In vitro* adhesion of N_2-fixing enteric bacteria to roots of grasses and cereals," *Appl. Environ. Microbiol.*, *49*, 1186–1190 (1985).

99. Haahtela, K., Waartiovaara, T., Sundman, V., and Skujins, J., "Root-associated N_2-fixation (acetylene reduction) by *Enterobacteriaceae* and *Azospirillum* strains in cold climate spodosols," *Appl. Environ. Microbiol.*, *41*, 203–206 (1981).

100. Haas, D., and Holloway, B., "Chromosome mobilisation by plasmid R68-45: a tool in *Pseudomonas* genetics," *Mol. Gen. Genet.*, *158*, 229–237 (1978).

101. Hadas, R., and Okon, Y., "Effect of *Azospirillum brasilense* inoculation on root morphology and respiration in tomato seedlings," *Biol. Fertil. Soils 5*, 241–247 (1987).

102. Hardy, R. W. F., Burns, R. C., and Holsten, R. D., "Application of the acetylene-ethylene reduction assay for measurement of nitrogen fixation," *Soil Biol. Biochem.*, *5*, 47–81 (1973).

103. Hartmann, A., and Burris, R. H., "Regulation of nitrogenase activity by oxygen in *Azospirillum brasilense* and *Azospirillum lipoferum*," *J. Bacteriol.*, *169*, 944–948 (1987).

104. Hartmann, A., Fu, H., and Burris, R. H., "Regulation of nitrogenase activity by ammonium chloride in *Azospirillum* spp.," *J. Bacteriol.*, *165*, 864–870 (1986).

105. Hartmann, A., and Kleiner, D., "Ammonium (Methylammonium) transport by *Azospirillum* spp.," *FEMS Microbiol. Lett.*, *15*, 65–67 (1982).

106. Hartmann, A., Singh, M., and Klingmüller, W., "Isolation and characterization of *Azospirillum* mutants excreting high amounts of indoleacetic acid," *Can. J. Microbiol.*, *29*, 916–923 (1983).

107. Hedden, P., MacMillan, J., and Phinney, B. O., "The metabolism of the gibberellins," *Ann. Rev. Plant Physiol.*, *29*, 149–192 (1978).

108. Heinrich, D., and Hess, D., "Chemotactic attraction of *Azospirillum lipoferum* by wheat roots and characterization of some attractants," *Can. J. Microbiol.*, *31*, 26–31 (1985).

109. Hirota, Y., Fujii, T., Sano, Y. and Iyama, S., "Nitrogen fixation in the rhizosphere of rice," *Nature*, *276*, 416–417 (1978).

110. Hochstein, L. I., and Tomlinson, G. A., "The enzymes associated with denitrification," *Ann. Rev. Microbiol.*, *42*, 231–261 (1988).

111. Jain, D. K., and Patriquin, D. G., "Characterization of a substance produced by *Azospirillum* which causes branching of wheat root hairs," *Can. J. Microbiol.*, *31*, 206–210 (1985).

112. Jara, P., Quiviger, B., Laurent, P., and Elmerich, C., "Isolation and genetic analysis of *Azospirillum brasilense* Nif mutants. *Can. J. Microbiol.*, *29*, 968–972 (1983).

113. Jeter, R. M., Sias, S. R., and Inraham, J. L., "Chromosomal location and function of genes affecting *Pseudomonas aeruginosa* nitrate assimilation," *J. Bacteriol.*, *157*, 673–677 (1984).

114. Johnston, A. W. B., Beynon, J. L., Buchanan-Wollaston, A. V., Setchell, S. M., Hirsch, P. R., and Beringer, J. E., "High frequency transfer of nodulation ability between strains and species of *Rhizobium*," *Nature 276*, 634–636 (1978).

115. Jones, R. L., and Varner, J. E., "The bioassay of gibberellins," *Planta 72*, 155–161 (1967).

116. Kado, S. T., and Liu, C. I., "Rapid procedure for detection and isolation of large and small plasmids," *J. Bacteriol.*, *145*, 1365–1373 (1981).

117. Kapulnik, Y., Okon, Y., and Henis, Y., "Changes of the root morphology of wheat caused by *Azospirillum* inoculation," *Can. J. Microbiol.*, *31*, 881–887 (1985).

118. Kefford, N. P., Brockwell, J., and Zwar, J. A., "The symbiotic synthesis of auxin by legumes and nodule bacteria and its role in nodule development," *Aust. J. Biol. Sci.*, *13*, 456–467 (1960).

119. Khammas, K. M., Ageron, E., Grimont, P. A. D., and Kaiser, P., "*Azospirillum irakense* sp. nov., a nitrogen-fixing bacterium associated with rice roots and rhizosphere soil," *Res. Microbiol.*, *140*, 679–693 (1989).

120. Kim, Y-M., Ahn, K-J., Beppu, T., and Uozumi T., "Nucleotide sequence of the *nifLA* operon of *Klebsiella oxytoca* NG13 and characterization of the gene products," *Mol. Gen. Genet.*, *205*, 253–259 (1986).

121. Kittell, B. L., Helinski, D. R., and Ditta, G. S., "Aromatic aminotransferase activity and indoleacetic acid production in *Rhizobium meliloti*," *J. Bacteriol.*, *171*, 5458–5466 (1989).

122. Klee, H., Montoya, A., Horodyski, F., Lichtenstein, C., Garfinkel, D., Fuller, S., Flores, C., Peschon, J., Nester, E., and Gordon, M., "Nucleotide sequence of the *tms* genes of the pTiA6NC octopine Ti plasmid: Two gene products involved in plant tumorigenesis," *Proc. Natl. Acad. Sci. USA*, *81*, 1728–1732 (1984).

123. Kleeberger, A., Castroph, H., and Klingmüller, W., "The rhizosphere microflora of wheat and barley with special reference to Gram-negative bacteria," *Arch. Microbiol.*, *136*, 306–311 (1983).

124. Kolb, W., and Martin, P., "Response of plant roots to inoculation with *Azospirillum brasilense* and to application of indoleacetic acid." In *Azospirillum III: genetics, physiology, ecology*, Klingmüller, W., ed. Berlin: Springer Verlag, 1985, pp. 215–221.

125. Kole, M. M., Page, W. J., and Altosaar, I., "Distribution of *Azotobacter* in eastern canadian soils and in association with plant rhizospheres," *Can. J. Microbiol.*, *34*, 815–817 (1988).

126. Kondorosi, A., Horvath, B., Göttfert, M., Putnoky, P., Rostas, K., Györgypal, Z., Kondorosi, E., Bachem, C., John, M., Schmidt, J., and Schell, J., "Identification and organization of *Rhizobium meliloti* genes relevant to the initiation and development of nodules." In *Nitrogen Fixation Research Progress*, Evans, H. J., Bottomley, P. J., and Newton, W. E., eds. Dordrecht: Martinus Nijhoff Publishers, 1985, pp. 73–78.

127. Krotzky, A., and Werner, D., "Nitrogen fixation in *Pseudomonas stutzeri*," *Arch. Microbiol.*, *147*, 48–57 (1987).

128. Kucey, R. M. N., "Alteration of size of wheat root systems and nitrogen fixation by associative nitrogen-fixing bacteria measured under field conditions," *Can. J. Microbiol.*, *34*, 735–739 (1988).

129. Kucey, R. M. N., "Plant growth-altering effects of *Azospirillum brasilense* and *Bacillus* C-11-25 on two wheat cultivars," *J. Appl. Bacteriol.*, *64*, 187–196 (1988).

130. Ladha, J. K., Barraquio, W. L., and Watanabe, I., "Isolation and identification of nitrogen-fixing *Enterobacter cloacae* and *Klebsiella planticola* associated with rice plants," *Can. J. Microbiol.*, *29*, 1301–1308 (1983).

131. Lalande, R., and Knowles, R., "Cytoplasmic content of *Azospirillum brasilense* Sp7 grown under aerobic and denitrifying conditions," *Can. J. Microbiol.*, *33*, 151–156 (1987).

132. Lamm, R. B., and Neyra, C. A., "Characterization and cyst production of *Azospirilla* isolated from selected grasses growing in New Jersey and New York," *Can. J. Microbiol.*, *27*, 1320–1325 (1981).

133. Lee, M., Breckenridge, C., and Knowles, R., "Effect of some culture conditions on the production of indole-3-acetic acid and a gibberellin-like substance by *Azotobacter vinelandii*," *Can. J. Microbiol.*, *16*, 1325–1330 (1970).

134. Leigh, J. A., Signer, E. R., and Walker, G. C., "Exopolysaccharide-deficient mutants of *Rhizobium meliloti* that form ineffective nodules," *Proc. Natl. Acad. Sci. USA*, *82*, 6231–6235 (1985).

135. Leong, S. A., and Expert, D., "Siderophores in plant-pathogen interactions," *Plant-Microbe Interactions, Molecular and Genetic Perspectives*, Kosuge, T., and Nester, E. W., eds. New York: McGraw-Hill, 1989, vol. 3, pp. 62–83.

136. Letham. D. S., and Palni, L. M. S., "The biosynthesis and metabolism of cytokinins," *Ann. Rev. Plant Physiol.*, *34*, 163–197 (1983).

137. Lieberman, M., and Kunishi, A. T., "Ethylene-forming systems in etiolated pea seedling and apple tissue," *Plant Physiol.*, *41*, 376–382 (1975).

138. Lifshitz, R., Kloepper, J. W., Scher, F. M., Tipping, E. M., and Lalibertä, M., "Nitrogen-fixing *Pseudomonas* isolated from roots of plants grown in the canadian high arctic," *Appl. Environ. Microbiol.*, *51*, 251–255 (1986).

139. Lima, E., Boddey, R. M., and Döbereiner, J., "Quantification of biological nitrogen fixation associated with sugar cane using ^{15}N aided nitrogen balance," *Soil Biol. Biochem.*, *19*, 246–252 (1987).

140. Lin, W., Okon, Y., and Hardy, H. W. F., "Enhanced mineral uptake by Zea mays and Sorgum bicolor roots inoculated with *Azospirillum brasilense*," *Appl. Environ. Microbiol.*, *45*, 1775–1779 (1983).

141. Long, S. R., "*Rhizobium*-legume nodulation: life together in the underground," *Cell 56*, 203–214 (1989).

142. Ludden, P. W., Okon, Y., and Burris, R. H., "The nitrogenase system of *Spirillum lipoferum*," *Biochem. J.*, *173*, 1001–1003 (1978).

143. Magalhaes, F. M., Baldani, J. I., Souto, S. M., Kuykendall, J. R., and Döbereiner, J., "A new acid-tolerant *Azospirillum* species," *An. Acad. Bras. Cien.*, *55*, 417–430 (1983).

144. Magie, A. R., Wilson, E. E., and Kosuge, T., "Indoleacetamide as an intermediate in the synthesis of indoleacetic acid in *Pseudomonas savastanoi*," *Science 141*, 1281–1282 (1963).

145. Malik, K. A., and Schlegel, H. G., "Chemolithoautotrophic growth of bacteria able to grow under N_2-fixing conditions," *FEMS Microbiol. Lett.*, *11*, 63–67 (1981).

146. Marocco, A., Bazzicalupo, M., and Perenzin, M. "Forage grasses inoculation with gentamycine and sulfaguanidine resistant mutants of *Azospirillum brasilense*," *Azospirillum II: Genetics, Physiology, Ecology*, Klingmüller, W., ed. Berlin: Springer Verlag, 1983, pp. 149–158.

147. Martin, G. B., Thomashow, M. F., and Chelm, B. K., "*Bradyrhizobium japonicum glnB*, a putative nitrogen-regulatory gene, is regulated by NtrC at tandem promoters," *J. Bacteriol.*, *171*, 5638–5645 (1989).

148. Martinez-Drets, G., Del Gallo, M., Burpee, C., and Burris, R. H., "Catabolism of carbohydrates and organic acids and expression of nitrogenase in *Azospirilla*," *J. Bacteriol.*, *159*, 80–85 (1984).

149. Martinez-Drets, G., Fabiano, E. and Cardona, A., "Carbohydrate catabolism in *Azospirillum amazonense*," *Appl. Environ. Microbiol.*, *50*, 183–185 (1985).

150. Maulik, P., and Ghosh, S., "NADPH/NADH-dependent cold-labile glutamate dehydrogenase in *Azospirillum brasilense*," *Eur. J. Biochem.*, *155*, 595–602 (1986).

151. McClung, C. R., and Patriquin, D. G., "Isolation of a nitrogen-fixing *Campylobacter* species from the roots of *Spartina alterniflora* Loisel," *Can. J. Microbiol.*, *26*, 881–886 (1980).

152. McClung, C. R., Van Berkum, P., Davis, R. E., and Sloger, C., "Enumeration and localization of N_2-fixing bacteria associated with roots of *Spartina alterniflora* Loisel," *Appl. Environ. Microbiol.*, *45*, 1914–1920 (1983).

153. Michiels, K., De Troch, P., Onyeocha, I., Van Gool, A., Elmerich, C., and Vanderleyden, J., "Plasmid localization and mapping of two *Azospirillum brasilense* loci that affect exopolysaccharide synthesis," *Plasmid 21*, 142–146 (1989).

154. Michiels, K., Vanderleyden, J., and Van Gool, A., "*Azospirillum*-plant root associations: A review," *Biol. Fertil. Soils*, *8*, 356–368 (1989).

155. Michiels, K. W., Vanderleyden, J., Van Gool, A. P., and Signer, E. R., "Isolation and characterization of *Azospirillum brasilense* loci that correct *Rhizobium meliloti exoB* and *exoC* mutants," *J. Bacteriol.*, *170*, 5401–5404 (1988).

156. Michiels, K., Vanderleyden, J., Van Gool, A., and Signer, E. R., "Isolation and properties of *Azospirillum lipoferum* and *Azospirillum brasilense* surface polysaccharide mutants," *Nitrogen Fixation with Non-Legumes*, F. A. Skinner, R. M. Boddey, and I. Fendrik, eds., Dordrecht: Kluwer Academic Press, 1989, pp. 189–195.

157. Michiels, K., Vanstockem, M., Vanderleyden, J., and Van Gool, A., "Stability of broad host range plasmids in *Azospirillum*, cloning of a 5.9 kb plasmid from *A. brasilense* R07," *Azospirillum III: Genetics, Physiology, Ecology*, Klingmüller, W., ed. Berlin: Springer Verlag, 1985, pp. 64–83.

158. Mishra, A. K., and Roy, P., "Mobilisation of *Spirillum lipoferum* chromosomal DNA by RP4," *Indian J. Microbiol.*, *21*, 60–62 (1981).

159. Mishra, A. K., Roy, P., and Bhattacharya, S., "Deoxynucleic acid-mediated transformation of *Spirillum lipoferum*" *J. Bacteriol.*, *137*, 1425–1427 (1979).

160. Moore, T. C., "Auxins," *Biochemistry and Physiology of Plant Hormones*, Moore, T. C., ed. Berlin: Springer Verlag, 1988, pp. 28–92.

161. Morgenstern, E., and Okon, Y., "Promotion of plant growth and NO_3^- and Rb^+ uptake in *Sorghum bicolor* × *Sorghum sudanense* inoculated with *Azospirillum brasilense*-Cd," *Arid Soil Res. Rehabil.*, *1*, 211–217 (1987).

162. Morris, R. O., "Genes specifying auxin and cytokinin biosynthesis in phytopathogens," *Ann. Rev. Plant Physiol.*, *37*, 509–538 (1986).

163. Nair, S. K., Jara, P., Quiviger, B., and Elmerich, C., "Recent developments in the genetics of nitrogen fixation in *Azospirillum*," *Azospirillum II: genetics, physiology, ecology*, Klingmüller, W., ed. Birkhauser Basel, EXS 48, pp. 29–38 (1983).

164. Nelson, L. M., and Knowles, R., "Effect of oxygen and nitrate on nitrogen fixation and denitrification by *Azospirillum brasilense* grown in continuous culture," *Can. J. Microbiol.*, *24*, 1395–1403 (1978).

165. Neuer, G., Kronenberg, A., and Bothe, H., "Denitrification and Nitrogen Fixation by *Azospirillum*. III. Properties of a wheat-*Azospirillum* association," *Arch. Microbiol.*, *141*, 364–370 (1985).

166. Neyra, C. A., and van Berkum, P., "Nitrate reduction and nitrogenase activity in *Spirillum lipoferum*," *Can. J. Microbiol.*, *23*, 306–310 (1977).

167. Nur, I., Okon, Y., and Henis, Y., "Comparative studies of nitrogen-fixing bacteria associated with grasses in Israel with *Azospirillum brasilense,*" *Can. J. Microbiol.*, *26*, 714–718 (1980).

168. Okon, Y., "*Azospirillum* as a potential inoculant for agriculture," *Trends in Biotechnology*, *3*, 223–228 (1985).

169. Okon, Y., Albrecht, S. L., and Burris, R. H., "Carbon and ammonia metabolism of *Spirillum lipoferum,*" *J. Bacteriol.*, *128*, 592–597 (1976).

170. Okon, Y., Albrecht, S. L., and Burris, R. H., "Methods for growing *Spirillum lipoferum* and for counting it in pure culture and in association with plants," *Appl. Environ. Microbiol.*, *33*, 85–88 (1977).

171. Okon, Y., Cakmakci, L., Nur, I., and Chet, I., "Aerotaxis and chemotaxis of *Azospirillum brasilense*: a note," *Microb. Ecol.*, *6*, 277–280 (1980).

172. Okon, Y., Fallik, E., Sarig, S., Yahalom, E., and Tal, S., "Plant growth promoting effects of *Azospirillum,*" *Nitrogen Fixation: Hundred Years After*. H. Bothe, F. J. de Bruijn, W. E. Newton, eds. Stuttgart: Gustav Fischer, 1988, pp. 741–746.

173. Okon, Y., and Hadar, Y., "Microbial inoculants as crop-yield enhancers," *CRC Crit. Rev. Biotech.*, *6*, 61–85 (1987).

174. Oliveira, R. G. B., and Drozdowicz, A., "Are *Azospirillum* bacteriocins produced and active in soil?" *Azospirillum IV: Genetics, Physiology, Ecology*, Klingmüller, W., ed. Berlin: Springer Verlag, 1988, pp. 101–108.

175. Onyeocha, I., Vieille, C., Zimmer, W., Baca, B. E., Flores, M., Palacios, P., and Elmerich, C., "Physical map and properties of a 90 MDa plasmid of *Azospirillum brasilense* Sp7," *Plasmid*, *23*, 169–182 (1990).

176. Pacovsky, R. S., "Diazotroph establishment and maintenance in the *Sorghum-Glomus-Azospirillum* association," *Can. J. Microbiol.*, *35*, 977–981 (1989).

177. Pacovsky, R. S., "Influence of inoculation with *Azospirillum brasilense* and *Glomus fasciculatum* on sorghum nutrition," *Nitrogen Fixation with Non-Legumes*. Skinner, F. A., Boddey, R. M., and Fendrik, I., eds. Dordrecht: Kluwer Academic Publishers, 1989, pp. 235–239.

178. Page, W. J., and Dale, P. L., "Stimulation of *Agrobacterium tumefaciens* growth by *Azotobacter vinelandii* ferrisiderophores," *Appl. Environ. Microbiol.*, *51*, 451–454 (1986).

179. Page, W. J., and Grant, G. A., "Partial repression of siderophore-mediated iron transport in *Azotobacter vinelandii* grown with mineral iron," *Can. J. Microbiol.*, *34*, 675–679 (1988).

180. Page, W. J., and von Tigerstrom, M., "Iron- and molybdenum-repressible outer membrane proteins in competent *Azotobacter vinelandii,*" *J. Bacteriol.*, *151*, 237–242 (1982).

181. Page, W. J., and von Tigerstrom, M., "Aminochelin, a catecholamine siderophore produced by *Azotobacter vinelandii,*" *J. Gen. Microbiol.*, *134*, 453–460 (1988).

182. Palm, C. J., Gaffney, T., and Kosuge, T., "Cotranscription of genes encoding indoleacetic acid production in *Pseudomonas syringae* subsp. *savastanoi,*" *J. Bacteriol.*, *171*, 1002–1009 (1989).

183. Papen, H., and Werner, D., "N$_2$-fixation in *Erwinia herbicola*" *Arch. Microbiol.*, *120*, 25–30 (1979).

184. Papen, H., and Werner, D., "Organic acid utilization, succinate excretion, encystation and oscillating nitrogenase activity in *Azospirillum brasilense* under microaerobic conditions," *Arch. Microbiol.*, *132*, 57–61 (1982).

185. Patel, U., Baxi, M. D., and Modi, V. V., "Evidence for the involvement of iron siderophore in the transport of molybdenum in cowpea *Rhizobium*," *Curr. Microbiol.*, *17*, 179–182 (1988).

186. Patriquin, D. G., and Döbereiner, J., "Light microscopy observations of tetrazolium-reducing bacteria in the endorhizosphere of maize and other grasses in Brasil," *Can. J. Microbiol.* *28*, 734–742 (1978).

187. Patriquin, D. G., Döbereiner, J., and Jain, D. K., "Sites and processes of association between diazotrophs and grasses," *Can. J. Microbiol.*, *29*, 900–915 (1983).

188. Pederson, W. L., Chakrabarty, K., Klucas, R. V., and Vivader, A. K., "Nitrogen fixation (acetylene reduction) associated with roots of winter wheat and sorghum in Nebraska," *Appl. Environ. Microbiol.*, *35*, 129–135 (1978).

189. Pedrosa, F. O., and Yates, M. G., "Regulation of nitrogen fixation (*nif*) genes of *Azospirillum brasilense* by *nifA* and *ntr* (*gln*) type gene products," *FEMS Microbiol. Lett.*, *23*, 95–101 (1984).

190. Pedrosa, F. O., Stephan, M., Döbereiner, J., and Yates, M. G., "Hydrogen-uptake hydrogenase activity in nitrogen-fixing *Azospirillum brasilense*," *J. Gen. Microbiol.*, *128*, 161–166 (1982).

191. Pereira, J. A. R., Cavalcante, V. A., Baldani, J. I., and Döbereiner, J., "Field inoculation of sorghum and rice with *Azospirillum* spp. and *Herbaspirillum seropedicae*," *Nitrogen fixation with non-legumes*, Skinner, F. A., Boddey, R. M., and Fendrik, I., eds. Dordrecht: Kluwer Academic Publishers, 1989, pp. 219–224.

192. Perroud, B., Bandhari, S. K., and Elmerich, C., "The *nifHDK* operon of *Azospirillum brasilense* Sp7," *Azospirillum III: Genetics, Physiology, Ecology*, Klingmüller, W., ed. Berlin: Springer Verlag, 1985, pp. 10–19.

193. Plazinski, J., Dart, P. J., and Rolfe, B., "Plasmid visualization and *nif* gene location in nitrogen-fixing *Azospirillum* strains," *J. Bacteriol.*, *155*, 1429–1433 (1983).

194. Polsinelli, M., Baldanzi, E., Bazzicalupo, M., and Gallori, E., "Transfer of plasmid pRD1 from *Escherichia coli* to *Azospirillum brasilense*," *Mol. Gen. Genet.*, *178*, 709–711 (1980).

195. Postgate, J. R., *The fundamentals of nitrogen fixation*, Cambridge: Cambridge Univ. Press, 1982.

196. Prakash, R. K., and Atherly, A., "Plasmids of *Rhizobium* and their role in symbiotic nitrogen fixation," *Int. Rev. Cytol.*, *104*, 1-24 (1986).

197. Punita, Jafri, S., Reddy, M. A., and Das, H. K., "Multiple chromosomes of *Azotobacter vinelandii*," *J. Bacteriol.*, *171*, 3133–3138 (1989).

198. Puvanesarajah, V., Schell, F. M., Stacey, G., Douglas, C. J. and Nester, E. W., "Role for 2-linked-beta-D-glucan in the virulence of *Agrobacterium tumefaciens*," *J. Bacteriol.*, *164*, 102–106 (1985).

199. Quiviger, B., Franche, C., Lutfalla, G., Rice, D., Haselkorn, R., and Elmerich, C., "Cloning of a nitrogen fixation (*nif*) gene cluster of *Azospirillum brasilense*," *Biochimie*, *64*, 495–502 (1982).

200. Rao, A. V., and Vankateswarlu, B., "Associative symbiosis of *Azospirillum lipoferum* with dicotyledonous succulent plants of the Indian desert," *Can. J. Microbiol.*, *28*, 778–782 (1982).

201. Ratti, S., Curti, B., Zanetti, G., and Galli, E., "Purification and characterization of glutamate synthase from *Azospirillum brasilense*," *J. Bacteriol.*, *163*, 724–729 (1985).

202. Reinhold, B., and Hurek, T., "Location of diazotrophs in the root interior with special attention to the kallar grass association," *Nitrogen Fixation with Non-Legumes*. Skinner, F. A., Boddey, R. M., and Fendrik, I., eds. Dordrecht: Kluwer Academic Publishers, 1989, pp. 209–218.

203. Reinhold, B., Hurek, T., and Fendrick, I., "Strain-specific chemotaxis of *Azospirillum* spp," *J. Bacteriol.*, *162*, 190–195 (1985).

204. Reinhold, B., Hurek, T., Fendrik, I., Pot, B., Gillis, M., Kersters, K., Thielemans, S., and De Ley, J., "*Azospirillum halopraeferens* sp. nov., a nitrogen-fixing organism associated with roots of Kallar grass (*Leptochloa fusca* (L.) Kunth)," *Int. J. Syst. Bacteriol.*, *37*, 43–51 (1987).

205. Reinhold, B., Hurek, T., Nieman, E. G., and Fendrik, I., "Close association of *Azospirillum* and diazotrophic rods with different root zone of Kallar grass," *Appl. Environ. Microbiol.*, *52*, 520–526 (1986).

206. Rennie, R. J., "Dinitrogen-fixing bacteria: computer assisted identification of soil isolates," *Can. J. Microbiol.*, *26*, 1275–1283 (1980).

207. Rennie, R. J., "A single medium for the isolation of acetylene-reducing (dinitrogen-fixing) bacteria from soils," *Can. J. Microbiol.*, *27*, 8–14 (1981).

208. Rennie, R. J., De Freitas, J. R., Ruschel, A. P., and Vose, P. B., "Isolation and identification of N_2-fixing bacteria associated with sugar cane (*Saccharum* sp.)," *Can J. Microbiol.*, *28*, 462–467 (1982).

209. Rennie, R. J., and Thomas, J. B., "N-15 determined effect of inoculation with N-fixing bacteria on nitrogen assimilation in Western Canadian wheats," *Plant Soil 100*, 213–223 (1985).

210. Reusch, R. N., and Sadoff, H. L., "Novel lipid components of the *Azotobacter vinelandii* cyst membrane," *Nature 302*, 268–270 (1983).

211. Reynders, L., and Vlassak, K., "Conversion of tryptophan to indoleacetic acid by *Azospirillum* sp," *Soil Biol. Biochem.*, *11*, 547–548 (1979).

212. Richard, C., "Bactériologie et épidémiologie des espèces du genre *Klebsiella*," *Bulletin Inst. Past.*, *80*, 127–145 (1982).

213. Riley, D., and Barber, S., "Effect of ammonium and nitrate fertilization on phosphorus uptake as related to root-induced pH changes at the root-soil interface," *Soil Sci. Soc. Am. Proc.*, *35*, 301–306 (1971).

214. Robson, R., Jones, R., Kennedy, C. K., Drummond, M., Ramos, J., Woodley, P. R., Wheeler, C., Chesshyre, J., and Postgate, J., "Aspects of genetics of *Azotobacter*," *Advances in Nitrogen Fixation Research*, Veeger, C., and Newton, W. E., eds. The Hague: M. Nijhoff & Junk Publishers 1984, pp. 643–651.

215. Roussos, S., Garcia, J. L., Rinaudo, G., and Gauthier, D., "Distribution de la microflore hétérotrophe aérobie et en particulier des bactéries dénitrifiantes et fixatrices d'azote dans la rhizosphère du riz," *Ann. Microbiol. (Inst. Pasteur)*, *131A*, 197–207 (1980).

216. Ruckdäschel, E., Kittell, B. L., Helinski, D. R., and Klingmüller, W., "Aromatic amino acid aminotransferases of *Azospirillum lipoferum* and their possible involvement in IAA biosynthesis," *Azospirillum IV: Genetics, Physiology, Ecology*, Klingmüller, W., ed. Berlin: Springer Verlag, 1988, pp. 49–53.

217. Sadasivan, L. and Neyra, C. A., "Flocculation in *Azospirillum brasilense* and *Azospirillum lipoferum*: exopolysaccharides and cyst formation," *J. Bacteriol.*, *163*, 716–723 (1985).

218. Sadasivan, L., and Neyra, C. A., "Cyst production and brown pigment formation in aging cultures of *Azospirillum brasilense* ATCC29145," *J. Bacteriol.*, *169*, 1670–1677 (1987).

219. Sadoff, H. L., "Encystment and germination in *Azotobacter vinelandii*," *Bacteriol. Rev.* *39*, 516–539 (1975).

220. Sadoff, H. L., Shimei, B., and Ellis, S., "Characterization of *Azotobacter vinelandii* deoxyribonucleic acid and folded chromosomes," *J. Bacteriol.*, *138*, 817–877 (1979).

221. Santero, E., Toukdarian, A., Humphrey, R., and Kennedy, C., "Identification and characterization of two nitrogen-fixation regulatory regions, *nifA* and *nfrX*, in *Azotobacter vinelandii* and *Azotobacter chroococcum*," *Mol. Microbiol.*, *2*, 303–314 (1988).

222. Saxena, B., Modi, M., and Modi, V. V., "Isolation and characterization of siderophores from *Azospirillum lipoferum* D-2," *J. Gen. Microbiol.*, *132*, 2219–2224 (1986).

223. Saxena, B., Vithlani, L., and Modi, V. V., "Siderophore-mediated transport of molybdenum in *Azospirillum lipoferum* strain D-2," *Curr.*, *Microbiol.*, *19*, 291–295 (1989).

224. Schank, S. C., Smith, R. L., Weiser, G. C., Zuberer, D. A., Bouton, J. H., Quesenberry, K. H., Tyler, M. E., Milam, J. R., and Littel, R. C., "Fluorescent antibody technique to identify *Azospirillum brasilense* associated with roots of grasses," *Soil Biol. Biochem.*, *11*, 287–295 (1979).

225. Schneider, E. A., and Wightman, F., "Metabolism of auxin in higher plants," *Ann. Rev. Plant Physiol.*, *25*, 487–513 (1974).

226. Schrank, I., Zaha, A., De Arujo, E. F., and Santos, D. S., "Construction of a gene library from *Azospirillum brasilense* and characterization of a recombinant containing the *nif* structural genes," *Brazil. J. Med. Biol. Res.*, *20*, 321–330 (1987).

227. Schwedock, J., and Long, S. R., "Nucleotide sequence and protein products of two new nodulation genes of *Rhizobium meliloti*, *nodP* and *nodQ*," *Mol. Plant-Microbe Interact.* *2*, 181–194 (1989).

228. Scott, D. B., Scott, C. A., and Döbereiner, J., "Nitrogenase activity and nitrate respiration in *Azospirillum* spp," *Arch. Microbiol.*, *121*, 141–145 (1979).

229. Sekiguchi, S., and Maruyama, Y., "Assimilatory reduction of nitrate in *Rhizobium meliloti*," *J. Basic Microbiol.*, *28*, 529–539 (1988).

230. Sekine, M., Ichikawa, T., Kuba, N., Kobayashi, M., Sakurai, A., and Syono, K., "Detection of IAA biosynthetic pathway from tryptophan via indole-3-acetamide in *Bradyrhizobium* spp.," *Plant Cell Physiol.*, *29*, 867–874 (1988).

231. Sekine, M., Watanabe, K., and Syono, K., "Molecular cloning of a gene for indole-3-aceteamid hydrolase from *Bradyrhizobium japonicum*," *J. Bacteriol.*, *171*, 1718–1724 (1989).

232. Sekine, M., Watanabe, K., and Syono, K., "Nucleotide sequence of a gene for indole-3-acetamide hydrolase from *Bradyrhizobium japonicum*," *Nucl. Acids Res. 17*, 6400 (1989).

233. Seldin, L., and Dubnau, D., "Deoxyribonucleic acid homology among *Bacillus polymixa*, *Bacillus macerans* and *Bacillus azotofixans*, and other nitrogen-fixing *Bacillus* strains," *Int. J. Syst. Bacteriol.*, *35*, 151–154 (1985).

234. Seldin, L., Van Elsas, J. D., and Penido, E. G. C., "*Bacillus azotofixans* sp. nov. a nitrogen-fixing species from Brazilian soil and grass roots," *Int. J. Syst. Bacteriol.*, *34*, 451–456 (1984).

235. Selveraj, G., and Iyer, V. N. "Suicide plasmid vehicles for insertion mutagenesis in *Rhizobium meliloti* and related bacteria," *J. Bacteriol.*, *158*, 580–589 (1984).

236. Simon, R., Priefer, U., and Pühler, A., "A broad host range mobilization system for *in vivo* genetic engineering: transposon mutagenesis in Gram-negative bacteria," *Bio/ Technology*, *1*, 784–791 (1983).

237. Singh, M., "Transfer of bacteriophage Mu and Transposon Tn5 into *Azospirillum*," *Azospirillum: Genetics, Physiology, Ecology*, Klingmüller, W., ed. Basel: Birkhäuser Verlag, 1982, EXS 42, pp. 35–43.

238. Singh, M., Kleeberger, A., and Klingmüller, W., "Location of nitrogen fixation (*nif*) genes on indigenous plasmids of *Enterobacter agglomerans*," *Mol. Gen. Genet.*, *190*, 373–378 (1983).

239. Singh, M., and Klingmüller, W., "Cloning of pEA3, a large plasmid of *Enterobacter agglomerans* containing nitrogenase structural genes," *Plant Soil*, *90*, 235–242 (1986).

240. Singh, M., and Klingmüller, W., "Transposon mutagenesis in *Azospirillum brasilense*: isolation of auxotrophic mutants and Nif⁻ mutants and molecular cloning of the mutagenized *nif* DNA," *Mol. Gen. Genet.*, *202*, 136–142 (1986).

241. Singh, M., Kreutzer, R., Acker, G., and Klingmüller, W., "Localization and physical mapping of a plasmid-borne 23 kb *nif* gene cluster from *Enterobacter agglomerans* showing homology to the entire *nif* gene cluster of *Klebsiella pneumoniae* M5a1," *Plasmid*, *19*, 1–12 (1988).

242. Singh, M., Tripathi, A. K., and Klingmüller, W., "Identification of a regulatory *nifA* type gene and physical mapping of cloned new *nif* regions of *Azospirillum brasilense*," *Mol. Gen. Genet.*, *219*, 235–240 (1989).

243. Singh, M. and Wenzel, W., "Detection and characterization of plasmids in *Azospirillum*," *Azospirillum: Genetics, Physiology, Ecology*, Klingmüller, W., ed. Basel: Birkhäuser Verlag, EXS 42, 1982, pp. 44–51.

244. Smit, G., Kijne, J. W., and Lugtenberg, B. J. J., "Correlation between extracellular fibrils and attachment of *Rhizobium leguminosarum* to pea root hair tips," *J. Bacteriol.*, *168*, 821–827 (1986).

245. Song, S.-D., Hartmann, A., and Burris, R. H., "Purification and properties of the nitrogenase of *Azospirillum amazonense*," *J. Bacteriol.*, *164*, 1271–1277 (1985).

246. Stephan, M. P., Zimmer, W., and Bothe, H., "Denitrification by *Azospirillum brasilense* Sp 7. II. Growth with nitrous oxide as respiratory electron acceptor," *Arch. Microbiol.*, *138*, 212–216 (1984).

247. Stewart, V., "Nitrate respiration in relation to facultative metabolism in Enterobacteria," *Microbiol. Rev.*, *52*, 190–232 (1988).

248. Stouthamer, A. H., "Dissimilatory reduction of oxidized nitrogen compounds," *Biology of anaerobic microorganisms*, Zehner, A. J. B., ed. New York: John Wiley & Sons, 1988, pp. 245–303.

249. Subba Rao, N. S., Tilak, K. V. B. R., and Singh, C. S., "Synergistic effect of vesicular-arbuscular mycorrhizas and *Azospirillum brasilense* on the growth of barley in pots," *Soil Biol. Biochem.*, *17*, 119–121 (1985).

250. Taller, B. J., and Wong, T.-Y., "Cytokinins in *Azotobacter vinelandii* culture medium," *Appl. Environ. Microbiol.*, *55*, 266–267 (1989).

251. Tang, Y. W., and Bonner, J., "The enzymatic inactivation of indoleacetic acid. I. Some characteristics of the enzyme contained in pea seedlings," *Arch. Biochem.*, *13*, 11–25 (1947).

252. Tarrand, J. J., Krieg, N. R., and Döbereiner, J., "A taxonomic study of the *Spirillum lipoferum* group with description a new genus, *Azospirillum* gen. nov. and two species,

Azospirillum lipoferum (Beijerinck) comb. nov. and *Azospirillum brasilense* sp. nov.," *Can. J. Microbiol.*, *24*, 967–980 (1978).

253. Thompson, B. J., Domingo, E., and Warner, R. C., "Pseudolysogeny conversion of *Azotobacter* phages," *Virology*, *102*, 267–277 (1980).

254. Tien, T. M., Diem, H. G., Gaskins, M. H., and Hubbel, D. H., "Polygalacturonic acid transeliminase production by *Azospirillum* species," *Can. J. Microbiol.*, *27*, 426–431 (1981).

255. Tien, T. M., Gaskins, M. H., and Hubbell, D. H., "Plant growth substances produced by *Azospirillum brasilense* and their effect on the growth of pearl millet (*Pennisetum amaricanum*)," *Appl. Environ. Microbiol.*, *37*, 1016–1024 (1979).

256. Tjepkema, J., "Nitrogenase activity in the rhizosphere of *Panicum vigatum*," *Soil Biol. Biochem.*, *7*, 179–180 (1975).

257. Umali-Garcia, M., Hubbell, D. H., and Gaskins, M. H., "Process of infection of *Panicum maximum* by *Spirillum lipoferum*," *Environmental Role of Nitrogen-Fixing Blue-Green Algae and Asymbiotic Bacteria*, *Ecol. Bull.*, *Stockholm 26*, 1978, pp. 373–379.

258. Umali-Garcia, M., Hubbell, D. H., Gaskins, D. H., and Dazzo, F. B., "Association of *Azospirillum* with grass roots," *Appl. Environ. Microbiol.*, *39*, 219–226 (1980).

259. Vairinhos, F., Wallace, W., Nicholas, D. J. D., "Simultaneous assimilation and denitrification of nitrate by *Bradyrhizobium japonicum*," *J. Gen. Microbiol.*, *135*, 189–194 (1989).

260. Van Berkum, P., and Bohlool, B. B., "Evaluation of nitrogen fixation by bacteria in association with roots of tropical grasses," *Microbiol. Rev.*, *44*, 491–517 (1980).

261. Van Rhijn, P., Vanstockem, M., Vanderleyden, J., and de Mot, R., "Isolation of behavioral mutants of *Azospirillum brasilense* by using Tn5 *lacZ*," *Appl. Environ. Microbiol.*, *56*, 990–996 (1990).

262. van't Riet, J., Stouthamer, A. H., and Planta, R. J., "Regulation of nitrate assimilation and nitrate respiration in *Aerobacter aerogenes*," *J. Bacteriol.*, *96*, 1455–1464 (1968).

263. Vande Broek, A., Van Gool, A., and Vanderleyden, J., "Electroporation of *Azospirillum brasilense* with plasmid DNA," *FEMS Microbiol. Lett.*, *61*, 177–182 (1989).

264. Vanstockem, M., Michiels, K., Vanderleyden, J., and Van Gool, A., "Transfer and random integration of Tn5 in *Azospirillum*," *Azospirillum III: Genetics, Physiology, Ecology*, Klingmüller, W., ed. Berlin: Springer Verlag, 1985, pp. 74–84.

265. Vanstockem, M., Michiels, K., Vanderleyden, J., and Van Gool, A., "Transposon mutagenesis of *Azospirillum brasilense* and *Azospirillum lipoferum*: physical analysis of Tn5 and Tn5-Mob insertions mutants," *Appl. Environ. Microbiol.*, *43*, 410–415 (1987).

266. Ventura, W., and Watanabe, I., "^{15}N dilution technique of assessing nitrogen fixation in association with rice," *Phillipp. J. Crop. Sci.*, *7*, 44–50 (1982).

267. Verreth, C., Cammue, B., Swinnen, P., Crombez, D., Michielsen, A., Michiels, K., Vang ool, A., and Vanderleyden, J., "Cloning and expression in *Escherichia coli* of the *Azospirillum brasilense* Sp7 gene encoding ampicillin resistance," *Appl. Environ. Microbiol.*, *55*, 2056–2060 (1989).

268. Vieille, C., and Elmerich, C., "Characterization in *Azospirillum brasilense* Sp7 of two plasmid genes homologous to *Rhizobium meliloti nodPQ*," *Mol. Plant-Microbe Interact.*, *3*, 389–400 (1990).

269. Vieille, C., Onyeocha, I., Galimand, M., and Elmerich, C., "Homology between plasmids of *Azospirillum brasilense* and *Azospirillum lipoferum*," *Nitrogen Fixation with*

Non-Legumes, Skinner, F. A., Boddey, R. M., and Fendrik, I., eds., Dordrecht: Kluwer Academic Publishers, 1989, pp. 165–172.

270. Von Bülow, J. F. W., and Döbereiner, J., "Potential for nitrogen fixation in maize genotypes in Brazil," *Proc. Natl. Acad. Sci. USA, 72*, 2389–2393 (1975).

271. Waelkens, F., Maris, M., Verreth, C., Vanderleyden, J., and Van Gool, A. P., "*Azospirillum* DNA shows homology with *Agrobacterium* chromosomal virulence genes," *FEMS Microbiol. Lett., 43*, 241–246 (1987).

272. Wang, T. L., Wood, E. A., and Brewin, N. J., "Growth regulators, *Rhizobium* and nodulation in peas. Indole-3-acetic acid from the culture medium of nodulating and non-nodulating strains of *R. leguminosarum*," *Planta 155*, 345–349 (1982).

273. Watanabe, I., So, R., Ladha, J. K., Katayama-Fujimura, Y., and Kuraishi, H., "A new nitrogen-fixing species of pseudomonad: *Pseudomonas diazotrophicus* sp. nov. isolated from the root of wetland rice," *Can. J. Microbiol., 33*, 670–678 (1987).

274. Weiler, E. W., Jourdan, P. S. and Conrad, W., "Levels of indole-3-acetic acid in intact and decapitated coleoptiles as determined by a specific and highly sensitive solid-phase enzyme immunoassay," *Planta, 153*, 561–571 (1981).

275. Wenzel, W., Singh, M., and Klingmüller, W., "Molecular cloning of nitrogen fixation genes from *Azospirillum*," *Azospirillum: Genetics, Physiology, Ecology*, Klingmüller, W., ed. Basel: Birkhäuser Verlag, 1983, EXS 48, pp. 39–46.

276. Westby, C. A., Cutshall, D. S., and Vigil, G. V., "Metabolism of various carbon sources by *Azospirillum brasilense*," *J. Bacteriol., 156*, 1369–1372 (1983).

277. Wetsby, C. A., Enderlin, C. S., Steinberg, N. A., Joseph, C. M., and Meeks, J. C., "Assimilation of $^{13}NH_4^+$ by *Azospirillum brasilense* grown under nitrogen limitation and excess," *J. Bacteriol., 169*, 4211–4214 (1987).

278. Wong, P. and Stenberg, N. E., "Characterization of *Azospirillum* isolated from nitrogen-fixing roots of haversted sorghum plants," *Appl. Environ. Microbiol., 38*, 1189–1191 (1979).

279. Wood, A. G., Menezes, E. M., Dykstra, C., and Duggan, D. E., "Methods to demonstrate the megaplasmids (or minichromosomes) in *Azospirillum*," *Azospirillum: Genetics, Physiology, Ecology*, Klingmüller, W., ed. Basel: Birkhäuser Verlag, EXS 42, 1982, pp. 18–34.

280. Wright, S. F., and Weaver, R. W., "Enumeration and identification of nitrogen-fixing bacteria from forage grass," *Appl. Environ. Microbiol., 42*, 97–101 (1981).

281. Yagoda-Shagam, J., Barton, L. L., Reed, W. P., and Chiovetti, R., "Fluorescein isothiocyanate-labeled lectin analysis of the surface of the nitrogen-fixing bacterium *Azospirillum brasilense* by flow cytometry," *Appl. Environ. Microbiol., 54*, 1831–1837 (1988).

282. Yamada, T., Palm, C. J., Brooks, B., and Kosuge, T., "Nucleotide sequences of *Pseudomonas savastanoi* indoleacetic acid genes show homology with *Agrobacterium tumefaciens* T-DNA," *Proc. Natl. Acad. Sci. USA, 82*, 6522–6526 (1985).

283. You, C. B., and Zhou, F., "Non-nodular endorhizospheric nitrogen fixation in wetland rice," *Can. J. Microbiol. 35*, 403–408 (1988).

284. Zafar, Y., Ashraf, M., and Malik, K. A., "Nitrogen fixation associated with roots of Kallar grass (*Leptochloa fusca* L. Kunth)," *Plant Soil, 90*, 93–106 (1986).

285. Zimmer, W., and Bothe, H., "The phytohormonal interactions between *Azospirillum* and wheat," *Plant Soil 110*, 239–247 (1988).

286. Zimmer, W., Danneberg, G., and Bothe, H., "Amperometric method for determining nitrous oxide in denitrification and nitrogenase-catalyzed nitrous oxide reduction," *Curr. Microbiol.*, *12*, 341–346 (1985).

287. Zimmer, W., Roeben, K., and Bothe, H., "An alternative explanation for plant growth promotion by bacteria of the genus *Azospirillum*," *Planta 176*, 333–342 (1988).

288. Zimmer, W., Stephan, M. P., and Bothe, H., "Denitrification by *Azospirillum brasilense* Sp7. I. Growth with nitrite as respiratory electron acceptor," *Arch. Microbiol.*, *138*, 206–211 (1984).

7

Actinorhizal Symbioses

Dwight D. Baker and Beth C. Mullin

I. Introduction

Research over the past decade has advanced the knowledge of actinorhizal symbioses, the symbiotic relationships between actinomycetes in the genus *Frankia* and a wide range of woody angiosperms. Symbioses are established when frankiae infect host plant roots and cause the formation of root nodules that are active in N_2 fixation. The ability to grow *Frankia* in pure culture has made it possible to study the physiology and genetics of *Frankia* independently of its host and to determine more accurately than before the host specificity and symbiotic effectivity of individual *Frankia* isolates. In this chapter we shall attempt to present enough background information to acquaint all readers with the basic aspects of actinorhizal symbioses. At the same time we shall present examples of the most current studies being undertaken on this important symbiotic system.

A. Actinorhizal Plants

The symbiotic host plants for *Frankia*, collectively called actinorhizal plants by consensus in 1978,[174] belong to 24 genera within eight plant families (Table 7-1). These host genera share no single known character that would identify them as uniquely symbiotic, nor are they closely related. The fact that the actinorhizal genera are more phylogenetically disparate than the symbiotic legumes, for example, suggests that the symbiotic relationship between them and *Frankia* occurred early in evolutionary time and that significant divergence has occurred since then. Alternatively, host plants may have developed the ability to participate in symbioses much later in a series of independent events.

Actinorhizal plant genera are distributed worldwide, and human activities over the past several centuries have contributed to this wide distribution. There are, however, some notable absences of actinorhizal plants in some

Table 7-1. Actinorhizal Host Plant Genera

Genus	Family	Numbers of Nodulated Species	Principal Geographic Distribution[a]
Allocasuarina	Casuarinaceae	58	Aus
Alnus	Betulaceae	42	NAm, SAm, Eur, NAs, SAs
Casuarina	Casuarinaceae	18	Aus, SAs, NAf, NAm, SAm
Ceanothus	Rhamnaceae	31	NAm
Cercocarpus	Rosaceae	4	NAm
Ceuthostoma	Casuarinaceae	2[b]	Aus
Chamaebatia	Rosaceae	1	NAm
Colletia	Rhamnaceae	3	Eur, NAf, SAm
Comptonia	Myricaceae	1	NAm
Coriaria	Coriariaceae	16	Aus, NAm, SAm, Eur
Cowania	Rosaceae	1	NAm
Datisca	Datiscaceae	2	NAm, SAs
Discaria	Rhamnaceae	5	SAm, Eur
Dryas	Rosaceae	3	NAm, Eur
Elaeagnus	Elaeagnaceae	35	NAs, NAm, Eur, SAs
Gymnostoma	Casuarinaceae	18	Aus
Hippophaë	Elaeagnaceae	2	Eur, NAs
Kentrothamnus	Rhamnaceae	1	SAm
Myrica	Myricaceae	28	SAf, NAm, SAm, Aus, SAs, NAs
Purshia	Rosaceae	2	NAm
Retanilla	Rhamnaceae	1	SAm
Shepherdia	Elaeagnaceae	2	NAm
Talguenea	Rhamnaceae	1	SAm
Trevoa	Rhamnaceae	2	SAm

[a]NAm = North America, SAm = South America, Eur = Europe, Aus = Australia and/or Oceania, NAf = northern Africa, SAf = southern Africa, NAs = northern Asia, SAs = southern Asia.

[b]Lack of adequate collections make inclusion of these taxa tentative.

localities and environments. With the exception of several species of *Myrica*, Africa is particularly lacking in native actinorhizal plants. Also, moist lowland tropical zones of South America and portions of Asia lack native actinorhizal species. Many actinorhizal genera can be characterized as inhabiting nutrient-poor sites in north or south temperate regions or similar environments at higher elevations in tropical or subtropical regions. Cold desert regions in North and South America are particularly rich in actinorhizal species of the Rosaceae and Rhamnaceae. It is likely that additional actinorhizal species remain to be discovered.

The most widely distributed actinorhizal host plants are *Alnus*, *Casuarina*, and *Elaeagnus*, which have been introduced by man into almost all continents. Although probably no particular efforts were made to introduce the symbiotic *Frankia* along with these host plants, numerous exotic actinorhizal species are observed to nodulate without inoculation. Enough difficulties with natural inoculation have been reported, however, to warrant the routine inoculation of introduced species.

The ability of many actinorhizal species to grow rapidly on poor soils has led to their widespread use in forestry, landscaping, soil stabilization, and the revegetation of disturbed sites. *Alnus* in temperate regions and *Casuarina* in more tropical regions are grown for timber, fuel, and pulp. *Alnus*, *Myrica*, and *Elaeagnus* species are interplanted with a wide range of non-N_2-fixing tree species and have been shown to increase growth of associated plants.[41] The reader is urged to consult several recent reviews on the present and potential utilization of actinorhizal plant species.[18,40,41,44,45,70,189]

B. *Frankia*

Frankia is the N_2-fixing microsymbiont of actinorhizal plants. It was first observed within root nodules by Woronin in 1866 (see Hawker and Fraymouth[66]). It is unlike most N_2-fixing bacteria in being multicellular and differentiated, but similar to some cyanobacteria in being filamentous. Taxonomically, it belongs to the Actinomycetales, an order of filamentous bacteria, and recently has been included, along with *Geodermatophilus* and *Blastococcus*, in the revised family Frankiaceae.[65] The latter two genera in this revised family are not symbiotic and do not fix N_2 but are similar to *Frankia* in morphology, development, and some genetic characteristics.

Original species designations within the genus *Frankia* that were based on cross-inoculation studies using crushed nodule inocula are no longer considered to be valid. New species designations have been proposed[84] but are not widely accepted although it is clear from DNA hybridization studies that distinct genetic species do exist among frankiae.[2,32,54] *Frankia* strains or isolates have been given letter and number designations,[9,88] but unfortunately no single system is universally used. In this chapter we have minimized referral to individual isolates and when referral is made the isolate name used is that found in the original description of that isolate.

II. Morphology of *Frankia*

Figure 7-1 shows a photomicrograph of the characteristic morphology of *Frankia* cultured *in vitro*. *Frankia* typically produces a vegetative mycelium composed of septate filaments. On this mycelium may be produced two distinct kinds of cells if physiological conditions are conducive.

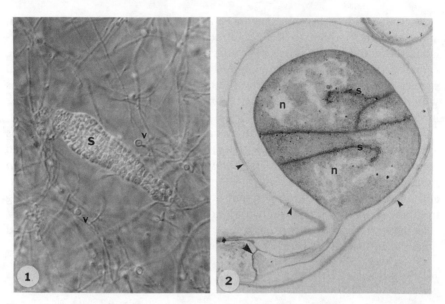

Figure 7-1. Nomarski interference optics light micrograph of a typical *Frankia* strain cultivated on a nitrogen-deficient medium. A sporangium (S) and numerous vesicles (v), the site for nitrogen fixation, are present on the vegetative mycelium.

Figure 7-2. Median longitudinal section through a *Frankia* vesicle. Incomplete septae (s) are present in the body of the vesicle. Nucleoid regions (n) are also frequently visible. A basal septum (large arrow) separates the vesicle from the subtending stalk cell, which in turn attaches it to the vegetative mycelium. The outer wall of the vesicle (small arrows), which in this preparation is widely separated from the vesicle membrane, shows its multilamellar nature. (Reprinted with permission from Lancelle et al.[87])

The first specialized cell is the vesicle, the site of N_2 fixation,[96,97] which is produced when substrate nitrogen is limiting.[168] The vesicle is an enlarged spherical cell attached to the substrate mycelium by a stalk cell (Figure 7-2). Within the cytoplasm of the vesicle, septations may occur that completely, or more usually incompletely, divide the cell into compartments.[87,110] The function of these septa is not understood, since fully functional vesicles have been observed without them.[110] Although vesicles traditionally have been thought to serve only one function, that of N_2 fixation, recent studies appear to indicate that the vesicle may also serve as a propagule.[134]

The second specialized cell produced by *Frankia* is the spore. Spores are produced in large numbers within amorphous sporangia (Figure 7-3).[16,89] Spores are neither ornamented nor motile and they show highly variable rates of germination, depending on culture conditions, bacterial strain, or presence of a suitable host.[178] Sporangia may be produced as terminal appendages of the vegetative mycelium or in some strains may occur as in-

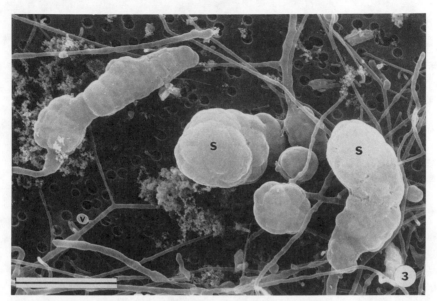

Figure 7-3. Scanning electron micrograph of *Frankia* illustrating the three-dimensional morphology of frankiae when cultured *in vitro*. Large sporangia (S) are easily distinguished from the vesicles (v). Note the variability in hyphal diameter of the mycelium. Bar = 10 μm. (Reprinted with permission from Baker and Seling.[14])

tercalary structures within the mycelium.[112] Unlike the majority of soil-borne aerobic actinomycetes, sporangia of *Frankia* are not aerial; that is, sporangia are borne not above the surface of the growth substrate, but within it.

The morphology of *Frankia* within host plant cells is similar to that observed *in vitro*. A vegetative mycelium is present in all infected host cells. Vesicles are produced in large numbers (Figure 7-4) when N_2 fixation is high, but the morphology of the vesicles may be significantly modified by the host plant. Four general vesicle morphologies have been reported[14] and these, combined with differences observed in vesicle anatomy and/or position within the host cell, result in characteristic structures within each host plant (e.g., Figure 7-5). Spherical vesicles not unlike those observed in pure cultures have been observed in the host plant genera *Alnus*, *Elaeagnus*, *Hippophaë*, and *Shepherdia*.[68,111] Obovate (flask- or pear-shaped) vesicles have been observed in *Ceanothus*[161] and other members of the Rhamnaceae,[34] as well as in *Dryas*, *Purshia*, and other members of the Rosaceae.[106,107] Lanceolate (fingerlike) vesicles have been observed in *Coriaria*, *Comptonia*, *Datisca*, and *Myrica*.[67,108,109] No recognizable vesicles have been observed in *Casuarina*[177] or *Allocasuarina*, but it is believed that an as yet unidentified specialized cell does exist in these symbioses for the localization of nitrogenase or that low O_2 concentrations within the host cell eliminate the need for a vesicle.[24]

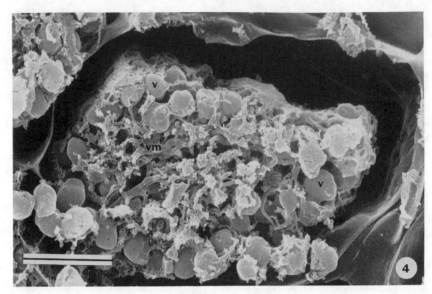

Figure 7-4. Scanning electron micrograph of *Frankia* within a root nodule of *Alnus*. Enlarged vesicles (v) are present on the vegetative mycelium (vm). The bacterium fills a majority of the cell. Bar = 10 μm. (Reprinted with permission from Baker and Seling.[14])

Sporangia have been observed within host cells of some actinorhizal plants (see the review by Torrey[170]), but these occur in only a minority of all the actinorhizal host plant genera. Nonetheless, the presence of sporangia within nodules of the genera *Alnus*, *Myrica*, and *Comptonia* has captured a significant amount of interest among *Frankia* researchers. The presence of spores within nodules is considered to be a genetic trait of the microsymbiont. The presence of sporangia in nodule cells has been correlated with a significant reduction in N_2 fixation capacity, but spore-positive strains (designated Sp^+) are more infective than the spore-negative (designated Sp^-) strains.[153,178,179,181] Sporangia and spores, when produced *in planta*, are very similar to those observed *in vitro*.

III. Isolation and Cultivation of *Frankia*

Attempts at isolating *Frankia* in pure culture date back to the late 1880s,[15] when interest in symbiotic microorganisms was high. Numerous scientists from the late 1880s to the mid-1900s claimed to have successfully isolated this filamentous bacterium, but none could demonstrate clearly the consistent reinfection of the host plant. Ernst Pommer, a German scientist, successfully isolated *Frankia* and for the first time demonstrated reasonable reinfection,[126] but the controversy within the scientific community at the time was so heated that his results were disregarded and his cultures lost.

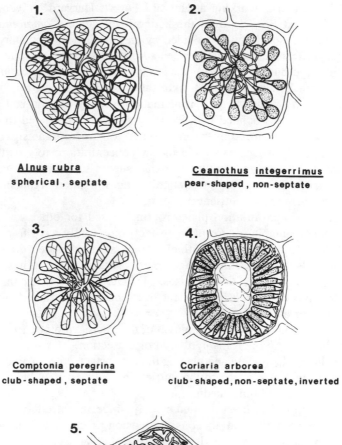

1.

Alnus rubra
spherical , septate

2.

Ceanothus integerrimus
pear-shaped , non-septate

3.

Comptonia peregrina
club-shaped , septate

4.

Coriaria arborea
club-shaped, non-septate, inverted

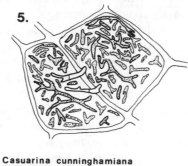

5.

Casuarina cunninghamiana
filamentous, non-vesiculate

Figure 7-5. Morphologies of *Frankia* vesicles observed in different actinorhizal host plants. Morphology is host controlled; the same bacterial strain will take different morphologies in different host plants. (Reprinted with permission from J. Torrey.[170])

In 1978, scientists working at Harvard Forest, Harvard University, successfully isolated *Frankia* from the actinorhizal genus *Comptonia*.[37] Confirmation of this isolation was provided by an independent laboratory.[83] From this one significant success have arisen additional successful isolations until at present large numbers of *Frankia* strains are held in collections around the world (see Lechevalier[88]). Despite the rapid advance in this area, isolation of *Frankia* strains is still problematic for many actinorhizal genera.[11]

The difficulties associated with isolating *Frankia* are related to their relatively slow growth in comparison to other soil microorganisms. *Frankia* also grows poorly when transferred in low concentrations to growth media. Therefore, to isolate and culture frankiae successfully, it is necessary to separate them effectively from contaminating soil microorganisms and to concentrate them in the nutrient medium.[11]

Use of the simple dilution plates routinely used for other soil microorganisms is not suitable for *Frankia*. In fact, only rarely have frankiae been isolated directly from soils. Routinely, frankiae are isolated directly from surface-sterilized root nodules using one of several methods. These methods may be categorized into three classes: (1) microdissection techniques, (2) serial filtration, or (3) density gradient fractionation (see Baker[11] for a thorough review of all methods).

Microdissection involves the removal of the microsymbiotic *Frankia* from root nodule cells by micromanipulation or dissection. The simplest version involves slicing the nodule and placing the cut surface onto an agar medium. After the bacterium grows into the nutrient medium, it may be subcultured and transferred to a liquid medium.

Serial filtration has been applied widely since its introduction by Benson.[19] In this technique an aqueous suspension of crushed root nodules is filtered through two or more nylon mesh filters. A large mesh filter (50 μm) removes large cellular debris from the suspension, but a smaller mesh filter (20 μm) collects the endosymbiont clusters that have been released from the root nodule cells during disruption. The collected frankiae are then transferred to agar media for colony isolation.

Isolation by density gradient centrifugation takes advantage of the density differences between frankiae and other microorganisms that contaminate nodule and soil samples. It has been used successfully for the isolation of *Frankia* from crushed nodule or soil suspensions and has been the only technique successful for isolating frankiae directly from soils.[13] In this procedure a crushed nodule or soil suspension is applied to a discontinuous sucrose gradient. *Frankia* becomes concentrated at an interface between layers of 2.5 M and 1.6 M sucrose.[11] The frankiae are collected from the density gradient and inoculated into agar media for colony isolation.

The choice of nutrient medium used for isolation attempts and cultivation is probably the most critical decision in assuring success. Frankiae are best isolated using a nitrogen-deficient medium so that there is selection against nondiazotrophs. Cultivation of frankiae may also be accomplished using similar media, but more rapid or consistent growth may be obtained using moderately complex media.[11] *Frankia* grows best in stationary liquid culture although gentle shaking may be of benefit for some strains. The most frequently used nitrogen-free minimal media are composed of a simple inorganic salts solution with sodium propionate as the sole carbon source (see Baker[11] for formulas). Complex media usually, although not always, contain some lipid-rich component such as lecithin or Tween that may facilitate the utilization of other carbon compounds.[90]

Despite isolation difficulties and slow growth, the majority of frankiae do not require unusual carbon compounds or vitamins. Organic acids have proved to be the best carbon sources for most frankiae, with succinate, pyruvate, and propionate being better than others.[31,92,139] Many sugars, both mono- and disaccharides, can be utilized; however, strains may differ greatly in their growth rate on different carbon sources.

Although *Frankia* strains have been routinely isolated and cultured for over a decade, infective strains have not been successfully isolated from all of the actinorhizal host plant genera. Infective strains currently do not exist in pure culture for any of the actinorhizal genera of the Rosaceae, Coriariaceae, Rhamnaceae, or Datiscaceae.[11] Numerous attempts have been made to isolate these microsymbionts and numerous *Frankia* strains have been cultivated. Unfortunately, those that have been cultivated have not had the ability to reinfect plants of their source host. Our understanding of infection physiology, initial plant-bacterial interactions, or strain diversity seems to be insufficient to addrress the problems of these recalcitrant microsymbionts.

IV. Genetics of *Frankia*

Frankia are fairly typical actinomycetes in terms of the overall physical properties of their genome. The G + C content of *Frankia* DNA (66–75%) is within the range of values reported for other actinomycetes,[4,5,54] and the genome molecular weights of *Frankia* strains ArI4 and EuI1 as determined by reassociation kinetic analysis are 8.3×10^9 and 6.0×10^9 daltons, respectively.[2] Analysis of 16S ribosomal RNA genes of two *Frankia* strains establishes a clear phylogenetic relationship between *Frankia* and the nonsymbiotic soil actinomycete *Geodermatophilus*.[56,65]

The uniform morphology and growth characteristics of *Frankia* strains in culture give no hint of the extent of genetic diversity that exists within the genus. Solution hybridization of total DNA from isolates from a broad range of host plants distributed across a wide geographical area results in hybridization values ranging from 97% to 0% sequence relatedness.[2,54] Solution hybridization of isolates from the single host species *Myrica pensylvanica*, all growing in the state of New Jersey, revealed the same extent of sequence divergence.[32] On an even smaller geographical scale, total protein patterns,[20,23,57] and restriction pattern analysis of total DNA and plasmid profiles[32,50,148] have revealed diversity in *Frankia* isolates from single host plants and in some cases from single nodule lobes. Despite the overall diversity among *Frankia* isolates, individual genotypes generally appear to be stable both in culture and in the soil environment.[103]

The symbiotic genes of *Frankia* have only recently been subject to study. Structural genes coding for the nitrogenase complex in *Frankia* are highly conserved and share a common evolutionary origin with nitrogenase genes from other diazotrophic bacteria. In all strains examined to date the *nifKDH* cluster is maintained in the free-living as well as symbiotic form of the bacteria and there is no evidence for gene rearrangement following the induction of nitrogenase activity.[119] DNA sequence analysis of *nifH* from an *Alnus* isolate[119] and a *Hippophaë* isolate[115] indicate that there is 93% nucleotide sequence similarity and 96% derived amino acid sequence similarity between these two strains. Comparison of these derived amino acid sequences to those from non-*Frankia* N_2-fixing bacteria show that the *Frankia* sequences are more similar to those from *Anabaena* than those of other Gram-positive N_2-fixing bacteria, raising the possibility that *nif* sequences in *Frankia* may have been transferred from *Anabaena* or that both *Anabaena* and *Frankia* received *nif* genes from a common source.

Through restriction mapping and hybridization with heterologous probes, three additional potential symbiotic genes have been mapped in *Frankia* strain ArI3. A DNA region containing *nifA*- and *nifB*-hybridizing sequences is present 4.5 kb away from *nifK*[149] and a region hybridizing to a pectate lyase probe is adjacent to *nifH*.[137,150] Other symbiotic genes may be accessible using heterologous gene probes as well; however, to date no *Rhizobium nod* genes have been found to hybridize reproducibly to *Frankia* DNA.

Many *Frankia* strains have been found to contain plasmids that vary in size from 8 kb to as much as 190 kb.[32,48,117,145,152] No functions have been ascribed to these plasmids, although in two cases plasmid DNA was found to hybridize to a *Klebsiella nif* gene probe.[51,146] *Frankia* isolates with identical DNA restriction patterns are found to harbor different plasmids, but no physiological differences between these strains have been detected.[31,32] Likewise, a clone of *Frankia* strain ArI3 regenerated from protoplasts and shown to be cured of its two plasmids appeared no different in cultural or symbiotic characteristics.[121]

Several factors have interfered with progress toward understanding the genetics of *Frankia* symbioses. Symbiotic mutants are not widely available for study because of the difficulty of generating such mutants, and a transformation system suitable for use with *Frankia* has not been identified or created. Until these tools are routinely available for the study of *Frankia*, progress will continue to be painfully slow. The reader is referred to three recent reviews on the genetics of actinorhizal symbioses.[103,117,150]

V. Host Specificity

The specificity of strains to nodulate certain host plants has been observed since the first *Frankia* strain was successfully isolated in pure culture. The affinities of pure-cultured strains differed significantly from relationships observed using suspensions of crushed root nodules. As a result, the taxonomy of *Frankia*, which had included specific epithets based on host specificity,[17] was invalidated.[89,90] Current belief among *Frankia* researchers is that host specificity should not be used to define taxa; however, there is a strong correlation between known host specificities of strains and genetic similarities.

From those *Frankia* strains that have been successfully isolated in pure culture, three or four major host specificity groups may be defined.[10,173] Host specificity group 1 includes *Frankia* strains capable of nodulating *Alnus*, *Comptonia*, and *Myrica*. Host specificity group 2 includes those strains capable of nodulating *Casuarina*, *Gymnostoma*, and some *Allocasuarina* species. Host specificity group 3 includes strains capable of nodulating *Elaeagnus*, *Hippophaë*, and *Shepherdia*. Complicating the definition of host specificity is the observation that species of *Myrica* and *Gymnostoma* are promiscuous; that is, they can be nodulated by strains from any of these three groups. Baker[10] defined a fourth host specificity group after observing that a number of *Frankia* strains, originally assigned to group 3, lacked the ability to nodulate the promiscuous genera. Although strains belonging to groups 3 and 4 must somehow be genetically distinct, recent studies have indicated that they are not so distinct from each other as they are from strains in groups 1 or 2.[113]

Additional host specificity groups probably will be defined in the future as infective *Frankia* strains are isolated from the Rhamnaceae, Rosaceae, Coriariaceae, and Datiscaceae. Because no currently isolated strains have the capability to nodulate host plants from these families, our understanding of host specificity among *Frankia* strains must be considered incomplete.

VI. Host Infection Processes

The infection processes by which *Frankia* initiates a symbiotic relationship with host plants have been observed to be of two general types, infection of root hairs and infection via intercellular penetration.[38,101] These symbiotic modes of entry into the host root are analogous to infections observed in the symbiotic legumes.[156,157] However, an additional entry mode has been observed in the legume-rhizobial symbioses, wound or crack entry, which has not been reported for actinorhizal symbioses.

A. Root Hair Infection

The deformation of host plant root hairs by *Frankia* was first observed several decades ago. That the deformed root hair was the site of entry for the bacterium was clarified by Taubert[162] and Angulo-Carmona.[6]

Root hair infection (RHI) in actinorhizal plants involves several distinct phases. The first phase is deformation and/or curling of the host plant root hair. Through some undefined chemical signal from *Frankia*, actively growing root hairs become deformed by branching, clubbing, or actual curling.[29] As the root hair becomes increasingly deformed, *Frankia* hyphae are frequently entrapped between segments of the root hair. It appears that in these zones of entrapment, *Frankia* hyphal branches initiate the digestion of the primary cell wall of the root hair. In response to the partial wall digestion, the host plant cell begins to build additional wall material around the site of digestion (Figure 7-6a,b). Continued digestion by the bacterium and concurrent wall building by the host plant create a tubular ingrowth within the root hair cell that has been termed the *encapsulation*. This structure is similar in function to the infection threads observed in the legume-rhizobial symbioses. Although it is convenient to refer to the bacterium as "within" the plant cells from the time of infection, in all respects and at all times the bacterium is exterior to the cell, that is, outside the plasma membrane and outside the encapsulation or cell wall.[110]

Prenodule initiation is the second phase in RHI after actual infection of the root hairs. In all root hair infected plants, a proliferation of cells in the hypodermis occurs at or very near the base of the infected root hair. The function of the prenodule is not well understood, but as the invading *Frankia* filaments grow into the root proper from the root hair, they branch and ramify into the newly formed cells of the prenodule. Vesicles may differentiate on the microsymbiont within the prenodule cells[101] and thus N_2 fixation may begin to benefit the plant immediately.

In the third phase of RHI the host plant initiates a nodule primordium from cells in the root pericycle. The nodule primordium initiated in response to *Frankia* infection is not significantly different from lateral root primordia, except perhaps in orientation to the root xylem poles.[36] The *Frankia* hyphae that have proliferated in the prenodule continue to proceed, accompanied by host-produced encapsulation and bacterial cell division, through the root cortical cells and toward the incipient nodule primordium. Upon entering the primordium, once again the *Frankia* filaments branch and invade numerous cells of the young nodule cortex.

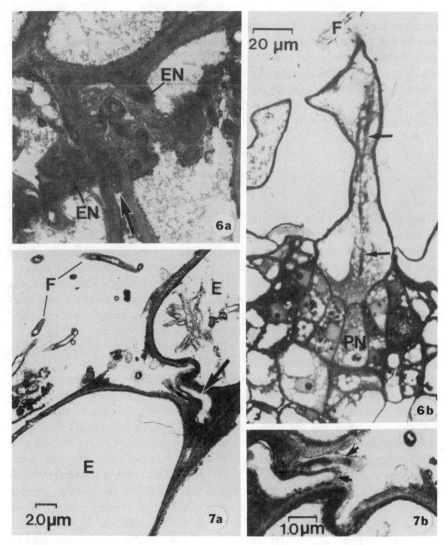

Figure 7-6. Details of the root hair mediated infection process in *Myrica*. (a) Major cell wall deposition and encapsulation (EN) of the microsymbiont (arrow) at the site of the infection in the root hair. (b) In early stages the bacterial hyphae (arrows) are observed within the infected root hair and the adjacent cortical cells of the root have divided to form a characteristic swelling called the prenodule (PN). (Reprinted with permission from Miller and Baker.[101])

Figure 7-7. Details of the intercellular penetration infection process in *Elaeagnus*. (a) Filaments of *Frankia* (F) are present on the surface of the root. Intercellular penetration occurs between adjacent epidermal cells (E) through the middle lamellar region (arrow). (b) Enlargement of the infection site. The bacterial hypha has entered between epidermal cells through an enzymatic digestion process (arrows). (Reprinted with permission from Miller and Baker.[101])

Almost immediately, *Frankia* begins to produce differentiated vesicles, which will be the site of N_2 fixation. In this, the fourth phase of RHI, the new nodule takes on the anatomy and physiology of a mature root nodule and significant N_2 fixation by the bacterium and nitrogen transfer to the host plant begins. The host plant nodule meristem will continue to divide and produce new cortical cells that will subsequently be invaded by *Frankia*. In this way, the nodule will grow in size to increase its functional capabilities as the plant grows.

B.   Intercellular Penetration

Until 1985 the root hair infection process was believed to be the sole mode of entry into the host root for *Frankia*. This was understandable because the majority of species studied intensively were infected via root hairs. Miller and Baker[100] described a unique mode of entry for the infection of *Elaeagnus*. They observed that *Elaeagnus*, as well as the other genera of the Elaeagnaceae, *Hippophaë* and *Shepherdia*, often failed to produce root hairs yet were well nodulated. Light microscopical observations also indicated that when root hairs were present they were not deformed.

Using the transmission electron microscope it is possible to follow the entry of *Frankia* into the root and the pattern of development for the formation of a new nodule. *Frankia* hyphae may be observed to grow between adjacent epidermal cells of the root (Figure 7-7a,b). *Frankia* filaments apparently digest away a portion of the middle lamellar region to gain entry. Once within the root, the bacterial filaments branch and ramify in the intercellular spaces between root cortical cells. In response to the presence of the bacteria, the host plant produces a darkly staining substance that is exuded into the intercellular spaces to surround the invading bacterium. This substance is digested by the bacterium as it continues to proliferate. Despite the fact that the bacterium remains intercellular to this point, differentiated vesicles may be observed, thus indicating the potential for N_2 fixation. A prenodule, typical of the RHI species, does not form in plants nodulated by intercellular penetration (IP).

A nodule primordium is initiated in the pericycle in a manner similar to that observed in RHI species. The *Frankia* hyphae grow toward this primordium and enter the newly forming nodule while remaining intercellular colonizers. At some point the hyphae within the nodule cortical zone branch and penetrate individual cells in a way similar to that observed in the infection of root hairs in the RHI process. At this point in development the bacterium is for the first time "within" host plant cells, though still remaining external to the host plasma membrane and a wall-like encapsulation. Numerous vesicles are produced and N_2 fixation commences.

Although IP processes have been reported only for the Elaeagnaceae, it is probable that other actinorhizal genera are infected via this mode of entry. Recent preliminary studies of *Ceanothus* and other members of the Rhamnaceae and Rosaceae suggest that numerous species may be infected by IP.[27]

VII. Nodule Growth and Development

Continued growth and development of the nodule after infection and initiation is similar in both RHI and IP host species. Each nodule lobe has an internal anatomy very similar to that of a lateral root, with a central cylinder of vascular tissue, a cortical region in which are found the infected cells, and a typical outer periderm layer. Most nodule lobes have a determinate meristem at the distal tip (Figure 7-8a). However, this meristem in some species (e.g., *Casuarina*, *Myrica*) may continue to grow and appears to escape the influence of the bacterium, forming an ageotropic "nodule root" (Figure 7-8b). Nodule roots are found primarily in plant species subject to periodic waterlogging where they are thought to play a functional role in transporting air to the root nodule lobes.[169]

Numerous times during each growing season, additional nodule lobe meristems are initiated, giving rise to a nodule composed of multiple lobes. In time this pattern of division gives rise to a coralloid and somewhat spherical structure that is characteristic of older nodules. In most actinorhizal species the root nodules are perennial, with the older portions of the nodule becoming lignified. In moderately old actinorhizal trees such as *Alnus* or *Casuarina* the size of these nodules may exceed 30 cm in diameter. In other actinorhizal species, particularly those native to arid regions, it appears that cycles of drought may induce the senescence of the nodules on a routine basis and nodules do not become much larger than a few centimeters in diameter.[79]

Nodule formation and growth are undoubtedly linked to altered phytohormone levels in the region of infected root tissue. Cytokinin and indole acetic acid occur in high amounts in *Alnus glutinosa* nodules,[69,190] and exogenously supplied cytokinin leads to the formation of numerous pseudonodules on uninfected *Alnus* and *Myrica* roots.[26] More recent studies have shown that both cytokinin and indole compounds are secreted by *Frankia* in culture,[28,160] and it is likely that these phytohormones have an effect on nodule growth and development.

VIII. Nodule Physiology

A. Carbon and Nitrogen

As the nodule develops, it becomes a sink for photosynthate.[188] In all actinorhizal plants examined to date, sucrose is the major form in which pho-

Figure 7-8. External morphologies of actinorhizal root nodules. (a) Typical coralloid root nodules of *Elaeagnus*, an example of the "*Alnus*-type" nodule. (b) Root nodule of *Myrica*, an example of the "*Myrica*-type" nodule. Nodule roots (nr) grow out of the tips of the nodule lobes in an ageotropic fashion.

tosynthate reaches the nodule. It is transported in the phloem and appears to be unloaded where phloem tissue comes in contact with infected and uninfected nodule cells.[73] Sucrose is taken into these cells and subsequently undergoes metabolism or is stored as starch. The photosynthate that reaches the nodule must be partitioned into three major physiological sinks. Some of it must be used for energy and carbon skeletons for nodule and endophyte growth, some must be used to generate reductants and to support oxidative phosphorylation for N_2 fixation, and some must be used by the plant to transport the fixed N_2 out of the nodule and throughout the plant.

The form of carbon that is supplied to symbiotic frankiae is not known. Studies on vesicle clusters isolated from nodule cells have shown that externally supplied sucrose, trehalose, maltose, glucose, and fructose are all able to increase respiratory rates in frankiae,[92] as are 6-phosphogluconate and glucose-6-phosphate.[183] Diurnal and seasonal fluctuations in the supply of photosynthate significantly affect rates of N_2 fixation in actinorhizal nodules.[1,125,136,186] In young nodules N_2 fixation rates are proportional to photosynthetic rates[42,61] and do not seem to be closely related to the levels of carbohydrate stored in nodules. Leaf area, light intensity, and weather conditions, especially temperature,[136] all affect rates of N_2 fixation in young nodulated plants.[76] In older nodules N_2 fixation may continue in the absence of photosynthesis, probably by utilizing carbohydrate stored in the nodules or adjacent root tissues.

Rates of N_2 fixation within actinorhizal nodules are also regulated by levels of combined nitrogen external to the nodule.[75,77] For example, within three days after treatment of *A. incana* seedlings with 20 mM NH$_4$Cl, nitrogenase activity of intact nodules has been shown to decrease to 10% of initial rates. This regulation does not appear to be due to a limited supply of photosynthate and appears to result rather from a decay of nitrogenase activity associated with vesicle senescence.

Although *Frankia* cells in culture can readily fix N_2, increased growth occurs when a reduced nitrogen source is provided. For most strains tested, ammonium ions are the preferred source of reduced nitrogen, although a number of other nitrogenous compounds may support growth as well.[139] A high-affinity ammonium transport system has been characterized in *Frankia* strain CpI1, but because it is inducible only under conditions of nitrogen starvation it is not likely to be involved in ammonium transport in the nodule where there are likely to be high levels of ammonium.[95] Glutamine synthetase II, which is present in *Frankia* filaments and in very low levels in vesicles,[52,114,175] is induced under conditions of nitrogen starvation. The combined action of these two inducible enzyme systems would enable *Frankia* to survive in soils very low in nitrogen and carbon. When nitrogen is not limited, glutamine synthetase I is active and is considered to be the major route for the assimilation of nitrogen.

Ammonium is very likely the form in which reduced nitrogen is transported from N_2-fixing vesicles to non-N_2-fixing hyphae and to the host cell cytoplasm. It is not clear yet what enzymes are responsible for ammonium assimilation in nodule cells. Both glutamine synthetase[71] and glutamate dehydrogenase[30] activities have been found in *Alnus* nodules and ^{13}NH$_4$$^+$ labeling has indicated that both may be involved in assimilating ammonium in *Alnus* nodules.[133] Analysis of the total amino acid composition of nodules and of xylem sap from cut stems of most *Alnus* species[30,58,99,133] and of *Casuarina equisetifolia*[185] clearly identify citrulline as the major amino acid in these species. In other actinorhizal plants studied, including *Alnus inokumai* and *Casuarina cunninghamiana*, predominant amino acids include arginine, asparagine, glutamine, or glutamate.[62,132,187] There is no evidence that any actinorhizal plants transport ureides, which are major nitrogen transport compounds in some legumes.[47]

B. Oxygen

Despite the extreme sensitivity of *Frankia* nitrogenase proteins to molecular O_2,[12,21,105] N_2 fixation by *Frankia* requires a high level of aerobic metabolism to generate the ATP and NADH necessary to reduce N_2. Contrary to the case found with legume nodules that are O_2-limited, actinorhizal nodules show maximum rates of N_2 fixation at atmospheric levels of O_2.[59,168] In fact, both free-living and symbiotic frankiae are able to fix N_2 at levels of O_2 ranging from 2 to 70 kPa O_2 provided that they have been preadapted to assay levels of O_2.[123,143,144] There appear to be at least two O_2 protection mechanisms in operation in *Frankia*. One is a rapid response to increased O_2 levels providing short-term protection and the second is a slower response providing long-term O_2 protection. In cultured as well as symbiotic frankiae, stepwise exposure to higher than ambient levels of O_2 results in immediate transient drops in nitrogenase activity followed by a recovery of activity.[141] It is proposed that at each stepwise increase in O_2, nitrogenase undergoes a conformational change that temporarily inactivates it and protects it from permanent damage. Increased respiratory rates stimulated by higher O_2 levels lower O_2 to permissible levels, and nitrogenase activity is resumed. In addition, the presence of superoxide dismutase and catalase within *Frankia* cells may provide protection for nitrogenase.[158]

Long-term protection from O_2 is provided by alterations in the vesicle wall. In free-living aerobic cultures of *Frankia* N_2 fixation occurs largely if not solely within spherical vesicles that are produced in response to low levels of reduced nitrogen.[98,114,163] The thickness of the vesicle envelope is regulated by the pO_2 of the culture medium, with wall thickness increasing with increasing levels of O_2.[123] The multilaminate vesicle envelope[172] has a lipid composition[86] differing from that of noninduced cultures and has an especially high amount of C_{22}–C_{26} polyhydroxy fatty acids or alcohols.[176] Although the structure of these long-chain compounds has not been determined, they may bear a resemblance to the C_{26}–C_{28} hydroxyl alcohols and fatty acids found in *Anabaena* heterocysts.[85] Recently, high levels of a triterpene hopanoid have been found in *Frankia* (A. Berry, personal communication) and it remains to be seen whether it is involved in vesicle function. The vesicle envelope is thought to provide a barrier to O_2 diffusion to protect the nitrogenase within. As O_2 levels in culture increase, new nitrogenase enzyme is synthesized, and thickening of the vesicle envelope occurs on existing as well as newly formed vesicles.[123] In free-living cultures grown at very low O_2 levels, N_2 fixation occurs in the absence of vesicle formation,[105] indicating that the formation of vesicles is not essential for N_2 fixation and is likely to be an O_2 protection mechanism allowing *Frankia* to fix N_2 over a very wide range of O_2 concentrations. This is in sharp contrast to rhizobia, which have no analogous intrinsic O_2 protection mechanism and is similar in nature to the O_2 protection mechanism of the heterocystous cyanobacteria.

Within nodule tissue the extent of vesicle formation and the shape of vesicles are controlled by the host plant. Figure 7-5 shows the variation in vesicle shape and anatomy that can exist within nodule cells. In *Alnus* nodules, vesicle shape is nearly spherical and vesicle envelope thickness appears to increase as concentrations of O_2 surrounding the nodule increase.[143] In *Alnus* there are no apparent host-mediated barriers to the diffusion of O_2 and infected cells are in contact with ambient levels of O_2 via numerous interconnected intercellular spaces.[165] In this case it is likely that the vesicle wall envelope provides the major barrier to O_2 diffusion.

The lack of endophyte vesicles in *Casuarina* nodules (see Figure 7-5) is not a characteristic of the endophyte, which does form vesicles in aerobic culture, but is likely to reflect the low O_2 levels within nodule tissue. Cytological analysis of *Casuarina* nodules indicates that infected host cell walls change in composition, becoming more hydrophobic and probably less permeable to O_2, upon penetration of the cells by *Frankia*.[25] Furthermore, infected cells are not in direct contact with air passages that connect outer and inner cortical cell layers in the nodules,[191] so it is unlikely that infected cells are at ambient O_2 levels. Studies of respiratory rates and nitrogenase activity in response to changes in temperature and pO_2 provide physiological evidence for the presence of a host-mediated barrier to O_2 in *Casuarina*.[167] However, intracellular hyphae in *Casuarina* nodules do possess a laminate surface layer that may be chemically similar to the multilaminate envelope seen in vesicles of *in vitro* cultured frankiae.[24] This layer may also be involved in the regulation of O_2 concentrations within symbiotic frankiae.

In *Myrica* species that are adapted to wetland habitats, vesicle envelope structure does not appear to change with changing pO_2 levels. Instead, nodule ventilation decreases with increasing levels of O_2 and few air spaces are found adjacent to infected cells. Nodule roots increase in length and in diameter with decreasing ambient O_2 levels, thus providing a mechanism to channel more O_2 to nodule cells.[144]

In yet another kind of O_2 response, *Coriaria* nodules are able to respond more quickly to changes in pO_2 by some mechanism that causes a rapid (5–30 min) increase in resistance above ambient O_2 levels and a rapid decrease in resistance below ambient O_2 levels.[140] In these nodules infected cells are surrounded by a dense suberized periderm whose thickness differs with the O_2 concentration at which the nodules are grown. The variable resistance barrier is thought to be a two-cell-layer gap between the endodermis and the internal periderm, adjacent to the single-nodule lenticel.

Hemoglobin has been found in the nodules of some species of actinorhizal plants but is undetectable in the nodules of others. It is present in the highest concentrations in the nodules of *Casuarina* and *Myrica*[166] and has been purified and characterized from *C. glauca*[7,55] and *M. gale*.[124] Hemoglobin from *C. glauca* has approximately 52% and 43% amino acid sequence similarity with hemoglobins from *Parasponia* and soybean, respectively, providing evidence that it is evolutionarily related to other globin proteins.[7,82] CO-reactive heme has been found in decreasing concentrations in nodules of *Alnus*, *Comptonia*, *Ceanothus*, *Hippophaë*, *Elaeagnus*, *Coriaria*, and *Purshia* but is not detectable in *Datisca*.[142,166] Whether the CO-reactive heme found in these actinorhizal nodules is hemoglobin remains to be established.

The presence of elevated levels of hemoglobin in *Casuarina* and *Myrica* can be correlated with nodule structure in these two species. In each case a barrier to O_2 diffusion surrounds infected nodule cells and lowers the O_2 concentration within infected cells. At lower O_2 levels hemoglobin may serve to facilitate the transport of O_2 to metabolically active frankiae. This role is consistent with the role hemoglobin is thought to play in legume nodules, where all nodules are found to have a barrier to O_2 diffusion to infected cells. Actinorhizal species, with lower levels of hemoglobin, have well-ventilated nodules and the only diffusion barrier to O_2 appears to be the vesicle envelope of the endophyte.

The finding of hemoglobin in the nodules of such evolutionarily diverse actinorhizal plants and in the roots of non-nodulating plants[33] has raised questions about the nonsymbiotic role of these plant globins. It has been suggested that a general role for plant globins may be that of an O_2 sensor responsible for shifts between aerobic and anaerobic metabolism.[7]

C. Effectiveness of Nitrogen Fixation

It has been shown that both partners in the actinorhizal symbiosis play a role in determining the overall effectiveness of N_2 fixation within a nodulated plant. Numerous studies have shown that inoculating different provenances or clones of *Alnus* with a single *Frankia* inoculum results in variation in growth and nodulation.[43,46,61,72,94] Likewise, uniform plants inoculated with different *Frankia* strains exhibited differences in performance.[43,46,63,94,116,127,138,159] In most cases it has not been possible to document the cause of the variation seen. In one study, however, differences in effectiveness of N_2 fixation, evidenced by growth and biomass production, were linked to the presence or absence of hydrogenase activity in nodules.[138] *Frankia* strains capable of expressing hydrogenase activity *in vitro* and *in planta* formed a more effective symbiosis with their host plants than did those without hydrogenase activity. The presence of an uptake hydrogenase that allows recycling of H_2 from nitrogenase-dependent H_2 evolution is common among *Frankia*.[21,22,128] Hydrogenase immunologically related to the large subunit of hydrogenase from *Alcaligenes* has been located in both filaments and vesicles of *Frankia* CpI1.[91]

It has been observed that, in general, nodules that contain sporulating *Frankia* (Sp$^+$ nodules) have lower rates of N_2 fixation than nodules lacking sporulating *Frankia* (Sp$^-$).[102,116,135] In both *Myrica* and *Comptonia*, it has been shown that endophytic vesicles and host cell cytoplasm rapidly senesce as sporangia of Sp$^+$ nodules develop, thus causing the lower rates of N_2 fixation observed.[181,182] Although the production of spores *in planta* is a stable characteristic of some *Frankia*, it is not known how different these strains are from the more common Sp$^-$ strains.

The reader is referred to three recent reviews on the physiology of actinorhizal symbioses.[73,142,169]

IX. Ecological Studies

Although found worldwide, actinorhizal plants are more numerous and more diverse in temperate regions, where they are frequently found on soils poor in nitrogen. They are pioneer species, able to colonize severely disturbed sites, and are thought to contribute to early succession in these sites by stabilizing the soil and building its nitrogen content. Many actinorhizal plants are mycorrhizal as well[129] and thus possess the capability of extracting other nutrients from the soil to further enhance their success on poor soils. The increased growth performance and C_2H_2 reduction rates due to tripartite or tetrapartite symbioses has been documented for several actinorhizal species.[39,130,131] As succession proceeds, actinorhizal plants become less numerous as non-N_2-fixing species are able to move into the area. However, in areas that remain less than optimum for many plant species (desert, chaparral, riparian), actinorhizal plants may remain dominant.

Volcanic sites provide an excellent opportunity to study the role of actinorhizal plants in early succession. In a recent study of the biological invasion of *Myrica faya* at a volcanic site in Hawaii, this actinorhizal species was found to fix 18.5 kg N ha^{-1} yr^{-1} and to increase significantly the amount of available nitrogen in the soil under the plants. Non-N_2-fixing species growing in this soil and benefiting from the companion plants accumulated greater biomass than those growing at locations not adjacent to *Myrica*.[184] The success of *Myrica* at this Hawaiian site was attributed to its N_2-fixing capabilities, its prolific seed production, its effective seed dispersal by exotic birds, and its rapid growth rates. The recent volcanic eruption of Mt. St. Helens in the state of Washington, U.S.A., has provided a similar opportunity to study the invasion of ash-covered mud deposits by *Alnus rubra*.[40]

In order for actinorhizal plants to be successful as pioneer species they must be nodulated by appropriate *Frankia* strains. *Frankia* seems to be present in many soils, even in the absence of host plants.[74,122,155] In most cases actinorhizal plants are nodulated wherever they are found. However, there are exceptions to this general observation. For example, using the ability of host plants to become nodulated as a means of assessing *Frankia* populations in the soil, no evidence was found for the presence of *Frankia* in peat soils from which the top layers had been previously harvested.[8]

Soils with the greatest ability to induce nodulation are not always those under host plants. Soil under the nonhost plant *Betula pendula* has been found to induce nodulation in *Alnus incana* to a greater extent than the soil under nearby *Alnus* trees.[153,154] In these same studies soil under nearby coniferous trees had a very low nodulation capacity, indicating a degree of specificity in the relationship between *Betula* soils and *Frankia*. Several explanations may be set forth to explain the high populations of *Frankia* in soils under birch. Root exudates or leaf litter of birch may supply a specific set of compounds favorable to the saprophytic growth of *Frankia*, although no stimulation of growth was found when *in vitro* cultured strains were supplemented with aqueous soil extracts.[154] Alternatively, an overall high level of microbial activity in soils under *Betula* may provide an enhanced environment for *Frankia* growth. Yet another explanation may be that the microbial composition of the soil under *Betula* stimulates nodulation by *Frankia*. It has been shown that microorganisms other than *Frankia* can enhance nodule formation by *Frankia*[80,81] and this could lead to the appearance of a larger *Frankia* population. As techniques are developed for directly assessing *Frankia* populations in the soil, it will be possible to measure independently nodulation efficiency and population density.

Despite the general presence of *Frankia* in the soil, nothing yet is known about the ability of *Frankia* to grow and compete in the soil. This is largely because reliable methods are not available for the direct isolation of *Frankia* from soil. Additionally, there is little information available about strain competition for nodulation of host plants. Genetic analysis of *Frankia* isolates indicates that there is genetic diversity within single localities.[20,23,32,57] Physiological analysis has shown that isolates differ significantly from one another in symbiotic properties.[53,102,116,127,138,180] New techniques are needed to be able to study the structure of *Frankia* populations in the soil and the effects that host and nonhost plants have on population dynamics.

X. Future Research Directions

There is a definite need to continue research on basic as well as practical aspects of actinorhizal symbioses. Physiological and biochemical studies of carbon and nitrogen metabolism in the nodule must be expanded and cytological observations of nodulation events need to be complemented by molecular analyses. The development of mutation selection and transformation systems for *Frankia* are essential to the systematic study of the genetics and the molecular biology of actinorhizal symbiosis. Likewise, the development of techniques for the rapid identification of *Frankia* in the soil and in nodules is essential for many ecological studies. Attempts to isolate frankiae from all host plant genera should continue and trials testing the effectiveness of various host plant and *Frankia* combinations need to be initiated for many actinorhizal species. Resulting superior symbiotic combinations will need to undergo field testing in a variety of habitats and should result in the increased utilization of actinorhizal plants in forestry, agriculture, land reclamation, and amenity planting.

Progress has been made in all of the areas discussed here. The ability to form and regenerate protoplasts from *Frankia* cultures is one step toward the development of a mutation selection and transformation system.[121,164] Another positive step is the demonstration that at least some isolated vesicles[134] and spores[178] (H. Berg, personal communication) have the ability to grow in culture. The study of *Frankia* plasmids[51,118,120] may lead to the construction of vectors for transformation, and the likelihood exists that *Frankia* strains have phage that could be developed into vectors (M. Lechevalier, personal communication).

Methods based on restriction pattern analysis[3,32,49,50,78] or restriction fragment length polymorphisms[119] have made it possible to distinguish reliably among isolates in culture and in some cases to predict some symbiotic characteristics of the isolates.[113] Direct hybridization of plasmid probes to DNA from crushed nodules has permitted the identification of plasmid-bearing strains within nodules[147] and polymerase chain reaction (PCR) amplification of strain-specific sequences in nodule DNA has permitted the identification of *Frankia* strains within individual nodules and has made it possible for competition studies between two *Frankia* strains to be undertaken.[151] Synthetic oligonucleotide probes complementary to variable regions of *Frankia* ribosomal RNA have been used to distinguish between strains grown in culture.[64] These probes also have been used to distinguish strains *in planta* from as little as 1 mg of nodule tissue. Preliminary studies have indicated that PCR amplification has made it possible to detect *Frankia* DNA in extracts of soils that have received additions of *Frankia* cells (A. Hilger and D. Myrold, personal communication), opening up the possibility for studies on the dynamics of *Frankia* populations in the soil.

Molecular studies have been initiated on host plant genes and proteins involved in symbiosis. Hemoglobin has been identified as a protein whose abundance increases in the nodules of some actinorhizal plants, and gel electrophoresis of total root and nodule proteins has shown the existence of other proteins that appear to be unique to nodules (A. Séguin and M. Lalonde, 2nd Int'l. Congress of Plant Molecular Biology, Jerusalem, 1988; H. Belefant and B. Mullin, unpublished). cDNA cloning of nodule RNA has been accomplished and should lead to the characterization of nodule-specific gene sequences (P. Twigg and B. Mullin, unpublished). Studies on non-nodulating plant species that are closely related to actinorhizal plants offer the opportunity for comparative biochemical and genetic analyses. These species may be the best candidate for attempts to broaden the host range of *Frankia*.[35,60,93]

References

1. Akkermans, A. D. L., and van Dijk, C., "The formation and nitrogen-fixing activity of the root nodules of *Alnus glutinosa* under field conditions," *Symbiotic Nitrogen Fixation in Plants*, Nutman, P. S., ed. Cambridge: Cambridge University Press, 1976, pp. 511–520.

2. An, C. S., Riggsby, W. S., and Mullin, B. C., "Relationships of *Frankia* based on deoxyribonucleic acid homology studies," *Int. J. Syst. Bacteriol. 35*, 140–146 (1985).

3. An, C. S., Riggsby, W. S., and Mullin, B. C., "Restriction pattern analysis of genomic DNA of *Frankia* isolates," *Plant and Soil 87*, 43–48 (1985).

4. An, C. S., Riggsby, W. S., and Mullin, B. C., "DNA relatedness of *Frankia* isolates ArI4 and EuI1 to other actinomycetes of cell wall type III," *The Actinomycetes 20*, 50–59 (1987).

5. An, C. S., Wills, J. W., Riggsby, W. S., and Mullin, B. C., "Deoxyribonucleic acid base composition of 12 *Frankia* isolates," *Can. J. Bot. 61*, 2859–2862 (1983).

6. Angulo-Carmona, A. F., "La formation des fixateurs d'azote chez nodules *Alnus glutinosa* (L.) Vill.," *Acta Bot. Neerl. 23*, 257–303 (1974).

7. Appleby, C. A., Bogusz, D., Dennis, E. S., and Peacock, W. J., "A role for hemoglobin in all plant roots," *Plant Cell and Environ. 11*, 359–367 (1988).

8. Arveby, A. S., and Huss-Danell, K., "Presence and dispersal of infective *Frankia* in peat and meadow soils in Sweden," *Biol. Fertil. Soils 6*, 39–44 (1988).

9. Baker, D., "Modification of catalog numbering system for *Frankia* strains and revision of strain numbers for DDB *Frankia* collection," *The Actinomycetes 20*, 85–88 (1987).

10. Baker, D. D., "Relationships among pure cultured strains of *Frankia* based on host specificity," *Physiol. Plant. 70*, 245–248 (1987).

11. Baker, D. D., "Methods for the isolation, culture, and characterization of the Frankiaceae: Soil actinomycetes and symbionts of actinorhizal plants," *Isolation of Biotechnological Organisms from Nature*, Labeda, D., ed. New York: McGraw-Hill, 1989, pp. 213–236.

12. Baker, D., and Huss-Danell, K., "Effects of oxygen and chloramphenicol on *Frankia* nitrogenase activity," *Arch. Microbiol. 144*, 233–236 (1986).

13. Baker, D., and O'Keefe, D., "A modified sucrose fractionation procedure for the isolation of frankiae from actinorhizal root nodules and soil samples," *Plant and Soil 78*, 23–28 (1984).

14. Baker, D., and Seling, E., "*Frankia*: New light on an actinomycete symbiont," *Biological, Biochemical and Biomedical Aspects of Actinomycetes*, Ortiz-Ortiz, L., Bojalil, L. F., and Yakoleff, V., eds. New York: Academic Press, 1984, pp. 563–574.

15. Baker, D., and Torrey, J. G., "The isolation and cultivation of actinomycetous root nodule endophytes," *Symbiotic Nitrogen Fixation in the Management of Temperate Forests*, Gordon, J. C., Wheeler, C. T., and Perry, D. A., eds. Corvallis: Forest Research Laboratory, Oregon State University, 1979, pp. 38–56.

16. Baker, D., and Torrey, J. G., "Characterization of an effective actinorhizal microsymbiont, *Frankia* sp. AvcI1 (Actinomycetales)," *Can. J. Microbiol. 26*, 1066–1071 (1980).

17. Becking, J. H., "Frankiaceae *fam. nov.* (Actinomycetales) with one new combination and six new species of the genus *Frankia* Brunchorst 1886, 174," *Int. J. Syst. Bacteriol. 20*, 201–220 (1970).

18. Benoit, L. F., and Berry, A. M., "Methods for production and use of actinorhizal plants in forestry, low maintenance landscapes and revegetation," *The Biology of Frankia and Actinorhizal Plants*, Schwintzer C. R., and Tjepkema, J. D., eds. New York: Academic Press, 1990, pp. 281–297.

19. Benson, D. R., "Isolation of *Frankia* strains from alder actinorhizal root nodules," *Appl. Environ. Microbiol. 44*, 461–465 (1982).

20. Benson, D. R., and Hanna, D., "*Frankia* diversity in an alder stand as estimated by sodium dodecyl sulfate-polyacrylamide gel electrophoresis of whole-cell proteins," *Can. J. Bot. 61*, 2919–2923 (1983).

21. Benson, D. R., Arp, D. J., and Burris, R. H., "Cell-free nitrogenase and hydrogenase from actinorhizal root nodules," *Science 205*, 688–689 (1979).

22. Benson, D. R., Arp, D. J., and Burris, R. H., "Hydrogenase in actinorhizal root nodules and root nodule homogenates," *J. Bacteriol. 142*, 138–144 (1980).

23. Benson, D. R., Buchholz, S. E., and Hanna, D. G., "Identification of *Frankia* strains by two-dimensional polyacrylamide gel electrophoresis," *Appl. Environ. Microbiol. 47*, 489–494 (1984).

24. Berg, H., and McDowell, L., "Endophyte differentiation in *Casuarina* actinorhizae," *Protoplasma 136*, 104–117 (1987).

25. Berg, H., and McDowell, L., "Cytochemistry of the wall of infected cells in *Casuarina* actinorhizae," *Can. J. Bot. 66*, 2038–2047 (1988).

26. Bermudez de Castro, F., Canizo, A., Costa, A., Miguel, C., and Rodriguez-Barrueco, C., "Cytokinins and nodulation of the non-legumes *Alnus glutinosa* and *Myrica gale*," *Recent Developments in Nitrogen Fixation*, Newton, W., Postgate, J. R., and Rodriguez-Barrueco, C., eds. London: Academic Press, 1977, pp. 439–450.

27. Berry, A. M., and Sunell, L. A., "The infection process and nodule development," *The Biology of Frankia and Actinorhizal Plants*, Schwintzer, C. R., and Tjepkema, J. D., eds. New York: Academic Press, 1990, pp. 61–81.

28. Berry, A. M., Kahn, R. K. S., and Booth, M. C., "Identification of indole compounds secreted by *Frankia* HFPArI3 in defined culture medium," *Plant and Soil 118*, 205–209 (1989).

29. Berry, A. M., McIntyre, L., and McCully, M. E., "Fine-structure of root hair infection leading to nodulation in the *Frankia-Alnus* symbiosis," *Can. J. Bot. 64*, 292–305 (1986).

30. Blom, J., Roelofsen, W., and Akkermans, A. D. L., "Assimilation of nitrogen in root nodules of alder (*Alnus glutinosa*)," *New Phytol. 89*, 321–326 (1981).

31. Bloom, R. A., Lechevalier, M. P., and Tate III, R. L., "Physiological, chemical, morphological and plant-infectivity characteristics of *Frankia* isolates from *Myrica pensylvanica*; correlation to DNA restriction patterns," *Appl. Environ. Microbiol. 55*(9), 2161–2166 (1989).

32. Bloom, R. A., Mullin, B. C., and Tate III, R. L., "DNA restriction patterns and solution hybridization studies of *Frankia* isolates from *Myrica pensylvanica* (Bayberry)," *Appl. Environ. Microbiol. 55*(9), 2155–2160 (1989).

33. Bogusz, D., Appleby, C. A., Landsmann, J., Dennis, E. S., Trinick, M. J., and Peacock, W. J., "Functioning haemoglobin genes in a non-nodulating plant," *Nature 331*, 178–180 (1988).

34. Bond, G., and Becking, J. H., "Root nodules in the genus *Colletia*," *New Phytol. 90*, 57–65 (1982).

35. Bousquet, J., Girouard, E., Strobeck, C., Dancik, B. P., and Lalonde, M., "Restriction fragment polymorphisms in the rDNA region among seven species of *Alnus* and *Betula papyrifera*," *Plant and Soil 118*, 231–240 (1989).

36. Callaham, D., and Torrey, J. G., "Prenodule formation and primary nodule development in roots of *Comptonia* (Myricaceae)," *Can. J. Bot. 55*, 2306–2318 (1977).

37. Callaham, D., Del Tredici, P., and Torrey, J. G., "Isolation and cultivation *in vitro* of the actinomycete causing root nodulation in *Comptonia*," *Science 199*, 899–902 (1978).

38. Callaham, D., Newcomb, W., Torrey, J. G., and Peterson, R. L., "Root hair infection in actinomycete-induced root nodule initiation in *Casuarina, Myrica*, and *Comptonia*," *Bot. Gaz. 140*(S), 1–9 (1979).

39. Chatarpaul, L., Chakravarty, P., and Subramanim, P., "Studies in tetrapartite symbioses. I. Role of ecto- and endomycorrhizal fungi and *Frankia* on the growth performance of *Alnus incana*," *Plant and Soil 118*, 145–150 (1989).

40. Dawson, J. O., "Actinorhizal plants: Their use in forestry and agriculture," *Outlook Agric. 15*(4), 202–208 (1986).

41. Dawson, J. O., "Interactions among actinorhizal plants and associated species," *The Biology of Frankia and Actinorhizal Plants*, Schwintzer, C. R., and Tjepkema, J. D., eds. New York: Academic Press, 1990, pp. 299–316.

42. Dawson, J. O., and Gordon, J. C., "Nitrogen fixation in relation to photosynthesis in *Alnus glutinosa*," *Bot. Gaz. 140*(S), 70–75 (1979).

43. Dawson, J. O., and Sun, S. H., "The effect of *Frankia* isolates from *Comptonia peregrina* and *Alnus crispa* on the growth of *Alnus glutinosa, A. cordata* and *A. incana* clones," *Can. J. Forest Res. 11*, 758–762 (1981).

44. Diem, H. G., and Dommergues, Y. R., "Current and potential uses and management of Casuarinaceae in the tropics and subtropics," *The Biology of Frankia and Actinorhizal Plants*," Schwintzer, C. R., and Tjepkema, J. D., eds. New York: Academic Press, 1990, pp. 317–342.

45. Diem, H. G., Duhoux, E., Simonet, P., and Dommergues, Y. R., "Actinorhizal symbiosis biotechnology: The present and the future," *Proceedings of the 8th International Biotechnology Symposium*, Durand, G., Bobichon, L., and Florent, J., eds. Paris: Société Francaise de Microbiologie, 1988, pp. 984–995.

46. Dillon, J. T., and Baker, D., "Variations in nitrogenase activity among pure-cultured *Frankia* strains tested on actinorhizal plants as an indication of symbiotic compatibility," *New Phytol. 92*, 215–219 (1982).

47. Dixon, R. O. D., and Wheeler, C. T., "Biochemical, physiological and environmental aspects of symbiotic nitrogen fixation," *Biological Nitrogen Fixation in Forest Ecosystems: Foundations and Applications*, Gordon, J. C., and Wheeler, C. T., eds. Dordrecht: Martinus Nijhoff/Dr. W. Junk, 1986, pp. 107–171.

48. Dobritsa, S. V., "Extrachromosomal circular DNAs in endosymbiont vesicles from *A. glutinosa* root nodules," *FEMS Microbiol. Lett. 15*, 87–91 (1982).

49. Dobritsa, S. V., "Restriction analysis of the *Frankia* spp. genome," *FEMS Microbiol. Lett. 29*, 123–128 (1985).

50. Dobritsa, S. V., and Stupar, O. S., "Genetic heterogeneity among *Frankia* isolates from root nodules of individual actinorhizal plants," *FEMS Microbiol. Lett. 58*, 287–292 (1989).

51. Dobritsa, S. V., and Tomashevsky, A. Yu., "Homology between structural nitrogenase genes (*nifHDK*) from *Klebsiella pneumoniae* and extrachromosomal DNAs from microsymbiont vesicles of *A. glutinosa*," *Biopolym. Cell 4*, 44 (1988).

52. Edmands, J., Noridge, N. A., and Benson, D. R., "The actinorhizal root-nodule symbiont *Frankia* sp. strain CpI1 has two glutamine synthetases," *Proc. Natl. Acad. Sci. USA 84*, 6126–6130 (1987).

53. Favilli, F., Margheri, M. C., Vagnoli, L., Bosco, M., and Balloni, W., "Infectiveness of several *Frankia* strains on *Alnus* spp. plants," *Nitrogen Fixation: Hundred Years After*, Bothe, H., de Bruijn, F. J., Newton, W. E., eds. Stuttgart: Gustav Fischer, 1988, p. 694.

54. Fernandez, M. P., Meugnier, H., Grimont, P.A.D., and Bardin, R., "Deoxyribonucleic acid relatedness among members of the genus *Frankia*," *Int. J. Syst. Bacteriol. 39*, 424–429 (1989).

55. Fleming, A. I., Wittenberg, J. B., Wittenberg, B. A., Dudman, W. F., and Appleby, C. A., "The purification, characterization and ligand-binding kinetics of hemoglobins from root nodules of the non-leguminous *Casuarina glauca-Frankia* symbiosis," *Biochem. Biophys. Acta 911*, 209–220 (1987).

56. Fox, G. E., and Stackebrandt, E., "The application of 16S rRNA cataloguing and 5S RNA sequencing in bacterial systematics," *Meth. Microbiol. 19*, 405–458 (1987).

57. Gardes, M., and Lalonde, M., "Identification and subgrouping of *Frankia* strains using sodium dodecyl sulfate polyacrylamide gel electrophoresis," *Physiol. Plant. 70*, 237–244 (1987).

58. Gardner, I. C., and Leaf, G., "Translocation of citrulline in *Alnus glutinosa*," *Plant Physiol. 35*, 948–950 (1960).

59. Gauthier, D. L., Diem, H. G., and Dommergues, Y., "*In vitro* nitrogen fixation by two actinomycete strains isolated from *Casuarina* nodules," *Appl. Environ. Microbiol. 41*, 306–308 (1981).

60. Giasson, L., and Lalonde, M., "Restriction pattern analysis of deoxyribonucleic acid isolated from callus and cell suspension of actinorhizal and nonactinorhizal Betulaceae," *Physiol. Plant. 70*, 304–310 (1987).

61. Gordon, J. C., and Wheeler, C. T., "Whole plant studies on photosynthesis and acetylene reduction in *Alnus glutinosa*," *New Phytol. 80*, 179–186 (1978).

62. Hafeez, F., Chaudhary, A. H., and Akkermans, A. D. L., "Physiological studies on N_2-fixing root nodules of *Datisca cannabina* L. and *Alnus nitida* Endl. from Himalaya region in Pakistan," *Plant and Soil 78*, 129–146 (1984).

63. Hahn, D., *16S rRNA as a Molecular Marker in Ecology of Frankia*, Ph.D. thesis, Department of Microbiology, Agricultural University, Wageningen, The Netherlands, 125 pp. (1990).

64. Hahn, D., Dorsch, M., Stackebrandt, E., and Akkermans, A. D. L., "Synthetic oligonucleotide probes for identification of *Frankia* strains," *Plant and Soil 118*, 211–219 (1989).

65. Hahn, D., Lechevalier, M. P., Fisher, A., and Stackebrandt, E., "Evidence for a close phylogenetic relationship between members of the genera *Frankia*, *Geodermatophilus*, and '*Blastococcus*' and emendation of the family Frankiaceae," *Syst. Appl. Microbiol. 11*, 236–242 (1989).

66. Hawker, L. E., and Fraymouth, J., "A reinvestigation of the root nodules of species of *Elaeagnus*, *Hippophaë*, *Alnus*, and *Myrica* with special reference to the morphology and life histories of the causative organisms," *J. Gen. Microbiol. 5*, 369–386 (1951).

67. Henry, M. F., "Cytologie ultrastructurale de l'endophyte présent dans les nodusités radiculaires de *Myrica gale* L.," *Bull. Soc. Bot. Fr 124*. 291–300 (1977).

68. Henry, M. F., "Etude ultrastructurale de l'endophyte présent dans les nodules racinaires d'*Elaeagnus angustifolia* L.," *Bull. Soc. Bot. Fr. 126*, 149–163 (1979).

69. Henson, I. E., and Wheeler, C. T., "Hormones in plants bearing nitrogen-fixing root nodules: Partial characterization of cytokinins from root nodules of *Alnus glutinosa* (L) Gaertn," *J. Exp. Bot. 28*(106), 1076–1086 (1977).

70. Hibbs, D. E., and Cromack, K., "Actinorhizal plants in the Pacific Northwest," *The Biology of Frankia and Actinorhizal Plants*," Schwintzer, C. R., and Tjepkema, J. D., eds. New York: Academic Press, 1990, pp. 343–363.

71. Hirel, B., Perrot-Rechenmann, C., Maudinas, B., and Gadal, P., "Glutamine synthetase in alder (*Alnus glutinosa*) root nodules. Purification, properties and cytoimmunochemical localization," *Physiol. Plant. 55*, 197–203 (1982).

72. Huss-Danell, K., "Nitrogen fixation and biomass production in clones of *Alnus incana*," *New Phytol. 85*, 503–511 (1980).

73. Huss-Danell, K., "The physiology of actinorhizal nodules," *The Biology of Frankia and Actinorhizal Plants*, Schwintzer, C. R., and Tjepkema, J. D., eds. New York: Academic Press, 1990, pp. 129–156.

74. Huss-Danell, K., and Frej, A. K., "Distribution of *Frankia* in soils from forest and afforestation sites in northern Sweden," *Plant and Soil 90*, 407–418 (1986).

75. Huss-Danell, K., and Hahlin, A. -S., "Nitrogenase activity decay and energy supply in *Frankia* after addition of ammonium to the host plant *Alnus incana*," *Physiol. Plant. 74*, 745–751 (1988).

76. Huss-Danell, K., Lundquist, P. -O., and Ekblad, A., "Growth and acetylene reduction acitivity by intact plants of *Alnus incana* under field conditions," *Plant and Soil 118*, 61–73 (1989).

77. Huss-Danell, K., Sellstedt, A., Flower-Ellis, A., and Sjöström, M., "Ammonium effects on function and structure of nitrogen-fixing root nodules of *Alnus incana* (L.) Moench.," *Planta 156*, 332–340 (1982)

78. Ide, P. I., "Optimization of conditions for restriction fragment length polymorphism analysis in the identification of *Frankia* strains," M.S. thesis, Department of Botany, The University of Tennessee, Knoxville, 23 pp. (1986).

79. Klemmedson, J. O., "Ecological importance of actinomycete-nodulated plants in the western United States," *Bot. Gaz. 140*(S), 91–96 (1979).

80. Knowlton, S., and Dawson, J. O., "Effects of *Pseudomonas cepacia* and cultural factors on the nodulation of *Alnus rubra* roots by *Frankia*," *Can. J. Bot. 61*, 2877–2882 (1983).

81. Knowlton, S., Berry, A., and Torrey, J. G., "Evidence that associated soil bacteria may influence root hair infection of actinorhizal plants by *Frankia*," *Can. J. Microbiol. 26*, 971–977 (1980).

82. Kortt, A. A., Inglis, A. S., Fleming, A. I., and Appleby, C. A., "Amino acid sequence of hemoglobin I from root nodules of the non-leguminous *Casuarina glauca-Frankia* symbiosis," *FEBS Lett. 231*(2), 341–346 (1988).

83. Lalonde, M., "Confirmation of the infectivity of a free-living actinomycete isolated from *Comptonia peregrina* root nodules by immunological and ultrastructural studies," *Can. J. Bot. 56*, 2621–2635 (1978).

84. Lalonde, M., Simon, L., Bousquet, J., and Séguin, A., "Advances in the taxonomy of *Frankia*: Recognition of species *alni* and *elaeagni* and novel subspecies *pommerii* and *vandijkii*," *Nitrogen Fixation: Hundred Years After*, Bothe, H., de Bruijn, F. J., and Newton, W. E., eds. Stuttgart: Gustav Fischer, 1988, pp. 671–680.

85. Lambein, F., and Wolk, C. P., "Structural studies on the glycolipids from the envelope of the heterocyst of *Anabaena cylindrica*," *Biochemistry 12*, 791–798 (1983).

86. Lamont, H. C., Silvester, W. B., and Torrey, J. G., "Nile red fluorescence demonstrates lipid in the envelope of vesicles from N_2-fixing cultures of *Frankia*," *Can. J. Microbiol. 34*, 656–660 (1988).

87. Lancelle, S. A., Torrey, J. G., Hepler, P. K., and Callaham, D. A., "Ultrastructure of freeze-substituted *Frankia* strain HFPCcI3, the actinomycete isolated from root nodules of *Casuarina cunninghamiana*," *Protoplasma 127*, 64–72 (1985).

88. Lechevalier, M. P., "Catalog of *Frankia* strains," *The Actinomycetes 19*, 131–162 (1985).

89. Lechevalier, M. P., and Lechevalier, H. A., "The taxonomic position of the actinomycetic endophytes," *Biological Nitrogen Fixation in the Management of Temperate Forests*, Gordon, J. C., Wheeler, C. T., and Perry, D. A., eds. Corvallis: Forest Research Laboratory, Oregon State University, 1979, pp. 111–121.

90. Lechevalier, M. P., and Lechevalier, H. A., "Taxonomy of *Frankia*," *Biological, Biochemical and Biomedical Aspects of Actinomycetes*, Ortiz-Ortiz, L., Bojalil, L. F., and Yakoleff, V., eds. New York: Academic Press, 1984, pp. 575–582.

91. Lindblad, P., and Sellstedt, A., "Immunogold localization of hydrogenase in free-living *Frankia* CpI1," *FEMS Microbiol. Lett. 60*, 311–316 (1989).

92. Lopez, M. F., Young, P., and Torrey, J. G., "A comparison of carbon source utilization for growth and nitrogenase activity in two *Frankia* isolates," *Can. J. Microbiol. 32*, 353–358 (1986).

93. MacKay, J., Séguin, A., and Lalonde, M., "Genetic transformations of 9 *in vitro* clones of *Alnus* and *Betula* by *Agrobacterium tumefaciens*," *Plant Cell Rep. 7*, 229–232 (1988).

94. MacKay, J., Simon, L., and Lalonde, M., "Effect of substrate nitrogen on the performance of *in vitro* propagated *Alnus glutinosa* clones inoculated with Sp^+ and sp^- *Frankia* strains," *Plant and Soil 103*, 21–31 (1987).

95. Mazzucco, C. E., and Benson, D. R., "^{14}C Methylammonium transport by *Frankia* sp. strain CpI1," *J. Bacteriol. 160*(2), 636–641 (1984).

96. Meesters, T. M., "Localization of nitrogenase in vesicles of *Frankia* sp. Cc1.17 by immunogold labelling on ultrathin cryosections," *Arch. Microbiol. 146*, 327–331 (1987).

97. Meesters, T. M., van Genesen, S. T., and Akkermans, A. D. L., "Growth, acetylene reduction activity and localization of nitrogenase in relation to vesicle formation in *Frankia* strains Cc1.17 and Cpl.2," *Arch. Microbiol. 143*, 137–142 (1985).

98. Meesters, T. M., van Vliet, M. W., and Akkermans, A. D. L., "Nitrogenase is restricted to the vesicles in *Frankia* strain EAN1pec," *Physiol. Plant. 70*, 267–271 (1987).

99. Miettinen, J. K., and Virtanen, A. I., "The free amino acids in the leaves, roots and root nodules of the alder (*Alnus*)," *Physiol. Plant. 5*, 540–557 (1952).

100. Miller, I. M., and Baker, D. D., "Initiation, development and structure of root nodules in *Elaeagnus angustifolia* L. (Elaeagnaceae)," *Protoplasma 128*, 107–119 (1985).

101. Miller, I. M., and Baker, D. D., "Nodulation of actinorhizal plants by *Frankia* strains capable of both root hair infection and intercellular penetration," *Protoplasma 131*, 82–91 (1986).

102. Monz, C. A., and Schwintzer, C. R., "The physiology of spore-negative and spore-positive nodules of *Myrica gale*," *Plant and Soil 118*, 75–87 (1989).

103. Mullin, B. C., and An, C. S., "The molecular genetics of *Frankia*," *The Biology of Frankia and Actinorhizal Plants*, Schwintzer, C. R., and Tjepkema, J. D., eds. New York: Academic Press, 1990, pp. 195–214.

104. Murray, M. A., Fontaine, M. S., and Tjepkema, J. D., "Oxygen protection of nitrogenase in *Frankia* sp. HFPArI3," *Arch. Microbiol. 139*, 162–166 (1984).

105. Murray, M. A., Zhang, Z., and Torrey, J. G., "Effect of O_2 on vesicle formation, acetylene reduction, and O_2-uptake kinetics in *Frankia* sp. HFPCcI3 isolated from *Casuarina cunninghamiana*," *Can. J. Microbiol. 31*, 804–809 (1985).

106. Newcomb, W., "Fine structure of the root nodules of *Dryas drummondii* Richards (Rosaceae)," *Can. J. Bot. 59*, 2500–2514 (1981).

107. Newcomb, W., and Heisey, R. M., "Ultrastructure of actinorhizal root nodules of *Chamaebatia foliolosa* (Rosaceae)," *Can. J. Bot. 62*, 1697–1707 (1984).

108. Newcomb, W., and Pankhurst, C., "Fine structure of actinorhizal root nodules of *Coriaria arborea* (Coriariaceae)," *New Zealand J. Bot. 20*, 93–103 (1982).

109. Newcomb, W., and Pankhurst, C., "Ultrastructure of actinorhizal root nodules of *Discaria toumatou* Raoul (Rhamnaceae)," *New Zealand J. Bot. 20*, 105–113 (1982).

110. Newcomb, W., and Wood, S. M., "Morphogenesis and fine structure of *Frankia* (Actinomycetales): The microsymbiont of nitrogen-fixing actinorhizal root nodules," *Int. Rev. Cytol. 109*, 1–88 (1987).

111. Newcomb, W., Baker, D., and Torrey, J. G., "Ontogeny and fine structure of effective root nodules of the autumn olive (*Elaeagnus umbellata*)," *Can. J. Bot. 65*, 80–94 (1987).

112. Newcomb, W., Callaham, D., Torrey, J. G., and Peterson, R. L., "Morphogenesis and fine structure of the actinomycetous endophyte of nitrogen-fixing root nodules of *Comptonia peregrina*," *Bot. Gaz. 140*(S), 22–34 (1979).

113. Nittayajarn, A., Mullin, B. C., and Baker, D. D., "Screening of symbiotic frankiae for host specificity by restriction fragment length polymorphism analysis," *Appl. Environ. Microbiol. 56*, 1172–1174 (1990).

114. Noridge, N. A., and Benson, D. R., "Isolation and nitrogen-fixing activity of *Frankia* sp. strain CpI1 vesicles," *J. Bacteriol. 166*, 301–305 (1986).

115. Normand, P., and Bousquet, J. "Phylogeny of nitrogenase sequences in *Frankia* and other nitrogen-fixing microorganisms," *J. Mol. Evol. 29*, 436–447 (1989).

116. Normand, P., and Lalonde, M., "Evaluation of *Frankia* strains isolated from provenances of two *Alnus* species," Can. J. Microbiol. 28, 1133–1142 (1982).

117. Normand, P., and Lalonde, M., "The genetics of actinorhizal *Frankia*: A review," *Plant and Soil 90*, 429–453 (1986).

118. Normand, P., Downie, J. A., Johnston, A. W. B., Kieser, T., and Lalonde, M., "Cloning of a multicopy plasmid from the actinorhizal nitrogen-fixing bacterium *Frankia* sp. and determination of its restriction map," *Gene 34*, 367–370 (1985).

119. Normand, P., Simonet, P., and Bardin, R., "Conservation of *nif* sequences in *Frankia*," *Mol. Gen. Genet. 213*, 238–246 (1988).

120. Normand, P., Simonet, P., Butour, J. L., Rosenberg, C., Moiroud, A., and Lalonde, M., "Plasmids in *Frankia* sp.," *J. Bacteriol. 155*, 32–35 (1983).

121. Normand, P., Simonet, P., Prin, Y., and Moiroud, A., "Formation and regeneration of *Frankia* protoplasts," *Physiol. Plant. 70*, 259–266 (1987).

122. Oremus, P. A. I., "Occurrence and infective potential of the endophyte of *Hippophaë rhamnoides* ssp. *rhamnoides* in costal sand-dune areas," *Plant and Soil 56*, 123–139 (1980).

123. Parsons, R., Silvester, W. B., Harris, S., Gruijters, W. T. M., and Bullivant, S., "*Frankia* vesicles provide inducible and absolute oxygen protection for nitrogenase," *Plant Physiol. 83*, 728–731 (1987).

124. Pathirana, M. S., "Occurrence of hemoglobin in the nitrogen-fixing root nodules of *Myrica gale*," M.S. thesis, University of Maine, Orono, 27 pp. (1989).

125. Pizelle, G., "Seasonal variations of the sexual reproductive growth and nitrogenase activity in mature *Alnus glutinosa*," *Plant and Soil 78*, 181–188 (1984).

126. Pommer, E. H., "Über die Isolierung des endophyten aus den Wurzelknöllchen *Alnus glutinosa* Gaertn. ünd über erfolgreiche Re-infektionsversuche." *Ber. Dtsch. Bot. Ges. 72*, 138–150 (1959).

127. Reddell, P., and Bowen, G. D., "*Frankia* source affects growth, nodulation and nitrogen fixation in *Casuarina* species," *New Phytol. 100*, 115–122 (1985).

128. Roelofsen, W., and Akkermans, A. D. L., "Uptake and evolution of H_2 and reduction of C_2H_2 by root nodules and nodule homogenates of *Alnus glutinosa*," *Plant and Soil 52*, 571–578 (1979).

129. Rose, S. L., "Mycorrhizal associations of some actinomycete nodulated nitrogen-fixing plants," *Can. J. Bot. 58*, 1449–1454 (1980).

130. Rose, S. L., and Youngberg, C. T., "Tripartite associations in snowbrush (*Ceanothus velutinus*): Effect of vesicular-arbuscular mycorrhizae on growth, nodulation and nitrogen fixation," *Can. J. Bot. 59*, 34–39 (1981).

131. Russo, R. O., "Evaluating alder-endophyte (*Alnus acuminata-Frankia*-Mycorrhizae) interactions. I. Acetylene reduction in seedlings inoculated with *Frankia* strain ArI3 and *Glomus intra-radices*, under three phosphorus levels," *Plant and Soil 118*, 151–155 (1989).

132. Schubert, K. R., "Products of biological nitrogen fixation in higher plants: Synthesis, transport, and metabolism" *Ann. Rev. Plant Physiol. 37*, 539–574 (1986).

133. Schubert, K. R., Coker III, G. T., and Firestone, R. B., "Ammonia assimilation in *Alnus glutinosa* and *Glycine max*," *Plant Physiol. 67*, 662–665 (1981).

134. Schultz, N. A., and Benson, D. R., "Developmental potential of *Frankia* vesicles," *J. Bacteriol. 171*, 6873–6877 (1989).

135. Schwintzer, C. R., "Spore-positive and spore-negative nodules," *The Biology of Frankia and Actinorhizal Plants*, Schwintzer, C. R., and Tjepkema, J. D., eds. New York: Academic Press, 1990, pp. 177–193.

136. Schwintzer, C. R., Berry, A. M., and Disney, L. D., "Seasonal patterns of root nodule growth, endophyte morphology, nitrogenase activity, and shoot development in *Myrica gale*," *Can. J. Bot. 60*, 746–757 (1982).

137. Séguin, A., and Lalonde, M. "Detection of pectolytic activity and *pel* homologous sequences in *Frankia*," *Plant and Soil 118*, 221–229 (1989).

138. Sellstedt, A., Huss-Danell, K., and Ahlqvist, A. -S., "Nitrogen fixation and biomass production in symbiosis between *Alnus incana* and *Frankia* strains with different hydrogen metabolism," *Physiol. Plant. 66*, 99–107 (1986).

139. Shipton, W. A., and Burggraaf, A. J. P. "A comparison of the requirements for various carbon and nitrogen sources and vitamins in some *Frankia* isolates," *Plant and Soil 69*, 149–161 (1982).

140. Silvester, W. B., and Harris, S. L., "Nodule structure and nitrogenase activity of *Coriaria arborea* in response to varying pO_2," *Plant and Soil 118*, 97–109 (1989).

141. Silvester, W. B., and Winship, L. J., "Transient responses of nitrogenase to acetylene and oxygen by actinorhizal nodules and cultured *Frankia*," *Plant Physiol. 92*, 480–486 (1990).

142. Silvester, W. B., Harris, S. L., and Tjepkema, J. D., "Oxygen regulation and hemoglobin," *The Biology of Frankia and Actinorhizal Plants*, Schwintzer, C. R., and Tjepkema, J. D., eds. New York: Academic Press, 1990, pp. 157–176.

143. Silvester, W. B., Silvester, J. K., and Torrey, J. G., "Adaptation of nitrogenase to varying oxygen tension and the role of the vesicle in root nodules of *Alnus incana* ssp. *rugosa*," *Can. J. Bot. 66*, 1772–1779 (1988).

144. Silvester, W. B., Whitbeck, J., Silvester, J. K., and Torrey, J. G., "Growth, nodule morphology and nitrogenase activity of *Myrica gale* with roots grown at various oxygen levels," *Can. J. Bot. 66*, 1762–1771 (1988).

145. Simonet, P., Capellano, A., Navarro, E., Bardin, R., and Moiroud, A., "An improved method for lysis of *Frankia* with achromopeptidase allows detection of new plasmids," *Can. J. Microbiol. 30*, 1292–1295 (1984).

146. Simonet, P., Haurat, J., Normand, P., Bardin, R., and Moiroud, A., "Localization of *nif* genes on a large plasmid in *Frankia* sp. strain ULQ0132105009," *Mol. Gen. Genet. 204*, 492–495 (1986).

147. Simonet, P., Thi Le, N., Teissier du Cros, E. T., and Bardin, R., "Identification of *Frankia* strains by direct DNA hybridization of crushed nodules," *Appl. Environ. Microbiol. 54*, 2500–2503 (1988).

148. Simonet, P., Thi Le, N., Moiroud, A., and Bardin, R., "Diversity of *Frankia* strains isolated from a single alder stand," *Plant and Soil 118*, 13–22 (1989).

149. Simonet, P., Normand, P., and Bardin, R., "Heterologous hybridization of *Frankia* DNA to *Rhizobium meliloti* and *Klebsiella pneumoniae nif* genes," *FEMS Microbiol. Lett. 55*, 141–146 (1988).

150. Simonet, P., Normand, P., Hirsch, A. M., and Akkermans, A. D. L., "The genetics of the *Frankia* actinorhizal symbiosis," *The Molecular Biology of Symbiotic Nitrogen Fixation*, Gresshoff, P. M., ed. Boco Raton, Fl: CRC Press, 1990, pp. 70–109.

151. Simonet, P., Normand, P., Moiroud, A., and Bardin, R., "Identification of *Frankia* strains in nodules by hybridization of polymerase chain reaction products with strain-specific oligonucleotide probes," *Arch. Microbiol. 153*, 235–240 (1990).

152. Simonet, P., Normand, P., Moiroud, A., and Lalonde, M., "Restriction enzyme digestion patterns of *Frankia* plasmids," *Plant and Soil 87*, 49–60 (1985).

153. Smolander, A., "*Frankia* populations in soils under different tree species with special emphasis on soils under *Betula pendula*," *Plant and Soil 121*, 1–10 (1990).

154. Smolander, A., and Sarsa, M. -L., "*Frankia* strains of soil under *Betula pendula*: Behavior in soil and in pure culture," *Plant and Soil 122*, 129–136 (1990).

155. Smolander, A., and Sundman, V., "*Frankia* in acid soils of forests devoid of actinorhizal plants," *Physiol. Plant. 70*, 297–303 (1987).

156. Sprent, J. I., "Which steps are essential for the formation of functional legume nodules?" *New Phytol. 111*, 129–153 (1989).

157. Sprent, J. I., and de Faria, S. M., "Mechanisms of infection of plants by nitrogen-fixing organisms," *Plant and Soil 110*, 157–165 (1988).

158. Steele, D. B., and Stowers, M. D., "Superoxide dismutase and catalase in *Frankia*," *Can. J. Microbiol. 32*, 409–413 (1986).

159. Steele, D. B., Ramirez, K., and Stowers, M. D., "Host plant growth response to inoculation with *Frankia*," *Plant and Soil 118*, 139–143 (1989).

160. Stevens, G. A., and Berry, A. M., "Cytokinin secretion by *Frankia* sp. HFPArI3 in defined medium. *Plant Physiol. 87*, 15–16 (1988).

161. Strand, R., and Laetsch, W. M., "Cell and endophyte structure of the nitrogen-fixing root nodules of *Ceanothus integerrimus* H. & A. I. Fine structure of the nodule and its endosymbiont," *Protoplasma 93*, 165–178 (1977).

162. Taubert, H., "Über den Infectionsvorgang und die Entwicklung der Knöllchen bei *Alnus glutinosa* Gaertn.," *Planta 48*, 135–156 (1956).

163. Tisa, L. S., and Ensign, J. C., "Isolation and nitrogenase activity of vesicles from *Frankia* sp. strain EAN1pec," *J. Bacteriol. 169*, 5054–5059 (1987).

164. Tisa, L. S., and Ensign, J. C., "Formation and regeneration of protoplasts of the actinorhizal nitrogen-fixing actinomycete *Frankia*," *Appl. Environ. Microbiol. 53*, 53–56 (1987).

165. Tjepkema, J. D., "Oxygen relations in leguminous and actinorhizal nodules," *Symbiotic Nitrogen Fixation in the Management of Temperate Forests*, Gordon, J. C., Wheeler, C. T., and Perry, D. A., eds. Corvallis: Forest Research Laboratory, Oregon State University, 1979, pp. 175–186.

166. Tjepkema, J. D. and Asa, D. J., "Total and CO-reactive heme content of actinorhizal nodules and the roots of some non-nodulating plants," *Plant and Soil 100*, 225–236 (1987).

167. Tjepkema, J. D., and Murray, M. A., "Respiration and nitrogenase activity in nodules of *Casuarina cunninghamiana* and cultures of *Frankia* sp. HFP020203: Effects of temperature and partial pressure of O_2," *Plant and Soil 118*, 111–118 (1989).

168. Tjepkema, J. D., Ormerod, W., and Torrey, J. G., "Vesicle formation and acetylene reduction activity in *Frankia* sp. CpI1 cultured in defined nutrient media," *Nature 287*, 633–635 (1980).

169. Tjepkema, J. D., Schwintzer, C. R., and Benson, D. R., "Physiology of actinorhizal nodules," *Ann. Rev. Plant Physiol. 37*, 209–232 (1986).

170. Torrey, J. G., "The site of nitrogenase in *Frankia* in free-living culture and in symbiosis," *Nitrogen Fixation Research Progress*, Evans, H. J., Bottomley, P. J., and Newton, W. E., eds. Dordrecht: Martinus Nijhoff, 1985, pp. 293–299.

171. Torrey, J. G., "Endophyte sporulation in root nodules of actinorhizal plants," *Physiol. Plant. 70*, 279–288 (1987).

172. Torrey, J. G., and Callaham, D., "Structural features of the vesicles of *Frankia* sp. CpI1 in culture," *Can. J. Microbiol. 28*, 749–757 (1982).

173. Torrey, J. G., and Racette, S., "Specificity among the Casuarinaceae in root nodulation by *Frankia*," *Plant and Soil 118*, 157–164 (1989).

174. Torrey, J. G., and Tjepkema, J. D., "Preface," *Bot. Gaz. 140*(S), i (1979).

175. Tsai, Y. -L., and Benson, D. R., "Physiological characteristics of glutamine synthetases I and II of *Frankia* sp. strain CpI1," *Arch. Microbiol. 152*, 382–386 (1989).

176. Tunlid, A., Schultz, N. A., Benson, D. R., Steele, D. B., and White, D. C., "Differences in fatty acid composition between vegetative cells and N_2-fixing vesicles of *Frankia* sp. strain CpI1," *Proc. Natl. Acad. Sci. USA 86*, 3399–3403 (1989).

177. Tyson, J. H., and Silver, W. S., "Relationships of ultrastructure to acetylene reduction (N_2 fixation) in root nodules of *Casuarina*," *Bot. Gaz. 140*(S), 44–48 (1979).

178. Tzean, S. S., and Torrey, J. G., "Spore germination and the life cycle of *Frankia in vitro*," *Can. J. Microbiol. 35*, 801–806 (1989).

179. van Dijk, C., "Ecological aspects of spore formation in the *Frankia-Alnus* symbiosis," Ph.D. thesis, State University, Leiden, The Netherlands, 154 pp. (1984).

180. VandenBosch, K. A., and Torrey, J. G., "Host endophyte interactions in effective and ineffective nodules induced by the endophyte of *Myrica gale*," *Can. J. Bot. 61*, 2898–2909 (1983).

181. VandenBosch, K. A., and Torrey, J. G., "Consequences of sporangial development for nodule function in root nodules of *Comptonia peregrina* and *Myrica gale*," *Plant Physiol. 76*, 556–560 (1984).

182. VandenBosch, K. A., and Torrey, J. G., "Development of endophytic *Frankia* sporangia in field- and laboratory-grown nodules of *Comptonia peregrina* and *Myrica gale*," *Am. J. Bot. 72*, 99–108 (1985).

183. Vikman, P-Å., and Huss-Danell, K., "Capacity for hexose respiration in symbiotic *Frankia* from *Alnus incana*," *Physiol. Plant. 70*, 349–354 (1987).

184. Vitousek, P. M., and Walker, L. R., "Biological invasion by *Myrica faya* in Hawaii: Plant demography, nitrogen fixation, ecosystem effects," *Ecol. Monogr. 59*(3), 247–265 (1989).

185. Walsh, K. B., Ng, B. H., and Chandler, G. E., "Effects of nitrogen nutrition on xylem sap composition of Casuarinaceae," *Plant and Soil 81*, 291–293 (1984).

186. Wheeler, C. T., "The diurnal fluctuation in nitrogen fixation in the nodules of *Alnus glutinosa* and *Myrica gale*," *New Phytol. 68*, 675–682 (1969).

187. Wheeler, C. T., and Bond, G., "The amino acids of non-legume root nodules," *Phytochemistry 9*, 705–708 (1970).

188. Wheeler, C. T., and Lawrie, A. C., "Nitrogen fixation in root nodules of alder and pea in relation to the supply of photosynthetic assimilates," *Symbiotic Nitrogen Fixation in Plants*, Nutman, P. S., ed. New York: Cambridge University Press, 1976, pp. 497–509.

189. Wheeler, C. T., and Miller, I. M., "Current and potential uses of actinorhizal plants in Europe," *The Biology of Frankia and Actinorhizal Plants*, Schwintzer, C. R., and Tjepkema, J. D., eds. New York: Academic Press, 1990, pp. 365–389.

190. Wheeler, C. T., Hensen, I. E., and McLaughlin, M. E., "Hormones in plants bearing actinomycete nodules," *Bot. Gaz. 140*(S), 52–57 (1979).

191. Zeng, S., Tjepkema, J. D., and Berg, R. H., "Gas diffusion pathway in nodules of *Casuarina cunninghamiana*," *Plant and Soil 118*, 119–123 (1989).

8
Ecology of *Bradyrhizobium* and *Rhizobium*

Peter J. Bottomley

I. Introduction

Interest in the ecology of rhizobia arose when scientists recognized that some of the idiosyncrasies of legume growth under field conditions might be linked to rhizobial behavior in soil. Although a large number of publications have accumulated and three extensive reviews have been devoted entirely to rhizobial ecology,[47,239,302] the topic is difficult to review. The subject matter is extremely diverse, the quality of the science is variable, and the findings can be repetitive, difficult to interpret, and inconsistent. The latter can be explained, in part, by our difficulty in establishing criteria with which findings of only regional or site-specific significance can be separated from those that have global significance. Despite the extensive bibliography of this chapter, the author wishes to draw the attention of the reader to other relevant reviews. Comprehensive articles on various aspects of the agronomic relationships between rhizobia and legumes[27,79,141,280] and on the phenomenon of competitive nodulation[109] have been published recently. Furthermore, the reader should be aware of pertinent information arising from research on the soil ecology of bacteria used as biological control agents of plant pathogens.[349,405]

II. Background to the Ecology of Rhizobia

A. *Inoculation Responses*

During the first half of the twentieth century, failure of legumes to achieve profitable yields was often due to the absence of symbiotically effective rhizobia in the soil.[66,114,124,272,388] Since treatment of seed with good-quality inoculants of appropriate rhizobia solved this problem and since subsequent plantings of the same legume at the same site did not usually respond to inoculation,[164,243] there seemed little doubt that rhizobia could persist in many soils after their introduction.

Although benefits to inoculation are still reported throughout the world, these situations can be predicted with no greater degree of certainty than they were 60 years ago.[129a] Inoculation responses are associated primarily with the first planting of a legume in soil having no prior history of the crop.[13,35,36,164,294,344,361] Inoculation responses are reported occasionally in situations where the population density of effective soil rhizobia is low and the level of mineral nitrogen is insufficient for adequate plant growth.[92,119,217,353]

Although indeterminate persistence of introduced rhizobia in fertile soils should be viewed as one of the success stories of twentieth-century agriculture, there is a concern that naturalized rhizobial populations will restrict future introductions of legumes from reaching their N_2-fixing potential. In the case of soybean [*Glycine max* (L.), Merrill], field studies show that many strains of *Bradyrhizobium japonicum* are incapable of supporting maximum yields on improved soybean cultivars.[2,56] Concern arises from the fact that strains belonging to certain serogroups of *B. japonicum* have become established as dominant soybean nodule occupants throughout the major production regions of the United States[90,147,206,403] and they cannot be displaced with better N_2-fixing strains when current inoculation technologies are practiced.[24,146,192,399]

B. Problem Environments

In contrast to growing legumes on highly productive soils, establishing them on marginal agricultural land raised complex problems. Low and inconsistent rainfall, extremes of temperature, and acidic soils of low nutrient status and poor water-holding capacity were major impediments to legume growth. Under such conditions the soil ecology of rhizobia was recognized to be an important issue. In southwest Australia, problems with the establishment and persistence of subclover (*Trifolium subterraneum* L.) on newly cleared lands were attributed to the lack of persistence of inoculant strains of *Rhizobium leguminosarum* bv. *trifolii*.[68-73] Lack of persistence of *Rhizobium meliloti* continues to plague attempts to establish and maintain annual *Medicago* spp. in the same region of the Australian continent.[178,179] Persistence of rhizobia in soil is a concern throughout many of the semiarid regions of the world where rhizobia of annual legumes encounter long periods of hot and dry soil conditions.[112,207,338]

Problems in subclover establishment were often associated with excess soil acidity[188,237,394] or were due to the presence of indigenous populations of *R. leguminosarum* bv. *trifolii* that were symbiotically ineffective on subclover[15,389] and occasionally competitive enough to prevent nodulation by inoculant strains.[174,182,197] During the past 30 years numerous reports from worldwide locations showed that indigenous soil rhizobial populations are suboptimally effective, relative to inoculant strains, on introduced forage legume species.[29,139,144,145,172,174,182,194,341,389]

C. Inconsistent Performance and Persistence of Inoculant Strains

Other phenomena that have confounded microbiologists include reduced representation of inoculant strains in nodules with each succeeding crop, and inconsistent inoculant success at different locations.[15,34,37,113,140,255,256,263,264,316,333,381] With the exception of the reports from southwest Australia,[178,302] there are no reports in the primary literature where inoculant strain decline was proven to be the direct cause of legumes failing to reach their yield potential in postestablishment years. In recent years, however, a decine in productivity of subclover and medic (annual *Medicago* spp.) pastures has been documented throughout many areas of southeast Australia. Mediocre effectiveness of nodule occupants and lack of persistence of rhizobia are only two of many factors that have been implicated in this complex phenomenon.[60,80,125]

D. Screening for Superior Inoculant Strains

In the search for inoculant strains with improved saprophytic and competitive traits, isolates of many *Rhizobium* and *Bradyrhizobium* species were identified that had superior tolerance of soil stresses such as low pH[76a,88,95,142,205,368,410,411] low phosphate,[17,18,64,229] extremes of temperature,[233,276] and desiccation.[26,131,160,247,248,292,352] Strains were screened for desirable phenotypes such as H_2-oxidizing ability,[151] bacteriocin-producing ability,[171] and superior nodulating competitiveness.[199,267,297,325,381] Despite these efforts to screen potential inoculants, only one example of large-scale success in the field has been reported. Distribution of inoculants containing acid-tolerant strains of *R. meliloti* has enabled annual medics (*Medicago polymorpha* L. and *M. murex* Willd.) to be successfully established in low-pH soils of the wheat belt of southwest Australia.[178,179]

Neither the presence of soil rhizobia of suboptimum effectiveness nor the use of inoculant strains lacking competitiveness and persistence has been disastrous to field legume production. However, several issues are in need of resolution, including the possibility of these phenomena confounding the interpretation of field experiments and restricting the ability of legumes to achieve their present, and future, N_2-fixing potential in production agriculture.

III. The Soil Habitat

A. Factors Affecting the Magnitude of Soil Populations of Rhizobia

i. Environmental Influences

Through use of the plant infection–soil dilution technique, in combination with most probable number (MPN) statistics, the density of rhizobial populations in soil is found to range from 10 to 10^7 per gram.[47,392] Although population density has been positively correlated with many factors, including organic matter, clay content, soil pH, base saturation, mean annual rainfall, irradiance, presence of legumes, or vigor of legume growth,[21,144,226,341,414] other studies have not shown any simple relationship between population density and specific environmental parameters.[139,169,400,401,412] Perhaps these discrepancies are related to the limitations of the labor-intensive MPN-technique, which imposes restrictions on the number of soil samples that can be analyzed and the time period over which samples are taken. In some studies substantial fluctuation in population density was revealed at the same site when soil samples were taken at intervals over a period of one or two years.[112,218,226,249,250,338]

Nonuniform distribution of rhizobia in soil can confound interpretation of environmental influences on population density. Wollum and Cassel[408] observed that the density of *B. japonicum* ranged between 10^2 and 10^7/g over short distances (0.3 m) in a field. Studies on *R. leguminosarum* bv. *trifolii* revealed that 90% or more of the total population either was located in the first 2-cm increment of the soil profile[322] or was associated with the desiccated nodule tissue of dead plants.[68]

Many legumes have the potential to nodulate with either *Rhizobium or Bradyrhizobium,*[107,111,190,225,370,392] yet there is virtually no information on the population densities of members of the two genera with the ability to nodulate the same host in the same soil.[107] In Chatel's classic work,[71,72] *Bradyrhizobium* sp. (*Lupinus*) was found to colonize sandy soils successfully in southwest Australia that were inhospitable to some but not all strains of *R. leguminosarum* bv. *trifolii*. Population of *Bradyrhizobium* spp. were detected in soil on low-rainfall sites in Hawaii where *Rhizobium* spp. were absent.[414]

ii. Presence of Host

In the case of *Rhizobium*, populations of different species and of biovars within a species can coexist at similar low densities in soil for many years in the absence of legumes.[218,289,290] Recent studies, however, show the impact that the presence of a specific host can have on disturbing the equilibrium between the soil population densities of closely related biovars within the same *Rhizobium* species. Populations of *R. leguminosarum* bv. *viciae* were 1000 times greater than of bv. *phaseoli* in pea (*Pisum sativum* L.) fields. In contrast, the relative population densities of the two biovars were reversed in soil recovered from neighboring bean [*Phaseolus vulgaris* (L.) Savi] fields.[218] Studies from the author's laboratory confirm this finding with populations of *R. leguminosarum* bvs. *trifolii* and *viciae*. In grazed subclover patures where vetch (*Vicia*) species were absent, *R. leguminosarum* bvs. *viciae* and *trifolii* coexisted at $<10^2$/g and $>10^6$/g, respectively. On an adjacent unimproved site where vetch species were the dominant legume cover and where clover (*Trifolium*) species were absent, the population densities of the two biovars were reversed (Bottomley, B, Strain, and Claycomb, unpubl. observations). Research is needed to evaluate whether the impact of the host is through a selective rhizosphere effect or is due to population release from senescing nodules. In the case of *Bradyrhizobium*, although *B. japonicum* persists in soil for many years in the absence of soybean,[44,200,400] there are no data on the dynamics of soil populations of *B. japonicum* relative to other bradyrhizobial populations of alternate host range.

B. Nonnodulating Rhizobia in Soil Populations

Although immunofluorescence allows researchers to determine the densities of soil-borne rhizobia without resort to the MPN technique, concern has been expressed about the proportion of cells that are viable and competent to nodulate within populations observed by immunofluorescence.[22,130,202,310] Using immunofluorescence, small, cell-like structures (1.0×0.2 μm) have been detected within soil-borne populations of *R. leguminosarum* bv. *trifolii*.[5,28] The density of the small, cell-like structures increased abruptly at the 15–20-cm depth increment under an undisturbed subclover pasture. Since there was no evidence for viability within the small cell subpopulation, the fluorescent structures probably represent cell "ghosts," or antigenic debris that had accumulated in the fine pores of the compacted clay layer as a result of downward water movement. Inaccessibility of these cells both to nutrients and to predators probably accounts for loss of viability and persistence.

Nonnodulating bacteria have been isolated that were antigenically related to, or showed partial DNA homology with, symbiotically competent biovars of *R. leguminosarum*.[186,358] Immunofluorescence counts of indigenous soil populations of *Bradyrhizobium* sp. (*Lotus*)[235] and *R. leguminosarum* bv. *trifolii*[28] have been reported to be significantly greater than population densities determined by the MPN procedure. Even though there is evidence that nonviable rhizobia cannot persist for significant lengths of time in soil,[24] estimates of population densities should be made with a combination of immunofluorescence and viability assays to assess whether the former method is overestimating the size of the population of viable, nodulation-competent rhizobia.

C. Relationship Between Rhizobia and the Soil Microbial Population

Information about the synecological relationship between rhizobia and the soil microbial population is almost nonexistent. The proportion that rhizobia contribute to the total soil bacterial population is difficult to ascertain, and values reported in the literature vary widely. Comparisons of epifluorescence counts of total bacteria with immunofluorescence counts of rhizobia show that *B. japonicum*[271] and *R. leguminosarum* bv. *trifolii*[28,30,101] comprise between 0.1% and 1.0% of the total bacterial population in bulk or rhizosphere soil. In studies where MPN values of rhizobia were compared with plate counts of total soil bacteria, the former contributed approximately 8% of the total bacteria in soil recovered from clover pastures and soybean fields,[247,250] and up to 25% of the total bacteria in clover rhizospheres.[336] While the plate count technique is known to underestimate the magnitude of viable soil bacteria,[30,238] enumeration by fluorescence microscopy suffers from the difficulty in determining what proportion of indigenous bacteria can be recovered from the soil matrix.[85,101,208,310,325]

Recent studies have shown that the residence time of microbial biomass can vary in different soils.[269,383] Cleyet-Marel and coworkers have shown the rate at which introduced populations of *B. japonicum* decline to equilibrium in soil can be described kinetically and that the rates differ between soils.[78,86] More research is needed to determine the residence time of rhizobial populations relative to other bacterial types under different soil conditions.

D. Influence of Soil Textural Properties on Rhizobia

Coarse-textured soils, prone to water deficits, acidification, and nutrient deficiencies, have been troublesome both for legume establishment and for maintaining productivity. Fine-textured, highly weathered acidic soils can also be problematical if they contain excesses of toxic elements such as aluminum and manganese, mineralogical fractions that possesses a high capacity and affinity to adsorb phosphate, and if they are prone to structural changes during wetting and drying cycles. Poorly-drained soils of high base saturation have not received the same attention, despite the difficulties usually encountered in cultivating legumes under these conditions. Bushby[47] emphasized that most ecological studies of rhizobia have centered on soils of the temperate regions with mineralogical signatures quite different from those of the highly weathered soils found in many regions of the tropics.

i. Rhizobia–Clay Interactions

Although the surface charge characteristics of *Rhizobium* and *Bradyrhizobium* are different, and can vary among strains of *Rhizobium*,[251,252] evidence for the surface charge characteristics of clay minerals directly influencing rhizobial behavior in field soil is circumstantial. An extracting solution containing partially hydrolyzed gelatin and ammonium phosphate (pH 8.2) was essential for recovering rhizobia introduced into soils of the Oxisol, Vertisol, and Andisol orders.[209] Although the authors suggested that rhizobia were difficult to recover because of strong electrostatic association with the mineral fraction, an alternate explanation might be that rhizobia were dispersed successfully from soil structure but were subsequently trapped within soil colloids as a result of anomalous flocculation behavior.[85,101]

ii. Clay Minerals and Nodulation

Circumstantial evidence for a role of clay minerals influencing the nodulating success of rhizobia arises from experiments on competitive nodulation between mixed strains of *R. leguminosarum* bv. *viciae* on lentils[260] (*Lens esculenta* Moench), *Rhizobium* sp. (*Leucaena*) on leucaena[270] [*Leucaena leucocephala* (Lam.) de Wit], and *Rhizobium* sp. (*Cicer*) on chickpea[361] (*Cicer arietinum* L.). Differences in the outcome of competition were seen when the experiments were conducted either in an Oxisol dominated by amorphous and variable-charge minerals or in less weathered soils containing predominantly constant-charge, crystalline clay minerals. In contrast, there was no influence of soil type on competitive nodulation among three *B. japonicum* strains when they were inoculated onto soybean grown in three different soils.[135]

E. Influence of Soil Structural Properties on Rhizobia

The relationship between soil structure and five specific aspects of the morphology, physiology, and ecology of rhizobia will be discussed.

i. Cell Size

Ozawa and Yamaguichi[296] reported that although 90% of indigenous rhizobia are tightly associated with soil particles, a long period of incubation was required before the majority of introduced rhizobia penetrated into the pores of soil aggregates. In a recent study, a greater percentage of introduced rhizobia became associated with soil particles ≥50 μm in diameter as the moisture content of soil was lowered.[311] These authors hypothesized that movement of rhizobia into soil aggregates was impeded until pores of a diameter similar to the length of a laboratory-grown rhizobia (≤3.0 μm) were vacated by water. With regard to the size of indigenous rhizobia, both *B. japonicum* and *R. leguminosarum* bv. *trifolii* have been shown to decrease in size after prolonged incubation in soil.[87,310] Morphologically atypical cells have been observed within indigenous soil populations of *B. japonicum*[23] and *R. leguminosarum* bv. *trifolii*.[28] Perhaps a small cell size confers an ecological advantage to rhizobia by allowing them to occupy small pores and avoid both matric water potential fluctuation and predation. On the other hand, occupancy of small pores might be disadvantageous, since nutrient accessibility will be limited, and access to a legume root will be restricted.

ii. Water Relations

Soil populations of *Bradyrhizobium* spp. have been shown to be more persistent than *Rhizobium* spp. in soils which endure seasonal moisture deficits.[71,72,414] It has been suggested that *Bradyrhizobium* enters dormancy more readily than *Rhizobium* because the former maintains a lower internal water content than the latter at low water potentials.[49] Appealing as this hypothesis may be, other studies have not provided supporting evidence.[160,247,248,303,352,380,413a] There is a need to address this controversy by conducting a systematic study of the water relations of well-characterized populations of both genera indigenous to the same soil.

iii. Response of Rhizobia to Fluctuations in Water Potential

Although rhizobia can survive prolonged periods of exposure to extremely low water potentials,[131] few studies have addressed either the efficiency of rhizobial growth at low matric potential[352] or growth recovery after a rapid water potential increase brought about by a rewetting process.[342] When polyethylene glycol was used to lower the matric water potential of the growth medium, field isolates representing two serotypes of *R. meliloti* were found to differ both in their ability to grow as matric potential was lowered to −1.5 MPa and in their growth recovery after "upshock" into conventional growth medium (−0.1 MPa).[50] Furthermore, nodulation by the upshock-sensitive strains was delayed relative to the insensitive strains after transfer from −1.0 MPa growth medium onto alfalfa seedlings. It is worth noting that although root hair infection of subclover does not occur at matric potentials lower than −0.3 MPa,[415] subclover seeds can germinate in soil at water potential as low as −1.2 MPa[266] (Leung and Bottomley, unpublished data). It remains to be determined whether rhizobia inhabiting dry surface soil can obtain nodulating advantage by regaining metabolic competence at water potentials sufficient for seed germination yet too low for root hair infection.

iv. Motility

Given the capacity of rhizobia to produce flagellae, numerous reports suggest that motility might be advntageous to rhizobia in colonizing legume roots and initiating the symbiosis.[9,31,53,121,268,359] Only one of the nodulation studies was carried out in a soil, and this soil was extremely coarse-textured.[268] The predominant macropore structure of that particular soil and the high moisture content maintained for plant growth represent a combination of soil characteristics that would facilitate bacterial movement to the root surface. The results of other soil studies cast doubt on the importance of motility for gaining access to the root surface under nonsaturated soil conditions. In soils of varying textures, movement of introduced *R. leguminosarum* bv. *trifolii* ceased when water-filled pores became discontinuous (−0.005 to −0.026 MPa), even though water-occupied pores were quite large (≤6.8 μm) relative to the size of a rhizobial cell.[148] Liu et al.[234] showed that a nonmotile mutant and the parent strain of *B. japonicum* colonized soybean rhizospheres equally well in a silt loam soil (usually characterized by poor structure and dominated by small pores).

At this point it is worth digressing to the real agricultural world. To avoid unnecessary soil compaction, the majority of legumes are planted in the field after surface soil has dried sufficiently to support wheel traffic. Under these nonsaturated conditions it is reasonable to speculate that motility confers no advantage and that only those rhizobia located on the periphery of aggregates will make contact with the emerging radicle and the primary nodulation sites. More research is needcd to gain a better understanding of microsite distribution of rhizobia in soil before a critical role for motility in nodulation of primary and secondary roots can be accepted.

v. Predation

Laboratory studies carried out with antibiotic and fungicide-resistant mutants of rhizobia introduced into soil show that protozoa and other microbial predators can cause a decline in the population density of introduced rhizobia.[67,91,143,165,177,228,314] It is worth commenting, however, that the population densities of rhizobia at which predators become ineffective are similar to values typically found in nonsterile field soil. Furthermore, since introduced rhizobia are distributed differently in soil from indigenous rhizobia,[296] the accessibility of both indigenous and introduced rhizobia to predators needs to be compared both in bulk soil and in rhizospheres.

F. Influence of Soil Chemical Properties on Rhizobia

i. The Soil Acidity Complex

As a result of their high nutrient content, legumes are demanding of the mineral resources of soil. Many soils of the world that can benefit from legume growth and N_2 fixation are often deficient in those minerals essential for legume growth. These soils are often inherently acidic (pH $\leq$ 5.5), or so weakly buffered as to be prone to acidification either as a result of legume growth or from applications of ammoniacal fertilizers to nonlegumes grown in the rotation.[106,246,315] Although application of lime can rectify the problem, the availability, cost, and mechanics of application can be serious impediments to liming.[77] Furthermore, highly weathered soils of low cation exchange capacity are easily overlimed, the consequences of which can be serious physical and chemical disorders.[330,343] Although the factor common to all acid soils is low pH, it cannot be overemphasized that substantial variation in texture, structure, clay mineralogy, organic matter content, activities of monomeric and polymeric ionic species of aluminum, and availability of Ca^{2+}, Mn^{2+}, Mo, and P make it extremely difficult to compare data from studies carried out in different acid soils. The data presented in Coventry et al.[80] exemplify this problem by providing examples of the different limitations to legume growth that were encountered in acid soils in Victoria, in southeast Australia.

ii. Influence of Acidity on Rhizobial Growth and Survival

Problems in legume establishment in acid soils have often been linked to the varying sensitivity of rhizobia to acidity. The subject has been extensively reviewed.[47,77,79,239,280] The identification of *Bradyrhizobium* strains that were capable of growth at low pH in the presence of soluble aluminum[95,205] stimulated numerous researchers to screen culture collections of other species for acid and aluminum-tolerant representatives.[76a,142,159,240,241,368,385,410,411] The logical extension of these efforts was to ask whether rhizobia indigenous to acid soils were successful colonizers because of their superior acid tolerance. Studies of indigenous rhizobia in acid soils have not produced results in support of this possibility.

In a survey of Hawaiian soils ranging from pH 4.9 to pH 7.9, the most dense populations of *Bradyrhizobium* spp. were recovered in Tropohumults of pH <5.0. The greater mean annual rainfall received by the more acidic soils had a positive influence on population density that overrode the negative impact, if any, of low soil pH.[414] No difference in acid tolerance was noted among isolates of *R. leguminosarum* bv. *trifolii* recovered across a soil pH transect.[413] The majority of the population of *R. leguminosarum* bv.

trifolii recovered from an extremely acid soil were no more acid tolerant than laboratory strains, and most of the population "avoided" low pH by colonizing the less acidic upper 2 cm of the soil profile.[322,323] Although liming an acid soil changed the outcome of nodule occupancy by four indigenous serotypes of *R. leguminosarum* bv. *trifolii* on subclover,[116,117] the soil population density of each of the four serotypes was unaffected by liming regardless of the influence of lime on nodulating abilities.[4]

iii. Influence of Acidity on Nodule Formation

Both formation and function of several legume–rhizobial symbioses have been shown to be sensitive to soil acidity. Indeed, the importance of pH to the infection process was recognized many years ago when small amounts of lime drilled with, or coated onto clover seed, were found to be beneficial to nodulation of legumes planted in acid soils.[3,237] The infection process of certain temperate legumes (*Trifolium, Pisum,* and *Medicago*) was found to be more sensitive to acidity than plant growth itself.[230,236,278] In addition, nodule formation[46,283] and the expression of *nod* genes by rhizobia[324] can be more sensitive to acidity or aluminum than the growth or survival of the rhizobia per se. Although both Al-sensitive and Al-tolerant strains of *Bradyrhizobium* sp. (*Vigna*) survived Al-rich acid soil conditions equally well,[159] the Al-sensitive strain nodulated cowpea [*Vigna unguiculata* (L.) Walp] less effectively in the acid soil than did the Al-tolerant strain.[161] In contrast, Graham et al.[142] found that an acid-tolerant strain of *R. leguminosarum* bv. *phaseoli* not only survived better in acid soil than an acid-sensitive strain but also gave better nodulation and bean [*Phaseolus vulgaris* (L.) Savi] yield.

iv. Influence of Acidity on Nodule Function

Although the growth of many legumes under nitrogen-fixing conditions is more sensitive to acidity and aluminum than when plants are grown with mineral forms of nitrogen,[11,12,105,277,283] there are exceptions where no differences were measured between growth dependent on symbiosis and mineral N.[232,282] Furthermore, differences in tolerance of acidity are documented between genera of legumes,[11,281] between species within a genus,[62,63,179] and between cultivars of a species.[187,208,293,385,386] Adaptability of legume species to acid, low-fertility soils is often revealed by their indifference to mineral nutrient supplements.[52,162] To the author's knowledge there are no published studies describing the behavior of rhizobia in the rhizospheres of both acid-tolerant and acid-sensitive legumes growing in acidic soils.

v. Influence of Phosphate

It has been documented that many legumes are inferior to other plant species in sequestering phosphate from soil.[59,201] As a consequence, many soils require supplementing with large quantities of phosphate fertilizer if legume production is to become an economically viable reality.[295] Since some legumes have a higher P requirement for symbiotic growth than when they are grown with fertilizer N,[65,184,351] the question arose of whether this phenomenon was linked to the P-sequestering characteristics of rhizobia.

Soybeans inoculated with P-deficient *B. japonicum* strain USDA 110 possessed lower nitrogenase activity and lower shoot N content than when inoculated with luxury P-grown bacteria.[274] Strains of *B. japonicum,*[64] *R. meliloti,*[17,18] and *R. leguminosarum* bv. *trifolii*[229] have been shown to differ in their ability to sequester low concentrations of inorganic P. In the case of *R. leguminosarum* bv. *trifolii,* strains that showed superior P-sequestering ability were members of an indigenous serotype that dominated nodules of subclover grown in a P-deficient acid soil.[4]

vi. Role of Phosphatases

The possibility exists that rhizobia utilize phosphatase enzymes to sequester soil organic P and circumvent any inadequacy of their inorganic P-uptake system. Soil isolates of *R. leguminosarum* bv. *trifolii* express acid phosphatase activity (optimum pH, 6.5) when grown under inorganic P-limited conditions.[229] In contrast, alkaline phosphatase (optimum pH 9.0) was detected in some strains of *Rhizobium,* and strains of *Bradyrhizobium* sp. produced neither alkaline nor acid phosphatase under similar growth conditions.[354] Further work is needed to resolve these discrepancies in P metabolism among rhizobia and to determine if these differences have any significance to soil-colonizing ability or nodule formation and function in P-deficient soils.

vii. Influence of Calcium

Under acid conditions higher levels of calcium are required for root hair infection than are needed for legume growth.[236,242,279] Although the availability of soil calcium is rarely considered to be a limiting factor for either plant or rhizobial growth, highly weathered soils of low base content can become depleted in exchangeable Ca after several years of intensive crop production.[81,82] Furthermore, calcium deficiency can be exacerbated by excesses of other cations. For example, subclover responds to calcium fertilization when grown in soils formed over serpentine, a parent material unusually high in magnesium relative to calcium.[198]

Although there is a low calcium requirement for growth of rhizobia under optimum laboratory conditions,[390] several recent findings suggest that a critical role for calcium in the ecology of rhizobia cannot be discounted. Supplemental calcium is essential for growth of *Rhizobium meliloti* both at low inorganic phosphate concentrations[17,18] and at low matric water potential created by a nonpermeating solute.[50] Acid-tolerant *R. meliloti* strains required higher levels of calcium for growth than did other *R. meliloti* strains.[291] Calcium plays a role in the synthesis of a surface adhesion protein important for attachment of *R. leguminosarum* bv. *viciae* to pea roots,[355,356] and, in the same biovar, calcium is involved in stabilizing oligomeric forms of outer-membrane proteins that are covalently linked to peptidoglycan.[99] Under acidic conditions supplemental calcium enhanced the induction of *nod* genes by 7,4′-dihydroxyflavone in *R. leguminosarum* bv. *trifolii*.[321,324]

IV. The Rhizosphere Habitat

A. *Selectivity of the Legume Rhizosphere*

In their book entitled *The Rhizosphere*, Curl and Truelove[89] commented: "The status of *Rhizobium* [in the rhizosphere] is difficult to assess from conflicting reports." The beneficial influence of legume roots upon the activity of soil microorganisms in their immediate vicinity was documented many years ago.[170] Subsequently, roots of leguminous plants were found to harbor higher concentrations of microorganisms in their vicinity than did roots of other plant species.[299,332,336] Japanese workers noted that legume rhizospheres stimulated growth of rhizobia more than other soil bacteria and that rhizobia proliferated more in rhizospheres of legumes than in those of monocots.[376,377,378] An extreme example of rhizosphere specificity was shown by *R. meliloti*, which only colonized the rhizospheres of *Medicago* spp. despite the presence of other plant species in the pasture community.[69,327]

B. *Nonselectivity of the Legume Rhizosphere*

Other studies, however, have provided no support for the argument that legume rhizospheres are particularly selective for rhizobia. Furthermore, they reveal the inconsistencies between studies. Under laboratory conditions, an individual strain of *B. japonicum* proliferated more extensively in a variety of nonhost rhizospheres than in the rhizosphere of the nodulating host, soybean.[304] In contrast, under field conditions, greater stimulation of indigenous *B. japonicum* occurred in soybean rhizospheres than in oat (*Avena sativa* L.) rhizospheres.[271] Although *Rhizobium leguminosarum* bv. *phaseoli* poorly colonized rhizospheres of soybean grown in the laboratory,[304] under field conditions the same biovar colonized the rhizosphere of soybean better than did indigenous *B. japonicum*.[325]

C. Complexity of Rhizosphere Studies

Technical limitations result in the rhizosphere being treated in an overly simplified manner. The complexity of rhizospheres fashioned in soil is revealed elegantly by the electron micrographs presented in Foster et al.[128] Despite the high numbers of bacteria in the rhizosphere, less than 10% of the actual root surface is colonized.[337] Root morphology and rate of root development are influenced by plant genotype, soil structure, mineral nutrient status of the soil, and water availability.[185,244] The frequency, length, density, and infectibility of root hairs can vary markedly among legume species.[231,287,288] Since soil bacteria (and presumably rhizobia) are not uniformly distributed on plant roots, and the patterns of infectibility of root hairs of soil-grown legumes remains to be established, research is needed to determine if these two phenomena infuence the outcome of competitive nodulation in soil.

D. The Inactive or Senescing Rhizosphere Habitat

Large populations of *B. japonicum* and *Bradyrhizobium* sp. (*Vigna*) were measured around senescing roots of field-grown soybean and black gram [*Vigna mungo* (L.) Hepper] and were attributed to senescing nodules.[48,271] The importance of nodule decay to establishing a soil population was pointed out by Kuykendall.[219] Even though an inoculant strain of *B. japonicum* only formed 1% of the nodules on soybean planted into a soil already containing an indigenous population of *B. japonicum*, this was sufficient to allow the inoculant to establish a resident soil population. The ability of disintegrating nodules to change the rhizobial subpopulation structure might be more exaggerated in undisturbed soils under pastures and minimum tillage systems. Soils managed in this manner contain a high organic carbon content and thereby have the potential to maintain larger rhizobial populations than conventionally tilled soils.

V. Intraplant Habitat

The intraplant habitat has received little attention, despite the possibility of significant microbial interactions. Since nodules of soil-grown legumes can be co-occupied by different strains of rhizobia,[101,260,271,318] co-occupancy can be of potential significance if the symbiotic effectiveness characteristics of the co-occupants differ. In studies of interstrain competition between a highly effective strain and a suboptimally effective strain of *R. leguminosarum* bv. *trifolii*, the former strain was found to occupy more than 50% of nodules on subclover cv. Woogenellup. The growth of the plants, however, resembled plants nodulated solely by the less effective strain.[102] The suboptimally effective plant phenotype could be explained by the fact that the majority of the nodules occupied by the effective strain were co-occupied by the strain of inferior effectiveness.

A second scenario can be envisaged where intraplant competition might have implications to field productivity. Prior to regrowth of forage legumes after mechanical harvest or animal grazing, meristematic activity of indeterminate nodules is stimulated, and new cortical cells are invaded by rhizobia existing within infection threads.[384] Competition between different strains within the infection threads could influence which of the strains form the majority of the new bacteroids and thereby have an impact upon N_2 fixation during regrowth. This aspect of rhizobial ecology also merits attention in those legume species where infection occurs intercortically and where rhizobia can remain in the infection thread even during their N_2-fixing state.[331,364]

VI. Composition of Soil Rhizobial Populations

Over a period of 50 years numerous studies have revealed the tremendous diversity of isolates that can be recovered from the same soil population of rhizobia (Table 8-1). Both serotyping and multilocus allozyme electrophoretic typing have revealed a similar number of different types (6–20) within each soil population, whereas other methods show much greater diversity. At this time no organized effort has been made to evaluate whether or not rhizobial subpopulation structure is similar in soils throughout the world.

A. Serotyping

Serotyping has a long history of use in population delineation, yet it has received criticism for the inability to discriminate between closely related but nonidentical strains. Although this criticism is valid, several research groups have shown consistently that only one or two serotypes of *B. japonicum*,[90,271] *R. leguminosarum* bv. *trifolii*,[4,100] and *R. leguminosarum* bv. *viciae*[245] dominate the root nodules of field-grown legumes despite a greater number of serotypes existing within those soil populations. Given that rhizobia within a serotype are not necessarily identical, it remains to be established if attributes that account for competitive nodulating ability are shared by members of the same serotype. Another concern with serology relates to the monopoly that immunofluorescence maintains as the only practical method for studying the autecology of indigenous rhizobia in soil. Despite the long and impressive record of this method, there is no concrete evidence that serotypes represent distinctly different ecotypes within a soil rhizobial population.

Table 8–1 Composition of Soil Rhizobial Populations

Method of Analysis	Rhizobium or Bradyrhizobium spp.	Number of Types in Soil Population	Plant Host	Location	Reference
Serology					
	R. meliloti	9	Medicago minima (L.) Bartal	New South Wales, Australia	181
	R. leguminosarum bv. trifolii	8	Trifolium glomeratum L.	N.S.W., Australia	181
	R. leguminosarum bv. trifolii	6	Trifolium repens L.	N.S.W., Australia	181
	R. leguminosarum bv. trifolii	7	T. glomeratum	N.S.W., Australia	313
	R. leguminosarum bv. trifolii	9	Trifolium spp.	Oregon, USA	379; Dashti and Bottomley, unpublished observation
	R. leguminosarum bv. trifolii	10	Trifolium spp.	California, USA	173
	R. leguminosarum bv. trifolii	10	T. repens	Otago, New Zealand	134
	R. leguminosarum bv. viciae	3	Pisum sativum L.	Washington, USA	33, 245
	Bradyrhizobium japonicum	6–9	Glycine max (L.) Merrill	USA	191
	B. japonicum	>6	G. max	Iowa, USA	90
	B. japonicum	7	G. max	Iowa, USA	147
	B. japonicum	>7	G. max	Maryland, USA	21

Table 8-1 Continued

Method of Analysis	*Rhizobium* or *Bradyrhizobium* spp.	Number of Types in Soil Population	Plant Host	Location	Reference
Multilocus allozyme electrophoresis					
	R. leguminosarum bv. *viciae*	10–15	*P. sativum*	Norfolk, UK	417–419
	R. leguminosarum bv. *viciae*	7	*P. sativum*	Roskilde, Denmark	123
	R. leguminosarum bv. *trifolii*	9	*T. repens*	Norfolk, UK	417
	R. leguminosarum bv. *trifolii*	3 to 10	*T. repens*	Wales, UK	157, 158
	R. leguminosarum bv. *phaseoli*	2	*Phaseolus vulgaris* (L.) Savi	Norfolk, UK	417
	R. leguminosarum bv. *phaseoli*	4	*P. vulgaris*	Hidalgo, Mexico	308
Protein profile patterns					
	B. japonicum	19	*G. max*	Wisconsin, USA	285
	B. japonicum	29	*G. max*	Wisconsin, USA	200
	R. leguminosarum bv. *trifolii*	14	*Trifolium subterraneum* L.	Oregon, USA	115
	R. meliloti	16	*Medicago sativa* L.	Oregon, USA	189

qqb,

309

Table 8-1 Continued

Method of Analysis	*Rhizobium* or *Bradyrhizobium* spp.	Number of Types in Soil Population	Plant Host	Location	Reference
Plasmid profile patterns					
	R. meliloti	13–17	*M. sativa*	Ontario, Canada	41
	R. meliloti	17	*M. sativa*	Berrenrath-Weiler, Fed. Rep. Ger.	42
	R. leguminosarum bv. *viciae*	18	*P. sativum*	Washington, USA	33
Intrinsic antibiotic resistance					
	R. leguminosarum bv. *viciae*	12	*P. sativum*	Co. Cork, Eire	267
	R. leguminosarum bv. *phaseoli*	54	*P. vulgaris*	Norfolk, UK	20
Phage typing					
	R. meliloti	51–65	*M. sativa*	Ontario, Canada	40
	R. meliloti	37–87	*M. sativa* *Melilotus alba* Desr.	Ontario, Canada	369

310

B. Multilocus Allozyme Electrophoretic Typing

Multilocus allozyme electrophoretic typing has been favored recently by some researchers, since it provides the opportunity to gain a better understanding of the genetic relatedness of subpopulations. The same electrophoretic types (ETs) of *R. leguminosarum* bvs. *trifolii* and *viciae* were recovered from nodules of *T. repens* L. and *Pisum sativum* planted into the same soil.[417] These findings confirmed, in a field setting, the close chromosomal relationship between pea and clover rhizobia that had previously been established only from studies of culture collection organisms.[84] Furthermore, isolates of bvs. *viciae* and *trifolii* with identical ETs were identified at different locations in the UK.[157,418] Although both of these findings were made 40 years ago with serological methods, there is a need to determine if a relationship exists between population delineation established by multilocus allozyme electrophoretic typing and by serotyping.[19]

C. Subpopulation Densities Within Soil Rhizobial Populations

The pioneering work in immunofluorescence by Schmidt's group at the University of Minnesota made it possible to determine the densities of subpopulations of indigenous soil rhizobia and to evaluate if the outcome of competitive nodulation was related to differences in subpopulation densities.[23,25,346,347] Neither bulk soil nor host or nonhost rhizosphere populations of three *B. japonicum* serogroups were significantly different from each other despite the overwhelming nodulating success of serogroup 123 on soybean.[120,271,319,320,326] In contrast with these findings, statistically significant differences exist between the densities of certain serotypes within *R. leguminosarum* bv. *trifolii* soil populations under permanent pasture.[4,28,101,379] Although reasons for the differences between the *Bradyrhizobium* and *Rhizobium* findings are not known, it is worth emphasizing that no evidence exists for believing that *Bradyrhizobium* and *Rhizobium* occupy the same ecological niche and behave similarly in soil.

In regard to the relationship between serotype population size and the outcome of competitive nodulation, findings from the author's laboratory on soil populations of *R. leguminosarum* bv. *trifolii* are in general agreement with those obtained on *B. japonicum*. No evidence was obtained for the outcome of competitive nodulation being directly related to serotype subpopulation densities.[4,101,379] Indeed, six of the nine serotypes of *R. leguminosarum* bv. *trifolii* found within this soil population are recovered consistently as minor nodule occupants, regardless of their population densities or of the clover species used as a host (Leung, Dashti, and Bottomley, unpublished data).

D. Physiological Activity of Soil Rhizobial Subpopulations

Many years have passed since the all-encompassing terms of *incursiveness* and *saprophytic competence* were introduced to describe the characteristics that account for the success of rhizobia in soil.[70,156] Although immunofluorescence allows subpopulation densities to be determined, conventional use of the method reveals nothing about the physiological competence of rhizobia in soil. Recently, immunofluorescence has been combined with a viability assay to study the response of rhizobia to added substrates.[30] Preliminary data from the author's laboratory have revealed the following: (1) the proportion of substrate-responsive cells differs between serotypes, (2) the rate of appearance of elongated cells differs between serotypes, and (3) substrate concentration and environmental conditions can influence both the rate of appearance and the final percentage of elongated cells in serotype populations.

Substrate responsiveness has been demonstrated recently among subpopulations of soil-borne indigenous *B. japonicum*[395,396] and of *R. leguminosarum* bv. *viciae*.[136] Serogroup 123 of *B. japonicum* outgrew serogroup 110 when soil was incubated under a H_2-containing microaerobic atmosphere.[396] These findings are of significance since previous laboratory studies showed the H_2-oxidizing phenotype was more prevalent among isolates of serogroup 110 of *B. japonicum*[61] than among members of serogroup 123.[206] These findings emphasize the healthy skepticism that should be maintained about concepts established with culture collection organisms until they are proven in natural populations of rhizobia in the soil environment. More creative research effort is needed to evaluate the responsiveness of members of soil and rhizosphere subpopulations to different nutrient sources under specific plant and soil conditions.

VII. Competitive Nodulation

No aspect of the ecology of rhizobia has received the attention that has been given to competitive nodulation between mixtures of rhizobia.[7,94,109] Numerous publications have shown that representatives both of simple mixtures of rhizobia and of complex indigenous soil rhizobial populations are distributed unequally in nodules and that biological and abiological factors can influence nodulating success (Table 8-2).

Date and Brockwell[94] summarized much of the information about competitive nodulation that was generated prior to 1976. In many cases, the outcome of nodule occupancy was extremely one-sided and was not related to the relative rhizosphere populations of different strains. In the majority of those cases the strain combinations included an effective and an ineffective strain with the former dominating the nodules.[223,309,328] The outcome of

Table 8–2   Variables Influencing Outcome of Competitive Nodulation on Legumes

Variable	Host	*Bradyrhizobium* or *Rhizobium* spp.	Reference
SOIL INFLUENCE			
Soil type or location	*Glycine max* (L.) Merrill	*B. japonicum*	54, 90, 147, 191
	Trifolium subterraneum L.	*R. leguminosarum* bv. *trifolii*	15, 34, 188, 333
	Pisum sativum L.	*R. leguminosarum* bv. *viciae*	8,267
	Cicer arietinum L.	*Rhizobium* sp. (*Cicer*)	13
	P. sativum	*R. leguminosarum* bv. *viciae*	245
	Leucaena leucocephala (Lam.) de Wit	*Rhizobium* sp. (*Leucaena*)	270
	Lens culinaris Medik. (prev. *L. esculenta* Moench)	*R. leguminosarum* bv. *viciae*	260
Soil temperature	*G. max*	*B. japonicum*	210, 221, 334*, 404
	Trifolium repens L.	*R. leguminosarum* bv. *trifolii*	154
Planting date	*G. max*	*B. japonicum*	57
Soil depth	*Prosopis glandulosa* L.	*Bradyrhizobium* and *Rhizobium* spp.	190
Soil pH	*G. max*	*B. japonicum*	90, 147
Lime amendment	*T. subterraneum* *T. repens*	*R. leguminosarum* bv. *trifolii*	4, 16, 117, 318
Phosphate amendment	*T. subterraneum*	*R. leguminosarum* bv. *trifolii*	4
	G. max	*B. japonicum*	166
pH change	*T. repens*	*R. leguminosarum* bv. *trifolii*	196*, 339*
	Phaseolus vulgaris (L.) Savi	*R. leguminosarum* bv. *phaseoli*	385*, 386*
PLANT INFLUENCE			
Intergenera	*P. sativum* *T. repens* *P. vulgaris*	*R. leguminosarum*	417
	P. sativum *L. culinaris*	*R. leguminosarum* bv. *viciae*	375
	Macroptilium atropurpureum (DC.) Urban *Stylosanthes guianensis* (Aubl.) Swartz	*Bradyrhizobium* sp.	129*
	Medicago sativa *Melilotus alba* Desr.	*R. meliloti*	369

Table 8–2   Continued

Variable	Host	Bradyrhizobium or Rhizobium spp.	Reference
Interspecies	*Medicago lupulina* L. *M. sativa*	*R. meliloti*	369
	T. subterraneum *Trifolium pratense* L.	*R. leguminosarum* bv. *trifolii*	329, 379
	T. subterraneum *T. repens*	*R. leguminosarum* bv. *trifolii*	254
Intercultivar	*T. subterraneum*	*R. leguminosarum* bv. *trifolii*	37, 101, 117
	T. repens	*R. leguminosarum* bv. *trifolii*	195, 265
	T. pratense	*R. leguminosarum* bv. *trifolii*	340*
	M. sativa	*R. meliloti*	40, 51
	P. vulgaris	*R. leguminosarum* bv. *phaseoli*	386*
	P. vulgaris	*R. leguminosarum* bv. *phaseoli*	168
	G. max	*B. japonicum*	55, 222
	G. max	*B. japonicum/R. fredii*	108, 263
	C. arietinum	*Rhizobium* sp. (*Cicer*)	361
Intracultivar	*T. repens*	*R. leguminosarum* bv. *trifolii*	155
	T. pratense	*R. leguminosarum* bv. *trifolii*	340*
	M. sativa	*R. meliloti*	38
MICROORGANISM INFLUENCE			
Bacteria (nonrhizobia)	*G. max*	*B. japonicum*	132*, 167
	M. sativa	*R. meliloti*	149*
Intrastrain	*G. max*	*B. japonicum*	275*
	M. sativa	*R. meliloti*	150*
Toxin or bacteriocin	*T. subterraneum*	*R. leguminosarum* bv. *trifolii*	171*, 372*, 373*
OTHER INFLUENCES			
Delayed inoculation	*G. max*	*B. japonicum*	213*, 215*
	T. subterraneum	*R. leguminosarum* bv. *trifolii*	345*
Plasmid removal	*Lotus pedunculatus* Cav.	*Rhizobium loti*	298*
	M. sativa	*R. meliloti*	39*
Plasmid exchange	*P. sativum*	*R. leguminosarum* bv. *viciae*	32*
Host genotype restriction	*T. subterraneum*	*R. leguminosarum* bv. *trifolii*	137,138
	P. sativum	*R. leguminosarum* bv *viciae*	407
	G. max	*B. japonicum*	83, 204, 257*

*Studies were not carried out in soil.

several other studies do not support the concept that an effective strain will always outcompete an ineffective strain regardless of population size.[6,102,129,193,284,393]

When differences in effectiveness among strains in the inoculant mixture are not extreme, most nonsoil studies show a relationship between nodulation success and the relative numbers of strain either in the inoculum or in the rhizosphere.[8,16,129,133,235,257,275] As mentioned earlier, immunofluorescence studies carried out in nonsterile soil containing indigenous populations of rhizobia have not confirmed that the outcome of competition is related to the relative numbers of different serotypes in rhizospheres. This discrepancy requires further attention and is often overlooked by many researchers of the competition phenomenon who have restricted their studies to simple mixtures of rhizobia under nonsoil conditions.

A. Issues to Be Resolved in Competitive Nodulation

A comprehensive understanding of the scientific basis behind competitive nodulation will depend upon refining methods for studying subpopulations in the rhizosphere. In addition, a closer liaison than currently exists will be required between microbial ecologists and those studying the molecular biology of symbiosis establishment. The reader's attention is drawn to the following observations made by the author as a result of reviewing the literature on this subject.

i. Numbers of Root Hairs Infected and Subpopulation Size

Although we are gaining a better understanding of rhizobial subpopulations in the rhizosphere, no attempts have been made to establish if a relationship exists between the number of root hairs infected and subpopulation size.

ii. Environmental Effects on Root Development

In studies of competitive nodulation, adequate attention has not always been given to the effect of environmental variables upon the structural development of the plant root, root hair infection characteristics, and nodule occupancy patterns along the root system. More consistent attention to these details can be found in recent publications.[261,312,397,418,422]

iii. Competitive Indicator' Strains

The outcome of competition between rhizobia in simple mixtures in nonsoil environments can be profoundly different from that observed in soil containing indigenous rhizobia.[101,102,214,381] For example, although there is no doubt that members of indigenous serogroup 123 outcompete indigenous serogroup 110 in many soils, conflicting data have arisen from competition experiments using simple mixtures of *B. japonicum* strain USDA 110 and strain USDA 123.[83,133,204,214] The same "USDA" strains of *B. japonicum* obtained from different culture collections can behave differently in competitive situations.[275] Similar discrepancies have arisen in the southeast region of the United States, where serogroup 31 is a dominant nodule occupant of soybean,[180] yet strain USDA 31 is poorly competitive against other strains in nonsoil studies.[133,357] To the author's knowledge, *R. leguminosarum* bv. *phaseoli* strain Kim 5 is the only strain of rhizobia that has been shown consistently to be a superior competitor in different field and laboratory studies.[16,199,305] Researchers need to make a serious effort to procure highly competitive individual isolates representing indigenous subpopulations that are competitive under field conditions. Along with this endeavor, we need to determine laboratory assay conditions under which competitive behavior resembles that observed under field soil conditions.

iv. Physiological State of Soil and Inoculant Rhizobia

Preexposure of one strain of rhizobia to seedlings can alter the pattern of competition with other strains applied later to the root system.[213,215,345] Other studies have not confirmed that superior competitive ability is related to the ability to rapidly infect root hairs and form nodules.[357,422] Discrepancies between competitive behavior under laboratory and field conditions might be an artifact of growing strains in rich medium. As a consequence, important differences in the physiological states of the strains in soil are overlooked. Studies from the author's laboratory show that the speed of nodulation by representatives of the most competitive indigenous serotypes of *R. leguminosarum* bv. *trifolii*[229] and *R. meliloti*[50] was not influenced by water stress or phosphate starvation. In contrast, nodulation by representatives of poorly competitive serotypes was delayed significantly after exposure to the two stresses. No differences in speed of nodulation were found when competitive and noncompetitive isolates were grown under nonstressed conditions.

v. Nodulation Patterns and N₂-Fixing Characteristics

In the case of soil-grown *V. mungo*,[48] *T. subterraneum*,[101] and *M. sativa*,[153] the proportion of nodules occupied by inoculant strains changes with plant age. This observation can be combined with the fact that the rate of N_2 fixation can vary over the life cycle of a legume and can vary among different cultivars of a species.[14,183] Since root nodules on secondary lateral roots can contribute significantly to the N economy of field-grown plants,[152,261,353a,409] careful attention must be given to the time when nodule occupancy is evaluated.

vi. Nodulation Success and Synergistic Effects

Uncertainty exists about what proportion of root nodules need to be occupied by a highly effective strain in order to measure a significant increase in the quantity of N_2 fixed. For example, although inoculation of red clover (*Trifolium pratense* L.) with single strain inoculants of *R. leguminosarum* bv. *trifolii* induced a similar yield response from field-grown plants, only 10% of nodules were occupied by one inoculant strain, whereas 85% were occupied by the other strain.[10] This issue can be further confounded by unpredictable synergistic effects on the yield of plants that have their nodules occupied by more than one strain.[10,102,133,253,307]

B. Molecular Biology in Relation to Competitive Nodulation

In recent years molecular biological approaches have generated a vast amount of knowledge about symbiosis establishment. Although most of this research has not focused directly on competitive nodulation, several recent publications may be relevant to this phenomenon.

i. Interstrain Variation in the Sequence and Control of nod Gene Products

Mutations in many of the *nod* genes result in a "delayed nodulation" phenotype,[97,98,175,176,365] or a destabilization of infection thread development.[58,406] Among rhizobia within the same host range group, it is possible that subtle differences existing in the structure or control of *nod* gene products influence the proportion of infected root hairs that abort. Evidence has been obtained recently for variation in the way that strains of *R. meliloti* control *nod* genes.[212]

ii. Expression of nod Genes by Nodulating and Nonnodulating Rhizobia in Soil Populations

Many *nodD* gene products can interact with flavonoids from either host or nonhost legumes and induce expression of *nod* genes.[363,416,421] Furthermore, the level of *nod* gene expression can influence whether nodulation is inhibited[211] or enhanced.[365] It was reported recently that nodulation of "Afghanistan" pea by *R. leguminosarum* bv. *viciae* strain Tom is blocked by nodulation-incompetent strain PF_2 through expression of early *nod* genes in the latter strain.[110] Perhaps the expression of *nod* genes in both nodulating and nonnodulating rhizobia in a rhizosphere results in synergistic effects that either interfere with or enhance nodulation by different subpopulations.

iii. Flavonoid Signatures and the nodD Gene Product

The *nodD* gene is a determinant of host range in some legumes and not in others.[362,363] Different spectra of flavonoid inducers are released from seedlings of different species.[363,420,421] The quantity (or quality) of flavonoid inducers released from seedlings of different cultivars of subclover is influenced by environmental pH.[321] Furthermore, the efficiency of flavonoid-dependent induction of *nod* genes in *R. leguminosarum* bv. *trifolii* is influenced both by environmental pH and by the individual strain into which the *nod* gene reporter system was inserted.[324] Further research is needed to evaluate how interactions between flavonoid signatures, environmental conditions at seed germination, and the chromosomal background of different rhizobial subpopulations might influence *nod* gene induction and competitive nodulation.

VIII. Agricultural Implications

Despite the advances made in our understanding of legume–rhizobial symbioses, little of this information has made an impact upon practical legume agronomy. In this section I attempt to address certain areas of rhizobial ecology that are in need of thorough examination and that may have applied significance.

A. Inoculant Technology

i. The Physiological State of Rhizobia in Inoculant Carriers

During the period between 1950 and 1970, advances were made in inoculant technology that resulted in the production of high densities of viable rhizobia in solid carriers.[45,96,335,367] We have no understanding of the physiological state of rhizobia in carrier materials, nor of the physiological transformations that rhizobia undergo after transfer to the seed. For example, it is known that peat inoculants benefit from a "curing" period, after which the survival of rhizobia on the seed is enhanced.[45,259] Although the benefits of enhanced survival are obvious, it is also possible the rhizobia now lack the ability to respond quickly to the germinating seed. A rapid response may be important if the inoculant strain is to compete successfully with indigenous soil rhizobia for nodulation sites. In this regard we can consider the following:

1. Neither lag nor stationary phase cells of *R. leguminosarum* bv. *trifolii* showed ability to induce *nod* genes in the presence of the flavonoid inducer 7,4′-dihydroxyflavone.[103]

2. Growth medium composition in general,[306] and excess ammonium specifically,[118] can inhibit the induction of *R. meliloti nod* genes by flavonoids.

3. Given the possibility that motility is important for rhizobia to colonize roots successfully, no attempts have been made to determine if peat-cured inoculant rhizobia possess flagellae.

4. Rhizobia can proliferate extensively in peat under conditions where the carrier was providing all nutrients essential for growth.[360] The high content of aromatic compounds (including flavonoids) in peat is of interest in this regard. Strains of *Bradyrhizobium* and *Rhizobium* grow on many aromatic compounds. with strains of *Rhizobium* being less versatile.[74,75,300,301] Since many flavonoids are antagonists of *nod* gene expression by more potent flavonoid inducers,[103,126,216,420] it is possible that rhizobia emerging from a peat carrier are "desensitized" to the specific *nod* inducers of the host as a result of exposure to a complex array of flavonoid compounds in the peat. Furthermore, if rhizobia emerge from the peat carrier with their aromatic degradative pathways induced, they may inadvertently modify the flavonoids released by the germinating seedling.

5. Poor survival of inoculant rhizobia after their application to the seed has been linked to seed coat toxicity.[258,366] Iron limitation was considered to be the causative factor of this phenomenon, since addition of FeEDTA to growth medium alleviated the toxic effect.[122] By gaining a better understanding of the physiological state of rhizobia in peat carriers, progress might be made in inoculant production and quality.

ii. Placement of Inoculant Strains

Although the nodule mass of some legumes is located primarily on the upper region of the taproot, other legume species nodulate more uniformly and indeterminately throughout their root system. The impact of lateral root nodules on N_2 fixation has been well documented.[152,261,353a,409] It is debatable whether or not a practical, economically viable technology can be developed to place inoculants strategically for interception of the secondary root system. Nevertheless, inoculant rhizobia have been successfully introduced into stands of alfalfa and birdsfoot trefoil (*Lotus corniculatus* L.) two to four weeks after seedling emergence and nodulation deficiencies were corrected.[330a] Moreover, when a *B. japonicum* inoculant was introduced in irrigation water 30 days after soybean emergence, a greater seed yield was obtained than was achieved with conventional seed inoculation.[76] It must be emphasized, however, that both of these studies were carried out in soil devoid of competing indigenous rhizobia. Perhaps future developments in fertilizer application and irrigation technologies hold the key to achieving this goal with legumes of particular rooting habits growing in soils of specific textural and structural properties.

iii.   Biological Control Strategies

Although some success has been achieved in the use of bacteria to control soil-borne plant pathogens,[203,349,405] only recently has progress been made with an analogous system in rhizobia. *R. leguminosarum* bv. *trifolii* strain T24 is known to inhibit growth and nodulation by other biovars of *R. leguminosarum* as a result of producing a unique peptide, trifolitoxin.[350,373] Since T24 is an ineffective N_2-fixing strain, the genes for both trifolitoxin production and resistance were cloned and expressed in a symbiotically effective strain of *R. leguminosarum* bv. *trifolii*.[371,374] As a result of acquiring these genes, the competitiveness of the recipient was enhanced.[372] This novel approach to the problem of competition should stimulate a search for other examples within the rhizobia.

B.   Intercropping

The availability of selective herbicides enables farmers to establish legumes under monoculture systems. Since the legume has no competition for available soil N, a significant portion of the total N assimilated by the crop is derived from the soil.[317] In contrast, pasture legumes are invariably grown in the presence of companion grasses, and, in many parts of the world, grain legumes are routinely intercropped with a nonlegume companion crop. Studies involving ^{15}N isotope dilution techniques show that legumes derive a higher percentage of N from N_2 fixation when grown in companion situations than they do under monoculture situations.[1,93,227,273] Nevertheless, opinions tend to be mixed upon the issue of whether legume yield is suppressed or unaffected under intercropping conditions. Given our limited understanding of rhizobia in the rhizosphere of monocultures, the author knows of no studies where the ecology of rhizobia has been compared in rhizospheres of monocultures and mixed-plant communities. Since many legumes compete poorly with cereals and grasses for soil resources,[59,201] plant competitiveness may be enhanced by focusing attention on the development and physiology of roots.

C.   Legume Breeding

Although this topic is to be covered elsewhere (Chapter 16), it is worth emphasizing that farmers base their choice of legume cultivar or species primarily upon such criteria as (1) disease resistance, (2) adaptation to the length of the growing season, (3) management of hygiene and harvest, (4) restrictions of food processing technology, and (5) marketability (e.g., size and color of pods and seeds). In the case of forage legumes, tolerance to grazing pressure, winter hardiness, percentage of hard seed produced, and levels of animal-damaging substances in the herbage are additional factors that influence the farmer's choice. As a result, N_2 fixation is not a phenotype to which breeders have devoted much attention. Since variety trials invariably occur at specific locations in a region, it is possible that cultivar performance is influenced by those members of a soil rhizobial community that nodulate the plants. Although several reports have shown that cultivars of a crop species nodulate with different members of a soil population,[40,51,55,117,222,369] the impact of this phenomenon upon N_2 fixation by cultivars under field conditions is unknown.

The tremendous diversity of rhizobia in soil populations might be insurance against cultivar selection having a negative impact on yield under field conditions. However, cultivar performance may be more seriously affected in soils where the population diversity of rhizobia is restricted. In this context, the diversity of the soil population of *R. leguminosarum* bv. *trifolii* was less in acid soils than in neutral soils.[157] If soil rhizobial populations were well characterized on all sites where extensive cultivar testing is undertaken, some of these concerns could be studied in a rationale manner.

D. Genetic Instability of Inoculant Strains

Numerous surveys have shown that the majority of rhizobia recovered from nodules of field-grown plants are generally less effective than the inoculant strains currently available.[29,139,144,323,382,389] The most popular hypthesis to explain this phenomenon suggests that the isolates represent indigenous rhizobia symbiotically effective with endemic legumes, but symbiotically "mismatched" with introduced agricultural varieties.[137,138] It is possible, however, that the results reflect the genetic instability of the effectiveness trait of an inoculant strain that is expressed after several years in the soil. Instability of symbiotic characteristics has been well documented over many years even under laboratory conditions.[127,139,163,220,224,275,286,382,387,388,391,402]

Obviously, environmentally induced change will be difficult to address with a strain that is inherently unstable. Passage of rhizobia through the host plant can influence the maintenance of host range determinants carried on plasmids.[104,365,398] If genetic change does occur in nodules, then one has to ponder (1) what percentage of the nodule occupants will be affected, (2) whether the variant(s) will return to the soil in a viable form, and (3) whether there is sufficient selective pressure in the nodule, or in the soil, to enrich for the variant relative to the large soil rhizobial population that did not nodulate the plant. The potency of nodules as a source of genetic variants will also depend on the number and size of nodules, the proportion of the nodule content that is viable, and the extent of development and viability of bacteroids.[262,374a,423]

E. Genetic Exchange Between Rhizobia in Soil and Nodules

Using restriction fragment length polymorphism, the same *sym* plasmid has been identified within chromosomally dissimilar isolates of both *R. leguminosarum* bvs. *trifolii* and *viciae* recovered from the same locations.[348,419] Furthermore, evidence was also obtained to suggest that one isolate possessed a "recombinant" *sym* plasmid in which *nif* genes had been transferred from one to another *sym* plasmid.[348] Broughton et al.[43] presented evidence that the *sym* plasmid of *R. leguminosarum* could be transferred to a non-nodulating mutant of *R. meliloti* in alfalfa (*Medicago sativa* L.) rhizospheres and thereby allow the plants to nodulate. Naturally occurring exchange of genes between rhizobia in soil probably depends upon the development of genetically distinct subpopulations in close proximity to each other. The question arises as to the nature of the microsites that could meet this criterion. Since nodules of soil-grown *Trifolium* and *Lens* can be extensively co-occupied by antigenically distinct strains,[4,101,260,318] the nodule seems a most logical site.

IX. Concluding Remarks

Although our understanding of rhizobial ecology is progressing at a slower rate than most of the other subdisciplines of biological N_2 fixation, I have several suggestions to make that might facilitate the resolution of this problem. Although rhizobial ecologists generally focus upon specific legume–rhizobial combinations and their idiosyncrasies under regional conditions, a concerted effort is needed to determine if rhizobial ecology can be transformed into a cosmopolitan science. To achieve this end, we must determine the population composition of rhizobia in soils located at major research centers throughout the world. Second, we need to expand our working knowledge of the ecology of these naturalized populations in different, albeit well-characterized soils under field conditions realistically managed for agricultural production. Third, wherever possible, we need to standardize our choice of biological materials, experimental designs, methods, and data analyses. If we can come to grips with these issues, then I believe it will become easier to separate those findings that are of limited comparative worth from those of more general significance. Too much of so-called rhizobial ecology is carried out either with laboratory strains under nonsoil conditions or in soil under conditions atypical of field situations. There are enough publications emphasizing the fact that findings in the laboratory bear little resemblance to field observations. We need to understand why! Although I believe that ecological research should be conducted primarily with natural soil populations, definitive proof of most hypotheses will only be obtained by subsequently studying representatives of these populations in simpler "artificial" systems. Model systems need to be developed in which the saprophytic and nodulating behavior of rhizobia in the soil environment can be accurately reproduced.

Acknowledgments

The author recognizes the financial support provided by the United States Department of Agriculture Competitive Research Grants Office and the Oregon Agricultural Experiment Station for his own findings presented in the manuscript. Special appreciation is given to Carlene Pelroy for processing this manuscript from drafts too numerous to count. My thanks go to Gary Stacey, Rick Weaver, Ed Schmidt, and Ben Bohlool, who provided constructive criticism during the final stages of preparation. This article reprsents Technical Paper No. 9157 of the Oregon Agricultural Experiment Station.

References

1. Abaidoo, R. C., van Kessel, C., "^{15}N-uptake N$_2$-fixation and rhizobial interstrain competition in soybean and bean, intercropped with maize," *Soil Biol. Biochem. 21*, 155–159 (1989).

2. Abel, G. H., and Erdman, L. W., "Response of Lee soybeans to different strains of *Rhizobium japonicum*," *Agron. J. 56*, 423 424 (1964).

3. Albrecht, W. A., and Poirot, E. M., "Fractional neutralization of soil acidity for the establishment of clover," *J. Am. Soc. Agron. 22*, 649–657 (1930).

4. Almendras, A. S., and Bottomley, P. J., "Influence of lime and phosphate on nodulation of soil-grown *Trifolium subterraneum* L. by indigenous *Rhizobium trifolii*," *Appl. Environ. Microbiol. 53*, 2090–2097 (1987).

5. Almendras, A. S., and Bottomley, P. J., "Cation and phosphate influences on the nodulating characteristics of indigenous serogroups of *Rhizobium trifolii* on soil grown *Trifolium subterraneum* L.," *Soil Biol. Biochem. 20*, 345–351 (1988).

6. Amarger, N., "Competition for nodule formation between effective and ineffective strains of *Rhizobium meliloti*," *Soil Biol. Biochem. 13*, 475–480 (1981).

7. Amarger, N., "Evaluation of competition in *Rhizobium* spp.," In: *Current Perspectives in Microbial Ecology* (Klug, M. J., and Reddy, C. A., eds.), Washington, D.C., Amer. Soc. Microbiol., 1984, pp. 300–305.

8. Amarger, N., and Lobreau, J. P., "Quantitative study of nodulation competitiveness in *Rhizobium* strains," *Appl. Environ. Microbiol. 44*, 583–588 (1982).

9. Ames, P., and Bergman, K., "Competitive advantage provided by bacterial motility in the formation of nodules by *Rhizobium meliloti*," *J. Bacteriol. 148*, 728–729 (1981).

10. Ames-Gottfred, N. P., and Christie, B. R., "Competition among strains of *Rhizobium leguminosarum* bv. *trifolii* and use of a diallel analysis in assessing competition," *Appl. Environ. Microbiol. 55*, 1599–1604 (1989).

11. Andrew, C. S., "Effect of calcium, pH and nitrogen on the growth and chemical composition of some tropical and temperate pasture legumes. I. Nodulation and growth," *Aust. J. Agric. Res. 27*, 611–623 (1976).

12. Andrew, C. W., Johnson, A. D., and Sandland, R. L., "Effect of aluminum on the growth and chemical composition of some tropical and temperate pasture legumes," *Aust. J. Agric. Res. 24*, 325–339 (1973).

13. Arsac, J. F., and Cleyet-Marel, J. C., "Serological and ecological studies of *Rhizobium* spp. (*Cicer arietinum* L.) by immunofluorescence and ELISA technique: Competitive ability for nodule formation between *Rhizobium* strains," *Plant and Soil 94*, 411–423 (1986).

14. Attewell, J., and Bliss, F. A., "Host plant characteristics of common bean lines selected using indirect measures of N_2 fixation," In: *Nitrogen Fixation Research Progress* (Evans, H. J., Bottomley, P. J., and Newton, W. E., eds.), Dordrecht, Martinus Nijhoff, 1985, pp. 3–9.

15. Baird, K. J., "Clover root-nodule bacteria in the New England region of New South Wales," *Aust. J. Agric. Res. 6*, 15–26 (1955).

16. Beattie, G. A., Clayton, M. K., and Handelsman, J., "Quantitative comparison of the laboratory and field competitiveness of *Rhizobium leguminosarum* biovar *phaseoli*," *Appl. Environ. Microbiol. 55*, 2755–2761 (1989).

17. Beck, D. P., and Munns, D. N., "Phosphate nutrition of *Rhizobium* spp.," *Appl. Environ. Microbiol. 47*, 278–282 (1984).

18. Beck, D. P., and Munns, D. N., "Effect of calcium on the phosphorus nutrition of *Rhizobium meliloti*," *Soil Sci. Soc. Am. J. 49*, 334–337 (1985).

19. Beltran, P., Musser, J. M., Helmuth, R., Farmer, J. J., Frerichs, W. M., Wachsmuth, I. K., Ferris, K., McWharter, A. C., Wells, J. G., Cravioto, A., and Selander, R. K., "Toward a population genetic analysis of *Salmonella*: Genetic diversity and relationships among strains of serotypes *S. choleraesius, S. derby, S. dublin, S. enteritidis, S. heidelberg, S. infantis, S. newport,* and *S. typhimurium*," *Proc. Natl. Acad. Sci. 85*, 7753–7757 (1988).

20. Beynon, J. L., and Josey, D. P., "Demonstration of heterogeneity in a natural population of *Rhizobium phaseoli* using variation in intrinsic antibiotic resistance," *J. Gen. Microbiol. 118*, 437–442 (1980).

21. Bezdicek, D. F., "Effect of soil factors on the distribution of *Rhizobium japonicum* serogroups," *Soil Sci. Soc. Am. Proc. 36*, 305–307 (1972).

22. Bezdicek, D. F., and Donaldson, M. D., "Flocculation of *Rhizobium* from soil colloids for enumeration by immunofluorescence," In: *Microbial Adhesion to Surfaces* (Berkeley, R. C. W., et al., eds.), Chichester, England, Ellis Horwood, 1980, pp. 297–305.

23. Bohlool, B. B., Schmidt, E. L., "Immunofluorescent detection of *Rhizobium japonicum* in soils," *Soil Sci. 110*, 229–236 (1970).

24. Bohlool, B. B., and Schmidt, E. L., "Persistence and competition aspects of *Rhizobium japonicum* observed in soil by immunofluorescence microscopy," *Soil Sci. Soc. Am. Proc. 37*, 561–564 (1973).

25. Bohlool, B. B., and Schmidt, E. L., "The immunofluorescence approach to microbial ecology," In: *Advances in Microbial Ecology,* vol. 4 (Alexander, M., ed.), New York, Plenum Press, 1980, pp. 203–241.

26. Boonkerd, N., and Weaver, R. W., "Survival of cowpea rhizobia in soil as affected by temperature and moisture," *Appl. Environ. Microbiol. 43,* 585–589 (1982).

27. Bottomley, P. J., and Brockwell, J., "Manipulation of rhizobial microflora for improving crop productivity and soil fertility in temperate regions: A critical assessment," In: *Microorganisms that Promote Plant Growth* (Dawson, J. D., and Dart, P. J., eds.), Dordrecht, Martinus Nijhoff, 1992, in press.

28. Bottomley, P. J., Dughri, M. H., "Population size and distribution of *Rhizobium leguminosarum* bv. *trifolii* in relation to total soil bacteria and soil depth," *Appl. Environ. Microbiol. 55,* 959–964 (1989).

29. Bottomley, P. J., and Jenkins, M. B., "Some characteristics of *Rhizobium meliloti* isolates from alfalfa fields in Oregon," *Soil Sci. Soc. Am. J. 47,* 1153–1157 (1983).

30. Bottomley, P. J., and Maggard, S. P., "Determination of viability within serotypes of a soil population of *Rhizobium leguminosarum* bv. *trifolii*," *Appl. Environ. Microbiol. 56,* 533–540 (1990).

31. Bowra, B. J., and Dilworth, M. J., "Motility and chemotaxis towards sugars in *Rhizobium leguminosarum*," *J. Gen. Microbiol. 126,* 231–235 (1981).

32. Brewin, N. J., Wood, E. A., and Young, J. P. W., "Contribution of the symbiotic plasmid to the competitiveness of *Rhizobium leguminosarum*," *J. Gen. Microbiol. 129,* 2973–2977 (1983).

33. Brockman, F. J., and Bezdicek, D. F., "Diversity within serogroups of *Rhizobium leguminosarum* bv. *viciae* in the Palouse region of Eastern Washington as indicated by plasmid profiles, intrinsic antibiotic resistance and topography," *Appl. Environ. Microbiol. 55,* 109–115 (1989).

34. Brockwell, J., Gault, R. R., Zorin, M., and Roberts, M. J., "Effects of environmental variables on the competition between inoculum strains and naturalized populations of *Rhizobium trifolii* for nodulation of *Trifolium subterraneum* L. and on rhizobia persistence in the soil," *Aust. J. Agric. Res. 33,* 803–815 (1982).

35. Brockwell, J., Holliday, R. A., Daoud, D. M., and Materon, L. A., "Symbiotic characteristics of a *Rhizobium*-specific annual medic, *Medicago rigidula* (L.) All.," *Soil Biol. Biochem. 20,* 593–600 (1988).

36. Brockwell, J., Roughley, R. J., and Herridge, D. F., "Population dynamics of *Rhizobium japonicum* strains used to inoculate three successive crops of soybean," *Aust. J. Agric. Res. 38,* 61–74 (1987).

37. Brockwell, J., Schwinghamer, E. A., and Gault, R. R., "Ecological studies of root-nodule bacteria introduced into field environments. V. A. critical examination of the stability of antigenic and streptomycin-resistance markers for identification of strains of *Rhizobium trifolii*," *Soil Biol. Biochem. 9,* 19–24 (1977).

38. Bromfield, E. S. P., "Variation in preference for *Rhizobium meliloti* within and between *Medicago sativa* cultivars grown in soil," *Appl. Environ. Microbiol. 48,* 1231–1236 (1984).

39. Bromfield, E. S. P., Lewis, D. M., and Barran, L. R., "Cryptic plasmid and rifampin resistance in *Rhizobium meliloti* influencing nodulation competitiveness," *J. Bacteriol. 164,* 410–413 (1985).

40. Bromfield, E. S. P., Sina, I. B., and Wolynetz, M. S., "Influence of location, host cultivar, and inoculation on the composition of naturalized populations of *Rhizobium meliloti* in *Medicago sativa* nodules," *Appl. Environ. Microbiol. 51,* 1077–1084 (1986).

41. Bromfield, E. S. P., Thurman, N. P., Whitwill, S. T., and Barran, L. R., "Plasmids and symbiotic effectiveness of representative phage types from two indigenous populations of *Rhizobium meliloti,*" *J. Gen. Microbiol. 133,* 3457–3466 (1987).

42. Broughton, W. J., Heycke, N., Priefer, U., Schneider, G-M., and Stanley, J., "Ecological genetics of *Rhizobium meliloti:* Diversity and competitive dominance," *FEMS Microbiol. Lett. 40,* 245–249 (1987).

43. Broughton, W. J., Samrey, U., and Stanley, J., "Ecological genetics of *Rhizobium meliloti* symbiotic plasmid transfer in the *Medicago sativa* rhizosphere," *FEMS Microbiol. Lett. 40,* 251–255 (1987).

44. Brunel, B., Cleyet-Marel, J.-C., Normand, P., and Bardin, R., "Stability of *Bradyrhizobium japonicum* inoculants after introduction into soil," *Appl. Environ. Microbiol. 54,* 2636–2642 (1988).

45. Burton, J. C., "Methods of inoculating seeds and their effect on survival of rhizobia," In: *Symbiotic Nitrogen Fixation in Plants* (Nutman, P. S., ed.), Cambridge, Cambridge University Press, 1975, pp. 175–189.

46. Bushby, H. V. A., "Changes in the numbers of antibiotic-resistant rhizobia in the soil and rhizosphere of field grown *Vigna mungo* cv. *Regur,*" *Soil Biol. Biochem. 13,* 241–245 (1981).

47. Bushby, H. V. A., "Ecology," In: *Nitrogen Fixation,* Vol. 2, *Rhizobium* (Broughton, W. J., ed.), Oxford, Clarendon Press, 1982, pp. 35–75.

48. Bushby, H. V. A., "Colonization of rhizospheres and nodulation of two *Vigna* species by rhizobia inoculated onto seed: Influence in soil," *Soil Biol. Biochem. 16,* 635–641 (1984).

49. Bushby, H. V. A., and Marshall, K. C., "Water status of rhizobia in relation to their susceptibility to desiccation and to their protection by montmorillonite," *J. Gen. Microbiol. 99,* 19–27 (1977).

50. Busse, M. D., and Bottomley, P. J., "Growth and nodulation responses of *Rhizobium meliloti* to water stress induced by permeating and nonpermeating solutes," *Appl. Environ. Microbiol. 55,* 2431–2436 (1989a).

51. Busse, M. D., and Bottomley, P. J., "Nitrogen-fixing characteristics of alfalfa cultivars nodulated by representatives of an indigenous *R. meliloti* serogroup," *Plant and Soil 117,* 255–262 (1989b).

52. Cadisch, G., Sylvester-Bradley, R., and Nosberger, J., "^{15}N-based estimation of nitrogen fixation by eight tropical legumes at two levels of P:K supply," *Field Crops Res. 22,* 191–194 (1989).

53. Caetano-Anolles, G., Wall, L. G., DeMichell, A. T., Macchi, E. M., Bauer, W. D., and Favelukes, G., "Role of motility and chemotaxis in efficiency of nodulation by *Rhizobium meliloti,*" *Plant Physiol. 86,* 1228–1235 (1988).

54. Caldwell, B. E., and Hartwig, E. E., "Serological distribution of soybean root nodule bacteria in soils of southeastern USA," *Agron. J. 62,* 621–623 (1970).

55. Caldwell, B. E., and Vest, G., "Nodulation interaction between soybean genotypes and serogroups of *Rhizobium japonicum,*" *Crop. Sci. 8,* 680–682 (1968).

56. Caldwell, B. E., and Vest, G., "Effect of *Rhizobium japonicum* strains on soybean yields," *Crop Sci. 10,* 19–21 (1970).

57. Caldwell, B. E., and Weber, D. F., "Distribution of *Rhizobium japonicum* serogroups in soybean nodules as affected by planting dates," *Argon. J. 62*, 12–14 (1970).

58. Canter Cremers, H. C. J., Spaink, H. P., Wifjes, A. H. M., Pees, E., Wijffelman, C. A., Okker, R. J. H., and Lugtenberg, B. J. J., "Additional nodulation genes on the *sym* plasmid of *Rhizobium leguminosarum* bv. *viciae*," *Plant Mol. Biol. 13*, 163–174 (1989).

59. Caradus, J. R., "Distinguishing between grass and legume species for efficiency of phosphorus use," *N.Z. J. Agric. Res. 23*, 75–81 (1980).

60. Carter, E. D., Wolfe, E. C., and Francis, C. M., "Problems of maintaining pastures in the cereal–livestock areas of southern Australia," *Proc. 2nd Aust. Agron. Conf.* (Norman, M. J., ed.), Wagga Wagga, 1982, pp. 68–87.

61. Carter, K. R., Jennings, N. T., Hanus, F. J., and Evans, H. J., "Hydrogen evolution and uptake by nodules of soybeans inoculated with different strains of *Rhizobium japonicum*," *Can. J. Microbiol. 24*, 307–311 (1978).

62. Carvalho de, M. M., Andrew, C. S., Edwards, D. G., and Asher, C. J., "Comparative performance of six *Stylosanthes* species in three acid soils," *Aust. J. Agric. Res. 31*, 61–76 (1980).

63. Carvalho de, M. M., Edwards, D. G., Andrew, C. S., and Asher, C. J., "Aluminum toxicity, nodulation, and growth of *Stylosanthes* species," *Agron. J. 73*, 261–265 (1981).

64. Cassman, K. G., Munns, D. N., and Beck, D. P., "Growth of *Rhizobium* strains at low concentrations of phosphate," *Soil Sci. Soc. Am. J. 45*, 520–523 (1981).

65. Cassman, K. G., Whitney, A. S., and Fox, R. L., "Phosphorus requirements of soybean and cowpea as affected by mode of N nutrition," *Agron. J. 73*, 17–22 (1981).

66. Cass-Smith, W. P., and Pittman, H. A. J., "The influence of methods of planting on the effective inoculation and establishment of subterranean clover," *J. Dep. Agric. W. Aust. 16*, 61–73 (1938).

67. Chao, W.-L., and Alexander, M., "Interaction between protozoa and *Rhizobium* in chemically amended soil," *Soil Sci. Soc. Am. J. 45*, 48–50 (1981).

68. Chatel, D. L., and Greenwood, R. M., "The location and distribution in soil of rhizobia under senesced annual legume pastures," *Soil Biol. Biochem. 5*, 799–808 (1973a).

69. Chatel, D. L., and Greenwood, R. M., "Differences between strains of *Rhizobium trifolii* in ability to colonize soil and plant roots in the absence of their specific host plants," *Soil Biol. Biochem. 5*, 809–813 (1973b).

70. Chatel, D. L., Greenwood, R. M., and Parker, C. A., "Saprophytic competence as an important character in the selection of *Rhizobium* for inoculation," *Trans. 9th Int. Congr. Soil Sci.*, Sydney, Angus and Robertson, 1968, Vol. 2, 65–73.

71. Chatel, D. L., and Parker, C. A,. "Survival of field-grown rhizobia over the dry summer period in western Australia," *Soil Biol. Biochem. 5*, 415–423 (1973a).

72. Chatel, D. L., and Parker, C. A., "The colonization of host–root and soil by rhizobia-1. Species and strain differences in the field," *Soil Biol. Biochem. 5*, 425–432 (1973b).

73. Chatel, D. L., Shipton, W. A., and Parker, C. A., "Establishment and persistence of *Rhizobium trifolii* in western Australian soils," *Soil Biol. Biochem. 5*, 815–824 (1973).

74. Chen, Y. P., Dilworth, M. J., and Glenn, A. R., "Degradation of mandelate and 4-hydroxymandelate by *Rhizobium leguminosarum* bv. *trifolii* TA1," *Arch. Microbiol. 151*, 520–523 (1989).

75. Chen, Y. P., Glenn, A. R., and Dilworth, M. J., "Uptake and oxidation of aromatic substrates by *Rhizobium leguminosarum* MNF 3841 and *Rhizobium trifolii* TA1," *FEMS Microbiol. Lett. 21,* 201–205 (1984).

76. Ciafardini, G., and Barbieri, C., "Effect of cover inoculation of soybean on production, N₂ fixation, and yield," *Agron. J. 79,* 645–648 (1987).

76a. Cooper, J. E., "Acid production, acid tolerance, and growth rate of *Lotus* rhizobia in laboratory media," *Soil Biol. Biochem. 14,* 127–137 (1982).

77. Cooper, J. E., Wood, M., and Holding, A. J., "The influence of soil acidity factors on rhizobia," In: *Temperate Legumes: Physiology, Genetics, and Nodulation* (Jones, D. G., and Davies, D. R., eds.), London, Pitman, 1983, pp. 319–335.

78. Corman, A., Crozat, Y., and Cleyet-Marel, J. C., "Modelling of survival kinetics of some *Bradyrhizobium japonicum* strains in soil," *Biol. Fertil. Soils 4,* 79–84 (1987).

79. Coventry, D. R., and Evans, J., "Symbiotic nitrogen fixation and soil acidity," In: *Soil Acidity and Plant Growth* (Robson, A. D., ed.), New York, Academic Press, 1989, pp. 103–137.

80. Coventry, D. R., Hirth, J. R., and Fung, K. K. H., "Nutritional restraints on subterranean clover grown on acid soils used for crop–pasture rotation," *Aust. J. Agric. Res. 38,* 163–176 (1987).

81. Coventry, D. R., Hirth, J. R., Reeves, T. G., and Burnett, V. F., "Growth and nitrogen fixation by subterranean clover in response to inoculation, molybdenum application and soil amendment with lime," *Soil Biol. Biochem. 17,* 791–796 (1985).

82. Coventry, D. R., Hirth, J. R., Reeves, T. G., and Jones, H. R., "Development of populations of *Rhizobium trifolii* and nodulation of subterranean clover following the cropping phase in crop–pasture rotations in south eastern Australia," *Soil Biol. Biochem. 17,* 17–22 (1985).

83. Cregan, P. B., and Keyser, H. H., "Host restriction of nodulation by *Bradyrhizobium japonicum* strain USDA 123 in soybean," *Crop. Sci. 26,* 911–916 (1986).

84. Crow, V. L., Jarvis, B. W. D., and Greenwood, R. M., "Deoxyribonucleic acid homologies among acid producing strains of *Rhizobium,*" *Int. J. Syst. Bacteriol. 31,* 152–172 (1981).

85. Crozat, Y., and Cleyet-Marel, J-C., "Problèmes méthodologiques posés par l'extraction et la récupération des bactéries telluriques pour leur quantification par immunofluorescence," *Agronomie 4,* 603–610 (1984).

86. Crozat, Y., Cleyet-Marel, J. C., and Corman, A., "Use of the fluorescent antibody technique to characterize equilibrium survival concentrations of *Bradyrhizobium japonicum* strains in soil," *Biol. Fertil. Soils 4,* 85–90 (1987).

87. Crozat, Y., Cleyet-Marel, J. C., Giraud, J. J., and Obaton, M., "Survival rates of *Rhizobium japonicum* populations introduced into different soils," *Soil Biol. Biochem. 14,* 401–405 (1982).

88. Cunningham, S. D., and Munns, D. N., "The correlation between extracellular polysaccharide production and acid tolerance in *Rhizobium,*" *Soil Sci. Soc. Am. J. 48,* 1273–1276 (1984).

89. Curl, E. A., and Truelove, B., *The Rhizosphere,* Berlin, Springer-Verlag, 1986, p. 288.

90. Damirgi, S. M., Frederick, L. R., and Anderson, I. C., "Serogroups of *Rhizobium japonicum* in soybean nodules as affected by soil type," *Agron. J. 59,* 10–12 (1967).

91. Danso, S. K. A., Keya, S. O., and Alexander, M., "Protozoa and the decline of *Rhizobium* populations added to soil," *Can. J. Microbiol. 21*, 884–895 (1975).

92. Danso, S. K. A., and Owiredu, J. D., "Competitiveness of introduced and indigenous cowpea *Bradyrhizobium* strains for nodule formation on cowpea (*Vigna unguiculata* (L.) Walp) in three soils," *Soil Biol. Biochem. 20*, 305–310 (1988).

93. Danso, S. K. A., Zapata, F., and Hardarson, G., "Nitrogen fixation in fababeans as affected by plant population density in sole or intercropped systems with barley," *Soil Biol. Biochem. 19*, 411–415 (1987).

94. Date, R. A., and Brockwell, J., "*Rhizobium* strain competition and host interaction for nodulation," In: *Plant Relations in Pastures* (Wilson, J. R., ed.), East Melbourne, CSIRO, 1978, pp. 206–216.

95. Date, R. A., and Halliday, J., "Selecting *Rhizobium* for acid, infertile soils of the tropics," *Nature (Lond.) 277*, 62–74 (1979).

96. Date, R. A., and Roughley, R. J., "Preparation of legume seed inoculants," In: *A Treatise on Dinitrogen Fixation* (Hardy, R. W. F., and Gibson, A. H., eds.), New York, Wiley, 1977, pp. 243–275.

97. Debelle, F., Rosenberg, C., Vasse, J., Mailett, F., Martinez, E., Denarie, J., and Truchet, G., "Assignment of symbiotic development phenotypes to common and specific nodulation (*nod*) genetic loci of *Rhizobium meliloti*," *J. Bacteriol. 168*, 1075–1086 (1986).

98. DeMaagd, R. A., Spaink, H. P., Pees, I., Mulders, I. H. M., Wifjes, A., Wijffelman, C. A., Okker, R. J. H., and Lugtenberg, B. J. J., "Localization and symbiotic function of a region on the *Rhizobium leguminosarum sym* plasmid pRL1JI responsible for a secreted flavonoid-inducible 50-kilodalton protein," *J. Bacteriol. 171*, 1151–1157 (1989).

99. De Maagd, R. A., Wientjes, F. B., and Lugtenberg, B. J. J., "Evidence for divalent cation (Ca^{2+})-stabilized oligomeric proteins and covalently bound protein–peptidoglycan complexes in the outer membrane of *Rhizobium leguminosarum*," *J. Bacteriol. 171*, 3989–3995 (1989).

100. Demezas, D. H., and Bottomley, P. J., "Identification of two dominant serotypes of *Rhizobium trifolii* in root nodules of uninoculated field-grown subclover," *Soil Sci. Soc. Am. J. 48*, 1067–1071 (1984).

101. Demezas, D. H., and Bottomley, P. J., "Autecology in rhizospheres and nodulating behavior of indigenous *Rhizobium trifolii*," *Appl. Environ. Microbiol. 52*, 1014–1019 (1986a).

102. Demezas, D. H., and Bottomley, P. J., "Interstrain competition between representatives of indigenous serotypes of *Rhizobium trifolii*," *Appl. Environ. Microbiol. 52*, 1020–1025 (1986b).

103. Djordjevic, M. A., Redmond, J. W., Batley, M., and Rolfe, B. G., "Clovers secrete specific phenolic compounds which either stimulate or repress *nod* gene expression in *Rhizobium trifolii*," *EMBO J. 6* 1173–1179 (1987).

104. Djordjevic, M. A., Zurkowski, W., and Rolfe, B. G., "Plasmids and stability of symbiotic properties of *Rhizobium trifolii*," *J. Bacteriol. 151*, 560–568 (1982).

105. Doerge, T. A., Bottomley, P. J., and Gardner, E. H., "Molybdenum limitations to alfalfa growth and nitrogen content on a moderately acid high-phosphorus soil," *Agron. J. 77*, 895–901 (1985).

106. Donald, C. M., and Williams, C. H., "Fertility and productivity of a podzolic soil as influenced by subterranean clover (*Trifolium subterraneum* L.) and superphosphate," *Aust. J. Agric. Res. 5*, 664–687 (1954).

107. Dowdle, S. F., and Bohlool, B. B., "Predominance of fast-growing *Rhizobium japonicum* in a soybean field in the People's Republic of China," *Appl. Environ. Microbiol. 50*, 1171–1176 (1985).

108. Dowdle, S. F., and Bohlool B. B., "Intra and inter-specific competition in *Rhizobium fredii* and *Bradyrhizobium japonicum* as indigenous and introduced organisms," *Can. J. Microbiol. 33*, 990–995 (1987).

109. Dowling, D. N., and Broughton, W. J., "Competition for nodulation of legumes," *Ann. Rev. Microbiol. 40*, 131–157 (1986).

110. Dowling, D. N., Stanley, J., and Broughton, W. J., "Competitive nodulation blocking of Afghanistan pea is determined by *nod* DABC and *nod* FE alleles in *Rhizobium leguminosarum*," *Mol. Gen. Genet. 216*, 170–174 (1989).

111. Dreyfus, B. L., and Dommergues, Y. R., "Nodulation of *Acacia* species by fast and slow growing tropical strains of *Rhizobium*," *Appl. Environ. Microbiol. 41*, 97–99 (1981).

112. Dudeja, S. S., and Khurana, A. L., "Persistence of *Bradyrhizobium* sp. (*Cajanus*) in a sandy loam," *Soil Biol. Biochem. 21*, 709–713 (1989).

113. Dudman, W. F., and Brockwell, J., "Ecological studies of root-nodule bacteria introduced into field environments. I. A survey of field performance of clover inoculants by gel immune diffusion serology," *Aust. J. Agric. Res. 9*, 739–747 (1968).

114. Duggar, J. F., "Nodulation of peanut plants as affected by variety, shelling of seed, and disinfection of seed," *J. Am. Soc. Agron. 27*, 286–288 (1935).

115. Dughri, M. H., and Bottomley, P. J., "Complementary methodologies to delineate the composition of *Rhizobium trifolii* populations in root nodules," *Soil Sci. Soc. Am. J. 47*, 939–945 (1983a).

116. Dughri, M. H., and Bottomley, P. J., "Effect of acidity on the composition of an indigenous soil population of *Rhizobium trifolii* found in nodules of *Trifolium subterraneum* L.," *Appl. Environ. Microbiol. 46*, 1207–1213 (1983b).

117. Dughri, M. H., and Bottomley, P. J., "Soil acidity and the composition of an indigenous population of *Rhizobium trifolii* in nodules of different cultivars of *Trifolium subterraneum* L.," *Soil Biol. Biochem. 16*, 405–411 (1984).

118. Dusha, T., Bakos, A., Kondorosi, A., de Bruijn, F. J., and Schell, J., "The *Rhizobium meliloti* early nodulation genes (*nod* ABC) are nitrogen-regulated: Isolation of a mutant strain with efficient nodulation capacity on alfalfa in the presence of ammonium," *Mol. Gen. Genet. 219*, 89–96 (1989).

119. Eardly, B. D., Hannaway, D. B., and Bottomley, P. J., "Nitrogen nutrition and yield of seedling alfalfa as affected by ammonium nitrate fertilization," *Agron, J. 77*, 57–62 (1985).

120. Ellis, W. R., Ham, G. E., and Schmidt, E. L., "Persistence and recovery of *Rhizobium japonicum* inoculum in a field soil," *Agron. J. 76*, 573–576 (1984).

121. El-Haloui, N. E., Oehin, D., and Tailliez, R., "Compétitivité pour l'infection entre souches de *Rhizobium meliloti:* Role de la mobilité," *Plant and Soil 95*, 337–344 (1986).

122. El-Zamik, F. I., and Wright, S. F., "Precautions in the use of yeast extract mannitol medium for evaluation of legume seed toxicity to *Rhizobium*," *Soil Biol. Biochem. 19*, 207–209 (1987).

123. Engvild, K. C., and Nielsen, G., "Strain identification in *Rhizobium* by starch gel electrophoresis of isoenzymes," *Plant and Soil 87*, 251–256 (1985).

124. Erdman, L. W. "New developments in legume inoculation," *Soil Sci. Soc. Am. J. 8,* 213–215 (1943).

125. Evans, J., Hochman, Z., O'Connor, G. E., and Osborne, G. J., "Soil acidity and *Rhizobium:* their effects on nodulation of subterranean clover on the slopes of southern New South Wales," *Aust. J. Agric. Res. 38,* 605–618 (1988).

126. Firmin, J. L., Wilson, J. E., Rossen, L., and Johnston, A. W. B., "Flavonoid activation of nodulation genes in *Rhizobium* reversed by other compounds present in plants," *Nature (Lond.) 324,* 90–92 (1986).

127. Flores, M., Gonzalez, V., Pardo, M. A., Leija, A., Martinez, E., Romero, D., Pinero, D., Davila, G., and Palacios, R., "Genomic instability in *Rhizobium phaseoli,*" *J. Bacteriol. 170,* 1191–1196 (1988).

128. Foster, R. C., Rovira, A. D., and Cock, T. W., *Ultrastructure of the Root–Soil Interface,* St. Paul, Amer. Phytopath. Soc., 1983, 157 pp.

129. Franco, A. A., and Vincent, J. M., "Competition amongst rhizobial strains for the colonization and nodulation of two tropical legumes," *Plant and Soil 45,* 27–48 (1976).

129a. Fred, E. B., Baldwin, I. L., and McCoy, E., *Root Nodule Bacteria and Leguminous Plants,* Madison, University of Wisconsin Press, 1932, 343 pp.

130. Frederickson, J. K., Bezdicek, D. F., Brockman, F. J., and Li, S. W., "Enumeration of Tn5 mutant bacteria in soil by using a most-probable-number-DNA hybridization procedure and antibiotic resistance," *Appl. Environ. Microbiol. 54,* 446–453 (1988).

131. Fuhrmann, J., Davey, C. B., and Wollum, A. G., "Desiccation tolerance of clover rhizobia in sterile soils," *Soil Sci. Soc. Am. J. 50,* 639–644 (1986).

132. Fuhrmann, J., and Wollum, A. G., "Nodulation competition among *Bradyrhizobium japonicum* strains as influenced by rhizosphere bacteria and iron availability," *Biol. Fert. Soils 7,* 108–112 (1989a).

133. Fuhrmann, J., and Wollum, A. G., "Symbiotic interactions between soybean and competing strains of *Bradyrhizobium japonicum,*" *Plant and Soil 119,* 139–145 (1989b).

134. Gaur, Y. D., and Lowther, W. L., "Distribution, symbiotic effectiveness, and fluorescent antibody reaction of naturalized populations of *Rhizobium trifolii* in Otago soils," *N.Z. J. Agric. Res. 23,* 529–532 (1980).

135. George, T., Bohlool, B. B., and Singleton, P. W., "*Bradyrhizobium japonicum*–environment interactions: Nodulation and interstrain competition in soils along an elevational transect," *Appl. Environ. Microbiol. 53,* 1113–1117 (1987).

136. Germida, J. J., "Growth of indigenous *Rhizobium leguminosarum* and *Rhizobium meliloti* in soils amended with organic nutrients," *Appl. Environ. Microbiol. 54,* 257–264 (1988).

137. Gibson, A. H., "Genetic control of strain-specific ineffective nodulation in *Trifolium subterraneum* L.," *Aust. J. Agric. Res. 15,* 37–49 (1964).

138. Gibson, A. H., "Nodulation failure in *Trifolium subterraneum* L. cv. Woogenellup (syn. Marrar)," *Aust. J. Agric. Res. 19,* 907–918 (1968).

139. Gibson, A. H., Curnow, B. C., Bergersen, F. J., Brockwell, J., and Robinson, A. C., "Studies of field populations of *Rhizobium:* Effectiveness of strains of *Rhizobium trifolii* associated with *Trifolium subterraneum* L. pastures in southeastern Australia," *Soil Biol. Biochem. 7,* 95–102 (1975).

140. Gibson, A. H., Date, R. A., and Brockwell, J., "A comparison of competitiveness and persistence amongst five strains of *Rhizobium trifolii,*" *Soil Biol. Biochem. 8,* 395–401 (1976).

141. Graham, P. H., and Chatel, D. L., "Agronomy," In: *Nitrogen Fixation,* Vol. 3, Legumes (Broughton, W. J., ed.), Oxford, Clarendon Press, 1983, pp. 56–98.

142. Graham, P. H., Viteri, S. E., Mackie, F., Vargas, A. T., and Palacios, A., "Variation in acid tolerance among strains of *Rhizobium phaseoli,*" *Field Crops Res. 5,* 121–128 (1982).

143. Habte, M., and Alexander, M., "Further evidence for the regulation of bacterial populations in soil by protozoa," *Arch. Microbiol. 113,* 181–183 (1977).

144. Hagedorn, C., "Effectiveness of *Rhizobium trifolii* populations associated with *Trifolium subterraneum* L. in southwest Oregon soils," *Soil Sci. Soc. Am. J. 42,* 447–451 (1978).

145. Hagedorn, C., Ardahl, A. H., and Materon, L. A., "Characteristics of *Rhizobium trifolii* populations associated with subclover in Mississippi soils," *Soil Sci. Soc. Am. J. 47,* 1148–1152 (1983).

146. Ham, G. E., Caldwell, V. B., and Johnson, H. W., "Evaluation of *Rhizobium japonicum* inoculants in soils containing naturalized populations of rhizobia," *Agron. J. 63,* 301–303 (1971).

147. Ham, G. E., Frederick, L. R., and Anderson, I. C., "Serogroups of *Rhizobium japonicum* in soybean nodules sampled in Iowa," *Agron. J. 63,* 69–72 (1971).

148. Hambdi, Y. A., "Soil–water tension and the movement of rhizobia," *Soil Biol. Biochem. 3,* 121–126 (1971).

149. Handelsman, J., and Brill, W. J., "*Erwinia herbicola* isolates from alfalfa plants may play a role in nodulation of alfalfa by *Rhizobium meliloti,*" *Appl. Environ. Microbiol. 49,* 818–821 (1985).

150. Handelsman, J., Ugalde, R. A., and Brill, W. J., "*Rhizobium meliloti* competitiveness and the alfalfa agglutinin," *J. Bacteriol. 157,* 703–707 (1984).

151. Hanus, F. J., Albrecht, S. L., Zablotowicz, R. M., Emerich, D. W., Russell, S. A., and Evans, H. J., "Yield and N content of soybean seed as influenced by *Rhizobium japonicum* inoculants possessing the hydrogenase characteristic," *Agron. J. 73,* 368–372 (1981).

152. Hardarson, G., Golbs, M., and Danso, S. K. A., "Nitrogen fixation in soybean (*Glycine max* L. Merrill) is affected by nodulation patterns," *Soil Biol. Biochem. 21,* 783–787 (1989).

153. Hardarson, G., Heichel, G. H., Barnes, J. K., and Vance, C. P., "Rhizobial strain preference of alfalfa populations selected for characteristics associated with N_2 fixation," *Crop Sci. 22,* 55–58 (1982).

154. Hardarson, G., and Jones, D. G., "Effect of temperature on competition amongst strains of *Rhizobium trifolii* for nodulation of two white clover varieties," *Ann. Appl. Biol. 92,* 229–236 (1979a).

155. Hardarson, G., and Jones, D. G., "The inheritance of preference for strains of *Rhizobium trifolii* by white clover (*Trifolium repens* L.)," *Ann. Appl. Biol. 92,* 329–333 (1979b).

156. Harris, J. R., "Rhizosphere relationships of subterranean clover. I. Interactions between strains of *Rhizobium trifolii,*" *Aust. J. Agric. Res. 5,* 247–270 (1954).

157. Harrison, S. P., Jones, D. G., and Young J. P. W., "*Rhizobium* population genetics: Genetic variation within and between populations from diverse locations," *J. Gen. Microbiol. 135,* 1061–1069 (1989).

158. Harrison, S. P., Young, J. P. W., and Jones, D. G., "*Rhizobium* population genetics: Effect of clover variety and inoculum dilution on the genetic diversity sampled from natural populations," *Plant and Soil 103*, 147–150 (1987).

159. Hartel, P. G., and Alexander, M., "Growth and survival of cowpea rhizobia in acid, aluminum-rich soils," *Soil Sci. Soc. Am. J. 47*, 502–506 (1983).

160. Hartel, P. G., and Alexander, M., "Role of extracellular polysaccharide production and clays in the desiccation tolerance of cowpea bradyrhizobia," *Soil Sci. Soc. Am. J. 50*, 1193–1198 (1986).

161. Hartel, P. G., Whelan, A. M., and Alexander, M., "Nodulation of cowpea and survival of cowpea rhizobia in acid, aluminum-rich soils," *Soil Sci. Soc. Am. J. 47*, 514–517 (1983).

162. Henzell, E. F., "The role of biological nitrogen fixation research in solving problems in tropical agriculture," *Plant and Soil 108*, 15–21 (1988).

163. Herridge, D. F., and Roughley, R. J., "Variation in colony characteristics and symbiotic effectiveness in *Rhizobium*," *J. Appl. Bacteriol. 38*, 19–29 (1975).

164. Herridge, D. F., Roughley, R. J., and Brockwell, J., "Low survival of *Rhizobium japonicum* inoculant leads to reduced nodulation, nitrogen fixation and yield of soybean in the current crop but not in the subsequent crop," *Aust. J. Agric. Res. 38*, 75–82 (1987).

165. Heynen, C. E., van Elsas, J. D., Kurkman, P. J., and van Veen, J. A., "Dynamics of *Rhizobium leguminosarum* biovar *trifolii* introduced into soil; the effect of bentonite clay on predation by protozoa," *Soil Biol. Biochem. 20*, 483–488 (1988).

166. Hicks, P. M., and Loynachan, T. E., "Phosphorus fertilization reduces vesicular-arbuscular mycorrhizal infection and changes nodule occupancy of field-grown soybean," *Agron. J. 79*, 841–844 (1987).

167. Hicks, P. M., and Loynachan, T. E., "Bacteria of the soybean rhizosphere and their effect on growth of *Bradyrhizobium japonicum*," *Soil Biol. Biochem. 21*, 561–566 (1989).

168. Hilali, A., Aurag, J., Molina, J. A. E., and Schmidt, E. L., "Competitiveness and persistence of strains of *Rhizobium phaseoli* introduced into a Moroccan sandy soil," *Biol. Fert. Soils 7*, 213–218 (1989).

169. Hiltbold, A. E., Patterson, R. M., and Reed, R. B., "Soil populations of *Rhizobium japonicum* in a cotton–corn–soybean rotation," *Soil Sci. Soc. Am. J. 49*, 343–348 (1985).

170. Hiltner, L., "Uber neuere Erfahrungen und Probleme auf dem gebiet der Bodenbakteriologie und unter besonderer Berucksichtigung der grundungung und Brache," *Arb. Dtsch Landwirt. Ges. 98*, 59–78 (1904).

171. Hodgson, A. L. M., Roberts, W. P., and Waid, J. S., "Regulated nodulation of *Trifolium subterraneum* L. inoculated with bacteriocin producing strains of *Rhizobium trifolii*," *Soil Biol. Biochem. 17*, 475–478 (1985).

172. Holding, A. J., and King, J., "The effectiveness of indigenous populations of *Rhizobium trifolii* in relation to soil factors," *Plant and Soil 18*, 191–198 (1963).

173. Holland, A. A., "Serological characteristics of certain root-nodule bacteria of legumes," *Anton van Leeuwenhoek 32*, 410–418 (1966).

174. Holland, A. A., "Competition between soil- and seed-borne *Rhizobium trifolii* in nodulation of introduced *Trifolium subterraneum*," *Plant and Soil 32*, 293–302 (1970).

175. Honma, M. A., and Ausubel, F. M., "*Rhizobium meliloti* has three functional copies of the *nod* D symbiotic regultory protein," *Proc. Natl. Acad. Sci. U.S.A. 84*, 8558–8562 (1987).

176. Horvath, B., Kondorosi, E., John, M., Schmidt, J., Torok, I., Gyorgypal, Z., Barabas, I., Wieneke, U., Schell, J., and Kondorosi, A., "Organization, structure and symbiotic function of *Rhizobium meliloti* nodulation genes determining host specificity for alfalfa," *Cell 46*, 335–343 (1986).

177. Hossain, A. K. M., and Alexander, M., "Enhancing soybean rhizosphere colonization by *Rhizobium japonicum*," *Appl. Environ, Microbiol. 48*, 468–472 (1984).

178. Howieson, J. G., and Ewing, M. A., "Acid tolerance in the *Rhizobium meliloti–Medicago* symbiosis," *Aust. J. Agric. Res. 37*, 55–64 (1986).

179. Howieson, J. G., and Ewing, M. A., "Annual species of *Medicago* differ greatly in their ability to nodulate on acid soils," *Aust. J. Agric. Res. 40*, 843–850 (1989).

180. Howle, P. K. W., Shipe, E. R., and Skipper, H. D., "Soybean specificity for *Bradyrhizobium japonicum* strain 110," *Agron. J. 79*, 595–598 (1987).

181. Hughes, D. Q., and Vincent, J. J., "Serological studies of the root-nodule bacteria. III. Tests of neighboring strains of the same species," *Proc. Linn. Soc. N.S.W. 67*, 142–152 (1942).

182. Ireland, J. A., and Vincent, J. M., "A quantitative study of competition for nodule formation," *Trans. 9th Int. Congr. Soil Sci.*, Vol. 2, Sydney, Angus and Robertson, 1968, pp. 85–93.

183. Israel, D. W., "Cultivar and *Rhizobium* strain effects on nitrogen fixation and remobilization by soybeans," *Agron. J. 73*, 509–516 (1981).

184. Israel, D. W., "Investigation of the role of phosphorus in symbiotic dinitrogen fixation," *Plant Physiol. 84*, 835–840 (1987).

185. Itoh, S., and Barber, S. A., "Phosphorus uptake by six plant species as related to root hairs," *Agron. J. 75*, 457–461 (1983).

186. Jarvis, B. W. D., Ward, L. J. H., and Slade, E. A., "Expression by soil bacteria of nodulation genes from *Rhizobium leguminosarum* bv. *trifolii*," *Appl. Environ. Microbiol. 55*, 1426–1434 (1989).

187. Jarvis, S. C., and Robson, A. D., "A comparison of the cation/anion balance of ten cultivars of *Trifolium subterraneum* L. and their effects on soil acidity," *Plant and Soil 75*, 235–243 (1983).

188. Jenkins, H. V., Vincent, J. M., and Waters, L. M., The root-nodule bacteria as factors in clover establishment in the red basaltic soils of the Lismore district, New South Wales. III. Field inoculation trials," *Aust. J. Agric. Res. 5*, 77–89 (1954).

189. Jenkins, M. B., and Bottomley, P. J., "Evidence for a strain of *Rhizobium meliloti* dominating the nodules of alfalfa," *Soil Sci. Soc. Am. J. 49*, 326–328 (1985).

190. Jenkins, M. B., Virginia, R. A., and Jarrell, W. M., "Rhizobial ecology of the woody legume mesquite (*Prosopsis glandulosa*) in the Sonoran desert," *Appl. Environ. Microbiol. 53*, 36–40 (1987).

191. Johnson, H. W., and Means, U. M., "Serological groups of *Rhizobium japonicum* recovered from nodules of soybeans (*Glycine max*) in field soils," *Agron. J. 55*, 269–271 (1963).

192. Johnson, H. W., Means, U. M., and Weber, C. R., "Competition for nodule sites between strains of *Rhizobium japonicum*," *Agron. J. 57*, 179–185 (1965).

193. Johnston, A. W. B., and Beringer, J. E., "Mixed inoculation with effective and ineffective strains of *Rhizobium leguminosarum*," *J. Appl. Bacteriol. 49*, 375–380 (1976).

194. Jones, D. G., "Symbiotic variation of *Rhizobium trifolii* with S.100 Nomark white clover (*Trifolium repens* L.)," *J. Sci. Fd. Agric. 14*, 740–743 (1963).

195. Jones, D. G., and Hardarson, G., "Variation within and between white clover varieties in their preference for strains of *Rhizobium trifolii*," *Ann. Appl. Biol. 92*, 221–228 (1979).

196. Jones, D. G., Morley, S. J., "The effect of pH on host plant 'preference' for strains of *Rhizobium trifolii* using fluorescent ELISA for strain identification," *Ann. Appl. Biol. 97*, 183–190 (1981).

197. Jones, M. B., Burton, J. C., and Vaughn, C. E., "Role of inoculation in establishing subclover in California annual grasslands," *Agron. J. 70*, 1081–1085 (1978).

198. Jones, M. B., Williams, W. A., and Ruckman, J. E., "Fertilization of *Trifolium subterraneum* L. growing on serpentine soils," *Soil Sci. Soc. Am. J. 41*, 87–89 (1977).

199. Josephson, K. L., and Pepper, I. L., "Competitiveness and effectiveness of strains of *Rhizobium phaseoli* isolated from the Sonoran desert," *Soil Biol. Biochem. 16*, 651–655 (1984).

200. Kamicker, B. J., and Brill, W. J., "Identification of *Bradyrhizobium japonicum* nodule isolates from Wisconsin soybean farms," *Appl. Environ. Microbiol. 51*, 487–492 (1986).

201. Keay, J., Biddiscombe, E. F., and Ozanne, P. G., "The comparative rates of phosphate absorption by eight annual pasture species," *Aust. J. Agric. Res. 21*, 33–44 (1970).

202. Kennedy, A. C., and Wollum, A. G., "Enumeration of *Bradyrhizobium japonicum* in soil subjected to high temperature: Comparison of plate count, most probable number and fluorescent antibody techniques," *Soil Biol. Biochem. 20*, 933–937 (1988).

203. Kerr, A., "Biological control of crown gall through production of agrocin 84," *Plant Dis. 64*, 25–30 (1980).

204. Keyser, H. H., and Cregan, P. B., "Nodulation and competition for nodulation of selected soybean genotypes among *Bradyrhizobium japonicum* serogroup 123 isolates," *Appl. Environ. Microbiol. 53*, 2631–2635 (1987).

205. Keyser, H. H., and Munns, D. N., "Tolerance of rhizobia to acidity, aluminum, and phosphate," *Soil Sci. Soc. Am. J. 43*, 519–523 (1979).

206. Keyser, H. H., Weber, D. F., and Uratsu, S. L., "*Rhizobium japonicum* serogroup and hydrogenase phenotype distribution in 12 states," *Appl. Environ. Microbiol. 47*, 613–615 (1984).

207. Khurana, A. L., and Dudeja, S. S., "Field population of rhizobia and response to inoculation, molybdenum and nitrogen fertilizer in pigeon pea," In: *Proc. Int. Workshop on Pigeonpea*, Vol. 2, ICRISAT Centre, Patancheru, AP, India, pp. 381–386 (1981).

208. Kim, M.-K., Edwards, D. G., and Asher, C. J., "Tolerance of *Trifolium subterraneum* cultivars to low pH," *Aust. J. Agric. Res. 36*, 569–578 (1985).

209. Kingsley, M. T., and Bohlool, B. B., "Release of *Rhizobium* spp. from tropical soils and recovery for immunofluorescence enumeration," *Appl. Environ. Microbiol. 42*, 241–248 (1981).

210. Kluson, R. A., Kenworthy, W. J., and Weber, D. F., "Soil temperature effects on competitiveness and growth of *Rhizobium japonicum* and on *Rhizobium*-induced chlorosis of soybeans," *Plant and Soil 95*, 201–207 (1986).

211. Knight, C. D., Rossen, L., Robertson, J. G., Wells, B., and Downie, J. A., "Nodulation inhibition by *Rhizobium leguminosarum* multicopy *nod* ABC genes and analysis of early stages of plant infection," *J. Bacteriol. 166,* 552–558 (1986).

212. Kondorosi, E., Gyuris, J., Schmidt, J., John, M., Duda, E., Hoffman, B., Schell, J., and Kondorosi, A., "Positive and negative control of *nod* gene expression in *Rhizobium meliloti* is required for optimal nodulation," *EMBO J. 8,* 1331–1340 (1989).

213. Kosslak, R. M., and Bohlool, B. B., "Suppression of nodule development of one side of a split-root system of soybeans caused by prior inoculation of the other side," *Plant Physiol. 75,* 125–130 (1984).

214. Kosslak, R. M., and Bohlool, B. B., "Influence of environmental factors in interstrain competition in *Rhizobium japonicum,*" *Appl. Environ. Microbiol. 49,* 1128–1133 (1985).

215. Kosslak, R. M., Bohlool, B. B., Dowdle, S., and Sadowsky, M. J., "Competition of *Rhizobium japonicum* strains in early stages of soybean nodulation," *Appl. Environ. Microbiol. 46,* 870–873 (1983).

216. Kosslak, R. M., Bookland, R., Barkeri, J., Paaren, H. E., and Appelbaum, E. R., "Induction of *Bradyrhizobium japonicum* common *nod* genes by isoflavones isolated from *Glycine max,*" *Proc. Natl. Acad. Sci. U.S.A. 84,* 7428–7432 (1987).

217. Kucey, R. M. N., "Responses of field beans (*Phaseolus vulgaris* L.) to levels of *Rhizobium leguminosarum* bv. *phaseoli* inoculation in soils containing effective *R. leguminosarum* bv. *phaseoli* populations," *Can. J. Plant Sci. 69,* 419–426 (1989).

218. Kucey. R. M. N., and Hynes, M. F., "Populations of *Rhizobium leguminosarum* bv. *phaseoli* and *viciae* in fields of bean or pea in rotation with nonlegumes," *Can. J. Microbiol. 35,* 661–667 (1989).

219. Kuykendall, L. D., "Influence of *Glycine max* nodulation on the persistence in soil of a genetically marked *Bradyrhizobium japonicum* strain," *Plant and Soil 116,* 275–277 (1989).

220. Kuykendall, L. D., and Elkan, G. H., "*Rhizobium japonicum* derivatives differing in nitrogen fixing efficiency and carbohydrate utilization," *Appl. Environ. Microbiol. 32,* 511–519 (1976).

221. Kvien, C. S., and Ham, G. E., "Effect of soil temperature and inoculum rate on the recovery of three introduced strains of *Rhizobium japonicum,*" *Agron. J. 77,* 484–489 (1985).

222. Kvien, C. S., Ham, G. E., and Lambert, J. W., "Recovery of introduced *Rhizobium japonicum* strains by soybean genotypes," *Agron. J. 73,* 900–905 (1981).

223. Labandera, C. A., and Vincent, J. M., "Competition between an introduced strain and native Uruguayan strains of *Rhizobium trifolii,*" *Plant and Soil 42,* 327–347 (1975a).

224. Labandera, C. A., and Vincent, J. M., "Loss of symbiotic capacity in commercially useful strains of *Rhizobium trifolii,*" *J. Appl. Bacteriol. 39,* 209–211 (1975b).

225. Lawrie, A. C., "Relationships among rhizobia from native Australian legumes," *Appl. Environ. Microbiol. 45,* 1822–1828 (1983).

226. Lawson, K. A., Barnet, Y. M., and McGilchrist, C. A., "Environmental factors influencing numbers of *Rhizobium leguminosarum* bv. *trifolii* and its bacteriophages in field soils," *Appl. Environ. Microbiol. 53,* 1125–1131 (1987).

227. Ledgard, S. F., Simpson, J. R., Freney, J. R., and Bergersen, F. J., "Field evaluation of ^{15}N techniques for estimating nitrogen fixation in legume–grass associations," *Aust. J. Agric. Res. 36,* 247–258 (1985).

228. Lennox, L. B., and Alexander, M., "Fungicide enhancement of nitrogen fixation and colonization of *Phaseolus vulgaris* by *Rhizobium phaseoli*," *Appl. Environ. Microbiol. 41*, 404–411 (1981).

229. Leung, K., and Bottomley, P. J., "Influence of phosphate on the growth and nodulation characteristics of *Rhizobium trifolii*," *Appl. Environ. Microbiol. 53*, 2098–2105 (1987).

230. Lie, T. A., "Effect of low pH on different phases of nodule formation in pea plants," *Plant and Soil 31*, 391–406 (1969).

231. Lim, G., "Studies on the physiology of nodule formation. VIII. The influence of the size of the rhizosphere population of nodule bacteria on root hair infection in clover," *Ann. Bot. 27*, 55–67 (1963).

232. Lindstrom, K., Sarsa, M.-L., Polkunen, J., and Kansanen, P., "Symbiotic nitrogen fixation of *Rhizobium* (*Galega*) in acid soils, and its survival in soil under acid and cold stress," *Plant and Soil 87*, 293–302 (1985).

233. Lipsanen, P., and Lindstrom, K., "Adaptation of red clover rhizobia to low temperatures," *Plant and Soil 92*, 55–62 (1986).

234. Liu, R., Tran, V. M., and Schmidt, E. L., "Nodulating competitiveness of a nonmotile Tn7 mutant of *Bradyrhizobium japonicum* in nonsterile soil," *Appl. Environ. Microbiol. 55*, 1895–1900 (1989).

235. Lochner, H. H., Strijdom, B. W., and Law, I. J., "Unaltered nodulation competitiveness of a strain of *Bradyrhizobium* sp. (*Lotus*) after a decade in soil," *Appl. Environ. Microbiol. 55*, 3000–3008 (1989).

236. Loneragan, J. F., and Dowling, E. F., "The interaction of calcium and hydrogen ions in the nodulation of subterranean clover," *Aust. J. Agric. Res. 9*, 464–472 (1958).

237. Loneragan, J. F., Meyer, D., Fawcett, R. G., and Anderson, A. J., "Lime pelleted clover seed for nodulation on acid soil," *J. Aust. Inst. Agric. Sci. 21*, 264–265 (1955).

238. Louw, H. A., and Webley, D. M., "The bacteriology of the root region of the oat plant grown under controlled pot culture conditions," *J. Appl. Bacteriol. 22*, 216–226 (1959).

239. Lowendorf, H. S., "Factors affecting survival of *Rhizobium* in soil," In: *Advances in Microbial Ecology*, vol. 4, (Alexander, M., ed.), New York, Plenum Press, 1980, pp. 87–124.

240. Lowendorf, H. S., and Alexander, M., "Selecting *Rhizobium meliloti* for inoculation of alfalfa planted in acid soils," *Soil Sci. Soc. Am. J. 47*, 935–938 (1983).

241. Lowendorf, H. S., Baya, A. M., and Alexander, M., "Survival of *Rhizobium* in acid soils," *Appl. Environ. Microbiol. 42*, 951–957 (1981).

242. Lowther, W. L., and Loneragan, J. F., "Calcium and nodulation in subterranean clover (*Trifolium subterraneum* L.)," *Plant Physiol. 43*, 1362–1366 (1968).

243. Lynch, D. L., and Sears, O. H., "The effect of inoculation upon yields of soybeans on treated and untreated soils," *Soil Sci. Soc. Am. Proc. 16*, 214–216 (1952).

244. Mackay, A. D., and Barber, S. A., "Effect of soil moisture and phosphate level on root hair growth of corn roots," *Plant and Soil 86*, 321–331 (1984).

245. Mahler, R. L., and Bezdicek, D. F., "Serogroup distribution of *Rhizobium leguminosarum* in peas in the Palouse of Eastern Washington," *Soil Sci. Soc. Am. J. 44*, 292–295 (1980).

246. Mahler, R. L., and McDole, R. E., "Effect of soil pH on crop yield in Northern Idaho," *Agron. J. 79*, 751–755 (1987).

247. Mahler, R. L., and Wollum, A. G., "Influence of water potential on the survival of rhizobia in a goldsboro loamy sand," *Soil Sci. Soc. Am. J. 44*, 988–992 (1980).

248. Mahler, R. L., and Wollum, A. G., "The influence of soil water potential and soil texture on the survival of *Rhizobium japonicum* and *Rhizobium leguminosarum* isolates in the soil," *Soil Sci. Soc. Am. J. 45*, 761–766 (1981a).

249. Mahler, R. L., and Wollum, A. G., "Seasonal variation of *Rhizobium trifolii* in clover pastures and cultivated fields in North Carolina," *Soil Sci. 132*, 240–246 (1981b).

250. Mahler, R. L., and Wollum, A. G., "Seasonal fluctuation of *Rhizobium japonicum* under a variety of field conditions in North Carolina," *Soil Sci. 134*, 317–324 (1982).

251. Marshall, K. C., "Electrophoretic properties of fast- and slow-growing species of *Rhizobium*," *Aust. J. Biol. Sci. 20*, 429–438 (1967).

252. Marshall, K. C., "Studies by microelectrophoretic and microscopic techniques of the sorption of illite and montmorillonite to rhizobia," *J. Gen. Microbiol. 56*, 301–306 (1969).

253. Martensson, A. M., Brutti, L., and Ljunggren, H., "Competition between strains of *Bradyrhizobium japonicum* for nodulation of soybeans at different nitrogen fertilizer levels," *Plant and Soil 117*, 219–225 (1989).

254. Masterson, C. L., and Sherwood, M. T., "Selection of *Rhizobium trifolii* strains by white and subterranean clovers," *Ir. J. Agric. Res. 13*, 91–99 (1974).

255. Materon, L. A., and Hagedorn, C., "Nodulation of crimson clover by introduced rhizobia in Mississippi soils," *Soil Sci. Soc. Am. J. 46*, 553–556 (1982).

256. Materon, L. A., and Hagedorn, C., "Competitiveness and symbiotic effectiveness of five strains of *Rhizobium trifolii* on red clover," *Soil Sci. Soc. Am. J. 47*, 491–495 (1983).

257. Materon, L. A., and Vincent, J. M., "Host specificity and interstrain competition with soybean rhizobia," *Field Crops Res. 3*, 215–224 (1980).

258. Materon, L. A., and Weaver, R. W., "Toxicity of arrowleaf clover seed to *Rhizobium trifolii*," *Agron. J. 76*, 471–473 (1984).

259. Materon, L. A., and Weaver, R. W., "Inoculant maturity influences survival of rhizobia on seed," *Appl. Environ. Microbiol. 49*, 465–467 (1985).

260. May, S. W., and Bohlool, B. B., "Competition among *Rhizobium leguminosarum* strains for nodulation of lentils (*Lens esculenta*)," *Appl. Environ. Microbiol. 45*, 960–965 (1983).

261. McDermott, T. R., and Graham, P. H., "*Bradyrhizobium japonicum* inoculant motility, nodule occupancy, and acetylene reduction in the soybean root system," *Appl. Environ. Microbiol. 55*, 2493–2498 (1989).

262. McDermott, T. R., Graham, P. H., and Brandwein, D. H., "Viability of *Bradyrhizobium japonicum* bacteroids," *Arch. Microbiol. 148*, 100–106 (1987).

263. McLoughlin, T. J., Alt, S. G., Owens, P. A., and Fetherston, C., "Competition for nodulation of field-grown soybeans by strains of *Rhizobium fredii*," *Can. J. Microbiol. 32*, 183–186 (1986).

264. McLoughlin, T. J., Bordeleau, L. M., and Dunican, L. K., "Competitiveness studies with *Rhizobium trifolii* in a field experiment," *J. Appl. Bacteriol. 56*, 131–135 (1984).

265. McLoughlin, T. J., and Dunican, L. K., "Competition studies with *Rhizobium trifolii* in laboratory experiments," *Plant and Soil 88*, 139–143 (1985).

266. McWilliam, J. R., Clements, R. J., and Dowling, P. M., "Some factors influencing the germination and early seedling development of pasture plants," *Aust. J. Agric. Res.* *21*, 19–32 (1969).

267. Meade, J., Higgins, P., and O'Gara, F., "Studies on the inoculation and competitiveness of a *Rhizobium leguminosarum* strain in soils containing indigenous rhizobia," *Appl. Environ. Microbiol.* *49*, 899–903 (1985).

268. Mellor, H. Y., Glenn, A. R., Arwas, R., and Dilworth, M. J., "Symbiotic and competitive properties of motility mutants of *Rhizobium trifolii* TA1," *Arch. Microbiol.* *148*, 34–39 (1987).

269. Merckx, R., den Hartog, A., and Van Veen, J. A., "Turnover of root-derived materials and related microbial biomass formation in soils of different textures," *Soil Biol. Biochem.* *17*, 565–569 (1985).

270. Moawad, H., and Bohlool, B. B., "Competition among *Rhizobium* spp. for nodulation of *Leucaena leucocephala* in two tropical soils," *Appl. Environ. Microbiol.* *48*, 5–9 (1984).

271. Moawad, H. A., Ellis, W. R., and Schmidt, E. L., "Rhizosphere response as a factor in competition among three serogroups of indigenous *Rhizobium japonicum* for nodulation of field-grown soybeans," *Appl. Environ. Microbiol.* *47*, 607–612 (1984).

272. Moodie, C. D., and Vandacevye, S. V., "Yield and nitrogen content of chickpea, *Cicer arietinum*, as affected by seed inoculation," *Soil Sci. Soc. Am. J.* *8*, 229–233 (1943).

273. Morris, D. R., and Weaver, R. W., "Competition for nitrogen-15 depleted ammonium nitrate and nitrogen fixation in arrowleaf clover–gulf ryegrass mixtures," *Soil Sci. Soc. Am. J.* *51*, 115–119 (1987).

274. Mullen, M. D., Israel, D. W., and Wollum, A. G., "Effects of *Bradyrhizobium japonicum* and soybean (*Glycine max* (L.) Merr.) phosphorus nutrition on nodulation and dinitrogen fixation," *Appl. Environ. Microbiol.* *54*, 2387–2392 (1988).

275. Mullen, M. D., and Wollum, A. G., "Variation among different cultures of *Bradyrhizobium japonicum* strains USDA 110 and 122," *Can. J. Microbiol.* *35*, 583–588 (1989).

276. Munevar, F., and Wollum, A. G., "Growth of *Rhizobium japonicum* strains at temperatures above 27°C," *Appl. Environ. Microbiol.* *42*, 272–276 (1981).

277. Munns, D. N., "Soil acidity and growth of a legume. I. Interactions of lime with nitrogen and phosphate on growth of *Medicago sativa L.* and *Trifolium subterraneum L.*," *Aust. J. Agric. Res. 16*, 733–741 (1965).

278. Munns, D. N., "Nodulation of *Medicago sativa* in solution culture. I. Acid-sensitive steps," *Plant and Soil 28*, 129–146 (1968).

279. Munns, D. N., "Nodulation of *Medicago sativa* in solution culture. V. Calcium and pH requirements during infection," *Plant and Soil 32*, 90–102 (1970).

280. Munns, D. N., "Acid tolerance in legumes and rhizobia," *Adv. Plant Nutr. 2*, 63–91 (1986).

281. Munns, D. N., Fox, R. L., and Koch, B. L., "Influence of lime on nitrogen fixation by tropical and temperate legumes," *Plant and Soil 46*, 591–601 (1977).

282. Munns, D. N., Hohenberg, T. S., Righetti, T. L., and Lauter, D. J., "Soil acidity tolerance of symbiotic and nitrogen-fertilized soybeans," *Agron. J. 73*, 407–410 (1981).

283. Munns, D. N., Keyser, H. H., Fogle, V. W., Hohenberg, J. S., Righetti, T. L., Lauter, D. L., Zaroug, M. G., Clarkin, K. L., and Whitacre, K. W., "Tolerance of soil acidity in symbioses of mung bean with rhizobia," *Agron. J. 71*, 256–260 (1979).

284. Nicol, H., and Thornton, H. G., "Competition between related strains of nodule bacteria and its influence on infection of the legume host," *Proc. Roy. Soc. Lond. 130*, 32–59 (1941).

285. Noel, K. D., and Brill, W. J., "Diversity and dynamics of indigenous *Rhizobium japonicum* populations," *Appl. Environ. Microbiol. 40*, 931–938 (1980).

286. Nutman, P. S., "Variation within strains of clover nodule bacteria in the size of nodule produced and in the effectivity of the symbiosis," *J. Bacteriol. 51*, 411–432 (1946).

287. Nutman, P. S., "Some observations on root-hair infection by nodule bacteria," *J. Exp. Bot. 10*, 250–263 (1959).

288. Nutman, P. S., "The relation between root hair infection by *Rhizobium* and nodulation in *Trifolium* and *Vicia*," *Proc. Roy. Soc. Lond. 165*, 122–137 (1962).

289. Nutman, P. S., and Hearne, R., "Persistence of nodule bacteria in soil under long-term cereal cultivation," *Rothamsted Experimental Station Report for 1979*, Part 2, pp. 77–90 (1980).

290. Nutman, P. S., and Ross, G. J. S., "*Rhizobium* in the soils of the Rothamsted and Woburn farms," *Rothamsted Experimental Report for 1970*, Part 2, pp. 148–167 (1970).

291. O'Hara, G. W., Goss, T. J., Dilworth, M. J., and Glenn, A. R., "Maintenance of intracellular pH and acid tolerance in *Rhizobium meliloti*," *Appl. Environ. Microbiol. 55*, 1870–1876 (1989).

292. Osa-Afiana, L. O., and Alexander, M., "Differences among cowpea rhizobia in tolerance to high temperature and desiccation in soil," *Appl. Environ. Microbiol. 43*, 435–439 (1982).

293. Osborne, G. J., Pratley, J. E., and Stewart, W. P., "The tolerance of subterranean clover (*Trifolium subterraneum* L.) to aluminum and manganese," *Field Crops. Res. 3*, 347–358 (1981).

294. Owiredu, J. D., and Danso, S. K. A., "Response of soybean [*Glycine max* (L.) Merrill] to *Bradyrhizobium japonicum* inoculation in three soils in Ghana," *Soil Biol. Biochem. 20*, 311–314 (1988).

295. Ozanne, P. G., Keay, J., and Biddiscombe, E. F., "The comparative applied phosphate requirements of eight annual pasture species," *Aust. J. Agric. Res. 20*, 809–818 (1969).

296. Ozawa, T., and Yamaguchi, M., "Fractionation and estimation of particle-attached and unattached *Bradyrhizobium japonicum* strains in soils," *Appl. Environ. Microbiol. 52*, 911–914 (1986).

297. Paau, A. S., "Improvement of *Rhizobium* inoculants," *Appl. Environ. Microbiol. 55*, 862–865 (1989).

298. Pankhurst, C. E., MacDonald, P. E., and Reeves, J. M., "Enhanced nitrogen fixation and competitiveness for nodulation of *Lotus pedunculatus* by a plasmid-curved derivative of *Rhizobium loti*," *J. Gen. Microbiol. 132*, 2321–2328 (1986).

299. Papavizas, G. C., and Davey, C. B., "Extent and nature of the rhizosphere of *Lupinus*," *Plant and Soil 14*, 215–236 (1961).

300. Parke, D., and Ornston, L. N., "Nutritional diversity of *Rhizobiaceae* revealed by auxanography," *J. Gen. Microbiol. 130*, 1743–1750 (1984).

301. Parke, D., and Ornston, L. N., "Enzymes of the β-ketoadipate pathway are inducible in *Rhizobium* and *Agrobacterium* spp. and constitutive in *Bradyrhizobium* spp," *J. Bacteriol. 165*, 288–292 (1986).

302. Parker, C. A., Trinick, M. J., and Chatel, D. L., "Rhizobia as soil and rhizosphere inhabitants," In: *A Treatise on Dinitrogen Fixation*, (Hardy, R. W. F., and Gibson, A. H., eds.), New York, Wiley, 1977, pp. 311–352.

303. Pena-Cabriales, J. J., and Alexander, M., "Survival of *Rhizobium* in soils undergoing drying," *Soil Sci. Soc. Am. J. 43*, 962–966 (1979).

304. Pena-Cabriales, J. J., and Alexander, M., "Growth of *Rhizobium* in unamended soil," *Soil Sci. Soc. Am. J. 47*, 81–84 (1983).

305. Pepper, I. L., Josephson, K. L., Nautiyal, C. S., and Bourque, D. P., "Strain identification of highly competitive bean rhizobia isolated from root nodules: Use of fluorescent antibodies, plasmid profiles and gene probes," *Soil Biol. Biochem. 21*, 749–753 (1989).

306. Peters, N. K., and Long, S. R., "Alfalfa root exudates and compounds which promote or inhibit induction of *Rhizobium meliloti* nodulation genes," *Plant Physiol. 88*, 396–400 (1988).

307. Peterson, E. A., Sirois, J. C., Berndt, W. C., and Miller, R. W., "Evaluation of competitive ability of *Rhizobium meliloti* strains for nodulation in alfalfa," *Can. J. Microbiol. 29*, 541–545 (1983).

308. Pinero, D., Martinez, E., and Selander, R. K., "Genetic diversity and relationships among isolates of *Rhizobium leguminosarum* bv. *phaseoli*," *Appl. Environ. Microbiol. 54*, 2825–2832 (1988).

309. Pinto, C. M., Yao, P. Y., and Vincent, J. M., "Nodulating competitiveness amongst strains of *Rhizobium meliloti* and *R. trifolii*," *Aust. J. Agric. Res. 25*, 317–329 (1974).

310. Postma, J., van Elsas, J. D., Govaert, J. M., and van Veen, J. A., "The dynamics of *Rhizobium leguminosarum* bv. *trifolii* introduced into soil as determined by immunofluorescence and selective plating techniques," *FEMS Microbiol. Ecol. 53*, 251–260 (1988).

311. Postma, J., van Veen, J. A., and Walter, S., "Influence of different initial soil moisture contents on the distribution and population dynamics of introduced *Rhizobium leguminosarum* bv. *trifolii*," *Soil Biol. Biochem. 21*, 437–442 (1989).

312. Pueppke, S. G., "Nodulation distribution on legume roots: Specificity and response to the presence of soil," *Soil Biol. Biochem. 18*, 601–606 (1986).

313. Purchase, H. F., and Vincent, J. M., "A detailed study of the field distribution of strains of clover nodule bacteria," *Proc. Linn. Soc. N.S.W. 74*, 227–236 (1949).

314. Ramirez, C., and Alexander, M., "Evidence suggesting protozoan predation on *Rhizobium* associated with germinating seeds and in the rhizosphere of beans (*Phaseolus vulgaris* L.)," *Appl. Environ. Microbiol. 40*, 492–499 (1980).

315. Rasmussen, P. E., and Rohde, C. R., "Soil acidification from ammonium-nitrogen fertilization in moldboard plow and stubble-mulch wheat-fallow tillage," *Soil Sci. Soc. Am. J. 53*, 119–122 (1989).

316. Read, M. P., "The establishment of serologically identifiable strains of *Rhizobium trifolii* in field soils in competition with native microflora," *J. Gen. Microbiol. 1*, 1–4 (1953).

317. Rennie, R. J., "Nitrogen fixation in agriculture in temperate regions," In: *Nitrogen Fixation Research Progress* (Evans, H. J., Bottomley, P. J., and Newton, W. E. eds.), Dordrecht, Martinus Nijhoff, 1985, pp. 659–664.

318. Renwick, A., and Jones, D. G., "The manipulation of white clover 'host preference' for strains of *Rhizobium trifolii* in an upland soil," *Ann. Appl. Biol. 108*, 291–302 (1986).

319. Reyes, V. G., and Schmidt, E. L., "Population densities of *Rhizobium japonicum* strain 123 estimated directly in soil and rhizospheres," *Appl. Environ. Microbiol. 37*, 854–858 (1979).

320. Reyes, V. G., and Schmidt, E. L., "Populations of *Rhizobium japonicum* associated with the surfaces of soil-grown roots," *Plant and Soil 61*, 71–80 (1981).

321. Richardson, A. E., Djordjevic, M. A., Rolfe, B. G., and Simpson, R. J., "Effects of pH, Ca and Al on the exudation from clover seedlings of compounds that induce the expression of nodulation genes in *Rhizobium trifolii*," *Plant and Soil 109*, 34–47 (1988).

322. Richardson, A. E., and Simpson, R. J., "Enumeration and distribution of *Rhizobium trifolii* under a subterranean clover-based pasture growing in an acid soil," *Soil Biol. Biochem. 20*, 431–438 (1988).

323. Richardson, A. E., and Simpson, R. J., "Acid-tolerance and symbiotic effectiveness of *Rhizobium trifolii* associated with a *Trifolium subterraneum* L.–based pasture growing in an acid soil," *Soil Biol. Biochem. 21*, 87–95 (1989).

324. Richardson, A. E., Simpson, R. J., Djordjevic, M. A., and Rolfe, B. G., "Expression of nodulation genes in *Rhizobium leguminosarum* bv. *trifolii* is affected by low pH and by Ca and Al ions," *Appl. Environ. Microbiol. 54*, 2541–2548 (1988).

325. Robert, F. M., and Schmidt, E. L., "Population changes and persistence of *Rhizobium phaseoli* in soil and rhizospheres," *Appl. Environ. Microbiol. 45*, 550–556 (1983).

326. Robert, F. M., and Schmidt, E. L., "Response to three indigenous serogroups of *Rhizobium japonicum* to the rhizosphere of pre-emergent seedlings of soybean," *Soil Biol. Biochem. 17*, 579–580 (1985).

327. Robinson, A. C., "The influence of host on soil and rhizosphere populations of clover and lucerne root nodule bacteria in the field," *J. Aust. Inst. Agric. Sci. 33*, 207–209 (1967).

328. Robinson, A. C., "Competition between effective and ineffective strains of *Rhizobium trifolii* in the nodulation of *Trifolium subterraneum*," *Aust. J. Agric. Res. 20*, 827–841 (1969a).

329. Robinson, A. C., "Host selection for effective *Rhizobium trifolii* by red clover and subterranean clover in the field," *Aust. J. Agric. Res. 20*, 1053–1060 (1969b).

330. Robson, A. D., *Soil Acidity and Plant Growth*, New York, Academic Press, 1989, p. 306.

330a. Rogers, D. D., Warren, R. D., and Chamblee, D. S., "Remedial postemergence legume inoculation with *Rhizobium*," *Agron. J. 74*, 613–619 (1982).

331. Rolfe, B. G., and Gresshoff, P. M., "Genetic analysis of legume nodule initiation," *Ann. Rev. Plant Physiol. Plant Mol. Biol. 39*, 297–319 (1988).

332. Rouatt, J. W., and Katznelson, H., "A study of the bacteria on the root surface and in the rhizosphere of crop plants," *J. Appl. Bacteriol. 24*, 164–171 (1961).

333. Roughley, R. J., Blowes, W. M., and Herridge, D. F., "Nodulation of *Trifolium subterraneum* L. by introduced rhizobia in competition with naturalized strains," *Soil Biol. Biochem. 8*, 403–407 (1976).

334. Roughley, R. J., Bromfield, E. S. P., Pulver, E. L., and Day, J. M., "Competition between species of *Rhizobium* for nodulation of *Glycine max*," *Soil Biol. Biochem. 17*, 467–470 (1980).

335. Roughley, R. J., and Vincent, J. M., "Growth and survival of *Rhizobium* spp. in peat culture," *J. Appl. Bacteriol. 30*, 362–376 (1967).

336. Rovira, A. D., "*Rhizobium* numbers in the rhizospheres of red clover and *Paspalum* in relation to soil treatment and the numbers of bacteria and fungi," *Aust. J. Agric. Res. 12*, 77–83 (1961).

337. Rovira, A. D., Newman, E. I., Bowen, H. J., and Campbell, R., "Quantitative assessment of the rhizoplane microflora by direct microscopy," *Soil Biol. Biochem. 6*, 211–216 (1974).

338. Rupela, O. P., Toomsan, B., Mittal, S., Dart, P. J., and Thompson, J. A., "Chickpea *Rhizobium* populations: Survey of influence of season, soil depth and cropping pattern," *Soil Biol. Biochem. 19*, 247-252 (1987).

339. Russell, P. E., and Jones, D. G., "Immunofluorescence studies of selection of strains of *R. trifolii* by S184 white clover (*T. repens* L.)," *Plant and Soil, 42*, 119–129 (1975a).

340. Russell, P. E., and Jones, D. G., "Variation in the selection of *Rhizobium trifolii* by varieties of red and white clover," *Soil Biol. Biochem. 7*, 15–18 (1975b).

341. Rys, G. J., and Bonish, P. M., "Effectiveness of *Rhizobium trifolii* populations associated with *Trifolium* species in Taranaki, New Zealand," *N.Z. J. Exp. Agric. 9*, 327–335 (1981).

342. Salema, M. P., Parker, C. A., Kidby, D. K., Chatel, D. L., and Armitage, T. M., "Rupture of nodule bacteria on drying and rehydration," *Soil Biol. Biochem. 14*, 15–22 (1982).

343. Sanchez, P. A., *Properties and Management of Soils in the Tropics*, New York, Wiley, 1976, p. 618.

344. Sanginga, N., Mulongoy, K., and Ayanaba, A., "Effectivity of indigenous rhizobia for nodulation and early nitrogen fixation with *Leucaena leucocephala* grown in Nigerian soils," *Soil Biol. Biochem. 21*, 231–235 (1989).

345. Sargent, L., Huang, S. Z., Rolfe, B. G., and Djordjevic, M. A., "Split-root assays using *Trifolium subterraneum* show that *Rhizobium* infection induces a systemic response that can inhibit nodulation of another invasive *Rhizobium* strain," *Appl. Environ. Microbiol. 53*, 1611–1619 (1987).

346. Schmidt, E. L., "Quantitative autecological study of microorganisms in soil by immunofluorescence," *Soil Sci. 118*, 141–149 (1974).

347. Schmidt, E. L., Bankole, R. O., and Bohlool, B. B., "Fluorescent-antibody approach to study of rhizobia in soil," *J. Bacteriol. 95*, 1987–1992 (1968).

348. Schofield, P. R., Gibson, A. H., Dudman, W. F., and Watson, J. M., "Evidence for genetic exchange and recombination of *Rhizobium* symbiotic plasmids in a soil population," *Appl. Environ. Microbiol. 53*, 2942–2947 (1987).

349. Schroth, M. N., and Hancock, J. G., "Selected topics in biological control," *Ann. Rev. Microbiol. 35*, 453–476 (1981).

350. Schwinghamer, E. A., and Belkengren, R. P., "Inhibition of rhizobia by a strain of *Rhizobium trifolii*: Some properties of the antibiotic and of the strain," *Arch. Mikrobiol. 64*, 130–145 (1968).

351. Singleton, P. W., Abdel-Magid, H. M., and Tavares, J. W., "Effect of phosphorus on the effectiveness of strains of *Rhizobium japonicum*," *Soil Sci. Soc. Am. J. 49*, 613–616 (1985).

352. Singleton, P. W., El Swaify, S. A., and Bohlool, B. B., "Effect of salinity on *Rhizobium* growth and survival," *Appl. Environ. Microbiol. 44*, 884-890 (1982).

353. Singleton, P. W., and Tavares, J. S., "Inoculation response of legumes in relation to the number and effectiveness of indigenous *Rhizobium* populations," *Appl. Environ. Microbiol. 51*, 1013–1018 (1986).

353a. Sloger, C., Bezdicek, D. F., Milberg, R., and Boonkerd, N., "Seasonal and diurnal variations in N_2 (C_2H_2)-fixing activity in field soybeans," In: *Nitrogen Fixation by Free-living Microorganisms* (Stewart, W. D. P., ed.), London, Cambridge University Press, 1975, pp. 271–284.

354. Smart, J. B., Robson, A. D., and Dilworth, M. J., "A continuous culture study of the phosphorus nutrition of *Rhizobium trifolii* WU95, *Rhizobium* NGR234 and *Bradyrhizobium* CB756," *Arch. Microbiol. 140*, 276–280 (1984).

355. Smit, G., Kijne, J. W., and Lugtenberg, B. J. J., "Involvement of both cellulose fibrils and a Ca^{2+}-dependent adhesin in the attachment of *Rhizobium leguminosarum* to pea root hair tips," *J. Bacteriol. 169*, 4294–4301 (1987).

356. Smit, G., Logman, T. J. J., Boerrigter, M. E. T. I., Kijne, J. W., and Lugtenberg, B. J. J., "Purification and partial characterization of the *Rhizobium leguminosarum* bv. *viciae* Ca^{2+}-dependent adhesin, which mediates the first step in attachment of cells of the family Rhizobiaceae to plant root hair tips," *J. Bacteriol. 171*, 4054–4062 (1989).

357. Smith, G. B., and Wollum, A. G., "Nodulation of *Glycine max* by six *Bradyrhizobium japonicum* strains with different competitive abilities," *Appl. Environ. Microbiol. 55*, 1957–1962 (1989).

358. Soberon-Chavez, G., and Najera, R., "Isolation from soil of *Rhizobium leguminosarum* lacking symbiotic information," *Can. J. Microbiol. 35*, 464–468 (1989).

359. Soby, S., and Bergman, K., "Motility and chemotaxis of *Rhizobium meliloti* in soil," *Appl. Environ. Microbiol. 46*, 995–998 (1983).

360. Somasegaran, P., and Halliday, J., "Dilution of liquid *Rhizobium* cultures to increase production capacity of inoculant plants," *Appl. Environ. Microbiol. 44*, 330–333 (1982).

361. Somasegaran, P., Hoben, H. J., and Gurgun, V., "Effect of inoculation rate, rhizobial strain competition and nitrogen fixation in chickpea," *Agron. J. 80*, 68–73 (1988).

362. Spaink, H. P., Okker, R. J. H., Wijffelman, C. A., Tak, T., Leentje Goosende Roo, E. P., van Brussel, A. A. N., and Lugtenberg, B. J. J., "Symbiotic properties of rhizobia containing a flavonoid-independent hybrid *nod* D product," *J. Bacteriol. 171*, 4045–4053 (1989).

363. Spaink, H. P., Wijffelman, C. A., Pees, E., Okker, R. J. H., and Lugtenberg, B. J. J., "*Rhizobium* nodulation gene *nod* D as a determinant in host specificity," *Nature* (Lond.) 328, 337–339 (1987).

364. Sprent, J. I., "Which steps are essential for formation of functional legume nodules?" *New Phytol. 111*, 129–153 (1989).

365. Surin, B. P., and Downie, J. A., "*Rhizobium leguminosarum* genes required for expression and transfer of host specific nodulation," *Plant Mol. Biol. 12*, 19–29 (1989).

366. Thompson, J. A., "Inhibition of nodule bacteria by an antibiotic from legume seed coats," *Nature (Lond.) 187*, 619–620 (1960).

367. Thompson, J. A., "Production and quality control of legume inoculants," In: *Methods for Evaluating Biological Nitrogen Fixation* (Bergersen, F. J., ed.), New York, Wiley, 1980, pp. 489–534.

368. Thornton, F. C., and Davey, C. G., "Acid tolerance of *Rhizobium trifolii* in culture media," *Soil Sci. Soc. Am. J. 47*, 496–501 (1983).

369. Thurman, N. P., and Bromfield, E. S. P., "Effect of variation within and between *Medicago* and *Melilotus* species on the composition and dynamics of indigenous populations of *Rhizobium meliloti*," *Soil Biol. Biochem. 20*, 31–38 (1988).

370. Trinick, M. J., "Biology," In: *Nitrogen Fixation, Vol. 2, Rhizobium* (Broughton, W. J., ed.), Oxford, Clarendon Press, 1982, pp. 77–146.

371. Triplett, E. W., "Isolation of genes involved in nodulation competitiveness from *Rhizobium leguminosarum* bv. *trifolii* T24," *Proc. Natl. Acad. Sci. U.S.A. 85*, 3810–3814 (1988).

372. Triplett, E. W., "Construction of a symbiotically effective strain of *Rhizobium leguminosarum* bv. *trifolii* with increased nodulation competitiveness," *Appl. Environ. Microbiol. 56*, 98–103 (1990).

373. Triplett, E. W., and Barta, T. M., "Trifolitoxin production and nodulation are necessary for the expression of superior nodulation competitiveness by *Rhizobium leguminosarum* bv. *trifolii* strain T24 on clover," *Plant Physiol. 85*, 335–342 (1987).

374. Triplett, E. W., Schink, M. J., and Noeldner, K. I., "Mapping and subcloning of the trifolitoxin production and resistance genes from *Rhizobium leguminosarum* bv. *trifolii* T24," *Mol. Plant Microbe. Interac. 2*, 202–208 (1989).

374a. Tsien, H. C., Cain, P. S., and Schmidt, E. L., "Viability of *Rhizobium* bacteriods," *Appl. Environ. Microbiol. 34*, 854–856 (1977).

375. Turco, R. F., and Bezdicek, D. F., "Diversity within two serogroups of *Rhizobium leguminosarum* native to soils in the palouse of eastern Washington," *Ann. Appl. Biol. 111*, 103–114 (1987).

376. Tuzimura, K., and Watanabe, I., "The growth of *Rhizobium* in the rhizosphere of the host plant," *Soil Sci. Plant Nutr. 8*, 19–24 (1962a).

377. Tuzimura, K., and Watanabe, I., "The effect of the rhizosphere of various plants on the growth of *Rhizobium*," *Soil Sci. Plant Nutr. 8*, 13–17 (1962b).

378. Tuzimura, K., and Watanabe, I., "Different growth and survival of *Rhizobium* species in the rhizosphere of various plants in different sorts of soil," *Soil Sci. Plant Nutr. 12*, 15–22 (1966).

379. Valdivia, B., Dughri, M. H., and Bottomley, P. J., "Antigenic and symbiotic characterization of indigenous *Rhizobium leguminosarum* bv. *trifolii* recovered from root nodules of *Trifolium pratense* L. sown into subterranean clover pasture soils," *Soil Biol. Biochem. 20*, 267–274 (1988).

380. Van Rensberg, H. J., and Strijdom, B. W., "Survival of fast- and slow-growing *Rhizobium* spp. under conditions of relatively mild desiccation," *Soil Biol. Biochem. 12*, 353–356 (1980).

381. Van Rensberg, H. J., and Strijdom, B. W., "Competitive abilities of *Rhizobium meliloti* strains considered to have potential as inoculants," *Appl. Environ. Microbiol. 44*, 98–106 (1982).

382. Van Rensburg, H. J., and Strijdom, B. W., "Effectiveness of *Rhizobium* strains used in inoculants after their introduction into soil," *Appl. Environ. Microbiol. 49*, 127–131 (1985).

383. Van Veen, J. A., Ladd, J. N., Martin, J. K., and Amato, M., "Turnover of carbon, nitrogen, and phosphorus through the microbial biomass in soils incubated with ^{14}C-, ^{15}N-, and ^{32}P-labelled bacterial cells," *Soil Biol. Biochem. 19*, 559–565 (1987).

384. Vance, C. P., Johnson, L. E. B., Halvorson, A. M., Heichel, G. H., and Barnes, D. K., "Histological and ultra-structural observations of *Medicago sativa* root nodule senescence after foliage removal," *Can. J. Bot. 58*, 295–309 (1980).

385. Vargas, A. A. T., and Graham, P. H., "*Phaseolus vulgaris* cultivar and *Rhizobium* strain variation in acid-pH tolerance and nodulation under acid conditions," *Field Crops Res. 19*, 91–101 (1988).

386. Vargas, A. A. T., and Graham, P. H., "Cultivar and pH effects on competition for nodule sites between isolates of *Rhizobium* in beans," *Plant and Soil 117*, 195–200 (1989).

387. Vincent, J. M., "Variation in the nitrogen fixing property of *Rhizobium trifolii*," *Nature (Lond.)* 153, 496–497 (1944).

388. Vincent, J. M., "The root-nodule bacteria of pasture legumes," *Proc. Linn. Soc. N.S.W. 79*, iv–xxxii (1954a).

389. Vincent, J. M., "The root-nodule bacteria as factors in clover establishment in the red basaltic soils of the Lismore district, New South Wales, I. A survey of 'native' strains," *Aust. J. Agric. Res. 5*, 55–60 (1954b).

390. Vincent, J. M., "Influence of calcium and magnesium on the growth of *Rhizobium*," *J. Gen. Microbiol. 28*, 653–658 (1962a).

391. Vincent, J. M., "Australian studies of the root-nodule bacteria. A review," *Proc. Linn. Soc. N.S.W. 87*, 8–38 (1962b).

392. Vincent, J. M., "Root-nodule symbioses with *Rhizobium*," In: *The Biology of Nitrogen Fixation* (Quispel, A., ed.), Amsterdam, North-Holland, 1974, pp. 265–341.

393. Vincent, J. M., and Waters, L. M., "The influence of the host on competition amongst clover root-nodule bacteria," *J. Gen. Microbiol. 9*, 357–370 (1953).

394. Vincent, J. M., and Waters, L. M., "The root-nodule bacteria as factors in clover establishment in the red basaltic soils of the Lismore district. New South Wales. II. Survival and success of inocula in laboratory trials," *Aust. J. Agric. Res. 5*, 61–76 (1954).

395. Viteri, S. E., and Schmidt, E. L., "Ecology of indigenous soil rhizobia: Response of *Bradyrhizobium japonicum* to readily available substrates," *Appl. Environ. Microbiol. 53*, 1872–1875 (1987).

396. Viteri, S. E., and Schmidt, E. L., "Chemoautotrophy as a strategy in the ecology of indigenous soil bradyrhizobia," *Soil Biol. Biochem. 21*, 461–463 (1989).

397. Wadisirisuk, P., Danso, S. K. A., Hardarson, G., and Bowen, G. D., "Influence of *Bradyrhizobium japonicum* location and movement on nodulation and nitrogen fixation in soybeans," *Appl. Environ. Microbiol. 55*, 1711–1716 (1989).

398. Wang, C. L., Beringer, J. E., and Hirsch, P. R., "Host plant effects on hybrids of *Rhizobium leguminosarum* biovars *viceae* and *trifolii*," *J. Gen. Microbiol. 132*, 2063–2070 (1986).

399. Weaver, R. W., and Frederick, L. R., "Effect of inoculum rate on competitive nodulation of *Glycine max* L. Merrill. II. Field studies," *Agron. J. 66*, 233–235 (1974).

400. Weaver, R. W., Frederick, L. R., and Dumenil, L. C., "Effect of soybean cropping and soil properties on numbers of *Rhizobium japonicum* in Iowa soils," *Soil Sci. 114*, 137–141 (1972).

401. Weaver, R. W., Morris, D. R., Boonkerd, N., and Sij, J., "Populations of *Brady-rhizobium japonicum* in fields cropped with soybean–rice rotations," *Soil Sci. Soc. Am. J. 51,* 90–92 (1987).

402. Weaver, R. W., and Wright, S. F., "Variability in effectiveness of rhizobia during culture and in nodules," *Appl. Environ. Microbiol. 53,* 2972–2974 (1987).

403. Weber, D. F., Keyser, H. H., and Uratsu, S. L., "Serological distribution of *Brady-rhizobium japonicum* from U.S. soybean production areas," *Agron. J. 81,* 786–789 (1989).

404. Weber, D. F., and Miller, V. L., "Effect of soil temperature on *Rhizobium japonicum* serogroup distribution in soybean nodules," *Agron. J. 64,* 796–798 (1972).

405. Weller, D. M., "Biological control of soilborne plant pathogens in the rhizosphere with bacteria," *Annu. Rev. Phytopathol. 26,* 379–407 (1988).

406. Wijffelman, C. A., Pees, E., van Brussel, A. A. N., Okker, R. J. H., and Lugtenberg, B. J. J., "Genetic and functional analysis of the nodulation region of the *Rhizobium leguminosarum sym* plasmid pRL1JI," *Arch. Microbiol. 143,* 225–232 (1985).

407. Winarno, R., and Lie, T. A., "Competition between *Rhizobium* strains in nodule formation interaction between nodulating and non-nodulating strains," *Plant and Soil 51,* 135–142 (1979).

408. Wollum, A. G., and Cassel, D. K., "Spatial variability of *Rhizobium japonicum* in two North Carolina soils," *Soil. Sci. Soc. Am. J. 48,* 1082–1086 (1984).

409. Wolyn, D. J., Attewell, D. J., Ludden, P. W., and Bliss, F. A., "Indirect measures of N_2 fixation in common bean (*Phaseolus vulgaris* L.) under field conditions: The role of lateral root nodules," *Plant and Soil 113,* 181–187 (1989).

410. Wood, M., and Cooper, J. E., "Screening clover and *Lotus* rhizobia for tolerance of acidity and aluminum," *Soil Biol. Biochem. 17,* 493–497 (1985).

411. Wood, M., and Cooper, J. E., "Acidity, aluminum, and multiplication of *Rhizobium trifolii:* Effects of temperature and carbon source," *Soil Biol. Biochem. 20,* 89–93 (1988).

412. Wood, M., Cooper, J. E., and Campbell, D. S., "A survey of clover and *Lotus* rhizobia in Northern Ireland pasture soils," *J. Soil. Sci. 36,* 357–365 (1985).

413. Wood, M., and Shepherd, G., "Characterization of *Rhizobium trifolii* isolated from soils of different pH," *Soil Biol. Biochem. 19,* 317–321 (1987).

413a. Woomer, P., and Bohlool, B. B., "Rhizobial ecology in tropical pasture systems," In: *Persistence of Forage Legumes* (Marten, G. C., et al., eds.), Madison, Amer. Soc. Agron. 1989, pp. 233–245.

414. Woomer, P., Singleton, P. W., and Bohlool, B. B., "Ecological indicators of native rhizobia in tropical soils," *Appl. Environ. Microbiol. 54,* 1112–1116 (1988).

415. Worrall, V. S., and Roughley, R. J., "The effect of moisture stress on infection of *Trifolium subterraneum* L. by *Rhizobium trifolii* Dang," *J. Exp. Bot. 27,* 1233–1241 (1976).

416. Yelton, M. M., Mulligan, J. T., and Long, S. R., "Expression of *Rhizobium meliloti nod* genes in *Rhizobium* and *Agrobacterium* backgrounds," *J. Bacteriol. 169,* 3094–3098 (1987).

417. Young, J. P. W., "*Rhizobium* population genetics: Enzyme polymorphism in isolates from peas, clover, beans, and lucerne grown at the same site," *J. Gen. Microbiol. 131,* 2399–2408 (1985).

418. Young, J. P. W., Demetriou, L., and Apte, R. G., *"Rhizobium* population genetics: Enzyme polymorphism in *Rhizobium leguminosarum* from plants and soil in a pea crop," *Appl. Environ. Microbiol. 53,* 397–402 (1987).

419. Young, J. P. W., and Wexler, M., *"Sym* plasmid and chromosomal genotypes are correlated in field populations of *Rhizobium leguminosarum," J. Gen. Microbiol. 134,* 2731–2739 (1988).

420. Zaat, S. A. J., Schripsema, J., Wijffelman, C. A., van Brussel, A. A. N., and Lugtenberg, B. B., "Analysis of the major inducers of the *Rhizobium nod* A promoter from *Vicia sativa* root exudate and their activity with different *nod* D genes," *Plant Mol. Biol. 13,* 175–188 (1989).

421. Zaat, S. A. J., Wijffelman, C. A., Mulders, I. H. M., van Brussel, A. A. N., and Lugtenberg, B. J. J., "Root exudates of various host plants of *Rhizobium leguminosarum* contain different sets of inducers of *Rhizobium* nodulation genes," *Plant Physiol. 86,* 1298–1303 (1988).

422. Zdor, R. E., and Pueppke, S. G., "Early infection and competition for nodulation of soybean by *Bradyrhizobium japonicum* 123 and 138," *Appl. Environ. Microbiol. 54,* 1996–2002 (1988).

423. Zhou, J. C., Tchan, Y. T., and Vincent, J. M., "Reproductive capacity of bacteroids in nodules of *Trifolium repens L.* and *Glycine max* (L.) Merr.," *Planta 163,* 473–482 (1985).

9

The Rhizobium Infection Process

Jan W. Kijne

I. Introduction

Bacteria of the genera *Rhizobium, Bradyrhizobium,* and *Azorhizobium* (collectively known as rhizobia) release mitogenic signal molecules that may induce cell divisions in the cortex of plant roots. Roots of leguminous plants (Fabaceae) and *Parasponia,* a woody member of the elm family, respond to these mitogens with root nodule formation as an ultimate result. During development of a nodule primordium (Figure 9-1), rhizobia migrate *intra*cellularly or *inter*cellularly toward the dividing root cells. Subsequently, the bacteria are taken up by these plant cells, either completely (in phagocytotic vesicles) or incompletely (in plasma membrane invaginations surrounded by cell wall material), and develop into N_2-fixing "organelles" maintained by the plant in return for ammonia (see Chapter 10). When rhizobial migration towards the nodule primordium is unsuccessful, plant cell divisions may continue and a so-called empty nodule may form.[91]

In general, plant cells do not ingest large particles by phagocytosis. Endocytotic uptake of rhizobia is one of the very few exceptions. Absence of phagocytosis in plant cells has been attributed to the presence of a cell wall and to cell turgor. Apparently, rhizobia are able to overcome these barriers. Interestingly, complete endocytosis of rhizobia is observed only in cells formed under the influence of rhizobial stimuli.

Only a limited set of host plants responds to mitogens released by a single rhizobial species or biovar. Consequently, nodulation is a host plant–specific process. This observation indicates that rhizobial mitogens are different from generally recognized cell division factors such as auxin and cytokinin, and that induction of a nodule primordium is different from plant hormone–triggered tumor and gall formation by *Agrobacterium* and *Pseudomonas savastanoi* bacteria, respectively. This does not preclude a contribution of plant hormones to induction of a nodule primordium. Allen and Allen[5] have shown that treatment with auxin-transport inhibitors induces nodule-like outgrowths

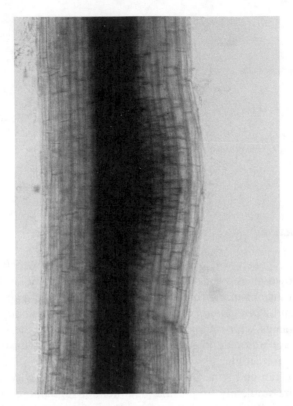

Figure 9-1. A nodule primordium in a *Vicia hirsuta* root. Root diameter = 0.4 cm. (Courtesy of Dr. A. A. N. van Brussel.)

on the roots of several plants, a phenomenon that has recently been studied at the molecular level by Hirsch et al.[105] Moreover, auxins and cytokinins induce cell divisions in pea root explants at sites similar to those of nodule primordium formation.[140] These observations suggest that rhizobial mitogens specifically influence the endogenous phytohormone balance in host plant roots, with induction of cell divisions as a result. The cell envelope of root cells of nonhost plants may be a barrier for reception of rhizobial signals: treatment with cell wall–degrading enzymes and polyethylene glycol renders some rice seedling roots responsive to rhizobial mitogenic activity.[6]

Entry of plant-associated bacteria into the intercellular space of plants is not exceptional. Plants are open organisms. Many openings between plant epidermal cells such as stomata or wounds resulting from emergence of lateral roots give direct access to the internal space of the plant body. However, only a few bacterial genera are able to colonize the plant's interior, most of them being plant-pathogenic species or pathovars, which penetrate plant cells only during or after plant cell death. In contrast, legume or *Parasponia* cells interacting with rhizobia normally are viable during and after endocytotic uptake of their symbiotic partners.

During or after specific response to a rhizobial mitogen, the host plant largely determines the mechanism of root invasion. One rhizobial strain infects *Vigna* roots intracellularly via root hairs and *Arachis* roots intercellularly via wounds.[195] Another strain infects *Parasponia* intercellularly via epidermal cracks and *Macroptilium* intracellularly via root hairs.[149] Also rhizobial uptake in root cells is host plant–determined. In *Parasponia* cells the last-mentioned *Rhizobium* strain is present in walled plasmamembrane invaginations (so-called fixation threads), whereas *Macroptilium* cells ingest the same bacteria via endocytotic vesicles. Furthermore, morphogenesis of a root nodule is under host control; for example, formation of collar-shaped nodules in *Lupinus angustifolius* and of cylindrical nodules in *Ornithopus sativus* can be induced by the same rhizobial strain.[120] Certain plants occasionally develop root nodules even in the absence of rhizobia,[219] which points at a spontaneous hormonal inbalance. These observations demonstrate that rhizobia may trigger infection- and nodulation-related processes but that the host plant determines their nature.

Many legumes, especially caesalpinioid genera, do not nodulate at all.[64] This may, for instance, be explained by the absence in these plants of cells able to recognize mitogens released by rhizobia present in the same habitat. But even the presence of infective rhizobia in the rhizosphere of a homologous host plant does not necessarily result in induction of a nodule primordium, root invasion, cell infection, or nodulation. Growth with a sufficient amount of combined nitrogen renders a legume resistant to nodulation (for a recent review, see Streeter[213]). Resistance is expressed at various steps of the infection process, depending on the symbiotic partners involved. Vice versa, the conditions under which rhizobia are grown determine their ability to invade the roots of the host plant.[128] Thus, both partners must be simultaneously conditioned for successful infection. Once a nodule primordium is induced, the plant (again) controls the number of root nodules by a systemic response that suppresses nodule formation in other parts of the root system.[29,123]

In this chapter, root hair infection by *Rhizobium leguminosarum* biovar *viciae*, the symbiont of *Pisum* (pea) and *Vicia* (vetch), will first be described in order to present a coherent story of one of the best-studied cases of nodulation. Next, it will be shown that root hair infection is only one of the possible mechanisms of root invasion by rhizobia. Then successive steps of the infection process will be discussed by comparing the infection process of pea and vetch with recent representative data on processes and molecules involved in other rhizobia–plant interactions, in search of common or specific features of nodulation. Host plant specificity will serve as a leading motive. The chapter takes as a starting point the excellent reviews on the infection process by Dart,[47,48,49] which marked the end of what might be called the premolecular period in the study of rhizobial infection. Since then, related reviews have been published by (in chronological order) Bauer,[14] Newcomb,[168] Dazzo and Gardiol,[56] Halverson and Stacey,[104] Downie and Johnston,[81] Rolfe and Gresshoff,[186] Djordjevic et al.,[76] Long,[145] and Sprent,[210] among others.

II. Infection by *Rhizobium leguminosarum* biovar *viciae*

A. *Outside the Root: Rhizosphere Interactions*

Successful root infection does not start at the root surface. Root infection by rhizobia is a multistep process, the success of which is partly determined by preinfection events in the rhizosphere. These events refer to conditioning of rhizobia for infection, activation of nodulation genes, targeting of the bacteria to appropriate sites at the root surface, and release of rhizobial stimuli.

Rhizobium leguminosarum biovar *viciae* (*Rl viciae*) bacteria are Gram-negative rods motile by means of peritrichous flagella. Preferred host plants of *Rl viciae* comprise species of the legumes *Pisum, Vicia, Lens,* and *Lathyrus.* Most *Rl viciae* strains are completely cross-nodulating within this group, the effectiveness being diverse. The bacteria, but also *Escherichia coli,* show positive chemotaxis towards pea root exudate.[96] A range of amino acids, sugars, and carboxylic acids is present in the exudate, and many of these were shown to be attractants. Apparently, chemotaxis is not a major factor in determination of host plant specificity.[96] The nodulation ability of flagella-less mutants of *Rl viciae* was found not to be affected; however, these mutants may be less competitive under field conditions.[204] After formation of lateral roots, a strong increase of rhizobial numbers on the pea root surface can be observed.[235] This increase may be correlated with the liberation of significant amounts of homoserine from the main root at the sites where laterals emerge.[234] Homoserine is an amino acid specific for *Pisum* and represents a preferred carbon and nitrogen source for *Rl viciae.*[236] Mutations abolishing homoserine catabolism have, however, no detectable effect on nodulation by *Rl viciae,*[87] suggesting that (specific) growth stimulation of rhizobia in the host plant rhizosphere is not necessarily required for nodulation. On the contrary, exponentially growing bacteria are less infective than growth-limited cells.[128] The absence, rather than the presence, of certain nutrients in the rhizosphere may be essential for successful infection.[127]

Many genes of *Rl viciae* that are involved in nodulation (*nod* genes) are located on a large Sym(biosis) plasmid. These genes encode common and host plant–specific nodulation functions and are organized in at least four

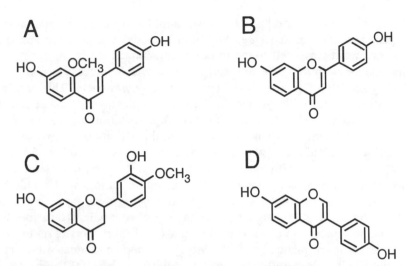

Figure 9-2. Examples of nodulation gene-inducing flavonoids. (A) 4,4'-dihydroxy-2'-methoxychalcone (alfalfa); (B) 7,4'-dihydroxyflavone (clover); (C) 7,3'-dihydroxy-4'-methoxyflavanone (vetch); (D) daidzein (soybean). For references, see text.

inducible operons. The expression of these operons requires the presence of a functional regulatory *nodD* gene as well as flavonoids exuded by the host plant root (Figure 9-2; see Chapters 14 and 17 and their references). Induction of *nod* genes appeared to start only a few minutes after addition of appropriate flavonoids.[248] The molecular mode of action of *nod* gene inducers in rhizobia has yet to be established. The flavanone inducer naringenin accumulates in the cytoplasmic membrane of *Rl viciae*,[181] in which it presumably activates the *nodD* gene product of the bacteria.[194,208] Different host plants of *Rl viciae* exude different flavonoid inducers.[249] Root exudate of *Vicia sativa* subsp. *nigra*, another host plant of *Rl viciae*, contains several inducers for the *nodA* promoter. Six of these are flavanones, including 3,5,7,3'-tetrahydroxy-4'-methoxyflavanone and 3,7,3'-trihydroxy-4'-methoxyflavanone.[251] Seed exudate of *P. sativum* contains four inducers, one of which was tentatively identified as apigenin-7-O-glucoside.[92] Interestingly, *P. sativum* also produces compounds of an isoflavone nature that inhibit the induction of *nod* genes of *Rl viciae*.[92] Production of specific *nod* gene inducers is, however, not restricted to host plants. Also, the nonhost *Trifolium repens* (white clover) exudes inducers able to interact with the *nodD* gene product of *Rl viciae*, whereas the *Medicago sativa* (alfalfa) seed flavone luteolin activates *nod* genes in concert with the *nodD* genes of either *R. meliloti, Rl trifolii,* or *Rl viciae*.[208] Some *nod* gene inducers are strong chemoattractants for *Rl viciae* with a low concentration threshold.[7] Loss of the Sym plasmid pRL1JI from the bacteria results in a decrease, but not elimination, of the chemotactic response. Addition of the flavanol kaempferol, an antagonist of *nod* gene induction in *Rl viciae,* did not influence the response to *nod* gene inducers, and it was itself found to be an attractant.[7]

Inoculation with *Rl viciae* cells causes a remarkable increase in *nod* gene–inducing activity of exudate of *V. sativa* subsp. *nigra* roots. This response is specifically induced by *Rl viciae;* it requires induction of *nod* genes, the presence of *nodD, nodABCIJ* and the host range–determining (*hsn*) *nodFEL* genes in the bacteria;[228,229] and it results from the activity of an extracellular heat-stable rhizobial factor with a molecular mass between 1000 and 10,000 Da.[229] In comparison with exudate of uninoculated plants, inoculated *Vicia* roots release several additional inducers including naringenin.[182] This rhizobial signal and/or similar signals also influence growth processes in the host root by causing an increase in the amount of root hairs (Hai), root hair deformation (Had), and development of thick and short roots (Tsr) in *Vicia sativa* subsp. *nigra* plants.[90,226,229,247] Moreover, *nod* gene–related factors of *Rl viciae* elicit expression of the early nodulin gene ENOD12 in root cells of *Pisum sativum* preparing for rhizobial infection[193] (see Chapter 15). Had and Tsr phenotypes, however, can also be induced by (1) culture supernatants of high-density populations of uninduced or Sym plasmid-cured rhizobia, or (2) low-density populations of rhizobia without *nodLEF* genes co-cultured with *Vicia* roots.[89,90,228] Apparently, a mixture of rhizobial *nod* gene–related and unrelated signal molecules is present in the rhizosphere of a host root. For *Rl viciae,* the chemical structure of either of these factors has yet to be established. Van Brussel et al.[227] reported that the minimal genetic requirements for induction of Had and Tsr in *Vicia* are *nodDABC,* those for root hair infection (Inf) are *nodEFDABC,* and those for nodule initiation (Noi) are *nodEFDABCIJ.* The latter set of *nod* genes is also sufficient to induce ENOD12 gene expression in pea.[193]

In addition to low-molecular-weight signals, *Rl viciae* secretes a Ca^{2+}-binding protein of M_r 50,000, encoded by *nodO*.[69,88] Synthesis of this protein is flavonoid-inducible and is required for nodulation of *Vicia sativa,* depending on the rhizobial chromosomal background.[67] The role of this protein in rhizobial infection has yet to be defined (see Chapter 14).

Induction of Tsr roots in *Vicia* by *nod* gene–related signals is suppressed by the ethylene inhibitor aminoethoxyvinylglycine (AVG) and can be mimicked by addition of the ethylene-releasing compound ethephon.[250] This suggests that one or more of these factors induce an increase in biosynthesis of the plant hormone ethylene in *Vicia* roots. In contrast, the root hair deformation response cannot be inhibited by AVG or induced by ethephon. Two hypotheses can be proposed to explain these observations: (1) the Tsr factor is different from the Had factor, or (2) both factors represent the same signal molecule, and physiological responses in the plant cells resulting in root hair deformation include ethylene biosynthesis as a secondary effect. In *Vicia sativa* subsp. *nigra,* the ethylene-mediated growth response of the roots is exceptional in that it results in inhibition of nodulation of the main root. Root cortical cells in *Rhizobium*- or ethephon-induced Tsr roots show hypertrophy in correlation with a rearrangement of microtubules (Van Spronsen et al., submitted), and root nodules are induced by *Rl viciae* only on lateral *V. sativa* roots at the sites of emergence. Treatment with AVG restores both normal root cell morphology and normal nodulation.[250] In other host plants, ethylene may be produced in less dramatic amounts. The possible relationship between ethylene production and synthesis of early nodulins has not yet been studied. However, preliminary data of Scheres et al.[192] suggest that expression of ENOD12 in pea root hairs can be induced by treatment with antiauxins such as 2,3,5-triiodobenzoic acid (TIBA) or *N*-(1-naphthyl)phthalamic acid (NPA). Taken together, these data demonstrate that soluble *nod* gene–related factors of low molecular mass influence the hormonal balance in host roots, root (cell) morphology, and flavonoid metabolism. The requirement of *hsn* genes for their production suggests that only host plants will recognize these signal molecules. This is consistent with the observation that *nod* gene–related Had factors active on *Vicia* produced by *Rl viciae* are inactive on the nonhost alfalfa.[59,90] Host plant–specific receptors for these signal molecules have yet to be discovered. It would be interesting to study their relationship to the set of taxonomic characteristics of the tribe Vicieae to which the host plants of *Rl viciae* belong. In addition to specific signal molecules, *Rl viciae* may produce minor amounts of the phytohormones auxin and cytokinin.[241,242] It is, however, still unknown whether phytohormone synthesis by rhizobia is involved in nodulation.

In summary, the host plant releases chemoattractants and *nod* gene–inducing signal molecules and, in return, the rhizobia produce (host-specific) signals that influence the growth and metabolism of root cells. One or more of these signal molecules is a specific mitogen, which probably acts by influencing the endogenous phytohormone balance.

B. Root Hairs, Target Cells for Infection

In host plants of *Rl viciae,* rhizobia migrate towards the nodule primordium by *intra*cellular passage (Figure 9-3). Primary target cells for infection by *Rl viciae* are young, growing root hairs. For these plants, wound entry or infection of epidermal cells has never been documented. Root hairs are tip-growing protuberances of plant root epidermal cells,[45] with high exocytotic and cell wall assembly activities at their top.[184] Epidermal cells forming a root hair are called trichoblasts. Differentiation of trichoblasts starts with preparation for cell division: (1) nuclei of trichoblasts enlarge and synthesize DNA in advance of root hair emergence,[46,165,220] (2) a strong topological

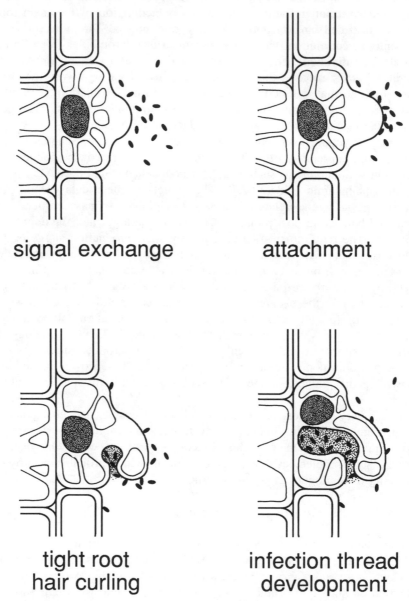

signal exchange

attachment

tight root
hair curling

infection thread
development

Figure 9-3. Early stages of root infection via a root hair by *Rhizobium* bacteria.

correlation exists between the site of root hair emergence and the cell division site in the foregoing root epidermal cytokinesis,[200] (3) root hair emergence is preceded by establishment of a microtubule array radiating from the nuclear envelope and by migration of the nucleus to the presumptive plane of cell division.[9] Next, trichoblasts are arrested in cytokinesis and, instead of a cell division site, a polar tip growth area is established, the future root hair tip. Endoplasmic microtubules connect the nucleus with this polarized area.[9] To enable turgor-driven root hair emergence, the outer periclinal wall of a trichoblast presumably is modified at the site of emergence and a polar mechanism of cell wall synthesis is established. During root hair emergence the nucleus usually migrates in step with the extending hair tip,[47,171] probably because nucleus and hair tip are connected by microtubules.[9,143]

Low-inoculum spot inoculation experiments with a nonmotile mutant of *Rl viciae* have shown that the majority of *Pisum* cells to be infected are localized in the root zone between the smallest emerging root hairs and the hairless root tip[72] and that these cells have yet to be formed at the time of inoculation. Root hair–inducing (Hai) signals produced by the bacteria may enhance or induce differentiation of at least part of the epidermal cells into trichoblasts. In pea, the zone of mature, nongrowing root hairs appeared to be almost completely resistant to infection.[70,72]

Attachment of *Rl viciae* to the root hair surface is considered to be the first step in the more intimate interactions between the bacteria and their primary target cells. In *Pisum,* tips of growing root hairs are preferred sites of attachment.[201] Consequently, attaching rhizobia interfere with the plant's cell wall assembly system. Attachment of *Rl viciae* to pea root hair tips is a two-step process.[205] In the first step, single rhizobial cells adhere to the surface of the root hair. In the second step, other rhizobia adhere to the root hair–bound bacterial cells, which may result in accumulation of rhizobia at the attachment site. The extent of accumulation depends on the number of rhizobia present in the rhizosphere. Different molecules appear to be involved in these two attachment steps. Smit et al.[205] discovered a rhizobial surface protein, designated as rhicadhesin, which probably is involved in binding of the bacteria to the root hair surface of plants. Rhicadhesin is a calcium-binding protein with a M_r of about 14 kDa as determined by SDS–polyacrylamide gel electrophoresis and appears to be common among Rhizobiaceae, including *Agrobacterium*. Sym plasmid-borne genes are not required for its synthesis, which is consistent with the observation that these genes do not play a role in root hair attachment of *Rl viciae* cells.[128,202] The plant receptor in rhicadhesin-mediated attachment has yet to be identified, but since rhicadhesin appeared to be involved in rhizobial attachment to various leguminous and nonleguminous plants, a common plant cell wall component may be envisaged.[205] In addition to rhicadhesin, physicochemical interactions between the surfaces of root hairs and rhizobia may contribute to the attachment process.[160]

Rhizobial cellulose fibrils mediate the accumulation of *Rl viciae* bacteria on pea root hair tips.[202] Fibril-negative mutants show normal nodulation properties; thus, fibril-mediated accumulation is not a prerequisite for successful root hair infection. Under appropriate culture conditions, rhizobial accumulation on pea root hair tips is accelerated by pea lectin molecules.[128] Lectins are sugar-binding (glyco)proteins other than enzymes or antibodies.[11] Pea lectin (Psl) is a 49 kDa protein that is usually described as a glucose–mannose binding lectin but that more specifically contains two high-affinity binding sites for oligosaccharides containing a trimannoside core with a neighboring fucosyl-α-1,6-N-acetylglucosamine group.[60,122] Psl is abundantly present in pea seeds, in which it constitutes a mixture of two isolectins, Psl1 and Psl2, and a small amount of the unprocessed Psl precursor.[232] In pea roots, lectin molecules similar to Psl2 and the Psl precursor are produced and subsequently secreted into the rhizosphere.[73] These molecules most probably are responsible for enhanced accumulation of infective *Rl viciae* cells on pea root hair tips.[72,128] Interestingly, carbon-limited *Rl viciae* cells are uninfective, they cannot be agglutinated by Psl, and, consequently, they do not show lectin-mediated accumulation. This may be explained by poor production of extracellular polysaccharides (discussed later). Prolonged presence in the pea rhizosphere restores the infectivity of these bacteria, suggesting that pea root exudate provides the rhizobia with a carbon source that promotes their ability to infect. Involvement of Sym plasmid-encoded gene products in lectin-enhanced accumulation could not be demonstrated.[128] This is consistent with the observations that several *Rhizobium* species are able to bind Psl[231] and that Psl can precipitate extracellular polysaccharides of various fast-growing *Rhizobium* species.[117]

Soon after binding of *Rl viciae* cells to a growing pea root hair, the root hair responds by curling.[101,169] The presence of living rhizobia at the root hair surface is a prerequisite for induction of tight ("marked") root hair curling.[246] In pea, tightly curled root hairs were found in small groups randomly distributed in the zone of emerging root hairs,[72] whereas other hairs showed various degrees of deformation. Haack[101] found 20–25% tightly curled pea root hairs upon inoculation with *Rl viciae* bacteria and even more upon inoculation with *Rl trifolii*. Other rhizobia, however, including *Rl phaseoli*, are unable to induce significant root hair curling in pea, which shows that induction of tight root hair curling has a high degree of host–plant specificity.

A computer simulation study of root hair curling by Van Batenburg et al.[225] predicted that tight curling would result from dominant induction of tip growth by attached rhizobia. Furthermore, simulation showed that excessive curling (e.g., *Rl trifolii* on pea) or hair deformation (e.g., *R. meliloti* on pea) may result from an inability of the rhizobia to initiate a proper infection site in the root hair curl. The common *nodABC* genes of rhizobia are minimally required for induction of root hair curling.[190] The hypothesis seems to be generally accepted that *nodABC* gene–related signal(s), causing root hair deformation in the *absence* of bacteria, induce hair curling when produced by surface-attached cells. Mutations in the genes *nodI, J, F, E, L, M,* and *N* result in excessive root hair curling and deformation, disturbed cell infection, and/or delayed nodulation.[34,79,80,197,207,245] These observations suggest that tight curling of root hairs induced by *Rl viciae* is a result of the concerted action of common and *hsn nod* gene products and explain the host–plant specificity of this process. Attachment of rhizobial cells at root hair tips may be responsible for the delivery of *nod* gene–related signals at the correct sites on the target cells.[146] In pea, however, infective rhizobia are able to attach to more than 95% of the growing root hairs,[128] whereas only 25% of these show tight curling.[101] This observation points at heterogeneity of the rhizobia with respect to production of curling factors and/or at heterogeneity of the root hairs with respect to signal reception.

The correlation between proper curling and proper cell infection is consistent with the computer simulation model. It is tempting to suggest that the rhizobial curling signal(s) is similar, if not identical, to one or more of the host-specific *nod* gene–related factors excreted by the bacteria after *nod* gene induction. This would add induction of tip growth in polarized root cells to the biological activities of these signals. Apparently, the curling signal of *Rl trifolii* is more closely related to the *Rl viciae* signal than are factors produced by other rhizobia. Inability of other rhizobia to induce marked curling of pea root hairs can also be explained by poor *nod* gene induction and/or by inadequate attachment.

Rhizobia entrapped in a root hair curl may be ingested by the hair cell. Cell infection may also occur at a contact site of two adjacent root hairs, sometimes with both hairs becoming infected. Infections in touching hairs amounted to about 4% of infections for pea.[101] This demonstrates that entrapment of rhizobia rather than hair curling is a prerequisite for ingestion. A detailed description at the ultrastructural level of root hair infection by *Rl viciae* is lacking. However, in view of the data concerning root hair infection by the closely related *Rl trifolii*[31] and pea root cortical cell infection by *Rl viciae*,[9] this process most probably includes (1) host cell wall degradation, and (2) invagination of the host plasma membrane. At the light-microscopic level, the first sign of cell infection is the development of a hyaline spot in the root hair curl. Localized cell wall degradation suggests the involvement of cell wall–degrading enzymes in the infection process. Pectinase, hemicellulase, and cellulase activities were found for *Rl viciae* by Hubbell et al.[114] and Martinez-Molina et al.[148] However, plant cells are also able to induce localized cell wall changes, root hair formation itself being a good example of this ability. Genesis of intercellular spaces at sites of fusion of a cell plate with the existing primary cell wall after cytokinesis requires a subtle and very localized mechanism of wall degradation.[116] The possibility that rhizobia activate the plant's own wall modification system is still open. Moreover, rhizobial signal molecules may interfere with the cell wall assembly mechanism at the root hair tip, resulting in local absence of cell wall material. Early reports on specific induction of host polygalacturonase by rhizobial factors[142] merit reexamination in light of the present state of knowledge.

Lectin is probably involved in the initiation and development of a rhizobial infection site. In pea, the pattern of Psl location entirely corresponds with the susceptibility of root epidermal cells to infection by *Rl viciae*.[72] Whereas the surface of the pea root tip is Psl-negative, lectin is present on the surface of epidermal cells at the time of their differentiation into trichoblasts. Trichoblasts with emerging root hair tips represent the majority of Psl-positive root epidermal cells (Figure 9-4) and are distributed around the root without preference for certain regions.[72] Older Psl-positive hairs are grouped opposite the (proto)xylematic poles of the central cylinder,[70] whereas the surface of the mature root hair zone and the zone between the root hairs and the cotyledons is characterized by absence of Psl and by an almost complete resistance to rhizobial infection. Extracellular Psl on the surface of pea roots is capable of binding specific saccharide receptor molecules.[72,74] Study of slightly plasmolized trichoblasts demonstrated that Psl predominantly is present at the external surface of the plasma membrane and that functional lectin molecules usually are absent from the root hair cell wall surface. These observations demonstrate that rhizobia may interact with both secreted and membrane-located Psl molecules.

Psl is encoded by a family of four genes, of which only one gene appears to be functional.[119] The nucleotide sequence of the *psl* gene has been reported by Gatehouse et al.[95] and Kaminski et al.[119] Transformation of white clover roots with a functional pea lectin gene confers upon these roots the ability to be nodulated by *Rl viciae*,[71] although nodulation is limited and delayed in time. In transformed white clover roots, Psl appeared to be present at sites similar to those on pea roots.[72,74] Because *Rl viciae* normally induces root hair curling in white clover but does not infect the cells, these results strongly suggest that Psl somehow contributes to the pea–*Rl viciae* host plant specificity barrier at the level of ingestion of the bacteria by root hair cells.

Initial uptake of rhizobia in a root hair is followed by the formation of a growing tubular invagination of the host plasma membrane, accompanied by deposition of cell wall material on subapical regions of the invaginating membrane. The resulting structure is called an infection thread and contains the invading rhizobia. Infection thread initiation coincides with an arrest of

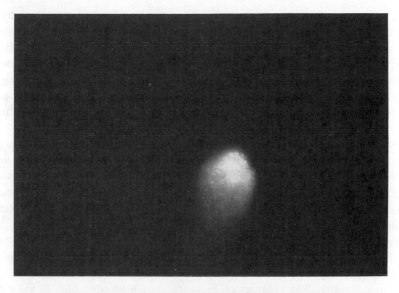

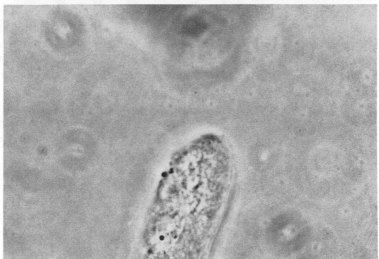

Figure 9-4. Tip of a growing pea root hair showing pea lectin as visualized by indirect immunofluorescence, with the corresponding phase-contrast image.[70]

tip growth of the root hair. In pea, infection threads are relatively broad, with a diameter of about 3–4 μm, showing several bacteria in a cross section. Cell divisions of rhizobia predominantly take place at the top of the thread. As with root hair growth, stable growth of an infection thread requires the formation of a persistent tip growth area. The hair nucleus migrates in step with the tip of the infection thread to the base of the root hair.

Apparently, the nucleus is uncoupled from the hair tip and recoupled to the advancing tip of the thread.[143] Some infection threads may branch or anastomose; others abort. Irregular growth of some threads may reflect instabilities in the interactions of the rhizobia with the host cell, resulting in unstable properties of the plasma membrane at the thread tip. Abortion of thread growth may result from a discontinuity in molecular communication between both partners and/or from rhizobial growth limitations in the thread tip.

Notably, a rhizobial infection thread in pea roots contains not only rhizobia but also a matrix whose origin is unclear. Electron micrographs demonstrate that the contact area between the host plasma membrane and the thread matrix is much larger than that between the membrane and the rhizobial surface. Apparently, the host cell ingests the growing matrix rather than the bacteria. This is consistent with the observation that host root hairs normally do not phagocytose individual *Rl viciae* cells. Bradley et al.[25] demonstrated that one component of the infection thread matrix in pea is a plant-derived glycoprotein.

Extracellular polysaccharides of *Rl viciae* may also contribute to the thread matrix. Rhizobia are Gram-negative bacteria with a cell envelope consisting of an outer membrane, a cytoplasmic membrane, and a periplasmic space in between. The bacteria may excrete extracellular material, the amount and composition of which depends on the nutritional conditions. Part of this material is more or less soluble, and it can be either recovered from the growth medium or easily washed off from the cells. Excreted material may also bind to the cell surface as a relatively insoluble bacterial capsule. Polysaccharides constitute an important part of the rhizobial surface components, including lipopolysaccharides (LPS) from the outer membrane, capsular polysaccharides (CPS) at the cell surface, and acidic exopolysaccharides (EPS) in the growth medium.[35] Furthermore, Rhizobiaceae are characterized by the production of cyclic glucan molecules, present in the periplasmic space and/or excreted into the growth medium (e.g., Dell et al.[66] and Miller et al.[159]). Each of these polysaccharides has a role in root infection via root hairs. However, the evidence primarily results from the study of the nodulation behavior of polysaccharide mutants, and pleiotropic effects may confuse the conclusions. Such genetic evidence suggests that rhizobial extracellular polysaccharides are involved in infection thread formation. *Rl viciae* secretes a high-molecular-weight acidic EPS that causes the slimy appearance of its colonies on solid media. Several nonmucoid EPS-minus mutants of *Rl viciae* fail to nodulate pea;[22,75] others, however, are infective. EPS consists of octasaccharide units containing galactose, glucuronic acid, and glucose in a molar ratio of 1:2:5 with pyruvate, acetate, and 3-hydroxybutyrate substituents.[174] *Rl viciae* EPS closely resembles the EPS of *Rl trifolii* but differs in that a higher number of galactose and glucose residues are substituted for by 3-hydroxybutyryl and acetyl groups, respectively. Interestingly, EPS structure may be influenced by *nod* genes. Transfer of the *hsn* genes *nodFERL* and, particularly, *nodMN* from *Rl trifolii* into *Rl viciae* changed the acylation pattern of EPS and resulted in a hybrid strain that, in contrast to the parent strain, exhibited efficient infection and nodulation of white clover, production of the *Rl trifolii*-type EPS, and an increased proportion of cells binding white clover lectin, trifoliin A.[175] Biovar-specific domains in EPS may contribute to the host–plant specificity of root hair infection. On the other hand, Sym plasmid–borne *nod* genes are not involved in determining either the glycosyl sequence of EPS or the location of noncarbohydrate substitutions.[175] Furthermore, it was shown that the NodE protein, and not the NodMN products, distinguishes the host range of nodulation of *Rl viciae* and *Rl trifolii*.[209]

Rather than EPS, LPS of *Rl viciae* has been reported to have biovar/ species-specific properties as visualized by its specific interaction with pea lectin.[117,118,243] LPS, a major constituent of the rhizobial outer membrane, contains three regions: lipid A, a core oligosaccharide, and an *O*-antigenic chain. The *O*-antigenic part is highly variable, even among strains of the same biovar/species;[36,252] moreover LPS on the surface of one cell may be very heterogeneous in composition. Mutants of *Rl viciae* completely lacking the *O*-antigen part of LPS are still able to attach to root hair tips[204] and to induce root hair curling, infection thread formation, and nodule initiation.[68] This implies that complete LPS (so-called LPS I) of *Rl viciae* is not required for early nodulation steps. Small biovar/species-specific structures may be present in LPS that as yet have not been detected. These structures, however, may also form part of other polysaccharides present as contaminants in LPS preparations. Certain capsular polysaccharides are thought to be attached to lipid A.[23] In the studies of Wolpert and Kamberger, LPS has been isolated from early-stationary-phase rhizobia grown in yeast–mannitol medium and, interestingly, in the absence of *nod* gene–inducing flavonoids. *Rl viciae* bacteria grown under these conditions are highly infective.[128]

Taken together, these observations accentuate the need for a more detailed analysis of EPS and LPS structures and for characterization of various EPS and LPS mutants. Mutual interactions between certain *hsn* genes and chromosomal genes encoding surface polysaccharide structures are still far from clear. For such studies, one should start from the notion that entrapped rhizobia may produce different surface components than rhizobia cultured under laboratory conditions, and that plant factors may change the size and composition of rhizobial surface polysaccharides.[206]

In summary, after attachment *Rl viciae* redirects tip growth in young root hairs resulting in development of an infection thread. Host-specific *nod* gene–related factors (released and/or associated with rhizobial surface polysaccharides), lectin, and thread matrix constituents are probably involved in tight hair curling and infection thread initiation.

C. Infection Thread Development

Observations on infection threads in pea root cortical cells indicate that persistent physical contact between *Rl viciae* cells and the host plasma membrane is not a prerequisite for continued infection thread growth.[9] Probably, extracellular signal molecules in the thread matrix continuously contribute to the interactions between the bacteria and the plasma membrane. These and/or other signal molecules also initiate the nodule primordium and condition the host cortical cells for future infection. It should be noted that in pea, infection thread formation precedes nodule initiation.

Approximately 48 hours after inoculation with infective *Rl viciae* cells, all pea root cortical cells in radial areas adjacent to an infected root hair respond with morphological changes. First, these highly vacuolated cells establish a nuclear-envelope (NE) radiating microtubule array that results in the formation of cytoplasmic strands that radiate from the nucleus.[9] Simultaneously, the orientation of the interphase microtubule array changes from a circumferential into a longitudinal position, an additional wall layer is deposited onto the original host cell wall, and the cells increase in size. The cell nuclei swell and migrate to the center of the cell. The induction of these morphological changes in pea cortical cells is accompanied by the expression of a host gene PsENOD12, encoding a secretory (hydroxy)proline-rich protein composed of repeating pentapeptides.[193] An ENOD12 transcript is also present in inoculated root hairs but is absent in root hairs from uninoculated pea seedlings. The ENOD12 protein may contribute to the formation of the additional wall layer and, later, to the formation of the infection thread wall. As already mentioned, expression of PsENOD12 can be induced by a soluble compound(s) produced by *Rl viciae* in the synthesis of which both common and host range *nod* genes appear to be involved (see Chapter 15).

In vacuolated plant cells, establishment of a NE-radiating microtubule array and swelling and migration of the nucleus to the cell center normally precede cell division. Indeed, the *inner* cortical cells in an infection area of pea roots enter mitosis after formation of a transvacuolar phragmosome within the plane of cell division. Particularly, cortical cells opposite a xylem pole are susceptible to induction of cell divisions.[140] The initial division of these cells is always anticlinal.[139,169] In contrast, *outer* cortical cells closer to the advancing infection thread do not divide and possibly become arrested in the pre-prophase of the cell cycle. In these cells, the NE-radiating microtubule array becomes anticlinically positioned in the center of the cell. The resulting transvacuolar structure is called a cytoplasmic bridge[9] (Figure 9-5). Apparently, the position of the cortical cells in the infection area deter-

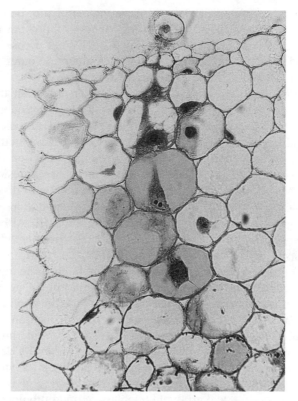

Figure 9-5. An infection area in a pea root, showing a
curled root hair and radially positioned cytoplasmic bridges.[9]

mines whether the nucleus-radiating strands fuse into a premitotic phragmosome (inner cortical cells) or into a cytoplasmic bridge (outer cortical
cells). Because these host cell responses occur in advance of the approaching
infection thread, activity of diffusible rhizobial signals or endogenous secondary signals must be assumed. Root infection *per se* appears to be redundant for a host response, since induction of cell divisions in pea root
cortical cells by *Rl viciae* bacteria unable to induce infection thread formation has been described.[134] Rhizobial signals involved in induction of host
cell divisions or in expression of PsENOD12 may be the same compounds.
Some of these compounds may be associated with rhizobial LPS, since it
was shown that LPS is able to replace cytokinin in inducing cortical cell
divisions in pea root explants.[126] Furthermore, plant signals are involved. A
plant factor from the root vascular tissue has been found to act in concert
with plant hormones for induction of cell divisions in inner cortical cells
opposite the (proto)xylematic poles in pea roots, similar to the situation with
nodule primordia.[140] Furthermore, Phillips[176] suggested that an inhibitor,

possibly abscisic acid, is translocated in the phloem from the cotyledons to the root and restricts cell divisions in the vicinity of phloem regions. Endogenous gradients of nutrients and growth regulators across the pea root cortex may determine the different behavior of inner and outer cortical cells in response to rhizobial stimuli (the gradient hypothesis[140]).

In pea roots, crossing of host cell walls by an infection thread occurs invariably opposite the axis of a cytoplasmic bridge.[9] Frequently, the site of fusion of the infection thread tip with the plasma membrane at the inner periclinal wall of an already infected cell does not coincide with the axis of the bridge in the adjacent cell. In these cases, the infection thread grows along the inner wall of the infected cell until it reaches a position opposite the bridge. At this site, both adjacent host walls gradually disappear and a pore is formed. Apparently, the site of cell wall degradation and infection thread passage is determined by the neighboring host cell. After penetration of the cell wall pore by the bacterial cells, the infection thread matrix contacts the plasma membrane of the neighboring cell. A tip-growth area is initiated, resulting in an invagination of the membrane. The diameter of the tubular membrane invagination usually exceeds that of the penetration pore and is similar to the diameter of the infection thread in previously infected cells. Infection thread initiation is accompanied by deposition of newly synthesized wall material in subapical regions of the invaginating membrane. At the very tip of a growing infection thread ultrastructurally detectable wall material is absent. Prominent features of the cytoplasm surrounding the thread tip are a well-developed rough endoplasmic reticulum, active dictyosomes, vesicles, mitochondria, and array of endoplasmic microtubules running from the host nucleus to the infection thread. These host organelles may be involved in membrane flow to the thread tip, in supply of matrix components and of precursors for thread-wall synthesis. The infection thread grows through the cytoplasmic bridge, passes the cell nucleus, and eventually reaches the inner periclinal wall of the cell. Infection of the neighboring cell proceeds in an identical way, as described earlier (Figure 9-6).

As mentioned earlier, formation of a nodule primordium in *Pisum sativum* begins with cell divisions in the inner cortex at some distance (about five cell layers) from the tip of the advancing infection thread.[139] New mitotic activity is initiated in neighboring cortical cells with the development of a centrifugally growing meristematic area as a result. The infection thread branches in the center of the expanding meristem, but meristematic cells and cells in the peripheral layers of the nodule primordium usually do not contribute to infection thread formation. This indicates that at this stage of the infection process rhizobial proliferation is confined to the intercellular space. However, infection thread formation is again initiated in swelling cells produced by the meristem.[125] In these cells, contact between the host plasma membrane and the thread matrix usually does not result in the establishment

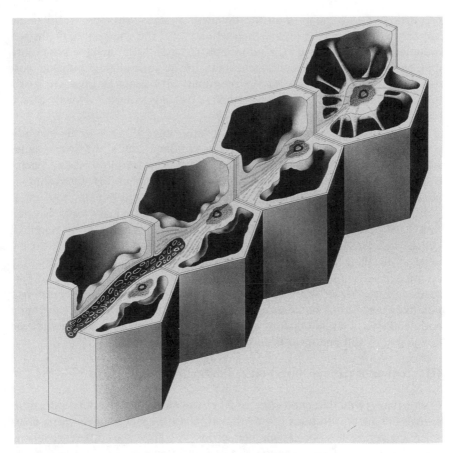

Figure 9-6. Three-dimensional representation of the cytoplasmic rearrangements prior to and during infection of pea root cortical cells by *Rhizobium*.[9]

of a persistent tip-growth area. Irregular thread structures are formed, including unwalled "droplets" or bulges of matrix material. From these bulges the pea cells ingest the *Rl viciae* bacteria but, remarkably, little, if any, matrix material.[125] As a result, the bacteria find their way into a cell compartment surrounded by a membrane of host origin, the symbiosome.[188] At this stage of nodulation, *Rl viciae* mutants with an altered LPS structure show symbiotic defects, a.o. impairment in release from the infection thread.[68] Chromosomal Tn5 mutants of *Rl viciae* lacking the O-antigenic polysaccharide of LPS either are resistant to phagocytotic uptake by the host cells[68] or degenerate prematurely shortly after ingestion.[178] Apparently, the O-antigenic polysaccharide is involved in the interactions with the host (plasma) membrane. A physical association between the symbiosome membrane and LPS from ingested *Rl viciae* cells has been demonstrated.[24] Physicochemical properties of LPS rather than the detailed chemical structure seem to be important for correct membrane interactions.[68] On the other hand, O-antigen mutants of *Rl viciae* have a pleiotropic phenotype,[68,178,204] and the possibility that other rhizobial molecules contribute to the interactions with the host membrane cannot be ruled out. Furthermore, LPS is not an invariant structure and may be subject to developmental changes during growth of the bacteria in the plant tissue.[230,244]

After uptake, *Rl viciae* cells divide several times along with the symbiosome membrane. Each bacterium remains confined singly in a host vesicle. Afterward, the bacteria adopt enlarged pleiomorph forms, so-called bacteroids, and start to reduce N_2 into ammonia (see Chapter 10). Interestingly, about half of the nodule cells do not ingest rhizobia and remain uninfected.

In summary, infection threads are formed in polarized root cells arrested in a stage before cytokinesis. Polarization is probably induced by rhizobial signals and closely resembles trichoblast differentiation. In addition to the vertical morphogenetical gradient that determines development of dividing epidermal cells into polarized trichoblasts, the rhizobia contribute to formation of a horizontal gradient with cell divisions in the inner cortex and polarized cells in the outer cortex. Apparently, cells produced by the young nodule meristem are not (sufficiently) polarized, with abortive infection thread development and endocytosis as results.

III.   Other Rhizobial Infections: Similarities, Differences

Comparison of the infection story of *Rl viciae* with other rhizobial infections reveals (1) general features of nodulation, and (2) specific phenomena characterizing certain host–bacterium interactions. A conspicuous difference among rhizobial infections concerns the site of root infection.

A.   Cracks, Cell Junctions, or Root Hairs

Bradyrhizobium parasponiae infects *Parasponia* (Ulmaceae), the only nonlegume plant genus known to form N_2-fixing root nodules with rhizobia.[217] Efficient nodulation is host plant–specific, since *Parasponia* is not readily nodulated with rhizobia from legumes. Upon inoculation, swollen multicellular root hairs are formed by the plant, and cell divisions are induced in outer cell layers of the root cortex. Continuous cell divisions cause development of callus-like bumps and separation of epidermal cells, especially at the basis of multicellular root hairs. As a result, the intercellular space of the root becomes (locally) continuous with the outer space ("crack formation"), and rhizobia invade the root ("crack entry"). Rhizobial proliferation in primary cell walls of dividing root cells suggests activity of cell wall–degrading enzymes. Extensive colonization of intercellular areas in the outer

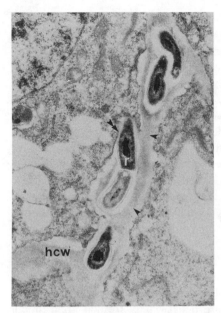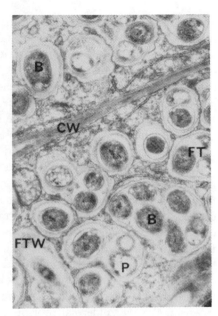

Figure 9-7. (a) A single file of intercellular rhizobia (r) in the cell wall (hcw, single arrows) of a *Parasponia rigida* root nodule. The double arrows indicate direct rhizobial contact with the host plasmamembrane.[131] (b) Cross-sectioned fixation threads (FT) in a *Parasponia rigida* root nodule. B: bacteroids; CW: host cell wall; FTW: fixation thread wall; P: poly-beta-hy-droxybutyrate.[177]

cortex may result in plant cell damage. However, proliferation of rhizobia between dividing cortical cells appears to be better controlled, and intracellular rhizobia in single files surrounded by plant cell wall material can be observed. Genuine infection threads are not formed, and rhizobia are not endocytotically released from the elaborate thread-like wall ingrowths.[131,215] Rhizobia within these ingrowths fix N_2, which led to use of the name *fixation threads* (Figure 9-7). Intercellular spread of rhizobia and formation of fixation threads have also been described for root nodules of many Caesalpinoideae-genera and of some Papilionideae like *Andira*.[61,62] *Parasponia* nodules contain a central vascular area surrounded by the symbiotic zone and resemble a modified lateral root. This type of nodulation is generally considered to represent a primitive rhizobial infection process (see Chapters 12 and 13), contrasting with the "sophisticated" nodulation process in pea (see earlier).

Other rhizobia–legume interactions result in intermediate nodulation types. *Arachis* (groundnut), a rather promiscuous host plant, and the more specialized *Stylosanthes,* a tropical forage legume, are nodulated from sites of lateral root emergence. Cell divisions are induced in the cortex of an emerging lateral root.[41,42] Growth of the young root tip causes separation of cortical and epidermal cells and enables entry and intercellular spread of rhizobia. Interestingly, *Arachis* produces root hairs only at these root junctions.[41] Upon inoculation, these hairs deform and curl to a certain extent. Large hypertrophic cells are situated at the root hair base, which become infected by rhizobia entrapped between these and neighboring cells. These infections however, are abortive. In *Stylosanthes,* such hypertrophic cells are lacking, and, characteristically, outer root cells penetrated by rhizobia die. In both genera, no infection threads are formed, and rhizobia colonize the root apoplast presumably by cell wall digestion or, in *Stylosanthes,* by progressive collapse of outer root cells. Eventually the bacteria are endocytotically ingested by dividing host cells from a modified irregular cell wall matrix into membrane-bound packets. Continuous host cell divisions result in development of a uniformly infected central nodule tissue. Also stem nodulation of *Sesbania rostrata* by *Azorhizobium caulinodans* starts at infection sites formed by lateral root primordia.[4]

In *Mimosa scabrella,* a tropical tree, rhizobial infection sites are junctions of epidermal cells.[63] The bacteria penetrate the radial walls and proliferate intercellularly. Next, a nodule primordium is induced in the inner cortex. Interestingly, *sub*epidermal root hairs are occasionally formed upon inoculation. Penetration of outer root cells starts from irregular cell wall ingrowths and usually results in an incompatible interaction like the situation in *Stylosanthes*. In contrast, the bacteria are "normally" ingested by cells of the developing nodule meristem and a healthy nodule is formed.

Junctions of differentiating trichoblasts are targets of rhizobial infection in the *Glycine* (soybean)-*Bradyrhizobium* symbiosis.[222] In the presence of rhizobia, root hairs curl shortly after emergence, thereby entrapping the bacteria in the cell junction. The rhizobia do not penetrate the walls between the trichoblasts, but initiate infection thread formation at the root hair base. Infection threads in soybean are relatively narrow, with a diameter of about 1 μm, and contain a single row of linearly aligned rhizobia (Figure 9-8). The rhizobial cell in the tip of the thread is in direct contact with the plasma membrane of the plant cell. The infection thread wall is also present at the tip of the thread, and only a limited amount of infection thread matrix material, if any, can be observed. Before infection thread formation, cell divisions are induced in the outer cortex of the root.[33] After passage through the root hair, the bacteria infect dividing cortical cells from sites of fusion of a young cell wall. Bacteria are endocytotically released directly from the tip of an infection thread,[98,189] simultaneously with the formation of infection-thread wall degradation vesicles.[189] Continuous cell proliferation eventually results in formation of a central nodule tissue containing about 50% uninfected cells.

These different infection processes, including the nodulation type of *Rl viciae*, have several essential features in common: (1) induction at a distance

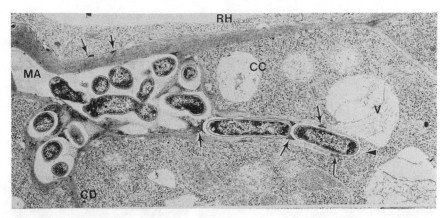

Figure 9-8. Development of a narrow infection thread in soybean. RH: root hair cell base; CC: subadjacent cortical cell; MA: intercellular matrix; CD: young cell wall; V: vacuole. The arrows indicate unidentified dark material in the walls and between the bacteria.[222]

of cell divisions in young root cells by rhizobia, (2) a compatible interaction of rhizobia with root cells either formed (nodule cells) or differentiating (trichoblasts) under rhizobial influence, and (3) no interaction or an incompatible interaction of rhizobia with differentiated root cells. Two different mechanisms of root invasion are apparent: (1) penetration of primary cell walls, and (2) formation of infection threads (i.e., tip-growing structures enabling passage *through* cells). The latter type of root invasion includes formation of narrow infection threads containing little thread matrix (e.g., soybean, bean) or broad matrix-rich threads (e.g., pea, clover, alfalfa). The broad type of infection thread is only formed in vacuolate polarized cells such as trichoblasts and trichoblast-like cortical cells. Interestingly, for formation of both types of infection threads the plane of plant cell division appears to be a determinative factor. Narrow threads start from junctions with a young primary cell wall, whereas broad threads start from a tip-growth area that develops in the plane of cell division.

Some (if not all?) rhizobia able to invade roots via infection thread formation apparently also have the ability to invade by crack entry. *Rl viciae* nodulates the heterologous subterranean clover, although the bacteria do not initiate infection thread formation in root hairs, hence do not induce tight curling of root hairs, and do not interact with clover lectin.[113] Nodules are formed at the sites of emergence of lateral roots. These nodules, however, do not fix N_2, and the nodule tissue senesces prematurely, like the situation with certain LPS mutants of *Rl viciae* in their own host plants.

Unfortunately, molecular data on rhizobial infection systems other than those of *R. leguminosarum, R. meliloti,* and *B. japonicum* (all root hair infections) are scanty or lacking. Existence of general molecular principles is demonstrated by the observation that the essential *nodABC* genes are present in "broad-infection-thread" rhizobia, in "narrow-infection-thread" rhizobia, and in "crack-entry/fixation-thread" rhizobia like *B. parasponiae*.[150] This observation most probably relates to the general property of rhizobia to induce cell divisions at a distance in host roots. Thus, the following discussion of similarities and differences in nodulation will necessarily be restricted to "broad-infection-thread" and "narrow-infection-thread" types of root hair infection. Evidently, more knowledge of other rhizobial infection processes is needed. Detailed study of the molecular biology of crack entry may well result in development of a better strategy for increasing the range of N_2-fixing systems than study of root hair infection. After all, *Parasponia* is a nonlegume, and the highly effective *Bradyrhizobium-Parasponia* symbiosis is capable of fixing up to a respectable 850 kg N ha^{-1} yr^{-1} under controlled conditions.[216]

B. Signal Exchange Before Cell Infection

As mentioned earlier, key features of nodulation by *Rl viciae* are (1) chemoattraction; (2) activation of *nod* genes; (3) production of rhizobial signals and surface polysaccharides; (4) attachment; (5) host responses: cortical cell divisions, root hair curling, and infection thread initiation; (6) infection thread growth; and (7) endocytotic release.

A chemotactic response to various sugars and amino acids is common among rhizobia.[99,147] For *R. meliloti,* chemotaxis genes were mapped on the chromosome.[254] Several *nod* gene-inducing flavonoids appear to be good chemoattractants,[3,28] and the *nodD* gene is involved in this response.[28] These observations are similar to the situation with the closely related *Agrobacterium tumefaciens,* the crown-gall bacterium. *Agrobacterium* is chemotactic to a range of sugars and amino acids[144] but is more sensitive to phenolic inducers of its virulence genes. This response requires the Ti plasmid-borne genes *VirA* and *VirG*[196] that, similar to *nodD* for rhizobia, encode the sensor-activator system for plant signal molecules. The molecular connection between chemotaxis and activation of nodulation/virulence genes is not yet understood. Electrotaxis may add to positioning of rhizobia at target sites for root infection.[156]

Motility of rhizobia appears to be an essential factor in competition for nodulation rather than an essential factor for nodulation itself. Following separate inoculations, motile and nonmotile strains of *Rhizobium* are equally successful in nodulation. However, upon inoculation with mixtures of strains, a motile rhizobial strain forms several times more nodules than its nonmotile derivative.[155]

NodD-dependent inducibility by plant flavonoids is generally observed for rhizobial *nod* genes, and various inducers have been identified (e.g., 7,4′-

dihydroxyflavone for *R. trifolii*,[183] luteolin and 4,4'-dihydroxy-2'-methox-ychalcone for *R. meliloti*,[153,172] and the isoflavones genistein and daidzein for *B. japonicum*.[10,124] The broad-host-range *Rhizobium* strain NGR234 responds to a broad spectrum of phenolic inducers, including flavones, flavanones, isoflavones, coumestans, coumarins, and hydroxybenzoic acids.[12] Interestingly, isoflavones are anti-inducers for *Rl viciae nod* genes, and *nod* inhibitors also have been found for other rhizobia (e.g., see ref. 173). For white clover roots, and possibly for other legumes, the amount and the site of release of inducers and anti-inducers appeared to vary over the root.[187] Biovar-specific induction of increased *nod*-gene-inducing activity in root exudate, as reported for *Vicia sativa* ssp. *nigra,* has not (yet) been reported for other host–rhizobia combinations. On the contrary, white clover roots show a similar response upon inoculation with nonhomologous *Rhizobium* strains.[187] For soybean, inoculation did not alter the ability of soybean root exudate to induce the *nodYABC* operon of *B. japonicum*.[214]

The observations on chemoattraction and *nod* gene activation by excreted legume root flavonoids strongly suggest that root infection by rhizobia is preceded by induction of *nod* genes in the rhizosphere. The minimal level of *nod* gene expression required for rhizobial infection of host roots remains to be determined. Notably, *nodD* gene products of different rhizobia recognize different sets of inducer molecules.[100,111,208] This interaction is involved in both the determination of host-specific nodulation, as has been shown for the *Rl trifolii*-red clover symbiosis,[208] as well as in optimizing the association with each of different host plants, as appears to be the case for *R. meliloti*.[110]

After activation of *nod* gene transcription by appropriate flavonoid inducers, rhizobia release symbiotic signal molecules. The major alfalfa-specific signal of *R. meliloti* has recently been identified as being a sulfated beta-1,4-tetrasaccharide of D-glucosamine in which three amino groups are acetylated and one is acylated with a C_{16} bis-unsaturated fatty acid (M_w 1102).[137] (Figure 9-9). The common *nodABC* genes as well as the *hsn nodH* and *Q* genes are involved in the production of this signal molecule, designated NodRm-1. A NodRm-1 solution in the range of $10^{-8}-10^{-11}M$ induces root hair deformation and branching on alfalfa but not on the heterologous host plant *Vicia*. Breaking of the glycosidic bonds results in a complete loss of Had and Hab activity. The significant root hair branching activity of NodRm-1 suggests that this signal is an inducer of tip growth and that it most probably is involved in tight curling of root hairs and in infection thread initiation if produced by attached rhizobia. NodRm-1 also appears to be a mitogen. If an acylated oligosaccharide structure turns out to be a general characteristic of host-specific rhizobial signals, legume lectins, characterized by specific sugar binding sites and conserved hydrophobic binding sites, may be good candidates for representing signal receptor molecules.[129] An

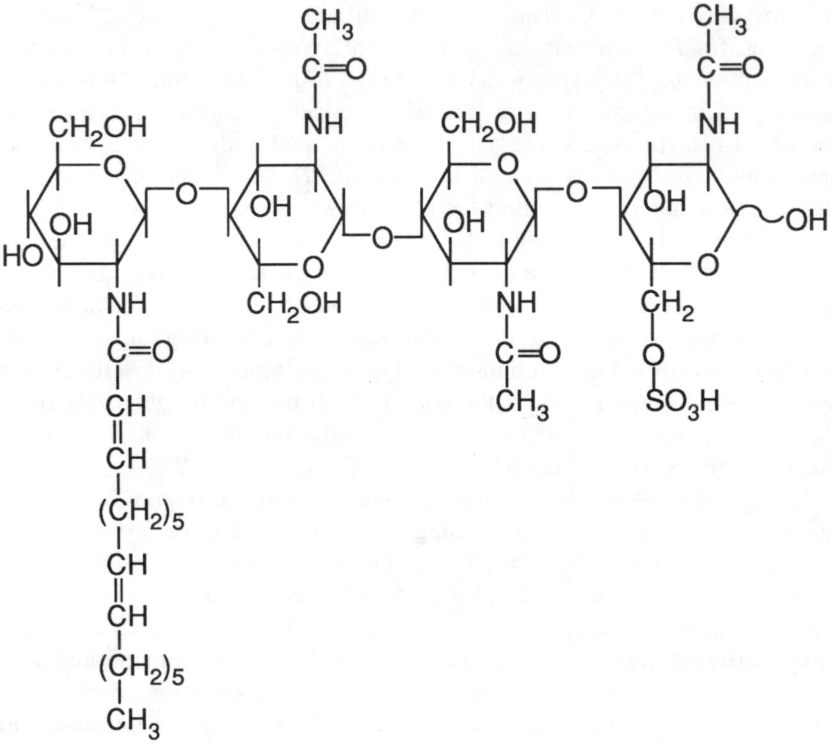

Figure 9-9. NodRm-1, an alfalfa-specific signal molecule produced and released by *Rhizobium meliloti*.[137]

alfalfa root lectin has not yet been isolated or localized. Now that NodRm-1 has been identified, the question of how rhizobia elicit morphogenetic responses in host plant cells may be answered in the near future.

These observations demonstrate that signal exchange is a general property of legume–rhizobia interactions and that molecular determination of host plant specificity of rhizobial infection starts outside the plant root.

C. Interactions at the Root Hair Surface

Root epidermal cells susceptible to infection by root-hair-infecting rhizobia are generally localized in the zone between the smallest emerging root hairs and the root tip of young seedlings, as shown for pea,[72] soybean,[17] *Vigna sinensis,* alfalfa, white clover and *Arachis hypogaea*,[18] *Glycine soja* and *Vigna unguiculata*.[179] This is the zone of trichoblast differentiation. In white clover, *Rl trifolii* induces branching of mature root hairs, thereby inducing additional infection sites in a root zone that is usually resistant to rhizobial infection.[20] The increase in number of legume root hairs upon inoculation with rhizobia may result from enhancement of trichoblast differentiation by rhizobial signals.

Rhizobia, like agrobacteria, are able to attach to the surface of both legume and nonlegume cells, including root hairs. Attachment of rhizobia to nonhost plants is commonly designated *nonspecific attachment*. The two-step model for rhizobial attachment—as proposed by Smit et al.[205] for *Rl viciae* (see earlier) and, in a comparable form, by Dazzo et al.[57] for *Rl trifolii*—is similar to the model proposed by Matthysse et al.[151] for attachment of *A. tumefaciens* to carrot tissue culture cells. Since rhicadhesin activity has been found for various rhizobia and the protein has also been isolated from *Agrobacterium*, rhicadhesin is a good candidate for mediating attachment of Rhizobiaceae to plant cells.[203] Like agrobacteria, various rhizobia produce cellulose fibrils. These fibrils are usually involved in bacterial floc formation. Cellulose-mediated accumulation of rhizobial cells at root hair tips may thus be considered as a special case of flocculation,[65,166] resulting in a firm cell-to-cell adhesion. For *B. japonicum*, fimbriae (pili) appear to play a similar role, as do cellulose fibrils for other rhizobia.[237–239] A correlation was found between presence of rhizobial flocs and successful infection of root hairs.[166] However, accumulation of rhizobia actually accentuates successful attachment of single cells to the root hair surface[203] and may only represent a secondary phenomenon resulting from heavy inoculation. Moreover, neither cellulose fibrils nor pili are essential for nodulation, since rhizobial mutants lacking either surface appendages show normal nodulation behavior on their respective host plants. Also for *Agrobacterium*, production of cellulose fibrils is not essential for establishing an intimate association with host plant cells.[152] Up to now, rhicadhesin-negative mutants of Rhizobiaceae have not been identified, and an absolute requirement of rhicadhesin for nodulation (either by root hair infection or by other ways) and/or plant tumor formation has not been shown (yet).

Additional involvement of root lectin in attachment may be a common phenomenon for root-hair-infecting rhizobia. Particularly *Rl trifolii*–clover root hair interactions have been extensively studied by Dazzo and coworkers (for reviews see refs. 55–58). Like pea lectin in *Pisum*, clover lectin (trifoliin A) is synthesized in seeds and roots of the plant and is excreted into the rhizosphere. In contrast to pea lectin, however, trifoliin A is present on the outer cell wall surface of growing root hairs, where it may mediate adhesion of adjacent clover root hair tips.[54] This observation explains differences in the results of attachment studies with clover and pea root hairs. For example, attachment of *Rl trifolii* cells to the surface of clover root hairs is strongly inhibited by the trifoliin A-hapten 2-deoxy-D-glucose,[51,255] whereas pea lectin-haptenic monosaccharides do not inhibit attachment of *Rl viciae* cells to the surface of pea root hairs. Attachment of *Rl trifolii* cells to clover root hairs shows a characteristic pattern that is biovar-specific, in contrast

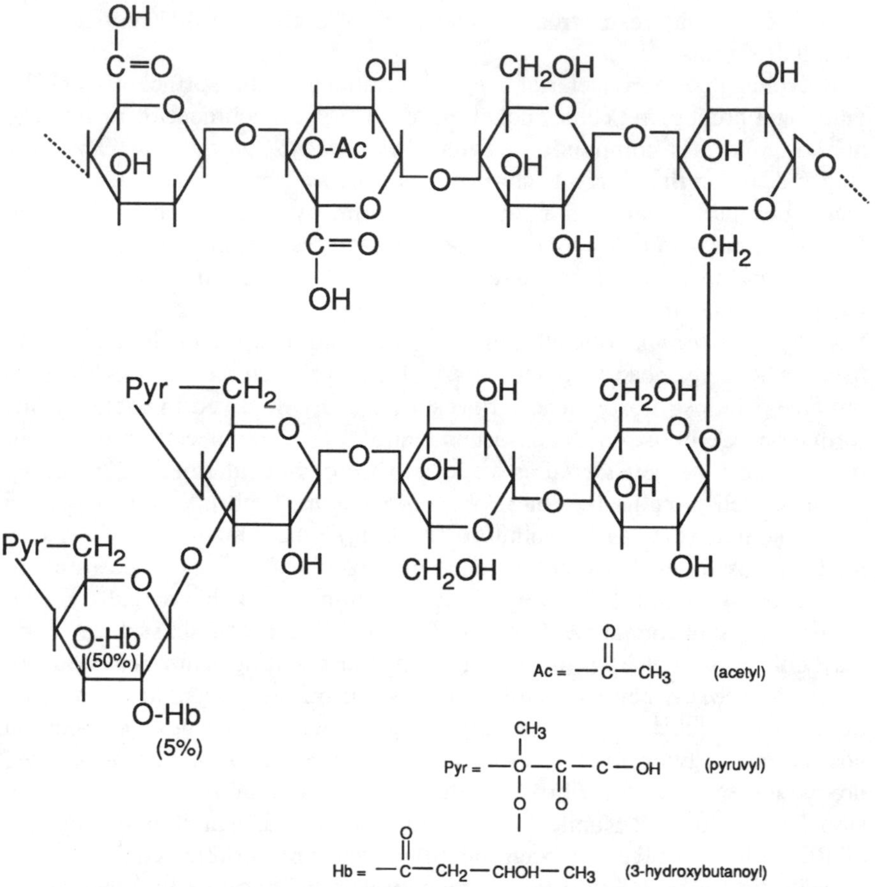

Figure 9-10. The octasaccharide unit of the trifoliin A-binding capsular polysaccharide of *Rhizobium leguminosarum* biovar *trifolii* strain 843. Fifty percent of the oligomers are acylated with D-3-hydroxybutanoyl groups, whereas about 5% of the oligomers bear a second D-3-hydroxybutanoyl group.[106]

to the situation with *Rl viciae* and pea root hairs. *Rl trifolii* bacteria produce several polysaccharide receptors for trifoliin A, which change with culture age.[53,199] Lectin binding sites are present in LPS[112] and in the acidic capsular polysaccharide (EPS/CPS)[106] (Figure 9-10) depending on the culture conditions. Genes essential for expression of lectin receptors on the bacterial surface are located on the *Rl trifolii* Sym plasmid (see also refs. 175 and 255). These observations show that lectin-mediated attachment of *Rl trifolii* cells, grown under appropriate culture conditions, is one of the features of the host-specific interaction with clover roots.

Lectin is also produced by and present on the surface of young soybean roots.[93,94,240] Soybean root lectin is similar, not identical, to the well-characterized soybean seed lectin SBA (or SBL), but it exhibits an identical carbohydrate specificity. *B. japonicum* has two major subspecies that differ in physiology, in DNA homology grouping, and in the nature of extracellular polysaccharide.[109] Binding of SBA appears to be specific for one of the *B. japonicum* subspecies (type I) and some related bacteria and is reversible by hapten inhibition with D-galactose or *N*-acetyl-D-galactosamine.[15,21] SBA-binding depends on the growth phase of the bacteria[19,161] and is significantly influenced by rhizosphere conditions.[16] Cells of this *B. japonicum* subspecies produce two forms of EPS with SBA-binding activity, a capsular heteropolysaccharide containing both terminal D-galactosyl and 4-O-methyl-D-galactosyl residues, and a similar but smaller diffusible form.[162,221] SBA does not bind to the outer membranes of the cells.[32,221] Attachment of *B. japonicum* cells to *Glycine soja* root hairs and epidermal cells is inhibited in the presence of SBA-haptenic monosaccharides.[211] In addition to SBA, soybean seeds contain a 4-O-methylglucuronic acid-specific lectin.[191] This lectin interacts with the other *B. japonicum* subspecies (type II), which secretes a soluble extracellular 4-O-methyl-D-glucurono-L-rhamnan[82] and which does not bind SBA.[78] Some varieties of soybean are preferentially nodulated by one of the *B. japonicum* subspecies,[30] and it remains to be examined if differences in lectin receptors are responsible for this preference. Furthermore, the influence of *nod* genes and *nod* gene induction on soybean lectin binding by *B. japonicum* cells has yet to be studied.

Attachment of rhizobia to alfalfa seedlings has been studied by using a low-inoculum attachment/colonization assay.[26,27] Various rhizobia are able to attach to alfalfa roots, but the homologous symbiont *R. meliloti* occupies specific colonization sites even in the presence of a large excess of other rhizobia. It has been suggested that rhizobial LPS, but not EPS and cyclic beta-1,2-glucan, is involved in specific adsorption to alfalfa roots.[130] The (so far undefined) specific colonization sites either are present at the time of inoculation or may be induced by signal molecules (e.g., NodRm-1) released by infective rhizobia. Differences in experimental approach hamper direct comparison of these results with the ones described earlier.

Early results of the study of lectin–rhizobia interactions led Bohlool and Schmidt[21] and Dazzo and Hubbell[50] to hypothesize that specific lectin-mediated binding of the homologous rhizobial partner to the root hair surface of the host plant is one of the essential factors in determination of host plant specificity in the root nodule symbiosis (the lectin recognition hypothesis). This recognition model predicts that host plants within one cross-inoculation group all produce similar lectins with unique sugar-binding properties to be complemented by specific sugar sequences at the surface of the homologous rhizobial partner. Indeed, available data indicate that different lectins in legumes are distributed following the cross-inoculation group classification.[233]

However, the lectin recognition hypothesis does not explain the results of attachment studies on the *Rl viciae*-pea symbiosis (see earlier). Moreover, under appropriate conditions *Rl viciae* bacteria abundantly attach to clover root hairs, whereas early lectin-hapten-insensitive attachment of *Rl trifolii* to clover root hairs has been reported[8,57] and discussed by Dazzo et al.[57] Furthermore, synthesis of extracellular pea lectin receptors is not limited to *Rl viciae* bacteria. The available data suggest a basic rhicadhesin-mediated mechanism for rhizobial attachment to a root hair surface. Root lectin may be additionally involved in attachment step 1 if present on the outer cell wall surface of target root hairs and/or in step 2 if secreted into the rhizosphere, provided that the rhizobia are appropriately conditioned. Indeed, "crucial events other than attachment are also needed to ensure successful infection of (clover) root hairs in this nitrogen-fixing symbiosis."[57]

Secreted root lectin may also increase the proportion of the rhizobial population that is capable of efficient nodulation.[102,103,212] Pretreatment with SBA or a galactose-binding lectin secreted by soybean roots enhances nodulation by low inocula of wild-type *B. japonicum* cells or by normal inocula of a mutant strain exhibiting a delayed-nodulation phenotype. Root exudates or D-galactose/N-acetylgalactosamine-binding lectins from other plants appeared to be inactive. Since the mutant strain shows normal root adsorption and root hair curling ability and chloramphenicol or rifampin prevented the lectin effect, it was suggested that soybean lectin on the root surface or in the rhizosphere induces a physiological response in bradyrhizobia that enhances an infection step beyond root hair curling.[103] This hypothesis corresponds with the results of Diaz et al.,[71] suggesting that root lectin is essential for root hair infection by rhizobia rather than for root adsorption.

On the other hand, pretreatment of legume roots with lectin-binding polysaccharides of rhizobia may enhance root hair infection.[2,58] Trifoliin A-binding CPS and LPS of *Rl trifolii* interact with clover root hairs, and addition of very low quantities of these polysaccharides results in enhancement of root hair infections by *Rl trifolii* in white clover, in contrast to addition of non-lectin-binding polysaccharides. This biovar-specific effect may somehow be related with the effect of similar quantities of LPS on biovar-specific colonization of alfalfa roots by *R. meliloti*.[30]

Taken together, these and other[133] observations strongly suggest a role of legume root lectin in specific interactions of rhizobia with its primary target cells. Lectin present on the outside of the cytoplasmic membrane of emerging root hairs may function as a receptor for extracellular rhizobial signals (e.g., NodRm1-like signals) or (fragments of) surface polysaccharides. Secreted lectin molecules may stimulate rhizobial infectability. A possible relationship between rhizobial surface polysaccharides and NodRm1-like signals merits further study. Moreover, to test the involvement of root lectin in the rhizobial infection process, host plants are required in which the amount and/or quality of root lectin is altered.

D. Broad or Narrow Infection Threads

Surface polysaccharides appear to play a crucial role in rhizobial interactions with root hairs and in development of either broad or narrow infection threads.

In *R. meliloti, ndv* mutants are defective in the production of cyclic beta-(1,2)-glucan, a periplasmic/extracellular ring-form oligosaccharide of about 20 glucose residues.[115] This glucan is supposed to play an essential role in hypoosmotic adaptation of wild-type bacteria (similar to the situation with *Agrobacterium*[157]), and its absence in *ndv* mutants is correlated with several cell surface defects that can be corrected by raising the osmolarity of the medium.[85] The mutants show a reduced ability to attach to and to induce infection threads in alfalfa roots, and elicit small pseudonodules.[84,86] It has been suggested that beta-(1,2)-glucan functions as a signal molecule active at the level of infection thread formation.[86] Exogenously supplied beta-(1,2)-glucan indeed inhances infection and nodulation of alfalfa[86] and clover.[1] However, symbiotic pseudorevertants of *ndv* mutants have been isolated that effectively nodulate alfalfa roots albeit still being defective in glucan production.[86] Whereas *Rhizobium* and *Agrobacterium* produce cyclic beta-(1,2)-glucan molecules, either neutral or with anionic substituents (e.g., refs. 13 and 158), *Bradyrhizobium* cells produce smaller oligoglucans, which are primarily linked by beta-1,6 and beta-1,3 glycosidic bonds.[159] Further study may provide firm evidence for a role of these characteristic cyclic glucans in rhizobial infection.

EPSs produced by different rhizobia differ in structure. For *R. leguminosarum, R. meliloti,* and *Bradyrhizobium* EPSs, these differences are obvious.[35] But EPSs of different biovars of *Rl leguminosarum,* initially reported to be identical,[154,185] also show significant differences most notably at the level of glycosyl substitution.[177] These results suggest that EPS structure may be one of the factors determining the host range of rhizobial infection; however, the role of (activated) *nod* genes in EPS synthesis has not yet been conclusively established. EPS-deficient mutants (*exo* mutants) of *Rl viciae, Rl trifolii, R. meliloti,* and *Rhizobium* NGR234 do not nodulate their host plants because of abortive infection thread formation.[22,40,43,75,135,167] Moreover, noncarbohydrate substituents appear to be important for normal nodulation (e.g., *R. meliloti* mutants, which secrete nonsuccinylated EPS induce empty alfalfa nodules with aborted infection threads[136]). On the other hand, production of acidic EPS per se seems to be more important than production of a particular type of EPS, since the symbiotic defect of EPS-deficient mutants of *R. meliloti* can be suppressed by production of another, structurally different acidic EPS that is normally cryptic.[97,253] Suppression of an EPS mutation by another rhizobial surface component may also explain effective nodulation of alfalfa by isolates of *R. meliloti*[41] apparently lacking the normal acidic EPS.[180] In some cases, addition of purified EPS partially restores nodulation ability of EPS mutants.[77]

Interestingly, acidic EPS does not seem to be essential for nodulation of bean and soybean.[22,75,121,132] Rae and Brewin (in preparation) suggested that this observation is correlated with the different process of infection thread formation in these plants (see earlier). The suggestion of Diebold and Noel[75] that production of rhizobial acidic EPS is necessary for indeterminate nodulation (formation of nodules with a persistent apical meristem, as found on pea, clover, and alfalfa) but is not essential for determinate nodulation (formation of round or oval nodules with decreasing meristematic activity, as found on bean and soybean) is consistent with this hypothesis. Requirement of EPS for broad rather than for narrow infection thread formation points at an essential, yet undefined, contribution of EPS to synthesis of a functional infection thread matrix.

In contrast to EPS, intact LPS (LPS I) appears to be required for formation of stable narrow infection threads.[170] This may be related to the direct contact of the bacteria with the host plasma membrane in these structures. Consistent with this suggestion, LPS I is not required for formation of broad infection threads[44,68,178] but appears to be essential at the stage of bacterial contact with the plasma membrane (i.e., during endocytotic uptake from the thread matrix[68]). Following this line of thought, narrow infection thread formation may be related to endocytotic uptake of single rhizobial cells. The observation of Roth and Stacey[189] that inside the top of infection threads in soybean nodules *de novo* membrane formation takes place in apposition to the bradyrhizobial outer membrane is consistent with this hypothesis. If uptake of rhizobia proceeds by a "membrane-zippering" mechanism, as suggested for receptor-mediated phagocytosis by macrophages, the distribution of endocytotic factors on the rhizobial surface may determine whether a bacterium is completely endocytozed or not. The rapidly accumulating knowledge of genetics and molecular structure of rhizobial LPS (e.g., see refs. 38, 39, 107, 108, 178, and 223) will contribute to definition of the role of LPS in rhizobial interactions with the cell surface of legumes.

In host plants developing narrow infection threads, cells responding to rhizobial mitogens are located in the outer cortex. Consequently, rhizobia in narrow infection threads migrate over relatively short distances in cells relatively rich in cytoplasm. Broad infection thread formation appears to be correlated with induction of a nodule primordium in the inner cortex. Apparently, a matrix type of infection thread is required for passage over larger distances through vacuolate host cells. In most cases, however, a detailed description of infection thread formation by wild-type and mutant rhizobia at the ultrastructural level is lacking. The intriguing relationship between cell infection and the plane of plant cell division especially needs further study.

E. Influence of Nitrate

The long-known inhibition of nodulation in the presence of relatively high amounts of combined nitrogen is still poorly understood at the molecular level.[213] Addition of at least 18 mM nitrate to the roots of alfalfa seedlings completely inhibits accumulation of *R. meliloti* cells on root hairs, root hair curling, infection thread development, and nodule formation.[218] This observation suggests that combined nitrogen somehow influences the signal-response system of both symbiotic partners. Since nitrate, in contrast to ammonium,[83] does not affect expression of the *nodABC* genes in *R. meliloti*, the influence of nitrate may be due to an effect on the host plant (e.g., regarding release of *nod* inducers or signal reception).

Also in white clover, nitrate (16 mM) or ammonium (1 mM) inhibits accumulation of *Rl trifolii* cells on root hairs, root hair deformation, root hair infection, and nodulation.[52] Since nitrate supply suppresses accumulation of clover lectin, trifoliin A, on the clover root surface and (indirectly) reduces the ability of trifoliin A to bind to *Rl trifolii* cells,[198] lectin apparently is a target for regulation of rhizobial infection by combined nitrogen. Synthesis of trifoliin A is not repressed in the presence of excess nitrate.[198] The mechanism by which nitrate influences lectin location and activity in clover is unknown. In pea, addition of 20 mM nitrate inhibits nodulation by *Rl viciae* but does not influence the presence of lectin on root hair tips and lectin-mediated accumulation of *Rl viciae* cells.[72] Apparently, the effect of nitrate supply on early rhizobial infection steps varies among different legumes.

Interestingly, addition of IAA suppresses some of the effects of nitrate supply on root hair curling, infection thread formation, and nodulation.[164,224] This observation suggests that the presence of combined nitrogen influences the phytohormone balance in legume roots. Such an influence will have pleiotropic effects in the plant, which may explain the large range of nodulation steps affected by combined nitrogen and the variability in response between different host plants.

IV. Concluding Remarks

Host plant specificity in the rhizobial root nodule symbiosis is determined early in the infection process (e.g., see ref. 138) and is expressed at three different levels: (1) rhizobia-induced nodules are confined to legumes and *Parasponia,* (2) individual rhizobial species or biovars preferably nodulate a limited set of legumes, and (3) some strains of a rhizobial species may nodulate a particular cultivar of a host plant species, whereas other strains of the same species cannot.[141] Molecular determinants of host plant specificity have now been identified for the following infection steps: *nod* gene activation, signal production by rhizobia, and interactions with host lectin. Each of these factors or a combination may explain host plant specificity at the species/biovar/strain levels (levels 2 and 3). But then, "Why legumes?"[210] Legumes are well-known producers of flavonoids, but flavonoid production and excretion is not limited to this plant family. Legumes produce a peculiar and homogeneous group of lectins,[233] but at present there are no indications why these rather than other types of plant lectins are suited to interact with rhizobia or rhizobial signals. Moreover, it is not yet known whether lectins are involved in crack entry infections. Do legumes offer the correct combination of factors? Are nitrogen stress, hormone balance, signal reception, lectin synthesis, and flavonoid release somehow related? Indications may be found by answering the question, "Why, of all Ulmaceae, *Parasponia*?"

Acknowledgments

Many stimulating discussions with my colleagues Nick Brewin, Clara Diaz, Ben Lugtenberg, Kees Recourt, Gerrit Smit, Herman Spaink, Gary Stacey, Ton van Brussel, Carel Wijffelman, and especially Bobby Bakhuizen are gratefully acknowledged.

References

1. Abe, M., Amemura, A., and Higashi, S., "Studies on cyclic beta-1,2-glucan obtained from the periplasmic space of *Rhizobium trifolii* cells," *Plant Soil 64*, 315–324 (1982).

2. Abe, M., Sherwood, J. E., Hollingsworth, R. I., and Dazzo F. B., "Stimulation of clover root hair infection by lectin-binding oligosaccharides from the capsular and extracellular polysaccharides of *Rhizobium trifolii*," *J. Bacteriol. 160*, 517–520 (1984).

3. Aguilar, J. J. M., Ashby, A. M., Richards, A. J. M., Loake, G. J., Watson, M. D., and Shaw, C. H., "Chemotaxis of *Rhizobium leguminosarum* biovar *phaseoli* towards flavonoid inducers of the symbiotic nodulation genes," *J. Gen. Microbiol. 134*, 2741–2746 (1988).

4. Alazard, D., Ndoye, I., and Dreyfus, B., "*Sesbania rostrata* and other stem-nodulated legumes." In: *Nitrogen Fixation: Hundred Years After,* Bothe, H., De Bruijn, F. J., and Newton, W. E., eds. Stuttgart: Gustav Fischer, 1988, pp. 765–769.

5. Allen, O. N., and Allen, E. K. *Proc. Acad. Sci. Bernice P. Bishop Museum Spec. Publ. 35*, 15–16 (1940).

6. Al-Mallah, M. K., Davey, M. R., and Cocking, E. C., "Formation of nodular structures on rice seedlings by rhizobia," *J. Exp. Bot. 40*, 473–478 (1989).

7. Armitage, J. P., Gallagher, A., and Johnston, A. W. B., "Comparison of the chemotactic behaviour of *Rhizobium leguminosarum* with and without the nodulation plasmid," *Mol. Microbiol. 2*, 743–748 (1988).

8. Badenoch-Jones, J., Flanders, D. J., and Rolfe, B. "Association of *Rhizobium* strains with roots of *Trifolium repens*," *Appl. Environ. Microbiol. 49*, 1511–1520 (1985).

9. Bakhuizen, R. "The plant cytoskeleton in the *Rhizobium*–legume symbiosis," Ph.D. thesis, Leiden University, The Netherlands, 1988.

10. Banfalvi, Z., Nieuwkoop, A., Schell, M., Besl, L., and Stacey, G., "Regulation of *nod* gene expression in *Bradyrhizobium japonicum*," *Mol. Gen. Genet. 214*, 420–424 (1988).

11. Barondes, S. H. "Bifunctional properties of lectins: Lectins redefined," *TIBS 13*, 480–484 (1988).

12. Bassam, B. J., Djordjevic, M. A., Redmond, J. W., Batley, M., and Rolfe, B. G., "Identification of a *nodD*-dependent locus in the *Rhizobium* strain NGR234 activated by phenolic factors secreted by soybeans and other legumes," *Mol. Plant-Microbe Interact. 1*, 161–168 (1988).

13. Batley, M., Redmond, J. W., Djordjevic, S. P., and Rolfe, B. G., "Characterisation of glycerophosphorylated cyclic beta-1,2-glucans from a fast-growing *Rhizobium* species, *Biochim. Biophys. Acta 901*, 119–126 (1987).

14. Bauer, W. D. "Infection of legumes by Rhizobia," *Annu. Rev. Plant Physiol. 32*, 407–449 (1981).

15. Bhuvaneswari, T. V., Pueppke, S. G., and Bauer, W. D. "Role of lectins in plant-microorganism interactions. I. Binding of soybean lectin to rhizobia," *Plant Physiol. 60*, 486–491 (1977).

16. Bhuvaneswari, T. V., and Bauer, W. D., "Role of lectins in plant–microorganism interactions. III. Influence of rhizosphere/rhizoplane culture conditions on the soybean lectin-binding properties of rhizobia," *Plant Physiol. 62*, 71–74 (1978).

17. Bhuvaneswari, T. V., Turgeon, B. G., and Bauer, W. D., "Early events in the infection of soybean (*Glycine max* (L.) Merr.) by *Rhizobium japonicum*," *Plant Physiol. 66*, 1027–1031 (1980).

18. Bhuvaneswari, T. V., Bhagwat, A. A., and Bauer W. D., "Transient susceptibility of root cells in four common legumes to nodulation by rhizobia," *Plant Physiol. 68*, 1144–1149 (1981).

19. Bhuvaneswari, T. V., Mills, K. K., Crist, D. K., Evans, W. R., and Bauer, W. D., "Effects of culture age on symbiotic infectivity of *Rhizobium japonicum*," *J. Bacteriol. 153*, 443–451 (1983).

20. Bhuvaneswari, T. V., and Solheim, B., "Root hair deformations in white clover/*Rhizobium trifolii* symbiosis," *Physiol. Plant. 68*, 1144–1149 (1985).

21. Bohlool, B. B., and Schmidt, E. L., "Lectins: A possible basis for specificity in the *Rhizobium*–legume root nodule symbiosis," *Science 185*, 269–271 (1974).

22. Borthakur, D., Barber, C. E., Lamb, J. W., Daniels, M. J., Downie, J. A., and Johnston, A. W. B. "A mutation that blocks exopolysaccharide synthesis prevents nodulation of peas by *Rhizobium leguminosarum* but not of beans by *Rhizobium phaseoli* and is corrected by cloned DNA from *Rhizobium* or the phytopathogen *Xanthomonas*," *Mol. Gen. Genet. 203*, 320–323 (1986).

23. Boulnois, G. J., and Jann, K., "Bacterial polysaccharide capsule synthesis, export and evolution of structural diversity," *Mol. Microbiol. 3*, 1819–1823 (1989).

24. Bradley, D. J., Butcher, G. W., Galfre, G., Wood, E. A., and Brewin, N. J., "Physical association between the peribacteroid membrane and lipopolysaccharide from the

bacteroid outer membrane in *Rhizobium*-infected pea root nodule cells," *J. Cell Sci.* 85, 47–61 (1986).

25. Bradley, D. J., Wood, E. A., Larkins, A. P., Galfre, G., Butcher, G. W., and Brewin, N. J., "Isolation of monoclonal antibodies reacting with peribacteroid membranes and other components of pea root nodules containing *Rhizobium leguminosarum,*" *Planta* 173, 149–160 (1988).

26. Caetano-Anolles, C. G., and Favelukes, G., "Quantitation of adsorption of rhizobia in low numbers to small legume roots," *Appl. Environ. Microbiol.* 52, 371–376 (1986).

27. Caetano-Anolles, C. G., and Favelukes, G., "Host–symbiont specificity expressed during early adsorption of *Rhizobium meliloti* to the root surface of alfalfa," *Appl. Environ. Microbiol.* 52, 377–382 (1986).

28. Caetano-Anolles, C. G., Crist-Estes, D. K., and Bauer, W. D., "Chemotaxis of *Rhizobium meliloti* to the plant flavone luteolin requires functional nodulation genes," *J. Bacteriol.* 170, 3164–3169 (1988).

29. Caetano-Anolles, G., Lagares, A., and Bauer, W. D., "*Rhizobium meliloti* exopolysaccharide mutants elicit feedback regulation of nodule formation in alfalfa," *Plant Physiol.* 91, 368–374 (1990).

30. Caldwell, B. E., and West, G., "Nodulation interactions between soybean genotypes and serogroups of *Rhizobium japonicum,*" *Crop Sci.* 8, 680–682 (1968).

31. Callaham, D. A., and Torrey, J. G., "The structural basis for infection of root hairs of *Trifolium repens* by *Rhizobium,*" *Can. J. Bot.* 59, 1647–1664 (1981).

32. Calvert, H. E., Lalonde, M., Bhuvaneswari, T. V., and Bauer, W. D., "Role of lectins in plant–microorganism interactions. IV. Ultrastructural localization of soybean lectin binding sites on *Rhizobium japonicum,*" *Can. J. Microbiol.* 24, 785–793 (1978).

33. Calvert, H. E., Pence, M. K., Pierce, M., Malik, N. S. A., and Bauer, W. D., "Anatomical analysis of the development and distribution of *Rhizobium* infections in soybean roots," *Can. J. Bot.* 62, 2375–2384 (1984).

34. Canter Cremers, H. C. J., Wijffelman, C. A., Pees, E., Rolfe, B., Djordjevic, M. A., and Lugtenberg, B. J. J., "Host-specific nodulation of plants of the pea cross-inoculation group is influenced by genes in fast growing *Rhizobium* downstream *nodC,*" *J. Plant Physiol.* 132, 398–404 (1988).

35. Carlson, R. W., "Surface chemistry." In: *Ecology of Nitrogen Fixation, Vol. 2, Rhizobium,* Broughton, W., ed. Oxford: Oxford University Press, 1982, pp. 199–234.

36. Carlson, R. W., "Heterogeneity of *Rhizobium* lipopolysaccharides," *J. Bacteriol.* 158, 1012–1017 (1984).

37. Carlson, R. W., Kalembassa, S., Turowski, D. A., Pachori, P., and Noel, K. D., "Characterization of the lipopolysaccharide from a *Rhizobium phaseoli* mutant that is defective in infection thread development," *J. Bacteriol.* 169, 4923–4928 (1987).

38. Carlson, R. W., Garci, F., Noel, D., and Hollingsworth, R., "The structures of the lipopolysaccharide core components from *Rhizobium leguminosarum* biovar *phaseoli* CE3 and two of its symbiotic mutants, CE109 and CE309," *Carbohydr. Res.* 195, 101–110 (1989).

39. Cava, J. R., Elias, P. M., Turowski, D. A., and Noel, K. D., "*Rhizobium leguminosarum* CFN42 genetic regions encoding lipopolysaccharide structures essential for complete nodule development on bean plants," *J. Bacteriol.* 171, 8–15 (1989).

40. Chakravorty, A. K., Zurkowski, W., Shine, J., and Rolfe, B. G., "Symbiotic nitrogen fixation: Molecular cloning of *Rhizobium* genes involved in exopolysaccharide synthesis and effective nodulation," *Mol. Gen. Genet. 192*, 459–465 (1982).

41. Chandler, M. R., "Some observations on infection of *Arachis hypogaea* L. by *Rhizobium*," *J. Exp. Bot. 29*, 749–755 (1978).

42. Chandler, M. R., Date, R. A., and Roughly, R. J., "Infection and root-nodule development in *Stylosanthes* species by *Rhizobium, J. Exp. Bot. 33*, 47–57 (1982).

43. Chen, H., Batley, M., Redmond, J., and Rolfe, B. G., "Alteration of the effective nodulation of a fast-growing broad host range *Rhizobium* due to changes in exopolysaccharide synthesis," *J. Plant Physiol. 120*, 331–349 (1985).

44. Clover, R. H., Kieber, J., and Signer, E. R., "Lipopolysaccharide mutants of *Rhizobium meliloti* are not defective in symbiosis," *J. Bacteriol. 171*, 3961–3967 (1989).

45. Cormack, R. G. H., "Development of root hairs in angiosperms," *Bot. Rev. 28*, 446–464 (1962).

46. Cutter, E. C., and Feldman, L. J., "Trichoblasts in *Hydrocharis*. II. Nucleic acids, proteins and consideration of cell growth in relation to endoploidy," *Am. J. Bot. 57*, 202–211 (1970).

47. Dart, P. J., "The infection process." In: *The Biology of Nitrogen Fixation*, Quispel, A., ed. Amsterdam: North-Holland, 1974, pp. 381–429.

48. Dart, P. J., "Legume root nodule initiation and development," In: *The Development and Function of Roots*, Torrey, J. G., and Clarkson, D. T., eds. London: Academic Press, 1975, pp. 467–506.

49. Dart, P. J., "Infection and development of leguminous nodules," In: *A Treatise on Dinitrogen Fixation, Section III, Biology*, Hardy, R. W. F., and Silver, W. S., eds. New York: Wiley, 1977, pp. 367–472.

50. Dazzo, F. B., and Hubbell, D. H., "Cross-reactive antigens and lectins as determinants of symbiotic specificity in the *Rhizobium*–clover association," *Appl. Microbiol. 30*, 1018–1033 (1975).

51. Dazzo, F. B., Napoli, C. A., and Hubbell, D. H., "Adsorption of bacteria to roots as related to host specificity in the *Rhizobium*–clover symbiosis," *Appl. Environ. Microbiol. 32*, 166–171 (1976).

52. Dazzo, F. B., and Brill, W. J. "Regulation by fixed nitrogen of host–symbiont recognition in the *Rhizobium*–clover symbiosis," *Plant Physiol. 62*, 18–21 (1978).

53. Dazzo, F. B., Urbano, M. R., and Brill, W. J., "Transient appearance of lectin receptors on *Rhizobium trifolii*," *Curr. Microbiol. 2*, 15–20 (1979).

54. Dazzo, F. B., Truchet, G. L., and Kijne, J. W., "Lectin involvement in root-hair tip adhesion as related to the *Rhizobium*–clover symbiosis," *Physiol. Plant. 56*, 143–147 (1982).

55. Dazzo, F. B., and Truchet, G. L., "Interactions of lectins and their saccharide receptors in the *Rhizobium*–legume symbiosis," *J. Membrane Biol. 73*, 1–16 (1983).

56. Dazzo, F. B., and Gardiol, A., "Host-specificity in Rhizobium–legume interactions." In: *Genes Involved in Microbe-Plant Interactions*, Verma, D. P. S., and Hohn, Th., eds. New York: Springer-Verlag, 1984, pp. 3–31.

57. Dazzo, F. B., Truchet, G. L., Sherwood, J. E., Hrabak, E. M., Abe, M., and Pankratz, S. H. "Specific phases of root hair attachment in the *Rhizobium trifolii*–clover symbiosis," *Appl. Environ. Microbiol. 48*, 1140–1150 (1984).

58. Dazzo, F. B., Hollingsworth, R. I., Sherwood, J. E., Abe, M., Hrabak, E. M., Gardiol, A. E., Pankratz, H. S., Smith, K. B., and Yang, H., "Recognition and infection of clover root hairs by *Rhizobium trifolii*." In: *Nitrogen Fixation Research Progress*, Evans, H. J., and Bottomley P. J., eds. Dordrecht: Martinus Nijhoff, 1985, pp. 239–245.

59. Debelle, F., Sharma, S. B., Rosenberg, C., Vasse, J., Maillet, F., Truchet, G., and Denarie, J., "Respective roles of common and host-specific *Rhizobium meliloti nod* genes in the control of lucerne infection." In: *Recognition in Microbe–Plant Symbiotic and Pathogenic Interactions*, Lugtenberg, B., ed. Berlin: Springer-Verlag, 1986, pp. 17–28.

60. Debray, H., Decout, D., Strecker, G., Spik, G., and Montreuil, J., "Specificity of twelve lectins towards oligosaccharides and glycopeptides related to N-glycosylproteins," *Eur. J. Biochem. 117*, 41–55 (1981).

61. De Faria, S. M., Sutherland, J. M., and Sprent, J. I., "A new type of infected cell in root nodules of *Andira* ssp. (Leguminosae)," *Plant Sci. 45*, 143–147 (1986).

62. De Faria, S. M., McInroy, S. G., and Sprent, J. I., "The occurrence of infected cells with persistent infection threads in legume root nodules," *Can. J. Bot. 65*, 553–558 (1987).

63. De Faria, S. M., Hay, H. T., and Sprent, J. I., "Entry of rhizobia into roots of *Mimosa scabrella* Bentham occurs between epidermal cells," *J. Gen. Microbiol. 134*, 2291–2296 (1988).

64. De Faria, S. M., Lewis, G. P., Sprent, J. I., and Sutherland, J. M., "Occurrence of nodulation in the Leguminosae," *New Phytol. 111*, 607–619 (1989).

65. Deinema, M. H., and Zevenhuizen, L. P. T. M., "Formation of cellulose fibrils by Gram-negative bacteria and their role in bacterial flocculation," *Arch. Microbiol. 78*, 42–57 (1971).

66. Dell, A., York, W. S., McNeil, M., Darvill, A. G., and Albersheim, P., "The cyclic structure of beta-D-(1,2)-linked D-glucans secreted by *Rhizobia* and *Agrobacteria*," *Carbohydr. Res. 117*, 185–200 (1983).

67. De Maagd, R. A., Spaink, H. P., Pees, E., Mulders, I. H. M., Wijfjes, A., Wijffelman, C. A., Okker, R. J. H., and Lugtenberg, B. J. J., "Localization and symbiotic function of a region on the *Rhizobium leguminosarum* Sym plasmid pRL1JI responsible for a secreted, flavonoid-inducible 50-kilodalton protein," *J. Bacteriol. 171*, 1151–1157 (1989).

68. De Maagd, R. A., Rao, A. S., Mulders, I. H. M., Goosen-de Roo, L., Van Loosdrecht, M. C. M., Wijffelman, C. A., and Lugtenberg, B. J. J., "Isolation and characterization of mutants of *Rhizobium leguminosarum* biovar *viciae* strain 248 with altered lipopolysaccharides: Possible role of surface charge or hydrophobicity in bacterial release from the infection thread," *J. Bacteriol. 171*, 1143–1150 (1989).

69. De Maagd, R. A., Wijfjes, A. H. M., Spaink, H. P., Ruiz-Sainz, J. E., Wijffelman, C. A., Okker, R. J. H., and Lugtenberg, B. J. J., "*nodO*, a new *nod* gene of the *Rhizobium leguminosarum* biovar *viciae* Sym plasmid pRLiJI, encodes a secreted protein," *J. Bacteriol. 171*, 6764–6770 (1989).

70. Diaz, C. L., Van Spronsen, P. C., Bakhuizen, R., Logman, G. J. J., Lugtenberg, E. J. J., and Kijne, J. W., "Correlation between infection by *Rhizobium leguminosarum* and lectin on the surface of *Pisum sativum* L. roots," *Planta 168*, 350–359 (1986).

71. Diaz, C. L., Melchers, L. S., Hooykaas, P. J. J., Lugtenberg, B. J. J., and Kijne, J. W., "Root lectin as a determinant of host–plant specificity in the *Rhizobium*–legume symbiosis. *Nature (Lond.) 338,* 579–581 (1989).

72. Diaz, C. L., "Root lectin as a determinant of host–plant specificity in the *Rhizobium*-legume symbiosis," Ph.D. thesis, Leiden University, The Netherlands, 1989.

73. Diaz, C. L., Hosselet, M., Logman, G. J. J., Van Driessche, E., Lugtenberg, B. J. J., and Kijne, J. W., "Distribution of glucose/mannose-specific isolectins in pea (*Pisum sativum* L.) seedlings," *Planta 181,* 451–461 (1990).

74. Diaz, C. L., Hooykaas, P. J. J., Lugtenberg, E. J. J., and Kijne, J. W., "Functional expression of pea lectin on root hair tips of transgenic white clover hairy roots." In: *Nitrogen Fixation: Achievements and Objectives,* Gresshoff, P. M., Newton, W. E., Roth, E. L., and Stacey, G., eds. New York: Chapman-Hall, 1990. p. 733.

75. Diebold, R., and Noel, K. D., "*Rhizobium leguminosarum* exopolysaccharide mutants: Biochemical and genetic analyses and symbiotic behavior on three hosts," *J. Bacteriol. 171,* 4821–4830 (1989).

76. Djordjevic, M. A., Gabriel, D. W., and Rolfe, B., "*Rhizobium,* the refined parasite of legumes," *Annu. Rev. Phytopathol. 25,* 145–168.

77. Djordjevic, S. P., Chen, H., Batley, M., Redmond, J. W., and Rolfe, B. G., "Nitrogen-fixing ability of exopolysaccharide synthesis mutants of *Rhizobium* sp. strain NGR234 and *Rhizobium trifolii* is restored by the addition of homologous exopolysaccharides," *J. Bacteriol. 169,* 53–60 (1987).

78. Dombrink-Kurtzman, M. A., Dick, Jr, W. E., Burton, K. A., Cadmus, M. C., and Slodki, M. E., "A soybean lectin having a 4-*O*-methylglucuronic acid specificity," *Biochem. Biophys. Res. Comm. 111,* 798–803 (1983).

79. Downie, J. A., Hombrecher, G., Ma, Q. S., Knight, C. D., Wells, B., and Johnston, A. W. B., "Cloned nodulation genes of *Rhizobium leguminosarum* determine host range specificity," *Mol. Gen. Genet. 190,* 359–365 (1983).

80. Downie, J. A., Knight, C. D., Johnston, A. W. B., and Rossen, L., "Identification of genes and gene products involved in the nodulation of peas by *Rhizobium leguminosarum,*" *Mol. Gen. Genet. 198,* 255–262 (1985).

81. Downie, J. A., and Johnston, A. W. B., "Nodulation of legumes by *Rhizobium,*" *Plant Cell. Environ. 11,* 403–412 (1988).

82. Dudman, W. F., "Structural studies on the extracellular polysaccharides of *Rhizobium japonicum* strains," *Carbohydr. Res. 66,* 9–23 (1978).

83. Dusha, I., Bakos, A., Kondorosi, A., De Bruijn, F. J., and Schell, J., "The *Rhizobium meliloti* early nodulation genes (*nodABC*) are nitrogen-regulated: Isolation of a mutant strain with efficient nodulation capacity on alfalfa in the presence of ammonium," *Mol. Gen. Genet. 219,* 89–96 (1989).

84. Dylan, T., Ielpi, L., Stanfield, S., Kashyap, L., Douglas, C., Yanowsky, M., Nester, E., Helinski, D. R., and Ditta, G., "*Rhizobium meliloti* genes required for nodule development are related to chromosomal virulence genes in *Agrobacterium tumefaciens,*" *Proc. Natl. Acad. Sci. USA 83,* 4403–4407 (1986).

85. Dylan, T., Helinski, D. R., and Ditta, G. S., "Hypoosmotic adaptation in *Rhizobium meliloti* requires β-(1,2)-glucan," *J. Bacteriol. 172,* 1400–1408 (1990).

86. Dylan, T., Nagpal, P., Helinski, D. R., and Ditta, G. S., "Symbiotic pseudorevertants of *Rhizobium meliloti ndv* mutants," *J. Bacteriol. 172,* 1409–1417 (1990).

87. Economou, A., Hawkins, F. K. L., and Johnston, A. W. B., "pRL1JI specifies the catabolism of L-homoserine and contains a gene, *rhi*, whose transcription is reduced in the presence of *nod* gene inducer molecules." In: *Nitrogen Fixation: Hundred Years After*, Bothe, H., De Bruijn, F. J., and Newton, W. E., eds. Stuttgart: Gustav Fischer, 1988, p. 462.

88. Economou, A., Hamilton, W. D. O., Johnston, A. W. B., and Downie, J. A., "The *Rhizobium* nodulation gene *nodO* encodes a Ca^{2+}-binding protein that is exported without N-terminal cleavage and is homologous to haemolysin and related proteins," *EMBO J. 9*, 349–354 (1990).

89. Faucher, C., Maillet, F., Vasse, J., Rosenberg, C., Van Brussel, A. A. N., Truchet, G., and Denarie, J., "*Rhizobium meliloti* host range *nod*H gene determines production of an alfalfa-specific extracellular signal," *J. Bacteriol. 170*, 5489–5499 (1988).

90. Faucher, C., Camut, S., Denarie, J., and Truchet, G., "The *nodH* and *nodQ* host range genes of *Rhizobium meliloti* behave as avirulence genes in *R. leguminosarum* bv. *viciae* and determine changes in the production of plant-specific extracellular signals," *Mol. Plant-Microbe Interact. 2*, 291–300 (1989).

91. Finan, T. M., Hirsch, A. M., Leigh, J. A., Johansen, E., Kuldau, G. A., Deegan, S., Walker, G. C., and Signer, E. R., "Symbiotic mutants of *Rhizobium meliloti* that uncouple plant from bacterial differentiation," *Cell 40*, 869–877 (1985).

92. Firmin, J. L., Wilson, K. E., Rossen, L., and Johnston, A. W. B., "Flavonoid activation of nodulation genes in *Rhizobium* reversed by other compounds present in plants," *Nature 324*, 90–92 (1986).

93. Gade, W., Jack, M. A., Dahl, J. B., Schmidt, E. L., and Wold, F., "The isolation and characterization of a root lectin from soybean (*Glycine max* (L.) cultivar Chippewa)," *J. Biol. Chem. 256*, 12905–12910 (1981).

94. Gade, W., Schmidt, E. L., and Wold, F., "Evidence for the existence of an intracellular root lectin in soybeans," *Planta 158*, 108–110 (1983).

95. Gatehouse, J. A., Bown, D., Evans, I. M., Gatehouse, L. N., Jobes, D., Preston, P., and Croy, R. R. D., "Sequence of the seed lectin gene from pea (*Pisum sativum* L.)," *Nucl. Acid Res. 15*, 7642 (1987).

96. Gaworzewska, E. T., and Carlile, M. J., "Positive chemotaxis of *Rhizobium leguminosarum* and other bacteria towards root exudates from legumes and other plants," *J. Gen. Microbiol. 128*, 1179–1188 (1982).

97. Glazebrook, J., and Walker, G. C., "A novel exopolysaccharide can function in place of the calcofluor-binding exopolysaccharide in nodulation of alfalfa by *Rhizobium meliloti*," *Cell 56*, 661–672 (1989).

98. Goodchild, D. J., and Bergersen, F. J., "Electron microscopy of the infection and subsequent development of soybean nodule cells," *J. Bacteriol. 92*, 204–213 (1966).

99. Gotz, R., Limmer, N., Ober, K., and Schmitt, R., "Motility and chemotaxis in two strains of *Rhizobium* with complex flagella," *J. Gen. Microbiol. 128*, 789–798 (1982).

100. Gyorgypal, Z., Iyer, N., and Kondorosi, A., "Three regulatory *nodD* alleles of divergent flavonoid specificity are involved in host-dependent nodulation by *Rhizobium meliloti*," *Mol. Gen. Genet. 212*, 85–92 (1988).

101. Haack, A., "Uber den Einfluss der Knöllchenbakterien auf die Wurzelhaare von Leguminosen und Nichtleguminosen," *Zentr. Bakteriol. Parasitenk. Abt. II 117*, 343–366 (1964).

102. Halverson, L. J., and Stacey, G., "Host recognition in the *Rhizobium*–soybean symbiosis. Evidence for the involvement of lectin in nodulation," *Plant Physiol. 77*, 621–625 (1985).

103. Halverson, L. J., and Stacey, G., "Effect of lectin on nodulation by wild-type *Bradyrhizobium japonicum* and a nodulation-defective mutant," *Appl. Environ. Microbiol. 51*, 753–760 (1986).

104. Halverson, L. J., and Stacey, G., "Signal exchange in plant–microbe interactions," *Microbiol. Rev. 50*, 193–211 (1986).

105. Hirsch, A. M., Bhuvaneswari, T. V., Torrey, J. G., and Bisseling, T., "Early nodulin genes are induced in alfalfa root outgrowths elicited by auxin transport inhibitors," *Proc. Natl. Acad. Sci. USA 86*, 1244–1248 (1989).

106. Hollingsworth, R. I., Dazzo, F. B., Hallenga, K., and Musselman, B., "The complete structure of the trifoliin A lectin-binding capsular polysaccharide of *Rhizobium trifolii* 843," *Carbohydr. Res. 172*, 97–112 (1988).

107. Hollingsworth, R. I., Carlson, R. W., Garcia, F., and Gage, D. A., "A new core tetrasaccharide component from the lipopolysaccharide of *Rhizobium trifolii* ANU843," *J. Biol. Chem. 264*, 9294–9299 (1989).

108. Hollingsworth, R. I., and Lill-Elghanian, D. A., "Isolation and characterization of the unusual lipopolysaccharide component, 2-amino-2-deoxy-2-N-(27-hydroxyoctaconasoyl)-3-O-(3-hydroxytetradecanoyl)-*gluco*-hexuronic acid, and its de-O-acylation product from the free lipid A of *Rhizobium trifolii* ANU843," *J. Biol. Chem. 264*, 14039–14042 (1989).

109. Hollis, A. B., Kloos, W. E., and Elkan, G. H., "DNA:DNA hybridization studies of *Rhizobium japonicum* and related *Rhizobiaceae*," *J. Gen. Microbiol. 123*, 215–222 (1981).

110. Honma, M. A., Asomaning, M., and Ausubel, F. M., "*Rhizobium meliloti nodD* genes mediate host-specific activation of *nodABC*," *J. Bacteriol. 172*, 901–911 (1990).

111. Horvath, B., Bachem, C. W. B., Schell, J., and Kondorosi, A., "Host-specific regulation of nodulation genes in *Rhizobium* is mediated by a plant-signal interacting with the *nodD* gene product," *EMBO J. 6*, 841–848 (1987).

112. Hrabak, E. M., Urbano, M. R., and Dazzo, F. B. "Growth-phase dependent immunodeterminants of *Rhizobium trifolii* lipopolysaccharide which bind Trifoliin A, a white clover lectin," *J. Bacteriol. 148*, 697–711 (1981).

113. Hrabak, E. M., Truchet, G. L., Dazzo, F. B., and Govers, F., "Characterization of the anomalous infection and nodulation of subterranean clover roots by *Rhizobium leguminosarum* 1020," *J. Gen. Microbiol. 131*, 3287–3302 (1985).

114. Hubbell, D. H., Morales, V. M., and Umali-Garcia, M., "Pectolytic enzymes in *Rhizobium*," *Appl. Environ. Microbiol. 35*, 210–213 (1978).

115. Ielpi, L., Dylan, T., Ditta, G., Helinski, D. R., and Stanfield, S. W., "The *ndvB* locus of *Rhizobium meliloti* encodes a 319 kDa protein involved in the production of beta-(1,2)-glucan," *J. Biol. Chem. 265*, 2843 (1990).

116. Jeffree, C. E., Dale, J. E., and Fry, S. C., "The genesis of intercellular spaces in developing leaves of *Phaseolus vulgaris* L.," *Protoplasma 132*, 90–98 (1986).

117. Kamberger, W., "An Ouchterlony double diffusion study on the interaction between legume lectins and rhizobial cell surface antigens," *Arch. Microbiol. 121*, 83–90 (1979).

118. Kamberger, W., "Role of cell surface polysaccharides in the *Rhizobium*-pea symbiosis," *FEMS Microbiol. Lett. 6*, 361–365 (1979).

119. Kaminski, P. A., Buffard, D., and Strosberg, A. D., "The pea lectin gene family contains only one functional lectin gene," *Plant Mol. Biol. 9*, 497–507 (1987).

120. Kidby, D. K., and Goodchild, D. J., "Host influence on the ultrastructure of root nodules of *Lupinus luteus* and *Ornithopus sativus, J. Gen. Microbiol. 45*, 147–152 (1966).

121. Ko, Y. H., and Gayda, R., "Nodule formation in soybeans by exopolysaccharide mutants of *Rhizobium fredii* USDA191," *J. Gen. Microbiol. 136*, 105–113 (1990).

122. Kornfeld, K., Reitman, M.L., and Kornfeld, R., "The carbohydrate-binding specificity of pea and lentil lectins. Fucose is an important determinant," *J. Biol. Chem. 256*, 6633–6640 (1981).

123. Kosslak, R. M., and Bohlool, B. B., "Suppression of nodule development of one side of a split-root system of soybeans caused by prior inoculation of the other side," *Plant Physiol. 75*, 125–130 (1984).

124. Kosslak, R. M., Bookland, R., Barkei, J., Paaren, H. E., and Appelbaum, E. R., "Induction of *Bradyrhizobium japonicum* common *nod* genes by isoflavones isolated from *Glycine max. Proc. Natl. Acad. Sci. 84*, 7428–7432 (1987).

125. Kijne, J. W., "The fine structure of pea root nodules. I. Vacuolar changes after endocytotic host cell infection by *Rhizobium leguminosarum,*" *Physiol. Plant Pathol. 5*, 75–79 (1975).

126. Kijne, J. W. Adhin, S. D., and Planqué, K., "Cell proliferation in pea root explants supplied with rhizobial LPS." In: *Cell Wall Biochemistry Related to Specificity in Host Plant–Pathogen interactions*, Raa, J., and Solheim, B., eds. Tromsö, Oslo, Bergen: Universitetsforlaget, 1976, pp. 415–416.

127. Kijne, J. W., Smit, G., Diaz, C. L., and Lugtenberg, B. J. J., "Attachment of *Rhizobium leguminosarum* to pea root hair tips," In: *Recognition in Microbe–Plant Symbiotic and Pathogenic Interactions*, Lugtenberg, B. J. J., ed. Berlin: Springer-Verlag, 1986, pp. 101–111.

128. Kijne, J. W., Smit, G., Diaz, C. L., and Lugtenberg, B. J. J., "Lectin enhanced accumulation of manganese-limited *Rhizobium leguminosarum* cells on pea root hair tips," *J. Bacteriol. 170*, 2994–3000 (1988).

129. Kijne, J. W., Diaz, C. L., and Lugtenberg, B. J. J., "Role of lectin in the pea-*Rhizobium* symbiosis." In: *Signal Molecules in Plants and Plant–Microbe interactions*, Lugtenberg, B. J. J., ed. Berlin: Springer-Verlag, 1989, pp. 351–358.

130. Lagares, A., and Favelukes, G., "Early recognition in the association of *Rhizobium meliloti* with alfalfa: Possible role of the lipopolysaccharide." In: *Nitrogen Fixation: Hundred Years After*, Bothe, H., De Bruijn, F. J., and Newton, W. E., eds. Stuttgart: Gustav Fischer, 1988, p. 477.

131. Lancelle, S. A., and Torrey, J. G., "Early development of *Rhizobium*-induced root nodules of *Parasponia rigida*. I. Infection and early nodule initiation," *Protoplasma 123*, 26–37 (1984).

132. Law, I. J., Yamamoto, Y., Mort, A. J., and Bauer, W. D., "Nodulation of soybean by *Rhizobium japonicum* mutants with altered capsule synthesis," *Planta 154*, 100–109 (1982).

133. Law, I. J., and Strijdom, B. W., "Role of lectins in the specific recognition of *Rhizobium* by *Lotononis bainesii*," *Plant Physiol. 74*, 779–785 (1984).

134. Le Gal, M. F., and Hobbs, S. L. A., "Cytological studies of the infection process in nodulating and non-nodulating pea gentoypes," *Can. J. Bot. 67*, 2435–2443 (1989).

135. Leigh, J. A., Signer, E. R., and Walker, G. C., "Exopolysaccharide deficient mutants of *Rhizobium meliloti* that form ineffective nodules," *Proc. Natl. Acad. Sci. USA 82,* 6231–6235 (1985).

136. Leigh, J. A., Reed, J. W., Hanks, A. M., Hirsch, A. M., and Walker, G. C., "*Rhizobium meliloti* mutants that fail to succinylate their calcofluor-binding exopolysaccharide are defective in nodule invasion," *Cell 51,* 579–587 (1987).

137. Lerouge, P., Roche, P., Faucher, C., Maillet, F., Truchet, G., Prome, J. C., and Dénarié, J., "Symbiotic host-specificity of *Rhizobium meliloti* is determined by a sulphated and acylated glucosamine oligosaccharide signal," *Nature 344,* 781–784 (1990).

138. Li, D., and Hubbell, D. H., "Infection thread formation as a basis of nodulation specificity in *Rhizobium*–strawberry clover associations," *Can. J. Microbiol. 15,* 1133–1136 (1969).

139. Libbenga, K. R., and Harkes, P. A. A., "Initial proliferation of cortical cells in the formation of root nodules in *Pisum sativum* L.," *Planta 114,* 17–28 (1973).

140. Libbenga, K. R., Van Iren, F., Bogers, R. J., and Schraag-Lamers, M. F., "The role of hormones and gradients in the initiation of cortex proliferation and nodule formation in *Pisum sativum* L.," *Planta 114,* 29–39 (1973).

141. Lie, T. A., "Symbiotic specialization in pea plants: The requirement of specific *Rhizobium* strains for peas from Afghanistan," *Ann. Appl. Biol. 88,* 462–465 (1978).

142. Ljunggren, H., and Fåhraeus, G. "Role of polygalacturonase in root hair invasion by nodule bacteria," *J. Gen. Microbiol. 26,* 521–528 (1961).

143. Lloyd, C. W., Pearce, K. J., Rawlins, D. J., Ridge, R. W., and Shaw, P. J., "Endoplasmic microtubules connect the advancing nucleus to the tip of legume root hairs, but F-actin is involved in basipetal migration," *Cell Motil. Cytoskel. 8,* 27–36 (1987).

144. Loake, G. J., Ashby, A. M., and Shaw, C. H., "Attraction of *Agrobacterium tumefaciens* C58C1 towards sugars involves a highly sensitive chemotaxis system," *J. Gen. Microbiol. 134,* 1427–1432 (1988).

145. Long, S. R., "Rhizobium–legume nodulation: Life together in the underground," *Cell 56,* 203–214 (1989).

146. Lugtenberg, B. J. J., Smit, G., Diaz, C. L., and Kijne, J. W., "Role of attachment of *Rhizobium leguminosarum* cells to pea root hair tips in targeting signals for early symbiotic steps." In: *Signal Molecules in Plants and Plant–Microbe Interactions,* Lugtenberg, B. J. J., ed. Berlin: Springer-Verlag, 1989, pp. 129–136.

147. Malek, W., "Chemotaxis in *Rhizobium meliloti* strain L5.30," *Arch. Microbiol. 152,* 611–612 (1989).

148. Martinez-Molina, E., Morales, V. M., and Hubbell, D. H., "Hydrolytic enzyme production by *Rhizobium,*" *Appl. Environ. Microbiol. 38,* 1186–1188 (1979).

149. Marvel, D. J., Kuldau, G., Hirsch, A., Richards, E., Torrey, J. G., and Ausubel, F. M., "Conservation of nodulation genes between *Rhizobium meliloti* and a slow-growing *Rhizobium* strain that nodulates a non-legume host. *Proc. Natl. Acad. Sci. USA 82,* 5841–5845 (1985).

150. Marvel, D. J., Torrey, J. G., and Ausubel, F. M., "*Rhizobium* symbiotic genes required for nodulation of legume and non-legume hosts," *Proc. Natl. Acad. Sci. USA 84,* 1319–1323 (1987).

151. Matthysse, A. G., Holmes, K. V., and Gurlitz, R. H. G., "Elaboration of cellulose fibrils by *Agrobacterium tumefaciens* during attachment to carrot cells," *J. Bacteriol. 145,* 583–595 (1981).

152. Matthysse, A. G., "Role of bacterial cellulose fibrils in *Agrobacterium tumefaciens* infections," *J. Bacteriol. 154*, 906–915 (1983).

153. Maxwell, C. A., Hartwig, U. A., Joseph, C. M., and Phillips, D. A., "A chalcone and two related flavonoids released from alfalfa roots induce *nod* genes of *Rhizobium meliloti*," *Plant Physiol. 91*, 842–847 (1989).

154. McNeil, M., Darvill, J., Darvill, A. G., Albersheim, P., Van Veen, R., Hooykaas, P., Schilperoort, R., and Dell, A., "The discernible structural features of the acidic polysaccharides secreted by different *Rhizobium* species are the same," *Carbohydr. Res. 146*, 307–326 (1986).

155. Mellor, H. Y., Glenn, A. R., Arwas, R., and Dilworth, M. J., "Symbiotic and competitive properties of motility mutants of *Rhizobium trifolii* TA1," *Arch. Microbiol. 148*, 34–39 (1987).

156. Miller, A. L., Raven, J. A., Sprent, J. I., and Weisenseel, M. H., "Endogenous ion currents traverse roots and root hairs of *Trifolium repens*," *Plant Cell Environ. 9*, 79–83 (1986).

157. Miller, K. J., Kennedy, E. P., and Reinhold, V. N., "Osmotic adaptation by Gram-negative bacteria: Possible role for periplasmic oligosaccharides," *Science 231*, 48–51 (1986).

158. Miller, K. J., Reinhold, V. J., Weissborn, A. C., and Kennedy, E. P., "Cyclic glucans produced by *Agrobacterium tumefaciens* are substituted with *sn*-1-phosphoglycerol residues. *Biochim. Biophys. Acta 901*, 112–118 (1987).

159. Miller, K. J., Gore, R. S., Johnson, R., Benesi, A. J., and Reinhold, V. N., "Cell-associated oligosaccharides of *Bradyrhizobium* ssp.," *J. Bacteriol. 172*, 136–142 (1990).

160. Morris, V. J., Brownsey, G. J., Harris, J. E., Gunning, A. P., Stevens, B. J. H., and Johnston, A. W. B., "Cation-dependent gelation of the acidic extracellular polysaccharides of *Rhizobium leguminosarum*: A non-specific mechanism for the attachment of bacteria to plant roots," *Carbohydr. Res. 191*, 315–320 (1989).

161. Mort, A. J., and Bauer, W. D., "Composition of the capsular and extracellular polysaccharides of *Rhizobium japonicum*: Changes with culture age and correlation with binding of soybean seed lectin to bacteria," *Plant Physiol. 66*, 158–163 (1980).

162. Mort, A. J., and Bauer, W. D., "Application of two new methods of cleavage of polysaccharides into specific oligosaccharide fragments. Structure of the capsular and extracellular polysaccharides of *Rhizobium japonicum* that bind soybean lectin," *J. Biol. Chem. 257*, 1870–1875 (1982).

163. Mulligan, J. T., and Long, S. R., "Induction of *Rhizobium meliloti nodC* expression by plant exudate requires *nodD*," *Proc. Natl. Acad. Sci. USA 82*, 6609–6613 (1985).

164. Munns, D. N., "Nodulation of *Medicago sativa* in solution culture. IV. Effects of indole-3-acetate in relation to acidity and nitrate," *Plant Soil 29*, 257–262 (1968).

165. Nagl, W., *Endopolyploidy and Polyteny in Differentiation and Evolution.* Amsterdam: North-Holland, 1978.

166. Napoli, C., Dazzo, F., and Hubbell, D., "Production of cellulose microfibrils by *Rhizobium*," *Appl. Microbiol. 30*, 123–131 (1975).

167. Napoli, C., and Albersheim, P., "*Rhizobium leguminosarum* mutants incapable of normal extracellular polysaccharide production," *J. Bacteriol. 141*, 1454–1456 (1980).

168. Newcomb, W., "Nodule morphogenesis." In: *International Review of Cytology, Suppl. 13*, Bourne, G. H., and Danielli, J. F., eds. New York: Academic Press, 1981, pp. 246–298.

169. Newcomb, W., Sipell, D., and Peterson, R. L., "The early morphogenesis of *Glycine max* and *Pisum sativum* root nodules," *Can. J. Bot. 57*, 2603–2616 (1979).

170. Noel, K. D., VandenBosch, K. A., and Kulpaca, B., "Mutations in *Rhizobium phaseoli* that lead to arrested development of infection threads," *J. Bacteriol. 168*, 1392–1401 (1986).

171. Nutman, P. S., Doncaster, C. C., and Dart, P. J., "Infection of clover by root nodule bacteria," black and white 16-mm optical sound track film available from the British Film Institute, London UK (1973).

172. Peters, N. K., Frost, J. W., and Long, S. R., "A plant flavone, luteolin, induces expression of *Rhizobium meliloti* nodulation genes," *Science 233*, 917–1008 (1986).

173. Peters, N. K., and Long, S. R., "*Rhizobium meliloti* nodulation gene inducers and inhibitors," *Plant Physiol. 88*, 396–400 (1988).

174. Philip-Hollingsworth, S., Hollingsworth, R. I., and Dazzo, F. B., "Host-range related structural features of the acidic extracellular polysaccharides of *Rhizobium trifolii* and *Rhizobium leguminosarum*," *J. Biol. Chem. 264*, 1461–1466 (1989).

175. Philip-Hollingsworth, S., Hollingsworth, R. I., Dazzo, F. B., Djordjevic, M. A., and Rolfe, B. G., "The effect of interspecies transfer of *Rhizobium* host-specific nodulation genes on acidic polysaccharide structure and *in situ* binding by host lectin," *J. Biol. Chem. 264*, 5710–5714 (1989).

176. Phillips, D. A., "A cotyledonary inhibitor of root nodulation in *Pisum sativum*," *Physiol. Plant. 25*, 482–487 (1971).

177. Price, G. D., Mohapatra, S. S., and Gresshoff, P. M., "Structure of nodules by *Rhizobium* strain ANU289 in the nonlegume *Parasponia* and the legume siratro (*Macroptilium atropurpureum*), *Bot. Gaz. 145*, 444–451 (1984).

178. Priefer, U. B., "Genes involved in lipopolysaccharide production and symbiosis are clustered on the chromosome of *Rhizobium leguminosarum* biovar *viciae* VF39," *J. Bacteriol. 171*, 6161–6168 (1989).

179. Pueppke, S. G., "*Rhizobium* infection threads in root hairs of *Glycine max* (L.) Merr., *Glycine soja* Sieb. & Zucc., and *Vigna unguiculata* (L.) Walp.," *Can. J. Microbiol. 29*, 69–76 (1982).

180. Putnoky, P., Grosskopf, E., Camlta, D. T., Kiss, G. B., and Kondorosi, A., "*Rhizobium fix* genes mediate at least two communication steps in symbiotic nodule development," *J. Cell Biol. 106*, 597–607 (1988).

181. Recourt, K., Van Brussel, A. A. N., Driessen, A. J., and Lugtenberg, B. J. J., "Accumulation of a *nod* gene inducer, the flavonoid naringenin, in the cytoplasmic membrane of *Rhizobium leguminosarum* biovar *viciae* is caused by the pH-dependent hydrophobicity of naringenin," *J. Bacteriol. 171*, 4370–4377 (1989).

182. Recourt, K., Van Brussel, A. A. N., Kijne, J. W., Schripsema, J., and Lugtenberg, B. J. J., "Inoculation of *Vicia sativa* subsp. *nigra* increases the number of *nod* gene inducing flavonoids released by the roots of the host plant." In: *Nitrogen Fixation: Achievements and Objectives,* Gresshoff, P. M., Newton, W. E., Roth, E. L., and Stacey, G., eds. New York: Chapman-Hall, 1990. p. 271.

183. Redmond, J. W., Batley, M., Djordjevic, M. A., Innes, R. W., Kuempel, P. L., and Rolfe, B. G., "Flavones induce expression of nodulation genes in *Rhizobium*," *Nature 323*, 632–634 (1986).

184. Ridge, R. W., "Freeze-substitution improves the ultrastructural preservation of legume root hairs," *Bot. Mag. Tokyo 101*, 427–441 (1988).

185. Robertsen, B. K., Aman, P., Darvill, A. G., McNeil, M., and Albersheim, P., "Host–symbiont interactions. V. The structure of acidic extracellular polysaccharides secreted by *Rhizobium leguminosarum* and *Rhizobium trifolii*," *Plant Physiol. 67*, 389–400 (1981).

186. Rolfe, B., and Gresshoff, P., "Genetic analysis of legume nodule initiation," *Annu. Rev. Plant Physiol. Plant Mol. Biol. 39*, 297–319 (1988).

187. Rolfe, B. G., Batley, M., Redmond, J. W., Richardson, A. E., Simpson, R. J., Bassam, B. J., Sargent, C. L., Weinman, J. J., Djordjevic, M. A., and Dazzo, F. B., "Phenolic compounds secreted by legumes." In: *Nitrogen Fixation: Hundred Years After*, Bothe, H., De Bruijn, F. J., and Newton, W. E., eds. Stuttgart: Gustav Fischer, 1988, pp. 405–409.

188. Roth, L. E., Jeon, K., Stacey, G., "Homology in endosymbiotic systems: The term 'symbiosome.'" In: *Molecular Genetics of Plant–Microbe Interactions*, Palacios, R., and Verma, D. P. S., eds. St. Paul: Amer. Phytopathol. Soc. Press, 1988, pp. 220–225.

189. Roth, L. E., and Stacey, G., "Bacterium release into host cells of nitrogen-fixing soybean nodules: The symbiosome membrane comes from three sources," *Eur. J. Cell Biol. 49*, 13–23.

190. Rossen, L., Johnston, A. W. B., and Downie, J. A., "DNA sequence of the *Rhizobium leguminosarum* nodulation genes *nodA, B*, and *C* required for root hair curling," *Nucl. Acid Res. 12*, 9497–9508 (1984).

191. Rutherford, W. M., Dick, Jr, W. E., Cavins, J. F., Dombrink-Kurtzman, M. A., and Slodki, M. E., "Isolation and characterization of a soybean lectin having a 4-*O*-methylglucuronic acid specificity," *Biochemistry 25*, 952–958 (1986).

192. Scheres, B., Van de Wiel, C., Zalensky, A., Hirsch, A., Van Kammen, A., and Bisseling, T., "Identification of *Rhizobium leguminosarum* genes and signal compounds involved in the induction of early nodulin gene expression." In: *Signal Molecules in Plants and Plant–Microbe Interactions*, Lutenberg, B. J. J., ed. Berlin: Springer-Verlag, 1989, pp. 367–377.

193. Scheres, B., Van de Wiel, C., Zalensky, A., Horvath, B., Spaink, H., Van Eck, H., Zwartkruis, F., Wolters, A.-M., Gloudemans, T., Van Kammen, A., and Bisseling, T. "The ENOD12 early nodulin is involved in the infection process during the pea-*Rhizobium leguminosarum* interaction," *Cell 60*, 281–294 (1990).

194. Schlaman, H. R. M., Spaink, H. P., Okker, R. J. H., and Lugtenberg, B. J. J., "Subcellular localization of the *nodD* gene product in *Rhizobium leguminosarum*," *J. Bacteriol. 171*, 4686–4693 (1989).

195. Sen, D., and Weaver, R. W., "A basis for different rates of N2-fixation by the same strain of *Rhizobium* in peanut and cowpea root nodules," *Plant Sci. Lett. 34*, 239–246 (1984).

196. Shaw, C. H., Ashby, A. M., Brown, A., Royal, C., and Loake, G. J., "*VirA & G* are the Ti-plasmid functions required for chemotaxis of *Agrobacterium tumefaciens* towards acetosyringone," *Molec. Microbiol. 2*, 413–418 (1988).

197. Shearman, C. A., Rossen, L., Johnston, A. W. B., and Downie, J. A., "The *Rhizobium leguminosarum* gene *nodF* encodes a polypeptide similar to acyl-carrier protein and is regulated by *nodD* plus a factor in pea root exudate," *EMBO J. 5*, 647–652 (1986).

198. Sherwood, J. E., Truchet, G. L., and Dazzo, F. B., "Effect of nitrate supply on the in-vivo synthesis and distribution of trifoliin A, a *Rhizobium trifolii*-binding lectin, in *Trifolium repens* seedlings," *Planta 162*, 540–547 (1984).

199. Sherwood, J. E., Vasse, J. M., Dazzo, F. B., and Truchet, G. L., "Development and Trifoliin A-binding ability of the capsule of *Rhizobium trifolii*," *J. Bacteriol. 159*, 145–152 (1984).

200. Sinnott, E. W., and Bloch, R., "Cell polarity and the differentiation of root hairs," *Proc. Natl. Acad. Sci. USA 25*, 55–58 (1939).

201. Smit, G., Kijne, J. W., and Lugtenberg, B. J. J., "Correlation between extracellular fibrils and attachment of *Rhizobium leguminosarum* to pea root hair tips," *J. Bacteriol. 168*, 821–827 (1986).

202. Smit, G., Kijne, J. W., and Lugtenberg, B. J. J., "Both cellulose fibrils and a Ca^{2+}-dependent adhesin are involved in the attachment of *Rhizobium leguminosarum* to pea root hair tips," *J. Bacteriol. 169*, 4294–4301 (1987).

203. Smit, G., "Adhesins from Rhizobiaceae and their role in plant–bacterium interactions," Ph.D. thesis, Leiden University, The Netherlands, 1988.

204. Smit, G., Kijne, J. W., and Lugtenberg, B. J. J., "Roles of flagella, lipopolysaccharide, and a Ca^{2+}-dependent cell surface protein in attachment of *Rhizobium leguminosarum* biovar *viciae* to pea root hair tips," *J. Bacteriol. 171*, 569–572 (1989).

205. Smit, G., Logman, T. J. J., Boerrigter, M. E. T. I., Kijne, J. W., and Lugtenberg, B. J. J., "Purification and partial characterization of the *Rhizobium leguminosarum* biovar *viciae* Ca^{2+}-dependent adhesin, which mediates the first step in attachment of cells of the family Rhizobiaceae to plant root hair tips," *J. Bacteriol. 171*, 4054–4062 (1989).

206. Solheim, B., and Fjellheim, K. E., "Rhizobial polysaccharide-degrading enzymes from roots of legumes," *Physiol. Plant. 62*, 11–17 (1984).

207. Spaink, H. P., Okker, R. J. H., Wijffelman, C. A., Pees, E., and Lugtenberg, B. J. J., "Promoters in the nodulation region of the *Rhizobium leguminosarum* Sym plasmid pRL1JI," *Plant Mol. Biol. 9*, 27–39 (1987).

208. Spaink, H. P., Wijffelman, C. A., Pees, E., Okker, R. J. H., and Lugtenberg, B. J. J., "*Rhizobium* nodulation gene *nodD* as a determinant of host specificity," *Nature 328*, 337–340 (1987).

209. Spaink, H. P., Weinman, J., Djordjevic, M. A., Wijffelman, C. A., Okker, R. H. J., and Lugtenberg, B. J. J., "Genetic analysis and cellular localization of the *Rhizobium* host-specificity determining NodE protein," *EMBO J. 8*, 2811–2818 (1989).

210. Sprent, J. I., "Which steps are essential for the formation of functional legume nodules?" *New Phytol. 111*, 129–153 (1989).

211. Stacey, G., Paau, A. S., and Brill, W. J., "Host recognition in the *Rhizobium*–soybean symbiosis," *Plant Physiol. 66*, 609–614 (1980).

212. Stacey, G., Halverson, L. J., Nieuwkoop, T., Banfalvi, Z., Schell, M. G., Gerhold, D., Deshmane, N., So, J. S., and Sirotkin, K. M., "Nodulation of soybean: *Bradyrhizobium japonicum* physiology and genetics." In: *Recognition in Microbe–Plant Symbiotic and Pathogenic Interactions*, Lugtenberg, B. J. J., ed. Berlin: Springer-Verlag, 1986, pp. 87–99.

213. Streeter, J., "Inhibition of legume nodule formation and N_2 fixation by nitrate," *CRC Crit. Rev. Plant Sci. 7*, 1–23 (1988).

214. Sutherland, T. D., Bassam, B. J., Schuller, L. J., and Gresshoff, P. M., "Early nodulation signals of the wild type and symbiotic mutants of soybean (*Glycine max*), *Mol. Plant Microbe Int. 3*, 122–128 (1990).

215. Trinick, M. J., "Structure of nitrogen-fixing nodules formed by *Rhizobium* on roots of *Parasponia andersonii* Planch," *Can. J. Microbiol. 25*, 565–578 (1979).

216. Trinick, M. J., "The effective rhizobium symbiosis with the nonlegume *Parasponia andersonii.*" In: *Current Perspectives in Nitrogen Fixation,* Gibson, A. H., and Newton, W. E., eds. Canberra: Australian Academy of Science, 1981, p. 480.

217. Trinick, M. J., and Hadobas, P. A., "Biology of the *Parasponia–Bradyrhizobium* symbiosis," *Plant Soil 110,* 177–185 (1988).

218. Truchet, G. L., and Dazzo, F. B., "Morphogenesis of lucerne root nodules incited by *Rhizobium meliloti* in the presence of combined nitrogen," *Planta 154,* 352–360 (1982).

219. Truchet, G., Barker, D. G., Camut, S., De Billy, F., Vasse, J., and Huguet, T., "Alfalfa nodulation in the absence of *Rhizobium,*" *Mol. Gen. Genet. 219,* 65–68 (1989).

220. Tschermak-Woess, E., and Hasitschka, G., "Uber Musterbildung in der Rhizodermis und Exodermis bei einigen Angiospermen und einer Polypodiacee," *Oesterr. Bot. Z. 100,* 646–651 (1953).

221. Tsien, H. C., and Schmidt, E. L., "Localization and partial characterization of soybean lectin-binding polysaccharide of *Rhizobium japonicum,*" *J. Bacteriol. 145,* 1063–1074 (1981).

222. Turgeon, B. G., and Bauer, W. D., "Ultrastructure of infection-thread development during infection of soybean by *Rhizobium japonicum,*" *Planta 163,* 328–349 (1985).

223. Urbanik-Sypniewska, T., Seydel, U., Greck, M., Weckesser, J., and Mayer, H., "Chemical studies on the lipopolysaccharide of *Rhizobium meliloti* 10406 and its lipid A region," *Arch. Microbiol. 152,* 527–532 (1989).

224. Valera, L., and Alexander, M., "Reversal of nitrate inhibition of nodulation by indolyl-3-acetic acid," *Nature 206,* 326 (1965).

225. Van Batenburg, F. H. D., Jonker, R., and Kijne, J. W., "*Rhizobium* induces marked root hair curling by redirection of tip growth, a computer simulation," *Physiol. Plant. 66,* 476–480 (1986).

226. Van Brussel, A. A. N., Zaat, S. A. J, Canter Cremers, H. C. J., Wijffelman, C. A., Pees, E., Tak, T., and Lugtenberg, B. J. J., "Role of plant root exudate and Sym plasmid-localized nodulation genes in the synthesis by *Rhizobium leguminosarum* of Tsr factor which causes thick and short roots on common vetch," *J. Bacteriol. 165,* 517–522 (1986).

227. Van Brussel, A. A. N., Pees, E., Spaink, H. P., Tak, T., Wijffelman, C. A., Okker, R. J. H., Truchet, G., and Lugtenberg, B. J. J., "Correlation between *Rhizobium leguminosarum nod* genes and nodulation phenotypes on *Vicia.*" In: *Nitrogen Fixation: Hundred Years After,* Bothe, H., De Bruijn, F. J., and Newton, W. E., eds. Stuttgart: Gustav Fischer, 1988, p. 483.

228. Van Brussel, A. A. N., "Symbiotic signals in early stages of the morphogenesis of *Rhizobium*-induced root nodules," *Symbiosis 9,* 135–146 (1990).

229. Van Brussel, A. A. N., Recourt, K., Pees, E., Spaink, H. P., Tak, T., Wijffelman. C. A., Kijne, J. W., and Lugtenberg, B. J. J., "An extracellular biovar-specific signal of *Rhizobium leguminosarum* biovar *viciae* induces increased *nod* gene-inducing activity in root exudate of *Vicia sativa* ssp. *nigra,*" *J. Bacteriol. 172,* 5394–5401 (1990).

230. VandenBosch, K. A., Brewin, N. J., and Kannenberg, E. L., "Developmental regulation of a *Rhizobium* cell surface antigen during growth of pea root nodules," *J. Bacteriol. 171,* 4537–4542 (1989).

231. Van der Schaal, I. A. M., Kijne, J. W., Diaz, C. L., and Van Iren, F., "Pea lectin binding by *Rhizobium*." In: *Lectins, Biology, Biochemistry, Clinical Biochemistry, Vol. 3,* Bøg-Hansen, T. C., and Spengler, G. A., eds. Berlin: Walter de Gruyter, 1983, pp. 531–538.

232. Van der Schaal, C. A. M., "Lectins and their possible involvement in the *Rhizobium*-leguminosae symbiosis," Ph.D. thesis, Leiden University, The Netherlands, 1983.

233. Van Driessche, E., "Structure and function of leguminous lectins." In: *Advances in Lectin Research. Vol. 1,* Franz, H., ed. Berlin: Springer-Verlag, 1988, pp. 73–134.

234. Van Egeraat, A. W. S. M., "Exudation of ninhydrin-positive compounds by pea seedling roots: A study of the sites of exudation and of the composition of the exudate," *Plant Soil 42,* 37–47 (1975).

235. Van Egeraat, A. W. S. M., "The growth of *Rhizobium leguminosarum* on the root surface and in the rhizospere of pea seedlings in relation to root exudates," *Plant Soil 42,* 367–379 (1975).

236. Van Egeraat, A. W. S. M., "The possible role of homoserine in the development of *Rhizobium leguminosarum* in the rhizosphere of pea seedlings," *Plant Soil 42,* 381–386 (1975).

237. Vesper, S. J., and Bauer, W. D., "Role of pili (fimbriae) in attachment of *Bradyrhizobium japonicum* to soybean roots," *Appl. Environ. Microbiol. 52,* 134–141 (1986).

238. Vesper, S. J., Malik, N. S. A., and Bauer, W. D., "Transposon mutants of *Bradyrhizobium japonicum* altered in attachment to host roots," *Appl. Environ. Microbiol. 53,* 1959–1961 (1987).

239. Vesper, S. J., and Bhuvaneswari, T. V., "Nodulation of soybean roots by an isolate of *Bradyrhizobium japonicum* with reduced firm attachment capability," *Arch. Microbiol. 150,* 15–19 (1988).

240. Vodkin, L. O., and Raikhel, N. V., "Soybean lectin and related proteins in seeds and roots of Le$^+$ and Le$^-$ soybean varieties," *Plant Physiol. 81,* 558–565 (1986).

241. Wang, T. L., Wood, E. A., and Brewin, N. J., "Growth regulators, *Rhizobium* and nodulation in peas. Indole-3-acetic acid from the culture medium of nodulating and non-nodulating strains of *R. leguminosarum*," *Planta 155,* 345–349 (1982).

242. Wang, T. L., Wood, E. A., and Brewin, N. J., "Growth regulators, *Rhizobium* and nodulation in peas. The cytokinin content of a wild-type and a Ti-plasmid containing strain of *R. leguminosarum*," *Planta 155,* 350–355 (1982).

243. Wolpert, J. S., and Albersheim, P., "Host–symbiont interactions. I. The lectins of legumes interact with the *O*-antigen-containing lipopolysaccharides of their symbiont *Rhizobia*," *Biochem. Biophys. Res. Comm. 70,* 729–737 (1976).

244. Wood, E. A., Butcher, G. W., Brewin, N. J., and Kannenberg, E. L., "Genetic derepression of a developmentally regulated lipopolysaccharide antigen from *Rhizobium leguminosarum* 3841," *J. Bacteriol. 171,* 4549–4555 (1989).

245. Wijffelman, C. A., Pees, E., Van Brussel, A. A. N., Okker, R. J. H., and Lugtenberg, B. J. J., "Genetic and functional analysis of the nodulation region of the *Rhizobium leguminosarum* Sym plasmid pRL1JI," *Arch. Microbiol. 143,* 225–232 (1985).

246. Yao, P. Y., and Vincent, J. M., "Host specificity in the root hair curling factor of *Rhizobium* spp.," *Aust. J. Biol. Sci. 22,* 413–423 (1969).

247. Zaat, S. A. J., Van Brussel, A. A. N., Tak, T., Pees, E., and Lugtenberg, B. J. J., "Flavonoids induce *Rhizobium leguminosarum* to produce *nodABC* gene-related factors

that cause thick, short roots and root hair responses on common vetch," *J. Bacteriol.* *169*, 3388–3391 (1987).

248. Zaat, S. A. J., Wijffelman, C. A., Spaink, H. P., Van Brussel, A. A. N., Okker, R. J. H., and Lugtenberg, B. J. J., "Induction of the *nodA* promoter of *Rhizobium leguminosarum* Sym plasmid pRL1JI by plant flavonones and flavones," *J. Bacteriol.* *169*, 198–204 (1987).

249. Zaat, S. A. J., Wijffelman, C. A., Mulders, I. H. M., Van Brussel, A. A. N., and Lugtenberg, B. J. J., "Root exudates of various host plants of *Rhizobium leguminosarum* contain different sets of inducers of *Rhizobium* nodulation genes," *Plant Physiol.* *86*, 1298–1303 (1988).

250. Zaat, S. A. J., Van Brussel, A. A. N., Tak, T., Lugtenberg, B. J. J., and Kijne, J. W., "The ethylene-inhibitor aminoethoxyvinylglycine restores normal nodulation by *Rhizobium leguminosarum* biovar *viciae* on *Vicia sativa* subsp. *nigra* by suppressing the 'thick and shorts roots' phenotype," *Planta 177*, 141–150 (1989).

251. Zaat, S. A. J., Schripsema, J., Wijffelman, C. A., Van Brussel, A. A. N., and Lugtenberg, B. J. J., "Analysis of the major inducers of the *Rhizobium nodA* promoter from *Vicia sativa* root exudate and their activity with different *nodD* genes," *Plant Mol. Biol. 13*, 175–188 (1989).

252. Zevenhuizen, L. P. T. M., Scholten-Koerselman, I., and Posthumus, M. A., "Lipopolysaccharides of *Rhizobium*," *Arch. Microbiol. 125*, 1–8 (1980).

253. Zhan, H., Levery, S. B., Lee, C. C., and Leigh, J. A., "A second exopolysaccharide of *Rhizobium meliloti* strain SU47 that can function in root nodule invasion," *Proc. Natl. Acad. Sci. USA 86*, 3055–3059 (1989).

254. Ziegler, R. J., Pierce, C., and Bergman, K., "Mapping and cloning of a *fla-che* region of the *Rhizobium meliloti* chromosome," *J. Bacteriol. 168*, 785–790 (1986).

255. Zurkowski, W., "Specific adsorption of bacteria to clover root hairs related to the presence of plasmid pWZ2 in cells of *Rhizobium trifolii*," *Microbios 27*, 27–32 (1980).

10

Physiology of Nitrogen-Fixing Legume Nodules: Compartments and Functions

Dietrich Werner

I. The Compartments

A. The Symbiosome

The term *symbiosome* proposed by Roth et al. includes the bacteroid, the symbiosome space, and the peribacteroid (symbiosome) membrane as a functional unit.[96] It is a very useful term, emphasizing the homology to other cell organelles, such as chloroplasts and mitochondria developed in evolution by independent endosymbiotic processes.[115]

i. The bacteroid

The differentiation of free-living cells of *Rhizobium* and *Bradyrhizobium* to bacteroids includes the changes summarized in Table 10-1. In the endosymbiotic stage, bacteria of the genus *Rhizobium, Bradyrhizobium, Sinorhizobium,* or *Azorhizobium* are called bacteroids. In size and shape, they can be similar to the free-living form, as in the genus *Bradyrhizobium* (Figure 10-1) or irregular and branched, as it is often found in biovars of the genus *Rhizobium.* DNA content in bacteroids per cell can be four to eight times larger than that in the free-living stage in *Rhizobium.*[13] However, this is due to the multicellular form of the bacteroids, which increase their size without cell division. In *Bradyrhizobium japonicum* it has been demonstrated that, per unit cell volume, the DNA content of both forms is about the same.[124,79] *In vitro* forms of bacteroids have been produced in soft agar cultures.[80] Between an upper layer and a bottom layer with rod-shaped cells was a narrow zone with pleomorphic bacteroidlike cells that included polyphosphate bodies and a polar distribution of PHB. In the nodules the bacteroid form is under host plant control, as demonstrated for one strain of *Rhizobium,* which in *Vigna mungo* formed rod-shaped bacteroids, whereas in *Arachis hypogaea* single spherical bacteroids were enclosed in peribacteroid membranes.[19]

Table 10-1   Differentiation of Bacteroids from Free-Living Cells of Rhizobium sp. and Bradyrhizobium sp. Modified from 115, 105

Process	Bacteroid State
Viability	Only slightly reduced
Osmotic sensitivity	Enhanced
EPS formation	Reduced
DNA content	Unchanged, related to the same cell size
Lipid metabolism	PHB accumulation, also in Fix⁻ strains of Bradyrhizobium
Electron transport	Special branch, without cytochromes c and aa_3
Hem-synthesis	Enhanced (probably for leghemoglobin)
Nitrogenase	up to 10% of the soluble protein
NifA-dependent processes	under oxygen control

Bacteroids from *Rhizobium* as well as from *Bradyrhizobium* strains can re-differentiate to free-living viable cells. Twenty to 90% of colony-forming bacteroids have been isolated from *Glycine max, Ornithopus sativus,* and *Phaseolus aureus.* Supplementation of the media with 0.2 or 0.3 *M* mannitol increases the percentage of colony-forming units.[104,112]

Biochemical and genetic evidence is increasing that *Bradyrhizobium japonicum* contains three different respiratory chains.[36] One chain is similar to the one in mitochondria:

$$2 \, H \rightarrow Q> \rightarrow FeS/bc_1 \rightarrow c \rightarrow aa_3 \rightarrow O_2$$

In bacteroids there is another specific branched electron transport pathway, leading from the bc_1 complex to the terminal oxidase, since mutations in cytochrome c and aa_3 were Fix⁺.[36] Cytochrome c oxidase purified from bacteroid membranes is apparently a glycoprotein.[60] Other bacteroid-specific cytochromes are P450, P420, c552, and c554.[3] These bacteroid-specific cytochromes are apparently different in *Rhizobium* and *Bradyrhizobium.* In bacteroids of *Rhizobium leguminosarum* biovar *viciae* no cytochrome P450 was found; instead a protein was found with a Soret peak at about 424 nm.[51] The somewhat larger O_2-tolerance of pea bacteroids compared to bacteroids from soybeans may be related to this protein.[3]

ii.   The Symbiosome Space

As can be seen on TEM sections (see Figure 10-1), the symbiosome space is apparently not empty but contains, besides the bacteroid, other components—with a much lower density, however, than the plant cytoplasm. Solid biochemical identification of components in this compartment requires that cross-contamination from the host cell cytoplasm, host cell mitochondria,

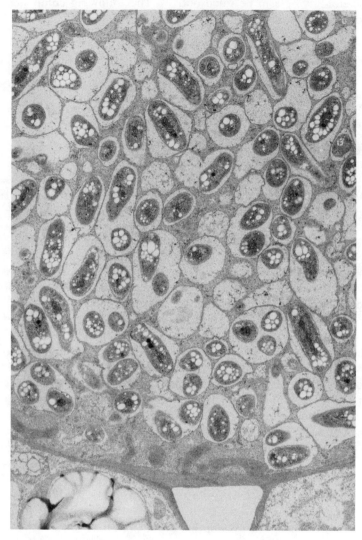

Figure 10-1. TEM section from an infected cell in nodules of *Glycine max* cv. Caloria with bacteroids from *Bradyrhizobium japonicum* 31-6-85. Stable symbiosomes fill the cell. Close to the cell wall a number of amyloplasts (without starch) are concentrated. (Original graph from PD Dr. E. Mörschel and C. Zimmermann) 13 600 × magnification.)

plastids, ER, and Golgi as well as contamination from bacteroid periplasm and bacteroid cytoplasm be ruled out.[69,48] The following components have been identified in this compartment:

Acid phosphatase[49]
α-mannosidase II[48]
Proteases[68]
Protease inhibitors[32]
α-glucosidase[49]
Aspartate aminotransferase[49]
Trehalase[64]

The symbiosome space can be considered as a special form of lytic compartment.[65] This is obvious in soybean nodules infected with *Bradyrhizobium japonicum* strain RH 31-Marburg, where lysis of bacteroids occurs in the vicinity of the host cell nucleus rather early after infection.[121] Alpha-mannosidase III is known to be a marker for the extracellular space, whereas the isoenzymes I and II are markers for the plant vacuome (one or several vacuoles). Since only isoenzyme II of α-mannosidase has been found in the symbiosome space, this means that the latter is different from the plant vacuome as well as from the extracellular space. How bacteroids in an effective symbiosis are stabilized in the symbiosome space in the presence of several hydrolases is not completely understood. Proteases could be inhibited by the presence of protease inhibitors.[32] Other hydrolases may be regulated by pH. A model for pH balance in symbiosis was proposed by Kannenberg and Brewin.[46] The space would be acidified by two transport activities through the peribacteroid membrane, the dicarboxylic acid transport and a vectorial protein ATPase. On the other side, actively N_2-fixing bacteroids would increase the pH by uptake and utilization of malate/succinate and by the excretion of ammonia into the symbiosome space. A consequence of this model would be a somewhat lower pH in *fix⁻* symbiosomes.

iii. The Peribacteroid (Symbiosome) Membrane (PBM)

The peribacteroid (symbiosome) membrane (PBM) is produced by the host plant by a membrane flow from the Golgi and/or directly from the ER. The fatty acid composition of the PBM in nodules of *Glycine max* is more similar to the ER than to the Golgi.[9] In this well-established pathway, it was shown by a number of independent methods that the microsymbiont has a decisive role in formation, stability, protein composition, and particle density of this membrane. This evidence follows.

1. The PBM from soybean nodules infected by mutant RH 31-Marburg of *Bradyrhizobium japonicum* has in the PF face a particle density of only 1200–1300 particles μm^{-2} compared to the wild type, with about 2200 particles in the PF face.[123]

2. A discrete set of particles in the size of 13–14 nm are missing in the EF face of the mutant-affected PBM compared to the wide type.[123]

3. Glycosylation of several PBM proteins is affected by the microsymbiont.[123]

4. Three other mutants of *Bradyrhizobium japonicum* are responsible for the loss of different specific PBM proteins.[71]

5. Different wild-type strains of *Bradyrhizobium japonicum* (strain 61-A-101 and strain 110*spc*4) change in the ratio of $18:n/16:0$ unsaturated to saturated fatty acids in the PBM significantly.[9]

6. The PBM in nodules of soybeans infected with *Bradyrhizobium japonicum* 61-A-24 shows early disintegration, followed by a significant accumulation of the phytoalexin glyceollin I in nodules to the same level as induced by a heavy infection with a phytopathogenic fungus.[122] The same type of hypersensitive reaction in glyceollin I accumulation can be produced with the *nifA* mutant 110*spc*4/A9 of *Bradyrhizobium japonicum*.[82] A hypersensitive response can also be produced in nodule cells with a wild-type *Bradyrhizobium japonicum* (nod[+] fix[+]) in combination with specific genotypes of *Glycine soja*[81] (Figure 10-2).

The conclusion is that there are at least four different types of signals (at least as many genes) in the microsymbiont that affect production, stability, and protein composition of the peribacteroid membrane.

Nodulin-24 and nodulin-26 have been identified to be PBM associated.[30] In the biosynthesis of the PBM a specific form of choline kinase (choline kinase II) is involved. In soybean it is present in significant activities only in nodules producing a stable peribacteroid membrane.[68] The occurrence of this isoenzyme is not related to N_2 fixation, since *fix*[−] nodules with a stable peribacteroid membrane (as produced by strain 110*spc*4/A3) also have choline kinase II.[70]

The following functions have been identified to be present in the PBM:

1. A H^+-ATPase of the host plasma membrane type[15] and probably a second ATPase of the vacuolar type,[6]
2. A calcium-dependent protein kinase activity,[7]
3. A dicarboxylate carrier,[20]
4. A magnesium-dependent pyrophosphatase.[8]

Summarizing the biosynthesis in the development of this membrane system, it is very interesting to compare it with the formation of protein bodies in

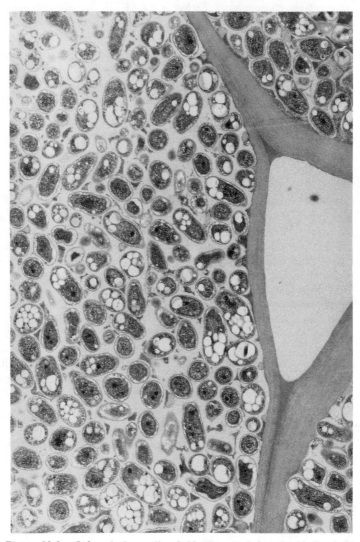

Figure 10-2. Infected plant cells of *Glycine soja* infected with *Bradyrhizobium japonicum* USDA 113 at the 29–33th d.p.i. showed massive cell wall thickening. Bacteria inside the plant cells accumulated large amount of PHB and had reached higher cell densities than in the earlier stages. They were no longer surrounded by a PBM or by any other intact plant structure, indicating that plant cells had undergone a hypersensitive reaction.[81] 13,600 × magnification.

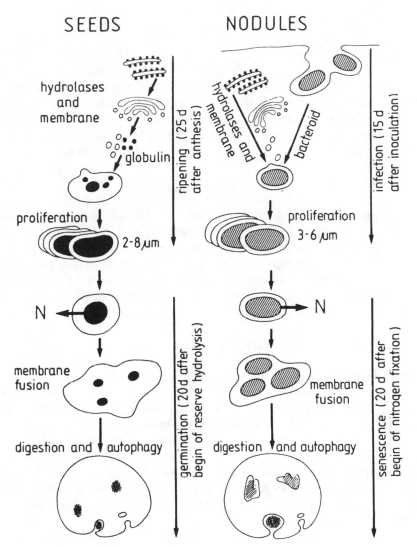

SEEDS NODULES

Figure 10-3. Schematic representation of the development cycle of protein bodies in soybean seeds/cotyledons compared to that of endosymbiotic *Bradyrhizobium*-containing organelles in soybean root nodules. N = fixed nitrogen. The time scale given is approximate and depends heavily upon external factors.[65]

seeds or cotyledons from seed legumes (Figure 10-3). In this respect the symbiosome can be considered to be homologous to protein bodies. There proteins are stored and finally exported, whereas in the symbiosome, nitrogen is fixed and finally exported. Membrane fusion, digestion, and autophagy in senescence are similar.

B.   The Host Cell

i.   The Cytoplasm

After infection the host cell cytoplasm becomes dense and granular in EM photography. The number of free ribosomes increases drastically in all but a few ineffective symbioses. The cytoplasm becomes red because of the appearance of leghemoglobin.[18,58] A host-cytoplasm-located galactosidase is stimulated at this time.[72] Its function, however, remains unknown.

ii.   The Nucleus

At a distinct stage of nodule development cell division in the cortex slows down and then stops. DNA synthesis continues and the host cell nucleus becomes polyploid, DNA levels in the bacteroid zone up to 32C (diploid = 2C) are reached.[63] Enlarged nuclei are to be seen in infected host cells, often with several nucleoli. Nuclei with crenated borders have been reported and thus probably seem to increase the surface area of the nuclear membrane to allow the passage of more mRNA into the cytoplasm.

iii.   Endoplasmic Reticulum (ER), Golgi

During the early stages of cell colonization, ER vesicles become more numerous throughout the host cell. During this time the ER-located enzymes choline phosphotransferase and GDP-DMP-mannosyltransferase are stimulated by 200% and 300%, respectively.[68] Direct connections between the ER and PBM are probably not significant, but a direct transfer of components by vesicle shuttle from the ER matrix to the symbiosome space still cannot be ruled out. The case of α-mannosidase II is an example. The symbiosome space-located isoenzyme II is synthesized on bound ribosomes of the rough ER[113] and remains in the plant vacuome,[48] but it has not been found in the Golgi.

The enzyme UDP-galactose-asialogalactofetuin-galactosyltransferase is found in both ER and Golgi[67] and provides subterminal galactose for glycoconjugates. This activity is stimulated 800% in effective symbioses, which is less than the 1600% stimulation recorded for the N-acetylgalactosaminyltransferase, which is completely Golgi-located[67] and provides terminal amino-sugar residues for glycoconjugates. Since PBM contains products of both the preceding enzymes, this provides biochemical support for the observations of Robertson and coworkers[90–92] that the Golgi is a major step in one of the pathways providing material for the PBM. Of the putative Golgi marker enzymes, glucan synthetase is not measurable in infected cells,[78] although IDPase is easily detectable. In transmission electron micrographs dictyosome bundles are often seen adjacent to peribacteroid membranes,[78] and freeze-fracture studies reveal continuities between Golgi and PBM in the form of vesicles.[90]

iv. Plastids and Mitochondria

Mitochondria in infected cells are cristae-rich through intensive folding of the inner mitochondrial membrane and move to the cell periphery.[116] This movement is probably associated with the need for O_2 for oxidative respiration in the mitochondria. The pO_2 inside the infected cells has been measured with microelectrodes to be $10nM$. Thus, mitochondrial respiration is O_2-limited,[88] and therefore their position at the plasma membrane probably reflects their "wandering" up the cellular O_2-gradient to the highest partial pressure. The biochemical modification in mitochondria of infected cells is more or less unresearched, which is astonishing, considering their central role in N_2 fixation. The limitation of O_2 in the infected cells means that mitochondrial metabolism is limited. This, in turn, restricts the supply of C_4 skeletons to the bacteroid, leading to energy limitation and lower N_2 fixation. N_2 fixation can be increased by 20–40% by doubling the atmospheric pO_2[120] (Figure 10-4). Thus, manipulations in host cell mitochondria may well lead to higher agricultural yields.

Plastids have been isolated,[16] but as with mitochondria, a thorough biochemical analysis has not yet been undertaken. In EM pictures plastids can often be seen closely associated with mitochondria.[118] This probably reflects the provision of carbon skeletons and energy by mitochondria for the plastid-located enzymes of NH_4^+ assimilation. O_2 consumption in isolated mitochondria from soybean nodules was about 180 μl O_2 min^{-1} mg^{-1} protein[85] with an estimation that about 80% of the mitochondria were from infected cells. No cyanide-insensitive respiration was noted. About the same activity was reported for mitochondria from cowpea nodules.[89] This means that, despite the low oxygen concentration in infected cells, the mitochondria can have a very active respiratory metabolism. The activity of the mitochondrial malate-dehydrogenase (MDH) in lupin nodules was estimated to be about 20% of the cytoplasmic MDH.[111]

C. The Nodules as New Plant Organs

Organization and development of the nodule meristem for indeterminate nodules are summarized in Figure 10-5. From top to bottom we can follow the development from the meristem the infection threads to the development of the symbiosomes, the mature symbiosomes with the N_2-fixation zone, and finally the senescence zone. The matrix glycoproteins, enod 12 and enod 2, are examples of proteins that apparently also have a function in plant

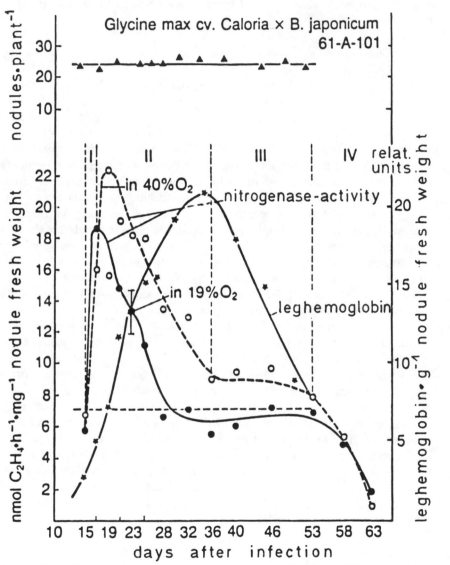

Figure 10-4. Nodule number, nitrogenase activity under 19% and 40% oxygen and leghemoglobin concentrations in nodules of *Glycine max* cv. Caloria infected with *Bradyrhizobium japonicum* 61-A-101, 15-63 d after infection. (Modified from ref. 115.)

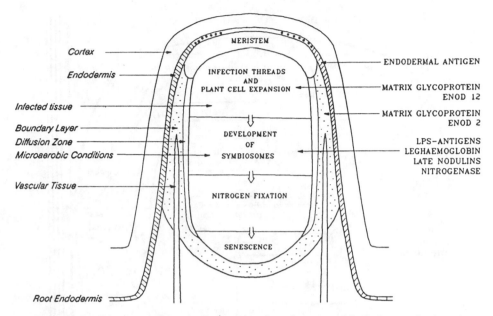

Figure 10-5. Diagrammatic longitudinal section through a pea nodule. Left, organization of cell layers; center, tissue invasion by *Rhizobium* sp. and the development of nitrogen fixation; right, the development of gene expression identified using antibody or cDNA probes.[46,75]

development in general. Compartments and functions for a determinate nodule are summarized in Figure 10-6.

II. The Functions

A. *Nitrogen Fixation*

The legume nodules can be considered as optimized organs for the functions of a N_2-fixing cell organelle (the symbiosome) just as the leaves of plants are the optimized organs for light-dependent CO_2 fixation by chloroplasts or other plastids.

N_2 fixation in legume nodules follows the following general equation:

$$N_2 + 8\,e^- + 8\,H^+ + 16\,MgATP \rightarrow 2NH_3 + H_2 + 16\,MgADP + 16\,Pi$$

i. *Rates of N_2 Fixation*

The rate of N_2 fixation in legumes depends on the host plant species (Table 10-2) and, in addition, on the microsymbiont and general factors, such as

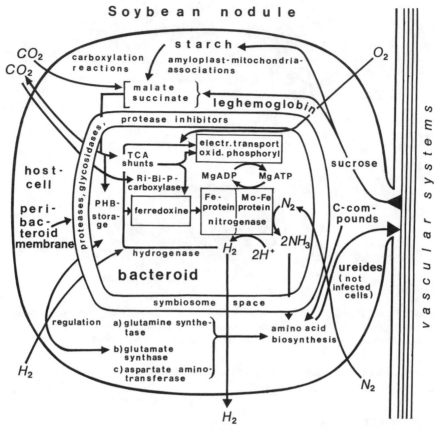

Figure 10-6. Scheme of the functions in a soybean-nodule. (Modified from refs. 115 and 27.)

soil type, temperature, macronutrients, micronutrients, and water regime. The agricultural legumes clover and alfalfa seem to have the highest average fixation rate, which is on the order of 250 kg N per hectare per year, whereas the reported data for soybeans, peanuts, and lentils are on the average only about 100 kg N. A much higher fixation rate was reported for *Sesbania rostrata,* the stem-nodulated tropical legume with figures of about 700 kg N ha^{-1} year^{-1}, combining the data reported for the dry season and the wet season.[53] For fast-growing N_2-fixing legume trees fixation rates of 80–150 kg N ha^{-1} year^{-1} were calculated.[74,93] A very interesting but hardly understood relationship apparently exists between the general growth rate of the host plant and the N_2-fixing capacity. Plants rapidly producing large vegetative and generative sinks (leaves, stems, and seeds) apparently have a large demand for nitrogen as well as for carbon from photosynthesis. Rapid N_2 fixation, on the other hand, can also enhance vegetative growth and seed

Table 10-2 N_2 *Fixation in Legumes*[86,77,53,74,93]

Legume	Fixation (kg N ha^{-1} year^{-1})	
	Minimum/Maximum	Average
Clover	45–670	250
Pea	50–500	150
Alfalfa	90–340	250
Lupin	140–200	150
Soybean	60–300	100
Peanut	50–150	100
Lentils	50–150	80
Broadbean	100–300	200
Sesbania rostrata	600–800	700
N_2-fixing trees	80–500	150

yield.[44] Seed yield in nodulated soybeans of 3000 kg/ha is comparable to plants fertilized in addition with 45–134 kg N/ha (Table 10-3). In terms of mean seed weight, the nodulated plants were superior to all plots fertilized with nitrate or urea. There is experimental evidence for alfalfa[22] as well as for soybeans[43] that the growth limitation of plants changes from source limitation to sink limitation. This means that during the later reproductive growth stages, the maximum photosynthetic activity can be achieved with only half maximum sunlight. At this stage the capacity of soybeans for nitrate uptake is already significantly reduced with less than 50% of its maximum capacity between development stages R5 and R6.[43] During this decline of nitrate reduction capacity an effective nodulated soybean plant can reach the maximum N_2-fixation capacity and thereby increase the growth of the plant and increase seed production as well. The inhibition of nodulation by nitrate, especially in soybeans, further complicates the interaction of nitrate assimilation and N_2 fixation. The genetic basis and the physiological implications of these effects have been studies intensively by P. Gresshoff and his co-workers.[21,23] Increased concentrations of nitrate from 0.5mM to 7.5 mM reduce the nodule number on wild-type soybeans of cultivar Bragg from 100 to 50 nodules per plant. In contrast, in the supernodulating mutant nts382 nodule number remains high, with up to 400 nodules per plant (Table 10-4). With a large number of nitrate-tolerant supernodulating soybean mutant lines it was established that, with one exception, in each line the same single gene is involved in this specific character.[23]

ii. Assimilation of Fixed Nitrogen

The N_2 fixed in the bacteroids is exported through the inner and outer membrane of the microsymbiont and through the peribacteroid membrane into

Table 10-3 Contribution by N_2 Fixation to Mean Seed Weight, Seed Yield, and N Content of Field-Grown Soybean[22]

Fertilizer N Added at	Rate of N Fertilizer	Mean Seed Weight		Increase in Mean Seed Weight	Seed Yield		Increase in Seed Yield	Increase Due to Seed Size	Total Seed N		Increase in Total Seed N
		Non-nodulated	Nodulated		Non-nodulated	Nodulated			Non-nodulated	Nodulated	
	kg ha⁻¹	mg		%	kg ha⁻¹		%	%	kg ha⁻¹		%
Planting	0	107	164	53.3	1300*	2920*	125.0	42.6	53	157	196.0
Planting	45	117	165	42.0	1747	3127	79.0	51.4	73	175	140.0
Planting	89	135	162	20.0	2307	2813	21.9	91.3	101	156	54.5
Planting	134	142	164	15.5	2380*	3040	27.7	56.0	108	172	59.3
R2	45	139	166	19.4	2153	2787	29.4	66.0	95	159	67.4
R2	89	158	170	7.6	2527	2913	15.3	49.7	115	164	42.6
R2	134	158	166	5.1	2840	3113	9.6	53.1	133	174	30.8
Avg.		136	165		2179	2959		58.7	97	165	

*Assuming 6.4% N, 1300, 2380, and 2920 kg of soybean seeds contain 83, 152, and 187 kg of N, respectively.

Table 10-4   The Effect of Applied Nitrate on Nodulation and Plant Dry Weight of Soybean cv. Bragg and Mutant nts383[a]

Genotype	Inoculum Dose, Cells per Pot	Nodule Number		Nodule Fresh Weight, g		Plant Dry Weight, g	
		0.5 mM KNO$_3$	7.5 mM KNO$_3$	0.5 mM KNO$_3$	7.5 mM KNO$_3$	0.5 mM KNO$_3$	7.5 mM KNO$_3$
cv. Bragg	10^9	107 ± 8	51 ± 10	1.07 ± 0.08	0.26 ± 0.03	2.89 ± 0.22	2.72 ± 0.35
nts382	10^9	304 ± 75	398 ± 59	1.91 ± 0.28	1.62 ± 0.16	1.34 ± 0.16	1.50 ± 0.08

[a]Plants were harvested at 50 days after sowing.[21]

the host plant cytosol. The pathways of ammonium assimilation in the host plant cytoplasm are summarized in Figure 10-7. The major enzymes involved catalyze the following reactions:

NH₃ Assimilation

Glutamine synthetase (ED 6.3.1.2):

$$2\ NH_3 + 2\ \text{L-glutamate} + 2\ ATP \rightarrow 2\ \text{L-glutamine} + 2\ ADP + 2\ Pi$$

Glutamate synthase (EC 2.6.1.53):

$$\text{2-oxoglutarate} + \text{L-glutamine} + NADH + H^+ \rightarrow 2\ \text{L-glutamate} + NAD^+$$

Asparate aminotransferase (EC 2.6.1.1):

$$\text{L-glutamate} + \text{oxaloacetate} \rightarrow \text{L-aspartate} + \text{2-oxoglutarate}$$

Asparagine synthetase (EC 6.3.5.4):

$$\text{L-aspartate} + \text{L-glutamine} + ATP \rightarrow \text{L-asparagine} + \text{L-glutamate} + AMP + PPi$$

Overall Reaction

$$2NH_3 + 3\ ATP + \text{oxaloacetate} + NADH + H^+ \rightarrow \text{L-asparagine} +$$
$$AMP + 2\ ADP + 2\ P_1 + NAD^+ + PPi$$

The overall costs for the fixation and assimilation of 1 mol N_2 is thereby 19 ATP. Compared to the 3 mol ATP per mol CO_2 fixed in C3 plant photosynthesis or 5 ATP in C4 plant photosynthesis, and to the assimilation of other inorganic anions, N_2 fixation is by far the most costly primary fixation process in terms of energy.

Glutamine synthetase (GS) is an octameric enzyme with a molecular weight of 860,000. Four different subunits (alpha, beta, gamma, and delta) have been identified in nodules of *Phaseolus vulgaris*.[10] The subunit delta is a plastidic form, the three others are cytosolic subunits. From these subunits two different nodule GS forms are synthesized, whereas roots contain only a single GS. Leaves also contain two isoenzymes, the cytosolic GS 1 and the chloroplast-associated form GS 2. Regulatory compounds for GS (e.g., GTP and AMP) affect both enzymes in the same way, whereas glucosamine-6-phosphate only inhibits GS 2.[54,37] Compared to roots, GS activity in nod-

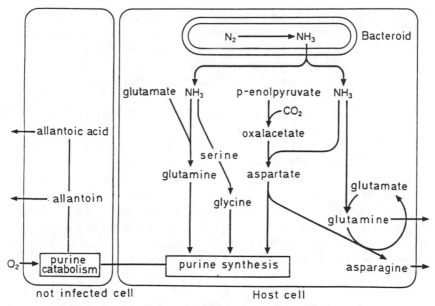

Figure 10-7. Pathways of N-compounds in infected and uninfected cells from ureide-exporting legume nodules. Modified from 25.

ules is about 15 times greater and represents more than 2% of the soluble nodule protein.[62] GS1 was expressed normally in nodules formed by *Brady-rhizobium* strains defective in nitrogenase activity, indicating that GS1 induction is not dependent on nitrogenase. The data on subunits of GS in soybean nodules are less clear than those for *Phaseolus* beans.

The next enzyme in the assimilation pathway is GOGAT; two NADH-dependent forms have been found in nodules of *Phaseolus vulgaris,* whereas in alfalfa nodules only one isozyme was detected.[2] In contrast to GS, GO-GAT seems to be a monomer with a high molecular mass of about 230 kDa. The K_m values for glutamine, 2-oxogultarate and NADH are 466 μM, 33 μM, and 42 μM^2, respectively. Expression of the nodule NADH-GOGAT in other plant organs is very low. Antibodies against the ferredoxin-dependent GOGAT do not cross-react with the NADH-dependent GOGAT, indicating that these enzyme are distinctly different. Besides the NADH-dependent GOGAT there is also a soybean nodule ferredoxin-dependent GOGAT that is readily recognized by antibodies against rice leaf ferredoxin-GOGAT.

The enzyme for formation of aspartate from glutamate is the aspartate aminotransferase (AAT). Two immunologically different forms have been detected in nodules of alfalfa (AAT-1 and AAT-2) with molecular masses of 42 and 40 kDa.[28] AAT form 2 is a dominant form in nodules. In roots about 60% of activity is present in AAT-1. Two forms of AAT have also been found in soybeans[97] and in peas.[5] AAT-2 is not a nodulin, but a nodule-enhanced protein with about a six-fold increase from roots to nodules on a fresh weight basis.[29] The nodule-enhanced form of AAT (synonymous with glutamate oxaloacetate transaminase/GOT) has much higher K_m values for glutamate (27 mM), oxaloacetate (0.14 mM), and aspartate (4.9 mM) as compared with the root form with 7.9 mM, 0.019 mM, and 0.84 mM for the three substrates.[5]

The central role of asparagine synthetase (EC 6.3.5.4) in amide-exporting nodules can be easily demonstrated by determination of amino acid concentrations in the cytosol.[106] The concentration of asparagine in the host plant cytosol was 44 mM. Ammonia concentration was 4.3 mM, glutamate concentration, 0.2 mM; and alanine concentration, 2.2 mM. The concentration of all other amino acids was below 1 mM. Asparagine synthetase activity increased 170-fold in nodules of soybeans compared to cortex tissue.[38] The apparent K_m values for the substrate aspartate, Mg-ATP, and glutamine are 0.24 mM, 0.076 mM, and 0.16 mM. The K_m for ammonium ions as a N donor is 40 times larger than that for glutamine.

In ureide-exporting legumes (e.g., soybeans and *Phaseolus* beans) glutamine, glycine, and aspartate are used for purine synthesis in plastids of the infected cells (see Figure 10-7). Purine nucleotide is transformed to xanthine and uric acid, which is transformed by peroxisomes in neighboring uninfected cells to allantoin. This is further transformed in the smooth ER of uninfected cells to allantoic acid.[72] The nodulin uricase II, located in the peroxisomes of the uninfected cells, is involved in this metabolism.[76] There is a strong correlation between N_2-fixation rates and the total N and ureides in soybeans. It is interesting that allantoin amino hydrolase responds to inhibition of N_2 fixation and urate transport by NH_4NO_3.[14]

B. Carbon Metabolism

i. Organic Substrates for Bacteroids

The carbon metabolic pathways in bacteroids of *Bradyrhizobium japonicum* are summarized in Figure 10-8. The major substrate entering from the host plant through the peribacteroid membrane into the symbiosome space, where it will be available for the bacteroid, is most likely to be malate and/or succinate. The evidence for this follows.

1. C4-dicarboxylate transport mutants of fast-growing rhizobia form ineffective (fix⁻) nodules.[94,17]

2. In *Bradyrhizobium japonicum,* a Tn5 mutant in dicarboxylic acid transport (succinate uptake system I reduced to about 10%, and the low-affinity uptake system II to about 50%) under free-living conditions had a N_2-fixing activity of only 4–7% of the wild type.[41,39]

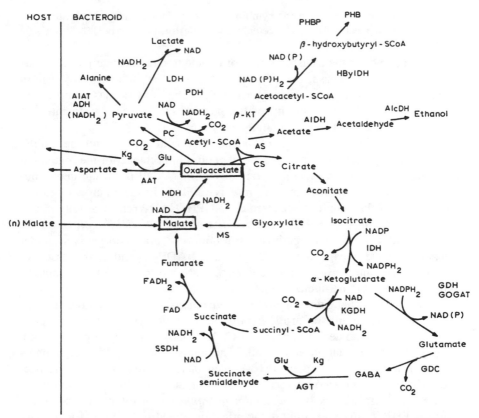

Figure 10-8. Carbon metabolic pathways in *Bradyrhizobium japonicum* bacteroids: (1) TCA cycle, CS, citrate synthase; IDH, NADP-isocitrate dehydrogenase; KGDH, α-ketoglutarate, dehydrogenase; MDH, malate dehydrogenase. (2) α-Ketoglutarate-glutamate shunt and δ-aminobutyrate pathway, GDH, glutamate dehydrogenase; GOGAT, glutamine: oxoglutarate aminotransferase; GDC, glutamate decarboxylase; AGT, δ-aminobutyric-glutamic transaminase; SSDH, succinate semialdehyde dehydrogenase; (3) Malate-aspartate shuttle, MDH, malate dehydrogenase; AAT, aspartate aminotransferase. (4) Modified dicarboxylic acid cycle, MS, malate synthase; MDH, malate dehydrogenase; PC, pyruvate carboxylase; PDH, pyruvate dehydrogenase. (5) Fermentation reactions, AS, acetyl-SCoA synthetase; AIDH, Aldehyde dehydrogenase; AlcDH, alcohol dehydrogenase; LDH, lactate dehydrogenase. (6) Poly-β-hydroxybutyrate shunt, β-KT, β-ketothiolase; HBylDH, β-hydroxybutyryl-SCoA reductase; PHBP, poly-β-hydroxybutyrate polymerase. (7) Other, AlAT, alanine aminotransferase; ADH, alanine dehydrogenase, Glu, glutamate, Kg, α-ketoglutarate. (Modified from ref. 61.)

3. Bacteroids take up dicarboxylic acids at a rate 30 to 50 times higher than sugars, and oxidize them very rapidly to CO_2.[98,99]

4. Malate and succinate are rapidly transported across the peribactroid membrane via a dicarboxylate carrier, whereas glucose and fructose uptake by symbiosomes is very slow.[20]

5. Malate is present in the symbiosome space in significant concentrations, meaning that it is actually available for the bactroids.[40]

A key enzyme for the transformation of malate to oxaloacetate in bacteroids is malate dehydrogenase, one of the few enzymes for carbon metabolism in bacteroids that has been studied in detail.[26] The enzyme has a molecular weight of 35 870. The activity of the purified enzyme is significantly reduced by addition of 2-oxoglutarate and by NADP.

The two following reactions of oxaloacetate with citrate synthase, on the one hand, the AAT, on the other, are of central importance for the regulation of carbon metabolism. The specific activity of NADP isocitrate dehydrogenase is reduced by 50% at 0.25 mM NADPH.[61] In the microaerobid environment of the nodule the bacteroid NADPH/NADP ratio was determined to be 2.7.[107] In *Azotobacter beijernickii* and in *Escherichia coli* the TCA cycle shifts from a cyclic pathway to a noncyclic branched pathway under anaerobid conditions, since citrate synthase, NADP-dependent isocitrate dehydrogenase and 2-oxoglutarate dehydrogenase are inhibited by accumulating reduced pyridine nucleotides.[45,1]

One striking characteristic of *Bradyrhizobium japonicum* carbon metabolism is the accumulation of PHB (poly-β-hydroxybutyrate). The activity of β-hydroxybutyrate dehydrogenase (E.C. 1.1.1.30) increases in parallel with the PHBA accumulation and parallels nitrogenase activity.[126] Tn5 mutants of *nifD* and *nifH* accumulate PHBA even faster.[35] Also in wild-type bacteroids PHBA can account for up to 50% of the dry weight of the microsymbiont.[50] Bacteroids of *Rhizobium leguminosarum* cv *viciae* can accumulate PHBA, however, only in the infection thread.[125] The specific activity of β-hydroxybutyrate dehydrogenase in effective (*fix*⁺) bacteroids was two to three times greater than that in ineffective (*fix*⁻) bacteroids.[118] The activity of the same enzyme in nitrogenase derepressed free-living cell of *Bradyrhizobium japonicum* also was about 50% greater than that in nitrogenase-repressed cells from liquid culture.[117]

Besides the oxaloacetate/aspartate pathway and the PHB pathway, the 2-oxoglutarate/glutamate pathway is apparently present in bacteroids, since a significant part of the TCA cycle compounds are converted to glutamate.[98]

ii. Host Cell Carbon Metabolism

In the host cell, as in the symbiosome, the microaerobic conditions (10 nM O$_2$) can affect the mitochondrial and cytosolic TCA cycles. As in flood-tolerant plants, the increased PEP-carboxylase/malate dehydrogenase pathway may be a response to microaerobic conditions. Those enzymes are increased 5- to 50-fold in nodules compared to roots in *Vicia faba*[55] and in *Pisum sativum*.[24] After ¹⁴CO$_2$ fixation in effective nodules, investigators found

80% of the ^{14}C label in malate and about 10% in citrate,[95] whereas the label in 2-oxoglutarate was very low. This indicates a low activity of isocitrate dehydrogenase in host cells.[47] The percentage of label in succinate after ^{14}CO$_2$ was supplied is rather low. Therefore, succinate dehydrogenase may be blocked as well, perhaps by the significant concentration of malonate in the cytosol of nodule host cells.[103] No malonate was detected in the symbiosome space, indicating that the PBM is a barrier for this competitive inhibitor of succinate dehydrogenase. Nevertheless, bacteroids of *Bradyrhizobium japonicum* are able to metabolize malonate. This activity is completely inhibited by 1 mM succinate, whereas the utilization of succinate is only reduced by 20-30% in the presence of a 100-fold excess of malonate. However, this capacity may be only important in senescent nodules and in bacteroids redifferentiating to free-living cells when the compartmentation by the peribacteroid membrane is lost.[119]

The main carbon source transported from the host plant into nodules is sucrose; there it is broken down by a catabolic form of sucrose synthase, a four-subunit enzyme that is identical to nodulin 100.[108] In addition, acid invertase (E.C. 3.2.1.26), δ-trehalase (E.C. 3.2.1.28), and maltase (E.C. 3.2.1.20) have been localized in the soybean nodule cytosol.[101] After conversion of UDP-glucose to glucose-6-phosphate, a significant percentage of the carbon source is used to form starch in the amyloplasts of infected and of noninfected cells. This is probably a carbon and energy buffer for N$_2$ fixation or for other stress periods when photosynthesis may be slightly reduced.[72] Several other enzymes leading from glucose-6-phosphate to acetyl CoA in the Emden-Meyerhof pathway and branching from intermediates to other compounds have been characterized. Phosphoenolpyruvate carboxylase has a Michaelis constant for PEP of 9.4×10^{-2} mM and for HCO$_3^-$ of 4.1×10^{-1} mM.[84] This means that rather high concentrations of HCO$_3^-$ have to be present. Pyruvate kinase from soybean nodule cytosol is inhibited by NH$_4^+$ in the presence of physiological concentrations of potassium and high PEP concentrations, whereas in the absence of potassium, ammonium ions activate the enzyme.[83] Rather high activities of malic enzyme (E.C. 1.1.1.40) have been found in the cytosol of lupin nodules with production of 4 μmol NADP $\times$ min^{-1} $\times$ g^{-1} fresh weight.[110] The affinity of the enzyme for pyruvate was several times lower than that for malate.

Estimates of the carbohydrate requirements for N$_2$ fixation by soybean nodules are summarized in Table 10-5. Adding the data to estimate the total requirement for N$_2$ fixation + H$_2$ evolution—for nodule maintenance, assimilation and transport of ammonia, and growth—indicates about 12 g carbohydrate per g N fixed.[102,87,33] In experiments using ^{11}CO$_2$ and ^{14}CO$_2$ labeling, translocation of assimilates in 3 to 4-week-old soybean plants from the leaves to the roots and thence to the nodules occurs within two to two

Table 10-5 Estimate of the Carbohydrate Requirement for N_2 Fixation by Glycine maxNodules[102,87,33]

Component	Basis for Estimate	g Carbohydrate/ g N Fixed
N_2 fixation + H_2 evolution	Respiration	7.29
Nodule "maintenance"	Respiration	2.68
Assimilation, transport of NH_4^+	Calculation	1.86
Growth	Calculation	0.26
Total		12.1 *

*A very similar estimate for total carbon cost (12.4) has been calculated by Gordon et al.[33] from entirely different data sets.

and a half hours.[31] With decreasing photosynthesis in the second half of the photoperiod, a larger percentage of the assimilates is translocated to the roots.

C. Oxygen Protection and Supply Mechanisms

i. Cork Layers

The outer cork layers of a legume nodule cover the outer cortex of this new plant organ.[11] This reduces gas diffusion significantly; however, it allows exchange through the lenticels. In addition, the incorporation of phlobaphenes into the cell wall of cork cells increases the resistance against pathogenic fungi.

ii. Diffusion Barriers

By inserting an O_2 microelectrode into soybean nodules, Tjepkema and Yocum[109] found a very sharp decline in O_2 concentration in the inner cortex. This decline was correlated with a layer of densely packed cells and a reduced number of intercellular spaces. In the outer part of the cortex the intercellular spaces amount to 1–5% of the tissue[57] with up to 10^5 air spaces per nodule in the bacteroid zone.[11] The O_2 diffusion barrier is regulated by the host plant in response to the carbon supply and the N_2 assimilation rate. Over a wide range of O_2 concentrations (from 21% to 80%) there was no damage to nitrogenase in the bacteroids. After changing from 80% oxygen incubation back to 21%, the assimilation rate was almost the same as that for plants incubated directly under air.[100] When the carbohydrate concentration in clover nodules was reduced by a direct treatment, the diffusion resistance increased within 24 hours to a maximum. At this time, however, the ability to control the resistance to oxygen diffusion was lost.[73] This is convincing evidence that oxygen concentration rather than carbohydrate availability limits nitrogenase activity in legumes. This fact apparently is true only for certain stages of nodule development (see Figure 10-8). With synchronized nodules (i.e., a constant number of nodules per plant), it can be shown that the O_2 enhancement effect on nitrogenase activity is only present in stages 2 and 3 of nodule development and not in stages 1 and 4. The increase of nitrogenase activity under 40% O_2 compared to 19% O_2 is essentially independent of the absolute activity of N_2 fixation.[120] The inhibition of N_2 fixation by nitrate also may be raised, to a certain extent, by increasing the external O_2 concentration.[114] N_2 fixation in legume nodules is reduced after incubation in an atmosphere without N_2 gas or by inhibition of the ammonium assimilation. The inhibition can be overcome by increasing the external O_2 concentration. The physiological result of this regulated O_2 diffusion barrier is a response to carbon supply from the plant as well as to N_2 fixation in the bacteroid. C, N, and O_2 metabolism are closely linked to each other.

The sensor mechanism by which changes in the carbon supply from the phloem or ammonia assimilation in the host cytosol are perceived is unknown.[57] Sucrose was proposed as an osmotically effective compound. However, it was shown that the sucrose pool in nodules of soybeans decreases when the diffusion resistance increases.[42] In this respect it might be interesting to look in more detail at inorganic components such as the potassium concentration in the water-filled diffusion barrier, since the potassium concentration in effective and ineffective nodules of soybeans is significantly different.[52] Another hypothesis concerns the ATP dependent conversion of sucrose to polysaccharides. It may be mediated by the adenylate energy charge, resulting in an O_2 effect on the available osmoticum sucrose.

iii. Leghemoglobin and Respiration

Leghemoglobin functions in the transport of O_2 and facilitates diffusion from the periphery of the infected cells to the bacteroid zone, where the apparent O_2 concentration is about 10 nM. Leghemoglobin constitutes up to 30% of the total soluble host plant protein in nodules and confers the red color. The four leghemoglobin genes in soybeans carry information of four different components produced: a, c1, c2, and c3. These are different in molecular weight, isoelectric point, and amino acid sequence. However, the immunological differences are small. Unfractionated FeII oxyleghemoglobin has a typical absorption maximum at 411 nm and two smaller maxima at 451 and 475 nm.[3] In deoxygenated leghemoglobin the 411-nm maximum shifts to 427 nm and the two smaller maxima are shifted to one maximum at 457 nm.[4] Leghemoglobin is detectable just before nitrogenase activity develops. Whether leghemoglobin is really a symbiotic protein and whether the globin part is host plant derived and the heme part is encoded by the microsymbiont are still questionable. In *Rhizobium meliloti* a mutation in δ-aminolevulinic acid synthesis induced production of white nodules,[59] whereas *Bradyrhizobium japonicum* with the same mutation induced fully effective nodules without leghemoglobin deficiency.[34] Nodules with highly active nitrogenase had a higher proportion of oxygenated leghemoglobin (30–48%) and a higher internal O_2 concentration (17–30 nM) compared to nodules with inhibited nitrogenase activity, which had only 11–17% oxygenated leghemoglobin and 5–8 nM internal oxygen concentration.[56] This O_2 concentration is sufficient to support the terminal oxidase in bacteroids, which has an apparent K_m of 5 nM. On the other hand, the K_m for the terminal oxidase in mitochondria of nodules is about 100 nM.[89] The often observed concentration of mitochondria (and amyloplasts) in the periphery of the infected cells may be a consequence of this.[72]

Abbreviations

AAT	Aspartate aminotransferase
ADP	Adenosine 5′-diphosphate
AMP	Adenosine 5′-monophosphate
Da	Dalton
d.p.i.	Days post-inoculation
EF	Exoplasmatic fracture face
EM	Electron microscopy
EPS	Exopolysaccharides
ER	Endoplasmic reticulum
GDP	Guanosine diphosphate
GOGAT	Glutamine oxoglutarate aminotransferase
GS	Glutamine synthetase
IDPase	Inosine diphosphatase
Km	Michaelis Constant
MDH	Malat-dehydrogenase
NAD	Nicotinamide adenine dinucleotide
NADH	Nicotinamide adenine dinucleotide—reduced form
Q	Ubichinon (coenzyme Q)
PBM	Peribacteroid membrane
PBS	Symbiosome space
PEP	Phosphoenol pyruvate

PF   Protoplasmatic fracture face

PHB   Poly-β-hydroxybutyrate

PHBA   Poly-β-hydroxybutyric acid

PPi   Inorganic phosphate

TEM   Transmission electron microscopy

UDP   Uridine-di-phosphate

References

1. Amarasingham, C. R., and Davis, B. D., "Regulation of α-ketoglutarate dehydrogenase formation in *Escherichia coli*," *Journal of Biological Chemistry, 240,* 3664–3667 (1965).

2. Anderson, M. P., Vance, C. P., Heichel, G. H., and Miller, S. S., "Purification and characterization of NADH-glutamate synthase from alfalfa root nodules," *Plant Physiology, 90,* 351–358 (1989).

3. Appleby, C. A., "Leghemoglobin and *Rhizobium* respiration," *Annual Review of Plant Physiology, 35,* 443–478 (1984).

4. Appleby, C. A., Bergersen, F. J., Preparation and experimental use of leghaemoglobin. In *Methods for Evaluating Nitrogen Fixation,* Bergersen, F. J., ed. Chichester: Wiley, 1980, p. 315–335.

5. Appels, M. A., *The Symbiosis Between Rhizobium leguminosarum and Pisum sativum: Regulation of the Nitrogenase Activity,* Landbouwuniversiteit te Wageningen, Proefschrift, 1989.

6. Bassarab, S., Mellor, R. B., and Werner, D., "Evidence for two types of Mg^{++}-ATPase in the peribacteroid membrane from *Glycine max* root nodules," *Endocytobiosis and Cell Research, 3,* 189–196 (1986).

7. Bassarab, S., and Werner, D., "Ca^{2+}-dependent protein kinase activity in the peribacteroid membrane from soybean root nodules," *Journal of Plant Physiology, 130,* 233–241 (1987).

8. Bassarab, S., and Werner, D., "Mg^{2+} dependent pyrophosphatase, a tonoplast enzyme in the peribacteroid membrane of *Glycine max* root nodules," *Symbiosis, 7,* 81–94 (1989).

9. Bassarab, S., Schenk, S. U., and Werner, D., "Fatty acid composition of the peribacteroid membrane and the ER in nodules of *Glycine max* varies after infection by different strains of the microsymbiont *Bradyrhizobium japonicum*," *Botanica Acta, 102,* 196–201 (1989).

10. Bennett, M. J., Lightfoot, D. A., and Cullimore, J. V., Studies on glutamine synthetase from root nodules of *Phaseolus vulgaris* L. In *Nitrogen Fixation: Hundred Years After,* Bothe, H., de Bruijn, F. J., and Newton, W. E., eds. Stuttgart: Gustav Fischer, 1988, p. 567.

11. Bergersen, F. J., and Goodchild, D. J., "Aeration pathways in soybean root nodules," *Australian Journal of Biological Sciences, 26,* 729–740 (1973).

12. Bergersen, F. J., *Root Nodules of Legumes: Structure and Functions,* Chichester: Research Studies Press, 1982.

13. Bisseling, T., van den Bos, R. C., van Kammen, A., van den Ploeg, M., van Duijn, P., and Houwers, A., "Cytofluorometrical determination of the DNA contents of bac-

teroids and corresponding broth-cultured *Rhizobium* bacteria," *Journal of General Microbiology, 101,* 79–84, (1977).

14. Blevins, D. G., Winkler, R. G., Rice, C. F., Walker, S., Lukaszweski, K., and Randall, D. D., Infection and development of leguminous nodules. In *Nitrogen Fixation: Hundred Years After,* Bothe, H., de Bruijn, F. J. and Newton, W. E., eds. Stuttgart: Gustav Fischer, 1988, p. 568.

15. Blumwald, E., Fortin, M. C., Rea, P. A., Verma D. P. S., and Poole, R. J., "Presence of host-plasma membrane type H^+-ATPase in the membrane envelope enclosing the bacteroids in soybean root nodules," *Plant Physiology, 78,* 665 (1985).

16. Boland, M. J., Hanks, J. F., Reynolds, P. H. S., Blevins, D. G., Tolbert, N. E., and Schubert, K. R., "Subcellular organization of ureide biogenesis from glycolytic intermediates and ammonium in nitrogen-fixing soybean nodules," *Planta, 155,* 45 (1982).

17. Bolton, E., Higgisson, B., Harrington, A., and O'Gara, F., "Dicarboxylic acid transport in *Rhizobium meliloti*: Isolation of mutants and cloning of dicarboxylic acid transport genes," *Archives of Microbiology, 144,* 142–146 (1986).

18. Brisson, N., and Verma, D. P. S., "Soybean leghemoglobin gene family: Normal, pseudo and truncated genes," *Proceedings of the National Academy of Sciences USA, 79,* 4055 (1982).

19. Dart, P. J., *A Treatise on Dinitrogen Fixation. Section III: Biology,* Hardy R. W. G., and Silver, W. S., eds. New York: Wiley, 1977, pp. 367–472.

20. Day, D. A., Yang, L.-J. O., and Udvandi, M. K, Nutrient exchange across the peribacteroid membrane of isolated symbiosomes. In *Nitrogen Fixation: Achievements and Objectives,* Gresshoff, P. M., Roth, L. E., Stacey, G., and Newton, W. E., eds. New York: Chapman & Hall, 1990, pp. 219–226.

21. Day, D. A., Carroll, B. J., Delves, A. C., and Gresshoff, P. M., "Relationship between autoregulation and nitrate inhibition of nodulation in soybeans," *Physiologia Plantarum, 75,* 37–42 (1989).

22. Deibert, E. J., Bijeriego, M., and Olson, R. A., "Utilization of ^{15}N fertilizer by nodulating and non-nodulating soybean isolines," *Agronomy Journal, 71,* 717–723 (1979).

23. Delves, A. C., Carroll, B. J., and Gresshoff, P. M., "Genetic analysis and complementation studies on a number of mutant supernodulating soybean lines," *Journal of Genetics, 67,* 1–8 (1988).

24. DeVries, G. E., In't Veld, P., and Kijne, J. W., "Production of organic acids in *Pisum sativum* root nodules as a result of oxygen stress," *Plant Science Letters, 20,* 115–123 (1980).

25. Dilworth, M., and Glenn, A., "How does a legume nodule work?" *Trends in Biological Sciences,* 519 (1984).

26. Emerich, D. W., Anthon, G. E., Hayes, R. R., Karr, D. B., Liang, R., Preston, G. G., Smith, M. T., and Waters, J. K., Metabolism of *Rhizobium*-leguminous plant nodules with an emphasis on bacteroid carbon metabolism. In *Nitrogen Fixation: Hundred Years After,* Bothe, H., de Bruijn, F. J., and Newton, W. E., eds. Stuttgart: Gustav Fischer, 1988, p. 531–546.

27. Evans, H. J., Harker, A. R., Papen, H., Russell, S. A., Hanus, F. J., and Zuber, M., "Physiology, biochemistry and genetics of the uptake hydrogenase in *Rhizobium*," *Annual Review of Microbiology, 41,* 335–361 (1987).

28. Farnham, M. W., Miller, S. S., Griffith, S. M., and Vance, C. P., "Aspartate aminotransferase in alfalfa root nodules," *Plant Physiology, 93,* 603–610 (1990).

29. Vance, C. P., Farnham, M. W., N. Degenhart, R. J. Larson, S. S. Miller, D. K. Barnes, and Bantt, J. S., Alfalfa root nodule aspartate aminotransferase (AAT): Biochemical importance and genetic control. In *Nitrogen Fixation: Achievements and Objectives*, Gresshoff, P. M., Roth, L. E., Stacey, G., and Newton, W. E., eds. New York: Chapman & Hall, 1990, pp. 693–699.

30. Fortin, M. G., Morrison, N. A., and Verma, D. P. S., "Nodulin 26, a PBM nodulin, as expressed independently of the development of the peribacteroid compartment," *Nucleic Acids Research, 15,* 813 (1987).

31. Fritz, R., Wieneke, J., and Führ, F., "Transportkinetik $\overline{11}$C-markierter Photosyntheseprodukte in Sojabohnen. I. Einfluβ von Tageszeit und Physiologischem Alter," *Angewandte Botanik, 60,* 317–327 (1986).

32. Garbers, C., Meckbach, R., Mellor, R. B., and Werner, D., "Protease (thermolysin) inhibition activity in the peribacteroid space of *Glycine max* root nodules," *Journal of Plant Physiology, 132,* 442–445 (1988).

33. Gordon, A. J., Ryle, G. J. A., Mitchell, D. F., and Powell, C. E., "The flux of ^{14}C-labelled photosynthate through soyabean root nodules during N_2-fixation," *Journal of Experimental Botany, 36,* 756–759 (1985).

34. Guerinot, M. L., and Chelm, B. K., "Bacterial δ-aminolevulinic acid synthase activity is not essential for leghemoglobin formation in the soybean/*Bradyrhizobium japonicum* symbiosis," *Proceedings of the National Academy of Sciences USA, 83,* 1837 (1986).

35. Hahn, H., Meyer, L., Studer, D., Regensburger, B., and Hennecke, H., "Insertion and deletion mutations within the nif region of *Rhizobium japonicum*," *Plant Molecular Biology, 3,* 158–168 (1984).

36. Hennecke, H., Bott, M., Ramseier, T., Thony-Meyer, L., Fischer, H.-M., Anthamatten, D., Kullik, I., and Thony, B., A genetic approach to analyze the critical rule of oxygen in bacteroid metabolism. In *Nitrogen Fixation: Achievements and Objectives*, Gresshoff, P. M., Roth, L. E., Stacey, G., and Newton, W. E., eds. New York: Chapman & Hall, 1990, pp. 293–300.

37. Hirel, B., and Gadal, P., "Glutamine synthase in rice. A comparative study of the enzymes from roots and leaves," *Plant Physiology, 66,* 619–623 (1980).

38. Huber, T. A., and Streeter, J. G., "Asparagine biosynthesis in soybean nodules," *Plant Physiology, 74,* 605–610 (1984).

39. Humbeck, C., and Werner, D., "Two succinate uptake systems in *Bradyrhizobium japonicum*," *Current Microbiology, 14,* 259–262 (1987).

40. Humbeck, C., and Werner, D., "Separation of malate and malonate pools by the peribacteroid membrane in soybean nodules," *Endocytobiosis and Cell Research, 4,* 185–196 (1987).

41. Humbeck, C., and Werner, D., "Delayed nodule development in a succinate transport mutant of *Bradyrhizobium japonicum*," *Journal of Plant Physiology, 134,* 276–283 (1989).

42. Hunt, S., King, B. J., Denison, R. F., Kouchi, H., Tajima, S., and Layzell, D. B., An osmotic mechanism for diffusion barrier regulation in soybean nodules. In *Nitrogen Fixation: Achievements and Objectives*, Gresshoff, P. M., Roth, L. E., Stacey, G., and Newton, W. E., eds. Chapman & Hall, 1990, 352.

43. Imsande, J., "Interrelationship between plant developmental stage, plant growth rate, nitrate utilization and nitrogen fixation in hydroponically grown soybean," *Journal of Experimental Botany, 39,* 775–785 (1988).

44. Imsande, J., "Rapid dinitrogen fixation during soybean pod fill enhances net photosynthetic output and seed yield: A new perspective," *Agronomy Journal, 81,* 549–556 (1989).

45. Jackson, F. A., and Dawes, E. A., "Regulation of the tricarboxylic acid cycle and poly-β-hydroxybutyrate metabolism in *Azotobacter beijerinkii* grown under nitrogen or oxygen limitation," *Journal of General Microbiology, 97,* 303–312 (1976).

46. Kannenberg, E. L., and Brewin, N. J., "Expression of a cell surface antigen from *Rhizobium leguminosarum* 3841 is regulated by oxygen and pH," *Journal of Bacteriology, 171,* 4543–4548 (1989).

47. King, B. J., Layzell, D. B., and Canvin, D. T., "The role of dark carbon dioxide fixation in root nodules of soybean," *Plant Physiology, 81,* 200–205 (1986).

48. Kinnback, A., Mellor, R. B., and Werner, D., "Alpha-mannosidase II isoenzyme in the peribacteroid space of *Glycine max* root nodules," *Journal of Experimental Botany, 38,* 1373–1377 (1987).

49. Kinnback, A., and Werner, D., "α-glucosidase and other enzyme activities in the peribacteroid space of soybean root nodules." In *Nitrogen Fixation: Achievements and Objectives,* Gresshoff, P. M., Roth, L. E., Stacey, G., and Newton, W. E., eds. New York: Chapman & Hall, 1990, p. 746.

50. Klucas, R. V., and Evans, H. J., "An electron donor system for nitrogenase-dependent acetylene reduction by extracts of soybean nodules," *Plant Physiology, 42,* 1458–1560 (1968).

51. Kretovich, W. L., Romanov, V. I., and Korolyov, A. V., "*Rhizobium leguminosarum* cytochromes (*Vicia faba*)," *Plant and Soil, 39,* 619–634 (1973).

52. Kuhlmann, K.-P., Stripf, R., Wätjen, W., Richter, F. W., Gloystein, F., and Werner, D., "Mineral composition of effective and ineffective nodules of *Glycine max* in comparison to roots: Characterization of developmental stages by differences in nitrogen, hydrogen, sulfur, molybdenum, potassium and calcium content," *Angewandte Botanik, 56,* 315–323 (1982).

53. Ladha, J. K., Pareek, R. P., So, R., and Becker, M., "Stem-nodule symbiosis and its unusual properties." In *Nitrogen Fixation: Achievements and Objectives,* Gresshoff, P. M., Roth, L. E., Stacey, G., and Newton, W. E., eds. New York: Chapman & Hall, 1990. p. 633–640.

54. Lara, M., Porta, H., Padilla, J., Folch, J., and Sanchez, F., "Heterogeneity of glutamine sythetase polypeptides in *Phaseolus vulgaris* L.," *Plant Physiology, 76,* 1019–1023 (1984).

55. Lawrie, A. C., and Wheeler, C. T., "Nitrogen fixation in the root nodules of *Vicia faba* L. in relation to the assimilation of carbon. II. The dark fixation of carbon dioxide," *New Phytologist, 74,* 437–445 (1975).

56. Layzell, D. B., Hunt, S., and Palmer, G. R., "Mechanism of nitrogenase inhibition in soybean nodules. Pulse-modulated spectroscopy indicates that nitrogenase activity is limited by O_2," *Plant Physiology, 92,* 1101–1107 (1990).

57. Layzell, D. B., Hunt, S., Moloney, A. H. M., Fernando, S. M., and Diaz del Castillo, L., Physiological, metabolic and developmental implications of CO_2 regulation in legume nodules. In *Nitrogen Fixation: Achievements and Objectives,* Gresshoff, P. M., Roth, L. E., Stacey, G., and Newton, W. E., eds. New York: Chapman & Hall, 1990, pp. 21–32.

58. Lee, J. S., and Verma, D. P. S., "Structure and chromosomal arrangement of leg-hemoglobin genes in kidney bean suggest divergence in soybean leghemoglobin gene loci following tetraploidization," *EMBO Journal, 3*, 2745 (1984).

59. Legon, S. A., Ditta, G. S., and Helsinki, D. R., "Heme biosynthesis in *Rhizobium*. Identification of a cloned gene coding for δ-aminolevulinic acid synthetase from *Rhizobium meliloti*," *Journal of Biological Chemistry, 257*, 8724 (1982).

60. Maier, R. J., Moshiri, F., Keefe, R. G., and Gabel, C., "Molecular analysis of terminal oxidases in electron-transport pathways of *Bradyrhizobium japonicum* and *Azotobacter vinelandii*." In *Nitrogen Fixation: Achievements and Objectives*, Gresshoff, P. M., Roth, L. E., Stacey, G., and Newton, W. E., eds. New York: Chapman & Hall, 1990, pp. 301–308.

61. McDermott, T. R., Griffith, S. M., Vance, C. P., and Graham, P. H., "Carbon metabolism in *Bradyrhizobium japonicum* bacteroids," *FEMS Microbiology Reviews, 63*, 327–340 (1989).

62. McParland, R. H., Guevara, J. G., Becker, R. R., and Evans, H. J., "The purification and properties of the glutamine synthetase from the cytosol of soya-bean root nodules," *Biochemistry Journal, 153*, 597–606 (1976).

63. Meijer, E. G. M., "Development of leguminous root nodules", In *Nitrogen Fixation 2: Rhizobium*, Broughton, E. J., ed. New York: Oxford University Press, 1982, pp. 312–331.

64. Mellor, R. B., "Distribution of trehalase in soybean root nodule cells: Implications for trehalose metabolism," *Journal of Plant Physiology, 133*, 173–177 (1988).

65. Mellor, R. B., "Bacteroids in the *Rhizobium*–legume symbiosis inhabit a plant internal lytic compartment: Implications for other microbial endosymbioses," *Journal of Experimental Botany, 40*, 831–839 (1989).

66. Mellor, R. B., Mörschel, E., and Werner, D., "Legume root response to symbiotic infection. Enzymes of the peribacteroid space," *Zeitschrift für Naturforschung, 39c*, 123–125 (1984).

67. Mellor, R. B., and Werner, D., Glycoconjugate interactions in soybean root nodules. In *Lectins*, Vol. IV, Bog-Hansen, T. C., and Breborowicz, J., eds. Berlin: Walter de Gruyter, 1985, pp. 268–276.

68. Mellor, R. B., Christensen, T. M. I. E., and Werner, D., "Choline kinase II is present only in nodules that synthesize stable peribacteroid membranes," *Proceedings of the National Academy of Sciences USA, 83*, 659–663 (1986).

69. Mellor, R. B., and Werner, D., "The fractionation of *Glycine max* root nodule cells: A methodological overview," *Endocytobiosis and Cell Research, 3*, 317–336 (1986).

70. Mellor, R. B., Thierfelder, H. Pausch, G., and Werner, D., "The occurrence of choline kinase II in the cytoplasm of soybean root nodules infected with various strains of *Bradyrhizobium japonicum*," *Journal of Plant Physiology, 128*, 169–172 (1987).

71. Mellor, R. B., Garbers, C., and Werner, D., "Peribacteroid membrane nodulin gene induction by *Bradyrhizobium japonicum* mutants," *Plant Molecular Biology, 12*, 307–315 (1989).

72. Mellor, R. B., and Werner, D., Legume nodule biochemistry and function. In *Molecular Biology of Symbiotic Nitrogen Fixation*, Gresshoff, P. M., ed. Boca Raton: CRC Press, 1990, pp. 111–129.

73. Minchin, F. R., Sheehy, J. E., and Minguez, M. I., "Characterization of the resistance to oxygen diffusion in legume nodules," *Annals of Botany, 55*, 53–60 (1985).

74. Moiroud, A., and Capellano, A., "Le robinier, *Robinia pseudoacacia* L., une espèce fixatrice d'azote intéressante?" *Annales des Sciences Forestières, 39,* 407–418 (1982).

75. Nap, J.-P., and Bisseling, T., Nodulin function and nodulin gene regulation in root nodule development. In *Molecular Biology of Symbiotic Nitrogen Fixation,* Gresshoff, P. M., ed. Boca Raton: CRC Press, 1990, pp. 181–229.

76. Nguyen, T., Zelechowska, M., Forster, V., Bergmann, H., and Verma, D. P. S., "Primary structure of the soybean nodulin −35 gene encoding uricase II localized in the peroxisomes of uninfected cells of nodules," *Proceedings of the National Academy of Sciences USA, 82,* 5040 (1985).

77. Nitragin Company, *Annual Report,* Milwaukee, Wisconsin, 1985.

78. Ostrowski, E. D., Mellor, R. B., and Werner, D., "The use of colloid gold labelling in the detection of plasma membrane from symbiotic and non-symbiotic *Glycine max* root cells," *Physiologia Plantarum, 66,* 270–276 (1986).

79. Paau, A. S., Oro, J., and Cowles, J. R., "DNA content of free-living rhizobia and bacteroids of various *Rhizobium*–legume associations," *Plant Physiology, 63,* 402–405 (1979).

80. Pankhurst, C. E., and Craig, A. S., "Effect of oxygen concentration, temperature and combined nitrogen on the morphology and nitrogenase activity of *Rhizobium* sp. strain 32H1 in agar culture," *Journal of General Microbiology, 106,* 207–219 (1978).

81. Parniske, M., Zimmermann, C., Cregan, P. B., and Werner, D., "Hypersensitive reaction of nodule cells in the *Glycine* sp./*Bradyrhizobium japonicum*-symbiosis occurs at the genotype-specific level," *Botanica Acta, 103,* 143–148 (1990).

82. Parniske, M., Fischer, H.-M., Hennecke, H., and Werner, D., "Accumulation of the phytoalexin glyceollin I in soybean nodules infected by a *Bradyrhizobium japonicum* nifA mutant," *Zeitschrift für Naturforschung, 46c,* 318–320 (1991).

83. Peterson, J. B., and Evans, H. J., "Properties of pyruvate kinase from soybean nodule cytosol," *Plant Physiology, 61,* 909–914 (1978).

84. Peterson, J. B., and Evans, H. J., "Phosphoenolpyruvate carboxylase from soybean nodule cytosol. Evidence for isoenzymes and kinetics of the most active component," *Biochimica et Biophysica Acta, 567,* 445–452 (1979).

85. Puppo, A., Dimitrijevic, L., and Rigaud, J., "O_2 consumption and superoxide dismutase content in purified mitochondria from soybean root nodules," *Plant Science, 50,* 3–11 (1987).

86. Quispel, A., "Frontiers of Research in symbiotic Nitrogen Fixation," *Frontiers of Research in Agriculture,* Roy, S. K., ed. Calcutta: Indian Statistical Institute, 1982, pp. 206–223.

87. Rainbird, R. M., Hitz, W. D., and Hardy, R. W. F., "Experimental determination of the respiration associated with soybean/*Rhizobium* nitrogenase function, nodule maintenance, and total nodule nitrogen fixation," *Plant Physiology, 75,* 49–53 (1984).

88. Rawsthorne, S., and LaRue, T. A., Respiration and oxidative phosphorylation of mitochondria from nodules of cowpea (*Vigna unguiculata* L.). In *Nitrogen Fixation Research Progress,* Evans, H. J., Bottomley, P. J., and Newton, W. E., eds. Dordrecht: Martinus Nijhoff, 1985, p. 351.

89. Rawsthorne, S., and LaRue, T. A., "Preparation and properties of mitochondria from cowpea nodules," *Plant Physiology, 81,* 1092–1096 (1986).

90. Robertson, J. G., Lyttleton, P., Bullivant, S., and Grayston, G. F., "Membranes of lupin root nodules. 1. The role of Golgi bodies in the biogenesis of infection threads and peribacteroid membranes," *Journal of Cell Science, 30,* 129–149 (1978).

91. Robertson, J. G., and Lyttleton, P., "Coated and smooth vesicles in the biogenesis of cell walls, plasma membranes, infection thread and peribacteroid membranes in root hairs and nodules of white clover," *Journal of Cell Science, 58,* 63–78 (1982).

92. Robertson, J. G., and Lyttleton, P., "Division of peribacteroid membranes in root nodules of white clover," *Journal of Cell Science, 69,* 147–157 (1984).

93. Röhm, M., personal communication, recalculated from *Fast Growing Trees and Nitrogen Fixing Trees,* Werner, D., and Müller, P., eds. Stuttgart: Gustav Fischer, 1990.

94. Ronson, C. W., Lyttleton, P., and Robertson, J. G., "C_4-dicarboxylate transport mutants of *Rhizobium trifolii* from ineffective nodules on *Trifolium repens,*" *Proceedings of the National Academy of Sciences USA, 78,* 4284–4288 (1981).

95. Rosendahl, L., Vance, C. P., and Pedersen, W. B., "Products of dark CO_2 fixation in pea root nodules. Support bacteroid metabolism," *Plant Physiology, 92* (1990).

96. Roth, L. E. Jeon, K., and Stacey, G., Homology in endosymbiotic systems: The term "symbiosome". In *Molecular Genetics of Plant–Microbe Interactions,* Palacios, R., and Verma, D. P. S., eds. St. Paul: The American Phytopathological Society Press. 1988, p. 220–225.

97. Ryan, E., Bodley, F., and Fottrell, P. F., "Purification and characterization of aspartate aminotransferases from soybean root nodules and *Rhizobium japonicum*" *Phytochemistry, 11,* 957–963 (1972).

98. Salminen, S. O., and Streeter, J. G., "Involvement of glutamate in the respiratory metabolism of *Bradyrhizobium japonicum* bacteroids," *Journal of Bacteriology, 169,* 495–499 (1987).

99. Salminen, S. O., and Streeter, J. G., "Uptake and metabolism of carbohydrates by *Bradyrhizobium japonicum* bacteroids," *Plant Physiology, 83,* 535–540 (1987).

100. Sheehy, J. E., Minchin, F. R., and Witty, J. F., "Biological control of the resistance to oxygen flux in nodules," *Annals of Botany, 52,* 565–571 (1983).

101. Streeter, J. G., "Enzymes of sucrose, maltose, and α, α-trehalose catabolism in soybean root nodules," *Planta, 155,* 112–115 (1982).

102. Streeter, J. G., and Salminen, S. O., Carbon metabolism in legume nodules. In *Nitrogen Fixation Research Progress,* Evans, H. J. Bottomley, P. J., and Newton, W. E., eds. Dordrecht: Martinus Nijhoff, 1985, pp. 277–283.

103. Stumpf, D. K., and Burris, R. H., "Organic acid contents of soybean: Age and source of nitrogen," *Plant Physiology, 68,* 989–991 (1981).

104. Sutton, W. D., Jepsen, N. M., and Shaw, B. D., "Changes in the number, viability, and amino-acid-incorporating activity of *Rhizobium* bacteroids during lupin nodule development," *Plant Physiology, 59,* 741–744 (1977).

105. Sutton, W. D., Pankhurst, C. E. and Craig, A. S., "The *Rhizobium* bacteroid state," *International Review of Cytology,* Suppl. 13, 149–177 (1981).

106. Ta, T.-C., Faris, M. A., and MacDowall, F. D. H., "Pathways of nitrogen metabolism in nodules of alfalfa (*Medicago sativa* L.)," *Plant Physiology, 80,* 1002–1005 (1986).

107. Tajima, S., Kouzai, K., and Kimura, I., NAD(P)/NAD(P)H ratio and energy charge in succinate degrading soybean nodule bacteroids. In *Nitrogen Fixation: Hundred Years After,* Bothe, H, de Bruijn, F. J., and Newton, W. E., eds. Stuttgart: Gustav Fischer, 1988, p. 564.

108. Thummler, F., and Verma, D. P. S., "Nodulin 100 of soybean is the subunit of sucrose synthase regulated by the availability of free heme in nodules," *Journal of Biological Chemistry, 262,* 14730–14736 (1987).

109. Tjepkema, J. D., and Yocum, C. S., "Measurement of oxygen partial pressure within soybean nodules by oxygen microelectrodes," *Planta, 119,* 351–360 (1974).

110. Tomaszewska, B., and Schramm, R. W., "Properties of NADP-malic enzyme from yellow lupin (*Lupinus luteus* L.) nodules," *Acta Physiologiae Plantarum, 10,* 63–72 (1988).

111. Tomaszewska, B., Mazurowa, H., and Schramm, R. W., "Activities of some enzymes of carbon metabolism connected with nitrogen fixation in the root nodules of the growing yellow lupin *Lupinus luteus* L.," *Acta Physiologiae Plantarum, 10,* 73–84 (1988).

112. Tsien, H. C., Cain, P. S., and Schmidt, E. L., "Viability of *Rhizobium* bacteroids," *Applied Environmental Microbiology, 34,* 854–856 (1977).

113. Van der Wilden, W., and Chrispeels, M. J., "Characterization of the isoenzymes of α-mannosidase located in the cell wall, protein bodies and endoplasmic reticulum of *Phaseolus vulgaris* cotyledons," *Plant Physiology, 71,* 82–87 (1983).

114. Vessey, J. K., Walsch, K. B., and Layzell, D. B., "Oxygen limitation of N_2-fixation in stem girdled and nitrate-treated soybean," *Physiologia Plantarum, 73,* 113–121 (1988).

115. Werner, D., *Symbioses of Plants and Microbes,* London: Chapman & Hall, 1991, 300 pp.

116. Werner, D., and Mörschel, E., "Differentiation of nodules of *Glycine max.* Ultrastructural studies of plant cells and bacteroids," *Planta, 141,* 169–177 (1978).

117. Werner, D., and Stripf, R., "Differentiation of *Rhizobium japoicum*. I. Enzymatic comparison of nitrogenase repressed and derepressed free living cells and of bacteroids," *Zeitschrift für Naturforschung, 33c,* 245–252 (1978).

118. Werner, D., Mörschel, E., Stripf, R., and Winchenbach, B., "Development of nodules of *Glycine max* infected with an ineffective strain of *Rhizobium japonicum*," *Planta, 147,* 320–329 (1980).

119. Werner, D., Dittrich, W., and Thierfelder, H., "Malonate and Krebs cycle intermediates utilization in the presence of other carbon sources by *Rhizobium japonicum* and soybean bacteroids," *Zeitschrift für Naturforschung, 37c,* 921–926 (1982).

120. Werner, D., and Krotzky, A., "Die symbiontische Stickstoff-Fixierung der Leguminosen," *Funktionelle Biologie und Medizin, 2,* 31–39 (1983).

121. Werner, D., Mörschel, E., Kort, R., Mellor, R. B., and Bassarab, S., "Lysis of bacteroids in the vicinity of the host cell nucleus in an ineffective (fix⁻) root nodule of soybean (*Glycine max*)," *Planta, 162,* 8–16 (1984).

122. Werner, D., Mellor, R. B., Hahn, M. G. and Grisebach, H., "Soybean root response to symbiotic infection. Glyceollin I accumulation in an ineffective type of soybean nodules with an early loss of the pertibacteroid membrane," *Zeitschrift für Naturforschung, 40c,* 179–181 (1985).

123. Werner, D., Mörschel, E., Garbers, C., Bassarab, S. and Mellor, R. B., "Particle density and protein composition of the peribacteroid membrane from soybean root nodules is affected by mutation in the microsymbiont *Bradyrhizobium japonicum*," *Planta, 174,* 263–279 (1988).

124. Wilcockson, J. and Werner, D., "On the DNA content of bacteroids of *Rhizobium japonicum*," *Zeitschrift für Naturfoschung, 34,* 793–796 (1979).

125. Wolff, A., Mörschel, R., Zimmermann, C., Parniske, M., Bassarab, S., Mellor, R. B. and Werner, D., *Physiological Limitations and the Genetic Improvement of Symbiotic Nitrogen Fixation*, (O,Gara, F., Manian, S. and Drevon, J. J., eds.) Dordrecht: Kluwer Academic Publishers, 1988, pp. 65–74.

126. Wong, P. P. and Evans, H. J., "Poly-β-hydroxybutyrate utilization by soybean (*Glycine max* Merr.) nodules and assessment of its role in maintenance of nitrogenase activity," *Plant Physiology, 47*, 750–755 (1971).

11

Hydrogen Cycling in Symbiotic Bacteria

Daniel J. Arp

I. Introduction

The concept of H_2 cycling in legumes arises as a result of the catalytic activities of two enzymes of the microbial symbiont—namely, nitrogenase and hydrogenase. Nitrogenase obligatorily produces H_2 during reduction of N_2, and hydrogenase functions to oxidize the H_2 produced by nitrogenase. This production of H_2 by one enzyme followed by the consumption of H_2 by another enzyme results in a cycling of H_2 (Figure 11–1). Because nitrogenase, by definition, is present in all N_2-fixing symbioses, all N_2-fixing symbioses produce H_2. In contrast, not all N_2-fixing symbioses express a hydrogenase to oxidize the H_2 produced by nitrogenase. Thus, not all symbioses are capable of H_2 cycling. Indeed, most legume root nodules that fix N_2 also evolve H_2, either due to insufficient hydrogenase to oxidize all the H_2 evolved or due to the complete absence of hydrogenase. The production of H_2 by nitrogenase indicates an inefficient use of the energy allocated to N_2 fixation. H_2 oxidation by hydrogenase has the potential to return to the system at least a portion of the energy expended in the production of H_2, thereby increasing the overall efficiency of N_2 fixation. H_2 oxidation may provide additional benefits as well. Therefore, considerable interest has focused and continues to focus on H_2 cycling in legumes.

This chapter will focus on the role of H_2 cycling in *(Brady)Rhizobium/* legume and other symbioses. However, generous use of examples from H_2 cycling in free-living microorganisms, both N_2-fixing and non-N_2-fixing, will be used as necessary to make various points and to reflect the current understanding and development of the field. A number of reviews of H_2 cycling have been published in the last few years and the reader is referred to these for additional information, further insight, and alternate interpretations.[3,37,44,77,93]

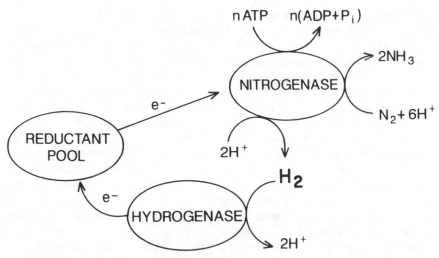

Figure 11–1. Scheme of H_2 cycling in N_2-fixing symbionts and free-living bacteria indicating the roles of hydrogenase and nitrogenase in this process.

II. H_2 Production by Nitrogenase

H_2 production by soybean (*Glycine max* L.) root nodules was first observed by Hoch, Little, and Burris[57] in 1957; they later[58] suggested that this process was catalyzed by nitrogenase. Evolution of H_2 by nitrogenase has since been confirmed with purified preparations of this enzyme from a number of microbial sources, both symbiotic and free-living. H_2 production has the same requirements as N_2 reduction (i.e., ATP and low potential reductant, and both components of nitrogenase are required to support activity[13,14]). In the absence of any other reducible substrate (e.g., N_2, C_2H_2), the flux of electrons through nitrogenase (and the rate of ATP hydrolysis) remains constant and all the reductant is used in the production of H_2. With mixtures of substrates, high electron flux rates (achieved by altering component ratio or ATP concentration) favor N_2 reduction, whereas low flux rates favor proton reduction.[16,52]

With N_2 as the substrate for nitrogenase, the flux of electrons is partitioned between protons and N_2. Increased pN_2 favors reduction of N_2.[102] However, extrapolation of the data to infinite pressures suggested that H_2 production would continue even at high pressures of N_2; that is, H_2 production would always accompany N_2 reduction. Furthermore, a limiting ratio of 1 N_2 reduced per H_2 produced was indicated by the data. This one-to-one ratio supported the idea that H_2 production by nitrogenase was an essential part of the reaction mechanism leading to the reduction of N_2, even though H_2 production did not require the presence of N_2. Definitive proof of this theory was lacking until Simpson and Burris[116] carried out nitrogenase reactions at very high N_2 pressures (up to 60 atm). Even at these high N_2 pressures, the ratio of N_2 reduced per H_2 produced remained near unity. As a result of these experiments, most proposed catalytic mechanisms for nitrogenase now take into account the need to evolve one H_2 for every N_2 reduced. H_2 production by nitrogenase is also unique in that it is the only electron transfer reaction by nitrogenase that is not inhibited by CO.[102] In the presence of kinetically saturating concentrations of CO, all the electron flux through nitrogenase is diverted to proton reduction to form H_2.

In addition to H_2 production by nitrogenase, H_2 interacts with nitrogenase in another way. In 1938, Wilson and colleagues[132] reported that when red clover (*Trifolium pratense* L.) plants were grown in the presence of H_2 as a diluent gas (to allow variation of the N_2 concentration), the plants did not grow as well as when argon was used as the diluent gas. This situation did not arise when the plants were grown in the presence of fixed nitrogen, thereby indicating that the N_2-fixing system was involved. They went on to determine that H_2 is a competitive inhibitor of N_2 reduction and determined a K_i of 0.14 atm H_2. This observation was later confirmed with purified nitrogenase; K_i's in the range of 0.11–0.19 atm H_2 have been measured.[100] H_2 inhibition of N_2 reduction is also related to the formation of HD by nitrogenase when exposed to D_2. Under turnover conditions in the presence of N_2 and D_2, HD is formed at the expense of D_2 consumption. An accounting of the electron flux reveals that electrons are diverted from N_2 reduction and H_2 formation (which still occurs in the presence of D_2) to the formation of HD.[15] The requirement for N_2 has led to a proposed mechanism whereby D_2 reacts with a reduction intermediate of N_2 to form two molecules of HD from one molecule of D_2. The additional electrons are supplied by nitrogenase. H_2 inhibition is a manifestation of this same catalytic activity of nitrogenase. H_2 inhibits nitrogenase by diverting electron flow from N_2 reduction to H_2 production. Although without the use of isotopes it is impossible to distinguish the H_2 produced as a consequence of N_2 reduction from the H_2 produced as a consequence of H_2 inhibition of N_2 reduction, it seems apparent that H_2 inhibition results in the formation of still more H_2.

It seems remarkable that an enzyme that functions physiologically to reduce atmospheric N_2 to ammonia should interact in such complex and fascinating ways with H_2. Whether this in some way reflects the evolutionary history of nitrogenase or is a natural consequence of the catalytic conditions (e.g., low potential reductant and protons poised on redox centers composed of transition metals) required to reduce N_2 is not known. More detailed discussions of nitrogenase catalysis may be found in the chapters by Yates and by Newton.

III. H_2 Oxidation in N_2-Fixing Systems

The ability to oxidize H_2 is widely distributed within the microbial kingdom. This capability is present in most of the major physiologic groups, including aerobes, anaerobes, methanogens, methylotrophs, autotrophs, phototrophs, cyanobacteria, and N_2-fixing microorganisms. The function of the hydrogenase varies with the physiology and metabolism of the microorganism. Among the N_2-fixing microorganisms, H_2 oxidation is also widespread, though by no means universal. H_2 oxidation, though first described in aerobic systems (both symbiotic and free-living)[99] also occurs in the phototrophs,[23] cyanobacteria,[59] methanotrophs,[23] and fermentative anaerobes.[22] For all N_2-fixing microorganisms that can oxidize H_2, the nitrogenase reaction provides one source of H_2. In some microorganisms (e.g., the clostridia) there are additional mechanisms of H_2 generation by the microorganism that are quantitatively more important than nitrogenase. In other cases (e.g., methanogens and chemoautotrophs) the major source of H_2 is the environment and nitrogenase is again of lesser importance. In the case of H_2 oxidation by aerobic N_2-fixing microorganisms—whether free-living or symbiotic—the major source of H_2 for oxidation by hydrogenase is that produced *in situ* by nitrogenase.

In the case of N_2-fixing symbioses, the ability to oxidize H_2 is associated with the microbial symbiont. Because two enzymes, nitrogenase and hydrogenase, are involved in the H_2 metabolism of root nodules, the H_2 status of a nodule is difficult to assess quantitatively. This led Schubert and Evans[110] to define the term *relative efficiency*, which relates the rate of H_2 evolution to the rate of electron flux through nitrogenase and is defined as follows:

Relative efficiency (RE) = 1 − (rate of H_2 evolution/rate of C_2H_2 reduction)

where the rate of acetylene reduction is taken as a measure of the rate of electron flux through nitrogenase. The value of RE gives a general indication of the efficiency of the N_2-fixing system with regard to the amount of H_2 evolved relative to the total electron flux through nitrogenase. The maximal value of 1 indicates no H_2 evolution and a high relative efficiency. A value of 0 (which has not been reported for a native system) would indicate that all the electron flux through nitrogenase is diverted to H_2 evolution. For a system lacking hydrogenase, the theoretical maximum for RE is 0.75, because, as described earlier, a minimum of 1 H_2 (requiring 2 e^-) must be produced for every N_2 reduced (requiring 6 e^- or 0.75 of the total e^- flux). The RE values are often lower than 0.75,[110] indicating that more of the electron flux through nitrogenase is directed toward H_2 production and that the ratio of H_2 produced to N_2 reduced is actually greater than 1. However, the presence of a hydrogenase complicates the interpretation of RE. Because the hydrogenase consumes H_2, the rate of H_2 evolution from the nodule is lowered. If all the H_2 produced by nitrogenase is consumed by hydrogenase, then a RE value of 1 is obtained. However, there is not always sufficient hydrogenase activity present to consume all the H_2 produced by nitrogenase.

Values between 0.75 and 1 indicate that some hydrogenase is present, though not enough to consume all the H_2. Values less than 0.75 do not exclude the presence of a hydrogenase, but if present, the activity is generally quite low. In one study of 19 legumes, RE values ranging from 0.20 to 0.99 were observed.[116] Typically, 30–60% of the energy consumed by nitrogenase will be used to produce H_2.[45,110] RE values for six actinorhizal systems included in the same survey ranged from 0.77 to 0.99.[110]

A number of surveys of root nodules of legumes and actinorhizal plants have been carried out to assess directly the distribution of H_2 oxidation capability in these symbioses. The results of a number of these surveys are summarized in Table 11–1. Clearly, the ability to oxidize H_2 is not the predominant phenotype. Although not apparent in the table, the *Bradyrhizobium* species that do express hydrogenase activity typically have sufficient activity to scavenge most, if not all, of the H_2 produced by nitrogenase. In contrast, for most of the *Rhizobium* species that express hydrogenase activity, the amount of activity is sufficient to recover only a small portion of the H_2 produced by nitrogenase. Minimisawa et al.[85] made the interesting observation that of 51 strains of *B. japonicum* examined, all those that produced the phytotoxin rhizobitoxine were HUP⁻. They also noted a correlation between the composition of the extracellular polysaccharide and the HUP phenotype.

The symbioses between actinorhizal plants and *Frankia* have a high occurrence of H_2 oxidation activity relative to the leguminous symbioses (see Table 11–1[10]). The significance of this is not clear but may be related to the fact that the extent of domestication of actinorhizal plants and *Frankia* is less than that of the *(Brady)Rhizobium*–legume symbioses. Perhaps selective pressures have been maintained in the actinorhizal symbioses that have been diminished in the legume symbioses.

IV. Benefits of H_2 Cycling

H_2 oxidation is generally considered a beneficial property of N_2-fixing microorganisms, whether in the symbiotic or free-living state.[28] Through H_2 oxidation, the H_2 generated by nitrogenase can be consumed and the reductant used in ways that are useful to the cells. The redox potential of H_2 oxidation is sufficiently low (-420 mV @ pH 7.0) that H_2 could be used as a reductant for all cellular processes. Therefore, those cells that express a hydrogenase are expected to carry out a more efficient fixation of N_2 by utilizing the H_2 produced by nitrogenase. This increased efficiency has been examined at the whole organism level in both free-living microorganisms and in legume–*Rhizobium* symbioses.

In free-living organisms, the most conclusive evidence that oxidation of the H_2 generated by nitrogenase actually benefits the microorganism has come

Table 11-1. *Distribution of HUP⁺ Phenotype among N₂-Fixing Symbioses*

Microbial Symbiont	Host Plant	No. of Isolates (I)[d] or Strains (S) Investigated	#Hup⁺	%Hup⁺	Ref.
Bradyrhizobium japonicum	*Glycine max*	32 (S)	6	19	21
Bradyrhizobium japonicum	*Glycine max*	1400 (I)	350	25	76
Bradyrhizobium sp. (Cowpea miscellany)	*Vigna unguiculata* *Vigna radiata* *Vigna mungo* *Cyanopsis tetragonoloba*	914 (I)	585– 640	64– 70	25
Bradyrhizobium sp. (Cowpea miscellany)	*Cicer arietinum*	611 (I)	0	0	25
Rhizobium sp. (Sesbania)	*Sesbania*	104 (I)	8	8	106
Rhizobium leguminosarum	*Pisum sativum*	108 (I)	14[a]	13	89
Rhizobium leguminosarum	*Pisum sativum*	15 (S)	6[a]	40	104
Rhizobium leguminosarum	*Pisum sativum*	23 (S)	19[c]	83	125
Rhizobium meliloti	*Medicago sativa*	19 (S)	0	0	105
Rhizobium trifolii	*Trifolium repens*	7 (S)	2[b]	30	105
Rhizobium astragalus	*Astragalus sinisca*	15 (I)	3	20	122
Frankia	*Alnus incana*	12 (I)	12	100	114

[a]Hydrogenase activity not high enough to recover all of the H₂ evolved by nitrogenase.

[b]Observed level of hydrogenase activity was very low.

[c]Hydrogenase activity high enough to recover all H₂ evolved by nitrogenase in only 3 stains.

[d]The designation of strain or isolate is taken from each reference. Different strains are presumed to be genetically distinct, while different isolates could, but need not be, genetically identical.

from continuous culture experiments with *Azotobacter chroococcum*.[133] Competition studies compared a hydrogenase–positive recombinant strain to a hydrogenase–negative mutant parent. The N_2-fixing continuous cultures were initiated with mixtures of cells of the two hydrogenase phenotypes, and the time required for domination by one or the other cell type was determined under different nutrient limitations. When limited for sucrose, the cultures were quickly dominated by the HUP^+ phenotype, even when the starting ratio favored the HUP^- phenotype by 100-fold. Phosphate limitation resulted in a moderate rate of domination by the HUP^+ phenotype. Ammonia-grown cultures (with no nitrogenase and therefore no H_2 production) showed no clear dominance pattern. These results clearly demonstrated that under appropriate conditions, the HUP^+ phenotype provides a competitive advantage to the bacteria. In contrast, Fe or O_2 limitations led to a slow domination by the HUP^- phenotype. This indicated that under certain conditions, hydrogenase may actually be disadvantageous.

To determine the effect of the HUP^+ phenotype of the rhizobial symbiont on the productivity of nodulated legumes, the total dry matter accumulation and the total N content of plants with and without the HUP^+ phenotype have been compared. The experiments consist of inoculating plants with HUP^+ and HUP^- rhizobia, growing the plants for a period of time under N_2-fixing conditions, and then harvesting and measuring the dry matter and N content of the resultant plants. In an early version of this type of experiment, HUP^+ strains of *Bradyrhizobium japonicum* were selected and compared to HUP^- strains.[1] The results gave higher N content (21%) and dry matter accumulation (14%) for HUP^+ strains. Similar results were reported with HUP^+ vs. HUP^- strains of *Rhizobium* on mungbean (*Vigna radiata* L.) plants[96] as well as with alder (*Alnus incana*) and various *Frankia* strains.[115] However, the strains selected for these studies were not likely to be isogenic other than the hydrogenase phenotype. Therefore, any differences in plant productivity could not be attributed specifically to the ability to oxidize H_2. In a more sophisticated version of the experiment with *B. japonicum* and soybeans, a revertible HUP^- mutant[72] was compared to a HUP^+ constructed strain that was isogenic with the HUP^- mutant (except for HUP).[43] Again, increased total dry matter content (9%) and N content (11%) were observed in plants inoculated with the HUP^+ *B. japonicum* constructed strain relative to plants inoculated with HUP^- *B. japonicum* mutant strain. Thus, the experimental evidence supports a beneficial role for hydrogenase in soybean root nodules. However, not all such studies have provided evidence of a benefit associated with H_2 cycling. For example, an increase in productivity of peas (*Pisum sativum* L.) inoculated with *Rhizobium leguminosarum* carrying the HUP-containing plasmid pRL6JI turned out not to be a result of the hydrogenase; rather, it was due to an as yet unidentified factor associated with the plasmid.[24] A survey of several *R. leguminosarum* strains also did not reveal a benefit to the HUP$^+$ phenotype.[89] As noted earlier, most pea root nodules have weak H$_2$ oxidation activity, and this may explain the apparent lack of benefit of a hydrogenase in this system. On the other hand, it may be that there are conditions where the presence of a hydrogenase simply does not provide any benefit. The presence or absence of a benefit may be related to such factors as energy coupling, nodule structure, or photosynthate partitioning.

In 1972, Dixon[29] published a seminal paper describing the occurrence and properties of hydrogenase in legume root nodules and offered three mechanisms by which the presence of a hydrogenase might increase the overall efficiency of biological N$_2$ fixation. A description of each postulated mechanism and a discussion of the evidence that has accumulated in support of each follows.

A. Postulated Mechanism 1

Oxidation of H$_2$ provides an additional source of reductant and energy to the N$_2$-fixing system that is not available to those systems without an H$_2$-oxidizing hydrogenase. If the N$_2$-fixing system is reductant or energy limited, then the additional source of reductant, H$_2$, should benefit the system and lead to an increased efficiency. This role of a hydrogenase has been likened to that of an afterburner for consuming the exhaust gases of an internal combustion engine. In both cases, the source of the fuel for the second step (the afterburner or hydrogenase) is generated by the inefficient use of energy by the first step (the internal combustion engine or nitrogenase). Neither the afterburner nor hydrogenase is provided with sufficient fuel to serve as a major source of energy, but both can make a significant contribution to the overall energy balance of the system in question.

Dixon[29] and Hyndman et al.[63] first provided evidence for esterification of inorganic phosphate coupled to H$_2$ oxidation in *R. leguminosarum* bacteroids and *Azotobacter vinelandii*, respectively. Emerich et al.[41] reported that steady-state levels of ATP were increased 20–40% by H$_2$ in bacteroids from H$_2$-oxidizing strains of *B. japonicum*. Thus, it is clear that H$_2$ oxidation can be coupled to ATP production. However, Nelson and Salminen[90] demonstrated that H$_2$ oxidation was strongly coupled to ATP formation in only 5 of 14 HUP$^+$ strains of *R. leguminosarum* taken from pea nodules. Therefore, H$_2$ oxidation is not always coupled to ATP production.

The low-potential reductant recovered upon oxidation of H$_2$ may be used in ways other than ATP production. For example, it has been shown that rates of acetylene reduction are increased in the presence of H$_2$ in bacteroids that express a hydrogenase.[41,103,107] Similar results were obtained with intact *Alnus rubra* nodules[88] and with free-living bacteria.[49] It seems likely that the electrons released upon the oxidation of H_2 can contribute to the reductant pool used by nitrogenase.

B. Postulated Mechanism 2

In aerobic N_2-fixing systems, H_2 oxidation is coupled, through the electron transport chain, to the reduction of O_2. Therefore, when H_2 is consumed, O_2 is also consumed and the *in situ* concentration of O_2 would be expected to be lower. Because nitrogenase is O_2 sensitive, this enzyme must be protected from exposure to O_2, even in organisms that require O_2 to support aerobic respiration. H_2 oxidation coupled to O_2 consumption may assist in the protection of nitrogenase from O_2. This process has been referred to as "respiratory protection." There are no N_2-fixing systems that require H_2 oxidation to protect nitrogenase from O_2 inactivation, and all these aerobic systems have other mechanisms that provide this protection. Therefore, while hydrogenase can contribute to the protection of nitrogenase from O_2 inactivation, it does not provide the primary mechanism for such protection.

Bacteroids isolated from legume nodules will reduce acetylene in the presence of an organic acid (e.g., succinate) and a low concentration of O_2. However, as the O_2 concentration is raised, the rate of acetylene reduction decreases (apparently due to inactivation of nitrogenase). When the bacteroids express a hydrogenase and the reaction mixture includes H_2, then the optimal concentration for O_2 is shifted to higher concentration.[41,103,107] This response was attributed to increased respiratory protection. Similar results were reported with cyanobacteria[112] and *Azotobacter*.[129] In each experiment, H_2 was added to the reaction mixture. Whether a similar protection would result from oxidation of the more limited supply of H_2 generated by nitrogenase is unknown. Respiratory protection has not been demonstrated in intact nodules, so it is not known if this additional mechanism for protection of nitrogenase from O_2 is significant in intact nodules. One must also consider the possibility that H_2 oxidation might be detrimental in a system that is O_2-limited rather than reductant-limited, especially if H_2 oxidation is not coupled bioenergetically as well as oxidation of carbon substrates. This could explain the results of Drevon et al.,[36] where HUP$^+$ soybeans were less productive than HUP$^-$ plants. It would also explain why HUP$^-$ *Azotobacter* dominate in O_2-limited chemostat competition experiments.[133]

C. Postulated Mechanism 3

Because N_2 reduction by nitrogenase is inhibited by H_2, H_2 oxidation may reduce any level of inhibition caused by H_2. Although this postulate is theoretically sound, it has not received as much attention in the literature as the first two. Dixon[29] suggested that it would be unlikely for H_2 to build up to inhibitory concentrations, given the apparent lack of a barrier to H_2 diffusion out of the nodule. However, he later argued,[31] based on models of H_2 diffusion through the nodular material and the rates of H_2 production, that inhibitory concentrations of H_2 could be reached. To answer this question definitively, two parameters must be measured. First, the K_i for H_2 for inhibition of N_2 reduction must be measured. This constant has now been determined for isolated, intact, soybean nodule bacteroids bathed in leghemoglobin and provided with succinate and an optimal concentration of O_2. The K_i value of 0.03 atm is the lowest K_i measured for either an intact system or a purified nitrogenase.[100] The low K_i indicates that a lower concentration of H_2 in the nodule than originally believed would significantly inhibit N_2 reduction. The second parameter needed for a complete assessment is to determine the H_2 and N_2 concentrations at various depths within the nodule. Although no such measurements have been made as yet, a sophisticated model of the gas diffusion through soybean nodules has indicated that a concentration of 114 μM (0.17 atm) H_2 could be reached in the infected cells of a soybean nodule and the concentration of N_2 would be 122 μM.[60] When the steady-state content of H_2 present in a HUP$^-$ nodule was measured and averaged over the entire nodule, a concentration of 0.02–0.03 atm H_2 was determined. This can be taken as a minimum concentration in the infected portion of the nodule. In contrast, HUP$^+$ nodules did not contain a measurable content of H_2. Taken together, these two parameters suggest that H_2 inhibition of N_2 reduction may be a major factor limiting the efficiency of N_2 fixation in legume nodules. Therefore, it now seems reasonable that a major benefit of H_2 cycling in nodules may be to relieve the inhibition of N_2 reduction by H_2.

V. Organismal Aspects of H_2 Oxidation

A. The Role of the Plant

The enzymes that interact directly with H_2 in N_2-fixing root nodules—namely, hydrogenase and nitrogenase—are found within the microbial symbiont of the nodule.[30] Thus, the role of the plant in H_2 cycling in nodules is indirect. Nonetheless, the plant has an important role in that it provides the reductant in the form of organic acids needed by the bacteroids to generate the ATP and reductant required by nitrogenase. This, in turn, will affect the rate of nitrogenase activity as well as the partitioning of electrons between N_2 and protons. The plant also has control over the concentration of O_2 that reaches the bacteroids. However, the mechanisms by which the plant communicates with the bacteroids, the extent to which it regulates the flow of carbon substrates or O_2, and the effect of these parameters on H_2 cycling by the bacteroids are not well understood.

B. Physiologic Aspects of H_2 Cycling in the Microbial Symbiont

In bacteroids, hydrogenase is an integral membrane protein; release from the membranes requires treatment with a detergent.[7] Little is known about how hydrogenase is integrated into the membrane of rhizobial species. In *Alcaligenes eutrophus*, immunocytochemical investigations of the intracellular location of the membrane-bound hydrogenase indicate that the enzyme can shift from a membrane location to the cytoplasm with changes in the growth conditions.[50] In *B. japonicum*, the sensitivity of hydrogenase to proteases depends on the redox status of the membranes.[86] While in O_2-oxidized membranes, hydrogenase was resistant to degradation by trypsin. In contrast, hydrogenase was readily degraded by trypsin in H_2-reduced membranes. Similar results were obtained with membrane-impermeant, protein-modifying reagents. The results indicate that hydrogenase undergoes a substantial redox-linked conformational change in the membranes. It is not known if hydrogenase is part of a larger complex in the membrane, as might be expected for a respiration-linked enzyme, or exists as an independent entity.

Because the H_2 oxidized by hydrogenase of bacteroids is generated *in situ*, a high affinity for H_2 would help ensure that the substrate is scavenged before leaving the nodule. In *B. japonicum* bacteroids, a remarkably low K_m for H_2 of 0.05 μM was determined.[42] This low K_m contributes to the ability of hydrogenase to recover nearly all the H_2 generated by nitrogenase. The electrons released upon oxidation of H_2 eventually reach O_2. A K_m value of approximately 10 μM was calculated for O_2 uptake coupled to H_2 oxidation plus endogenous respiration.[40] Hydrogenase itself is not capable of transferring electrons from H_2 to O_2 directly. Numerous studies have indicated the role of the electron transport chain in the transfer of electrons from H_2 and O_2 in *B. japonicum*. The results have been thoroughly discussed in two recent reviews[44,93] and are only briefly considered here. Inhibitors of the electron transport chain also inhibit H_2 oxidation coupled to O_2, but not to artificial *e*-acceptors.[38,92,94] Oxidation of H_2 leads to reduction of cytochromes, ubiquinone, and a flavin-containing component of the electron transport chain.[38,92,94,95] In the case of *B. japonicum*, evidence has accumulated that a unique cytochrome (component 559-H_2) is involved in the transfer of electrons from H_2 to O_2,[39] but the position of this cytochrome in the electron transport chain is not known. The immediate acceptor of electrons from hydrogenase has not yet been identified in species of *Rhizobium*, *Bradyrhizobium*, *Azotobacter*, or other N_2-fixing bacteria with membrane-associated hydrogenases.

As discussed earlier, H_2 oxidation can be coupled to production of ATP. Detailed studies have not been carried out to determine P/O coupling ratios with H_2 as *e*-donor. It seems clear that H_2 oxidation supports phosphorylation through the generation of a proton gradient. Uncouplers of oxidative phosphorylation such as CCCP (carbonyl cyanide m-chlorophenyl hydrazone) stimulate H_2-supported respiration but decrease the ATP content of *R. japonicum* bacteroids.[41] Again, detailed studies of proton translocation are lacking. The coupling most likely occurs at one or more of the traditional coupling sites in the electron transport chain. However, the question arises as to whether hydrogenase itself is a coupling site through scalar production of protons, given that protons are the product of H_2 oxidation. This would only require that the protons formed upon oxidation of H_2 be released to the periplasmic face of the inner membrane. In spheroplasts of *Paracoccus denitrificans,* proton ejection linked to H_2 oxidation was demonstrated when O_2 was the *e*-acceptor.[32] However, with benzyl viologen as *e*-acceptor (which accepts electrons directly from hydrogenase), no acidification of the medium was observed, suggesting that protons produced upon oxidation of H_2 are released into the cytoplasmic space.

3. Chemolithoautotrophy

A number of bacteria are capable of growth with H_2 as the sole energy source and CO_2 as the sole carbon source. However, this form of chemoautotrophic growth had not been associated with the N_2-fixing bacteria that enter into a symbiosis with legumes. With the establishment of methods for expressing hydrogenase in free-living cultures of *B. japonicum*[79] and the characterization of the conditions for expression of ribulose bisphosphate carboxylase,[117] it became apparent that these bacteria also had the two critical enzymes for chemoautotrophic growth. Further experiments revealed that at least some strains of Hup$^+$ *B. japonicum*[55,71] and other species of *Rhizobium* and *Bradyrhizobium*[124] are capable of growth, albeit slow growth, with H_2 as the sole energy source and CO_2 as the virtual sole carbon source. This metabolic diversity may contribute to the ability of Hup$^+$ species of *Rhizobium* and *Bradyrhizobium* to survive in the highly competitive soil environment.

Autotrophic cultures of *B. japonicum* have proved useful in studies of electron transport coupled to H_2 oxidation, investigation of hydrogenase, and other aspects of H_2 metabolism. In another example, autotrophic cultures were used to demonstrate that selenium increases hydrogenase expression. Furthermore, selenium copurified with the hydrogenase, which indicates that selenium is a constituent of this enzyme.[12] However, the role of selenium in this enzyme is not understood.

VI. Molecular Characterization of *Bradyrhizobium* Hydrogenase

Hydrogenases have been isolated from a large number of microorganisms and they comprise a diverse group of enzymes with regard to their bio-

Table 11-2. Properties of the NiFe Hydrogenase of Bradyrhizobium japonicum

Physiological function:	To oxidize the H_2 produced by nitrogenase
Intracellular location:	Cytoplasmic membrane
Physical properties:	
Molecular weight:	100,000
Subunit composition:	$(\alpha, \beta : 65,000; 35,000$ MW$)$
Metal content:	Ni, Fe (1 Ni : 10–11 Fe)
Catalytic properties of purified enzyme:	
Reactions catalyzed:	H_2 oxidation, H_2 evolution, isotope exchange
Specific activity:	60–70 μmol H_2 oxidized/min · mg protein
K_m for H_2:	1–2 μM
Electron acceptors:	Native: unknown
	Artificial: methylene blue, benzyl viologen, others
Inhibitors:	CO, C_2H_2, O_2
Inactivators:	O_2, cyanide

chemical properties.[2,3,17,37,46,59,93,127] The largest single class of hydrogenases is made up of those that have an $\alpha\beta$-dimeric subunit structure and contain Ni and Fe (as nonheme Fe). There is a remarkable level of immunologic cross-reactivity within this group of hydrogenases,[66] given that representative enzymes have been isolated from aerobes, phototrophs, facultative anaerobes, desulfovibrians, and N_2-fixing microorganisms. The hydrogenases of the aerobic, N_2-fixing microorganisms A. vinelandii and B. japonicum (Table 11–2) belong to this group.

A. Physical Properties

The hydrogenase of B. japonicum has been purified to homogeneity from bacteroids,[4,7] derepressed cells,[119] and chemoautotrophic cells[56] under anaerobic and aerobic conditions. The enzyme has a native molecular weight near 100,000 and consists of two subunits with molecular weights near 65,000 and 35,000. The purified enzyme contains Ni and Fe in a ratio of 1:10–11.[7] The iron is present in the form of Fe sulfur centers, as evidenced by a broad absorption spectrum in the visible region from 300 to 600 nm[4] and the presence of a complex EPR spectrum with g values <2.0 (Figure 11–2). The EPR signal is similar to that for the reduced state of the particulate hydrogenase from A. eutrophus.[107] Although the exact number and type of redox centers remains uncertain, by analogy with the well-characterized hydrogenase from Desulfovibrio gigas (which is also a Ni- and Fe-containing dimeric hydrogenase), a redox inventory consisting of 2 Fe_4S_4 centers, 1 Fe_3S_4 center, and a Ni center would seem reasonable.[46]

Hydrogenase from B. japonicum does exhibit some anomalous physical properties that have not yet received adequate explanation. For example, the apparent molecular weight derived from gel permeation experiments is 65,000.[7]

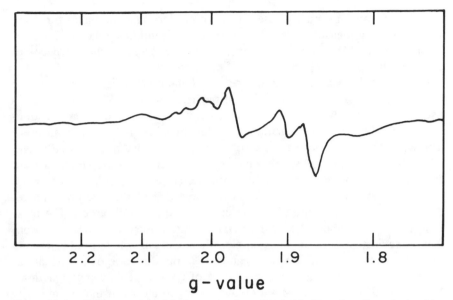

Figure 11–2. EPR spectrum of hydrogenase from *Bradyrhizobium japonicum* bacteroids. The protein was purified under anaerobic conditions and the spectrum was recorded while the hydrogenase was reduced with dithionite. $T = 5.1$ K; microwave power = 5 mW; microwave frequency = 9.41 gHz. (L. Seefeldt, G. Jensen, P. Stephens, D. Arp, unpublished results.)

The reason for this low estimate is not clear (the actual molecular weight is near 100,000) but may indicate a shape that deviates substantially from the spherical shape of the globular proteins used as molecular weight standards. The protein also exhibits a remarkable resistance to denaturation in the presence of SDS when in the reduced and active state.[113] In fact, the dimeric, active form of the enzyme can be electrophoresed into an acrylamide gel containing SDS without loss of activity. In contrast, when they are oxidized (e.g., by exposure to O_2), the subunits dissociate readily in the presence of SDS.

B. Catalytic Properties

The hydrogenases expressed in the endosymbionts of legume and actinorhizal root nodules function to oxidize H_2. Therefore, it is not surprising that they are most efficient at oxidizing H_2 *in vitro*. The K_m for H_2 of purified hydrogenase is near 1 μM, which is among the lowest values reported for hydrogenases in general.[3,7,67] The specific activity of homogeneous enzyme is near 60 μmol H_2 oxidized/min $\times$ mg protein.[4,56,119] Although this is on the low end of the reported specific activities for hydrogenases in general, it is comparable to other hydrogenases that function primarily to scavenge the H_2 generated by nitrogenase. If one considers that the value for kcat/ K_m (which is a measure of catalytic efficiency[130] and has a theoretical maximum near 10^9 M^{-1} sec^{-1}) is near 1×10^8 M^{-1} sec^{-1}, then it becomes apparent that this is indeed a very efficient enzyme.

Given that the natural electron acceptor is unknown, it is necessary to use artificial acceptors in studies with the purified enzyme. Methylene blue is used most frequently, but others (e.g., benzyl viologen, dichlorophenolindophenol, phenazine methosulfate, and ferricyanide) have also been used. The K_m's associated with these electron acceptors vary over a wide range, while the V_{max} values fall within a narrow range.[5] In addition to H_2 oxidation, *B. japonicum* and *A. vinelandii* hydrogenases can be made to evolve H_2 under appropriate conditions.[6,7,67] In the presence of the low-potential reductant, methyl viologen, a low rate of H_2 evolution is observed. The maximal rate is only about 2% of the maximal rate observed for H_2 oxidation. This is consistent with the physiologic role of the enzyme to oxidize H_2 and indicates a strong commitment to catalysis in the direction of H_2 oxidation. Hydrogenases in general are also capable of an isotope exchange reaction.[2] Where the isotopic composition of the hydrogen atoms in the H_2 and H_2O are different, hydrogenase will catalyze the equilibration of the mixture. Thus, in the presence of H_2 and D_2O, hydrogenase will catalyze the formation of HD, D_2, and HDO. This reaction occurs in the absence of an electron mediator. For *B. japonicum* hydrogenase,[6] exchange is slow compared to H_2 oxidation and is comparable to the rates of H_2 evolution. This probably reflects a common rate-limiting step between H_2 evolution and isotope exchange.

A number of inhibitors of *B. japonicum* and *A. vinelandii* hydrogenases have been characterized. CO is a rapid-equilibrium, reversible inhibitor that is competitive with H_2 and noncompetitive versus the electron acceptor.[5] This result suggests that it binds to the same site as H_2. Smith et al.[118] demonstrated inhibition of *A. chroococcum* hydrogenase by acetylene. Later studies revealed that acetylene is an active-site-directed, slow-binding, reversible inhibitor of *A. vinelandii* and *B. japonicum* hydrogenases.[61] The binding is competitive versus H_2. In fact, the amount of H_2 commonly found as a contaminant in acetylene[62] is sufficient to decrease greatly the rate of acetylene binding to hydrogenase. Exposure of *B. japonicum*[5] and *A. vinelandii*[112] hydrogenase to O_2 results in multiple effects. On short time exposures to low concentrations, O_2 is a reversible inhibitor of H_2 oxidation. The inhibition is uncompetitive versus H_2 and noncompetitive versus the electron acceptor. This indicates that O_2 must bind subsequent to H_2 binding, but at a site other than the electron acceptor site. O_2 can also lead to an irreversible inactivation of hydrogenase. For *B. japonicum* hydrogenase, the half-time for this inactivation is about 70 min in air.[7] H_2 can protect *A. vinelandii* hydrogenase from inactivation by O_2.[112] A final effect of O_2 occurs when hydrogenase is still associated with the membranes where hydrogenase can be oxidized to a stable but inactive state. This inactive state has allowed the hydrogenase to be purified under aerobic conditions.[119] In the case of *A. vinelandii* hydrogenase,[111] cyanide was characterized as an inactivator of hydrogenase. Cyanide binds to the oxidized (e.g., by incubation in the presence of H_2 and O_2) form of hydrogenase, which results in irreversible inactivation. No inactivation or inhibition by cyanide occurs when hydrogenase is in an active, reduced state.

Our knowledge of the hydrogenase from *B. japonicum* can be coupled with information derived from studies of other NiFe hydrogenases to formulate the following model of the catalytic mechanism of these hydrogenases. The kinetic data support a two-site ping pong mechanism in which H_2 binds to and is oxidized at one site on the enzyme followed by transfer of electrons through the enzyme to an electron acceptor. A reasonable site for the interaction of H_2 with the enzyme is at the Ni redox site. Oxidation of H_2 would lead to formation of a proton and a Ni hydride. Electrons are then passed from the Ni hydride through the enzyme via the FeS centers to the electron acceptor binding site. Acetylene and CO inhibit by competing with H_2 for binding to the Ni, whereas O_2 probably binds to and oxidizes one or more of the FeS centers.

Although many studies of hydrogenase require purified enzyme, it is important to consider the effect of purification on the catalytic properties of hydrogenase. For example, the apparent K_m for H_2 increases from 0.5 μM when measured in intact bacteroids to near 1 μM for the purified enzyme.[3] The enzyme is relatively insensitive to O_2 when in the intact cells but becomes sensitive upon purification. The pH optimum for oxidation of H_2 changes upon solubilization of hydrogenase from the membranes. Because of this tendency for catalytic properties of hydrogenase to change upon purification, one must be cautious in extrapolating observations with the purified hydrogenase to the situation in the intact organism.

VII. Genetics Associated with Hydrogenase

The genes essential for H_2 oxidation in *Rhizobium* and *Bradyrhizobium* have been designated *hup*, for *h*ydrogen *up*take, and the phenotype for H_2 oxidation is HUP$^+$. Initial studies directed at understanding *hup* genes focused on the generation and selection of mutant strains that lacked the H_2 oxidation phenotype (HUP$^-$).[78,82] The selections were made easier by the establishment of methods to express hydrogenase in the free-living state[79] and to directly couple H_2 oxidation to reduction of an artificial electron acceptor in single colonies on culture plates.[54] This procedure selected for mutants that were specifically blocked in ability to oxidize H_2, and it ruled out the selection of less specific mutants such as those involved in electron transport.

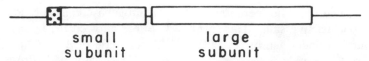

small
subunit

large
subunit

Figure 11–3. Arrangement of the structural genes for hydrogenase in *Bradyrhizobium japonicum*, *Azotobacter chroococcum*, *Rhodobacter capsulatus*, and other Ni- and Fe-containing hydrogenases. The stippled area indicates the putative leader sequence.

The *hup* genes in *B. japonicum* do not appear to be associated with a plasmid.[20] The H_2 uptake positive strains of *B. japonicum* examined did not possess any discernible plasmids (the procedure would not have detected plasmids of molecular weight greater than 280×10^6), while most strains with little or no H_2 uptake activity did possess plasmids. In contrast, the gene for hydrogenase is located on a sym plasmid in *R. leguminosarum*.[26,97] A large plasmid (270×10^6 Da) also encodes the H_2 uptake capability of *A. eutrophus*.[47]

Several cosmids capable of complementing HUP⁻ strains have been isolated from a *B. japonicum* gene bank.[19,53,68] One cosmid, pHU52, appeared to encode all essential HUP determinants.[68] When pHU52 was transferred into an originally HUP⁻ wild-type strain of *B. japonicum*, the strain was converted to the HUP⁺ phenotype. Although the number of specific genes and their organization has not yet been determined for the *hup* locus of *B. japonicum* (or of any NiFe hydrogenase), it is clear that this locus is considerably larger than the approximately 2.5 kilobases (kb) required to encode the structural polypeptides. From the effect of a series of transposon insertions on HUP phenotype, the *hup* locus spans at least 16 kb of DNA.[53] Transfer of subclones from pHU52 into an *Escherichia coli* maxicell expression system when coupled to the lambda promoter resulted in expression of proteins that cross-reacted with antibody raised against the purified *B. japonicum* hydrogenase.[134] One of these subclones (5.9 kb) was shown to contain the structural genes for the two subunits of hydrogenase. These genes have now been sequenced,[108] as have the analogous counterparts from a number of other microorganisms that express Ni- and Fe-containing hydrogenases.

The structural genes for *B. japonicum* hydrogenase are encoded in a single operon.[108] There is a putative leader peptide of 46 amino acids prior to the start site of the protein (which was determined by translating the DNA sequence to the corresponding amino acid sequence and comparing it to the N terminal amino acid sequence of the isolated protein). The small subunit sequence follows immediately after the leader sequence. In the case of *B. japonicum*, 32 nucleotides separate the small subunit gene from the large subunit gene. This arrangement of the structural genes (Figure 11–3) is similar to that reported in *Rhodobacter capsulatus*,[70] *Desulfovibrio baculatus*,[83] and *D. gigas*[74] (all Ni–Fe hydrogenases) and appears to be conserved in *R*.

leguminosarum as well.[73] However, this arrangement is distinct from that reported for *Desulfovibrio vulgaris* hydrogenase[128] (an "Fe-only" hydrogenase).

A comparison of the sequences of the genes coding for the hydrogenase structural genes from *B. japonicum*,[108] *D. gigas*,[74] *R. capsulatus*,[70] *A. chroococcum* (small subunit),[124] and *D. baculatus*[83] reveals considerable homology. This sequence homology even extends to a hydrogenase gene from the archaebacteria *Methanobacterium thermoautotrophicum*.[101] Much of the homology is found in the location of conserved cysteines, which are the likely metal binding sites in the protein, and their flanking amino acids.

The cosmid pHU52, which contains the hydrogenase genome, was used to transform HUP⁻ strains of *B. japonicum, Rhizobium trifolii*, and *Rhizobium meliloti*.[69] The strains could then be induced to express hydrogenase activity in the free-living state. When used as inocula on soybeans, clover, and alfalfa, the transformants gave rise to bacteroids with hydrogenase activity. However, the activity was very low, indicating that the HUP genome had not been stably integrated into the chromosome of the recipient. Given the potential of H_2 oxidation to increase N_2 fixation efficiency, it is essential to establish methods to transfer and stably integrate into the chromosome of the recipient the genome for hydrogenase.

VIII. Regulation of Hydrogenase Expression

The level of hydrogenase activity and protein expressed by *B. japonicum* in the free-living state is regulated by the cell in response to the environment. In the case of free-living cultures, the principal factors that influence hydrogenase expression are H_2, O_2, and carbon sources.[79,80,84] Induction of hydrogenase activity requires that the O_2 partial pressure be kept low; maximal levels of activity were observed at 1% O_2.[79] Expression of hydrogenase activity is sensitive to O_2 in most autotrophs and aerobic N_2-fixing microorganisms, although the O_2 partial pressure at which inhibition occurs is variable.[131] Mutants have been isolated that are hypersensitive to O_2 with regard to hydrogenase expression,[81] as well as mutants that are tolerant of high concentrations of O_2 and that express hydrogenase constitutively.[84] The mechanism of O_2 regulation is not understood.

Although it was first reported that reduced carbon sources repressed the synthesis of hydrogenase,[80,84] it was later shown that this was due to a depletion of O_2 and alteration of the pH coupled to carbon substrate oxidations.[126] Hydrogenase expression is regulated by carbon sources in some bacteria, but no general model has emerged.[18,48]

In the case of *B. japonicum*, it is not clear if H_2 is required for induction of hydrogenase activity. In one case, no hydrogenase activity was observed unless H_2 was included in the gas phase during hydrogenase expression.[80]

However, a later report[51] concluded that H_2 was not required for hydrogenase expression in cells derepressed for nitrogenase synthesis (H_2 evolution by nitrogenase was blocked with C_2H_2). H_2 is required for expression of hydrogenase in *Rhizobium* ORS571.[27] H_2 concentration clearly affects the final level of hydrogenase activity observed in *Alcaligenes latus*.[35] A low level of hydrogenase activity was observed when 0.001% H_2 was included in the gas phase during expression and the level of hydrogenase activity continued to increase with increasing H_2 up to 30% H_2 in the gas phase. Furthermore, hydrogenase activity varied by a factor of over 100 in response to changes in the H_2 concentration. The mechanism of this remarkable regulation is unknown. On the other hand, *A. eutrophus* particulate hydrogenase is expressed at high levels in the absence of H_2 and appears to depend on the redox potential of the cell.[48]

Hydrogenase activity can be expressed in the absence of nitrogenase activity in free-living cultures of *B. japonicum*.[80,84] This indicates that expression of these physiologically related enzymes is not necessarily coordinate. Nonetheless, conditions that support expression of nitrogenase activity seem to result in expression of hydrogenase as well[51] (even in the absence of H_2), suggesting that expression of hydrogenase may be coordinate with nitrogenase expression (even though the converse is not true). In this regard, it is interesting to note that mutants deficient in hydrogenase and nitrogenase activity have been isolated.[87] An involvement of DNA gyrase in hydrogenase expression has been proposed.[91] Cyclic nucleotides may also play a role in the regulation of hydrogenase expression.[75]

With the realization that the hydrogenases found in legume symbioses are Ni-containing enzymes, the possibility arose that Ni might regulate the expression of hydrogenase. It was shown that depletion of the derepression medium of Ni resulted in an inability of the cells to express hydrogenase activity.[65] For *B. japonicum*,[121] and *A. latus*,[34] it was further shown that when cells were deprived of Ni and then exposed to hydrogenase expression conditions, no protein accumulated that cross-reacted with antibody raised against the hydrogenase. This indicated either that Ni was required for synthesis of hydrogenase (regulation at the transcriptional or translational level) or that hydrogenase protein synthesized in the absence of Ni was rapidly degraded and did not accumulate to immunologically detectable levels. In *Anabaena*, depletion of Ni resulted in the absence of soluble hydrogenase activity, but normal levels of protein were detected immunologically.[99]

From the viewpoint of N_2-fixing symbioses, the more interesting aspects of the regulation of hydrogenase expression are those that can be attributed to the host. Dixon[29] demonstrated that a strain of *R. leguminosarum* that expressed hydrogenase activity in pea nodules expressed little or no hydrogenase in nodules of other hosts. Keyser et al.[64] reported similar results with *B. japonicum* strains on soybean and cowpea hosts. Although two strains expressed hydrogenase activity only on the cowpea host, one strain was HUP[+] on both hosts. Bedmar et al.[8] demonstrated that within a single species of peas, different cultivars produced nodules with significantly different levels of hydrogenase activity from a single strain of *R. leguminosarum,* even though nitrogenase activity was similar. Shoot–root grafting studies revealed that a transmissible shoot factor was involved in regulating the level of hydrogenase in the nodules.[9]

Based on results with free-living *B. japonicum,* it is tempting to speculate that the same factors that are needed to express hydrogenase in the free-living state (e.g., low O_2 and possibly the presence of H_2) are also responsible for regulating hydrogenase expression in the nodule (i.e. where pO_2 is low and where nitrogenase produces H_2). However, this simple model does not explain the host effect on hydrogenase expression. Furthermore, Lambert et al.[69] showed that conjugal transfer of cosmid pHU52 conferred hydrogenase activity to HUP[-] hosts in both the free-living and symbiotic states. However, transfer of cosmid pHU1 (which lacked an additional 5.5-kb EcoR1 fragment found on the right-hand side of pHU52) conferred hydrogenase activity in nodules but did not confer hydrogenase activity in the free-living state. This further indicates that regulation of hydrogenase expression in the symbiotic state is distinct from that in the free-living state. The details of host regulation of hydrogenase expression have not been sorted out at the molecular level, even though they are essential to efforts to exploit the beneficial aspects of hydrogenase activity in legumes.

IX. Conclusions

H_2 production by nitrogenase has been identified as one of the factors that limits the efficiency of N_2 fixation. It is largely the potential of H_2 cycling to lessen the impact of this reduction in efficiency that has driven research efforts in this area. During the last 10 years, we have made fundamental advances in many aspects of H_2 cycling in legumes and free-living bacteria. For example, the structural genes for hydrogenase have now been sequenced, the hydrogenase protein has been purified to homogeneity, the pathway of electron flow from H_2 to O_2 has been largely elucidated, and the benefit associated with H_2 cycling has been quantified. Yet in each of these areas, significant gaps in our knowledge still exist. For example, what additional genes are found in the *hup* locus and what are their functions? How is the hydrogenase protein integrated structurally and functionally into the membrane? What is the immediate acceptor of electrons from hydrogenase? Why are benefits from H_2 cycling not always apparent? Other areas requiring additional work include the role of the host in regulating expression of hydrogenase, the mechanisms by which hydrogenase improves the efficiency of N_2 fixation, and the establishment of methods to stably integrate the *hup* genes into the chromosome of recipient bacteria. Further research efforts should continue to provide answers to these and other questions. Ultimately, we hope to be in a position to apply our knowledge of H_2 cycling to ensure that all agronomically important N_2-fixing symbioses express a H_2-cycling system that benefits the plant.

Acknowledgments

Research on H_2 cycling in the author's laboratory has been funded by grants from USDA, DOE, NSF, and the California Agricultural Experiment Station.

References

1. Albrecht, S. L., Maier, R., Hanus, F. J., Russell, S. A., Emerich, D. W., and Evans, H. J., "Hydrogenase in *Rhizobium japonicum* increases nitrogen fixation by nodulated soybeans," *Science 203,* 1255–1257 (1979).

2. Adams, M. W. W., Mortenson, L. E., and Chen, J. S., "Hydrogenase," *Biochim. Biophys. Acta 594,* 105–176 (1981).

3. Arp, D. J., "H_2-oxidizing Hydrogenases of Aerobic N_2-fixing Microorganisms," in *Nitrogen Fixation and CO_2 Metabolism,* Ludden, P. W., and Burris, J. E. New York: Elsevier, 1985, pp. 121–131.

4. Arp, D. J., "*Rhizobium japonicum* hydrogenase: Purification to homogeneity from soybean nodules, and molecular characterization," *Arch. Biochem. Biophys. 237,* 504–512 (1985).

5. Arp, D. J., and Burris, R. H., "Kinetic mechanism of the hydrogen-oxidizing hydrogenase from soybean nodule bacteroids," *Biochemistry 20,* 2234–2240 (1981).

6. Arp, D. J., and Burris, R. H., "Isotope exchange and discrimination by the H_2-oxidizing hydrogenase from soybean root nodules," *Biochim. Biophys. Acta 700,* 7–15 (1982).

7. Arp, D. J., and Burris, R. H., "Purification and properties of the particulate hydrogenase from the bacteroids of soybean root nodules," *Biochim. Biophys. Acta 570,* 221–230 (1979).

8. Bedmar, E. J., Edie, S. A., and Phillips, D. A., "Host plant cultivar effects on hydrogen evolution by *Rhizobium leguminosarum,*" *Plant Physiol. 72,* 1011–1015 (1983).

9. Bedmar, E. J., and Phillips, D. A., "A transmissible plant shoot factor promotes uptake hydrogenase activity in *Rhizobium* symbionts," *Plant Physiol. 75,* 629–633 (1984).

10. Benson, D. R., Arp, D. J., and Burris, R. H., "Hydrogenase in actinorhizal root nodules and root nodule homogenates," *J. Bacteriol. 142,* 138–144 (1980).

11. Bothe, H., Tennigkeit, J., Eisbrenner, G., and Yates, M. G., "The hydrogenase–nitrogenase relationship in the blue-green alga *Anabaena cylindrica,*" *Planta 133,* 237–242 (1977).

12. Boursier, R., Hanus, R. J., Papen, H., Becker, M. M., Russell, S. A., and Evans, H. J., "Selenium increases hydrogenase expression in autotrophically cultured *Bradyrhizobium japonicum* and is a constituent of the purified enzyme," *J. Bacteriol. 170,* 5594–5600 (1988).

13. Bulen, W. A., Burns, R. C., and LeComte, J. R., "Nitrogen fixation: Hydrosulfite as electron donor with cell-free preparations of *Azotobacter vinelandii* and *Rhodospirillum rubrum*," *Proc. Natl. Acad. Sci. USA 53*, 532–539 (1965).

14. Bulen, W. A., and LeComte, J. R., "The nitrogenase system from *Azotobacter:* Two-enzyme requirement for N_2 reduction, ATP-dependent H_2 evolution, and ATP hydrolysis," *Proc. Natl. Acad. Sci. USA 56*, 979–986 (1966).

15. Burgess, B. K., Wherland, S., Stiefel, E. L., and Newton, W. E., "HD formation by nitrogenase: A probe for N_2 reduction intermediates," in *Molybdenum Chemistry of Biological Significance*, Newton, W. E., and Otsuka, S. New York: Plenum, 1980, pp. 73–84.

16. Burris, R. H., and Hageman, R. V., "Electron partitioning from dinitrogenase to substrate and the kinetics of ATP utilization," in *Molybdenum Chemistry of Biological Significance*, Newton, W. E., and Otsuka, S. New York: Plenum Press, 1980, pp. 23–37.

17. Cammack, R., Fernandez, V. M., and Schneider, K., "Nickel in hydrogenases from sulfate-reducing, photosynthetic, and hydrogen-oxidizing bacteria," *The Bioinorganic Chemistry of Nickel*, Lancaster, J. R., Jr. New York: VCH, 1988, pp. 167–190.

18. Cangelosi, G. A., and Wheelis, M. L., "Regulation by molecular oxygen and organic substrates of hydrogenase synthesis in *Alcaligenes eutrophus*," *J. Bacteriol. 159*, 138–144 (1984).

19. Cantrell, M. A., Haugland, R. A., and Evans, H. J., "Construction of a *Rhizobium japonicum* gene bank and use in isolation of a hydrogen uptake gene," *Proc. Natl. Acad. Sci. USA 80*, 181–185 (1983).

20. Cantrell, M. A., Hickok, R. E., and Evans, H. J., "Identification and characterization of plasmids in hydrogen uptake positive and hydrogen uptake negative strains of *Rhizobium japonicum*," *Arch. Microbiol. 131*, 102–106 (1982).

21. Carter, K. R., Jennings, N. T., Hanus, J., and Evans, H. J., "Hydrogen evolution and uptake by nodules of soybeans inoculated with different strains of *Rhizobium japonicum*," *Can. J. Microbiol. 24*, 307–311 (1978).

22. Chen, J. H., and Blanchard, D. K., "Isolation and properties of a unidirectional H_2-oxidizing hydrogenase from the strictly anaerobic N_2-fixing bacterium *Clostridium* W5," *Biochem. Biophys. Res. Commun. 84*, 1144–1150 (1978).

23. Chen, Y. P., and Yoch, D. C., "Regulation of 2 nickel-requiring (inducible and constitutive) hydrogenases and their coupling to nitrogenase in *Methylosinus trichosporium* OB36," *J. Bacteriol. 169*, 4778–4783 (1987).

24. Cunningham, S. D., Kapulnik, Y., Brewin, N. J., and Phillips, D. A., "Uptake hydrogenase activity determined by plasmid pRL6JI in *Rhizobium leguminosarum* does not increase symbiotic nitrogen fixation," *Appl. Environ. Microbiol. 50*, 791–794 (1985).

25. Dadarwal, K. R., Sindhu, S. S., and Batra, R., "Ecology of Hup$^+$ *Rhizobium* strains of cow pea miscellany: Native frequency and competence," *Arch. Microbiol. 141*, 255–259 (1985).

26. DeJong, T. M., Brewin, N. J., Johnston, A. W. B., and Phillips, D. A., "Improvement of symbiotic properties in *Rhizobium leguminosarum* by plasmid transfer," *J. Gen. Microbiol. 128*, 1829–1838 (1982).

27. De Vries, W., Stam, H., and Stouthamer, A. H., "Hydrogen oxidation and nitrogen fixation in rhizobia, with special attention focused on strain ORS 571," *Antonie van Leeuwenhoek 50*, 505–524 (1984).

28. Dixon, R. O. D., "Hydrogenases and efficiency of nitrogen fixation in aerobes," *Nature 262*, 173 (1976).

29. Dixon, R. O. D., "Hydrogenase in legume root nodule bacteroids: Occurrence and properties," *Arch. Mikrobiol. 85*, 193–201 (1972).

30. Dixon, R. O. D., "Hydrogenase in pea root nodule bacteroids," *Arch. Mikrobiol. 62*, 272–283 (1968).

31. Dixon, R. O. D., Blunden, E. A. G., and Searl, J. W., "Intercellular space and hydrogen diffusion in pea and lupin root nodules," *Plant Sci. Lett. 23*, 109–116 (1981).

32. Doussiere, J., Porte, F., and Vignais, P. M., "Orientation of hydrogenase in the plasma membrane of *Paracoccus Denitrificans*," *FEBS Lett. 114*, 291–294 (1980).

34. Doyle, C. M., and Arp, D. J., "Nickel affects expression of the nickel-containing hydrogenase of *Alcaligenes latus*," *J. Bacteriol. 170*, 3891–3896 (1988).

35. Doyle, C. M., and Arp, D. J., "Regulation of H_2 oxidation activity and hydrogenase protein levels by H_2, O_2, and carbon substrates in *Alcaligenes latus*," *J. Bacteriol. 169*, 4463–4468 (1987).

36. Drevon, J. J., Kalia, V. C., Heckmann, M. O., and Salsac, L., "Influence of hydrogenase on the growth of *Glycine* and *Vigna spp.*," *Nitrogen Fixation Res. Prog.*, Evans, H. J., Bottomley, P. J., and Newton, W. E. Dordrecht: Martinus Nijhoff, 1985, p. 358.

37. Eisbrenner, G., and Evans, H. J., "Aspects of hydrogen metabolism in nitrogen-fixing legumes and other plant–microbe associations," *Ann. Rev. Plant Physiol. 34*, 105–136 (1983).

38. Eisbrenner, G., and Evans, H. J., "Carriers in electron transport from molecular hydrogen to oxygen in *Rhizobium japonicum* bacteroids," *J. Bacteriol. 149*, 1005–1012 (1982).

39. Eisbrenner, G., and Evans, H. J., "Spectral evidence for a component involved in hydrogen metabolism of soybean nodule bacteroids," *Plant Physiol. 70*, 1667–1672 (1982).

40. Emerich, D. W., Albrecht, S. L., Russell, S. A., Ching, T., and Evans, H. J., "Oxyleghemoglobin-mediated hydrogen oxidation by *Rhizobium japonicum* USDA 122 DES bacteroids," *Plant Physiol. 65*, 605–609 (1980).

41. Emerich, D. W., Ruiz-Argueso, T., Ching, T. M., and Evans, H. J., "Hydrogen-dependent nitrogenase activity and ATP formation in *Rhizobium japonicum* bacteroids," *J. Bacteriol. 137*, 153–160 (1979).

42. Emerich, D. W., Ruiz-Argueso, T., Russell, S. A., and Evans, H. J., "Investigation of the H_2 oxidation system in *Rhizobium japonicum* 122 DES nodule bacteroids," *Plant Physiol. 66*, 1061–1066 (1980).

43. Evans, H. J., Hanus, F. J., Haugland, R. A., Cantrell, M. A., Xu, L. S., Russell, S. A., Lambert G. R., and Harker, A. R., "Hydrogen recycling in nodules affects nitrogen fixation and growth of soybeans," *World Soybean Research Conference III: Proceedings*, Shibles, R. Boulder: Westview Press, 1985, pp. 935–942.

44. Evans, H. J., Harker, A. R., Papen, H., Russell, S. A., Hanus, F. J., and Zuber, M., "Physiology, biochemistry, and genetics of the uptake hydrogenase in rhizobia," *Ann. Rev. Microbiol. 41*, 335–361 (1987).

45. Evans, H. J., Purohit, K., Cantrell, M. A., Eisbrenner, G., Sterling, A. R., Hanus, F. J., and Lepo, J. E., "Hydrogen losses and hydrogenases in nitrogen-fixing organ-

isms," *Current Perspectives in N₂ Fixation*, Gibson, A. H., Newton, and W. E. New York: Elsevier/North Holland, 1981, pp. 84–96.

46. Fauque, G., Peck, H. D., Jr., Moura, J. J. G., Huynh, B. H., Berlier, Y., Der-Vartanian, D. V., Teixeria, M., Przybyla, A. E., Lespinat, P. A., Moura, I., and LeGall, J., "The three classes of hydrogenases from sulfate-reducing bacteria of the genus *Desulfovibrio*," *FEMS Microbiol. Rev. 54*, 299–344 (1988).

47. Friedrich, B., Heine, E., Finck, A., and Friedrich, C. G., "Nickel requirement for active hydrogenase formation in *Alcaligenes eutrophus*," *J. Bacteriol. 145*, 1144–1149 (1981).

48. Friedrich, C. G., "Derepression of hydrogenase during limitation of electron donors and derepression of ribulosebisphosphate carboxylase during carbon limitation of *Alcaligenes eutrophus*," *J. Bacteriol. 149*, 203–210 (1982).

49. Fu, C., and Knowles, R., "H₂ supports nitrogenase activity in carbon-starved *Azospirillum lipoferum* and *A. amazonense*," *Can. J. Microbiol. 34*, 825–829 (1988).

50. Gerberding, H., and Mayer, F., "Localization of the membrane-bound hydrogenase in *Alcaligenes eutrophus* by electron microscopic immunocytochemistry," *FEMS Microbiol. Lett. 50*, 265–270 (1988).

51. Graham, L. A., Stults, L. W., and Maier, R. J., "Nitrogenase–hydrogenase relationships in *Rhizobium japonicum*," *Arch. Microbiol. 140*, 243–246 (1984).

52. Hageman, R. V., Burris, R. H., "Electron allocation to alternative substrates of *Azotobacter* nitrogenase is controlled by the electron flux through dinitrogenase," *Biochim. Biophys. Acta 591*, 63–75 (1980).

53. Haugland, R. A., Cantrell, M. A., Beaty, J. S., Hanus, F. J., Russell, S. A., and Evans, H. J., "Characterization of *Rhizobium japonicum* hydrogen uptake genes," *J. Bacteriol. 159*, 1006–1012 (1984).

54. Haugland, R. A., Hanus, F. J., Cantrell, M. A., and Evans, H. J., "Rapid colony screening method for identifying hydrogenase activity in *Rhizobium japonicum*," *Appl. Environ. Microbiol. 45*, 892–897 (1983).

55. Hanus, F. J., Maier, R. J., and Evans, H. J., "Autotrophic growth of H₂-uptake-positive strains of *Rhizobium japonicum* in an atmosphere supplied with hydrogen gas," *Proc. Natl. Acad. Sci. USA 76*, 1788–1792 (1979).

56. Harker, A. R., Xu, L. S., Hanus, F. J., and Evans, H. J., "Some properties of the nickel-containing hydrogenase of chemolithotrophically grown *Rhizobium japonicum*," *J. Bacteriol. 159*, 850–856 (1984).

57. Hoch, G. E., Little, H. N., and Burris, R. H., "Hydrogen evolution from soybean nodules," *Nature 179*, 430–431 (1957).

58. Hoch, G. E., Schneider, K. C., and Burris, R. H., "Hydrogen evolution and exchange, and conversion of N₂O to N₂ by soybean root nodules," *Biochim. Biophys. Acta 37*, 273–279 (1960).

59. Houchins, J. P., "The physiology and biochemistry of hydrogen metabolism in *Cyanobacteria*," *Biochim. Biophys. Acta 768*, 227–255 (1984).

60. Hunt, S., Gaito, S. T., and Layzell, D. B., "Model of gas exchange and diffusion in legume nodules," *Planta 73*, 128–141 (1988).

61. Hyman, M. R., and Arp, D. J., "Acetylene is an active-site-directed, slow-binding, reversible inhibitor of *Azotobacter vinelandii* hydrogenase," *Biochemistry 26*, 6447–6454 (1987).

62. Hyman, M. R., and Arp, D. J., "Quantification and removal of some contaminating gases from acetylene used to study gas-utilizing enzymes and microorganisms," *Appl. Environ. Microbiol. 53,* 298–303 (1987).

63. Hyndman, L. A., Burris, R. H., and Wilson, P. W., "Properties of hydrogenase from *Azotobacter vinelandii,*" *J. Bacteriol. 65,* 522–531 (1953).

64. Keyser, H. H., vanBerkum, P., and Weber, D. F., "A comparative study of the physiology of symbioses formed by *Rhizobium japonicum* with *Glycine max, Vigna unguiculata,* and *Macroptilium atropurpurem,*" *Plant Physiol. 70,* 1626–1630 (1982).

65. Klucas, R. V., Hanus, F. J., Russell, S. A., and Evans, H. J., "Nickel: A micronutrient element for hydrogen-dependent growth of *Rhizobium japonicum* and for expression of urease activity in soybean leaves," *Proc. Natl. Acad. Sci. USA 80,* 2253–2257 (1983).

66. Kovacs, K. L., Seefeldt, L. C., Tigyi, G., Doyle, C. M., Mortenson, L. E., and Arp, D. J., "Immunological relationship among hydrogenases," *J. Bacteriol. 171,* 430–435 (1989).

67. Kow, Y. W., and Burris, R. H. "Purification and properties of membrane-bound hydrogenase from *Azotobacter vinelandii,*" *J. Bacteriol. 159,* 564–569.

68. Lambert, G. R., Cantrell, M. A., Hanus, F. J., Russell, S. A., Haddad, K. R., and Evans, H. J., "Intra- and interspecies transfer and expression of *Rhizobium japonicum* hydrogen uptake genes and autotrophic growth capability," *Proc. Natl. Acad. Sci. USA 82,* 3232–3236 (1985).

69. Lambert, G. R., Harker, A. R., Cantrell, M. A., Hanus, F. J., Russell, S. A., Haugland, R. A., and Evans, H. J., "Symbiotic expression of cosmid-borne *Bradyrhizobium japonicum* hydrogenase genes," *Appl. Environ. Microbiol. 53,* 422–428 (1987).

70. Leclerc, M., Colbeau, A., Cauvin, B., Vignais, P. M., "Cloning and sequencing of the genes encoding the large and the small subunits of the H_2 uptake hydrogenase (*hup*) of *Rhodobacter capsulatus,*" *Mol. Gen. Genet. 214,* 97–107 (1988).

71. Lepo, J. E., Hanus, F. J., and Evans, H. J., "Chemoautotrophic growth of hydrogen-uptake-positive strains of *Rhizobium japonicum,*" *J. Bacteriol. 141,* 664–670 (1980).

72. Lepo, J. E., Hickok, R. E., Cantrell, M. A., Russell, S. A., and Evans, H. J., "Revertible hydrogen uptake-deficient mutants of *Rhizobium japonicum,*" *J. Bacteriol. 146,* 614–620 (1981).

73. Leyva, A., Palacios, J. M., and Ruiz-Argueso, T., "Conserved plasmid hydrogen-uptake (*hup*)-specific sequences with Hup$^+$ *Rhizobium leguminosarum* strains," *Appl. Environ. Microbiol. 53,* 2539–2542 (1987).

74. Li, C., Peck, H. D., LeGall, J., and Przybyla, A. E., "Cloning, characterization, and sequencing of the genes encoding the large and small subunits of the periplasmic [NiFe]hydrogenase of *Desulfovibrio gigas,*" *DNA 6,* 539–551 (1987).

75. Lim, S. T., and Shanmugam, K. T., "Regulation of hydrogen utilisation in *Rhizobium japonicum* by cyclic AMP," *Biochim. Biophys. Acta 584,* 479–492 (1979).

76. Lim, S. T., Uratsu, S. L., Weber, D. F., and Keyser, H. H., "Hydrogen uptake (hydrogenase) activity of *Rhizobium japonicum,*" in *Basic Life Sciences: Vol. 17, Genetic Engineering of Symbiotic Nitrogen Fixation and Conservation of Fixed Nitrogen,* Lyans, J. M., Valentine, R. C., Phillips, D. A., Rains, D. W., and Huffaker, R. C., eds. New York: Plenum, 1981, pp. 159–171.

77. Maier, R. J., "Biochemistry, regulation, and genetics of hydrogen oxidation in *Rhizobium,*" *CRC Crit. Rev. Biotech. 3,* 17–38 (1986).

78. Maier, R. J., "*Rhizobium japonicum* mutant strains unable to grow chemoautotrophically with H₂," *J. Bacteriol. 145*, 533–540 (1981).

79. Maier, R. J., Campbell, N. E. R., Hanus, F. J., Simpson, F. B., Russell, S. A., and Evans, H. J., "Expression of hydrogenase activity in free-living *Rhizobium japonicum*," *Proc. Natl. Acad. Sci. USA 75*, 3258–3262 (1978).

80. Maier, R. J., Hanus, F. J., and Evans, H. J., "Regulation of hydrogenase in *Rhizobium japonicum*," *J. Bacteriol. 137*, 824–829 (1979).

81. Maier, R. J., and Merberg, D. M., "*Rhizobium japonicum* mutants that are hypersensitive to repression of H₂ uptake by oxygen," *J. Bacteriol. 150*, 161–167 (1982).

82. Maier, R. J., Postgate, J. R., and Evans, H. J., "*Rhizobium japonicum* mutants unable to use hydrogen," *Nature 276*, 494–495 (1978).

83. Menon, N. K., Peck, H. D., Jr., LeGall, J., Przybyla, A. E., "Cloning and sequencing of the genes encoding the large and small subunits of the periplasmic (NiFeSe) hydrogenase of *Desulfovibrio baculatus*," *J. Bacteriol. 169*, 5401–5407 (1987).

84. Merberg, D., O'Hara, E. B., and Maier, R. J., "Regulation of hydrogenase in *Rhizobium japonicum*: Analysis of mutants altered in regulation by carbon substrates and oxygen," *J. Bacteriol. 156*, 1236–1242 (1983).

85. Minamisawa, K., "Comparison of extracellular polysaccharide composition, rhizobitoxine production, and hydrogenase phenotype among various strains of *Bradyrhizobium japonicum*," *Plant Cell Physiol. 30*, 877–884 (1989).

86. Moshiri, F., and Maier, R. J., "Conformational changes in the membrane-bound hydrogenase of *Bradyrhizobium japonicum*," *J. Biol. Chem., 263*, 17809–17816 (1988).

87. Moshiri, F., Stults, L., Novak, P., and Maier, R. J., "Nif⁻Hup⁻ Mutants of *Rhizobium japonicum*," *J. Bacteriol. 155*, 926–929 (1983).

88. Murry, M. A., and Lopez, M. F., "Interaction between hydrogenase, nitrogenase, and respiratory activities in a *Frankia* isolate from *Alnus rubra*," *Can. J. Microbiol. 35*, 636–641 (1989).

89. Nelson, L. M., and Child, J. J., "Nitrogen fixation and hydrogen metabolism by *Rhizobium leguminosarum* isolates in pea root nodules," *Can. J. Microbiol. 27*, 1028–1034 (1981).

90. Nelson, L. M., and Salminen, S. O., "Uptake hydrogenase activity and ATP formation in *Rhizobium leguminosarum* bacteroids," *J. Bacteriol. 151*, 989–995 (1982).

91. Novak, P. D., and Maier, R. J., "Inhibition of hydrogenase synthesis by DNA gyrase inhibitors in *Bradyrhizobium japonicum*," *J. Bacteriol. 169*, 2708–2712 (1987).

92. O'Brian, M. R., and Maier, R. J., "Electron transport components involved in hydrogen oxidation in free-living *Rhizobium japonicum*," *J. Bacteriol. 152*, 422–430 (1982).

93. O'Brian, M. R., and Maier, R. J., "Hydrogen metabolism in *Rhizobium*: Energetics, regulation, enzymology and genetics," *Adv. Microb. Physiol. 29*, 1–52 (1988).

94. O'Brian, M. R., and Maier, R. J., "Involvement of cytochromes and flavoprotein in hydrogen oxidation in *Rhizobium japonicum* bacteroids," *J. Bacteriol. 155*, 481–487 (1983).

95. O'Brian, M. R., and Maier, R. J., "Role of ubiquinone in hydrogen-dependent electron transport in *Rhizobium japonicum*," *J. Bacteriol. 161*, 775–777 (1985).

96. Pahwa, K., and Dogra, R. C., "H₂-recycling system in mungbean *Rhizobium* in relation to N₂-fixation," *Arch. Microbiol. 129*, 380–383 (1981).

97. Palacios, J. M., Leyva, A., and Ruiz-Argueso, T., "Generation and characterization of hydrogenase deficient mutants of *Rhizobium leguminosaum,*" *J. Plant Physiol. 132,* 412–416 (1988).

98. Peters, G. A., Evans, W. R., and Toia, R. E., Jr., "*Azolla–Anabaena azollae* relationship," *J. Plant Physiol. 58,* 119–126 (1976).

99. Phelps, A. S., and Wilson, P. W., "Occurrence of hydrogenase in nitrogen-fixing organisms," *Soc. Exp. Biol. Med. 47,* 473–378 (1941).

100. Rasche, M. E., and Arp, D. J., "Hydrogen inhibition of nitrogen reduction by nitrogenase in isolated soybean nodule bacteroids," *Plant Physiol. 91,* 663–668 (1989).

101. Reeve, J. N., Beckler, G. S., Cram, D. S., Hamilton, P. T., Brown, B. W., Krzycki, A. F., Alex, L., Orme-Johnson, W. H., and Walsh, C. T., "A hydrogenase-linked gene in *Methanobacterium thermoautotrophicum* strain H encodes a polyferredoxin," *Proc. Natl. Acad. Sci. USA 86,* 3031–3035 (1989).

102. Rivera-Ortiz, J. M., and Burris, R. H., "Interactions among substrates and inhibitors of nitrogenase," *J. Bacteriol. 123,* 537–545 (1975).

103. Ruiz-Argueso, T., Emerich, D. W., and Evans, H. J., "Hydrogenase system in legume nodules: A mechanism of providing nitrogenase with energy and protection from oxygen damage," *Biochem. Biophys. Res. Commun. 86,* 259–264 (1979).

104. Ruiz-Argueso, T., Hanus, J., and Evans, H. J., "Hydrogen production and uptake by pea nodules as affected by strains of *Rhizobium leguminosarum,*" *Arch. Microbiol. 116,* 113–118 (1978).

105. Ruiz-Argueso, T., Maier, R. J., and Evans, H. J., "Hydrogen evolution from alfalfa and clover nodules and hydrogen uptake by free-living *Rhizobium meliloti,*" *Appl. Environ. Microbiol. 73,* 583–587 (1979).

106. Saini, I., Chander Dogra, R., and Nagpal, P., "Uptake hydrogenase in fast-growing strains of *Rhizobium* sp. (*Sesbania*) in relation to nitrogen fixation," *J. Appl. Bacteriol. 62,* 449–452 (1987).

107. Salminen, S. O., and Nelson, L. M., "Role of uptake hydrogenase in providing reductant for nitrogenase in *Rhizobium leguminosarum* bacteroids," *Biochim. Biophys. Acta 764,* 132–137 (1984).

108. Sayavedra-Soto, L. A., Powell, G. K., Evans, H. J., and Morris, R. O., "Nucleotide sequence of the genetic loci encoding subunits of *Bradyrhizobium japonicum* uptake hydrogenase," *Proc. Natl. Acad. Sci. USA 85,* 8395–8399 (1988).

109. Schneider, K., Patil, D. S., and Cammack, R., "ESR properties of membrane-bound hydrogenase from aerobic hydrogen bacteria," *Biochim. Biophys. Acta 748,* 353 (1983).

110. Schubert, K. R., and Evans, H. J., "Hydrogen evolution: A major factor affecting the efficiency of nitrogen fixation in nodulated symbionts," *Proc. Nat. Acad. Sci. USA 73,* 1207–1211 (1976).

111. Seefeldt, L. C., and Arp, D. J., "Cyanide inactivation of hydrogenase from *Azotobacter vinelandii,*" *J. Bacteriol. 171,* 3298–3303 (1989).

112. Seefeldt, L. C., and Arp, D. J., "Oxygen effects on the nickel- and iron-containing hydrogenase from *Azotobacter vinelandii,*" *Biochemistry 28,* 1588–1596 (1989).

113. Seefeldt, L. C., and Arp, D. J., "Redox-dependent subunit dissociation of *Azotobacter vinelandii* hydrogenase in the presence of sodium dodecyl sulfate," *J. Biol. Chem. 262,* 16816–16821 (1987).

114. Sellstedt, A., "Occurrence and activity of hydrogenase in symbiotic *Frankia* from field-collected *Alnus incana,*" *Physiol. Plant. 75,* 304–308 (1989).

115. Sellstedt, A., Huss-Danell, K., and Ahlqvist, A. S., "Nitrogen fixation and production in symbioses between *Alnus incana* and *Frankia* strains with different hydrogen metabolism," *Physiol. Plant. 66*, 99–107 (1986).

116. Simpson, F. B., and Burris, R. H., "A nitrogen pressure of 50 atmospheres does not prevent evolution of hydrogen by nitrogenase," *Science 224*, 1095–1097 (1984).

117. Simpson, F. B., Maier, R. J., and Evans, H. J., "Hydrogen-stimulated CO_2 fixation and coordinate induction of hydrogenase and ribulosebisphosphate carboxylase in a H_2-uptake positive strain of *Rhizobium japonicum*," *Arch. Microbiol. 123*, 1–8 (1979).

118. Smith, L. A., Hill, S., and Yates, M. G., "Inhibition by acetylene of conventional hydrogenase in nitrogen-fixing bacteria," *Nature 262*, 209–210 (1976).

119. Stults, L. W., Moshiri, F., and Maier, R. J., "Aerobic purification of hydrogenase from *Rhizobium japonicum* by affinity chromatography," *J. Bacteriol. 166*, 795–800 (1986).

120. Stults, L. W., O'Hara, E. B., and Maier, R. J., "Nickel is a component of hydrogenase in *Rhizobium japonicum*," *J. Bacteriol. 159*, 153–158 (1984).

121. Stults, L. W., Sray, W. A., and Maier, R. J., "Regulation of hydrogenase biosynthesis by nickel in *Bradyrhizobium japonicum*," *Arch. Microbiol. 146*, 280–283 (1986).

122. Sun, J.-H., Chen, B.-J., Wang, H., and Song, H.-Y., "Hydrogen-recycling system increasing nitrogen fixation in chinese milk vetch *Rhizobium*," *Kexue Tongbao 33*, 1723–1729 (1988).

123. Tibelius, K. H., Robson, R. L., and Yates, M. G., "Cloning and characterization of hydrogenase genes from *Azotobacter chroococcum*," *Mol. Gen. Genet. 206*, 285–290 (1987).

124. Tilak, K. V. B. R., Schneider, K., and Schlegel, H. G., "Autotrophic growth of strains of *Rhizobium* and properties of isolated hydrogenase," *Curr. Microbiol. 10*, 49–52 (1984).

125. Truelsen, T. A., and Wyndaele, R., "Recycling efficiency in hydrogenase uptake positive strains of *Rhizobium leguminosarum*," *Physiol. Plant. 62*, 45–50 (1984).

126. Van Berkum, P., and Maier, R. J., "Lack of carbon substrate repression of uptake hydrogenase activity in *Bradyrhizobium japonicum* SR," *J. Bacteriol. 170*, 1962–1964 (1988).

127. Vignais, P. M., Colbeau, A., Willison, J. C., and Jouanneau, Y., "Hydrogenase, nitrogenase, and hydrogen metabolism in the photosynthetic bacteria," *Adv. Microb. Physiol. 26*, 155–234 (1985).

128. Voordouw, G., and Brenner, S., "Nucleotide sequence of the gene encoding the hydrogenase from *Desulfovibrio vulgaris*," *Eur. J. Biochem. 148*, 515–520 (1985).

129. Walker, C. C., and Yates, M. G., "The hydrogen cycle in nitrogen-fixing *Azotobacter chroococcum*," *Biochimie 60*, 225–231 (1978).

130. Walsh, C., *Enzymatic Reaction Mechanisms*. San Francisco: Freeman, 1977, pp. 33–35.

131. Wilde, E., and Schlegel, H. G., "Oxygen tolerance of strictly aerobic hydrogen-oxidizing bacteria," *Antonie van Leeuwenhoek 48*, 131–143 (1982).

132. Wilson, P. W., Umbreit, W. W., and Lee, S. B., "Mechanism of symbiotic nitrogen fixation," *Biochem. J. 32*, 2084–2095 (1938).

133. Yates, M. G., and Campbell, F. O., "The effect of nutrient limitation on the competition between an H_2-uptake hydrogenase positive (Hup$^+$) recombinant strain of *Azo-*

tobacter chroococcum and the Hup⁻ mutant parent in mixed populations," *J. Gen. Microbiol. 135*, 221–226 (1989).

134. Zuber, M., Harker, A. R., Sultana, M. A., and Evans, H. J., "Cloning and expression of *Bradyrhizobium japonicum* uptake hydrogenase structural genes in *Escherichia coli*," *Proc. Natl. Acad. Sci. USA 83*, 7668–7672 (1986).

12

Evolution of Nitrogen-Fixing Symbioses

Janet I. Sprent and *John A. Raven*

I. Introduction

Symbiotic systems on land are now thought to be quantitatively the most important source of biologically fixed N_2. However, this is unlikely always to have been the case. To set the scene for the evolution of symbiotic systems, it is therefore appropriate first to consider the evolution of biological N_2 fixation per se in the light of the development of the present environment. This chapter will thus begin at the beginning and attempt to unfold a possible route by which we arrive at modern N_2-fixing symbioses. Although the emphasis will be on legumes, it is pertinent to consider why these are currently more widespread than other symbiotic systems. Possible symbioses involving animals will not be considered (but see Ref. 90).

II. Evolution of the Present Atmosphere

Figure 12-1 outlines the way in which the modern atmosphere may have evolved, concentrating on features necessary for the evolution of the major metabolic pathways (respiration, photosynthesis, N_2 fixation) relevant to this book. Although there is considerable discussion as to the detail of many aspects (e.g., when oxidized nitrogen compounds first became available), the overall picture is more or less agreed.

The earliest organisms probably had a wide variety of metabolizable substances at their disposal. It has been argued that this period—3.9 to 3.5 Ga BP (3.9 to 3.5 × 10^9 years before present)—was the first of those when major diversification of microorganisms occurred.[48] Shortly after this, during major sedimentation of C compounds,[75,90] organisms became carbon (relative to N) deficient. Pressures then favored evolution of carbon dioxide fixation.[90] An elegant proposal as to how chemoautotrophy may have evolved first has recently been published.[105] Evidence from atomic fossils[75] is consistent with the early formation of the carboxylating enzyme ribulose bis-

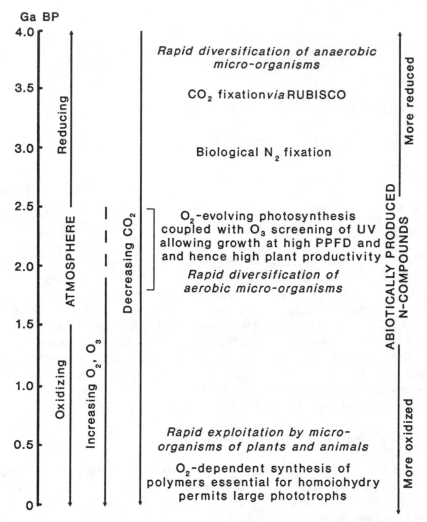

Figure 12-1. Possible sequences of events during the evolution of the modern atmosphere and types of organism–metabolic processes. (After Raven and Sprent[75] and Knoll and Bauld.[48])

phosphate carboxylase-oxygenase (RUBISCO). The ability to fix carbon dioxide in due course led to a nitrogen limitation and the scene was set for the evolution of N_2 fixation. Note that organisms at this time are assumed to have a C:N ratio similar to extant microbes, namely in the range of 4–11 (the higher values often being associated with the production of extracellular polysaccharides). Whether early nitrogenase proteins had a requirement for Mo, V (or other metals), or neither is an intriguing question, but one that we feel unable to address with our present knowledge.

The next step was the evolution of oxygenic photosynthesis, probably by cyanobacteria. This opened up many more possibilities and some problems, such as the need, in oxic environments, to protect nitrogenase from O_2. It allowed a second major burst of prokaryotic evolution.[48] At this stage the environment almost certainly had (not necessarily in the same habitats) all the inorganic N compounds found in the extant N-cycle, from ammonia through N_2 gas to nitrate. Thus, probably included in the additional microbial diversity were microorganisms carrying out all the processes of the nitrogen cycle. There is no *a priori* reason to suggest that denitrification should not have been present, especially since many extant N_2-fixing bacteria can also denitrify. Evidence from 5S rRNA sequences that, for example, *Paracoccus denitrificans* ATCC 13543 is closely related to *Rhodopseudomonas sphaeroides* and *Achromobacter cycloclastes* IAM 1013 is closely related to *Agrobacterium tumefaciens* (which in turn is closely related to *Rhizobium*)[61] is consistent with the suggestion of an evolutionary connection between the ability to reduce N_2 gas and oxidized nitrogen (in this case by *Nitrobacter winogradskyi, Nitrosolobus multiformis*).[109] Further discussion of possible early "biochemical experimentation" with nitrogen oxides in the environment can be found elsewhere.[98] Such "biochemical experimentation" with nitrogen oxides and nitrate could have preceded the buildup of significant O_2 in the atmosphere and ocean, or even the evolution of oxygenic photosynthesis, since nitrogen oxides and nitrate could have been generated in the absence of O_2 from H_2O vapor and N_2 under the influence of electrical discharges.[113]

How early ozone appeared in the atmosphere is not universally agreed, but once the atmospheric level of O_2 rose significantly, ozone levels were also likely to have increased. By screening out harmful UV radiation an ozone layer would have allowed organisms to grow at high photon flux densities and hence higher rates of plant productivity could be achieved. Later, when plants began to colonize land, they needed polymers such as lignin, suberin, and cutin for strengthening and protection against desiccation. These are synthesized by oxygen-dependent processes[75] that may have evolved initially as oxygen-protective mechanisms.[34] Large phototrophs then became possible. During the evolution of different groups of land plants the O_2 concentration of the atmosphere may have oscillated considerably (Figure 12-2).[9] Prior to the evolution of vascular plants, macroscopic animals radiated rapidly in the sea; arthropods radiated on land in parallel with the vascular plants.[8] Together, these plants and animals provided niches allowing a third major diversification of microorganisms.[48] Symbiotic N_2-fixing species formed an important component of these microorganisms.

III. Evolution of Nitrogen-Fixing Organisms

If N_2 fixation evolved prior to oxygenic photosynthesis, oxygen sensitivity of nitrogenase was not then a problem. Whether or not nitrogenase evolved

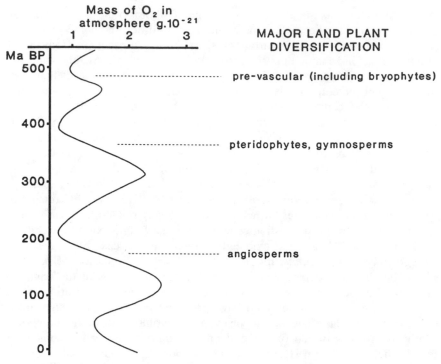

Figure 12-2. Hypothetical relationship between possible fluctuations in atmospheric oxygen levels and rapid species diversification in groups of plants containing N_2-fixing symbionts. (After Gottlieb[34] and references therein.)

for purposes such as detoxification of cyanide has been discussed elsewhere[71] and does not alter the arguments for selection pressures leading to successful N_2-fixing organisms. It is, however, interesting to note that one of the candidates for an ancestor of rhizobia is a precursor of *Azorhizobium* (see Section VI.A), the extant form of which can assimilate cyanide-N into cell N.[94] The range of bacteria that can fix N_2 has been discussed extensively elsewhere in this book. However, it is pertinent in the present context to summarize some of the evolutionary relationships among them, as suggested by molecular clock observations (see also Chapter 1). Leaving aside the archaebacteria and with the exception of a very few green sulfur and green non-sulfur bacteria and *Heliobacterium*,[31,109] all known eubacterial N_2-fixing genera fall into one of a cluster of groups (Figure 12-3); see also ref. 91). However, many genera have not yet been examined for N_2-fixing potential. The Proteobacteria have been discussed in detail in Chapter 1; symbiotic forms will be discussed later. Evidence from cyanobacteria (see also Chapter 4) suggests that nonheterocystous forms (nine genera considered, of which five have N_2-fixing species) pre-date heterocystous forms (seven genera consid-

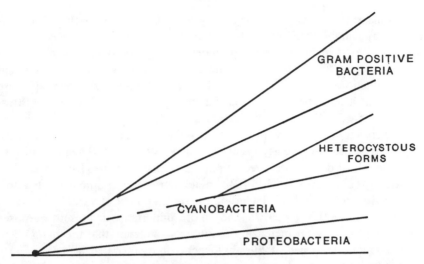

Figure 12-3. Possible separation of major eubacterial groups containing N₂-fixing members. The closed circle is the hypothetical origin and the dotted line indicates that the scale does not necessarily relate to that of the continuous lines.

ered).[32] Although fossilized structures resembling heterocysts have been observed, it is now thought that these may belong to ancestors of extant *Chloroflexus*-like organisms, which are considerably older than cyanobacteria.[32] Current evidence is more consistent with the evolution of heterocysts after oxygenic photosynthesis, as a mechanism for protection of nitrogenase.

The only known symbiotic Gram-positive genus is *Frankia* (see Chapter 7), an actinomycete of comparatively recent origin.

IV. Evolution of Host Plants

If we take all types of N₂-fixing symbiosis, we see that they are almost exclusively terrestrial. The occasional symbiotic diatoms, littoral-infralittoral lichens, and (more debatably aquatic as far as inorganic C source is concerned), the water fern *Azolla* are the known exceptions. All have cyanobacterial symbionts, although diatoms may also house other N₂-fixing eubacteria.[91]

Evidence for the timing of the events leading to evolution of extant plants is of two types. First, the fossil record. This is based upon large numbers of records of (usually) incomplete plants and of necessity is restricted to plants from habitats where fossilization was possible (i.e., mainly swamps and other wet areas). Second, an increasing amount of "molecular clock" data is being published. This relates to relatively few but well-characterized species. Whereas the fossil time scale is quite well defined, the molecular clock time scale is less so. However, the molecular approach is not constrained by suitable habitats for fossilization. Further, we now have atomic fossil data to help bridge the gap, most notably for our purposes on the carbon isotopes $^{13}C/^{12}C$ that enable the assessment of evidence for particular photosynthetic pathways.[75,86] This section will attempt a synthesis of these aspects, but it must be stressed that much is speculative. The major difference between the fossil and molecular clock lines of argument lies in the time when angiosperms evolved. Fossil evidence[97] is generally considered to be consistent with an early to mid-Cretaceous (about 150 Ma BP) origin. Molecular clock data in some recent publications puts values from more than 200 Ma[110] to even over 300 Ma BP,[53] much closer to the time when features associated with life on land, including the development of vascular systems, occurred (400–500 Ma BP). Perhaps more important for present purposes is that there is now an increasing amount of evidence that the mono- and dicotyledonous branches of the angiosperm tree separated at a very early stage.[10] No monocotyledonous plant is known to have a true symbiotic N_2-fixing structure. Regardless of exactly when this separation occurred, the atomic fossil data suggest that the ancestral forms had C_3 photosynthesis with RUBISCO as the primary carboxylating enzyme. The carbon isotope values suggest that the C_4 pathway (with PEP carboxylase as the primary carboxylating enzyme) is much more recent.[75] The change from C_3 to C_4 involves comparatively few genes and probably occurred several times in both monocotyledons and dicotyledons.[86] Molecular clock data support the isotope evidence, for example, suggesting a relatively recent separation of ancestors of barley (*Hordeum vulgare* L.) (C_3) and maize (*Zea mays* L. (C_4).[114] In view of the advantages of C_4 metabolism in warm dry areas, why are no legumes (or indeed other symbiotic N_2-fixing plants) known to be C_4? Many, including most acacias, occur naturally in communities also containing C_4 grasses. However, C_4 trees are very rare and have only evolved from less woody C_4 plants on islands with a restricted tree flora.[64]

Before considering the different types of symbiosis, grouped according to microsymbiont, we shall examine the selection pressures in favor of plants acquiring a N_2-fixing partner. In all cases that have been tested experimentally, combined nitrogen inhibits the setting up and/or operation of the symbiosis. This circumstantial evidence of the costliness of reducing N_2 gas suggests that the acquisition of an endosymbiont occurred when combined nitrogen limited plant productivity. We should stress here that limitation of nutrients generally (including water) are best withstood by the production of a good root system, possibly augmented with mycorrhizas (which have been found early in the fossil plant record).[96] To make nodules requires a considerable investment of nitrogen that otherwise could have been used to make roots.[90] The early fossil record of mycorrhizas (e.g., in cycads[96]) in the Triassic, ~230 Ma BP, and possibly in the Devonian, suggests that nutrients may indeed have been limiting at the time when vascular plants evolved.

Plants use products of photosynthesis very effectively in producing polymers such as cellulose, lignin, cutin, and suberin to give strength, transport systems, and protection against desiccation. This allows them to function as whole organisms at C:N ratios greatly in excess for those of animals and microorganisms.[90] Had plants not produced these N-free polymers or developed vacuolate cells,[75] they may have had an even greater need for N at the time of their major radiation. Even so, when plants began extensively to colonize land, they may have needed more nitrogen than was available abiotically. Extant free-living N_2-fixing organisms would have been confined to wet areas or wet episodes. By developing a symbiotic relationship, plants gained access to a controllable supply of fixed N and the microorganisms to an environment that was protected from desiccation (especially in homoiohydric plants) and could provide fixed carbon together with all those nutrients available in soil and taken up by absorbing organs such as roots.

We now return to the putative changes in atmospheric O_2 content since about 550 Ma BP[9] and the suggestion by Gottlieb[34] that the peaks of O_2 are related to diversification of major plant groups (see Figure 12-2).

The O_2 values were derived by Budyko et al.[9] from a consideration of crustal C, Fe, and S contents and the redox state of these elements. We first consider the changes in terrestrially derived (continental crust) reduced C (Figure 15 of Budyko et al.[9]), which are essentially antiparallel to those of O_2. The maximum rate of increase in quantity of sedimentary reduced C derived from photosynthesis on land can be used, with the C:N ratio in the sediments, to estimate minimal requirements for combined N input from abiotic plus biotic N_2 fixation. Maximum rates of reduced C accumulation in the Cambrian–Ordovician, the Devonian–lower Carboniferous, and the Triassic–Cretaceous, are 5.45 10^{12} g C per year (i.e., some 10^{-4} of the current net primary production rate on land). With a C:N mass ratio in the sediment reduced C of ≤100 (relatively unmodified plant structural material) a net input of combined nitrogen of 5.45 10^{10} g N per year over the whole productive land surface is required. Is this beyond the capacity of abiotic N_2 fixation in these periods? The present-day abiotic input from lightning acting on 21 kPa O_2 and 78 kPa N_2 yields 2–3 10^{12} g N per year worldwide[95] (i.e., ~10^{12} g N per year over the land surface of the earth). Even with minimum estimates (~4 kPa) of O_2 partial pressure by Budyko et al.[9] [i.e., those obtaining at the start of the episodes of net organic (reduced) C accumulation], a linear relation of abiotic N_2 fixation to O_2 partial pressure would still be adequate to provide the combined N related to reduced C storage over these three periods (each of several tens of millions of years duration). Of course, combined nitrogen inputs would also have been needed for any net short-term organic C and N storage in live biomass and surface deposits of C and N, and to offset denitrification losses. We have no way of estimating these requirements on the basis of the sedimentary record save to say that they will have continued even when the sedimentary organic pool was *decreasing*. Furthermore, the correlation noted by Budyko et al.[9] and by Gottlieb[34] between major diversification episodes for land plants and increases in atmospheric O_2 (and increases in reduced C sedimentation) may not be a directly causal relation. Diversification of taxa does not necessarily lead to increased biomass, although it is a likely outcome of the evolution of the vascular and, later, the tree habit, and the increased range of habitats available to plants as life forms diversified. Thus, although it is likely that the major phases of plant diversification were correlated with a need for increased combined N input rates, the sedimentary record cannot be cited as direct evidence for such a need.

The other aspect of the computation by Budyko et al.[9] and the speculations of Gottlieb[34] that is relevant to N_2 fixation is the O_2 content of the atmosphere. This could be related to abiotic N_2 fixation (increased rate with increased O_2 partial pressure in the 4–35-kPa range indicated in Figure 34 of Budyko et al.[9] after normalization of the quoted atmospheric O_2 mass over the last 550 Ma to the current O_2 mass and O_2 partial pressure of 21 kPa) and to biological N_2 fixation (increased potential for inhibition by O_2 as the partial pressure increases in the range 4–35 kPa). Before drawing too many conclusions as to the effect of these quoted variations in O_2 partial pressure on N_2 fixation, it should be noted that the upper and lower limits are at or beyond the extremes that seem plausible on the basis of the fossil record of the Phanerozoic. The *upper* limit of 35 kPa might cause a fire risk incompatible with the extent of vegetation revealed by the fossil record,[12] although many extant photosynthetic cells of plants operate with at least 35 kPa O_2 present in the light[74,76] (Raven, in preparation) when they are not relying on CO_2 diffusion from the atmosphere to RUBISCO. The *lower* limit of 4 kPa could pose problems for O_2 supply to large animals and plants (Raven, in preparation) based on the size of organisms in the past and responses of similarly sized organisms today to changed O_2 partial pressures.

In conclusion, and without putting too much emphasis on the work of Budyko et al.[9] and Gottlieb[34] in view of somewhat contradictory conclusions based on C and S stable isotope mass balances in the Phanerozoic,[79] we suggest that there were three episodes of particularly high demand for N_2 fixation associated with diversification of elements of the land flora. The first two would be of prevascular land plants (including bryophytes?) in the Cambrian–Ordovician, and of the pteridophytes and gymnosperms in the Devonian–Carboniferous. Extant N_2-fixing symbioses involving these plants all have cyanobacterial endosymbionts. The third episode, in the late Mesozoic, is associated with angiosperms that, with the sole exception of *Gun-*

Table 12-1. *Symbioses Involving Cyanobacteria*

Classification of Host	Host Genera	Cyanobacterial Genera
Bacillariophyceae	*Chaetoceros* *Rhizosolenia*	Not known but heterocystous
Anthocerophyta	*Anthoceros* and some others	*Nostoc*
Hepaticophyta	*Blasia* *Cavicularia* *Porella*	*Nostoc*
Pteridophyta	*Azolla*	*Nostoc* but usually called *Anabaena*
Cycadophyta	All nine extant genera	*Nostoc (Anabaena, Calothrix)*
Angiospermae	*Gunnera*	*Nostoc*
Lichenised fungi	≥18 lichen genera	*Nostoc, Scytonema, Stigonema, Calothrix, Dichothrix?* others

nera, have rhizobia or *Frankia* as endosymbionts. We now consider in turn these three types of symbiosis.

V. Cyanobacterial Symbioses

Cyanobacterial symbioses are characterized by a very wide range of host organisms but a restricted range of endosymbionts (Table 12-1). From an evolutionary point of view, the lichens are a particularly interesting group and probably the most ancient of the N_2-fixing symbioses. They are unique in having a nonphotosynthetic host—a fungus, most commonly a member of the Ascomycotina. About 21% of all fungi may form associations with one or more photobionts[39]: all those containing cyanobacteria that have been tested fix N_2. It has been suggested that the earliest lichens were terricolous associations between ancestors of the modern Ascomycotina and cyanobacteria.[39,40] The number of partners in lichens (≥ one mycobiont; ≥ one photobiont), the wide variation in arrangement of cells of the partners, and many other lichen features suggest a continuing process of both evolution and devolution. These and other fascinating aspects of lichens are discussed by Hawksworth.[39,40]

Other cyanobacterial symbionts have photosynthetic host plants. Nearly all the endosymbionts are *Nostoc* spp; none lacks heterocysts (see Chapter 4). Since the main function of heterocysts is to protect nitrogenase from oxygen and since location inside a plant can provide a low O_2 environment, why are no nonheterocystous cyanobacteria known to form symbioses? Answers to this question are largely speculative, but possibilities include (1) heterocystous forms were more likely to have been found in habitats where

Table 12-2. Possible Evolutionary Trends in Cyanobacterial Symbioses

	PRIMITIVE ———> ADVANCED			
Host:	Anthoceros Blasia	Azolla	Cycads	Gunnera
Location of Symbiont:	Leaf pockets	Leaf pockets	Root cortex intercellular	Glands at the base of petioles, intracellular
Access to Light:	Good	Good	Poor to zero	Low to zero

the host plants were evolving, (2) the location inside the plant is not sufficiently low in O_2 (most likely when cyanobacteria are close to photosynthetic cells as in *Azolla*, bryophytes, and lichens) to permit efficient operation of nitrogenase in vegetative cells. Evidence on this is very sketchy; a counterargument is that the vegetative cells are not photosynthesizing (at least in the N_2-fixing stages), and therefore there is no internally generated O_2.

Leaving out the diatoms (Bacillariophyceae) about which little is known and the lichens, we can trace a number of developments in cyanobacterial symbioses as we proceed from lower plant gametophytic hosts to higher plant, sporophytic hosts[54,65,90,91] (Table 12-2), observable in extant forms. Can we deduce anything from these about ancestral forms? The vegetative structures of *Anthoceros* and *Blasia* are very similar. *Nostoc* invades (by motile hormogonia whose formation may be induced by host exudates[10]) pockets in the thallus. This invasion is partly at least under host control but requires the motile stage and water to allow entry. The host exerts further control in that the symbiotic *Nostoc* has repressed photosynthesis, has enhanced heterocyst development, and exports most of its fixed N for assimilation by the host. Many looser associations between bryophytes and N_2-fixing cyanobacteria are known.[54] In the early days of colonization of land these and more advanced symbioses may have been very common. Many habitats currently occupied by bryophytes are less likely to be N limited. As world biomass increased and the shade required by many bryophytes was largely provided by vegetation, there may have been sufficient N cycling and stem flow to sustain them. In other words, the need for symbiotic N_2 fixation may have been transitory in evolutionary terms.

Azolla is unique in many ways, including being the only fern known to house a nitrogen-fixing symbiont. The aquatic habitats in which it thrives are generally low in N. Although the endophyte is housed in leaf pockets, these are closed at maturity and their internal structure is highly specialized[65] but not to the extent of excluding other microorganisms.[91] The plant "maintains" a culture of *Anabaena* in an apical cup made from young leaves. There is a sophisticated system for ensuring that an inoculum survives the resting stage ready to infect young sporophytes.

However, as with leaf cavities in *Anthoceros,* coralloid roots in cycads and leaf glands in *Gunnera,* the anatomic structures accommodating the cyanobacteria are also formed in their absence.[65] With rare known exceptions,[100] nodules on legumes and nonlegumes are only formed in response to the presence of the endophyte. However, this does not mean that in evolutionary terms the endophyte-housing structures were not developed to support a need for N_2 fixation, enabling the host to colonize areas low in combined N. Where, as in the case of ferns, evolution began in swampy areas, carbon compounds may have been plentiful, and free-living (autotrophic and heterotrophic) organisms may have fixed sufficient N_2 for the ecosystem (as they may do now in some rice paddy fields[91]). Fossil evidence of *Azolla* is much more recent than that of many other non-N_2-fixing ferns, such as the related *Salvinia,* and of cycads.[90] Thus, we cannot exclude the possibility that the symbiosis with *Anabaena* was a single event related to a particular need.

Conifers may have evolved in the comparatively dry Permian (250 Ma BP) and have numerous adaptations to dry conditions. No conifer is known to house N_2-fixing symbionts, yet all of their relatives, the cycads, do. There is evidence that the latter evolved in the late Carboniferous (280–300 Ma BP) and abounded in the Mesozoic (as did conifers), becoming less abundant in the Cretaceous. The fact that they have retained motile spermatozoids suggests that they probably did not evolve in dry areas, although some extant forms are very drought, and even fire, resistant.[91] The *Nostoc* or other genus is located inside the root, but between cells, although the mode of infection is not understood.[91] However it occurs, the plant is apparently able to exert some kind of selectivity, since only one type of cyanobacterium is found in any one coralloid root. Whether or not other bacteria are present in healthy symbioses is a matter of discussion, with the balance of opinion being in favor of their absence. This tendency toward maintaining an endophyte in more or less pure culture and exerting some type of control on its metabolism can be considered an advanced symbiotic character. Overall, the cycads appear to have been the most uniformly successful symbioses involving cyanobacteria and plants. We postulate that the symbiotic habitat was acquired early. Extant cycads from all continents have been found consistently to be symbiotic. Indeed, there is more consistency (100%!) in N_2 fixation among cycads than even papilionoid legumes.

Cycads evolved at a time when there were (probably) several lines of development toward leaves.[97] Leaves (rather than photosynthetic stems) are obviously advantageous to phototrophs, but because of their high levels of protein for light harvesting and carbon dioxide fixation (relative to structural and transport material), they are rich in N. One reason cycads may have been so successful is that their symbiotic N_2 fixation allowed them to develop more effective photosynthetic structures. Present–day cycads are often found as understory plants in forests where they may fix significant amounts of N_2. In parts of Australia, for example, the genus *Macrozamia* fills a niche in some eucalypt forests that in other areas is occupied by woody nodulated legumes such as species of *Acacia* and *Daviesia*.

Like *Azolla*, *Gunnera* is an anomaly, albeit a very interesting one. It is the only symbiosis involving cyanobacteria where the endophyte is intracellular. In this respect it has similarities with legume and nonlegume nodules. The Gunneraceae are a comparatively old angiosperm family. This fact, taken with the universal occurrence of symbiosis in extant species of *Gunnera*, suggests that the symbiosis might have arisen early.[90] Whether the particular leaf glands of *Gunnera*, through which infection occurs and which are at or near ground level, favored an initially casual association that later developed into a complex symbiosis is not known. Recent evidence that *Gunnera* may be dependent on N_2 fixed by *Nostoc*, because it has a reduced capacity to assimilate nitrate, suggests that this may be so.[63]

One feature common to all cyanobacterial symbioses, whether the endophyte has access to light or not, is that, when N_2 fixation is taking place, photosynthesis is suppressed. Although heterocyst frequency is increased, vegetative cells still occur and are dark blue-green. Fully differentiated heterocysts appear unable to divide (unlike *Frankia* vesicles; see later), and so vegetative cells will be necessary to maintain the symbiotic strains *ex planta*. However, to retain the photosynthetic apparatus appears rather profligate. As far as is known, infection processes of the type found in legumes and other actinorhizal plants do not occur (although how *Nostoc* invades *Gunnera* cells is not fully understood). Cyanobacteria as a whole are very well adapted to phototrophy. Unlike rhizobia and *Frankia*, they appear not to be closely related to organisms that may be parasitic or saprophytic [and hence have potential for extracellular enzyme (hydrolytic) production]. Perhaps it is not surprising that, in spite of many "attempts," they are not, on a global scale, as successful in symbiosis as are the nodule-forming organisms.

VI. Symbioses between Legumes and Rhizobia

A. Evolution of Rhizobia

As discussed in Chapter 1, the α subgroup of the Proteobacteria contains a number of N_2-fixing genera, including all the Gram-negative symbiotic forms. If we look at the possible relationships among this group, some interesting evolutionary points emerge (Figure 12-4). We shall consider these from the top of the figure downward. The genera bracketed as group A include two capable of growing on fixed nitrogen in their own free-living state (*Beijerinckia* and *Azorhizobium*). The ability of *Azorhizobium* to assimilate the product

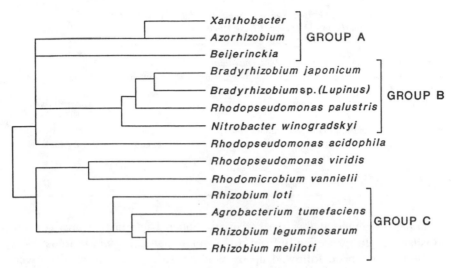

Figure 12-4. Relations between extant rhizobia and other bacteria. (See also Chapter 1.)

(ammonium) of its own N_2 fixation is diagnostic: other rhizobia cannot do this (see Chapter 1). Further, *Azorhizobium* is only distantly related to other known rhizobia. The structure (DNA sequence) and action (control by O_2 and combined nitrogen) of the regulatory gene *nifA* in *Azorhizobium* have features in common with both other rhizobia and with *Klebsiella pneumoniae* (in the γ group of Proteobacteria).[73] Group B contains both bradyrhizobia, photosynthetic organisms, and a nitrite-oxidizing species. Group C includes rhizobia *sensu stricto,* photosynthetic species, and the plant tumor-inducing agrobacteria. Thus, there appear to be three distinct types of rhizobia—others may yet be added, especially if *Photorhizobium*[24] is properly established. The ancestor of all groups may have been a free-living organism that could live chemo- and/or photoautotrophically as well as fix N_2, and it may have predated the evolution of higher plants.[112]

Table 12-3 summarizes the major features of the hypothetical ancestor that have been modified or retained to give modern rhizobia. It is consistent with the observation that some extant rhizobia contain RUBISCO and may be able to live autotrophically on CO_2 and H_2.[37,51,52,85] It is likely that considerable modification to these suggestions will be made as more information becomes available. Numerous questions remain, such as how the differences in location of symbiotic genes (chromosomal or plasmid) arise. Fortunately, as we study a wider range of rhizobia, particularly from the tropics and from more primitive legumes, and as we acquire more comparative data on the structure of key molecules, the prospects for answering some of these questions look promising. Further, the identification of bacteria by gene probes as well as phenotypic characters will be important.

Table 12-3.   Features of Extant Rhizobia Seen in Relation to a Hypothetical
Photosynthetic, Free-Living, Nitrogen-Fixing Ancestor

Losses
Light-harvesting pigments (except *Photorhizobium*)
Ability to grow autotrophically (except some strains of *B. japonicum* and *R. meliloti* which contain RUBISCO)
Control of N fixation by ammonia and oxygen (some aspects retained by *Azorhizobium*)
Gains
Ability to infect certain plants and induce nodule formation
Adaptations to existence in plant cytosolic environment; for example, control by high [K⁺] low pO₂
No change?
Ability to produce phytohormones

Young[111] suggested that the soil may contain bacteria that are indistinguishable electrophoretically from rhizobia but that lack *sym* plasmids. This
suggestion has been followed up by workers such as Jarvis et al.,[45] who
point out that studying only those rhizobia isolated from nodules might obscure relationships with other organisms. Generalizing this argument, Coplin[16]
notes that plant pathologists tend to ignore the needs of pathogens (and symbionts) to survive and/or grow *ex planta*. Plasmids that are essential for
host invasion may be useless in a saprophytic context. It is argued that the
varied requirements of pathogens and symbionts to survive in hosts and *ex
planta* argues for some stability in plasmid as well as chromosomal DNA.
Regardless of the narrowness or broadness of host range of individual strains
of rhizobia, the preceding arguments indicate a generalist strategy—retain-
ing the potential both for symbiotic and saprophytic existence. We have
argued elsewhere[90] that if it is a selective advantage for rhizobia to retain
the ability to nodulate host plants and to fix N_2, then it must be possible for
bacteria from active nodules to be released in a viable form into soil. It was
suggested that in determinate nodules such as those of soybean, where development of infected cells is more or less synchronized, bacteroids are not
greatly enlarged compared with free-living forms and should retain viability.
Evidence quoted then and later[93] in support of this has been reinforced.[46]
Conversely, in indeterminate nodules, especially where mature bacteroids
become greatly enlarged and pleomorphic, they lose viability. However, these
nodules always have a developmental spectrum of rhizobia, including undifferentiated viable ones. Again, recent evidence supports that quoted in
earlier papers.[50] The α subgroup of the Proteobacteria also includes *Azospirillum*. This and other N_2-fixing genera may be found inside plant roots,
particularly in members of the Gramineae, but they do not induce nodules.
Azospirilla appear to have features in common with rhizobia. For example,
A. brasilense can produce exopolysaccharides (EPS) that can substitute for
those lacking in certain mutants of *Rhizobium meliloti*.[57] Since EPS are an

essential part of the infection of some but not all[21] hosts by rhizobia, it will be interesting to see if they have a role to play in *Azospirillum* infection, about which virtually nothing is known (see also Chapter 6). The fact that several DNA fragments in azospirilla are homologous with *Rhizobium nod* DNA sequences[56] poses a number of interesting questions.

B. Legumes

In considering the evolution of specificity between legumes and rhizobia, Young and Johnston[112] have given an excellent account of the taxonomic problems of both host and rhizobia. They point out the virtues of letting rhizobia sort out the taxonomy of the hosts! One of the instances they cite is the papilionoid legume tribe Phaseoleae. We shall extend some of their arguments to illustrate what may be considered recent evolution, before considering the possible situation when legumes first evolved.

Phaseolus vulgaris L. probably rivals *Brassica oleracea* L. in having a wide range of cultivars used to provide food in many parts of the world. As a dry bean it forms much of the staple diet of South America and now also large areas of Africa (e.g, Kenya, Tanzania).[1] In the United States navy beans are canned in enormous quantities and are exported to many other countries. Additionally, french beans are a widely grown green vegetable. One could argue that evolution has been encouraged by plant breeders. Although interactions between host plants and rhizobia are considered elsewhere (see Chapter 16), it is worth making some particular points here, since wherever *Phaseolus vulgaris* grows it appears to have nodulation problems. The host plant is essentially of tropical–subtropical origin, and yet it can nodulate with a diversity of rhizobia, including some closely related to strains from temperate legumes such as *Pisum* and *Vicia* (see Chapter 1). In more northern latitudes the symbiosis may be limited by low temperatures (probably via the host), and in some parts of South America and Africa it may be limited by high temperatures (probably via the endophyte). However, it has recently been discovered (M. Hungria and A. A. Franco, personal communication) that many rhizobia isolated from trees can effectively nodulate tropical cultivars of *P. vulgaris* and that some of the resulting symbioses are heat tolerant. We cannot exclude the possibility that our current ideas on interactions between legumes and rhizobia are confounded by the breeders of host cultivars and possibly by the producers of legume inoculants. As Young and Johnston[112] point out (with respect to soybean), there are hazards of drawing broad evolutionary conclusions from studies of highly bred crop species! Fortunately, we are now obtaining more information from natural environments, particularly from tropical areas where the Leguminosae are thought to have originated. We now see that many of the dogmas (e.g., root hair infection, release of bacteria from infection threads being essential) may

Table 12-4.   Genera of Caesalpinioid Legumes with Well-Confirmed Nodulating Species[27]

Genus	Tribe	Group
Campsiandra	Caesalpinieae	Peltophorum
Melanoxylon	Caesalpinieae	Peltophorum
Moldenhawrea	Caesalpinieae	Peltophorum
Dimorphandra	Caesalpinieae	Dimorphandra
Erythrophleum	Caesalpinieae	Dimorphandra
Sclerolobium	Caesalpinieae	Sclerolobium
Tachigali	Caesalpinieae	Sclerolobium
Chamaecrista	Cassieae	—

be features of the more advanced, largely temperate legumes of agricultural significance.

Although the symbioses between legumes and rhizobia requires many interactions, we shall concentrate on the host plant, since this appears to be the major partner determining the mode of infection and the final structure of the symbiotic organ, the nodule. Further, there are a number of instances where the same strain nodulates members of different legume subfamilies[88] and even nonlegumes like *Parasponia*. For a detailed analysis, particularly of the bacterial aspects of the infection process, readers are referred to the excellent review of Young and Johnston[112] and to other chapters in the present volume.

i.   Subfamily Caesalpiniodeae

Legumes appear to have evolved fairly early in the evolution of the angiosperms. Fossil evidence suggests that the Caesalpinioideae is the oldest of the three subfamilies (the other two being the Mimosoideae and the Papilionoideae, following the taxonomic scheme of Polhill and Raven[68]). The current position with respect to nodulation in all three subfamilies has been assessed in the light of recent host taxonomic revisions.[27] Since the majority of the Mimosoideae and Papilionoideae can nodulate, and it is generally agreed that these two subfamilies evolved from the Caesalpinioideae, it is reasonable to suppose that their caesalpinoid ancestor also nodulated. Because nodulation is less common in this subfamily, we shall begin by considering in which groups it does occur. Over half the genera and more than 80% of all caesalpinioid species have yet to be examined for nodulation. However, some members of each tribe have been examined, allowing tentative conclusions to be drawn. In doing this, only fully confirmed positive reports of nodulation will be considered, and of these most have been studied at the structural level in our laboratory. We prefer to err on the cautious side at this stage. The genera that we feel confident have nodulated species are listed in Table 12-4. Either all other reports of nodulated genera have not been independently confirmed or the number of negative reports (often for the same species) exceeds the number of positive reports.[27]

The tribe Caesalpinieae contains 47 genera, arranged in eight groups.[68] Nodulated genera occur in only three of these. First, the Sclerolobium group of three genera—*Sclerolobium, Tachigali,* and *Diptychandra.* There are no reports on the latter genus. It would be very interesting to know whether, like its fellows, it can nodulate. The Peltophorum group has three nodulating genera, five genera for which there is no information, three genera with negative reports, and two genera with unconfirmed positive reports. Of the 10 genera in the Pterogyne group, only two have confirmed nodulation, and there is one unconfirmed report, one negative group, and six with no information. The Caesalpinieae, in common with much of the subfamily, present great taxonomic difficulties.[106] Watson,[106] who has made a great contribution to this area, reflects that "it may well be that evolution has bequeathed us an inherently messy situation."

Taking the parsimonious approach, there is no evidence that the genera in the Caesalpinieae that nodulate do not have a common ancestor. A systematic search for nodulation in those genera for which there are no reports would be extremely helpful.

The genus *Cassia,* consisting of over 500 species, is now generally divided into three—*Cassia, Chamaecrista,* and *Senna.* American and African cassias have been reclassified and work is under way for those of other regions. Nearly half of the reclassified species have been included in *Chamaecrista:* Of these, forty-three have been reported to nodulate, with two negative reports and the rest awaiting examination. There are no confirmed reports of nodulation in the other two genera. This would seem to establish the value of nodulation as a taxonomic character, even though it was not the basis for subdivision of *Cassia.* In discussing their reasons for giving generic status to the subgenera of *Cassia,* Irwin and Barnaby[44] support Bentham's original views and suggest that "these syndromes of differential characters must have survived from a remote past, antecedent to dispersal around the world of the three major cassioid types, antecedent certainly to the evolution of most species known today, and perhaps antecedent to distant separation of the continents." Again, being parsimonious, could *Chamaecrista* have diverged from a common nodulated caesalpinioid ancestor? We do not know. Certainly the current taxonomic position is not inconsistent with nodulation evolving more than once. Some molecular clock studies on selected genera would help greatly to try to decide among these various possibilities, answers to which have more than intrinsic interest, since they would help to define why some legumes apparently cannot nodulate.

A further point suggesting close relationships between extant nodulated caesalpinioid legumes arises from studies of nodule structure.[29] All those that have been examined have bacteria retained within infection threads (persistent infection threads) throughout the active life of the nodule. This property is shared only by a few papilionoid legumes and *Parasponia* (Ulmaceae, Chapter 13).

Considering the Leguminosae as a whole, the most archaic genera are thought to be *Gleditsia* (honey locust), *Gymnocladus, Ceratonia* (carob), *Zenia,* and *Cercis*[69] (Judas tree); none of these has been confirmed as nodulating. On the other hand, the base groups from which the main radiation occurred may have included the Sclerobium and Dimorphandra groups, where nodulation is definitely found. These suggestions are consistent with nodulation first occurring in the Caesalpinioideae, at some stage that may have preceded separation of continents, since nodule structure in extant genera is remarkably uniform. This would put the origin earlier in the Cretaceous than the fossil record indicates and possibly even in the late Jurassic, consistent with some of the molecular clock data on the separation of monocotyledonous and dicotyledonous plants (see Section IV) and with the pressure for N_2 fixation suggested by the atmospheric O_2 data.

ii.   Subfamily Mimosoideae

Over 90% of the species examined from the subfamily Mimosoideae nodulate. Where nodule structure has been examined, no evidence of persistent infection threads has been found (ref. 29 and unpublished data from our laboratory), unlike the situation in the Papilioniodeae (see the next section). The tribe Parkieae is considered to be the most likely link with the Caesalpinioideae. It has two genera, *Parkia* and *Pentaclethra*. The latter is thought to be closely related to *Dimorphandra*[23] on a variety of vegetative and reproductive characters. The nodule structure of these two genera is different, *Pentaclethra* not having the persistent infection threads found in *Dimorphandra.*[29] *Parkia* is interesting in that some species appear to nodulate and others do not. Of the nonnodulating groups of the tribe Mimoseae, Dinizia and Xylia have affinities with *Pentaclethra* or *Dimorphandra,* and thus may be primitive. Some members of the Newtonia group and the monotypic Filleopsis group have similarities with the nodulated caesalpinioid genus *Erythrophleum.* In the tribe Ingeae, all genera examined can nodulate, although some may also have nonnodulating species. *Acacia* (tribe Acacieae) also has a few well-documented nonnodulating species. Spontaneous as well as induced nonnodulating genotypes of crop legumes are found, so it is possible that nonnodulating species of otherwise consistently nodulating genera are secondary. Loss of nodulating ability in whole groups of genera appears less likely. The interface between mimosoid and caesalpinioid legumes is rather broad, with links involving nodulation particularly unclear, since the nodulating genus thought to link the two subfamilies, *Dimorphandra,* has nodules unlike those of studied mimosoid species.

In the Leguminosae generally, some species appear never to have root hairs, others produce them erratically[25] (in time and space), and yet others produce hairs in a way that elementary botany textbooks would have us believe is typical. Obviously, absence of hairs precludes an infection pathway involving them. In some cases, where nodules are associated with lateral roots, infection is via wounds.[92] However, at least in several mimosoid legumes, nodules are found on plants lacking hairs and not associated with lateral roots. For one of these, *Mimosa scabrella* Benth. (an important fuel wood and general purpose tree), an infection pathway between epidermal cells has been established.[26] Whether such pathways have any taxonomic and evolutionary significance is not yet clear.

iii. Subfamily Papilionoideae

In the subfamily Papilionoideae, 97% of the species examined were nodulated. While there are occasional negative reports in most tribes, the only consistent ones are from a small number of tribes toward the probable base of the subfamily; some have other features of taxonomic–evolutionary interest.

The tribe Swartzieae is usually considered to be at the base of the Papilionoideae and to form a link with the Caesalpinioideae, in which it was previously included. It contains 12 genera, of which only *Swartzia* is known to nodulate (six have not been examined). Nodules on *Swartzia* do not contain the persistent infection threads characteristic of caesalpinioid nodules.[29] The tribe Sophoreae is generally agreed to be heterogeneous.[67] There are several genera for which there are no reliable reports of nodulation, but it would be unwise to draw any conclusions from these at present. Few studies have yet been made on their structure, but some at least lack persistent infection threads (Sutherland and Sprent, unpublished data). Two of the three genera of the Dipterygeae appear not to nodulate; there is no report for *Taralea*.

The tribes Dalbergieae and Millettieae (formerly Tephrosieae) both contain genera in which nodules have persistent infection threads.[29,92] For other reasons these genera do not fall entirely happily within their currently allotted tribes. We believe this additional fact argues strongly for their forming a direct link with the Caesalpinoideae. The Dalbergieae also contain several genera that appear not to nodulate. Apart from these and the genera with persistent infection threads, all nodules examined in the Dalbergieae are of the aeschynomenoid type as defined by Corby in his extensive work on the systematic value of legume nodules.[17,18]

C. Evolution of Infection Mechanisms

Before infection of plants by rhizobia can occur, compatible organisms must come together in a more or less specific way. Evolution of specificity per se has been discussed recently[112] and will not be considered in detail here. Similarly, host specificity forms a major part of Chapter 9. We shall thus consider only aspects germane to the evolution of symbiotic systems. More details of certain arguments can be found elsewhere,[93] where some of them were first developed.

Although, on morphological grounds, Corby[17,18] regarded the aeschynomenoid type of nodule as advanced, on grounds of infection processes we have regarded it as having a primitive origin.[93] Our definition of an aeschynomenoid nodule includes Corby's morphologic characters (generally small and oblate, lacking lenticels and occurring in the axils of fine roots) but adds the vital anatomic character that the infected tissue is uniform (i.e., lacks uninfected interstitial cells). From these characters we infer that the type of infection process described for *Arachis* (peanut, groundnut) and *Stylosanthes* occurs.[13,14] Although varying in detail, this essentially takes place through wounds where lateral roots emerge (Chapter 9), hence the association with fine roots. Nodules on the stems of *Aeschynomene* and *Sesbania* are associated with adventitious root initials and have a generally similar infection pathway.[101,104] The aeschynomenoid type of nodule is common in the legume tribes Aeschynomeneae (8 of the 25 genera have been examined[93] and Dalbergieae (at least six genera, including some important timber genera such as *Dalbergia*). *Sesbania* is the only genus so far known to have two types of nodule, stem nodules that are aeschynomenoid and root nodules that are rounded when young but later become indeterminate: These root nodules are formed following hair infections and contain interstitial cells in the infected region.[38,62] Stem and root nodules on *Aeschynomene* are similar in structure and may have the same endophyte, which, at least in some cases, may be the putative *Photorhizobium*.[24] Those on *Sesbania rostrata* may be formed by different genera of rhizobia—stem nodules containing *Azorhizobium*. Taxonomically, *Sesbania* has been included in the Aeschynomeneae, but more recently transferred to the Robinieae.[70] Thus, both the simple type of infection and the primitive character of some of the rhizobia involved are consistent with the argument that wound infection evolved at an early stage. If legumes evolved in swampy areas, we even have the possibility that stem nodules preceded root nodules, since extant species of *Aeschynomene* and *Sesbania* show suppression of root nodulation in favor of stem nodulation when grown under flooded conditions. Other aeschynomenoid nodules on roots that have been studied in sufficient detail (*Arachis* and *Stylosanthes*) are infected by *Bradyrhizobium*. Perhaps nodulation began with photosynthetic ancestral rhizobia infecting stems and then, under drier conditions, migrating to roots. This migration could be accompanied by loss of ability to photosynthesize and to obtain carbon from the host plant and in due course to the complex control systems found in extant nodules from plants such as soybean (see Chapter 10).

Although a wound infection has obvious similarities to *Agrobacterium,* it has been impossible with rhizobia, unlike agrobacteria, to obtain infection with artificial wounds (e.g., made with a scalpel). We have suggested that the processes of cell wall breakdown that accompany the emergence of lateral roots may predispose them to infection by rhizobia.[89] Progress of rhizobia within root tissues following wound infection is intercellular as well as through intercellular spaces. In some cases root cortical cells may be invaded, hypertrophy, and die. In due course one or a few cells toward the inner cortex become invaded, the rhizobia being surrounded at this stage with an amorphous zoogloea thought to be of host origin. Organized infection threads do not occur. Such invaded cells divide repeatedly to form the uniform infected tissue typical of an aeschynomenoid nodule. At maturity, bacteria are enclosed by peribacteroid membranes, which, in the case of *Arachis,* may have various inclusions on either side.[5] Cell division does not precede invasion as it does in, for example, soybean. Since the same *Bradyrhizobium* strain can nodulate *Arachis* via wounds and *Vigna* (nodules similar to soybean) via hairs,[82] differences in infection and nodule development must be host controlled. There is some evidence that wound infections may occur in other species, such as the aquatic *Neptunia plena* Benth. (Mimosoideae; James and Sprent, unpub.), but with formation of infection threads as nodules develop and indeterminate nodule growth. In this case infected zones contain interstitial cells. The transition from an intercellular infection at the initial stage to an infection thread stage is thought to occur as shown in Figure 12-5. In another legume, *Mimosa scabrella,* from the subfamily Mimosoideae, infection was found to occur between intact epidermal cells.[26] Rhizobia subsequently passed between cells and through intercellular spaces, but in due course infection threads were formed and nodules had a broadly similar structure to those of *Neptunia.* We suspect that, at least in legumes without root hairs and with nodules not associated with lateral roots, this newly described type of infection may be common. Finally, we have the root hair type of infection found in advanced legumes, such as clovers, soybean, and peas. In caesalpinoid genera with persistent infection threads, the only type of infection process described so far (in *Dimorphandra*) is by root hairs.[25,28] However, we cannot exclude the possibility that other types of infection may occur, not involving infection threads. *Parasponia* has such a system, where infection occurs through an epidermis ruptured by cortical cells proliferating in response to surface-colonizing rhizobia (see Chapter 13).

Thus, in different extant nodules we have almost all possible combinations, from no infection threads at all (*Arachis*) to infection threads from the first stages of nodulation throughout the entire active period (*Dimorphandra*). Further, even in advanced nodules such as soybean with a very well-defined sequence of events in nodulation, stress such as salinity may

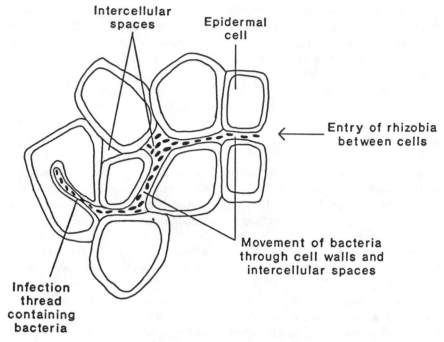

Figure 12-5. Possible sequence of events leading from an intercellular infection to an infection thread. After cells are invaded by infection threads bacteria may be released (as in *Neptunia plena*) or possibly retained in a branched thread system (as in *Parasponia*).

lead to major modifications, including intercellular rhizobia (E. K. James. personal communication). Thus, the structural evidence is entirely consistent with the suggestion[103] that there is a continuum between intercellular spaces and the infection thread matrix. Elegant labeling with monoclonal antibodies showed that glycoproteins with common epitopes could be found in intercellular spaces, the matrix of infection threads, and even nonlegume cells such as carrot suspension cultures.[103] It was thus suggested that infection threads are really an extension of the intercellular space system, providing tunnels through which rhizobia cross root cortices. Although one may agree entirely with this suggestion, there is a further problem to be addressed: How do infection threads overcome the turgor of the host plant cells?[22] The initial problem, when the hair wall or cross wall between cortical cells is breached, is overcome by the pressure of neighboring cells (or parts of the same cell in the case of some curled root hairs). However, to push into a new cell requires something to push against. The thread wall, which has a different chemical composition to the primary wall of root cortical cells,[41] if cemented to the primary wall, could function in this way.[89] The consequence of these features is that infection threads are essential if rhizobia are to pass through, rather than between, cells (passing through cells is a highly organized process; see Chapter 9). This in turn means that they are a necessary part of root hair infections and, if primitive nodulated legumes lacked root hairs, must be an advanced feature, albeit one found in extant members of all three subfamilies.

Recent evidence that the Mimosoideae and Papilionoideae were established earlier than once thought and evidence that some mimosoid fossils have features in common with extant nonnodulating genera leave open the question of whether nodulation occurred more than once during nodule evolution. Figure 12-6 summarizes some possible evolutionary sequences based mainly on nodule characters.

D. *Agrobacterium—A Red Herring?*

It has been argued here and elsewhere that rhizobia differ from other endosymbiotic nitrogen-fixing organisms in that they induce the formation of outgrowths (nodules) that they subsequently inhabit. Their taxonomic relationship with *Agrobacterium* (see Figure 12-3) suggests that outgrowth formation may be similar in the two types of infection. However, agrobacteria incite galls by transmitting phytohormone-coding DNA to the host, where it is incorporated into the genome. Indeed, in the case of crown gall (*A. tumefaciens*), transformed cells will continue to divide in the absence of a pathogen, an apparently pointless procedure. Other plant pathogens such as *Pseudomonas syringae* (in the γ group of the Proteobacteria) may induce tumors as a result of phytohormones that they themselves produce. The related species *Ps. aeruginosa,* containing various DNA fragments from *R. trifolii,* showed ability in some cases to deform and/or branch clover root hairs and form pseudonodules.[66] Recent work (see also Chapters 9 and 15) has shown that auxin transport inhibitors can elicit pseudonodules on alfalfa (lucerne) roots and induce some early nodulin genes.[43] It is significant that these pseudonodules have an outer cortex with some differentiation of vascular tissue. Additionally, *nod*A and *nod*B genes of *R. meliloti* are claimed to lead to generation of cytokininlike compounds, which stimulate mitosis (to varying extents) in alfalfa, soybean, carrot, barley, and wheat protoplasts.[78,80] One of the genes that was induced by auxin transport inhibitors was ENOD2. This has also been found to be expressed in alfalfa plants that "nodulated" in the absence of *Rhizobium.*[100] The authors used the term *nodules* rather than *pseudonodules* because they were anatomically nodulelike with cortex, peripheral vascular system, and endodermis. Thus, it appears that some legumes have an inherent ability to nodulate if suitably triggered. Rhizobia, possibly via phytohormones, may usually provide the trigger (probably after they in turn are triggered by host exudates). The fact that cell division factors formed by *R. meliloti*[80] could stimulate division in

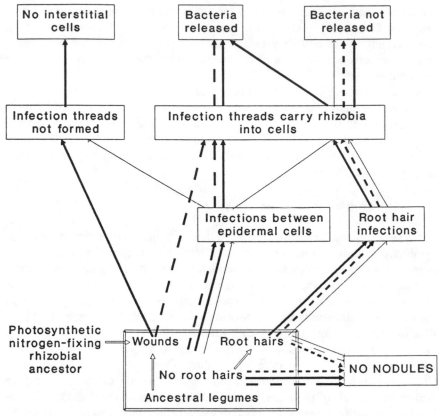

Figure 12-6. Possible sequences of events leading to extant nodule forms. All pathways indicated by bold lines have been demonstrated; faint lines remain to be confirmed. Pathways indicated by continuous lines are for papilionoid species; long dashes, for mimosoid, and short dashes, for caesalpinioid species. The papilionoid pathway leading from an intercellular infection via infection threads to released bacteria is based on unpublished data of Cordeiro and Sprent for *Lonchocarpus muelhbergianus*. References to other pathways may be found in the text.

monocotyledonous plant protoplasts (barley and wheat) may relate to the "nodular structures" recently reported on rice as a result of combined treatment with rhizobia, cell wall degrading enzymes, and an osmoticum.[2] These structures, however, lack an organized anatomy, and the rhizobia were either intercellular or in dying cells; they did not show detectable nitrogenase activity.

We may thus speculate that it is not so much the formation of nodules that is the problem (i.e., to this extent *Agrobacterium* is a red herring), but the introduction of rhizobia into them and then the fixation of N_2.

E. Parasponia—The Enigma

The discovery of nodules on *Parasponia* (see Chapter 13) revolutionized many of the concepts of the interaction between legumes and rhizobia. It remains the only genus outside the Leguminosae known to be nodulated by rhizobia. Its nodules are rootlike, rather than stemlike, confirming the major role of the host in controlling nodule structure. The retention of bacteria within infection threads is now known not to be a unique feature.[29] However, the precise nature of the thread wall may be crucial for activity, which may be transitory and associated with the presence of hemoglobin.[98] The intercellular movement of rhizobia following initial infection is also very similar to that of some legumes and nonlegumes. This suggests that it may be the nature of the cell wall of a particular host that makes it amenable to penetration by microorganisms, including rhizobia and *Frankia*,[88] a point that will be extended in Section VIII.

VII. Evolution of Actinorhizal Plants

In spite of much effort by many people, no unifying theme has emerged about these symbioses from the host taxonomic side. This section makes no further attempt, but considers what we may learn from actinorhizas about the minimum requirements of a prokaryotic–higher plant N_2-fixing symbiosis.

Although we know relatively little about the detailed molecular genetics and biochemistry of nonlegume nodules, we know a lot about their structure. Studies have now been made on all known nodulated genera[60,91] (see Chapter 7). It is clear that as a group they have adopted a range of solutions to one of the most fundamental problems, that of oxygen protection of nitrogenase (see Chapter 7). At one extreme we may have the *Casuarina* system, where insufficient O_2 reaches the endophyte to require production of thick-walled vesicles; instead hemoglobin may be necessary to ensure that sufficient O_2 reaches the endophyte. At the other extreme we have relatively free movement of O_2, with nitrogenase being protected largely by vesicle walls. A particularly interesting case is that of *Coriaria*,[84] where the extent of suberization in host cells varies with O_2 supply. Sophisticated variable diffusion resistances, as found in advanced legumes (see Chapter 10), appear rare.

It is time for the legume researchers to take a leaf out of the nonlegume researchers' book and grow a wider range of nodulated plants under a range of O_2 conditions to see whether they all behave in a soybean–pea–cloverlike fashion. Some active caesalpinoid and mimosoid nodules appear much less pink than those on herbaceous species. They may have structural features (such as cork[87]) or walls of persistent infection threads that make a contribution to the control of O_2 diffusion.

It was argued earlier (see Section VI.A) that, if nitrogenase was a selective advantage for rhizobia, then some viability must be retained by rhizobia in active nodules. Applying the same arguments to *Frankia,* we immediately face the problem that we know virtually nothing about how the organism survives, let alone grows (if it does) in soil. On the information to date there is an inverse relationship between spore-forming nodules and effectivity. *Frankia* vesicles and cyanobacterial heterocysts have much in common, in particular the development of lipid layers in walls as an oxygen-protection device.[102] However, unlike fully differentiated heterocysts, *Frankia* vesicles may be able to regenerate vegetative hyphae[81] and thus become possible candidates for perennation from active nodules.

Like legumes, actinorhizal plants have both root hair and epidermal infection mechanisms; as yet, wound infections have not been described. The intercellular passage of rhizobia[26] and that of *Frankia* during epidermal infections show striking similarities.[58] As in legumes, the infection pathway is host controlled.[59]

There are no known actinorhizal plants with stem nodules and no known legume actinorhizas. Unlike legumes, where nodules have many stemlike attributes,[88] nonlegume nodules are clearly modified roots. Their endophyte may have evolved later than rhizobia.[91] It can apparently survive for long periods in the soil, possibly growing there, since potential actinorhizal species, planted in soil not known ever to have supported actinorhizal plants, almost invariably nodulate. When the symbiosis evolved is largely a matter of speculation. The scattered taxonomic nature of the host genera suggests that it may have evolved several times, in response to local selective pressures such as nitrogen deficiency following retreat of glaciers,[83] a view supported by fossil evidence of nodules.[4] The possibility of several initial symbiotic events, taken with the wide variation in nodule structure of extant nonlegume nodules, suggests that it may be possible to "construct" further symbioses in the foreseeable future.

VIII. Future Evolution

This section considers how studies of a variety of N_2-fixing symbioses may be used to produce new symbioses. We shall not enter the discussion of whether it would be better to develop new symbiotic systems or to introduce *nif* directly into eukaryotic host genomes.[90]

First, we discuss microbial aspects, including both potential endosymbionts and the possible role of other soil organisms.

Cyanobacteria, being endosymbiotic with a wide variety of different types of plants, are obvious candidates for trying to synthesize new symbioses. Heterocystous forms have their own oxygen protection system and therefore would need fewer control systems in the plant. Various attempts have been made to introduce cyanobacteria (as well as other N_2-fixing organisms) into plant protoplasts. For example, cells of *Anabaena variabilis* have been incorporated into tobacco protoplasts;[55] however, the protoplasts would not divide and in due course ejected the *Anabaena* cells. More success was obtained using tobacco callus and using *A. variabilis*.[35] During the growth of the callus, *Anabaena* cells were included in intercellular spaces and were subsequently found in this location in shoots regenerated from the callus, where they produced heterocysts and reduced acetylene. However, more internal cyanobacteria were present in chlorophyll-deficient than in fully green shoots. There was no evidence that cyanobacteria could break down cell walls to effect invasion.

Frankia has only relatively recently been obtained in pure culture, where it usually grows very slowly. This could present difficulties in trying to obtain the type of cohabitation described earlier for tobacco and *Anabaena*. As an attempt to hasten advances in study of *Frankia*, rather than for obtaining new symbioses, protoplast fusions between *Frankia* and the fast-growing actinomycete *Streptomyces* have been reported;[72] attempts to use these hybrids to nodulate *Alnus rubra* met with limited success.

Numerous successful attempts have been made to grow rhizobia in association with callus tissue from both legumes[15] and nonlegume plants.[77] How stable these are is not clear, and there has been no reproducible evidence on intracellular location of bacteria. Similarly, as noted earlier, rhizobia were not incorporated into live cells of the nodulelike structures on rice.[2] Is it necessary to have intracellular endosymbionts? The problem is really one of accommodating sufficient N_2-fixing cells to provide the needs of an entire plant. In a relatively simple symbiosis such as *Azolla–Anabaena*, the endosymbiont may account for 16% of total plant protein and clearly is adequate.[65] Cycads are relatively slow growing and the endophyte cavities in the coralloid roots may occupy around 40% of the cross-sectional area (J. I. Sprent, unpub.). This is at least comparable with the proportion of nodule area occupied by rhizobia or *Frankia*. Even allowing for the fact that only heterocysts fix N_2, the cycad symbiosis probably has sufficient N_2-fixing capacity for the whole plant. The same is unlikely to be true of a system where the endosymbiont occupies only intercellular spaces of the type found in callus or in normal mesophytic tissue. This is one reason that it is difficult to see how *Azospirillum* and similar bacteria could fix large amounts of N_2 in those parts of the plant where they are found internally— there may be insufficient bacterial cells present. To achieve sufficient fixation in a plant-controllable fashion, intracellular (in the sense of within the bounds of host-cell walls) location of endophytes in a special structure such as a nodule makes a great deal of sense. It allows a greatly increased area for exchange of metabolites between symbionts as well as accommodating more endosymbiont cells. The latter may be especially important, since, for example, N_2 fixation in legume nodules may be limited by the number of bacteroids present.[49]

Thus, we need to consider how to construct new nodules and to put endosymbionts into their cells. Making nodules may be relatively simple, for reasons indicated in Section VI.D. It is probably a matter of finding the correct hormone balance for the plant in question, allowing for the hormones produced by the potential endophyte. (In soils there is the complicating factor of phytohormones produced by other organisms.) Getting the endosymbionts into the nodules is likely to be more difficult. It is now clear that many of the early interactions between symbionts occur at their surfaces. Plant cell walls are known to vary greatly at the structural and at the metabolic levels. Both proteins[11] (including glycoproteins) and oligosaccharides[30] are essential parts not only of cell growth but of reactions to environmental stimuli, including pathogens and symbionts. Enhanced production of proline-rich or hydroxyproline-rich wall polymers in specific stages of nodule morphogenesis (e.g., those coded by ENOD2 and ENOD12[78,107]) and the linking of these to physiologic functions such as the oxygen diffusion barrier and morphologic processes such as lateral root initiation[47] suggest that a detailed knowledge of cell wall composition is a required precursor to planned (rather than serendipitous) production of new symbioses.

Many of the compounds (nodulins) previously thought to be nodule-specific are now known to be nodule-enhanced, with their activity normally being at a lower level in nonnodule tissue (see Chapter 15). In some cases the genes for nodulins may be widespread in plants. This has been shown for hemoglobin genes, which from their genetic structure (including the presence of introns) appear to have evolved early.[3] The observation that hemoglobin genes may be expressed at a low level in nonnodule tissue and in *Trema*, a nonlegume genus closely related to *Parasponia*,[7] suggests that, should it be desirable for engineered symbioses, hemoglobin expression should not present an insoluble problem. Evidence of homology between nodulin-26 (a plant-derived component of the peribacteroid membrane) and transporters in both bovine lens fiber and *Escherichia coli*[6] further suggests that the means for molecular communication between symbiotic partners are potentially available.

In genetically engineered host plants, it may be necessary also to take into account the regulation of defense genes. Soybeans, for example, have multiple copies of the gene for chalcone synthase that are differentially regulated by UV light, elicitors from fungi such as *Phytophthora megasperma*, rhizobia, and agrobacteria.[108] It may be necessary to add copies of particular genes of this type to potential hosts. In this context the recent use of *Agrobacterium rhizogenes* to effect transgenic roots and root nodules[36] holds great promise.

Difficult decisions arise in deciding which host plants to target. Because of their importance as a source of food, N_2-fixing cereals have been a dream for many. Clearly, some progress has been made in obtaining outgrowths on rice.[2] However, because of the early evolutionary divergence of monocots and dicots (see Section IV) and the numerous anatomic differences between roots of extant monocots and dicots, obtaining intracellular rhizobia may be more difficult. If we knew how *Azospirillum* obtains entry to grasses, this could become a better potential alternative to rhizobia. Within the dicots, is it better to attempt to nodulate nonnodulating genera of legumes, to try to add to *Parasponia* as a nonlegume nodulating with rhizobia, or to extend the range of *Frankia*-induced nodules? Two recent reviews have posed the question "Why legumes?"[89,112] In neither were overriding reasons for legumes found. The only major known feature unique to legume nodules is their stemlike anatomy. The observation that the product of ENOD12 produced during infection of peas by rhizobia is also strongly expressed in certain stem tissues,[78] and the suggestion that the ancestral legume nodules may have been on stems (see Section VI.C) underwrites this "special" nature of legumes. However, stemlike anatomy is not a prerequisite for successful nodulation with rhizobia, as *Parasponia* nodules are rootlike.

If successful invasion of host cells is achieved, probably using intercellular infection processes, then the problem of maintaining the symbiosis arises. Natural symbioses vary greatly in longevity, although individual N_2-fixing nodule cells probably have a maximum active life span of a few weeks. Possible interactions between metabolic pathways of symbionts and control of cell division during development of various symbionts have recently been discussed.[42] Individual research groups attempting to design new symbioses will each decide on their own priorities as to the best protocol to be adopted. With molecular techniques now available, the goal should be within reach, especially if we consider some of the simpler ways in which nature herself has evolved successful systems. Manuscript completed March 1990.

References

1. Adams, M.W., Coyne, D.P., Davies, H.J.C., Graham, P.H., and Frances, C.A., "Common bean (*Phaeseolus vulgaris* L.)," in *Grain Legume Crops*, R.J. Summerfield and E.H. Roberts, eds. London: Collins, pp. 433–476.

2. Al-Mallah, M.K., Davey, M.R., and Cocking, E.C., "Formation of nodular structures on rice seedlings by *Rhizobia*," *J. Exp. Bot. 40*, 473–478 (1989).

3. Appleby, C.A., Bogusz, D., Dennis, E.S., and Peacock, W.J., "A role of haemoglobin in all plant roots?" *Plant Cell Environ. 11*, 359–367 (1988).

4. Baker, D., and Miller, N.G., "The ultrastructural evidence for the existence of actinorhizal symbioses in the late pleistocene," *Can. J. Bot. 58*, 1612–1620 (1980).

5. Baker, M.E., and Saier, M.H., "A common ancestor for bovine lens fiber major intrinsic protein, soybean nodulin-26 protein, and E. coli glycerol facilitator," Cell 60, 185–186 (1990).

6. Bal, A.K., Hameed, S., and Jayaram, S., "Ultrastructural characteristics of the host–symbiont interface in nitrogen-fixing peanut nodules," Protoplasma 150, 19–26 (1989).

7. Bogusz, F., Appleby, C.A., Landsmann, J., Dennis, E.S., Trinick, M.J., and Peacock, W.J., "Haemoglobin genes in non-nodulating plants, and their expression in nonsymbiotic tissue," Nature 331, 178–180 (1988).

8. Briggs, D.E.G., and Crowther, P.R., eds., Palaeobiology: A Synthesis. Oxford: Blackwell, 1990.

9. Budyko, M.I., Ronov, A.B., and Yanshin, A.L., History of the Earth's Atmosphere. Berlin: Springer-Verlag, 1985.

10. Campbell, E.L., and Meeks, J.C., "Characteristics of hormogonia formation by symbiotic Nostoc spp. in response to the presence of Anthoceros punctatus on the extracellular products," Appl. Environ. Microbiol. 55, 125–131 (1989).

11. Cassab, G.I., and Varner, J.E., "Cell wall protein," Ann. Rev. Plant Physiol. Plant Mol. Biol. 39, 321–353 (1988).

12. Chaloner, W.G., "Fossil charcoal as an indication of palaeoatmospheric oxygen level," J. Geol. Soc. 146, 171–174 (1989).

13. Chandler, M.R., "Some observations on infection of Arachis hypogaea L. by Rhizobium," J. Exp. Bot. 29, 749–755 (1978).

14. Chandler, M.R., Date, R.A., and Roughley, R.J., "Infection and root nodule development in Stylosanthes species by Rhizobium," J. Exp. Bot. 33, 47–57 (1982).

15. Child, J.J., and LaRue, T.A., "A simple technique for the establishment of nitrogenase in soybean callus culture," Plant Physiol. 53, 88–90 (1974).

16. Coplin, D.L., "Plasmids and their role in the evolution of plant pathogenic bacteria," Ann. Rev. Phytopathol. 27, 187–212 (1989).

17. Corby, H.D.L., "The systematic value of leguminous root nodules," in Advances in Legume Systematics, R.M. Polhill and P.H. Raven, eds. Kew: Royal Botanic Gardens, 1981, pp. 657–669.

18. Corby, H.D.L., "Types of rhizobial nodules and their distribution among Leguminosae," Kirkia 13, 53–123 (1988).

19. Crepet, W.L., and Taylor, D.W., "The diversification of the Leguminosae: First fossil evidence of the Mimosoideae and Papilionoideae," Science 228, 1087–1089 (1985).

20. Crepet, W.L. and Taylor, D.W., "Primitive mimosoid flowers from the paleocene-eocene and their systematic and evolutionary implications," Am. J. Bot. 73, 548–563 (1986).

21. Diebold, R., and Noel, K.D., "Rhizobium leguminosarum exopolysaccharide mutants: Biochemical and genetic analyses and symbiotic behaviour on three hosts," J. Bacteriol. 171, 4821–4830 (1989).

22. Dixon, R.O.D., "Rhizobia (with particular reference to relationships with host plants)," Ann. Rev. Microbiol. 23, 137–158 (1969).

23. Elias, T.S., "Parkieae (Wight & Arn.) Benth. (1842)," in Advances in Legume Systematics, R.M. Polhill and P.H. Raven, eds. Kew: Royal Botanic Gardens, 1981, p. 153.

24. Ellis, J.M., Eardly, B.D., Hungria, M., Hardy, R.W.F., Bizzo, N.W., and Eaglesham, A.R., "Photosynthetic N$_2$-fixing *Rhizobium*," in *12th American Symbiotic Nitrogen Fixation Conference Proceedings*, Ames, Iowa, 1989.

25. Faria, S.M. de, *Occurrence, Infection Pathways and Structure of Root Nodules from Woody Species of the Leguminosae*, Ph.D thesis, University of Dundee, Scotland, 1988.

26. Faria, S.M. de, Hay, G.T., and Sprent, J.I., "Entry of rhizobia into roots of *Mimosa scabrella* Bentham occurs between epidermal cells," *J. Gen. Microbiol. 134*, 2291–2296 (1988).

27. Faria, S.M. de, Lewis, G.P., Sprent, J.I., and Sutherland, J.M., "Occurrence of nodulation in the Leguminosae," *New Phytol. 111*, 607–619 (1989).

28. Faria, S.M. de, McInroy, S.G., Rowell, P. and Sprent, J.I., "Some properties of "primitive" legume nodules," in *Nitrogen Fixation: Hundred Years After*, H. Bothe, F.J. de Bruijn, and W.E. Newton, eds. Stuttgart: Gustav Fischer, 1988, p. 524.

29. Faria, S.M. de, McInroy, S., and Sprent, J.I., "The occurrence of infected cells, with persistent infection threads, in legume root nodules," *Can. J. Bot. 65*, 533–538 (1987).

30. Fry, S.C., *The Growing Plant Cell Wall: Chemical and Metabolic Analyses*. Harlow, Essex: Longman, 1988.

31. Gest, H., and Favinger, J.L., "*Heliobacterium chlorum*, an anoxygenic brownish-green photosynthetic bacterium containing a 'new' form of bacteriochlorophyll," *Arch. Microbiol. 136*, 11–16 (1983).

32. Giovannoni, S.J., Turner, S., Olsen, G.J., Burns, S., Lane, D.J., and Pace, N.R., "Evolutionary relationships among cyanobacteria and green chloroplasts," *J. Bacteriol. 170*, 3584–3592 (1988).

33. Gober, J.W., and Kashket, E.R., "K$^+$ regulates bacteroid-associated functions of *Bradyrhizobium*," *Proc. Natl. Acad. Sci. 84*, 4650–4654 (1987).

34. Gottlieb, O.R., "The role of oxygen in phytochemical evolution towards diversity," *Phytochemistry 28*, 2545–2558 (1989).

35. Gusev, M.V., Korzhenevskaya, T.G., Pyvovarova, Baulina, O.I., and Butenko, R.G., "Introduction of a nitrogen-fixing cyanobacterium into tobacco regenerates," *Planta 167*, 1–8 (1986).

36. Hansen, J., Jorgensen, J.-E., Stougaard, J., and Marcker, K.A., "Hairy roots—a short cut to transgenic root nodules," *Plant Cell Rep. 8*, 12–15 (1989).

37. Hanus, F.J., Maier, R.J., and Evans, H.J., "Autotrophic growth of H$_2$-uptake-positive strains of *Rhizobium japonicum* in an atmosphere supplied with hydrogen gas," *Proc. Natl. Acad. Sci. USA 76*, 1788–1792 (1979).

38. Harris, J.O., Allen, E.K., and Allen, O.N., "Morphological development of nodules on *Sesbania grandiflora* Poir with reference to the origin of nodule rootlets," *Am. J. Bot. 36*, 651–661 (1949).

39. Hawksworth, D.L., "The variety of fungal–algal symbioses, their evolutionary significance, and the nature of lichens," *Bot. J. Linn. Soc. 96*, 3–20 (1988).

40. Hawksworth, D.L., "Co-evolution of fungi and algae in lichen association," in *Coevolution of Fungi with Plants and Animals*, K.A. Pirozynski and D.L. Hawksworth, eds. London: Academic, 1988, pp. 125–148.

41. Higashi, S., Kushiyama, K., and Abe, M., "Electron microscopic observations of infection threads in driselase treated nodules of *Astragalus sinicus*," *Can. J. Microbiol. 32*, 947–952 (1987).

42. Hill, D.J., "The control of the cell cycle in microbial symbionts," *New Phytol. 112*, 175–184 (1989).

43. Hirsch, A.M., Bhuvaneswari, T.V., Torrey, J.G., and Bisseling, T., "Early nodulin genes are induced in alfalfa root outgrowths elicited by auxin transport inhibitors," *Proc. Natl. Acad. Sci. USA 86*, 1244–1248 (1989).

44. Irwin, H.S., and Barneby, R.C., "Cassieae Bronn (1822)," in *Advances in Legume Systematics*, R.M. Polhill and P.H. Raven, eds. Kew: Royal Botanic Gardens, 1981, pp. 97–106.

45. Jarvis, B.D.W., Ward, L.J.H., and Slade, E.A., "Expression by soil bacteria of nodulation genes from *Rhizobium leguminosarum* biovar *trifolii*," *Appl. Environ. Microbiol. 55*, 1426–1434 (1989).

46. Karr, D.B., and Emerich, D.W., "Uniformity of the microsymbiont population from soybean nodules with respect to buoyant density," *Plant. Physiol. 86*, 693–699 (1988).

47. Keller, B., and Lamb, C.J., "Specific expression of a novel cell wall hydroxyproline-rich glycoprotein gene in lateral root initiation," *Genes Develop. 3*, 1639–1646 (1989).

48. Knoll, A.H., and Bauld, J., "The evolution of ecological tolerances in prokaryotes," *Trans. Roy. Soc. Edinb. Earth Sci. 80*, 209–223 (1989).

49. Lin, J., Walsh, K.B., Canvin, D.T., and Layzell, D.B., "Structural and physiological basis for effectivity of soybean nodules formed by fast-growing and slow-growing bacteria," *Can. J. Bot. 66*, 526–534 (1988).

50. McRae, D.G., Miller, R.W., and Berndt, W.B., "Viability of alfalfa nodule bacteroids isolated by density gradient centrifugation," *Symbiosis 7*, 67–80 (1989).

51. Manian, S.S., Gumbleton, R., Buckley, A.M., and O'Gara, F., "Nitrogen fixation and carbon dioxide assimilation in *Rhizobium japonicum*," *Appl. Environ. Microbiol. 48*, 276–279 (1984).

52. Manian, S.S., and O'Gara, F., "Derepression of ribulose bisphosphate carboxylase activity in *Rhizobium meliloti*," *FEMS Microbiol. Lett. 14*, 95–99 (1982).

53. Martin, W., Gierl, A., and Saedler, H., "Molecular evidence for pre-cretaceous angiosperm origins," *Nature 339*, 46–48 (1989).

54. Meeks, J.C., "Cyanobacterial–bryophyte associations," in *CRC Handbook of Symbiotic Cyanobacteria*, R.N. Rai, ed. 1989, pp. 43–63.

55. Meeks, J. C., Malmberg, R. L., and Wolk, C. P., "Uptake of auxotrophic cells of a heterocyst-forming cyanobacterium by tobacco protoplasts and the fate of their associations," *Planta 139*, 55–60 (1978).

56. Michiels, K., Vanderleyden, J., and Gool, A. van, "*Azospirillum*—plant root associations: A review," *Biol. Fert. Soils 8*, 356–368 (1989).

57. Michiels, K., Vanderleyden, J., Gool, A. van, and Signer, E. R., "Isolation and characterization of *Azospirillum brasilense* loci that correct *Rhizobium meliloti exoB* and *exoC* mutations," *J. Bacteriol. 170*, 5401–5404 (1988).

58. Miller, I. M., and Baker, B. D., "The initiation, development and structure of root nodules in *Eleagnus augustifolia* L. (Eleagnaceae)," *Protoplasma 128*, 107–119 (1985).

59. Miller, I. M., and Baker, B. D., "Nodulation of actinorhizal plants by *Frankia* strains capable of both root hair infection and intercellular penetration," *Protoplasma 131*, 82–91 (1986).

60. Newcomb, W., and Wood, S. M., "Morphogenesis and fine structure of *Frankia* (Actinomycetales): The microsymbiont of nitrogen-fixing actinorhizal root nodules," *Int. Rev. Cytol. 109*, 1–88 (1987).

61. Ohkubo, S., Iwasaki, H., Hori, H., and Osawa, S., "Evolutionary relationship of denitrifying bacteria as deduced from 5S rRNA sequences," *J. Biochem. 100*, 1261–1267 (1986).

62. Olsson, J. E., and Rolfe, B. G., "Stem and root nodulation of the tropical legume *Sesbania rostrata* by *Rhizobium* strains ORS-571 and WE7, *J. Plant. Physiol. 121*, 199–210.

63. Osborne, B. A., "Comparison of photosynthesis and productivity of *Gunnera tinctoria* Milina (Mirbel) with and without the phycobiont *Nostoc punctiforme* L.," *Plant Cell Environ. 12*, 941–946 (1989).

64. Pearcy, R. W., "The light environment and growth of C_3 and C_4 tree species in the understory of a Hawaiian forest," *Oecologia 58*, 19–25 (1983).

65. Peters, G. A., and Meeks, J. C., "The *Azolla–Anabaena* symbiosis: basic biology," *Ann. Rev. Plant Physiol. Plant Mol. Biol. 40*, 193–210 (1989).

66. Plazinski, J., and Rolfe, B. G., "Sym plasmid genes of *Rhizobium trifolii* expressed in *Lignobacter* and *Pseudomonas* strains," *J. Bacteriol. 163*, 1261–1269 (1985).

67. Polhill, R. M., "Sophoreae Sprengel (1818)," in *Advances in Legume Systematics*, R. M. Polhill and P. H. Raven, eds., Kew: Royal Botanic Gardens, 1981, pp. 213–230.

68. Polhill, R. M., and Raven, P. H., eds., *Advances in Legume Systematics*, Part I. Kew: Royal Botanic Gardens, 1981.

69. Polhill, R. M., Raven, P. H., and Stirton, C. H., "Evolution and systematics of the Leguminosae," in *Advances in Legume Systematics*, R. M. Polhill and P. H. Raven, eds. Kew: Royal Botanic Gardens, 1981, pp. 1–26.

70. Polhill, R. M., and Sousa, M., "Robinieae (Benth.) Hutch. (1964)," in *Advances in Legume Systematics*, R. M. Polhill and P. H. Raven, eds. Kew: Royal Botanic Gardens, 1981, pp. 283–288.

71. Postgate, J. R., *The Fundamentals of Nitrogen Fixation*. Cambridge: Cambridge University Press, 1982.

72. Prakash, R. K., and Cummings, B., "Creation of novel nitrogen-fixing actinomycetes by protoplast fusion of *Frankia* with streptomyces," *Plant Molec. Biol. 10*, 281–289, (1988).

73. Ratet, P., Pawlowski, K., Schell, J., and Bruijn, F. J. de, "The *Azorhizobium caulinodans* nitrogen fixation regulatory gene, *nif* A, is controlled by the cellular nitrogen and oxygen status," *Mol. Microbiol. 3*, 825–838 (1989).

74. Raven, J. A. "Ribulose bisphosphate carboxylase activity in terrestrial plants: significance of O_2 and CO_2 diffusion," *Curr. Adv. Plant Sci. 9*, 579–590 (1977).

75. Raven, J. A., and Sprent, J. I., "Photography, diazotrophy and paleoatmospheres, biological catalysts and the H, C, N and O cycles," *J. Geol. Soc. 146*, 161–170 (1989).

76. Samish, Y. B., "Oxygen build-up in photosynthesizing leaves and canopies is small," *Photosynthetica 9*, 372–375 (1975).

77. Schetter, C., and Hess, D., "Nitrogenase activity in *in vitro* associations between callus tissue of non-leguminous horticultural plants and *Rhizobium*," *Plant Sci. Lett. 9*, 1–5 (1977).

78. Scheres, B., Wiel, C. Van de, Zalensky, A., Horvath, B., Spaink, H., Eck, H. van, Zwartkruis, F., Wolters, A.-M., Gloudemans, T., Kammen, A. van, and Bisseling, T., "The ENOD12 gene product is involved in the infection process during the pea–rhizobium interaction," *Cell 60*, 281–294 (1990).

79. Schidlowski, M., "A 3,800-million year isotopic record of life from carbon in sedimentary rocks," *Nature 333*, 313–318 (1988).

80. Schmidt, J., Wingender, R., John, M., Wieneke, U., and Schell, J., "*Rhizobium meliloti nod*A and *nod*B genes are involved in generating compounds that stimulate mitosis of plant cells," *Proc. Natl. Acad. Sci. USA 85*, 8578–8582 (1988).

81. Schultz, N. A., and Benson, D. R., "Developmental potential of *Frankia* vesicles," *J. Bacteriol. 171*, 6873–6877 (1989).

82. Sen, D., and Weaver, R. W., "A basis of different rates of N_2-fixation by some strains of *Rhizobium* in peanut and cowpea root nodules," *Plant Sci. Lett. 34*, 239–246 (1984).

83. Silvester, W. B., "Dinitrogen fixation by plant associations excluding legumes," in *A Treatise on Dinitrogen Fixation*, R. W. F. Hardy and A. H. Gibson, eds. New York: Wiley Interscience, 1977, pp. 141–190.

84. Silvester, W. B., and Harris, S. L., "Nodule structure and nitrogenase activity of *Coriaria arborea* in response to varying pO_2," *Plant Soil 118*, 119–123 (1989).

85. Simpson, F. B., Maier, R. J., and Evans, H. J., "Hydrogen stimulated CO_2 fixation and co-ordinate induction of hydrogenase and ribulose bisphosphate carboxylase in a H_2-uptake positive strain of *Rhizobium japonicum*," *Arch. Microbiol. 123*, 1–8 (1979).

86. Spicer, R. A., "Physiological characteristics of land plants in relation to environment through time," *Trans. Roy. Soc. Edinb. Earth Sci. 80*, 321–329 (1989).

87. Sprent, J. I., "Nitrogen fixation in arid environments," in *Plants for Arid Lands*, G. E. Wickens, J. R. Goodin, and D. V. Field, eds. London: George Allen & Unwin, 1985, pp. 215–219.

88. Sprent, J. I., "Which steps are essential for the formation of functional legume nodules?" *New Phytol. 111*, 129–153 (1989).

89. Sprent, J. I., and Faria, S. M. de, "Mechanisms of infection of plants by nitrogen fixing organisms," *Plant Soil 110*, 157–163 (1988).

90. Sprent, J. I., and Raven, J. A., "Evolution of nitrogen fixing symbioses," *Proc. Roy. Soc. Edinb. B 85*, 215–237 (1985).

91. Sprent, J. I., and Sprent, P., *Nitrogen Fixing Organisms: Pure and Applied Aspects*, London: Chapman and Hall, 1990.

92. Sprent, J. I., Sutherland, J. M., and Faria, S. M. de, "Some aspects of the biology of nitrogen-fixing organisms," *Phil. Trans. Roy. Soc. Lond. B 317*, 111–129 (1987).

93. Sprent, J. I., Sutherland, J. M., and Faria, S. M. de, "Structure and function of root nodules from woody legumes," in *Advances in Legume Biology*, Vol. 29, C. H. Stirton and J. L. Zarucchi, eds. Missouri: Monog. Syst. Bot., Missouri Bot. Gard., 1989, pp. 559–578.

94. Stam, H., Stouthamer, A. H., and Verseveld, H. W. van, "Hydrogen metabolism and energy costs of nitrogen fixation," *FEMS Microbiol. Rev. 46*, 73–92 (1987).

95. Stedman, D. M., and Shetter, R. E., "The global budget of atmospheric nitrogen species," in *Trace Atmospheric Constituents: Properties, Transformations and Fates*, S. E. Schwartz, ed. New York: Wiley, 1983, pp. 411–454.

96. Stubblefield, S. P., and Taylor, T. N., "Recent advances in palaeomycology," *New Phytol. 108*, 3–25 (1988).

97. Thomas, B. A., and Spicer, R. A., *The Evolution and Palaeobiology of Land Plants*, London: Croom Helm, 1987.

98. Towe, K. M., "Habitability of early earth; clues from the physiology of nitrogen fixation and photosynthesis," *Origins of Life 15*, 235–250 (1985).

99. Trinick, M. J., Goodchild, D. J., and Miller, C., "Localization of bacteria and hemoglobin in root nodules of *Parasponia andersonii* containing both *Bradyrhizobium* strains and *Rhizobium leguminosarum* biovar *trifolii*," *Appl. Environ. Microbiol. 55*, 2046–2055 (1989).

100. Truchet, G., Barker, D. G., Camut, S., Billy, F. de, Vasse, J., and Huguet, T., "Alfalfa nodulation in the absence of *Rhizobium*," *Mol. Gen. Genet. 219*, 65–68 (1989).

101. Tsien, H. C., Dreyfus, B. I., and Schmidt, E. L., "Initial stages in the morphogenesis of nitrogen-fixing stem nodules of *Sesbania rostrata*," *J. Bacteriol. 156*, 888–897 (1983).

102. Tunlid, A., Schultz, N. A., Benson, D. R., Steele, D. B., and White, D. C., "Differences in fatty acid composition between vegetative cells and N_2-fixing vesicles of *Frankia* sp. strain CpI1," *Proc. Natl. Acad. Sci. USA 86*, 3399–3403 (1989).

103. VandenBosch, K. A., Bradley, D. J., Knox, J. P., Perotto, S., Butcher, G. W., and Brewin, N. J., "Common components of the infection thread matrix and the intercellular space identified by immunocytochemical analysis of pea nodules and uninfected roots," *EMBO J. 8*, 335–342 (1989).

104. Vaughn, K. C., and Elmore, C. D., "Ultrastructural characterization of nitrogen-fixing stem nodules in *Aeschynomene indica*," *Cytobios 42*, 49–62 (1985).

105. Wächtershäuser, G., "Evolution of the first metabolic cycles," *Proc. Nat. Acad. Sci. USA 87*, 200–204 (1990).

106. Watson, L., "An automated system of generic descriptions for Caesalpinioideae, and its application to classification and key making," in *Advances in Legume Systematics*, R. M. Polhill and P. H. Raven, eds. Kew: Royal Botanic Gardens, 1981, pp. 65–80.

107. Wiel, C. Van de, Scheres, B., Franssen, H., Lierop, M.-J. van, Lammeren, A. van, Kammen, A. van, and Bisseling, T., "The early nodulin transcript ENOD2 is located in the nodule parenchyma (inner cortex) of pea and soybean root nodules," *EMBO J. 9*, 1–7, (1990).

108. Wingender, R., Röhrig, H., Höricke, C., Wing, D., and Schell, J., "Differential regulation of soybean chalcone synthase genes in plant defence, symbiosis and upon environmental stimuli," *Mol. Gen. Genet. 218*, 315–322 (1989).

109. Woese, C. R., "Bacterial evolution," *Microbiol. Rev. 51*, 221–271 (1987).

110. Wolfe, K. H., Gouy, M., Yang, Y.-W., Sharp, P. M., and Li, W.-H., "Date of the monocot–dicot divergence estimated from chloroplast DNA sequence data," *Proc. Natl. Acad. Sci. USA 86*, 6201–6205 (1989).

111. Young, J. P. W., "*Rhizobium* population genetics: Enzyme polymorphism in isolates from peas, clover, beans and lucerne grown at the same site," *J. Gen. Microbiol. 131*, 2399–2408 (1985).

112. Young, J. P. W., and Johnston, A. W. B., "The evolution of specificity in the legume–rhizobium symbiosis," *Trends Ecol. Evol. 4*, 341–349 (1989).

113. Yung, Y. L., and McElroy, M. B., "Fixation of nitrogen in the prebiotic atmosphere," *Science 203*, 1002–1004 (1979).

114. Zurawski, G., and Clegg, M. T., "Evolution of higher plant chloroplast DNA-encoded genes: Implications for structure-function and phylogenetic studies," *Ann. Rev. Plant Physiol. 38*, 391–418 (1987).

Notes Added in Proof

1. The suggestion that angiosperms originated at least 3–9 Ma BP (p. 464) on the basis of molecular clock data[53] has been reasonably challenged by other investigators on the bases of both molecular (Li, N, H. *et al,* 1989) and palaeontological (Crane *et al.* 1989) data, and defended by Martin *et al* (1989). *References:* Crane, P. R., Donoghue, M. J., Doyle, J. A. and Fries, E. M. *"Angiosperm origins." Nature, 342,* 131 (1989); Li, W-H., Gouy, M., Wolfe, K. H. and Sharp, P. M. "Angiosperm origins," *Nature, 342,* 141–2 (1989); Martin, W., Gierl, A. and Saedler, H., "Angiosperm origins," *Nature, 342,* 132, (1989).

2. The atmospheric O_2 values since 550 Ma BP (pp. 465–6) deduced by Budyke *et al*[9] and used by Gottlieb[34] need reconsideration in relation to the more recent analysis of Berner and Canfield (1989). This later work, which is likely to yield better estimates of atmospheric O_2 contents since 550 Ma BP, only yields one major peak of O_2 occurring in the late carboniferous and reaching 45 K Pa. Over the rest of the last 550 Ma the estimated O_2 partial pressure was between 29 K Pa and 12 K Pa, i.e. within ±9 K Pa of the extant 21 K Pa. *Reference:* Berner, R. A. and Canfield, D. E., "A new model for atmospheric oxygen over paleontological time," *Am, J, Sci., 289,* 333–361 (1989).

3. Raven, in preparation (p. 466) is now in press as Raven (1991). *Reference:* Raven, J. A., "Long-term functioning of enucleate sieve elements: possible mechanisms of damage avoidance and damage repair," *Plant Cell Environm., 14,* in press (1991).

4. The role of photosynthesis in cyanobacterial symbioses is further discussed by Raven *et al,* (1991). *Reference:* Raven, J. A., Sprent, J. I. and Parsons, R., "Cyanobacteria in symbioses with other phototrophs: the role of photosynthesis," *Br. Phycol. J., 26,* 95 (1991).

5. The suggestion that the Aeschynomeneae are a comparatively early branch of the Papilionoideae (p. 478) is supported by recent molecular (cpDNA) data. *Reference:* Lavin, M., Doyle, J. J. and Palmer, J. D. "Evolutionary significance of the loss of the chloroplast—DNA inverted repeat in the Leguminosae sub-family Papilionoideae. *Evolution, 44,* 390–402 (1990).

13

The *Rhizobium* Symbiosis of the Nonlegume *Parasponia*

Jan-Hendrik Becking

I. Introduction

In 1909, during the 10th Congress of the Netherlands Indies Agricultural Syndicate, discussing green manure applications, the Dutch forester S. P. Ham[44] drew attention to the fact that the tree called *anggrung* (Javanese) or *kuraj* (Sundanese) on Java was *Parasponia parviflora* Miquel, belonged to the Ulmaceae, occurred in the mountainous regions of Java, and bore root nodules that probably possessed N_2-fixing capacity. Although the latter supposition was explicitly mentioned, this observation was not further followed or verified.

In 1935 the herbarium-botanist E. W. Clason[18] made a vegetation survey of the upper regions of Mount Kelud (2050 m) in East Java (Upper Badak region), where the vegetation had been repeatedly destroyed by eruptions of this volcano in the last century. The existing vegetation on the new ash-layered soil was observed to be mainly *Parasponia parviflora,* being a typically forest-forming species on these sites. Junghuhn,[51] in 1844, had collected *Parasponia parviflora* on Mt. Kelud, and from his specimens Miquel,[65] in 1851, described this species. Clason[18] also described the abundant occurrence of root nodules on these *Parasponia* plants and surmised its N_2-fixing capacity, which apparently aided its adaptation as a pioneer vegetation on these virgin soils. Extensive herbarium material, including root nodules, of *Parasponia* parviflora collected by Clason still is present in the Herbarium Bogoriense at Bogor, West Java, Indonesia.

Independently and unaware of both previous observations, Trinick,[98] in 1973, observed root nodules on *Trema aspera*. This was actually a misidentification later amended to *Trema cannabina*[105] and finally classified as *Parasponia rugosa*.[1] The genus *Trema* is closely related to *Parasponia,* and in habit and in leaf shape is very similar in morphologic appearance, so that sometimes even botanists have confused these genera. Trinick[98] found the nodulated *Parasponia rugosa* as a weed growing between the rows of tea in the Pangia district of Papua, New Guinea. He was the first to isolate a *Rhizobium* strain from the root nodules and to confirm N_2-fixation ability by performing acetylene reduction tests.

II. Plant Species and Relationships

A. Taxonomy

The genus *Parasponia* belongs to the family Ulmaceae (order: Urticales), a family with about 15 genera and around 200 spp. of worldwide distribution.[87]

Field examination for root nodules in other related genera of the Ulmaceae on Java (Indonesia)—that is, the genera *Celtis* (*C. cinnamomea, C. philippinensis, C. rigescens, C. sumatrana, C. tetandra*, and *C. wightii*), *Gironniera* (*G. subequalis* and *G. supidata*), *Ulmus* (*U. parviflora*), and *Trema* [*T. cannabina* (= *T. amboinensis*), *T. guinensis, T. micrantha, T. orientalis*, and *T. tomentosa*] by Akkermans et al.,[2] and by Becking[11] (and unpublished data)—revealed the absence of root nodulation in all members of these genera. Moreover, many Urticaceae species were tested [i.e., representatives of the genera *Boehmeria, Cypholophus, Debregeasia, Elatostema, Dendrocnide, Laportea, Lecanthus, Leucosyke, Nothocnide, Pilea, Pipturus*, and *Villebrunea* by Akkermans et al.[2] and by Becking (unpublished results)]. All these species showed no root nodulation.

Apparently, *Parasponia* is the sole nonleguminous plant genus that possesses root nodules and a *Bradyrhizobium–Rhizobium* symbiosis. Greenhouse pot experiments and glass tube laboratory experiments under aseptic growth conditions and with surface-sterilized seeds performed with the *Parasponia* rhizobium endophyte and *Parasponia* and *Trema* plants confirmed that only *Parasponia* plants became nodulated and all *Trema* seedlings remained nonnodulated (Becking,[11,12] and unpublished results).

Therefore, any claim of nodulation of representatives of *Trema* is likely to be an error, particularly because in the field *Parasponia* and *Trema* are very difficult to distinguish and are often confused. Members of both genera have very similarly shaped and sized urticaceous leaves with 3-nerves at the base and approximately the same tree habit (Figures 13–1 and 13–2). However, in detail both genera can be easily distinguished by their inflorescences. The arrangement of the perianth, which manifests itself most clearly in the male flower, is induplicate-valvate (folded inward and margins do not overlap) in *Trema* and imbricate (overlapping) in *Parasponia*. It also can be observed in female flowers, but it presents itself there less prominently (see Figures 13–1A and 13–2A). The inflorescences in *Parasponia* and *Trema* are either ♂ or ♀ (very rarely ♂♀), and both types are usually borne on separate vegetative branches. They have a branched panicle or thyrse. Pan-

Figure 13–1. *Trema orientalis* (L.) Bl. (A) Twig with female flowers. (B) Twig with female inflorescences (to the left and center), fruits and a male inflorescence (to the right). (C) Stem with ripe (black) fruits and one unripe fruit (to the left).
Bar marker in A = 0.5 cm; in B = 2.0 cm; in C = 1.0 cm.

Figure 13–2. Parasponia parviflora Miq. (A) Twig with female flowers. Note the short stalks of the inflorescens (in contrast to *Trema*) and the urticoid leaves with 3-nerves at the base (as in *Trema*). The figure shows the intrapetiolar stipules (see S) in *Parasponia,* which ends in long, bifid apices (S*). (B) Twig with ovoid (in *Trema* spherical) ripe fruits, which in nature are pink–red.
Bar marker in A and B = 1 cm.

icle branching is usually more prominent in *Trema,* and therefore these inflorescences stand farther from the stem (see Figure 13–1) than those of *Parasponia* (see Figure 13–2).

In addition, both genera can be distinguished by their stipules, which are small, lateral, and free in *Trema* and large, connate (united), and intrapetiolar in *Parasponia* (see S in Figure 13–2A). Since the stipules readily fall off in *Parasponia* species, they can be best observed at the innovations of the stem tips and on the terminal bud. The large (6–9-mm-long) connate stipules of *Parasponia parviflora* are bicarinate and end in long-bifid apices (see Figure 13–2A, asterisk).

The fruit of *Parasponia* is a fleshy drupe and ovoid in shape. It is more round and spherical in *Trema*. Whereas ripe *Parasponia* fruits become pink-red, those of *Trema* are generally black (Figures 13–1B, 13–1C, and 2B). The only exception is *Trema cannabina,* which develops deep-orange or red fruits. Fruiting in *Parasponia* and *Trema* occurs in rather young plants. Also under greenhouse conditions, fruiting appears rather rapidly, as both species bear fruits within a year when they are only about 30–40 cm high.

The number of chromosomes of *Parasponia parviflora* has been determined to be $n = 10$ (Becking[12]). In the root nodules some plant cells are apparently diploid or polyploid, as chromosome numbers of $n = 20$ or more have been counted (Becking, unpublished results). In *Trema* species chromosome numbers are variable. Cytogenetic studies have indicated that in *Trema politoria* from India it is $n = 10 + B$ (Mehra,[63] and Mehra and Gill[64]); in *T. orientalis, $n = 18$ (Arora[7]) or $n = 20$ (Gajapathy,[42] Hsu[47]) or $n = 10$ (Hans[45]); and in *T. tomentosa* (cited as *T. amboinensis*), $n = 10$ or 80 (Hans[45]).

B. Distribution

Parasponia species occur from southern Sumatra (Lampungs), Java, Philippines, Celebes (Sulawesi), eastward to New Guinea and the Pacific Islands (Polynesia and Melanesia). Representatives of this genus do not occur in India, the east coast of the Asiatic continent, the Malay Peninsula, and Borneo. Thus, *Parasponia* has a typical southeast Asia and Pacific Island distribution. In contrast, the 10–15 existing *Trema* species have a much more worldwide distribution, occurring in Central and South America, tropical and South Africa, the Asiatic mainland (India, Burma, Vietnam, southern China), through southeast Asia toward the Pacific Islands and northern Australia.

Within the genus *Parasponia,* five species are recognized (i.e., *P. andersonii, P. melastomatifolia, P. parviflora, P. rigida,* and *P. rugosa*). The species *P. rigida* and *P. melastomatifolia* occur exclusively on New Guinea; *P. andersonii,* on New Guinea and further eastward on Polynesia (Tahiti) and Melanesia (Fiji Islands and New Caledonia); and *P. rugosa,* on Java (Mt. Kelud and Mt. Lamongan), the Lesser Sunda Islands (Bali), Philippines, Celebes, the Moluccas, and New Guinea. *Parasponia parviflora* has the most westward distribution and at the same time the most restricted area, occurring only on Java [west Java (e.g., Mt. Gede–Pangrango, Mt. Salak) and east Java (Mt. Kelud)] and in southern Sumatra (Lampungs) at a few localities.

Figure 13–3. Germinating seeds and seedlings of *Parasponia parviflora*. (A) Seeds in the process of germinating, bar marker = 1 mm. (B) Seedlings germinated in soil. (C) Germinated plant with seed coat still on cotyledon leaf.
Bar marker in B = 1 cm; in C = 0.5 cm.

C. Propagation

i. From Seeds

The fruit of *Parasponia* is an ovoid drupe; the pericarp is fleshy and fibrous and ripe fruit is pink. The endocarp (i.e., seed) is hard, stony, and dark brown. In *P. parviflora* the fruits average 3.9 (3.2–4.2) mm in length and 2.7 (2.2–2.9) mm in width, weighing an average 13.0 (10.2–16.8) mg. The round seeds are very small, measuring merely 1.3–1.5 mm in diameter and weighing only on average 1.03 (0.9–1.5) mg (Figure 13–3A).

Under natural conditions the ripe fruit falls to the ground, where the fleshy pericarp is readily decomposed under the tropical conditions of high temperature and humidity. *Parasponia parviflora* often grows in thickets along river banks, and the fruits that drop into the water are transported by the rapidly flowing mountain currents and rivulets. In this case, the pericarp is soon removed by abrasion against stones and other objects. Among these biotypes, *Parasponia* plants and seedlings are often found in temporary dry riverbeds, growing in the sand between the stones.

Since *Parasponia* fruits are fleshy, they are also readily dispersed by various aspects of frugivorous birds and arboreous mammals. On Java I observed that the fruit is frequently eaten by species of *Zosterops* (*Z. palpebrosa* and *Z. gallio*) and *Lophozosterops* (*L. javanica*), which visit the trees in large flocks, and by various *Pycnonotus* species (e.g., *P. aurigaster* and *P. bimaculatus*).

Under greenhouse and laboratory conditions *Parasponia* seeds usually germinate poorly. It also takes a long period when the fleshy endocarp has been previously removed. In an experiment in Indonesia I fed caged *Zosterops* species with *Parasponia parviflora* fruits still present on the twigs. The seeds, which pass the intestinal tract undamaged, can be recovered from the droppings. These seeds showed much better germination (~40–60%) than untreated seeds or intact fruits. This observation prompted us to give *Parasponia* and *Trema* seeds a pretreatment with diluted acid before sowing. After various preliminary trials a treatment of soaking the seeds for 24 h in a solution of 4 ml of 37% HCl in 250 ml water gave very satisfactory results. The treated seeds were freed from acid by thoroughly rinsing on a plastic sieve in running tapwater for about 10 min. Subsequently, the seeds were planted in small trays (16 × 22 cm, high 5 cm) with soil and kept in a greenhouse (22° C day, 20° C night). After such a treatment the first seeds germinated after 2 weeks, but usually germination was somewhat protracted (also under normal, natural conditions) and extended over 1–3 months (see Figure 13–3). However, in this period at least 70–80% of the seeds germinated.

For obtaining sterile plants in test tubes, the acid-treated seeds, after thorough washing in tapwater, were surface sterilized by shaking in 7–10% sodium hypochlorite for 10–20 min and then were washed with sterile distilled water. Such axenic plants were usually cultivated on agar slants on a N-free mineral medium (only in the beginning a small amount of combined N was added) in wide glass tubes (2 cm diameter, length about 22 cm) plugged with cotton wool.

ii. From Cuttings

In contrast to general opinion, *Parasponia* can very easily be propagated from cuttings. Relatively young shoots of *Parasponia* were cut in pieces about 10–15 cm in length, and the basal end of the cuttings was dusted with the commercial preparation Verapon 0.5%, which contains naphthylacetic acid hydrazid 0.5%. Other commercial rooting powders, such as Rhizopon A (containing 3-indolelacetic acid 0.5%) and Stim-Root (containing 3-indolebutyric acid 0.4% or 0.8%) also give rise to adventitious roots, but are less successful.

After dusting, the basal ends of the cuttings were placed in humus-rich pot soil in small plantlet trays covered under glass and kept at high relative humidity (80–90%) at a temperature of about 20° C (day 22° C, night 20° C). Within 4–6 weeks such cuttings produced advantitious roots that were nodulated very easily with an appropriate *Parasponia* rhizobium strain. The success rate of such vegetative propagation is 60–80%. The only disadvantage is that *Parasponia* plants have the habit of somewhat drooping twigs, and in a number of these rooted cuttings this habit of somewhat negative geotropy was retained. But after further development most of such plants produced a more upright growth.

Vegetative propagation of *Parasponia* was first claimed by Trinick and Galbraith[105] to be difficult, but our results are confirmed by Gresshoff et al.,[43] who reported a similar outcome (about 80% of the cuttings being successful) with *Parasponia rigida*.

D. Nodulation Frequency in the Field

So far at all sites with *Parasponia* vegetation (*P. parviflora* and *P. rugosa*) examined on Java (West, Central, and East Java), the *Parasponia* plants proved to be nodulated. However, nodulation was found to be more abundant on the younger plants, and particularly on plants growing on exposed sites.

III. *Rhizobium* Classification

Root-nodule bacteria of the former genus *Rhizobium* (Bergey's *Manual of Determinative Bacteriology,* 8th ed., 1974) are currently classified into three genera: *Rhizobium* ("fast growers"), *Bradyrhizobium* ("slow growers"), and *Azorhizobium* ("stem-nodulating" species). The original classification, mainly based on cross-inoculation groups,[41] has been changed on the basis of genetic analysis. Moreover, some former species have been fused into one species with several biovars (*Rhizobium leguminosarum* biovar *trifolii,* biovar *phaseoli,* and biovar *viceae,*[50] whereas some new *Rhizobium* species have only been recognized recently (e.g., *R. loti, R. fredii, R. galegae*).

In addition to a slower growth rate (colony formation on agar: about 7 days in *Bradyrhizobium* and 3 days in *Rhizobium*), *Bradyrhizobium* species distinguish themselves from the genus *Rhizobium* by some physiologic characteristics, such as no acid but alkaline production in culture media, and in cell morphology being usually monotrichous, in contrast to peritrichous flagellation in *Rhizobium*. Further, they show a far lower frequency of genetic exchange and a higher intrinsic antibiotic resistance.[85] Finally, there is no evidence for plasmids carrying symbiotic genes in *Bradyrhizobium*. DNA homology studies and ribosomal RNA analyses have shown that *Rhizobium* and *Bradyrhizobium* are not closely related. *Rhizobium* and *Agrobacterium* are more closely related to each other than either is to *Bradyrhizobium*.[46,48,114] Ribosomal RNA sequence analysis showed that *Bradyrhizobium* is a member of the *Rhodopseudomonas palustris* rRNA branch and therefore is closely related to this taxon.[48] *Rhodopseudomonas palustris* is a representative of the phototrophic family Rhodospirillaceae, the purple nonsulfur bacteria.

The genus *Azorhizobium*[36] is somewhat intermediate between *Bradyrhizobium* and *Rhizobium* species, since it (strain ORS571) resembles the "fast-growing" rhizobia (*Rhizobium* spp.) in growth rate (generation time c. 90 min), but many of its physiologic characteristics are more typical for the "slow-growing" group (*Bradyrhizobium* spp.).[31-34] Moreover, *Azorhizobium* strain ORS571 does not appear to harbor any (symbiotic) plasmids like *Bradyrhizobium*. *Bradyrhizobium* and *Azorhizobium* have the common characteristic that they are able to fix N_2 asymbiotically.[35] A ribosomal RNA sequence analysis has revealed that *Azorhizobium* ORS571 belongs to the *Rhodopseudomonas* rRNA branch and therefore, like *Bradyrhizobium*, is related to *Rhodopseudomonas palustris*.[48]

A. *Parasponia: Bradyrhizobium–Rhizobium Strains*

Rhizobium isolates from *Parasponia* root nodules are usually "slow-growing" strains and therefore can be classified as belonging to the genus *Bradyrhizobium*.[78,81,108] A large number of different *Parasponia–Bradyrhizobium* strains are known up to now. The strains most often cited in the literature are NGR231 and CP241 (from *P. rugosa,* Trinick[101]) and CP272, CP283, and CP284 (from *P. andersonii,* Trinick[101]), RP501 (from *P. parviflora,* Tjepkema and Cartica[95]), and ANU298, which is a streptomycin-resistant derivative of strain CP283 from *P. andersonii* (used by Scott et al.[80] and Shine et al.[83]).

Also some "fast-growing" rhizobia (i.e., representatives of the genus *Rhizobium*) isolated from tropical legumes can nodulate *Parasponia*. The best-known strains are NGR234 isolated from *Lablab purpureum* nodules and a strain NGR71 obtained from *Acacia farnesiana* nodules. The strain ANU240 used by Rolfe et al.[76] is a spontaneous streptomycin-resistant mutant of the *Lablab* strain NGR234. Although the previously mentioned strains produced nodules on *Parasponia,* Trinick and Galbraith[106] showed that these nodules were ineffective in N_2 fixation. However, some strains from *Leucaena leucocephala* (NGR7, NGR89, etc.) and *Mimosa* spp. (NGR191, NGR189) were able to nodulate *Parasponia*. In addition, the nodules showed some, although low, nitrogenase activity. Nodulation was mentioned by Trinick and Galbraith[106] to be completely absent in *Parasponia* plants on inoculation with fast-growing rhizobia of temperate legumes, such as the *R. legumi-nosarum* biovars *phaseoli* and *trifolii,* and with *R. meliloti.*

Becking[12] isolated 22 strains of *Bradyrhizobium* from *Parasponia par-viflora* root nodules (i.e., the strains Pp 1A, 1B, 2, 4A, 4B, 6, 7A, 8, 19, 25, 26, 31, 50, 51, 52, 53, 54, M1A, M1b, M2A, M2B, and M3A). Most of these strains were isolated from field-collected root nodules from various provenances (Mt. Gede, Mt. Salak, etc.) in western Java. However, some strains also were obtained from spontaneously nodulated *Parasponia* plants grown in greenhouse soil, in which previously no tropical legume had been cultivated. This greenhouse soil was a humus-rich pot soil commercially fabricated (mixed) and distributed in 25-kg plastic bags for floriculture (Tri-MF No. 26 pot soil, Trio B. V., Westerhaar, The Netherlands). The isolation procedure of our *Parasponia*–rhizobium isolates was the normal technique of surface sterilization of root nodules[111] followed by plating on yeast extract, mannitol 1%, agar plates.

Growth test of the isolated strains showed that they belonged to the "slow-growing" rhizobium group (i.e., *Bradyrhizobium*), as rough estimates indicated a generation time of about 6 h or more and that they produced alkalinity in a mineral salt-manitol medium. A more precise investigation of the growth requirements and flagellation arrangement of the latter strains has not yet been made.

B. Cross-inoculation Experiments

We performed cross-inoculation experiments with *Parasponia–Bradyrhizobium* strains and various other *Rhizobium* ("fast-growing") strains on *Parasponia parviflora* and legume plant material. Similar cross-inoculation experiments performed by other authors,[108,109] but with *Parasponia andersonii*, will be discussed.

In our experiments (Becking[12]) sterile plant material of *Parasponia parviflora* and various legumes was obtained from surface-sterilized seeds using a treatment of ethanol, detergent, and Na-hypochlorite. After the germination of the seeds on mineral agar plates at 27° C, the seedlings were aseptically transferred to test tubes containing agar slants with mineral agar containing a small amount of combined N (30 mg NH_4NO_3 per liter), plugged with cotton wool. For the small-seeded *Parasponia* and the smaller-seeded legumes, glass tubes of about 22 cm in length and 2 cm in diameter were used. For the larger-seeded legume species, such as *Arachis, Lupinus, Vigna,* and *Macroptilium* spp., 1-liter milk bottles (height 27 cm, diameter 8.5 cm) or large glass tubes (length 47 cm, diameter 5.5 cm) were employed for cultivation. Each test was performed in 5–10 replicates. In a number of cases with the larger leguminous plants pot experiments with perlite also were performed; cross-contamination was avoided by keeping the plants as far as possible under aseptic conditions.

In these cross-inoculation experiments the following distinction can be made: (1) *Parasponia* seedlings inoculated with *Parasponia–Bradyrhizobium* isolates or legume–rhizobium isolates and (2) leguminous plants inoculated with *Parasponia–Bradyrhizobium* isolates.

i. Parasponia–Bradyrhizobium Isolates on Parasponia Plants

All our *Parasponia–Bradyrhizobium* isolates (so-called Pp strains) produced root nodulation on *Parasponia parviflora* (its original host) in one or more tubes of the 10 replicates. In a few strains all test tubes produced positive root nodulation, but usually root nodulation varied between 80% and 90%. In a few strains root nodule formation was much less, as only one or two of the tubes (10% and 20%) were nodulated. This variability apparently is caused by a different degree of "root-cortex penetration ability" or by the genetic heterogeneity of the *Parasponia* plant material. These *Parasponia* plants raised from seeds are certainly genetically heterogenous, as also is evident from pot experiments in which plants showed a great variability in leaf shape and growth habit as mature plants grown under the same greenhouse conditions.

In addition, the symbiosis itself showed different degrees of effectiveness. Visual ratings of growth of the plants in the test tubes and the perlite cultures suggested that these *Bradyrhizobium* strains were, in their symbiosis, effective, partly ineffective, or completely ineffective. These observations were corroborated by the variable nitrogenase activities of the root nodules of the intact plants, as they varied from very effective to nil.

ii. Legume Rhizobium Isolates on Parasponia Plants

Many rhizobia strains available in the *Rhizobium* collection of the Laboratory of Microbiology, Agricultural University, Wageningen, were tested for their ability to produce root nodulation in *Parasponia*. These strains were mainly temperate fast-growing strains belonging to the genus *Rhizobium*, but the collection also contained some *Bradyrhizobium* strains. The strains will be presented according to the numbering codes used for the *Rhizobium* Collection of the Laboratory of Microbiology, Agricultural University, Wageningen; its classification has been adapted based on current information.

a. Rhizobium leguminosarum. The fast-growing strains can be divided into three biovares.

1. Biovar Viceae. Twelve strains of *Biovar viceae* (R4, R5, R7, R10, R72–R77, R80, and R82) were tested, ranging from very effective strains [i.e., including the highly effective strain R7 (=PRE) (on *Pisum sativum*)] to ineffective strains [e.g., the ineffective strain R10 (=S313)]. Remarkably enough, five strains (i.e., 41% of all strains tested) produced root nodulation on *Parasponia parviflora*. Among the strains producing root nodules were the ineffective strain R10 (on *Pisum sativum*), but also highly effective strains such as R7. Root nodulation was most prominently developed on the lateral roots. The nodular structures developed varied a great deal in appearance. Often they resembled more or less calluslike outgrowths [so-called pseudo–root nodules, Figure 13–4A), which contained the *Rhizobium* endophyte, as evident from some nodules examined internally.

2. *Biovar phaseoli.* Of seven strains (R20–R25 and R124) tested, only one strain (R22) produced nodular structures on *Parasponia parviflora* roots. The other strains only occasionally caused a local thickening of the taproot or lateral roots. The presence of rhizobia could be ascertained in the nodular tissue of the *Parasponia* produced by strain R22.

3. *Biovar trifolii.* Twenty strains (R27–R34, R36–R39, R51, R52, R62, R63, R70, R71, R85, and R86) were tested, including the highly effective strains R38 (=K8) and R86 (=S459), and the ineffective strain R30 (=Coryn). Twelve strains, including the effective strains R38 and R86, but also the ineffective strain R30, produced root nodulation in the *Parasponia* seedlings. Thus, the ability to produce root nodulation on *Parasponia* is certainly not related to the effectiveness of the strain on its original host (clover). As observed in the other *R. leguminosarum* biovars, the formation of root nodules was often very delayed, as it usually took 3 weeks to 2 months before distinct root nodules could be seen by the naked eye. Moreover, root nodulation did not occur in all replicate test tubes. In about half of the nodulation-positive strains only one or two test tubes of the series of 5–10 replicates showed root nodules, but with strains R36, R51, and R63 all replicate tubes were positive. These nodules formed on *Parasponia* had the normal nodule structure, and inside the nodular tissue the presence of the *Rhizobium* endophyte was observed.

b. *Rhizobium meliloti.* Eleven strains (R12–R17, R132, R134–R136, and R140) were tested. However, no strain produced normal root nodules, but again with most strains a kind of pseudo–root nodule was formed (see Figures 13–4B and 13–4C). The pseudonodules were usually free from rhizobium bacteria, but in a few cases the original strains could be reisolated from the root nodule.

c. *Bradyrhizobium japonicum.* The 10 strains of this species (R87–R90, R95, R109, R114, RCC705, RCC707, and RCC711) tested were negative in respect to normal root nodule formation. However, many strains produced a kind of callus tissue along the roots and in some of these pseudo–root nodules the presence of the rhizobium-endophyte could be detected (see Figures 13–4D and 13–4E).

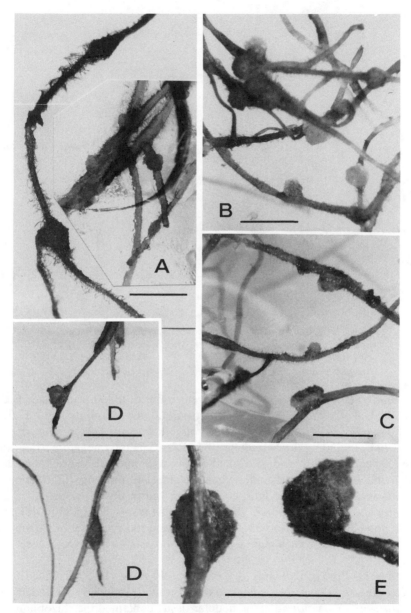

Figure 13-4. Abnormal root nodule formation on *Parasponia parviflora* by various rhizobium strains from legume origin. (A) *Rhizobium leguminosarum* biovar *viceae*. (B) and (C) *Rhizobium meliloti*. (D) and (E) *Bradyrhizobium japonicum*. Bar marker in A − D = 2 mm; in E = 1 mm.

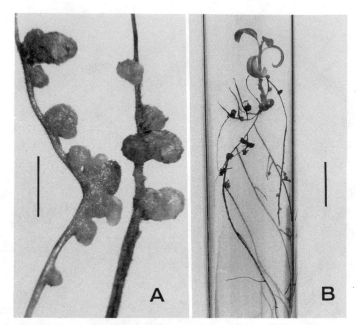

Figure 13–5. Root nodules produced by *Bradyrhizobium* strains on *Parasponia parviflora.* (A) *Arachis-Bradyrhizobium* strain R141. (B) *Lupinus-Bradyrhizobium* strain R61.
Bar marker in A = 2 mm; in B = 1 cm.

d. Bradyrhizobium lupini. Of the seven strains of this species tested [i.e., R18, R61, R92, R130, R131, R138 (from Lupine, *Lupinus* spp., Figure 13–5B) and R26 (from Serradella, *Ornithopus* spp.)], six produced root nodulation, including the Serradella strain R26. The only negative strain was R18. Especially with strain R61 and R131, root nodulation was abundant and found in all or nearly all replicate tests. The presence of rhizobium bacteria within the root nodules could be proved in nodule slices.

e. Arachis–Bradyrhizobium. The two strains tested (R14 and R142) produced good root nodulation on *Parasponia,* and the presence of rhizobium bacteria within the root nodule tissue could be established (see Figure 13–5A).

f. Albizzia–Bradyrhizobium. The *Rhizobium* strain (R139) isolated from *Albizia distachya* (syn. *A. lophantra*) of Australia, produced root nodulation in *Parasponia parviflora* and the nodular tissue contained the rhizobium endophyte.

g. Nitrogen Fixation Capacity

All the previously mentioned associations of legume-*Rhizobium–Bradyrhizobium* strains and *Parasponia parviflora* seedlings, producing either normal root nodules or so-called pseudo–root nodules, proved to be ineffective with regard to N_2 fixation, as none of these structures showed nitrogenase activity when tested with the sensitive C_2H_2-reduction assay (see later).

h.   Experiments by other Authors

Trinick[99] mentioned that *Bradyrhizobium–Rhizobium* strains of some tropical legumes such as *Vigna unguiculata, Pueraria phaseoloides, Macroptilium atropurpureum, M. lathyroides, Cassia mimosoides,* and *Glycine max* could produce root nodulation in *Parasponia rugosa ("Trema cannabina")*.

In more detailed experiments with *Parasponia andersonii* and a large collection of *Bradyrhizobium–Rhizobium* isolates of tropical legumes, available in Australian culture collections (CSIRO, Univ. of Western Australia, and standard Australian cultures), nodulation was observed on *Parasponia* with 34 of the 39 slow-growing rhizobia (*Bradyrhizobium* spp.) tested.[104,106] However, the nitrogenase activity of the nodulated plants varied widely. Effective associations were obtained with some, but not all isolates from *Cajanus cajan, Centrosema pubescens, Crotalaria anagyroides, Flemingia congesta, Inocarpus fragiferum, Macroptilium lathyroides, Phaseolus calcaratus, Stizolobium deeringianum,* and *Stylosanthese gracilis.* Ineffective nodules developed with rhizobia from *Albizia stipulata, Cassia mimosoides, Desmodium microphyllum, Pterocarpus indicus, Uraria lagopoides, Dolichos africanum,* and the wide–host range cowpea *Rhizobium* CB756. Isolates of *Bradyrhizobium japonicum* and *B. lupini* also were tested and produced nodulation, but the association proved to be ineffective. The latter result is in agreement with our observations.[12]

Moreover, according to Trinick and Galbraith,[105] all fast-growing strains of rhizobia (*Rhizobium* spp.) obtained from *Lablab purpureus, Acacia farnesiana, Leucaena leucocephala,* and *Mimosa invisa* produced root nodules on *Parasponia andersonii* but had only a low level of nitrogenase activity when tested with the C_2H_4-reduction assay in detached root nodules. Nodulation of *Parasponia* was also obtained by the rhizobia (*Bradyrhizobium– Rhizobium*) of the *Sesbania* group,[108] but again with low nitrogenase activity. Subsequently, Trinick and Hadobas[109] extended the list of tropical legumes that produce nodulation on *Parasponia (P. andersonii)* to isolate from *Crotalaria paulina* and *Macrotyloma africanum.* In all cases, these strains could fully nodulate *Macroptilium atropurpureum* but were only partially effective (10–20%) in *Parasponia.* Temperate fast-growing rhizobia such as *R. phaseoli, R. leguminosarum, R. trifolii* (all now currently combined in *R. leguminosarum* biovars) were observed to be unable to produce root nodulation in *Parasponia.*[106] This result is contradictory to our observations.[12] However, later, Trinick and Hadobas[108] confirmed our results with *R. leguminosarum* biovar *trifolii* strains after more prolonged exposure, as it takes 1.5–2 months before nodules become visible. Apparently, nodulation of *Parasponia* by the usual tropical rhizobia is not delayed to such an extent.

iii. Nodulation of Legumes with Isolates from Parasponia

In view of the previously mentioned results on nodulation of *Parasponia* plants by slow-growing rhizobia (*Bradyrhizobium* spp.) strains and fast-growing rhizobia (*Rhizobium* spp.) strains, it was worthwhile to test the reverse combinations.

Six *Parasponia parviflora* isolates from our collection (Pp 1A, Pp 5A, Pp 8, Pp 10, and Pp 54) were tested in five replicates on temperate legumes (*Trifolium repens, T. pratense, Pisum sativum, Medicago sativa, Lupinus luteus,* and *L. angustifolius*) and, on tropical legumes (*Arachis hypogaea* (Peanut), *Cajanus cajan* (Pigeon pea), *Phaseolus mungo* (Mungbean), *P. vulgaris* var. Pinto (Pintobean), *P. aureus, P. luneatus, P. calcaratus, P. acutifolius, Glycine max var.* Altoma, *G. max* var. Calori, *G. max* var. Early-Akita, *Macroptilium atropurpureum, M. lathyroides, M. bracteatum, M. martii, Vigna unguiculata* var. *sesquipedalis, V. unguiculata* var. Black Eye, *V. unguiculata* var. Remshorn, and *Stylosanthes guinensis*). Since *Parasponia* isolates may show delayed nodulation on legumes, the final examination was 2 months after inoculation with the isolates. In all species of the temperate legumes tested, no root nodulation developed. The tropical legumes showed root nodulation, but it varied greatly with the species concerned. In *Arachis hypogaea* no root nodules were formed, but at the base of the lateral roots calluslike outgrowths were produced along the root surface (Figure 13–6C). All six *Phaseolus* species tested were negative for nodulation, except *P. vulgaris* var. Pinto, where abnormal root nodules were formed (see Figure 13–6A). A closer examination showed that some parts of these nodules were endophyte-free, but other sections of the root nodule distinctly contained the rhizobial endophyte. With regard to the observed nodulation of the bean group by *Parasponia* isolates, it is worth mentioning that Lange[60] observed that *P. vulgaris* formed nodules with many slow-growing rhizobia (*Bradyrhizobium* spp.) obtained from native legumes of western Australia. Abundant and apparently normal nodules were developed on *Macroptilium atropurpureum* and to a lesser extent on *M. lathyroides,* but the two other *Macroptilium* species (*M. bracteatum* and *M. martii*) were negative. On *M. bracteatum* only some swollen lateral roots were produced. The normal *Macroptilium* root nodules produced by the *Parasponia–Bradyrhizobium* contained rhizobia and were effective in N_2 fixation. Excellent and effective nodulation and N_2 fixation also were obtained on *Vigna unguiculata* in its various varieties (see Figure 13–6B). In *Cajanus cajan* (Pigeon pea) and in the tropical soil-cover legume *Stylosanthes guinensis* root

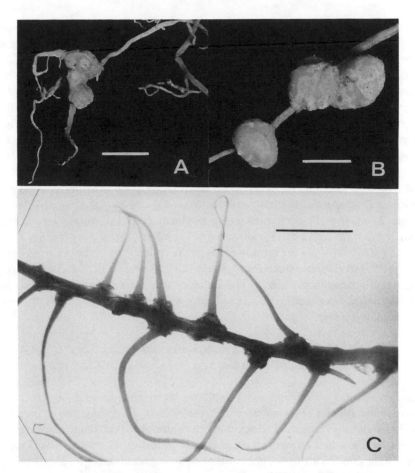

Figure 13–6. Root nodules produced by *Parasponia–Bradyrhizobium* on legumes. (A) on *Phaseolus vulgaris* var. *pinto* (Pintobean), ineffective nodules; (B) on *Vigna unguiculata* (Cowpea), effective nodules; (C) *Arachis hypogaea* (Peanut) abnormal, ineffective nodular structures.
Bar marker in A − C = 4 mm.

nodulation was entirely absent. It must be mentioned that in our experiments not all the *Bradyrhizobium* isolates from *Parasponia* gave the same outcome in the various tests and that positive results were not always present in all replicates. Nevertheless, the conclusion can be drawn that the various legume isolates nodulate *Parasponia* plants more readily than the *Parasponia* isolates nodulate legumes. Moreover, *Parasponia* is more selective and compatible with its own isolates than with legume isolates. The same is true with regard to effectiveness.

Trinick and Galbraith[106] observed that the highly effective *Parasponia andersonii* isolate CP283 produced greater plant top dry weight on legumes than the corresponding control strains from *Macroptilium lathyroides, Lablab purpureus, Fleminga congesta,* and all *Vigna unguiculata* subspecies. When, however, *Parasponia* strain CP283 was compared with other rhizobia from tropical legumes on *M. atropurpureum,* it was intermediate in effectiveness. Many other strains of rhizobia isolated from *Parasponia* were less successful in nodulating legumes. They were able to nodulate *Vigna unguiculata* spp., *M. atropurpureum,* and sometimes *M. lathyroides* and *Flemingia congesta,* but the associations were less effective and often no N_2 was fixed. Although the *Leucaena–Minosa* group of rhizobia was able to nodulate *Parasponia,* rhizobia isolates from *Parasponia* were unable to nodulate *Leucaena.* Plants representing soybean (*Glycine max*), lupine (*Lupinus angustifolius*), pea (*Pisum sativum*), clover (*Trifolium pratense*), and medic (*Medicago sativa*) groups were not nodulated by rhizobia from Parasponia.[106] Two isolates from *Parasponia* (CP283 and NGR231) ineffectively nodulated *Phaseolus vulgaris.* The latter observation is consistent with our observation of nodulation of *P. vulgaris* var. Pinto (tropical!) with some *Parasponia* isolates.[12]

IV. Time Taken to Nodulate *Parasponia* and Dual Occupancy of Nodules

All *Parasponia–Bradyrhizobium* isolates were able to nodulate the four *Parasponia* species (*P. andersonii, P. parviflora, P. rigida,* and *P. rugosa*) in 7–14 days (Trinick,[101] Becking[12] and unpublished data, Trinick and Hadomas[108]), when grown in axenic culture in test tubes on agar slants under good light conditions in a light cabinet or in a greenhouse, at 25–28° C. In contrast to this, Trinick and Hadomas[108] showed that *Bradyrhizobium* strains from tropical legumes took 31–74 days to form nodules on *Parasponia.* Because of this delay, *Parsponia* selects from a mixed soil population specific strains that are able to induce early nodulation. Hence, in the competition studies between legume and *Parasponia* isolates, *Parasponia* selects specifically for its own strains.

On the other hand, Trinick and Hadobas[109] showed that *Bradyrhizobium* strains isolated from tropical legumes were slow to nodulate *Parasponia andersonii* (25–55 days) compared to their relatively rapid nodulation of *Macroptilium atropurpureum* (8–15 days). The competitive abilities of *Bradyrhizobium* strains are thus primarily governed by the time required to nodulate plants, as early nodulation is favorable in competition. Thus, in general *Parasponia* isolates nodulate *Parasponia* plants more rapidly than they nodulate tropical legumes. The only exception is *Parasponia* strain CP283, which nodulates legumes rapidly but produced delayed nodulation

(30–50 days) on *Parsponia andersonii*. For this reason Trinick and Hadobas[109] suggested that this *Parasponia* isolate may be of legume origin.

From the previously mentioned observations it is clear that in a mixed culture of *Parasponia* and legume isolates, the isolates from *Parasponia* always dominate in the root nodules of *Parasponia*. With mixed cultures of *Parasponia–Bradyrhizobium* isolates, dual and triple occupancy of *Parasponia* nodules often could be detected by differences in streptomycin sensitivity among the strains and by serologic techniques.[109] Only one legume isolate (NGR86, from *M. atropurpureum*) could simultaneously occupy one third of the nodules produced by the *Parasponia* isolates, when mixed inocula were applied. Moreover, among the *Parasponia* isolates the competitiveness for nodulation on *P. andersonii* was not associated with effectiveness of N_2 fixation. A highly effective strain may be a poor competitor and a least effective strain may be a good competitor and thus able to dominate in the root nodules.[109]

V. *In Vitro* Expression of Nitrogenase Activity in *Parasponia–Bradyrhizobium*

Bradyrhizobium species are usually able to fix N_2 asymbiotically and some are even able to grow chemoautotrophically with CO_2 and H_2.[55,70] Therefore, it could be expected that *Parasponia–Bradyrhizobium* isolates would exhibit the same property. Mohapatra et al.[67] demonstrated that *Parasponia–Bradyrhizobium* strain ANU289 was able to fix N_2 (C_2H_2 reduction assay) under *in vitro* conditions on agar and in liquid stationary culture. The nitrogenase activity was 40–70 nmol C_2H_4 per mg protein after 7 days incubation. The strain was able to give effective nodulation on *Macroptilium atropurpureum* (Siratro) as well in *Parasponia andersonii* and *P. rugosa*. The O_2 requirement for developing nitrogenase activity *in vitro* was between 0 and 10% O_2 in the gas phase, and nitrogenase showed increased sensitivity to oxygen repression at 20% O_2.

VI. Nodule Structure and Cytology

A. *Infection and Early Nodule Initiation*

The first information on rhizobium entrance and early infection in *Parasponia* is from Trinick.[103] He described that an infection thread like that in legumes is formed inside a root hair, and by passage through the root hair the bacteria can invade cortical cells of the host plant. The further development to a mature nodule was, however, more like the actinorhizal type (*Alnus*-type) of root nodule formation, because the root nodule is actually a modified lateral root initiated by a root primordium arising in the pericycle.

In contrast to this, Lancelle and Torrey[56] found no infection threads in root hairs, but instead an invasion of the rhizobium endophyte through the intercellular spaces of the epidermis into the root cortex. The earliest sign of root nodule initiation is described by the presence of clumps of multicellular root hairs and an initiation of cell divisions in the outer cortex always subjacent to the multicellular root hairs. After this, the second step is the infection of the host cells themselves by the rhizoidal cells and the formation of typical intracellular persistent infection threads, which probably mark the beginning of a true symbiotic relationship in these nodules. The further nodule morphogenesis was observed by Lancelle and Torrey[57] to be essentially the same as the actinorhizal associations, including prenodule formation and the initiation of modified lateral roots that form the nodule lobe primordia of the later coralloid root nodules.

Bender et al.[14] carried out a time course study of the earliest stages of infection in *Parasponia andersonii*. An infection zone formed close behind the growing tip of the taproot was similar to that found in legumes. Here, as in legumes, bacteria colonize the root surface; however, they erode the mucilage layer of the primary cell wall of the epidermal cells within 24 h after inoculation. Four days after inoculation a swelling of the root beneath the area of erosion of the root surface was visible, and this was followed by a rupture of the overlying epidermal layer. The removal of the epidermis and the emergence of exposed cells provided the site through which bacteria enter the root and colonize the cortex.

Our observations with *Parasponia parviflora* are more in agreement with those of Bender et al.[14] than with those of the two first–mentioned studies. The first step of root nodule initiation in *Parasponia* is a strong proliferation of the rhizobia bacteria at the root surface (Figure 13–7). This colonization of the root surface probably gives rise to the production of signal substances causing a rejuvenation and active mitotic activity of the cortical tissue directly underlying the colonization site. A production of a cytokinin compound by the *Parasponia*-nodule rhizobium has been demonstrated by Sturtevant and Taller[92]; it was known earlier for *Bradyrhizobium japonicum* and *R. leguminosarum*.[72,73] An erosion of the mucilage layer covering the primary cell wall of the root surface as mentioned by Bender et al.[14] could not be observed by us with certainty, but may be plausible, since the rhizobia bacteria need substrate for growth. Because of the rapid meristematic proliferation of the cortical cells within the root, which is possibly elicited by the rhizobia bacteria on the root surface, an epidermal dome is formed at the outside of the root at the site of infection. Further increase of this dome causes a rupture of the epidermal cells at the root surface in a tangential longitudinal direction. In consequence of this rupture, the epidermal cells often curl upward at their free ends (Figure 13–8). Through the raised fissure or wound the endophyte invades the exposed cortical tissue and the bacteria

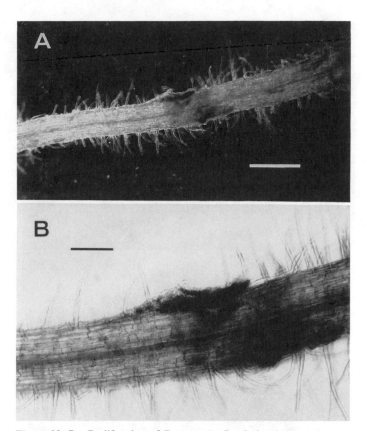

Figure 13-7. Proliferation of *Parasponia–Bradyrhizobium* on the root surface of *Parasponia parviflora*. (B) is a magnification of (A). Living, unstained preparation.
Bar marker in A = 200 μm; in B = 100 μm.

spread rapidly by proliferation through the intercellular spaces toward the cortical cells, which display active meristematic activity in the underlying layers (Figure 13–9). Often one or more centers of meristematic activity may develop at root nodule initiation, and these regions are invaded separately from the outside by the nonencapsulated endophyte present in the intercellular spaces. From electron micrographs, it is evident that within the intracellular spaces the rhizobium endophyte is not protected by cell wall deposits, but the cells are freely embedded in a polysaccharide slime matrix probably of bacterial origin (Figure 13–10).

Actual penetration of the endophyte intracellularly, like that of certain legumes, is accompanied by the formation of an infection thread. The infection thread wall is formed as an ingrowth of the host cell wall by the formation of new cell wall components. Therefore, the infection thread is

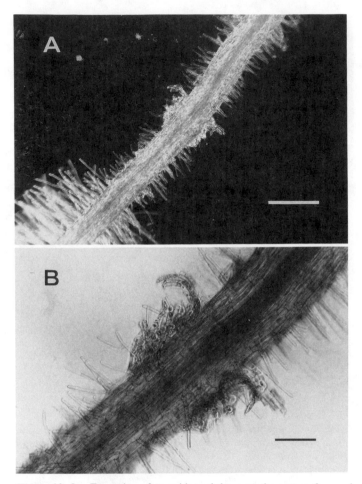

Figure 13–8. Formation of an epidermal dome on the root surface and rupture of the epidermal cells in tangential longitudinal direction with a curling up of the epidermal cells at their free ends. Living, unstained preparation.
Bar marker in A = 300 μm; in B = 100 μm.

always relatively thick-walled, being formed by secondary cell wall material of host origin (Figure 13–11). Moreover, in consequence of this, infection threads always showed a continuity with the secondary cell wall deposits of the host. The infection thread probably serves to avoid direct contact between host cytoplasm and endophyte in the early stages of the association, to prevent antagonistic reactions.

Nodule development in *Parasponia* is definitely not associated with root hair curling or root hair deformation at the site of infection. Moreover, it is

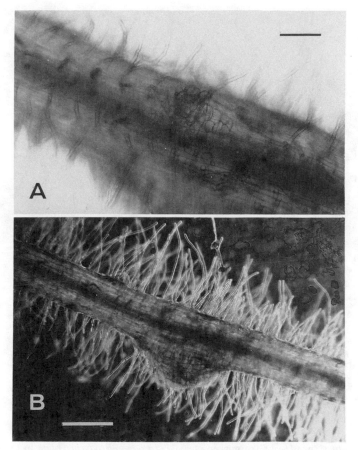

Figure 13–9. Penetration of *Parasponia–Bradyrhizobium* cells through the intercellular spaces of the wound fissure into the underlying cortical cells and the gradual development of a root nodule. Living, unstained preparation.
Bar marker in A = 100 μm; in B = 300 μm.

independent of the presence of root hairs, because nodule initiation in *Parasponia* may occur on bare or nearly bare root surfaces. The deformation or swelling of root hair tips, sometimes observed by us, also is a phenomenon uncorrelated with root nodule induction; rather, it is an effect of the production of auxins or other growth hormones by the endophyte.

B. Nodule Structure and Development

The structure of *Parasponia* (*Trema cannabina* var. *scabra* = *P. rugosa*) root nodules was originally described by Trinick and Galbraith[105] to have

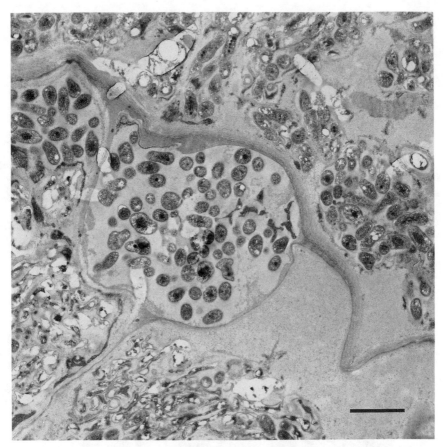

Figure 13–10. Invasion of the noncapsulated *Parasponia–Bradyrhizobium* in the intercellular spaces and their penetration in uncapsulated form intracellularly into adjacent host cells. Transmission electron micrograph; bar marker = 2 μm.

the overall features of the actinorhizal root nodules of a nonlegume rather than a legume root nodule. Moreover, evidence was obtained that the rhizobial cells within the bacterial zone are not released from so-called infection threads, but these structures remained intact and finally filled the host cells completely. These infection threads possessed active N_2-fixation activity. Only a very small proportion of the intracellular infection threads were observed to release rhizobial cells into the host cytoplasm.[105]

In a light and transmission electron microscopic study of the root nodules of *Parasponia parviflora* we observed persistent threadlike structures within infected host cells (Figure 13–12). It was shown that these structures contained in cross-section a row or a number of rows of rhizobial cells (Figure 13–13). When these structures were freed from the host cell in cell squashes,

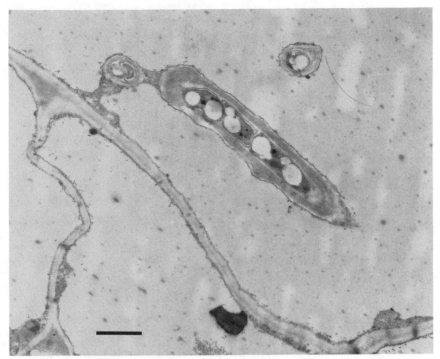

Figure 13–11. Intracellular penetration of an infection thread containing *Parasponia–Brady-rhizobium* bacteria. Note the thick, lignified walls of the infection thread and the lining up of the bacterial cells within the infection thread in a single row.
Transmission electron micrograph; bar marker = 1 μm.

they exhibited independent structures of variable length and thickness, usually 40–50 μm long and 1.3–3.0 μm thick (Figure 13–14). Occasionally, threadlike structures up to 540–1034 μm have been measured (Becking[11] and unpublished data).

In a second morphologic study, but with *Parasponia andersonii,* Trinick[100] confirmed that *Parasponia* root nodules resemble actinorhizal root nodules; that is, they are modified lateral roots having an apical meristem and a central vascular bundle surrounded by an endophyte-infected zone. Two types of infection threads were recognized by light and electron microscopy. One is a darkly stained or electron-dense structure, the other is a lightly stained and apparently thinner-walled structure of varying wall thickness occurring only intracellularly and often occupying the whole host cell content. It was concluded that the rhizobia cells were never released from the infection threads into the host cytoplasm, in contrast to earlier observations with *P. rugosa.*[105] The persistence of both types of infection threads was proved in serial sections.[100]

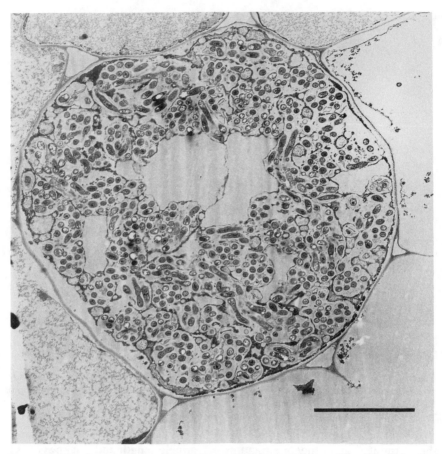

Figure 13–12. Overview of an infected host cell of *Parasponia parviflora*. The living host cell is completely filled with the so-called fixation threads. The latter structures are, according to the present author, peribacteroid membranes that have incorporated cell wall constituents. Transmission electron micrograph; bar marker = 10 μm.

A comparative morphologic study of root nodules formed by the same *Parasponia–Bradyrhizobium* strain (ANU289) on *Parasponia rigida* and on a legume *Macroptilium atropurpureum* (Siratro) was conducted by Price et al.[75] The nonlegume root nodule is described as a modified lateral root with multiple regions of apical meristematic activity; the nodule is indeterminate and has a centralized vascular system. The bacteroids are not released from membrane-bound infection threads, which appear to develop into lightly staining, thin-walled threads containing numerous loosely packed bacteroids. For these structures the name *fixation threads* is proposed, because the bacteroids within the threads showed active nitrogenase activity.[75] In contrast to the nonlegume nodule, the legume nodule was found to be a

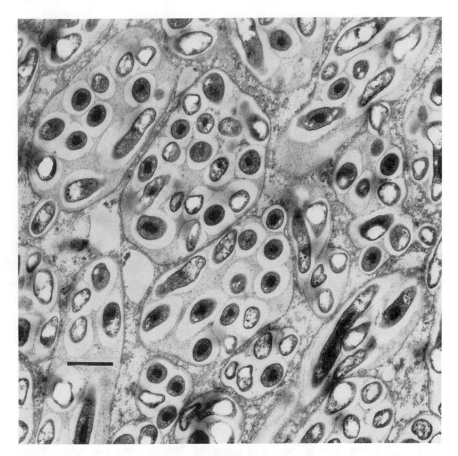

Figure 13–13. Cross sections of the so-called fixation threads at higher magnification. As evident, these "sacs" have a double membrane, and within the "sacs" the bacteria are imbedded in a polysaccharide matrix.
Transmission electron micrograph; bar marker = 1 μm.

spherical determinate nodule with a peripheral vascular system. The invaded region contained hypertrophied host cells, in which bacteroids developed, which were packaged in a peribacteroid membrane.

Smith et al.[86] investigated by histochemical methods the differences in composition and structure of the two types of infection threads. Both types of infection threads were observed to contain pectic and cellulosic components. However both types responded differently on cellulase and pectinase treatment as these enzymes produced no change in the infection threads, but they dissolved the fixation threads to some degree. Moreover, the fixation threads contained lipids and a suberinlike compound, whereas the invasive infection threads were lignified as well as suberized. Becking,[11] in his 1979

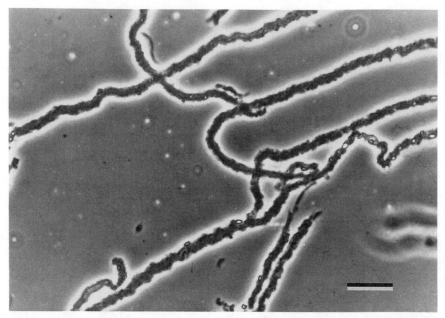

Figure 13–14. Freed peribacteroid "sacs" in squashes of *Parasponia parviflora* nodules demonstrating that these sacs are independent structures. The larger filamentous structures are composed of many individual sacs, as these structures easily stick together. The smaller sacs are probably the more simple structures.
Transmission electron micrograph; bar marker = 10 μm.

paper, refrained from calling the intracellular threads observed in sections of *Parasponia parviflora* root nodules, infection threads. In the previously mentioned article I only indicated them as "intracellular threadlike structures," because I had another opinion of their function. Additional light, transmission, and scanning electron microscopic work on *Parasponia parviflora* root nodules (Becking, unpublished data) have only strengthened my view, which clearly deviates from the current opinions on infection and fixation threads in root nodules of *Parasponia*[75,86,100,105] and some legumes.[28,29,90]

The two types of infection threads are, according to my opinion, completely different structures. The so-called fixation thread cannot be called a form of an infection thread. The observed solid, thick-walled thread structures (Figures 13–11 and 13–15) are the true infection threads. They originate in the intercellular spaces as newly formed ingrowths of the host cell wall, and as in legumes, they accompany the bacteria on their way through the root nodule. The infection thread wall, as in legumes,[69,90] clearly is of plant origin; in staining reactions they fuse completely with the host cell wall. These infection threads, starting from intercellular spaces containing

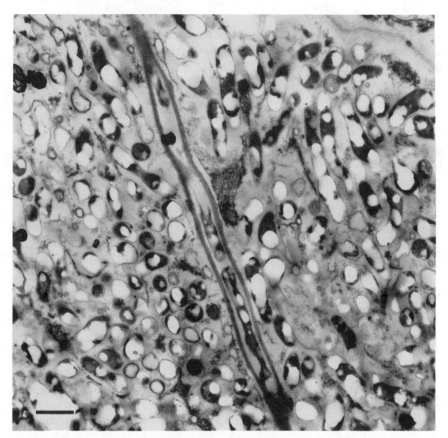

Figure 13–15. A true (invasive) infection thread in a senescent host cell of *Parasponia parviflora*. Note the degraded peribacteroid membranes of the sacs liberating the bacteria, and the persistent thick-walled infection thread with a single row of bacteria.
Transmission electron micrograph; bar marker = 1 μm.

rhizobia bacteria, traverse the nodule in all directions intracellularly and intercellularly, bringing the bacteria to the infection zone(s). These true invasive infection threads are rather persistent in the root nodule tissue, as they also can be observed in senescent nodule parts. Their lignification probably prevents their enzymatic degradation and thus protects the bacteria inside the threads from lysis (see Figure 13–15). These true infection threads are continuously connected with each other, and in *Parasponia* the rhizobium bacteria inside the infection thread are always lined up in a single row (see Figure 13–15).

The other type of infection thread is generally called a fixation thread. This thread type is clearly thin-walled and is confined to mature host cells. These structures are formed or proliferate within the living host cell, and usually the cell is completely packed with such structures. According to our observations, these threads are discontinuous from each other (i.e., they always are separate, independent structures). Moreover, they are unable to traverse the primary host cell walls. We could not confirm Trinick's[100] observation that the lightly stained infection threads (i.e., the fixation threads) can pass through host cell walls and are interconnected with the thick-walled infection threads. In our opinion the intracellular fixation threads represent membrane-bound structures identical to the peribacteroid membrane that enclose a few to numerous bacteroid cells in some legume associations. This membrane has been demonstrated to be of plant origin, but it may incorporate elements of the rhizobial surface.[17] In *Parasponia* these membrane envelopes, enclosing a number or several rows of bacteria, often are stretched in one direction, so that they form threadlike structures (see Figures 13–13 and 13–14). In *Parasponia* root nodules these membranes contain far more plant-originating material than is found in the peribacteroid membranes of legumes. In *Parasponia* the membranes may include cell wall–like deposits such as polysaccharides, suberinelike compounds, and lipids.[86] Since these membrane envelopes enclosing bacteroids are mainly plant determined, it is clear that the same *Parasponia–Bradyrhizobium* strain inoculated on legumes (*Vigna unguiculata* spp., cowpea) will produce in these nodules only thin-walled membrane envelopes (Figure 13–16) characteristic of legumes.

According to our observations (Becking, in preparation), the rhizobial cells escape from the true invasive infection threads to invade very young host nodule cells. At this stage the rhizobium cells are not yet surrounded by a peribacteroid membrane, but this membrane envelope develops gradually around the bacterial cells and is further consolidated by cell wall material simultaneously with the change of bacteria to bacteroids. In their formation, membrane envelopes often enclose bundles or rows of rhizobial cells. In a mature *Parasponia* host cell, these membrane envelopes have reached their maximal development (Figure 13–17) and concomitantly the bacteroids become very active in N_2 fixation. In addition, the electron micrographs show the presence of numerous poly-β-hydroxybutyrate globules within the bacteroids, which is an indication that these cells are nitrogen rather than carbon limited. As at the very young infection stages, and in senescence of the *Parasponia* root nodules, the peribacteroid membranes may be absent (or disintegrating in senescent tissue) from the host cytoplasm (see Figure 13–15). The released bacteria and the membrane structures become available for resorption by the plant. As is clear from the electron micrographs (see Figure 13–15), the true invasive infection thread is far more resistant to these activities than all other structures within the host cell.

Root nodule morphogenesis studies of root nodules of *Parasponia* have indicated that they show closer resemblance to root nodules induced by *Frankia* species on nonlegumes than to *Rhizobium*-induced root nodules of legumes.

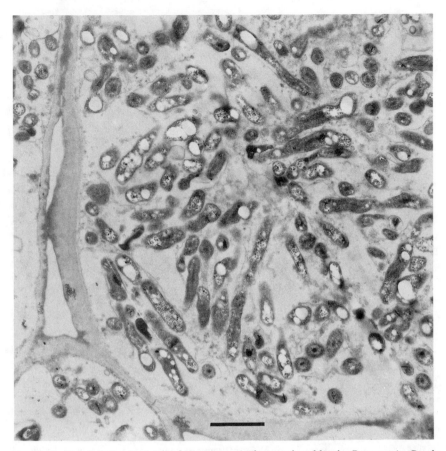

Figure 13–16. Infected host cell of *Vigna unguiculata* produced by the *Parasponia–Brady-rhizobium* endophyte. Visible are the thin peribacteroid envelopes enclosing a few to many rhizobium cells.
Transmission electron micrograph; bar marker = 2 μm.

The nonlegume actinorhizal root nodules are modified lateral roots with multiple regions of apical meristematic activity. The elongate, cylindrical nodule is indeterminate and has a centralized vascular system. In contrast to this, legume nodules have a peripheral vascular system incorporated into the root nodule cortex. Particularly, tropical legumes often show a nodule morphology that is spherical, determinate, and has peripheral vascular strands. Such a type of nodule also is developed on the stems of *Sesbania* spp.[25,34,37,38] These nodules develop from infections at stem sites, where adventitious, preformed root primordia may emerge. The vascular bundle of the protruding root primordium is incorporated into the nodule, but in subsequent growth

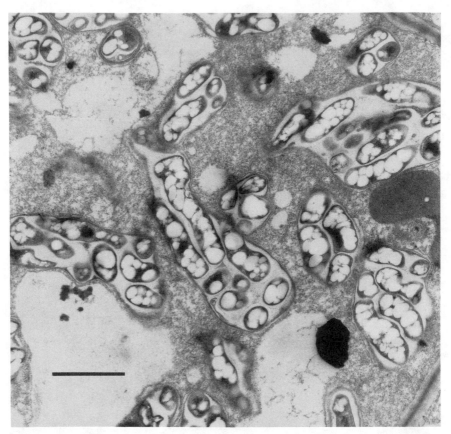

Figure 13–17. Mature *Parasponia parviflora* infected host cell. The bacteroids are in groups enclosed by a double membrane envelope of which the inner one is consolidated with cell wall material. The membrane sacs are surrounded by active cytoplasm and numerous vacuoles. The bacteroids contain numerous poly-β-hydroxybutyrate globules, indicating that they are nitrogen limited rather than carbon limited.
Transmission electron micrograph; bar marker = 2 μm.

the central vascular bundle becomes arrested and gives rise to peripheral vascular strands, as are found in "typical" tropical legume nodules.

According to our observations, the *Parasponia* root nodule morphology and morphogenesis distinguish them from actinorhizal root nodules in several points. They appear to have a more rhizobial structure than recognized up to now. They probably are related as well to typical determinate nodules of some typical legumes as to the indeterminate nodules of some tropical woody legumes. Nodules of the latter type have an apical nodule meristem and therefore exhibit indeterminate growth. Moreover, they are perennial. Unfortunately, knowledge on the morphology and morphogenesis of the pe-

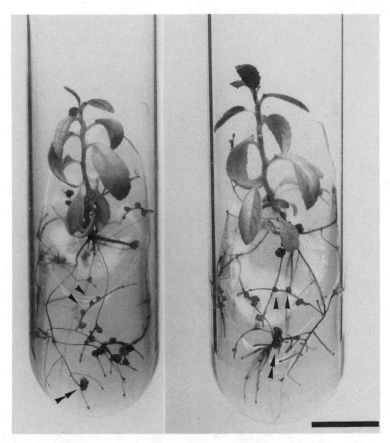

Figure 13–18. *Parasponia parviflora* seedlings on agar culture showing numerous globose nodules of determinate growth (arrowheads) and a few elongated nodules with indeterminate growth (double arrowhead).
Bar marker = 1 cm.

rennial nodules of woody legumes is scanty, except for an old publication by Spratt.[88] The nodules of some tropical woody legumes, especially those in the subfamilies Ceasalpinioideae and Papilionoideae, may be important for comparison with *Parasponia* nodules.

Our observations have indicated that *Parasponia* plants may develop two types of nodules: smaller root nodules with a determinate growth pattern (Figure 13–18) and larger root nodules with an indeterminate growth pattern. Both nodule types can be found on the same plant (Figure 13–19). Although not usual, they can be initiated by the same *Bradyrhizobium* strain. The smaller *Parasponia* root nodules may emerge from a kind of wound callus and then often show a reduced development of the central vascular

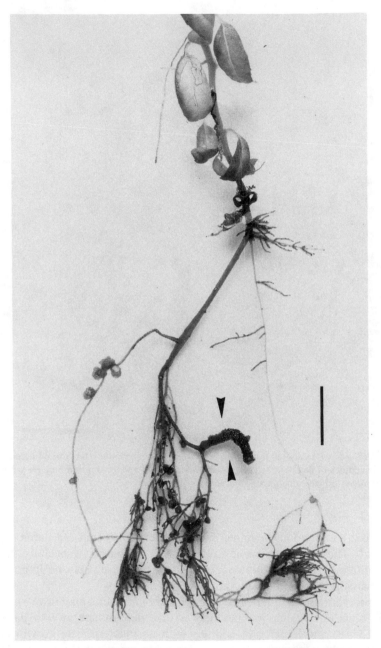

Figure 13–19. Parasponia parviflora seedling at more advanced stage of growth showing numerous determinate globose nodules and one large cylindrical nodule with indeterminate growth (see arrowhead).
Bar marker = 1 cm.

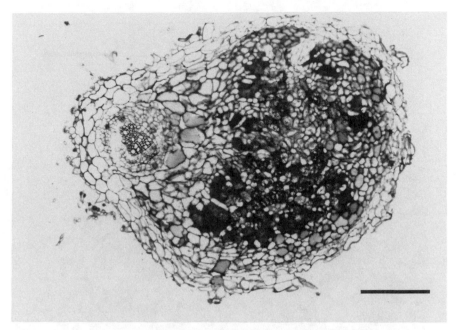

Figure 13–20. Transverse section of a determinate *Parasponia* nodule perpendicular to the supporting root. It shows the vascular bundle of the supporting root, the nodule epidermis and cortical cells, but also a more central zone of infected cells. The dark-stained cells at the periphery of the nodule are host cells filled with tannin deposits.
Light micrograph: Erythrosin-stained preparation.
Bar marker = 200 μm.

strand (Figure 13–20). They show arrested growth, and because the meristematic region is more spherical, they give rise to oblate or spheroid nodules. Such round nodules are found in *Arachis* (Aeschynomenoid type[19,20]) and in *Desmodium* (Desmodioid type[19,20]). The formation of a "prenodule" in the sense of that described by Becking[9] for actinorhizal symbiosis is absent. These nodules form a more or less calluslike outgrowth resembling true determinate legume root nodules. Their internal organization also is reminiscent of legume root nodules, as uninfected cells, interstital cells, hypotrophic cells containing the rhizobium endophyte, and cells with tannin deposits are regularly arranged (Figure 13–21).

The other type of root nodule resembles the nodules of tropical legumes with an apical meristem and therefore indeterminate growth; it produces a basically enlongated cylindrical root nodule with a potential bifurcation and branching. These nodules are always perennial and are found, for instance, in tropical legumes such as *Crotalaria* and *Acacia* species (Crotalarioid type[19,20]). This type of *Parasponia* root nodule (Figure 13–22) differs from the indeterminate tropical legume nodules because its vascular system is cen-

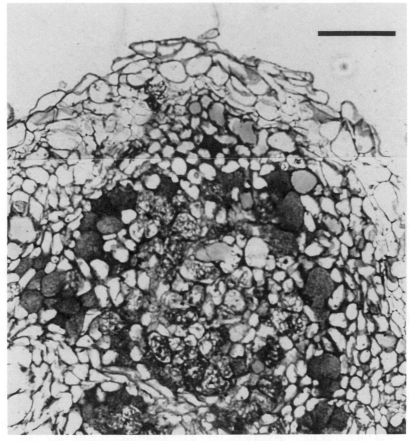

Figure 13–21. Detail of a transverse section of a globose *Parasponia* root nodule with determinate growth. The picture gives a part of the outer section of the root nodule with the epidermis, a row of cells containing tannin deposits and central host cells filled with the so-called fixation threads interspersed with host cells without bacteria.
Light micrograph; Erythrosin-stained preparation.
Bar marker = 100 µm.

tral, rather than peripheral as in legumes. Moreover, its morphogenesis is different from actinorhizal root nodules, as a clear prenodule is not formed. The meristematic proliferation of cells within the cortex is at least partly incorporated into the developing nodule. The apical meristem and the central vascular tissue of this indeterminate nodule is derived from a lateral or adventitious root originating in the pericycle of the main root. It may be from a preformed, dormant root primordium, but it is also possible that it is initiated later by signals from the developing root nodule. The latter may be possible, since in some cases we could not find a dormant root primordium

Figure 13–22. Nodulated *Parasponia parviflora* plant and nodule types of indeterminate nodules. (A) Young *Parasponia* plant with small cylindrical determinate nodules. (B) Large (older) coralloid *Parasponia* nodule showing large lenticels at the surface. (C) Enlarged nodule cluster showing new nodular tissue development at the nodule tips after a period of arrested growth.
Bar marker in A = 4.0 cm; in B = 1.0 cm; in C = 0.5 cm.

in the very early stages at sites of nodule initiation, but in later developmental stages it always was present. In these indeterminate root nodules of *Parasponia* the vascular system can be bifurcated or branched several times, and in this way it gives rise to the coralloid root nodules that we observe in field and pot experiments (see Figure 13–22). Perennial coralloid nodules or branched cluster nodules are known for many tropical woody legumes (e.g., in *Swartzia trinitensis* of the Caesalpinioideae[3]).

As shown in *Parasponia*, the two types of root nodules may occur on a single plant (see Figure 13–19). Two types of nodules on the same plant have been reported for lupines (*Lupinus* species), where collar-shaped or girdling nodules are formed on older, thicker roots, and elongate, cylindrical nodules are formed on the finer roots.[22,23,88] In relation to *Parasponia*, it is of special interest that two types of root nodules have also been reported for the woody legume *Prosopis glandulosa* (Mesquite, Mimosoideae), where typical elongate nodules and smaller spherical nodules resembling those of cowpea occur together on the same plant. The spherical root nodules were, however, shown to be ineffective.[112] In our *Parasponia* experiments both nodule types contain the so-called infection threads in the host cells, and both types have the capacity to fix atmospheric N_2 (see Figure 13–21). In general, the nitrogenase activity of *Parasponia* nodules is very variable (see Section VI). Particularly the small spherical nodules had variable and often low nitrogenase activity, sometimes even being ineffective, whereas the cylindrical indeterminate nodules showed overall a much higher activity. It is not clear whether the oblate–spherical nodules are ephemeral, but because of the poor N_2-fixing capacity of the root nodules, these *Parasponia* plants often developed N-deficiency symptoms (yellow leaves and stunted growth; see Figure 13–18) and shortened life time.

De Faria et al.[27-29] recently reported the presence of persistent fixation threads in the root nodules of *Andira* spp. (Leguminosae, subfamily Papilionoideae), which in morphologic appearance are very reminiscent of the persistent fixation threads found in *Parasponia* root nodules. In addition, these persistent fixation threads have been reported for a number of other genera of woody legumes belonging to the Caesalpinioideae.[28] According to us, the described cytology of the cells is very similar to that observed in *Parasponia* root nodules; besides the persistent fixation threads, true infection threads spreading from the intercellular spaces have been observed. As in *Parasponia*, we consider the fixation threads of *Andira* and related woody legumes to be a special type of peribacteroid membrane supplemented with cell wall material. In conclusion, any relationship of nodule structure and morphogenesis of *Parasponia* must be sought in the nodule development of the woody legumes rather than in the actinorhizal symbioses of *Frankia* with nonlegumes.

VI. Physiology of the Symbiosis

A. *Nitrogen Fixation of Parasponia–Bradyrhizobium in Parasponia*

The N_2-fixation ability of *Parasponia* root nodules is well established. Field experiments on Java, where *Parasponia parviflora* was tested under ambient conditions (24–25° C), indicated N_2-fixation rates varying from 0.7 to 15.2 μmol C_2H_2 g^{-1} fresh weight nodules h^{-1} as measured directly after nodule detachment. Our figure[11] for detached field nodules of 0.7–0.8 μmol C_2H_4 g^{-1} fresh weight (or 3.0–4.1 μmol C_2H_4 g^{-1} dry weight) nodule tissue h^{-1} is about 20 times lower than the maximal figure of 15.2 μmol C_2H_4 g^{-1}

fresh weight nodules h^{-1} presented by Akkermans et al.[2] for the same *Parasponia* species tested under similar conditions. The latter value is, however, consistent with N_2-fixation values of 11.5–11.7 µmol C_2H_4 g^{-1} nodule fresh weight h^{-1} found by Trinick[98] with intact plants of *Parasponia rugosa* (*"Trema canabina"*) in pot experiments with Leonard jar assemblies. However, in the latter experiments nodule detachment reduced the N_2-fixation activity of the root nodules to about half (Table 13–1). The great variability in nitrogenase activity is probably caused by age and size differences of the nodules tested. Small (~3 mm) young nodules have the highest activity, and we tested the larger field nodules about 1.5–2.0 cm in diameter (see Figure 13–22). As in actinorhizal root nodules,[10] most of the nitrogenase activity is located in the nodule section below the nodule tip and a little more downward, which nodule part contains host cells completely filled with mature bacteroids (so-called fixation threads). Trinick[100] measured the highest nitrogenase activities for this region as compared to other nodule parts; the nodule base had a nitrogenase activity of either zero or very low. However, the maximal nitrogenase activities as measured in 2-mm transverse segments of this most active region were rather low i.e., 20.1–24.0 nmol C_2H_4 min^{-1} g^{-1} fresh wt. nodules (=1.2–1.4 µmol C_2H_4 h^{-1} g^{-1} fresh wt.) as compared to other N_2-fixation values found for intact plants or *Parasponia* root nodules immediately after detachment (see Table 13–1).

Parasponia plants can be cultivated under various culture conditions. We grew *Parasponia parviflora* plants in water culture, in perlite or soil culture, and in glass tubes on agar slants.[12] Nitrogen fixation values of the nodules from these plants showed a great variability, caused probably by the rhizobium strain used as inoculum as well as by the plant environmental conditions. In water culture the nitrogenase value of the *Parasponia* root nodules was generally very low, averaging 0.5 ± 0.2 µmol C_2H_4 g^{-1} fresh weight nodules h^{-1}. But this result reflects the high water content of these nodules (~90–91%), since on a dry weight basis the mean value is 5.6 ± 2.2 µmol C_2H_4 g^{-1} dry weight nodules h^{-1}. On perlite culture the *Parasponia* plants produced good to reasonable growth, and testing the N_2-fixation activity of detached root nodules of 6-, 8-, and 9-month-old plants yielded average values of 5.1, 6.4, and 3.5 µmol C_2H_4 g^{-1} fresh weight h^{-1}, respectively.[12] However, the standard deviation of the nitrogenase activity of the individual nodules in the various separate series and replicates was substantial. For instance, in the 8-month-old plants a nitrogenase activity as high as 25.9 and 28.2 µmol C_2H_4 g^{-1} fresh weight nodule h^{-1} has been measured. The nitrogenase activity of the nodules of 9-month-old plants in another experimental series was less variable, and excluding some ineffectively nodulated plants, N_2-fixation rates of 9.7–11.0 µmol C_2H_4 g^{-1} fresh weight nodule h^{-1} were found.

Table 13–1. Nitrogen Fixation (C2H4) of Parasponia spp. Root Nodules from Various Sources

Species	Culture Condition	Nodule Size (Length) mm	μmol C$_2$H$_4$ g^{-1} Nodule Fresh Wt h^{-1}	μmol C$_2$H$_4$ g^{-1} Nodule Dry Wt h^{-1}	Author
P. rugosa	Greenhouse/pot experiments				
	Intact plants	—	11.5–11.7	—	Trinick[98]
	Detached nodules		4.6	—	Trinick[98]
P. parviflora	Field-collected detached nodules	3	15.2 ± 0.7	—	Akkermans et al.[2]
		10–30	1.8 ± 0.8	—	
P. parviflora	Field-collected detached nodules	15–20	0.7–0.8	3.1–4.1	Becking[11]
P. andersonii	Greenhouse/light cabinet. Intact plants in glass tube agar culture				
	with Bradyrhizobium strains of varying effectiveness	—	1.13–5.90[a]	—	Trinick and Galbraith[106]
	with Rhizobium strains	—	0.64–0.79[a]	—	Trinick and Galbraith[106]
	in detached nodules	—	0.27–0.52[a]	—	Trinick and Galbraith[106]
P. parviflora	Greenhouse/light cabinet				
	Intact pot plants in soil	3–8	10.6–12.8 av. 11.9 ± 1.1	55.8–67.3[b] av. 62.4 ± 6.0	Becking[13]
	Intact glass tube agar culture	2–3	1.5–5.2 av. 3.7 ± 1.4	8.8–30.5[b] av. 21.9 ± 8.1	Becking[13]
	Cultures inoculated with Bradyrhizobium strains of various effectiveness				

[a]Calculated from nmol C$_2$H$_4$ g^{-1} fresh weight nodule min^{-1} values.
[b]Calculated from data that these nodules contain 81% water in soil culture and 83% in agar culture.

Apart from genomic variability of the *Parasponia* plants (seed propagation), which may cause differences in nitrogenase activity, the effectiveness of the rhizobium isolates may influence this variability. Tests on five different *Parasponia–Bradyrhizobium* strains on 6-month-old *Parasponia parviflora* seedlings grown on perlite gave N_2-fixation values of the root nodules from about 9.0 μmol C_2H_4 g^{-1} fresh weight nodule h^{-1} to considerably lower values.[12]

Soil cultures of *Parasponia* generally gave plants with the best and most homogeneous nitrogenase activities, especially when intact plants were tested. Nitrogenase values of approximately 11.9 ± 1.1 μmol C_2H_4 g^{-1} fresh weight nodule h^{-1} have been measured, but detachment of these root nodules usually reduced the nitrogenase activity to about 2.0–2.9 μmol C_2H_4 g^{-1} fresh weight nodule h^{-1}.[12]

Plants cultivated in closed test tubes on mineral–agar slants provide a favorable test material under aseptic conditions for evaluating effectiveness of various *Bradyrhizobium* and *Rhizobium* strains on *Parasponia*. Moreover, these cultures are handy for nitrogenase activity measurements *in situ*; the cotton plug is replaced by a rubber stopper (Suba seal), and an appropriate volume of C_2H_2 is added to the gas atmosphere. After the experiment the plant can be harvested and the nodule weight determined. Experiments with 2.5-month-old seedlings, 1.5 month after inoculation, indicated nitrogenase activities as measured in the intact plant of 3.6–5.2 μmol C_2H_4 g^{-1} fresh weight nodules h^{-1}, and in another series values of 1.5–5.2 μmol C_2H_4 fresh weight nodules h^{-1} (see Table 13–1). However, individual root nodules of these plants, detached and tested separately, gave very variable results. Some of these nodules showed a nitrogenase activity of 0.7–1.6 μmol C_2H_4, indicating poor activity; some had no activity at all; and other nodules had an activity of about 4.9–15.0 μmol g^{-1} fresh weight nodule h^{-1}, indicating good activity. Although the individual plants were inoculated with a single rhizobium isolate, the nitrogenase activity of the nodules obtained from the same plant varied considerably.

The explanation for these results may be that the plant material is heterogenous (seed propagation), the nodule age may be different, the bacteria strains may show differences in effectiveness, and the nodule type produced (see earlier discussion) may influence the outcome of the nitrogenase activity.

In similar glass tube experiments but with *Parasponia andersonii*, Trinick[101] also observed a great variability in nitrogenase activity among the various plants and strains tested. Moreover, many strains proved to be ineffective. According to his findings the highest nitrogenase activity on a plant or nodule weight basis did not necessarily correspond with the most effective strains of rhizobia expressed on a plant top dry weight basis, possibly because of irregular nodule degeneration and/or variation in peak enzyme activity and its duration. Otherwise highly effective strains often had lower nodule weight than less effective strains, whereas other highly effective strains had higher nodule mass. The rates of C_2H_2 reduction also varied greatly among the strains. The highest values given were 90.5 nmol C_2H_4 g^{-1} fresh weight nodule min^{-1} and 171 nmol C_2H_4 g^{-1} fresh weight nodule min^{-1}, which correspond to values of 5.4 μmol and 10.3 μmol C_2H_4 g^{-1} fresh weight nodule h^{-1}, respectively. These values are of the same order of magnitude as our results.[12]

B. Nitrogen Fixation of Parasponia–Bradyrhizobium in Legumes

Trinick and Galbraith[106] tested 15 isolates of *Parasponia* for nodulation and N_2 fixation effectiveness in legumes. Only one strain from *Parasponia* (CP283) produced greater plant top dry weight than the corresponding control strain on *Macroptilium lathyrioides, Lablab purpureus, Flemingia congesta,* and all *Vigna* species. However, when CP283 was compared with 11 rhizobia from tropical legumes on *Macroptilium atropurpureum,* it was intermediate in effectiveness. The other 14 strains of rhizobia isolated from *Parasponia* were less successful. They were able to nodulate *Vigna* spp., *Macroptilium atropurpureum,* and sometimes *M. lathyrioides* and *Flemingia congesta,* but the associations were far less effective and often supported no N_2 fixation.

Becking[12] studied the effectiveness of several *Parasponia* isolates on *Vigna unguiculata* in perlite culture and in sterilized soil and on *Parasponia parviflora,* its original host plant. The nodules of 2-month-old *Vigna unguiculata* plants in perlite culture had an average nitrogenase activity of 14.3 μmol C_2H_4 g^{-1} fresh weight root nodule h^{-1}, whereas the *Parasponia* plants of the same age had a mean activity of 3.2 μmol C_2H_4 g^{-1} fresh weight nodule h^{-1}. Soil cultures showed the same general pattern. The nodules of 3-month-old *Vigna* plants had an activity of 10.4 μmol C_2H_4 g^{-1} fresh weight nodule h^{-1}, in contrast to *Parasponia* nodules of plants of the same age with 2.0 μmol C_2H_4 g^{-1} fresh weight nodule h^{-1}. Relatively effective rhizobium isolates of *Parasponia* on *Parasponia parviflora* (e.g., strains Pp31) always were less effective in N_2 fixation on *Parasponia* in comparison with legumes (e.g., *Vigna* spp.).[12]

Similar results are reported by Price et al.[75]; they found that *Parasponia–Bradyrhizobium* isolate ANU289 produced N_2-fixing nodules on both *Macroptilium atropurpureum* (Siratro) and *Parasponia rigida.* Fixation rates as measured in detached, moistened nodules with the C_2H_2 reduction assay showed on average in siratro an activity of 16.0 and in Parasponia 8.9 μmol $C_2H_4 \cdot$ g^{-1} fresh weight nodule h^{-1}. In this case the *Macroptilium* nodules were measured 6 weeks after inoculation and the *Parasponia* root nodules

Table 13–2. Hydrogen Evolution and Relative Efficiency of Parasponia parviflora Root Nodules

Root Nodules	H_2 Evolution μmol H_2 g^{-1} Fresh Weight h^{-1}	C_2H_2 Reduction μmol C_2H_4 g^{-1} Fresh Weight h^{-1}	Relative efficiency (a)*	(b)*
Soil culture				
(Ar:O_2)	4.470 ± 0.152	7.2 ± 1.4	0.96	0.97
(Air)	0.187 ± 0.049	2.1 ± 0.1		
Water culture				
(Ar:O_2)	0.53 ± 0.02	0.399 ± 0.120	0.72	0.62
(Air)	0.15 ± 0.01	0.420 ± 0.06		

*(a) RE = 1 − (H_2 evolved in air ÷ H_2 evolved in argon).

(b) RE = 1 − (H_2 evolved in air ÷ C_2H_2 reduced).

were measured after 17 weeks. Such *Parasponia* nodules were considered to be of similar developmental, but not chronological age.

C. Hydrogen Production and Relative Efficiency of Nitrogen Fixation in Parasponia

Nitrogenase-dependent H_2 evolution and relative efficiency of N_2 fixation were studied in detached root nodules of 5-month-old *Parasponia* plants grown in soil or in water culture. The two types of nodules were chosen, as previous tests have revealed (see foregoing) significant differences in N_2-fixation rates in the two types of nodules. Hydrogen production was monitored under two conditions—in air and in an atmosphere of argon (A) with 20% (v/v O_2).[13] For calculation of the relative efficiency, the formula suggested by Schubert and Evans[77] has been used.

The results presented in Table 13–2 show that the relative efficiency of soil-grown root nodules was high (96% and 97% for the two equations) and that the relative efficiency of water culture–grown nodules was distinctly lower, with 72% and 62% relative efficiency.

D. $^{15}N_2$-Fixation Experiments with Parasponia

Since all N_2-fixation studies with *Parasponia* are based on the indirect C_2H_2 reduction assay, direct isotope $^{15}N_2$ studies for confirmation of N_2 fixation are desirable, and such data can supply the conversion ratio for $C_2H_4:N_2$. This factor is indispensable for the assessment of the N_2-fixation capacity of *Parasponia* in field studies.

Detached soil-grown and water-culture-grown *Parasponia parviflora* nodules were exposed to an atmosphere of N_2 (10%, v/v), argon (70%), and

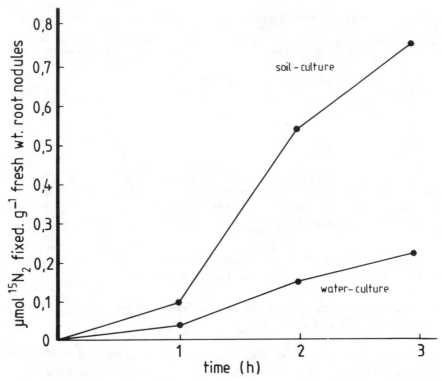

Figure 13–23. Time course of $^{15}N_2$ fixation in detached root nodules of *Parasponia parviflora* plants grown in soil or water culture.

O_2 (20%), in which the N_2 gas was labeled with 60 atom % ^{15}N. The results presented in Figure 13–23 show that, as in the C_2H_2-reduction tests, the soil culture root nodules were two to three times more active in N_2 fixation than the water culture–grown *Parasponia* root nodules.

The nitrogenase activity of *Parasponia* root nodules was earlier observed to be very variable and nodule dependent. Therefore, for an accurate determination of the conversion factor, both the C_2H_2 reduction and the $^{15}N_2$ fixation should be measured in the same material under identical conditions (25° C) and as quickly as possible in succession. Because exposure to C_2H_2 may induce inhibitory effects on subsequent N_2 fixation, the $^{15}N_2$ tests were performed before the C_2H_2 reduction assay. For the C_2H_2 reduction tests, 10% C_2H_2, v/v in air was used, whereas for the $^{15}N_2$ experiment a synthetic gas mixture was employed (see earlier) containing N_2 with 50 atom % ^{15}N excess. For comparison, effective root nodules of *Pisum sativum* were included in the test. Table 13–3 shows the mean conversion ratio $C_2H_4:N_2$ was 6.7 in *Parasponia*, whereas in *Pisum* root nodules it was 10.0.

Table 13–3. $^{15}N_2$ *Fixation and Determination of Conversion Factor* C_2H_4/N_2 *of Parasponia parviflora Root Nodules*

Species	No. Exp.	$\%^{15}N$ in Sample	N-total Sample (mg)	Fresh Weight Sample (g)	μmol $C_2H_4 \ h^{-1} \ g^{-1}$ Fresh Weight	μmol $N_2 \ h^{-1} \ g^{-1}$ Fresh Weight	Conversion Ratio C_2H_4/N_2
Parasponia parviflora	1	0.40	7.17	1.146	1.130	0.143	7.9
	2	0.38	11.88	1.968	0.209	0.052	4.0
	3	0.43	10.51	1.898	1.937	0.245	7.9
	4	0.45	9.39	1.650	3.062	0.333	9.2
	5	0.38	11.68	1.945	0.651	0.147	4.4
							av. 6.7 ± 2.3
*Pisum sativum**	6	0.56	2.70	0.681	4.693	0.544	8.6
	7	0.45	2.63	0.751	2.336	0.205	11.4
							av. 10.0 ± 2.0

**Pisum sativum* root nodules included for reference.

Note: The experiments were conducted in an atmosphere of Ar 70%, O_2 20%, N_2 10%. The nitrogen gas contained 50 atom $\%^{15}N$. Exposure periods for $^{15}N_2$ experiments 60 min, for the C_2H_2 experiments 30 min. Natural atom $\%^{15}N = 0.3680$. The C_2H_2 assays were conducted on nodules after the measurements of the $^{15}N_2$ fixation.

E.   Natural ^{15}N Abundance of Parasponia Root Nodules

The ^{15}N abundance of fixed N in the aerial portion of many plants was found to reflect closely the atmospheric N_2 within 2 $\delta^{15}N$, but nodules of soybean and some other legumes were observed to be enriched in ^{15}N compared to other tissues.[52,82,113] The magnitude of the enrichment was strongly correlated with the effectiveness of the symbiosis in soybeans[53] and some other nodulated legumes.[112] Evidently, some isotopic fractionation associated with N_2 fixation or assimilation of fixed N within the nodule causes the nodule to become uniquely enriched in ^{15}N compared to the rest of the plant.

To determine whether a similar relationship in ^{15}N enrichment exists in *Parasponia*, we measured the ^{15}N abundance (the per mil ^{15}N abundance compared to atmospheric N_2) of leaves and nodules of 6-month-old *Parasponia parviflora* plants grown in soil. The plants were cultivated in pots in a climatic chamber at a constant temperature of 27° C and at a light intensity of 30,000 lux (12 h light and 12 h dark).

Table 13–4 shows that the nitrogen content of the *Parasponia* nodules was about two times higher than that of the leaves. The $\delta^{15}N$ abundance compared to atmospheric N_2 was observed in the root nodules to be significantly higher than in the leaves. The $\delta^{15}N$ value of the leaves was 0.25, whereas that of the nodules was 5.20. This outcome is consistent with the $\delta^{15}N$ abundance values of effective N_2 fixation in legumes.[53,112]

Table 13–4. The Natural ^{15}N Abundance and $\delta^{15}N$ Determinations of Leaves and Nodules of Parasponia parviflora Grown in Soil

Material	No. of samples	% N	^{15}N atom %	$\delta^{15}N$ 29‰
Parasponia parviflora				
Leaves	5	2.75 ± 0.21	0.3663	0.25
Nodules	8	4.01 ± 0.04	0.3681	5.20

F. Oxygen Relations and Hemoglobin in Parasponia Root Nodules

With infiltration studies Tjepkema and Cartica[95] demonstrated that intercellular spaces are lacking in the inner layers of the nodule cortex of Parasponia rigida. Oxygen measurements with a microelectrode indicated an O_2 diffusion barrier in the inner cortex, since the O_2 partial pressure in the zone of the infected cells was much lower than that in the cortex. Also measurements of nitrogenase activity and O_2 uptake as a function of temperature and partial pressure of O_2 were consistent with diffusion limitations of O_2 uptake by the inner cortex. In contrast, actinorhizal root nodules have an interlacing network of air spaces within the cortex, which India ink infiltrations shows are directly connected to the soil atmosphere.[93] Here the mechanical diffusion barrier for O_2 is the multilaminate vesicle envelope of the Frankia endophyte.[84,96,97,117] At that time, Parasponia nodules were presumed to lack hemoglobin compounds,[21] but the observed energy usage for N_2 fixation was found to be similar to that in legume root nodules. This indicated that O_2 regulation in legume and Parasponia is very similar but differs from O_2 regulation in actinorhizal root nodules.[93,117,118]

Hemoglobins were demonstrated later in Parasponia root nodules by Appleby et al.,[6] and older claims[24] of the existence of hemoglobins in actinorhizal root nodules were confirmed.[94] The Parasponia hemoglobin was found to be different from cowpea hemoglobin, although both could be induced by the same Bradyrhizobium strain (CP283). This indicates that the plant has control over the type of hemoglobin that is produced in the nodules and that the bacterial strain has less influence. The purified Parasponia hemoglobin was observed by Appleby et al.[6] to contain only one subunit type of about 21,000 D in Parasponia andersonii, whereas Gresshoff et al.[43] using related techniques isolated from Parasponia rigida (infected with ANU289) a hemoglobin of molecular weight ~19,000.

In experiments with Parasponia andersonii, Trinick[102] found that the pO_2 optimum for nitrogenase was narrow in contrast to that in legumes but similar to that of actinorhizal systems. The nitrogenase activity fell rapidly above and below 0.2 atm O_2; restoration of nitrogenase (C_2H_2) activity after inhibition by 0.4 and 0.9 atm pO_2 was slow.

Table 13–5.   *Effect of O_2 Concentration on the Nitrogenase (C_2H_2) Activity of Excised Parasponia parviflora Root Nodules*

O_2 Concentration	μmol C_2H_4 g^{-1} Fresh Weight h^{-1} in $(Ar + O_2)$*	μmol C_2H_4 g^{-1} Fresh Weight h^{-1} in $(Air)^\dagger$	% $(Ar + O_2) \div (Air)$	Mean % $(Ar + O_2) \div (Air)$
3%	0.087	4.167	2.09	
	0.157	6.983	2.25	1.95 ± 0.39
	0.093	6.147	1.51	
10%	0.690	2.830	24.38	
	0.429	2.329	18.42	21.40 ± 4.22
20%	3.877	4.222	91.83	
	3.264	6.285	51.93	80.76 ± 25.19
	3.178	3.226	98.51	
30%	14.841	8.458	175.47	
	15.286	5.923	258.08	216.77 ± 58.42
40%	0.024	5.654	0.42	
	10.357	4.602	225.05	91.32 ± 118.28
	1.182	2.437	48.50	

*$(Ar + 10\%$ (v/v) $C_2H_2) + $ variable concentrations of O_2.

$^\dagger(Air + 10\%$ (v/v) $C_2H_2)$.

Each experiment lasted 1.5 hour [i.e., in succession 30 min (air), 30 min ($Ar + O_2$), and 30 min (air)].

Our two types of experiments with *Parasponia parviflora* on N_2 fixation at different partial pressures of O_2, however, gave different results. Because of the variable nitrogenase activity of *Parasponia* root nodules, the definitive measurement at a particular pO_2 pressure was made after and before measurements under standard conditions (i.e., nitrogenase activity in air with 0.1 atm C_2H_2). By this method, the relative effect of the pO_2 as compared to measurements under standard conditions can be determined. All partial O_2 pressures were checked with a Packard Model 427 gas chromatograph equipped with a thermal conductivity detector (Katharometer). Table 13–5 shows that the maximal nitrogenase (C_2H_2) activity of *Parasponia* root nodules was observed at a pO_2 of 0.30 atm, and at a pO_2 of 0.40 atm the nitrogenase activity was reduced to about the same value as that found in air (pO_2 0.21 atm). At lower partial pressures of O_2 (e.g., 0.03 and 0.10 atm O_2), the nitrogenase activity was markedly reduced.

In another type of experiment, detached root nodules of 9-month-old *Parasponia parviflora* plants (8 months after inoculation) were tested in 40– ml serum bottles containing 1.0–1.5 g fresh weight nodule tissue at 24° C. Nodules were exposed to O_2 concentrations of 0.045, 0.09, 0.14, 0.18, 0.27, and 0.36 atm O_2 (checked by gas chromatography). Although the N_2-fixation values show a great variability in the individual tests (see dots in Figure 13–

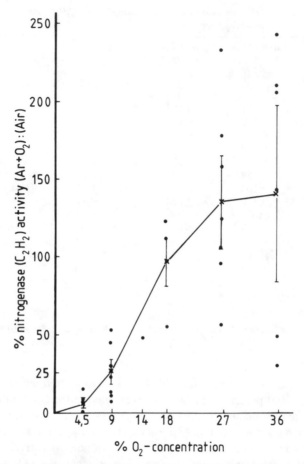

Figure 13–24. Effect of partial O_2 pressure on N_2 fixation (C_2H_2 reduction) of detached root nodules of *Parasponia parviflora* plants grown in soil culture.

24), which is also reflected in the large standard deviations, the results indicate that nitrogenase (C_2H_2) activity of the *Parasponia* nodules increases to a pO_2 of 0.36 atm. However, above a pO_2 of 0.40 atm, the nitrogenase is gradually inhibited, as shown in Table 13–5.

To have an insight into the energy requirement of the nitrogenase complex, we studied the O_2 production and the CO inhibition in relation to nitrogenase (C_2H_2) activity. As evident from Table 13–6, the CO_2:C_2H_4 ratio in relation to normal air (0.21 atm O_2) is increased at a pO_2 of 0.10 atm and decreased at a pO_2 of 0.40 atm. At a pO_2 of 0.20 atm the ratio of CO_2:C_2H_4 was 10.6. Table 13–7 gives the time course of the O_2 consumption and CO production in relation to N_2 fixation during three consecutive hours. It shows that the respiratory quotient CO_2/O_2 during the first 2 h was

Table 13–6. *Effect of Oxygen Concentration on Respiration and N_2 Fixation of Parasponia parviflora Root Nodules**

Oxygen Partial Pressure (% atm)	CO_2 Production μmol g^{-1} h^{-1}	N_2 Fixation μmol C_2H_4 g^{-1} Fresh Weight h^{-1}	Ratio CO_2/C_2H_4
10	14.0	0.73	19.2
20	22.5	2.12	10.6
40	33.2	4.12	8.1

*Detached root nodules of 6-month-old plants in soil culture.

Table 13–7. *Time Course of Respiration and Nitrogen Fixation of Excised Parasponia parviflora Root Nodules in Air**

Time Interval (min)	O_2 Consumption (μmol O_2 g^{-1} Fresh Weight h^{-1})	CO_2 Production (μmol CO_2 g^{-1} Fresh Weight h^{-1})	Nitrogen Fixation (μmol C_2H_4 g^{-1} Fresh Weight h^{-1})	Respiratory Quotient (RQ)	CO_2/C_2H_4
0–60	4.5	8.5	0.8	1.9	10.6
60–120	6.1	9.2	0.7	1.5	13.1
120–180	9.6	8.7	0.7	0.9	12.4

*Root nodules of 18-month-old plants in soil culture.

Table 13–8. *Effect of CO on the Nitrogenase Activity of Parasponia parviflora Root Nodules**

CO % in Air	Nitrogen Fixation (μmol C_2H_4 g^{-1} Fresh Weight Root Nodule h^{-1})	% Activity	% Inhibition
0	3.75 ± 0.37	100	—
0.01	4.31 ± 2.23	115	−15
0.1	1.28 ± 0.65	34	66
1.0	0.23 ± 0.18	6	94

*Before exposure to CO, the nodule material was exposed to C_2H_2 (10% v/v), for 1 h to determine the actual nitrogenase activity (as reference).

1.5–1.9, whereas the ratio $CO_2:C_2H_4$ during this period, as in the previous experiment, was 10.6–13.1.

Application of CO strongly inhibits the nitrogenase activity of *Parasponia*. At a partial pressure of CO of 0.001 atm the inhibition is 66%, whereas at a CO pressure of 0.01 atm this inhibition is 94% (Table 13–8). CO inhibits nitrogenase directly and, in addition, may have some blocking effect on the bacteroid cytochrome system. As shown by Appleby et al.,[5] *Para-*

sponia–Bradyrhizobium bacteroids (strain NGR231) have a high concentration of cytochrome *aa*3.

VII. Genetic Aspects

A genetic study of the *Parasponia–Bradyrhizobium* symbiosis is important, as it may clarify the molecular basis of this symbiosis. As this is the only known nonlegume with a rhizobium symbiosis, investigations may offer a clue to possible extensions of this symbiosis to other nonlegumes of agricultural importance.

Such attempts have been made by Al-Mallach et al.[4] in allegedly producing nodular structures on rice seedlings by *Rhizobium–Bradyrhizobium* strains. The nodules formed, however, were abnormal and non-N_2 fixing.

Information on the molecular genetics of *Parasponia* and the *Parasponia Bradyrhizobium–Rhizobium* isolates has focused, with respect to the bacterium, on the genes involved in the nodulation process and the genes playing a role in N_2 fixation; with regard to the plant, it has focused on genes playing a role in the hemoglobin synthesis.

A. Bacterial Genes

i. Nodulation Genes

Attempts have been made to identify in *Parasponia* rhizobia, structural and functional homologues of nodulation (*nod*) genes previously identified in legume-nodulating rhizobia and to determine whether they are required for *Parasponia* nodulation. Additional attempts have been made to identify and transfer regions of DNA from *Parasponia* nodulating strains to a rhizobium strain that normally fails to nodulate *Parasponia* and to examine whether this transfer has conferred nodulation ability.

Genetic studies of *Rhizobium* have revealed that over 26 genes are involved in the nodule induction process. A number of these, the so-called common *nod* genes, are functionally interchangeable among species, as mutations in them can be complemented by a homologous gene from another species. In addition, DNA sequence analyses have shown that there is a substantial structural conservation within their coding regions. This conservation has made it possible to identify in some rhizobium strains that can nodulate *Parasponia* the *nodA, nodB, nodC,* and *nodD* genes.[62,78] This indicates that nonlegumes and legumes are able to respond to the same bacterial signal(s). To illustrate, Marvel et al.[61] isolated a region of DNA of *Parasponia–Bradyrhizobium* RP501. This cloned *Parasponia–Bradyrhizobium* RP501 nodulation locus hybridized to DNA fragments carrying the *R. meliloti nodABC* genes. Subsequent hybridization analyses of this region have indicated that it contains the structural homologues of *nodA, nodB,* and *nodD,* in addition to the *nodC* gene. On transfer to a *Rhizobium meliloti* mutant containing transposon Tn5 in the *nodC* gene, this DNA fragment was able to complement the mutation and allow the strain to nodulate its original host, alfalfa, but not siratro, cowpeas, or *Parasponia.* Moreover, a strain carrying a Tn5 insertion adjacent to the three evolutionary conserved *nodABC* genes vigorously nodulated RP501 legume hosts but was incapable of nodulating *Parasponia;* this may have identified a nonlegume-specific nodulation function.

Australian workers studying *Parasponia* often used strain NGR234 and the related strain ANU240, which is a spontaneous streptomycin-resistant mutant of NGR234. These organisms are not *Bradyrhizobium* species, but fast-growing *Rhizobium* strains. These strains contain plasmids, in contrast to true *Bradyrhizobium* species. The rhizobium strain NGR234 was originally isolated from the legume *Lablab purpureus* and was later found to nodulate *Parasponia.* Heat curing of a Sym plasmid in this strain showed that the genes involved in nodulation of both legumes and *Parasponia,* as well as the nitrogenase genes, reside on this plasmid.[68]

In experiments with *Rhizobium* strain NGR234, Bender et al.[15] found that the narrow host range of *Rhizobium leguminosarum* biovar *trifolii* ANU843 can be extended from clovers to *Parasponia* by the addition of the *nodD1* gene from *Rhizobium* NGR234, indicating that the *nodD1* gene is the key determinant to its extension of host range.

Structural homologues to the *nodI* and *nodJ* genes, first identified in *R. leguminosarum,*[39] also have been identified by DNA sequence analysis in *Parasponia–Bradyrhizobium.*[79] In addition to the *nod* genes *nodD, nodKABC,* and *nodIJ,* there is one other region of the *Parasponia–Bradyrhizobium* ANU289 (a streptomycin-resistant derivative of strain CP283) genome carrying structural homologues of *Rhizobium nod* genes. This region carries sequences homologous to the *nodL, nodM,* and *nodN* genes of *R. leguminosarum* and *R. trifolii.*[79] In *Parasponia* it is likely that the "host-specific" reactions are mainly mediated by the *nodD* gene, as in the nodulation of legume plants. Therefore, the *Parasponia* nodulation process probably has the same genetic mechanism in the bacterium as in legume nodulation, although the specificity for *Parasponia* in its recognition by specific plant-derived *nod* gene inducers is not yet clarified.

ii. Nitrogen Fixation Genes

DNA sequences carrying *Klebsiella pneumoniae nif* genes have been used as hybridization probes for *Parasponia–Bradyrhizobium* homologues. These studies have indicated that *Parasponia* rhizobium contains homologues of the *nif* structural genes *nifH, nifD,* and *nifK.* The *nifH* gene is transcribed on a separate operon to *nifDK,* and a *nifE* homologue has been identified immediately next to *nifK*.[80,116] DNA sequence analysis revealed a substantial amount of amino acid sequence homology between these genes and those from other N_2-fixing prokaryotes.[115,116] As in *Bradyrhizobium japonicum,* but unlike other N_2-fixing organisms, the iron–protein subunit of nitrogenase (*nifH*) is encoded on an operon separated from other components of the nitrogenase enzyme complex. Thus, being unlinked, it is possible that it may be independently regulated and therefore noncoordinately expressed.[80]

B. Plant Genes

i. Hemoglobin Genes

Leghemoglobin is a symbiotically induced oxygen carrier protein present in large amounts in legume nodules. The protein is well characterized, and genes encoding it have been cloned from many legumes. Although initially the presence or absence of a hemoglobin compound in *Parasponia* was controversial,[21] later Appleby et al.[6] established its occurrence in *Parasponia* root nodules. It was suggested that *Parasponia* hemoglobin, as in legumes, plays a role as an O_2-carrier protein and thus is a necessary part of the plant's N_2-fixation symbiosis. It should be added that hemoglobinlike compounds also have been discovered and characterized in the actinorhizal symbioses of nonlegumes such as *Casuarina* species.[24,40,94]

Kortt et al.[54] have determined the amino acid sequence of the hemoglobin (I) from *Parasponia andersonii* and have shown it contains 40% sequence homology with leghemoglobins from soybean and lupin. The isolated *Parasponia* hemoglobin had three introns located in identical positions to the introns found in legume hemoglobin genes. The *Parasponia* sequences were shown to cross-hybridize with sequences in closely related nonnodulating species like *Trema tomentosa.* Comparison of the *Trema* gene with the *Parasponia* gene have revealed four exons and three introns in identical positions. The *Trema* and *Parasponia* exon sequences have 93% nucleotide similarity, and there is approximately 80% similarity in the introns. *Parasponia* hemoglobin has 30 amino acid residues common to all plant hemoglobins whose structures are known.[16] The homologous region of the *Trema* genome has been isolated and shown to be a functional hemoglobin gene transcribed in a tissue-specific fashion in roots.[16] This strengthened the opinion that the hemoglobin genes are ubiquitous in the plant kingdom and that the gene is transcribed in the roots of all nonnodulated plants. In view of the widespread occurrence of hemoglobin genes in plants, a common evolutionary origin of legume and nonlegume hemoglobins has been suggested.[58] It also has been suggested that they may have some cryptic function in nonsymbiotic tissue (i.e., it is possible that the hemoglobins play a role in the respiratory metabolism of root cells of all plant species).[59]

VIII. Ecological Aspects

Nearly pure *Parasponia* stands (*P. parviflora* and *P. rugosa*) of about the same age have been found on volcanic ash soils deposited by nineteenth-century and more recent eruptions (1835 and later) of Mt. Kelud, East Java.[18,51] These trees occurring in dense stands were not taller than 6–7.5 m and had a trunk diameter of about 15–22 cm. They were sparsely mixed with *Acacia montana* and some other small trees and had an undervegetation of *Saccharum spontaneum* and tree ferns. The tree also was found in other regions of Java[66,110] and Bali[30] around erosion gullies and on heavily eroded, agglomerated gravel or sand areas supporting only sparse covers of grasses and mosses. Here single *Parasponia* plants grew among rocks and gravel.

We found *Parasponia parviflora* particularly abundant on the lower slopes of the mountains of West Java (Mt. Gede and Mt. Salak) occurring with *Trema orientalis* in secondary forest close to virgin forest and forming a canopy above *Saccharum* grass vegetation. It also grows in mountainous regions along streams in exposed riverbeds between stones and gravel. In Papua New Guinea (Pangia district) *Parasponia rugosa* and *P. andersonii* were found on recently cultivated or exposed soils (e.g., between rows of tea plants).[98]

It is clear that *Parasponia* species are typical pioneer plants forming a natural succession in covering bare, nitrogen-poor eroded soils. No field data are available for nitrogen increments in natural habitats based on stand biomass and N_2-fixing capacity per area; however, some greenhouse and pot experiments have been reported. Trinick[101] indicated that under greenhouse conditions the N_2-fixing activity of the root nodules was sufficient to allow young trees to grow 45 cm per month. Further growth was rapid, and after another 6 months of growth in containers the 2-m plants weighed 400 g dry. Growth studies also suggested that plants grown 50 cm apart under glasshouse conditions could fix between the age of 4 and 10 months at a rate of about 850 kg N ha^{-1} year^{-1} with a biomass production of 30 tons dry matter ha^{-1} year^{-1} [103,104,108]

IX. Conclusions

Parasponia, a woody member of the elm family (Ulmaceae), is the only nonleguminous genus whose members are known to form an effective symbiosis with *Bradyrhizobium* and some other *Rhizobium* species. All five *Parasponia* species known (*P. andersonii, P. melastomatifolia, P. parviflora, P. rigida,* and *P. rugosa*) appear to be nodulated. Their distribution is primary from southeast Asia eastward to the Pacific Islands. Southern Sumatra and West Java mark the western boundary of distribution.

Parasponia can be nodulated with either *Bradyrhizobium* or *Rhizobium* species (i.e., by both slow- and fast-growing rhizobium strains). *Parasponia–Bradyrhizobium* isolates also will cross–inoculate on legumes, and particularly effective nodules are produced on siratro (*Macroptilium atropurpureum* and allied species) and on several varieties of cowpea (*Vigna unguiculata*). Other tropical legumes and temperate legumes usually are not nodulated by *Parasponia–Bradyrhizobium* isolates. Rhizobia from temperate legumes, such as clover (*Trifolium repens*), may produce nodules on *Parasponia* after sufficient exposure time, but these nodules usually are less effective or completely ineffective. The time for nodulation is variable, and a typical *Bradyrhizobium* strain isolated from *Parasponia* root nodules takes 7–14 days to nodulate its host plant *Parasponia,* whereas the nodulation of *Parasponia* plants with *Bradyrhizobium* or *Rhizobium* strains of legume origin is usually delayed to one or several months. In competition studies it was shown that *Bradyrhizobium* strains of *Parasponia* origin nearly always dominate in the nodules produced on *Parasponia.* This probably is because early nodulation of *Parasponia* by its own rhizobial strain confers an advantage. Nevertheless, under particular conditions of strain and environment, dual occupancy of *Bradyrhizobium* and *Rhizobium* strains in a single *Parasponia* root nodule has been observed.

The N_2-fixation capacity of *Parasponia* root nodules varies, depending on strain and culture conditions of the *Parasponia* plants (soil, perlite, water culture, mineral–agar culture). Nodules on a single plant produced by the same *Bradyrhizobium* strain may show variation in nitrogenase activity. In general, soil-borne nodules have the highest activity. The maximal nitrogenase activity measured in *Parasponia* root nodules is usually about 13–15 μmol C_2H_4 g^{-1} fresh weight nodule tissue h^{-1}. This is lower than that of most legumes. Occasionally, nitrogenase activities of 26–28 μmol C_2H_4 g^{-1} fresh weight nodules h^{-1} have been measured.

The nodule structure of *Parasponia* root nodules is comparable to that of actinorhizal nodules of nonlegumes; that is, the nodules are indeterminate with an apical meristem and central vasicular tissue. Nodule initiation in *Parasponia* occurs by entry of rhizobia through a wound or crack into the cortex. According to the author, *Parasponia* root nodule development is more legumelike than has been recognized so far. Unlike actinorhizal root nodules, a pseudonodule is not formed. The cortical cells are rejuvenated by mitotic cell divisions in one or more centers, and their further growth is rhizobial-like in appearance.

Intracellular infection threads and fixation threads have been observed in mature infected host cells. I do not regard the fixation threads to be a kind of infection thread, but the fixation threads are peribacteroid membranes enclosing a few to many bacterial cells (bacteroids). In *Parasponia* root nodules these mature membranes contain more cell wall deposits than usually found in legumes. On legumes (e.g., *Vigna* spp.) the same *Parasponia–Bradyrhizobium* strain produces similar envelope structures containing many cells, but with thinner membrane boundaries. An additional argument, that the so-called fixation threads are membrane envelopes, is that they always appear as separate, independent structures of variable length when observed in squashes. True invasive infection threads are always continuous and characterized by persistent, thick, and highly lignified walls.

Two types of root nodules can be induced on *Parasponia* roots, sometimes by the same strain. One type is the elongated and branched nodular structure containing a central vascular bundle and an apical meristem exhibiting indeterminate growth. These root nodules are perennial. Such nodule structures are the usual nodule type found in field studies. The other type has an arrested or reduced central vascular strand and a more spherical meristem. They are unbranched, and by their lateral expansion (spherical meristem) these root nodules become oblate to round in form and resemble closely the spherical nodules on cowpea and related species. However, they are not true determinate legume nodules, as they lack a peripheral closed vascular system in the cortex as found in legume nodules. Growth of these *Parasponia* nodules is time restricted and therefore certainly determinate. Moreover, these nodules are ephemeral, as the nodule tissue degenerates more quickly than the tissues of the branched indeterminate nodules.

Various studies have shown that determinate nodules are commonly formed on the roots of tropical legumes (*Vigna* spp., *Glycine max, Phaseolus vulgaris* spp., *Sebania rostrata*, etc.), whereas indeterminate nodules generally occur on temperate legumes (*Pisum sativum, Trifolium repens, Medicago sativa*, etc.). The morphology of the nodule seems to be primarily plant controlled, since broad host range rhizobia are known to form one type of nodule on one host and the other type on another host. In *Parasponia*, however, both nodule types occur on the same plant and sometimes also may be induced by the same rhizobial strain. Nodule morphology is correlated with the nitrogen metabolism of nodules. In indeterminate nodules, nitrogen is transported as amides (mainly asparagine), while determinate nodules transport fixed nitrogen as ureides.[71] One theory[91] implies that the division in nodular types is related to the limited solubility of ureides as agents of transport of fixed nitrogen. Therefore, the determinate type of nodule is found predominantly in tropical regions with high temperatures, whereas the amides glutamine and asparagine, being several times more soluble in water than ureides, are of special importance in temperate climates.

Parasponia is a tropical mountain (temperate) plant. The transport of nitrogen in the xylem bleeding sap has not been investigated, so there is no information whether the root nodules are ureide- or amide-exporting. However, Becking[12] analyzed the amino acid composition of indeterminate root nodules of soil and water culture–grown *Parasponia* plants. In *Parasponia*, aspartic and glutamic acid and the amides glutamine and asparagine were present in the nodules in high concentration. Actinorhizal root nodules (*Alnus* and *Casuarina*) analyzed simultaneously showed similar composition. The *Casuarina* root nodules in addition were high in proline.[12]

Both types of root nodules on *Parasponia* develop bacteroids with nitrogenase activity. In the opinion of the author, the internal structure and development of *Parasponia* root nodules are more related to legume nodules than to actinorhizal root nodules of nonlegumes. They are similar to root nodules of tropical legumes; the perennial root nodules of woody legumes of the Papilionoideae and Caesalpinioideae resemble closely the indeterminate nodules of *Parasponia*. In addition, the internal organization of the latter legume nodules (genus *Andira*, Papilionoideae) show structures very reminiscent of the so-called infection threads and fixation threads observed in infected *Parasponia* host cells.[28,29] The present author considers these structures to be like those in *Parasponia* nodules true invasive infection threads, whereas the "fixation threads" are independent membrane envelopes (peribacteroid membranes) consolidated with cell wall constituents.

Three conserved nodulation genes—*nodA*, *nodB*, and *nodC*—are required for nodulation of legumes by several *Rhizobium* and *Bradyrhizobium* species. The same genes and *nodD* have been identified in *Bradyrhizobium—Rhizobium* strains as genes required for nodulation of *Parasponia* and some legumes. A strain (RP501) carrying a Tn5 insertion adjacent to the three conserved *nodABC* genes nodulated RP501 legume hosts but was incapable of nodulating *Parasponia*, thus possibly identifying a nonlegume nodulation function. In a strain (NGR234 from *Lablab purpureus*) that could effectively nodulate *Parasponia*, the *nodD1* has been identified as the key determinant in the extension of its host range to the nonlegume *Parasponia*. Studies have indicated that the *Parasponia–Bradyrhizobium/Rhizobium* bacteria contain the homologues of the *nif* structural genes *nifH*, *nifD*, and *nifK*, and a considerable amount of amino acid sequence homology exists between these genes and those of other nitrogen-fixing organisms. Hemoglobin compounds have been demonstrated in *Parasponia* root nodules. Investigations have identified the hemoglobin genes in *Parasponia*, as well as in the related nonnodulated species such as *Trema*. It has been suggested that the hemoglobins of legumes and nonlegumes have a common evolutionary origin and that the gene is ubiquitous in plant species, probably playing a common function in the respiratory metabolism of root cells.

REFERENCES

1. Akkermans, A. D. L., Abdulkadir, S., and Trinick, M. J., "Nitrogen-fixing root nodules in Ulmaceae," *Nature (Lond.) 274*, 190 (1978).

2. Akkermans, A. D. L., Abdulkadir, S., and Trinick, M. J., "N$_2$-fixing root nodules in Ulmaceae: *Parasponia* or (and) *Trema* spp.," *Plant and Soil 49*, 711–715 (1978).

3. Allen, O. N., and Allen, E. K., *The Leguminosae: A Source Book of Characteristics, Uses, and Nodulation*. University of Wisconsin Press, Madison, 1981, p. 812.

4. Al-Mallah, M., Davey, M. R., and Cocking, E. C., "Formation of nodular structures on rice seedlings by rhizobia," *J. Exp. Bot. 40*, 473–478 (1989).

5. Appleby, C. A., Bergersen, F. J., Ching, T. M., Gibson, A. H., Gresshoff, P. M., and Trinick, M. J., "Cytochromes of Rhizobia from Parasponia, pea and soybean nodules, and from nitrogen fixing continuous cultures," in *Current Perspectives in Nitrogen Fixation*, A. Gibson and W. Newton, eds., Australian Academy of Sciences, Canberra, 1981, p. 368.

6. Appleby, C. A., Tjepkema, J. D., and Trinick, M. J., "Hemoglobin in a nonleguminous plant, *Parasponia:* Possible genetic origin and function in nitrogen fixation," *Science 220*, 951–953 (1983).

7. Arora, C. M., "New chromosome report," *Bull. Bot. Surv. India 2*, 305 (1960).

8. Baird, L. M., Virginia, R. A., and Webster, B. D., "Development of root nodules in the woody legume, *Prosopis glandulosa* Torr.," *Bot. Gaz. 146*, 39–43 (1985).

9. Becking, J. H., "Root nodules in non-legumes," in *The Development and Function of Roots*, J. G. Torrey and D. T. Clarkson, eds., Academic Press, London, 1975, pp. 507–566.

10. Becking, J. H., "Endophyte and association establishment in non-leguminous nitrogen-fixing plants," in *Recent Developments in Nitrogen Fixation*, W. Newton, J. R. Postgate, and C. Rodriguez-Barrueco, eds., Academic Press, London, 1977, pp. 551–567.

11. Becking, J. H., "Root–nodule symbiosis between *Rhizobium* and *Parasponia* (Ulmaceae)," *Plant and Soil 51*, 289–296 (1979).

12. Becking, J. H., "The *Parasponia parviflora*–Rhizobium symbiosis. Host specificity, growth and nitrogen fixation under various conditions," *Plant and Soil 75*, 309–342 (1983).

13. Becking, J. H., "The *Parasponia parviflora*–Rhizobium symbiosis. Isotopic nitrogen fixation, hydrogen evolution and nitrogen-fixation efficiency, and oxygen relations," *Plant and Soil 75*, 343–360 (1983).

14. Bender, G. L., Nayudu, M., Goydych, W., and Rolfe, B. G., "Early infection events in the nodulation of the non-legume *Parasponia andersonii* by *Bradyrhizobium*," *Plant Sci. 51*, 285–293 (1987).

15. Bender, G. L., Nayudu, M., Le Strange, K. K., and Rolfe, B. G., "The *nodD1* gene from *Rhizobium* strain NGR234 is a key determinant in the extension of host range to the nonlegume *Parasponia*," *Mol. Plant-Microbe Interactions 1*, 259–266 (1988).

16. Bogusz, D., Appleby, E. A., Landsmann, J., Dennis, E. S., Trinick, M. J., and Peacock, W. J., "Functioning haemoglobin genes in non-nodulating plants," *Nature (Lond.) 331*, 178–180 (1988).

17. Bradley, D. J., Butcher, G. W., Galfre, G., Wood, E. A., and Brewin, N. J., "Physical association between peribacteroid membrane and lipo-polysaccharide from the bacteroid outer membrane in *Rhizobium*-infected pea root nodule cells," *J. Cell. Sci. 85*, 47–61 (1986).

18. Clason, E. W., "The vegetation of the Upper-Badak region of Mount Kelut (East Java)," *Bull. Jardin Bot.*, Ser. III, *13*, 509–518, Tables 5, 6 (1935).

19. Corby, H. D. L., "The systematic value of Leguminous root nodules," in *Advances in Legume Systematics,* R. M. Polhill and P. H. Raven, eds., Royal Botanic Gardens, Kew, 1981, pp. 657–669.

20. Corby, H. D. L., Polhill, R. M., and Sprent, J. I., "Taxonomy," in *Nitrogen Fixation, Vol. 3: Legumes,* W. J. Broughton, ed., Clarendon Press, Oxford, 1983, pp. 1–35.

21. Coventry, D. R., Trinick, M. J., and Appleby, C. A., "A search for a leghaemoglobin-like compound in root nodules of *Trema cannabina* Lour," *Biochim. Biophys. Acta 420,* 105–111 (1976).

22. Dart, P. J., "Legume root nodule initiation and development," in *The Development and Function of Roots,* J. G. Torrey and D. T. Clarkson, eds., Academic Press, London, 1975, pp. 467–506.

23. Dart, P. J., "Infection and development of leguminous nodules," in *A Treatise on Dinitrogen Fixation,* Vol. 3, R. W. F. Hardy and W. S. Silver, eds., Wiley, New York, 1977, pp. 367–472.

24. Davenport, H. E., "Haemoglobin in the root nodules of *Casuarina cunninghamiana,*" *Nature (Lond.), 186,* 653–654.

25. De Bruijn, F. J., "The unusual symbiosis between the diazotrophic stem-nodulating bacterium *Azorhizobium caulinodans,*" in *Plant–Microbe Interactions: Molecular and Genetic Perspectives,* Vol. 3, T. Kosuge and E. W. Webster, eds., McGraw-Hill, New York, 1989, pp. 457–504.

26. De Bruijn, F. J., Pawlowski, K., Ratet, P., Hilgert, U., Wong, C. H., Schneider, M., Meyer, Z. A., H. and Schell, J., "Molecular genetics of nitrogen fixation by *Azorhizobium caulinodans* ORS571, the diazotrophic stem nodulating symbiont of *Sesbania rostrata,*" in *Nitrogen Fixation: Hundred Years After,* H. Bothe, F. J. de Bruijn and W. E. Newton, eds., Gustav Fischer, Stuttgart, 1988, pp. 351–355.

27. De Faria, S. M., McInroy, S. G., Rowell, P., and Sprent, J. I., "Some properties of "primitive" legume nodules." in *Nitrogen Fixation: Hundred Years After,* H. Bothe, F. J. de Bruijn, and W. E. Newton, eds., Gustav Fischer, Stuttgart, 1988, p. 524.

28. De Faria, S. M., McInroy, S. G., and Sprent, J. I., "The occurrence of infected cells, with persistent infection threads, in legume root nodules," *Can. J. Bot. 65,* 553–558 (1987).

29. De Faria, S. M., Sutherland, J. M., and Sprent, J. I., "A new type of infected cell in root nodules of *Andira* spp. (Leguminosae)," *Plant Sci. 45,* 143–147 (1986).

30. De Voogd, C. N. A., "Botanische aanteekeningen van de Kleine Soenda eilanden (VIII). De Batoer of Bali," *Trop. Natuur 29,* 37–53 (1940).

31. Dreyfus, B. L., *La symbiose entre Rhizobium et Sesbania rostrata légumineuse à nodules caulinaires,* Doctoral thesis, Université Paris VII, 1982.

32. Dreyfus, B. L., Alazard, D., and Dommergues, Y. R., "Stem-nodulating rhizobia," in *Current Perspectives of Microbiological Ecology,* M. G. Klug and C. E. Reddy, eds., American Society of Microbiology, Washington, 1984, pp. 161–169.

33. Dreyfus, B. L., and Dommergues, Y. R., "Nitrogen-fixing nodules induced by *Rhizobium* on the stem of the tropical legume *Sesbania rostrata,*" *FEMS Microbiol. Lett. 10,* 313–317 (1981).

34. Dreyfus, B. L., and Dommergues, Y. R., "Stem nodules on the tropical legume *Sesbania rostrata,*" in *Current Perspectives in Nitrogen Fixation,* A. H. Gibson and W. E. Newton, eds., Australian Academy of Science, Canberra, 1981, p. 471.

35. Dreyfus, B. L., Elmerich, C., and Dommergues, Y. R., "Free-living *Rhizobium* strain able to grow under N₂ as a sole nitrogen source," *Appl. Environ. Microbiol. 45*, 711–713 (1983).

36. Dreyfus, B. L., Garcia, J. L., and Gillis, M., "Characterization of *Azorhizobium caulinodans* gen. nov., sp. nov., a stem-nodulating nitrogen-fixing bacterium isolated from *Sesbania rostrata*," *Int. J. Syst. Bact. 38*, 89–98 (1988).

37. Duhoux, E., "Ontogenese des nodules caulinaires du *Sesbania rostrata* (légumineuse)," *Can. J. Bot. 62*, 982–994 (1984).

38. Duhoux, E., and Dreyfus, B., "Nature des sites d'infection par le *Rhizobium* de la tige de la légumineuse, *Sesbania rostrata* Brem.," *C. R. Acad. Sci. (Paris), 294*, 407–411 (1982).

39. Evans, I. J., and Downie, J. A., "The *nodI* gene product of *Rhizobium leguminosarum* is closely related to ATP-binding bacterial transport proteins; nucleotide sequences analysis of the *nodI* and *nodJ* genes," *Gene 43*, 95–101 (1986).

40. Fleming, A. I., Wittenberg, J. B., Wittenberg, B. A., Dudman, W. F., and Appleby, C. A., "The purification, characterization and ligand-binding kinetics of hemoglobins from root nodules of the non-leguminous *Casuarina glauca–Frankia* symbiosis," *Biochem. Biophys. Acta 911*, 209–220 (1987).

41. Fred, E. B., Baldwin, I. L., and McCoy, E., *Root Nodule Bacteria and Leguminous Plants*, University of Wisconsin Studies in Science, No. 5, Madison, 1932, p. 343.

42. Gajapathy, C., "Cytological studies in some Indian medicinal plants," *Bull. Bot. Surv. India, 3*, 49–51 (1961).

43. Gresshoff, P. M., Newton, S., Mohapatra, S. S., Scott, K. F., Howitt, S., Price, G. D., Bender, G. L., Shine, J., and Rolfe, B. G., "Symbiotic nitrogen fixation involving *Rhizobium* and the non-legume *Parasponia*," in *Advances in Nitrogen Fixation Research*, C. Veeger and W. E. Newton, eds., Proc. 5th Int. Symp. on Nitrogen Fixation, Aug.–Sept. 1983, Noordwijkerhout, The Netherlands, 1984, pp. 483–489.

44. Ham, S. P., "Note on *Parasponia* ('Anggroeng')," *Nederlandsch-Indisch Landbouw-Syndicaat, Handelingen 10e Congres*, Bandung, Java, N.O.I., Part 2, Verslag 2e Aflev. *Groene Bemesting 26* (1909).

45. Hans, A. S., "Polyploidy in Trema (Ulmaceae)," *Cytologia 36*, 341–345 (1971).

46. Hennecke, H., Kaluza, K., Thöny, B., Fuhrmann, M., Ludwig, W., and Stackebrandt, E., "Concurrent evolution of nitrogenase genes and 16S rRNA in *Rhizobium* species and other nitrogen fixing bacteria," *Arch. Microbiol. 142*, 342–348 (1985).

47. Hsu, C.-C., "Preliminary chromosome studies on the vascular plants of Taiwan (I)," *Taiwania 13*, 117–129 (1967).

48. Jarvis, B. D. W., Gillis, M., and De Ley, J., "Intra- and intergeneric similarities between ribosomal ribonucleic acid cistrons of *Rhizobium* and *Bradyrhizobium* species and some related bacteria," *Int. J. Syst. Bacteriol. 36*, 129–138 (1986).

49. Jordan, D. C., "Transfer of *Rhizobium japonicum* Buchanan 1980 to *Bradyrhizobium* gen. nov., a genus of slow-growing root nodule bacteria from leguminous plants," *Int. J. Syst. Bacteriol. 32*, 136–139 (1982).

50. Jordan, D. C., "Family III. Rhizobiaceae Conn 1938, 321," in *Bergey's Manual of Systematic Bacteriology*, Vol. 1, N. R. Krieg and J. G. Holt, eds., Williams & Wilkins, Baltimore, 1984, pp. 234–244.

51. Junghuhn, F., *Java, zijne Gedaante, zijn Plantentooi en zijn inwendige Bouw*, Vol. 2, C. W. Mieling, 's-Gravenhage, The Netherlands, 1853, pp. 671, 549.

52. Kohl, D. H., and Shearer, G., "Isotopic fractionation associated with symbiotic N_2 fixation and uptake of NO_2^- by plants," *Plant Physiol.* 66, 51–56 (1980).

53. Kohl, D. H., Shearer, G., and Harper, J., "Estimates of N_2 fixation based on differences in the natural abundance of ^{15}N in nodulating and nonnodulating isolines of soybeans," *Plant Physiol.* 66, 61–65 (1980).

54. Kortt, A. A., Burns, J. E., Trinick, M. J., and Appleby, C. A., "The amino acid sequence of hemoglobin I from *Parasponia andersonii,* a nonleguminous plant," *FEBS Lett.* 180, 55–60 (1985).

55. Kurz, W. G. W., and LaRue, T. A., "Nitrogenase activity in rhizobia in absence of plant host," *Nature (Lond.),* 256, 407–408 (1975).

56. Lancelle, S. A., and Torrey, J. G., "Early development of *Rhizobium*–induced root nodules of *Parasponia rigida*. I. Infection and early nodule initiation," *Protoplasma* 123, 26–37 (1984).

57. Lancelle, S. A., and Torrey, J. G., "Early development of *Rhizobium*–induced root nodules of *Parasponia rigida*. II. Nodule morphogenesis and symbiotic development," *Can. J. Bot.* 63, 25–35 (1985).

58. Landsmann, J., Dennis, E. A., Higgins, T. J. V., Appleby, C. A., Kortt, A. A., and Peacock, W. J., "Common evolutionary origin of legume and non-legume plant haemoglobins," *Nature (Lond.)* 324, 166–168 (1986).

59. Landsmann, J., Llewellyn, D., Dennis, E. S., and Peacock, W. J., "Organ regulated expression of the *Parasponia andersonii* haemoglobin gene in transgenic tobacco plants." *Mol. Gen. Genet.* 214, 68–73 (1988).

60. Lange, R. T., "Nodule bacteria associated with the indigenous Leguminosae of southwestern Australia," *J. Gen. Microbiol.* 61, 351–359 (1961).

61. Marvel, D. J., Kuldau, G., Hirsch, A., Richards, E., Torrey, J. G., and Ausubel, F. M., "Conservation of nodulation genes between *Rhizobium meliloti* and a slow-growing *Rhizobium* strain that nodulates a nonlegume host," *Proc. Natl. Acad. Sci. USA 82,* 5841–5845 (1985).

62. Marvel, D. J., Torrey, J. G., and Ausubel, F. M., "*Rhizobium* symbiotic genes required for nodulation of legume and nonlegume hosts," *Proc. Natl. Acad. Sci. USA* 84, 1319–1323 (1987).

63. Mehra, P. N., "Cytogenetical studies of hardwoods," *Nucleus 15,* 64–83 (1972).

64. Mehra, P. N., and Gill, B. S., "Cytological studies in Ulmaceae, Moraceae, and Urticaceae," *J. Arnold Arbor.* 55, 663–677 (1974).

65. Miquel, F. A. W., *Plantae Junghuhnianae,* H. R. de Breuk, Leiden, 1851, pp. 522, 68–69.

66. Miquel, F. A. W., *Flora van Nederlands-Indië,* Vol. 1, Afd. 2, p. 704, Plate XVI, 218, C. G. van der Post, Amsterdam, 1859.

67. Mohapatra, S. S., Bender, G. L., Shine, J., Rolfe, B. G., and Gresshoff, P. M., "In vitro expression of nitrogenase activity in *Parasponia–Rhizobium* strain ANU289," *Arch. Microbiol.* 134, 12–16, (1983).

68. Morrison, N. A., Hau, C. Y., Trinick, M. J., Shine, J., and Rolfe, B. G., "Heat curing of a Sym plasmid in a fast-growing *Rhizobium* sp. that is able to nodulate legumes and the nonlegume *Parasponia* sp.," *J. Bacteriol.* 153, 527–531 (1983).

69. Newcomb, W., "Nodule morphogenesis and differentiation," *Int. Rev. Cytol. Suppl.* 13, 247–297 (1981).

70. Pangan, J. D., Child, J. J., Scowcroft, W. R., and Gibson, A. H., "Nitrogen fixation by *Rhizobium* cultures on defined medium," *Nature (Lond.), 256,* 406–407 (1975).

71. Pate, J. S., and Atkins, C. A., "Nitrogen uptake, transport, and utilization," in *Nitrogen Fixation, Vol. 3: Legumes,* W. J. Broughton, ed., Clarendon Press, Oxford, 1983, pp. 245–298.

72. Phillips, D. A., and Torrey, J. G., "Cytokinin production by *Rhizobium japonicum,*" *Physiol. Plant. 23,* 1057–1063 (1970).

73. Philips, D. A., and Torrey, J. G., "Studies on cytokinin production by *Rhizobium,*" *Plant Physiol. 49,* 11–15, (1972).

74. Polhill, R. M., "The Papilionoideae," in *Advances in Legume Systematics,* Part I, R. M. Polhill and P. H. Raven, eds., Royal Botanic Gardens, Kew, 1981, pp. 191–204.

75. Price, G. D., Mohapatra, S. S., and Gresshoff, P. M., "Structure of nodules formed by *Rhizobium* strain ANU289 in the nonlegume *Parasponia* and the legume Siratro (*Macroptilium atropurpureum*)," *Bot. Gaz. 145,* 444–451 (1984).

76. Rolfe, B. G., Djordjevic, M. A., Morrison, N. A., Plazinski, J., Bender, G. L., Ridge, R., Zurkowski, S., Tellam, J. T., Gresshoff, P. M., and Shine, J., "Genetic analysis of the symbiotic regions in *Rhizobium trifolii* and *Rhizobium parasponia,*" in *Molecular Genetics of the Bacteria–Plant Interaction,* A. Pühler, ed., Springer-Verlag, Berlin, 1983, pp. 188–203.

77. Schubert, K. R., and Evans, H. J., "Hydrogen evolution: A major factor affecting the efficiency of nitrogen fixation in nodulated symbionts," *Proc. Natl. Acad. Sci. USA 73,* 1207–1211 (1976).

78. Scott, K. F., "Conserved nodulation genes from the non-legume symbiont *Bradyrhizobium* sp. (*Parasponia*)," *Nucleic Acids Res. 14,* 2905–2929 (1986).

79. Scott, K. F., and Bender, G. L., "The *Parasponia–Bradyrhizobium* symbiosis," in *Molecular Biology of Symbiotic Nitrogen Fixation,* P. M. Greshoff, ed., CRC Press, Boca Raton, Florida, 1990, 231–251.

80. Scott, K. F., Rolfe, B. G., and Shine, J., "Nitrogenase structural genes are unlinked in the nonlegume," *DNA 2,* 141–148 (1983).

81. Scott, K. F., Saad, M., Price, G. D., Gresshoff, P. M., Kane, H., and Chua, K. Y., "Conserved nodulation genes are obligatory for nonlegume nodulation," in *Molecular Genetics and Plant–Microbe Interactions,* Martinus Nijhoff, Dordrecht, 1987, pp. 238–240.

82. Shearer, G., Feldman, L., Bryan, B. A., Skeeters, J. L., Kohl, D. H., Amarger, N., Mariotti, F., and Mariotti, A., "[15]N abundance of nodules as an indicator of N metabolism in N_2-fixing plants," *Plant Physiol. 70,* 465–468 (1982).

83. Shine, J., Scott, K. F., Fellows, F. F., Djordjevic, M. A., Schofield, P., Watson, J. M., and Rolfe, B. G., "Molecular anatomy of the symbiotic region in *Rhizobium trifolii* and *Rhizobium parasponia,*" in *Molecular Genetics of the Bacterial–Plant Interactions,* A. Pühler, ed., Springer-Verlag, Berlin, 1983, pp. 204–209.

84. Silvester, W. B., Silvester, J. K., and Torrey, J. G., "Adaption of nitrogenase to varying oxygen tension and the role of the vesicle in root nodules of *Alnus incana* spp. *rugosa,*" *Can. J. Bot. 66,* 1772–1779 (1988).

85. Simon, R., and Priefer, U., "Vector technology of relevance to nitrogen fixation research," in *The Molecular Biology of Symbiotic Nitrogen Fixation,* P. Gresshoff, ed., CRC Press, Boca Raton, Florida, 1990, pp. 13–49.

86. Smith, C. A., Skvirsky, R. C., and Hirsch, A. M., "Histochemical evidence for the presence of a suberinlike compound in *Rhizobium*–induced nodules of the nonlegume *Parasponia rigida*," *Can. J. Bot. 64*, 1474–1483 (1986).

87. Soepadmo, E., "Ulmaceae," in *Flora Malesiana*, Ser. I., Vol. 8, C. G. G. J. van Steenis, ed., 1977, pp. 31–76.

88. Spratt, E. R., "A comparative account of the root-nodules of the Leguminosae," *Ann. Bot. (Lond.) 33*, 189–199, Plate 8, 9 (1919).

89. Sprent, J. I., "Functional evolution in some Papilionoid root nodules," in *Advances of Legume Systematics*, R. M. Polhill and P. H. Raven, eds., Royal Botanic Gardens, Kew, U.K., 1981, pp. 671–676.

90. Sprent, J. I., and de Faria, S. M., "Mechanisms of infection of plants by nitrogen fixing organisms," *Plant and Soil 110* (spec. vol.), 157–165 (1988).

91. Sprent, J. I., and Embrapa, E., "Root nodule anatomy, type of export product and evolutionary origin of some Leguminosae," *Plant Cell Environ. 3*, 35–43 (1980).

92. Sturtevant, D. B., and Taller, B. J., "Cytokinin production by a *Parasponia* nodule bacterium," *Abst. 89th Annual Meeting Am. Soc. Microbiol.*, Vol. 89, New Orleans, Louisiana, 1989, p. 300.

93. Tjepkema, J. D., "Oxygen regulation and energy consumption in the root nodules of legumes, *Parasponia*, and actinorhizal plants," in *Current Perspectives in Nitrogen Fixation*, A. H. Gibson and W. E. Newton, eds., Australian Academy of Sciences, Canberra, 1981, p. 368.

94. Tjepkema, J. D., "Hemoglobins in the nitrogen-fixing root nodules of actinorhizal plants," *Can. J. Bot. 61*, 2924–2929 (1983).

95. Tjepkema, J. D., and Cartica, R. J., "Diffusion limitation of oxygen uptake and nitrogenase activity in the root nodules of *Parasponia rigida* Merr. and Perry," *Plant Physiol. 69*, 728–733 (1982).

96. Tjepkema, J. D., Ormerod, W., and Torrey, J. G., "Vesicle formation and acetylene reduction activity in *Frankia* CpII cultured in defined nutrient media," *Nature (Lond.) 187*, 633–635 (1980).

97. Torrey, J. G., and Callaham, D., "Structural features of the vesicle of *Frankia* sp. CpII in culture," *Can. J. Microbiol. 28*, 749–757 (1982).

98. Trinick, M. J., "Symbiosis between *Rhizobium* and the Non-legume, *Trema aspera*," *Nature (Lond.) 244*, 459–460 (1973).

99. Trinick, M. J., "Rhizobium symbiosis with a non-legume," in *Proc. 1st Int. Symposium on Nitrogen Fixation*, W. E. Newton and C. J. Nyman, eds., Washington State University Press, Pullman, 1976, pp. 507–517.

100. Trinick, M. J., "Structure of nitrogen-fixing nodules formed by *Rhizobium* on roots of *Parasponia andersonii*," *Can. J. Microbiol. 25*, 565–578 (1979).

101. Trinick, M. J., "Growth of *Parasponia* in agar tube culture and symbiotic effectiveness of isolates from *Parasponia* spp.," *New Phytol. 85*, 37–45 (1980).

102. Trinick, M. J., "Effects of oxygen, temperature and other factors on the reduction of acetylene by root nodules formed by *Rhizobium* on *Parasponia andersonii* Planch," *New Phytol. 86*, 27–38 (1980).

103. Trinick, M. J., "The non-legume *Rhizobium* association," in *Current Perspectives in Nitrogen Fixation*, A. H. Gibson and W. E. Newton, eds., Australian Academy of Sciences, Canberra, 1981, p. 255.

104. Trinick, M. J., "The effective *Rhizobium* symbiosis with the non-legume *Parasponia andersonii*," in *Current Perspectives in Nitrogen Fixation*, A. H. Gibson and W. E. Newton, eds., Australian Academy of Sciences, Canberra, 1981, p. 480.

105. Trinick, M. J., and Galbraith, J., "Structure of root nodules formed by *Rhizobium* on the non-legume *Trema cannabina* var. *scabra*," *Arch. Microbiol. 108*, 159–166 (1976).

106. Trinick, M. J., and Galbraith, J., "The *Rhizobium* requirements of the non-legume *Parasponia* in relationship to the cross-inoculation group concept of legumes," *New Phytol. 86*, 17–26 (1980).

107. Trinick, M. J., Goodchild, D. J., and Miller, C., "Localization of bacteria and hemoglobin in root nodules of *Parasponia andersonii* containing both *Bradyrhizobium* strains and *Rhizobium leguminosarum* biovar. *trifolii*," *Appl. Environ. Microbiol. 55*, 2046–2055 (1989).

108. Trinick, M. J., and Hadobas, P. A., "Biology of the *Parasponia–Bradyrhizobium* symbiosis," *Plant and Soil 110*, (spec. vol.) 177–185 (1988).

109. Trinick, M. J., and Hadobas, P. A., "Competition by *Bradyrhizobium* strains for nodulation of the nonlegume *Parasponia andersonii*," *Appl. Environ. Microbiol. 55*, 1242–1248 (1989).

110. Van Steenis, C. G. G. J. (with artists Hamzah, A., and Toha, M., colored plates), *The Mountain Flora of Java*, E. J. Brill, Leiden, 1972.

111. Vincent, J. H., *A Manual for the Practical Study of Root Nodule Bacteria*, Blackwell, Oxford, p. 164.

112. Virginia, R. A., Baird, L. M., La Favre, J. S., Jarrell, W. M., Bryan, B. A., and Shearer, G., "Nitrogen fixation efficiency, natural ^{15}N abundance, and morphology of mesquite (*Prosopis glandulosa*) root nodules," *Plant and Soil 79*, 273–284 (1984).

113. Virginia, R. A., and Delwiche, C. C., "Natural ^{15}N abundance of presumed N_2-fixing and non-N_2-fixing plants from selected ecosystems," *Oecologia* (Berlin) *54*, 317–325 (1982).

114. Wedlock, N. D., and Jarvis, B. D. W., "DNA homologies between *Rhizobium fredii*, rhizobia, that nodulate *Galega* sp., and other *Rhizobium* and *Bradyrhizobium* species," *Int. J. Bacteriol. 36*, 550–558 (1986).

115. Weinman, J. J., Ph.D. thesis, Australian National University, Canberra, 1986.

116. Weinman, J. J., Fellows, F. F., Gresshoff, P. M., Shine, J., and Scott, K. F., "Structural analysis of the genes encoding the molybdenum-iron protein of nitrogenase in *Parasponia rhizobium* strain ANU289," *Nucleic Acids Res., 12*, 8329–8344 (1984).

117. Winship, L. J., and Tjepkema, J. D., "The role of diffusion in oxygen protection of nitrogenase in nodules of *Alnus rubra*," *Can. J. Bot. 61*, 2930–2936 (1983).

118. Winship, L. J., and Tjepkema, J. D., "Nitrogen fixation and respiration by root nodules of *Alnus rubra* Bong. Effects of temperature and oxygen concentration," *Plant and Soil 87*, 91–107 (1985).

14

Genetic Analysis of *Rhizobium* Nodulation

Sharon R. Long

I. Introduction

The formation of nodules on host plants by *Rhizobium* provides the location and the physiological environment for symbiotic nitrogen fixation. Genetic and molecular biological studies of nodule formation have begun to reveal some of the mechanisms involved and provide the prospect for manipulation of nodulation and of the consequent nitrogen fixation in nodules.

Nodulation genetics is a fast–moving field. The details of gene maps, gene product function, and regulation set down in this review will thus be incomplete from the vantage of a few year's time. Therefore, this review has the larger goal of providing some perspective on the unsolved questions, guidelines for studies in new systems, and a framework for how genetics relates to other disciplines. These larger issues will continue to be important as our understanding expands and intensifies in future years.

II. Nodulation as a Phenotype

Before nodulation begins, the two partners–*Rhizobium,* the microsymbiont, and a host plant, the macrosymbiont—each exist as free-living organisms. Early rhizosphere association of the two organisms is probably important for a number of reasons, including nutrition and chemotaxis.[184] The role of exuded plant compounds in stimulating symbiotic gene expression is now well documented for *in vitro* studies, as discussed later; a role in field conditions as well is indicated by study of different genetic lines of alfalfa.[109]

The complex cellular and developmental events occurring during nodulation are described by Kijne in this volume. In general, the steps that lead to the emergence of an infected, visible nodule structure are controlled by *Rhizobium* nodulation (*nod*) genes, and mutants with defects in the morphogenetic steps are termed Nod⁻ mutants.[13] This group includes some mutants with problematic phenotypes: They may result in formation of an in-

Table 14-1

Gene Name	Phenotype in Mutant
nod, nol	Failure to synthesize signal molecule(s)?; no nodule formation, nodule delay, or host range change
ndv, exo	Surface polysaccharide or metabolic changes; empty nodules form
fix	Infected nodules form, no N_2 fixation; may get some N_2 fixation in free-living cells
nif	Infected nodules form, no symbiotic N_2 fixation; no N_2 fixation in free-living cells

fected nodule, but infrequently, or on the incorrect host plant. Useful working phenotypes have been proposed for this somewhat complex set of mutants, including Hsn (host specificity of nodulation[101]) or Nod^d (nodulation delay).

A related group of mutants, including a number associated with surface defects, lead to the formation of nodulelike structures devoid of bacteria, or in which bacteria are not effectively released from infection threads into plant cells; this phenotype is termed nodule development defective (Ndv^-).[142] Among the genes in which mutations cause Ndv^- symbiotic phenotypes are *exo, ndv, lps,* and some metabolic pathway genes.

The third general symbiotic phenotype is characterized by structurally normal nodules, invaded by bacteria within infection threads, that nonetheless do not fix N_2; these are termed Fix^-. In these nodules bacteria may or may not be released into target cells. Such a phenotype may occur because of a defect in known *nif* genes, defined by relationship to *nif* genes essential for free-living organisms, or by a requirement for nitrogen fixation in free-living cells. The Fix^- phenotype also arises from a defect in *fix* genes, which encode functions apparently specific to symbiotic fixation or bacteroid differentiation in various *Rhizobium, Bradyrhizobium,* and *Azorhizobium* systems.

III. General Background: Approaches to Nodulation Genetics

The genetic exploration of *Rhizobium* has been greatly advanced by the use of recombinant DNA and transposable elements and by improvement in physical analysis of very large plasmids. This section includes a brief sketch of recent advances, and a discussion of some further experiments needed to make genetic analysis more complete.

A. Available Molecular and Genetic Techniques

Analysis of symbiotic genetics and molecular biology uses many of the same cloning and physical analysis techniques used in other systems, but also relies on continually improving techniques designed to overcome *Rhizobium*'s technical difficulties. These have been reviewed elsewhere,[80,111,126,129,145] and only a limited sketch can be included here. Mutagenesis relies to a great extent on the use of transposons, especially Tn*5* derivatives. Among these are several with resistances such as gentamycin and spectinomycin, or oxytetracycline, replacing the original that encodes Neomycin/Kanamycin resistance via the NPTII gene.[46,80] Another Tn*5* derivative includes the origin of transfer for inc-P plasmids; this has been used to mobilize otherwise nontransferable plasmids[168] and to create *Hfr* strains with origins at defined locations on a Rhizobium chromosome.[31,80,144]

Transfer of cloned genes can be accomplished with broad host range vectors, prominently those in the inc-P and inc-Q incompatibility groups. In brief, these vectors have been used both for functional complementation studies of cloned genes and for studies of gene regulation via cloned gene fusions, cloned *trans*-acting factors, and combinations of both (reviewed by Martinez et al.[129]; for examples see refs. 5, 54, 157, 169, 172, and 175). In studies of gene regulation, fusions to reporter genes are widely used, and both *lacZ* (β-galactosidase) and *uidA* (β-glucuronidase) gene fusions have succeeded in *Rhizobium* (for examples see refs. 138, 139, and 580).

The sophistication of the genetics available in various *Rhizobium* species differs widely, and as a rule, versatile genetic approaches—combining the ability to manipulate native and recombinant plasmids with conjugational mapping techniques, transducing phage, and so forth—have been of great help in understanding nodulation at the molecular level. Thus, improvements in genetics for each bacterial species may be required as an accompaniment to the push for more nodulation studies.

B. Location, Linkage, and Definition of Nod Genes

i. Plasmids in Rhizobium

Studies of *Rhizobium* have revealed the presence of very large plasmids.[145] In the case of *R. meliloti* these are referred to as "megaplasmids."[103,117,155] Indeed, their molecular size is over 1000 kb (1200 and 1500 kb[21]). In *R. leguminosarum* plasmids vary in size.[145] One study found *R. leguminosarum* bv. *trifolii* pSym to vary from about 50 to over 600 kb.[92] Plasmids bearing *nod, nif, exo,* or other symbiotic genes are referred to as pSym. More generally, plasmids are designated by the species name and strain number, followed by a lower–case letter designation starting with "a" for the smallest plasmid.

Symbiotic genes are located on plasmids in several *Rhizobium* species, as indicated by tests such as physical hybridization of *nif, nod,* and other genes to plasmids, and restoration or alteration of symbiotic properties by plasmid transfer.[105,129,145] Specific descriptions of individual plasmid-linked *nod* genes follow.

The symbiotic genes are not always all carried on the same plasmid. For example, exo-polysaccharide (EPS) synthesis genes are on pSymb in *R. meliloti,* whereas *nod* and *nif* genes are on pSyma.[73,103] Other plasmid properties such as transmissibility do not necessarily correlate with presence of symbiotic genes.[44,111,145]

Some symbiosis plasmids have been isolated and restriction mapped; examples are several plasmids from *Rhizobium leguminosarum* (see ref. 145 for detailed references). The large size of the *R. meliloti* megaplasmids precludes this, but genetic methods have made it possible to carry out fine structure maps. For example, Charles and Finan[31] used the general transducing phage φM12 to construct a genetic map of the plasmid pRmeSU47b, which bears exopolysaccharide genes and some metabolic genes such as dicarboxylate transport (*dct*) and thiamine synthesis (*thi*) genes. A feature of this analysis was the use of three transposable elements, each with a different resistance to establish linked overlapping transduction segments.

In *Bradyrhizobium,* large plasmids are found in some strains, but these do not appear to carry *sym* genes.[95] The stem nodulation strain *Azorhizobium caulinodans* has not been observed to contain a large plasmid.[183] In these and other cases,[191] all symbiotic genes thus are probably on the chromosome, and methods for chromosomal mapping are important. This will also be true for pSym-containing *Rhizobium,* in which some chromosomal genes are essential for full symbiotic function. In *B. japonicum,* reiterated insertion sequences were used to characterize and order deletions, which provides an acceleration to single–gene studies.[91] Conjugation and transduction mapping of chromosomes has been carried out in numerous *Rhizobium* strains (for reviews, see refs. 126 and 129; see also Osteras et al.[143a]), and the extension of this approach to other *Rhizobium* and to *Bradyrhizobium* and *Azorhizobium* will be useful for symbiotic studies. Ultimately, transducing phage linkage maps such as that established by Charles and Finan for pRmeSU47b would be appropriate to construct for the chromosomes of all *Rhizobium, Bradyrhizobium,* and *Azorhizobium* strains.

ii. Perspective on Gene Localization

Studies of nodulation genetics are largely concerned at present with defined genes and regions, but the decision to narrow study to these genes is worth reexamining. Specifically, the finding that *Rhizobium leguminosarum* bv. *viciae* plasmid transfer could confer rescue of symbiotic defects and/or change in host range (reviewed by Kondorosi and Johnston[111]) had a profound effect on the direction of nodulation gene research. Thus, after it was established that transfer of pRL1JI upon a plasmid cured *R. leguminosarum* bv. *viciae* background could restore nodulation, researchers peppered that plasmid with transposable element insertions to find individual symbiotic gene regions and also established that either of two relatively small cosmid-borne cloned segments could replace the intact plasmid in restoring host-specific nodulation.[20,127] Parallel work was carried out in *R. leguminosarum* bv. *trifolii*.[48,51,162] In *R. meliloti,* comparable studies have been somewhat hampered by the lack of a completely plasmid-cured strain; but studies of strains with large pSym deletions similarly led to the conclusion that *nod* genes are clustered in a fairly small region.[40,112]

The currently defined canon of "essential nodulation genes," described in more detail later, has emerged from the intensive analysis of the DNA segments defined by such early studies. Without a doubt, some similar essential nodulation genes will be found in other *Rhizobium, Bradyrhizobium,* and *Azorhizobium* strains.[183,186] However, it is possible that other essential nodulation genes are present elsewhere in *Rhizobium,* such as in the chromosome. The strategies described earlier would not have detected such genes, in that the recipient strains were also *Rhizobium* and therefore bore the same set of chromosomal genes. Also, some genes with subtle effects may have been missed in the initial screens. Indeed, more recent studies in *R. meliloti* indicate that genes affecting nodulation efficiency exist outside this region (see refs. 113, 146, and 151; J. Dénarié, personal communications). More *R. leguminosarum* bv. *viciae* genes with possible nodulation roles are also being described.[24,43,66]

A further complication is that *nod* genes in some species are now being shown to be dispersed. In *Bradyrhizobium japonicum,* one *nod* gene cluster is linked to *nifA,* but other genes are distant.[81] In *R. leguminosarum* bv. *phaseoli, nodABC* are not linked to other *nod* genes.[35,38,148,515]

Therefore, especially for researchers planning to study new *Rhizobium* species, it should be emphasized that random mutagenesis and phenotypic testing in each new species or strain should be used to supplement the search for *nod* or other genes homologous to those currently known. A search for genes co-regulated with known *nod* genes may be one useful approach.[9,113,160] Current work on broad host range strains,[1,11,19,81,122,140] as well as studies on mutations of defined *nod* genes,[39,51,101] suggests that several plant hosts should be included in the phenotype assays.

IV. Nodulation Genes and Plant Phenotypes

Rhizobium meliloti and the biovars of *Rhizobium leguminosarum* have been extensively studied at the molecular genetic level. Substantial information is now also available for *Bradyrhizobium japonicum.* These species serve to illustrate the identification of genes by a combination of molecular genetic techniques and phenotypic testing (see maps in Figure 14–1). Identification

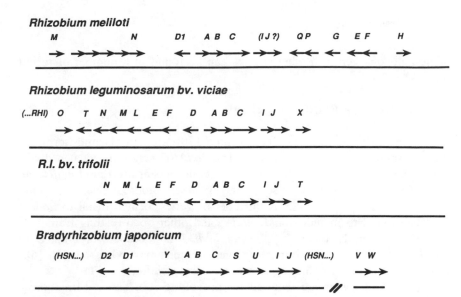

Rhizobium meliloti

M   N   D1   A B C   (IJ?)   QP   G E F   H

Rhizobium leguminosarum bv. viciae

(...RHI) O   T N ML E F D A B C I J   X

R.l. bv. trifolii

N ML E F D A B C I J   T

Bradyrhizobium japonicum

(HSN...) D2 D1   Y A B C S U I J (HSN...)   V W

Figure 14-1. Genes in the nodulation clusters of *Rhizobium* and *Bradyrhizobium* (see text for further details). *R. meliloti* nodulation genes are located on pSym-a, and are grouped in several apparent operons; as shown here, *nif* genes would be at the right side of the map. On the left (furthest from *nif*), open reading frames equivalent to *nodM* and *nodN* are separated by four open reading frames whose protein products are as yet uncharacterized (E. Kondorosi, personal communication). The regulatory *nodD1* gene is located divergently from *nodABC*; whether *nodI* and *nodJ* are located downstream of *nodD* is inferred from complementation studies and partial sequence, but has not been shown by complete sequencing and protein analysis. Another cluster includes *nodFE*, followed by *nodG* and *nodPQ*. It has not been demonstrated that these five genes are grouped in an operon, although some evidence suggests inducibility. The *nodH* gene is divergent from *nodFE*. Further regulatory genes such as *syrM*, *nodD3* and *nodD2* are also located on pSym-a.

The two biovars of *Rhizobium leguminosarum* shown here have a number of similarities in pSym-located *nod* gene organization. The common *nod* gene cluster includes *nodABCIJH;* downstream of this, certain strains of bv *viciae* have the broad host range gene *nodX*, while in bv. *trifolii* this position is occupied by *nodT*. In contrast, bv *viciae* carries *nodT* downstream of *nodFELMN*, a cluster apparently conserved in both biovars. Also in bv. *viceae*, the convergently transcribed *nodO* gene is located, adjacent to a series of rhizosphere inducible ("rhi") genes (not shown). What genes are located downstream of *nodN* in bv. *trifolii* is not known. Another difference between the two biovars is the relative location of *nif*, which would be at the left of the biovar *viciae* map shown here, and at the right of the bv. *trifolii nod* cluster. A third *Rhizobium leguminosarum* strain, biovar *phaseoli*, has a number of distinct features to its *nod* gene organization, that are currently being studied.

The *nod* genes of *Bradyrhizobium japonicum* are probably located on the chromosome, and several gene locations have so far been identified (see also chapter by Barbour et al., for further details). In one cluster (left side of map shown here) that is linked to *nifA* and to several *fix* genes, there are two tandem *nodD* genes located divergently from a cluster that includes a novel gene, *nodY*, just upstream from the *nodABC* genes; two other genes, *nodS* and *nodU*, intervene between *nodC* and *nodIJ*. Two other novel genes have been identified in this cluster; *nolA*, downstream from the two *nodD* genes, and *nodZ*, a host range gene downstream from *nodJ*. A completely different location loosely linked to the structure *nif* genes includes two *nod* genes, *nodVW*, that are probably regulatory. A third *nod* gene location distantly linked to *nifH* is inferred from deletion analysis.

of homologous genes in other species can be found in a number of recent reports.[1,2,3,4,33,35,86,87,117,122,130,148,149,183]

A. Nod *Genes of* Rhizobium meliloti

The *R. meliloti nod* genes are located on a 1200 kb sym plasmid (termed a megaplasmid) in at least two strains. The *nodABC* genes are located divergently expressed from *nodD1*, and in a second cluster are found *nodH* and a divergently oriented set of genes, *nodFE, nodG,* and *nodPQ*.[30,41,68,77,101,107,112,164,181] A third set of genes located distal to *nodD1* has been found to include *nodM, nodN,* and other open reading frames.[113] *NodD2* is distantly linked,[83,99,146] and *nodD3* is more closely linked to *nodH*.[61,99,139,159] At least two more *nod* gene regions may exist (see refs. 158, 151, and 146; and J. Dénarié, personal communication).

The *nodABC* genes are required for provoking root hair deformation—without these genes, no deformation or curling occurs at all in the host—and for stimulating early cell divisions in the inner cortex of the host plant.[40,59,60,112] Function of these genes is conserved across *Rhizobium* and *Bradyrhizobium* species.[50,78,112,112,131,] Whether *nodI* and *nodJ* follow *nodABC* is discussed later, under *R. leguminosarum.*

The *nodFE* genes of *R. meliloti* affect the timing and degree of nodulation and affect the type of root hair deformation on alfalfa.[40,101,180] These mutant defects are not restored by the cloned *nodFE* gene regions of other *Rhizobium,* despite the fact that the genes are very similar at the nucleotide level.[39,101] One possibility that this suggests is that the *nodFE* products may carry out a similar reaction on slightly different substrates.

The *R. meliloti nodH* gene is required for nodulation of alfalfa but not for nodulation of *Melilotus albus,* another plant in the *R. meliloti* "cross–inoculation group."[40,101,143,180] This is in part, though, a function of environmental conditions in that *nodH* strains nodulate either plant better if roots are shielded from light.[143] The *nodH* gene is, according to hybridization analysis, not generally found in other *Rhizobium* species,[153] but the broad host range strain NGR234 has a gene set capable of functionally complementing a *nodH* mutant for alfalfa nodulation.[122] The presence of *nodH* or *nodF* and *nodE* genes in other *Rhizobium* species inhibits their nodulation of their usual host.[39]

Mutations in the *nodG* gene have a very slight phenotype in SU47[180] but appear more visibly in the Rm41 strain background, where *nodG* mutants have a detectable delay in nodulation.[101] Mutations in *nodP* and *nodQ* show a delay in nodulation in SU47 backgrounds[30,40,163a]: The significance of this phenotype needs to be reevaluated in light of the discovery that there are very highly conserved second copies of *nodP* and *nodQ* located elsewhere in the *R. meliloti* genome.[163a]

B.   Nod *Genes of* R. leguminosarum

The *R. leguminosarum* biovars *viciae* and *trifolii* show some differences in gene arrangement. The *nod* genes are linked to *nif*, but the particulars are different for the two biovars, and both differ from *R. meliloti*. The *nodABC* genes are clustered,[156,163] and it has been shown in *R. l. viciae* that they are followed by two more genes, *nodI* and *nodJ*[69]; these five genes probably form an operon, together with *nodX*, found in some strains of *R. l. viciae*.[36] As is true for other bacteria, the *R. leguminosarum nodABC* genes are required for root hair curling and for nodule formation.[156,162,187] A single *nodD* gene is adjacent to *nodABC*.[162,163,167,176]

Nodulation of pea does not require *nodIJ*, but these are required for vetch nodulation.[24,25,69] Conservation of *nodIJ* in *R. l. trifolli* has been established by DNA sequence analysis,[179] and in *R. meliloti* is postulated, but presently this is only inferred from partial sequence[107] and complementation data.[25,187]; existence of the entire operon in *R. meliloti* has not been demonstrated yet.

In contrast to *R. meliloti*, *R. leguminosarum* shows *nodFE* downstream of *nodD*.[163,167] The action of *nodFE*[162,167] in host range has been extensively studied by genetic tests in the two *R. leguminosarum* biovars: mutations in this region in *R. l. trifolii* diminish the nodulation of that strain on clover, its usual host, and extend the nodulation ability to include a usual nonhost, pea.[51] Cloned genes have been manipulated to produce hybrid proteins of *R. l. viciae* and *R. l. trifolii nodE*, and particular protein regions that confer preferential nodulation of pea (or vetch) vs. clover have been delimited.[174]

In the two *R. leguminosarum* biovars, other genes are found that differ between the two biovars or among strains within a single biovar. These include *nodLMN*, *nodO*, and *nodX*. The *nodL*[178] gene is expressed in the *nodFE* operon; *nodMN* forms a distinct operon.[178] The *nodL* locus appears most needed for nodulation of peas, *Lens*, and *Lathyrus*; *nodM* and *nodN* mutations have a slight effect on nodulation of *Vicia*. The *nodO* gene appears to be a functional backup system for *nodFE*,[58] although the sequences are very different[179]: *nodO* is a secreted, Ca^{2+}-binding protein.[66] The *nodO* gene on its own appears required in some backgrounds for nodulation of *Vicia sativa*.[43] In *R. l. trifolii*, the *nodT* gene has been identified downstream of *nodIJ*,[179] while protein and sequence studies show it to be downstream of *nodM* in *R. l. viciae*,[24,179] in which it is required for nodulation of vetch.

The *nodX* gene is required for *R. l. viciae* to form nodules on certain cultivars of the host *Pisum sativum*, and this is conditioned by a single gene in the host plant.[88] A *nodX* gene sequence[36] is not found in all *R. l. viciae* strains. In strains without *nodX*, the DNA segment between the downstream end of *nodJ* and the downstream end of the linked *nif* gene cluster is roughly the same length, but the DNA sequence does not encode a protein.[36] The "single gene in bacterium and single gene in host" pattern is superficially reminiscent of the gene-for-gene pattern of cultivar–specific avirulence (resistance) in plant–pathogen interactions. There may be a positive ligand-receptor in each case, leading to triggering of a hypersensitive response in the parasite–host interaction and to the triggering of a positive host signal in the case of the symbiosis.[47]

Additional genes have recently been identified in the region of the *nod* cluster. A number of *rhi* (rhizosphere expressed) genes are located on the distal side of *nodO* from the common *nod* genes[66]; the roles of the genes and their products are not known.

C.   Nod *Genes of* Bradyrhizobium japonicum

In *Bradyrhizobium japonicum*, the *nod* genes are located in a separate cluster from the structural *nif* genes, although they appear to be roughly linked to the regulatory *nifA* gene. There are two *nodD* genes linked to but divergent from an operon containing at least eight genes, *nodYABCSUIJ*.[82,84,141,171] Thus, *Bradyrhizobium* contains a unique gene, *nodY,* at the beginning of the operon, and *nodS* and *nodU* are inserted in a position between the location of *nodC* and *nodI* in other species.[8,82,84] The *Bradyrhizobium nodABC* genes are required for nodulation as in *Rhizobium*.[141,131,84] The *nodY, nodS,* and *nodU* genes appear not to be required for nodulation of soybean, the most commonly studied host of *Bradyrhizobium japonicum,* but may be required for other symbiotic properties.[82] Two new and apparently unlinked genes that affect certain plant hosts have been designated *nodV* and *nodW.*[81] The sequences of these genes suggest they are regulators, possibly of as yet unidentified symbiotic genes.

Functions and organization of *B. japonicum nod* genes have interesting parallels in other organisms. In *Bradyrhizobium* sp. parasponia, a small gene designated *nodK* is located just upstream of *nodA,* as is *nodY* in *B. japonicum.*[165] However, the two genes are very distinct, adding to the mystery of what their functions might be. Genes very homologous to *nodS* and *nodU* are found in a broad–host range bacterium, *Rhizobium* sp. NGR234, and therefore they may represent genes needed for other hosts.[121] A combinatorial basis for host range in promiscuous bacteria is suggested by analysis of the symbiotic plasmid of NGR234, which has three general regions— HsnI, HsnII, and HsnIII—involved in nodulation of various plants.[122,140]

D.   *Perspective on Rhizobium Nod Genes*

i.   *Categories of nod Genes*

It can be seen from the maps and the preceding description that nodulation genes fall into several general categories: First, genes, such as *nodABC,* that are conserved both structurally and functionally across the fast–growing *Rhizobium*; it can be inferred that these are carrying out the same function in each symbiosis. However, such a conclusion must be considered tentative until an actual biochemical mechanism for gene action is demonstrated.

A second group of genes, such as *nodFE* or *nodIJ*, is conserved in terms of nucleotide sequence but appears to have preferential effects on, or to be required for, one host plant compared to others. One theoretical model that might account for this would be that such gene products are themselves signals whose conserved domains interact with common functions of bacterium and or plant and whose divergent regions have interactions with some plants and not with others. A second possibility is that such genes encode enzymes that carry out a conserved reaction (formation of a particular linkage, for example) with slightly different substrates according to the particular "allele" or species source of the gene.

Finally, a number of *nod* genes are unique to some *Rhizobium* species or biovars, and thus by inference carry out completely different functions in the various symbioses. The actions of these will need to be worked out for each system, but clues to mechanism already available (see later) indicate that at least some species-specific genes, such as *nodH* and *nodPQ,* may function in sequence with the common nodulation genes. This observation may provide a head start on identification of their biochemical functions.

ii. Future Studies in New Symbiotic Bacteria

A retrospective look at the progress in nodulation genetics in these three representative bacterial species shows several lessons important for future studies on other species. First, to find nodulation genes within a region, it helps to use as large a DNA segment as possible as a target for mutagenesis by transposons and to saturate for mutagenesis if one wishes to find all possible phenotypes. In addition, it is desirable to map and test all transposon insertions, not merely to determine map positions for those with a detectable phenotype. By this means, it is possible to show what DNA regions are not involved in nodulation, as well as those that are involved. For example, *nodD1* and *nodP* in *R. meliloti* were found in studies that used large fragments and intensive mutagenesis but were not observed in parallel studies carried out on the same organisms.[107,180]

Another lesson for future research is the importance of testing phenotype on more than one host plant and if possible in more than one bacterial strain background. *NodH* of *R. meliloti* was found to be required for nodulation of *Medicago* (alfalfa) but not of sweetclover, *Melilotus*.[101] By means of testing bacterial extracts on more than one plant, Faucher et al.[70] found the effect of *nodH* is to make a supernatant factor switch from effectiveness on alfalfa root hairs to effectiveness on vetch, according to the presence or absence of *nodH*.[71] The role of *nodFE* for host range in *R. leguminosarum* bv. *trifolii* was dramatically apparent when *nodFE* strains were tested for nodulation both on clover (the usual host) and on pea.[51] Similarly, the role of *nodIJ* in *R. leguminosarum* bv *viciae* has also been demonstrated by its requirement for nodulation on vetch, although not apparently required for pea.[25,69]

The elementary lessons of molecular genetics also should be emphasized. A number of studies on symbiotic mutants have revealed the presence of native insertion sequences in various *Rhizobium* species, thus the first lesson: Never assume that a transposon is the cause of a mutation until it has been reconstructed by marker exchange or transduction and has been complemented by the appropriate wild type clone. A converse variation on this theme constitutes the second lesson: In *Rhizobium* studies, do not assume that a gene has no essential function, just because an insertion in that gene has no symbiotic phenotype. The gene might be duplicated—at least!—and the redundant copies sufficient to allow function.[148] In *R. meliloti*, *nodD* and *nodPQ* are examples of this.[153,163a] Third, it is also now apparent that non-homologous genes may serve as back-up systems for each other: The functional redundancy of *nodFE* and *nodO* in *R. l. viciae* exemplifies this.[58]

A final special caution that was important to *Rhizobium* genetics long before molecular biology came along should also be repeated: Beware of mixed nodules and of revertants in the nodule population. This also applies to genes carried on recombinant plasmids in *Rhizobium*, since these can be lost with consequent changes in phenotype and in growth rate within the emerging nodule.

V. Regulation

The accelerating pace of work on nodulation gene mechanism (the analysis of *Rhizobium* factors that cause plant effects, for example) has been made possible by discoveries relating to nodulation gene regulation (for reviews see refs. 57, 115, 125, and 154). Over a number of years, various physiologic and developmental studies had indicated the likelihood that preexposure of plants and bacteria to each other had effects on their subsequent nodulation efficiency or on the production of factors (reviewed by Long[124]). Direct examination of nodulation gene expression, through use of reporter gene fusions to *nod* operons, or by RNA analysis, has shown that the *nod* genes are typically not expressed in free-living *Rhizobium* cultures, although a very low basal level of expression can sometimes be observed, that varies with culture age. The exception is the *nodD* gene, which in most bacteria is constitutively expressed.

Expression of the *nod* genes is caused by a combination of plant signals and bacterial regulatory elements. In several *Rhizobium* systems it has been shown that small signal molecules from the host plant cause induction of *nod* gene expression in the homologous *Rhizobium* or *Bradyrhizobium* strains.[85,100,106,116,137,157] In alfalfa, luteolin (3',4',5,7-tetrahydroxyflavone) and 4,4'-methoxychalcone have been characterized as prominent inducers, along with others that are somewhat less active, produced by alfalfa seeds and roots.[138,144,132,93,94] It is interesting that alfalfa roots secrete a different spectrum of flavonoids than the germinating seeds.[132] In clover, the roots produce 4',7-dihydroxyflavone and other inducing compounds, along with inhibitors.[49,150] In soybean, daidzein (4',7-dihydroxyisoflavone) and other isoflavones are produced.[116] *R. l.* bv. *viciae* also respond to flavonoids, including inducers such as naringenin and various inhibitors.[74,192]

For bacteria to induce genes in response to the plant signal, they must have a working copy of the *nodD* gene product. Ths protein acts as a positive regulator of the inducible *nod* gene operons.[57,125,126,189] By sequence homology, NodD is a member of a growing family of identified positive regulators referred to as the LysR–NodD family.[96] These proteins display fairly strong homology to each other, and rather little homology to other regulators with the exception of suggestive homology to a Cro-like helix-turn-helix motif.[167]

In many *Rhizobium* species, there exist multiple copies of *nodD*. These studies begin with the demonstration of DNA-segment homology between a known *nodD* gene and other fragments within the genome of a bacterium.[153] In *R. meliloti, R. fredii,* and *B. japonicum*, genetic analyses have shown that the multiple copies are functional: For example, in *R. meliloti,* as long as one of the three *nodD* copies is wild type, the strain is Nod$^+$.[98,99] Only when all three copies are inactivated by insertion does the strain lose nodulation phenotype. One of the *nodD* genes, *nodD3*, activates *nod* gene expression independently of flavonoids.[139] In *B. japonicum,* there may be other *nodD* genes in addition to those already found, for a double mutant of *nodD1* and *nodD2* is still Nod$^+$.[84] An alternative explanation is that there may be other regulators besides *nodD*. In *R. fredii,* the two *nodD*'s appear to have distinct functions,[2] and in *R. l.* bv. *phaseoli* two of three *nodD*'s appear to be flavonoid-independent activators.[37,38]

The existence of multiple *nodD*'s raises the question of their function and significance in the symbiosis. There is evidence that some of the *nodD* diversity relates to the interaction with various inducers,[90] as discussed later. In addition, it is evident that numerous other regulatory genes may be involved in controlling the activity of the multiple *nodD*'s. In *R. meliloti,* the *nodD3* gene is dependent for activity on a linked gene, *syrM*, whose sequence shows it to be highly homologous to the *nodD* family.[8a,128] Since *syrM* is highly expressed within nodules but not in cultured *Rhizobium* cells[166] (and J. Swanson, S. R. Long, manuscript submitted for publication), special circuits may exist for the control of nodulation genes at later stages in the symbiosis.

The regulation of the various *nodD* genes shows interesting differences within and between bacterial species. In most *Rhizobium, nodD* is constitutive[137,157]; however, in *B. japonicum, nodD1* expression shows inducibility by isoflavones.[8] The *nodD* gene of *R. leguminosarum* bv. *viciae* displays negative autoregulation of its own expression,[157] but *nodD1* of *R. meliloti* appears not to be autoregulatory.[137] Instead, strains of *R. meliloti* have a separate locus that down-regulates expression of the *nodD1* gene.[114] The *nodD3* gene product of *R. meliloti* is capable of activating *nod*-box dependent and *nod* gene expression independently of inducer.[139] However, the expression of *nodD3* is controlled by another locus, *syrM*,[8a,139] and one or both genes may be regulated by symbiotic and environmental signals (ref. 61 and J. Swanson and S. R. Long, submitted). The control of expression for regulatory genes is clearly variable enough to merit extensive study in each individual symbiotic system.

In addition to the *trans*–acting protein, NodD, *nod* gene regulation involves *cis*-acting elements referred to as *nod* boxes.[157] These are highly conserved sequences found upstream of most nodulation genes; they have an unusual length, almost 50 bases, and the pattern of conservation is also striking. The transcripts of *R. meliloti nodA, nodF,* and *nodH* have been mapped[77,139] and shown to initiate about 26–28 bases downstream (3′) of the end of the *nod* box; the position in *R. leguminosarum* bv. *viciae nodA* is comparable.[173]

The molecular mechanism of *nod* gene regulation is not well understood. For example, it is currently not known which form of RNA polymerase holoenzyme is involved in *nod* box promoter recognition.[75] The binding of nodD to *nod* box promoters has been shown for several systems,[76,97,114] and it has been shown that the footprint of NodD corresponds well with the *nod* box DNA.[76,114] However, it is not known what the role of inducer is: NodD–DNA binding is inducer-independent.

Despite the current lack of biochemical evidence for the inducer's role in NodD-mediated promoter activation, genetic evidence is strong that some type of direct interaction occurs between NodD proteins and inducer molecules. Spaink et al.[175] determined that the ability of a strain to respond strongly or weakly to structurally different inducers (with different hydroxylation patterns or different skeletons, for example) was changed if different alleles of the *nodD* gene were placed in that strain. Mutational changes in *nodD* also alter response to inducer,[22,23] and the *nodD1* and *nodD2* genes of *R. meliloti* show differential response to diverse plants and compounds.[90,98,139] This implies a structurally sensitive binding of NodD protein and an inducer. Broad host range *Rhizobium* strains have a NodD that responds to a wider variety of inducer structures than does the NodD of narrow-host-range *Rhizobium*, such as *R. meliloti* or *R. leguminosarum* bv. *trifolii*.[3,12,100] Mutations or recombinant variations in *nodD* genes that alter the tolerance of the NodD protein for various inducers also have an effect on the *in vivo* symbiotic performance of the *Rhizobium* for various plant hosts, indicating that inducer–NodD interactions are in some cases a first critical level of host–*Rhizobium* specificity.[133,173] However, no evidence rules out a direct role for additional proteins, currently unknown, in the response to inducers and/or in other aspects of NodD-mediated expression.

A.   Perspective on Gene Expression

The circuit comprising plant inducers and *Rhizobium* NodD's provides a satisfying first approximation of how *nod* genes are regulated. Among the generalizations from current studies that should be applied to other systems are the need to look out for multiple *nodD* genes, the importance of characterizing the native inducers synthesized by various plant hosts, and importance of looking both for activating and autoregulatory (negative) properties of NodD protein.

The complexities that may exist alongside the simple primary circuit are also becoming apparent, and these suggest further studies that should be carried out on each *Rhizobium* system. Indeed, there may be significant differences to be found, along with new universal principles, in the comparative study of various *Rhizobium*, *Bradyrhizobium*, and *Azorhizobium* systems. In the following section, I would like to point out some important unknowns.

Beginning with the inducers, it is essential to examine several questions. As indicated already, it is important to define the spectrum of inducers actually synthesized and secreted by each plant—keeping in mind that these may vary during development as in alfalfa[132]—and to define the way in which the various inducers interact with various NodD's. Other questions also remain. What is the total set of effects that occur in *Rhizobium* cells responding to inducer? Are all of them dependent on NodD, or are their responses to flavonoid inducers outside the known NodD–*nod* box circuit? Recent work on alfalfa indicates a NodD-independent effect of luteolin and certain other flavonoids on the rate of *R. meliloti* growth (D. Phillips, personal communication).

Circuits also come into view as the multiple *nodD* gene systems are considered. In *R. meliloti* strain 41, a repressor controls expression of *nodD1*,[114] and in strain SU47 *syrM* probably activates expression of *nodD3* and of the exopolysaccharide regulatory gene *syrA* (ref. 139 and J. Swanson and S. R. Long, unpublished results). What other genes are involved in control of *nodD* genes themselves and control of inducible *nod* box–type promoters? One may expect, in all *Rhizobium* systems, to find genes controlling the relationship of *nod* gene expression to other cell activities such as nutrition and growth. The active search should continue for other loci that qualitatively or quantitatively control *nod* gene expression. In *R. meliloti,* a chromosomal locus that down-regulates *nod* gene expression in high ammonia concentration has been described,[61] and a locus required for activity of all NodD's, and possibly for expression of some *nodD* genes, is also located on the chromosome (J. Ogawa and S. Long). These genes are probably a small subset of those that will be known within a few years. *Bradyrhizobium, Azorhizobium,* and *Rhizobium* systems may differ sufficiently that independent genetic studies of each system will be necessary to find the relevant components. The finding of new putative regulators *nodV* and *nodW* and of effective nodulation in a NodD⁻ *Bradyrhizobium japonicum*[81,84] may prefigure the characterization of such regulatory differences.

Circuits and multiple genes needed for *nod* gene regulation should be considered with respect to the subject of *nod* promoter mechanism. Several very specific questions remain unanswered: What form of RNA polymerase (i.e., which sigma factor) recognizes *nod* box promoters? One would predict that formation of the actively transcribing complex would involve NodD, inducer, RNA polymerase, and promoter DNA. But are there other proteins involved? No genetic or biochemical data rule this out at present. Such proteins may be sought biochemically and through genetic experiments such as those described earlier for circuitry studies.

Finally, regulatory studies will be incomplete until they examine levels of control beyond transcription initiation. Transcription termination and transcript stability, translational initiation, translational coupling, and protein stability may all be subject to specific control in nodulation systems. Several *nod* gene open-reading frames show overlap sequence with a neighbor, which is associated with translational coupling. This may be tested by creating nonpolar mutations that alter the translational efficiency of the upstream gene and observing the rate of translation of the downstream gene.

The existence of another level of regulation is suggested by the observation that *nod* operons show extraordinarily long transcript leaders, about 200 bases between transcript initiation and translation start site.[77,139,173] Inverted repeats found in these leaders, or other structural or sequence features, may affect transcript timing, termination, or stability.[77] Examination of this and other molecular behaviors will be necessary before a full picture of *nod* regulation is available.

VI. Genetics of Surface Determinants

The outer surfaces of bacterial cells are complex. The complexity is all the more problematic in *Rhizobium* and *Bradyrhizobium,* where a number of species biovars and natural isolates need to be analyzed with respect to how surface components relate to successful host-specific nodulation. Genetics is an essential companion to chemical analysis in a subject area such as this: Through genetic analysis it is possible to distinguish truly significant variation in carbohydrate apart from fortuitous and irrelevant differences from one species, strain, or biovar, to another.

Numerous lines of evidence show a role for various extracellular or surface carbohydrates in nodulation. The recent role of genetics in study of each of these components will be summarized in this section.

A. Exopolysaccharides

Extracellular (exo⁻) polysaccharides (EPS) can be recovered from culture fluid of bacterial cells and purified for chemical study.[26] Several cases show a genetic link between the abundance and/or structure of EPS and symbiotic properties.

Genes for production of extracellular polysaccharides have been extensively characterized in *Rhizobium meliloti,* and new studies are emerging in other *Rhizobium* species. The major research concerns have been the identification of genes for polysaccharide synthesis, accompanied by chemical analysis of normal and mutant polysaccharide structure and analysis of symbiotic phenotype, and the study of gene regulation.

i. R. meliloti exo Genes and Phenotypes

The screen for mutants in *R. meliloti* was assisted by the observation that this species normally produces an acidic extracellular (exo-) polysaccharide now termed EPS I that binds fluorescent dyes such as Calcofluor or Cellofluor,[119] laundry whiteners that have long been used in plant cell wall studies because they fluoresce when bound to beta-glucans such as cellulose. Mutants that do not synthesize the *R. meliloti* polysaccharide have been grouped into seven complementation groups, and to these have been added mutants that do not add some of the acidic side groups to the octasaccharide repeat unit and regulatory mutants with altered expression of EPS I.[53,118,119,123,136] Altogether, at least 12 complementation groups of EPS I– related *exo* genes are known, and almost all of these are linked on the larger (1500 kb) megaplasmid pSymb in *R. meliloti.*[73] In addition, at least six complementation groups exist for genes affecting synthesis of a recently discovered second (noncalcofluor binding) exopolysaccharide termed *EPS II.*[80] Although EPS II is structurally different from EPS I, it substitutes phenotypically on most host plants.

As it had previously been proposed that extracellular polysaccharides were the basis for host recognition in the *Rhizobium* system, it was an important landmark that the Exo- mutants in *R. meliloti* were found to have completely normal host range and instead to have a different defect: They are able to curl root hairs and in many cases to initiate invasion, but they cannot carry out infection. The few infection threads that form, abort early, and although nodules are formed by plant cell division, the plant tissues are devoid of bacteria and are termed *empty nodules*.[72,136]

ii. *R. leguminosarum* and Other Rhizobium Exopolysaccharide Genes

Nonmucoid strains found following a screen of mutant *R. leguminosarum* bv. *phaseoli* strains led to identification of cloned genes for and mutations in *pss* (polysaccharide synthesis) genes.[15,16,17] Because these were chromosomal genes, they could be mutated by Tn*5* in both *R. leguminosarum* bv. *phaseoli* and *R. l.* bv. *viciae*. The phenotypes were surprisingly different, in that a *pss*::Tn*5* mutant of *R. l.* bv. *phaseoli* was Nod$^+$ Fix$^+$ on bean, but *pss*::Tn*5* of *R. l.* bv. *viciae* was Nod on pea.[15] A preliminary study had also indicated polysaccharides to be necessary for *R. leguminosarum* bv. *trifolii* infection of clovers.[45] This indicates that EPS may be essential for symbiosis in one system (e.g., pea) but redundant in another (e.g., bean). Even in a single system, though, the relationship of genotype to phenotype may be complicated. While the reduction of EPS synthesis by a *pss*::Tn*5* mutation leaves symbiosis relatively unaffected, reduction by EPS by overexpression of the inhibitory *psi* gene causes a symbiotic defect.[16,17]

Rhizobium broad–host range strain NGR nodulates a variety of legumes. Mutants chosen for altered mucoid phenotype displayed normal symbiotic phenotype on some hosts but ineffective nodulations on others.[32] Using one such mutant to generate an R-prime, Chen et al. define at least five linked loci and two more unlinked genes encoding functions for production of EPS. Some of the Tn*5* insertion mutations initially appeared to be (now assigned to *exoY*) dominant to wild type when carried on an R-prime; this was subsequently found to be a regulatory effect from linked gene *exoX* that is inhibitory when overexpressed high copy relative to copy number of wild-type *exoY*.

iii. Regulation

Relatively little is currently known about the regulation of polysaccharide genes. Regulatory loci have been identified in *R. meliloti, R. leguminosarum* bv. *phaseoli,* and *Rhizobium* sp. NGR234. The EPS-I in *R. meliloti* is regulated in response to nitrogen in the medium, and mutants that derepress EPS expression in the presence of N have been localized to the *exoR* gene; a second gene, *exoS,* also appears to be a negative regulator of *exo* gene expression.[53] EPS-related genes are expressed by bacteria within nodules.[110]

In *R. phaseoli,* the *psi* gene product has a negative effect on the synthesis of polysaccharides, and *psi* in turn is repressed by the *psr* regulatory gene.[16] A similar observation has been made in *Rhizobium* sp. NGR234: *exoX* appears to repress exopolysaccharide synthesis, but the presence of *exoY* keeps *exoX* in balance (probably as a repressor) and permits *exo* expression to continue.[89] Similar regulatory relationships probably exist in *R. meliloti* (G. Walker, personal communication).

It might be predicted that at some level, *exo* genes should share regulatory controls with other *sym* genes. The relationship to nitrogen described earlier is particularly interesting in that light! A connection of nodulation and exopolysaccharide gene regulation is implied by the observation that a *R. meliloti* regulatory genes, *syrM,* up-regulates *nod* gene expression and production of polysaccharide capsule.[8,139] Further work on *exo–nod* relationships in numerous species is needed to determine how *Rhizobium* may coordinate these two symbiotic functions.

B. Neutral Glucans

Another category of extracellular carbohydrate is the beta 1,2 glucan, a cyclic neutral molecule found in *Agrobacterium* as well as *Rhizobium* and *Bradyrhizobium.*[10,26,134] The glucan molecules vary in degree of polymerization among various *Rhizobium* and *Agrobacterium* and may carry anatomic moieties such as phosphoglycerol.[10,134,135] Two genes that are required in *R. meliloti* for the export and synthesis of this molecule are *ndvA* and *ndvB* (for nodule development). *R. meliloti* with mutations in these genes are similar to Exo- bacteria in that they provoke formation of empty nodules.[63,79,102] These two genes are homologous to the *chvA* and *chvB* genes of *Agrobacterium tumefaciens.*[63] The smaller gene, *ndvA,* encodes a protein that shows some homology to Hly B protein (hemolysin).[177] The *ndvB* gene is extraordinarily large, over 10 kb long, encoding a protein of 319 kDa.[104] This gene product correlates with a very large membrane protein shown to be associated with nucleotide sugars.[193]

The role of the 1,2 glucan in nodulation is not completely clear. There are indications that it is essential for physiologic adjustment of free–living bacteria to osmotic stress,[62,135] and this may be important for growth, cell division, and regulation within the infection thread. *R. meliloti, ndvA, ndvB* mutants have pleiotropic defects, including decreased motility and loss of phage sensitivity, and these are partly corrected by growing the *ndv* mutants in a high–osmolarity environment. Second-site suppressors of *ndv*-caused motility defects are not fully corrected for symbiotic function, and *ndv* pseudorevertants selected for restored symbiotic function do not show a regaining of β-1,2-glucan synthesis.[62,64] Thus, it is possible that the *ndv* requirement for symbiosis does not act solely through B-glucan involvement.

C. Lipopolysaccharide (LPS)

The third major extracellular polysaccharide that has been genetically studied in *Rhizobium* and *Bradyrhizobium* is lipopolysaccharide.

As with other Gram negatives, lipopolysaccharide (LPS) in *Rhizobium* is a component of the outer membrane and consists of a lipid moiety (lipid A), which is part of the outer leaflet of the outer membrane, and polysaccharde whose inner core (lipid-attached) portion is conserved and whose outer O-chain) part is variable. Among the screens used to find *Rhizobium* strains mutant for LPS are colony morphology and reactivity with phage and/or with monoclonal antibodies.[18,28,29,42,107a,147]

LPS does appear to be required for some, but possibly not all, effective symbioses. Mutants lacking the O-antigen chain are defective for symbiosis (they form empty nodules) in *R. leguminosarum* bv. *phaseoli*.[28,29,42] On the other hand, an *R. meliloti* LPS mutant was effective in symbiosis with alfalfa.[34] LPS-defective mutants of some *Rhizobium* are symbiotically normal, but such mutants in other cases make empty nodules or show other defects; this indicates that LPS sometimes has structural information that in other systems may be carried by other molecules. This is also supported by the finding that a certain type of LPS can phenotypically suppress the symbiotic defect of *R. meliloti* Exo$^-$ mutants.[188]

In addition to analysis of mutants and the clones corresponding to mutated genes, some genetic clues are available from study of diverse strains and/or from construction of hybrid strains with new, different pSym. One of the first lessons that can be gleaned from this work is that there is not necessarily just one effective LPS structure for a particular biovar. For example, strains of *R. leguminosarum* bv. *trifolii* display more than one structure of LPS outer polysaccharide.[27]

These complications point out the value of a detailed genetic dissection of LPS synthesis in individual *Rhizobium* and *Bradyrhizobium*. Analysis of *R. leguminosarum* bv. *phaseoli* has shown that at least five genetic regions are required for LPS synthesis, as shown by complementation of LPS mutants in that strain.[18,29] By further complementation tests with 2 mutants of *R. l.* bv. *trifolii*, it was shown that at least one biochemical step for O-antigen must be common to the two different biovars. Another *R. l.* bv. *trifolii* mutant appeared to have a defect in a step specific to synthesis of a strain-specific structure, because the *R. l.* bv. *phaseoli* genes that restored LPS synthesis to that *R. l.* bv. *trifolii* mutant did so in such a way that the resulting LPS resembled *R. l. phaseoli,* not *R. l. trifolii*. Significantly, the *trifolii* LPS mutant, which was now complemented so that it synthesized *R. l. phaseoli* LPS, was nevertheless restored to normal clover nodulation. This, together with the finding of diverse LPS structures within biovars, suggests that substantial variation is tolerated for LPS in symbiosis.

The continued definition of genes, and characterization of associated enzymatic activity through determination of LPS structures, will be necessary before complementation analysis can reveal exactly how LPS may relate to symbiotic effectiveness, and in particular which LPS structural features, if any, relate to host specificity.

A further complication may lie in the fact that the structure of LPS itself may change during symbiotic development.[170,190] Therefore, analysis of LPS genetically caused structural defects that are or are not associated with symbiotic defects should be interpreted with caution until it is determined what differences exist in the symbiotic form of LPS. Finally, this discovery of developmental regulation of LPS provides a further impetus for characterizing LPS genes: The expression of these genes may be symbiotically regulated and may tie in with global controls for bacteroid differentiation.[190]

D. *Perspective on Extracellular Polysaccharides*

What do the various extracellular polysaccharides do? Several lines of evidence indicate that EPS molecules act as signals and may act also as physical lubrication or defense during the second stage of nodulation, when the bacterium is penetrating into the plant within infection threads. Djordevic et al.[52] showed that the intact large–molecular–weight polysaccharide of *Rhizobium* sp. NGR234 can restore function to Exo⁻ bacteria. J. Leigh (personal communication) and G. Walker (personal communication) have found that not the intact EPS, but the octasaccharide repeat units of the *R. meliloti* polysaccharide have this property. These results argue for the direct molecular involvement of EPS itself in symbiosis.

It could be argued that the polysaccharide is simply an osmoticum for the bacterial colony within the infection thread. However, it has been pointed out that mutant polysaccharides lacking one of the modifying acidic groups are ineffective at nodulation. One interpretation of this is that structural features of the polysaccharide are important and that a specific plant receptor exists. Whether such receptors correlate with lectins previously proposed to bind bacteria will need to be tested. The exact nature of the bacterial ligand for such lectins has continued to be debated. Another interpretation of the structure function data is that structural features of EPS are needed for bacterial processing of the EPS in order to produce appropriate oligomers.

In examining the overall literature on genetics of *Rhizobium* cellular polysaccharides, several cautions and opportunities are apparent. The temptation to correlate primary structure with function (invasiveness, host specificity, etc.) should be resisted during early stages of study. Three genetic studies that suggest the need for caution are the Glazebrook and Walker discovery of a cryptic second EPS,[80] the finding by Dylan et al. that symbiotic pseudo-revertants of *ndv* mutants are not restored for beta-glucan production,[64] and the report by Williams et al. that an EPS mutant phenotype can be suppressed by a second-site mutation affecting LPS.[188] These examples also remind us of what a powerful tool genetics can be, in its ability to uncover hidden and/or interacting components through second–site mutations.

Generalizing from this specific comment, it is apparent that understanding polysaccharides in nodulation will require detailed genetic study within each *Rhizobium* species or biovar, followed by analysis of hybrid strains with new gene combination, and accompanied by structural characterization of intermediates and final polysaccharide products. Such combined genetic–biochemical study will be an aid to the search for plant binding and receptors.

VII. Mechanism

The *nod* genes other than *nodD* and probably *nodVW* are assumed to participate in the mechanism of invasion. On first principles, several kinds of role can be postulated: First, the *nod*-gene–encoded proteins could themselves be secreted to act as signals, or enzymes, at the plant target cell. Second, *nod* proteins could be enzymes modifying the bacterial cell surface to facilitate bacterial cell–plant cell interaction. Third, *nod* proteins could be enzymes that synthesize signals that act on plant cells.

All these functions remain possible in the abstract, and the discovery that the secreted NodO protein can partly substitute phenotypically for the membrane–bound (and probably enzymatic) NodE protein should encourage researchers to keep an open mind about how various *nod* gene products may function. Since a great deal of active current research is focused on the signal-synthesis model for *nod* gene product action, this will be summarized here.

E. Nod *Gene Factors*

The chapter by Kijne in this book describes the study of factors, produced by *Rhizobium,* that have biological effects on growing root hairs and roots of host plants. The demonstration that these factors are related to nodulation has come most convincingly from genetics study, in that it can be shown that the presence of the factors in the supernatants of bacterial cultures requires the genetic presence of *nod* genes in the strain and also requires that the bacteria be induced with plant extract or pure flavonoid inducers. For example, van Brussel et al. showed that such a supernatant factor caused alterations in overall root growth (*t*hick *s*hort *r*oots (tsr) factor) and in root hair development, and that this factor correlated with the genetic presence and with the induction by flavonoids of *nod* genes.[182] *R. meliloti* factors

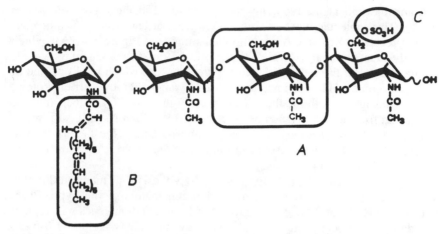

Figure 14-2. Structure of the *Rhizobium meliloti*-produced factor, NodRm-1. The compex molecule includes several distinct components, whose synthesis must be accounted for by study of *nod* gene-encoded functions. As other structures are discovered, in *R. meliloti* and other bacteria, the number of functions to be accounted for will expand.

(A) is a bcta-1,4-linked N-acetylglucosamine residue. Structural studies of NodRm-1 indicate it to be a tetramer of this modified sugar; ongoing studies in several organisms suggest that other sizes of oligomer may also be found (E. M. Atkinson, K. Faull and S. R. Long, unpublished observations; H. Spaink and B. Lugtenberg, N. Campbell and W. Broughton, and J. Denarie, personal communications). Among the functions to be demonstrated will be the formation of the glucosamine and the addition of the acetyl group, and the formation of the beta-1,4-linkage. If the backbond is characteristic of all Nod factors, then the common *nod* genes are appealing candidates to encode the basic linkage functions.

(B) is an N-acyl linked fatty acid; the structure of fatty acids in other Nod factors has not yet been determined. Among the synthetic processes that need to be studied are, the formation of the fatty acid-acyl donor species; control of the hydrocarbon chain size and state of reduction; and donation of the acyl group to the modified sugar. Another important question will be whether this occurs before or after linkage of the sugar subunits.

(C) is a sulfate, linked in NodRm-1 to the reducing and sugar subunit. The participation of *nod* gene encoded functions in sulfate metabolism has been demonstrated by Schwedock and Long[164] (1990), who found that the *nodPQ* encoded enzymes activate sulfate to form APS, and by Lerouge et al., who found that a strain mutated in *nodH* fails to form the sulfated NodRm-1 structure, and instead secretes a tetramer lacking sulfate. This combination of mutant analysis and direct biochemical study may be a useful approach to deduction of function for other *nod* genes.

causing root hair deformation on specific hosts[70,71] and causing cell division nonspecifically in plant protoplasts[161] have also been described.

The structure has been determined for one such factor produced by *Rhizobium meliloti*.[120] The molecule has been designated NodRm-1 and is secreted by Nod[+] cells that have been induced with the flavonoid, luteolin. Its structure is shown in Figure 14–2. It is a somewhat complex molecule, consisting of a tetramer of β-1, 4-linked *N*-acetyl glucosamines, of which the unit at the nonreducing end bears an *N*-acyl substitution of a C-16 fatty acid and the unit at the reducing end is modified at the 6-position with a sulfate. This molecule must be constructed from at least three types of metabolic precursor: first, the N-acetylglucosamine (see Figure 14–2A), which probably must be charged via a nucleotide before formation of glycoside linkages; second, the fatty acid (see Figure 14–2B), which modifies one of the glucosamines and which may be carried by an ACP (acyl carrier protein) prior to linkage; and third, the sulfate (see Figure 14–2C), which may be transferred from a carrier (such as APS or PAPS), via a specific sulfotransferase. Acetyl modifications have been found on some Nod factors as well (H. Spaink, J. Dénarié, personal communications; E.M. Atkinson and S. Long, unpublished observations).

The structural complexity supports the hypothesis that many nodulation genes encode enzymes synthesizing nodulation factors of this type. Proof that the NodRm-1 or 2 factor is synthesized by nodulation genes will require elucidation of the complete synthetic pathway, ultimately including *in vitro* synthesis of the factor(s) by the purified enzymes encoded by *nod* genes. Recombinant DNA studies will greatly facilitate certain aspects of this research. Single *nod* genes can be overexpressed as a means to purifying the *nod* gene proteins; these can be tested directly for enzymatic function and can be used to raise antibodies for localization of protein within wild–type cells, and for inhibition of *in vitro* synthesis, as a means of proving enzyme function.

Recent progress studying the NodRm-1 factor illustrates a possible path to identifying function for *nod* genes. The presence of all *R. meliloti nod* genes, including *nodQ* and *nodH,* is necessary for production of alfalfa-specific supernatant factors.[70,71] Supernatants of a *nodH* mutant, for example, lose hair deformation (Had) activity on alfalfa but now gain Had for vetch (*Vicia sativa*). In such strains, NodRm-1 is synthesized.[120] When *nodH* is mutated, a different molecule, NodRm-2, is synthesized; this lacks the sulfate at the reducing-end sugar moiety, and this purified molecule also shows the altered host specificity characteristic of supernatants of *nodH* mutants.[152] This suggests that the *nodH* product is involved in the transfer of sulfate to a *Vicia*-active tetrasaccharide. The *nodQ* gene product is also implicated genetically in fully active production of the sulfated NodRm-1.[70] The homology of *nodP* to *E. coli* DNA[163a] has made it possible to identify *nodP* and *nodQ* as homologous to *cysD* and *cysN*, which encode the *E. coli* ATP sulfurylase that synthesizes APS[164]; the cloned *nodPQ* genes display an ATP sulfurylaselike activity *in vitro*. Thus, *nodPQ* and *nodH* can be inferred to participate in the synthesis of an activated sulfate carrier and in the transfer of a sulfate to a modified oligosaccharide signal molecule.

F. Nod *Gene Mechanism*

Functions for other *nod* genes have been proposed by several researchers (J. Denarie; J. A. Downie; E. Kondorosi and A. Kondorosi; H. Spaink; and the author of this review and colleagues). The homology of *nodM* to *glmS* suggests that *nodM* may be responsible for synthesis of glucosamine, a precursor to the oligosaccharide. *NodL* has homology to *lacA* and might carry out an acetylation reaction, perhaps of glucosamine.[55] Because *nodF* and *nodE* are homologous to ACP and to a β-ketoadipate synthase, these may function in providing the fatty acid for transfer to a sugar residue. The *nodABC* genes have few helpful homologies, although a slight similarity of *nodC* to yeast chitin synthase (E. M. Atkinson, B. Rushing, S. R. Long, unpublished) supports the proposal that this gene product might be involved in polymerization or other reactions of *N*-acetylglucosamine. The *nodABC* genes may act together to synthesize the oligosaccharide backbone of signal molecules of the NodRm-1 type, or family. Whether this backbone is synthesized in discrete lengths of a few residues or polymerized as a long chain and subsequently cleaved will need to be tested directly.

i. Clues from Nucleotide Sequence Homology and Localization

The sequences of various nodulation genes have been determined by a number of laboratories. Comparisons of these sequences with others in the data base have provided suggestions of possible functions of possible substrates for a few *nod* genes. Some are homologous to known regulatory genes: *nodD* (as discussed earlier), *nodV*, and *nodW*. Possible enzyme relationships are cited in the following table:

gene	homology and key reference
nodF	acyl carrier protein[167]
nodE	ketoadipate synthetase[14]
nodG	various dehydrogenases[6,41,77,101]
nodI	ATPase membrane transport[69]
nodL	*lacA*-lac acetylase[55]
nodM	*glmS*-glucosamine synthetase (J. A. Downie, E. Kondorosi, personal communications)
nodO	*hly*-hemolysin[65]
nodP, nodQ	*cysD, cysN*-ATP sulfurylase[164]

Besides specific homologies, sequence information can also suggest possible membrane localization by examining predicted open reading frames for hydrophobic sections of protein. The following *nod* gene products are predicted to be membrane associated: NodC[107,156,181]; NodI, NodJ[69]; NodE[167]; and NodT.[179] In two cases—NodC[161] and NodE[174]—biochemical study has confirmed membrane localization. On other genes, some information is available. The Köhn group reported NodA and NodB both to be cytoplasmic,[161] although others have found biochemical and microscopic evidence that NodA is associated with the inner (cytoplasmic) membrane.[67,108] The *nodO* gene product was biochemically demonstrated to be secreted,[43,65,66] and *nodT* protein may be periplasmic.[179]

G. Future Genetic Studies Related to Mechanism

Genetic studies will continue to be needed both for in-depth analysis of currently known *nod* factors, such as NodRm-1, and for exploration of other components—if any—of the *Rhizobium* nodulation mechanism. As *nod* factor structures are determined, it will become apparent which central metabolic pathways may be necessary for providing precursors to *nod* genes for factor synthesis. Is there one locus encoding each of the critical enzymes, or more? How is expression controlled for these loci? Are symbiotic signals involved? Metabolic genetics will be an essential route to finding the genes for enzymes antecedent to the pathway directed by the *nod* genes themselves.

Summary

Two themes should be emphasized in final perspective on nodulation genetics.

The first theme is the importance of using genetics in all its power. The essence of experimental genetics is to find inheritable variation and to use logic and further genetic analysis to understand the underlying process. The first phase of nodulation genetics has consisted almost exclusively of making null mutations (insertions or deletions) and cloning and sequencing the corresponding genes. The next phase may well consist of further, more sophisticated combinations of genetics and biochemical analysis: the use of second–site mutations to help prove interacting components, and temperature–sensitive mutant alleles, used along with antibody inhibition to demonstrate enzymatic function of particular *nod* gene products. Behind all genetic studies of *nod* genes should be the caution that there may be other functions involved besides those that synthesize signals such as NodRm-1. One should always design genetic strategies capable of detecting unexpected results.

The second theme is the necessity of studying regulation at all its levels. The ability to regulate is one of the most essential of all biological attributes. Therefore, by pursuing regulation closely, one inevitably gets to the very heart of the process, to the driving forces and the constraints. Regulation must be thought of in the broadest sense, however. There are many levels of regulation, of which transcription initiation is only one. More sophisticated, biochemically detailed study will be needed before we understand more than the first level of symbiotic gene regulation. Detailed understanding of *nod, exo, ndv,* and *lps* gene control will certainly be accompanied by a greatly enriched sense of how nodulation is driven in developmental time and over evolution.

REFERENCES

1. Appelbaum, E. R., Johansen, E., and Chartrain, N., "Symbiotic mutants of USDA-191, a fast-growing *Rhizobium* that nodulates soybeans," *Mol. Gen. Genet. 201,* 454–461 (1985).

2. Appelbaum, E. R., Thompson, D. V., Idler, K., and Chartrain, N., "*Rhizobium japonicum* USDA 191 has two *nodD* genes that differ in primary structure and function," *J. Bacteriol. 170,* 12–20 (1988).

3. Bachem, C. W. B., Banfalvi, Z., Kondorosi, E., Schell, J., and Kondorosi, A., "Identification of host range determinants in the *Rhizobium* species MPIK-3030," *Mol. Gen. Genet. 203,* 42–48 (1986).

4. Bachem, C. W. B., Kondorosi, E., Banfalvi, Z., Horvath, B., Kondorosi, A., and Schell, J., "Identification and cloning of nodulation genes from the wide host range *Rhizobium* strain MPIK-3030," *Mol. Gen. Genet. 199,* 271–278 (1985).

5. Badgasarian, M., Lurz, R., Ruckert, B., Franklin, F. C. H., Bagdasarian, M. M., Frey, J., and Timmis, K. N., "Specific-purpose plasmid cloning vectors. II. Broad host range, high copy number, RSF1010-derived vectors, and a host-vector system for gene cloning in *Pseudomonas,*" *Gene 16,* 237–247 (1981).

6. Baker, M. E., "Human placental 17 beta-hydroxysteroid dehydrogenase is homologous to NodG protein of *Rhizobium meliloti,*" *Mol. Endocrinol. 3,* 881–884 (1989).

7. Banfalvi, Z., Kondorosi, E., and Kondorosi, A. "*Rhizobium meliloti* carries two megaplasmids," *Plasmid 13,* 129–138 (1985).

8. Banfalvi, Z., Nieuwkoop, A., Schell, M., Besl, L., and Stacey, G., "Regulation of *nod* gene expression in *Bradyrhizobium japonicum, Mol. Gen. Genet. 214,* 420–424 (1988).

8a. Barnett, M. J. and S. R. Long. 1990. DNA sequence and translational product of a new nodulation-regulatory locus: SyrM has sequence similarity to NodD proteins. J. Bacteriol. 172: 3695–3700.

9. Bassam, B. J., Djordevic, M. A., Redmond, J. W., Batley, M., and Rolfe, B. G., "Identification of a *nodD*-dependent locus in the *Rhizobium* strain NGR234 activated by phenolic factors secreted by soybeans and other legumes," *Mol. Plant-Microbe Interact. 1,* 161–168 (1988).

10. Batley, M., Redmond, J. W., Djordjevic, S. P., and Rolfe, B. G., "Characterization of glycerophosphorylated cyclic beta-1,2-glucans from a fast-growing *Rhizobium* species," *Biochim. Biophys. Acta 901,* 119–126 (1987).

11. Bender, G. L., Goydych, W., Rolfe, B. G., and Nayudu, M., "The role of *Rhizobium* conserved and host specific nodulation genes in the infection of the nonlegume *Parasponia andersonii,*" *Mol. Gen. Genet. 210,* 299–306 (1987).

12. Bender, G. L., Nayudu, M., Le Strange, K. K., and Rolfe, B. G., "The *nodD1* gene from *Rhizobium* strain NGR234 is a key determinant in the extension of host range to the nonlegume *parasponia,*" *Mol. Plant-Microbe Interact. 1,* 259–266 (1988).

13. Beringer, J. E., Brewin, N. J., and Johnston, A. W. B., "The genetic analysis of *Rhizobium* in relation to symbiotic nitrogen fixation," *Heredity 45*, 161–186 (1980).

14. Bibb, M. J., Biro, S., Motamedi, H., Collins, J. F., and Hutchinson, C. R., "Analysis of the nucleotide sequence of the *Streptomyces glaucescens* tcmI genes provides key information about the enzymology of polyketide antibiotic biosynthesis," *EMBO J. 8*, 2727–2736 (1989).

15. Borthakur, D., Barber, C. E., Lamb, J. W., Daniels, M. J., Downie, J. A., and Johnston, A. W. B., "A mutation that blocks exopolysaccharide synthesis prevents nodulation of peas by *Rhizobium-leguminosarum* but not of beans by *Rhizobium-phaseoli* and is corrected by cloned DNA from the phytopathogen *Xanthomonas*," *Mol. Gen. Genet. 203*, 320–323 (1986).

16. Borthaku, D., Barker, R. F., Latchford, J. W., Rossen, L., and Johnston, A. W. B., "Analysis of *pss* genes of *Rhizobium leguminosarum* required for exopolysaccharide synthesis and nodulation of peas: Their primary structure and their interaction with *psi* and other nodulation genes," *Mol. Gen. Genet. 213*, 155–162 (1988).

17. Borthakur, D., Downie, J. A., Johnston, A. W. B., and Lamb, J. W., "Psi a plasmid-linked *Rhizobium-phaseoli* gene that inhibits exopolysaccharide production and which is required for symbiotic nitrogen fixation," *Mol. Gen. Genet. 200*, 278–282 (1985).

18. Brink, B. A., Miller, J., Carlson, R. W., and Noel, K. D., "Expression of *Rhizobium leguminosarum* CFN42 genes for lipopolysaccharide in strains derived from different *R. leguminosarum* soil isolates," *J. Bacteriol. 172*, 548–555 (1990).

19. Broughton, W. J., Wong, C. H., Lewin, A., Samrey, U., Myint, H., Meyer, Z.-A., Dowling, D. N., and Simon, R., "Identification of *Rhizobium* plasmid sequences involved in recognition of *Psophocarpus, Vigna* and other legumes," *J. Cell Biol. 102*, 1173–1182 (1986).

20. Buchanan-Wollaston, A. V., Beringer, J. E., Brewin, N. J., Hirsch, P. R., and Johnston, A. W. B., "Isolation of symbiotically defective mutants in *Rhizobium-leguminosarum* by insertion of the transposon Tn-5 into a transmissible plasmid," *Mol. Gen. Genet. 178*, 185–190 (1980).

21. Burkhardt, B., Schillik, D., and Pühler, A., "Physical characterization of *Rhizobium meliloti* megaplasmids," *Plasmid 17*, 13–25 (1987).

22. Burn, J. E., Hamilton, W. D., Wootton, J. C., and Johnston, A. W. B., "Single and multiple mutations affecting properties of the regulatory gene *nodD* of *Rhizobium*," *Mol. Microbiol. 3*, 1567–1577 (1989).

23. Burn, J., Rossen, L., and Johnston, A. W. B., "Four classes of mutations in the *nodD* gene of *Rhizobium leguminosarum* biovar *viciae* that affect its ability to autoregulate and/or activate other *nod* genes in the presence of flavonoid inducers," *Genes Div. 1*, 456–464 (1987).

24. Canter Cremers, H. C. J., Spaink, H. P., Wijfjes, A. H. M., Pees, E., Wijffelman, C. A., Okker, R. J. H., and Lugtenberg, B. J. J., "Additional nodulation genes on the Sym plasmid of *R. leguminosarum* biovar *viciae*," *Plant Mol. Biol. 13*, 163–174 (1990).

25. Canter Cremers, H. C. J., Wijffelman, C. A., Pees, E., Rolfe, B. G., Djordjevic, M. A., and Lugtenberg, B. J. J., "Host specific nodulation of plants of the pea cross-inoculation group is influenced by genes in fast growing *Rhizobium* downstream *nodC*," *J. Plant Physiol. 132*, 398–404 (1988).

26. Carlson, R. W., "Surface chemistry," in Broughton, W. (ed.), *Ecology of Nitrogen Fixation*, Vol. 2, Oxford: Oxford University Press, 1982, pp. 199–234.

27. Carlson, R. W., "The heterogeneity of *Rhizobium* lipopolysaccharides," *J. Bacteriol.* *158*, 1012–1017 (1984).

28. Carlson, R. W., Kalembasa, S., Turowski, D., Pachori, P., and Noel, K. D., "Characterization of the lipopolysaccharide from a *Rhizobium phaseoli* mutant that is defective in infection thread development," *J. Bacteriol.* *169*, 4923–4928 (1987).

29. Cava, J. R., Elias, P. M., Turowski, D. A., and Noel, K. D., "*Rhizobium leguminosarum* CFN42 genetic regions encoding lipopolysaccharide structures essential for complete nodule development on bean plants," *J. Bacteriol.* *171*, 8–15 (1989).

30. Cervantes, E., Sharma, S. B., Maillet, F., Vasse, J., Truchet, G., and Rosenberg, C., "The *Rhizobium meliloti* host-range *nodQ* gene encodes a protein which shares homology with translation elongation and initiation factors," *Mol. Microbiol.* *3*, 745–755 (1989).

31. Charles, T. C., and Finan, T. M., "Genetic map of *Rhizobium meliloti* megaplasmid pRmeSU47b," *J. Bacteriol.* *172*, 2469–2476 (1990).

32. Chen, H., Batley, M., Redmond, J., and Rolfe, B. G., "Alteration of the effective nodulation properties of a fast-growing broad host range *Rhizobium* due to changes in exopolysaccharide synthesis," *J. Plant Physiol.* *120*, 331–350 (1985).

33. Chua, K.-Y., Pankhurst, C. E., Macdonald, P. E., Hopcroft, D. H., Jarvis, B. D. W., and Scott, D. B., "Isolation and characterization of transposon Tn5-induced symbiotic mutants of *Rhizobium-loti*," *J. Bacteriol.* *162*, 335–343 (1985).

34. Clover, R., Keiber, J., and Signer, E. R., "Lipopolysaccharide mutants of *Rhizobium meliloti* are not defective in symbiosis," *J. Bacteriol.* *171*, 3961–3967 (1989).

35. Davila, G., Brom, S., Flores, M., Girard, M. L., Gonzalez, V., Louzada, M., Martinez, J., Martinez, E., Palacios, R., Piñero, D., Romero, D., and Valdez, A. M., "The symbiotic genome of *Rhizobium phaseoli*," in Verma, D. P. S., and R. Palacios (eds.), *Molecular Plant–Microbe Interactions*. St. Paul: APS Press, 1988, pp. 187–191.

36. Davis, E. O., Evans, I. J., and Johnston, A. W. B., "Identification of *nodX*, a gene that allows *Rhizobium leguminosarum* biovar *viciae* strain TOM to nodulate Afghanistan peas," *Mol. Gen. Genet.* *212*, 531–535 (1988).

37. Davis, E. O., and Johnston, A. W. B., "Regulatory functions of the three *nodD* genes of *Rhizobium leguminosarum* biovar *phaseoli*," *Mol. Microbiol.* *4*, 933–941 (1990).

38. Davis, E. O., and Johnston, A. W. B., "Analysis of three *nodD* genes in *Rhizobium leguminosarum* biovar *phaseoli*; *nodD1* is preceded by *nolE*, a gene whose product is selected from the cytoplasm," *Mol. Microbiol.* *4*, 921–932 (1990).

39. Debellé, F., Maillet, F., Vasse, J., Rosenberg, C., De Billy, F., Truchet, G., Dénarié, J., and Ausubel, F. M., "Interference between *Rhizobium meliloti* and *Rhizobium trifolii* nodulation genes: Genetic basis of *R. meliloti* dominance. *J. Bacteriol.* *170*, 5718–5727 (1988).

40. Debellé, F., Rosenberg, C., Vasse, J., Maillet, F., Martinez, E., Dénarié, J., and Truchet, G., "Assignment of symbiotic developmental phenotypes to common and specific nodulation (*nod*) genetic loci of *Rhizobium meliloti*," *J. Bacteriol.* *168*, 1075–1086 (1986).

41. Debellé, F., and Sharma, S. B., "Nucleotide sequence of *Rhizobium meliloti* RCR2011 genes involved in host specificity of nodulation," *Nucleic Acids Res.* *14*, 7453–7472 (1986).

42. De Maagd, R. A., Rao, A. S., Mulders, I. H. M., Goosen-de Roo, L., Van Loosdrecht, M. C. M., Wijffelman, C. A., and Lugtenberg, B. J. J., "Isolation and characterization of mutants of *Rhizobium leguminosarum* bv. *viciae* 248 with altered lipopolysaccharides: Possible role of surface charge or hydrophobicity in bacterial release from the infection thread," *J. Bacteriol. 171*, 1143–1150 (1989).

43. De Maagd, R. A., Spaink, H. P., Rees, E., Mulders, I. H. M., Wijfjes, A., Wijffelman, C. A., Okker, R. J. H., and Lugtenberg, B. J. J., "Localization and symbiotic function of a region on the *Rhizobium leguminosarum* sym plasmid pRL1JI responsible for a secreted, flavonoid-inducible 50-kilodalton protein," *J. Bacteriol. 171*, 1151–1157 (1989).

44. Dénarié, J., Boistard, P., Casse-Delbart, F., Atherley, A. G., Berry, J. O., and Russell, P., "Indigenous plasmids of *Rhizobium*," in Giles, K., and A. Atherly (eds.), *Biology of the Rhizobiacea*. New York: Academic Press, 1981, p. 225.

44a. Derylo, M., Skorupska, A., Bednara, J., and Lorkiewicz, Z., "*Rhizobium-trifolii* mutants deficient in exopolysaccharide production," *Physiol. Plant 66*, 699–704 (1986).

45. Deshmane, N., and Stacey, G., "Identification of *Bradyrhizobium nod* genes involved in host-specific nodulation," *J. Bacteriol. 171*, 3324–3330 (1989).

46. De Vos, G. F., Walker, G. C., and Signer, E. R., "Genetic manipulations in *Rhizobium meliloti* utilizing two new transposon Tn5 derivatives," *Mol. Gen. Genet. 204*, 485–491 (1986).

47. Djordevic, M. A., Gabriel, D. W., and Rolfe, B. G., "*Rhizobium*—the refined parasite of legumes," *Ann. Rev. Phytopathol. 25*, 145–168 (1987).

48. Djordjevic, M. A., Innes, R. W., Wijffelman, C. A., Schofield, P. R., and Rolfe, B. G., "Nodulation of specific legumes is controlled by several distinct loci in *Rhizobium-trifolii*," *Plant Mol. Biol. 6*, 389–402 (1986).

49. Djordjevic, M. A., Redmond, J. W., Batley, M., and Rolfe, B. G., "Clovers secrete specific phenolic compounds which either stimulate or repress *nod* gene expression in *Rhizobium-trifolii*," *EMBO J. 6*, 1173–1180 (1987).

50. Djordjevic, M. A., Schofield, P. R., Ridge, R. W., Morrison, N. A., Bassam, B. J., Plazinski, J., Watson, J. M., and Rolfe, B. G., "*Rhizobium* nodulation genes involved in root hair curling (Hac) are functionally conserved," *Plant Mol. Biol. 4*, 147–160 (1985a).

51. Djordjevic, M. A., Schofield, P. R., and Rolfe, B. G., "Tn5 mutagenesis of *Rhizobium trifolii* host-specific nodulation genes result in mutants with altered host-range ability," *Mol. Gen. Genet. 200*, 463–471 (1985b).

52. Djordjevic, S. P., Chen, H., Batley, M., Redmond, J. W., and Rolfe, B. G., "Nitrogen fixation ability of exopolysaccharide synthesis mutants of *Rhizobium-SP* strain NGR234 and *Rhizobium-trifolii* is restored by the addition of homologous exopolysaccharides," *J. Bacteriol. 169*, 53–60 (1987).

53. Doherty, D., Leigh, J. A., Glazebrook, J., and Walker, G. C., "*Rhizobium meliloti* mutants that overproduce the *R. meliloti* acidic calcofluor-binding exopolysaccharide," *J. Bacteriol. 170*, 4249–4256 (1988).

54. Donnelly, D. F., Birkenhead, K., and O'Gara, F., "Stability of IncQ and IncP-1 vector plasmids in *Rhizobium* spp.," *FEMS Microbiol. Lett. 42*, 141–145 (1987).

55. Downie, J. A., "The *nodL* gene from *Rhizobium leguminosarum* is homologous to the acetyl transferase encoded by *lacA* and *cysE*," *Mol. Microbiol. 3*, 1649–1651 (1989).

56. Downie, J. A., Economou, A., Hamilton, W. D. O., and Johnston, A. W. B., "Exported protein encoded by *nod* genes from *Rhizobium leguminosarum* bv. *viciae*," in Gresshoff, P., L. E. Roth, G. Stacey, and W. E. Newton (eds.), *Nitrogen Fixation: Achievements and Objectives*. New York: Chapman Hall, 1990, pp. 201–206.

57. Downie, J. A., and Johnston, A. W. B., "Nodulation of legumes by *Rhizobium*," *Plant Cell Environ. 11*, 403–412 (1988).

58. Downie, J. A., and Surin, B. P., "Either of two *nod* gene loci can complement the nodulation defect of a *nod* deletion mutant of *Rhizobium leguminosarum* bv. *viciae*," *Mol. Gen. Genet. 222*, 81–86.

59. Dudley, M. E., Jacobs, T. W., and Long, S. R., "Microscopic studies of cell divisions induced in alfalfa roots by *Rhizobium meliloti*," *Planta 171*, 289–301 (1987).

60. Dudley, M. E., and Long, S. R., "A non-nodulating alfalfa mutant displays neither root hair curling nor early cell division in response to *Rhizobium meliloti*," *Plant Cell 1*, 65–72 (1989).

61. Dusha, I., Bakos, A., Kondorosi, A., De Bruijn, F. J., and Schell, J., "The *Rhizobium meliloti* early nodulation genes (*nodABC*) are nitrogen-related: Isolation of a mutant strain with efficient nodulation capacity on alfalfa in the presence of ammonium," *Mol. Gen. Genet. 219*, 89–96 (1989).

62. Dylan, T., Helinski, D. R., and Ditta, G. S., "Hypoosmotic adaptation in *Rhizobium meliloti* requires beta-$(1 \rightarrow 2)$-glucan," *J. Bacteriol. 172*, 1400–1408 (1990).

63. Dylan, T., Ielpi, L., Stanfield, S., Kashyap, L., Douglas, C., Yanofsky, M., Nester, E., Helinski, D. R., and Ditta, G., "*Rhizobium meliloti* genes required for nodule development are related to chromosomal virulence genes in *Agrobacterium tumefaciens*," *Proc. Natl. Acad. Sci. USA 83*, 4403–4407 (1986).

64. Dylan, T., Nagpal, P., Helinski, D. R., and Ditta, G. S., "Symbiotic pseudorevertants of *Rhizobium meliloti ndv* mutants," *J. Bacteriol. 172*, 1409–1417 (1990).

65. Economou, A., Hamilton, W. D. O., Johnston, A. W. B., and Downie, J. A., "The *Rhizobium* nodulation gene *nodO* encodes a Ca^{2+}-binding protein that is exported without N-terminal cleavage and is homologous to haemolysin and related proteins," *EMBO J. 9*, 349–354 (1990).

66. Economou, A., Hawkins, F. K. L., Downie, J. A., and Johnston, A. W. B., "Transcription of *rhiA*, a gene on a *Rhizobium leguminosarum* bv. *viciae* Sym plasmid, requires *rhiR* and is repressed by flavanoids the induce *nod* genes," *Mol. Microbiol. 3*, 87–93 (1989).

67. Egelhoff, T. T., "Molecular analysis of *Rhizobium meliloti* nodulation genes and gene products," Ph.D. dissertation, Stanford University, Stanford, CA, 1987.

68. Egelhoff, T. T., Fisher, R. F., Jacobs, T. W., Mulligan, J. T., and Long, S. R., "Nucleotide sequence of *Rhizobium meliloti* 1021 nodulation genes: *nodD* is read divergently from *nodABC*," *DNA 4*, 241–248 (1985).

69. Evans, I. J., and Downie, J. A., "The *nodI* gene product of *Rhizobium leguminosarum* is closely related to the ATP-bindinng bacterial transport proteins: Nucleotide sequence analysis of the *nodI* and *nodJ* genes," *Gene 43*, 95–102 (1986).

70. Faucher, C., Camut, S., Dénarié, J., and Truchet, G., "The *nodH* and *nodQ* host range genes of *Rhizobium meliloti* behave as avirulence genes in *R. leguminosarum* bv. *viciae* and determine changes in the production of plant-specific extracellular signals," *Mol. Plant-Microbe Interact. 2*, 291–300 (1989).

71. Faucher, C., Maillet, F., Vasse, J., Rosenberg, C., Van Brussel, A. A. N., Truchet, G., and Denarie, J., *"Rhizobium meliloti* host range *nodH* gene determines production of an alfalfa-specific extracellular signal," *J. Bacteriol. 170,* 5489–5499 (1988).

72. Finan, T. M., Hirsch, A. M., Leigh, J. A., Johansen, E., Kuldau, G. A., Deegan, S., Walker, G. C., and Signer, E. R., "Symbiotic mutants of *Rhizobium meliloti* that uncouple plant from bacterial differentiation," *Cell 40,* 869–877 (1985).

73. Finan, T. M., Kunkel, B., De Vos, G. F., and Signer, E. R., "Second symbiotic megaplasmid in *Rhizobium meliloti* carrying exopolysaccharide and thiamine synthesis genes," *J. Bacteriol. 167,* 66–72 (1986).

74. Firmin, J. L., Wilson, K. E., Rossen, L., and Johnston, A. W. B., "Flavonoid activation of nodulation genes in *Rhizobium* reversed by other compounds present in plants," *Nature 324,* 90–92 (1986).

75. Fisher, R. F., Brierley, H. L., Mulligan, J. T., and Long, S. R., "Transcription of *Rhizobium meliloti* nodulation genes: Identification of a *nodD* transcription initiation site *in vitro* and in *vivo," J. Biol. Chem. 262,* 6849–6855 (1987).

76. Fisher, R., and Long, S. R., "DNA footprint analysis of the transcriptional activator proteins NodD1 and NodD3 on inducible *nod* gene promoters," *J. Bacteriol. 171,* 5492–5502 (1989).

77. Fisher, R. F., Swanson, J., Mulligan, J. T., and Long, S. R., "Extended region of nodulation genes in *Rhizobium meliloti* 1021. II. Nucleotide sequence, transcription start sites, and protein products," *Genetics 117,* 191–201 . (1987b).

78. Fisher, R. F., Tu, J. K., and Long, S. R., "Conserved nodulation genes in *R. meliloti* and *R. trifolii," Appl. Environ. Microbiol. 49,* 1432–1435 (1985).

79. Geremia, R. A., Cavaignac, S., Zorreguieta, A., Toro, N., Olivares, J., and Ugalde, R. A., "A *Rhizobium meliloti* mutant that forms ineffective pseudonodules in alfalfa produces exopolysaccharide but fails to form beta $-(1 \rightarrow 2)$ glucan," *J. Bacteriol. 169,* 880–884 (1987).

80. Glazebrook, J., and Walker, G. C., "A novel exopolysaccharide can function in place of the calcofluor-binding exopolysaccharide in nodulation of alfalfa by *Rhizobium meliloti," Cell 56,* 661–672 (1989).

81. Göttfert, M., Grob, P., and Hennecke, H., "Proposed regulatory pathway encoded by the *nodV* and *nodW* genes, determinants of host specificity in *Bradyrhizobium japonicum," Proc. Natl. Acad. Sci. USA 87,* 2680–2684 (1990).

82. Göttfert, M., Hitz, S., and Hennecke, H., "Identification of *nodS* and *nodU*, two inducible genes inserted between the *Bradyrhizobium japonicum nodYABC* and *nodIJ* genes," *Mol. Plant-Microbe Interact. 3,* 308–316 (1990).

83. Göttfert, M., Horvath, B., Kondorosi, E., Putnoky, P., Rodriguez-Quinones, F., and Kondorosi, A., "At least two *nodD* genes are necessary for efficient nodulation of alfalfa by *Rhizobium meliloti," J. Mol. Biol. 191,* 411–420 (1986).

84. Göttfert, M., Lamb, J. W., Gasser, R., Semenza, J., and Hennecke, H., Mutational analysis of the *Bradyrhizobium japonicum* common *nod* genes and further *nod* box-linked genomic DNA regions," *Mol. Gen. Genet. 215,* 407–415 (1989).

85. Göttfert, M., Weber, J., and Hennecke, H., "Induction of a *nodA-lacZ* fusion in *Bradyrhizobium japonicum* by an isoflavone," *J. Plant Physiol. 132,* 394–397 (1988).

86. Goethals, K., Gao, M., Tomekpe, K., Van Montagu, M., and Holsters, M., "Common *nodABC* genes in *Nod* locus 1 of *Azorhizobium caulinodans:* Nucleotide sequence and plant-inducible expression," *Mol. Gen. Genet. 219,* 289–298 (1989).

87. Goethals, K., Van Den Eede, G., Van Montagu, M., and Holsters, M., "Identification and characterization of a functional *nodD* gene in *Azorhizobium caulinodans* ORS571," *J. Bacteriol. 172*, 2658–2666 (1990).

88. Goetz, R., Evans, I. J., Downie, J. A., and Johnston, A. W. B., "Identification of the host-range DNA which allows *Rhizobium-leguminosarum* strain TOM to nodulate cultivar afghanistan peas," *Mol. Gen. Genet. 201*, 296–300 (1985).

89. Gray, J. X., Djordjevic, M. A., and Rolfe, B. G., "Two genes that regulate exopolysaccharide production in *Rhizobium* sp. strain NGR234: DNA sequences and resultant phenotypes," *J. Bacteriol. 172*, 193–203 (1990).

90. Györgypal, Z., Iyer, N., and Kondorosi, A., "Three regulatory *nodD* alleles of diverged flavonoid-specificity are involved in host-dependent nodulation by *Rhizobium meliloti*," *Mol. Gen. Genet. 212*, 85–92 (1988).

91. Hahn, M., and Hennecke, H., "Mapping of a *Bradyrhizobium japonicum* DNA region carrying genes for symbiosis and an asymmetric accumulation of reiterated sequences," *Appl. Environ. Microbiol. 53*, 2247–2252 (1987).

92. Harrison, S. P., Jones, D. G., Schünmann, P. H. D., Forster, J. W., and Young, J. P., "Variation in *Rhizobium leguminosarum* biovar *trifolii* sym plasmids and the association with effectiveness of nitrogen fixation," *J. Gen. Microbiol. 134*, 2721–2730 (1988).

93. Hartwig, U. A., Maxwell, C. A., Joseph, C. M., and Phillips, D. A., "Interactions among flavonoid *nod* gene inducers released from alfalfa seeds and roots," *Plant Physiol. 91*, 1138–1142 (1989).

94. Hartwig, U. A., Maxwell, C. A., Joseph, C. M., and Phillips, D. A., "Chrysoeriol and luteolin released from alfalfa seeds induce *nod* genes in *Rhizobium meliloti*," *Plant Physiol. 92*, 116–122 (1990).

95. Haugland, R. A., Cantrell, M. A., Beaty, J. S., Hanus, F. J., Russell, S. A., and Evans, H. J., "Characterization of *Rhizobium japonicum* hydrogen uptake genes," *J. Bacteriol. 159*, 1006–1012 (1984).

96. Henikoff, S., Haughn, G. W., Calvo, J. M., and Wallace, J. C., "A large family of bacterial activator proteins," *Proc. Natl. Acad. Sci. USA 85*, 6602–6606 (1988).

97. Hong, G.-F., Burn, J. E., and Johnston, A. W., "Evidence that DNA involved in the expression of nodulation (*nod*) genes in *Rhizobium* binds to the product of the regulatory gene *nodD*," *Nucleic Acids Res. 15*, 9677–9690 (1987).

98. Honma, M. A., Asomaning, M., and Ausubel, F. M., "*Rhizobium meliloti nodD* genes mediate host-specific activation of *nodABC*," *J. Bacteriol. 172*, 901–911 (1990).

99. Honma, M. A., and Ausubel, F. M., "*Rhizobium meliloti* has three functional copies of the *nodD* symbiotic regulatory gene," *Proc. Natl. Acad. Sci. USA 84*, 8558–8562 (1987).

100. Horvath, B., Bachem, C. W. B., Schell, J., and Kondorosi, A., "Host-specific regulation of nodulation genes in *Rhizobium* is mediated by a plant-signal interacting with the *nodD* gene product," *EMBO J. 6*, 841–848 (1987).

101. Horvath, B., Kondorosi, E., John, M., Schmidt, J., Torok, I., Györgypal, L. Z., Barabas, I., Wieneke, U., Schell, J., and Kondorosi, A., "Organization structure and symbiotic function of *Rhizobium meliloti* nodulation genes determining host specificity for alfalfa," *Cell 46*, 335–344 (1986).

102. Hoying, J. B., Behm, S. M., and Lang-Unnasch, N., "Cloning and characterization of *Rhizobium meliloti* loci required for symbiotic root nodule invasion," *Mol. Plant-Microbe Interact. 3*, 18–27 (1990).

103. Hynes, M. F., Simon, R., Müller, P., Niehaus, K., Labes, M., and Pühler, A., "The two megaplasmids of *Rhizobium meliloti* are involved in the effective nodulation of alfalfa," *Mol. Gen. Genet. 202,* 356–362 (1986).

104. Ielpi, L., Dylan, T., Ditta, G. S., Helinski, D. R., and Stanfield, S. W., "The *ndvB* locus of *Rhizobium meliloti* encodes a 319-kDa protein involved in the production of beta-(1 → 2)-glucan," *J. Biol. Chem. 265,* 2843–2851 (1990).

105. Innes, R. W., Hirose, M. A., and Kuempel, P. L., "Induction of nitrogen fixing nodules on clover requires only 32 kilobase pairs of DNA from the *Rhizobium trifolii* symbiosis plasmid," *J. Bacteriol. 170,* 3793–3802 (1988).

106. Innes, R. W., Kuempel, P. L., Plazinski, J., Canter-Cremers, H., Rolfe, B. G., and Djordjevic, M. A., "Plant factors induce expression of nodulation and host-range genes in *Rhizobium trifolii,*" *Mol. Gen. Genet. 201,* 426–432 (1985).

107. Jacobs, T. W., Egelhoff, T. T., and Long, S. R., "Physical and genetic map of a *Rhizobium meliloti* nodulation gene region and nucleotide sequence of *nodC,*" *J. Bacteriol. 162,* 469–476 (1985).

107a. Johansen, E., Finan, T. M., Gefter, M. L., and Signer, E. R., "Monoclonal antibodies to *Rhizobium meliloti* and surface mutants insensitive to them," *J. Bacteriol. 160,* 454–457 (1984).

108. Johnson, D., Roth, L. E., and Stacey, G., "Immunogold localization of the NodC and NodA proteins of *Rhizobium meliloti,*" *J. Bacteriol. 171,* 4583–4588 (1989).

109. Kapulnik, Y., Joseph, C. M., and Phillips, D. A., "Flavone limitations to root nodulation and symbiotic nitrogen fixation in alfalfa," *Plant Physiol. 84,* 1193–1196 (1987).

110. Keller, M., Müller, P., Simon, R., and Pühler, A., "*Rhizobium meliloti* genes for exopolysaccharide synthesis and nodule infection located on megaplasmid 2 are actively transcribed during symbiosis," *Mol. Plant-Microbe Interactions, 1,* 267–2744 (1988).

111. Kondorosi, A., and Johnston, A. W. B., "The genetics of *Rhizobium,*" in Giles, K. L., and A. G. Atherly (eds.), *Biology of the Rhizobiaceae.* New York: Academic Press, 1981, p. 191.

112. Kondorosi, E., Banfalvi, Z., and Kondorosi, A., "Physical and genetic analysis of a symbiotic region of *Rhizobium meliloti:* Identification of nodulation genes," *Mol. Gen. Genet. 193,* 445–452 (1984).

113. Kondorosi, E., Györgypal, Z., Dusha, I., Baev, N., Pierre, M., Hoffmann, B., Himmelbach, A., Banfalvi, Z., and Kondorosi, A., "*Rhizobium meliloti* nodulation genes and their regulation," in Gresshoff, P., L. E. Roth, G. Stacey, and W. E. Newton (eds.), *Nitrogen Fixation: Achievements and Objectives.* New York: Chapman Hall, 1990, pp. 207–213.

114. Kondorosi, E., Gyuris, J., Schmidt, J., John, M., Duda, E., Hoffmann, B., Schell, J., and Kondorosi, A., "Positive and negative control of *nod* gene expression in *Rhizobium meliloti* is required for optimal nodulation," *EMBO J. 8,* 1331–1340 (1989).

115. Kondorosi, E., and Kondorosi, A., "Nodule induction on plant roots by *Rhizobium,*" *TIBS 11,* 296–299 (1986).

116. Kosslak, R., Bookland, R., Barkei, J., Paaren, H. E., and Appelbaum, E. R., "Induction of *Bradyrhizobium japonicum* common *nod* genes by isoflavones isolated from *Glycine max,*" *Proc. Natl. Acad. Sci. USA 84,* 7428–7432 (1987).

117. Lamb, J. W., Downie, J. A., and Johnston, A. W. B., "Cloning of the nodulation *nod* genes of *Rhizobium phaseoli* and their homology to *Rhizobium leguminosarum nod* DNA," *Gene 34,* 235–242 (1985).

118. Leigh, J. A., Reed, J. W., Hanks, J. F., Hirsch, A. M., and Walker, G. C., *"Rhizobium meliloti* mutants that fail to succinylate their calcofluor-binding exopolysaccharide are defective in nodule invasion," *Cell 51,* 579–587 (1987).

119. Leigh, J. A., Signer, E. R., and Walker, G. C., "Exopolysaccharide-deficient mutants of *Rhizobium meliloti* that form ineffective nodules," *Proc. Natl. Acad. Sci. USA 82,* 6231–6235 (1985).

120. Lerouge, P., Roche, P., Faucher, C., Maillet, F., Truchet, G., Promée, J. C., and Dénarié, J., "Symbiotic host-specificity of *Rhizobium meliloti* is determined by a sulphated and acylated glucosamine oligosaccharide signal," *Nature 344,* 781–784 (1990).

121. Lewin, A., Cervantes, E., Chee-Hoong, W., and Broughton, W. J., *"nodSU,* two new *nod* genes of the broad host range *Rhizobium* strain NGR234 encode host-specific nodulation of the tropical tree *Leucaena leucocephala," Mol. Plant-Microbe Interact., 3*: 317–326 (1990).

122. Lewin, A., Rosenberg, C., Meyer, H., Wong, C. H., Nelson, L., Manen, J.-F., Stanley, J., Dowling, D. N., Denarie, J., and Broughton, W. J., "Multiple host-specificity loci of the broad host-range *Rhizobium*-SP NGR234 selected using the widely compatible legume *Vigna unguiculata," Plant Mol. Biol. 8,* 447–459 (1987).

123. Long, S., Reed, J. W., Himawan, J., and Walker, G. C., "Genetic analysis of a cluster of genes required for synthesis of the calcofluor-binding exopolysaccharide of *Rhizobium meliloti," J. Bacteriol. 170,* 4239–4248 (1988).

124. Long, S. R., "Genetics of *Rhizobium* nodulation," in Kosuge, T., and E. Nester (eds.), *Plant-Microbe Interactions.* New York: Macmillan, 1984, p. 265.

125. Long, S. R., *"Rhizobium*-legume nodulation: Life together in the underground," *Cell 56,* 203–214 (1989).

126. Long, S. R., *"Rhizobium* genetics," *Ann. Rev. Genet. 23,* 483–506 (1989).

127. Ma, Q.-S., Johnston, A. W. B., Hombrecher, G., and Downie, J. A., "Molecular genetics of mutants of *Rhizobium leguminosarum* which fail to fix nitrogen," *Mol. Gen. Genet. 187,* 166–171 (1982).

128. Maillet, F., Debellé, F., and Dénarié, J., "Role of the *nodD* and *syrM* genes in the activation of the regulatory gene *nodD3,* and of the common and host specific *nod* genes of *Rhizobium meliloti,"* *Molec. Microbiol., 4*: 1975–1984 (1990).

129. Martinez, E., Romero, D., and Palacios, R., "The *Rhizobium* genome," *CRC Crit. Rev. Plant Sci. 9,* 59–93 (1990).

130. Marvel, D. J., Kuldau, G., Hirsch, A., Richards, E., Torrey, J. G., and Ausubel, F. M., "Conservation of nodulation genes between *Rhizobium meliloti* and a slow-growing *Rhizobium* strain that nodulates a nonlegume host," *Proc. Natl. Acad. Sci. USA 82,* 5841–5845 (1985).

131. Marvel, D. J., Torrey, J. G., and Ausubel, F. M., *"Rhizobium* symbiotic genes required for nodulation of legume and nonlegume hosts," *Proc. Natl. Acad. Sci. USA 84,* 1319–1323 (1987).

132. Maxwell, C. A., Hartwig, U. A., Joseph, C. M., and Phillips, D. A., "A chalcone and two related flavonoids released from alfalfa roots induce *nod* genes of *Rhizobium meliloti," Plant Physiol. 91,* 842–847 (1989).

133. McIver, J., Djordjevic, M. A., Weinman, J. J., Bender, G. L., and Rolfe, B. G., "Extension of host range of *Rhizobium leguminosarum* bv. *trifolii* caused by point mutations in *nodD* that result in alterations in regulatory function and recognition of inducer molecules," *Mol. Plant-Microbe Interact. 2,* 97–106 (1989).

134. Miller, K. J., Gore, R. S., Johnson, R., Benesi, A. J., and Reinhold, V. N., "Cell-associated oligosaccharides of *Bradyrhizobium* spp.," *J. Bacteriol. 172*, 136–142 (1990).

135. Miller, K. J., Kennedy, E. P., and Reinhold, V. N., "Osmotic adaptation by gram-negative bacteria: Possible role for periplasmic oligosaccharides," *Science 231*, 48–51 (1986).

136. Müller, P., Hynes, M., Kapp, D., Niehaus, K., and Pühler, A., "Two classes of *Rhizobium meliloti* infection mutants differ in exopolysaccharide production and in coinoculation properties with nodulation mutants," *Mol. Gen. Genet. 211*, 17–26 (1988).

137. Mulligan, J. T., and Long, S. R., "Induction of *Rhizobium meliloti nodC* expression by plant exudate requires *nodD*," *Proc. Natl. Acad. Sci. USA 82*, 6609–6613 (1985).

138. Mulligan, J. T., and Long, S. R., "*nodD* enhances induction of *nodC* by plant exudate," in Evans, H. J., P. J. Bottomley, and W. E. Newton (eds.), *Nitrogen Fixation Research Progress*. Dordrecht: Martinus Nijhoff, 1985, p. 122.

139. Mulligan, J. T., and Long, S. R., "A family of activator genes regulates expression of *Rhizobium meliloti* nodulation genes," *Genetics 122*, 7–18 (1989).

140. Nayudu, M., and Rolfe, B. G., "Analysis of R-primes demonstrates that genes for broad host range nodulation of *Rhizobium* strain NGR-234 are dispersed on the Sym plasmid," *Mol. Gen. Genet. 206*, 326–337 (1987).

141. Nieuwkoop, A. J., Banfalvi, Z., Deshmane, N., Gerhold, D., Schell, M. G., Sirotkin, K. M., and Stacey, G., "A locus encoding host range is linked to the common nodulation genes of *Bradyrhizobium japonicum*," *J. Bacteriol. 169*, 2631–2638 (1987).

142. Noel, K. D., Vandenbosch, K. A., and Kulpaca, B., "Mutations in *Rhizobium-phaseoli* that lead to arrested development of infection threads," *J. Bacteriol. 168*, 1392–1401 (1986).

143. Ogawa, J., Brierley, H., and Long, S. R., "Analysis of *Rhizobium meliloti* nodulation mutant WL131: Novel insertion sequence ISRm 3 in *nodG* and altered *nodH* protein product." *J. Bacteriol.*, *173*: 3060–3065 (1991).

143a. Osteras, M., Stanley, J., Broughton, W. J., and Dowling, D. N., "A chromosomal genetic map of *Rhizobium* sp. NGR234 generated with Tn5-mob," *Mol. Gen. Genet. 220*, 157–160 (1989).

144. Peters, N. K., Frost, J. W., and Long, S. R., "A plant flavone, luteolin, induces expression of *Rhizobium meliloti* nodulation genes," *Science 233*, 917–1008 (1986).

145. Prakash, R. K., and Atherly, A. G., "Plasmids of *Rhizobium* and their role in symbiotic nitrogen fixations," in Prakash, R. K., and A. G. Atherly (eds.), *International Review of Cytology* (Academic Press) *104*, 1–24 (1986).

146. Putnoky, P., and Kondorosi, A., "Two gene clusters of *Rhizobium meliloti* code for early essential nodulation functions and a third influences nodulation efficiency," *J. Bacteriol. 167*, 881–887 (1986).

147. Puvanesarajah, V., Schell, F. M., Gerhold, D., and Stacey, G., "Cell surface polysaccharides from *Bradyrhizobium japonicum* and a non-nodulating mutant," *J. Bacteriol. 169*, 137–141 (1987).

148. Quinto, C., Martinez, J., Cevallos, M. A., Davalos, A., and Peralta, Y., "Genomic organization of nodulation genes in *Rhizobium phaseoli*," in Verma, D. P. S., and N. Brisson (eds.), *Molecular Genetics of Plant-Microbe Interactions*. Dordrecht: Martinus Nijhoff, 1987, p. 214.

149. Ramakrishnan, N., Prakash, R. K., Shantharam, S., Duteau, N. M., and Atherly, A. G., "Molecular cloning and expression of *Rhizobium fredii* USDA193 nodulation genes extension of host range for nodulation," *J. Bacteriol. 168,* 1087–1095 (1986).

150. Redmond, J. W., Batley, M., Djordjevic, M. A., Innes, R. W., Kuempel, P. L., and Rolfe, B. G., "Flavones induce expression of nodulation genes in *Rhizobium," Nature 323,* 632–634 (1986).

151. Renalier, M.-H., Batut, J., Ghai, J., Terzaghi, B., Gherardi, M., David, M., Garnerone, A.-M., Vasse, J., Truchet, G., Huguet, T., and Boistard, P., "A new symbiotic cluster on the pSym megaplasmid of *Rhizobium meliloti* 2011 carries a functional *fix* gene repeat and a *nod* locus," *J. Bacteriol. 169,* 2231–2238 (1987).

152. Roche, P., Lerouge, P., Faucher, C., Promé, J.-C., Truchet, G., and Dénarié, J., "*Rhizobium meliloti* extra cellular *nod* signals," in Gresshoff, P., L. E. Roth, G. Stacey, and W. E. Newton (eds.), *Nitrogen Fixation: Achievements and Objectives.* New York: Chapman Hall, 1990.

153. Rodriguez-Quinones, F., Banfalvi, Z., Murphy, P., and Kondorosi, A., "Interspecies homology of nodulation genes in *Rhizobium," Plant Mol. Biol. 8,* 61–75 (1987).

154. Rolfe, B., and Gresshoff, P., "Genetic analysis of legume nodule initiation," *Annu. Rev. Plant Physiol. Plant Mol. Biol. 39,* 297–319 (1988).

155. Rosenberg, C., Boistard, P., Dénarié, J., and Casse-Delbart, F., "Genes controlling early and late functions in symbiosis are located on a megaplasmid in *Rhizobium meliloti," Mol. Gen. Genet. 184,* 326–333 (1981).

156. Rossen, L., Johnston, A. W. B., and Downie, J. A., "DNA sequence of the *Rhizobium leguminsarum* nodulation genes *nodAB* and *C* required for root hair curling," *Nucleic Acids Res. 12,* 9497–9508 (1984).

157. Rossen, L., Shearman, C. A., Johnston, A. W. B., and Downie, J. A., "The *nodD* gene of *Rhizobium leguminosarum* is autoregulatory and in the presence of plant exudate induces the *nod A, B, C* genes," *EMBO J. 4,* 3369–3373 (1985).

158. Rostas, K., Kondorosi, E., Horvath, B., Simoncsits, A., and Kondorosi, A., "Conservation of extended promoter regions of nodulation genes in *Rhizobium," Proc. Natl. Acad. Sci. USA 83,* 1757–1761 (1986).

159. Rushing, B., Yelton, M. M., and Long, S. R., "Genetic and physical analysis of the *nodD3* region of *Rhizobium meliloti," Nucleic Acids Res., 19*: 921–927 (1991).

160. Sadowsky, M. J., Olson, E. R., Foster, V. E., Kosslak, R. M., and Verma, D. P. S., "Two host-inducible genes of *Rhizobium fredii* and characterization of the inducing compound," *J. Bacteriol. 170,* 171–178 (1988).

161. Schmidt, J., Wingender, R., John, M., Wieneke, U., and Schell, J., "*Rhizobium meliloti nodA* and *nodB* genes are involved in generating compounds that stimulate mitosis of plant cells," *Proc. Natl. Acad. Sci. USA 85,* 8578–8582 (1988).

162. Schofield, P. R., Ridge, R. W., Rolfe, B. G., Shine, J., and Watson, J. M., "Host-specific nodulation is encoded on a 14kb DNA fragment in *Rhizobium trifolii," Plant Mol. Biol. 3,* 3–11 (1984).

163. Schofield, P. R., and Watson, J. M., "DNA sequence of *Rhizobium trifolii* nodulation genes reveals a reiterated and potentially regulatory sequence preceding *nodABC* and *nodFE," Nucleic Acids Res. 14,* 2891–2903 (1986).

163a. Schwedock, J., and Long, S. R., "Nucleotide sequence and protein products of two new nodulation genes of *Rhizobium meliloti, nodP* and *nodQ," Mol. Plant-Microbe Interactions 2,* 181–194 (1989).

164. Schwedock, J. and Long, S. R., ATP sulphurylase activity of the nodP and nodQ gene products of *Rhizobium meliloti*. Nature *348*, 644–646.

165. Scott, K. F., "Conserved nodulation genes from the non-legume symbiont *Bradyrhizobium* sp. (*Parasponia*)," *Nucleic Acids Res. 14*, 2905–2919 (1986).

166. Sharma, S. B., and Signer, E. R., "Temporal and spatial regulation of the symbiotic genes of *Rhizobium meliloti* in planta revealed by transposon Tn5-*gus*A," *Genes Develop. 4*, 344–356 (1990).

167. Shearman, C. A., Rossen, L., Johnston, A. W. B., and Downie, J. A., "The *Rhizobium leguminosarum* nodulation gene *nodF* encodes a polypeptide similar to acyl-carrier protein and is regulated by *nodD* plus a factor in pea root exudate," *EMBO J. 5*, 647–652 (1986).

168. Simon, R., "High frequency mobilization of gram negative bacterial replicons by the *in vivo* constructed Tn5 mob transposon," *Mol. Gen. Genet. 196*, 413–420 (1984).

169. Simon, R., Priefer, U., and Pühler, A., "A broad host range mobilization system for *in vivo* genetic engineering: Transposon mutagenesis in gram negative bacteria," *Biotechnology 1*, 784–791 (1983).

170. Sindhu, S. S., Brewin, N. J., and Kannenberg, E. L., "Immunochemical analysis of lipopolysaccharides from free-living and endosymbiotic forms of *Rhizobium leguminosarum*," *J. Bacteriol. 172*, 1804–1813 (1990).

171. So, J.-S., Hodgson, A. L. M., Haugland, R., Leavitt, M., Banfalvi, Z., Nieuwkoop, A. J., and Stacey, G., "Transposon-induced symbiotic mutants of *Bradyrhizobium japonicum:* Isolation of two gene regions essential for nodulation," *Mol. Gen. Genet. 207*, 15–23 (1987).

172. Spaink, H. P., Okker, R. J. H., Wijffelman, C. A., Pees, E., and Lugtenberg, B. J. J., "Promoters in the nodulation region of the *Rhizobium leguminosarum* Sym plasmid pRL1JI," *Plant Mol. Biol. 9*, 29–37 (1987).

173. Spaink, H. P., Okker, R. J. H., Wijffelman, C. A., Tak, T., Goosen-de Roo, L., Pees, E., Van Brussel, A. A. N., and Lugtenberg, B. J. J., "Symbiotic properties of rhizobia containing a flavonoid-independent hybrid *nodD* product," *J. Bacteriol. 171*, 4045–4053 (1989).

174. Spaink, H. P., Weinman, J., Djordjevic, M. A., Wijffelman, C. A., Okker, R. J. H., and Lugtenberg, B. J. J., "Genetic analysis and cellular localization of the *Rhizobium* host specificity-determining NodE protein," *EMBO J. 8*, 2811–2818. (1989).

175. Spaink, H. P., Wijffelman, C. A., Pees, E., Okker, R. J. H., and Lugtenberg, B. J. J., "*Rhizobium* nodulation gene *nodD* as a determinant of host specificity," *Nature 328*, 337–340 (1987).

176. Squartini, A., Van Veen, R. J. M., Regensburg-Tuink, T., Hooykaas, P. J. J., and Nuti, M. P., "Identification and characterization of the *nodD* gene in *Rhizobium leguminosarum* strain 1001," *Mol. Plant-Microbe Interact. 1*, 145–149 (1988).

177. Stanfield, S., Ielpi, L., O'Brochta, D., Helinski, D. R., and Ditta, G. S., "The *ndvA* gene product of *Rhizobium meliloti* is required for β-(1-2)glucan production and has homology to the ATP binding export protein H1yB," *J. Bacteriol. 170*, 3523–3530 (1988).

178. Surin, B. P., and Downie, J. A., "Characterization of the *Rhizobium leguminosarum* genes *nodLMN* involved in efficient host specific nodulation," *Mol. Microbiol. 2*, 173–183 (1988).

179. Surin, B. P., Watson, J. M., Hamilton, W. D. O., Economou, A., and Downie, J. A., "Molecular characterization of the nodulation gene, *nodT*, from two biovars of *Rhizobium leguminosarum*," *Mol. Microbiol. 4*, 245–252 (1990).

180. Swanson, J., Tu, J. K., Ogawa, J. M., Sanga, R., Fisher, R., and Long, S. R., "Extended region of nodulation genes in *Rhizobium meliloti* 1021. I. Phenotypes of Tn*5* insertion mutants," *Genetics 117*, 181–189 (1987).

181. Török, I., Kondorosi, E., Stepkowski, T., Posfai, J., and Kondorosi, A., "Nucleotide sequence of *Rhizobium meliloti* nodulation genes," *Nucleic Acids Res. 12*, 9509–9524 (1984).

182. Van Brussel, A. A. N., Zaat, S. A. J., Canter Cremers, H. C. J., Wijffelman, C. A., Pees, E., Tak, T., and Lugtenberg, B. J. J., "Role of plant root exudate and sym plasmid-localized nodulation genes in the synthesis by *Rhizobium-leguminosarum* of TSR factor which causes thick and short roots on common vetch," *J. Bacteriol. 165*, 517–522 (1986).

183. Van Den Eede, G., Dreyfus, B., Goethals, K., Van Montagu, M., and Holsters, M., "Identification and cloning of nodulation genes from the stem-nodulating bacterium ORS571," *Mol. Gen. Genet. 206*, 291–299 (1987).

184. Vincent, J. M., *Biology of Nitrogen Fixation*, Amsterdam: North-Holland, 1974, p. 265.

185. Vincent, J. M., *Nitrogen Fixation*, Vol. II. Baltimore: University Park Press, 1980, p. 103.

186. Ward, L. J. H., Rockman, E. S., Ball, P., Jarvis, B. D. W., and Scott, D. B., "Isolation and characterization of a *Rhizobium loti* gene required for effective nodulation of *Lotus pedunculatus*," *Mol. Plant-Microbe Interact. 2*, 224–232 (1989).

187. Wijffelman, C. A., Pees, E., Van Brussel, A. A. N., Okker, R. J. H., and Lugtenberg, B. J. J., "Genetic and functional analysis of the nodulation region of the *Rhizobium-leguminosarum* sym plasmid pRL-1JI," *Arch. Microbiol. 143*, 225–232 (1985).

188. Williams, M. N. V., Hollingsworth, R. I., Klein, S., and Signer, E. R., "The symbiotic defect of *Rhizobium meliloti* exopolysaccharide mutants is suppressed by *IpsZ*⁺, a gene involved in lipopolysaccharide biosynthesis," *J. Bacteriol. 172*, 2622–2632 (1990).

189. Winsor, B. A. T., "A nod at differentiation: The *nodD* gene product and initiation of *Rhizobium* nodulation," *Trends Genet. 5*, 199–201 (1989).

190. Wood, E. A., Butcher, G. W., Brewin, N. J., and Kannenberg, E. L., "Genetic derepression of a developmentally regulated lipopolysaccharide antigen from *Rhizobium leguminosarum* 3841," *J. Bacteriol. 171*, 4545–4555 (1980).

191. Young, J. P. W., and Johnston, A. W. B., "The evolution of specificity in the legume–rhizobium symbiosis," *Trends Evol. Ecol. 4*, 331–349 (1989).

192. Zaat, S. A. J., Wijffelman, C. A., Spaink, H. P., Van Brussel, A. A. N., Okker, R. J. H., and Lugtenberg, G. J. J., "Induction of the *nodA* promoter of *Rhizobium leguminosarum* Sym plasmid PRL1JI by plant flavanones and flavones," *J. Bacteriol. 169*, 198–204 (1987).

193. Zorreguieta, A., Geremia, R. A., Cavaignac, S., Cangelosi, G. A., Nester, E. W., and Ugalde, R. A., "Identification of the product of an *Agrobacterium tumefaciens* chromosomal virulence gene," *Mol. Plant-Microbe Interact. 1*, 121–127 (1988).

15

Nodulins in Root Nodule Development

Henk J. Franssen, Jan-Peter Nap,
and *Ton Bisseling*

I. Introduction

Nodule formation on the roots of leguminous plants has been well analyzed morphologically.[9,24,46,73,88] Based on these analyses, root nodule formation has been divided into three distinct steps: (1) the preinfection stage, (2) infection and nodule organogenesis, and (3) nodule function and maintenance.[126] During the preinfection stage (brady)rhizobia interact with growing root hairs and induce root hair curling, thereby trapping the bacteria within a "pocket."[115,116] During the second stage the bacteria penetrate the plant cell wall by partly dissolving it,[18,96,97,116] and invade the root hair cell and the root cortex through the infection thread. Meanwhile, independently from the infection process,[74,89] some cells of the root cortex dedifferentiate. Cortical cells start dividing at several places and at these sites the nodule primordia are formed.[19] Infection threads grow toward these primordia, and upon contact rhizobia are released from the tips of the infection threads into the cytoplasm of the plant cells. This release is an endocytotic process,[8] in which the bacteria become surrounded by the peribacteroid membrane.[99] After release, bacteria and peribacteroid membranes divide in a coordinated fashion to fill the host cytoplasm.[98] During the third stage the bacteria differentiate into pleiomorphic bacteroids that fix N_2.

There are two main categories of legume nodules, determinate and indeterminate,[20] that differ in overall morphology and in the presence of an apical meristem. In general, temperate legumes such as *Pisum, Vicia, Trifolium,* and *Medicago* develop indeterminate nodules, and determinate nodules occur on the roots of tropical legumes such as *Glycine, Phaseolus*, and *Vigna*. Indeterminate and determinate nodules are composed of similar tissues, all formed from the nodule meristem. The central tissue contains infected and uninfected cells and is surrounded by several other tissues. The outermost tissue is the nodule cortex, formerly named the outer cortex.[87–89,119] The nodule cortex is separated from the nodule parenchyma, previously named the inner cortex[87–89,119] by an endodermis. Vascular bundles are located in the nodule parenchyma. Indeterminate nodules are characterized by a persistent apical meristem and, therefore, can be divided into distinct zones that differ in developmental stage.[87] The most distal zone to the root is the apical meristem, which is bounded by a more proximal zone (the infection zone) containing elongating plant cells that become infected by *Rhizobium*. In the early symbiotic zone, host cells differentiate into infected and uninfected cells, whereas infected cells fully packed with bacteroids together with uninfected cells are found in the late symbiotic zone. In this late symbiotic zone, nitrogen fixation and ammonia assimilation occur. The most proximal zone in a mature indeterminate nodule is the senescent region where both plant cells and bacteroids degenerate.

In contrast to indeterminate nodules, determinate nodules do not have a persistent meristem.[89] After release of (brady)rhizobia from the infection thread, the infected cells continue to divide until about one week after the onset of N_2 fixation. When mitotic activity has ceased, increase in nodule size is caused by cell expansion. As a consequence, the developmental phases in a determinate nodule are separated temporally rather than spatially.

The formation of both determinate and indeterminate nodules is accompanied by nodule–specific expression of plant genes, the nodulin genes.[70,122] Nodulin genes are, by definition, expressed exclusively during the development of the symbiosis, consequently, they are expressed neither in uninfected roots nor in other parts of the host plant.

Nodulin genes are differentially expressed during nodule development.[40,44,46,48,71,106] The majority of nodulin genes are first expressed around the onset of N_2 fixation and are named late nodulin genes. Nodulin genes that are expressed at earlier stages of development are named early nodulin genes. Early nodulin genes are expressed when the plant become infected and the nodule structure is being formed. Plant genes specifically expressed in the preinfection stage cannot be called nodulin genes according to the definition because they are not expressed in a nodule. However, since the expression of these genes is linked to root nodule formation, we will also refer to these as early nodulin genes.[43]

Nodulins have been identified by comparing *in vitro* translation products from nodule and root mRNA and by identifying cDNA clones by differential screening. Early nodulin cDNA clones are designated by ENOD, while late nodulin cDNA clones are best indicated by NOD. The names of both NOD and ENOD clones should be preceded by the plant genus (in upper case) and species (in lower case) initials. When obtained as protein or *in vitro* translation product, nodulins are indicated by the letter N, followed by the plant genus and species initials in lower case and the molecular weight as determined by SDS polyacrylamide gel electrophoresis (SDS-PAGE). In case a biochemical function can be assigned to the nodulin, addition of the prefix N to the name of the protein is recommended.[86,122]

In this chapter, we shall discuss the nodulins involved in the different stages of nodule formation, their possible functions, the regulation of nodulin gene expression, and the communication between the macro- and microsymbiont in relation to the induction of nodulin gene expression.

II. Nodulins in the Development of Nodules

A. Early Nodulins Involved in the Preinfection Stage

In the preinfection stage, bacteria attach to the root hair surface and the root hairs deform and curl. Recently, changes in root hair gene expression upon interaction with *Rhizobium* bacteria have been described.[43] In pea root hairs, two genes, RH-42 and RH-44, have been identified whose expression appears to be influenced by the interaction of the root hairs with *Rhizobium*. The products of these genes have been detected by *in vitro* translation of root hair mRNA followed by two-dimensional gel electrophoresis (2D-PAGE) of the translation products. The expression of one of these genes, named RH-42 because it encodes a 42–kDa protein, is induced in root hairs upon inoculation of the plants with *Rhizobium*. This gene is not expressed in root hairs of uninoculated plants or in developing nodules. The expression of the other gene, RH-44, is markedly enhanced in root hairs upon inoculation, but low levels of the RH-44 gene product are detected in uninoculated roots and in nodules. Curling and deformation are the only processes unique to the root hair, because formation and growth of the infection thread also occur in the developing root nodule. The pattern of expression of the RH-42 and RH-44 genes suggests, therefore, that their gene products are involved in curling and deformation of the root hair.

B. Early Nodulins Involved in the Infection Process

The cytologic aspects of the infection process in pea have been studied extensively by Newcomb[87–89] and Bakhuizen.[7] They have shown that after entrapment of rhizobia in a curled root hair a local lesion of the root hair cell wall is induced. Subsequently, the host plant deposits cell wall–like material around these lesions and an infection thread is formed. This infection thread grows toward the root inner cortical cell layers. The outer cortical cells become prepared for infection thread growth before the infection thread penetrates these cells. They form vacuolar cytoplasmic bridges apparently via a reorientation of the microtubular cytoskeleton connected with transvacuolar migration of nuclei. The cytoplasmic bridges then serve as pathways for the growing infection threads. In the root cortical cells that become prepared for infection thread growth an additional cell wall is formed.

These processes are accompanied by the expression of two nodulin genes for which the cDNA clones pPsENOD12 and pPsENOD5 were isolated from a pea nodule cDNA library.[102,103]

In situ hybridization studies revealed that the PsENOD12 gene is induced in root hairs, root cortical cells, and nodule cells containing growing infection threads. Interestingly, this early nodulin gene is expressed not only in cells containing growing infection threads, but also within several cell layers in front of the infection thread tip (Figure 15–1), where cells are preparing for infection thread growth.

The PsENOD12 early nodulin contains a putative signal peptide at the N-terminus, and the rest of the protein is composed of two repeating pentapeptides, each containing two prolines. This structure recalls the structure of a hydroxyproline-rich cell wall protein,[5] and therefore PsENOD12 is likely a cell wall component. Because the PsENOD12 gene is expressed in cells already containing, and in cells preparing for the growing infection thread, the PsENOD12 protein may be part of the infection thread and/or of the additional cell wall of the root cortical cell.

The spatial distribution of the PsENOD5 mRNA in infected pea roots is strikingly different from that of the PsENOD12 transcript. Whereas the PsENOD12 mRNA is present within several root cortical cell layers in front of the infection thread tip, the PsENOD5 gene is expressed only in cells containing the growing infection thread. The PsENOD5 protein also is a proline-rich early nodulin, but it does not show a repetitive structure as PsENOD12 and PsENOD2 (see Section II.C.iii). PsENOD5 is rich in Pro, Ala, Gly, and Ser residues, and therefore this protein may be related to the arabinogalactan proteins that are plasma membrane components.[64] Thus, PsENOD5 may be a component of the infection thread plasma membrane.

C.   Early Nodulins Involved in
Nodule Organogenesis

i.   Primordium and Meristem

Nodule organogenesis starts with *Rhizobium*-induced cell divisions in the root cortex. The mitotic activity induced by *Rhizobium* results in the formation of a primordium. Infection threads reach the primordium, and bacteria are released into the plant cells about 5 days after inoculation of pea plants. At this stage of development a meristem is formed at the distal site of the primordium. The meristematic cells can be distinguished morphologically from the primordium cells as being smaller in size and containing smaller vacuoles. Concomitantly with the formation of the apical meristem,

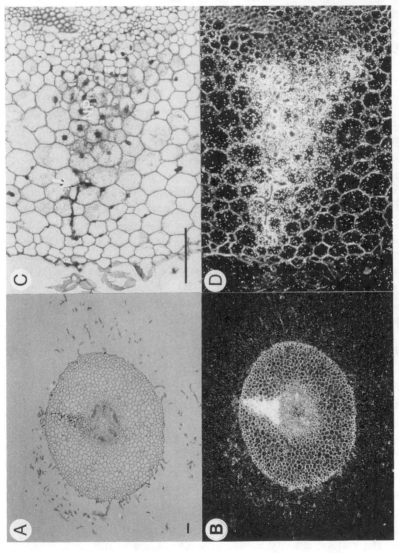

Figure 15–1. Localization by *in situ* hybridization of ENOD12 transcript in pea root segments of a 6–day old pea seedling, 3 days after inoculation. A, B, C, and D are corresponding photographs. A and C are bright file micrographs. B and D are dark field micrographs. In the dark field micrographs, silver grains representing hybridization are visible on white spots. The infection thread is indicated with an arrow and the nodule primordium is denoted by NP. (The figure is provided by C. van de Wiel.)

cells at the proximal site of the primordium differentiate into the different nodule tissues. At later stages of development, the apical meristem adds cells to these nodule tissues.

Two lines of evidence suggest that the infection-related PsENOD12 early nodulin is also involved in nodule primordium formation. By 48 hours after inoculation of pea seedlings, the infection threads have reached the outer cortical cells of the root, and cell division is induced in the root inner cortical cells. At this stage of development, the PsENOD12 gene is expressed both at the infection site and in the dividing inner cortical cells. These two sites are spatially separated by the root cortex cells, which do not express the PsENOD12 gene.[102] This observation suggests that the PsENOD12 gene is involved in both the infection process and the formation of a nodule primordium. The role of the PsENOD12 gene in the latter process is supported by the expression of the PsENOD12 gene in nodulelike structures formed by the auxin transport inhibitor naphtylphtalamic acid (NPA).[104] Since infection events do not occur in the NPA–formed nodulelike structures, it is plausible that PsENOD12 is involved in nodule primordium formation.

In the meristematic cells the PsENOD12 gene is not expressed. At present, no nodulins have been identified in nodule meristems. Therefore, it is unclear whether the nodule meristem is distinguished from other plant meristems at the molecular level.

ii. Central Tissue

The most conspicious nodule tissue is the central tissue, containing both the infected and uninfected cell types. The infected cells harbor the bacteroids and, consequently, the N_2 fixation occurs here. In determinate nodules, the uninfected cells form a network,[87-89] by which the assimilation products— allantoins and ureides—are likely transported to the nodule vascular system. The final stages of ammonia assimilation in a determinate nodule occur in the uninfected cells. In an indeterminate nodule, in which the fixed nitrogen is mainly transported as amides, no specific role has been assigned to the uninfected cell type. Tissues of indeterminate nodules are of graded age, with the youngest cells located near the apical meristem and the oldest near the root attachment point. The expression of a certain gene during development of a cell type can therefore be studied in a single longitudinal section of a nodule. By hybridizing longitudinal sections of an indeterminate pea nodule with different pea early nodulin gene probes, it was shown that the formation of the infected cell type involves the sequential induction of early nodulin genes.[103]

The PsENOD12 gene is expressed in all cells of the invasion zone directly adjacent to the apical meristem, where active infection thread growth occurs. This expression is consistent with a function of PsENOD12 in the infection process. In older parts of the nodule, the PsENOD12 transcript is not detectable. The infection process–related early nodulin transcript PsENOD5 also occurs in the invasion zone of the pea nodule. Although the PsENOD12 and PsENOD5 mRNAs are both present in the invasion zone of the nodule, a striking difference is observed with respect to their spatial distribution in this zone. Whereas PsENOD12 mRNA is present in all cells of the invasion zone, PsENOD5 mRNA is only detected in a limited number of cells. Also in contrast to PsENOD12 mRNA, PsENOD5 mRNA is also present is the infected cells of the early symbiotic zone.[103] Therefore, the expression of the PsENOD5 gene found in the invasion zone may reflect a molecular differentiation into infected cells, before infected and uninfected cells can be distinguished cytologically. In the early symbiotic zone, nodule cells elongate, rhizobia proliferate, but the infection threads cease growth. Consequently, the PsENOD5 protein is functional during both infection thread growth and bacterial proliferation. It could be part of the peribacteroid and infection thread membrane. In the early symbiotic zone, where PsENOD5 gene expression is maximal, the transcripts of the early nodulin genes PsENOD3 and PsENOD14 are first detectable. The PsENOD3 and PsENOD14 genes are exclusively expressed in the infected cells.[103] The PsENOD3 and PsENOD14 polypeptides have a molecular mass of 6 kDa and are 55% homologous. These early nodulins have a putative N-terminal signal peptide and they contain four cysteine residues with a spatial distribution resembling that of metal-binding proteins. Therefore, PsENOD3 and PsENOD14 may be metal-binding proteins—for instance, involved in Fe or Mo transport to the bacteroids—that require a supply of these metal ions for the synthesis of the enzyme nitrogenase.[103]

Like the PsENOD12 and PsENOD5 genes, the PsENOD3 and PsENOD14 genes are transiently expressed during development of the nodule. In the late symbiotic zone the amount of the PsENOD3 and PsENOD14 mRNAs decreases. In the infected cells of the late symbiotic zone, late nodulin transcripts like leghemoglobin mRNA are present.[103]

In indeterminate nodules, no specific functions have been assigned to the uninfected cells, but it has been shown that the pea early nodulin Nps-40′ is specifically located in this cell type.[120] Nps-40′ is an early nodulin that has been identified by comparing *in vitro* translation products from nodule and root RNA. Using antibodies directed against Nps-40′, it was shown that this early nodulin is structurally similar to the *Vicia sativa* early nodulin Nvs-40 and to the early nodulin Nms-30 of alfalfa.

iii. Nodule Parenchyma

The nodule parenchyma was previously named the nodule inner cortex.[119] The nodule parenchyma is located inside the endodermis of the nodule, but no other plant cortical tissue can be found at such a position. The morphology of the nodule parenchyma is different from that of cortical tissues because of the smaller intercellular spaces.[113,128] Also the expression of the early nodulin gene ENOD2 in pea,[119] soybean,[119] and alfalfa[121] nodule parenchyma and GmENOD13 in soybean nodule parenchyma[121] distinguishes this tissue from cortical tissues, and suggests that it will have a specific function in symbiosis.

The groups of Witty[128] and Tjepkema[113] have shown that nodule parenchyma has an important role in regulating the free O_2 concentration in the nodule. Witty and Tjepkema were able to prove that the low O_2 concentration in the nodule is achieved by the high O_2 consumption rate of the rhizobia combined with a diffusion barrier in the nodule parenchyma.[128] The few and small intercellular spaces in the nodule parenchyma contribute to the formation of the O_2 diffusion barrier.[113,128] The early nodulin ENOD2 is composed of two repeating pentapeptides containing two proline residues each. As with PsENOD12, this structure suggests that this early nodulin is a cell wall component. The soybean early nodulin protein GmENOD13 is 50% homologous with GmENOD2 and has a similar repetitive nature.[37,43]

Since the cell wall is a major factor in determining cell morphology, it is likely that early nodulins like ENOD2 and GmENOD13 contribute to the special morphology of the nodule parenchyma and consequently to the formation of the O_2 diffusion barrier.

D. Late Nodulins Involved in Nodule Function and Maintenance

The process of actual N_2 fixation starts when all cell types of the nodule have obtained their specific morphologic characteristics. Around the onset of N_2 fixation, the expression of a set of nodulin genes, the late nodulin genes, first becomes detectable. Late nodulins have been identified in nodules of all leguminous plants studied so far, including alfalfa[66]; bean[93]; clover[56]; lupine[110], pea[11]; soybean[4]; *Vicia*[84]; and the tropical legume *Sesbania*.[24] The start of late nodulin gene expression in the symbiotic interaction process coincides more or less with the beginning of N_2 fixation. Consequently, late nodulins will likely have functions related to the symbiotic N_2–fixation process, such as assimilation, transport of metabolites across the peribacteroid membrane, or creation of the physiologic conditions required for N_2 fixation in the bacteroids.

i. Metabolic Nodulins

In root nodules of several legume species, late nodulins have been identified with a specific role in the carbon, nitrogen, and oxygen metabolism of the nodule. These nodulins are referred to as metabolic nodulins. Leghemoglobin (Lb) is the most abundant late nodulin, comprising almost 20% of the total soluble nodule proteins, and is an O_2 carrier in the infected cells of the nodule that facilitate oxygen diffusion toward the bacteroids.[47,107,127] The Lbs are encoded by more than one gene. In the soybean genome, for example, four genes and several pseudogenes have been characterized.[13,15,16] On the protein level four major Lbs and several minor components representing modified major Lbs have been detected.[39] Slight differences have been observed in the time course of synthesis of the different Lbs during nodule development on both soybean[77,123] and pea.[117] The functional significance of the differential expression of the genes is unclear.

In soybean nodules Ngm-35 is the 33–kDa subunit of n-uricase[10] and is the second most abundant protein in the cytoplasm of soybean nodules. This nodule-specific uricase has no homology with the two-enzyme system that is responsible for the uricase activity in uninoculated roots of soybean. The n-uricase is the product of a totally different gene from the root and leaf uricase. Southern blot hybridizations of soybean genomic DNA using n-uricase cDNA clones as a probe indicated several DNA fragments homologous with Ngm-35 sequences, suggesting the existence of a small number of genes.[91] Uricase is a key enzyme in the ureide biosynthetic pathway of determinant nodules[70] and is specifically located in the peroxisomes of the uninfected cells.[10,91] Sequence analysis of the soybean n-uricase cDNA clone did not reveal a signal peptide, so information for transport into peroxisomes must reside in the protein itself. Two hydrophobic domains in the amino acid sequence may facilitate translocation across the peroxisomal membrane.[91]

Glutamine synthetase (GS) catalyzes the first step in ammonium assimilation in determinate and indeterminate nodules. GS is an octameric protein with a heterooligomeric structure. In alfalfa, a nodulin gene coding for a GS subunit has been identified.[28] In French bean, it has long been claimed that GS subunit genes are nodulins.[21,33,67,125] However, expression of the supposed n-GS gene has recently been detected in other plant organs,[33] with the notable exception of roots. In soybean, the occurrence of n-GS is controversial,[52,53,105] whereas in pea no n-GS has been detected.[112]

With respect to nodule carbon metabolism it has been shown that the subunit of the sucrose synthase that acts in soybean nodules is nodulin Ngm-100.[111]

In soybean nodules the nitrogen fixed in the bacteroid is exported as ureide, allantoin, and allantoic acid. In ureide metabolism proline is synthesized in high rates. The high amount of free proline has been reported to have an osmoregulatory function in the nodule. The enzyme pyroline-5-carboxylate reductase (P5CR) has a key role in proline biogenesis. Nodule P5CR has been found in the plant cytosol, and its activity is substantially higher than that reported for other plant tissues.[64A]. Hence, there is a possibility that also in the ureide metabolism in the nodule, nodule–specific forms of regular enzymes are involved, as is true for many of the enzymes of the carbon and nitrogen metabolic pathway.

Besides the preceding described late nodulins, several enzymes have been detected in root nodules of legumes that differ in physical, kinetic, or immunochemical properties from corresponding enzymes in roots. Examples are phosphoenolpyruvate carboxylase,[25] choline kinase,[80] xanthine dehydrogenase,[90] purine nucleosidase,[69] and malate dehydrogenase.[2] The nodule–specific forms of these different enzymes might be true nodulins. However, it is also possible that these nodule–specific forms are the result of modifications of root enzymes.

ii. Peribacteroid Membrane Nodulins

Peribacteroid membranes surround bacteroids in infected cells. This membrane develops from the plasma membrane, once the bacteria are released from the infection thread by endocytosis. Later in development, biogenesis of the peribacteroid membrane involves the coalescence of Golgi vesicles.[79,99,114,124] Both of these observations come from immunologic studies in which antibodies directed against antigens in the pea peribacteroid membrane also recognized antigens in plasma membranes and Golgi vesicle membranes.[118] Nevertheless, both biochemical and molecular studies have shown that the peribacteroid membrane contains unique proteins. In soybean root nodules, a nodule–specific isoform of choline-kinase I[81] and H^+ ATPase,[12] and several other nodulins have been identified that are localized in the peribacteroid membrane but do not occur in other membranes. Since the peribacteroid membrane is the physical and metabolic interface between the invading *Rhizobium* and its eukaryotic partner, nodulins may function in the exchange of metabolites between the symbionts. Currently, peribacteroid membrane nodulins have only been described in soybean. Nodulin Ngm-24 is exclusively located in the peribacteroid membrane. Secondary structure analysis suggests that this nodulin resides on the surface of the peribacteroid membrane, facing the peribacteroid space. Therefore, it is not likely that Ngm-24 functions as a transmembrane transport molecule.[32,34] Another peribacteroid membrane nodulin, Ngm-26,[32] probably is a transmembrane protein exposed both to the peribacteroid space and to the plant cytoplasm. The amino acid sequence of Ngm-26 shows a significant homology to the bovine lens fiber major intrinsic protein MIP and *E. coli* glycerol facilitator.[6,101] The latter protein is the only known pore-type protein in the *E. coli* cytoplasmic membrane and has a function in transport of small molecules across membranes. Therefore, Ngm-26 may have a role in small molecule transport across the peribacteroid membrane.

The peribacteroid membrane nodulin Ngm-23[57] is a member of a small family of nodulins, of which five members have been characterized.[57,100] Nucleotide sequence analysis of cDNA clones and of corresponding genes revealed that the coding sequences of these nodulins contain two regions with 70–90% homology, interrupted by a third region that is unique to each nodulin.[100] The two conserved regions are centered around four cysteine residues, the spatial distribution of which suggests that these nodulins may be metal-binding polypeptides. Although all these nodulins possess a potential hydrophobic signal sequence suggesting they may be associated with membranes, it was shown by using antiserum raised against soluble nodule–specific proteins that at least one of them is located in the cytosol of nodule cells.[57]

III. Nodulin Existence

By definition,[122] nodulins are plant-genome–encoded proteins that are found only in root nodules and not in any other parts of the host plant. Consequently, nodulin genes are plant genes exclusively expressed during the development of the symbiosis. In laboratory practice, however, it is virtually impossible to screen all plant organs and tissues, so nodulin genes have usually been identified by comparing gene expression in uninoculated roots and nodules.[11,46,66,71,105,106] Therefore, none of the nodulin genes discussed earlier have ever been formally proven to be genuine nodulin genes in the narrowest sense of the definition. Theoretically, nodulins may not even exist.

In the few cases where plant organs other than the uninoculated root were screened for expression of a presumed nodulin gene, results have been remarkable. A low–level expression of the early nodulin gene PsENOD12 has been demonstrated in stems and flowers.[102] Two PsENOD12 genes are present in the pea genome, both of which are expressed in nodule, stems, and flowers.[50] Detailed analyses of the expression pattern of the gene for the n-GS subunit in bean showed this gene to be expressed, albeit at a low level, in stems, petioles, and cotyledons.[33] Low levels of n-uricase are found in soybean callus and, occasionally, in uninoculated roots.[68]

In all these cases, the genes involved can no longer be formally considered as true nodulin genes. Even for the most undisputed nodulin leghemoglobin, it has been speculated that Lb gene expression, although never detected, may occur in tissues other than the root nodule. This hypothesis is supported by the fact that a single hemoglobin gene highly homologous to leghemoglobin genes is expressed in roots as well as in nodules.[65] Recently, the peribacteroid membrane nodulin Ngm-26 was found to show a substantial homology to the major intrinsic protein of the bovine lens fiber membrane and to the E. coli glycerol facilitator membrane protein.[6,101] This homology might indicate that Ngm-26 is a member of a gene family of which Ngm-

26 represents a nodulin, while other members are also expressed in symbiotic tissues. However, it could also mean that peribacteroid membrane nodulin genes are also transcribed in nonsymbiotic tissues.

In all cases described earlier, the occurrence of nodulins or nodulin gene expression in nonnodule tissue has been demonstrated by relatively insensitive techniques through protein or RNA transfer blot analysis, or *in vitro* translation of mRNA. In general, over 25,000 genes are likely expressed in each plant organ,[62,63] only a few thousand of which are expressed at a sufficient level to be detected by these methods. It can be anticipated that by applying more sensitive detection methods, like the polymerase chain reaction, more presumed nodulin genes will be found expressed at other positions in the plant and, therefore, will cease to be nodulins *stricto senso*. This implies that it will become important to rename several of the known nodulin genes in the future. Until that time, it seems unavoidable that the term *nodulin* will be used less accurately.

IV. Regulation of Nodulin Gene Expression

The molecular basis of tissue-specific gene expression is a major subject of research in plant[72] and animal systems.[51] Studies on both plant and animal organ-specific gene expression have shown the complexity of the regulation of gene expression. In the promoter region of genes, *cis*-acting sequences like tissue-specificity elements and enhancer and silencer sequences are present. To these region–specific proteins, so-called *trans*-acting factors bind either singly or in complex with other proteins, and apparently control gene expression.

Nodulin genes are highly regulated both spatially and temporally during nodule development, and the mechanisms underlying regulation of nodulin gene expression will be as complex as in other cases of organ-specific gene expression. The ultimate control over nodule specificity of gene expression will thus be determined by the interaction of *cis*-regulatory sequences and *trans*-acting factors. However, specific stimuli will be essential to elicit nodulin gene expression. In this respect, signal molecules from *Rhizobium* are likely to initiate nodulin gene expression, although it has also been suggested that physiologic conditions within the nodule play a role in the regulation of nodulin gene expression. In this section, we shall discuss the *cis*-regulatory sequences and *trans*-acting factors governing nodulin gene expression, the physiologic conditions in the nodule that might contribute to nodulin gene expression, and the *Rhizobium* genes coding for signal molecules eliciting nodulin gene expression.

A. Cis- *and* Trans-*Regulatory Elements*

The importance of the promoter region for nodule specific gene expression has been demonstrated by the transgenic approach.[23,60,61,108,109] The nodule–

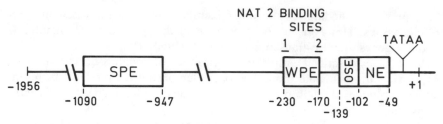

Figure 15–2. Organization of regulatory DNA sequences in the 2–kb 5' region of the soybean leghemoglobin *Lbc3* gene. Binding sites for the NAT2 *trans*-acting factor are indicated by 1 and 2. SPE = strong positive element, WPE = weak positive element, OSE = organ-specific element, NE = negative element. For a functional description of the various elements, see text. (This figure is kindly provided by J. Stougaard, Aarhus, Denmark.)

specific and developmentally correct expression in transgenic leguminous plants of a chimeric gene, consisting of the promoter of nodulin genes fused to the chloramphenicol acetyltransferase coding sequence, proved that at least for the investigated soybean genes for Lbc3 and Ngm-23 and the sesbania Lb gene the 5'-upstream region carries all required *cis*-regulatory sequences. Furthermore, this result shows that all relevant *trans*-acting factors are conserved in different legume species. By 5'-deletion analysis, several *cis*-regulatory elements have been identified in the 5' upstream region of the soybean Lbc3 and Ngm-23 genes.[60,61] The positions of these elements for the Lbc3 gene are shown in Figure 15–2. In addition to an element absolutely essential for nodule-specific expression (OSE), several elements have been identified that influence the level, but not the organ specificity, of Lbc3 gene expression. Certain elements have a strong (SPE) or a weak positive (WPE) effect on the expression level, whereas others have a negative effect on the expression level (NE) of the gene. Similar *cis*-regulatory elements have been identified in the promoter of a *Sesbania* Lb gene.[23]

By comparing the nucleotide sequences of different nodulin genes, several consensus sequences in the promoter region have been identified. Mauro et al.[78] described three different consensus sequences present in front of Lbc3, Ngm-23, and Ngm-24. However, none of these consensus sequences coincide with one of the *cis*-regulatory elements in Lbc3 or Ngm-23. Sandal et al.[100] have identified two additional consensus sequences that are present in the promoter regions of the genes encoding Lbc3 and the peribacteroid membrane proteins, Ngm-21, Ngm-22, Ngm-23, Ngm-44, and Ngm-24. All genes containing these sequences are expressed in the infected cell and are first expressed at the same step in nodule development. The conservation of the consensus sequences in the 5' flanking regions of these nodulin genes might suggest that they are involved in the activation of these genes by the same transcription factor.

In the Lbc3 gene and Ngm-23 gene promoter the consensus sequences described by Sandal et al.[100] are located within the OSE, suggesting that these sequences might be directing nodule–specific expression. The sequences are not present in the promoter region of the early nodulin gene GmENOD2[37A] or in the *n*-uricase encoding gene, which are expressed in the nodule parenchyma,[119] or the uninfected cell,[91] respectively. The significance of the consensus sequences with regard to nodule–specific, cell type, or signal–specific expression is not clear at this stage.

Different *trans*-acting factors may interact with the different *cis*-elements. However, at present no *trans*-acting factors binding specifically to these motifs have been identified. One nodule–specific protein from soybean (NAT2) has been identified that binds to AT-rich motifs in the 5′ region of Lbc3,[58] but deletion analyses of this promoter have shown that the AT-rich motifs determine neither the nodule specificity nor the level of the expression of the Lbc3 gene. Furthermore, proteins have also been identified that bind to the same AT-rich motifs in soybean roots and leaves. All these DNA-binding proteins have a relationship to the high-mobility-group proteins (HMG).[58] HMGs have been identified in several animal systems and are rather basic proteins that bind to DNA. However, they do not appear to have a role in tissue–specific gene expression. Apparently, no *trans*-acting factors that determine nodulin gene expression have yet been identified.

B. Physiologic Conditions Regulating Nodulin Gene Expression

The physiologic conditions within the root nodule are often claimed to play a role in the regulation of nodulin gene activity. In particular, O_2, ammonium, and heme have been implicated in the regulation of nodulin gene expression. However, the experimental data available to support these claims are not convincing.

Late nodulin genes are first expressed just before the onset of N_2 fixation and the actively dividing rhizobia create a low O_2 environment within the nodule. This low O_2 concentration may act upon late nodulin gene expression, notably Lb gene expression. Lb keeps the free O_2 concentration low, thus allowing the extremely O_2–sensitive nitrogenase enzyme to function. In an attempt to mimic the conditions for nodulin gene induction, pea roots have been grown under different O_2 concentrations.[47] None of the genes that are expressed at enhanced levels in pea roots under these low O_2 concentrations were found to be nodulins.

O_2 has been postulated to be a factor regulating *n*-uricase gene expression in soybean root nodules. The specific activity of *n*-uricase increases at least a 150-fold in nodules. The specific activity of *n*-uricase in callus tissue was enhanced threefold by lowering the O_2 concentration,[68] but the physiologic significance of this observation seems limited in comparison to the increase in activity in nodules. Thus, the role of oxygen in inducing *n*-uricase gene expression remains doubtful. On the other hand, although the oxygen concentration in the nodule may not be important for plant gene expression, the *nifA* gene of both *R. meliloti*[27] and *B. japonicum*[32] is induced in free-living bacteria when the O_2 concentration is reduced to microaerobic levels.

Heme has been shown to regulate the expression in yeast of a chimaeric gene consisting of the 5'-flanking region of the soybean Lbc3 gene and the coding sequence of the neomycin phosphotransferase gene. However, this regulation is not transcriptional, but at the posttranscriptional level.[59] The significance of any heme-mediated mechanism *in vivo* is unknown. It has also been shown that in the presence of free heme the nodulin sucrose synthase dissociates rapidly *in vitro,*[111] thereby decreasing the enzyme activity. Although heme may thus influence the activity of nodulins, there is no evidence that heme is involved in the induction of nodulin gene expression.

Ammonia has been implicated in the regulation of expression of GS genes in soybean,[53] but the GS genes investigated appeared not to be nodulin genes. Three lines of evidence strongly suggest that ammonia is not involved in the induction of nodulin gene expression: In normal nodule development, all nodulin genes are expressed before nitrogen fixation starts,[44,46] all nodulin genes are expressed in nodules induced by *Rhizobium* and *Bradyrhizobium nif* and *fix* mutants,[44,46] and on the roots of cowpea (*Vigna unguiculata*) grown in an argon–oxygen environment containing negligible amounts of N_2 gas, nodules could be induced by *Rhizobium* in which both nitrogenase activity and leghemoglobin are detectable.[3] Neither the availability of N_2 nor the product of nitrogenase activity, ammonia, is thus essential for the induction of nodulin gene expression.

C. Rhizobial Signals

Nodulin gene expression does not appear to be induced by alterations in the physiologic conditions of the nodule. It is therefore conceivable that it is the *Rhizobium* that delivers signals for inducing nodulin gene expression. A relatively low number of *Rhizobium* genes, the nodulation (*nod*) genes, is essential for the induction of the first steps of nodule formation (i.e., root hair deformation and curling), the infection process, and the induction of cortical cell divisions. The *nod* genes can be roughly divided into two groups—four common *nod* genes (*nodDABC,* which are structurally and functionally conserved in all *Rhizobium* species) and several host-specific *nod* genes that determine the host specificity of the bacterium.

Nodulation gene expression is under control of nodD, which might act as a transcriptional activator in *Rhizobium* when the bacteria are grown in the presence of flavonoids. Upon induction of the *nod* genes, *Rhizobium* secretes low–molecular–weight compounds that cause root hair deformations and the transcription of the PsENOD12, PsENOD5, and RH-44 genes.[43,55,102]

By comparison with HPLC chromatograms of the low–molecular–weight compounds produced by wild-type *Rhizobium meliloti* and different *nod* mutants, Faucher et al.[29,30] concluded that host–specific *nod*-gene products of *R. meliloti* modify the signal molecule that is formed by the activity of the *nod*ABC genes. Therefore, the induction of *nod* genes results in the production of a variety of closely related molecules, each of which might induce different steps in nodule formation.

The determination of exactly which of the low–molecular–weight compounds induces a specific step in nodule development is becoming a topical subject, therefore, assays have to be developed. It was found that root hair deformation can be induced in the absence of the bacterium by a compound secreted by *Rhizobium* grown *ex planta*.[129] Therefore, a bioassay based on the induction of root hair deformation could be applied to the purification of the root hair deformation factor. At the moment, it has not been possible to induce infection thread formation by applying soluble compounds. Therefore, it is very unlikely that bioassays can be developed to purify the factors inducing the infection process. However, the expression of the early nodulin genes PsENOD5 and PsENOD12 can be induced by soluble factors. As described earlier, these two early nodulins are in some way involved in the infection process. Therefore, assays based on the induction of these early nodulin genes can be used to identify and purify compounds involved in eliciting the infection process.

Apart from the *nod* genes, mutational analyses indicated that the rhizobial genes involved in exopolysaccharide synthesis are also essential for the infection process. Polysaccharides or oligosaccharide fragments might be essential for the induction of the infection process.[31,94] Alternatively, surface components like the exopolysaccharides might circumvent a defense response.[85,86] In the latter case, the exopolysaccharides will have a passive role in forming nodules rather than inducing steps in nodule development.

Although the compounds secreted by *Rhizobium* upon induction of the *nod* genes have now been purified, the mechanism by which they induce nodule formation is still unclear. However, clues about the mechanism by which the *Rhizobium* signal molecules induce nodule development are obtained from studies on nodulelike structures formed by auxin transport inhibitors. Naphtylphtalamic acid (NPA) and triiodobenzoic acid (TIBA) are compounds that are thought to block polar transport of auxin in plants. By the early 1950s it was already known that these compounds induce the formation of nodulelike outgrowths on roots of several plant species. Legumes turned out to be particularly susceptible to these compounds[1] and cytologic analyses show that these structures resemble *Rhizobium*-induced nodules in several aspects.[54] More recently, the resemblance was shown also in molecular terms, since the alfalfa and pea ENOD2 genes and the early nodulin gene PsENOD12 and Nms-30 are expressed in NPA-formed nodulelike structures in pea and alfalfa, respectively.[54,104] Although it is unclear how NPA and TIBA act, it can be concluded that root nodule formation is induced by disturbance of the phytohormone balance. Hence, upon activation of the bacterial *nod* genes, products are formed that interfere with the phytohormone balance in roots. This conclusion is supported by studies of Cooper and Long,[76] who showed that mutations in common *nod* genes of *R. meliloti* can be complemented with the zeatin gene of the T-DNA of *Agrobacterium* that is required for cytokinin production.

In the NPA/TIBA-formed nodulelike structures, the ENOD2 transcript is found by *in situ* hybridization studies in a peripheral tissue resembling the ENOD2 gene expressing nodule parenchyma of *Rhizobium*-induced nodules. Therefore, NPA/TIBA induces the formation of a nodule primordium that subsequently differentiates into a tissue resembling the nodule parenchyma. If NPA is able to induce a cascade of events, *Rhizobium* signals may also do this.

Although several early nodulin genes are expressed, in the NPA-induced, nodulelike structures, none of the late nodulins transcripts have ever been detected in these structures.[54] This strongly suggests that, besides the signals mimicked by NPA, other compounds secreted by *Rhizobium* are essential for the latter steps in development. At the moment it is unclear which *Rhizobium* genes are involved in triggering late nodulin gene expression, but good candidates are the genes involved in lipopolysaccharide (LPS) and capsular polysaccharide (CPS) synthesis, because in many cases mutations in these genes lead to nodules in which development stops shortly after release of the bacterium from the infection thread.[41,92,95] However, the *nod* genes still might have a role in the induction of late nodulin gene expression.[85]

V. Concluding Remarks

The formation of N_2-fixing nodules on roots of leguminous plants is a multistep process in which nodulin genes are expressed during the different stages of root nodule formation. In indeterminate nodules successive developmental stages are spatially separated, and it is shown that *Rhizobium* sp. causes induction of nodulin gene expression at various time points of development. The division of nodulin genes into early and late nodulins should therefore not be regarded as reflecting the existence of two main time points at which *Rhizobium* sp. induces nodulin gene expression. Rather, it represents an arbitrary division within a set of genes induced by the bacterium in a sequential manner. Early and late nodulin genes mark specific stages of nodule development. In addition to nodulin gene expression, the expression of specific (*Brady*)*rhizobium* genes might also be used to characterize specific stages of nodule development. Recently, the nodule cells in which the bacteria express the structural *nif* genes were identified by *in situ* hybridization. When the spatial distribution of more nodulin mRNAs and bacterial transcripts has been determined, a more detailed description of the steps involved in nodule development will become available. Consequently, this might lead to a redefinition of the zones now used.

The central tissue of the nodule contains the cells harboring the N_2-fixing bacteroids. Not surprisingly, the vast majority of nodulin genes are expressed in the infected cell zone. However, nodulin genes are also expressed in other nodule tissues. At least two early nodulin genes are expressed in the nodule parenchyma ("inner cortex"), and recently it was shown that the soybean early nodulin ENOD40 is located in the pericycle of the nodule vascular bundle.[38] This strongly suggests that, in addition to the central tissue, other nodule tissues will have a specific function in the symbiotic N_2–fixation process. The typical characteristics of these tissues may be major determinants of the efficiency of N_2 fixation as discussed for the nodule parenchyma.

The multistep nature of root nodule formation indicates that specific signals from *Rhizobium* sp. may be involved at several stages in inducing the next step(s) in nodule development. At present only signals inducing root hair deformation have been purified and studied. This deformation factor can elicit root hair deformation, and a bioassay could aid in efforts to purify this compound. However, it is very unlikely that bioassays can be developed to evaluate factors triggering the infection process or different subsequent steps of nodule development. Therefore, the availability of cloned nodulin genes that are markers for specific steps in nodule development will be essential in efforts to purify bacterial factors that induce these steps. The assay based on the induction of the ENOD12 gene in root hairs of pea seedlings shows that in principle these assays can be developed.

Auxin transport inhibitors can induce nodulelike structures on the roots of legumes that on a morphologic and molecular level resemble *Rhizobium*-induced nodules. Moreover, mutations in the bacterial *nod* genes can partly be complemented with the zeatin gene that is involved in cytokinin production. Both observations strongly suggest that *Rhizobium* sp. factors inducing root nodule formation will resemble normal plant signals or that *Rhizobium* sp. compounds will interfere with normal signal transduction pathways. Therefore, characterization of bacterial signals inducing steps in root nodule formation might increase our insight into signal molecules and signal transduction pathways used in normal plant development.

The discovery that presumed nodulin genes like the *n*-GS gene of *Phaseolus vulgaris* are used in nonsymbiotic developmental programs suggests that several other genes that are still supposed to be nodule specific will be used in other parts of the plant. Therefore, in addition to true nodule-specific genes, root nodule formation will involve the induction of genes that are not expressed in roots but that are used in other parts of the plant. The mechanisms by which *Rhizobium* is able to induce the expression of these sets of genes in a temporal and spatial controlled manner are not yet completely understood. However, in addition to the process of symbiotic N_2 fixation, the unraveling of these mechanisms is one of the most appealing reasons to study root nodule development.

Acknowledgments

We thank all colleagues who have communicated results prior to official publication; Roger Pennell, for critical reading of the manuscript; and Marie-José van Iersel and Gré Heitkönig, for secretarial assistance, H. J. F. was financially supported by a grant from the Netherlands Organization for the Advancement of Pure Research.

References

1. Allen, E. K., Allen, O. N., and Newman, A. S., "Pseudonodulation of leguminous plants induced by 2-bromo-3,5-dichlorobenzoic acid," *Am. J. Bot. 40*, 429–435 (1953).

2. Appels, M. A., and Haaker, H., "Identification of cytoplasmic nodule-associated forms of malate dehydrogenase involved in the symbiosis between *Rhizobium leguminosarum* and *Pisum sativum*," *Eur. J. Biochem. 171*, 515–522 (1987).

3. Atkins, C. A., Shelp, B. J., Kno, J., Peoples, M. B., and Pate, J. S., "Nitrogen nutrition and the development and senescence of nodules on cowpea seedlings," *Planta 162*, 316–326 (1984).

4. Auger, S., and Verma, D. P. S., "Induction and expression of nodule-specific host genes in effective and ineffective root nodules of soybean," *Biochemistry 20*, 1300–1306 (1981).

5. Averyhardt-Fullard, V., Datta, K., and Marcus, A., "A hydroxyproline-rich protein in the soybean cell wall," *Proc. Natl. Acad. Sci. USA 85*, 1083–1085 (1988).

6. Baker, M. E., and Saier, M. H., "A common ancestor for bovine lens fiber major intrinsic protein, soybean nodulin-26 protein, and *E. coli* glycerol facilitator," *Cell 60*, 185–186 (1990).

7. Bakhuizen, R., "The plant cytoskeleton in the *Rhizobium*–legume symbiosis," Thesis, State University of Leiden, Leiden, The Netherlands, 1988.

8. Bassett, B., Goodman, R. N., and Novacky, A., "Ultrastructure of soybean nodules. Release of rhizobia from the infection thread," *Can. J. Microbiol. 23*, 573–582 (1977).

9. Bauer, W. D., "Infection of legumes by rhizobia," *Ann. Rev. Plant Physiol. 32*, 407–449 (1981).

10. Bergmann, H., Preddie, E., and Verma, D. P. S., "Nodulin-35: A subunit of specific uricase (uricase II) induced and localized in the uninfected cells of soybean nodules," *EMBO J. 2*, 2333–2339 (1983).

11. Bisseling, T., Been, C., Klugkist, J., Van Kammen, A., and Nadler, K., "Nodule-specific host proteins in effective and ineffective root nodules of *Pisum sativum*," *EMBO J. 2*, 961–966 (1983).

12. Blumenwald, E., Fortin, M. G., Rea, P. A., Verma, D. P. S., and Poole, R. J., "Presence of host-plasma membrane type H$^+$-ATPase in the membrane envelope enclosing the bacteroids in soybean root nodules," *Plant Physiol. 78*, 665–672 (1985).

13. Bojsen, K., Abildsten, D., Jensen, E., Paludan, K., and Marcker, K. A., "The chromosomal arrangement of six soybean leghemoglobin genes," *EMBO J. 2*, 1165–1168 (1983).

14. Brewin, N. J., Robertson, J. G., Wood, E. A., Wells, B., Larkins, A. P., Galfre, G., and Butcher, G. W., "Monoclonal antibodies to antigens in the peribacteroid membrane from *Rhizobium*-induced root nodules of pea crossreact with the plasma membranes and Golgi bodies," *EMBO J. 4*, 605–611 (1985).

15. Brisson, N., and Verma, D. P. S., "Soybean leghemoglobin gene family: Normal, pseudo, and truncated genes," *Proc. Natl. Acad. Sci. USA 79*, 4055–4059 (1982).

16. Brisson, N., Pombo-Gentile, A., and Verma, D. P. S., "Organization and expression of leghaemoglobin genes," *Can. J. Biochem. 60*, 272–278 (1982).

17. Brown, G. G., Lee, J. S., Brisson, N., and Verma, D. P. S., "The evolution of a plant globin gene family," *J. Mol. Evol. 21*, 19–32 (1984).

18. Callaham, D. A., and Torrey, J. G., "The structural basis for infection of root hairs of *Trifolium repens* by *Rhizobium*," *Can. J. Bot. 59*, 1647–1664 (1981).

19. Calvert, H. E., Pence, M. K., Pierce, M., Malik, N. S. A., and Bauer, W. D., "Anatomical analysis of the development and distribution of *Rhizobium* infections in soybean roots," *Can. J. Bot. 62*, 2375–2384 (1984).

20. Corby, H. D. L., Polhill, R. M., and Sprent, J. I., "Taxonomy." In Broughton, W. J. ed., *Nitrogen Fixation, Vol. 3: Legumes,* Clarendon Press, Oxford, 1983, pp. 1–35.

21. Cullimore, J. V., Lara, M., Lea, P. J., and Miflin, B. J., "Purification and properties of two forms of glutamine synthetase from the plant fraction of *Phaseolus* root nodules, *Planta 157*, 245–253 (1983).

22. Dart, P. J., "Infection and development of leguminous nodules." In Hardy, R. W. F., and Silver, W. S. (eds.), *A Treatise on Dinitrogen Fixation*, Vol. III, Wiley, New York, 1977, pp. 367–472.

23. De Bruijn, F. J., Felix, G., Grunenberg, B., Hoffman, H. J., Metz, B., Ratet, P., Simons-Schreier, A., Szabodos, L., Welters, P. and Schell, J., "Regulation of plant genes specifically induced in nitrogen-fixing nodules: *cis*-acting elements and *trans*-acting factors in leghemoglobin gene expression," *Plant Molec. Biol. 13*, 319–325 (1989).

24. De Lajudie, P., and Huguet, T., "Plant gene expression during effective and ineffective nodule development of the tropical stem-nodulated legume *Sesbania rostrata*," *Plant Molec. Biol. 10*, 537–548 (1988).

25. Deroche, M.-E., Carrayol, E., and Jolivet, E., "Phosphoenolpyruvate carboxylase in legume nodules," *Physiol. Veg. 21*, 1075–1081 (1983).

26. Dickstein, R., Bisseling, T., Reinhold, V. N., and Ausubel, F. M., "Expression of nodule-specific genes in alfalfa root nodules blocked at an early stage of development" *Genes Develop. 2* (6), 677–687 (1989).

27. Ditta, G., Virts, E., Palomares, A., and Kim, C. H., "The *nif* A gene of *Rhizobium meliloti* is oxygen regulated," *J. Bacteriol. 169*, 3217–3223 (1987).

28. Dunn, M. K., Dickstein, R., Feinbaum, R., Burnett, B. K., Peterman, T. K., Thoidis, G., Goodman, H. M., and Ausubel, F. M., "Developmental regulation of nodule-specific genes in alfalfa root nodules," *Mol. Plant-Microbe Interact. 1,* 66–74 (1988).

29. Faucher, C., Camut, S., Dénarié, J., and Truchet, G., "The *nod*H and *nod*Q host range genes of *Rhizobium meliloti* behave as avirulence genes in *Rhizobium legumi-nosarum* bv. *viciae* and determine changes in production of plant-specific extracellular signals," *Mol. Plant-Microbe Interact. 2,* 291–300 (1989).

30. Faucher, C., Leronge, P., Roche, P., Rosenberg, C., Debellé, F., Vane, J., Cervantes, E., Shanna, S. B., Truchet, G., Primé, J. C., and Dénarié, J., "The common *nod*ABC genes and the *nod*H and *nod*Q host-range genes of *Rhizobium meliloti* determine the production of low molecular weight extracellular signals," In Lugtenberg, B. J. J. (ed.), *Signal Molecules in Plants and Plant–Microbe Interactions.* NATO ASI Series, 1989, pp. 379–387.

31. Finan, T. M., Hirsch, A. M., Leigh, J. A., Johansen, E., Kuldau, G. A., Deegan, S., Walker, G. C., and Signer, E. R., "Symbiotic mutants of *Rhizobium meliloti* that uncouple plant from bacterial differentiation," *Cell 40,* 869–877 (1985).

32. Fischer, H. M., and Hennecke, H., "Direct response of *Bradyrhizobium japonicum* nifA-mediated nif gene regulation to cellular oxygen status," *Mol. Gen. Genet. 209,* 621–626 (1987).

33. Forde, B. G., Day, H. M., Turton, J. F., Wen-Jun, S., Cullimore, J. V., and Oliver, J. E., "Two glutamine synthase genes from *Phaseolus vulgaris* L. display contrasting developmental and spatial patterns of expression in transgenic *Lotus corniculatus* plants," *Plant Cell 1,* 391–401 (1989).

34. Fortin, M. G., Morrison, N. A., and Verma, D. P. S., "Nodulin-26, a peribacteroid membrane nodulin is expressed independently of the development of the peribacteroid compartment," *Nucleic Acids Res. 15,* 813–824 (1987).

35. Fortin, M. G., Zelechowska, M., and Verma, D. P. S., "Specific targeting of membrane nodulins to the bacteroid-enclosing compartment in soybean nodules," *EMBO J. 4,* 3041–3046 (1985).

36. Franssen, H. J., Nap, J. P., Gloudemans, T., Stiekema, W., Van Dam, H., Govers, F., Louwerse, J., Van Kammen, A., and Bisseling, T., "Characterization of cDNA for nodulin-75 of soybean: A gene product involved in early stages of root nodule development," *Proc. Natl. Acad. Sci. USA 84,* 4495–4499 (1987).

37. Franssen, H. J., Scheres, B., Van de Wiel, C., and Bisseling, T., "Characterization of soybean (hydroxy)proline-rich proteins," In Palacios, R., and Verma, D.P.S. (eds.), Molecular Genetics of Plant-Microbe Interactions 1988. APS Press, St. Paul 1988, pp. 321–326.

37a. Franssen, H. J., Thompson, D. V. Idler, K., Kormelink, R., Van Kammen, A., and Bisseling, T., "Nucleotide sequence of two soybean ENOD2 early nodulin genes encoding Ngm-75," *Plant Mol. Biol. 14,* 103–106 (1989).

38. Franssen, H. J., personal communication.

39. Fuchsman, W. H., and Appleby, C. A., "Separation and determination of the relative concentrations of the homogeneous components of soybean leghemoglobin by isoelectric focusing," *Biochim. Biophys. Acta 579,* 314–324 (1979).

40. Fuller, F., and Verma, D. P. S., "Appearance and accumulation of nodulin mRNAs and their relationship to the effectiveness of root nodules," *Plant Mol. Biol. 3,* 21–28 (1984).

41. Gardiol, A. E., Hollingsworth, R. P., and Dazzo, F. B., "Alterations of surface properties in a Tn5 mutant strain of *Rhizobium trifolii*," *J. Bacteriol. 169*, 1161–1167 (1987).

42. Gloudemans, T., "Plant gene expression in early stages of *Rhizobium*–legume symbiosis." Thesis, Agricultural University, Wageningen, The Netherlands, 1988.

43. Gloudemans, T., Bhuvaneswari, T. V., Moerman, M., Van Brussel, A. A. N., Van Kammen, A., and Bisseling, T., "Involvement of *Rhizobium leguminosarum* nodulation genes on gene expression in pea root hairs," *Plant Mol. Biol. 12*, 157–167 (1989).

44. Gloudemans, T., De Vries, S. C., Bussink, H.-J., Malik, N. S. A., Franssen, H. J., Louwerse, J., and Bisseling, T., "Nodulin gene expression during soybean (*Glycine max*) nodule development," *Plant Mol. Biol. 8*, 395–403 (1987).

45. Goodchild, D. J., "The ultrastructure of root nodules in relation to nitrogen fixation." In Bourne, G. H., Danielli, J. F., and Jeon, U. W. (eds.), *Studies in Ultra Structure*, Academic Press, New York, 1977, pp. 235–288.

46. Govers, F., Gloudemans, T., Moerman, M., Van Kammen, A., and Bisseling, T., "Expression of plant genes during the development of pea root nodules," *EMBO J. 4*, 861–867 (1985).

47. Govers, F., Moerman, M., Hooymans, J., Van Kammen, A., and Bisseling, T., "Microaerobisis is not involved in the induction of pea nodulin gene expression," *Planta 169*, 513–517 (1986).

48. Govers, F., Nap, J. P., Moerman, M., Franssen, H. J., Van Kammen, A., and Bisseling, T., "cDNA cloning and developmental expression of pea nodulin genes," *Plant Mol. Biol. 8*, 425–435 (1987).

49. Govers, F., Franssen, H., Pieterse, C., Wilmer, J., and Bisseling, T., "Function and regulation of the early nodulin gene ENOD2." In Grierson, P., Zycett, G. (eds.), *Genetic Engineering of Crop Plants*, Butterworth Scientific, London, 1990, pp. 250–269.

50. Govers, F., personal communication.

51. Grierson, D., and Covey, S. N., *Plant Molecular Biology*, 2nd ed., Blackie, London, 1988.

52. Henson, I. E., and Wheeler, C. T., "Hormones in plants bearing nitrogen-fixing root nodules: The distribution of cytokinins in *Vicia faba* L.," *New Phytol. 76*, 433–439 (1976).

53. Hirel, B., Bouet, C., King, B., Layzell, D., Jacobs, F., and Verma, D. P. S., "Glutamine synthetase genes are regulated by ammonia provided externally or by symbiotic nitrogen fixation," *EMBO J. 6*, 1167–1171 (1987).

54. Hirsch, A. M., Bhuvaneswari, T. V., Torrey, J.-G., and Bisseling, T., "Early nodulin genes are induced in alfalfa root outgrowths elicited by auxin transport inhibitors," *Proc. Natl. Acad. Sci. USA 86*, 1244–1248 (1989).

55. Horvath, B., personal communication.

56. Hrabak, E. M., Truchet, G. L., Dazzo, F. B., and Govers, F., "Characterization of the anomalous infection and nodulation of subterranean clover roots by *Rhizobium leguminosarum* 1020," *J. Gen. Microbiol. 131*, 3287–3302 (1985).

57. Jacobs, F. A., Zhang, M., Fortin, M. G., and Verma, D. P. S., "Several nodulins of soybean share structural domains but differ in their subcellular locations," *Nucleic Acids Res. 15*, 1271–1280 (1987).

58. Jacobsen, K., Laursen, N. B., Jensen, E. Ø, Marcker, A., Poulsen, C., and Marcker, K. A., "HMG1-like proteins from leaf and nodule nuclei interact with different AT motifs in soybean nodulin promoters," *Plant Cell 2*, 85–94 (1990).

59. Jensen, E., Marcker, K. A., and Villadsen, I. S., "Heme regulates the expression in *Saccharomyces cerevisiae* of chimaeric genes containing 5'-flanking soybean leghemoglobin sequences," *EMBO J. 5*, 843–847 (1986).

60. Jensen, E. Ø, Marcker, K. A., Schell, J., and De Bruijn, F. J., "Interaction of a nodule specific, trans-acting factor with distinct DNA elements in the soybean leghaemoglobin LbC3 5' upstream region," *EMBO J. 7*, 1265–1271 (1988).

61. Jørgensen, J.-E., Stougaard, J., Marcker, A., and Marcker, K. A., "Root nodule specific gene regulation: Analysis of the soybean nodulin N23 gene promoter in heterologous symbiotic systems," *Nucleic Acids Res. 16*, 39–50 (1988).

62. Kamalay, J. C., and Goldberg, R. B., "Organ-specific nuclear RNAs in tobacco," *Proc. Natl. Acad. Sci. USA 81*, 2801–2805 (1984).

63. Kamalay, J. C., and Goldberg, R. B., "Regulation of structural gene expression in tobacco," *Cell 19*, 935–946 (1980).

64. Knox, J. P., Day, S., and Roberts, K., "A set of cell surface glycoproteins forms an early marker of cell position, but not cell type, in the root apical meristem of *Daucus carota* L.," *Development 106*, 47–56 (1989).

64a. Kohl, D. H., Schubert, K. R., Carter, M. B., Hagedoorn, C. H., and Shearer, G., "Proline metabolism in N_2–fixing root nodules: Energy transfer and regulation of purine biosynthesis," *Proc. Natl. Acad. Sci. USA 85*, 2036–2040 (1988).

65. Landsmann, J., Dennis, E. S., Higgins, T. J. V., Appleby, C. A., Kortt, A. A., and Peacock, W. J., "Common evolutionary origin of legume and non-legume plant haemoglobins," *Nature 324*, 166–168 (1986).

66. Lang-Unnasch, N., and Ausubel, F. M., "Nodule-specific polypeptides from effective alfalfa root nodules and from ineffective nodules lacking nitrogenase," *Plant Physiol. 77*, 833–839 (1985).

67. Lara, M., Cullimore, J. V., Lea, P. J., Miflin, B. J., Johnston, A. W. B., and Lamb, J. W., "Appearance of a novel form of plant glutamine synthetase during nodule development in *Phaseolus vulgaris* L.," *Planta 157*, 254–258 (1983).

68. Larsen, K., and Jochimsen, B., "Expression of nodule-specific uricase in soybean callus tissue is regulated by oxygen," *EMBO J. 5*, 15–19 (1986).

69. Larsen, K., and Jochimsen, B., "Expression of two enzymes involved in ureide formation in soybean regulated by oxygen." In Verma, D. P. S., and Brisson, N., (eds.), *Molecular Genetics of Plant–Microbe Interactions*, Nijhoff, Dordrecht, 1987, pp. 133–137.

70. Legocki, R. P., and Verma, D. P. S., "A nodule-specific plant protein (Nodulin-35) from soybean," *Science 205*, 190–193 (1979).

71. Legocki, R. P., and Verma, D. P. S., "Identification of "nodule-specific" host proteins (nodulins) involved in the development of *Rhizobium*–legume symbiosis," *Cell 20*, 153–163 (1980).

72. Lewin, B., *Genes, Vol. IV*, Wiley, New York, 1989.

73. Libbenga, K. R., and Harkes, P. A. A., "Initial proliferation of cortical cells in the formation of root nodules in *Pisum sativum* L.," *Planta 114*, 17–28 (1973).

74. Libbenga, K. R., and Bogers, R. J., "Root nodule morphogenesis." In Quispel, A. (ed.), *The Biology of Nitrogen Fixation*, North-Holland, Amsterdam, 1974, pp. 430–472.

75. Long, S. R., Peters, N. K., Mulligan, J. T., Dudley, M. E., and Fisher, R. F., "Genetic analysis of *Rhizobium*–plant interactions." In: Lugtenberg, B. (ed.), *Recognition*

in *Microbe–Plant Symbiotic and Pathogenic Interactions,* Springer-Verlag, Berlin, 1986, pp. 1–15.

76. Long, S. R., "*Rhizobium*–legume nodulation: Life together in the underground," *Cell 56*, 203–214 (1989).

77. Marcker, A., Lund, M., Jensen, E., and Marcker, K. A., "Transcription of the soybean leghemoglobin genes during nodule development," *EMBO J. 3*, 1691–1695 (1984).

78. Mauro, V. P., Nguyen, T., Katinakis, P., and Verma, D. P. S., "Primary structure of the soybean nodulin-23 gene and potential regulatory elements in the 5'-flanking region of nodulin and leghemoglobin genes," *Nucleic Acids Res. 13*, 239–249 (1985).

79. Mellor, R. B., and Werner, D., "Peribacteroid membrane biogenesis in mature legume root nodules," *Symbiosis 3*, 75–100 (1987).

80. Mellor, R. B., Christensen, T. M. I. E., Bassarab, S., and Werner, D., "Phospholipid transfer from ER to the peribacteroid membrane in soybean nodules," *Z. Naturforsch. 40c*, 73–79 (1985).

81. Mellor, R. B., Christensen, T. M. I. E., and Werner, D., "Choline kinase II is present only in nodules that synthesize stable peribacteroid membranes," *Proc. Natl. Acad. Sci. USA 83*, 659–663 (1986).

82. Metz, B. A., Welters, P., Hoffman, H. J., Jensen, E. Ø, Schell, J., and De Bruijn, F., "Primary structure and promoter analysis of leghemoglobin genes of the stem-nodulated tropical legume *Sesbania rostrata:* Conserved coding sequences *cis*-elements and *trans*-acting factors," *Mol. Gen. Genet. 214*, 181–191 (1988).

83. Miflin, B. J., and Lea, P. J., "Ammonia assimilation." In *The Biochemistry of Plants,* Miflin, B. J. (ed.), Academic Press, New York, 1980, pp. 169–202.

84. Moerman, M., Nap, J. P., Govers, F., Schilperoort, R., Van Kammen, A., and Bisseling, T., "*Rhizobium nod* genes are involved in the induction of two early nodulin genes in *Vicia sativa* root nodules," *Plant Mol. Biol. 9*, 171–179 (1987).

85. Nap, J. P., Moerman, M., Van Kammen, A., Govers, F., Gloudemans, T., Franssen, H. J., and Bisseling, T., "Early nodulins in root nodule development." In Verma, D. P. S., and Brisson, N. (eds.), *Molecular Genetics of Plant–Microbe Interactions,* Nijhoff, Dordrecht, 1987, pp. 96–101.

86. Nap, J. P., and Bisseling, T., "Nodulin function and nodulin gene regulation in root nodule development." In Gresshoff, P. M. (ed.), *The Molecular Biology of Symbiotic Nitrogen Fixation,* CRC Press, Florida, 1989, pp. 181–229.

87. Newcomb, W., "A correlated light and electron microscopic study of symbiotic growth and differentiation in *Pisum sativum* root nodules," *Can. J. Bot. 54*, 2163–2186 (1976).

88. Newcomb, W., "Nodule morphogenesis and differentiation." In Giles, K. L., and Atherley, A. G., (eds.), *Biology of the Rhizobiaceae,* Int. Rev. Cytol., Academic Press, New York, 1981, pp. 247–297.

89. Newcomb, W., Sippel, D., and Peterson, R. L., "The early morphogenesis of *Glycine max* and *Pisum sativum* root nodules," *Can. J. Bot. 57*, 2603–2616 (1979).

90. Nguyen, J., Machal, L., Vidal, J., Perrot-Rechenmann, C., and Gadal, P., "Immunochemical studies on xanthine dehydrogenase of soybean root nodules. Ontogenic changes in the level of enzyme and immunocytochemical localization," *Planta 167*, 190–195 (1986).

91. Nguyen, T., Zelechowska, M., Foster, V., Bergmann, H., and Verma, D. P. S., "Primary structure of the soybean nodulin-35 gene encoding uricase II localized in the

peroxisomes of uninfected cells off nodules," *Proc. Natl. Acad. Sci. USA 82*, 5040–5044 (1985).

92. Noel, K. D., Van den Bosch, K. A., and Kulpaca, B., "Mutations in *Rhizobium phaseoli* that lead to arrested development of infection threads," *J. Bacteriol. 168*, 1392–1401 (1986).

93. Padilla, J. E., Campos, F., Lara, M., and Sanchez, F., "Expression of major nodulins from bacterial-conditioned symbiosis in *Phaseolus vulgaris:* An approach to developmental plant responses during nodule ontogeny." In Palacios, R. (ed.), *Proceedings of the 4th Int. Symp. on Molecular Genetics of Plant Microbe Interactions,* Acapulco, Mexico, 1988.

94. Puhler, A., Hynes, M. F., Kapp, D., Muller, P., and Neihaus, K., "Infection mutants of *Rhizobium meliloti* are altered in acidic exopolysaccharide production." In Lugtenberg, B., (ed.), *Recognition in Microbe–Plant Symbiotic and Pathogenic Interactions,* Springer-Verlag, Berlin, 1986, pp. 29–37.

95. Puvanesarajah, V., Schell, F. M., Gerhold, D., and Stacey, G., "Cell surface polysaccharides from *Bradyrhizobium japonicum* and a non-nodulatingg mutant," *J. Bacteriol. 169*, 137–141 (1987).

96. Ridge, R. W., and Rolfe, B. G., "*Rhizobium* sp. degradation of legume root hair cell wall at the site of infection thread origin," *Appl. Environ. Microbiol. 50*, 717–720 (1985).

97. Ridge, R. W., and Rolfe, B. G., "Sequence of events during the infection of the tropical legume *Macroptilium atropurpureum* Urb. by the broad-host-range, fast-growing *Rhizobium* ANU240," *J. Plant Physiol. 122*, 121–137 (1986).

98. Robertson, J. G., and Lyttleton, P., "Division of peribacteroid membranes in root nodules of white clover," *J. Cell Sci. 69*, 147–157 (1984).

99. Robertson, J. G., Lyttleton, P., Bullivant, S., and Grayston, G. F., "Membranes in lupin root nodules. I. The role of Golgi bodies in the biogenesis of infection threads and peribacteroid membranes," *J. Cell Sci. 30*, 129–149 (1978).

100. Sandal, N. N., Bojsen, K., and Marcker, K. A., "A small family of nodule specific genes from soybean," *Nucleic Acids Res. 15*, 1507–1519 (1987).

101. Sandal, N. N., and Marcker, K. A., "Soybean nodulin-26 is homologous to the major intrinsic protein of the bovine lens fiber membrane," *Nucleic Acids Res. 19*, 9347 (1988).

102. Scheres, B., Van de Wiel, C., Zalensky, A., Horvath, B., Spaink, H., Van Eck, H., Zwartkruis, F., Wolters, A. M., Gloudemans, T., Van Kammen, A., and Bisseling, T., "The ENOD12 gene product is involved in the infection process durring pea–*Rhizobium* interaction," *Cell 60*, 281–294 (1990).

103. Scheres, B., Van Engelen, F., Van der Knaap, E., Van de Wiel, C., Van Kammen, A., and Bisseling, T., "Sequential induction of nodulin gene expression in the developing pea nodule," The Plant Cell 2, 687–700 (1990).

104. Scheres, B., personal communication.

105. Sengupta-Gopalan, C., and Pitas, J., "Expression of nodule specific glutamine synthetase genes during nodule development in soybeans," *Plant Mol. Biol. 7*, 189–199 (1986).

106. Sengupta-Gopalan, C., Pitas, J. W., Thompson, D. V., and Hoffman, L. M., "Expression of host genes during nodule development in soybeans," *Mol. Gen. Genet. 203*, 410–420 (1986).

107. Sheehy, J. E., Minchin, F. R., and Witty, J. F., "Control of nitrogen fixation in a legume nodule: An analysis of the role of oxygen diffusion in relation to nodule structure," *Ann. Bot. 55,* 549–562 (1985).

108. Stougaard, J., Petersen, T. E., and Marcker, K. A., "Expression of a complete soybean leghemoglobin gene in root nodules of transgenic *Lotus corniculatus,*" *Proc. Natl. Acad. Sci. USA 84,* 5754–5757 (1987).

109. Stougaard, J., Sandal, N. N., Grøn, A., Köhle, A., and Marcker, K. A., "5' analysis of the soybean leghemoglobin lbc3 gene: Regulatory elements required for promoter activity and organ specificity," *EMBO J. 6,* 3565–3569 (1987).

110. Strózycki, P., Koniecznyu, A., and Legocki, A. B., "Identification and synthesis *in vitro* of plant-specific proteins in yellow lupin root nodules," *Acta Biochim. Polin. 32,* 27–34 (1985).

111. Thummler, F., and Verma, D. P. S., "Nodulin-100 of soybean is the subunit of sucrose synthase regulated by the availability of free heme in nodules," *J. Biol. Chem. 262,* 14730 (1987).

112. Tingey, S. V., Walker, E. L., and Coruzzi, G. M., "Glutamine synthetase genes of pea encode distinct polypeptides which are differentially expressed in leaves, roots and nodules," *EMBO J. 6,* 1–9 (1987).

113. Tjepkema, J. D., and Yocum, C. S., "Measurement of oxygen partial pressure within soybean nodules by oxygen microelectrodes," *Planta 119,* 351–360 (1974).

114. Tu, J. C., "Structural similarity of the membrane envelopes of rhizobial bacteroids and the host plasma membrane as revealed by freeze-fracturing," *J. Bacteriol. 122,* 691–694 (1975).

115. Turgeon, B. G., and Bauer, W. D., "Early events in the infection of soybean by *Rhizobium japonicum.* Time course and cytology of the initial infection process," *Can. J. Bot. 60,* 152–161 (1982).

116. Turgeon, B. G., and Bauer, W. D., "Ultrastructure of infection-thread development during the infection of soybean by *Rhizobium japonicum,*" *Planta 163,* 328–349 (1985).

117. Uheda, E., and Syono, K., "Physiological role of leghaemoglobin heterogeneity in pea root nodule development," *Plant Cell Physiol. 23,* 75–84 (1982).

118. Van den Bosch, K. A., Bradley, D. J., Knox, J. P., Perotto, S., Butcher, G. W., and Brewin, N. J., "Common components of the infection thread matrix and the intercellular space identified by immunocytochemical analysis of pea nodules and uninfected roots," *EMBO J. 8,* 335–342 (1989).

119. Van de Wiel, C., Scheres, B., Franssen, H., Van Lierop, M. J., Van Lammeren, A., Van Kammen, A., and Bisseling, T., "The early nodulin transcript, ENOD2 is located in the nodule parenchyma (inner cortex) of pea and soybean root nodules," *EMBO J. 9,* 1–7 (1990).

120. Van de Wiel, C., Thesis, A histochemical study of root nodule development, Agricultural University, Wageningen, The Netherlands.

121. Van de Wiel, C., and Hirsch, A., personal communication.

122. Van Kammen, A., "Suggested nomenclature for plant genes involved in nodulation and symbiosis," *Plant Mol. Biol. Rep. 2,* 43–45 (1984).

123. Verma, D. P. S., Ball, S., Guerin, C., and Wanamaker, L., "Leghemoglobin biosynthesis in soybean root nodules. Characterization of the nascent and released peptides and the relative rate of synthesis of the major leghemoglobins," *Biochemistry 18,* 476–483 (1979).

124. Verma, D. P. S., Kazazian, V., Zogbi, V., and Bal, A. K., "Isolation and characterization of membrane envelope enclosing the bacteroids in soybean root nodules," *J. Cell Biol. 78,* 919–936 (1978).

125. Verma, D. P. S., Fortin, M. G., Stanley, J., Mauro, V. P., Purohit, S., and Morrison, N., "Nodulins and nodulin genes of *Glycine max,* a perspective," *Plant Mol. Biol. 7,* 51–61 (1986).

126. Vincent, J. M., Factors controlling the legume–*Rhizobium* symbiosis. In Newton, W. E., and Orme-Johnson, W. H. (eds.), *Nitrogen Fixation, Vol. II,* University Park Press, Baltimore, 1980, pp. 103–129.

127. Wittenberg, J. B., Bergersen, F. J., Appleby, C. A., and Turner, G. L., "Facilitated oxygen diffusion. The role of leghemoglobin in nitrogen fixation by bacteroids isolated from soybean root nodules," *J. Biol. Chem. 249,* 4057–4066 (1974).

128. Witty, J. F., Minchin, F. R., Shøt, L., and Sheehy, J. E., *Nitrogen Fixation and Oxygen in Legume Root Nodules.* Oxford surveys of plant and cellular biology, 3, 1986, pp. 275–315.

129. Zaat, S. A. J., Van Brussel, A. A. N., Tak, T., Pees, E., and Lugtenberg, B. J. J., "Flavonoids induce *Rhizobium leguminosarum* to produce *nod*DABC gene-related factors that cause thick, short roots and root hair responses on common vetch," *J. Bacteriol. 169,* 3388–3391 (1987).

16

Plant Genetics of Symbiotic Nitrogen Fixation

Donald A. Phillips and *Larry R. Teuber*

I. Introduction

Many genetically controlled traits in legumes affect the final phenotype of symbiotic N_2 fixation in the *Rhizobium*–legume association. Some genes alter root nodule formation; others influence physiological functions within the mature nodule. Understanding the nature of specific genetic traits that affect symbiotic N_2 fixation will clarify the underlying biological events of the symbiosis and in some cases may contribute to the agronomic management of N_2-fixing legumes. This review summarizes information currently available on simple and complex host plant genetic traits that affect the *Rhizobium*–legume association and indicates how those factors contribute to our understanding and practical use of the symbiosis. Nodulin gene expression (see Chapter 15) will not be discussed, but one must recognize that many of the phenomena outlined here probably are associated with expression of presently unrecognized nodulins.

II. Single-Gene Traits

Many simply inherited plant genes that affect root nodule formation (Table 16–1) and function (Table 16–2) have been described. Simplified phenotypic groupings of those genes contribute to organizing our partial knowledge of their biochemical functions. When these phenotypes are completely understood, categories, such as "nonnodulating" and "ineffective," undoubtedly will be associated with many different biochemical causes.

A. Genes Affecting Nodule Formation

The earliest published work showing a genetic effect of the host legume on root nodule phenotype reported a recessive allele responsible for nonnodulation in red clover (*Trifolium pratense* L.).[72] Since that time many naturally

Table 16-1. Root Nodule Formation Phenotypes in Legumes*

Species	Gene	Phenotype	Reference
Arachis hypogaea L.	—	Nod⁻ in double recessive	32, 71
Cicer arietinum L.	rn_1	Nod⁻	16
	rn_2, rn_3	Nod⁻, temperature dependent	16
Glycine max (L.) Merr.	rj_1	Nod⁻, strain-dependent	10, 22, 113
	nts	Nod⁺⁺ ± NO_3^-, shoot controlled	8, 9
	—	Nod⁺⁺ ± NO_3^-	35
Medicago sativa L.	nn_1, nn_2	Nod⁻ in double recessive	82
Melilotus alba annua Desr.	$sym_a, sym_b, sym_c, sym_d, sym_e$	Nod⁻	55, 60
	$nlsn$	Nod⁺⁺ ± NO_3^-	80
Phaseolus vulgaris L.	sym_1, sym_5	Nod⁻, temperature dependent	54–56, 58, 61
	sym_2, sym_4, sym_6	Nod⁻, strain-dependent	46, 56, 62, 63
Pisum sativum L.	$sym_7, sym_8, sym_9, sym_{10}, sym_{11}, sym_{14}$	Nod⁻	58
	sym_{12}	delayed nodulation in soil, Nod⁻ in strain tests, originally line K5	48, 90
	$sym_{15}, sym_{16}, sym_{17}$	few nodules, short lateral roots	57
	sym_{18}	Nod⁻, strain-dependent	60
	sym_{19}	Nod⁻, isolated as line K24	60, 90
	nod_3	Nod⁺⁺ ± NO_3^-, root controlled	49
	C	Controls nodule number	79
	brz	few or no nodules, bronze leaves, isolated as E107	58, 60
Trifolium pratense L.	—	Nod⁺⁺ ± NO_3^-, shoot controlled	23
	r	Nod⁻	72
Vicia faba L.	sym_1	Nod⁻; Nod⁺Fix⁻; strain dependent	24

*Nod⁻ plants lacked nodules under the test conditions. Nod⁺⁺ plants had hypernodulation. Fix⁻ nodules were ineffective with the rhizobial strain used.

Table 16-2.   Root Nodule Function Phenotypes in Legumes

Species	Gene	Phenotype	Reference
Cicer arientinum L.	rn_4	Fix$^-$	15
	rn_5	Poorly effective	15
Glycine max (L.) Merr.	Rj_2, Rj_3, Rj_4	Fix$^-$, strain dependent	7, 108, 109
	—	Recessive Fix$^+$ with *R. fredii*, Fix$^-$ with *B. japonicum* USDA 123	21
Medicago sativa L.	$in_1, in_2, in_3,$ in_4, in_5	Fix$^-$	82
Pisum sativum L.	sym_{13}	Fix$^-$	58
	sym_{20}	Fix$^-$	60, 89
	—	Dominant allele increases H_2 uptake	85
	—	Fix$^-$, seven separate alleles	23
Trifolium pratense L.	*i, ie, n, d*	Fix$^-$	4, 74–76
T. incarnatum L.	rt_1	Fix$^-$	99
T. subterraneum L.	—	Fix$^-$ strain dependent	31
Vicia faba L.	sym_1	Nod$^-$; Nod$^+$Fix$^-$; strain dependent	24

occurring genes or induced mutations that prevent root nodule formation have been reported in various legumes (see Table 16–1). Some genes are specific for certain rhizobial strains [e.g., rj_1 in soybean (*Glycine max* L. Merr.);[22] sym_2 in pea[62] (*Pisum sativum* L.)]; others show temperature-dependent phenotypes [e.g., rn_3 in chickpea (*Cicer arietinum* L.);[16] sym_1 in pea[61]]. One complicating factor is that not all nodule phenotypes can be separated easily into nonnodulating and ineffective categories. For example, swellings on "Afghanistan" pea roots, which are associated with sym_2,[46] have some features similar to ineffective nodules.[17] Another problem is that in at least one case (e.g., the sym_2 pea locus), interactions with modifying genes can complicate the interpretation.[58,79] Although some nonnodulating phenotypes have been characterized morphologically at the cellular level [e.g., *nod*49 in soybean[65] and nn_1, nn_2 in alfalfa (*Medicago sativa* L.)[25]], as yet none has been fully explained at the biochemical level. Observations such as the capacity of some rhizobitoxine-producing strains of *Bradyrhizobium japonicum* to nodulate rj_1rj_1 soybeans[22] or the observed modification of a 66-kDa protein in four of five sym_5sym_5 pea mutants[29] may lead to molecular explanations. In other cases, however, possible biochemical explanations have not been supported by experimental data. For example, the nonnodulating phenotype in alfalfa controlled by the presence of both nn_1 and nn_2[82] is not associated with an inability of the plant to produce *nod*-gene-inducing *flavonoids* required for rhizobial infection.[81] Shoot-root grafting experiments involving wild-type and nonnodulating lines show that generally the nodulation phenotype is not controlled by the shoot phenotype.[17,19,23,54,90] One exception is the K24 pea mutant in which both the root and the shoot phenotype are involved in the nonnodulating trait.[90] The total number of loci conferring nonnodulating phenotypes is not known for any legume, but mutagenesis programs in pea have identified nearly 20 separate loci (see Table 16–1), and more Nod⁻ mutants remain to be characterized.[60] Thus, easily 25–30 loci may be capable of producing Nod⁻ peas.

A recently described phenotype that may clarify root nodule development as well as increase agronomic yields in the nitrate-tolerant symbiosis (Nts) trait, which allows a legume to produce abundant nodules in the presence and absence of nitrate [*nts* in soybean;[8,9] an undesignated locus and *nod*$_3$ in pea;[23,49] *ntsn* in common bean (*Phaseolus vulgaris* L.)[80]]. Data from root-shoot grafting studies in soybean[19] and pea[23] indicate that phenotypes produced by the *nts* genes in those species are controlled by the shoot. In contrast, the similar phenotype produced in pea by *nod*$_3$ is controlled by the root.[90] Complementation studies with a number of soybean *nts* mutants indicate that all were mutated at the same locus,[18] which is similar to the number in pea.[23] Anatomical studies of roots in normal soybean show that *B. japonicum* induces many clusters of subepidermal cell divisions, which are suppressed before they form root nodule meristems, while the nts382 mutant line allows many more of the centers of mitotic activity to form nodules.[65] Such data are consistent with the observation that wild-type shoots suppress nodulation when grafted on to mutant *nts* roots[19] and with data suggesting that extracts of inoculated parental soybeans suppress nodulation in the mutant *nts* roots.[36] The possibility that an autoregulatory compound is produced in wild-type soybean shoots after exposure to rhizobia is discussed more completely in a recent review.[92]

B. Genes Affecting Nodule Function

Many Fix⁻ phenotypes (see Table 16-2), like some Nod⁻ phenotypes (see Table 16-1), depend on the particular rhizobial strain present. At this time no specific biochemical models are available to explain the Fix⁻ phenotypes. The demonstration in pea that mutations at seven different loci can cause Nod⁺Fix⁻ plants[23] indicates that several explanations will be necessary when all interactions are understood.

A critical observer might question whether simple categories of Fix⁺ and Fix⁻ obscure the complex nature of the *Rhizobium*–legume interaction. In those cases where quantitative data on nodule function have been reported (e.g., growth and N accumulation under N_2-dependent conditions;[31] or $^3H_2$-uptake activities[85]) intermediate phenotypes are evident. Whether those intermediate phenotypes represent the effects of minor genetic factors or simply other sources of variation in the experiments remains to be determined. Nevertheless, most workers would acknowledge that a complete understanding of the symbiosis will require a more precise quantitative characterization of nodule activity than simple allocation to Fix$^+$ or Fix$^-$ categories. As more extensive tests are made for host plant genetic effects on physiological functions—such as assimilation of NH_3, reduction of CO_2, transport of specific metabolites through the peribacteroid membrane, and H_2 metabolism—new sources of variation and greater scientific insight may result.

III. Complex Genetic Effects on Nitrogen Fixation

Many investigators have used quantitative genetic tools to test if agriculturally important legumes can be selected for increased symbiotic N_2 fixation (Table 16-3). The concepts on which those studies are based will be examined briefly here, and successful attempts to increase N_2 fixation in common bean and alfalfa will be discussed. In many cases the results obtained in these studies were influenced by the presence of specific rhizobial strains. Thus, one should consider information presented here in conjunction with results discussed under Sections IV and V.A.

One goal of quantitative genetics is to identify genetic differences responsible for phenotypic variation. Frequently, this goal is achieved through the application of statistical methods. Several approaches begin with a standard analysis of variance. Assigning sources of variation to genetic causes (plant or bacterial), to environmental factors, to interactions between those sources, or to experimental error allows one to quantify the variability of those factors in a particular experiment. If the same experiment includes controlled sources of genetic variation, such as two parent cultivars and their progenies, one can use the analysis to estimate the heritability of a trait (e.g., C_2H_2-reduction activity, ARA). Heritability can be expressed as broad sense (H_b^2), narrow sense (h_n^2), and realized (h_r^2). Broad-sense and narrow-sense heritability estimates are theoretical determinations that have predictive value in assessing the potential for genetic gain through selection. Realized heritability estimates are measures of an actual response to selection.

Estimates of H_b^2 and h_n^2 are derived from experimentally determined estimates of genotypic and phenotypic variances. Genetic variance (V_G) is composed of additive (V_A) and nonadditive (dominant and epistatic) variances, whereas phenotypic variance (V_P) includes genetic, genotype X environment, and error variances. By definition, $H_b^2 = V_G/V_P$, and $h_n^2 = V_A/V_P$. Realized heritability is determined as the selection response divided by the selection differential. All types of gene action have an additive genetic variance component. The size of that component depends on the average effect of an allele and the frequency of the allele in a population. Since most plant breeding methods are designed to use additive genetic variance, narrow-sense heritability estimates usually have the most predictive value. Some mating designs (e.g., diallel) provide estimates of general combining ability

Table 16–3. *Associations Observed Between Complex Genetic Traits Related to Symbiotic N_2 Fixation and Allied Traits**

Species	Measured Trait	Genetic Estimate	Allied Traits	Reference
Arachis hypogaea L.	ARA	H_b^2: 0.60 to 0.82	Shoot mass, fruit mass	1
Desmodium intortum (Mill.) Urb.	nodule mass	—	Total shoot N	47
D. sandwicense E. Mey	ARA	—	Shoot mass	88
Glycine max (L.) Merr.	seed N from N_2	H_b^2: 0.53 to 0.60	Seed yield, shoot mass	93
Medicago sativa L.	ARA	h_r^2: 0.78	Nodulation, shoot mass & N%	26
	ARA	—	Nodule GOGAT, PEPC; shoot mass	5
	Total fresh weight	H_b^2:0.10	—	68
	ARA	—	Nodule number, shoot mass, fibrous roots	97
	ARA	GCA/SCA: 0.45 to 3.41	Stem and nodule number	104
	Total N	GCA/SCA: 1.50 to 5.50	—	104
	Forage mass and %N	mass h_r^2: 0.05 to 0.40 %N h_r^2: −0.08 to 0.78	Forage quality	53, 106
	ARA	h_r^2: 0.01 to 0.25	Nodule number and mass, shoot mass, fibrous roots	20, 106
	nodule mass	h_r^2: 0.17 to 0.45	ARA, shoot mass, fibrous roots, nodule number	110

Table 16–3. Continued

Species	Measured Trait	Genetic Estimate	Allied Traits	Reference
Phaseolus vulgaris L.	ARA	—	Total N, total N_2 fixed	2
	ARA	—	Total mass, nodule mass	94
Pisum sativum L.	ARA	H_b^2: 0.80; h_n^2: 0.76	Shoot mass and N	42, 43
Psophocarpus tetragonolobus (L.) D.C.	ARA	H_b^2: 0.48 to 0.58; h_n^2: 0.55	—	48
	Nodule mass	H_b^2: 0.22 to 0.25; h_n^2: 0.20	—	48
Trifolium pratense L.	ARA	—	Nodule number, flowering	77
T. incarnatum L.	ARA	H_b^2: 0.91; h_r^2: 0.63	Shoot mass	100, 102
T. repens L.	Nodule volume	H_b^2: 0.06 to 0.48	Nodule number	51
	Nodule number	H_b^2: 0.03 to 0.50	—	51
	ARA	—	Nodule number and mass	70
Vicia faba L.	Nodule number and mass	—	Total mass, shoot N	27
Vigna unguiculata (L.) Walp	Nodule mass	—	Shoot N	34

*ARA—C_2H_2-reduction activity; GCA—general combining ability; GOGAT—glutamate synthase activity; H_b^2—broadsense heritability; h_n^2—narrow-sense heritability; h_r^2—realized heritability; PEPC—phosphoenolpyruvate carboxylase activity; SCA—specific combining ability.

(GCA) and specific combining ability (SCA). When epistatic genetic variance is assumed to be zero, which is a common assumption, GCA gives an estimate of V_A, whereas SCA estimates dominance variance. The GCA/SCA ratio therefore provides an estimate of the proportion of the genetic variance available for selection in a breeding program. All measures of heritability must be viewed with caution, because although high heritabilities suggest that selection is possible, genetic gain depends also on the intensity of selection and the V_P for that trait.

Several general points about the observations reported in Table 16-3 deserve comment. First, ARA, an indirect and short-term measure of nitrogenase activity, is heritable in various plant species using different methodologies in both controlled environments[26,48,100,102,104,110] and field conditions.[1] Thus, valid concerns that ARA seldom permits an accurate quantitative measure (as opposed to estimate) of actual N_2 reduction do not bar the use of this technique as a tool for plant breeders. Second, other traits that are easier to measure than ARA, such as shoot weight[42] or a visual score of root nodule mass or number, also were heritable and generally correlated with N_2 fixation.[27,34,47,48,51,110] Third, although many of the observations reported in Table 16-3 suggest that genetic variation in legumes can be used to increase N_2 fixation, extending that knowledge to understand the biological basis of the observed phenotype(s) or using it to produce cultivars with enhanced N_2 fixation can require considerable work and insight. Despite those problems, several groups have used quantitative genetic information from Table 16-3 to enhance N_2 fixation in bean[2,103] and alfalfa.[83,110]

In bean, the inbred backcross line method was used to increase N_2 fixation in the low N_2-fixing cultivar Sanilac but not in the moderate N_2-fixing cultivar Porrillo Sintetico.[2,103] That technique transferred desirable ARA traits from a donor parent, Puebla 152, into a Sanilac-like genetic background by using Sanilac as a recurrent parent and selecting for high ARA and seed yield. Lines that reduced much more N_2 than Sanilac under field conditions were produced, and the increases apparently were not directly related to differences in growth habit or maturity of the two parents (Table 16-4). The amount of N_2-derived N in lines 24-17 and 24-21 did not differ significantly from the midparent value, but line 24-21 maintained the growth habit and maturity date of the Sanilac parent. Thus, the results show clearly that genes required for good N_2-fixation traits can be transferred into a cultivar lacking those characteristics. The biological basis of the increased N_2 fixation in these bean lines is not clear from the data available. Although much of the increase in N_2 fixation by line 24-17 may be associated with the longer growth period relative to Sanilac, the same explanation cannot be given for line 24-21. Because both lines 24-17 and 24-21 are still genetically quite different from Sanilac at the molecular level, it may be difficult to identify a single morphological, physiological, or biochemical factor responsible for

Table 16-4. Nitrogen Parameters in Common Bean Lines Selected for Increased N_2 Fixation *

Bean Line	Seed N	Total N	N_2- Derived N	Maturity Date	Growth Habit
	---------------	mg N/plant	---------------	days	
Puebla 152	1083	1223	674	114	Vine
Sanilac	636	739	46	96	Bush
24-17	898	1073	475	107	Vine
24-21	842	987	306	93	Bush
Midparent (calculated)	860	981	360	105	
LSD (0.05)	172	177	136		

*Two lines, 24-17 and 24-21, were produced from Puebla 152 and Sanilac by the inbred backcross line method using Sanilac as the recurrent parent for two cycles. N_2 fixation was estimated under field conditions with the ^{15}N-dilution method. (Adapted from *Crop Science* 28: 773–778.[103])

the increased N_2 fixation. Whether this new potential for increasing N_2 fixation in beans will produce major agricultural benefits remains to be determined, but differences among bean cultivars for N_2-fixation capacity[91,95] indicate that cultivars in many bean market classes could be improved for this trait.

Heritable genotypic variation for ARA in alfalfa inoculated with a mixture of *R. meliloti* strains was first reported by Seetin and Barnes[97] in 1977, and its correlated association with actual N_2 fixation was established the following year by Duhigg et al.[26] This correlation was exploited by Viands et al.[110] to develop alfalfa populations that fixed significantly more N_2 under controlled conditions. In that study, two cycles of selection for increased ARA, shoot dry weight, root nodule mass, and fibrous roots all produced plants with shoots containing more N from N_2 reduction than the original population under controlled conditions. Field tests with those populations selected for N_2 fixation failed to show any significant effect of breeding on forage yield, N content, or N_2 fixed.[3] Whether that result was caused by selection for a favorable legume–*Rhizobium* interaction that was not expressed under field conditions or by other complex changes in environmental variables between the glasshouse and the field is unknown. As yet, no molecular or biochemical basis for the increases in N_2 fixation under controlled conditions has been reported. However, a positive contribution, of phosphoenolypyruvate carboxylase (PEPC) and glutamate synthase (GOGAT) in the nodule cytosol is suggested by changes in total plant N and root N concentration of populations selected directly for high and low activities of these enzymes.[50] Any such success from plant selection in alfalfa normally rests on increasing the frequency of favorable alleles in the population, because this species is an outcrossing tetraploid and every plant is a separate genotype.

A second research program designed to increase forage yield and quality in alfalfa by enhancing N_2 fixation also was stimulated by the initial demonstrations that ARA was heritable.[26,97] That effort, a continuing collaboration between the authors of this chapter, involves selection for integrative growth and N assimilation traits. The goal has been to produce alfalfa populations with improved N_2-fixation capacity and to use those materials to identify individual traits affecting N_2 fixation at the same time the plants are tested for potential agronomic benefits. Initial tests demonstrated the existence of genotypic variation in forage N concentration when plants were grown under N_2-dependent conditions with a single *Rhizobium* strain or on 8 mM NH_4NO_3.[87] That fact was used with three different starting materials to develop alfalfa populations with significant increases in forage dry weight (+17 to 54%) and N concentration (+5 to 9%) under controlled conditions.[106,107] When one improved population, HP32, was tested with other strains of *R. meliloti* and various concentrations of combined N, it still accumulated more total C and N than the original Hairy Peruvian starting material.[83] Direct measures of forage quality in plants grown under controlled conditions showed that trait was not impaired by selection for yield.[20] Attempts to identify fundamental differences between the original and the improved plant populations showed that in Hairy Peruvian alfalfa the number of root nodules and stems was increased by the selection protocol.[53] Although such phenotypic changes may be related to a decreased autoregulatory control of root nodule initiation similar to that reported for other legumes (see Table 16-1), other data suggest that the response may be associated with a limitation in production of flavonoid nodulation signals by alfalfa.[52] New information on the nature and diversity of flavonoid *nod*-gene-inducing compounds released by alfalfa seeds[39] and roots[66] should facilitate a direct examination of the latter possibility. Field trials over several years at four locations with materials selected from Moapa 69 indicate that the increased forage yield measured in controlled conditions[107] is not observed in the field. After 3 years, however, there is no evidence of a significant decrease in forage yield (L. R. Teuber and D. A. Phillips, unpublished data), and the selected populations continue to express the increase in forage N concentration observed under controlled conditions. Data from ^{15}N-dilution and N-difference analyses at one location indicate the increase in forage N concentration is a result of enhanced N_2 fixation (Table 16-5). If these promising results are confirmed throughout the 5-year test, the improved alfalfa populations will be important germplasm for alfalfa growers and breeders.

At this time it cannot be claimed that quantitative genetic techniques have clarified our understanding of symbiotic N_2 fixation. There are indications that statistical methods based in that discipline have helped produce plant

Table 16-5. *Effects of Selection for Increased N₂ Fixation on Forage N Content in Alfalfa**

Alfalfa Population	Forage Yield	Forage N Concentration	Total Forage N	Forage N from N₂	
				N Difference	¹⁵N Dilution
	kg ha⁻¹	g kg⁻¹	---------------------- kg ha⁻¹ ----------------------		
Moapa 69	3,330	32.4	107	37.3	58.1
Moapa 69-33	3,330	34.7	115	45.3	68.2
CUF101	3,400	32.1	109	39.3	63.3
MN IN-Sar	2,090	33.1	69.7	0.0	0.0
LSD (0.05)	141	1.4	6	5.5	5.7
CV (%)	7.4	7.3	9.6	24.3	14.1

*Moapa 69-33 was produced from Moapa 69 by three cycles of selection for increased forage yield and N concentration. Data represent means of six field harvests during the first year of production at Davis, CA. An ineffectively nodulating alfalfa population, MN IN-Sar, was used as a baseline control for calculating N₂ fixation by the N-difference and ¹⁵N-dilution methods. CUF101 served as an agronomic control. (D. A. Phillips and L. R. Teuber, unpublished results).

materials with increased N₂ fixation, but specific genes responsible for those changes have not yet been identified. The benefits of using the legume–*Rhizobium* symbiosis as a paradigm for testing quantitative genetic techniques should not be overlooked by theoretical geneticists. The increasing availability of specific rhizobial and plant genes that affect the symbiosis offers important tools for developing and testing quantitative genetic theory at the population level.

IV. Legume X *Rhizobium* Interactions

Many investigators have examined the effect of genetic components on N₂ fixation by fitting experimental data to a fixed-effects analysis-of-variance model. In such analyses a factor termed the legume X *Rhizobium* interaction can be separated from the influences of plant and bacterial genotypes. In practical terms, the presence of such an interaction can indicate that as plant cultivars and bacterial strains are tested in different combinations, no one cultivar or strain is always associated with the greatest N₂ fixation. A survey of the literature shows that legume X *Rhizobium* interactions have been quantified by this method in at least five species of legumes (Table 16-6). The phenomenon was described by other methods in red clover.[73] These interactions can seldom be explained at the basic biological level, but they cause marked differences in N₂ fixation and plant growth. Thus, an awareness of the phenomenon is important for plant breeders attempting to select for increased N₂ fixation.

Table 16-6. *Contribution of Interactions Between Legume Cultivars and Rhizobial Strains to Total Variance in Studies with Multiple Strains and Cultivars*

Species	Interaction Effect	Reference
	% of total phenotype variance	
Medicago sativa L.	Total N: 66%	5
	Shoot mass: 53%	28
	Total N: 41%	104
	ARA: 37%, 66%	105
Pisum sativum L.	Total N: 12%; total mass: 15%	98
	ARA: 5%	45
Trifolium incarnatum L.	Shoot mass: 6%; ARA: 13%	101
T. repens L.	Shoot mass: 23%	67
Vicia faba L.	Shoot mass: 74%	69

Plant breeders have debated whether to select for increased N_2 fixation in the presence of a single rhizobial strain or with a mixture of strains. Those who advocate a mixed-strain inoculum suggest that selection in the presence of multiple rhizobial genotypes will produce the best host–strain combinations.[3,37] Those advocating the use of a single-strain inoculum believe that because the most competitive rhizobia are not necessarily the most effective at reducing N_2, the presence of multiple rhizobial genotypes may confound the selection of plant genes favoring N_2 fixation.[86,101] Certainly the degree to which less effective strains from a mixed population nodulate genetically desirable hosts will slow the rate of genetic gain when selecting for host genes that favor increased N_2 fixation.[100] In the case of alfalfa, germplasm selected for increased N_2 fixation in the presence of one *R. meliloti* strain stimulated greater N_2 fixation by other strains.[83] Thus, a major difficulty that remains to be addressed by those who select plants in the presence of mixed rhizobial inocula is how their plants that show enhanced N_2 fixation under controlled conditions will interact with various populations of indigenous rhizobia that will be encountered in soils at different sites.[30]

Ultimately, selection procedures must be based on the genetic control of traits. The large size of the legume X *Rhizobium* interaction in *Vicia faba* L. was used to argue that most of the genetic variation was under nonadditive genetic control.[69] One conclusion from that study was that programs designed to increase N_2 fixation should simultaneously select both the plant and the *Rhizobium*. That conclusion was weakened by the fact that the type of genetic variance controlling the *Vicia* X *Rhizobium* interactions could not be determined by the analysis used. Other studies showed that significant additive and nonadditive genetic variance is present for several traits related to N_2 fixation.[45,104] Thus, some investigators conclude that both legumes and rhizobia should be selected independently.[45] Because rhizobia that nodulate one plant effectively are more likely to perform similarly on related genotypes than on unrelated plants,[38,67] legume selection with one rhizobial strain is a reasonable approach to increasing N_2 fixation. It follows that relative rankings of strains in closely related plant materials will not change markedly. Thus, alfalfa plants selected for N_2 fixation with a single strain of *Rhizobium* also improved symbiotic performance of several other *Rhizobium* strains.[83] Although that result shows that selecting plant genotypes with one bacterial strain is effective, it does not necessarily invalidate a more laborious simultaneous selection of both symbionts.

The practical problems and opportunities resulting from the legume X *Rhizobium* interaction are especially apparent in the case of soybean. Indigenous *B. japonicum* strains frequently are only moderately effective at fixing N_2, but they prevent superior inoculated strains from forming root nodules. Thus, the advantages of using plant genes to block indigenous strains and permit nodulation by desirable rhizobial strains is obvious. The rj_1 allele is one plant gene that prevents root nodule formation by many indigenous strains.[22] A more complex host plant effect is evident in several primitive soybean introductions [*Glycine max* (L.) Merr.] that restrict, but do not prevent, effective root nodule formation by *B. japonicum* strains in the 123 scrocluster.[11,13,14] A possibly related, but more extreme, host plant effect has been observed with the primitive *G. max* cultivar Peking[21] and the wild soybean (*Glycine soja* Sieb. and Zucc.),[12] which produce Fix^- nodules with *B. japonicum* USDA123 and Fix^+ nodules with *R. fredii*. Other "promiscuous" *G. max* lines that apparently form Fix^+ nodules with both *B. japonicum* and *Bradyrhizobium* sp. organisms indigenous to African soils are known.[59] Genetic relationships among these phenotypes have not yet been reported. It may be possible, however, to use the alleles involved to control indigenous rhizobia while supplying more effective strains as inoculants.

V. Agronomic Contributions of Plant Symbiosis Genes

A. *Optimizing Breeding Methodologies*

A successful plant improvement program requires that the breeder select phenotypes under conditions permitting the separation of environmental and genetic effects. Thus, breeding efforts designed to improve N_2 fixation must consider several factors in developing a selection protocol. Evidence suggests that the most important factors affecting selection for N_2 fixation are the following: (1) general environmental conditions and developmental stage at which the plant is evaluated, (2) the rhizobial genotype(s) present (i.e., a single- versus multiple-strain inoculum), (3) the N source, (4) the selection criteria, (5) the selection method, and (6) the breeding method.

Environmental factors—such as temperature, photoperiod, and mineral nutrient availability—affect plant growth and chemical composition.[112] If one supplies a constant, uniform, nonlimiting environment in every selection cycle, genotypic differences will be expressed more fully as phenotypic variation, when plants are evaluated at a specific time or developmental stage. Researchers selecting for N_2 fixation have evaluated phenotypes after planting[78] or clipping[3] and at specified physiologic stages.[87,106] All methods identified phenotypic differences with a genetic basis.

Although many investigators favor using an inoculum that contains a single rhizobial strain, others support the use of a mixed-strain inoculum. The issues surrounding that debate were discussed adequately in Section IV and will not be mentioned further.

The significant amount of soil N available in most agricultural soils means that legumes have the opportunity to utilize both N_2 and combined N. Plant genotypes in a single species, however, can vary in their capacity to use those forms of N as sole N sources,[87] and combined N can affect the phenotypic variation associated with concurrent or subsequent N_2 fixation.[96] As a result of these and other observations, some workers have discussed the advantages of selecting for N_2 fixation in the presence of combined N.[44,91,96] Tests that compared genetic gains produced by selecting alfalfa for forage dry weight and N concentration under N_2- followed by NH_4NO_3-dependent conditions with the progress on either N source alone showed that sequential selection on the two N sources was superior to selection on either individual N source.[107] Results of both theoretical and actual selection experiments indicate that genetic gains in forage dry weight and N concentration are maximum for alfalfa fertilized with approximately 33 kg N/ha under field conditions (L. R. Teuber et al., unpublished results).

In determining selection criteria, plant breeders must understand how the desired traits are interrelated and how selection may affect other plant characteristics. Many plant traits are correlated with measures of N_2 fixation (see Table 16-3). For example, shoot N concentration is negatively correlated with both N_2 fixation and shoot weight.[42] Such inverse relationships among advantageous traits point to the desirability of multiple-trait selection methods.[2,3,33,106] Simulation experiments have predicted that single-trait selection for either forage N concentration or forage dry weight would produce a 1–10% decrease in the unselected trait.[107] An application of independent culling levels to the two traits resolved that problem successfully and indicates the value of this technique for selecting physiological characteristics. An alternative method using a base index would be theoretically as effective in this case, but using that technique would require the additional labor of measuring each trait on every plant. Independent culling levels permit the rejection of many plants before they are assessed for all traits.

Choice of a plant breeding method depends on the type and amount of genetic variance present in the germplasm to be improved. Most breeding methods utilize additive genetic variance. The information available indicates that in out-crossing species additive genetic effects influencing N_2 fixation are sufficiently large for intrapopulation breeding methods to be effective.[42,78,88,97,104] Thus, at present there is insufficient evidence to merit breeding programs designed for development of hybrid varieties. Successful selection programs have used phenotypic recurrent selection in out-crossing species[3,78,106,107] and the inbred backcross line method in a self-pollinating crop.[2,103] Two research groups have used strain-cross technologies that involve crosses between unrelated plant lines that were previously selected for N_2 fixation.[3,41,78] Those reports provide no information about the performance of crosses between the unrelated germplasm sources prior to selection, so one cannot eliminate the possibility that the observed results were produced by uncontrolled heterotic factors associated with gametic phase disequilibrium.

A physiological trait like N_2 fixation is controlled by complex biochemical processes. Because those interacting factors may compensate for changes in a selected trait, small improvements in a single biochemical process may have little effect on whole plant yield.[64] Thus, when determining progress in a breeding program, it is important to document improvement in the trait for which selection has been conducted under the original growing conditions. Evaluation can then be extended to new environments, and other responses of the selected plants can be examined. Finally, it must be emphasized that selection for increased N_2 fixation can be only one component of any comprehensive plant improvement program.[3,33]

B. Direct Uses of Genetically Altered Plant Materials

Some plant materials already developed for enhanced symbiotic N_2 fixation and presumably more in the future will contribute directly to improving agronomic and horticultural plant cultivars. Results reported for common bean[2,103] (Table 16-4) indicate that a plant line with a seed yield similar to the recurrent parent Sanilac and fixing significantly more N_2 under field conditions is now available. This transference of good N_2-fixation traits to a line with poor symbiotic characteristics must be distinguished from attempts to enhance good fixation capabilities by mutating basic biological functions in excellent symbiotic partners such as Bragg[8,9] or Williams[35] soybean. Both approaches can contribute to increasing agronomic N_2 fixation, but the mutational method offers an added benefit of helping to understand symbiotic processes as it improves them.

In soybean, plant genes that determine which rhizobial strains form nodules are available (see Section IV). Whether plant breeders can identify or develop particular plant genotypes that increase N_2 fixation in diverse field conditions remains to be determined. There may be specific regions where one rhizobial symbiont prevails [e.g., *B. japonicum* serocluster 123 in the

U.S. Midwest, or *B*. sp. (cowpea) in tropical Africa] and a single plant genotype can be used to prevent unfavorable indigenous rhizobia from forming nodules. Another agronomic strategy that might be explored more fully is the possibility of growing mixtures of nodulating and nonnodulating plant genotypes. Initial data for soybean suggest that such mixtures allow a more efficient use of soil N and N_2 resources,[6] but whether more N_2 is fixed will required additional complex tests in which soil N content is analyzed over time.

Attempts to enhance N_2 fixation in forage legumes may also contribute agronomically improved plant materials. Although early methods that increased forage yield and N_2 fixation under controlled conditions[110] produced no significant improvement under field conditions,[3] other related techniques that also work in controlled environments[83,106,107] are producing modest but significant increases in forage N concentration and N_2-derived N content in the field without a decrease in forage yield (see Table 16-5). Selection directly in the field may produce plants better adapted for the complex environmental conditions found there. Thus far alfalfa populations produced by selection for increased forage yield and N concentration in the field show the desired phenotypes under controlled conditions, but no field results are yet available for these plant materials (L. R. Teuber and D. A. Phillips, unpublished results).

Another intriguing approach to using N_2 fixation more effectively in agricultural systems has been an attempt to develop alfalfa plants that conserve soil N by fixing more N_2 and storing it in roots and crowns.[40,41] The concept is that instead of growing a dormant alfalfa plant for several years in the U.S. Midwest, one might grow a nondormant type, typical of the U.S. Southwest, for one season. The nondormant types produce significant forage growth under Midwest conditions after the normal Midwestern cultivars decrease growth rates in response to shorter photoperiods and cool temperatures. The nondormant cultivars will not survive the Midwestern winter, but they can be plowed into the soil as a green manure several weeks after the last harvest. The cultivar Nitro was developed and tested at four locations for this purpose.[40,41] The agronomic value of the concept is evident in the fact that, when averaged across the four test sites, three nondormant cultivars fixed 45% more N_2 (P < 0.05) than two dormant cultivars between the third hay harvest and the final plowdown date.[41] Nitro fixed 15% more N_2 (P < 0.05) than the nondormant cultivars at one location but increases at the other three sites were not significant. Because no genetic control was included to test if the strain-cross method used to produce Nitro was responsible for its performance, one cannot determine if the selection protocol actually increased N_2 fixation.

One additional use of genetically altered legumes is the benefit that Nod mutants of agronomic species offer agronomists who estimate N_2 fixation under field conditions. Such mutants overcome some, but not all, of the criticisms leveled against nonleguminous reference crops that are used to estimate N_2 fixation by N-difference,[15]N-dilution, or [15]N A-value techniques.[111] The Nod⁻ legumes, however, do offer one type of baseline against which differences between N_2 fixation by other plants in the experiment can be calibrated.[84]

VI. Conclusions

Plant genes show both simple and complex effects on the *Rhizobium*–legume symbiosis. Defining the biochemical bases of mutant plant phenotypes already available will help clarify the basic biology of this symbiosis. Results indicate that breeding agronomic plants for increased N_2 fixation is possible, but poorly understood interactions between legumes and rhizobia can complicate such efforts.

Acknowledgments

The authors thank Drs. P. B. Cregan and T. A. LaRue for helpful comments during the preparation of this manuscript. Funds from the U.S. Department of Agriculture Competitive Grants Office and the National Science Foundation supported portions of the authors' research programs described in this review.

REFERENCES

1. Arrendell, S., Wynne, J. C., Elkan, G. H., and Isleib, T. G., "Variation for nitrogen fixation among progenies of a Virginia X Spanish peanut cross," *Crop. Sci. 25*, 865–869 (1985).

2. Attewell, J., and Bliss, F. A., "Host plant characteristics of common bean lines selected using indirect measures of N_2 fixation," in *Nitrogen Fixation Research Progress*, Evans, H. J., et al., eds. Dordrecht: Martinus Nijhoff, 1985, pp. 3–9.

3. Barnes, D. K., Heichel, G. H., Vance, C. P., and Ellis, W. R., "A multiple-trait breeding program for improving the symbiosis of N_2 fixation between *Medicago sativa* L. and *Rhizobium meliloti*," *Plant Soil 82*, 303–314 (1984).

4. Bergersen, F. J., and Nutman, P. S., "Symbiotic effectiveness in nodulated red clover. IV. The influence of the host factors i_1 and *ie* upon nodule structure and cytology," *Heredity 11*, 175–184 (1957).

5. Burton, J. C., and Wilson, P. W., "Host plant specificity among the Medicago in association with root-nodule bacteria," *Soil Sci. 47*, 293–303 (1939).

6. Burton, J. W., Brim, C. A., and Rawlings, J. O., "Performance of nonnodulating and nodulating soybean isolines in mixed culture and nodulating cultivars," *Crop Sci. 23*, 469–473 (1983).

7. Caldwell, B. E., "Inheritance of a strain-specific ineffective nodulation in soybeans," *Crop Sci. 6,* 427–428 (1966).

8. Carroll, B. J., McNeil, D. L., and Gresshoff, P. M., "A supernodulation and nitrate-tolerant symbiotic (*nts*) soybean mutant," *Plant Physiol. 78,* 34–40 (1985).

9. Carroll, B. J., McNeil, D. L., and Gresshoff, P. M., "Isolation and properties of soybean [*Glycine max* (L.) Merr.] mutants that nodulate in the presence of high nitrate concentrations," *Proc. Natl. Acad. Sci. USA 82,* 4162–4166 (1985).

10. Carroll, B. J., McNeil, D. L., and Gresshoff, P. M., "Mutagenesis of soybean (*Glycine max* [L.] Merr.) and the isolation of nonnodulating mutants," *Plant Sci. 47,* 109–119 (1986).

11. Cregan, P. B., and Keyser, H. H., "Host restriction of nodulation by *Bradyrhizobium japonicum* strain USDA 123 in soybean," *Crop Sci. 26,* 911–916 (1986).

12. Cregan, P. B., and Keyser, H. H., "Influence of *Glycine* spp. on competitiveness of *Bradyrhizobium japonicum* and *Rhizobium fredii,*" *Appl. Environ. Microbiol. 54,* 803–808 (1988).

13. Cregan, P. B., Keyser, H. H., and Sadowsky, M. J., "Host plant effects on nodulation and competitiveness of the *Bradyrhizobium japonicum* serotype strains constituting serocluster 123," *Appl. Environ. Microbiol. 55,* 2532–2536 (1989).

14. Cregan, P. B., Keyser, H. H., and Sadowsky, M. J., "Soybean genotype restricting nodulation of a previously unrestricted serocluster 123 Bradyrhizobia," *Crop Sci. 29,* 307–312 (1989).

15. Davis, T. M., "Two genes that confer ineffective nodulation in chickpea (*Cicer arietinum* L.)," *J. Hered. 79,* 476–478 (1988).

16. Davis, T. M., Foster, K. W., and Phillips, D. A., "Inheritance and expression of three genes controlling root nodule formation in chickpea," *Crop Sci. 26,* 719–723 (1986).

17. Degenhardt, T. L., LaRue, T. A., and Paul, E. A., "Investigation of a non-nodulating cultivar of *Pisum sativum,*" *Can. J. Bot. 54,* 1633–1636 (1976).

18. Delves, A. C., Carroll, B. J., and Gresshoff, P. M., "Genetic analysis and complementation studies on a number of mutant supernodulating soybean lines," *J. Genet. 67,* 1–8 (1988).

19. Delves, A. C., Mathews, A., Day, D. A., Carter, A. S., Carroll, B. J., and Gresshoff, P. M., "Regulation of the soybean–*Rhizobium* nodule symbiosis by shoot and root factors," *Plant Physiol. 82,* 588–590 (1986).

20. Demment, M. W., Teuber, L. R., Borque, D. P., and Phillips, D. A., "Changes in forage quality of improved alfalfa populations," *Crop Sci. 26,* 1137–1143 (1986).

21. Devine, T. E., "Inheritance of soybean nodulation response with a fast-growing strain of *Rhizobium,*" *J. Hered. 75,* 359–361 (1984).

22. Devine, T. E., and Weber, D. F., "Genetic specificity of nodulation," *Euphytica 26,* 527–535 (1977).

23. Duc, G., and Messager, A., "Mutagenesis of pea (*Pisum sativum* L.) and the isolation of mutants for nodulation and nitrogen fixation," *Plant Sci. 60,* 207–213 (1989).

24. Duc, G., and Picard, J., "Note on the presence of the sym-1 gene in *Vicia faba* hampering its symbiosis with *Rhizobium leguminosarum,*" *Euphytica 35,* 61–64 (1986).

25. Dudley, M. E., and Long, S. R., "A non-nodulating alfalfa mutant displays neither root hair curling nor early cell division in response to *Rhizobium meliloti,*" *Plant Cell 1,* 65–72 (1989).

26. Duhigg, P., Melton, B., and Baltensperger, A., "Selection of acetylene reduction rates in 'Mesilla' alfalfa," *Crop Sci. 18*, 813–816 (1978).

27. El-Sherbeeny, M. H., Lawes, D. A., and Mytton, L. R., "Symbiotic variability in *Vicia faba*. 2. Genetic variation in *Vicia faba*," *Euphytica 26*, 377–383 (1977).

28. Erdman, L. W., and Means, U. M., "Strain variation of *Rhizobium meliloti* on three varieties of *Medicago sativa*," *Agron. J. 45*, 625–629 (1953).

29. Fearn, J. C., and LaRue, T. A., "An altered constitutive peptide in *sym* 5 mutants of *Pisum sativum* (L.)," *Plant Mol. Biol., 14*, 207–216 (1990).

30. Gibson, A. H., "Genetic variation in the effectiveness of nodulation of lucerne varieties," *Aust. J. Agric. Res. 13*, 388–399 (1962).

31. Gibson, A. H., "Genetic control of strain-specific ineffective nodulation in *Trifolium subterraneum* L.," *Aust. J. Agric. Res. 15*, 37–49 (1964).

32. Gorbet, D. W., and Burton, J. C., "A non-nodulating peanut," *Crop Sci. 19*, 727–728 (1979).

33. Graham, P. H., and Temple, S. R., "Selection for improved nitrogen fixation in *Glycine max* (L.) Merr. and *Phaseolus vulgaris* L.," *Plant Soil 82*, 315–327 (1984).

34. Graham, R. A., and Scott, T. W., "Varietal characteristics and nitrogen fixation in cowpea," *Trop. Agric. 60*, 269–271 (1983).

35. Gremaud, M. F., and Harper, J. E., "Selection and initial characterization of partially nitrate tolerant mutants of soybean," *Plant Physiol. 89*, 169–173 (1989).

36. Gresshoff, P. M., Krotzky, A., Mathews, A., Day, D. A., Schuller, K. A., Olsson, J., Delves, A. C., and Carroll, B. J., "Suppression of the symbiotic supernodulation symptoms of soybeans," *J. Plant Physiol. 132*, 419–423 (1988).

37. Hardarson, G., Heichel, G. H., Barnes, D. K., and Vance, C. P., "Rhizobial strain preference of alfalfa populations selected for characteristics associated with N_2 fixation," *Crop Sci. 22*, 55–58 (1982).

38. Hardarson, G., and Jones, D. G., "The inheritance of preference for strains of *Rhizobium trifolii* by white clover (*Trifolium repens*)," *Ann. Appl. Biol. 92*, 329–333 (1979).

39. Hartwig, U. A., Maxwell, C. A., Joseph, C. M., and Phillips, D. A., "Chrysoeriol and luteolin released from alfalfa seeds induce *nod* genes in *Rhizobium meliloti*," *Plant Physiol. 92*, 116–122 (1990).

40. Heichel, G. H., and Barnes, D. K., "Opportunities for meeting crop nitrogen needs from symbiotic nitrogen fixation," in *Organic Farming: Current Technology and Its Role in a Sustainable Agriculture*. ASA Special Publ. No. 46, Bezdicek, D. F., et al., eds. Madison: Amer. Soc. Agronomy, 1984, pp. 49–59.

41. Heichel, G. H., Barnes, D. K., Vance, C. P., and Sheaffer, C. C., "Dinitrogen fixation technologies for alfalfa improvement," *J. Prod. Agric. 2*, 24–32 (1989).

42. Hobbs, S. L. A., and Mahon, J. D., "Heritability of $N_2(C_2H_2)$ fixation rates and related characters in peas (*Pisum sativum* L.)," *Can. J. Plant Sci. 62*, 265–276 (1982).

43. Hobbs, S. L. A., and Mahon, J. D., "Variability, heritability, and relationship to yield of physiological characters in peas," *Crop Sci. 22*, 773–779 (1982).

44. Hobbs, S. L. A., and Mahon, J. D., "Effects of pea (*Pisum sativum*) genotypes and *Rhizobium leguminosarum* strains on $N_2(C_2H_2)$ fixation and growth," *Can. J. Bot. 60*, 2594–2600 (1982).

45. Hobbs, S. L. A., and Mahon, J. D., "Variability and interaction in the *Pisum sativum* L.–*Rhizobium leguminosarum* symbiosis," *Can. J. Plant Sci. 63*, 591–599 (1983).

46. Holl, F. B., "Host plant control of the inheritance of dinitrogen fixation in the *Pisum–Rhizobium* symbiosis," *Euphytica 24*, 676–770 (1975).

47. Imrie, B. C., "The use of agar tube culture for early selection for nodulation of *Desmondium intortum*," *Euphytica 24*, 625–631 (1975).

48. Iruthayathas, E. E., Vlassak, L., and Laeremans, R., "Inheritance of nodulation and N_2 fixation in winged beans," *J. Hered. 76*, 237–242 (1985).

49. Jacobsen, E., "Modification of symbiotic interaction of pea (*Pisum sativum* L.) and *Rhizobium leguminosarum* by induced mutations," *Plant Soil 82*, 427–438 (1984).

50. Jessen, D. L., Barnes, D. K., and Vance, C. P., "Bidirectional selection in alfalfa for activity of nodule nitrogen and carbon-assimilating enzymes," *Crop Sci. 28*, 18–22 (1988).

51. Jones, D. G., and Burrows, A. C., "Breeding for increased nodule tissue in white clover (*Trifolium repens* L.)," *J. Agric. Sci. 71*, 73–79 (1968).

52. Kapulnik, Y., Joseph, C. M., and Phillips, D. A., "Flavone limitations to root nodulation and symbiotic nitrogen fixation in alfalfa," *Plant Physiol. 84*, 1193–1196 (1987).

53. Kapulnik, Y., Teuber, L. R., and Phillips, D. A., "Seedling development as a component of increased yield in an improved alfalfa population," *Crop Sci. 26*, 770–775 (1986).

54. Kneen, B. E., and LaRue, T. A., "Nodulation resistant mutant of *Pisum sativum* (L.)," *J. Hered. 75*, 238–240 (1984).

55. Kneen, B. E., and LaRue, T. A., "Induced symbiosis mutants of pea (*Pisum sativum*) and sweetclover (*Melilotus alba annua*)," *Plant Sci. 58*, 177–182 (1988).

56. Kneen, B. E., LaRue, T. A., and Weedin, N., "Genes reported to affect symbiotic nitrogen fixation by peas," *Pisum News Lett. 16*, 31–34 (1984).

57. Kneen, B. E., LaRue, T. A., and Weedin, N., "Mutants defective in symbiotic nitrogen fixation," *Pisum News Lett. 21*, 31 (1989).

58. Kneen, B. E., Vam Vikites, D., and LaRue, T. A., "Induced symbiosis mutants of *Pisum sativum*," in *Molecular Genetics of Plant-Microbe Interactions*, Verma, D. P. S., and Brisson, N., eds. Dordrecht: Martinus Nijhoff, 1987, pp. 79–84.

59. Kueneman, E. A., Root, W. R., Dashiell, K. E., and Hohenberg, J., "Breeding soybeans for the tropics capable of nodulating effectively with indigenous *Rhizobium* spp.," *Plant Soil 82*, 387–396 (1984).

60. LaRue, T. A., personal communication, 1989.

61. Lie, T. A., "Temperature-dependent root-nodule formation in pea cv. Iran," *Plant Soil 34*, 751–752 (1971).

62. Lie, T. A., "Host genes in *Pisum sativum* L. conferring resistance to European *Rhizobium leguminosarum* strains," *Plant Soil 82*, 415–425 (1984).

63. Lie, T. A., and Timmermans, P. C. J. M., "Host-genetic control of nitrogen fixation in the legume–*Rhizobium* symbiosis: Complication in the genetic analysis due to maternal effects," *Plant Soil 75*, 449–453 (1983).

64. Mahon, J. D., "Limitations to the use of physiological variability in plant breeding," *Can. J. Plant Sci. 63*, 11–21 (1983).

65. Mathews, A., Carroll, B. J., and Gresshoff, P. M., "Development of *Bradyrhizobium* infections in a supernodulating and non-nodulating mutants of soybean (*Glycine max* [L.] Merrill)," *Protoplasma 150*, 40–47 (1989).

66. Maxwell, C. A., Hartwig, U. A., Joseph, C. M., and Phillips, D. A., "A chalcone and two related flavonoids released from alfalfa roots induce *nod* genes of *Rhizobium meliloti*," *Plant Physiol. 90*, 842–847 (1989).

67. Mytton, L. R., "Plant genotype × rhizobium strain interactions in white clover," *Ann. Appl. Biol. 80*, 103–107 (1975).

68. Mytton, L. R., Brockwell, J., and Gibson, A. H., "The potential for breeding an improved lucerne–*Rhizobium* symbiosis. 1. Assessment of genetic variation," *Euphytica 33*, 401–410 (1984).

69. Mytton, L. R., El-Sherbeeny, M. H., and Lawes, D. A., "Symbiotic variability in *Vica faba*. 3. Genetic effects of host plant, rhizobium strain and of host × strain interaction," *Euphytica 26*, 785–791 (1977).

70. Mytton, L. R., and Rys, G. J., "The potential for breeding white clover (*Trifolium repens* L.) with improved nodulation and nitrogen fixation when grown with combined nitrogen. 2. Assessment of genetic variation in *Trifolium repens*," *Plant Soil 88*, 197–211 (1985).

71. Nigam, S. N., Nambiar, P. T. C., Dwivedi, S. L., Gibbons, R. W., and Dart, P. J., "Genetics of nonnodulation in groundnut (*Arachis hypogaea* L.). Studies with single and mixed *Rhizobium* strains," *Euphytica 31*, 691–693 (1982).

72. Nutman, P. S., "Nuclear and cytoplasmic inheritance of resistance to infection by nodule bacteria in red clover," *Heredity 3*, 263–291 (1949).

73. Nutman, P. S., "Symbiotic effectiveness in nodulated red clover. I. Variation in host and in bacteria," *Heredity 8*, 35–46 (1954).

74. Nutman, P. S., "Symbiotic effectiveness in nodulated red clover. II. A major gene for ineffectiveness in the host," *Heredity 8*, 47–60 (1954).

75. Nutman, P. S., "Symbiotic effectiveness in nodulated red clover. III. Further studies on inheritance of ineffectiveness in the host," *Heredity 11*, 157–173 (1957).

76. Nutman, P. S., "Symbiotic effectiveness in nodulated red clover. V. The *n* and *d* factors for ineffectiveness," *Heredity 23*, 537–551 (1968).

77. Nutman, P. S., "Improved nitrogen fixation in legumes by plant breeding; the relevance of host selection experiments in red clover (*Trifolium pratense* L.) and subterranean clover (*T. subterraneum* L.)," *Plant Soil 82*, 285–301 (1984).

78. Nutman, P. S., and Riley, J., "Breeding of nodulated red clover (*Trifolium pratense*) for high yield," *Ann. Appl. Biol. 98*, 319–331 (1981).

79. Ohlendorf, H., "Untersuchungen zur Vererbung der Resistenz von *Pisum sativum* gegen *Rhizobium leguminosarum* Stamm *311d*," *Z. Pflanzenzuchtg. 91*, 13–24 (1983).

80. Park, S. J., and Buttery, B. R., "Nodulation mutants of white bean (*Phaseolus vulgaris* L.) induced by ethyl-methane sulphonate," *Can. J. Plant Sci. 68*, 199–202 (1988).

81. Peters, N. K., and Long, S. R., "Alfalfa root exudates and compounds which promote or inhibit induction of *Rhizobium meliloti* nodulation genes," *Plant Physiol. 88*, 396–400 (1988).

82. Peterson, M. A., and Barnes, D. K., "Inheritance of ineffective nodulation and non-nodulation traits in alfalfa," *Crop Sci. 21*, 611–616 (1981).

83. Phillips, D. A., Cunningham, S. D., Bedmar, E. J., Sweeney, T. C., and Teuber, L. R., "Nitrogen assimilation in an improved alfalfa population," *Crop Sci. 25,* 1011–1015 (1985).

84. Phillips, D. A., Jones, M. B., and Foster, K. W., "Advantages of the nitrogen-15 dilution technique for field measurements of symbiotic dinitrogen fixation in legumes," in *Field Measurement of Dinitrogen Fixation and Denitrification,* Hauck, R. D., and Weaver, R. W., eds., Madison: Amer. Soc. Agronomy, 1986, pp. 11–21.

85. Phillips, D. A., Kapulnik, Y., Bedmar, E. J., and Joseph, C. M., "Development and partial characterization of nearly isogenic pea lines (*Pisum sativum* L.) that alter uptake hydrogenase activity in symbiotic *Rhizobium,*" *Plant Physiol. 92,* 983–989 (1990).

86. Phillips, D. A., and Teuber, L. R., "Genetic improvement of symbiotic nitrogen fixation in legumes," in *Nitrogen Fixation Research Progress,* Evans, H. J., et al., eds. Dordrecht: Martinus Nijhoff, 1985, pp. 11–17.

87. Phillips, D. A., Teuber, L. R., and Jue, S. S., "Variation among alfalfa genotypes for reduced nitrogen concentration," *Crop Sci. 22,* 606–610 (1982).

88. Pinchbeck, B. R., Hardin, R. T., Cook, F. D., and Kennedy, I. R., "Genetic studies of symbiotic nitrogen fixation in Spanish clover," *Can. J. Plant Sci. 60,* 509–518 (1980).

89. Postma, J. G., Jager, D., Jacobsen, E., and Feenstra, W. J., Characterization of a non-fixing mutant of pea (*Pisum sativum* L.), *Nitrogen Fixation: Hundred Years After,* Bothe, H., et al., eds. Stuttgart: Gustav Fischer, 1988, p. 640.

90. Postma, J. G., Jacobsen, E., and Feenstra, W. J., "Three pea mutants with an altered nodulation studied by genetic analysis and grafting," *J. Plant Physiol. 132,* 424–430 (1988).

91. Rennie, R. J., and Kemp, G. A., "N_2-Fixation in field beans quantified by ^{15}N isotope dilution: II. Effect of cultivars of beans," *Agron. J. 75,* 645–649 (1983).

92. Rolfe, B. G., and Gresshoff, P. M., "Genetic analysis of legume nodule initiation," *Ann. Rev. Plant Physiol. Plant Mol. Biol. 39,* 297–319 (1988).

93. Ronis, D. H., Sammons, D. J., Kenworthy, W. J., and Meisinger, J. J., "Heritability of total and fixed N content of the seed in two soybean populations," *Crop Sci. 25,* 1–4 (1985).

94. Rosas, J. C., and Bliss, F. A., "Host plant traits associated with estimates of nodulation and nitrogen fixation in common bean," *Hort. Sci. 21,* 287–289 (1986).

95. Ruschel, A. P., Vose, P. B., Matsui, E., Victoria, R. L., and Saito, S. M. T., "Field evaluation of N_2–fixation and N-utilization by *Phaseolus* bean varieties determined by ^{15}N isotope dilution," *Plant Soil 65,* 397–407 (1982).

96. Rys, G. J., and Mytton, L. R., "The potential for breeding white clover (*Trifolium repens* L.) with improved nodulation and nitrogen fixation when grown with combined nitrogen. 1. Effects of different amounts of nitrate nitrogen on phenotypic variation," *Plant Soil 88,* 181–195 (1985).

97. Seetin, M. W., and Barnes, D. K., "Variation among alfalfa genotypes for rate of acetylene reduction," *Crop Sci. 17,* 783–787 (1977).

98. Skøt, K., "Cultivar and *Rhizobium* strain effects on the symbiotic performance of pea (*Pisum sativum*)," *Physiol. Plant 59,* 585–589 (1983).

99. Smith, G. R., and Knight, W. E., "Inheritance of ineffective nodulation in crimson clover," *Crop Sci. 24,* 601–604 (1984).

100. Smith, G. R., Knight, W. E., and Peterson, H. H., "Variation among inbred lines of crimson clover for N_2 fixation (C_2H_2) efficiency," *Crop Sci. 22,* 716–719 (1982).

101. Smith, G. R., Knight, W. E., Peterson, H. L., and Hagedorn, C., "The effect of *Rhizobium trifolii* strains and crimson clover genotypes on N_2 fixation," *Crop Sci. 22,* 970–972 (1982).

102. Smith, G. R., Knight, W. E., and Peterson, H. L., "The inheritance of N_2 fixation efficiency in crimson clover," *Crop Sci. 22,* 1091–1094 (1982).

103. St. Clair, D. A., Wolyn, D. J., DuBois, J., Burris, R. H., and Bliss, F. A., "Field comparison of dinitrogen fixation determined with nitrogen-15-depleted and nitrogen-15-enriched ammonium sulfate in selected inbred backcross lines of common bean," *Crop Sci. 28,* 773–778 (1988).

104. Tan, G. Y., "Genetic variation for acetylene reduction rate and other characters in alfalfa," *Crop Sci. 21,* 485–488 (1981).

105. Tan, G. Y., and Tan, W. K., "Interaction between alfalfa cultivars and *Rhizobium* strains for nitrogen fixation," *Theor. Appl. Genet. 71,* 724–729 (1986).

106. Teuber, L. R., Levin, R. P., Sweeney, T. C., and Phillips, D. A., "Selection for N concentration and forage yield in alfalfa," *Crop Sci. 24,* 553–558 (1984).

107. Teuber, L. R., and Phillips, D. A., "Influences of selection method and nitrogen environment on breeding alfalfa for increased forage yield and quality," *Crop Sci. 28,* 599–604 (1988).

108. Vest, G., "Rj_3—a gene conditioning ineffective nodulation in soybean," *Crop Sci. 10,* 34–35 (1970).

109. Vest, G., and Caldwell, B. E., "Rj_4—a gene conditioning ineffective nodulation in soybean," *Crop Sci. 12,* 692–693 (1972).

110. Viands, D. R., Barnes, D. K., and Heichel, G. H., *Nitrogen Fixation in Alfalfa,* USDA Tech. Bull. No. 1643, 1981.

111. Vose, P. B., and Victoria, R. L., "Re-examination of the limitations of nitrogen-15 isotope dilution technique for the field measurement of dinitrogen fixation," in *Field Measurement of Dinitrogen Fixation and Denitrification,* Hauck, R. D., and Weaver, R. W., eds., Madison: Amer. Soc. Agronomy, 1986, pp. 23–41.

112. Whitehouse, W. W., and Griffith, W. K., "Soil fertility and fertilization of forages," in *Forages: The Science of Grassland Agriculture,* 3rd ed., Heath, M. E., et al., eds. Ames: Iowa State University Press, 1973, pp. 403–424.

113. Williams, L. F., and Lynch, D. L., "Inheritance of a non-nodulating character in the soybean," *Agron. J. 46,* 28–29 (1954).

17

Molecular Genetics of
Bradyrhizobium Symbioses

W. Mark Barbour, Shui-Ping Wang, and Gary Stacey

I. Introduction

The bacterial family Rhizobiaceae contains three genera—*Rhizobium*, *Bradyrhizobium*, and *Azorhizobium*—that have the ability to form N_2-fixing symbioses in specialized root nodules of leguminous plants[51,116] (see Chapter 1). The genus *Bradyrhizobium* has only one named species, *B. japonicum*, with other strains lumped together in a miscellaneous group[116] (see Chapter 1). These latter strains are referred to as *B*. sp., followed by the plant species in parenthesis. The bradyrhizobia are differentiated by slow growth (>8-hour generation time), an alkaline reaction on yeast extract mannitol medium, an inability to use disaccharides as a carbon source, the presence of one to two polar or subpolar flagella, a high G + C content (63–66%), the ability to grow autotrophically on H_2 as an energy source, and the ability to induce nitrogenase *ex planta*.[116]

Each species of *Bradyrhizobium*, *Rhizobium*, or *Azorhizobium* has a defined symbiotic host range. Generally, *Bradyrhizobium* spp. have a broader host range and nodulate tropical and subtropical legumes, whereas *Rhizobium* spp. have a more restricted host range and nodulate temperate legumes.[116] These organisms are vital for nitrogen cycling in nature and are agronomically important, as they provide fixed nitrogen for crop plants, alleviating the need for exogenously added fertilizer. *Bradyrhizobium* spp. nodulate agronomically important crops, including soybean [*Glycine max* (L.) Merr.], peanut (*Arachis hypogaea* L.), cowpea [*Vigna unguiculata* (L.) Walp.], mungbean [*Vigna radiata* (L.) Wilczek] and pigeon pea [*Cajanus cajun* (L.) Millsp.]. Additionally, *B*. sp. (*Parasponia*), which nodulates the woody dicot *Parasponia*, is the only rhizobia known to form an effective (N_2-fixing) symbiosis with a nonlegume[215] (see Chapter 13).

A goal of symbiotic N_2-fixation research is the expansion of the range of crops that can be effectively nodulated. The broad host range of bradyrhizobia offers a system for understanding the molecular basis for host range determination. Unfortunately, the bradyrhizobia have proven difficult to study genetically owing mostly to their slow growth rate, high intrinsic resistance to many antibiotics (such as kanamycin, ampicillin, tetracycline, and chloramphenicol), and relative lack of efficient genetic transfer systems. Also, as will be discussed later, symbiotically important genes in *Bradyrhizobium* spp. are somewhat dispersed on the chromosome in comparison to the clustered genes found on the Sym plasmids of *Rhizobium* spp.

Despite the difficulties experienced in studying *Bradyrhizobium*, there has been substantial recent progress that has uncovered several unique features of *Bradyrhizobium* molecular genetics. These features include the presence of unique nodulation genes and absence of other genes commonly found in *Rhizobium* spp. In addition, *Bradyrhizobium* spp. show significant differences in the regulation of both nodulation and N_2-fixation genes.

II. Genes Involved in Symbiosis

Molecular genetic studies have identified a number of symbiotic genes in *Bradyrhizobium*. These genes have been identified in many cases by DNA hybridization using known gene probes from *Rhizobium* species (e.g., refs. 158, 179, 188) or with *nif* genes from *Klebsiella pneumoniae* (e.g., ref. 97). Other genes have been identified by transposon mutagenesis (e.g., ref. 196), by deletion analysis (e.g., ref. 90), and by DNA sequencing (e.g., refs. 80, 154). The three main classes of symbiotic genes are the *nod, fix,* and *nif* genes as defined by the symbiotic phenotype of a mutation in each gene. A Nod mutant forms no nodules or nodulates more slowly than the wild type, whereas a Fix mutant forms nodules unable to fix N_2. The discrimination between *nod* and *fix* genes is not always clear. For example, mutants that nodulate slowly and form partially developed nodules with no or reduced N_2 fixation could be classified as Fix or Nod mutants. *Nif* genes are responsible for the synthesis and functioning of the nitrogenase enzyme complex and are classified according to homology with *nif* genes of nonsymbiotic bacteria (e.g., *K. pneumoniae*).

Many genes have been characterized that are required for or affect symbiosis at some level but that are not *nif, fix,* or *nod* genes. For example, genes encoding such functions as exopolysaccharide synthesis (e.g., ref. 131), lipopolysaccharide synthesis (e.g., refs. 24, 28, 156), heme synthesis (e.g., ref. 128), dicarboxylic acid transport (e.g., ref. 174), uptake hydrogenase (e.g., ref. 23), or nitrogen regulation may influence symbiosis (reviewed in ref. 88) but also serve functions in free-living cells. The nature of some of these functions will be discussed. It is likely that many of the *fix* and *nod* genes will be renamed once their specific biochemical function is elucidated.

III. Organization and Function of Nodulation Genes

A. Common Nodulation Genes

Much of this review will focus on the symbiotic genetics of the soybean symbiont, *Bradyrhizobium japonicum* strain USDA 110, as this is the best-characterized strain. The majority of the symbiotic genes described in USDA 110 are clustered in two areas of the genome (Figure 17-1). These two clusters have recently been shown, using R-prime analysis, to be separated by 500–1000 kb (D. Kuykendahl, personal communication).

Two general classes of nodulation genes have been described. The so-called common nodulation genes are found in all *Rhizobium* and *Bradyrhizobium* isolates studied and are essential for the earliest steps in the nodulation process. In many cases the common *nod* genes are functionally interchangeable between *Rhizobium* and *Bradyrhizobium* species (cf. refs. 48, 73, 77, 121, 145, 154). Additionally, genes have been identified that enable the bacteria to nodulate specific hosts and that are termed host specificity genes (e.g., refs. 43, 48, 49, 112, 154). Not all nodulation genes are easily classified in this way as they may exhibit characteristics of both groups (e.g., *nodD*, refs. 9, 81, 89, 110, 111, 150, 200).

All the nodulation genes thus far described in *Bradyrhizobium* are apparently located on the chromosome. Typically, *Rhizobium* strains harbor large symbiotic plasmids that carry *nod, nif,* and *fix* genes. Although large plasmids have been demonstrated in some bradyrhizobia, no symbiotic genes have been localized on these plasmids (e.g., ref. 147).

Much of the research focus on the genetics of nodulation has been directed to characterizing the organization, sequence, regulation, and function of the common *nod* genes—*nodA, nodB, nodC,* and *nodD*. The *nodABC* genes are essential for at least two early steps in the nodulation process—root hair curling and cortical cell division (reviewed in refs. 45, 133). These genes were first identified in *R. meliloti* (e.g., refs. 113, 121, 134, 214) but were subsequently found in all *Rhizobium* and *Bradyrhizobium* species (e.g., refs. 45, 133, 158, 175, 179, 188). Mutations in any of these genes results in a Nod⁻ phenotype on all hosts. Furthermore, it was shown that an *Agrobacterium* strain harboring the *R. meliloti nodDABC* genes could form nodules on alfalfa (*Medicago sativa* L.).[105]

A considerable amount of evidence now indicates that the *nodABC* genes' products are involved in the synthesis of a factor that induces the characteristic early nodulation responses by the host (see Chapter 14 for a detailed discussion). Recently, Lerouge et al.[129] reported on the chemical structure of a root hair curling factor produced by *R. meliloti*. This factor, NodRm-1, was identified as N-acyl-tri-N-acetyl-β-1, 4-D-glucosamine tetrasacchar-

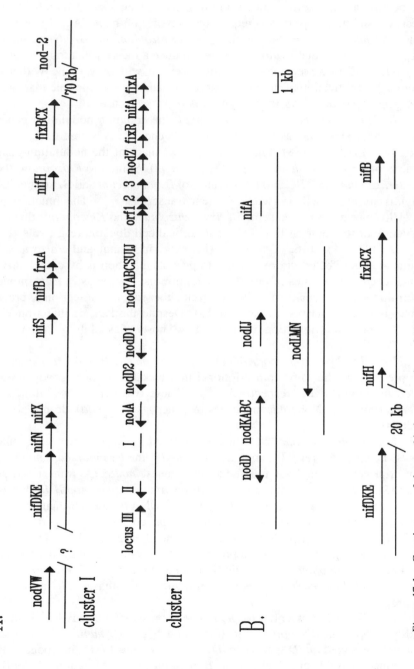

Figure 17-1. Genetic maps of the symbiotic gene regions of (A) *B. japonicum*; (B) *B.* sp. (*Parasponia*). The arrows designate transcriptional units and the approximate length of each transcript.

ide bearing a sulfate group on carbon 6 of the reducing sugar moiety. Synthesis of NodRm-1 appeared to require the presence of the *nodA* and *C* genes. Since the *nodABC* genes are "common" to *Rhizobium* and *Bradyrhizobium* species, it is likely that *Bradyrhizobium* produce a factor structurally related to NodRm-1. However, proof of this assertion will only come with the chemical identification of the factors produced and the specific role that each *nod* gene's product plays in the synthesis of these factors.

In *B. japonicum* and *B.* sp. (*Parasponia*) the common nodulation genes *nodA, nodB,* and *nodC* as well as other *nod* genes are clustered (e.g., refs. 126, 145, 158, 179, 188). Figure 17–1 shows maps of the nodulation gene regions of *B. japonicum* and *B.* sp. (*Parasponia*). Similar to *Rhizobium*, the *Bradyrhizobium nodABC* and *nodD* genes of *B. japonicum* and *B.* sp. (*Parasponia*) are expressed on two, divergent transcripts.[154,188] The function of *nodABC* appears to be the same in *Bradyrhizobium* and *Rhizobium;* that is, induction of root hair curling and plant cortical cell division (e.g., refs. 46, 48, 52, 121, 145, 146, 154, 175). However, in peanut and *Parasponia,* plants are not infected via root hairs. In peanut, infection is by crack entry, which occurs at the sites of emerging lateral roots,[30] whereas in *Parasponia* a third infection mechanism, different from either root hair infection or crack entry, is used (described in Chapter 13). Despite the lack of infection via curled root hairs, infection of both these hosts was shown to require *nodABC*.[146,158]

Unlike *Rhizobium* spp., *nodABC* of *B. japonicum* and *B.* sp. (*Parasponia*) are preceded in the same transcriptional unit by an additional gene, designated *nodY* and *nodK,* respectively.[154,188] These two genes show little sequence similarity. Nonpolar mutations in either *nodK* or *nodY* do not affect nodulation.

In *Rhizobium* spp., *nodABC* are cotranscribed along with two additional genes, *nodIJ* (e.g., ref. 60). In addition to *nodY,* the *B. japonicum nodABC* transcript contains at least four additional genes, *nodSUIJ,* transcribed as *nodYABCSUIJ*.[80,82,154] Despite their location, mutations in *nodSUIJ* do not significantly affect nodulation.[80] Recently,[136] DNA sequence analysis has revealed other genes downstream of *nodJ* that could also be part of the *nodYABCSUIJ* transcript (i.e., ORF1, 2, and 3; Figure 17-1). DNA sequencing indicates that *nodIJ* are present in *B.* sp. (*Parasponia*) in a similar position to *B. japonicum*.[189] In addition, an approximately 2-kb gap exists between *nodC* and *nodI* in *B. parasponia,* suggesting that *nodSU* may also be present.[189]

The *nodD* gene is essential for *nod* gene expression and may be found in multiple copies in *Rhizobium* spp. (e.g., *nodD*$_{1-3}$ in *R. meliloti*[89,111]). Two tandemly arranged *nodD* genes (*nodD*$_1$, *nodD*$_2$; Figure 17-1), independently transcribed, have been identified in *B. japonicum*.[4,82] Mutation of *nodD*$_1$ resulted in only a slight delay in nodulation of soybean.[82,154] Mutation of

$nodD_2$ and a mutation of both $nodD_1$ and $nodD_2$ resulted in a longer delay in nodule formation.[82] Remarkably, a $nodD_1$ mutation resulted in an almost complete loss of inducibility of the common *nod* genes.[8] The fact that $nodD_1$ is required for *nod* gene induction but is not essential for nodulation seems paradoxical. One explanation is that even a very low level of *nod* gene transcription is sufficient for nodulation under the conditions used in this study.

The *B*. sp. (*Parasponia*) ANU289 or RP501 *nodD* gene is also apparently not required for nodulation, as Tn5 mutations within the gene did not affect nodulation on legume or nonlegume hosts.[146,189] Not surprisingly, Southern hybridization of ANU289 genomic DNA revealed two putative *nodD* loci in this strain.[189] Therefore, multiple *nodD* genes may be common in *Bradyrhizobium*.

B. Host Specificity Genes

In addition to the common *nod* genes, additional genes are involved in determining the specific host plant(s) nodulated. Examples are the *nodEFGHQ* genes of *R. meliloti* (e.g., refs. 41, 61, 62, 72, 112, 187, 207) and the *nodLMNO* genes of *R. leguminosarum* (e.g., refs. 21, 22, 39, 49, 50, 56, 193, 198). The biochemical functions of these genes is currently unknown. However, recent evidence indicates that the host specificity genes of *R. meliloti* may be involved in specifically modifying the factor produced in response to *nodABC* (e.g., refs. 7, 29, 62). For example, NodH has been suggested to act as a sulfate transferase that modifies a compound NodRm-2 to make NodRm-1.[29] Although less developed, current models suggest that host specificity in *R. leguminosarum* is similarly determined by the synthesis of a NodRm-2-like compound that is subject to specific modification by host specificity genes (e.g., *nodEF*, see ref. 201).

A number of genes have been identified in *B. japonicum* that appear to be involved in host range determination. The first such gene characterized was *nodZ*, which is located downstream of *nodJ*[154] (see Figure 17-1). The *nodZ* gene was identified by hybridization to a host specificity region from *R*. sp. MPIK3030 and was shown to be required for nodulation of siratro (*Macroptilium atropurpureum* Urb.), but not soybean.[154,203] Siratro is the normal host for *R*. sp. MPIK3030. The *nodZ* gene is unlike other *nod* genes in that it is apparently regulated by NifA, similar to *nif* and *fix* genes[203] (see discussion later). Consistent with this observation, the *nodZ* gene is expressed in soybean bacteroids (Schell and Stacey, unpublished). Although apparently not essential for nodulation of soybean, NodZ may be required for bacterial growth and development in siratro.

Deshmane and Stacey[43] reported the identification of additional host specificity genes in *B. japonicum*. Three loci were identified by transposon mutagenesis within a 5.3-kb *Eco*R1 fragment downstream of $nodD_2$ (loci I, II, and III; Figure 17-1). Mutations in locus I or II resulted in delayed nodulation on all hosts tested, whereas a locus III mutation resulted in a Nod⁻ phenotype on soybean but did not affect nodulation of siratro or *Glycine soja*. Thus, locus III represents a soybean-specific nodulation gene. The nodulation host range of the locus III mutant resembles the wild-type host range of *R. fredii* strains. Indeed, it was shown that locus III DNA hybridized to all *B. japonicum* strains tested but not to any *R. fredii* strains. Mutations in locus III resulted in an inability to induce soybean root cortical cell division, although root hair curling did occur.[43] Therefore, one possibility is that the locus III region encodes functions similar to that proposed for the *nodH* gene of *R. meliloti;* that is, a role in modifying the NodABC factor produced by the bacteria.

Hahn and Hennecke[90] identified two regions (*nod*-1 and *nod*-2) essential for soybean nodulation by constructing deletions adjacent to cluster I (see Figure 17-1). Mutation of the *nod*-1 region resulted in reduced nodulation of soybean and no nodulation of three other hosts—siratro, cowpea, and mungbean. Therefore, the *nod*-1 region appears to encode host-specific nodulation functions. The *nod*-1 region was recently sequenced and two genes, *nodV* and *nodW* were identified.[79] Sequence comparisons revealed that the NodV and NodW proteins are members of a family of two-component regulatory systems. These regulatory systems are involved in sensing environmental stimuli (reviewed in refs. 204–205). NodV has similarity to the membrane-associated sensor subfamily, a few members of which have been shown to be protein kinases.[204–205] NodW has similarity to the transcriptional regulatory protein subfamily. A model was proposed that NodV senses a stimulus and activates NodW (perhaps via phosphorylation), which in turn regulates the transcription of genes important for nodulation of specific plant hosts.[79] Gottfert and Hennecke[79] pointed out the similarity between the phenotypes of mutations in *nodVW* and *nodZ*; that is, both are essential for siratro nodulation, but not for soybean. Therefore, *nodZ* was suggested as an obvious target for regulation by NodVW. This hypothesis has yet to be tested.

Hybridization studies with *B. japonicum* genomic DNA have revealed no significant homology with the well-characterized host specificity genes of *Rhizobium* spp. (e.g., *nodEFGH* from *R. meliloti* or *nodFELMN* from *R. leguminosarum*). However, recently Scott and Bender[189] reported that DNA encoding the *nodLMN* genes of *R. leguminosarum* biovar *viciae* hybridized to genomic DNA from *B. parasponia* ANU289. The hybridizing region was cloned but has not been characterized further.

In addition to the preceding report, Marvel et al.[146] reported that a Tn5 insertion downstream of *nodC* in *B.* sp. (*Parasponia*) resulted in a Nod⁻ phenotype on *Parasponia* but in a Nod⁺ phenotype on cowpea and siratro.

This locus is roughly analogous in location to the *nodZ* gene in *B. japonicum*.

C. Genotype-Specific Nodulation Genes

In soybean production areas of the midwestern United States, members of *B. japonicum* serocluster 123 are the dominant indigenous competitors for soybean nodulation.[58,92,152] These strains are often less effective at fixing N_2 than commercially available strains and therefore a method to supplant these indigenous, competitive strains within the nodules could have agronomic benefits. Cregan and colleagues at the USDA Beltsville Laboratory of Nitrogen Fixation initiated a research program to identify soybean genotypes that would exclude nodulation by members of the 123 serocluster and therefore could theoretically favor nodulation by the competing inoculant strain. Cregan and Keyser[32] identified several soybean genotypes that restricted nodulation and reduced the competitiveness of *B. japonicum* strain USDA123. These genotypes were nodulated normally by other, non-123 serogroup strains (e.g., USDA110). Two of the genotypes, Plant Introduction 377578 and PI 371607, were subsequently shown to differentially restrict nodulation by 20 different serocluster 123 isolates.[119] Subsequent work has found other soybean genotypes that will restrict members of serocluster 123 or other strains in serogroups 127 or 129[33,34] (Cregan and Sadowsky, personal communication). In the case of nodulation restriction of serocluster 123 strains and those of serogroup 127 or 129, a single, dominant plant gene appears to be involved in each case (Cregan, personal communication).

Recently, Sadowsky et al.[181] have identified a single, dominant bacterial gene that is essential for nodulation of the serocluster 123 restrictive soybean genotypes. This gene has been termed *nolA* and is an example of a genotype-specific nodulation gene (GSN) that determines infection of specific plant genotypes within a given legume species. NolA shows sequence similarity to the MerR protein, a transcriptional regulatory protein required for the expression of mercury resistance genes (e.g., ref. 95), especially within the helix-turn-helix DNA binding motif. Therefore, *nolA* may encode a DNA-binding, transcriptional-regulatory protein. The *nolA* gene of *B. japonicum* appears to represent the bacterial counterpart of a gene-for-gene interaction system involved in soybean nodulation. The fact that other soybean genes are likely to exist that determine nodulation restriction of different *B. japonicum* strains[33,34] (see earlier) suggests that other bacterial genes may also exist that are essential for nodulation of these specific plant genotypes. If this conjecture is correct, then *B. japonicum*–soybean nodulation could be in part determined by a "race-specific" genetic system analogous to that found in many pathogen–plant relationships (cf., refs. 91, 118). Genotype-specific nodulation genes have been reported in a few other cases but have not been extensively studied. The first report of such a gene was in *R. leguminosarum* bv. *viciae* strain TOM, where a single gene, *nodX,* was identified that allowed this strain to nodulate Afghanistan pea.[37,84,108,130] *nodX* is induced by pea root exudate or by the flavones eriodictyol and hesperitin. Studies of genetic crosses between European genotypes and Afghanistan pea revealed that nodulation restriction of bv. *viciae* strains was determined by a single recessive plant gene.[107] Therefore, a gene-for-gene interaction appears to exist in which a single dominant gene in *R. leguminosarum* bv. *viciae* strain TOM determines nodulation of Afghanistan pea as controlled by a single recessive plant gene.[45] The biochemical function of NodX is unknown.

In a different study, Pueppke and colleagues found that *R. fredii* strain USDA257 nodulates soybean cultivar Peking normally but cannot nodulate cultivar McCall.[102,103] Heron et al.[101] isolated five mutants of strain USDA257 in which the host range was extended to include McCall. These mutants are interesting in that they appear to identify bacterial genes that act negatively to restrict nodulation on specific genotypes of soybean. The bacterial genes identified have not been characterized. In addition, the plant genetic character that determines this host restriction phenotype has not been defined.

Table 17-1 summarizes the *Bradyrhizobium* nodulation genes and their proposed functions.

IV. Regulation of Nodulation Gene Expression

Nodulation is a complex process with the potential for bacterial gene regulation at several points during the infection process. In *Rhizobium* spp., the *nodD* gene product is constitutively expressed and is essential for the expression of the other *nod* genes. NodD is a member of a large family of prokaryotic regulatory proteins examples of which include *E. coli* LysR, IlvY, CysB, *Salmonella typhimurium* MetR, and *Enterobacter cloacae* AmpR.[96] In *Rhizobium,* with the exception of the *nodD* gene, nodulation genes are only expressed when the bacteria are exposed to the host plant or host plant exudates/extracts (reviewed in refs. 91, 132, 133). The *nod* gene–inducing compounds have been purified from several plants and in all cases found to be flavonoid chemicals (e.g., refs. 8, 47, 83, 93, 123, 148, 164, 177, 223, 224). The mechanism by which these compounds induce *nod* gene expression is not completely understood. The current model indicates that NodD interacts directly with the inducing compound and stimulates transcription. The evidence in favor of this model is that the specificity for the inducing flavonoid is determined by the primary sequence of NodD.[20,109,150,199,200] Indeed, this property of NodD in part determines host specificity, since different *Rhizobium* species show differential responses to particular flavonoids, as determined by their specific NodD (e.g.. ref. 133).

Table 17-1. Nodulation Genes of Bradyrhizobium *and Their Functions.*

Gene	Species	Proposed Function
nodABC	*B. japonicum*[179] *B.* sp. *(Parasponia)*[145,188]	Synthesis of a low-molecular-weight factor essential for root hair curling and cortical cell division
nodD₁	*B. japonicum*[154] *B.* sp. *(Parasponia)*[146,189]	Positive transcriptional regulator required for *nod* gene expression
nodD₂	*B. japonicum*[82] *B.* sp. *(Parasponia)*[189]	Unknown function; mutation causes delayed nodulation
nodI	*B. japonicum*[80,154] *B.* sp. *(Parasponia)*[189]	Unknown function; sequence similarity to ATP-dependent transport proteins
nodJ	*B. japonicum*[80,154] *B.* sp. *(Parasponia)*[189]	Unknown function; membrane location
nodK	*B.* sp. *(Parasponia)*[188]	Unknown function
nodLMN	*B.* sp. *(Parasponia)*[189]	Found by hybridization; host range determination in *Rhizobium*
nodSU	*B. japonicum*[80]	Unknown function
nodV	*B. japonicum*[79]	Host range determination; sequence similarity to membrane sensor family
nodW	*B. japonicum*[79]	Host range determination; sequence similarity to transcriptional regulatory family
nodY	*B. japonicum*[154]	Unknown function
nodZ	*B. japonicum*[154,203]	Host range determination
nolA	*B. japonicum*[181]	Genotype specific nodulation

In fact, the purpose of multiple *nodD* genes within a single *Rhizobium–Bradyrhizobium* may be to allow the interaction with flavonoid inducers coming from different members of the host range (e.g., refs. 81, 89, 110, 111). NodD has the ability to bind to the promoter of *nod* genes in the absence of inducer (e.g., refs. 70, 71, 109, 122). Therefore, one possibility is that binding of the inducer may induce a conformational change in NodD essential for the stimulation of transcription. The specific binding site of NodD upstream of the *nod* genes has been identified by DNA footprint analysis[71,122] and is, in general, well conserved upstream of *nod* genes in all species of *Rhizobium–Bradyrhizobium* analyzed (e.g., refs. 82, 154, 178, 197). This sequence has been termed the *nod* box and is approximately 47 base pairs in length. Deletion or mutation of the *nod* box sequence abolishes the ability of NodD to bind[20,109] or to stimulate transcription.[109,178,197]

The regulation of the *nodY*, *nodA*, and *nodC* genes of *B. japonicum* has been studied by the use of translational *lacZ* fusions to these genes.[8,83,123] The expression of these genes requires an active NodD₁ and is dependent on the addition of isoflavones (e.g., genistein and daidzein, compounds present in soybean seed and root exudates). A well-conserved *nod* box sequence was found 5′ of the *nodY* gene.[154] A *nod* box sequence was also found in an analogous position 5′ of the *nodK* genes of *B*. sp. (*Parasponia*).[188] However, the regulation of the *nod* genes of *B*. sp. (*Parasponia*) has not been studied in detail. These characteristics of the *nod* genes of *Bradyrhizobium* suggest that their regulation is similar to that of the better-studied *Rhizobium*. However, in *Rhizobium* species, isoflavones have been reported to be inhibitors of *nod* gene expression.[177]

A more striking difference between *Rhizobium* and *Bradyrhizobium* deals with the regulation of the *nodD₁* gene. Translational *lacZ* fusions to the *nodD₁* gene of *B. japonicum* indicate that this gene is not constitutively expressed to an appreciable level and its expression is dependent on the presence of plant-produced isoflavones.[8] The expression of *nodD₁* is also dependent on a functional NodD1 protein; therefore, expression is autoregulated. More recently, Smit et al.[195] reported the isolation of compounds excreted by soybean roots that specifically induced *nodD₁* expression; for example, these compounds did not induce a *nodY-lacZ* fusion. Therefore, the plant appears to have the potential to induce *nodD₁* expression without inducing other *nod* genes. Whether this occurs naturally and what role this ability may play during infection is presently unknown. The compounds that specifically induce *nodD₁* expression in *B. japonicum* have recently been identified as glycosylated derivatives of genistein and daidzein (Smit, Carlson, and Stacey, unpublished).

The fact that *nodD₁* is inducible by isoflavones led to an examination of the promoter sequence of this gene and the identification of a divergent *nod* box–like sequence.[8] This sequence was postulated to be the binding site for nodD that allowed *nodD₁* induction. Consistent with this hypothesis was the recent demonstration that deletion of this sequence resulted in the inability to induce *nodD₁* expression upon addition of genistein, daidzein, or soybean seed exudate.[219]

The reader should not be left with the idea that there are no inducible *nodD* genes in *Rhizobium* species. Recently, Johnston et al.[37a,115] reported that the *nodD₁* gene of *R. leguminosarum* bv. *phaseoli* was induced by isoflavones, a result very similar to the situation in *B. japonicum*. In addition, the *nodD₃* gene of *R. meliloti* is also positively regulated with expression dependent on the synthesis of a novel regulatory protein, SyrM[135,153] (see Chapter 14). These studies indicate that the regulation of *nodD* expression can be very complex. It is presently unclear what benefit the organism gains by this complex regulation. A model for the regulation of *nodD₁* and the *nodYABC* transcript in *B. japonicum* is shown in Figure 17-2.

In addition to the *nodD₁* and *nodYABCSUIJ* transcripts, several other operons require the presence of isoflavones for expression. These include *nolA*,[181] locus II and III,[43] and orf3 downstream of *nodJ*[136] (see Figure 17-1). The *nodYABCSUIJ* operon has been shown to be inducible by 50-fold with soybean seed extract using a *nodY–lacZ* fusion.[8] However, all the other induc-

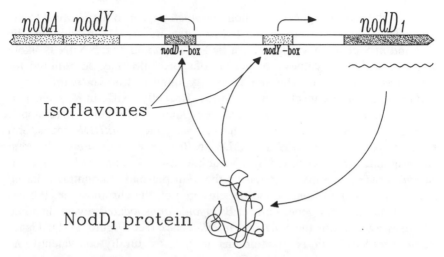

Figure 17-2. A model for regulation of *B. japonicum* nodulation genes. Shaded regions indicate gene coding sequences or *nod* boxes (i.e., NodD-box and NodY-box). Arrows indicate transcriptional start sites and direction. A low level of NodD is formed constitutively. In the presence of the inducer (i.e., specific isoflavones), NodD binds at both *nod*-boxes and activates transcription of the *nodD₁*, and *nodYABCSUIJ* operons. The figure is not drawn to scale.

ible loci described are induced from two- to 10-fold. This low level of induction has also been reported for *nod* genes in *Rhizobium* spp. (e.g., refs. 37, 162, 182, 225).

Not all *nod* genes in *Bradyrhizobium* appear to be preceded by a recognizable *nod* box. For example, a recognizable *nod* box has not been found upstream of *nolA*.[181] In this regard, the previous report of Sadowsky et al.[182] showed isoflavone-inducible genes in *R. fredii,* a soybean symbiont, that lacked a *nod* box sequence; this may indicate that multiple mechanisms control host-inducible genes in soybean rhizobia.

V. Organization and Function of the *Nif–Fix* Genes

As indicated previously, *nif* genes are homologous, both structurally and functionally, to genes in free-living N₂-fixing organisms, such as *K. pneumoniae*. In *K. pneumoniae*, the *nif* gene cluster is comprised of at least 20 genes (see Chapters 20 and 21). A few genes in *Rhizobium* and *Bradyrhizobium* have been identified that correspond to *K. pneumoniae nif* genes. These include *nifA*, which encodes a transcriptional regulatory protein (reviewed in ref. 88); *nifHDK*, the structural genes of the nitrogenase enzyme complex (e.g., refs. 31, 75, 97, 167, 168, 180, 190, 191, 213); and nifB[17,74,85,176] and nifE genes,[54,157] both of which appear to be involved in the biosynthesis or processing of the FeMo cofactor.[163,172,192] In addition, the nifN and nifS genes have been identified in B. japonicum.[54] NifN is likely involved in the biosynthesis of FeMo cofactor; NifS may be required for nitrogenase maturation. The nifX gene was found just downstream of nifN in B. japonicum (Hennecke, unpublished). Recently, NifX in K. pneumoniae has been proposed to act as a negative regulator of nif gene expression.[78]

In Klebsiella and Rhizobium, such as R. meliloti, the nifHDK genes form a single operon, while nifH and nifDK in Bradyrhizobium are in different transcriptional units separated by several kilobases.[2,67] In Rhizobium species the nif genes are tightly clustered on the Sym plasmid. In contrast, the nif genes in Bradyrhizobium are loosely clustered on the chromosome. For example, the regulatory gene, nifA, of B. japonicum is located several hundred kilobases away from the nifH and nifDK genes (see Figure 17-1). Genes required for symbiotic N₂ fixation that are not structurally equivalent to K. pneumoniae nif genes are called fix genes. As in the case of nif genes, the fix genes in Bradyrhizobium strains are loosely clustered on the chromosome. For example, the fixBC genes are located in symbiotic cluster I and the fixA gene is located in the cluster II region (see Figure 17-1), whereas in R. meliloti the fixABC genes form a single operon. The biochemical functions of a few of the fix genes in Bradyrhizobium have been deduced from studies of the corresponding genes in other bacteria. The fixABC genes have been found in Rhizobium and Bradyrhizobium,[53,74,85,86,126,186] and proposed to function in electron transport to nitrogenase. FixC was reported to contain a signal sequence for membrane insertion.[53] In B. japonicum, fixABC were shown to be essential for N₂ fixation in planta and in vitro under microaerobic, free-living conditions.[86] The fixLJ genes were first reported in R. meliloti[36] and were shown to be the transcriptional regulators of nifA and fixK expression. These genes control nifA transcription in response to the ambient O₂ concentration.[40] These genes were recently found in B. japonicum.[3] FixLJ show sequence homology to the same family of two-component regulatory protein systems as referrerd to earlier in the case of NodVW. FixL is homologous to the family of membrane sensor proteins, whereas FixJ shows homology to the family of transcriptional regulatory proteins. Sequence analysis of the B. japonicum fixLJ genes revealed that they are probably cotranscribed and are highly homologous (60%) to their counterparts in R. meliloti.[3]

The fixR gene has only been identified in B. japonicum and is found within the same operon as nifA.[210] Recently,[98] Hennecke reported that FixR shows sequence similarity to a NAD-dependent dehydrogenase, suggesting that this protein may play a redox role in the cell. In K. pneumoniae, nifL proceeds nifA within the same transcript and negates nifA function under aerobic conditions (reviewed in ref. 88). There is no sequence similarity between NifL and FixR and no evidence to suggest that they play similar roles within the cell.

Nif and *fix* genes in bradyrhizobia other than *B. japonicum* are less well characterized. The *nifH* and *nifDK* genes have been cloned from *B.* sp. (*Parasponia*)[191] and *B.* sp. (cowpea).[222] Similar to *B. japonicum,* the *nifH* gene of *B.* sp. (*Parasponia*) is transcribed on a separate operon located approximately 20 kb from *nifDK*.[191] In addition, *nifE* was localized immediately downstream of *nifDK*.[189] The *nifB* gene was also identified but linkage to other symbiotic genes has not been reported.[189] The *nifA* gene has been identified near the nodulation gene cluster in *B.* sp. (*Parasponia*),[189] again analogous to the situation in *B. japonicum*. Only three *fix* genes (i.e., *fixBCX*, Figure 17-1) have been identified in *B.* sp. (*Parasponia*).[189] These genes are very similar in their location, relative organization, and DNA sequence to other *Bradyrhizobium* species (cf. refs. 74, 87).

In addition to the named genes in *B. japonicum* and other *Bradyrhizobium* species, several additional *fix* loci have been identified by mutagenesis. For example, Fix⁻ mutants defective in heme metabolism (e.g., refs, 159, 169) and molybdenum metabolism[142] have been reported. Additional Fix⁻ mutants have been isolated, but the corresponding genes have yet to be characterized (e.g., refs. 155, 170, 196).

As indicated earlier during the discussion of *nod* genes, the distinction between a Fix⁻ and Nod⁻ phenotype is sometimes difficult to make. For example, a hypothetical mutation that affected growth of the bacteria within the plant could be judged Nod⁻ if the phenotype was exhibited very early during infection and Fix⁻ if exhibited later. Surprisingly, *nifA* is an example of a gene that appears to display both Nod and Fix character (e.g., refs. 65, 104, 184, 194). As discussed later in detail, NifA is required for the expression of the other *nif* and fix genes. A NifA⁻ mutant of *R. meliloti* forms nodules on alfalfa that are altered in ultrastructure,[104,194] suggesting a possible role for NifA in nodule development. The effect of a *nifA* mutation on nodule development is more severe in *B. japonicum,* where the later stages of nodule development are severely impaired.[65] Although plant nodule–specific genes (nodulins, see Chapter 15) appear to be expressed in these nodules, the level of leghemoglobin and a few other nodulins is extremely reduced.[206] This is not commonly found in nodules induced by other Fix⁻ mutants. These results indicate that NifA may play a role in the normal bacterial induction of nodule development and nodulin expression.

Table 17-2 lists the *nif* and *fix* genes reported to date in *Bradyrhizobium* and information concerning their possible biochemical functions.

VI. Regulation of nif–fix Gene Expression in Bradyrhizobium

Information concerning the transcriptional regulation *nif* and *fix* genes largely comes from the study of only two bacterial species, *K. pneumoniae* and *R.*

Table 17-2. Symbiotic Genes, nif and fix, Found in Bradyrhizobia

Gene	Species	Proposed Function
nifA	B. japonicum[65] B. sp. (Parasponia)[189]	Transcriptional regulator of nif and fix expression
nifH	B. japonicum[1,75,97,117] B. sp. (cowpea)[222] B. sp. (Parasponia)[191]	Structural gene for dinitrogenase reductase
nifDK	B. japonicum[1,75,97,117] B. sp. (cowpea)[222] B. sp. (Parasponia)[191]	Structural genes for dinitrogenase
nifB	B. japonicum[74,99] B. sp. (Parasponia)[189]	FeMoco synthesis
nifE	B. japonicum[54,99] B. sp. (Parasponia)[189]	FeMoco synthesis
nifN		FeMoco synthesis
nifS	B. japonicum[54,99]	Maturation of nitrogenase
fixA	B. japonicum[74,86,126]	Electron transport to nitrogenase
fixBC	B. japonicum[74,86] B. sp. (Parasponia)[189]	Electron transport to nitrogenase
fixLJ	B. japonicum[3]	Sequence similarity to family of two-component regulatory proteins (see ref. 204)
fixR	B. japonicum[98,210]	Function unknown; sequence similarity to dehydrogenases

meliloti (reviewed in refs. 88, 133; see Chapter 21). Briefly, in *K. pneumoniae*, the regulatory operon, *nifLA*, is activated by NtrC during ammonia starvation, and, under low O_2 concentration, NifA in turn activates all other *nif* genes. The *nifL* gene encodes a negative regulatory protein that negates NifA function in the presence of O_2 or intermediate levels of bound nitrogen. In *Rhizobium* and *Bradyrhizobium* spp., NifA plays a similar role in the regulation of *nif* and *fix* genes. However, no *nifL* counterpart has been reported in *Rhizobium* or *Bradyrhizobium*. Activation of *nif* and/or *fix* genes by NifA is dependent on another gene, *rpoN* (also called *glnF*, *ntrA rpoE*), which encodes an alternative sigma factor, σ.[54] Unlike the situation in *K. pneumoniae*, *R. meliloti nifA* expression does not require NtrC.[208] Rather, *nifA* expression is controlled via the products of the *fixLJ* operon.[36,40] FixLJ controls *nifA* expression in response to microaerobiosis. Under low O_2 conditions, FixLJ mediated *nifA* expression, whereas FixLJ is responsible for *nifA* repression in the presence of high O_2.[40]

All NtrA-dependent promoters have the $-24/-12$ conserved sequence (GG-N_9-TGC), which is the site recognized by σ,[54] the *rpoN* product (reviewed in ref. 211). In addition, some *nif–fix* promoters are preceded by one or two copies of an upstream activator sequence (UAS; 5'-TGT-N_{10}-

ACA-3′). This sequence is not essential for expression of these genes but is responsible for strong enhancement of their transcription. The UAS sequence has been shown in *K. pneumoniae* by deletion analysis and DNA footprinting to be the binding site for NifA.[15,16,100]

The *R. meliloti nifA* gene has its own promoter. However, transcription can also be initiated at the upstream *fixA* promoter, which itself is NifA-dependent.[11,44,120] Thus, the *nifA* gene in *R. meliloti* is autoregulated by a read-through mechanism from the *fixABCX* operon situated just upstream of the *nifA* gene.[11,120,218]

The *B. japonicum nif–fix* gene regulation system is the only well-studied system among *Bradyrhizobium* species and has several unique features in comparison to *K. pneumoniae* and *R. meliloti*. In *B. japonicum*, NifA also plays a central role in *nif–fix* gene regulation. NifA-dependent activation of *nif* and *fix* genes in *B. japonicum* also involves the recognition of promoters by the alternative sigma factor, σ.[54] Recently, two separate, strongly conserved *rpoN* (*ntrA*) genes were identified and sequenced.[69] A sequence comparison of RpoN1 and RpoN2 revealed the presence of an extra stretch of 48 amino acids in RpoN2. Mutagenesis of either *rpoN* gene gave a Nod⁺Fix⁺ phenotype, whereas a *rpoN1–rpoN2* double mutant had a Nod⁺Fix⁻ phenotype. These results suggest that both *rpoN* genes are active and can functionally complement one another.

Consistent with the requirement for σ,[54] all *Bradyrhizobium nif–fix* promoters that have been examined have the typical −24 and −12 consensus sequence as discussed earlier (reviewed in ref. 211). In addition, some also have the upstream activator sequence (UAS) located at various distances from their transcriptional start sites. The NifA-dependent *nif–fix* genes are not expressed under high O_2 concentration but are induced microaerobically, through a mechanism mediated by NifA.[65,86a] Unlike *nifA* of *R. meliloti*, the *B. japonicum nifA* gene does not have its own promoter. It is cotranscribed with *fixR*. The *fixRnifA* promoter is a typical σ⁵⁴-dependent −24/−12 activating promoter.[209, 210] It also has a UAS that that is located 50–86 bp upstream of the transcriptional start; however, this UAS may not bind NifA but another, as yet, unidentified regulatory protein.[209, 210] Unlike other N_2-fixing organisms, *nifA* is expressed under aerobic conditions.[209] Indeed, the *fixRnifA* promoter is active in free-living *B. japonicum* cultures grown aerobically, microaerobically and anaerobically, as well as in the symbiotic state. The activation mechanisms of the *fixRnifA* promoter under the different O_2 levels are not the same. The data suggest that under aerobic conditions, *fixRnifA* expression is dependent on the presence of a novel regulatory protein that requires the UAS sequence.[209,210] The NifA protein that is produced under these conditions is inactive. Under microaerobic conditions and *in planta*, *fixRnifA* expression is enhanced by the presence of active NifA. Therefore, the expression of the *B. japonicum fixRnifA* operon is dually controlled—by NifA-independent activation and by NifA-dependent autoregulation in response to cellular O_2 status.[209] As indicated previously, active NifA is required for *nif–fix* gene expression in *B. japonicum*. The conundrum presented by these data is why an apparently inactive NifA protein is expressed under aerobic conditions. One possibility is that NifA may act as an O_2-controlled repressor under these conditions. However, there is currently no evidence to support this assertion.

The *B. japonicum* NifA is O_2 sensitive (e.g., refs. 68, 124), its activity in *E. coli* being drastically and irreversibly reduced within minutes of exposure to O_2.[124] Comparison of the amino acid sequence of *B. japonicum* NifA to that from *K. pneumoniae* revealed the presence of a rhizobia-specific protein domain containing two conserved cysteine residues in a Cys–X_4–Cys motif. Fischer et al.[66] examined the role of these two cys residues and two additional residues at the end of the central domain in *B. japonicum* NifA by substituting them individually with serine via site-directed mutagenesis. The resultant mutant proteins were inactive. Changes of residues in the four amino acid spacer between the cysteines had not effect on NifA activity. It was further demonstrated that *B. japonicum* NifA activity was very sensitive to the presence of chelating agents in the growth medium, whereas *K. pneumoniae* NifA was not.[66] Therefore, the cysteine residues were proposed to be involved in metal binding. The NifA-bound metal ions could be Fe^{2+} or Fe^{3+}, the oxidation state of which would be influenced by the ambient O_2 concentration. A similar Cys–X_4–Cys motif has been found in the NifA proteins from *Rhizobium* spp. (see Chapter 21) and therefore may be a common feature of rhizobia.

As pointed out in the introduction, some strains of *Bradyrhizobium japonicum* possess the ability to fix N_2 *ex planta* under microaerobic conditions, although they do not grow under these conditions. However, activation of *nif* expression does not appear to involve NtrC under these conditions (e.g., ref. 208). Rather, *ex planta* expression of the *nif* genes probably involves NifA, which is active under microaerobic conditions.

As indicated previously, *nifA* expression in *R. meliloti* is regulated by FixLJ in response to ambient O_2 concentration.[36,40] Recently, Anthamatten and Hennecke[3] cloned the *B. japonicum fixLJ* genes by hybridization to a *fixLJ* gene probe from *R. meliloti*. DNA sequencing of this region indicated that *fixLJ* are organized in one operon. A nonpolar deletion in *fixL* resulted in a 90% reduction of N_2-fixation activity, the same phenotype as a *fixJ* insertion mutant. By monitoring the activity of a plasmid-encoded *fixR-lacZ* fusion, it was demonstrated that *fixRnifA* expression was independent of *fixLJ* under aerobic conditions.[3] Therefore, it is unlikely that FixJ is the previously unidentified regulator of *nifA* expression. Although FixLJ do not appear to be involved in *fixRnifA* regulation, the data clearly indicate that they have an important role in N_2 fixation. Additional experiments will likely identify *nif–fix* genes that are regulated by FixLJ. A model summarizing the regulation of the *nif–fix* genes in *B. japonicum* is shown in Figure 17-3.

VII. Other Genes Related to Symbiotic Nitrogen Fixation in Bradyrhizobium

A. *exo and lps*

In addition to those genes described previously, a variety of genes have been described that influence the symbiosis of *Bradyrhizobium* spp. with their host plants. Examples include genes that influence the cell surface polysaccharides of rhizobia. There is considerable precedence for an important role for cell surface polysaccharides in nodulation by both *Rhizobium* and *Bradyrhizobium* species. In *R. meliloti,* 12 complementation groups (*exo*) have been identified that affect the synthesis of exopolysaccharide (EPS) and are required for infection of alfalfa.[131] Similarly, mutations affecting lipopolysaccharide (LPS) synthesis in *R. leguminosarum* bv. *viciae* results in the formation of Fix⁻ nodules (e.g., refs. 28, 38, 165). In *B. japonicum*, EPS mutants appear to nodulate normally (e.g., ref. 127), whereas mutations affecting LPS synthesis result in defects in nodulation (e.g., refs. 140, 166, 202).

B. *hup*

Hydrogen metabolism in *B. japonicum* has been extensively studied (reviewed in refs. 57, 139; see Chapter 11). H_2 is produced normally during the reduction of N_2 by nitrogenase.[18,19,106] Some rhizobia possess an uptake hydrogenase that has the potential to return to the system at least a portion of the energy lost in the production of H_2 by nitrogenase. The presence of an uptake hydrogenase has been shown to increase the overall efficiency of nitrogen fixation by *B. japonicum* (reviewed in ref. 57). Hydrogenase isolated from *B. japonicum* consists of two subunits and contains nickel and iron (as nonheme iron).[5] Two genes designated *hup* encoding the hydrogenase were cloned and sequenced[23,94,185] and shown to be cotranscribed. The expression of hydrogenase in *B. japonicum* was shown to be O_2 regulated, and dependent on the availability of H_2 and carbon sources.[141,151] The mechanism of O_2 regulation of hydrogenase is still unknown.

C. *gln*

The utilization of fixed nitrogen requires the enzymes for nitrogen assimilation. Glutamine synthetase (GS) is the key enzyme for ammonia assimilation and is encoded by the gene *glnA*. The regulation of *glnA* and asso-

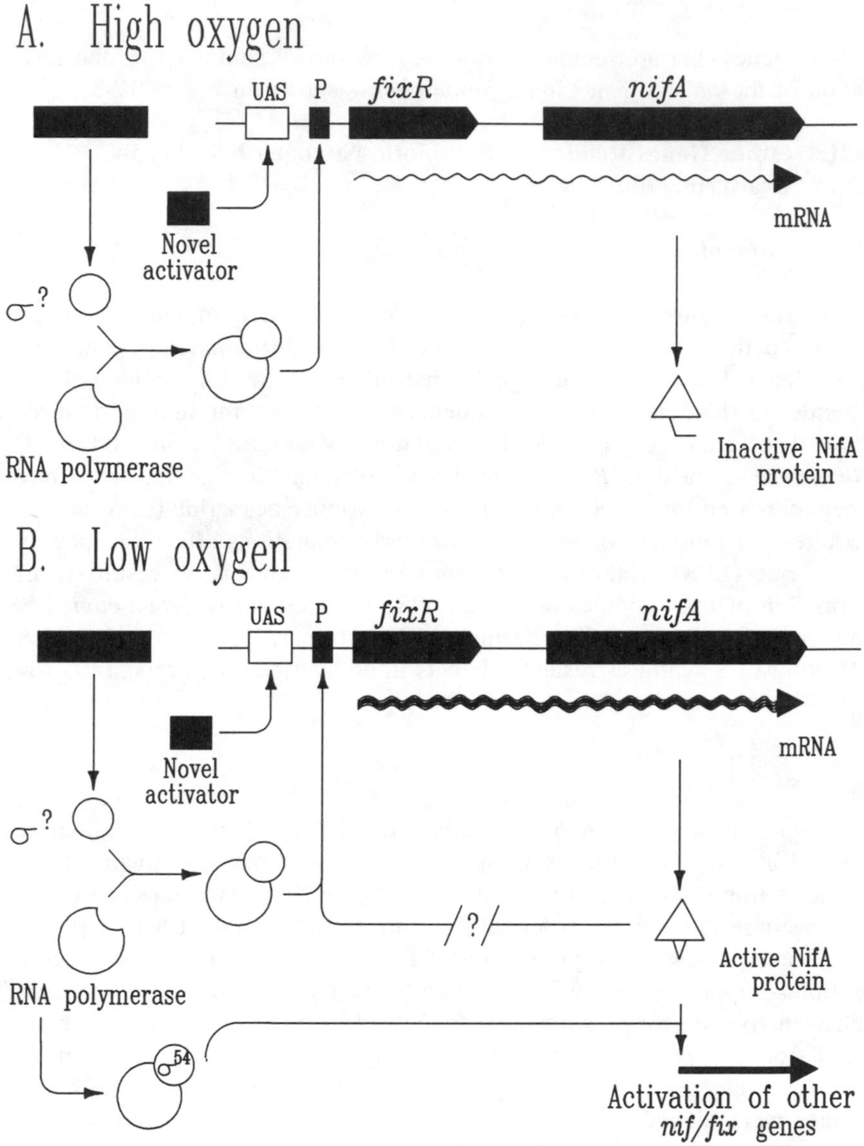

Figure 17-3. A model for regulation of *B. japonicum nif/fix* genes. Models for regulation under high O_2 (A) and low O_2 (B) are shown. Under high O_2, RNA polymerase with an associated σ factor (perhaps $σ^{54}$) and a proposed novel regulatory protein bind to the promoter (P) and upstream UAS sequence, respectively, resulting in the expression of the *fixRnifA* operon. An apparently inactive NifA is made under these conditions. Under low O_2, an active NifA protein is made, resulting in an increase in the expression of the *fixRnifA* operon. It has not been unequivocally shown that NifA acts directly on the *nifA* promoter. NifA also acts under these conditions to stimulate the expression of other *nif* and *fix* genes. The figure is not drawn to scale.

ciated genes has been extensively studied in the *Enterobacteriaceae* (reviewed in refs. 125, 138, 171). Free-living *B. japonicum* cells assimilate ammonia primarily by the coordinate activity of glutamine synthetase (GS) and glutamate synthase (GOGAT) (e.g., refs. 14, 217). However, GS activity decreases in bacteroids in concert with the derepression of nitrogenase activity (e.g., refs. 14, 216). The regulation of nitrogen assimilation pathways is therefore an integral part of the development of an effective symbiosis. Rhizobia differ from the *Enterobacteriaceae* in possessing multiple genes for GS (e.g., refs 35, 42, 63). For example, *B. japonicum* possesses at least two genes for GS, termed *glnA* and *glnII* (e.g., refs. 25–27, 35). The *B. japonicum glnA* gene is homologous to the *E. coli glnA*; however, unlike *E. coli*, transcription is not affected by the cellular nitrogen level. In contrast, *glnII* expression is nitrogen regulated.[25,27] Under aerobic, nitrogen-limited growth conditions, *glnII* expression requires active NtrC. However, under microaerobic conditions *glnII* expression is independent of NtrC.[143] In addition to the genes for glutamine synthetase, the *glnB* gene has been cloned and sequenced from *B. japonicum*.[144] GlnB encodes the P_{II} protein, a key component of the cascade system that controls glutamine synthetase enzyme activity in response to the carbon–nitrogen balance of the cell (reviewed in refs. 88, 125, 137, 138, 171). The *glnB* gene of *B. japonicum* was shown to be regulated by NtrC and NtrA both in free-living and symbiotic cells.[144]

D. *dct and Associated Genes*

The carbon sources used by bacteroids for N_2 fixation are now commonly assumed to be dicarboxylic acids (e.g., succinate, malate, and fumarate). Dicarboxylic acids are the most effective substrates for respiration and N_2 fixation by isolated soybean bacteroids (e.g., ref. 10). *Rhizobium* strains that are defective in succinate dehydrogenase[76] or malate dehydrogenase[220] activity were shown to be incapable of N_2 fixation. The bacteroids apparently depend on their host plants to provide dicarboxylic acids to fuel the reduction of N_2 to ammonia.[6,13,59,64,174,221] Correspondingly, *Rhizobium* mutants defective in dicarboxylic acid transport (*dct*) form ineffective nodules on their respective host plants (e.g., refs. 64, 174, 221). Genes for dicarboxylic acid transport (*dct*) have been found in *R. leguminosarum* bv. *viciae* (e.g., refs. 114, 173) and in *R. meliloti* (e.g., refs. 13, 59, 114). Evidence suggests that succinate, malate, and fumarate are transported via a common system in both cultured cells and bacteroids of *B. japonicum*.[149,183] Birkenhead et al.[12] reported that conjugation of the *R. meliloti dct* genes into *B. japonicum* increased the growth rate of the bacteria, as well as the succinate uptake and nitrogenase activity of free-living cells under microaerobic conditions.[12] These results indicate that an increase in the flow of carbon and energy sources may increase symbiotic N_2 fixation. Specific *dct* genes have yet to be isolated from *Bradyrhizobium* spp.

E. Others

A gene designated *frxA* was identified in *B. japonicum* and shown to be cotranscribed with the *nifB* gene.[55] Sequence comparisons suggest that *frxA* encodes a ferredoxin-like protein and therefore may take part in redox reactions within the cell. Several laboratories are now working to elucidate the pathway of electron transport within *B. japonicum* (e.g., refs. 160, 169, 212). For example, genes encoding cytochrome bc_1[212] and cytochrome oxidase[161] have been cloned. These studies are directed toward elucidating the pathway of electron flow to nitrogenase. In addition, Maier et al.[142] reported the identification of genes involved in molybdate transport, a function likely important to efficient N_2 fixation. The genetic tools developed in *B. japonicum* for the isolation of *nod*, *fix*, and *nif* genes will undoubtedly be used in the future to examine more broadly the molecular genetics of *Bradyrhizobium* species.

VIII. Conclusions

The study of *Bradyrhizobium* symbioses has revealed unique features that make these systems useful for the understanding of host specificity and symbiotic gene regulation. The molecular genetics of *Bradyrhizobium japonicum* and *B.* sp. (*Parasponia*) have been examined to some detail. As shown in Figure 17-1, current information suggests that the symbiotic genes of these two species are similarly arranged. Whether this is generally true of *Bradyrhizobium* spp. can only be ascertained by examination of several other strains. The broad host range of bradyrhizobia includes important agricultural crops and the nonlegume *Parasponia,* the only non-legume-rhizobia symbiosis known. As detailed earlier, numerous host-specific nodulation genes have been identified in *B. japonicum*. Further research will eventually lead to a molecular description of the host-specific nodulation mechansims used by this important group of rhizobia. An immediate question to be addressed is whether the host specificity genes of *Bradyrhizobium* spp. function in a similar manner to the genes from *Rhizobium* spp. Current information suggests that similarities are likely; howeverr, other information (e.g., *nodZ*, *nolA*) suggests that differences will also be found.

The study of *nod* gene regulation in *B. japonicum* has revealed several unique features. These include the induction of *nod* genes by isoflavones, the host-inducible expression of $nodD_1$, and the functionality of a divergent *nod* box sequence within the *nodD1* promoter. Likewise, *nif–fix* gene regulation in *B. japonicum* differs somewhat from that described in *Rhizobium*.

The regulatory gene *nifA* is cotranscribed with the *fixR* gene and is expressed under aerobic conditions (in an apparently inactive form). Two different mechanisms for the expression of *nifA* under aerobic and microaerobic (or symbiotic) conditions have been described.

Several *Bradyrhizobium* symbiotic genes have been described that have not been found in *Rhizobium*. These include *nodY, nodK, nodZ, nodV, nodW, nolA, nifS,* and *fixR*. It will be interesting to see how many of these genes will subsequently be identified in *Rhizobium* spp.

The agronomic importance of crops nodulated by *Bradyrhizobium* spp. provides sufficient justification for the study of the symbiotic genetics of these bacteria. However, further justification can be taken from the discovery of the unique and interesting characteristics of *Bradyrhizobium* symbioses. Information gained from the study of *Rhizobium* will continue to be applied to the investigation of *Bradyrhizobium* symbioses but will not be sufficient for a complete understanding of these important systems. The study of only a few *Rhizobium* and *Bradyrhizobium* species has revealed surprising diversity in organization, function, and regulation. As indicated in Chapter 12, those systems presently under study do not reflect the breadth of diversity that exists in rhizobia–legume symbioses. Therefore, it is likely that other differences will be found as a wider sampling of rhizobial species is examined. These comparative studies could be extremely valuable in formulating general principles that apply to rhizobial systems and perhaps to other plant–microbe interactions.

Acknowledgments

The authors wish to express their gratitude to Michael Gottfert, Hans-Martin Fischer, Hauke Hennecke, and Sharon Long for their comments on the chapter. Recent results presented from the authors' laboratory were supported by Public Health Service grants GM33494-05 and GM40183-02 from the National Institutes of Health and grant 62-600-1636 from the U.S. Department of Agriculture.

References

1. Adams, T.H., and B.K. Chelm, "The *nifH* and *nifDK* promoter regions from *Rhizobium japonicum* share structural homologies with each other and with nitrogen-regulated promoters from other organisms," *J. Mol. Appl. Genet.* 2:392–405 (1984).

2. Adams, T.H., C.R. McClung, and B.K. Chelm, "Physical organization of the *Bradyrhizobium japonicum* nitrogenase gene region," *J. Bacteriol. 159*:857–862 (1984).

3. Anthamatten, D. and H. Hennecke. Identification and functional analysis of *fixLJ*-like genes in *Bradyrhizobium japonicum*. p. 508 *In* P.M. Gresshoff, L.E. Roth, G. Stacey, and W.E. Newton (eds.), *Nitrogen Fixation: Achievements and Objectives,* Chapman and Hall, New York, 1990.

4. Appelbaum, E., "The *Rhizobium/Bradyrhizobium*–legume symbiosis," pp. 131–158. In P. M. Gresshoff (ed.), *Molecular Biology of Nitrogen Fixation,* CRC Press, Boca Raton, FL, 1989.

5. Arp, D.J., and R.H. Burris, "Purification and properties of the particulate hydrogenase from the bacteroids of soybean root nodules," *Biochim. Biophys. Acta 570*:221–230 (1979).

6. Arwas, R., I.A. McKay, F.R.P. Rowney, M.J. Dilworth, and A.R. Glenn, "Properties of organic acid utilization mutants of *Rhizobium leguminosarum* strain 300," *J. Gen. Microbiol. 131*:2059–2066 (1985).

7. Banfalvi, Z., and A. Kondorosi, "Production of root hair deformation factors by *Rhizobium meliloti* nodulation genes in *Escherichia coli:* HsnD (NodH) is involved in the plant host-specific modification of the NodABC factor," *Plant Mol. Biol. 13*:1–12 (1989).

8. Banfalvi, Z., A. J. Nieuwkoop, M. G. Schell, L. Besl, and G. Stacey, "Regulation of nod gene expression in *Bradyrhizobium japonicum*," *Mol. Gen. Genet. 214*:420–424 (1988).

9. Bender, G.L., M. Nayudu, K.K. Le Strange, and B.G. Rolfe, "The *nodD*1 gene from *Rhizobium* strain NGR234 is a key determinant in the extension of host range to the nonlegume *Parasponia*," *Mol. Plant-Microbe Interact. 1*:259–266 (1988).

10. Bergersen, F.J., and G.L. Turner, "Nitrogen fixation by the bacteroid fraction of breis of soybean root nodules," *Biochim. Biophys. Acta 141*:507–515 (1967).

11. Better, M., G. Ditta, and D.R. Helinski, "Deletion analysis of *Rhizobium meliloti* symbiotic promoters," *EMBO J. 4*:2419–2424 (1985).

12. Birkenhead, K., S.S. Manian, and F. O'Gara, "Dicarboxylic acid transport in *Bradyrhizobium japonicum:* Use of *Rhizobium meliloti dct* gene(s) to enhance nitrogen fixation" *J. Bacteriol. 170:*184–189 (1988).

13. Bolton, E., B. Higgisson, A. Harrington, and F. O'Gara, "Dicarboxylic acid transport in *Rhizobium meliloti:* Isolation of mutants and cloning of dicarboxylic acid transport genes," *Arch. Microbiol. 144*:142–146 (1986).

14. Brown, C.M., and M.J. Dilworth. "Ammonia assimilation by *Rhizobium* cultures and bacteroids," *J. Gen. Microbiol. 86*:39–48 (1975).

15. Buck, M., W. Cannon, and J. Woodcock, "Mutational analysis of upstream sequences required for transcriptional activation of the *Klebsiella pneumoniae nifH* promoter," *Nucleic Acids Res. 15*:9945–9956 (1987).

16. Buck, M., S. Miller, M. Drummond, and R. Dixon, "Upstream activator sequences are present in the promoters of nitrogen fixation genes," *Nature 320*:374–378 (1986).

17. Buikema, W. J., J. A. Klingensmith, S. L. Gibbons, and F. M. Ausubel, "Conservation of structure and location of *Rhizobium meliloti* and *Klebsiella pneumoniae nifB* genes," *J. Bacteriol. 169*:1120–1126 (1987).

18. Bulen, W. A., R. C. Burns, and J. R. LeComte, "Nitrogen fixation: Hydrosulfite as electron donor with cell free preparations of *Azotobacter vinelandii* and *Rhodosprillum rubrum*," *Proc. Natl. Acad. Sci. (USA) 53:*532–539 (1965).

19. Bulen, W.A., and J.R. LeComte, "The nitrogenase system from *Azotobacter*: Two-enzyme requirement for N_2 reduction, ATP-dependent H_2 evolution, and ATP hydrolysis," *Proc. Natl. Acad. Sci. (USA) 56*:979–986 (1966).

20. Burn, J.E., W.D. Hamilton, J.C. Wootton, and A.W.B. Johnston, "Single and multiple mutations affecting properties of the regulatory gene *nodD* of *Rhizobium*," *Mol. Microbiol.* 3:1567–1577 (1989).

21. Canter Cremers, H.C.J., H.P. Spaink, A.H.M. Wijfjes, E. Pees, C.A. Wijffelman, R.J.H. Okker, and B.J.J. Lugtenberg, "Additional nodulation genes on the sym plasmid of *Rhizobium leguminosarum* biovar *viciae*," *Plant Mol. Biol.* 13:163–174 (1989).

22. Canter Cremers, H.C.J., C.A. Wijffelman, E. Pees, B.G. Rolfe, M.A. Djordjevic, and B.J.J. Lugtenberg, "Host specific nodulation of plants of the pea cross-inoculation group is influenced by genes in fast growing *Rhizobium* downstream of *nodC*," *J. Plant Physiol.* 132:398–404 (1988).

23. Cantrell, M.A., R.A. Haugland, and H.J. Evans, "Construction of a *Rhizobium japonicum* gene bank and use in isolation of a hydrogen uptake gene," *Proc. Natl. Acad. Sci. (USA)* 80:181–185 (1983).

24. Carlson, R.W., S. Kalembasa, D. Turowski, P. Pachori, and K.D. Noel, "Characterization of the lipopolysaccharide from a *Rhizobium phaseoli* mutant that is defective in infection thread development," *J. Bacteriol.* 169:4923–4928 (1987).

25. Carlson, T.A., and B.K. Chelm, "Apparent eucaryotic origin of glutamine synthetase II from the bacterium *Bradyrhizobium japonicum*," *Nature* 322:568–570 (1986).

26. Carlson, T.A., M.L. Guerinot, and B.K. Chelm, "Characterization of the gene encoding glutamine synthetase I (*glnA*) from *Bradyrhizobium japonicum*," *J. Bacteriol.* 162:698–703 (1985).

27. Carlson, T.A., G.B. Martin, and B.K. Chelm, "Differential transcription of the two glutamine synthetase genes of *Bradyrhizobium japonicum*," *J. Bacteriol.* 169:5861–5866 (1987).

28. Cava, J.R., H. Tao, and K.D. Noel, "Mapping of complementation groups within a *Rhizobium leguminosarum* Cfn42 chromosomal region required for lipopolysaccharide synthesis," *Mol. Gen. Genet.* 221:125–128 (1990).

29. Cervantes, E., S. B. Sharma, F. Maillet, J. Vasse, G. Truchet, and C. Rosenberg, "The *Rhizobium meliloti* host range *nodQ* gene encodes a protein which shares homology with translation elongation and initiation factors," *Mol. Microbiol.* 3:745–755 (1989).

30. Chandler, M.R., "Some observations on infection of *Arachis hypogea* L. by *Rhizobium*," *J. Exp. Bot.* 29:749–755 (1978).

31. Corbin, D., L. Barran, and G. Ditta, Organization and expression of *Rhizobium meliloti* nitrogen fixation genes," *Proc. Natl. Acad. Sci. (USA)* 80:3005–3009 (1983).

32. Cregan, P.B., and H.H. Keyser, "Host restriction of nodulation by *Bradyrhizobium japonicum* strain USDA123 in soybean," *Crop Sci* 26:911–916 (1986).

33. Cregan, P.B., H.H. Keyser, and M.J. Sadowsky, "Host plant effects on nodulation and competitiveness of the *Bradyrhizobium japonicum* serotype strains constituting serocluster-123," *Appl. Env. Microbiol.*" 55:2532–2536 (1989).

34. Cregan, P.B., H.H. Keyser, and M.J. Sadowsky, "Soybean genotype restricting nodulation of a previously unrestricted serocluster 123 *Bradyrhizobium*," *Crop Sci* 29:307–312 (1989).

35. Darrow, R.A., and R.R. Knotts, "Two forms of glutamine synthetase in free-living root-nodule bacteria," *Biochem. Biophys. Res. Comm.* 78:554–559 (1977).

36. David, M., M.L. Daveran, J. Batut, A. Dedieu, O. Domergue, J. Ghai, C. Hertig, P. Boistard, and D. Kahn, "Cascade regulation of nif gene expression in *Rhizobium meliloti*," *Cell 541*:671–683 (1988).

37. Davis, E.O., I.J. Evans, and A.W.B. Johnston, "Indentification of *nodX*, a gene that allows *Rhizobium leguminosarum* biovar *viciae* strain TOM to nodulate Afghanistan peas," *Mol. Gen. Genet. 212*:531–5350 (1988).

37a. Davis, E.O., and A.W.B. Johnston, "Regulatory functions of the three *nodD* genes of *Rhizobium leguminosarum* biovar *phaseoli*," *Mol. Microbiol. 4*:933–941 (1990).

38. de Maagd, R.A., A.S. Rao, I.H.M. Mulders, L. Goosen-de Roo, M.C.M. Van Loosdrecht, C.A. Wijffelman, and B.J.J. Lugtenberg, "Isolation and characterization of mutants of *Rhizobium leguminosarum* biovar *viciae* strain 248 with altered lipopolysaccharides: Possible role of surface charge or hydrophobicity in bacterial release from the infection thread," *J. Bacteriol. 171*:1143–1150 (1989).

39. de Maagd, R.A., A.H.M. Wijfjes, H.P. Spaink, J.E. Ruiz-Sainz, C.A. Wijffelman, R.J.H. Okker, and B.J.J. Lugtenberg, "*nodO*, a new nod gene of the *Rhizobium leguminosarum* biovar *viciae* Sym plasmid pRL1JI, encodes a secreted protein," *J. Bacteriol. 171*:6764–6770 (1989).

40. de Philip, P., J. Batut, and P. Boistard, "*Rhizobium meliloti* FixL is an oxygen sensor and regulates *R. meliloti nifA* and *fixK* genes differently in *Escherichia coli*," *J. Bacteriol. 172*:4255–4262 (1990).

41. Debelle, F., C. Rosenberg, J. Vasse, F. Maillet, E. Martinez, J. Denarie, and G. Truchet, "Assignment of symbiotic developmental phenotypes to common and specific nodulation (*nod*) genetic loci of *Rhizobium meliloti*," *J. Bacteriol. 168*:1075–1086 (1986).

42. Debruijn, F.J., S. Rossbach, M. Schneider, P. Ratet, S. Messmer, W.W. Szeto, F.M. Ausubel, and J. Schell, "*Rhizobium meliloti* 1021 has three differentially regulated loci involved in glutamine biosynthesis, none of which is essential for symbiotic nitrogen fixation," *J. Bacteriol. 171*:1673–1682 (1989).

43. Deshmane, N., and G. Stacey, "Identification of *Bradyrhizobium nod* genes involved in host-specific nodulation," *J. Bacteriol. 171*:3324–3330 (1989).

44. Ditta, G., E. Virts, A. Palomares, and K. Choong-Hyun, "The *nifA* gene of *Rhizobium meliloti* is oxygen regulated," *J. Bacteriol. 169*:3217–3223 (1987).

45. Djordjevic, M.A., D.W. Gabriel, and B.G. Rolfe, "Rhizobium—the refined parasite of legumes," *Annu. Rev. Phytopathol. 25*:145–168 (1987).

46. Djordjevic, M.A., R.W. Innes, C.A. Wijffelman, P.R. Schofield, and B.G. Rolfe, "Nodulation of specific legumes is controlled by several distinct loci in *Rhizobium trifolii*," *Plant Mol. Biol. 6*:389–394 (1986).

47. Djordjevic, M.A., J.W. Redmond, M. Batley, and B.G. Rolfe, "Clovers secrete specific phenolic compounds which either stimulate or repress *nod* gene expression in *Rhizobium trifolii*," *EMBO J. 6*:1173–1179 (1987).

48. Djordjevic, M.A., P.R. Schofield, and B.G. Rolfe, "Tn5 mutagenesis of *Rhizobium trifolii* host-specific genes result in mutants with altered host range ability," *Mol. Gen. Genet. 200*:463–471 (1985).

49. Downie, J.A., G. Hombrecher, Q.-S. Ma, C.D. Knight, B. Wells, and A.W. B. Johnston, "Cloned nodulation genes of *Rhizobium leguminosarum* determine host-range specificity," *Mol. Gen. Genet. 190*:359–365 (1983).

50. Downie, J.A., C.D. Knight, A.W.B. Johnston, and L. Rossen, "Identification of genes and gene products involved in the nodulation of peas by *Rhizobium leguminosarum*," *Mol. Gen. Genet. 198*:591–597 (1985).

51. Dreyfus, B., J.L. Garcia, and M. Gillis, "Characterization of *Azorhizobium caulino-dans* gen., nov., sp. nov., a stem-nodulating nitrogen-fixing bacterium isolated from *Sesbania rostrata*," *Int. J. Syst. Bacteriol. 38*:89–98 (1988).

52. Dudley, M.E., T.W. Jacobs, and S.R. Long, "Microscopic studies of cell divisions induced in alfalfa roots by *Rhizobium meliloti*," *Planta 171*:289–301 (1987).

53. Earl, C.D., C.W. Ronson, and F.M. Ausubel, "Genetic and structural analysis of the *Rhizobium meliloti fixA, fixB, fixC*, and *fixX* genes," *J. Bacteriol. 169*:1127–1136 (1987).

54. Ebeling, S., M. Hahn, H.M. Fischer, and H. Hennecke, "Identification of *nifE-*, *nifN-*, and *nifS*-like genes in *Bradyrhizobium japonicum*," *Mol. Gen. Genet. 207*:503–508 (1987).

55. Ebeling, S., J.D. Noti, and H. Hennecke, "Identification of a new *Bradyrhizobium japonicum* gene (*frxA*) encoding a ferredoxin-like protein," *J. Bacteriol. 170*:1999–2001 (1988).

56. Economou, A., W.D.O. Hamilton, A.W.B. Johnston, and J.A. Downie, "The *Rhizobium* nodulation gene *nodO* encodes a Ca2+-binding protein that is exported without N-terminal cleavage and is homologous to haemolysin and related proteins," *EMBO J. 9*:349–354 (1990).

57. Eisbrenner, G., and H.J. Evans, "Aspects of hydrogen metabolism in nitrogen-fixing legumes and other plant-microbe associations," *Ann. Rev. Plant. Physiol. 34*:105–136 (1983).

58. Ellis, W.R., G.E. Ham, and E.L. Schmidt, "Persistence and recovery of *Rhizobium japonicum* inoculum in a field soil," *Agron. J. 76*:573–576 (1984).

59. Englke, T., H. Jagadish, and A. Puhler, "Biochemical and genetic analysis of *Rhizobium meliloti dctA* mutants defective in C4-dicarboxylic acid transport," *J. Gen. Microbiol. 133*:3019–3029 (1987).

60. Evans, I.J., and J.A. Downie, "The *nodI* gene product of *Rhizobium leguminosarum* is closely related to ATP-binding bacterial transport proteins; nucleotide sequence analysis of the *nodI* and *nodJ* genes," *Gene 43*:95–101 (1986).

61. Faucher, C., S. Camut, J. Denarie, and G. Truchet, "The *nodH* and *nodQ* host range genes of *Rhizobium meliloti* behave as avirulence genes in *R. leguminosarum* bv. *viciae* and determine changes in the production of plant-specific extracellular signals," *Mol. Plant-Microbe Interact. 2*:291–300 (1989).

62. Faucher, C., F. Maillet, J. Vasse, C. Rosenberg, A.A.N. van Brussel, G. Truchet, and J. Denarie, "*Rhizobium meliloti* host range *nodH* gene determines production of an alfalfa-specific extracellular signal," *J. Bacteriol. 170*:5489–5499 (1988).

63. Filser, M.M.K., C. Moscatelli, A. Lambertz, E. Vincze, M. Guida, G. Salzano, and M. Iaccarino, "Characterization and cloning of two *Rhizobium leguminosarum* genes coding for glutamine synthetase activities," *J. Gen. Microbiol. 132*:2561–2569 (1986).

64. Finan, T.M., A.M. Hirsch, J.A. Leigh, E. Johansen, A. Kuldau, J.M. Wood, and D.C. Jordan, "Symbiotic properties of C4-dicarboxylic acid transport mutants of *Rhizobium leguminosarum*," *J. Bacteriol. 154*:1403–1413 (1983).

65. Fischer, H.-M., A. Alverez-Morales, and H. Hennecke, "The pleiotropic natures of symbiotic regulatory mutants: *Bradyrhizobium japonicum nifA* gene is involved in control of *nif* gene expression and formation of determinant symbiosis," *EMBO J. 5*:1165–1173 (1986).

66. Fischer, H.-M., T. Bruderer, and H. Hennecke, "Essential and non-essential domains in the *Bradyrhizobium japonicum* NifA protein: identification of indispensible cysteine

residues potentially involved in redox reactivity and/or metal binding," *Nucleic Acids Res. 16*:2207–2224 (1988).

67. Fischer, H.-M., and H. Hennecke, "Linkage map of the *Rhizobium japonicum nifH* and *nifDK* operons encoding the polypeptides of the nitrogenase enzyme complex," *Mol. Gen. Genet. 196*:537–540 (1984).

68. Fischer, H.-M., and H. Hennecke, "Direct response of *Bradyrhizobium japonicum nifA*-mediated gene regulation to cellular oxygen status," *Mol. Gen. Genet. 209*:621–626 (1987).

69. Fischer, H.-M., I. Kullik, B. Herzog, H. Knobel, S. Fritsche, T. Bruderer, and H. Hennecke, "Two functional homologs of the σ^{54} gene (*rpoN*, *ntrA*) in *Bradyrhizobium japonicum*. p. 533 In P.M. Gresshoff, L.E. Roth, G. Stacey, and W.E. Newton (eds.), *Nitrogen Fixation: Achievements and Objectives*. Chapman and Hall, New York, 1990.

70. Fisher, R.F., T.T. Egelhoff, J.T. Mulligan, and S.R. Long, "Specific binding of proteins from *Rhizobium meliloti* cell-free extracts containing NodD to DNA sequences upstream of inducible nodulation genes," *Genes Devel. 2*:282–293 (1988).

71. Fisher, R.F., and S.R. Long, "DNA footprint analysis of the transcriptional activator proteins NodD1 and NodD3 on inducible *nod* gene promoters," *J. Bacteriol. 171*:5492–5502 (1989).

72. Fisher, R.F., J.A. Swanson, J.T. Mulligan, and S.R. Long, "Extended region of nodulation genes in *Rhizobium meliloti* 1021. II. Nucleotide sequence, transcription start sites and protein products," *Genetics 117*:191–201 (1987).

73. Fisher, R.F., J.K. Tu, and S.R. Long, "Conserved nodulation genes in *R. meliloti* and *R. trifolii*," *Appl. Environ. Microbiol. 49*:1432–1435 (1985).

74. Fuhrmann, M., H.-M. Fischer, and H. Hennecke, "Mapping of *Rhizobium japonicum nifB-*, *fixBC-*, and *fixA*-like genes and identification of the *fixA* promoter," *Mol. Gen. Genet. 199*:315–322 (1985).

75. Fuhrmann, M., and H. Hennecke, "*Rhizobium japonicum* nitrogenase Fe protein gene (*nifH*)," *J. Bacteriol. 158*:1005–1011 (1984).

76. Gardiol, A., A. Arias, C. Cervenansky, and G. Martinez-Drets, "Succinate dehydrogenase mutant of *Rhizobium meliloti*," *J. Bacteriol. 151*:1621–1623 (1982).

77. Goethals, K., G. van den Eede, M. Van Montagu, and M. Holsters, "Identification and characterization of a functional *nodD* gene in *Azorhizobium caulinodans* ORS571," *J. Bacteriol. 172*:2658–2666 (1990).

78. Gosink, M.M., N.M. Franklin, and G.P. Roberts, "The product of the *Klebsiella pneumoniae nifX* gene is a negative regulator of the nitrogen fixation (*nif*) regulon," *J. Bacteriol. 172*:1441–1447 (1990).

79. Gottfert, M., P. Grob, and H. Hennecke, "Proposed regulatory pathway encoded by the *nodV* and *nodW* genes, determinants of host specificity in *Bradyrhizobium japonicum*," *Proc. Natl. Acad. Sci. (USA) 87*:2680–2684 (1990).

80 Gottfert, M., S. Hitz, and H. Hennecke, "Identification of *nodS* and *nodU*, two inducible genes inserted between the *Bradyrhizobium japonicum nodYABC* and *nodIJ* genes," *Mol. Plant-Microbe Interact. 3*:308–316 (1990).

81. Gottfert, M., B. Horvath, E. Kondorosi, P. Putnoky, F. Rodriquez-Quinones, and A. Kondorosi, "At least two *nodD* genes are necessary for efficient nodulation of alfalfa by *Rhizobium meliloti*," *J. Mol. Biol. 191*:411–420 (1986).

82. Gottfert, M., J.W. Lamb, R. Gasser, J. Semenza, and H. Hennecke, "Mutational analysis of the *Bradyrhizobium japonicum* common *nod* genes and further *nod* box-linked genomic DNA regions," *Mol. Gen. Genet. 215*:407–415 (1989).

83. Gottfert, M., J. Weber, and H. Hennecke, "Induction of a *nodA–lacZ* fusion in *Bradyrhizobium japonicum* by an isoflavone," *J. Plant Physiol. 132*:394–397 (1988).

84. Gotz, R., I.J. Evans, J.A. Downie, and A.W.B. Johnston, "Identification of the host-range DNA which allows *Rhizobium leguminosarum* strain TOM to nodulate cv. Afghanistan peas," *Mol. Gen. Genet. 201*:296–300 (1985).

85. Gronger, P., S.S. Manian, H. Reilander, M. O'Connell, U.B. Priefer, and A. Puhler, "Organization and partial sequence of a DNA region of the *Rhizobium leguminosarum* symbiotic plasmid pRL6JI containing the genes *fixABC*, *nifA*, *nifB* and a novel open reading frame," *Nucleic Acids Res. 15*:31–49. (1987).

86. Gubler, M., and H. Hennecke, "*fixA, B*, and *C* genes are essential for symbiotic and free-living microaerobic nitrogen fixation," *FEBS Lett. 200*:186–192 (1986).

86a. Gubler, M., and H. Hennecke, "Regulation of the *fixA* gene and *fixBC* operon in *Bradyrhizobium japonicum*," *J. Bacteriol. 170*:1205–1214 (1988).

87. Gubler, M., T. Zurcher, and H. Hennecke, "The *Bradyrhizobium japonicum fixBCX* operon: Identification of *fixX* and of a 5′ mRNA region affecting the level of the *fixBCX* transcript," *Mol. Microbiol. 3*:141–148 (1989).

88. Gussin, G.N., C.W. Ronson, and F.M. Ausubel, "Regulation of nitrogen fixation genes," *Ann. Rev. Genet. 20*:567–591 (1986).

89. Gyorgypal, Z., N. Iyer, and A. Kondorosi, "Three regulatory *nodD* alleles of divergent flavonoid-specificity are involved in host-dependent nodulation by *Rhizobium meliloti*," *Mol. Gen. Genet. 212*:85–92 (1988).

90. Hahn, M., and H. Hennecke, "Cloning and mapping of a novel nodulation region from *Bradyrhizobium japonicum* by genetic complementation of a deletion mutant," *Appl. Environ. Microbiol. 54*:55–61 (1988).

91. Halverson, L. J., and G. Stacey, "Signal exchange in plant-microbe interactions," *Microbiol. Rev. 50*:193–225 (1986).

92. Ham, G.E., V.B. Caldwell, and H.W. Johnson, "Evaluation of *Rhizobium japonicum* inoculants in soils containing naturalized populations of rhizobia," *Agron. J. 63*:301–303 (1971).

93. Hartwig, U.A., C.A. Maxwell, C.M. Joseph, and D.A. Phillips, "Chrysoeriol and luteolin released from alfalfa seeds induce *nod* genes in *Rhizobium meliloti*," *Plant Physiol. 92*:116–122 (1990).

94. Haugland, R.A., M.A. Cantrell, J.S. Beaty, F.J. Hanus, S.A. Russell, and H.J. Evans, "Characterization of *Rhizobium japonicum* hydrogen uptake genes," *J. Bacteriol. 159*:1006–1012 (1984).

95. Helmann, J.D., Y. Wang, I. Mahler, and C.T. Walsh, "Homologous metalloregulatory proteins from both Gram-positive and Gram-negative bacteria control transcription of mercury resistance operons," *J. Bacteriol. 171*:222–229 (1989).

96. Henikoff, S., G.W. Haughn, J.M. Calvo, and J.C. Wallace, "A large family of bacterial activator proteins," *Proc. Natl. Acad. Sci. (USA) 85*:6602–6606 (1989).

97. Hennecke, H., "Recombinant plasmids carrying nitrogen fixation genes from *Rhizobium japonicum*," *Nature 291*:354–355 (1981).

98. Hennecke, H., "A genetic approach to analyze the critical role of oxygen in bacteroid metabolism. pp. 293–300. In P.M. Gresshoff, L.E. Roth, G. Stacey, and W.E. New-

ton (eds.), *Nitrogen Fixation: Achievements and Objectives*, Chapman and Hall, New York, 1990.

99. Hennecke, H., H.-M. Fischer, S. Ebeling, M. Gubler, B. Thony, M. Gottfert, J. Lamb, M. Hahn, T. Ramseier, B. Regensburger, A. Alvarez-Morales, and D. Studer, "*nif*, *fix*, and *nod* gene clusters in *Bradyrhizobium japonicum* and *nifA*-mediated control of symbiotic nitrogen fixation," pp. 191–196. In D. P. S. Verma, and N. Brisson (eds.), *Molecular Genetics of Plant–Microbe Interactions*, Martinus Nijhoff, Dordrecht, 1987.

100. Hennecke, H., H.-M. Fischer, M. Gubler, B. Thony, D. Anthamatten, I. Kullik, S. Ebeling, S. Fritsche, and T. Zurcher, "Regulation of *nif* and *fix* genes in *Bradyrhizobium japonicum* occurs by a cascade of two consecutive steps of which the second one is oxygen sensitive," pp. 339–344. In H. Bothe, F.J. de Bruijn, and W.E. Newton (eds.), *Nitrogen fixation: Hundred Years After*. Gustav Fischer, Stuttgart.

101. Heron, D.S., T. Ersek, H.B. Krishnan, and S.G. Pueppke, "Nodulation mutants of *Rhizobium fredii* USDA257." *Mol. Plant-Microbe Interact.* 2:4–10 (1989).

102. Heron, D.S., and S.G. Pueppke, "Mode of infection, nodulation specificity, and indigenous plasmids of 11 fast-growing *Rhizobium japonicum* strains," *J. Bacteriol.* 160:1061–1066 (1984).

103. Heron, D.S., and S.G. Pueppke, "Regulation of nodulation in the soybean–*Rhizobium* symbiosis. Strain and cultivar variability," *Plant Physiol.* 84:1391–1396 (1987).

104. Hirsch, A.M., M. Bang, and F.M. Ausubel, "Ultrastructural analysis of ineffective alfalfa nodules formed by *nif*: : Tn5 mutants of *Rhizobium meliloti*," *J. Bacteriol.* 155:367–380 (1983).

105. Hirsch, A. M., D. Drake, T.W. Jacobs, and S. R. Long, "Nodules are induced on alfalfa roots by *Agrobacterium tumefaciens* and *Rhizobium trifolii* containing small segments of the *Rhizobium meliloti* nodulation region," *J. Bacteriol.* 161:223–230 (1985).

106. Hoch, G.E., H.N. Little, and R.H. Burris, "Hydrogen evolution from soybean nodules," *Nature* 179:430–431 (1957).

107. Holl, F.B., "Host plant control of the inheritance of dinitrogen fixation in the *Pisum–Rhizobium* symbiosis," *Euphytica* 24:767–770 (1975).

108. Hombrecher, G., R. Gotz, N.J. Dibb, J.A. Downie, A.W.B. Johnston, N.J. Brewin, "Cloning and mutagenesis of nodulation genes from *Rhizobium leguminosarum* TOM, a strain with extended host range," *Mol. Gen. Genet.* 194:293–298 (1984).

109. Hong, G.-F., J.E. Burn, and A.W.B. Johnston, "Evidence that DNA involved in the expression of nodulation (*nod*) genes in *Rhizobium* binds to the product of the regulatory gene *nodD*," *Nucleic Acids Res.* 15:9677–9698 (1987).

110. Honma, M.A., M. Asomaning, and F.M. Ausubel, "*Rhizobium meliloti nodD* genes mediate host-specific activation of *nodABC*," *J. Bacteriol.* 172:901–911 (1990).

111. Honma, M.A., and F.M. Ausubel, "*Rhizobium meliloti* has three functional copies of the *nodD* symbiotic regulatory gene," *Proc. Natl. Acad. Sci. (USA)* 84:8558–8562 (1987).

112. Horvath, B., E. Kondorosi, M. John, J. Schmidt, I. Torok, Z. Gyorgypal, I. Barabas, U. Wieneke, J. Schell, and A. Kondorosi, "Organization, structure, and symbiotic function of *Rhizobium meliloti* nodulation genes determining host specificity for alfalfa," *Cell* 46:335–343 (1986).

113. Jacobs, T.W., T.T. Egelhoff, and S.R. Long, "Physical and genetic map of a *Rhizobium meliloti* nodulation gene region and nucleotide sequence of *nodC*," *J. Bacteriol.* 162:469–476 (1985).

114. Jiang, J., B.H. Gu, L.M. Albright, and B.T. Nixon, "Conservation between coding and regulatory elements of *Rhizobium meliloti* and *Rhizobium leguminosarum dct* genes," *J. Bacteriol. 171*:5244–5253 (1989).

115. Johnston, A.W.B., J.W. Latchford, E.O. Davis, A. Economou, and J.A. Downie, "Five genes on the sym plasmid of *Rhizobium leguminosarum* whose products are located in the bacterial membrane or periplasm,"pp. 311–318. In B.J.J. Lugtenberg (ed.), *Signal Molecules in Plants and Plant–Microbe Interactions*, Springer-Verlag, Berlin, 1989.

116. Jordan, D.C., "Transfer of *Rhizobium japonicum* to *Bradyrhizobium*, slow-growing root nodule bacterium from leguminous plants," *Int. J. Syst. Bacteriol. 32*:136–139 (1982).

117. Kaluza, K., M. Fuhrmann, M. Hahn, B. Regensburger, and H. Hennecke, "In *Rhizobium japonicum* the nitrogenase genes *nifH* and *nifDK* are separated," *J. Bacteriol. 155*:915–918 (1983).

118. Keen, N.T., and B. Staskawicz, "Host range determinants in plant pathogens and symbionts," *Ann. Rev. Microbiol. 42*:421–440 (1988).

119. Keyser, H.H., and P.B. Cregan, "Nodulation and competition for nodulation of selected soybean genotypes among *Bradyrhizobium japonicum* serogroup 123 isolates," *Appl. Environ. Microbiol. 53*:2631–2635 (1987).

120. Kim, C.-H., D.R. Helinski, and G. Ditta, "Overlapping transcription of the *nifA* regulatory gene in *Rhizobium meliloti*," *Gene 50*:141–148 (1986).

121. Kondorosi, A., E. Kondorosi, C.E. Pankhurst, W.J. Broughton, and Z. Banfalvi, "Physical and genetic analysis of a symbiotic region of *Rhizobium meliloti:* Identification of nodulation genes," *Mol. Gen. Genet. 193*:445–452 (1984).

122. Kondorosi, E., J. Gyuris, J. Schmidt, M. John, E. Duda, B. Hoffman, J. Schell, and A. Kondorosi, "Positive and negative control of *nod* gene expression in *Rhizobium meliloti* is required for optimal nodulation," *EMBO J. 8*:1331–1340 (1989).

123. Kosslak, R.M., R. Bookland, J. Barkei, H.E. Paaren, and E.R. Appelbaum, "Induction of *Bradyrhizobium japonicum* common *nod* genes by isoflavones isolated from *Glycine max.*," *Proc. Natl. Acad. Sci. (USA) 84*:7428–7432 (1987).

124. Kullik, I., H. Hennecke, and H.-M. Fischer, "Inhibition of *Bradyrhizobium japonicum nifA*-dependent *nif* gene activation by oxygen occurs at the NifA protein level and is irreversible," *Arch. Microbiol. 151*:191–197 (1989).

125. Kustu, S., K. Sei, and J. Keener, "Nitrogen regulation in enteric bacteria," pp. 139–154. In I.R. Booth and C.F. Higgins (eds.), *Regulation of Gene Expression—25 Years On*. Cambridge University Press, Cambridge.

126. Lamb, J.W., and H. Hennecke, "In *Bradyrhizobium japonicum* the common nodulation genes, *nodABC*, are linked to *nifA* and *fixA*," *Mol. Gen. Genet. 202*:512–517 (1986).

127. Law, I.J., Y. Yamamoto, A.J. Mort, and W.D. Bauer, "Nodulation of soybean by *Rhizobium japonicum* mutants with altered capsule synthesis," *Planta 154*:100–109 (1982).

128. Leong, S.A., G.S. Ditta, and D.R. Helinski, "Heme biosynthesis in *Rhizobium*. Identification of a cloned gene coding for delta-aminolevulinic acid synthetase from *Rhizobium meliloti*," *J. Biol. Chem. 257*:8724–8730 (1982).

129. Lerouge, P., P. Roche, C. Faucher, F. Maillet, G. Truchet, J.C. Prome, and J. Denarie, "Symbiotic host-specificity of *Rhizobium meliloti* is determined by a sulphated and acylated glucosamine oligosaccharide signal," *Nature 344*:781–784 (1990).

130. Lie, T.A., "Symbiotic specialization in pea plants: The requirement of specific *Rhizobium* strains for peas from Afghanistan," *Ann. Appl. Biol. 88*:462–465 (1978).

131. Long, S., J.W. Reed, J. Himawan, and G.C. Walker, "Genetic analysis of a cluster of genes required for synthesis of the calcofluor-binding exopolysaccharide of *Rhizobium meliloti*," *J. Bacteriol. 170*:4239–4248 (1988).

132. Long, S.R., "*Rhizobium*–legume nodulation: Life together in the underground," *Cell 56*:203–214 (1989).

133. Long, S.R., "*Rhizobium* genetics," *Ann. Rev. Genet. 23*:483–506 (1989).

134. Long, S.R., W. Buikema, and F.M. Ausubel, "Cloning of *Rhizobium meliloti* nodulation genes by direct complementation of Nod-mutants," *Nature 298*:485–488 (1982).

135. Long, S.R., J. Schwedock, T. Egelhoff, M. Yelton, J. Mulligan, M. Barnett, B. Rushing, and R. Fisher, "Nodulation genes and their regulation in *Rhizobium meliloti*," pp. 145–151. In B.J.J. Lugtenberg (ed.), *Signal Molecules in Plants and Plant–Microbe Interactions*, Springer-Verlag, Berlin, 1989.

136. Luka, S., and G. Stacey, "Characterization of a new host-inducible genetic locus in *Bradyrhizobium japonicum*," p. 557. In P.M. Gresshoff, L.E. Roth, G. Stacey, and W.E. Newton (eds.), *Nitrogen Fixation: Achievements and Objectives*, Chapman and Hall, New York, 1990.

137. Magasanik, B., "Genetic control of nitrogen assimilation in bacteria," *Ann. Rev. Genet. 16*:135–168 (1982).

138. Magasanik, B., "Reversible phosphorylation of an enhancer binding protein regulates the transcription of bacterial nitrogen utilization genes," *Trends Biochem. Sci. 13*:475–479 (1988).

139. Maier, R.J., "Biochemistry, regulation, and genetics of hydrogen oxidation in *Rhizobium*," *CRC Crit. Rev. Biotechnol. 3*:17–38 (1986).

140. Maier, R.J., and W.J. Brill, "Involvement of *Rhizobium japonicum* O-antigen in soybean nodulation," *J. Bacteriol. 133*:1295–1299 (1978).

141. Maier, R.J., N.E.R. Campbell, F.J. Hanus, F.B. Simpson, S.A. Russell, and H.J. Evans, "Expression of hydrogenase activity in free-living *Rhizobium japonicum*," *Proc. Natl. Acad. Sci. (USA) 75*:3258–3262 (1978).

142. Maier, R.J., N.E.R. Graham, R.G. Keefe, T. Pihl, and E. Smith, "*Bradyrhizobium japonicum* mutants defective in nitrogen fixation and molybdenum metabolism," *J. Bacteriol. 169*:2548–2554 (1987).

143. Martin, G.B., K.A. Chapman, and B.K. Chelm, "Role of the *Bradyrhizobium japonicum ntrC* gene product in differential regulation of the glutamine synthetase II gene (*glnII*)," *J. Bacteriol. 170*:5452–5459 (1988).

144. Martin, G.B., M.F. Thomashow, and B.K. Chelm, *Bradyrhizobium japonicum glnB*, a putative nitrogen-regulatory gene, is regulated by Ntrc at TANDEM promoters," *J. Bacteriol. 171*:5638–5645 (1989).

145. Marvel, D.J., G. Kuldau, A. Hirsch, E. Richards, J.G. Torrey, and F.M. Ausubel, "Conservation of nodulation genes between *Rhizobium meliloti* and a slow-growing *Rhizobium* strain that nodulates a nonlegume host," *Proc. Natl. Acad. Sci. (USA) 82*:5841–5845 (1985).

146. Marvel, D.J., J.G. Torrey, and F.M. Ausubel, *Rhizobium* symbiotic genes required for nodulation of legume and nonlegume hosts," *Proc. Natl. Acad. Sci. (USA) 84*:1319–1323 (1987).

147. Masterson, R.V., P.R. Russell, and A.G. Atherly, "Nitrogen fixation (*nif*) genes and large plasmids of *Rhizobium japonicum,*" *J. Bacteriol. 152*: 928–931 (1982).

148. Maxwell, C.A., U.A. Hartwig, C.M. Joseph, and D.A. Phillips, "A chalcone and two related flavonoids released from alfalfa roots induce *nod* genes of *Rhizobium meliloti,*" *Plant Physiol. 91*:842–847 (1989).

149. McAllister, C.F., and J.E. Lepo, "Succinate transport by free-livinig forms of *Rhizobium japonicum,*" *J. Bacteriol. 153*:1155–1162 (1983).

150. McIver, J., M.A. Djordjevic, J.J. Weinman, G.L. Bender, and B.G. Rolfe, "Extension of host range of *Rhizobium leguminosarum* biovar *trifolii* caused by point mutations in *nodD* which result in alterations in regulatory function and recognition of inducer molecules," *Mol. Plant-Microbe Interact. 2*:97–106.

151. Merberg, D., E.B. O'Hara, and R.J. Maier, "Regulation of hydrogenase in *Rhizobium japonicum*: Analysis of mutants altered in regulation by carbon substrates and oxygen," *J. Bacteriol. 156*:1236–1242 (1983).

152. Moawad, H.A., W.R. Ellis, and E.L. Schmidt, "Rhizosphere response as a factor in competition among three serogroups of indigenous *Rhizobium japonicum* for nodulation of field grown soybeans," *Appl. Environ. Microbiol. 47*:607–612 (1984).

153. Mulligan, J.T., and S.R. Long, "A family of activator genes regulates expression of *Rhizobium meliloti* nodulation genes," *Genetics 122*:7–18 (1989).

154. Nieuwkoop, A.J., Z. Banfalvi, N. Deshmane, D. Gerhold, M.G. Schell, K.M. Sirotkin, and G. Stacey, "A locus encoding host range is linked to the common nodulation genes of *Bradyrhizobium japonicum,*" *J. Bacteriol. 169*:2631–2638 (1987).

155. Noel, K.D., G. Stacey, S. Tandon, L.E. Silver, and W.J. Brill, "Mutants of *Rhizobium japonicum* defective in symbiotic nitrogen fixation," *J. Bacteriol. 152*:485–494 (1982).

156. Noel, K.D., K.A. VandenBosch, and B. Kulpaca, "Mutations in *Rhizobium phaseoli* that lead to arrested development of infection threads," *J. Bacteriol. 168*:1392–1401 (1986).

157. Norel, F., N. Desnoues, and C. Elmerich, "Characterization of DNA sequences homologous to *Klebsiella pneumoniae nifH, D, K,* and *E* in the tropical *Rhizobium* ORS571," *Mol. Gen. Genet. 199*:352–356 (1985).

158. Noti, J.D., B. Dudas, and A.A. Szalay, "Isolation and characterization of nodulation genes from *Bradyrhizobium* sp. (Vigna) strain IRc78," *Proc. Natl. Acad. Sci. (USA) 82*:7379–7383 (1985).

159. O'Brian, M.R., P.M. Kirschbom, and R.J. Maier, "Bacterial heme synthesis is required for expression of the leghemoglobin holoprotein but not the apoprotein in soybean root nodules," *Proc. Natl. Acad. Sci. (USA) 84*:8390–8393 (1987).

160. O'Brian, M.R., P.M. Kirschbom, and R.J. Maier, "Tn5-induced cytochrome mutants of *Bradyrhizobium japonicum*: Effects of the mutations on cells grown symbiotically and in culture," *J. Bacteriol. 169*:1089–1094 (1987).

161. O'Brian, M.R., and R.J. Maier, "Isolation of cytochrome aa₃ from *Bradyrhizobium japonicum,*" *Proc. Natl. Acad. Sci. (USA) 84*:3219–3223 (1987).

162. Olson, E.R., M.J. Sadowsky, and D.P.S. Verma, "Identification of genes involved in *Rhizobium*–legume symbiosis by Mu-dI (kan,lac)-generated transcriptional fusions," *Biotechnology 3*:145–149 (1985).

163. Paustian, T.D., V.K. Shah, and G.P. Roberts, "Purification and characterization of the *nifN* and *nifE* gene products from *Azotobacter vinelandii* mutant UW-45," *Proc. Natl. Acad. Sci. (USA) 86*:6082–6086 (1989).

164. Peters, N.K., J.W. Frost, and S.R. Long, "A plant flavone, luteolin, induces expression of *Rhizobium meliloti* nodulation genes," *Science 233*:977–980 (1986).

165. Priefer, U.B., "Genes involved in lipopolysaccharide production and symbiosis are clustered on the chromosome of *Rhizobium leguminosarum* biovar *viciae* Vf39," *J. Bacteriol. 171*:6161–6168 (1989).

166. Puvanesarajah, V., F.M. Schell, D. Gerhold, and G. Stacey, "Cell surface polysaccharides from *Bradyrhizobium japonicum* and a nonnodulating mutant," *J. Bacteriol. 169*:137–141 (1987).

167. Quinto, C., H. de la Vega, M. Flores, L. Fernandez, T. Ballado, G. Soberon, and R. Palacios, "Reiteration of nitrogen fixation gene sequences in *Rhizobium phaseoli*," *Nature 299*:724–726 (1982).

168. Quinto, C., H. de la Vega, M. Flores, J. Leemans, M.A. Cevallos, M.A. Pardo, R. Azpiroz, M.L. Girard, E. Calva, and R. Palacios, "Nitrogenase reductase: A functional multigene family in *Rhizobium phaseoli*," *Proc. Natl. Acad. Sci. (USA) 82*:1170–1174 (1985).

169. Ramseier, T.M., B. Kaluza, D. Studer, T. Gloudemans, T. Bisseling, P.M. Jordan, R.M. Jones, M. Zuber, and H. Hennecke, "Cloning of a DNA region from *Bradyrhizobium japonicum* encoding pleiotropic functions in heme metabolism and respiration," *Arch. Microbiol. 151*:203–212 (1989).

170. Regensburger, B., L. Meyer, M. Filser, J. Weber, D. Studer, J.W. Lamb, H.M. Fischer, M. Hahn, and H. Hennecke "*Bradyrhizobium japonicum* mutants defective in root-nodule bacteroid development and nitrogen fixation," *Arch. Microbiol. 144*:355–366 (1986).

171. Reitzer, L.J., and B. Magasanik, "Ammonia assimilation and the biosynthesis of glutamine, glutamate, aspartate, asparagine, L-alanine, and D-alanine, pp. 302–320. In F.C. Neidhardt (ed.), *Escherichia coli* and *Salmonella typhimurium: Cellular and Molecular Biology*. American Society for Microbiology, Washington, D.C., 1987.

172. Roberts, G.P., and W.J. Brill, "Gene–product relationships of the *nif* regulon of *Klebsiella pneumoniae*," *J. Bacteriol. 144*:210–216 (1980).

173. Ronson, C.W., P.M. Astwood, and J.A. Downie, "Molecular cloning and genetic organization of C4-dicarboxylate transport genes from *Rhizobium leguminosarum*," *J. Bacteriol. 160*:903–909 (1984).

174. Ronson, C.W., P. Lyttleton, and J.G. Robertson, "C4-dicarboxylate transport mutants of *Rhizobium trifolii* form ineffective nodules on *Trifolium repens*," *Proc. Natl. Acad. Sci. (USA) 78*:4284–4288 (1981).

175. Rossen, L., A.W.B. Johnston, and J.A. Downie, "DNA sequence of the *Rhizobium leguminosarum* nodulation genes *nodAB* and *C* required for root hair curling," *Nucleic Acids Res. 12*:9497–9508 (1984).

176. Rossen, L., Q.S. Ma, E.A. Mudd, A.W.B. Johnston, and J.A. Downie, "Identification and DNA sequence of *fixZ*, a *nifB*-like gene from *Rhizobium leguminosarum*," *Nucleic Acids Res. 12*:7123–7134 (1984).

177. Rossen, L., C.A. Shearman, A.W.B. Johnston, and J.A. Downie, "The *nodD* gene of *Rhizobium leguminosarum* is autoregulatory and in the presence of plant exudate induces the *nodABC* genes," *EMBO J. 4*:3369–3373 (1985).

178. Rostas, K., E. Kondorosi, B. Horvath, A. Simoncsits, and A. Kondorosi, "Conservation of extended promoter regions of nodulation genes in rhizobia," *Proc. Natl. Acad. Sci. (USA) 83*:1757–1761 (1985).

179. Russell, P., M.G. Schell, K.K. Nelson, L.J. Halverson, K.M. Sirotkin, and G. Stacey, "Isolation and characterization of the DNA region encoding nodulation functions in *Bradyrhizobium japonicum," J. Bacteriol. 164*:1301–1308 (1985).

180. Ruvkin, G.B., and F.M. Ausubel, "Interspecies homology of nitrogenase genes," *Proc. Natl, Acad. Sci. (USA)* 77:191–194 (1980).

181. Sadowsky, M.J., P.B. Cregan, M. Gottfert, A. Sharma, D. Gerhold, F. Rodriguez-Quinones, H.H. Keyser, H. Hennecke, and G. Stacey, "The *Bradyrhizobium japonicum nolA* gene and its involvement in the genotype-specific nodulation of soybeans," *Proc. Natl. Acad. Sci. (USA) 88*, 637–641 (1991).

182. Sadowsky, M.J., E.R. Olson, V.E. Foster, R.M. Kosslak, and D.P.S. Verma, "Two host-inducible genes of *Rhizobium fredii* and characterization of the inducing compound," *J. Bacteriol. 170*:171–178 (1988).

183. San Francisco, M.J.D., and G.R. Jacobson, "Uptake of succinate and malate in cultured cells and bacteroids of two slow-growing species of *Rhizobium," J. Gen. Microbiol. 131*:765–773 (1985).

184. Sanjuan, J., and J. Olivares, "Implication of *nifA* in regulation of genes located on a *Rhizobium meliloti* cryptic plasmid that affect nodulation efficiency," *J. Bacteriol. 171*:4154–4161 (1989).

185. Sayavedra-Soto, L.A., G.K. Powell, H.J. Evans, and R.O. Morris, "Nucleotide sequence of the genetic loci encoding subunits of *Bradyrhizobium japonicum* uptake hydrogenase," *Proc. Natl. Acad. Sci. (USA) 85*:8395–8399 (1988).

186. Schetgens, R.M.P., J.G.J. Hontelez, R.C. van den Bos, and A. Van Kammen, "Identification and phenotypical characterization of a cluster of *fix* genes, including a *nif* regulatory gene, from *Rhizobium leguminosarum* PRE," *Mol. Gen. Genet. 200*:368–3740 (1985).

187. Schwedock, J., and S.R. Long, "Nucleotide sequence and protein products of two new nodulation genes of *Rhizobium meliloti, nodP* and *nodO," Mol. Plant-Microbe Interact. 2*:181–194 (1989).

188. Scott, K.F., "Conserved nodulation genes from the non-legume symbiont *Bradyrhizobium* sp. (Parasponia)," *Nucl. Acids Res. 14*:2905–2910 (1986).

189. Scott, K.F., and G.L. Bender, "The *Parasponia–Bradyrhizobium* symbiosis, pp. 231–251. In P.M. Gresshoff (ed.), *Molecular Biology of Symbiotic Nitrogen Fixation*, CRC Press, Boca Raton, FL, 1990.

190. Scott, K.F., B.G. Rolfe, and J. Shine, "Biological nitrogen fixation: Primary structure of the *Rhizobium trifolii* iron protein gene," *DNA 2*:149–155 (1983).

191. Scott, K.F., B.G. Rolfe, and J. Shine, "Nitrogenase structural genes are unlinked in the nonlegume symbiont *Parasponia* rhizobium," *DNA 2*:149–148 (1983).

192. Shah, V.K., J. Imperial, R.A. Ugalde, P.W. Ludden, and W.J. Brill, "In vitro synthesis of the iron–molybdenum cofactor of nitrogenase," *Proc. Natl. Acad. Sci. (USA) 83*:1636–1640 (1986).

193. Shearman, C.A., L. Rossen, A.W.B. Johnston, and J.A. Downie, "The *Rhizobium leguminosarum* nodulation gene *nodF* encodes a polypeptide similar to acyl-carrier protein and is regulated by *nodD* plus a factor in pea root exudate," *EMBO J. 5*:647–652 (1986).

194. Shen, S.C., S.P. Wang, G.Q. Yu, and J.B. Zhu, "Expression of the nodulation and nitrogen fixation genes in *Rhizobium meliloti," Genome 31*:354–360 (1989).

195. Smit. G., V. Puvanesarajah, R.W. Carlson, and G. Stacey, "*Bradyrhizobium japonicum nodD* can specifically be induced by soybean seed extract compounds which do not induce the *nodYABC* genes." p. 274. In P. Gresshoff, E. Roth, G. Stacey, and W.E. Newton (eds.), *Nitrogen Fixation: Achievements and Objectives*, Chapman and Hall, New York, 1990.

196. So, J.-S. A.L.M. Hodgson, R. Haugland, M. Leavitt, Z. Banfalvi, A.J. Nieuwkoop, and G. Stacey, "Transposon-induced symbiotic mutants of *Bradyrhizobium japonicum:* Isolation of two gene regions essential for nodulation," *Mol. Gen. Genet. 207*:15–23 (1987).

197. Spaink, H.P., R.J.H. Okker, C.A. Wijffelman, E. Pees, and B.J.J. Lugtenberg "Promoters in the nodulation region of the *Rhizobium leguminosarum* Sym plasmid pRL1JI," *Plant Mol. Biol. 9*:27–39 (1987).

198. Spaink, H.P., J. Weinman, M.A. Djordjevic, C.A. Wijffelman, R.J.H. Okker, and B.J.J. Lugtenberg, "Genetic analysis and cellular localization of the *Rhizobium* host specificity-determining NodE protein," *EMBO J. 8*:2811–2818 (1989).

199. Spaink, H.P., C.A. Wijffelman, R.J.H. Okker, and B.J.J. Lugtenberg, "Localization of functional regions of the *Rhizobium nodD* product using hybrid *nodD* genes," *Plant Mol. Biol. 12*:59–73 (1989).

200. Spaink, H.P., C.A. Wijffleman, W. Pees, R.J.H. Okker, and B.J.J. Lugtenberg, "*Rhizobium* nodulation gene *nodD* as a determinant of host specificity," *Nature 328*:337–340 (1987).

201. Stacey, G., "Compilation of the *nod, fix,* and *nif* genes of rhizobia and information concerning their function." p. 239–244. In P. Gresshoff, E. Roth, G. Stacey, and W.E. Newton (eds.), *Nitrogen Fixation: Achievements and Objectives*, Chapman and Hall, New York, 1990.

202. Stacey, G., L.A. Pocratsky, and V. Puvanesarajah, "Bacteriophage that can distinguish between wild-type *Rhizobium japonicum* and a non-nodulating mutant," *Appl. Env. Microbiol. 48*:68–72 (1984).

203. Stacey, G., M.G. Schell, and N. Deshmane, "Determinants of host specificity in the *Bradyrhizobium japonicum*–soybean symbiosis," pp. 394–399. In B.J.J. Lugtenberg (ed.), *Signal Molecules in Plants and Plant–Microbe Interactions*, Springer-Verlag, Berlin, 1989.

204. Stock, J.B., A.J. Ninfa, and A.M. Stock, "Protein phosphorylation and regulation of adaptive responses in bacteria," *Microbiol. Rev. 53*:450–490 (1989).

205. Stock, J.B., A.M. Stock, and J.M. Mottonen, "Signal transduction in bacteria," *Nature 344*:395–400 (1990).

206. Studer, D., T. Gloudemans, H.J. Franssen, H.-M. Fischer, T. Bisseling, and H. Hennecke, "In the *Bradyrhizobium japonicum*–soybean symbiosis the bacterial nitrogen fixation regulatory gene (*nifA*) is also involved in control of nodule-specific host plant gene expression, *Eur. J. Cell Biol. 45*:177–184 (1987).

207. Swanson, J.A., J.K. Tu, J. Ogawa, R. Sanga, R.F. Fisher, and S.R. Long, "Extended region of nodulation genes in *Rhizobium meliloti* 1021. I. Phenotypes of Tn*5* insertion mutants," *Genetics 117*:181–189 (1987).

208. Szeto, W.W., and F. Cannon, "An *ntrC* homologue in *B. japonicum*," pp. 150–254. In D.P.S. Verma and N. Brisson (eds.), *Molecular Genetics of Plant–Microbe Interaction*, Martinus Nijhoff, Dordrecht.

209. Thony, B., D. Anthamatten, and H. Hennecke, "Dual control of the *Bradyrhizobium japonicum* symbiotic nitrogen fixation regulatory operon *fixR nifA:* Analysis of cis- and trans- acting elements, *J. Bacteriol. 171*:4162–4169 (1989).

210. Thony, B., H.-M. Fischer, D. Anthamatten, T. Bruderer, and H. Hennecke, "The symbiotic nitrogen fixation regulatory operon (*fixRnifA*) of *Bradyrhizobium japonicum* is expressed aerobically and is subject to a novel, *nifA*-independent type of activation," *Nucl. Acids Res. 15*:8479–8499 (1987).

211. Thony, B., and H. Hennecke, "The $-24/-12$ Promoter Comes of Age," *FEMS Microbiol. Rev. 63*:341–357 (1989).

212. Thony-Meyer, L., D. Stax, and H. Hennecke, "An unusual gene cluster for the cytochrome bc1 complex in *Bradyrhizobium japonicum* and its requirement for effective root nodule symbiosis," *Cell 57*:683–697 (1989).

213. Torok, I., and A. Kondorosi, "Nucleotide sequence of the *R. meliloti* nitrogenase reductase (*nifH*) gene," *Nucleic Acids Res. 9*:5117–5123 (1981).

214. Torok, I., E. Kondorosi, T. Stepkowski, J. Postfai, and A. Kondorosi, "Nucleotide sequence of *Rhizobium meliloti* nodulation genes," *Nucleic Acids Res. 12*:9509–9524 (1984).

215. Trinick, M.J., "Symbiosis between *Rhizobium* and the nonlegume, *Trema aspera,*" *Nature 244*:459–461 (1973).

216. Upchurch, R.G., and G.H. Elkan, "Ammonia assimilation in *Rhizobium japonicum* colonial derivatives differing in nitrogen-fixing efficiency," *J. Gen. Microbiol. 104*:219–225 (1978).

217. Vairinhos, F., B. Bhandari, and D.J.D. Nicholas, "Glutamine synthetase, glutamate synthase and glutamate dehydrogenase in *Rhizobium japonicum* strains grown in cultures and in bacteroids from nodules of *Glycine max,*" *Planta 159*:207–215 (1983).

218. Virts, E.L., S.W. Stanfield, D.R. Helinski, and G.S. Ditta, "Common regulatory elements control symbiotic and microaerobic induction of *nifA* in *Rhozobium meliloti,*" *Proc. Natl. Acad. Sci (USA) 85*:3062–3065 (1988).

219. Wang, S.-P., and G. Stacey, "A divergent *nod* box sequence is essential for *nodD*1 induction in *B. japonicum.*" p. 600. In P.M. Gresshoff, E. Roth, G. Stacey, and W.E. Newton (eds.), *Nitrogen Fixation: Achievements and Objectives,* Chapman and Hall, New York, 1990.

220. Waters, J.K., A.A. Preston, R.T. Liang, and D.W. Emerich, "Malate dehydrogenase-deficient mutants of *Rhizobium japonicum,*" p. 220. In H.J. Evans, P.J. Bottomley, and W.E. Newton (eds.), *Nitrogen Fixation Research Progress,* Martinus Nijhoff, Dordrecht, 1985.

221. Watson, R.J., Y.-K. Chan, R. Wheatcroft, A.-F. Yang, and S. Han, "*Rhizobium meliloti* genes required for C4-dicarboxylate transport and symbiotic nitrogen fixation are located on a megaplasmid," *J. Bacteriol. 170*:927–934 (1988).

222. Yun, A.C., and A.A. Szalay, "Structural genes of dinitrogenase and dinitrogenase reductase are transcribed from two separate promoters in the broad host range cowpea *Rhizobium* strain IRc78," *Proc. Natl. Acad. Sci. (USA) 81*:7358–7362 (1984).

223. Zaat, S.A.J., A.A.N. van Brussel, T. Tak, E. Pees, and B.J.J. Lugtenberg, "Flavonoids induce *Rhizobium leguminosarum* to produce *nodDABC* gene-related factors that cause thick, short roots and root hair responses on common vetch," *J. Bacteriol. 169*:3388–3391 (1987).

224. Zaat, S.A.J., C.A. Wijffelman, I.M. Mulders, A.A.N. van Brussel, and B.J.J. Lugtenberg, "Root exudates of various host plants of *Rhizobium leguminosarum* contain different sets of inducers of *Rhizobium* nodulation genes," *Plant Physiol.* 86:1298–1303 (1988).

225. Zaat, S.A.J., C.A. Wijffelman, H.P. Spaink, A.A.N. van Brussel, R.J.H. Okker, and B.J.J. Lugtenberg, "Induction of the *nodA* promoter of *Rhizobium leguminosarum* Sym plasmid pRL1JI by plant flavanones and flavones," *J. Bacteriol.* 169:198–204 (1987).

18

The Enzymology of Molybdenum-Dependent Nitrogen Fixation

M. G. Yates

1. Introduction

This chapter deals exclusively with the structure, function, and mechanism of molybdenum nitrogenases, topics that have received exhaustive attention over the last 30 years, since the first report of cell-free extracts from *Clostridium pasteurianum* capable of fixing N_2 appeared.[16] Mo nitrogenases, without exception, consist of two proteins that, in the presence of MgATP, an electron source, and anaerobic conditions, combine to reduce N_2 to ammonia with the concomitant production of H_2.[142]

$$N_2 + 8H^+ + 8e^- \rightarrow 2NH_3 + H_2 \qquad (1)$$

In the absence of an added reducible substrate nitrogenase reduces protons to H_2.

These proteins are both oxygen-sensitive, iron-sulfur proteins, the largest of which contains molybdenum in an iron-molybdenum cofactor, called FeMoco, which can be isolated in an active form. It is widely accepted that this cofactor is the site of N_2 binding and reduction and that electrons are transferred from the smaller, Fe protein to the MoFe protein in an ATP-dependent reaction. This is the simplest description of nitrogenase activity. This chapter describes present knowledge of the structure and function of these two proteins, the cofactor, FeMoco and the mechanism of N_2 reduction. Individual proteins are referred to by the initials of their parent organism and the suffix 1 or 2; for example, Kp1 and Kp2 are, respectively, the MoFe and Fe proteins of *Klebsiella pneumoniae* nitrogenase. The last major reviews of these topics were by Thorneley and Lowe,[161] Lowe et al.,[101] Orme-Johnson,[122] and Eady.[24] This review concentrates mainly on the literature produced after those reviews.

II. The Fe Proteins

The Fe proteins are dimers (α_2) of identical subunits with total molecular weights ranging from 58,000 to 72,000. Estimated specific activities of the purified proteins, with sodium dithionite as the electron donor and an optimum level of MoFe protein, range from less than 1000 to a maximum of 2500 nmoles of H_2 mg protein^{-1} min^{-1} for purified Bp2, the Fe protein of *Bacillus polymyxa*[34] and 2692 nmoles C_2H_2 mg protein^{-1} min^{-1} for purified Av2.[120] These contrast with *in vivo* estimations for Av2 of 6250 nmoles C_2H_2 mg protein^{-1} min^{-1} [84] and for *Rhodopseudomonas capsulata*, Rc2 of 5000 nmoles C_2H_2 mg protein^{-1} min^{-1}.[78] These differences may reflect the different rate of electron transfer from the biological and chemical electron donors to the Fe protein, but they may also be due to underestimation of the Fe protein concentration *in vivo* or, alternatively, to inactive protein in the *in vitro* preparations. Proteins from different sources have been analyzed by amino acid or DNA sequencing and show a high degree of homology among species, whether derived from aerobes, anaerobes, facultative anaerobes, or cyanobacteria.[101] Lowe et al.[101] pointed out that the homologies between the proteins were far higher than between their respective genes: as much as 70% compared with 30% in some cases, indicating that although DNA sequences may have evolved considerably, the protein sequences had not. They concluded that the enzymic activities severely constrained the primary structures of these proteins. However, despite their similarities, crucial differences exist between these Fe proteins that prevent some enzymically active, interspecies cross-reactions with MoFe proteins; for example, Cp2, the Fe protein from *Clostridium pasteurianum*, forms an enzymically inactive complex with Av1, the MoFe protein from *Azotobacter vinelandii*.[35]

A. Metal Content

All purified Fe proteins are considered to contain one [4Fe–4S] cluster per dimer. This conclusion is based on the results of four analytical approaches made in the 1970s:

1. *Mossbauer spectroscopy:* Smith and Lang[145] showed that the Mossbauer spectrum of reduced Kp2 was similar in form and temperature behavior to those of 4Fe ferredoxins.

2. *Cluster displacement techniques:* the Fe–S cluster from Cp2 was displaced by hexamethylphosphoramide–thiophenol treatment and shown to be similar by EPR spectroscopy to Fe–S clusters extracted from 4Fe ferredoxins.[124]

3. *Linear electric field effects:* these measured the shift in g values of EPR centers subjected to strong electric fields, showing that Cp2 contained a [4Fe–4S] center.[124]

4. *Measurement of magnetic circular dichroism:* the results for reduced Av2, the *A. vinelandii* Fe protein, and Kp2 were found to be quantitatively comparable to the result for a [4Fe–4S] ferredoxin.[150]

On the assumption that the Fe–S cluster is crucial to the Fe proteins role, all this evidence posed the obvious question: How is a single [4Fe–4S] cluster bound between two identical subunits? Earlier speculations that the cluster was sited between two mirror image faces of the subunits was substantiated by the work of Hausinger and Howard.[58] The cysteinyl residues of Av2 were labeled by alkylation with [2–¹⁴C]iodoacetic acid and radioactive peaks were counted in a peptide map based on the known amino acid sequence.[57] They showed that cysteinyl residues 97 and 132 were more heavily labeled than the other five cysteines and that the presence of MgATP or MgADP plus α,α'-dipyridyl, which removed the Fe, increased this label to more than 0.8 mol of the residue per mol of subunit. They proposed that these residues bound the [4Fe–4S] cluster in a symmetrical arrangement between the two subunits. Recently, Howard et al.[71] changed the conserved cysteine residues 38, 85, 97, 132, and 184 of Av2 to serine by site-directed mutagenesis and showed that replacement of either cys-97 or cys-132 resulted in total loss of activity, whereas replacement of any one of the other three cysteine residues left approximately 10% activity. These findings gave rise to the conclusion that cysteine residues 97 and 132 bound the [4Fe–4S] cluster and replacement of either meant that a functional [4Fe–4S] cluster could not be incorporated into Av2 apoprotein.

Hausinger and Howard[58] also suggested that cys-85 may be involved in MgATP binding because MgATP partly protected this residue against carboxy methylation whereas MgADP did not. A third suggestion was that the cys residues 38, 85, and 151 were close to the Fe–S center, whereas residues 5 and 184 were distant and closer to the outer surfaces of the subunits.

B. EPR Data

Two observations about the Fe protein are not compatible with the preceding scheme. These are (1) the $S = 1/2$ spin EPR spectra, normally associated with dithionite-reduced Fe proteins, integrate to far less than an ideal value of one electron per mole and (2) dye-oxidized Fe proteins frequently reduce with a stoichiometry of two electrons per mole. Neither of these observations is consistent with a single [4Fe–4S] cluster operating between oxidations states 2^- and 3^-. Rather, they suggest that there are two redox centers in the protein. Lowe[96] obtained computer-simulated EPR spectra of both Kp2 and MgATP-bound Kp2 that indicated the presence of two interacting paramagnetic centers sufficiently close to quench EPR signals.

Braaksma et al.[9,10] and Haaker et al.[49] provided evidence to support this suggestion by (1) showing that the EPR spectra of Av2 integrated to one electron spin per mole in the presence of 50% ethylene glycol and to 1.38 electron spins per mole when MgADP was also included and (2) isolating Av2 with as high as 8.8 Fe per mole. Moreover, they observed a broad correlation between Fe content and specific activity between three and nine Fe atoms per mole Av2. This led to the conclusion that Av2 contained at least two interacting [4Fe–4S] clusters. However, later data[50] demonstrated that the high Fe content of Av2 was linked to oxygen shock treatment and that Fe protein, containing four Fe atoms per mole, was as active as protein with higher Fe. In addition, only four Fe atoms per mole were released by MgATP plus the chelator bathophenanthroline disulfonate, with either 8Fe or 4Fe proteins. Furthermore, Hagen et al[54] analyzed EPR spectra of "oxygen shock"–derived Av2 and found it identical to that of Fe protein from "unshocked" cells, with an integration of 0.2 electrons per mole. They disagreed with Lowe's[96] view of interacting spins, but claimed that the EPR signal was consistent with a [4Fe–4S] cluster under g strain. They concluded that Av2 had to have more than one [4Fe–4S] cluster.

This enigma was apparently resolved when three groups of investigators reported the presence of an EPR species with spin $S = 3/2$ in Fe proteins, which was EPR-silent at $g = 1.94$. Lindahl et al.[91] and Watt and Mc-Donald[173] observed a broad $g = 4$ or 5, $S = 3/2$ signal in dithionite-reduced Av2, which, together with the $S = 1/2$, $g = 1.94$ signal integrated maximally to 0.84[173] or 1.02[91] spins per mole. Lindahl et al. noted that 50% ethylene glycol treatment yielded predominantly the $g = 1.94$ form, whereas 0.4 M urea produced the $g = 5$ form. Hagen et al.[55] also obtained similar signals with purified Av2, Kp2, and Ac2 (*Azotobacter chroococcum* Fe protein) and showed that the ethylene glycol–nucleotide treatment of Av2 partly converted the $S = 3/2$ to the $S = 1/2$ spin state. Although it is difficult to quantify such broad signals (described tentatively by Watt and McDonald as a broad signal at $g = 4$ but more specifically by Hagen et al.[55] as a "complex signal" with peaks at g values of 5.9, 4.8, 4.3, and 3.4), this second signal indicates the presence of two conformers in Fe proteins of spin state $S = 1/2$ and $S = 3/2$ respectively, which together account for the spin integration of a single [4Fe–4S] cluster. Morgan et al.[116] later confirmed the presence of the $S = 3/2$ form in Av2. This removes one of the reasons for postulating a second, interacting EPR center. However, doubt was cast on the validity of the $S = 3/2$ spin state by Meyer et al.,[112] whose proton NMR studies of Cp2 implied that the high-spin ground states arise from [4Fe–4S] cluster distortion upon freezing.

C. Electron Counting

The second reason for postulating a "two-cluster" Fe protein is not so easily explained away. Yates et al.[182] and Thorneley et al.[162] observed a biphasic dithionite–mediated reduction of phenazine methosulfate–oxidized Ac2 in the presence of MgATP, a very rapid phase utilizing one electron per mole of Ac2, which corresponded with the full development of the $S = 1/2$ spin EPR signal, and a slow phase utilizing a second electron. Any inactivation of the Ac2 reduced the size of this rapid phase, which implied that the slow phase was due to interaction of sodium dithionite with inactive protein. The specific activity of 1400 nmoles C_2H_2 reduced mg^{-1} protein min^{-1} was probably consistent with the protein being half-active and, hence, the simplest conclusion from these data was that fully active Ac2 would transfer two electrons per mole. Walker and Mortenson[169] had previously titrated Cp2 of specific activity: 2300 nmoles C_2H_2 min^{-1} mg $protein^{-1}$ with dye and ferricyanide and observed a stoichiometry of 1.4 to 2 electrons per mole. Braaksma et al.[9] monitored the EPR spectrum during oxidation and reduction of Av2 ± MgATP and, using Nernst equation fits, obtained a value of $n = 2$, implying a two-electron shift.

The preceding observations are balanced by those indicating a one-electron transfer per mole of Fe protein during the redox cycle. Zumft et al.[182] used EPR monitoring, as did Braaksma et al.[9] but observed $n = 1.02$ (±0.02) for the oxidation–reduction of Cp2 (±MgATP). Ljones and Burris[94] oxidized Cp2 physiologically with a minimal amount of Cp1 plus MgATP and obtained a value on rereduction of one electron per mole. Watt et al.[176] using indigo-sulfonate– or methylene blue–oxidized Av2, the reduction of which was measured coulometrically or electrospectrophotometrically, also found one electron per mole of Av2. Morgan et al.[116] investigated the redox properties of both the $S = 3/2$ and $S = 1/2$ spin states of Av2 and came to the same conclusion. Thorneley and Lowe,[159] in their kinetic analysis of the Fe protein cycle, used physiologically oxidized Kp2 and they also assumed a one-electron transfer. Ashby and Thorneley[7] later investigated the reduction by sodium dithionite of indigo-carmine dye–oxidized Kp2 ± MgADP and observed a rapid single-electron transfer slowed approximately 30-fold by MgADP. Ljones and Burris[94] argued that dye-oxidized Fe protein could be damaged to produce nonphysiological redox centers. Anderson and Howard[1] showed that $Av2_{ox}$ produced a [2Fe–2S] center on treatment with ATP and α,α'-dipyridyl—a production of two such centers would account for the two electrons per mole, although, presumably, in inactive protein. Pagani et al.[126] reported that oxidized Kp2 treated with MgATP and α,α'-bipyridyl produced a 2Fe Kp2 that could be partly reactivated by treatment with the sulfur transferase rhodanese. Thus, the present opinion is that the one-electron transfer per mole of Fe protein is enzymically relevant. However, the presence of the second electron in some instances has not been satisfactorily explained.

III. The MoFe Proteins

Eady and Smith[31] described the purification and properties, including metal and acid-labile sulfur contents, of the nitrogenase proteins that have been used as the main sources for biochemical investigations: *K. pneumoniae*, *C. pasteurianum*, and *Azotobacter*. Metal contents varied widely from 7.8 to 38 g atoms Fe per mole and from 1 to 2 g atoms of Mo per mole of protein. Molecular weights ranged from 200,000 to 240,000 and the subunits were approximately 50,000 and 60,000 respectively. Reported specific activities of these MoFe proteins ranged from 780 to 3613 (this value was nmoles H_2 produced; Burgess et al.[14] pointed out that 10% acetylene, commonly used in the gas phase, did not suppress H_2 production completely with purified proteins, and ethylene production, therefore, underestimated specific activity), or the extraordinarily high value of 6100 nmoles C_2H_2 reduced mg^{-1} protein min^{-1} for Av1 reported by Nicholas and Deering.[120]

At one time it was considered that MoFe proteins may contain only one subunit type, but Kennedy et al.[80] showed that division into one or two bands occurred in SDS gels, depending on the type of SDS used. The later discovery that two genes (*nifK*, *nifD*) coded for these subunits confirmed their findings. Hybridization experiments showed that *nifD* genes were highly conserved.

Ioannidis and Buck[75] compared the derived amino acid sequences of seven *nifD* genes of *K. pneumoniae*, *A. vinelandii*, Rhizobium cowpea species, *Rhizobium* sp., *Parasponium*, *Bradyrhizobium japonicum*, *Anabaena* 7120, and *Clostridium pasteurianum*. Homology was high among the first six (67–72%) and lowest between *K. pneumoniae* and *C. pasteurianum* NifD proteins (44%). Regions of least homology were at the N- and C-terminal ends. Five cysteine residues (at positions 63, 89, 155, 184, and 275 in *K. pneumoniae* NifD) were conserved. Holland et al.[65] compared NifK proteins; these were similar along most of their sequences except between residues 225–300 (*K. pneumoniae* NifK) where deletions or insertions occurred. NifD proteins possessed regions of low similarity between residues 225 and 340 whereas the NifD were homologous to NifK subunits between residues 120 and 180. NifK proteins have three conserved cysteines: residues 69, 94, and 154 of Kp NifK. Arnold et al.[5] sequenced the whole *nif* region of *K. pneumoniae* and reported several differences in their NifK sequence from that of Holland et al.[65] but confirmed the NifD sequence reported by Ioannidis and Buck.[75]

A. Arrangement of the Metal Clusters

High-specific-activity MoFe proteins usually contain high metal contents: 24–36 Fe and 1–2 Mo per mole (240,000 Da). The consensus opinion on the metal distribution in these proteins was described by Lowe et al.:[101] they contain two FeMoco molecules of empirical structure $MoFe_{6-8}S_{4-10}$, also known as M clusters, and four probable [4Fe–4S] groups known as P clusters. The arrangement of these clusters within the four subunits is not known, although current research, both crystallographic and site-directed mutagenesis, is aimed at determining this.

i. Crystallography

Single crystals of Av1, Kp1, and Cp1 (the MoFe protein of *C. pasteurianum* nitrogenase) have been obtained giving diffraction patterns to 2–4 Å,[177,180] but these have not been adequately analyzed to indicate the locations of the metal clusters. However, single-crystal EXAFS of Cp1 indicate a nonlinear Fe–Mo–Fe configuration in FeMoco.[39]

ii. Site-Directed Mutagenesis

Brigle et al.[11] constructed six mutations in *A. vinelandii* NifD involving cysteine to serine changes at residues 154, 183, and 275, glutamine to glutamic acid at residues 151 and 191, and aspartic to glutamic acid at residue 161. The first three mutations yielded very low (0.2–6% of wild-type) Av1 activity, whereas the other mutations yielded moderate (20–75% of wild-type) activities. Dean et al.[25] extended this survey to cover all the conserved cysteines (to serine) in both subunits. All mutants had severely restricted diazotrophic growth except substitution 153 in the β subunit, which remained unaffected. This cysteine clearly has a different role from the others, but the authors maintain that it could still be involved in metal binding, since oxygen will substitute for sulfur as a metal ligand.

Kent et al.[82] reported similar investigations of Kp1 subunits in which the cysteines were changed to alanine rather than serine. Mutations in the α subunit at positions 63, 89, 155, and 275 and in the β subunit at positions 69, 94, and 152 all resulted in loss of diazotrophic growth, Kp1 activity, and the EPR signal. Cysteine-184 replacement greatly diminished but did not completely eliminate growth or Kp1 activity. This activity parallels the position 183 mutation in the Av1 α subunit. On the other hand, the differences in activity of the respective position 153 mutations emphasize the different response when alanine, rather than serine, replaced cysteine and support the suggestion that this cysteine is involved in cluster binding. Kent et al.[82] confirmed that substituting serine for cysteine-152 of the β subunit retained activity. They also showed that an extract prepared from the mutant with the α-subunit cysteine-275 replaced by alanine complemented extracts of a mutant unable to synthesize the iron–molybdenum cofactor. This evidence indicated that cysteine-275 is involved in cofactor binding.

Scott et al.[137] have advanced this approach a stage further. They noticed that residues 150–170 of the *nifE* gene product, which is believed to be involved in FeMoco assembly, were homologous to residues 180–201 of NifD, the α subunit. Substitution of the α-subunit residues histidine-195 by asparagine and glutamine-191 by lysine yielded Nif⁻ mutants that reduced C₂H₂ to C₂H₄ and ethane and changed the characteristic EPR spectrum of nitrogenase. The authors argued cogently that this ethane-producing phenotype was not due to expression of either the vanadium or the alternative nitrogenase, which also produce ethane from C₂H₂, but due solely to the introduced mutations. Thus, the α-subunit residues his-195, gln-191, and cys-275 all seem to play a role in FeMoco binding. Whether the β-subunit cys-153 binds cofactor or P cluster is not known.

A different approach was adopted by Govezensky and Zamir,[47] who studied the effect of single-residue replacements in NifD on the aggregation of Kp1. Native, nondenaturing electrophoresis of Kp1 on acrylamide gels yielded one major form (*a*) and two, slower-moving, variants (*b*, *c*). Changing Gly-186 to Asp, Gly-195 to Glu, Ser-443 to Pro, and Gly-455 to Asp all produced Nif⁻ mutants in which the slowest-migrating band *c* was the predominant form. This band was also the predominant form in NifE and NifN mutants, which lack FeMoco. Thus, these residues may be involved in cofactor binding. Mutations in the N terminus of NifD, which were also Nif⁻ did not affect the electrophoretic pattern of Kp1.

Kent et al.[81] also studied the electrophoretic mobilities of Kp1, observing four forms, three corresponding to those of Govezensky and Zamir and a fourth that was probably an O₂-damaged form of the predominant band (form *a*). They analyzed 19 mutants, 11 Nif⁻ and 8 Nif⁺ or Nif±; 9 of which were described by Kent et al.[82] The abundance of form *a* broadly correlated with the Nif activity expressed by the mutant; in the Nif⁻ mutants the most abundant forms were *c* and a fast-moving form *e*. Both these forms contained the α and β subunits of Kp1. Addition of FeMoco to a NifB⁻ mutant Nif proteins produced form *a* from form *c*; a similar experiment with the Nif⁻ Cys-275 mutant showed only a slight change in mobility of form *c*, consistent with the evidence[82] that Cys-275 is necessary for binding FeMoco. This work also indicated that Cys-89 (α subunit) and Cys-94 (β subunit) were involved in subunit interactions. These results are summarized in Table 18-1.

iii. Chemical Approach to Cluster Binding

Experimental chemical approaches, together with x-ray crystallography, once suitable reference points are established, will help to resolve the tertiary and quaternary structures of the MoFe protein. A third experimental approach has arisen because all the metal ions in the MoFe protein cannot possibly be bound by conserved cysteines: since the ratio is 32 Fe and 2 Mo to 16 conserved cysteines, there would only be enough to bind the P clusters and none to bind FeMoco, or else the metal ions would be insufficient to provide four Cys ligands to each P cluster. This third approach involves the study of how well amino acids other than cysteine bind to Fe–S clusters. Evans and Leigh[37] showed that serine, tyrosine, and proline but not tryptophan nor threonine bound to $(Me_4N)_2[Fe_4S_4(SBu^t)_4]$ in MeCN to displace t-butylthiol and yield, for example, $(Me_4N)_2[Fe_4S_4(methyl-L-serinate)_4]$. Such a residue could (and presumably does when serine replaces Cys-153 in the β subunit) serve as a ligand for Fe in the Nif clusters. Another possible Fe-binding residue is aspartate, which probably bound the new [4Fe–4S] cluster formed when ferredoxin III from *Desulfovibrio africanus* was treated with Fe^{2+}.[46]

iv. Individual Subunits

Another approach to the problem of determining the location of the metal clusters is to delete either the *nifK* or the *nifD* gene and derepress the strains to produce the single subunit. Robinson et al.[128] isolated such mutants of *A. vinelandii* and obtained instant nitrogenase activity upon mixing crude extracts, which indicated that the metal clusters were *in situ* and did not require time for assembly. It suggests that the clusters are associated with individual subunits and not shared between them, as is probably the case in the Fe protein. Unfortunately, the subunits are difficult to purify, but Robinson et al.[129] solubilized the β subunit with ethylene glycol, achieved a partial purification by anion exchange, and showed that the subunit formed a tetramer that contained iron.

B. The P Clusters

The consensus opinion is that P clusters are unusual [4Fe–4S] clusters that are distinct from the Fe associated with FeMoco; for example, the MCD of oxidized Kp1 from a NifB⁻ mutant of *K. pneumoniae*, which lacks FeMoco, was identical to that of oxidized Kp1 from the wild type.[130] They show a transient $g = 1.93$ EPR signal characteristic of $[4Fe–4S]^{1+}$ during thionine oxidation of Kp1, which suggests that the dithionite-reduced species are $[4Fe–4S]^0$,[144] and also a signal that is induced by solid thionine[86] (discussed later). The $[4Fe–4S]^0$ state has not been obtained in model compounds, which supports the suspicion that the ligands are unconventional—that is, amino acid residues other than cysteine bind the Fe atoms.[101]

The belief in the existence of P clusters is based on the Mossbauer spectra of the dithionite-reduced protein and extrusion data, which indicated that

Table 18-1 Site-Directed Mutagenesis: Effect of Single Amino Acid Residue Changes in NifD and NifK Proteins

Organism	Mutation	Nif Phenotype Growth on N_2[1]	Electrophoretic Form of Nif[3]	Normal EPR Signal Intensity[4]	Possible Role of Normal Residue[5]	Ref.
K. pneumoniae	None	++++	a	+		82
	NifDK Δ	−	None	−		81,82
	NifD 63 C → A	−	e	−		81,82
	NifD 89 C → A	−	e	−		81,82
	NifD 155 C → A	−	e	−		81,82
	NifD 155 C → S	−	e	+/−		81,82
	NifD 184 C → A	+	b,c	WB		81,82
	NifD 275 C → A	−	c	−	CB	81,82
	NifK 69 C → A	−	e	WB		81,82
	NifK 94 C → A	−	e	+		81,82
	NifK 112 C → A	++++	a	−		81,82
	NifK 152 C → A	−	e	+		81,82
	NifK 152 C → S	++	a	+	MB	81,82
	NifD 65 Y → F	++	a	+		82
	NifD 276 Y → F	++++	a	+		82
	NifD 281 Y → F	++++	a	+		82
	NifD 354 Y → F	++++	a	+		82
	Double Mutations					
	NifD 89 C → A + NifK 94 C → A	+	a	+/−		82
	NifD 155 C → A + NifK 94 C → A	−	e	−		82
	NifD 275 C → A + NifK 152 C + A	−	e	WB		82
	NifD 275 C + A + NifK 152 C → S	−	e	WB		82
	NifD 85 G → R	−	a			47

Organism / Mutation	Activity[1]	Electrophoretic form[3]	[4]	[5]	Ref.
NifD 121 E → K	−	a		CB	47
NifD 186	−	c		CB	47
NifD 195 G → E	−	c		CB	47
NifD 443 S → P	−	c		CB	47
NifD 445 G → D	−	c		CB	47
NifD 62 C → S	−	ND			25
NifD 88 C → S	−	ND			25
NifD 154 C → S	−	ND	−		11
NifD 183 C → S	±	ND			11
NifD 275 C → S	−	ND			11
NifK 70 C → S	−	ND			25
NifK 95 C → S	−	ND			25
NifK 153 C → S	+	ND		MB	25
NifD 83 H → N	+	ND			25
NifD 151 Q → E	+	ND			11
NifD 159 I → V	+	ND			25
NifD 160 G → A	+	ND			25
NifD 161 D → E	+	ND			11
NifD 191 Q → E	−[2]	ND		CB	11
NifD 271 N → D	+	ND			25
NifD 276 Y → S	+	ND			25
NifD 277 R → H	−	ND			25
NifD 279 M → I	+	ND			25
NifD 280 N → D	+	ND			25
NifD 191 Q → K	−[2]	ND	+	CB	137
NifD 195 H → N	−[2]	ND	WB	CB	137

A. vinelandii

[1] Kent et al. used +++; other authors used either + or the text to describe wild-type activity.

[2] These mutants were Nif⁻ but reduced acetylene.

[3] The electrophoretic form of Nif is based upon its mobility on native gel electrophoresis: *a* is the wild-type form; *b* and *c* are slow-moving forms, whereas *e* is a fast-moving form with respect to *a*.

[4] WB = weak and broad.

[5] CB = cofactor binding; MB = metal binding. ND not determined.

90% of the non-cofactor Fe was extracted as [4Fe–4S] clusters.[85] The Mossbauer spectra have been described many times in original papers and reviews.[101,122,143] The Mossbauer spectrum of dithionite-reduced Kp1 shows four Fe species (M_4, M_5, M_6, and S or Fe^{2+}, D, M^N, and S, depending on the nomenclature used). M_4 and M_5 were considered to be derived from the P clusters and M_6 from FeMoco and S remained unassigned, although it may be associated with nonparamagnetic Fe in FeMoco.[98] Both M_4 and M_5 are high-spin ferrous ions distinguishable by the degree of quadrupole splitting, and each P cluster contained one M_4 and three M_5 Fe atoms.[83,128] However, Mossbauer studies on hybrid Kp1 containing [57]Fe-enriched P clusters and [56]Fe-FeMoco, caused McLean et al.[107] to dispute this view. Species S appeared to be part of the P clusters rather than a separate cluster type or associated with FeMoco. They suggested that Kp1 contained two slightly different types of P clusters: one made of $M_4:M_5$ (1:3) and the other of $M_4:M_5:S$ (1:2:1). Whether these are associated with the different subunit types is not known.

The spin state of the Fe in P clusters reflects the complexity of the ligand environment and the redox potential (E_m) of the cluster. Mossbauer data defined the spin state of the P clusters of Av1 and Cp1 as between $S = 3/2$ and $S = 9/2$[72] but MCD of Kp1 indicated that the P clusters were in spin states $S = 5/2$ or $S = 7/2$.[130] Hagen et al.[55] challenged the conventional P cluster concept on the basis of their EPR studies with Av1, Ac1 (the MoFe protein of *A. chroococcum* nitrogenase), and Kp1 proteins, oxidized with solid thionine. They observed excited-state EPR signals with g values of 10.4, 5.8, and 5.5. The size and temperature dependence of the $g = 10.4$ peak suggested a spin state $S = 7/2$ associated with the P clusters and a P cluster: FeMoco ratio of 1. They proposed that either the P clusters were [8Fe–8S] clusters or only two of the conventional [4Fe–4S] clusters were oxidized by thionine to the $S = 7/2$ spin state, the others remaining frozen after a one-electron oxidation. Lindahl et al.[93] repeated these observations upon Av1 and Kp1 with soluble and solid thionine. Oxidation by soluble thionine yielded one species (P_{ox}), whereas oxidation by a solid thionine suspension yielded two main half-integer spins species that they designated P_{ox} and P_{ox} ($S = 7/2$). Mossbauer data indicated that these states were isoelectronic since their isomer shifts were approximately equal. Moreover, when the solid thionine was removed by anaerobic gel chromatography, the P_{ox} ($S = 7/2$) reverted to P_{ox}, causing the authors to argue that P_{ox} ($S = 7/2$) was caused by a conformational change induced by interaction with solid thionine. They disagreed with both of the alternative proposals offered by Hagen et al.;[56] the first (the [8Fe–8S] clusters) on the grounds of incomplete quantification (that their samples contained an unknown percentage of P_{ox} $S = 7/2$) and with the second model on the basis that it requires the formation of an EPR silent state with zero or integer spin. The Mossbauer data of Lindahl et al.[93] showed that the solid-thionine-oxidized P clusters all had half-integer electronic spin. Obviously, this $S = 7/2$, P cluster phenomenon needs further investigation with other oxidizing agents.

i. Redox Titrations of MoFe Proteins

The simplest redox scenario of the MoFe protein is that it contains six redox centers of two types: four P clusters and two cofactors. Redox titrations indicated that six electrons could be removed from Av1 by a mild oxidant without loss of activity. Coulimetric titration of this dye-oxidized Av1 with sodium dithionite indicated two reduction regions at -290 and -480 mV, each requiring three electrons; the EPR-active cofactor was first reduced followed by the P clusters.[171] Careful oxidation of reduced protein indicated that P clusters are oxidized first, followed by cofactor.[122] More detailed analyses[36,174] revealed a three-electron reduction of P centers, two electrons for the cofactors and one electron for an unknown center. Later experiments[175] showed that Av1 can lose up to 9 or 12 electrons when oxidized by ferricyanide, dichlorophenolindophenol, or cytochrome c. These higher oxidation states develop an EPR signal at $g = 1.99$ and 1.89, also observed by Smith et al.,[146,147] but their physiological significance remains unknown.

Smith et al.[144,148] observed a different EPR spectrum with g values of 2.05, 1.95, and 1.81 when Kp1 was oxidized by low levels of ferricyanide. Rapid-freeze experiments showed a diminution of the cofactor EPR signal concomitant with the disappearance of the new signal. The authors argued that the new signal was due to intramolecular electron transfer to the P clusters in the $[4Fe-4S]^{2+}$ level to yield a $[4Fe-4S]^{1+}$ state and that this is evidence that P clusters are in the $[4Fe-4S]^{0}$ oxidation state in dithionite-reduced Kp1. There is some evidence that P clusters can be oxidized to this P^{1+} oxidation state during turnover and thus may act as an electron buffer, supplying cofactor with electrons to reduce substrate. This suggestion can be reconciled with Watt's observations that in dye-oxidized Av1, the cofactor is reduced before the P clusters[174] only if the cofactor, or cofactor–substrate complex, achieves a more reduced state than the P clusters during turnover. It follows that P clusters may not be involved in the reduction of the highest oxidation state of the MoFe protein during turnover and, therefore, may not be on the direct electron transfer pathway to the cofactor. The alternative possibility is that formation of the active complex involves a conformational change such that P clusters are reduced before the cofactor. Watt's data are also inconsistent with the four P cluster hypothesis, unless one of these P clusters is the "unknown redox center." This is unlikely since the order of oxidation in Watt's experiments is P clusters–cofactor–unknown center. Thus, both the function and number of P clusters remains an enigma.

C. The Molybdenum Cofactor

The structure of FeMoco is perhaps the most intensively researched unknown in N_2 fixation. Both Lowe et al.[101] and Orme-Johnson[122] in their respective reviews concurred on the empirical formula of $MoFe_6S_{4-9}Fe_{1-2}$, the six irons being paramagnetic and the one to two irons diamagnetic. Loss of these diamagnetic irons apparently destroyed activity.[139] However, new purification methods[110,179] obtained active FeMoco with an MoFe ratio of 1:6, but it was not reported whether all these Fe atoms were paramagnetic. Lough et al.[95] reported a rapid method for isolating *A. vinelandii* FeMoco, but the low Fe:Mo ratio indicated that it was impure. Since the preceding reviews, numerous reports have appeared on cofactor structure and properties.[2-4,19,24,32,40,41,42,44,45,61,95,107,110,118,119,151,163,164,168,179]

i. Spectroscopic Techniques

Numerous spectroscopic techniques have been employed to study FeMoco and a selection of these studies since 1985, together with the conclusions drawn, is listed in Table 18-2. The EXAFS data agree broadly on the Mo–Fe, Mo–S, Fe–Fe, and Fe–S bond distances in the first coordination spheres but not on the number of atoms within those spheres. Nor does everyone agree on the number of soft ligands (C, O, N) in close proximity. Only Arber et al.[3] found evidence for a second Fe coordination sphere. Clearly, EXAFS data on nitrogenase are difficult to interpret and more-sophisticated theory and technology is necessary. The same criticism can be made about ENDOR (electron nuclear double resonance) spectroscopy and ESE (electron spin echo) spectroscopy. All the conclusions drawn in Table 18-2 are tentative because of the complexity of FeMoco, and further refinement of these techniques is urgent. Nevertheless, some important probabilities have emerged from these studies: (1) The two Fe coordination spheres are at 2.64 and 3.67Å in Fe EXAFS. (2) The similarities of the environments of the Fe atoms in isolated FeMoco compared with the dissimilarities when FeMoco is associated with the apoproteins. However, it must be emphasized that Mössbauer and ENDOR measure different parameters and, for direct comparison, ENDOR studies of cofactor are necessary. (3) The multiplicity of forms the isolated FeMoco can adopt indicates that relatively strong constraints must be placed on it in the holoprotein. (4) Single-crystal, polarized x-ray absorption spectroscopy of Cp1 indicates the presence of a nonlinear Fe–Mo–Fe configuration consistent with $MoFe_3$, tetrahedral, or $MoFe_4$, square pyramidal, structures. (5) H^1 ENDOR observed proton exchange with D_2O, suggesting liganded OH^- or H_2O as possible sites for displacement by reducible substrates.

Other spectroscopic techniques, including NMR, MCD, FTIR (Fourier transform infrared), redox, and electrochemical titrations have all contributed to knowledge about FeMoco. Several potential model compounds have been proposed (e.g., Eldredge et al.[33]), but the key to its inorganic core remains elusive.

ii. FeMoco Contains Homocitrate

Mutations in the *nifV* gene of *K. pneumoniae* altered the substrate-reducing pattern of nitrogenase.[106,108] This NifV⁻ substrate-reducing pattern was obtained when isolated FeMoco from NifV⁻ nitrogenase complemented extracts of NifB⁻ (FeMoco-less) mutants *in vitro*. FeMoco from the wild-type *K. pneumoniae* restored the wild-type activity pattern to NifB⁻ extracts.[59] This was strong evidence that FeMoco contained the N_2-binding site and the active center of nitrogenase.

Hoover et al.[70] reported that the *nifV* gene coded for a low-molecular-weight factor and, in two subsequent publications, showed that homocitrate accumulated in NifV⁺ *K. pneumoniae* and that homocitrate cured the NifV⁻ phenotype.[68,69] Shah et al.[140] devised an *in vitro* system for FeMoco biosynthesis. This required the protein products of, at least, the *nifB*, *nifE*, and *nifN* genes of *K. pneumoniae* together with molybdate and ATP. When FeMoco was synthesized *in vitro* in the presence of tritium-labeled homocitrate, the label was incorporated into the isolated nitrogenase; homocitrate was identified by NMR spectroscopy and found to be present in nitrogenase in a 1:1 molar ratio with Mo.[67] In a further interesting development of this work, Imperial et al.[73] determined the critical structural features of homocitrate necessary for efficient nitrogenase functioning (i.e., the preferential reduction of N_2 rather than H^+). These were the hydroxyl group, the 1 and 2 carboxyl groups, and the R configuration of the chiral center. For instance, homoisocitrate and isocitrate substitution favored high levels of proton reduction. Indeed, of 26 alternative homocitrate analogs tested, all supported very inefficient N_2 reduction compared to H_2 production and decreased the yield of both products. Clearly, the homocitrate requirement is specific: Hoover et al. suggested that homocitrate may dictate the structure of FeMoco and be involved in substrate binding or electron transfer. Kinetic studies are now needed on these analog nitrogenases. Presumably, both FeVaco and the third alternative nitrogenase cofactor of *A. vinelandii* contain homocitrate, since the *nifV* gene is common to all three systems.

D. MgATP- and MgADP-Binding to Nitrogenase Proteins

MgATP is a mandatory requirement for nitrogenase activity whereas MgADP is an inhibitor. That both nucleotides bind to the Fe protein is widely ac-

Table 18-2 Recent Analyses of the MoFe Protein and FeMoco by Spectroscopic Techniques

Techniques	Sample	Conclusion	Ref.
Mo EXAFS	MoFe protein	4–5 S at 2.37 Å 3–4 Fe at 2.67 Å 2(C, N or O) at 2.1 Å	101
Mo EXAFS	FeMoco	3–5 S at 2.37 Å 3–4 Fe at 2.67 Å 2–3 (C, O, N) at 2.1 Å	101
Mo EXAFS	Av1	3 S at 2.36 Å 2.8 Fe at 2.68 Å 3 O at 2.09 Å	122
Mo EXAFS	Crystalline Cp1	4-S at 2.35 Å 2–4 Fe at 2.68 Å The Fe–Mo–Fe configuration is nonlinear	39
Mo EXAFS	Kp1 NifV⁻ Kp1	4–5 S at 2.37 Å 2–4 Fe at 2.69 Å X (O, N) at 2.18 Å	32 109
Mo EXAFS	Cp1 Av1 FeMoco	3–5 S at 2.37 Å 3–4 Fe at 2.68 Å 2–3 (O, N) at 2.10 Å	18
Fe EXAFS	Av1	3.9 S at 2.2 Å	122

Method	Sample	Description	Reference
Fe EXAFS	FeMoco	2–5 S (Fe) at 2.25 Å 1.4–3.2 Fe at 2.66 Å 0.3–0.5 Mo at 2.76 Å 0–2 (O, N, C) at 1.81 Å	101
Fe EXAFS	FeMoco	3 S at 2.2 Å 2.2 Fe at 2.64 Å 1.2 Fe at 3.67 Å 0.8 Mo at 2.7 Å	3 41,42
S, K, and Mo L XANES	FeMoco	Improves X-ray absorption edge spectra	61
Mo ENDOR	Kp1 NifV Kp1	Kp1 and NifVKp1 differ in non-S ligands near Mo	109
Mo and H ENDOR	Av1	FeMoco contains 1 Mo as Mo^{II} or Mo^{VI}	63
Fe ENDOR	Av1	No two Fe sites in FeMoco have the same environment	64 164
Fe, Mo, and S ENDOR	Av1, Kp1, Cp1	FeMoco's contains a single Mo, probably Mo IV. Av1 and Kp1 ENDOR are similar and differ from Cp1 H^1 ENDOR indicated interacting protons exchangeable in D$_2$O	168
Mossbauer	Hybrid Kp1	MoFe contains two pairs of P clusters	107
Mossbauer	FeMoco	All Fe atoms have a similar environment	118
EPR	FeMoco	Different forms of FeMoco exist	119

Table 18-2 (Continued)

Techniques	Sample	Conclusion	Ref.
ESE	Cp1 FeMoco	N is directly coordinated to the MoFe center	151
H, Fe, Mo, N ENDOR	Kp1	1. N is directly coupled to Mo	163,164
	NifV⁻ Kp1	2. NifV Kp1 differs from Kp1 in ligand binding to 1 Fe in FeMoco-	
		3. NifV cofactor contains an exchangeable proton	

cepted.[101,117] MgATP changes the EPR spectrum of the reduced Fe protein and also makes it more O_2-sensitive and more accessible to SH-group reagents and to Fe chelators. Both the MoFe protein and MgADP prevent this last accessibility. The first indications that MgATP bound to nitrogenase proteins were by Bui and Mortenson,[12] who found that MgATP bound to Cp2 but not to Cp1 and by Biggins and Kelly,[8] who reported that MgATP bound to both Kp1 and Kp2 and also to O_2-damaged Kp2.

Since then, numerous methods—equilibrium or flow dialysis, gel equilibrium or gel filtration, susceptibility to chelating agents, stopped-flow pre-steady-state kinetics, and circular dichroism spectroscopy—have been used to determine MgATP or MgADP binding to nitrogenase proteins (Table 18-3). In addition, NMR evaluation of Mn binding indicated four possible Mg- (and hence Mg-nucleotide-) binding sites to Kp1.[83]

Most of the data indicate that the Fe protein binds two MgATP and two MgADP, probably at the same two sites, since MgADP has been shown to inhibit MgATP binding competitively.[20,164] The calculated dissociation constants vary considerably between samples and between the methods employed (see Table 18-3). Cordewener et al.[20] observed that the dissociation constants of both nucleotides with Av2 were affected by the specific activities of the samples used, although not in any simple manner, and, consequently, by the amount of inactive protein present, which also binds nucleotides. Another difficulty was emphasized by Watt et al.,[176] who claimed that Av2 was heterogeneous in the presence of sodium dithionite, which made direct measurement of binding constants meaningless. They, instead, constructed thermodynamic cycles using the binding constant data of $Av2_{ox}$/ 2MgATP, measured by either equilibrium or flow dialysis, UV-visible or circular dichroism spectroscopy, to calculate a composite binding constant of $2.2 \times 10^6 \ M^{-2}$ for $Av2_{red}$/2MgATP.

Despite these reservations about the validity of binding measurements and the wide variation in the reported binding constants, the results of Cordewener et al.[20,21] and Watt et al.[176] on the Fe proteins and Miller and Eady[114] on the MoFe proteins, Kp1 and Ac1, all suggest that the tightest binding occurs between MgADP and the oxidized proteins and the weakest binding occurs between MgATP and the reduced proteins. Lindahl et al.[92] observed that purified Av2 samples were contaminated by MgADP, presumably another indication that MgADP binds tightly to the Fe protein: apyrase treatment, which hydrolyzes both ATP and ADP to AMP, sharpened the EPR spectrum of isolated "native Av2" to become "resolved Av2." Such tight binding might have been the reason for T'so and Burris'[165] observation (see Table 18-3) that Cp2 contained only one MgADP-binding site and why the binding of MgADP to $Kp1_{ox}$ was not observed earlier (R. R. Eady, personal communication).

Table 18-3 Data on Nucleotide Binding to the Nitrogenase Proteins

Source	Nucleotide	Sites	Binding: K_D	Method	Reference
The Fe Protein					
$Cp2_{red}$	MgATP	2	17 μM	Gel equilibration	165
	MgADP	1	5 μM		
$Cp2_{red}$	MgATP		50 μM	Gel equilibration	34
$Cp2_{red}$	MgATP		83 μM	Gel equilibration	53
$Av2_{red}$	MgATP	2	430 μM;220 μM	Chelation by bathophenanthroline	53
$Av2_{ox}$	MgATP	2	<5 μM	Circular dichroism	149
	MgADP	2	<5 μM	Circular dichroism	
$Cp2_{ox}$	MgATP	2	<5 μM	Circular dichroism	
	MgADP	2	<5 μM	Circular dichroism	
$Av2_{red}$	MgATP	1	560 μM	Flow dialysis	21
$Av2_{ox}$		2	290 μM	Flow dialysis	
$Av2_{red}$	MgATP	2	220 μM;1710 μM	Flow dialysis	20
	MgADP	2	91 μM;44 μM		
$Av2_{ox}$	MgATP	2	49 μM;180 μM		
	MgADP	2	24 μM;39 μM		
$Kp2_{ox}$	MgADP	2	$4 \times 10^{10}\ M^{-2}$	Stopped-flow kinetics	7

	Nucleotide			Method	Ref.
Av2$_{red}$	MgATP		$4 \times 10^7\ M^{-2}$	Flow dialysis	176
				Dialysis	
The Mo Fe protein					
Cp1$_{red}$	MgATP	None		Gel equilibration	165
Kp1$_{red}$	MgATP	4	600 μM	Gel filtration	115
Av1$_{red}$	MgATP	None		Flow dialysis	21
Av1$_{red}$	MgADP	None		Flow dialysis	20
Av1$_{ox}$	MgADP	2	Tight	Gel filtration	114
Av1$_{red}$	MgADP	2	Tight	Gel filtration	114
Av1$_{ox}$	MgATP		Weak		
Av1$_{red}$	MgATP		None		
Fe protein[+]	MgATP		400 μM	Stopped-flow kinetics	152
MoFe complex	MgADP		20 μM	Stopped-flow kinetics	156
Fe + VFe protein complex	MgATP		230 μM	Stopped-flow kinetics	155
Fe + VFe protein complex	MgADP		30 μM	Stopped-flow kinetics	155

There are some data, however, that are not consistent with this rule of thumb that MgADP binds more tightly to Av2 than does MgATP. Lindahl et al.[92] perturbed dithionite-reduced Av2 with either ethylene glycol or hexamethylphosphoramide (12–15%) or, alternatively, urea (0.4 M). Subsequent addition of MgATP changed the EPR spectrum to closely resemble that of native Av2/MgATP, whereas adding MgADP had little effect on the EPR spectrum of the perturbed native form. This implies that MgATP binds more strongly than MgADP, at least to induce the conformation that alters the EPR spectrum. Perhaps the experiments of Lindahl et al.[92] should be repeated in a different order by assessing the effect of perturbing solvents on the EPR spectra of the preformed $Av2_{red}$/MgADP and $Av2_{red}$/MgATP states. This information might be relevant to the understanding of the regulation of nitrogenase activity by MgADP *in vivo*.

Nitrogenase ATPase activity belongs to the combined, but not to the individual, nitrogenase proteins, which suggests that the active site may be shared. This, in turn, gives rise to the possibility that MgATP may bridge the two component proteins. Attempts to measure direct binding of MgATP to the MoFe protein, have, by and large, been unrewarding. Only one report described four low-affinity MgATP-binding sites on reduced Kp1,[115] and Kimber et al.[83] inferred four Mg^{2+}-binding sites from NMR studies of Kp1. Miller and Eady[114] observed tight binding of 2MgADP to either dye-oxidized or dithionite-reduced Ac1, but only weak binding of MgATP to $Ac1_{ox}$ and none to $Ac1_{red}$. Other reports were also negative (see Table 18-3). However, Robson[131] analyzed Nif HDK-derived sequences and found three regions in NifH proteins and one in NifK proteins that were analogous to nucleotide-binding domains in ATPases. Region β in the NifH proteins contained the Cys-85 residue identified by Hausinger and Howard[58] as MgATP-binding. These regions need further investigation by site-directed mutagenesis.

IV.   The Mechanism of Nitrogenase Action

A.   The Lowe/Thorneley Model

The electron transfer sequence of N_2 fixation, electron donor → Fe protein → MoFe protein → N_2, was derived mainly from EPR and Mossbauer studies of the isolated nitrogenase proteins together with EPR of the steady state of the enzyme during turnover.[125,145,146,147,183] The next stage in the search for the mechanism of nitrogenase action, the kinetic model derived from pre-steady-state kinetic studies of *K. pneumoniae* nitrogenase activity by Lowe and Thorneley[99,100] and Thorneley and Lowe,[159,160] has been reviewed comprehensively[101,122,161] and will be described only briefly here. It comprises two main processes: the Fe and the MoFe protein cycles (Figures 18-

1 and 18-2). The Fe protein cycle is a redox cycle during which an electron is transferred to the MoFe protein. It consists of five steps:

1. $Kp2_{ox}(MgADP)_2$ is reduced to $Kp2_{red}(MgADP)_2$.[7]
2. MgADP is replaced by MgATP.
3. $Kp2_{red}(MgATP)_2$ complexes with Kp1.
4. MgATP hydrolysis is followed by electron transfer to Kp1.
5. Dissociation of the complex into $Kp1_{red} + Kp2_{ox}(MgADP)_2$ occurs.

N_2 reduction to $2NH_3 + H_2$ is an eight-electron process, the Fe protein cycle has to operate eight times, during which the MoFe protein is progressively reduced, binds and reduces substrates in turn, and finally returns to its highest oxidation state, E_0. Since each Fe protein cycle is a single electron transfer, the MoFe protein cycle involves one FeMoco, associated with an $\alpha\beta$ dimer of Kp1, rather than an $\alpha_2\beta_2$ tetramer. The scheme was devised to explain the lag and burst phases of product appearance during substrate reduction and the slowness of the overall process. For example, H_2 production under argon shows a lag phase, the length of which depends on the ratio of component proteins.[52,99] H_2 production under N_2 shows the same lag phase, followed by a burst of H_2 production, before attaining a steady rate. The lag was explained by assuming two slow reduction steps before H_2 was released[52,99] and the burst was due to rapid replacement of H_2 by N_2 binding. Determination of the rate constants of the individual reactions led to the following conclusions: (1) The rate-limiting step was the protein–protein dissociation of the complex $Kp2_{ox}(MgADP)_2 Kp1_{red}$ ($k = 6.4$ sec^{-1}) rather than ATP hydrolysis, electron transfer, or substrate reduction. (2) The dissociation rate of the $Kp2_{ox}(MgADP)_2Kp1_{red}$ complex is independent of the oxidation state of the reduced Kp1 species. (3) Substrate binding and product release could only occur with uncomplexed Kp1. (4) N_2 bound to a more reduced form of Kp1 (probably E_3) than that which could release H_2. (5) The hydrazido group ($=N-NH_2$) bound intermediate,[157] which releases hydrazine on quenching with acid or alkali is probably bound to oxidation state E_4. (6) One molecule of NH_3 may be released at oxidation state E_5, which could leave a readily reducible bound nitrido group ($E\equiv N$), which on further reduction, under physiological or assay conditions, yields the second NH_3 molecule and restores the original oxidation state E_0. (7) If the reducing power is low, then oxidation state E_2H_2 will lose H_2 and revert to state E_0. This is prevented by sufficient reductant, MgATP and a molar excess of the Fe protein over the MoFe protein. Given these conditions, then the reduction of $Kp2_{ox}(MgADP)_2$, the rate of MgATP-MgADP exchange on the reduced protein (200 sec^{-1},[156,7] and the rate of $Kp2_{red}(MgATP)_2Kp1$ complex formation will all be considerably faster than the rate-limiting complex dis-

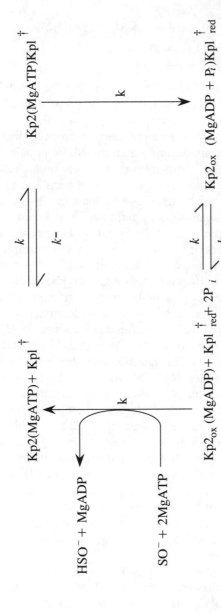

Figure 18-1. The Fe protein cycle: the oxidation and reduction of the Fe protein of *K. pneumoniae*.[99,101,161]

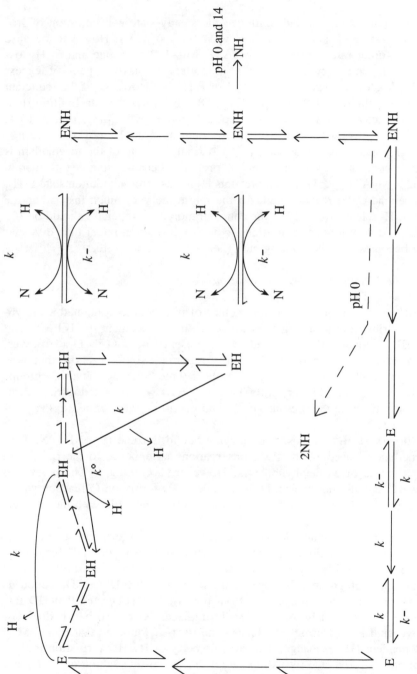

Figure 18-2. The MoFe protein cycle for nitrogenase indicating oxidation states of H_2 release and N_2 reduction and release.[101,161]

sociation step. This would minimize the steady-state concentration of free Kp1 at state E_2H_2 and decrease the wasteful loss of H_2. However, because the first-order rate constants of MgATP/MgADP exchange and of H_2 loss from E_2H_2 are very similar, it is important to have a pool of excess $Kp2_{red}(MgATP)_2$ to complex the Kp1 of E_2H_2. (8) However, if the reductant is low and the reduction of $Kp2_{ox}(MgADP)_2$ becomes the rate-limiting step, then the reformation of complex $Kp2_{ox}(MgADP)_2KP1_{red}$ (k_3, Figure 18-1) could also prevent H_2 evolution. This would have the benefit of "freezing" the E_2H_2 state and saving energy. (9) A final requisite of the mechanism is a high concentration of nitrogenase proteins, because complex formation [$Kp2_{red}(MgATP)_2Kp1$], which prevents H_2 release from oxidation state E_2H_2, is a second-order reaction and, for the rate to be very much faster than the rate of H_2 release from E_2H_2, the concentration of nitrogenase should be 100 μM.[161] At very low concentrations, complex formation rather than dissociation becomes the rate-limiting step (hence the dilution effect[158]).

B. The Mechanism of HD Formation

Chatt[17] postulated that the active center of nitrogenase contained a molybdenum trihydrido grouping to account for the replacement of H_2 by N_2 and the HD formation that occurred in the presence of D_2O or D_2. However, Burgess et al.[13,15] and Wherland et al.[178] reported two observations that were not compatible with Chatt's suggestion of three equivalent hydrido groups at the active site: (1) Very little T_2O or (THO) was obtained during nitrogenase activity in the presence of T_2 and (2) no D_2 was formed in the presence of HD.

Burgess et al.[15] proposed an enzyme-bound diazene (E—NH=NH) intermediate to account for these observations as opposed to the postulated free diimide of earlier hypotheses. However, according to Thorneley and Lowe,[161] this did not explain N_2-independent HD formation (a reaction whose existence has been questioned[76,87]), nor was it compatible with the competitive inhibition of N_2 fixation by H_2.

Figure 18-3 accounts for many facts relating to HD formation and accepts the suggested stoichiometry that one electron in six supports HD formation. It depicts the active center as a metal dihydro grouping with a proton bound to an adjacent group. This proton interacts with the HD or D_2 bound as $E_3^-HD(H^+)$ and thus releases HD but not D_2. It is suggested that HD formation is enhanced by N_2 because D_2 replaces N_2 faster than it does H_2. However, Li and Burris[87] and Jensen and Burris[76] found no evidence for N_2-independent HD exchange. The stoichiometry of H_2:HD formation in their experiments predicted by extrapolation that at infinite saturation of N_2 and D_2 the H_2:2HD ratio would approach 1, as predicted by the Cleland scheme[48] (Figure 18-4) rather than the H_2:6HD predicted by the Guth and Burris

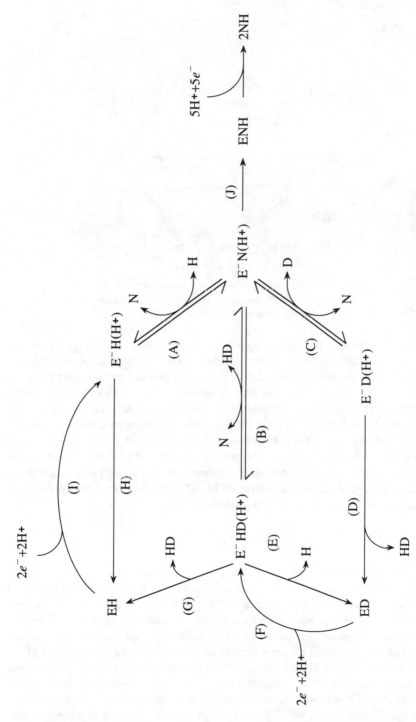

Figure 18-3. Mechanism for N_2-catalyzed HD formation that gives no D_2 in the presence of N_2 and HD and no T^+ from T_2.[100]

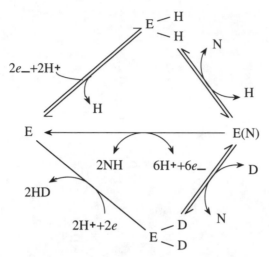

Figure 18-4. Mechanism of nitrogenase-catalyzed H_2 evolution, N_2 reduction, and inhibition of NH_3 formation by $H_2(D_2)$ as suggested by Cleland.[48]

scheme,[48] although later data measured at 50 atm $H_2 + N_2$ or $D_2 + N_2$ support the Guth and Burris stoichiometry (R. H. Burris, personal communication). Lowe and Thorneley[100] predicted that either ratio was possible depending on the overall component protein concentration and ratio, or other variables. Regardless of which scheme is correct (the Lowe–Thorneley scheme allows for N_2-independent HD exchange, whereas those of Guth and Burris[48] and of Cleland[48] accept the experimental evidence that N_2-independent HD formation does not exist), it is currently considered that the active center contains a metal-dihydro grouping plus one other H atom but not a metal–trihydro grouping. The interaction of metal-hydrido groups and protons to release H_2 has been studied by Henderson.[62]

C. MgATP Hydrolysis and Electron Transfer

Eady et al.[30] investigated the pre-steady-state kinetics of electron transfer between Kp2 and Kp1 and concomitant MgATP hydrolysis. They concluded that the two activities occurred simultaneously. Hageman et al.[53] investigated Av nitrogenase and came to a similar conclusion. However, Cordewener et al.[23] observed an initial burst of reductant-independent ATPase activity with dye-oxidized proteins similar to that in reductant-dependent ATPase activity. This clearly showed that the initial burst of MgATP hydrolysis was not dependent on electron transport and led the authors to propose that MgATP hydrolysis must precede electron transfer and, presumably, must enable it to occur. Thorneley et al.[154] provided evidence to support this suggestion, by following MgATP hydrolysis by stopped-flow microcalorimetry at 6° C. Under such conditions MgATP hydrolysis was approximately three times faster than electron transfer between Kp2 and Kp1. Cordewener et al.[23] attempted to determine the rate-limiting step of MgATP hydrolysis by measuring the off rate of MgADP or P_i from oxidized $Kp2_{ox}$ or $Kp2_{ox}$ + $Kp1_{ox}$. Rapid gel filtration indicated a slow off rate of one MgATP or MgADP but the off rate of the other nucleotide was too rapid for the technique. P_i binding and release rates were faster than those of the nucleotide and, therefore, not rate-limiting. Cordewener et al. made the interesting suggestion that the burst of ATP hydrolysis was due to the hydrolysis of both MgATP molecules binding to the two binding sites of the Fe protein, but, subsequently, only one, low-affinity, site was actively binding and hydrolyzing MgATP while the other was occupied by MgADP. Unfortunately, they did not test this hypothesis by preincubating and gel filtering the proteins with MgADP before testing for the burst of MgATP hydrolysis. MgADP bound with lower affinity to the reduced protein, consistent with the rapid dissociation rate calculated by Thorneley and Cornish-Bowden.[156] Cordewener et al.[22] reported kinetic studies on MgATP hydrolysis and C_2H_2 reduction by reduced and oxidized proteins and showed that MgATP hydrolysis and nitrogenase activity were proportional to the Av2 activity, but that $O_2{}^-$ damage to Av1 uncoupled MgATP hydrolysis from substrate reduction. They claimed that reductant-independent MgATPase activity was not due to a futile cycle of electron transfer, as suggested by Orme-Johnson and Davis,[123] since dye-oxidized proteins hydrolyze MgATP. The authors also argued that reactivation of the $Av2_{ox}(MgATP)_2$ complex was the rate-limiting step in reductant-independent ATPase activity and suggested that this might be so in the reductant-dependent activity. However, Ashby and Thorneley[7] emphasized that reduction of $Kp2_{ox}(MgADP)_2$ occurred before the nucleotides exchange.

V. Alternative Substrates of Nitrogenase

Rivera-Ortiz and Burris[126] described four types of interactions between alternative substrates and inhibitors and nitrogenase: competitive, noncompetitive, unclassified, and negative inhibition. A significant contribution of the Lowe–Thorneley mechanism of N_2 reduction is the realization that noncompetitive inhibition kinetics may reflect binding to different oxidation states of the enzyme active site during the MoFe protein cycle, rather than to different sites on the enzyme.

A. N_2O Reduction

Three publications have dealt with N_2O reduction in recent years.[77,88,90] N_2O inhibits N_2 fixation competitively, presumably because it binds to the same site or oxidation state as N_2.[127] Jensen and Burris[77] confirmed this competitive inhibition with purified proteins, and also that the products were N_2 and H_2O, and discovered that if the N_2O concentration was low, the N_2 formed was further reduced to NH_3. At infinite N_2O concentration, H_2 production (by either N_2 binding or, presumably, intrinsic H_2 production) and HD production were completely inhibited. Presumably, the metal dihydrido group was used to produce H_2O, which suggested to Vaughn and Burgess[167] that N_2O binds through its oxygen to the metal. H_2 inhibited N_2O-dependent N_2 reduction but not its production. Liang and Burris[88] expanded these observations and showed that N_2 and N_2O were mutually competitive and competitive against C_2H_2, whereas C_2H_2 was noncompetitive against both N_2 and N_2O. This latter situation was explained by assuming C_2H_2 bound to two forms of the enzyme, one of which also bound N_2 and N_2O. If this is the case, it is difficult to understand the noncompetitive kinetics unless the affinity of C_2H_2 for the N_2-binding form is lower than for the C_2H_2-binding form. It must be emphasized that steady-state kinetics of substrate reductions by nitrogenase are difficult to interpret because substrates compete for electrons as well as binding sites and the component protein ratio is also a significant variable. Liang and Burris[90] also investigated the NifV^- nitrogenase of *K. pneumoniae*, which differs from the wild type by favoring H_2 production over N_2 fixation *in vivo* and by catalyzing H_2 production, which is partly inhibited by CO. The NifV^- enzyme diverted 51% of the electron flux to H^+ reduction at 3 atm of N_2 and the authors calculated that this would only diminish to 47% at infinite N_2. Moreover, N_2O failed to abolish H_2 production, which prompted the speculation that NifV^- nitrogenase had more than one H_2-evolving form.

B. Cyanide and Methylisocyanide

In a comprehensive investigation of the steady-state kinetics of cyanide reduction by nitrogenase, Li et al.[86] showed that CN^- was a reversible inhibitor of total electron flow, which actually enhanced MgATP hydrolysis. They decided that HCN, on the other hand, was a substrate, yielding both CH_3NH_2, in a four-electron reduction, and CH_4 plus NH_3, in a six-electron reduction (Table 18-4). However, it is difficult to envisage how HCN can be a substrate, because cyanide usually binds to metals through the carbon atom (R. L. Richards, personal communication). Lowe et al.[98] devised a scheme to resolve this difficulty, which will be described later. Electron balance studies showed that the NH_3/CH_4 product ratio was larger than 1, which led Li et al.[86] to suggest that "excess NH_3" was released by a two-electron reduction mechanism, with formaldehyde, which was not detected, as a second product. C_2H_2 diminished, and N_2O completely abolished, this "excess" NH_3 formation. These results implied that C_2H_2 or N_2O bound simultaneously

Table 18-4. *Reactions Catalyzed by Nitrogenase*

	Reaction	Ref.[a]
1.	$N_2 + 6H^+ + 6e^- \rightarrow 2NH_3$	88
2.	$HCN + 6H^+ + 6e^- \rightarrow CH_4 + NH_3$	98
3.	$HCN + 4H^+ + 4e^- \rightarrow CH_3NH_2$	98
4.	$CH_3NC + 6H^+ + 6e^- \rightarrow CH_4 + CH_3NH_2$	133
	$CH_3NC + 4H^+ + 4e^+ \rightarrow CH_3NHCH_3$	19
5.	$HN_3 + 6H^+ + 6e^- \rightarrow N_2H_4 + NH_3$	132
6.	$N_3^- + 3H^+ + 2e^- \rightarrow N_2 + NH_3$	132
7.	$N_2O + 2H^+ + 2e^- \rightarrow N_2 + H_2O$	88,90
8.	$C_2H_2 + 2H^+ + 2e^- \rightarrow C_2H_4$	97,38
9.	$3CH{=}CHCH_2 + 6H^+ + 2e^- \rightarrow CH_2CH_2CH_2 + 2CH_3CH{=}CH_2$	43
10.	$2H^+ + 2e^- \rightarrow H_2$	88
11.	$CH_2{-}N{=}N + 6H^+ + 6e^- \rightarrow CH_3NH_2 + NH_3$	105
12.	$CH_2{-}N{=}N + 8H^+ + 8e^- \rightarrow CH_4 + 2NH_3$	105
13.	$NO_2^- + 7H^+ + 6e^- \rightarrow NH_3 + 2H_2O$	167
14.	$N{\equiv}CNH_2 + 6H^+ + 6e^- \rightarrow CH_3NH_2 + NH_3$	113
15.	$N{\equiv}CNH_2 + 8H^+ + 8e^- \rightarrow CH_4 + 2NH_3$	113
16.	$C_2H_4 + 2H^+ + 2e^- \rightarrow C_2H_6$	6
17.	$N_2H_4 + 2H^+ + 2e^- \rightarrow 2NH_3$	15
Catalytic activities associated with individual nitrogenase proteins:		
		Ref.
Av1	$H_2 \rightarrow 2H^+ + 2e^-$	171
Av2	$S_2O_4^{2-} \rightarrow S_2O_3^{2-} + SO_2^{2-}$	104
Stoichiometric reaction of the Fe protein with O_2:		
Kp2	$O_2 + 2H^+ + 2e^- \rightarrow H_2O_2$	153

[a] The references are the most recent on the alternative substrates.

with cyanide and promoted a six-electron reduction by preventing the release of NH_3 from the higher oxidation state of the MoFe protein cycle. This was a somewhat surprising result since cyanide, apparently, binds to a higher oxidation state than that of proton release, probably E_0 of the Lowe and Thorneley model, whereas both C_2H_2 and N_2O bind to lower oxidation states and, therefore, might be expected to inhibit the six- and four-electron reductions more than the two-electron reduction. A possible explanation is that C_2H_2- or N_2O binding, together with the release of "excess" NH_3, all occur at E_3 and mutually interfere. This might explain why, if N_2O bound, it apparently was not reduced. Perhaps the experiment should be repeated with $^{15}N_2O$.

The evidence of Li et al.[86] that prompted them to assert that cyanide bound to a more highly oxidized state of the MoFe protein cycle than the state that releases H_2 was the observation that a high Av2/Av1 ratio favored H_2 release rather than cyanide reduction, the opposite to that which occurs when N_2 is the substrate. Lowe et al.[98] strengthened this conclusion when they showed that a high total protein concentration favored N_2 reduction over H_2

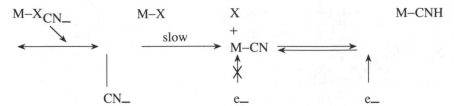

Figure 18-5. Possible mechanism for cyanide binding to nitrogenase involving a slow ligand replacement step, the inhibition of total electron flow by CN⁻, and proton-induced reduction.[98]

release, whereas H_2 release was favored over substrate reduction with cyanide. Pre-steady-state kinetics, however, showed that the H_2 release time lag was the same with or without cyanide, but an unprecedented 3-sec lag occurred before CH_4 was released. Lowe et al.[98] proposed that this slow release was due to a slow, cyanide-induced modification of the binding site, necessary before either inhibition can occur or product is released (Figure 18-5). The figure also explains how CN⁻ can be both inhibitor and substrate and how it has to be protonated before reduction. Such a modified binding site might have prevented N_2O reduction in the experiment of Li et al.[86]

Rubinson et al.[133] reported that methylisocyanide reduction by nitrogenase was analogous to HCN reduction, including the following similarities: (1) Both inhibited total electron flux through nitrogenase, an effect that was decreased by other substrates or CO. (2) Both uncoupled MgATP hydrolysis from electron flow. (3) An increased Av2/Av1 ratio favored H_2 production over CH_3NC or cyanide reduction, indicating that both substrates bind at a redox state of the MoFe protein cycle more oxidized than that which releases H_2. (4) Electron balance studies show a product imbalance: in the case of CH_3NC the CH_3NH_2/CH_4 ratio was larger than 1. This implies a two-electron reduction process analogous to cyanide reduction:

$$CH_3NC \xrightarrow{2H+} CH_3NHCH \xrightarrow{H_2O} CH_3NH_2 + HCHO$$

(5) CH_3NC also can be reduced by either six or four electrons (see Table 18-4).

These similarities are striking, but there are two differences between cyanide and methylisocyanide with respect to nitrogenase action: (1) cyanide enhances MgATP hydrolysis whereas CH_3NC does not; (2) only CO or azide relieves inhibition of total electron flow by CN⁻, whereas inhibition by CH_3NC is relieved by CO, azide, N_2O, N_2, or C_2H_2. Rubinson et al.[133] argued that these differences indicated that CN⁻ and CH_3NC bound to different sites. However, it is surely possible that the bound inhibitor–enzyme complexes are sufficiently different, even if bound at the same site or the same oxidation state, to effect the observed differences. It is noteworthy that the affinity for CN^- (K_i = 27 μM) was greater than that for CH_3NC (K_i = 158 μM).

Both cyanide and CH_3NC bind to metals through the carbon atom. Kelly et al.[79] examined isocyanide reduction in the presence of D_2O and observed that CD_4 was formed. This was good evidence that CH_3NC bound nitrogenase through the isocyanide carbon. Conradson et al.[19] attempted to get evidence of CN^- and CH_3NC binding to FeMoco by ^{19}F-NMR and EXAFS. Although both species bound to FeMoco, neither bound to the Mo, according to EXAFS, nor were they removed by CO or azide, which reverse their inhibition of total electron flux. This leaves the option that CN^- and CH_3NC bound to the Fe of the cofactor and such binding may be nonspecific. Such a conclusion is consistent with the evidence of Biggins and Kelly,[8] who found that [14]CN^- bound to Kp1 and Kp2 and deduced that this binding was nonspecific. Conradson et al.[19] also found no evidence that C_2H_2, N_2, azide, or CO bound to the molybdenum of FeMoco and speculated that cofactor was inaccessible to substrate binding except under turnover conditions. However, Smith et al.[147] observed that C_2H_2 altered the EPR spectrum of Kp1, which suggested that C_2H_2 could bind to Kp1-bound cofactor.

C. Azide

Schöllhorn and Burris[135] first demonstrated azide reduction by nitrogenase, apparently in a two-electron reduction to N_2 plus NH_3. Dilworth and Thorneley[26] confirmed this observation, using purified Kp nitrogenase proteins, but also observed that hydrazine was a third reaction product. They also showed that a stoichiometric excess of NH_3 was produced and, to account for this, proposed that azide could be reduced by three mechanisms in two-, six- and eight-electron steps to yield N_2 + NH_3, N_2H_4 + NH_3, and NH_3, respectively, with N_3^- as the substrate in each case (see Table 18-4). Rubinson et al.[132] confirmed the production of both hydrazine and "excess" ammonia when azide was reduced but disagreed about the nature of the substrate and the source of "excess" NH_3. They observed that the concentration of N_2H_4 produced was directly proportional to the HN_3 but not to the N_3^- concentration. The "excess" NH_3 concentration, on the other hand, correlated with the N_3^- concentration and not with HN_3. In fact, either HN_3 or D_2, both known inhibitors of N_2 fixation (and D_2 inhibits no other substrate reduction), inhibited "excess" NH_3 production. In the case of inhibition by D_2 the drop in NH_3 production exactly matched the rise in N_2.

Rubinson et al.[132] also addressed the question of how the "excess" NH_3 arose from N_2 reduction, given that the amount of N_2 produced, if distributed in the gas phase, would be well below the K_s for N_2. They argued that N_2 must be produced close to the N_2-binding site in a receptive state (i.e., state E_3 of the Lowe–Thorneley model). This makes an interesting comparison with N_2O reduction when the N_2 produced is also reduced further, though not at infinite N_2O levels.

There is a close similarity between the reactions:

$$N_3^- + 3H^+ + 2e^- \rightarrow N_2 + NH_3$$
$$N_2O + 2H^+ + 2e^- \rightarrow N_2 + H_2O$$

which are both catalyzed by nitrogenase. Rubinson et al.[132] proposed the following mechanism for N_3^- reduction.

$$E + N_3^- \rightarrow E{-}N{-}N{\equiv}N^- \xrightarrow{3H^+} E{-}NH_2 + N_2$$

and, by analogy, Vaughn and Burgess[167] proposed a similar mechanism for N_2O reduction:

$$E + {}^-O{-}^+N{\equiv}N \rightarrow E{-}O{-}N{\equiv}N: \xrightarrow{2H^+} E + H_2O + N_2$$

Both mechanisms involve N_2 release from the enzyme. A further similarity is that the N_2 produced can be reduced to NH_3.[77,132] N_2O is a competitive inhibitor of N_2 reduction and presumably binds to the same oxidation state of the Lowe–Thorneley mechanism. Reduction of N_2O followed by substrate release would restore the same oxidation level, immediately available for N_2 binding. N_3^-, on the other hand, is apparently not a competitive inhibitor of N_2 reduction, and presumably the oxidation state of the MoFe protein on product release is unlikely to be the same as that which binds N_2. If the Lowe–Thorneley mechanism is correct, that substrates bind to and are released only by the free MoFe protein, then in the case of N_3^- reduction, a slow reduction cycle to produce the correct binding state would have to occur, which could allow dispersal of N_2. If, on the other hand, the mechanism was such that the terminal N of the bound azide produced NH_3, rather than the enzyme-bound N, and the N_2 remained bound until the enzyme–N–N complex was reduced to the right oxidation state to produce NH_3, then dispersal would not occur. Such a mechanism would be consistent with Dilworth and Thorneley's observation that at high pH, N_3^- would abolish H_2 production.[26]

A similar mechanism could be invoked for N_2O reduction; that is, the terminal N rather than O could bind to the enzyme. Little is known about N_2O binding to metal complexes; Tuan and Hoffman[166] calculated that N-linked complexes are likely to be more stable than O-linked complexes.

A third important difference of opinion between Dilworth and Thorneley[26] and Rubinson et al.[132] concerned H_2 production in the presence of azide. The earlier authors' data indicated that at high pH and infinite azide concentration H_2 production would be completely inhibited. Rubinson et al. disagreed. Their data indicated that even at a high azide concentration (25 mM) and Av2/Av1 = 8 at pH 7.6, 35% of the electrons still went to produce H_2. This, they argued, made azide reduction different from that of C_2H_2, cyanide, or CH_3NC and closely resembled N_2 reduction. However, their data indicated that the percentage electron allocation to H_2 decreased with the Av2/Av1 ratio. This is in clear contrast to what happens during N_2 fixation and resembles cyanide or isocyanide reduction. Whether or not H_2 production is inhibited is important, because it indicates to which oxidation state, in the Lowe–Thorneley mechanism, a substrate binds. If a substrate abolishes H_2 production by nitrogenase, then it probably binds to a more oxidized state of the MoFe protein cycle than that of H_2 release and, therefore, prevents the latter from happening, rather than simply competing with H^+ for the electron source. It is possible to envisage, given an infinite substrate concentration and a high Fe–MoFe protein ratio, that any substrate other than N_2 should be able to outcompete H^+ reduction. However, if a substrate promotes H_2 release as the Fe/MoFe protein ratio increases, then it is mandatory that it bind to a higher oxidation state than that of H_2 release. In the present case, if Dilworth and Thorneley's predictions are correct that H_2 release is abolished at infinite azide concentrations and if "excess" NH_3 is produced in that condition, this argues either in favor of the eight-electron reduction of azide to NH_3 or that the two nitrogen atoms produced by the two-electron reduction remain bound to the enzyme and, therefore, do not need to displace H_2 before further reduction.

D. Nitrite

Vaughn and Burgess[167] reported that nitrite was a substrate for nitrogenase and was reduced in a six-electron process to NH_3. No intermediates were reported. They suggested a mechanism involving an N-bound NO_2^- that was primarily reduced to a nitrosyl-bound (E—N=O) intermediate that was then reduced to NH_3 + H_2O. Nitrite also inactivated the Fe protein, a process that was enhanced by ATP. Meyer[111] had also observed this inactivation of the Fe protein by NO_2^- and concluded that NO, which was produced nonenzymically in the reaction, was the destructive agent. Vaughn and Burgess, however, observed that levels of NO higher than anything present in the assay mixture did not damage the Fe protein, although even higher NO concentrations did. They concluded that NO_2^- was the damaging agent.

E. Cyanamide

Miller and Eady[113] reported that cyanamide was reduced by both Kp and Ac nitrogenases in a six- or eight-electron process to yield methylamine and ammonia or methane and ammonia, respectively (see Table 18-4). The vanadium nitrogenase of *A. chroococcum* also reduced cyanamide but with a higher K_m and lower V_{max} than the molybdenum enzyme. H_2 production, in the presence of cyanamide, increased at Kp2/Kp1 ratios above 3, analogous to the situation with cyanide as the substrate. This suggested that cyanamide also bound to a more oxidized form of the enzyme than that at which H_2 was released.

F. Cyclopropene

Perhaps the most interesting feature of cyclopropene reduction by nitrogenase is its unique product pattern: a cyclopropane/propene ratio of $1:2$.[102] Recently, Gemoets et al.[43] compared reduction of cyclopropene by Kp nitrogenase and NifV⁻Kp nitrogenase, which lacks the homocitrate moiety of FeMoco. The product ratio with the latter enzyme was cyclopropane–propene, $1:1.4$. Isolated FeMoco reduced cyclopropene with borohydride but only to give cyclopropane and very little propene (McKenna et al.[103]). The results suggest that the unique ratio of products is a property of the intact cofactor constrained by the ligand surround of the apoprotein.

G. Diazirine

McKenna et al.,[105] following their observation that cyclopropene was reduced by nitrogenase, reported that the cyclic compound diazirine (CH_2N_2) was also reduced by the enzyme (see Table 18-4), like cyanamide, in six- or eight-electron steps to yield methylamine or methane and ammonia. H_2 was also produced. The absence of HD exchange indicated that the methane was not formed via a bound methyldiazine intermediate. Orme-Johnson[122] reviewed diazirine reduction in some detail.

H. Acetylene and Ethylene

Lowe et al.[97] reported four transient EPR signals during Kp nitrogenase turnover that arose from Fe–S clusters in the MoFe protein. One of these, a sharp axial signal at $g = 1.925$, 2.000, and 2.000 was observed only in the presence of C_2H_4 and is, as yet, the only indication of a nitrogenase–product complex. However, $^{13}C_2H_2$ and $^{13}C_2H_4$ addition provided no evidence that acetylene or ethylene bound to Fe–S clusters and the preceding EPR signal was later shown to be from FeMoco.[59] C_2H_2 was shown to bind to two sites (or oxidation states) and C_2H_4 to one. Ashby et al.[6] later showed that high levels of C_2H_4 inhibited total electron flux and the ethylene was reduced to ethane at about 1% of the total electron flux, provided the component protein ratios were extreme, that is, at Kp2/Kp1 > 5 or high Kp1/Kp2 ratios, but not at Kp2/Kp1 around 1. In a recent study, Fisher et al.[38] have shown that (1) C_2H_2 inhibits total electron flux at high component protein concentrations and (2) C_2H_4 is released at oxidation state E_3 of the MoFe protein cycle.

VI. Reactions Catalyzed by Single Nitrogenase Proteins

A. H_2 Oxidation by Av1

Wang and Watt[170] observed that purified and crystalline samples of Av1 possessed uptake hydrogenase activity in that they catalyzed T_2 uptake in the presence of methylene blue. The four-electron oxidized state of Av1 was most active; the dithionite-reduced and six-electron oxidized states were less so. Critical observations that distinguished this activity from Av uptake hydrogenase were (1) no response to the inhibitors CO or C_2H_2, (2) no H_2-producing activity with reduced methyl viologen, and (3) the ability to use O_2 as an electron acceptor, which uptake hydrogenase can only effect via the respiratory chain. The authors noted that H_2 protected Av1 from O_2 damage. This interesting observation has not been followed up or confirmed; it needs reinvestigating using *Azotobacter* mutants that lack the conventional uptake hydrogenase.

B. Dithionite Oxidation by Av2

McKenna et al.[104] reported that the dithionite-removing ability of nitrogenase Fe proteins (Av2) was due to the protein acting as a dismutase yielding mainly $S_2O_3^{2-}$ and SO_3^{2-}. This was followed by the oxidation of Av2 without significant loss of activity. MgATP accelerated the dismutase activity.

C. Stoichiometric O_2 Reduction by Ac2 and Kp2

Thorneley and Ashby[153] studied the oxidation of Ac2 and Kp2 together with their MgATP and MgADP complexes, by O_2 and H_2O_2. Both proteins, either free or complexed to nucleotides, reduced O_2 without becoming inactivated, provided the molar excess of protein was fourfold. At higher O_2 levels the proteins were oxygen damaged. The authors described the nondamaging O_2 uptake as auto-protection by analogy with respiratory or conformational protection. The fastest-reacting species with excess O_2 was Ac2(MgADP)$_2$, with a second-order rate constant of 1.5×10^{-5} M^{-1} sec^{-1}. The reaction rates correlated closely with the midpoint potentials of the reacting species. The reaction rates when the proteins, rather than O_2, were in excess were slightly slower but of a similar order. These reactions were biphasic and the kinetics suggested an intermediate such as H_2O_2. Hydrogen peroxide slowly oxidized both proteins, prompting the suggestion that catalase probably operated in conjunction with the Fe protein during autoprotection. Additional aspects of this mechanism are discussed in Chapter 2.

VII. Conclusions

Clearly, the most pressing need in nitrogenase biochemistry is to unravel the structure of FeMoco and how it is constrained within the apo-protein. To this end, the establishment of a method of *in vitro* synthesis of cofactor and the discovery that it contains homocitrate are major advances. We now know that the lack of homocitrate (NifV$^-$) or its replacement by analogs causes different substrate responses. This is also true of the isolation of the cofactor from the apo-protein, indicating that the binding constraints within the holo-protein are of paramount importance. Site-directed mutagenesis has been used successfully to identify some of the binding residues, and no doubt more will be discovered in the near future. Evidence is accumulating that cofactor is bound by the α subunit of the MoFe protein, but the β subunit has been shown to contain iron and to undergo immediate biochemical complementation with the α subunit *in vitro*, suggesting that both subunits can be formed intact in separate mutants. In time these separate subunits will be purified and analyzed for their metal-cluster contents, and perhaps we will discover whether cofactor can bind only to the α subunit or to both. Amino acid–metal cluster chemistry will help to indicate likely binding residues. Thus, we can expect steady progress on the MoFe apo-protein. Progress on understanding the chemical nature of the cofactor will depend on more refined spectroscopic techniques (the emergency of ENDOR is marked), but most of all, it will depend on whether pure and analyzable crystals are obtained.

Perhaps the major breakthrough in the Fe protein biochemistry is the discovery of the $S = 3/2$ spin EPR signal, which, together with the $S = 1/2$ spin state EPR, yields an electron integration of one spin per mole. Much effort has been spent on investigating nucleotide binding to both proteins, with a consensus agreement that two molecules of MgATP or MgADP bind to the Fe protein, but with little agreement on nucleotide binding to the MoFe protein. Perhaps site-directed mutagenesis will in time be used to address which regions of the Fe protein are important in nucleotide binding.

The Lowe–Thorneley model is the most comprehensive analysis of nitrogenase function, and its value lies in the ability of its predictions to be tested experimentally. One of its tenets is that it presumes interaction between one Fe protein and the α, β dimer of the MoFe protein containing one FeMoco molecule. One would expect, therefore, an *in vivo* ratio of at least two Fe proteins per MoFe protein $\alpha_2\beta_2$ tetramer. As observed *in vitro*, a high ratio of Fe/MoFe protein and a high overall concentration of proteins encourages

N_2 reduction over H_2 production. However, at least three publications record the *in vivo* Fe protein/MoFe protein ratio as low as between 1.4 and 1:1.[27,78,84] A second interesting development is an interpretation of crystallographic data on Av1: that the two cofactors are close together, sufficiently close for N_2 to bridge the Mo atoms.[140] Whether the Lowe–Thorneley model will require modification in view of these preliminary findings and interpretation remains to be seen. One has to bear in mind that the Lowe–Thorneley model was devised using sodium dithionite as the reductant. Reduced flavodoxin transfers electrons faster than SO_2^- to the Fe protein.[51,133,181] No doubt, in time, the pre-steady-state and steady-state kinetics of nitrogenase activity with its biological electron donor will be investigated. The future for biochemical research into the molybdenum nitrogenase seems almost spoiled for choice.

Acknowledgments

I thank Drs. Susan Hill, Roger Thorneley, Ray Richards, and, in particular, David Lowe for useful discussions, Barry Smith and David Lowe for reading the manuscript, and Mrs. R. J. Foote for typing.

References

1. Anderson, G.L., and J.B.Howard, "Reactions with oxidized iron protein of *Azotobacter vinelandii* nitrogenase: Formation of a 2Fe center," *Biochemistry* 23:2118–2122 (1984)

2. Arber, J.M., B.R. Dobson, R.R. Eady, A.C. Flood, C.D. Garner, C.A. Gormal, S.S. Hasnain, B.E. Smith, and P. Stephens, "X-ray absorption spectroscopy of the VFe protein of *Azotobacter chroococcum* and the iron molybdenum cofactor of *Klebsiella pneumoniae*," *Reueil der Travaux Chimique des Pays-Bas* 106:237 (1987).

3. Arber, J.M., A.C. Flood, F.D.Garner, C.A. Gormal, S.S. Hasnain, and B. E. Smith. "Iron-edge X-ray absorption spectroscopy of the iron molybdenum cofactor of nitrogenase from *Klebsiella pneumoniae*," *Biochem. J.* 252:421–425 (1988).

4. Arber, J.M., A.C. Flood, C.D. Garner, S.S. Hasnain, and B.E. Smith, "X-ray absorption spectroscopy of $Cu–M_6–S$ and Fe–Mo–S systems of biological relevance," *J. Phys.* 47:C8–1159–1163 (1986).

5. Arnold, W., A. Rump, W. Klipp, U.B. Priefer, and A. Pühler. "Nucleotide sequence of a 24,206 base pair DNA fragment carrying the entire nitrogen fixation gene cluster of *Klebsiella pneumoniae*," *J. Mol. Biol.* 203:715–738 (1988).

6. Ashby, G.A., M.J. Dilworth, and R.N.F. Thorneley. "*Klebsiella pneumoniae* nitrogenase: Inhibition of hydrogen evolution by ethylene and the reduction of ethylene to ethane," *Biochem. J.* 247:547–554 (1987).

7. Ashby, G.A., and R.N.F. Thorneley, "Nitrogenase of *Klebsiella pneumoniae*. Kinetic studies on the Fe protein involving reduction by sodium dithionite, the binding of MgADP and a conformation change that alters the reactivity of the 4Fe–4S centre," *Biochem. J.* 246:455–465 (1987).

8. Biggins, D.R., and M. Kelley, "Interaction of nitrogenase from *Klebsiella pneumoniae* with ATP or cyanide," *Biochim. Biophys. Acta 205*:288–299 (1970).

9. Braaksma, A., H. Haaker, H. Grande, and C. Veeger. "The effect of the redox potential on the activity of the nitrogenase and on the Fe protein of *Azotobacter vinelandii*," *Eur. J. Biochem. 121*:483–491 (1982).

10. Braaksma, A., H. Haaker, and C. Veeger, "Fully active Fe protein of the nitrogenase from *Azotobacter vinelandii* contains at least eight iron atoms and eight sulphide atoms per molecule," *Eur. J. Biochem. 133*:71–76 (1983).

11. Brigle, K.E., R.A. Setterquist, D.R. Dean, J.J. Cantwell, M.C. Weiss, and W.E. Newton, "Site-directed mutagenesis of the nitrogenase MoFe protein of *Azotobacter vinelandii*," *Proc. Natl. Acad. Sci. USA 84*(20):7066–7069 (1987).

12. Bui, P.T., and L.E. Mortenson, "Mechanism of the enzymic reduction of N_2: The binding of adenosine 5'-triphosphate and cyanide to the N_2-reducing system," *Proc. Natl. Acad. Sci. USA 61*:1021–1027 (1968).

13. Burgess, B.K., J.L. Corbin, J.F. Rubinson, J.-G. Li, M.J. Dilworth, and W.E. Newton, "Nitrogenase Reactivity," in C. Veeger and W.E. Newton (eds.), *Advances in Nitrogen Fixation*. Proceedings of the 5th International Symposium on Nitrogen Fixation, Noordwijkerhout, Aug. 28–Sep. 3, 1983, Nijhoff/Junk, The Hague, 1984, p. 146.

14. Burgess, B.K., D.B. Jacobs, and E.I. Stiefel, "Large scale purification of high activity *Azotobacter vinelandii* nitrogenase," *Biochim. Biophys. Acta 614*:196–209 (1980).

15. Burgess, B.K., S. Wherland, W.E. Newton, and E.I. Stiefel, "Nitrogenase reactivity: Insight into the nitrogen-fixing process through hydrogen-inhibition and HD-forming reactions," *Biochemistry 20*:5140–5146 (1981).

16. Carnahan, J.E., L.E. Mortenson, J.F. Mower, and J.E. Castle, "Nitrogen fixation in cell-free extracts of *Clostridium pasteurianum*," *Biochim. Biophys. Acta 44*:520–535 (1960).

17. Chatt, J., "Chemistry relevant to the biological fixation of nitrogen," in W.D.P. Stewart and J.R. Gallon (eds.), *Proc. Phytochem. Soc. Eur.*, Academic Press, London, 1980, pp. 1–18.

18. Conradson, S.D., B.K. Burgess, W.E. Newton, L.E. Mortenson, and K.O. Hodgson. "Structural studies of the molybdenum site in the MoFe protein and its FeMo cofactor by EXAFS," *J. Am. Chem. Soc. 109*:7507–7515 (1987).

19. Conradson, S.D., B.K. Burgess, S.A. Vaughn, A.L. Roe, B. Hedman, K.G. Hodgson, and R.H. Holm, "Cyanide and methylisocyanide binding to the isolated iron-molybdenum cofactor of nitrogenase," *J. Biol. Chem. 264*(27):15967–15974 (1989).

20. Cordewener, J., H. Haaker, P. van Ewijk, and C. Veeger, "Properties of the MgATP and MgADP binding sites on the Fe protein of nitrogenase from *Azotobacter vinelandii*," *Eur. J. Biochem. 148*:499–508 (1985).

21. Cordewener, J., H. Haaker, and C. Veeger, "Binding of MgATP to the nitrogenase proteins from *Azotobacter vinelandii*," *Eur. J. Biochem. 132*:47–54 (1983).

22. Cordewener, J., M. Krüse-Wolters, H. Wassink, H. Haaker, and C. Veeger, "The role of MgADP hydrolysis in nitrogenase catalysis," *Eur. J. Biochem. 172*:739–745 (1988).

23. Cordewener, J., A. Ten Asbroek, H. Wassink, R.R. Eady, H. Haaker, and C. Veeger, "Binding of ADP and orthophosphate during the ATPase reaction of nitrogenase," *Eur. J. Biochem. 162*:265–270 (1987).

24. Day, E.P., T.A. Kent, P.A. Lindahl, E. Munck, W.H. Orme-Johnson, H. Roder, and A. Roy. "Squid measurements of metalloprotein magnetisation. New methods applied to the nitrogenase proteins," *Biophys. J.* 837–854 (1987).

25. Dean, D.R., K.E. Brigle, H.D. May, and W.E. Newton. "Site-directed mutagenesis of the nitrogenase MoFe protein," in H. Bothe, F.J. de Bruijn, and W.E. Newton (eds.), *Nitrogen Fixation: Hundred Years After*. Proceedings of the 7th International Congress on Nitrogen Fixation, Köln, Mar. 13–20, Gustav Fischer, Stuttgart, 1988, pp. 107–113.

26. Dilworth, M.J., and R.N.F. Thorneley, "Nitrogenase of *Klebsiella pneumoniae:* Hydrazine is a product of azide reduction," *Biochem. J.* 193:971–983 (1981).

27. Dingler, C., J. Kuhla, H. Wassink, and J. Oelze, "Levels and activities of nitrogenase proteins in *Azotobacter vinelandii* grown at different dissolved oxygen concentration," *J. Bacteriol.* 170:2148–2151 (1988).

28. Dunham, W.R., W.R. Hagen, A. Braaksma, H.J. Grande, and H. Haaker, "The importance of quantitative Mossbauer spectroscopy of MoFe protein from *Azotobacter vinelandii,*" *Eur. J. Biochem.* 146:497–501 (1985).

29. Eady, R.R., "Enzymology of free-living diazotrophs," in W.J. Broughton and S. Pühler (eds.), *Nitrogen Fixation, Vol. 4: Molecular Biology,* Clarendon Press, Oxford, 1986, pp. 1–49.

30. Eady, R.R., D.J. Lowe, and R.N.F. Thorneley, "Nitrogenase of *Klebsiella pneumoniae:* A pre-steady-state burst of ATP hydrolysis is coupled to electron transfer between the component proteins," *FEBS Lett.* 95:211–213 (1978).

31. Eady, R.R., and B.E. Smith, "Physicochemical properties of nitrogenase and its components," in R.W.F. Hardy, F. Bottomley, and R.C. Burns (eds.), *A Treatise on Dinitrogen Fixation,* New York, 1979, pp. 399–490.

32. Eidsness, M.K., A.M. Flank, B.E. Smith, A.C. Flood, C.D. Garner, and S.P. Cramer, "EXAFS of *Klebsiella pneumoniae* nitrogenase MoFe protein from wild type and nifV mutant strains," *J. Am. Chem. Soc.* 108:2746–2748 (1986).

33. Eldredge, P.A., R.F. Bryan, S. Ekkehard, and B.A. Averell, "The [MoFe$_6$S$_6$(CO)$_{16}$]$^{2-}$ ion. A new model for the FeMo cofactor of nitrogenase," *J. Am. Chem. Soc.* 110:5573–5574 (1988).

34. Emerich, D.W., and R.H. Burris, "Nitrogenase from *Bacillus polymyxa:* Purification and properties of the component proteins." *Biochem. Biophys. Acta* 536:172–183 (1978).

35. Emerich, D.W., T. Ljones, and R.H. Burris, "Nitrogenase: Properties of the catalytically inactive complex between the *Azotobacter vinelandii* MoFe protein and the *Clostridium pasteurianum* Fe protein," *Biochem. Biophys. Acta* 527:359–369 (1978).

36. Euler, W.B., J. Martinsen, J.W. McDonald, G.D. Watt, and Z.-C. Wang, "Double integration and titration of the electron paramagnetic resonance signal in the molybdenum iron protein of *Azotobacter vinelandii,*" *Biochemistry* 23:3021–3024 (1984).

37. Evans, D.J., and G.J. Leigh, "Serinato-, tyrosinato- and prolinato-derivatives of Fe$_4$S$_4$ clusters," *J. Chem. Soc. Chem. Commun.* 395–396 (1988).

38. Fisher, K., R.N.F. Thorneley, and D.J. Lowe, "Nitrogenase of *Klebsiella pneumoniae.* Mechanism of acetylene reduction," *Biochem. J.,* 272:621–625, 1990.

39. Flank, A.M., M. Weiniger, L.E. Mortenson, and S.P. Cramer, "Single crystal EXAFS of nitrogenase," *J. Am. Chem. Soc.* 108:1049–1055 (1986).

40. Frank, P., S.F. Gheller, W.E. Newton, and K.O. Hodgson, "Purification and spectroscopic characteristics in N-methyl formamide of the Azotobacter vinelandii Fe-Mo cofactor," Biochem. Biophys. Res. Commun. 163:746–754 (1989).

41. Garner, C.D., J.M. Arber, I. Harvey, S.S. Hasnain, R.R. Eady, B. E. Smith, E. Boer, and R. Wever. "Characterization of molybdenum and vanadium centres in enzymes by X-ray absorption spectroscopy," Polyhedron 8:1649–1652 (1989).

42. Garner, C.D., J.M. Arber, S.S. Hasnain, B.R. Dobson, R.R. Eady and B. E. Smith, "X-ray absorption spectroscopic studies of the catalytic centres of nitrogenases," Physica B 158:74–77 (1989).

43. Gemoets, J.P., J. Bravo, C.E. McKenna, G.J. Leigh, and B.E. Smith, "Reduction of cyclopropene by NifV⁻ and wild-type nitrogenases from Klebsiella pneumoniae," Biochem. J. 258:487–491 (1989).

44. George, G.N., R.E. Bare, H. Jin, E.I. Stiefel, and R.C. Prince, "EPR spectroscopic studies on the molybdenum–iron site of nitrogenase from Clostridium pasteurianum," Biochem. J . 262:349–352 (1989).

45. George, G.N., W.E. Cleland, J.H. Enemark, B.E. Smith, C.A. Kipke, S. Roberts, and S. P. Cramer, "L-edge spectroscopy of molybdenum compounds and enzymes," in Stanford Synchroton Radiation Laboratory Activity Report, S. Robinson and K. Cantwell (eds.), 1988, p. 145a.

46. George, S.J., F.A. Armstrong, C.E. Hatchikian, and A.J. Thomson, "Electrochemical and spectroscopic characterization of the conversion of the 7Fe into the 8Fe form of ferrodoxin III from Desulfovibrio africanus. Identification of a [4Fe–4S] cluster with a non-cysteine ligand," Biochem. J. 264:275–284 (1989).

47. Govezensky, D., and A. Zamir, "Structure function relationship in the α subunit of Klebsiella pneumoniae nitrogenase MoFe protein from analysis of nifD mutants," J. Bacteriol. 171:5729–5735 (1989).

48. Guth, J.H., and R.H. Burris, "Inhibition of nitrogenase-catalysed NH_3 formation by H_2," Biochemistry 22:5111–5222 (1983).

49. Haaker, H., A. Braaksma, J. Cordewener, J. Klugkist, H. Wassink, H. Grande, R. R. Eady, and C. Veeger, "Iron-sulfide content and ATP-binding properties of nitrogenase component II from Azotobacter vinelandii," in C. Veeger and W. E. Newton (eds.) Advances in Nitrogen Fixation Research. Nijhoff/Junk, The Hague, 1984, pp. 123–131.

50. Haaker, H., J. Cordewener, A.T. Asbroek, H. Wassink, R.R. Eady, and C. Veeger, "The role of Fe protein in nitrogenase catalysis," in Evans, H.J., P.J. Bottomley, and W.E. Newton (eds.) Nitrogen Fixation Research Progress, Martinus Nijhoff, Dordrecht, 1985, pp. 567–576.

51. Hageman, R.V., and R.H. Burris. "Kinetic studies on electron transfer and interaction between nitrogenase components from Azotobacter vinelandii," Biochemistry 17:4117–4124 (1978).

52. Hageman, R.V., and R.H. Burris. "Electron allocation to alternative substrates of Azotobacter nitrogenase is controlled by the electron flux through dinitrogenase," Biochem. Biophys. Acta 591:63–75 (1980).

53. Hageman, R.V., W.H. Orme-Johnson, and R.H. Burris, "Role of magnesium adenosine 5'-triphosphate in the hydrogen evolution reaction catalyzed by nitrogenase from Azotobacter vinelandii," Biochemistry 19:2333–2342 (1980).

54. Hagen, W.R., W.R. Dunham, A. Braaksma, and H. Haaker, "On the prosthetic group(s) of component II from nitrogenase. EPR of the Fe protein from *Azotobacter vinelandii,*" *FEBS Lett. 187*:146–150 (1985).

55. Hagen, W.R., R.R. Eady, W.R. Dunham, and H. Haaker, "A novel 3/2 EPR Signal associated with native Fe protein of nitrogenase," *FEBS Lett. 189*:250–254 (1985).

56. Hagen, W.R., H. Wassink, R.R. Eady, B.E. Smith, and H. Haaker, "Quantitative EPR of an S = 7/2 system in thionine-oxidized MoFe proteins of nitrogenase. A redefinition of the P-cluster concept," *Eur. J. Biochem. 169*:457–465 (1987).

57. Hausinger, R.P., and J.B. Howard, "The amino acid sequences of the nitrogenase-iron protein from *Azotobacter vinelandii,*" *J. Biol. Chem. 257*:2483–2490 (1982).

58. Hausinger, R.P., and J.B. Howard, "Thiol reactivity of the nitrogenase Fe protein from *Azotobacter vinelandii,*" *J. Biol. Chem. 258*:13486–13492 (1983).

59. Hawkes, T.R., P.A. McLean, and B.E. Smith, "Nitrogenase of *nifV* mutants of *Klebsiella pneumoniae* contains an altered form of the iron-molybdenum cofactor," *Biochem. J. 217*:317–321 (1984).

60. Hawkes, T.R., and B.E. Smith, "Purification and characterisation of the inactive MoFe protein (Nif B⁻ Kp1) of the nitrogenase from *nifB* mutants of *Klebsiella pneumoniae,*" *Biochem. J. 209*:43–50 (1983).

61. Hedman, B., P. Frank, S.F. Cheller, A.Z. Roe, W.E. Newton, and K.O. Hodgson, "New structural insight into the iron molybdenum cofactor from *Azotobacter vinelandii* nitrogenase through sulphur K and molybdenum L X-ray absorption edge studies," *J. Am. Chem. Soc. 110*:3798–3805 (1988).

62. Henderson, R.A., "The detection and reactivity of [MoH₄(η²-H₂)(Ph₂PCH₂CH₂PPh₂)₂]²⁺," *J. Chem. Soc., Chem. Commun.* 1670–1672 (1987).

63. Hoffman, B. M., J. E. Roberts, and W. H. Orme-Johnson, "⁹⁵Mo and ¹H ENDOR spectroscopy of the nitrogenase MoFe protein," *J. Am. Chem. Soc. 104*:860–862 (1982).

64. Hoffman, B. M., R. A. Venters, and J. E. Roberts, "⁵⁷Fe ENDOR of the nitrogenase MoFe protein," *J. Am. Chem. Soc. 104*:4711–4712 (1982).

65. Holland, D., A. Zilberstein, D. Govezensky, D. Solomon, and A. Zamir, "Nitrogenase MoFe protein subunits from *Klebsiella pneumoniae* expressed in foreign hosts. Characteristics and interactions," *J. Biol. Chem. 262*:8814–8820 (1987).

66. Hoover, T. R., J. Imperial, J. Liang, P. W. Ludden, and V. K. Shah, "Dinitrogenase with altered substrate specificity results from the use of homocitrate analogues for *in vitro* synethesis of the iron–molybdenum cofactor," *Biochemistry USA. 27*:3647–3651 (1988).

67. Hoover, T. R., J. Imperial, P. W. Ludden, and V. K. Shah, "Homocitrate is a component of the iron-molybdenum cofactor of nitrogenase." *Biochemistry USA 28*:2768–2771 (1989).

68. Hoover, T. R., J. Imperial, P. W. Ludden, and V. K. Shah, "Homocitrate cures the NifV⁻ phenotype in *Klebsiella pneumoniae,*" *J. Bacteriol. 170*:1978–1979 (1988).

69. Hoover, T. R., A. D. Robertson, R. L. Cerny, R. N. Hayes, J. Imperial, V. K. Shah, and P. W. Ludden. "Identification of the V factor needed for synthesis of the iron–molybdenum cofactor of nitrogenase as homocitrate," *Nature 329*:855–857 (1987).

70. Hoover, T. R., V. K. Shah, G. P. Roberts, and P. W. Ludden, "*nifV*-dependent, low molecular-weight factor required for *in vitro* synthesis of iron-molybdenum cofactor of nitrogenase," *J. Bacteriol. 167*:999–1003 (1986).

71. Howard, J. B., R. Davis, B. Moldenhauer, V. L. Cash, and D. Dean, "FeS cluster ligands are the only cysteines required for nitrogenase Fe protein activities," *J. Biol. Chem. 264*:11270–11274 (1989).

72. Huyhn, B. H., M. T. Henzl, J. A. Christner, R. Zimmerman, W. H. Orme-Johnson, and E. Munck, "Mossbauer studies of the MoFe protein from *Clostridium pasteurianum* W5," *Biochim. Biophys. Acta 623*:124–138 (1980).

73. Imperial, J., T. R. Hoover, M. S. Madden, P. W. Ludden, and V. K. Shah, "Substrate reduction properties of dinitrogenase activated *in vitro* are dependent upon the presence of homocitrate or its analogues during iron–molybdenum cofactor synthesis." *Biochemistry USA 28*:7796–7799 (1989).

74. Imperial, J., V. K. Shah, R. A. Ugalde, P. W. Ludden, and W. J. Brill, "Iron–molybdenum cofactor synthesis in *Azotobacter vinelandii* nif⁻ mutants," *J. Bacteriol. 169*:1784–1786 (1987).

75. Ioannidis, I., and M. Buck, "Nucleotide sequence of the *Klebsiella pneumoniae nifD* gene and the predicted amino acid sequence of the α subunit of nitrogenase MoFe protein," *Biochem. J. 247*:287–291 (1987).

76. Jensen, B. B., and R. H. Burris, "Effect of high pN_2 and high pD_2 on NH_3 production, H_2 evolution and HD formation by nitrogenase," *Biochemistry 24*:1141–1147 (1985).

77. Jensen, B. B., and R. H. Burris, "N_2O as a substrate and as a competitive inhibitor of nitrogenase," *Biochemistry 25*:1083–1088 (1986).

78. Jouanneau, Y., B. Wong, and P. M. Vignais, "Stimulation by light of nitrogenase synthesis in cells of *Rhodopseudomonas capsulata* growing in N-limited continuous cultures," *Biochim. Biophys. Acta 808*:149–155 (1985).

79. Kelly, M., R. L. Richards, and J. R. Postgate, "Reduction of cyanide and isocyanide by nitrogenase of *Azotobacter chroococcum*," *Biochem. J. 102*:1c–3c (1967).

80. Kennedy, C. K., R. R. Eady, E. Kondorosi, and D. K. Rekosh, "The molybdenum–iron protein of *Klebsiella pneumoniae* nitrogenase. Evidence for non-identical subunits from peptide mapping," *Biochem. J. 155*:383–389 (1976).

81. Kent, H. M., M. Baines, C. Gormal, B. E. Smith, and M. Buck, "Analysis of site-directed mutations in the α- and β-subunits of *Klebsiella pneumoniae* nitrogenase," *Mol. Microbiol.* (in press), 1990, *4*:1497–1504.

82. Kent, H. M., I. Ioannides, C. Gormal, B. E. Smith, and M. Buck, "Site-directed mutagenesis of Kp nitrogenase. Effects of modifying conserved cysteine residues in the α and β subunits," *Biochem. J. 264*:257–264 (1989).

83. Kimber, S. J., E. O. Bishop, and B. E. Smith, "Evidence on the role of ATP in the mechanism of nitrogenase from proton NMR relaxation studies on metal and nucleotide binding to the molybdenum–iron protein," *Biochim. Biophys. Acta 705*:383–395 (1982).

84. Klugkist, J., H. Haaker, H. Wassink, and C. Veeger, "The catalytic activity of nitrogenase in intact *Azotobacter vinelandii* cells," *Eur. J. Biochem. 146*:509–515 (1985).

85. Kurtz, D. M., R. S. MacMillan, B. K. Burgess, L. E. Mortenson, and R. H. Holm, "Identification of iron–sulfur centers in the iron–molybdenum proteins of nitrogenase," *Proc. Natl. Acad. Sci. USA 76*:4986–4989 (1979).

86. Li, J.-G., B. K. Burgess, and J. L. Corbin, "Nitrogenase reactivity: Cyanide as a substrate and inhibitor," *Biochemistry 21*:4393–4402 (1982).

87. Li, J. L., and R. H. Burris, "Influence of pN_2 and pD_2 on HD formation by various nitrogenases," *Biochemistry 22*:4472–4480 (1983).

88. Liang, J., and R. H. Burris, "Interactions among N_2, N_2O and C_2H_2 as substrates and inhibitors of nitrogenase from *Azotobacter vinelandii*," *Biochemistry* 27:6726–6732 (1988).

89. Liang, J., and R. H. Burris, "Hydrogen burst associated with nitrogenase catalyzed reactions," *Proc. Natl. Acad. Sci. USA* 85:9446–9450 (1988).

90. Liang, J., and R. H. Burris, "N_2O reduction and HD formation by nitrogenase from a Nif V⁻ mutant of *Klebsiella pneumoniae*," *J. Bacteriol.* 171:3176–3180 (1989).

91. Lindahl, P. A., E. P. Day, T. A. Kent, W. H. Orme-Johnson, and E. Munck, "Mossbauer, EPR and magnetization studies of the Azotobacter Fe protein. Evidence for a $[4Fe-4S]^{1+}$ cluster with spin S = 3/2," *J. Biol. Chem.* 260:11160–11173 (1985).

92. Lindahl, P. A., N. J. Gorelick, E. Munck, and W. H. Orme-Johnson, "EPR and Mossbauer studies of nucleotide-bound nitrogenase iron protein from *Azotobacter vinelandii*," *J. Biol. Chem.* 262:14945–14954 (1987).

93. Lindahl, P. A., V. Papaefthymiou, W. H. Orme-Johnson, and E. Munck, "Mossbauer studies of solid thionine-oxidized MoFe protein of nitrogenase," *J. Biol. Chem.* 263:19412–19418 (1988).

94. Ljones, T., and R. H. Burris, "Nitrogenase: The reaction between the Fe protein and bathophenanthroline-disulfonate as a probe for interactions with MgATP," *Biochemistry* 17:1866–1872 (1978).

95. Lough, S. M., D. B. Jacobs, S. F. Perker, J. W. McDonald, and G. D. Watt, "A facile method of isolation for the iron–molybdenum cofactor of nitrogenase and bacterial ferretin from extracts of *A. vinelandii*," *Inorg. Chem. Acta* 151:227–232 (1989).

96. Lowe, D. J., "Simulation of the electron paramagnetic resonance spectrum of the iron protein of nitrogenase: A prediction of the existence of a second paramagnetic center," *Biochem. J.* 175:955–957 (1978).

97. Lowe, D. J., R. R. Eady, and R. N. F. Thorneley, "Electron paramagnetic studies on the nitrogenase of *Klebsiella pneumoniae*: Evidence for acetylene and ethylene-nitrogenase transient complexes," *Biochem. J.* 173:277–290 (1978).

98. Lowe, D. J., K. Fisher, R. N. F. Thorneley, S. W. Vaughn, and B. K. Burgess. "Kinetics and mechanism of the reaction of cyanide with molybdenum nitrogenase from *Azotobacter vinelandii*," *Biochemistry* 28:8460–8466 (1989).

99. Lowe, D. J., and R. N. F. Thorneley, "The mechanism of *Klebsiella pneumoniae* nitrogenase action: Pre-steady-state kinetics of H_2 formation," *Biochem. J.* 224:877–886 (1984).

100. Lowe, D. J., and R. N. F. Thorneley, "The mechanism of *Klebsiella pneumoniae* nitrogenase action: The determination of rate constants required for the simulation of the kinetics of N_2 reduction and H_2 evolution," *Biochem. J.* 224:895–901 (1984).

101. Lowe, D. J., R. N. F. Thorneley, and B. E. Smith, "Nitrogenase," in P. Harrison (ed.), *Metalloproteins. Part 1: Metal Proteins with Redox Roles*, Verlag-Chemie, Basel, 1985, pp. 207–249.

102. McKenna, C. E., and C. W. Huang, "*In vivo* reduction of cyclopropene by *Azotobacter vinelandii* nitrogenase," *Nature* 280:609–611 (1979).

103. McKenna, C. E., J. B. Jones, H. Eran, and C. W. Huang, "Reduction of cyclopropene as criterion of active-site homology between nitrogenase and its FeMo cofactor," *Nature* 280:611–612 (1979).

104. McKenna, C. E., R. Menard, C. J. Dao, P. H. Stephens, and M.-C. McKenna, "Self oxidation of dithionite-AV2 solutions: The phenomenon defined and explained," in H.

Bothe, F. J. de Bruijn, and W. E. Newton (eds.), *Nitrogen Fixation: Hundred Years After,* Gustav Fischer, Stuttgart, 1988, p. 131.

105. McKenna, C. E., P. J. Stephens, H. Eran, G. M. Luo, F. X. Zhang, M. Ding, and H. T. Nguyen, "Substrate interactions with nitrogenase and its FeMo cofactor: Chemical and spectroscopic investigations," in C. Veeger and W. E. Newton (eds.), *Advances in Nitrogen Fixation Research,* Nijhoff/Junk, The Hague, 1984, p. 115.

106. McLean, P. A., and R. A. Dixon, "Requirement of *nifV* gene for production of wild-type nitrogenase enzyme in *Klebsiella pneumoniae,*" *Nature* 292:655–656 (1981).

107. McLean, P. A., V. Papaefthymiou, W. H. Orme-Johnson, and E. Munck, "Isotopic hybrids of nitrogenase. Mossbauer study of MoFe protein with selective ^{57}Fe enrichment of the P clusters." *J. Biol. Chem.* 262:12900–12903 (1987).

108. McLean, P. A., B. E. Smith, and R. A. Dixon, "Nitrogenase of *Klebsiella pneumoniae nifV* mutants: Investigation of the novel carbon monoxide-sensitivity of hydrogen evolution by the mutant enzyme." *Biochem. J.* 211:589–597 (1983).

109. McLean, P. A., A. E. True, M. A. Nelson, S. Chapman, M. R. Godfrey, B. K. Teo, W. H. Orme-Johnson, and B. M. Hoffman, "On the difference between iron–molybdenum cofactor of wild type and *nifV* mutant molybdenum iron proteins of *Klebsiella pneumoniae*: ENDOR, EXAFS and EPR evidence," *J. Am. Chem. Soc.* 109:943–945 (1987).

110. McLean, P. A., D. A. Wink, S. K. Chapman, A. B. Hickman, D. M. McKillop, and W. H. Orme-Johnson. "A new method for extraction of iron molybdenum cofactor (FeMoco) from nitrogenase absorbed to DEAE cellulose. 1. Effects of anions, cations and pre-extraction treatments," *Biochemistry* 28:9402–9406 (1989).

111. Meyer, J., "Comparison of carbon monoxide, nitric oxide and nitrite as inhibitors of the nitrogenase from *Clostridium pasteurianum,*" *Arch. Biochim. Biophys.* 210:246–256 (1981).

112. Meyer, J., J. Gaillard, and J. M. Moulis, "The high spin states of nitrogenase iron proteins. A ^1H-NMR investigation of Cp2," in Bothe, H., F. J. de Bruijn, and W. E. Newton (eds.), *Nitrogen Fixation: Hundred Years After,* Gustav Fischer, Stuttgart, 1988, p. 132.

113. Miller, R. W., and R. R. Eady, "Cyanamide: A new substrate for nitrogenase," *Biochim. Biophys. Acta* 952:290–296 (1988).

114. Miller, R. W., and R. R. Eady, "Molybdenum nitrogenase of *Azotobacter chroococcum:* Tight binding of MgADP to the MoFe protein," *Biochem. J.* 263:725–729 (1989).

115. Miller, R. W., R. L. Robson, M. G. Yates, and R. R. Eady, "Catalysis of exchange of terminal phosphate groups of ATP and ADP by purified nitrogenase proteins," *Can. J. Biochem.* 58:542–548 (1980).

116. Morgan, T. V., R. C. Prince, and L. E. Mortenson, "Electrochemical titration of the S = 3/2 and S = 1/2 states of the iron protein of nitrogenase," *FEBS Lett.* 206:4015–4018 (1986).

117. Mortenson, L. E., "ATP and nitrogen fixation," in W. R. Ulrich, R. J. Aparacio, P. J. Syrrett, and F. Cartillo (eds.), *Inorganic Nitrogen Metabolism,* Springer-Verlag, Berlin, *Heidelberg, New York,* 1987, pp. 165–172.

118. Newton, W. E., S. F. Gheller, R. H. Sands, and W. R. Dunham, "Mössbauer spectroscopy applied to the oxidized and semi-reduced states of the iron molybdenum cofactor of nitrogenase," *Biochem. Biophys. Res. Commun.* 162:882–891 (1989).

119. Newton, W. E., S. F. Gheller, B. J. Feldman, W. R. Dunham, and F. A. Schultz, "Isolated iron–molybdenum cofactor exists in multiple forms in its oxidized and semi-reduced states," *J. Biol. Chem. 264*:1924–1927 (1989).

120. Nicholas, D. J. D., and J. V. Deering, "Repression, derepression and activation of nitrogenase in *Azotobacter vinelandii*," *Aust. J. Biol. Sci. 29*:147–161 (1976).

121. Orme-Johnson, W. J., "Biochemistry of nitrogenase," in A. Hollaender, R. H. Burris, P. R. Day, R. W. F. Hardy, D. R. Helinski, M. R. Lamborg, L. Owens, and R. C. Valentine (eds.), *Genetic Engineering for Nitrogen Fixation,* Plenum New York, 1977, pp. 317–332.

122. Orme-Johnson, W. H., "Molecular basis of biological nitrogen fixation," *Ann. Rev. Biophys. Chem. 14*:419–459 (1985).

123. Orme-Johnson, W. H., and L. C. Davis, "Current topics and problems in the enzymology of nitrogenase," in W. Lovenberg (ed.), *Iron Sulfur Proteins,* Academic Press, New York, 1976, pp. 15–60.

124. Orme-Johnson, W. H., L. C. Davis, M. T. Henzl, B. A. Averill, N. R. Orme-Johnson, E. Munck, and R. Zimmerman, "Components and pathways in biological nitrogen fixation," in W. E. Newton, J. R. Postgate, and C. Rodriguez-Barrueco (eds.), *Recent Developments in Nitrogen Fixation,* Academic Press, London, 1977, pp. 131–178.

125. Orme-Johnson, W. H., W. D. Hamilton, T. Ljones, M. Y. W. T'so, R. H. Burris, V. K. Shah, and W. J. Brill, "Electron paramagnetic resonance of nitrogenase and nitrogenase components from *Clostridium pasteurianum* W5 and *Azotobacter vinelandii* O. P," *Proc. Natl. Acad. Sci. USA 69*:3142–3145 (1972).

126. Pagani, S., M. Eldridge, and R. R. Eady, "Nitrogenase of *Klebsiella pneumoniae*: Rhodanese-catalysed restoration of activity of the inactive 2Fe species of the Fe protein," *Biochem. J. 244*:485–488 (1987).

127. Rivera-Ortiz, J. M., and R. H. Burris, "Interactions among substrates and inhibitors of nitrogenase," *J. Bacteriol. 123*:537–545 (1975).

128. Robinson, A. C., B. K. Burgess, and D. R. Dean, "Activity reconstitution and accumulation of nitrogenase components in *Azotobacter vinelandii* mutant strains containing defined deletions within the nitrogenase structural gene cluster," *J. Bacteriol. 166*:180–186 (1986).

129. Robinson, A., J. Li, D. Chun, D. R. Dean, and B. K. Burgess, "Characterization of the subunits of the iron molybdenum of nitrogenase from *Azotobacter vinelandii*," in H. Bothe, F. J. de Bruijn, and W. E. Newton (eds.), *Nitrogen Fixation: Hundred Years After,* Gustav Fischer, Stuttgart, 1988, p. 136.

130. Robinson, A. E., A. J. M. Richards, A. J. Thomson, T. R. Hawkes, and B. E. Smith, "Low temperature magnetic circular dichroism spectroscopy of the iron molybdenum cofactor and the complementary cofactor-less MoFe protein of *Klebsiella pneumoniae* nitrogenase," *Biochem. J. 219*:495–503 (1984).

131. Robson, R. L., "Identification of possible adenine nucleotide binding sites in nitrogenase Fe and MoFe proteins by amino acid sequence comparison," *FEBS Lett. 173*:394–398 (1984).

132. Rubinson, J. F., B. K. Burgess, J. L. Corbin, and M. J. Dilworth, "Nitrogenase activity: Azide reduction," *Biochemistry 24*:273–283 (1985).

133. Rubinson, J. F., J. L. Corbin, and B. K. Burgess, "Nitrogenase reactivity: Methyl isocyanide as a substrate and inhibitor," *Biochemistry 22*:6260–6268 (1983).

134. Scherings, G., H. Haaker, and C. Veeger, "Regulation of nitrogen fixation by Fe-S protein II in *Azotobacter vinelandii*," *Eur. J. Biochem. 77*:621–630 (1977).

135. Schöllhorn, R., and R. H. Burris, "Reduction of azide by the N_2-fixing enzyme system," *Proc. Natl. Acad. Sci. USA 57*:1317–1323 (1967).

136. Schaltz, F. A., S. F. Gheller, and W. E. Newton, "Iron molybdenum cofactor of nitrogenase: Electrochemical determination of the electron stoichiometry of the oxidized semireduced couple," *Biochem. Biophys. Res. Commun. 152*(2):629–635 (1988).

137. Scott, D. J., H. D. May, W. E. Newton, K. E. Brigle, and D. R. Dean, "Role for the nitrogenase MoFe protein in FeMo cofactor binding and catalysis," *Nature 343*:188–190 (1990).

138. Shah, V. K., and W. J. Brill, "Isolation of an iron–molybdenum cofactor from nitrogenase," *Proc. Natl. Acad. Sci. USA 74*:3249–3253 (1977).

139. Shah, V. K., and W. J. Brill, "Isolation of a molybdenum–iron cluster from nitrogenase," *Proc. Natl. Acad. Sci. USA 78*:3438–3340 (1981).

140. Shah, V. K., J. Imperial, R. A. Ugalde, P. W. Ludden, W. J. Brill, "*In vitro* synthesis of the iron–molybdenum cofactor of nitrogenase," *Proc. Natl. Acad. Sci. USA 83*:1636–1640 (1986).

141. Shilov, A. E., "Catalytic reduction of dinitrogen in protic media: Chemical model of nitrogenase," *J. Mol. Catal. 41*:221–234 (1987).

142. Simpson, F. B., and R. H. Burris, "A nitrogen pressure of 50 atmospheres does not prevent evolution of hydrogen by nitrogenase," *Science 224*:1095–1097 (1984).

143. Smith, B. E., "Reactions and physicochemical properties of the Nitrogenase MoFe proteins," in A. Müller and W. E. Newton (eds.), *Nitrogen Fixation: The Chemical-Biochemical-Genetic Interface,* Plenum, New York, 1983, pp. 23–62.

144. Smith, B. E., G.-X. Chen, R. A. Dixon, T. R. Hawkes, D. J. Lowe, P. A. McLean, M. J. O'Donnell, and J. R. Postgate, "The properties and interactions of the metal atoms in the MoFe protein of nitrogenase," *Chem. Scripta 21*:57–60 (1983).

145. Smith, B. E., and G. Lang, "Mossbauer spectroscopy of the nitrogenase proteins from *Klebsiella pneumoniae:* Structural assignments and mechanistic conclusions," *Biochem. J. 137*:169–180 (1974).

146. Smith, B. E., D. J. Lowe, and R. C. Bray, "Nitrogenase of *Klebsiella pneumoniae:* Electron paramagnetic resonance studies on the catalytic mechanism," *Biochem. J. 130*:641–643 (1972).

147. Smith, B. E., D. J. Lowe, and R. C. Bray, "Studies by electron paramagnetic resonance on the catalytic mechanism of nitrogenase of *Klebsiella pneumoniae*," *Biochem. J. 135*:331–341 (1973).

148. Smith, B. E., D. J. Lowe, G.-X. Chen, M. J. O'Donnell, and T. R. Hawkes, "Evidence on intramolecular electron transfer in the MoFe protein of nitrogenase from *Klebsiella pneumoniae* from rapid freeze electron paramagnetic resonance studies of its oxidation by ferricyanide," *Biochem. J. 209*:207–213 (1983).

149. Stephens, P. J., C. E. McKenna, M.-C. McKenna, H. T. Nguyen, and D. J. Lowe, "Circular dichroism and magnetic circular dichroism of nitrogenase proteins," in Chien Ho (ed.), *Electron Transport and Oxygen Utilization,* Elsevier, Amsterdam, 1983, pp. 405–409.

150. Stephens, P. J., C. E. McKenna, B. E. Smith, H. T. Nguyen, M.-C. McKenna, A. J. Thomas, F. Devlin, and J. B. Jones, "Circular dichroism and magnetic circular dichroism of nitrogenase proteins," *Proc. Natl. Acad. Sci. USA 76*:2585–2590 (1979).

151. Thomann, H., T. V. Morgan, H. Jin, S. J. N. Burgmayer, R. E. Bare, and E. I. Stiefel, "Protein nitrogen coordination to the FeMo center of nitrogenase from *Clostridium pasteurianum*," *J. Am. Chem. Soc. 109*:7913–7914 (1987).

152. Thorneley, R. N. F., "Nitrogenase of *Klebsiella pneumoniae:* A stopped-flow study of magnesium-adenosine triphosphate-induced electron transfer between the component proteins," *Biochem. J. 145*:391–396 (1975).

153. Thorneley, R. N. F., and G. A. Ashby, "Oxidation of nitrogenase iron protein by dioxygen without inactivation could contribute to high respiration rates of Azotobacter species and facilitate nitrogen fixation in other aerobic environments," *Biochem. J. 261*:181–187 (1989).

154. Thorneley, R. N. F., G. A. Ashby, J. V. Howarth, N. C. Millar, and H. Gutfreund, "A transient kinetic study of the nitrogenase of *Klebsiella pneumoniae* by stopped-flow calorimetry," *Biochem. J. 264*:657–661 (1989).

155. Thorneley, R. N. F., N. H. J. Bergström, R. R. Eady, and D. J. Lowe, "Vanadium nitrogenase of *Azotobacter chroococcum:* MgATP-dependent electron transfer within the protein complex," *Biochem. J. 257*:789–794 (1989).

156. Thorneley, R. N. F., and A. Cornish-Bowden, "Kinetics of nitrogenase of *Klebsiella pneumoniae:* Heterotropic interactions between magnesium-adenosine 5'-diphosphate and magnesium-adenosine 5'-triphosphate," *Biochem. J. 165*:255–262 (1977).

157. Thorneley, R. N. F., R. R. Eady, and D. J. Lowe, "Biological nitrogen fixation by way of an enzyme-bound dinitrogen-hydride intermediate," *Nature 272*:557–558 (1978).

158. Thorneley, R. N. F., R. R. Eady, and M. G. Yates, "Nitrogenases of *Klebsiella pneumoniae* and *Azotobacter chroococcum.* Complex formation between the component proteins," *Biochim. Biophys. Acta 403*:269–284 (1975).

159. Thorneley, R. N. F., and D. J. Lowe, "The mechanism of *Klebsiella pneumoniae* nitrogenase action: Pre-steady-state kinetics of an enzyme-bound intermediate in N_2 reduction and of NH_3 formation," *Biochem. J. 224*:887–894 (1984).

160. Thorneley, R. N. F., and D. J. Lowe, "The mechanism of *Klebsiella pneumoniae* nitrogenase action: Simulation of the dependence of the H_2-evolution rate on component-protein concentration and ratio and sodium dithionite concentration," *Biochem. J. 224*:903–909 (1984).

161. Thorneley, R. N. F., and D. J. Lowe, "Kinetics and mechanism of the nitrogenase enzyme system," in T. G. Spiro (ed.), *Molybdenum Enzymes*, Wiley, New York, 1985, pp. 222–284.

162. Thorneley, R. N. F., M. G. Yates, and D. J. Lowe, "Nitrogenase of *Azotobacter chroococcum:* Kinetics of the reduction of oxidized iron-protein by sodium dithionite," *Biochem. J. 155*:137–144 (1976).

163. True, A. E., P. McLean, M. J. Nelson, W. H. Orme-Johnson, and B. M. Hoffman, "Comparison of wild-type and *nifV* mutant molybdenum–iron proteins of nitrogenase from *Klebsiella pneumoniae* by ENDOR spectroscopy," *J. Am. Chem. Soc. 112*:651–657 (1990).

164. True, A. E., M. J. Nelson, R. A. Venters, W. H. Orme-Johnson, and B. M. Hoffman, "^{57}Fe hyperfine coupling tensors of the FeMo cluster in *Azotobacter vinelandii* MoFe protein: Determination by polycrystalline ENDOR spectroscopy," *J. Am. Chem. Soc. 110*:1935–1943 (1988).

165. T'so, M. Y. W., and R. H. Burris, "The binding of ATP and ADP by nitrogenase components from *Clostridium pasteurianum*," *Biochim. Biophys. Acta 309*:263–270 (1973).

166. Tuan, D. F.-T., and R. Hoffman, "N vs O linkage and σ vs π Bonding in Transition metal complexes of N_2O and NCO^-," *Inorg. Chem. 24*:871–876 (1985).

167. Vaughn, S. A., and B. K. Burgess, "Nitrite, a new substrate for nitrogenase," *Biochemistry 28*:419–424 (1989).

168. Venters, R. A., M. J. Nelson, P. A. McLean, A. E. True, M. A. Levy, B. M. Hoffman, and W. H. Orme-Johnson, "ENDOR of the resting state of nitrogenase molybdenum iron proteins from *Azotobacter vinelandii*, *Klebsiella pneumoniae* and *Clostridium pasteurianum*. ^1H, ^{57}Fe ^{95}Mo and ^{33}S Studies," *J. Am. Chem. Soc. 108*:3487–3498 (1986).

169. Walker, G. A., and L. E. Mortenson, "Effect of magnesium adenosine 5'-triphosphate on the accesibility of the iron of clostridial azoferredoxin, a component of nitrogenase," *Biochemistry 13*:2382–2388 (1974).

170. Wang, Z.-C., and G. D. Watt, "H_2 uptake activity of the MoFe protein component of *Azotobacter vinelandii* nitrogenase," *Proc. Natl. Acad. Sci. USA 81*:376–379 (1984).

171. Watt, G. D., A. Burns, and D. L. Tennent, "Stoichiometry and spectral properties of the MoFe cofactor and noncofactor redox centres in the MoFe protein of nitrogenase from *Azotobacter vinelandii*," *Biochemistry 20*:7272–7277 (1981).

172. Watt, G. D., A. Burns, S. Lough, and D. L. Tennent, "Redox and spectroscopic properties of oxidized MoFe protein from *Azotobacter vinelandii*," *Biochemistry 21*:4926–4932 (1980).

173. Watt, G. D., and J. W. Macdonald, "Electron paramagnetic resonance spectrum of the iron protein of nitrogenase: Existence of a $g = 4$ spectral component and its effect on spin quantization," *Biochemistry 24*:7226–7231 (1985).

174. Watt, G. D., and Z.-C. Wang, "Further redox reactions of metal clusters in the molybdenum–iron protein of *Azotobacter vinelandii* nitrogenase," *Biochemistry 25*:5196–5202 (1986).

175. Watt, G. D., and Z.-C. Wang, "Protein interaction with and higher oxidation states of the nitrogenase MoFe protein from *Azotobacter vinelandii*," *Biochemistry USA 28*:1844–1850 (1989).

176. Watt, G. D., Z.-C. Wang, and R. R. Knotts, "Redox reactions of nucleotide binding to the iron protein of *Azotobacter vinelandii*," *Biochemistry 25*:8156–8162 (1986).

177. Weinenger, M. S., and L. E. Mortenson, "Crystallographic properties of the MoFe proteins of nitrogenase from *Clostridium pasteurianum* and *Azotobacter vinelandii*," *Proc. Natl. Acad. Sci. USA 79*:378–380 (1982).

178. Wherland, S., B. K. Burgess, E. I. Steifel, and W. E. Newton, "Nitrogenase activity: Effect of component ratio on electron flow and distribution during nitrogen fixation," *Biochemistry 20*:5132–5140 (1981).

179. Wink, D. A., P. A. McLean, A. B. Hickman, and W. H. Orme-Johnson, "A new method for extraction of iron–molybdenum cofactor (FeMoco) from nitrogenase absorbed to DEAE cellulose. 2. Solubilization of FeMoco in a wide range of organic solvents," *Biochemistry 28*:9407–9412 (1989).

180. Yamane, T., M. S. Weinenger, L. E. Mortenson, and M. G. Rossman, "Molecular symmetry of the MoFe protein of nitrogenase. Structural homology/nitrogen fixation/X-ray crystallography," *J. Biol. Chem. 257*:1221–1223 (1982).

181. Yates, M. G., "Electron transport to nitrogenase in *Azotobacter chroococcum*. Azotobacter flavodoxin hydroquinone as an electron donor," *FEBS Lett. 27*:63–67 (1972).

182. Yates, M. G., R. N. F. Thorneley, and D. J. Lowe, "Nitrogenase of *Azotobacter chroococcum:* Inhibition by ADP of the reduction of oxidized Fe protein by sodium dithionite," *FEBS Lett. 60*:89–93 (1975).

183. Zimmerman, R., E. Münck, W. J. Brill, V. K. Shah, M. T. Henzl, J. Rawlings, and W. H. Orme-Johnson, "Nitrogenase X: Mössbauer and EPR studies on reversibly oxidized MoFe protein from *Azotobacter vinelandii* O.P. Nature of the iron centers," *Biochim. Biophys. Acta 537*:185–207 (1978).

184. Zumft, W. G., L. E. Mortenson, and G. Palmer, "Electron paramagnetic resonance studies on nitrogenase. Investigation of the oxidation reduction behaviour of azoferredoxin and molybdoferredoxin with potentiometric and rapid-freeze techniques," *Eur. J. Biochem. 46*:525–535 (1974).

19

Alternative Nitrogen Fixation Systems

Paul E. Bishop and R. Premakumar

I. Introduction

The process of N_2 fixation has been studied for many years in the aerobic soil bacterium *Azotobacter vinelandii*, but it was not realized until recently that this diazotroph harbors three genetically distinct nitrogenases.[23,48,61] One of these enzyme complexes is the well-characterized conventional molybdenum-containing nitrogenase (nitrogenase-1). Nitrogenase-1 is only expressed in medium containing molybdenum (Mo) and is composed of two components: dinitrogenase reductase-1 (also called component II or Fe protein) and dinitrogenase-1 (also designated component I or MoFe protein). Dinitrogenase reductase-1, which serves as an electron donor to dinitrogenase-1, is a dimer of two identical subunits with an M_r of approximately 60,000. A single [4Fe–4S] cluster is bridged between the two subunits.[36,40] Dinitrogenase-1, with an M_r of about 220,000, is a tetramer that is made up of two pairs of nonidentical subunits (α and β). Dinitrogenase-1 contains two types of metal centers that are involved in the redox reactions of the N_2 reduction process: P centers that might be organized as four unusual 4Fe–4S clusters[27,55] and two identical FeMo cofactors (FeMoco), which are almost certainly the sites for N_2 binding and reduction.[78]

Nitrogenase-2 is a vanadium-containing enzyme complex that is synthesized in N-free medium lacking Mo but containing vanadium (V).[37,38,68] This enzyme complex consists of two components: dinitrogenase reductase-2, a dimer of two identical subunits, and dinitrogenase-2, which is now thought to be a hexamer (M_r of about 240,000) of two dissimilar pairs of large subunits (α and β) and a pair of small subunits (δ).[70] Dinitrogenase reductase-2 has an M_r of about 62,000 and contains four Fe atoms and four acid-labile sulfide groups per dimer.[30,38] Dinitrogenase-2 contains two V atoms, 23 Fe atoms, and 20 acid-labile sulfide groups per molecule.[29] A cofactor (FeVaco) analogous to FeMoco has been extracted from dinitrogenase-2 using *N*-methylformamide.[79]

Nitrogenase-3 does not appear to contain either Mo or V and is made under Mo- and V-deficient conditions.[23] This enzyme is composed of two components: dinitrogenase reductase-3 and dinitrogenase-3. Dinitrogenase reductase-3 is a dimer (M_r of approximately 65,000) of two identical subunits while dinitrogenase-3 is a tetramer (M_r of about 216,000) composed of two dissimilar pairs of subunits (α and β). However, dinitrogenase-3 may actually be a hexamer, since the structural gene operon for nitrogenase-3 contains an open reading frame (ORF) that potentially encodes a protein similar to the δ subunit for dinitrogenase-2.[48] Dinitrogenase reductase-3 contains four Fe atoms and four acid-labile sulfide groups per dimer. Dinitrogenase-3 contains approximately 24 Fe atoms and 18 acid-labile sulfide groups per molecule; however, it lacks significant amounts of Mo or V. It is also interesting to note that dinitrogenase-3 can be isolated in at least two active configurations: $\alpha_2\beta_2$ and $\alpha_1\beta_2$.

The existence of nitrogenase-2 and nitrogenase-3 was unknown prior to 1980. Until that time it was generally thought that Mo was absolutely required for N_2 fixation even though scattered reports indicated that some diazotrophs could grow slowly in N-free medium lacking Mo. These low rates of N_2 fixation were usually attributed to the incorporation of trace amounts of contaminating Mo into nitrogenase. The early work of Bortels[14] on *Azotobacter* species established that adding small amounts of Mo to growth media enhanced N_2 fixation. Bortels[16] also showed that low concentrations of V stimulated the growth of *Azotobacter* species under diazotrophic conditions. For many years it seemed possible that V could substitute for Mo in dinitrogenase-1 since many of the chemical properties of these two metals are similar.[21,54] By the early to mid-1970s it appeared that stimulation by V might be explained by other hypotheses such as the incorporation of V into dinitrogenase-1 with consequent stabilization of the enzyme and a more effective utilization of the small amount of Mo found in Mo-starved cells.[5]

In the early 1980s evidence was presented indicating that *A. vinelandii* contained at least two nitrogenase systems: the conventional Mo-containing nitrogenase system and an alternative nitrogenase system expressed in the absence of Mo. This evidence was built around the core observation that Nif$^-$ (unable to fix N_2) mutant strains underwent phenotypic reversal (i.e., Nif$^-$ to Nif$^+$) under conditions of Mo deprivation. These reports[9,10,60,64] were received with skepticism because they challenged the long-held belief that Mo was absolutely required for N_2 fixation and that nitrogenases were essentially the same regardless of their source. The latter notion was further bolstered by the results of Southern blot experiments by Ruvkun and Ausubel,[71] which indicated that some of the structural genes encoding nitrogenases from diverse diazotrophic organisms were highly conserved at the nucleotide sequence level.

Since the Nif$^-$ strains of *A. vinelandii* that were first used to demonstrate phenotypic reversal probably contained point mutations, it was considered possible that phenotypic reversal was due to increased leakiness of the mutant phenotypes under conditions of Mo starvation and not to derepression of an alternative N_2-fixation system. This possibility was ruled out when mutant strains carrying deletions in the structural genes for nitrogenase-1 were shown to undergo phenotypic reversal under Mo-deficient conditions.[8,11,67] Strains with deletions in the structural genes (*nifHDK*) also facilitated the isolation of nitrogenase-2 from *A. chroococcum*[68] and nitrogenase-2 and nitrogenase-3 from *A. vinelandii*.[23,37,38]

In order to set the discovery of an alternative N_2-fixation system in a historical context, the next section describes early findings on the metal requirements for N_2 fixation by *Azotobacter* species.

II. Early Studies on Metal Requirements for N_2 Fixation by Azotobacter

In 1901 Beijerinck[4] reported finding a bacterium that outgrew all other microorganisms in N-free medium. This bacterium was named *Azotobacter chroococcum* after it was established that it was able to fix atmospheric nitrogen. The discovery of a free-living soil microorganism with N_2-fixation capabilities gave rise to speculation about the contribution of *Azotobacter* to soil fertility. Subsequently, factors potentially leading to an increased number of azotobacter cells in the soil and to a stimulation of N_2 fixation by this organism were studied. It was observed that pure cultures of *A. chroococcum* lost their excellent N_2-fixation capabilities after transfer in minimal medium. This lower level of N_2 fixation was reversed upon the addition of sterilized soil or aqueous soil extracts to the growth medium. Kremieniewski[53] showed that this stimulatory effect could also be achieved by adding humic acid preparations, and for many years the stimulatory agents were thought to be the organic constituents of soil extracts. However, in 1929 Bortels[12] suggested that the observed increases in N_2 fixation were due to the inorganic components of soil extracts and not the organic material.

In 1930, Bortels[14] published the results of a search for elements stimulatory to N_2 fixation by *A. chroococcum*. Tungsten (W) was the first element tested; however, only a small stimulatory effect was observed with this metal. Next, the effect of Mo and uranium on diazotrophic growth was examined. Though uranium in its radioactive form had been previously postulated as a possible catalyst for N_2 fixation, it did not stimulate N_2 fixation. The addition of molybdate (0.0005%), on the other hand, caused a threefold increase in N_2-dependent growth over controls to which no Mo had been added. In another report, Bortels[13] demonstrated that vanadium (V) compounds were almost as stimulatory as Mo salts. This result was also confirmed by Burk and Horner.[20]

In 1936, Bortels[16] reported the results of quantitative studies on the effect of addition of Na_2MoO_4, $NaVO_3$, and Na_2WO_4 to cultures of *A. chroococcum* and *A. vinelandii*. As little as 20 nM Na_2MoO_4 increased the amount of N fixed three- to fourfold over controls without added Mo, and optimum concentrations of Na_2MoO_4 were determined to be around 1 µM for *A. chroococcum* and 0.1 µM for *A. vinelandii*. Up to 26 times as much N_2 was fixed during a six-day incubation period in the presence of Mo as compared to controls in which Mo was not added. Concentrations of $NaVO_3$ as low as 80 nM stimulated N_2 fixation by *A. chroococcum*, and *A. vinelandii* responded to concentrations as low as 8 nM. Optimal concentrations for both organisms were observed to be approximately 1 µM $NaVO_3$. Tungsten exhibited only a weak stimulatory effect. However, this metal was found to increase N_2 fixation by cells growing under conditions where Mo was present in suboptimal amounts or where only V was present.

Bortels and scientists following him focused their attention on Mo and its role in biological N_2 fixation. There are, however, a few observations described in Bortels' publications that are not entirely consistent with the view that Mo is absolutely essential for N_2 fixation. The degree to which Mo or V stimulated N_2 fixation was strain dependent. One strain responded to the addition of 50 µM Na_2MoO_4 by only a 1.8-fold increase in the amount of N_2 fixed as compared to the control without added Mo. The response to V was almost the same as for Mo. However, the strain described in the 1936 publication[16] showed a 16-fold increase in the amount of N_2 fixed in the presence of 50 µM Na_2MoO_4 and a ninefold increase in the presence of 7 µM $NaVO_3$ as compared to control cultures. Before Bortels published his observations on the role of Mo in N_2 fixation, *Azotobacter* had been studied under N_2-fixing conditions in media to which Mo had not been added. As previously mentioned, a decrease in N_2 fixation was observed in these media after several sequential transfers; however, some growth under nitrogen-fixing conditions was always observed and *A. vinelandii*, in particular, grew well under these conditions. The argument that this diazotrophic growth was due to contaminating Mo in the growth medium cannot be refuted easily, especially considering what is now known about the Mo-storage capacity of *Azotobacter*.[62] On the other hand, it appears that carry-over of Mo through the inoculum and contamination of the growth medium must have been minimal, since both Bortels[14,16] and Burk and Horner[20] reported that nanomolar concentrations of added Mo exhibited a stimulatory effect and that saturation was reached in the range of 0.1–1 µM Na_2MoO_4.

The question of Mo or V contamination of the growth medium arises again in connection with the results of further experiments carried out by Bortels,[15] where the effects of different Fe and Mo or V concentrations on

N_2 fixation by *A. chroococcum* and *A. vinelandii* were tested. The conclusion drawn from these experiments was that *A. chroococcum* was virtually unable to fix N_2 in the absence of added Mo, but that *A. vinelandii,* in contrast, exhibited N_2 fixation under Mo-deficient conditions provided that sufficient Fe was included in the medium.

Several years later, Horner et al.[41] tested the effect of Mo and V on two strains of *A. vinelandii,* seven strains of *A. chroococcum,* and one strain of *A. agile.* Molybdenum and V stimulated N_2 fixation by all these strains and the maximum amount of N_2 fixed was observed in the presence of Mo. One of the *A. vinelandii* strains and all of the *A. chroococcum* strains tested grew poorly under Mo-deficient conditions. The *A. agile* strain and the second *A. vinelandii* strain, on the other hand, fixed 70.8% and 59%, respectively, of the N_2 reduced in the Mo-sufficient medium. Therefore, the authors questioned "whether Mo should be considered essential for such organisms."

Esposito and Wilson[32] reexamined the trace metal requirement of *A. vinelandii* using the 8-hydroxyquinoline extraction technique for media purification. In addition, deionized distilled water was employed and the glassware was treated with an EDTA solution. Despite these precautions, *A. vinelandii* strain O, the parent strain of strain OP (the most commonly used laboratory strain), fixed N_2 in Mo-deficient medium at about 25% of the amount fixed under Mo-sufficient conditions. Vanadium addition increased the amount of N_2 fixed slightly above that observed in Mo-deficient medium. From these results the authors nevertheless concluded that Mo was essential for N_2 fixation by *A. vinelandii* strain O and that V could not substitute for Mo.

Becking[3] focused his studies on the strain-dependent correlation of the effect of Mo and V on N_2 fixation by members of the family Azotobacteriaceae. Molybdenum stimulated N_2 fixation by all strains tested. *A. vinelandii* strains required the least amount of Mo (0.0004 ppm or 4.2 nM) for half-maximal nitrogen fixation. The addition of 0.002 ppm (21 nM) sufficed to achieve half-maximal N_2 fixation in *A. agile,* whereas *A. chroococcum* required a concentration of 0.05 ppm (0.52 μM). *Beijerinckia* species showed a variable Mo requirement with concentrations ranging between 0.004 ppm (41.7 nM) and 0.043 ppm (0.45 μM). In the absence of added Mo, some N_2 fixation was always observed, though in most cases the amount of N_2 fixed was less than 10% of that fixed in the presence of 1.6 μM Na_2MoO_4. However, selected strains of *A. vinelandii, A. agile,* and *B. mobile* were able to fix 18–53% of the N_2 fixed in the presence of Mo when cultured in Mo-deficient medium. Most *A. vinelandii* and *A. chroococcum* strains responded to V with an increase in N_2 fixation. Of the 20 *A. agile* strains tested, only one was able to utilize V in a way that benefited N_2 fixation. Vanadium did not affect N_2 fixation by any of the *Beijerinckia* strains tested.

III. Biochemical Properties of Alternative Mo-Independent Nitrogenases

A. An alternative Nitrogenase (Nitrogenase-2) that Contains Vanadium

The nitrogenase component containing V (the so-called V-nitrogenase) was partially purified from V-grown *A. vinelandii* cells[5,21,22,54] almost 20 years ago. Burns et al.[22] showed that "V-nitrogenase," when compared to conventional dinitrogenase-1, exhibited different catalytic properties. These include (1) a lower affinity for C_2H_2; (2) different product ratios from acrylonitrile reduction; (3) different percent electron allocations to substrates (N_2, C_2H_2, H^+, and CH_2CHCN); and (4) a different K_i for CO inhibition. Physical and chemical differences between "V-nitrogenase" and dinitrogenase-1 were also found; for example, "V-nitrogenase" did not exhibit an electron paramagnetic resonance (EPR) signal at $g = 3.67$ and "V-nitrogenase" contained two thirds less Fe than dinitrogenase-1. Some of the physical and chemical differences were attributed to loss of metals upon fractionation on DEAE cellulose; for example, "V-nitrogenase" contained neither Mo nor V beyond background levels after purification on DEAE cellulose. From these results Burns et al.[22] postulated that V was actively involved in catalysis since the turnover number calculated per Mo was ≥ 2230 and the number per V was calculated as ≥ 112, which is comparable to 196, the turnover number per Mo in dinitrogenase-1. However, Benemann et al.[5] proposed that the V effect was due to the incorporation of V into the nitrogenase-1 complex with a consequent stabilization of the enzyme and a more efficient utilization of the small amount of Mo found in Mo-starved cells.

In 1982 it was reported[10] that mutants of *A. vinelandii* carrying lesions of the structural genes of nitrogenase-1 were able to respond to the presence of V in the medium. The pattern of NH_4^+-repressible proteins from extracts of these mutants on two-dimensional gels was the same as for the wild type grown in N-free medium containing 5 μM V_2O_5. Based on these results it was proposed that the "V-nitrogenase" purified by Burns et al.[22] was actually an alternative dinitrogenase that might be maximally derepressed in the presence of V due to the exclusion of even trace amounts of Mo. Another study showed that the addition of increasing concentrations of V to cultures of Nif mutants of *A. vinelandii* led to increased specific activities of nitrogenase in these cultures.[60]

More recently, Mo-independent V-containing nitrogenases (nitrogenase-2) have been purified from W-tolerant *nifHDK* deletion mutants of *A. chroococcum*[29,30,68] and *A. vinelandii*.[37,38] The properties of nitrogenase-2 are summarized in Tables 19-1, 19-2, and 19-3. From Table 19-1 it can be seen that C_2H_2 is a poor substrate for the enzyme; a high proportion of the elec-

Table 19-1. Comparison of Specific Activities of Dinitrogenase-1 and Dinitrogenase-2[a]

Specific Activity[b] (nmol product/min/ mg protein)	Dinitrogenase-2		Dinitrogenase-1 A. vinelandii[23]
	A. chroococcum[29]	A. vinelandii[37]	
NH$_4^+$ formation	350	660	1270
C$_2$H$_4$ from C$_2$H$_2$	608	220	2000
C$_2$H$_6$ from C$_2$H$_2$	0.74	N.D.[c]	N.D.
H$_2$ evolution			
under N$_2$ atm	928	N.D.	867
under Ar atm	1348	1400	3520
under 0.1 atm	998	N.D.	303
C$_2$H$_2$ plus 0.9 atm Ar			

[a]Data were complied from references cited with each strain.

[b]Specific activities obtained in a homologous assay in which the corresponding complementary dinitrogenase reductase was used.

[c]N.D. denotes "not determined."

trons are allocated to the reduction of H$^+$ to H$_2$. Even with N$_2$ as a substrate, about half of the electron flux was shunted into H$_2$ evolution. From a kinetic study of electron transfer from dinitrogenase reductase-2 to dinitrogenase-2 by stopped-flow spectrophotometry, it has been proposed that this difference in reactivity between dinitrogenase-1 and dinitrogenase-2 for the substrates N$_2$ and H$^+$ resides at the substrate-binding center of dinitrogenase-2.[82] Another interesting feature of nitrogenase-2 is its ability to reduce C$_2$H$_2$ to ethane as well as to C$_2$H$_4$.[25,26] This ability to reduce C$_2$H$_2$ to ethane, not shared by nitrogenase-1, may serve as a basis for distinguishing alternative nitrogenases from conventional nitrogenases.[25] Nitrogenase-2 was also observed to be more effective in N$_2$ reduction at lower temperatures than nitrogenase-1. The component responsible for this relatively good performance at low temperatures (down to 5° C) was found to be dinitrogenase reductase-2.[56] It was speculated that this low-temperature effectiveness might be one of the reasons for the persistence of this less-efficient nitrogenase in azotobacters.[56]

1. Dinitrogenase Reductase-2

The physicochemical properties of dinitrogenase reductase-2, purified from A. chroococcum[30,68] and from A. vinelandii[38,63] along with those of A. vinelandii dinitrogenase reductase-1 are given in Table 19-2. It can be seen that the properties of dinitrogenase reductase-2 are very similar to those of dinitrogenase reductase-1. This similarity extends to the ability to form active complexes with dinitrogenase-1 of either system.[30,38,68] Dinitrogenase reductase-1 and dinitrogenase reductase-2 from A. vinelandii have also been shown to be antigenically similar.[38,63] These enzymes are homodimers with

Table 19-2. Comparison of the Physicochemical Properties of Dinitrogenase Reductase-1 and Dinitrogenase Reductase-2[a]

Properties	Dinitrogenase Reductase-2		Dinitrogenase Reductase-1
	A. chroococcum[30,68]	*A. vinelandii*[38,63]	*A. vinelandii*[19,23,50,59,63,77]
Specific activity (nmol product/min/ mg protein)			
H_2 evolution			
under Ar	1211	N.D.[b]	2210
	1107[c]		
under N_2	648	N.D.	796
under 0.1 atm C_2H_2	435	N.D.	128
NH_4^+ formation	337	N.D.	1030
C_2H_4 from C_2H_2	341	1100–1600[c]	1720
	999[c]	870[c]	1500
M_r	61,000	63,000	67,000
		67,000	
Subunit structure	γ_2	γ_2	γ_2
Subunit M_r	32,000	31,000	34,000
		34,000	
Fe content (g atom mol^{-1})	3.7 ± 0.2	3.6 ± 0.4	3.7 ± 0.2
Acid-labile sulfide (g atom mol^{-1})	3.92 ± 0.31	4.4 ± 0.26	3.8 ± 0.21
EPR (*g* values)	2.035,1.961	2.05,1.96	2.05,1.96
	1.892	1.88	1.88

[a]Data were compiled from references cited with each strain.

[b]N.D. denotes "not determined."

[c]Specific activity calculated from an assay in which dinitrogenase reductase-2 was complemented with dinitrogenase-1 from the same organism.

subunit M_rs ranging from 31,000 to 34,000 (see Table 19-2). Each of the dinitrogenase reductases has four Fe and four acid-labile sulfides per mole of protein. These Fe and S^{2-} ions are arranged in a single [4Fe–4S] cluster that is shared between the two subunits of dinitrogenase reductase-1.[40] The same arrangement probably exists in dinitrogenase reductase-2 as well. Dinitrogenase reductase-2 is O_2 labile, with the enzyme from *A. chroococcum* having a $t_{1/2}$ of 36 sec in air.[30]

At low temperatures dithionite-reduced dinitrogenase reductase-2 exhibited EPR signals that are consistent with the presence of $S = 1/2$ and $S = 3/2$ spin states.[30,39] These signals are similar to those reported for other dinitrogenase reductases. The presence of MgATP or MgADP caused changes in line shape indicating conformational alterations in dinitrogenase reductase-2 from *A. vinelandii*.[38]

Table 19-3. Comparison of the Physicochemical Properties of Dinitrogenase-1 and Dinitrogeanse-2[a]

Properties	Dinitrogenase-2		Dinitrogenase-1
	A. chroococcum[28,29]	A. vinelandii[37]	A. vinelandii[23,76]
Molecular mass (M_r)	210,000	200,000	221,000
Subunit structure	$\alpha_2\beta_2\delta_2$	$\alpha_2\beta_2(\delta_2)^b$	$\alpha_2\beta_2$
Subunit M_r	50,000	52,000	54,000
	55,000	55,000	58,000
	13,000	N.D.[c]	
Metal and S^{2-} content (g atom mol^{-1})			
V	2 ± 0.3	0.7 ± 0.3	N.D.
Mo	0.06	<0.05	2
Fe	19 ± 2	9 ± 2	33
S^{2-}	21 ± 1	N.D.[c]	27
EPR (g values)	5.6, 4.35	5.8, 5.4,	4.32, 3.65
	3.77, 1.93	4.34, 2.04	2.01
		1.93	

[a]Data were compiled from references cited with each strain.

[b]The probable existence of δ subunits is based on genetic evidence.

[c]N.D. denotes "not determined."

2. Dinitrogenase-2

Dinitrogenase-2 from A. *chroococcum* is a hexamer with an M_r of 238,000 and is composed of three types of subunits with M_rs of 50,000, 55,000, and 14,000.[28,61] The structural genes encoding these subunits have been sequenced.[70] The third subunit (M_r 14,000) is encoded by a new gene, *vnfG*, which is located between the *vnfD* and *vnfK* structural genes. It seems likely that dinitrogenase-2 from A. *vinelandii* also has a third subunit since the sequence of the structural genes shows *vnfG* between *vnfD* and *vnfK* as in A. *chroococcum*.

Properties of dinitrogenase-2 (summarized in Table 19-3) are similar to dinitrogenase-1 with the exception that Mo is replaced by V. Dinitrogenases-1 and -2 are able to form active complexes with dinitrogenase reductase of either system.[29] The EPR studies of dithionite-reduced dinitrogenase-2 from either A. *chroococcum*[29] or A. *vinelandii*[37,57] reveal that this protein contains an EPR center with an $S = 1/2$ spin system in addition to the center with an $S = 3/2$ spin system as found in MoFe proteins (ascribed to a FeMoco center). The presence of a signal at $g = 3.77$ in the EPR spectrum of A. *chroococcum* dinitrogenase-2, but not in the EPR spectrum of A. *vinelandii* dinitrogenase-2, may indicate some difference in the environment of the cofactor between these two dinitrogenases. A recent study of dinitrogenase-2 from *A. vinelandii* by the passage EPR technique indicated the presence of a third spin center (S3) that had a broad, structureless absorption envelope extending from approximately 2800 to 5500 G with a peak near 3500 G.[39] It was speculated that S3 arises from a paramagnetic site in the enzyme coupled to a metal center.[39] Low-temperature MCD and EXAFS studies of dinitrogenase-2 provide strong evidence for the presence of P clusters that are similar to P clusters found in MoFe proteins and of a cofactor containing Fe and V.[1,2,34,35,58] The presence of the cofactor (FeVaco) was confirmed by the *N*-methyl formamide extraction of acid-precipitated dinitrogenase-2 from *A. chroococcum*.[79] The brown solution, containing V, Fe, and S^{2-} in the ratio of 1:6:5, exhibited an EPR spectrum with *g* values at 4.5, 3.6, and 2.0, which is characteristic of an $S = 3/2$ spin system. This preparation also activated purified cofactorless dinitrogenase-1 complemented with dinitrogenase reductase-1. The reconstituted system catalyzed the reduction of H^+ to H_2 and the reduction of C_2H_2 to C_2H_4 and ethane (which is characteristic of nitrogenase-2).

B. An Alternative Nitrogenase (Nitrogenase-3) that Does not Appear to Contain Molybdenum or Vanadium

An alternative nitrogenase that does not seem to contain either Mo or V has recently been purified from *A. vinelandii*.[23] The physicochemical properties of this enzyme are summarized in Table 19-4. This second alternative nitrogenase (nitrogenase-3) is composed of two components. The dinitrogenase component (dinitrogenase-3) is composed of two different subunits that assemble into at least two active configurations: $\alpha_2\beta_2$ and $\alpha_1\beta_2$. The Fe and acid-labile sulfide content for the two subunit configurations of dinitrogenase-3 are 24 Fe atoms and 18 S^{2-} groups per M_r of 216,000 ($\alpha_2\beta_2$) and 11 Fe atoms and 9 S^{2-} groups per M_r of 158,000 ($\alpha_1\beta_2$). The dinitrogenase reductase component (dinitrogenase reductase-3) is composed of two protein subunits of identical M_r (32,500) and contains four Fe and four acid-labile S^{2-} groups per M_r of 65,000. The protein subunits of nitrogenase-3 comigrated with the four Mo- and NH_4^+-repressible proteins originally observed in cell extracts of tungstate-tolerant Nif$^+$ pseudorevertants and in cell extracts of Mo-starved wild-type cells.[9] Thus, nitrogenase-3 appears to be the nitrogenase hypothesized to exist in the alternative N_2 fixation system first described in 1980.[9] The pattern of substrate reduction efficiency resembles the pattern shown by nitrogenase-2 and is $H^+ > N_2 > C_2H_2$.

One characteristic of nitrogenase-3 (also shared by nitrogenase-2) is the high rate of H_2 evolution relative to NH_3 or C_2H_4 formation regardless of the overall electron flux. In contrast to nitrogenase-1, nitrogenase-3 catalyzes relatively high rates of H_2 evolution even in the presence of C_2H_2.[23]

Table 19-4. Physicochemical Properties of the Components of Nitrogenase-3 from A. vinelandii[a]

Properties	Dinitrogenase-3	Dinitrogenase Reductase-3
Specific activity[b] (nmol product/ min/mg protein		
NH$_4^+$ formation	38	29
H$_2$ evolution		
under N$_2$ atm	213	488
under Ar atm	253	503
under 0.1 atm of C$_2$H$_2$ plus 0.9 atm of Ar	202	394
C$_2$H$_4$ from C$_2$H$_2$	28	22
Molecular mass (M_r)[c]	216,000	65,000
Subunit structure	$\alpha_2\beta_2(\delta_2)$[d]	γ_2
	58,000	32,500
Subunit M_r	50,000	
Metal[e] and S^{2-} content (g atom mol^{-1})		<0.007
Mo	0.126	
	<0.046	
	<0.010	0.008
V	<0.010	
Fe	24	3.5
S^{2-}	17.6	4.0

[a] Data compiled from ref. 23.

[b] Specific activities were obtained in a homologous assay in which the corresponding complementary component was used.

[c] Calculation based on proposed subunit composition and subunit M_r determined by SDS-PAGE.

[d] (δ_2) indicates the probable existence of a third subunit based on genetic evidence.

[e] Determined by neutron activation analysis; where two separate determinations were made, both values are shown.

Cell-free extracts containing nitrogenase-3 catalyzed the reduction of C$_2$H$_2$ to ethane, a property shared with nitrogenase-2 (D. Dalton, R. Premakumar, and P. Bishop, unpublished observations). The specific activities obtained with purified nitrogenase-3 using several different substrates were extemely low as compared to those obtained with nitrogenase-1 (see Table 19-4). Several possible explanations for these low activities are: (1) nitrogenase-3 may be an intrinsically less-efficient enzyme, (2) optimum conditions for enzymatic activity may not have been found, and (3) nitrogenase-3 may require an additional unknown protein(s) or trace element (that is lost during purification) for maximal activity. With regard to the third possibility, this protein(s) could be the product of anfG (an open reading frame (ORF) ORF located between *anfD* and *anfK*) and/or either of the products of the ORFs located 3' to *anfK*.[48]

Dinitrogenase reductase-1 and dinitrogenase reductase-2 from *A. vinelandii* showed negligible activity in a complementation reaction with dinitrogenase-3 (0.8 nmol C_2H_4/min/mg protein).[23] Dinitrogenase reductase-3 was also not very effective in complementing dinitrogenase-1 in the acetylene reduction assay since the specific activity of dinitrogenase-1 in this reaction was only 0.8% of that with dinitrogenase reductase-1.[63] Unlike *Clostridium pasteurianum* dinitrogenase reductase-1, which does not exhibit activity in heterocomplexes and is a potent inhibitor of homocomplexes,[31] dinitrogenase reductase-3 did not inhibit the activity of dinitrogenase-1 and dinitrogenase reductase-1,[63] Peptide mapping showed that dinitrogenase reductase-1 shares roughly 50% amino acid sequence similarity with dinitrogenase reductase-3. There is also antigenic similarity between the two dinitrogenase reductases.[63]

IV. Genetics of Alternative Nitrogen-Fixation Systems

A. *nif* Genes Shared by the Three Nitrogenase Systems in Azotobacter vinelandii

Two *nif* genes are known to be shared by the three nitrogenase systems. One of these is *nifB*, since NifB⁻ mutants (strains CA30 and UW45) are unable to grow under Mo-sufficient, Mo-deficient plus V, and Mo-deficient diazotrophic conditions.[45,47] In the wild type a *nifB*-hybridizing transcript (4 kb in size) is also observed under all three diazotrophic conditions.[45] Interestingly, these transcripts also include *nifQ*, which is not required for growth in Mo-deficient plus V or Mo-deficient N-free medium. Although the role that NifQ (the *nifQ* gene product) might play during growth under the latter conditions is unknown, it has been speculated that NifQ could make the expression of nitrogenase-2 and nitrogenase-3 less sensitive to repression by trace amounts of Mo. This speculation is founded on the observation that *nifQ* mutations (and mutations polar onto *nifQ*) cause strains to exhibit lag periods longer than those seen with NifQ⁺ strains during diazotrophic growth.[45] Since NifB is known to be required for FeMoco synthesis, it is probably safe to assume that this gene product plays an analogous role in the formation of FeVaco (the V-containing cofactor of dinitrogenase-2[79]). In the case of dinitrogenase-3, NifB must be involved in some function other than one that relates to Mo or V. The other *nif* gene shared by the three nitrogenase systems is *nifM*. NifM is required for maturation of dinitrogenase reductase-1 and a NifM⁻ mutant (strain MV21) was unable to grow diazotrophically either in the presence or absence of Mo or V.[51] At least two regulatory genes have been found that are required for growth under Mo-sufficient or Mo-deficient conditions; *nfrX*[72] and *ntrD* (A. Bali and C. Kennedy, personal communication). Other genes that are shared among the three N_2-fixation systems will undoubtedly be identified after systematic testing of available Nif⁻ strains[42] for diazotrophic growth under Mo-deficient conditions in the presence and absence of V.

B. Genes Encoding Nitrogenase-2

The structural genes encoding dinitrogenase-2 and dinitrogenase reductase-2 have been cloned, sequenced, and mutagenized for both *A. chroococcum*[69,70] and *A. vinelandii*[65] (R. Joerger, T. Loveless, R. Pau, L. Mitchenall, B. Simon, and P. Bishop, unpublished results). These genes have been designated as *vnf* (*v*anadium *n*itrogen *f*ixation).[70] In contrast to the single operon (*nifHDK*) encoding the subunits for nitrogenase-1, the genes encoding the nitrogenase-2 proteins are split between two operons (Figure 19-1). *vnfH* encodes the dinitrogenase reductase-2 subunits and is part of a two-gene operon. The ORF 3' to *vnfH* encodes a ferredoxinlike protein that has not been ascribed a function. *vnfH* is preceded by a potential promoter sequence that would be predicted to interact with core RNA polymerase containing the sigma 54 factor (*ntrA*, *rpoN*, or *glnF* gene product). The *vnfDGK* operon, located 1.0 kbp (*A. vinelandii*) or 2.5 kbp (*A. chroococcum*) downstream from the *vnfH-Fd* operon, encodes the subunits for dinitrogenase-2. *vnfD* encodes the α subunit and *vnfK* encodes the β subunit. In *A. vinelandii* the 1.0-kbp region between the *vnfH-Fd* and *vnfDGK* operons does not appear to contain any identifiable ORFs (R. Joerger, T. Loveless, R. Pau, L. Mitchenall, B. Simon, and P. Bishop, unpublished results).

The third subunit, δ (M_r of 13,274), is encoded by *vnfG*.[70] This gene does not have a counterpart in the nitrogenase-1 system (system 1 in Figure 19-1); however, it does share some sequence similarity to *anfG*, an ORF located between *anfD* and *anfK* (Figure 19-1). Whether or not the δ subunit is required for full activity of dinitrogenase-2 is presently unknown. The *vnfDGK* genes appear to be cotranscribed and transcription is NH_4^+-repressible. The largest and most abundant transcript is 3.4 kb in length and the two transcripts present in lesser amounts are 1.9 and 1.7 kb.[70]

The removal of a 1.4-kbp *BglII* fragment, which spans all of *vnfG* and the 3' and 5' ends of *vnfD* and *vnfK*, from the genomes of both *A. chroococcum*[70] and *A. vinelandii*[61] results in deletion strains that lack dinitrogenase-2. When this deletion was transferred to *A. chroococcum* strain MCD1155 (carrying a deletion of the structural genes for nitrogenase-1), the resulting double deletion strain was unable to grow under any N_2-fixing condition. This result indicates that *A. chroococcum* does not contain a third nitrogenase.[70] A similar double-deletion strain of *A. vinelandii* (strain RP206) grew in N-free media lacking Mo.[61] This finding provided genetic evidence

System 1

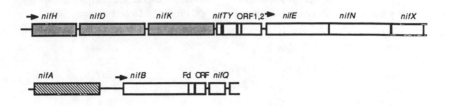

System 2

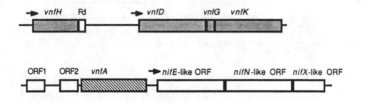

System 3

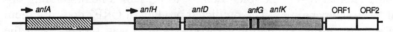

Figure 19-1. Organization of some genes involved in the three nitrogenase systems of *A. vinelandii*. Systems 1, 2, and 3 refer to genes required for nitrogenases 1, 2, and 3, respectively. Arrows indicate promoters and direction of transcription.

for the expression of a third nitrogenase (nitrogenase-3) in *A. vinelandii* that lacks Mo and V and also supported previously published results describing the isolation and partial characterization of nitrogenase-3.[23]

C. Genes Encoding Nitrogenase-3

The structural genes encoding nitrogenase-3 in *A. vinelandii* have been cloned, sequenced, and mutagenized.[48] These genes have been designated *anf* (*al*ternative *n*itrogen *f*ixation) and they are organized in a single operon, *anfHDGK*, ORF1, ORF2 (see Figure 19-1). The subunits of dinitrogenase reductase-3 are encoded by *anfH* and the α and β subunits of dinitrogenase-

Table 19-5. Sequence Comparisons Between nif, vnf, and anf Genes of A. vinelandii[a]

Gene Comparisons	Percent Nucleotide Sequence Identity	Percent Amino Acid Sequence Identity of Predicted Gene Products
nifH × vnfH	88.5	91.0
nifH × anfH	69.3	62.8
vnfH × anfH	70.1	63.5
nifD × vnfD	52.6	33.0
nifD × anfD	49.6	32.7
vnfD × anfD	65.8	54.4
vnfG × anfG	55.9	39.8
nifK × vnfK	51.5	31.1
nifK × anfK	50.7	32.1
vnfK × anfK	69.8	57.4
nifD × nifK	—	19.8
vnfD × vnfK	—	24.5
anfD × anfK	—	21.5

[a]The values for vnfH, vnfD, vnfG, and vnfK are derived from unpublished sequence data (R. Joerger, T. Loveless, R. Pau, L. Mitchenall, B. Simon, and P. Bishop). The source of sequence data for nif and anf genes are refs. 42 and 48, respectively.

3 are encoded by anfD and anfK, respectively. anfG probably encodes a third subunit (δ) for dinitrogenase-3. The anfHDGK operon is preceded by a potential promoter sequence that would interact with RNA polymerase containing sigma 54. The predicted protein products of the two ORFs that are located 3' to anfK (see Figure 19-1) do not show overall similarity to any known nif gene products. However, the predicted ORF1 product contains some sequence identity to the NH_2-terminal part of dinitrogenase reductase and another region exhibits identity to presumed heme-binding domains of P-450 cytochromes. The predicted product of ORF2 does not seem to show any interesting similarity with other amino acid sequences in the Bionet data base.[48] Deletions plus insertions placed in several regions of the anfHDGK operon resulted in Anf⁻ mutants that were unable to grow in N-free, Mo-deficient medium. However, growth in media containing Mo or V was normal.[48] The absence of nitrogenase-3 proteins in these Anf⁻ mutants was also confirmed by two-dimensional gel electrophoresis.[48] Transcription from the anfHDGK, ORF1, ORF2 operon results in NH_4^+-repressible transcripts that are 6.0, 4.3, and 2.6 kb in length (R. Premakumar, M. Jacobson, and P. Bishop, unpublished results).

D. Sequence Comparisons Between Structural Genes

In Table 19-5 sequence comparisons are shown for the structural genes and their presumed protein products. Overall sequence identity is greater for nifH, vnfH, and anfH and their products (dinitrogenase reductases) than for the genes encoding the three dinitrogenases. Five cysteine residues are conserved across the three dinitrogenase reductase proteins, and two of these conserved residues appear to serve as ligands for the Fe–S center, which is bound symmetrically between the dinitrogenase reductase-1 subunits.[48] A motif characteristic of nucleotide-binding domains (Gly–X–Gly–XX–Gly) is also present in the three gene products.[48] The percent identity between *nifH* and *vnfH* is quite high (88.5%) and suggests that these two genes may have diverged relatively recently (in evolutionary time) from an ancestral gene, or that functional constraints are very similar for the two dinitrogenase reductases. In contrast, *anfH* seems to be more distantly related to *nifH* and *vnfH* (about 70% identity). This correlates well with the inability of dinitrogenase reductase-3 to yield high nitrogenase activity in a complementation assay with dinitrogenase-1. In contrast, dinitrogenase reductase-2 gives high activity in the same assay.[23]

The *vnfDK* and *anfDK* genes are more similar to each other than either set of genes is to *nifDK*, with *vnfK* and *anfK* sharing slightly more identity (69.8%) than *vnfD* and *anfD* (65.8%). The products of *nifDK*, *vnfDK*, and *anfDK* contain cysteine and histidine residues that are conserved in nearly all dinitrogenase proteins that have thus far been examined. These highly conserved Cys and His residues are thought to be coordinating ligands for Fe–S centers. Some of these Cys residues are essential for dinitrogenase activity as demonstrated by site-directed mutagenesis experiments.[17]

As previously mentioned, *vnfG* and *anfG* show identity with respect to both nucleotide sequence (55.9%) and amino acid sequences of the presumed products (39.8%). This may indicate that these gene products function as subunits of their respective dinitrogenases in a similar fashion.

Finally, the percent identity between amino acid sequences of the α and β subunits of each dinitrogenase (see Table 19-5) suggests that the genes encoding these two subunits may have evolved from a common ancestral gene, as previously suggested for Mo-containing dinitrogenases.[81] Based on these identity comparisons, it has been speculated that *nifD* and *nifK* may have diverged somewhat earlier during evolution than the genes encoding the subunits of dinitrogenase-2 and -3.[48,70]

An observation that may relate to the occurrence of alternative nitrogenases in other organisms is the finding that predicted products of *nifH*-like genes from two very different diazotrophs exhibit a high degree of identity with the *anfH* product (Table 19-6). One of these is NifH3 (81.7% identity) from the obligate anaerobe *Clostridium pasteurianum* and the other is NifH1 (72.4% identity) from the thermophilic archaebacterium, *Methanococcus thermolithotrophicus*. It remains to be seen, however, whether these identities signify functional similarity to the nitrogenase-3 system. In the case of *nifH3* from *C. pasteurianum,* no evidence for transcription could be found under Mo-sufficient diazotrophic conditions; however, transcription under

Table 19-6. Sequence Comparisons of Predicted Products of nif, vnf, and anf Genes from A. vinelandii and of nif Genes from Methanococcus thermolithotrophicus and Clostridium pasteurianum[a]

Gene Product Comparison	Percent Amino Acid Identity
Av NifH × Cp NifH3	63.2
Av NifH × Mt NifH1	63.4
Av VnfH × Cp NifH3	62.3
Av VnfH × Mt NifH1	63.0
Av AnfH × Cp NifH3	81.7
Av AnfH × Mt NifH1	72.4
Cp NifH3 × Mt NifH1	70.3
Av NifD × Mt NifD	40.4
Av VnfD × Mt NifD	41.5
Av AnfD × Mt NifD	38.3

[a]Source of sequence: *A. vinelandii* (AV) (see footnote, Table 19-5); *Clostridium pasteurianum* (Cp);[84] and *Methanococcus thermolithotrophicus* (Mt)[80].

Mo-deficient conditions was not examined.[84] Although the predicted product of *nifH1* from *M. thermolithotrophicus* shows a fairly high degree of identity (72.4%) with the *anfH* product, it is clear that the *M. thermolithotrophicus* *nifD* product shows much less identity (38.3%) with the *anfD* product; therefore, it can be concluded that the degree of identity between amino acid sequences of dinitrogenase reductase proteins does not necessarily correlate with the percent identity observed for the α and β subunits of different dinitrogenases (also see Table 19-5). This higher degree of variability is also observed with the conventional Mo-containing dinitrogenases.[27]

E. Predicted Properties of the Structural Gene Products

The molecular weights and pIs as calculated from the predicted structural gene products are presented in Table 19-7. These values are generally in good agreement with those obtained using electrophoretic techniques.

F. Nonstructural Genes

Recently, two *nifA*-like ORFs were identified in DNA cloned from *A. vinelandii*. One of these ORFs was recognized in the DNA sequence flanking the Tn5 insertion carried by a Vnf⁻ mutant (strain CA46) that is unable to express nitrogenase-2 when derepressed in N-free medium containing V.[46,47] Since this mutant synthesizes both nitrogenase-1 and nitrogenase-3 under Mo-deficient N-free conditions in the presence or absence of V, the *nifA*-like ORF was designated *vnfA*.[46] The other *nifA*-like ORF was located approximately 700 bp upstream from *anfH* (see Figure 19-1). A mutant (strain

Table 19-7. *Predicted Molecular Weights and pIs of the Nitrogenase Structural Gene Products from A. vinelandii*[a]

Gene	Total Number of aa	Predicted Products Calculated Molecular Weight	Estimated pI
vnfH	290	31,026	4.58
nifH	290	31,514	4.48
anfH	275	29,883	4.72
vnfD	474	53,874	5.94
nifD	492	55,285	5.93
anfD	518	58,409	6.34
vnfG	113	13,337	4.83
anfG	132	15,342	5.00
vnfK	475	52,772	5.45
nifK	523	59,455	5.98
anfK	462	51,176	4.95

[a]The sources of the sequence data from which the values presented in this table were derived are listed in the footnote for Table 19-5.

CA66), carrying a deletion plus insertion in this *nifA*-like ORF, synthesized only nitrogenase-2 proteins after derepression in Mo-deficient media with or without V. Thus, this ORF was designated *anfA*.[46] The highest degree of similarity between the predicted products of *nifA*, *vnfA*, and *anfA* is in the C-terminal half of the proteins where the potential RNA polymerase–sigma 54 interaction sites and ATP-binding domains are located. A potential DNA-binding domain[33] that includes a helix-turn-helix motif is present in both predicted products of *vnfA* and *anfA*.[46]

nifENX-like ORFs are located immediately downstream from *vnfA*.[85] Preliminary results with a strain containing Tn5-*lacZ* inserted in the 3'-terminal end of the *nifN*-like ORF indicate that the *nifN*- and/or *nifX*-like ORF is required for diazotrophic growth in Mo-deficient media with or without V. However, in the presence of Mo this strain shows wild-type growth (E. Wolfinger and P. Bishop, unpublished results). These results imply that these *nifENX*-like genes are required for functional nitrogenase-2 and nitrogenase-3 but not nitrogenase-1.

V. Regulation of Expression of the Alternative N_2-Fixation Systems

Knowledge of how nitrogenase-2 and nitrogenase-3 are regulated at the level of gene expression is rather rudimentary. Transcription of genes involved in systems 2 and 3 appears to be initiated at potential promoter sites that conform to the RNA polymerase–sigma 54 recognition sequence [CT<u>GG</u>-N_8-TT<u>GC</u>A; where the underlined nucleotides are invariant[7]]. Such sites are found in the 5' noncoding region of all *vnf* and *anf* operons examined to date except for *vnfA* where one potential promoter is situated within the 3' end of ORF2 (see Figure 19-1) and another is located in the region between ORF1 and ORF2. As might be expected from these observations, *ntrA* is required for all three nitrogenase systems.[83] Expression of nitrogenase-1 in *A. vinelandii* is activated by NifA.[6] This gene product is thought to recognize upstream activating sequences (UAS) [TGT-N_{10}-ACA[18]]. A characteristic feature of alternative system operons is the apparent lack of this UAS. This is not particularly surprising since NifA is not required for expression of nitrogenase-2 or nitrogenase-3.[46] VnfA and AnfA may function as activators for *vnf* and *anf* operons, respectively, and the binding sites for these proteins may be quite different from the binding site for NifA.

The factors that regulate expression and activity of these activator proteins from *A. vinelandii* are not yet known. However, it is clear that ammonium and metals are involved in the regulation process. Early on it was observed that proteins attributed to alternative nitrogenases were absent in *A. vinelandii* cells grown in the presence of NH_4^+ or Mo.[9,10] Nitrogenase-2 (previously called N_2ase B_1) was present in cells grown in Mo-deficient medium containing V, whereas cells grown in Mo-deficient medium in the absence of V expressed nitrogenase-3 (formerly designated N_2ase B_2). Dinitrogenase reductase-2 was also expressed in cells grown in Mo-deficient medium even in the absence of V.[10,63] In another study[43] *nifH*-hybridizing transcripts from the *nifHDK* operon were found to be undectable in *A. vinelandii* cells derepressed under Mo-deprived conditions. Rather, under these conditions, a different set of *nifH*-hybridizing transcripts was observed that probably originated from the *vnfH-Fd* operon. Although the details of how the expression of nitrogenase-2 and nitrogenase-3 is modulated by NH_4^+ and metals remain obscure, progress on these regulatory aspects can be expected to be rapid now that transcriptional *lacZ* fusions, which are integrated into the *A. vinelandii* genome, have been constructed for *vnfH*, *vnfD*, and *anfH* (C. Kennedy, personal communication, and our unpublished results). Experiments with an *anfH-lacZ* fusion strain indicate that transcription initiated from the *anfH* promoter is repressed by NH_4^+, Mo, and V. Furthermore, experiments with this *lacZ* fusion strain indicate that dinitrogenase reductase-2 is required for *in vivo* transcription of the *anfHDGK* operon (R. Joerger, E. Wolfinger, and P. Bishop, unpublished results). This may be one reason why dinitrogenase reductase-2 is always present under Mo-deficient conditions (as noted previously) even though dinitrogenase-2 is absent and dinitrogenase reductase-2 is unable to biochemically complement dinitrogenase-3.[23]

A new gene, *ntrD*, is required for growth on nitrate and for diazotrophic growth under Mo-sufficient and Mo-deficient conditions. NtrD$^-$ mutants are blocked in transcription from the *anfH* promoter, but not from the *vnfH* and *vnfD* promoters (A. Bali and C. Kennedy, personal communication). Another regulatory gene, *nfrX*, is necessary for growth under both Mo-sufficient and Mo-deficient conditions, but not in the presence of V.[72] Thus, at least four gene products (AnfA, dinitrogenase reductase-2, NtrD, and NfrX) may be involved in the regulation of nitrogenase-3.

VI. The Occurrence of Alternative Nitrogenases in Diazotrophs Other than the Azotobacteriaceae

Evidence is accumulating which suggests that diazotrophs other than the azotobacters contain alternative nitrogenase systems. Recently, the cyanobacterium *Anabaena variabilis* has been reported to possess an alternative V-containing nitrogenase.[52] Multiple *nifH*-like sequences have also been identified in several cyanobacteria species; these include *Anabaena*,[66] *Calothrix*,[49] and *Nostoc commune*.[24] These reiterated *nifH*-like sequences may indicate the presence of alternative nitrogenases. Along these lines, it is interesting that *Calothrix antarctica* has been reported[73] to fix N_2 in Mo-deficient medium provided cobalt was present. The archaebacterium *Methanosarcina barkeri* has been shown to require either Mo or V for diazotrophic growth; thus, this diazotroph may synthesize a V-containing enzyme.[74] *Rhodobacter capsulatus* has been shown to have multiple copies of *nif*-like sequences that are ordinarily silent. However, at least one of these copies is capable of activation by mutation.[75] This observation is reminiscent of the situation in *A. vinelandii* where Nif^+ pseudorevertants can be readily isolated.[9] Recent results with a *nifHDK* deletion strain of *R. capsulatus* unequivocally demonstrate that this organism contains an alternative nitrogenase that is optimally expressed in medium that is nearly Mo-free (<0.05 ppb Mo) (K. Schneider and W. Klipp, personal communication). The partially purified dinitrogenase contained 20 g atoms of Fe per mole protein and negligible amounts of Mo, V, W, and Re. These findings represent the first definitive evidence that alternative nitrogenases occur outside of the Azotobacteriaceae. Multiple *nifH*-like sequences in *C. pasteurianum*[84] are likely to have a functional significance (e.g., *nifH3*) since this organism is able to fix significant quantities of N_2 in Mo-deficient medium and N_2 fixation is stimulated by V in addition to Mo.[44] It is also interesting to note that this diazotroph reduces C_2H_2 to ethane in Mo-deficient medium in the presence and absence of V.[25]

Although it is convenient to classify nitrogenases according to their metal content, it is also possible that some Mo-containing nitrogenases share more similarity with either nitrogenase-2 or nitrogenase-3 than with nitrogenase-1. Souillard and Sibold[80] have raised this possibility by suggesting that methanogens may have an unusual Mo-containing nitrogenase that has biochemical properties similar to those of nitrogenase-3. A finding that may support this notion is the isolation of a Mo-containing nitrogenase from *Xanthobacter autotrophicus* that contains a small subunit ($M_r = 12,300$) that could be similar to the δ subunits of dinitrogenase-2 and dinitrogenase-3 (K. Schneider, S. Köppe, F. Hagen, and A. Müller, personal communication). The enzyme is also able to reduce C_2H_2 to ethane (another indicator that this nitrogenase may resemble the alternative nitrogenases).

VII. Future Directions

With the availability of nucleotide sequence data for the structural genes encoding the alternative nitrogenases, it is now possible to construct specific hybridization probes that could prove useful in the screening of diazotrophs for alternative system genes. This assumes, of course, that the *vnf* and *anf* genes of *A. vinelandii* will have some conserved sequences that are common to *vnf* and *anf* genes of other diazotrophs but not to *nif* genes. Another possibility is the use of degenerate oligonucleotides as polymerase chain reaction (PCR) primers flanking an intragenic region that is highly variable between the *nifH*, *vnfH*, and *anfH* genes. These primers could be used to amplify this particular sequence from genomic DNA of diverse diazotrophs. The PCR products could then be cloned and sequenced to determine whether distinct multiple *nifH*-like genes exist in the genomic DNA of a given diazotroph. This technique has been utilized to analyze a portion of a *nifH* gene from the marine cyanobacterium *Trichodesmium thiebautii*.[86]

Central to the understanding of why alternative nitrogenases exist is the role that these enzymes play under natural conditions. One possible factor is low temperature, as suggested by the finding that low temperature favors N_2 reduction by nitrogenase-2.[56] This possibility is supported by the observation that β-galactosidase expression was not affected by Mo in *vnfH-lacZ* and *vnfD-lacZ* fusion strains at 10° C, whereas at 30° C, expression was considerably diminished by the presence of Mo (C. Kennedy, personal communication). Thus, nitrogenase-2 could be ideally suited for low-temperature diazotrophy. A more obvious role for Mo-independent nitrogenases might be N_2 fixation in low-pH soils with high iron oxide contents where Mo is known to be biologically limiting. Such soils are common to the southeastern part of the United States and to many tropical regions of the world. Other Mo-deficient environments might be locations where Mo is removed by other organisms or where the natural abundance is low. The challenge will be to devise assays to detect alternative nitrogenase activity. One assay that holds promise is the reduction of C_2H_2 to ethane by alternative nitrogenases but not by conventional Mo-containing nitrogenases.[25]

Another approach could be to test natural environments for the potential to express active alternative nitrogenases. Thus, environmental samples (e.g., soil extracts, sea water, and lake water) could be tested for the ability to support N_2 fixation by single nitrogenase strains of *A. vinelandii* (i.e., double-deletion strains inactivated for two of the three nitrogenases). The effect of environmental samples on expression of the three nitrogenase structural gene operons could also be tested using *nifH-*, *vnfD-*, and *anfH-lacZ* fusion strains.

In conclusion, the study of alternative nitrogenase systems is still in its infancy and many fundamental questions remain unanswered. For example, the importance of these nitrogenases to the global nitrogen cycle remains an open question, as does the functional role that these enzymes play in the natural environment. A thorough study of these nitrogenases at the biochemical and biophysical level will undoubtedly add to our general understanding of how nitrogenases catalyze the reduction of N_2 to ammonia. Our knowledge concerning regulation by metals should be aided by studying the details of how Mo and V regulate the expression of the three nitrogenases in *A. vinelandii*.

Finally, knowledge of alternative nitrogenases may find applications in agriculture. For example, if alternative N_2-fixation systems are found in the agronomically important rhizobia, it is possible that these systems can be genetically manipulated in order to improve the effectiveness of these microsymbionts.

References

1. Arber, J.M., Dobson, B.R., Eady, R.R., Hasnain, S.S., Garner, C.D., Matsushita, T., Nomura, M., and Smith, B.E., "Vanadium K-edge X-ray absorption spectroscopy of the functioning and thionine-oxidized forms of the VFe-protein of the vanadium nitrogenase from *Azotobacter chroococcum*," *Biochem. J. 258*, 733–737 (1989).

2. Arber, J.M., Dobson, B.R., Eady, R.R., Stevens, P., Hasnain, S.S., Garner, C.D., and Smith, B.E., "Vanadium K-edge X-ray absorption spectrum of the VFe protein of the vanadium nitrogenase of *Azotobacter chroococcum*," *Nature (Lond.) 325*, 372–374 (1987).

3. Becking, J.H., "Species differences in molybdenum and vanadium requirements and combined nitrogen utilization by *Azotobacteriaceae*," *Plant and Soil 16*, 171–201 (1962).

4. Beijerinck, M.W., "Über oligonitrophile Mikroben," *Zentralbl. Bakteriol. Parasitenkd. Infektionskr. II Abt. 7*, 561–582 (1901).

5. Benemann, J.R., McKenna, C.E., Lie, R.F., Traylor, T.G., and Kamen, M.D., "The vanadium effect in nitrogen fixation by *Azotobacter*," *Biochim. Biophys. Acta 264*, 25–38 (1972).

6. Bennett, L.T., Cannon, F., and Dean, D.R., "Nucleotide sequence and mutagenesis of the *nifA* gene from *Azotobacter vinelandii*," *Mol. Microbiol. 2*, 315–321 (1988).

7. Beynon, J., Cannon, M., Buchanan-Wollaston, V., and Cannon, F., "The *nif* promoters of *Klebsiella pneumoniae* have a characteristic primary structure," *Cell 34*, 665–671 (1983).

8. Bishop, P.E., Hawkins, M.E., and Eady, R.R., "Nitrogen fixation in Mo-deficient continuous culture by a strain of *Azotobacter vinelandii* carrying a deletion of the structural genes for nitrogenase (*nifHDK*)," *Biochem. J. 238,* 437–442 (1986).

9. Bishop, P.E., Jarlenski, D.M.L., and Hetherington, D.R., "Evidence for an alternative nitrogen fixation system in *Azotobacter vinelandii*," *Proc. Natl. Acad. Sci. (USA) 77,* 7342–7346 (1980).

10. Bishop, P.E., Jarlenski, D.M.L., and Hetherington, D.R., "Expression of an alternative nitrogen fixation system in *Azotobacter vinelandii*," *J. Bacteriol. 150,* 1244–1251 (1982).

11. Bishop, P.E., Premakumar, R., Dean, D.R., Jacobson, M.R., Chisnell, J.R., Rizzo, T.M., and Kopczynski, J., "Nitrogen fixation by *Azotobacter vinelandii* strains having deletions in structural genes for nitrogenase," *Science 232,* 92–94 (1986).

12. Bortels, H., "Biokatalyse und Reaktionsempfindlickeit bei niederen und höheren Pflanzen," *Angew. Bot. 11,* 285–332 (1929).

13. Bortels H., "Kurze Notiz über die Katalyse der biologischen Stickstoffbindung," *Zentralbl. Bakteriol. Parasitenkd. Infektionskr. II. Abt. 87,* 476–477 (1933).

14. Bortels H., "Molybdän als Katalysator bei der biologischen Stickstoffbindung." *Arch. Mikrobiol. 1,* 333–342 (1930).

15. Bortels, H., "Über die Wirkung von Agar sowie Eisen, Molybdän, Mangan und anderen Spurenelementen in stickstofffreier Nährlösung auf *Azotobacter*," *Zentralbl. Bakteriol. Parasitenkd. Infektionskr. II. Abt. 100,* 373–393 (1939).

16. Bortels, H., "Weitere Untersuchungen über die Bedeutung von Molybdän, Vanadium, Wolfram und andere Erdaschenstoffe für stickstoffbindende und andere Mikroorganismen," *Zentralbl. Bakteriol. Parasitenkd. Infektionskr. II. Abt. 95,* 193–218 (1936).

17. Brigle, K.E., Setterquist, R.A., Dean, D.R., Cantwell, J.S., Weiss, M.C., and Newton, W.E., "Site-directed mutagenesis of the nitrogenase MoFe protein of *Azotobacter vinelandii*," *Proc. Natl. Acad. Sci. (USA) 84,* 7066–7069 (1987).

18. Buck, M., Miller, S., Drummond, M., and Dixon, R., "Upstream activator sequences are present in the promoters of nitrogen fixation genes," *Nature (Lond.) 320,* 374–378 (1986).

19. Bulen, W.A., Nitrogenase from *Azotobacter vinelandii* and reactions affecting mechanistic interpretations, in *Proc. 1st Symp. Nitrogen Fixation,* Vol. 1, eds. Newton, W.E., and Nyman, C.J. Washington State University Press, 1976, pp. 177–186.

20. Burk, D., and Horner, C.K., "The specific catalytic role of molybdenum and vanadium in nitrogen fixation and amide utilization by *Azotobacter*," *Trans. III. Int. Congr. Soil Sci.,* Vol. 1, Oxford, 1935, pp. 152–155.

21. Burns, R.C., Fuchsman, W.H., and Hardy, R.W.F., "Nitrogenase from vanadium-grown *Azotobacter:* Isolation, characteristics, and mechanistic implications," *Biochem. Biophys. Res. Commun. 42,* 353–358 (1971).

22. Burns, R.C., Stasny, J.T., and Hardy, R.W.F., "Isolation and charcteristics of a modified nitrogenase from *Azotobacter vinelandii* including "vanadium-Fe" protein from cells grown on medium enriched in vanadium, in *Proc. 1st Symp. Nitrogen Fixation,* Vol. 1, eds. Newton, W.E., and Nyman, C.J., Washington State University Press, 1976, pp. 196–207.

23. Chisnell, J.R., Premakumar, R., and Bishop, P.E., "Purification of a second alternative nitrogenase from a *nifHDK* deletion strain of *Azotobacter vinelandii*," *J. Bacteriol. 170,* 27–33 (1988).

24. Defrancesco, N., and Potts, M., "Cloning of *nifHD* from *Nostoc commune* UTEX 584 and of a flanking region homologous to part of the *Azotobacter vinelandii nifU* gene," *J. Bacteriol. 170,* 3297–3300 (1988).

25. Dilworth, M.J., Eady, R.R., Robson, R.L., and Miller R.W., "Ethane formation from acetylene as a potential test for vanadium nitrogenase *in vivo*," *Nature (Lond.), 327,* 167–168 (1987).

26. Dilworth, M.J., Eady, R.R., and Eldridge, M.E., "The vanadium nitrogenase of *Azotobacter chroococcum:* Reduction of acetylene and ethylene to ethane," *Biochem. J. 249,* 745–751 (1988).

27. Eady, R.R., "Enzymology in free-living diazotrophs," in *Nitrogen Fixation,* Vol. 4, eds. Broughton, W.J., and Pühler, A., Oxford, Clarendon Press, 1986, pp. 1–49.

28. Eady, R.R., "The vanadium nitrogenase of *Azotobacter*," *Polyhedron 8,* 1695–1700 (1989).

29. Eady, R.R., Robson, R.L., Richardson, T.H., Miller, R.W., and Hawkins, M., "The vanadium nitrogenase of *Azotobacter chroococcum*: Purification and properties of the VFe protein," *Biochem. J. 244,* 197–207 (1987).

30. Eady, R.R., Richardson, T.H., Miller, R.W., Hawkins, M., and Lowe, D.J., "The vanadium nitrogenase of *Azotobacter chroococcum*: Purification and properties of the Fe protein," *Biochem. J. 256,* 189–196 (1988).

31. Emerich, D.W., and Burris, R.H., "Interactions of heterologous nitrogenase components that generate catalytically inactive complexes," *Proc. Natl. Acad. Sci. (USA) 73,* 4369–4373 (1976).

32. Esposito, R.G., and Wilson, P.W., "Trace metal requirement of *Azotobacter*," *Proc. Soc. Exp. Med. 93,* 564–567 (1956).

33. Fischer, M.-H., Bruderer, T., and Hennecke, H., "Essential and non-essential domains in the *Bradyrhizobium japonicum* NifA protein: Identification of indispensable cysteine residues involved in redox reactivity and/or metal binding," *Nucleic Acids Res. 16,* 2207–2224 (1988).

34. Garner, C.D., Arber, J.M., Harvey, I., Hasnain, S.S., Eady, R.R., Smith, B.E., Boer, E.D., and Wever, R., "Vanadium K-edge X-ray absorption spectroscopy of the functioning and thionine-oxidized forms of the VFe protein of the vanadium nitrogenase from *Azotobacter chroococcum*," *Polyhedron 8,* 1649–1652 (1989).

35. George, G.N., Coyle, C.L., Hales, B.J., and Cramer, S.P., "X-ray absorption of *Azotobacter* vanadium nitrogenase," *J. Am. Chem. Soc. 110.* 4057–4059 (1988).

36. Gillum, W.O., Mortenson, L.E., Chen, J.S., and Holm, R.H., "Quantitative extrusion of the Fe_4S_4 cores of active sites of ferredoxins and the hydrogenase of *Clostridium pasteurianum*," *J. Am. Chem. Soc. 99,* 584–595 (1977).

37. Hales, B.J., Case, E.E., Morningstar, J.E., Dzeda, M.F., and Mauterer, L.A., "Isolation of a new vanadium-containing nitrogenase from *Azotobacter vinelandii*," *Biochemistry 25,* 7251–7255 (1986).

38. Hales, B.J., Langosch, D.J., and Case, E.E., "Isolation and characterization of a second nitrogenase Fe-protein from *Azotobacter vinelandii*," *J. Biol. Chem. 261,* 15301–15306 (1986).

39. Hales, B. J., True, A.E., and Hoffman, B.M., "Detection of a new signal in the EPR spectrum of vanadium nitrogenase from *Azotobacter vinelandii*," *J. Am. Chem. Soc. 111,* 8519–8520 (1989).

40. Hausinger, R.P., and Howard, J.B., "Thiol reactivity of the nitrogenase Fe-protein from *Azotobacter vinelandii*," *J. Biol. Chem. 258*, 13486–13492 (1983).

41. Horner, C.K., Burk, D., Allison, F.E., and Sherman, M.S., "Nitrogen fixation by *Azotobacter* as influenced by molybdenum and vanadium," *J. Agric. Res. 65*, 173–193 (1942).

42. Jacobson, M.R., Brigle, K.E., Bennett, L.T., Setterquist, R.A., Wilson, M.S., Cash, V.L., Beynon, J., Newton, W.E., and Dean, D.R., "Physical and genetic map of the major *nif* gene cluster from *Azotobacter vinelandii*," *J. Bacteriol. 171*, 1017–1027 (1989).

43. Jacobson, M.R., Premakumar, R., and Bishop, P.E., "Transriptional regulation of nitrogen fixation by molybdenum in *Azotobacter vinelandii*," *J. Bacteriol. 167*, 480–486 (1986).

44. Jensen, H.L., "The influence of molybdenum and vanadium on nitrogen fixation by *Clostridium butyricum* and related organisms," *Proc. Linn. Soc. N.S. Wales 72*, 73–86 (1947).

45. Joerger, R.D., and Bishop, P.E., "Nucleotide sequence and genetic analysis of the *nifB-nifQ* region from *Azotobacter vinelandii*," *J. Bacteriol. 170*, 1475–1487 (1988).

46. Joerger, R.D., Jacobson, M.R., and Bishop, P.E., "Two *nifA*-like genes required for expression of alternative nitrogenases of alternative nitrogenasess by *Azotobacter vinelandii*," *J. Bacteriol. 171*, 3258–3267 (1989).

47. Joerger, R.D., Premakumar, R., and Bishop, P.E., "Tn5-induced mutants of *Azotobacter vinelandii* affected in nitrogen fixation under Mo-deficient and Mo-sufficient conditions," *J. Bacteriol. 168*, 673–682 (1986).

48. Joerger, R.D., Jacobson, M.R., Premakumar, R., Wolfinger, E.D., and Bishop, P.E., "Nucleotide sequence and mutational analysis of the structural genes (*anfHDGK*) for the second alternative nitrogenase from *Azotobacter vinelandii*," *J. Bacteriol. 171*, 1075–1086 (1989).

49. Kallas, T., Rebiere, M.-C., Rippka, R., and Tandeau de Marsac, N., "The structural *nif* genes of the cyanobacteria *Gloeothece* sp. and *Calothrix* sp. share homology with those of *Anabaena* sp., but the *Gloeothece* genes have a different arrangement," *J. Bacteriol. 155*, 427–431 (1983).

50. Kennedy, C., Eady, R.R., Kondorosi, E., and Klavans Rekosh, D., "The molybdenum-iron protein of *Klebsiella pneumoniae* nitrogenase. Evidence for non-identical subunits from peptide mapping," *Biochem. J. 155*, 383–389 (1976).

51. Kennedy, C., Gamal, R., Humphrey, R., Ramos, J., Brigle, K., and Dean, D., "The *nifH*, and *nifN* genes of *Azotobacter vinelandii*: Characterization by Tn5 mutagenesis and isolation from pLAFR1 gene banks," *Mol. Gen. Genet. 205*, 318–325 (1986).

52. Kentemich, T., Danneberg, G., Hundeshagen, B., and Bothe, H., "Evidence for the occurrence of the alternative, vanadium-containing nitrogenase in the cyanobacterium *Anabaena variabilis*," *FEMS Microbiol. Lett. 51*, 19–24 (1988).

53. Kremieniewski, S., "Untersuchungen über *Azotobacter chroococcum*." *Zentralbl. Bakteriol Parasitenkd. Infektionskr. II. Abt. 23*, 161–173 (1909).

54. McKenna, C. E., Benemann, J.R., and Traylor, T.G., "A vanadium containing nitrogenase preparation: Implications for the role of molybdenum in nitrogen fixation," *Biochem. Biophys. Res. Commun. 41*, 1501–1508 (1970).

55. McLean, P.A., Papaefthymiou, V., Münck, E., and Orme-Johnson, W.H.," Use of isotopic hybrids of the MoFe protein to study the mechanism of nitrogenase catalysis,"

in *Nitrogen Fixation: Hundred Years After*, eds. Bothe, H., deBruijn, F.J., and Newton, W.E., Gustav Fischer, Stuttgart, 1988, pp. 101–106.

56. Miller, R.W., and Eady, R.R., "Molybdenum and vanadium nitrogenases of *Azotobacter chroococcum:* Low temperature favours N_2 reduction by vanadium nitrogenase," *Biochem. J. 256*, 429–432 (1988).

57. Morningstar, J.E., and Hales, B.J., "Electron paramagnetic resonance study of the vanadium-iron protein of nitrogenase from *Azotobacter vinelandii*," *J. Am. Chem. Soc. 109*, 6854–6855 (1987).

58. Morningstar, J.E., Johnson, M.K., Case, E.E., and Hales, B.J., "Characterization of the metal clusters in the nitrogenase molybdenum-iron and vanadium-iron proteins of *Azotobacter vinelandii* using magnetic circular dichroism spectroscopy," *Biochemistry 26*, 1795–1800 (1987).

59. Münck, E., Rhodes, H., Orme-Johnson, W.H., Davis, L.C., Brill, W.J., and Shah, V.K., "Nitrogenase VIII. Mossbauer and EPR Spectroscopy. The MoFe protein component from *Azotobacter vinelandii* OP," *Biochim. Biophys. Acta 400*, 32–53 (1975).

60. Page, W.J., and Collinson, S.K., "Molybdenum enhancement of nitrogen fixation in a Mo-starved *Azotobacter vinelandii* Nif⁻ mutant," *Can. J. Microbiol. 28*, 1173–1180 (1982).

61. Pau, R.N., Mitchenall, L.A., and Robson, R.L., "Genetic evidence for an *Azotobacter vinelandii* nitrogenase lacking molybdenum and vanadium," *J. Bacteriol. 171*, 124–129 (1989).

62. Pienkos, P.T., Shah, V.K., and Brill, W.J., "Molybdenum in nitrogenase," in *Molybdenum and Molybdenum-Containing Enzymes*, ed. Coughlan, M., Pergamon Press, Elmsford, N.Y., 1980, pp. 387–401.

63. Premakumar, R., Chisnell, J.R., and Bishop, P.E., "A comparison of the three dinitrogenase reductases expressed by *Azotobacter vinelandii*," *Can. J. Microbiol. 35*, 344–348 (1989).

64. Premakumar, R., Lemos, E.M., and Bishop, P.E., "Evidence for two dinitrogenase reductases under regulatory control by molybdenum in *Azotobacter vinelandii*," *Biochim. Biophys. Acta 797*, 64–70 (1984).

65. Raina, R., Reddy, M.A., Ghosal, D., and Das, H.K., "Characterization of the gene for the Fe-protein of the vanadium dependent alternative nitrogenase of *Azotobacter vinelandii* and construction of a Tn5 mutant," *Mol. Gen. Genet. 214*, 121–127 (1988).

66. Rice, D., Mazur, B.J., and Haselkorn, R., "Isolation and physical mapping of nitrogen fixation genes from the cyanobacterium *Anabaena* 7120, " *J. Biol. Chem. 257*, 13157–13163 (1982).

67. Robson, R.L., "Nitrogen fixation in strains of *Azotobacter chroococcum* bearing deletions of a cluster of genes coding for nitrogenase," *Arch. Microbiol. 146*, 74–79 (1986).

68. Robson, R.L., Eady, R.R., Richardson, T.H., Miller, R.W., Hawkins, M., and Postgate, J.R., "The alternative nitrogenase of *Azotobacter chroococcum* is a vanadium enzyme," *Nature (Lond.) 322*, 388–390 (1986).

69. Robson, R.L., Woodley, P.R., and Jones, R., "Second gene (*nifH**) coding for a nitrogenase iron-protein in *Azotobacter chroococcum* is adjacent to a gene coding for a ferredoxin-like protein," *EMBO J. 5*, 1159–1163 (1986).

70. Robson, R.L., Woodley, P.R., Pau, R.N., and Eady, R.R., "Structural genes for the vanadium nitrogenase from *Azotobacter chroococcum*," *EMBO J. 8*, 1217–1224 (1989).

71. Ruvkun, G.B., and Ausubel, F.M., "Interspecies homology of nitrogenase genes," *Proc. Natl. Acad. Sci. (USA) 77*, 191–195 (1980).

72. Santero, E., Toukdarian, A., Humphrey, R., and Kennedy, C., "Identification and characterization of two nitrogen fixation regulatory genes, *nifA* and *nfrX*, in *Azotobacter vinelandii* and *Azotobacter chroococcum*," *Mol. Microbiol. 2*, 303–314 (1988).

73. Saubert, S., and Strijdom, B.W., "The response of certain nitrogen-fixing microorganisms to cobalt in the presence and absence of molybdenum," *S. Afr. Agric. Sci. 11*, 769–774 (1968).

74. Scherer, P., "Vanadium and molybdenum requirement for the fixation of molecular nitrogen by two *Methanosarcina* strains," *Arch. Microbiol. 151*, 44–48 (1989).

75. Scolnik, P.A., and Haselkorn, R., "Activation of extra copies of genes coding for nitrogenase in *Rhodopseudomonas capsulata*," *Nature (Lond.) 307*, 289–292 (1984).

76. Shah, V.K., and Brill, W.J., "Nitrogenase IV. Simple method of purification to homogeneity of nitrogenase components from *Azotobacter vinelandii*," *Biochim. Biophys. Acta 305*, 445–454 (1973).

77. Shah, V.K., and Brill, W.J., "Isolation of an iron–molybdenum cofactor from nitrogenase," *Proc. Natl. Acad. Sci. (USA) 74*, 3269–3253 (1977).

78. Smith, B.E., Bishop, P.E., Dixon, R.A., Eady, R.R., Filler, W.A., Lowe, D.J., Richards, A.J.M., Thomson, A.J., Thorneley, R.N.F., anf Postgate, J.R., The iron–molybdenum cofactor of nitrogenase. In *Nitrogen Fixation Research Progress*, eds. Evans, H.J., Bottomley, P.J., and Newton, W.E., Martinus Nijhoff, Dordrecht, 1985, pp. 597–603.

79. Smith, B.E., Eady, R.R., Lowe, D.J. and Gormal, C., "The vanadium–iron protein of vanadium nitrogenase from *Azotobacter chroococcum* contains an iron–vanadium cofactor," *Biochem. J. 250*, 299–302 (1988).

80. Souillard, N., and Sibold, L., "Primary structure, functional organization and expression of nitrogenase structural genes of thermophilic archaebacterium *Methanococcus thermolithotrophicus*," *Mol. Microbiol. 3*, 441–552 (1989).

81. Thöny, B., Kaluza, K., and Hennecke, H., "Structural and functional homology between the α and β subunits of the nitrogenase MoFe protein as revealed by sequencing the *Rhizobium japonicum nifK* gene," *Mol. Gen. Genet. 198*, 441–448 (1985).

82. Thorneley, R.N.F., Bergstrom, N.H.J., Eady, R.R., and Lowe, D.J., "Vanadium nitrogenase of *Azotobacter chroococcum:* MgATP-dependent electron transfer within the protein complex," *Biochem. J. 257*, 789–794 (1989).

83. Toukdarian, A., and Kennedy, C. "Regulation of nitrogen metabolism in *Azotobacter vinelandii:* Isolation of *ntr* and *glnA* genes and construction of *ntr* mutants," *EMBO J. 5*, 399–407 (1986).

84. Wang, S.-Z., Chen, J.-S., and Johnson, J.L., "The presence of five *nifH*-like sequences in *Clostridium pasteurianum:* Sequence divergence and transcriptional properties," *Nucleic Acids Res. 16*, 439–454 (1988).

85. Wolfinger, E.D., Pau, R.N., and Bishop, P.E., "Multiple *nifE*- and *nifN*-like genes in *Azotobacter vinelandii*," *Ann. Mtg. Am. Soc. Microbiol.* 185 (Abstract H-97) (1989).

86. Zehr, J.P., McReynolds, L.A., "Use of degenerate oligonucleotides for amplification of the *nifH* gene from the marine cyanobacterium *Trichodesmium thiebautii*," *Appl. Environ. Microbiol. 55*, 2522–2526 (1989).

20

Biochemical Genetics of Nitrogenase

Dennis R. Dean and Marty R. Jacobson

I. Introduction

Biological nitrogen fixation is catalyzed by nitrogenase. This entity is a complex metalloenzyme composed of two separable components designated Fe protein and MoFe protein. Both component proteins are required for catalysis. The Fe protein serves as a specific ATP-binding, one-electron reductant of the MoFe protein, whereas the site for substrate binding and reduction is almost certainly located upon the MoFe protein. During catalysis, Fe protein probably binds two MgATP molecules, is reduced, associates with the MoFe protein, and donates a single electron to the MoFe protein in a reaction coupled to ATP hydrolysis and component protein dissociation. The catalytic reduction of N_2 is usually indicated by the following equation:

$$N_2 + 8e^- + 8H^+ + 16\,MgATP \xrightarrow[\text{nitrogenase}]{} 2NH_3 + H_2 + 16\,MgADP + 16Pi$$

Because multiple electrons are required for N_2 reduction, multiple cycles of component protein association and dissociation are required for nitrogenase turnover. As shown in the preceding equation, eight protons and eight reducing equivalents are consumed and a single H_2 molecule evolved per N_2 reduced. An anaerobic environment is also required for nitrogenase catalysis and both component proteins are extremely O_2 labile. Ferredoxins or flavodoxin are probable physiologic electron donors to nitrogenase *in vivo,* whereas dithionite is usually used as the source for reducing equivalents *in vitro.*[30] Nitrogenase is also known to be a promiscuous enzyme capable of reducing substrates other than the apparent physiologic ones, N_2 and H+.[48,117,139,156,163,168,212,250] Most notable among these is C_2H_2, which nitrogenase can reduce by two electrons to yield C_2H_4.[48,212] This feature is particularly useful in assay of nitrogenase because C_2H_2 and C_2H_4 are easily separated and quantitated by gas chromatography. Hydrogen evolution appears to be integral to nitrogenase catalysis,[225] and some organisms are apparently able to use an uptake hydrogenase to recapture a portion of the energy otherwise lost through this process (see Chapter 11 in this volume). A detailed discussion of the catalytic properties of nitrogenase is presented in Chapter 18 of this volume.

Native Fe protein is an approximately 60,000-D homodimer. A single [4Fe–4S] cluster is believed to be symmetrically bridged between the Fe protein subunits. This redox center is probably the immediate electron donor to the MoFe protein. The MoFe protein is an $\alpha_2\beta_2$ protein of about 220,000 D and contains two Mo atoms and about 32 Fe and 32 sulfide atoms per native molecule. About 16 Fe atoms can be extruded from each MoFe protein in the form of four [4Fe–4S] clusters (also designated as P clusters) by treatment of the native protein with thiols in a denaturing organic solvent. Although nothing is known about the function of the P clusters in relation to nitrogenase catalysis, they are capable of reversible oxidation and reduction and, as such, it seems likely they are involved in some aspect of the accumulation and intramolecular delivery of multiple electrons to the substrate reduction site. All or most of the remaining Fe and both Mo atoms are contained within two identical FeMo-cofactors. As will be discussed in a later section, there is considerable evidence indicating that FeMo-cofactor is located near or is part of the substrate binding and reduction site.

Other N_2-fixing complexes that are structurally related to but genetically distinct from the Mo-dependent nitrogenase discussed earlier have recently been identified in certain azotobacters. These other systems have catalytic components that are analogous to the Fe protein and the MoFe protein.[119,205,206] A major difference in these various N_2-fixing systems appears to reside within the metal compositions of their respective cofactor species (see, for example, ref. 227). Interestingly, the various N_2-fixing systems appear to share some gene products that are required for the maturation of their respective catalytic components.[118,131] These and other features of nitrogenases that apparently do not contain Mo, referred to as "alternative nitrogenases," are presented in Chapter 10 of this volume and will not be discussed further here.

The primary products of the genes encoding the nitrogenase structural components (*nifH*, Fe protein subunit; *nifD*, MoFe protein α-subunit; *nifK*, MoFe protein β-subunit) are not catalytically competent. Rather, immature nitrogenase components are processed to active forms through the action of products of certain other N_2-fixation-specific (*nif*) gene products. For example, it is known that at least six *nif*-specific gene products are required for the formation of FeMo-cofactor or insertion of FeMo-cofactor into the apo-MoFe protein, and at least one *nif*-specific gene product is required for accumulation of active Fe protein. Thus, the enormous complexity concerning the structure and catalytic mechanism of nitrogenase is also reflected in the organization, structures, and functions of those genes and their products required for its formation and activation. There are a number of fundamental, and as yet unanswered, questions attached to N_2-fixation research. For example, what are the structures and specific functions of the metalloclusters contained within the nitrogenase component proteins and how are these prosthetic groups organized within the individual nitrogenase component proteins? Also, what are the biochemical processes required for the synthesis and maintenance of a catalytically competent nitrogenase and what are the specific contributions that the individual *nif*-specific gene products make to such processes?

In this chapter we discuss aspects concerning the structure and function of nitrogenase and its associated *nif*-specific gene products. In particular, we concentrate upon the comparative analysis of the nitrogenase component proteins and *nif*-specific and *nif*-associated genes and their products as revealed by DNA sequence analysis. These features are discussed in relation to what is known or suspected concerning the composition and the biophysical properties of the various metalloclusters contained within the respective nitrogenase component proteins. Also, we discuss gene- and site-directed mutagenesis strategies and biochemical reconstitution studies that have been applied toward the analysis of nitrogenase components and other *nif*-specific genes and their products. Historical aspects concerning the biochemical, physiologic, and genetic analyses of biological N_2 fixation are presented in the prefatory chapter.

II. Diazotrophs and Biochemical Nomenclature

Although N_2 fixation is carried out only by certain Eubacteria and Archaebacteria, the capacity for N_2 fixation is dispersed widely among a phylogenetically diverse group of these organisms. Organisms capable of growth using N_2 as their only nitrogen source are generically referred to as diazotrophs. That nitrogenases from phylogenetically diverse diazotrophic organisms are similar in structure and mechanism was indicated by experiments of Detroy et al.,[47] who showed that complementary nitrogenase components isolated from different organisms can be mixed to form heterologous, catalytically competent enzymes. As will be discussed in this chapter, such structural similarities in nitrogenase components from different species was later confirmed by detailed heterologous reconstitution studies[18,39,61–63,129,172,231,245] and numerous DNA sequence analyses (see later discussion).

Various nomenclatures have been used to designate the nitrogenase Fe protein and MoFe protein components. For example, Fe protein = component II = dinitrogenase reductase, and MoFe protein = component I = dinitrogenase. These different nomenclatures have a variety of origins and each nomenclature makes sense in the context for which it is used. Fe protein and MoFe protein refer to the metallocompositions of the respective components, whereas component I and component II refer to the elution sequence of the MoFe protein and Fe protein, respectively, during column chromatography.[31] The dinitrogenase reductase and dinitrogenase nomenclature was proposed by Hageman and Burris[85,86] on the basis of their observation that Fe protein (dinitrogenase reductase) serves as a one-electron reductant of the MoFe protein (dinitrogenase), which likely provides the site for substrate binding and reduction. Extensive heterologous-component protein-mixing experiments, such as those pioneered by Detroy et al.,[47] also demanded the introduction of a nomenclature that distinguishes not only a particular component protein but also the source of that protein. The Sussex group proposed a convenient shorthand for this purpose.[56] In this nomenclature, first letters of the genus and species designations are combined to indicate the component protein source, followed by 1 or 2 to indicate the MoFe protein (component I) or the Fe protein (component II), respectively. For example, Fe protein from *Azotobacter vinelandii* is indicated as Av2. All the nomenclatures described earlier are widely used in the published literature and are sometimes interchangeably used in a single publication.

III. Electron Transport to Nitrogenase

A source of reducing equivalents of sufficiently low potential is required for regeneration of reduced Fe protein during nitrogenase turnover. Both flavodoxin and ferredoxins have been shown capable of serving this function *in vitro*.[10,11,21,45,249,252,274–276] *Klebsiella pneumoniae,* however, is the only organism for which the electron transfer pathway to nitrogenase has been established by complementary genetic and biochemical approaches.[44,95,174,199,221] Two gene products, a flavodoxin (*nifF* gene product) and a pyruvate–flavodoxin oxidoreductase (*nifJ* gene product) constitute an unbranched *nif*-specific electron transport system that couples the oxidation of pyruvate to reduction of the Fe protein. Strains bearing mutations in either *nifF* or *nifJ* are deficient in *in vivo* nitrogenase activity, but extracts from such strains exhibit substantial nitrogenase activity *in vitro* when dithionite is used as an artificial electron donor.[95,200] Both flavodoxin and pyruvate–flavodoxin oxidoreductase were purified on the basis of their ability to restore activity to crude extracts of the complementary mutant strain when pyruvate was used as the ultimate electron donor. Shah and colleagues[221] used preparations of *K. pneumoniae* flavodoxin, pyruvate flavodoxin oxidoreductase, Fe protein, and MoFe protein plus pyruvate and an ATP-regenerating source to couple the oxidation of pyruvate to C_2H_2 reduction *in vitro*. This reaction occurred at a stoichiometric ratio of pyruvate molecules consumed per C_2H_4 molecule produced, which is consistent with the two electrons available from pyruvate oxidation and the two electrons required for reduction of C_2H_2 to C_2H_4. These results established that in *K. pneumoniae*, only the *nifF* and *nifJ* gene products are required for electron transport to nitrogenase. Spectral measurements have indicated that pyruvate–flavodoxin oxidoreductase decarboxylates pyruvate to form acetyl-CoA and CO_2 and couples this reaction to the reduction of semireduced flavodoxin to the hydroquinone form ($2e^-$ reduced state).[45,221,274] During nitrogenase turnover flavodoxin is thus shuttled between the semiquinone and hydroquinone states by a reduction step coupled to oxidation of pyruvate and an oxidation step coupled to reduction of Fe protein.

Drummond[51] has developed a tertiary structure model for the *K. pneumoniae* flavodoxin, and the closely related *A. vinelandii* flavodoxin, by comparing their amino acid sequences to the partial sequence of the crystallographically determined structure of the long-chain flavodoxin from *Anacystis nidulans*. Although such models are likely to provide only an approximate structure, they are important because they provide a rational basis for amino acid substitution studies that can be used to probe functional features of the molecule. Drummond's model indicates a concentration of acidic residues flanking flavodoxin's proposed FMN pocket. Such charged residues could be important for the orientation of specific redox partners during electron transfer.[51] Similarly, a concentration of positively charged residues located near the amino end of *nif*-specific flavodoxins could indicate a docking region important for the protein–protein interactions that are likely to occur prior to or during electron transfer. Drummond points out, however, that because the Fe protein is capable of being reduced by structurally unrelated and fundamentally different redox partners (e.g., flavodoxin or ferredoxin), the redox potential of the electron donor might well play the more important role in this process. This idea is supported by the likely near surface location of Fe protein's Fe–S cluster, as was indicated by Fe protein alkylation studies of Hausinger and Howard.[91] Drummond[51] also proposed that Tyr-101 within the *K. pneumoniae* flavodoxin plays a major role in determining its redox potentials. This assignment is based on analogous aromatic groups (Trp residues) within the *A. nidulans* and *Clostridium* MP flavodoxins that are located near, or interface with, the flavin moiety.

Purification of the *nifJ* gene product was reported by three different research groups.[19,221,251] All groups agree on the subunit composition of pyruvate–flavodoxin oxidoreductase, a dimer of identical $M_r = 120,000$ subunits, but disagree on the Fe–S composition of the native enzyme. The estimate of Wahl and Orme-Johnson[251] is supported by primary amino acid sequence information deduced from the *nifJ* gene sequence[4,33] that indicates the presence of two, closely spaced, canonical ferredoxinlike [4Fe–4S] cluster-binding sequences per subunit. The proposed Fe–S cluster-binding amino acid sequence arrangement within the deduced primary sequence and spectral data obtained by Wahl and Orme-Johnson[251] indicates the presence of four clusters of the [4Fe–4S] type per holoenzyme. The presence of two such clusters per subunit is consistent with the oxidation of pyruvate, yielding two reducing equivalents, and the stoichiometric relationship of pyruvate oxidation and C_2H_2 reduction. Nevertheless, it is difficult to reconcile coupling of two $1e^-$ transfers (per subunit), which appears to occur in pyruvate-directed reduction of the pyruvate–flavodoxin oxidoreductase Fe–S centers, to the $1e^-$ electron transfer of flavodoxin from the semiquinone to the hydroquinone state. This situation indicates either that a fully ($2e^-$) reduced pyruvate flavodoxin–oxidoreductase subunit reduces two flavodoxin molecules simultaneously or that two $1e^-$ transfers occur sequentially.

Possible pathways for electron transport to nitrogenase in other diazotrophic organisms remain to be established. For example, no gene analogous to the K. pneumoniae nifJ product has been found in the azotobacters. This finding, albeit a negative result and therefore inconclusive, is in line with observations of Shah et al.[221] They compared the relative efficiency of flavodoxins isolated from A. vinelandii and K. pneumoniae in in vitro pyruvate-dependent, pyruvate–flavodoxin oxidoreductases-driven C_2H_2 reduction assays. Substitution of the A. vinelandii flavodoxin for the K. pneumoniae flavodoxin in this system was much less efficient, although fully reduced flavodoxin from either source was equally efficient in driving nitrogenase catalysis. One interpretation of this result is that the inefficiency of A. vinelandii flavodoxin in the heterologous system resides in its relative inability to interact with the pyruvate–flavodoxin oxidoreductase from K. pneumoniae. Such inefficiency could, however, be interpreted several different ways, only one being that A. vinelandii does not have a nif-specific pyruvate–flavodoxin oxidoreductase.

The primary structure of the flavodoxin isolated from A. vinelandii[13,239] shares a high level of sequence identity when compared to K. pneumoniae flavodoxin.[4,50] Furthermore, the gene encoding the A. vinelandii flavodoxin (nifF) is located within the major nif cluster and is preceded by canonical nif-specific regulatory element sequences.[13,114] The A. vinelandii flavodoxin is not, however, absolutely nif-specific as it accumulates at a low level even under conditions that completely repress accumulation of the nitrogenase structural components.[10,13] Also, unlike the K. pneumoniae nifF product, the A. vinelandii nifF-encoded product is not essential for N_2 fixation.[13] Thus, although the nifF product from A. vinelandii is likely to have some function in electron transport to nitrogenase, there must be another physiological electron donor as well. One candidate is the A. vinelandii fdxA gene product that encodes the 7Fe-ferredoxin, ferredoxin-I. Reduced ferredoxin-I is capable of supplying reducing equivalents to nitrogenase in vitro, but mutant strains of A. vinelandii deleted for fdxA or deleted for both fdxA and nifF remain capable of diazotrophic growth.[151,166] Moreover, fdxA is apparently expressed constitutively and its coding region is not preceded by canonical *nif* regulatory element sequences.[166] Thus, the true physiologic role of fer-redoxin-I in N_2 fixation remains obscure. Klugkist et al.[136] have reported the presence of three flavodoxins in *A. vinelandii* ATCC 478, only one of which appears to be *nif* specific. The genetic studies described earlier were per-formed using *A. vinelandii* OP, for which only one flavodoxin, the *nifF* gene product, has been identified.

Haaker and Klugkist[83] have proposed a model in which electron transport to nitrogenase in the obligately aerobic diazotroph *A. vinelandii* is coupled with electron transport to O_2. In this model, a NADPH dehydrogenase pres-ent in the cytoplasmic membrane binds NADPH, which then specifically donates both of its electrons to two different redox centers. One electron is transferred to a redox center operating at -160 mV, which in turn donates its electron to a respiratory chain component with O_2 being the ultimate acceptor. The other electron can theoretically be transferred from NADPH to a redox center operating at a potential of -500 mV. Two *nif*-specific membrane-bound proteins are suggested as potential participants in this pro-cess, possibly coupling electron flow to other carriers such as flavodoxin or ferredoxin. This model provides a compelling explanation for a number of other observations reported by this group.[84,135] Nevertheless, proof of the model awaits identification of the genetic determinants of the proposed elec-tron transport pathway as well as biochemical characterization of the indi-vidual components.

Speculation on the nature of genes encoding products that participate in N_2-fixation-specific electron transport in aerobic and microaerobic diazo-trophic organisms has also emerged from the deduced sequence of the *fixABCX* products and mutational analyses of these genes in certain organisms.[57,82] The "*fix*" mnemonic indicates, by agreed convention, those genes associated with N_2 fixation in rhizobial species but apparently not having structural homologues in the facultative anaerobe *K. pneumoniae*. The major rationale for proposing that *fixABC* products are involved in electron transport to ni-trogenase is that homologues to the *fixABC* genes have been identified only in organisms capable of aerobic or microaerobic N_2 fixation. For example, homologues to the *Rhizobium meliloti fixABC* genes[57] are present in *Brady-rhizobium japonicum*,[82] *Rhizobium leguminosarum*,[81] *Azorhizobium cauli-nodans*,[127] *A. vinelandii*[82] (Wientjens, unpublished), and *Azotobacter chroo-coccum*.[66] Also, another gene called *fixX* is cotranscribed with the *fixBC* cluster in *B. japonicum*[82] and *fixABC* gene clusters in *R. leguminosarum*,[81] *Rhizobium trifolii*,[108] and *R. meliloti*.[55,57] The *fixX* gene encodes a ferredoxin (see Figure 20-1). Nevertheless, there is no direct biochemical or genetic evidence that supports a role for any *fix* gene product in electron transport to nitrogenase. In fact, evidence that was interpreted to challenge this view was reported by Kaminski et al.[127] They found that crude extracts of *A. caulinodans* ORS571 *fixA* or *fixC* mutants did not support high levels of

```
          10        20        30        40        50        60        70        80        90       100

Rc fdx    MPTVAYTRGGAEYTPVVLMKIDEQKCIGCGRCFKVCGGRDVMSLHGLTEDGQVVAPGTDEWDEVEDEIVKKVMALTGAENCIGCGACARVCPSECQTHAALS

Rt fixX   MKAIVKRRVEDKLYQNRYLVDPGRPHISVRKHLFPTPNLIALTQVCPAKCYQLNDRRQVIIVSDGCLEGCTCNVLCGPDGDIEWTYPRGGFGVLFKFG

Rm fixX   MKTAIAERIEDKLYQNRYLVDAGRPHITVRPHRSPSLNLLALTRVCPAKCYELNETGQVEVTADGCMEGCTCRVLCEANGDVEWSYPRGGFGVLFKFG

Rc fdxN   MAMKIDPELCTSCGDCEPVCPTNAIAPKKGVYVINADT-CTECEGEHDLPQC-VNACMTDNCINPAA

Av fdx    MALKIVESCVNCWACVDVCPSEAISLAGPHFEISASK-CTECDGDYAEKQC-ASICPVEGAILLADGTPANPPGSLTGIPPERLAEAMREIQAR

An fdxN   MAYTITSQCISCKLCSSVCPTGAIKIAENGQHWIDSELCTNCVDTVYTVPQCKAGCPTCDGCVKVPSDYWEGWFANYNRVIAKLTKQDYWERWFNCYSQKFSEQLQKHQGEILGV

Rm fdxN   MAFKIIASQCTQCGACEFECPRGAVNFKGEKYVIDPTK-CNECKGGFDTQQC-ASVCPVSNTCVPA

Bj frxA   MPFKIIASQCTSCSACEPLCPNVAISEKGGNFVIEAAK-CSELRGHFDEPQC-AAACPVDQTCVVDRALPRYQAPV

Av fdxV   MANAIDGYECTVCGDCEPVCPTGSIVFRDDHYAIEADS-CNECTDVGETR-C-LGVCPVDLCIQPLDD

Ac fdxV   MAMAIDGYECTVCGDCKPVCPTGSIVLQGGIYVIDADS-CNECADLGEPR-C-LGVCPVDFCIQPLDD
```

Figure 20-1. Primary sequences of ferredoxinlike gene products associated with nitrogen fixation in various organisms. Arrowheads indicate clustered Cys residues. For those polypeptides having similar organizations of clustered Cys residues the Cys residues are aligned. Rc, *Rhodobacter capsulatus*; Rt, *R. trifolii*; Rm, *R. meliloti*; Av, *A. vinelandii*; An, *Anabaena* 7120; Bj, *B. japonicum*; Ac, *A. chroococcum*. The *fdxV* genotypic designation indicates ferredoxinlike genes associated with the vanadium nitrogenases in *A. vinelandii* and *A. chroococcum*. This designation is added here for clarity but is not used in the published literature.

C_2H_2 reduction activity *in vitro* even when dithionite was used as an artificial electron donor. If these mutant strains were defective in their physiologic electron transport pathway, the use of dithionite as the terminal electron donor should theoretically circumvent such defects in *in vitro* assays. Nevertheless, the specific consequences that eliminating electron transport to nitrogenase *in vivo* might have upon the catalytic stability of the component proteins in this organism are not known, and, therefore, such negative results cannot be considered absolutely conclusive. Kaminski et al.[27] did present evidence that *A. caulinodans fixC* mutants are primarily impaired in Fe protein activity, although whether this is a direct or indirect effect is not yet known. Specific aspects concerning the physiology and genetics of N_2 fixation in associative N_2-fixing bacteria are presented in Chapter 6 of this volume.

In addition to the ferredoxin encoded by *fixX*, several other ferredoxins whose expression appears to be *nif*-regulated have also been identified in a variety of diazotrophic organisms.[59,118,134,164,169,211] However, only one of these ferredoxins has been shown, by using insertional inactivation of the gene, to be required for N_2 fixation.[134] A comparison of some N_2 fixation-related ferredoxinlike gene products identified from various organisms is shown in Figure 20-1. A [2Fe–2S] ferredoxin capable of donating electrons to nitrogenase *in vitro* has also been isolated from *Anabaena* and characterized.[21] The gene encoding this ferredoxin has been isolated and its expression shown to be *nif*-specific.[20]

IV. Fe Protein

The redox active center of the Fe protein is believed to be an Fe–S cluster of the cubane [4Fe–4S] type. A simple model is that a typical [4Fe–4S] cluster within the Fe protein is responsible for single electron delivery to the MoFe protein. The actual situation, however, appears more complicated because the Fe protein's characteristic $S = \frac{1}{2}$ EPR signal integrates only to approximately 0.2 electrons per molecule. This apparent paradox was addressed by several groups who identified an additional EPR signal ($S = \frac{3}{2}$) associated with isolated Fe protein.[87,88,141,167,264] The sum of both EPR signals is 1 spin/mole. These results can be reconciled if it is considered that the Fe–S cluster exists within isolated Fe protein in two forms, one having the well-characterized $S = \frac{1}{2}$ state and the other existing in a novel $S = \frac{3}{2}$ spin state. Meyer and his colleagues[162] have used ^1H NMR to analyze Fe protein isolated from *Clostridium pasteurianum*. The ^1H NMR spectra obtained in these experiments were interpreted as arising from a single spin state ladder at room temperature, similar to that occurring in other [4Fe–4S] proteins, assuming only an $S = \frac{1}{2}$ spin state at low temperature. It was therefore inferred that the $S = \frac{3}{2}$ spin state resulted from interactions between the

[4Fe–4S] active site and its protein and solvent environment in the frozen state.

Fe protein is the obligate electron donor to the MoFe protein and binding of MgATP to Fe protein and its subsequent hydrolysis appears essential to that process. Although there is evidence that two molecules of MgATP bind per Fe protein and both ATPs are consumed during electron delivery, the relationship of MgATP hydrolysis, electron transfer, and component protein dissociation is not well understood[27,109,116,142,224,243,246,253,260,261,266] Neverthe- less, a number of biochemical and biophysical features of isolated Fe protein are known to change as a consequence of MgATP binding and such prop- erties provide some clues concerning the role of MgATP in nitrogenase ca- talysis. Binding of MgATP to isolated Fe protein results in (1) a consider- ably lowered redox potential (more negative) of the Fe–S cluster,[167,279] (2) elicitation of a more axial EPR spectrum lineshape,[144,180,229,279] (3) alterations in the circular dichroic and magnetic circular dichroic spectra,[235] (4) in- creased susceptibility of the Fe–S cluster iron molecules to removal by che- lation,[143,253] (5) a decrease in the ^{57}Fe quadropole splittings in Mossbauer spectra,[141] (6) disruption of Fe protein crystals,[195] (7) alterations in the reac- tivity of certain Fe protein thiol groups,[91] and (8) alteration of the ^{1}H NMR spectra.[162] Taken together, these features indicate that MgATP binding causes a conformational change in the Fe protein resulting in a lowered redox po- tential of the Fe–S cluster and in its greater accessibility to solvent and external reagents. Such interpretations are consistent with a role for the Fe protein in physical interaction with the MoFe protein and electron transfer from the Fe protein's reduced Fe–S cluster to a redox site located upon the MoFe protein.

Hausinger and Howard[91] correlated *in vitro* alkylation rates of specific Fe protein cysteinyl residues to the concomitant destruction of the Fe–S cluster as a method to develop a cluster-binding model. On the basis of these data, they suggested that the Fe–S cluster is symmetrically bound between the two identical Fe protein subunits through mercaptide ligands provided by cysteinyl residues 98 and 133 (numbers refer to the amino acid sequence of the *A. vinelandii* Fe protein in Figure 20-2). A twofold symmetry of the Fe– S cluster environment was also indicated by ^{1}H NMR spectroscopy,[162] and this observation was interpreted to support the Hausinger–Howard model. As we will discuss later, there is now biochemical-genetic data that also support, but do not prove, this model.

The *nifH* gene, encoding Fe protein, has been isolated and the corre- sponding nucleotide sequences determined for the following organisms: *K. pneumoniae*,[214,236] *A. vinelandii*,[24] *C. pasteurianum*,[36] *R. trifolii*,[215] *Rhizo- bium* sp. (*Parasponium*),[216] *R. meliloti*,[244] *B. japonicum*,[72] *Rhodobacter cap- sulatus*,[121] *Rhizobium* ORS571,[175] *Rhizobium phaseoli*,[191] *Rhizobium* strain ANU240,[8] *Anabaena* 7120,[161] *Thiobacillus ferrooxidans*,[188] *Azospirillum brasilense* Sp7,[42,68] *Frankia* strain HRN18a,[176] *Frankia* strain Ar13,[177] *Methanobacterium ivanovii,*[232] *Methanococcus voltae,*[233] and *Methanococcus thermolithotrophicus.*[232] In certain organisms *nifH* exists as a member of a gene family. This was first shown in the case of *Anabaena* 7120 where two unlinked gene sequences exhibited heterologous hybridization to a *K. pneumoniae nifH*-specific gene probe.[197] In *A. vinelandii* there are two other genes whose products bear high sequence identity when compared to the *nifH* product sequence.[119,192] There is now considerable genetic and biochemical evidence that in *A. vinelandii,* the products of these genes participate in alternative N_2-fixation systems (see Chapter 19). In *R. phaseoli* there are three *nifH* genes having identical polypeptide sequences, and none of these is individually required for diazotrophic growth.[191] In *Rhizobium* ORS571 there are two *nifH* genes encoding slightly nonidentical, but functional, Fe proteins.[175] In the case of *C. pasteurianum,* there are six *nifH*-like sequences having from 68–100% sequence identity when compared to the primary structure of isolated *C. pasteurianum* Fe protein,[257] whose sequence was determined by peptide sequencing methods.[238] Although the significance of multiple *nifH*-like genes present in *C. pasteurianum* is not yet known, most of them appear to be transcribed under N_2-fixing conditions. The most peculiar situation encountered so far was the identification of a *nifH*-like sequence located within a photosynthetic gene cluster in *R. capsulatus.*[94] A gene whose potential product is similar in sequence to the *nifH*-like gene from *R. capsulatus* is also encoded within and expressed from the chloroplast genome of *Marchantia polymorpha.*[73,179] The identification of multiple *nifH*-like genes in some bacteria and a *nifH*-like gene in a eukaryote suggests the interesting possibility that Fe protein–like entities could have reductase functions unrelated to nitrogenase catalysis.

In Figure 20-2 we compare Fe protein primary sequences, in most cases deduced from DNA sequence data, for the following organisms: *Anabaena* 7120, *T. ferrooxidans, R. meliloti, R. trifolii, B. japonicum, R. capsulatus, K. pneumoniae, A. vinelandii,* and *C. pasteurianum.* Examination of the aligned Fe protein sequences reveals features common to all or most of the respective polypeptides. Among all diazotrophic Eubacteria whose Fe protein sequences are known, there are five conserved cysteinyl residues: Cys-39, Cys-86, Cys-98, Cys-133, and Cys-185 (numbers refer to the *A. vinelandii* sequence). In contrast, there are two methanogens having Fe proteins that do not share the complete set of conserved cysteinyl residues found in Eubacteria. In *M. thermolithotrophicus* and *M. voltae* the residue corresponding to Cys-185 found in all Eubacteria is a Tyr residue.[232,233] There is very strong sequence identity among all Fe protein sequences in regions surrounding interspecifically conserved residues Cys-39, Cys-98, Cys-133, and Cys-185 (in Eubacteria). There is also fairly strong sequence identity surrounding Cys-86, but this conservation is not as striking as in the other

```
                      10        20        30       ▼40        50        60
An    TDENIRQIAFYGKGGIGKSTTSQNTLAAMAEMGQRIMIVGCDPKADSTRLMLHSKAQTTVLHLAAER
Tf   AMSDKLRQIAFYGKGGIGKSTTSQKHLAALAEMGQKILIVGCDPKADSTRLILHSKAQDTVLSLAAEA
Rm      AALRQIAFYGKGGIGKSTTSQNTLAALVDLGQKILIVGCDPKADSTRLILNAKAQDTVLHLAATE
Rt      AALRQIAFYGKGGIGKSTTSQNTLAALVELGQKILIVGCDPKADSTRLILNSKAQGTVLDLAATK
Bj      ASLRQIAFYGKGGIGKSTTSQNTLAALAEMGQKILIVGCDPKADSTRLILHAKAQDTILSLAASA
Rc      GKLRQIAFYGKGGIGKSTTSQNTLAALVEMGQKILIVGCDPKADSTRLILNTKLQDTVLHLAAEV
Kp       TMRQCAIYGKGGIGKSTTTQNLVAALAEMGKKVMIVGCDPKADSTRLILHAKAQNTIMEMAAEV
Av       AMRQCAIYGKGGIGKSTTTQNLVAALAEMGKKVMIVGCDPKADSTRLILHSKAQNTIMEMAAEA
Cp         RQVAIYGKGGIGKSTTTQNLTSGLHAMGKTIMVVGCDPKADSTRLLLGGLAQKSVLDTLREE

          70        80     ▼  90      ▼ 100       110       120       130 ▼
An    GAVEDLELHEVMLTGFRGVKCVESGGPEPGVGCAGRGIITAINFLEENGAYQD-LDFVSYDVLGDVVC
Tf    GSVEDLELEDVMKVGYRDIRCVESGGPEPGVGCAGRGVITSINFLEENGAYD-GANYVSYDVLGDVVC
Rm    GSVEDLELEDVLKVGYRGIKCVESGGPEPGVGCAGRGVITSINFLEENGAYNDV-DYVSYDVLGDVVC
Rt    GSVEDLELGDVLKTGYGGIKCVESGGPEPGVGCAGRGVITSINFLEENGAYDDV-DYVSYDVLGDVVC
Bj    GSVEDLELEDVMKVGYQDIRCVESGGPEPGVGCAGRGVITSINFLEENGAYENI-DYVSYDVLGDVVC
Rc    GSVEDLEVEDVVKIGYKGIKCTEAGGPEPGVGCAGRGVITAINFLEENGAYDDV-DYVSYDVLGDVVC
Kp    GSVEDLELEDVLQIGYGDVRCAESGGPEPGVGCAGRGVITAINFLEEEGAYEDDLDFVFYDVLGDVVC
Av    GTVEDLELEDVLKAGYGGVKCVESGGPEPGVGCAGRGVITAINFLEEEGAYEDDLDFVFYDVLGDVVC
Cp    G--EDVELDSILKEGYGGIRCVESGGPEPGVGCAGRGIITSINMLEQLGAYTDDLDYVFYDVLGDVVC

            140       150       160        170       180  ▼    190      200
An    GGFAMPIREGKAQEIYIVTSGEMMAMYAANNIAR GILKYAHSGGVRLGGLICNSRKVDREDELIMNLA
Tf    GGFAMPIRKQ-AQEIYIVMSGEMMAMYAANNISK GVLKYANSGGVRLGGLICNERQTDKELELAEALA
Rm    GGFAMPIRENKAQEIYIVMSGEMMALYAANNIAK GILKYAHAGGVRLGGLICNERHTDRELDLAEALA
Rt    GGFAMPIRENKAQEIYIVMSGEMMALYAANNIAR GILKYASAGSVRLGGLICNERQTDRELDLAEALA
Bj    GGFAMPIRENKAQEIYIVMSGEMMAMYAANNISK GILKYANSGGVRLGGLICNERQTDKELELAEALA
Rc    GGFAMPIRENKAQEIYIVMSGEMMALYAANNIAK GILKYANSGGVRLGGLICNERKTDRELELAEALA
Kp    GGFAMPIRENKAQEIYIVCSGEMMAMYAANNISK GIVKYAKSGKVRLGGLICNSRQTDREDELIIALA
Av    GGFAMPIRENKAQEIYIVCSGEMMAMYAANNISK GIVKYANSGSVRLGGLICNSRNTDREDELIIALA
Cp    GGFAMPIREGKAQEIYIVASGEMMALYAANNISK GIQKYAKSGGVRLGGIICNSRKVANEYELLDAFA
```

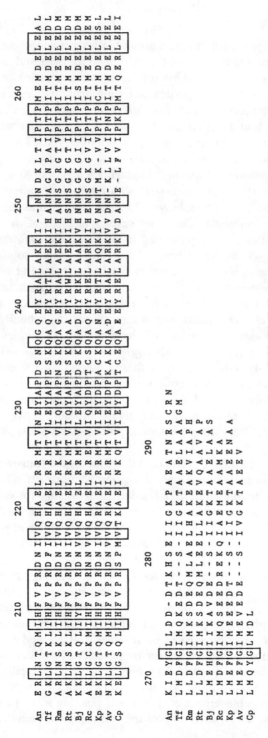

Figure 20-2. Alignment of Fe protein primary sequences. An, *Anabaena* 7120; Tf, *T. ferrooxidans*; Rm, *R. meliloti*; Rt, *R. trifolii*; Bj, *B. japonicum*, Rc, *R. capsulaus*; Kp, *K. pneumoniae*; Av, *A. vinelandii*; Cp, *C. pasteurianum*. Numbering is according to the *A. vinelandii* sequence beginning from the initiating methionine (which is not included in the figure). Residues that are conserved in all Fe proteins shown in the figure are boxed. The five conserved Cys residues are indicated by arrowheads.

cases. Strong sequence identity located around at least two of the conserved Cys residues was expected because of the likelihood for cysteinyl mercaptide ligation to the Fe–S cluster. A striking interspecies conservation in the folding pattern for Fe proteins is also indicated by a remarkable conservation of residues, which are frequently located within reverse turns. For example, of the 28 Gly residues found within the *A. vinelandii* Fe protein primary sequence, 24 are found at the corresponding positions in most known eubacterial Fe protein sequences (see Figure 20-2).

As previously mentioned, Hausinger and Howard proposed a model in which a single [4Fe–4S] cluster is bound with twofold symmetry between the identical subunits.[91] Cys-98 and Cys-133 were identified as the probable four ligands (two per subunit). There are several features of the cysteinyl ligand environment that can be predicted if the Hausinger–Howard model is correct. If the Fe–S cluster is bridged between the two subunits, it is expected that the appropriate Cys residues would be located at or near reverse turns within regions of fairly high hydrophobicity. Such an arrangement could accommodate requirements for a near-surface location and proper orientation of Fe–S cluster ligands within their respective subunits, and the potential for subunit–subunit contact of these regions. Residues that immediately flank Cys-98 (Pro–Gly–Val–Gly–Cys98–Ala–Gly) and Cys-133 (Val–Val–Cys133–Gly–Gly–Phe) meet these criteria. Also, the conservation and near location of charged amino acids flanking these regions (e.g., Glu-93, Arg-101, Asp-129, and Arg-140) could be important for the formation of salt bridges between identical subunits or could provide sites for ionic interactions necessary for component protein association or electron transfer.

There are also biochemical–genetic data that support the Hausinger–Howard model. In collaboration with Howard's group we used site-directed mutagenesis techniques to substitute Ser residues individually for each of the five interspecifically conserved Cys residues within the *A. vinelandii* Fe protein sequence.[105] Diazotrophic growth properties and component protein activities of the respective mutant strains revealed that of the five interspecifically conserved Cys residues, only Cys-98 and Cys-133 were absolutely required—namely, substitution of Ser for Cys-98 or Cys-133, respectively, eliminated diazotrophic growth and Fe protein activity. Ser substitutions for Cys-39, Cys-86, or Cys-185 eliminated neither diazotrophic growth nor the Fe protein activity of the respective mutant strains. Nevertheless, an important structural or functional role for residues Cys-39 and Cys-86 was indicated because substitution of Ser for these residues resulted in lowered diazotrophic growth rates and lowered Fe protein activity or stability in crude extracts. The ability to substitute Ser for Cys-185 with no apparent effect is consistent with the replacement of Tyr at this position in Fe protein from certain methanogens. Thus, there is probably no functional requirement for a thiol group at the residue 185 position. Although these results are consistent with the Hausinger–Howard model, they certainly do not prove it. For example, the potential for an oxygen replacement of a thiol ligand with consequent retention of biological activity, although probably remote in this case, has not been eliminated. Also, the possibility that the protein is rearranged to accommodate a different ligand coordination, when the normal ligand is denied, has not been explored in this case. Such a cluster-driven protein rearrangement has, in fact, been demonstrated in amino acid substitution experiments involving the known cysteinyl ligands to ferredoxin-I from *A. vinelandii*.[152] A similar conclusion was also inferred from site-directed substitution studies of proposed cysteinyl thiol donors to the [4Fe–4S] cluster present in glutamine phosphoribosylpyrophosphate amidotransferase from *B. subtilis*.[150]

In *Rhodospirillum rubrum* and several other diazotrophic organisms, nitrogenase activity is physiologically regulated by reversible mono-ADP ribosylation. This topic is discussed at length in Chapter 3 of this volume. ADP-ribosylation, which results in inactivation of the Fe protein, occurs under physiologic conditions where N_2 fixation is either unnecessary or futile. The biochemical counterparts involved in ADP-ribosylation (an ADP-ribosyltransferase[146]) and removal (ADP-ribose glycohydrolase[209]) have been purified from *R. rubrum* and characterized. Also, the genes encoding these enzymes have been cloned and their nucleotide sequences determined.[71] Ludden and his associates have shown that the Fe protein ADP-ribosylation site corresponds to the *A. vinelandii* Arg-101 residue[187] and that Fe proteins isolated from organisms that are not known to be ADP-ribosylated *in vivo* can serve as ADP-ribosylation substrates *in vitro*.[146] For example, Fe protein isolated from *A. vinelandii* or *C. pasteurianum* can be ADP-ribosylated *in vitro* by treatment with purified ADP-ribosyltransferase. Interestingly, only one of the two Arg-101 residues present within a homodimeric Fe protein molecule is modified.[187] This observation makes sense in light of the Hausinger–Howard model, which implies a close juxtaposition of the respective Arg-101 residues from complementary subunits. Thus, ADP-ribosylation of one Arg-101 residue could cause steric hindrance of the modification of the remaining Arg-101 residue. Potential long-range effects could be involved as well. Other than removal of the initiating Met residue, ADP-ribosylation of the Fe protein in certain organisms, and the presence of an Fe–S cluster within native Fe protein, there is no evidence for covalent, posttranslational modification of Fe protein *in vivo*. This conclusion comes from a comparison of Fe protein primary sequences for *A. vinelandii* and *C. pasteurianum* determined by chemical methods[90,238] to sequences deduced from the respective *nifH* gene sequences.[24,36]

Davis and his colleagues[35] have used an undirected mutagenesis approach to implicate an important role for the Arg-101 residue in electron transfer.

These investigators have been determining the nucleotide sequences of *K. pneumoniae nifH* genes from a series of mutants previously identified as having point mutations in that locus. To date, six mutant alleles have been sequenced. Among these is a mutant strain having His-101 substituted for Arg-101. As might be expected, Fe protein from this mutant strain does not serve as a substrate for ADP-ribosylation.[145] The altered His-101 Fe protein is incapable of catalyzing reduction of protons or C_2H_2 in the presence of MoFe protein, but it will catalyze the hydrolysis of MgATP under the same conditions. Moreover, the His-101 Fe protein has a number of properties characteristic of the normal Fe protein; namely, it contains a functional [4Fe–4S] cluster and it is capable of undergoing the MgATP-dependent conformational change.[145] These data suggest that the His-101 Fe protein is able to associate with MoFe protein but improper orientation of the respective complexes prevents productive electron transfer. Lowery et al.[145] point out that elimination of some, but not all, functions of the Fe protein resulting from a single amino acid substitution indicate the probability of at least two, and probably several, dynamic interactions between Fe protein and MoFe protein during electron transfer.

Using site-directed mutagenesis of the *nifH* gene from *A. vinelandii*, we have confirmed that substitution of His for Arg-101 results in elimination of the diazotrophic growth capability of the resultant mutant (Cash, Howard, and Dean, unpublished). We have also placed a number of other substitutions at the Arg-101 position. Some substitutions, for example, Tyr-101 and Leu-101, permit formation of at least a partially active Fe protein because such mutant strains remain capable of diazotrophic growth, albeit at a reduced rate. These results indicate that a positively charged residue at the Arg-101 position is neither sufficient nor necessary for Fe protein activity. These results are not necessarily incompatible with the conclusions of Lowery et al.[145] concerning the overall functionality of the Arg-101 residue. Perhaps the most valuable feature that has emerged from Fe protein Arg-101 substitution experiments and the ADP-ribosylation phenomenon is that these approaches have now given investigators the opportunity to physically separate specific events that occur during component protein association and electron transfer.

Interspecies Fe protein sequence conservations and comparison of Fe protein primary sequences to known ATP-binding proteins led Robson[204] to identify three potential ATP-binding sites upon the Fe protein. One of the assigned ATP sites is truly convincing. There is a region of 11 amino acids located near the N-terminus that is conserved in all known Fe protein sequences (see Figure 20-2). Within this interspecifically conserved region there is a canonical ATP-binding sequence having the motif Gly–X–Gly–X–X–Gly–Lys–Ser, which is also present in ATP-dependent protein kinases. Jones et al.[123] used the three-dimensional structure of lactate dehydrogenase to model an Fe protein nucleotide binding fold that includes the preceding ATP-binding motif. Although no site-directed amino acid substitution experiments have yet been reported for residues within the proposed nucleotide binding domain of the Fe protein, such studies will undoubtedly play a major role in unraveling the mechanism of nitrogenase catalysis.

Not only can similarities in primary sequences of proteins having homologous functions provide insight concerning functional domains, but certain differences among the primary sequences of such proteins can be valuable as well. Consideration of the differences in interspecific Fe protein sequences is particularly useful in light of the numerous *in vitro,* heterologous-component protein-mixing experiments reported previously. For example, although most heterologous Fe protein and MoFe protein complexes are catalytically active, a reconstituted mixture of Fe protein isolated from *C. pasteurianum* (Cp2) and MoFe protein isolated from *A. vinelandii* (Av1) is ineffective in substrate reduction.[61,63,64] This heterologous mixture results in the formation of a tight, inactive complex. Differences between the primary amino acid sequences of Cp2 and Av2 have been used in attempts to explain this phenomenon and to assign a potential component protein interaction site located upon the Fe protein.[36,236] A comparison of the primary sequences of Av2 and Cp2 in Figure 20-2 reveals a 65% sequence identity between these proteins. The most striking differences are located at their respective carboxyl ends where Av2 is elongated by 13 residues when compared to Cp2. These 13 amino acids include five negatively charged residues and one positively charged residue. It has been suggested that these differences in size and charge density located at the carboxyl ends of the respective Fe protein sequences could account for formation of the tight, ineffective complex in Av1–Cp2 reconstitution experiments.[36,236]

Some insight concerning the formation of the tight, ineffective complex in heterologous Av1–Cp2 reconstitution experiments was also obtained by Murrell et al.[173] They found that ADP-ribosylation of Cp2 prevents its irreversible complexation with Av1. Again this points to an important role for the Fe protein Arg-101 residue, a near neighbor of a proposed Fe–S cluster ligand (Cys-98), in proper component protein interaction or in electron transfer. Interestingly, once the complex is formed, Cp2 no longer serves as a substrate for ADP-ribosylation,[173] indicating that Arg-101 becomes buried during component protein association. This result was not unexpected in light of the biochemical properties of the heterologous, inactive Av1–Cp2 complex.[64] When complexed with the Av1, Cp2's Fe–S cluster becomes resistant to Fe chelators, even in the presence of MgATP. Nevertheless, MgATP remains capable of binding to such a complex and does induce a conformational change. These results were interpreted to indicate that Fe protein's MgATP-binding sites remain accessible to nucleotides during component protein complexation, whereas the Fe–S cluster becomes sandwiched between the component proteins during this event.[64]

Another approach used to study the nature of nitrogenase component protein interaction has involved component protein cross-linking studies. Willing et al.[268] have shown that treatment of a mixture of A. vinelandii Fe protein and MoFe protein with a reagent capable of covalently linking amino and carboxyl groups results only in equimolar cross-linking of an Fe protein subunit with a MoFe protein β-subunit. This cross-linking was inhibited by elevated salt concentrations, independent of the presence of adenine nucleotides, and resulted in the loss of enzyme activity. These results imply that at least one of the Fe protein subunits and a MoFe protein β-subunit interact or are in very close proximity at some point during nitrogenase turnover, and that formation of a salt bridge could be important during docking of the component proteins. Identification of the specific residues that are cross-linked and the effects of substituting such residues with other amino acids, using directed mutagenesis techniques, should clarify the mechanistic significance of component protein cross-linking studies.

V. MoFe Protein

The MoFe protein is considerably more complex than the Fe protein in both its structure and its metallocomposition. It has a native M_r of about 220,000 and contains approximately 32 Fe, 32 sulfides, and two Mo atoms per native molecule. Kennedy and co-workers[130] used denaturing SDS polyacrylamide gel electrophoresis of purified MoFe protein and peptide-mapping techniques to show that the MoFe protein is composed of two non-identical subunits. These subunits were later resolved and isolated by ion exchange chromatography.[148,237] The individual subunits are present in native MoFe protein as an $\alpha_2\beta_2$ tetramer.

About 16 Fe atoms can be extruded from each MoFe protein molecule as four [4Fe–4S] clusters by treatment of the native protein with thiols in a denaturing organic solvent.[137] These Fe–S clusters have been described by their redox properties and their Mossbauer and circular and magnetic circular dichroic spectroscopic features[54,65,107,158,171,194,228,230,235,255,262,263,265,278] (see Chapter 18). Orme-Johnson, Munck, and their colleagues have used the nomenclature "P" to describe the Fe–S clusters.[278] This designation comes from the conclusion that the Fe–S clusters are ligated by amino acid R groups and are therefore covalently protein-bound. Trivial designations—Fe^{2+}, D, and S—have also been used to indicate the nature of the iron atoms present in P clusters, as described by Mossbauer spectroscopy. Although P clusters are extruded in the form of four [4Fe–4S] clusters,[137] there is no compelling evidence that indicates they are actually present within the native protein in that form. For example, larger clusters could rearrange to form [4Fe–4S] clusters upon their extrusion.[128] Indeed, the EPR spectrum of thionine–oxidizcd MoFe protein has recently been interpreted to indicate that the P clusters are actually organized into two 8Fe clusters or are organized into two pairs of inequivalent 4Fe clusters.[89] Mossbauer studies using MoFe protein samples having P clusters selectively enriched with ^{57}Fe were also interpreted to indicate that P clusters are organized into two pairs of slightly inequivalent clusters.[158] There is no definitive evidence concerning the biosynthesis or molecular structure of the P clusters nor is there any direct experimental evidence concerning their possible function in nitrogenase catalysis. The nature of the P clusters is reviewed in more detail in Chapter 8.

All of the remaining Fe atoms contained within a MoFe protein molecule can be extruded in the form of two identical FeMo-cofactors by acid–base treatment followed by extraction with the chaotropic solvent *N*-methylformamide (NMF)[218] There is compelling evidence that FeMo-cofactor provides, or is part of, the actual substrate-binding and reduction site. This evidence is as follows. Certain mutant strains that are incapable of synthesizing FeMo-cofactor produce an inactive and EPR-silent MoFe protein.[218] Such inactive MoFe protein can be reconstituted by the addition of FeMo-cofactor that has been extrated from native MoFe protein. MoFe protein reconstituted in this way regains not only its enzymatic activity but also the biologically unique $S = {}^3/_2$ EPR signal characteristic of semireduced MoFe protein. Isolated FeMo-cofactor also exhibits an $S = {}^3/_2$ EPR signal similar in g value to MoFe protein but it has a considerably broadened lineshape. Furthermore, certain mutant strains, those having a defective *nifV* gene (see later discussion), produce an altered form of FeMo-cofactor.[92] MoFe proteins from such mutants exhibit dramatic changes in their substrate reduction properties. Finally, MoFe proteins from mutant strains having substitutions for certain amino acid residues targeted as providing or being located near FeMo-cofactor-binding sites also exhibit altered substrate reduction properties and a changed EPR spectrum.[213] These features are described in more detail in later sections of this chapter.

It seems obvious that the metalloclusters contained within the nitrogenase MoFe protein are redox centers involved in the acceptance, storage, and intramolecular delivery of electrons to substrate during nitrogenase catalysis. Consequently, the structure and reactivity of these metalloclusters is of considerable importance concerning the elucidation of the molecular mechanism of biological N_2 fixation. The most direct approach would appear to involve a complete biophysical and chemical description of the individual metallocluster types in both their protein-bound and extracted forms. However, the complexity of the problem has confounded such attempts, which, to date, have not resulted in an understanding of the molecular structure of MoFe protein's metalloclusters. A major problem arises from the large number of metal atoms involved. Also, because the absolute number of Fe atoms contained in each fully active MoFe protein molecule is not known and because the percentage of fully active MoFe protein in any given preparation is not known, quantitative spectroscopic analyses are subject to certain assumptions concerning these values. Such uncertainties are, of course, also related to the extreme O_2 lability of the MoFe protein and its metalloclusters. In the case of FeMo-cofactor, it is possible to extract FeMo-cofactor from native MoFe protein that can be used to reconstitute cofactorless MoFe protein.[218] However, the low volatility and chemical instability of NMF, the organic solvent most frequently used to extract FeMo-cofactor, have denied efforts to crystallize and solve FeMo-cofactor's molecular structure. New methods for the extraction of FeMo-cofactor into organic solvents other than NMF provide some promise for overcoming these difficulties.[159,271] Details concerning the numerous analytical techniques that have been used to characterize isolated FeMo-cofactor are discussed in Chapters 18 and 22 of this volume and will not be discussed further here.

Recently, a biochemical-genetic approach for analysis of the metalloclusters in their MoFe protein-bound forms has been initiated by us and by the Sussex group. The basic rationale of this approach involves specific substitution of amino acid residues targeted as direct ligands to or as being located near individual metallocluster types. The approach has two overall goals. First, a description of the catalytic and spectroscopic consequences of such substitutions should provide substantial information concerning the distribution and functionality of particular prosthetic group types and should provide insight concerning the contribution of the polypeptide environment to those functions. Second, by alteration of the specific ligands binding the metallocluster types to the MoFe protein, solvents more amenable than NMF to crystallographic and analytical techniques might be used directly for extraction of FeMo-cofactor from its protein matrix. Strategies that have been used to formulate a rational approach for the site-directed substitution of specific residues within the MoFe protein are discussed later.

Specific amino acids contained within the MoFe protein sequence can be targeted as potential metallocluster ligands by considering that solvents required for extrusion of the individual cluster types are likely to duplicate the functional groups of the particular cluster ligand. The spectroscopic properties and the reactivity of the metalloclusters within or extruded from the MoFe protein can also provide clues in this regard. Cluster extrusion requirements,[137,218] spectroscopic studies (see, for example, refs. 158 and 228) and biochemical reconstitution studies[93,185,218] all suggest that FeMo-cofactor is not physically associated with the P clusters. Thus, assignment of specific metallocluster domains is somewhat simplified because independent polypeptide environments can be assigned for the two cluster types.

Fe–S clusters that are bound to proteins via cysteinyl thiol linkages—for example, ferredoxin [4Fe–4S] clusters—are quantitatively extruded by unfolding the protein in an organic solvent in the presence of excess thiols.[181] Thus, extrusion of the P clusters by this method indicates a probability for mainly cysteinyl mercaptide ligands to the P clusters.[137] Mossbauer data were used to assign two major environments for the four iron atoms proposed to be contained in each P cluster.[228,278] These sites were labeled D and Fe^{2+} and were originally proposed to exist in a ratio of 3:1 in each cluster.[278] More recently, the three D sites in two of the P clusters were proposed to be organized into slightly inequivalent sites: two D sites plus one S site.[158] One interpretation of the Mossbauer data is that there are four P clusters differentiated into two pairs of nearly identical four Fe-containing entities. Nevertheless, the spectroscopic data have not eliminated the possibility for the existence of only two identical 8Fe clusters, a possibility also previously suggested.[89] In any event, the distribution of the Fe species contained within P clusters into 12 nearly identical sites (10 D + 2 S) and an additional four other sites ($4Fe^{2+}$) provides a convenient starting point for consideration of potential cluster coordinating residues located within the MoFe protein. Extrusion of FeMo-cofactor into NMF suggests a probability of N-ligands to FeMo-cofactor.[218] This possibility is also strengthened by electron spin echo experiments performed on MoFe protein[241] and FT-IR analysis of extracted FeMo-cofactor.[254] Reaction of isolated FeMo-cofactor with a single thiolate per Mo atom also suggests a probability for at least one cysteinyl mercaptide ligand to FeMo-cofactor.[32]

A number of research groups have tried to gain structural insights concerning the MoFe protein by interspecies comparison of MoFe protein α- and β-subunit primary sequences deduced from the corresponding gene sequences, the inference being that structural features critical to MoFe protein function will be reflected in conservation of interspecific primary sequences. Sources for MoFe protein primary sequences deduced from gene sequence data include *K. pneumoniae*[4,97,113,214,234] *A. vinelandii*,[24,96,114] *Anabaena* 7120,[78,138,154] *T. ferrooxidans*,[193] *B. japonicum*,[126,242] *C. pasteurianum*,[256,258] cowpea *Bradyrhizobium* sp. (cowpea) strain IRC78, *Rhizobium* cp. (*Parasponium*) strain ANU289,[267] and *R. meliloti* (β-subunit, cited in Ref. 2). In the case of the *Anabaena* 7120 *nifD* gene, a developmentally regulated chromosomal rearrangement is required for formation of the intact MoFe protein α-subunit[78] (see also Chapter 4). In Figure 20-3 the deduced MoFe protein α-subunit sequences from *C. pasteurianum, K. pneumoniae, Anabaena* 7120, *A. vinelandii,* and *B. japonicum* are aligned, as are the respective MoFe protein β-subunits from these species. These comparisons reveal strong interspecies sequence conservations. A striking interspecies conservation in the folding of the individual subunits is also suggested by a remarkable conservation of residues that are frequently located within re-

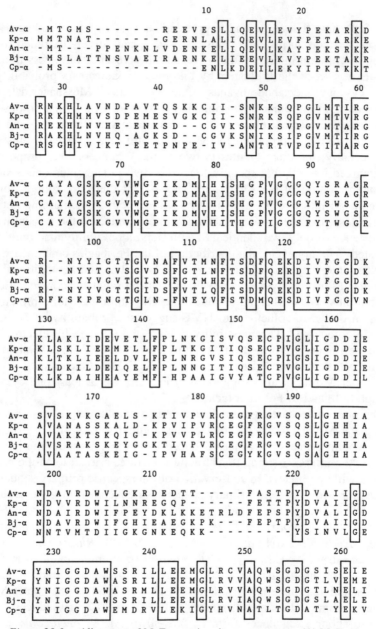

Figure 20-3. Alignment of MoFe protein primary sequences. (A) MoFe protein α-subunit alignments; (B) MoFe protein β-subunit alignments. Av, *A. vinelandii;* Kp, *K. pneumoniae;* An, *Anabaena* 7120; Bj, *B. japonicum;* Cp, *C. pasteurianum.* Numbering is according to the *A. vinelandii* sequence. Residues conserved in all sequences shown are boxed.

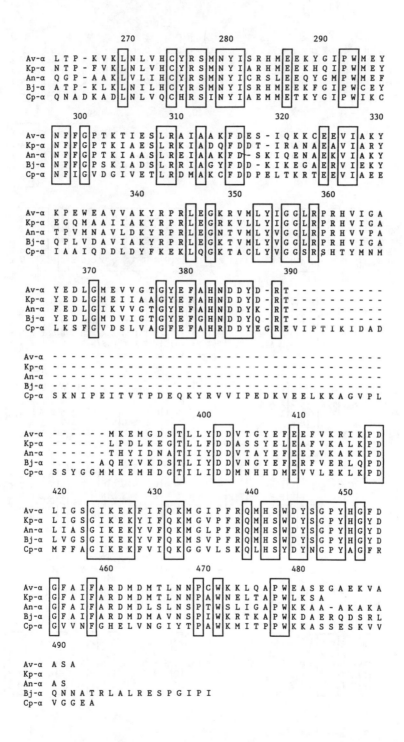

```
                270              280              290
Av-α  L T P - K V K L N L V H C Y R S M N Y I S R H M E E K Y G I P W M E Y
Kp-α  N T P - F V K L N L V H C Y R S M N Y I A R H M E E K H Q I P W M E Y
An-α  Q G P - A A K L L V L I H C Y R S M N Y I C R S L E E Q Y G M P W M E F
Bj-α  A T P - K L K L N I L H C Y R S M N Y I S R H M E E K F G I P W C E Y
Cp-α  Q N A D K A D L N L V Q C H R S I N Y I A E M M E T K Y G I P W I K C

          300              310              320              330
Av-α  N F F G P T K T I E S L R A I A A K F D E S - I Q K K C E E V I A K Y
Kp-α  N F F G P T K I A E S L R K I A D Q F D D T - I R A N A E A V I A R Y
An-α  N F F G P T K I A A S L R E I A A K F D - S K I Q E N A E K V I A K Y
Bj-α  N F F G P S K I A D S L R R I A G Y F D D - K I K E G A E R V I E K Y
Cp-α  N F I G V D G I V E T L R D M A K C F D D P E L T K R T E E V I A E E

                340              350              360
Av-α  K P E W E A V V A K Y R P R L E G K R V M L Y I G G L R P R H V I G A
Kp-α  E G Q M A A I I A K Y R P R L E G G R K V L L Y I G G L R P R H V I G A
An-α  T P V M N A V L D K Y R P R L E G N T V M L Y V G G L R P R H V V P A
Bj-α  Q P L V D A V I A K Y R P R L E G K T V M L Y V G G L R P R H V I G A
Cp-α  I A A I Q D D L D Y F K E K L Q G K T A C L Y V G G S R S H T Y M N M

          370              380              390
Av-α  Y E D L G M E V V G T G Y E F A H N D D Y D - R T - - - - - - - - - - -
Kp-α  Y E D L G M E I I A A G Y E F A H N D D Y D - R T - - - - - - - - - - -
An-α  F E D L G I K V V G T G Y E F A H N D D Y K - R T - - - - - - - - - - -
Bj-α  Y E D L G M D V I G T G Y E F G H N D D Y Q - R T - - - - - - - - - - -
Cp-α  L K S F G V D S L V A G F E F A H R D D Y E G R E V I P T I K I D A D

Av-α  - - - - - - - - - - - - - - - - - - - - - - - - - - - - - - - - - - - -
Kp-α  - - - - - - - - - - - - - - - - - - - - - - - - - - - - - - - - - - - -
An-α  - - - - - - - - - - - - - - - - - - - - - - - - - - - - - - - - - - - -
Bj-α  - - - - - - - - - - - - - - - - - - - - - - - - - - - - - - - - - - - -
Cp-α  S K N I P E I T V T P D E Q K Y R V V I P E D K V E E L K K A G V P L

                        400              410
Av-α  - - - - - - M K E M G D S T L L Y D D V T G Y E F E E F V K R I K P D
Kp-α  - - - - - - L P D L K E G T L L F D D A S S Y E L E A F V K A L K P D
An-α  - - - - - - T H Y I D N A T I I Y D D V T A Y E F E E F V K A K K P D
Bj-α  - - - - - A Q H Y V K D S T L I Y D D V N G Y E F E R F V E R L Q P D
Cp-α  S S Y G G M M K E M H D G T I L I D D M N H H D M E V V L E K L K P D

      420              430              440              450
Av-α  L I G S G I K E K F I F Q K M G I P F R Q M H S W D Y S G P Y H G F D
Kp-α  L I G S G I K E K Y I F Q K M G V P F R Q M H S W D Y S G P Y H G Y D
An-α  L I A S G I K E K Y V F Q K M G L P F R Q M H S W D Y S G P Y H G Y D
Bj-α  L V G S G I K E K Y V F Q K M S V P F R Q M H S W D Y S G P Y H G Y D
Cp-α  M F F A G I K E K F V I Q K G G V L S K Q L H S Y D Y N G P Y A G F R

          460              470              480
Av-α  G F A I F A R D M D M T L N N P C W K K L Q A P W E A S E G A E K V A
Kp-α  G F A I F A R D M D M T L N N P A W N E L T A P W L K S A
An-α  G F A I F A R D M D L S L N S P T W S L I G A P W K K A A - A K A K A
Bj-α  G F A I F A R D M D M A V N S P I W K R T K A P W K D A E R Q D S R L
Cp-α  G V V N F G H E L V N G I Y T P A W K M I T P P W K K A S S E S K V V

      490
Av-α  A S A
Kp-α
An-α  A S
Bj-α  Q N N A T R L A L R E S P G I P I
Cp-α  V G G E A
```

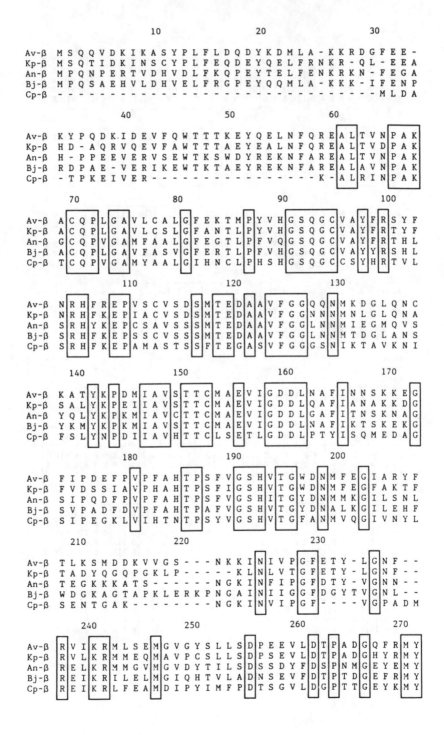

```
                10                  20                  30
Av-β  M S Q Q V D K I K A S Y P L F L D Q D Y K D M L A - K K R D G F E E -
Kp-β  M S Q T I D K I N S C Y P L F E Q D E Y Q E L F R N K R - Q L - E E A
An-β  M P Q N P E R T V D H V D L F K Q P E Y T E L F E N K R K N - F E G A
Bj-β  M P Q S A E H V L D H V E L F R G P E Y Q Q M L A - K K K - I F E N P
Cp-β  - - - - - - - - - - - - - - - - - - - - - - - - - - - - - - M L D A

                40                  50                  60
Av-β  K Y P Q D K.I D E V F Q W T T T K E Y Q E L N F Q R E A L T V N P A K
Kp-β  H D - A Q R V Q E V F A W T T T A E Y E A L N F Q R E A L T V D P A K
An-β  H - P P E E V E R V S E W T K S W D Y R E K N F A R E A L T V N P A K
Bj-β  R D P A E - V E R I K E W T K T A E Y R E K N F A R E A L A V N P A K
Cp-β  - T P K E I V E R - - - - - - - - - - - - - - - - K - A L R I N P A K

                70                  80                  90                  100
Av-β  A C Q P L G A V L C A L G F E K T M P Y V H G S Q G C V A Y F R S Y F
Kp-β  A C Q P L G A V L C S L G F A N T L P Y V H G S Q G C V A Y F R T Y F
An-β  G C Q P V G A M F A A L G F E G T L P F V Q G S Q G C V A Y F R T H L
Bj-β  A C Q P L G A V F A S V G F E R T L P F V H G S Q G C V A Y Y R S H L
Cp-β  T C Q P V G A M Y A A L G I H N C L P H S H G S Q G C C S Y H R T V L

                110                 120                 130
Av-β  N R H F R E P V S C V S D S M T E D A A V F G G Q Q N M K D G L Q N C
Kp-β  N R H F K E P I A C V S D S M T E D A A V F G G N N N M N L G L Q N A
An-β  S R H Y K E P C S A V S S S M T E D A A V F G G L N N M I E G M Q V S
Bj-β  S R H F K E P S S C V S S S M T E D A A V F G G L N N M T D G L A N S
Cp-β  S R H F K E P A M A S T S S F T E G A S V F G G G S N I K T A V K N I

                140                 150                 160                 170
Av-β  K A T Y K P D M I A V S T T C M A E V I G D D L N A F I N N S K K E G
Kp-β  S A L Y K P E I I A V S T T C M A E V I G D D L Q A F I A N A K K D G
An-β  Y Q L Y K P K M I A V C T T C M A E V I G D D L G A F I T N S K N A G
Bj-β  Y K M Y K P K M I A V S T T C M A E V I G D D L N A F I K T S K E K G
Cp-β  F S L Y N P D I I A V H T T C L S E T L G D D L P T Y I S Q M E D A G

                180                 190                 200
Av-β  F I P D E F P V P F A H T P S F V G S H V T G W D N M F E G I A R Y F
Kp-β  F V D S S I A V P H A H T P S F I G S H V T G W D N M F E G F A K T F
An-β  S I P Q D F P V P F A H T P S F V G S H I T G Y D N M M K G I L S N L
Bj-β  S V P A D F D V P F A H T P A F V G S H V T G Y D N A L K G I L E H F
Cp-β  S I P E G K L V I H T N T P S Y V G S H V T G F A N M V Q G I V N Y L

                210                 220                 230
Av-β  T L K S M D D K V V G S - - - N K K I N I V P G F E T Y - L G N F - -
Kp-β  T A D Y Q G Q P G K L P - - - - - K L N L V T G F E T Y - L G N F - -
An-β  T E G K K K A T S - - - - - N G K I N F I P G F D T Y - V G N N - -
Bj-β  W D G K A G T A P K L E R K P N G A I N I I G G F D G Y T V G N L - -
Cp-β  S E N T G A K - - - - - - - - N G K I N V I P G F - - - - V G P A D M

                240                 250                 260                 270
Av-β  R V I K R M L S E M G V G Y S L L S D P E E V L D T P A D G Q F R M Y
Kp-β  R V L K R M M E Q M A V P C S L L S D P S E V L D T P A D G H Y R M Y
An-β  R E L K R M M G V M G V D Y T I L S D S S S D Y F D S P N M G E Y E M Y
Bj-β  R E I K R I L E L M G I Q H T V L A D N S E V F D T P T D G E F R M Y
Cp-β  R E I K R L F E A M D I P Y I M F P D T S G V L D G P T T G E Y K M Y
```

786

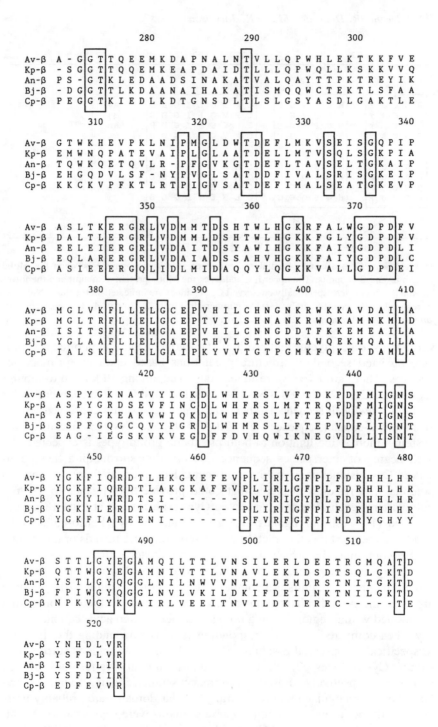

```
                    280             290             300
Av-β  A - G G T T Q E E M K D A P N A L N T V L L Q P W H L E K T K K F V E
Kp-β  - S G G T T Q Q E E M K E A P D A I D T L L L Q P W Q L L K S K K V V Q
An-β  P S - G T K L E D A A D S I N A K A T V A L Q A Y T T P K T R E Y I K
Bj-β  - D G G T T L K D A A N A I H A K A T I S M Q Q W C T E K T L S F A A
Cp-β  P E G G T K I E D L K D T G N S D L T L S L G S Y A S D L G A K T L E

                    310             320             330             340
Av-β  G T W K H E V P K L N I P M G L D W T D E F L M K V S E I S G Q P I P
Kp-β  E M W N Q P A T E V A I P L G L A A T D E L L M T V S Q L S G K P I A
An-β  T Q W K Q E T Q V L R - P F G V K G T D E F L T A V S E L T G K A I P
Bj-β  E H G Q D V L S F - N Y P V G L S A T D D F I V A L S R I S G K E I P
Cp-β  K K C K V P F K T L R T P I G V S A T D E F I M A L S E A T G K E V P

                    350             360             370
Av-β  A S L T K E R G R L V D M M T D S H T W L H G K R F A L W G D P D F V
Kp-β  D A L T L E R G R L V D M M L D S H T W L H G K K F G L Y G D P D F V
An-β  E E L E I E R G R L V D A I T D S Y A W I H G K K F A I Y G D P D L I
Bj-β  E Q L A R E R G R L V D A I A D S S A H V H G K K F A I Y G D P D L C
Cp-β  A S I E E E R G Q L I D L M I D A Q Q Y L Q G K K V A L L G D P D E I

                    380             390             400             410
Av-β  M G L V K F L L E L G C E P V H I L C H N G N K R W K K A V D A I L A
Kp-β  M G L T R F L L E L G C E P T V I L S H N A N K R W Q K A M N K M L D
An-β  I S I T S F L L E M G A E P V H I L C N N G D D T F K K E M E A I L A
Bj-β  Y G L A A F L L E L G A E P T H V L S T N G N K A W Q E K M Q A L L A
Cp-β  I A L S K F I I E L G A I P K Y V V T G T P G M K F Q K E I D A M L A

                    420             430             440
Av-β  A S P Y G K N A T V Y I G K D L W H L R S L V F T D K P D F M I G N S
Kp-β  A S P Y G R D S E V F I N C D L W H F R S L M F T R Q P D F M I G N S
An-β  A S P F G K E A K V W I Q K D L W H F R S L L F T E P V D F F I G N S
Bj-β  S S P F G Q G C Q V Y P G R D L W H M R S L L F T E P V D F L I G N T
Cp-β  E A G - I E G S K V K V E G D F F D V H Q W I K N E G V D L L I S N T

                    450             460             470             480
Av-β  Y G K F I Q R D T L H K G K E F E V P L I R I G F P I F D R H H L H R
Kp-β  Y C K F I Q R D T L A K G K A F E V P L I R L G F P L F D R H H L H R
An-β  Y G K Y L W R D T S I - - - - - - - - P M V R I G Y P L F D R H H L H R
Bj-β  Y G K Y L E R D T A T - - - - - - - - P L I R I G F P I F D R H H H H R
Cp-β  Y G K F I A R E E N I - - - - - - - - P F V R F G F P I M D R Y G H Y Y

                    490             500             510
Av-β  S T T L G Y E G A M Q I L T T L V N S I L E R L D E E T R G M Q A T D
Kp-β  Q T T W G Y E G A M N I V T T L V N A V L E K L D S D T S Q L G K T T D
An-β  Y S T L G Y Q G G L N I L N W V V N T L L D E M D R S T N I T G K T T D
Bj-β  F P I W G Y Q G G L N V L V K I L D K I F D E I D N K T N I L G K T T D
Cp-β  N P K V G Y K G A I R L V E E I T N V I L D K I E R E C - - - - - T E

                    520
Av-β  Y N H D L V R
Kp-β  Y S F D L V R
An-β  I S F D L I R
Bj-β  Y S F D I I R
Cp-β  E D F E V V R
```

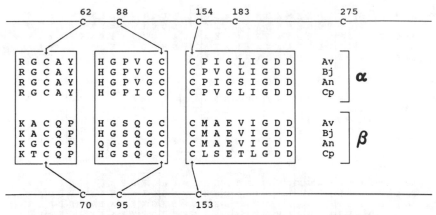

Figure 20-4. Comparison of spatial and sequence identity recognized between the MoFe protein α- and β-subunits. The upper line indicates the MoFe protein α-subunit and the lower line the β-subunit. Interspecifically conserved Cys residues are indicated on the lines. Numbers correspond to the conserved Cys residue numbers in Figure 20-3. Av, *A. vinelandii;* Bj, *B. japonicum;* An, *Anabaena* 7120; Cp, *C. pasteurianum.*

verse turns (e.g., glycine and proline). This conclusion is supported by computer-assisted secondary structure predictions based upon the primary sequences. There are five Cys residues conserved among all known α-subunit sequences: Cys-62, Cys-88, Cys-154, Cys-183, and Cys-275 (numbers refer to the *A. vinelandii* sequence). There are three Cys residues conserved among all known β-subunit sequences: Cys-70, Cys-95, and Cys-153. Because some cysteinyl mercaptide ligation to both prosthetic group types is expected, the high degree of interspecies sequence conservation surrounding these conserved Cys residues is probably significant. However, as there are only 16 potential Cys residues available that could provide thiol ligands to the metalloclusters (excluding non-conserved Cys residues) and at least 18 appear required (16 P cluster ligands and a minimum of two FeMo-cofactor ligands), there must be other metallocluster-coordinating ligands or the P cluster Fe atoms must be arranged in some novel cluster form.

Primary sequence conservations are also observed when various MoFe protein α- and β-subunits are compared to each other (Figure 20-4). There are three interspecifically conserved Cys residues within the α-subunit that are located within regions having both spatial and amino acid sequence identity when compared to the corresponding regions surrounding the three interspecifically conserved β-subunit Cys residues. These regions surround α-subunit Cys residues 62, 88, and 154, and β-subunit Cys residues 70, 95, and 153, respectively. Based on primary sequence data, Lammers and Haselkorn[138] recognized that structurally similar domains are probably present in both the α- and β-subunits. These domains were suggested to encompass the first three conserved Cys residues within the α-subunit and the three

conserved Cys residues in the β-subunit. A low level of structural relatedness between the MoFe protein α- and β-subunits was also previously recognized by amino acid composition comparisons[148] and x-ray diffraction
analysis.[273] Hennecke and his colleagues[126] suggested that regions conserved
at both the interspecies and intersubunit level could represent structural prosthetic group-binding features common to both subunits of the MoFe protein
and that an evolutionary relationship exists between the α- and β-subunits.
This proposal makes sense if the P cluster organization model (i.e., two
pairs of nearly equivalent 4Fe clusters) is considered correct. For example,
if there are two pairs of nearly equivalent P clusters there must be four
structurally similar domains within the MoFe protein that accommodate these
clusters. Because there are no repeated sequences within either subunit, we
suggest that P clusters are bound to those regions of the respective α- and
β-subunits that share obvious sequence identity to each other.

As noted by Wang et al.,[258] the predicted reverse turns located immediately following the first conserved Cys residues and immediately preceding
the second conserved Cys residues of both the α- and β-subunits indicate a
near surface subunit location for these residues. The overall hydrophobic
nature of the regions located between the predicted reverse turns in both
subunits also suggests these regions could represent subunit–subunit contact
sites. Because pairs of P clusters might interact during electron accumulation
and intramolecular delivery, it seems reasonable to suggest that intersubunit
contact sites might be located near the individual P cluster domains. Such
placement also accommodates the possibility that the four proposed P clusters are actually arranged into two larger clusters bridged between the α-
and β-subunits. There is, in fact, some evidence that regions located near
the MoFe protein α-subunit Cys-88 region and the MoFe protein β-subunit
Cys-95 region might interface with each other. Substitution by Ala for either
of the *K. pneumoniae* residues that correspond to the *A. vinelandii* α-subunit
Cys-88 or β-subunit Cys-95 residues (see Figure 20-3) eliminates MoFe protein activity. However, substitution by Ala for both residues restores a low
level of activity (Kent, Baines, Gormal, Smith and Buck, unpublished). In
an analogous type of experiment (Dean, Newton, Brigle, unpublished), we
substituted Asn for the *A. vinelandii* α-subunit His-83 or the β-subunit His-
90 residue. Neither substitution resulted in a drastic alteration in the diazotrophic growth capability of the respective mutant. However, a combination of both substitutions eliminated diazotrophic growth.

Recently, site-directed amino acid substitution studies of proposed P cluster domains have been initiated.[26,41,132,226] Results of these preliminary studies are surprising. In collaboration with Newton's group we have shown
that, in *A. vinelandii*, individual substitution of Ser for α-subunit residues
Cys-62, Cys-88, Cys-154, or β-subunit residues Cys-70 and Cys-95 nearly
eliminates nitrogenase activity and diazotrophic growth of the respective mu-

tant strains.[41] Similarly, Kent et al.[132] have substituted Ala, and in some cases Ser, for the corresponding residues from *K. pneumoniae,* and these mutants display the same phenotype. Such results are consistent with the proposal that these Cys residues provide essential P cluster thiol ligands. However, more recently we have found (Setterquist, Newton, and Dean, unpublished) mutant strains having Thr or Asp substituted for α-subunit residue Cys-88 remain capable of diazotrophic growth, albeit at a reduced rate. Also, for both *A. vinelandii* and *K. pneumoniae,* substitution of β-subunit residue Cys-153 with Ser does not significantly reduce diazotrophic growth rates.[41,132] Such results demonstrate that α-subunit Cys-154 and β-subunit Cys-153 cannot have strict functional equivalence because replacement of α-subunit Cys-154 with Ser eliminates both nitrogenase activity and diazotrophic growth. There are at least three interpretations for these observations: (1) α-subunit Cys-88 and β-subunit Cys-153 are not P cluster ligands; (2) partial disruption of P clusters by certain substitutions does not eliminate catalytic activity, (3) removal of certain P cluster ligands causes a rearrangement of the protein, which results in formation of a different P cluster ligand set. This latter possibility has precedent in site-directed mutagenesis studies involving ferredoxin I from *A. vinelandii.*[152] Consideration of the possibility for normal P cluster ligands other than Cys, as well as potential substitution of P cluster thiol ligands in certain mutant strains, is particularly relevant, because evidence for coordination of His[240] and Asp[76] to Fe–S clusters has been reported. Also, Evans and Leigh[67] have demonstrated the potential for Tyr, Ser, and Pro as biological Fe–S cluster ligands because these amino acids are able to coordinate synthetic Fe–S clusters.

If regions targeted as potential candidates for P cluster domains are not considered as potential FeMo-cofactor environments, the remaining conserved Cys residues available for FeMo-cofactor binding are α-subunit residues Cys-183 and Cys-275. Regions surrounding both of these residues are reasonable candidates for FeMo-cofactor domains because they also contain conserved amino acids that might function as N-donor ligands. Residues His-195 and His-196 are most notable in this regard because of the previously published evidence for His ligation to FeMo-cofactor.[241]

The MoFe protein subunits are not required for FeMo-cofactor biosynthesis (see later discussion on FeMo-cofactor biosynthesis). Thus, it was suggested that certain FeMo-cofactor biosynthetic gene products might share primary sequence identity when compared to the MoFe protein subunits.[25,40] Such comparisons were used to obtain information concerning the identification of potential FeMo-cofactor-binding domains within the MoFe protein. For example, it was suggested that products of the *nifE* and *nifN* genes form a scaffold upon which FeMo-cofactor is assembled.[25] Indeed, the *nifE* and *nifN* products bear considerable sequence identity when compared to the MoFe protein α- and β-subunits, respectively (Figure 20-5).[2,4,23,25,40,165,217,259]

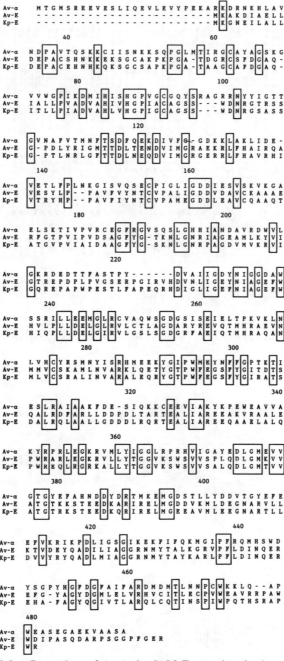

Figure 20-5. Comparison of *A. vinelandii* MoFe protein subunits to the *nifE* and *nifN* product sequences from *A. vinelandii* and *K. pneumoniae*. (A) Comparison of the MoFe protein α-subunit to *nifE* gene products; (B) comparison of the MoFe protein β-subunit to *nifN* gene products. Conserved sequences are boxed. Numbers refer to *A. vinelandii* MoFe protein subunit sequences.

791

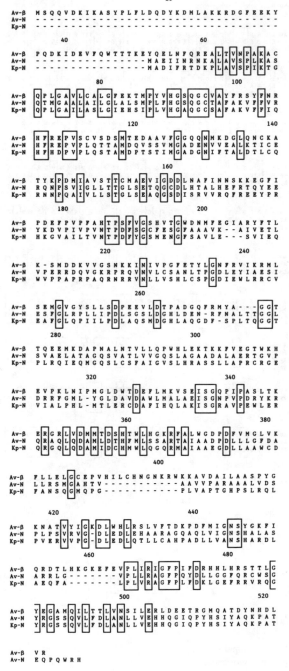

```
                          20
Av-β  M S Q Q V D K I K A S Y P L F L D Q D Y K D M L A K K R D G F E E K Y
Av-N  - - - - - - - - - - - - - - - - - - - - - - - - - - - - - - - - - - - -
Kp-N  - - - - - - - - - - - - - - - - - - - - - - - - - - - - - - - - - - - -

          40                                    60
Av-β  P Q D K I D E V F Q W T T T K E Y Q E L N F Q R E A L T V N P A K A C
Av-N  - - - - - - - - - - - - - - - - M A E I I N R N K A L A V S P L K A S
Kp-N  - - - - - - - - - - - - - - - M A D I F R T D K P L A V S P I K T G

               80                                    100
Av-β  Q P L G A V L C A L G F E K T M P Y V H G S Q G C V A Y F R S Y F N R
Av-N  Q T M G A A L A I L G L A L S M P L F H G S Q G C T A F A K V F F V R
Kp-N  Q P L G A I L A S L G I E H S I P L V H G A Q G C S A F A K V F F I Q

                         120                                    140
Av-β  H F R E P V S C V S D S M T E D A A V F G G Q Q N M K D G L Q N C K A
Av-N  H F R E P V P L Q T T A M D Q V S S V M G A D E N V V E A L K T I C E
Kp-N  H F H D P V P L Q S T A M D P T S T I M G A D G N I F T A L D T L C Q

                                   160
Av-β  T Y K P D M I A V S T T C M A E V I G D D L N A F I N N S K K E G F I
Av-N  R Q N P S V I G L L T T G L S E T Q G C D L H T A L H E F R T Q Y E E
Kp-N  R N N P Q A I V L L S T G L S E A Q G S D I S R V V R Q F R E E Y P R

          180                                    200
Av-β  P D E F P V P F A H T P S F V G S H V T G W D N M F E G I A R Y F T L
Av-N  Y K D V P I V P V N T P D F S G C F E S G F A A A V K - - A I V E T L
Kp-N  H K G V A I L T V N T P D F Y G S M E N G F S A V L E - - - S V I E Q

                         220                                    240
Av-β  K - S M D D K V V G S N K K I N I V P G F E T Y L G N F R V I K R M L
Av-N  V P E R R D Q V G K R P R Q V N V L C S A N L T P G D L E Y I A E S I
Kp-N  W V P P A P R P A Q R N R R V N L L V S H L C S P G D I E W L R R C V

                                   260
Av-β  S E M G V G Y S L L S D P E E V L D T P A D G Q F R M Y A - - - G G T
Av-N  E S F G L R P L L I P D L S G S L D G H L D E N - R F N A L T T G G L
Kp-N  E A F G L Q P I I L P D L A Q S M D G H L A Q G D F - S P L T Q G G T

          280                                    300
Av-β  T Q E E M K D A P N A L N T V L L Q P W H L E K T K K F V E G T W K H
Av-N  S V A E L A T A G Q S V A T L V V G Q S L A G A A D A L A E R T G V P
Kp-N  P L R Q I E Q M G Q S L C S F A I G V S L H R A S S L L A P R C R G E

                    320                                    340
Av-β  E V P K L N I P M G L D W T D E F L M K V S E I S G Q P I P A S L T K
Av-N  D R R F G M L - Y G L D A V D A W L M A L A E I S G N P V P D R Y K R
Kp-N  V I A L P H L - M T L E R C D A F I H Q L A K I S G R A V P E W L E R

                                   360                                    380
Av-β  E R G R L V D M M T D S H T W L H G K R F A L W G D P D F V M G L V K
Av-N  Q R A Q L Q D A M L D T H F M L S S A R T A I A A D P D L L L G F D A
Kp-N  Q R G Q L Q D A M I D C H M W L Q G Q R M A I A A E G D L L A A W C D

                                        400
Av-β  F L L E L G C E P V H I L C H N G N K R W K K A V D A I L A A S P Y G
Av-N  L L R S M G A H T V - - - - - - - - - - - A A V V P A R A A A L V D S
Kp-N  F A N S Q G M Q P G - - - - - - - - - - - P L V A P T G H P S L R Q L

          420                                    440
Av-β  K N A T V Y I G K D L W H L R S L V F T D K P D F M I G N S Y G K F I
Av-N  P L P S V R V G - D L E D L E H A A R A G Q A Q L V I G N S H A L A S
Kp-N  P V E R V V P G - D L E D L Q T L L C A H P A D L L V A N S H A R D L

                    460                                    480
Av-β  Q R D T L H K G K E F E V P L I R I G F P I F D R H H L H R S T T L G
Av-N  A R R L G - - - - - - - V P L L R A G F P Q Y D L L G G F Q R C W S G
Kp-N  A E Q F A - - - - - - - L P L V R A G F P L F D K L G E F R R V R Q G

                              500                                    520
Av-β  Y E G A M Q I L T T L V N S I L E R L D E E T R G M Q A T D Y N H D L
Av-N  Y R G S S Q V L F D L A N L L V E H H Q G I Q P Y H S I Y A Q K P A T
Kp-N  Y R G S S Q V L F D L A N L L V E H H Q G I Q P Y H S I Y A Q K P A T


Av-β  V R
Av-N  E Q P Q W R H
```

Also, the *nifEN* products isolated from an *A. vinelandii nifB* mutant were shown to form a tetrameric complex analogous to the MoFe protein.[185] Because FeMo-cofactor must escape from the *nifEN* product biosynthetic complex during maturation of the MoFe protein, their respective FeMo-cofactor-binding sites are likely to be structurally similar but functionally inequivalent. Comparison of the MoFe protein α-subunit regions targeted as providing potential FeMo-cofactor domains to *nifE* product sequences provide considerable support for this idea. This comparison is schematically shown in Figure 20-6.

Klipp and co-workers[4,165] proposed an entirely different FeMo-cofactor-binding region, also based on comparisons of MoFe protein α- and β-subunits to the *nifE* and *nifN* products, respectively. They suggested that a conserved His–Gly–X–X–Gly–Cys motif present in the MoFe protein α- and β-subunits and in the *nifE* and *nifN* gene products (see Figures 20-3 and 20-4) could provide intersubunit FeMo-cofactor-binding sites. These regions correspond to our proposed P cluster domains. Our interpretation of the conservation of the His–Gly–X–X–Gly–Cys motif present in the *nifE* and *nifN* products is that P-like clusters might also be present in the *nifEN* products tetrameric complex.[25] Indeed, the *nifEN* products complex isolated from the *A. vinelandii nifB* mutant does appear to have an Fe–S cluster, although the nature, function, and integrity of this cluster within the complex, as isolated, is not yet known.[185] The presence of one or more Fe–S clusters within the *nifEN* products complex could be involved in electron transfer reactions required for maturation of FeMo-cofactor or required for escape of FeMo-cofactor from the *nifEN* product complex. Site-directed mutagenesis studies have shown that, in *A. vinelandii*, substitution of Ser for the conserved *nifE* product Cys residue that corresponds to the MoFe protein α-subunit Cys-154 residue, reduces but does not abolish diazotrophic growth.[270] In contrast, substitution of Ser for the *nifE* product Cys residue that corresponds to the MoFe protein α-subunit Cys-275 residue does abolish diazotrophic growth.

There is now reasonable biochemical-genetic evidence supporting the idea that both of the MoFe protein α-subunit residues Cys-183 and Cys-275 are located at or near the FeMo-cofactor site within the MoFe protein. An important experiment concerning the possibility that the MoFe protein α-subunit residue Cys-275 provides an essential thiol ligand to FeMo-cofactor was reported by the Sussex group.[132,226] They found that substitution of Ala for Cys-275 in *K. pneumoniae* dramatically increased the pool of FeMo-cofactor in crude extracts of the mutant strain. This was shown by mixing extracts of the Ala-275 mutant strain with an extract of a *nifB* mutant (*nifB* mutants accumulate a MoFe protein devoid of FeMo-cofactor). Also, the elevated pool of the FeMo-cofactor in extracts of the Ala-275 mutant was detected by EPR spectroscopy. The simplest interpretation of these results is that, in the absence of the proposed thiol ligand (Cys-275), FeMo-cofactor is only

MoFe Protein α-subunit

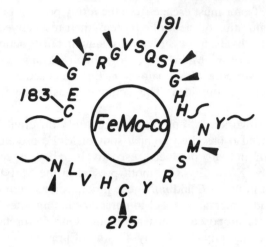

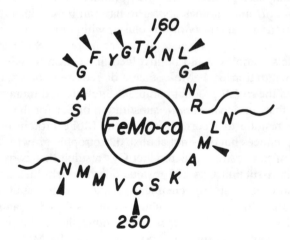

nifE Product

Figure 20-6. Schematic representation of residues conserved within the *A. vinelandii* MoFe protein α-subunit and the *A. vinelandii* nifE product proposed to provide FeMo-factor-binding or FeMo-cofactor intermediate-binding domains. Arrowheads indicate residues conserved within both the MoFe protein α-subunit and the *nifE* product.

loosely bound to the altered MoFe protein and is thus available for reconstitution of wild-type apo-MoFe protein. Support for this hypothesis comes from the observation that MoFe protein from the Ala-275 mutant exhibits a native electrophoretic mobility characteristic of apo-MoFe protein (Kent, Baines, Gormal, Smith, and Buck, unpublished). Conradson et al.[38] have used fluorine-19 chemical shifts as a probe of the reactivity of isolated FeMo-cofactor. Based on these studies they suggest that, if Cys-275 does provide a thiol ligand to FeMo-cofactor, formation and breaking of such a linkage could be in dynamic equilibrium and related to substrate binding or reduction.

The approach we have taken to analyze a potential role for the region located near α-subunit residue Cys-183 in FeMo-cofactor-binding has involved the substitution of certain residues within this region that could provide N-donor ligands to FeMo-cofactor.[213] Residues that were substituted include Gln-191 (substituted with Lys) and His-195 (substituted with Asn). In both cases the choice for substituting amino acid was guided by the corresponding *nifE* gene product sequence (see, for example, Figure 20-6). In this way, we hoped to alter the functionality of a potential FeMo-cofactor environment without imparting severe global alterations to the polypeptide structure. Altered nitrogenases from both mutant strains were able to reduce C_2H_2 to C_2H_6, a property not exhibited by native nitrogenase. Also, MoFe protein from both mutant strains exhibited an EPR spectrum altered in lineshape and g values when compared to wild type. These results provide strong evidence that the region spanning Gln-191 to His-195 includes a FeMo-cofactor-binding domain. They also demonstrate that FeMo-cofactor is associated with the α-subunit and that the catalytic activity of nitrogenase is influenced not only by the structure of FeMo-cofactor but also by its polypeptide environment. That this latter conclusion is correct was first indicated by experiments of Smith et al.[227] who showed that reconstitution of apo-Mo-nitrogenase (i.e., cofactorless MoFe protein) with the V-containing cofactor (FeVaco) produced a species that exhibited substrate reduction patterns different from either Mo-nitrogenase or V-nitrogenase. In our current FeMo-cofactor-binding model we propose that FeMo-cofactor is predominantly held within the MoFe protein through the α-subunit, and major binding domains are located near residues Cys-183 and Cys-275 (see Figure 20-6). However, a possible role for the β-subunit in binding FeMo-cofactor has not been eliminated, and we believe this possibility also deserves serious consideration.

Another approach used to probe structural and functional properties of individual amino acid residues within the nitrogenase MoFe protein α-subunit has involved undirected mutagenesis. Govensky and Zamir[80] identified certain amino acid substitutions that were incorporated into the MoFe protein α-subunit of *K. pneumoniae* as a result of chemical mutagenesis. The position and nature of these substitutions were correlated to the assembly and native electrophoretic mobility of the altered MoFe proteins. Conclusions that Govensky and Zamir have drawn from this independent approach are largely in agreement with those conclusions made from sequence comparisons and directed substitution studies described earlier.

VI. Genetic Organization of *nif* Genes

The primary products of the nitrogenase structural genes are not capable of catalyzing N_2 fixation. Rather, the products of a number of other *nif*-specific genes are required for the maturation of the nitrogenase structural components. Also, as we have already discussed, certain *nif*-specific gene products are required for electron transport to nitrogenase. Finally, there are also *nif*-specific regulatory elements that are responsible for activation or inhibition of *nif* gene expression under the appropriate physiologic conditions. Structural and functional aspects of the products of these genes from *K. pneumoniae*, *nifA* (positive regulatory element), and *nifL* (negative regulatory element), and regulatory elements involved in controlling *nif* gene expression in other organisms are discussed in Chapter 22 of this volume.

Early work concerning the identification of genes whose products are involved in N_2 fixation centered mainly on the facultative anaerobe *K. pneumoniae*. This choice was based on the fact that *K. pneumoniae* was the only diazotrophic organism amenable to application of the classical bacterial genetic manipulations developed for *E. coli*. By about 1980, 17 *nif*-specific genes had been identified in *K. pneumoniae*. The identification of these genes was accomplished by genetic complementation experiments, biochemical reconstitution studies, and *in vivo* DNA-directed gene expression studies.[49,60,104,149,160,190,198,199,200] Nucleotide sequence data later revealed the presence of three additional *nif*-specific genes within the *K. pneumoniae nif* cluster.[4,17,183] Thus, in *K. pneumoniae* there are a total of 20 *nif*-specific genes. Nucleotide sequences are now available for all the following genes: *nifJ*,[4,33] *nifH*,[4,214,236] *nifD*,[4,113,214] *nifK*,[4,97,234] *nifT*,[4,17] *nifY*,[4,17] *nifE*,[4,217] *nifN*,[4,217] *nifX*,[4,17] *nifU*,[4,15] *nifS*,[4,15] *nifV*,[4,15] *nifW*,[4,17,183] *nifZ*,[4,183] *nifM*,[4,183] *nifF*,[4,50] *nifL*,[4,53] *nifA*,[4,29,52] *nifB*,[4,28] and *nifQ*.[4,28] As shown in Figure 20-7, these 20 genes are organized into eight transcriptional units. At least two of the transcriptional units, those contained within the *nifUSVWZM* and *nifM* gene clusters, appear to overlap.[16,183,222]

Mazur, Rice, and Haselkorn,[155] and Ruvkun and Ausubel[208] recognized that interspecies conservation of nitrogenase primary sequences, as deduced from heterologous component reconstitution experiments, were also likely to be reflected in their respective coding sequences. Thus, these investigators were able to use heterologous hybridization techniques successfully as a means to identify and clone nitrogenase structural components from a variety of diazotrophic organisms. This was accomplished using cloned *nifHDK* genes from *K. pneumoniae*[34,189] as the heterologous probe. This approach, also coupled with DNA sequence analyses, has resulted in isolation of *nif*-specific gene clusters from numerous diazotrophic organisms. In Figure 20-7 we compare the present status of *nif* gene mapping studies for *K. pneumoniae*, *A. vinelandii*, *R. capsulatus*, *B. japonicum*, and *C. pasteurianum*. The organization of the *nif* genes of *Anabaena* 7120 is described in Chapter 4. Various aspects of the structure and function of *nif*-specific genes as revealed by DNA sequence analysis and biochemical studies are discussed later and are briefly summarized in Table 20-1.

VII. *nifM*

The *nifM* gene has been identified only in *K. pneumoniae*,[4,183] *A. vinelandii*,[114,115,131] and *A. chroococcum*.[66] Complementation of a *K. pneumoniae* *nifM* point mutation with cloned *A. chroococcum*[66] or *A. vinelandii*[131] *nifM* DNA was used to demonstrate that the *nifM* product probably has an identical function both in the azotobacters and in *K. pneumoniae*. Sequence comparisons of *nifM* gene products from *A. vinelandii* and *K. pneumoniae*, however, revealed only a very low level of interspecies sequence identity between them, and this sequence identity is largely confined to the C-terminal third of the respective polypeptides.[4,114,183] Such comparisons suggest that the active portion of the *nifM* product is likely to be located near the C-terminal region of the polypeptide. The low level of sequence identity when *nifM* products from *A. vinelandii* and *K. pneumoniae* are compared probably indicates why a homolog to *nifM* has not yet been identified by heterologous hybridization experiments involving other organisms.

Roberts et al.[200] analyzed several *K. pneumoniae* *nifM* point mutants for nitrogenase component protein activities and found that such strains are primarily deficient in Fe protein activity. *A. vinelandii* *nifM* mutants are deficient in Fe protein activity as well[115] (Table 20-2). As a method to examine the possible functions of *nif*-specific gene products in relation to Fe protein maturation, Howard et al.[106] used a binary plasmid expression system to direct the synthesis of the *K. pneumoniae* Fe protein in combination with expression of various other *nif* genes. *E. coli* served as the host for these experiments and *nif* gene expression was performed under anaerobic conditions. Expression of Fe protein in the absence of *nifM* gene expression resulted in accumulation of no active Fe protein and only very low Fe protein polypeptide accumulation. However, when *nifM* was expressed simultaneously with Fe protein, an active Fe protein whose properties were indistinguishable from the native Fe protein was obtained. This activity could not be stimulated by the presence of other *nif*-specific genes. These results were also independently confirmed by Paul and Merrick,[184] who employed

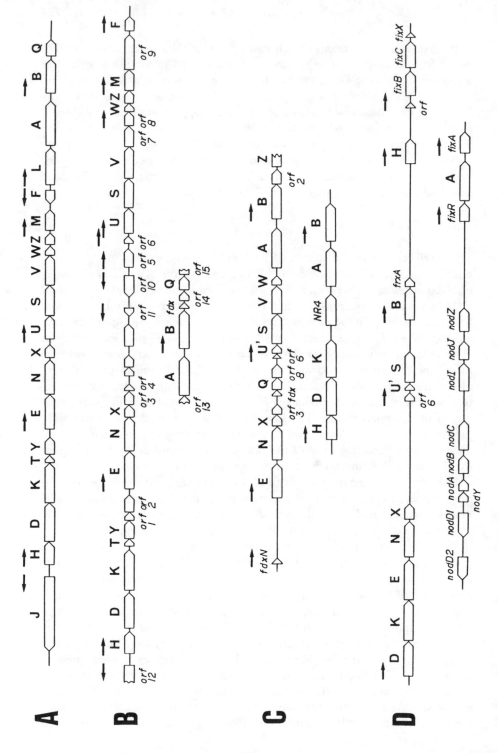

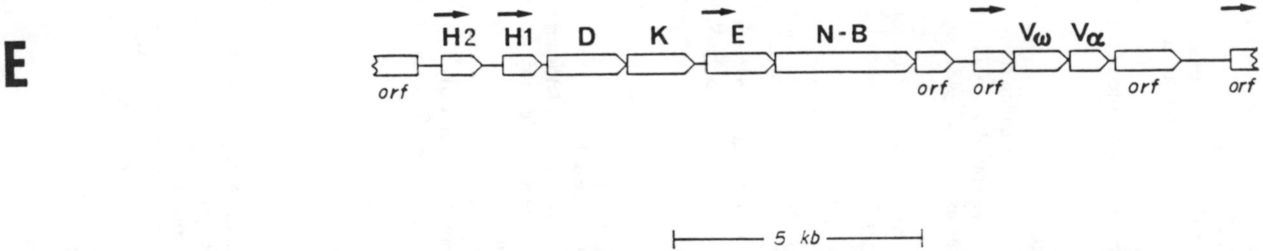

Figure 20-7. Comparison of the physical organizations of *nif*, *fix*, and "*nif*-associated" genes from (A) *K. pneumoniae*, (B) *A. vinelandii*, (C) *R. capsulatus*, (D) *B. japonicum*, (E) *C. pasteurianum*. Arrows indicate the position and direction for known or proposed transcription initiation sites. The numbering of the orfs (open reading frames) is arbitrary. Orfs from different organisms having the same numerical designation have a high level of primary sequence identity. U' indicates an incomplete *nifU* gene sequence when compared to the corresponding *K. pneumoniae* or *A. vinelandii* sequence. Notice that in *C. pasteurianum* the *nifN* and *nifB* genes are fused and the *nifV* gene is split into two genes. Orf13 from *A. vinelandii* is probably *nifL* (Kennedy, unpublished). For information concerning the development of the *K. pneumoniae* gene map see Arnold et al.[4] and references cited therein. For information concerning development of the *A. vinelandii* gene map see Jacobson et al.,[114] Bennett et al.,[12] and Joerger and Bishop[118] and references cited therein. A *fixABC* gene cluster from *A. vinelandii* has also been cloned and characterized (Wientjens, unpublished). Also, a nearly identical organization for most *nif* genes from *A. chroococcum* when compared to *A. vinelandii* has been reported (Evans et al.,[66] and Jones et al.[124]). Information concerning the organization of the *R. capsulatus nif* genes can be found in Ahombo et al.,[3,6,7] Jones and Haselkorn,[121,122] Klipp et al.,[133] Moreno-Vivian et al.,[164,165] Masepohl et al.,[153] Schatt et al.,[211] and Willison et al.[269] Unpublished data obtained from Masepohl and Klipp were also used to assemble the *R. capsulatus* gene map presented in this figure. Information concerning the organization of the *B. japonicum nif* genes can be found in Adams et al.,[1] Ebeling et al.,[58,59] Fischer et al.[70] Fuhrmann and Hennecke,[72] Kaluza and Hennecke,[126] Noti et al.,[178] and Thony et al.[242] Unpublished data obtained from Ebeling and Hennecke were also used to assemble the *B. japonicum nif* gene map presented in this figure. Information concerning the organization of the *C. pasteurianum nif* genes can be found in Chen et al.[36] and Wang et al.[256,255,258,259] Unpublished data obtained from Wang, Johnson, and Chen were also used to assemble the *C. pasteurianum* gene map presented in this figure.

Table 20-1. *nif-Specific Genes and Their Products' Known or Proposed Functions*

Gene	Product and Known or Proposed Function
nifH	Fe protein subunit, product forms homodimer ($M_r \approx 60,000$). A single [4Fe–4S] cluster is believed to be bridged between the identical subunits. Binds two molecules of MgATP and reduces the MoFe protein by single electron deliveries. Fe protein is also required for FeMo-cofactor biosynthesis.
nifD	MoFe protein α-subunit, forms $\alpha_2\beta_2$ tetramer ($M_r \approx 220,000$) with β-subunit. Holoprotein contains 2FeMo-cofactor molecules and four [4Fe] or two [8Fe] clusters called P clusters. FeMo-cofactor is the site of substrate reduction.
nifK	MoFe protein β-subunit.
nifF	Flavodoxin, physiologic reductant of the Fe protein.
nifJ	Pyruvate–flavodoxin–oxidoreductase, couples the oxidation of pyruvate to the reduction of flavodoxin. Composed of two identical $M_r \approx 120,000$ subunits. The holoenzyme probably contains four [4Fe–4S] centers.
nifM	Required for activation of the Fe protein.
nifU	Function not known. Appears to be involved primarily in stabilization of the Fe protein. Might have a function in protection or recycling of oxygen labile Fe–S centers.
nifS	Function not known. Appears to be involved primarily in stabilization of the Fe protein. Might have a function in protection or recycling of oxygen labile Fe-S centers.
nifV	Probably encodes a homocitrate synthase. Deduced gene product exhibits a fair level of primary sequence identity when compared to the *S. typhimurium leuA* gene product. Homocitrate is an organic component of FeMo-cofactor.
nifE	Required for FeMo-cofactor biosynthesis. Forms an $\alpha_2\beta_2$ tetramer ($M_r \approx 210,000$) with the *nifN* gene product. Exhibits a fair level of sequence identity when compared to the MoFe protein α-subunit.
nifN	Required for FeMo-cofactor biosynthesis. Forms an $\alpha_2\beta_2$ tetramer with the *nifE* gene product. Exhibits a fair level of sequence identity when compared to the MoFe protein β-subunit.
nifB	Required for FeMo-cofactor biosynthesis.
nifQ	Involved in FeMo-cofactor biosynthesis, probably at an early step. The $nifQ^-$ phenotype can be suppressed by elevated levels of Mo, cysteine, or cystine.
nifW	Function not known, required for full activity of the MoFe protein.
nifZ	Function not known, required for full activity of the MoFe protein. Might have a function related to FeMo-cofactor formation or insertion.
nifA	Positive regulatory element.
nifL	Negative regulatory element.
nifX	Specific function not known, probably a negative regulatory element.
nifT	Function unknown, not required for diazotrophic growth.
nifY	Function unknown, not required for diazotrophic growth.

*Table 20-2. Nitrogenase Component Protein Activities in **A**. vinelandii Mutants Having Defined Deletions Within nif-Specific Genes*

Strain	Doubling Time	Fe Protein Sp Activity	MoFe Protein Sp Activity	FeMo-cofactor Reconstitution
Wild-type	2.5	45.7	52.2	43.4
▲ nifH	NG	0.0	0.2	22.6
▲ nifD	NG	30.0	0.0	ND
▲ nifK	NG	57.0	0.0	ND
▲ nifE	NG	30.4	0.7	15.1
▲ nifN	NG	28.0	0.7	18.2
▲ nifU	NG	2.5	15.4	14.8
▲ nifS	15.5	3.5	13.1	13.3
▲ nifUS	NG	0.5	4.1	4.3
▲ nifV	9.0	47.3	5.9	13.4
▲ nifW	5.4	40.9	24.4	21.0
▲ nifZ	5.3	48.4	17.3	16.0
▲ nifM	NG	0.1	15.0	19.0
▲ nifZM	NG	0.1	0.5	3.7
▲ nifB	NG	28.6	0.2	17.8

Doubling time is expressed in hours. NG indicates no detectable growth. Component protein activities were determined in the presence of saturating levels of the complementary component. Activities are expressed as nanomoles acetylene reduced per minute per milligram crude extract. FeMo-cofactor reconstitutions indicate the maximum MoFe protein activity determined in the presence of saturating levels of Fe protein and isolated FeMo-cofactor. Values reported in this table are taken from the published literature, except for the ▲ nifB strain. Activities for this mutant are from our unpublished data.

a nearly identical approach but used more specific gene constructs. These results were interpreted to indicate that only *nifM* product is *directly* required for activation of Fe protein.

What is the function of *nifM* product in Fe protein maturation? One suggested possibility is that the *nifM* product is required for insertion of the Fe–S center into an apo-Fe protein.[106] Along this line of thought, Pagani et al.[182] suggested that *nifM* product could have a role analogous to the rhodenase-catalyzed reactivation of Fe–S-containing proteins whose metal clusters have been leached from their polypeptide matrices. On the basis of site-directed amino acid substitution studies involving the proposed Fe protein Fe–S cluster cysteinyl ligands, Howard et al.[105] suggested that *nifM* might not be directly responsible for Fe–S cluster insertion. Nevertheless, there is no compelling evidence for or against a role for the *nifM* product in formation of Fe protein's Fe–S cluster at this time. Berman et al.[14] have shown that the *nifM* product is not required for formation of the homodimeric Fe protein. Evidence for this was also based upon expression of Fe protein in a foreign host using recombinant plasmid constructs.

VIII. *nifU* and *nifS*

Genes corresponding to the *K. pneumoniae nifU* gene[4,15,160,200] have been identified in *A. chroococcum*,[66] *A. vinelandii*,[15,114,115] *Anabaena* 7120,[170] *A. brasilense* Sp7,[74] and *Nostoc commune*.[43] Potential gene products having limited sequence identity when compared to *nifU* products from the preceding organisms were also identified in *R. capsulatus* (Masepohl and Klipp, unpublished) and *B. japonicum* (Ebeling and Hennecke, unpublished). These latter *nifU*-like products are much smaller than the corresponding *nifU* products from *K. pneumoniae, A. vinelandii,* and *Anabaena* 7120, having sequence identity corresponding only to the C-terminal regions of the other *nifU* products. The *nifU*-like product from *B. japonicum* exhibits only very weak sequence identity when compared to other *nifU* products. Alignment of the deduced amino acid sequences of the *K. pneumoniae, A. vinelandii,* and *Anabaena* 7120 *nifU* products shows regions of high sequence conservation in the N-terminal half and in the C-terminal quarter of the respective polypeptides (Figure 20-8). There is relatively low sequence identity among *nifU* gene products stretching for about 60 residues between these conserved domains. There are nine conserved cysteinyl residues contained within all three *nifU* products, and these residues are found in regions of fairly high primary sequence conservation. These cysteinyl residues are not organized into any known metallocluster-binding motif, so their significance remains unknown. There is a Cys–X–X–Cys motif that is repeated twice within the *nifU* product sequences and there is some low, but probably significant, sequence conservation when these two regions are compared to each other. Such internal sequence conservation could indicate the presence of two equivalent domains within the *nifU* product.

On the basis of complementation experiments, as well as Mu reversion studies, Roberts and Brill[199] suggested that *nifU* might not be essential for N_2 fixation in *K. pneumoniae*. In contrast, Merrick and co-workers[160] presented evidence that *nifU* does represent a discrete complementation group within *K. pneumoniae*. Nonpolar insertional inactivation of the *nifU*-like gene from *R. capsulatus*, which precedes *nifS* in this organism, does not eliminate diazotrophic growth (Masepohl and Klipp, unpublished). However, because a number of *nif* genes appear to be reiterated in this organism (e.g., *nifA* and *nifB*[133]), results with the *R. capasulatus nifU*-like mutant cannot be considered unequivocal. An in-frame deletion located within the *nifU* gene from *A. vinelandii* does eliminate diazotrophic growth[114,115] (see Table 20-2). The *A. vinelandii nifU* deletion strain exhibits about a 20-fold reduction in Fe protein activity and about a threefold reduction in MoFe protein activity (see Table 20-2). Thus, it appears that, in *A. vinelandii,* the *nifU* product is required primarily for maturation or catalytic stability of the Fe protein. Whether or not the effect that deleting *nifU* has upon Fe protein and MoFe protein

```
        ** *  **   *  * *                  * * ***  ▼******   *        *
Kp  NifU  MWNYSEKVKDHFFNPRNARVVDNAN------AVGDVGSLSCGDALRLMLRVDPQSEIIEE   54
Av  NifU  MWDYSEKVKEHFYNPKNAGAVEGAN------AIGDVGSLSCGDALRLTLKVDPETDVILD   54
An  NifU  MWDYTDKVLELFYDPKNQGVIEENGEPGVKVATGEVGSIACGDALRLHIKVEVESDKIVD   60


        *****▼ ***********  *  ** **    *  ***** ***  ***▼****  ***
Kp  NifU  AGFQTFGCGSAIASSSALTELIIGHTLAEAGQITNQQIADYLDGLPPEKMHCSVMGQEAL  114
Av  NifU  AGFQTFGCGSAIASSSALTEMVKGLTLDEALKISNQDIADYLDGLPPEKMHCSVMGREAL  114
An  NifU  SRFQTFGCTSAIASSSALTEMIKGLTLDEALKVSNKDIADYLGGLPEAKMHCSVMGQEAL  120


        **  *  **      ** *  ▼  ▼* *  *     *      * *    * **  *****▼
Kp  NifU  RAAIANFRGESL--EEEHDEGKLICKCFGVDEGHIRRAVQNNGLTTLAEVINYTKAGGGC  172
Av  NifU  QAAVANYRGETI--EDDHEEGALICKCFAVDEVMVRDTIRANKLSTVEDVTNYTKAGGGC  172
An  NifU  EAAIYNYRGIPLAAHDEDDEGALVCTCFGVSENKVRRIVIENDLTDAEQVTNYIKAGGGC  182
Rc  NifU'                                                            MR    2


Kp  NifU  TSCHEKIELALAEILA----------QQPQTTPAVASG---------KDPHWQSVVD---  210
Av  NifU  SACHEAIERVLTEELAARGEVFVAAPIKAKKKVKVLAPEPAPAPVAEAPAAAPKLSNLQR  232
An  NifU  GSCLAKIDDIIKDVKENKAATNLNNKGGSKPTNIPNSGQKRPLTNVQKIALIQKVLDEEV  240
Rc  NifU'  DMQDDDTKSPAPPPAAAAAARRAAGQAAPDASALRDRFAKLAQADTPEAATDAAAAADDE   62


        **      *  **   *  *      *  *  *  *  ▼   ▼       **
Kp  NifU  -----TI-AELRPHIQADGGDMALLSVTNHQVTVSLSGSCSGCMMTDMTLAWLQQ-KLME  263
Av  NifU  IRRIETVLAAIRPTLQRDKGDVELIDVDGKNVYVKLTGACTGCQMPSMTLGGIQQ-RLIE  291
An  NifU  ----------RPVLIADGGDVELYDVDGDIVKVVLQGACGSCSSSTATLKIAIESRLRD  289
Rc  NifU'  VTRIRALIDEMRPTFRRDGGDIELVRVEGAKVIVHLSGACAGCMLAGQTLYGVQK-RITD  121


              *
Kp  NifU  RTGCYMEVVAA                                                 274
Av  NifU  ELGEFVKVIPVSAAAHAQMEV                                       312
An  NifU  RINPSLVVEAV                                                 300
Rc  NifU'  VLGRPFRVIPDIRH                                             135
```

Figure 20-8. Alignment of *nifU* gene products from *K. pneumoniae* (Kp), *A. vinelandii* (Av), *Anabaena* 7120 (An), and the *nifU*-like product from *R. capsulatus* (Rc). Alignments of the complete *nifU* gene products are taken from Mulligan and Haselkorn.[170] Residues conserved in all species are indicated by * and conserved cys residues are indicated by ▼.

activity in *A. vinelandii* crude extracts is a direct one is not clear (see Table 20-2). As mentioned earlier, studies involving the expression of the *K. pneumoniae* Fe protein and *nifM* product in *E. coli* appear to indicate that *only* the *nifM* product is directly required for activation of Fe protein.[106,184] If *nifU* is required for Fe protein maturation or stability in *A. vinelandii,* such involvement could be related to the metabolism of the inorganic Fe or sulfide required for formation of Fe–S clusters or the protection of such clusters from oxidative inactivation once they are formed. Such roles for the *nifU* product could account for the observed reduction in both Fe protein and MoFe protein activities in the absence of *nifU* product in *A. vinelandii.* An essential role in N$_2$ fixation for the *nifU* product in *A. vinelandii,* but perhaps not in certain other organisms, could be related to the fact that *A. vinelandii* is an obligate aerobe.

Genes corresponding to the *K. pneumoniae nifS* gene[4,15,199,200] have been identified in *A. chroococcum*,[66] *A. vinelandi*,[15,114,115] *Anabaena* 7120,[170] *A. brasilense* Sp7,[74] *R. capsulatus* (Masepohl and Klipp, unpublished), and *B. japonicum*.[58] Alignment of the available primary sequences of the deduced *nifS* gene products from these organisms reveals a high level of sequence identity throughout. Nevertheless, neither the deduced *nifS* product sequences themselves nor the interspecies alignments have provided any significant insight concerning the potential function of the *nifS* product.

nifS mutants have been isolated from *K. pneumoniae*,[160,199,200] *A. vinelandii*,[114,115] *B. japonicum*,[58] and *R. capsulatus* (Masepohl and Klipp, unpublished). A nonpolar insertion mutation within *nifS* eliminates diazotrophic growth in *R. capsulatus*, whereas *nifS* mutants from *K. pneumoniae, B. japonicum*, or *A. vinelandii* are severely reduced in their diazotrophic growth capabilities. Biochemical defects resulting from inactivation of *nifS* have been examined for both *K. pneumoniae* and *A. vinelandii nifS* mutants. Roberts et al.[200] reported that, in *K. pneumoniae*, elimination of *nifS* function results in a drastic reduction of both Fe protein and MoFe protein activities. Because the effect was more pronounced upon Fe protein activity, these investigators suggested that the primary target for *nifS* function might reside in Fe protein maturation. Component protein activities present in *K. pneumoniae nifS* mutants are qualitatively the same as those obtained for an *A. vinelandii nifS* deletion mutant, the difference being that the *A. vinelandii nifS* deletion strain has higher levels of both component protein activities[115] (see Table 20-2). Nevertheless, like *K. pneumoniae nifS* mutants, the most pronounced consequence of deleting *nifS* in *A. vinelandii* appears to be in the accumulation of active Fe protein. It is difficult to reconcile these data with the conclusion that only *nifM* product is required for maturation and stability of the Fe protein. However, it should be noted that experiments involving inactivation of a gene in its native chromosome are fundamentally different from experiments involving high-level expression of genes in foreign hosts. For example, if the function of the *nifS* product involves protection of the component proteins or their metalloclusters from oxidative inactivation, the very high level of expression of Fe protein in *E. coli* under strictly anaerobic conditions could obviate the requirement for such a function. Thus the high production of active Fe protein itself might serve to remove potentially damaging intracellular oxidants and thereby mask a requirement for *nifS* product under these conditions.

Although individual deletion of the *A. vinelandii nifU* or *nifS* genes results in similar reductions of both Fe protein and MoFe protein activities, it seems unlikely their products are involved in exactly the same physiologic function. This was shown by deletion of both *nifU* and *nifS* from *A. vinelandii*, which resulted in almost complete elimination of Fe protein activity and a drastic reduction in MoFe protein activity[115] (see Table 20-2). Again, these results point to an important physiologic role for both the *nifU* and *nifS* products in accumulation or maintenance of active Fe protein in *A. vinelandii*. Whether the effect upon MoFe protein arises indirectly from accumulation of inactive Fe protein is not clear. The lowered accumulation of MoFe protein activity in *nifU*-, *nifS*-, and *nifUS*-deletion strains does not, however, result from a defect in FeMo-cofactor biosynthesis[115] (see Table 20-2). Because it is clear that only *nifM* product is absolutely required for Fe protein *activation,* we favor the idea that *nifU* and *nifS* products are most likely to be involved in physiologic stabilization of Fe protein, and probably MoFe protein, either directly or indirectly, once the active species are formed.

IX. *nifW* and *nifZ*

Homologues to the *pneumoniae nifW* and *nifZ* genes[4,17,183] have been identified in *A. vinelandii*,[114] *A. chroococcum* (Evans and Robson, unpublished), and *R. capsulatus* (Masepohl and Klipp, unpublished). There is nothing remarkable about the deduced primary sequences of *nifW* or *nifZ* products, nor have interspecies alignments of these polypeptides provided any clues concerning their functions.

Paul and Merrick[184] examined the accumulation of component protein activities in nitrogenase-derepressing *K. pneumoniae nifW* and *nifZ* insertion mutants. Under the conditions examined, neither the *nifW* nor the *nifZ* product was required for diazotrophic growth. Such mutant strains did, however, exhibit lowered MoFe protein activity, and it was therefore concluded that both *nifW* and *nifZ* products are primarily required for the full formation or activation of MoFe protein. Support for this conclusion was reached on the basis of similar studies involving *A. vinelandii nifW* and *nifZ* deletion mutants[115] (see Table 20-2). It is not yet known whether *nifW* and *nifZ* products are involved in the same function. Similar reductions in MoFe protein activities in *A. vinelandii nifW* and *nifZ* deletion strains indicate that *nifW* and *nifZ* products could be involved in the same function. Also, expression of *nifW* and *nifZ* products appear to be translationally coupled (on the basis of DNA sequence data) in both *A. vinelandii* and *K. pneumoniae*. However, reduction in the accumulation of MoFe protein activity appears to be more pronounced in *nifZ* mutants than in *nifW* mutants. Also, *nifW* and *nifZ* are not cotranscribed in *R. capsulatus* (Masepohl and Klipp, unpublished).

An additional clue concerning the possible function of the *nifZ* product was obtained by construction of an *A. vinelandii* mutant strain deleted for both *nifM* and *nifZ*.[115] Such a strain contains negligible Fe protein activity and only very low MoFe protein activity (see Table 20-2). Addition of purified FeMo-cofactor to crude extracts prepared from the nitrogenase-derepressed *nifMZ* deletion strain results in significant reconstitution of MoFe protein activity (see Table 20-2). These results indicate that in the absence of *nifM* product the *nifZ* product is required for FeMo-cofactor biosynthesis or insertion. This observation is in line with the suggestion of Paul and Merrick[184] that *nifW* and *nifZ* might catalyze reactions that can occur non-enzymatically, although at a low rate, or that other *nif*-specific gene products might substitute for *nifW* or *nifZ* product functions. On the basis of FeMo-cofactor reconstitution studies using the *A. vinelandii nifMZ* deletion strain, we conclude that at least the *nifZ* product has some function related to formation of FeMo-cofactor or insertion of FeMo-cofactor into the immature MoFe protein.

X. *nifQ*

Genes corresponding to the *K. pneumoniae nifQ* gene[4,28,149] have been identified in *A. vinelandii*[118] and *R. capsulatus*.[164] Imperial et al.[111] have shown that *nifQ* mutants of *K. pneumoniae* are deficient in FeMo-cofactor biosynthesis and that this *nifQ* phenotype can be suppressed by the addition of elevated levels of Mo to the growth medium. Suppression of the *nifQ* phenotype by elevated levels of Mo was also confirmed in *A. vinelandii*[118] and *R. capsulatus*[164] *nifQ* mutants. The function of the *nifQ* product does not appear to reside in the accumulation of Mo but rather in an initial step in the FeMo-cofactor biosynthetic pathway. This conclusion is based on the observation that *nifQ* mutants are not defective in another Mo-requiring biosynthetic pathway, Mo-cofactor biosynthesis.[112] Moreover, Mo uptake kinetics measured by ^{99}Mo incorporation rates were unchanged in *nifQ* mutants.[111] Ugalde et al.[248] reported that supplementation of cystine or cysteine to the growth medium spared *K. pneumoniae nifQ* mutants the elevated Mo requirement for physiologic formation of active nitrogenase. This observation was interpreted to indicate that *nifQ* product function probably involves formation of a Mo- and S-containing FeMo-cofactor precursor. Also, because either high levels of Mo, cysteine, or cystine can spare the *nifQ* phenotype, it was suggested that the reaction catalyzed by the *nifQ* product might occur nonenzymatically but at a low rate.

Alignment of *nifQ* product primary sequences, deduced from the gene sequences, for *K. pneumoniae*,[4,28] *A. vinelandii*,[118] and *R. capsulatus*[164] shows a clustering of four Cys residues located near the C-terminus and arranged in the pattern Cys–X_4–Cys–X–X–Cys–X_5–Cys. Such clustering of potential metal-binding thiol groups lends support to the proposal that the *nifQ* product could catalyze a reaaction involving coordination of Mo, S, and Fe. Location of the Cys-rich region very near the C-terminus and interspersion of a relatively high number of hydrophilic amino acids in this region suggest a probable near surface location of a prosthetic group if a FeMo-cofactor

```
         30                40                50              60
          ▼                                 ▼       ▼       ▼              ▼

Kp NifB  A A H P C Y S R H G H H R F A R M H L P V A P A C N L Q C N Y C N R K F D C S
Av NifB  Q N H P C Y S E E A H H Y F A R M H V A V A P A C N I Q C H Y C N R K Y D C A
Rc NifB  K D H P C Y S E E A H H H F A R M H V S V A P A C N I Q C N Y C N R K Y D C S
Bj NifB  K N H P C Y S E D A H H H Y A R M H V A V A P A C N I Q C N Y C N R K Y D C A
Rm NifB  K N H P C F S E E A H H Y F A R M H V A V A P A C N I Q C N Y C N R K Y D C A
Rl NifB  K D H P C F S E L A H H Y F A R M H V D V A P A C N I Q C N Y C N R K Y D C T
An NifB  A K H P C Y S E E A H H H Y A R M H V A V A P A C N I Q C N Y C N R K Y D C A
Cp "NifB" C T H P C Y G D N A H K - F A R M H I P V A P S C N I S C N Y C N R K Y D C T
```

Figure 20-9. Comparison of Cys-rich regions from *nifB* gene product sequences. Numbers refer to the *A. vinelandii nifB* product sequence. Residues conserved in all species are boxed. Numbers refer to the *K. pneumoniae nifB* gene product sequence, which contains a total of 468 residues. Conserved Cys residues are indicated by arrowheads. The source of *nifB* product sequences are as follows: Kp, *K. pneumoniae*; Av, *A. vinelandii*; Bj, *B. japonicum*; Rm, *R. meliloti*; R1, *R. leguminosarum*; An, *Anabaena* 7120; Cp, *C. pasteurianum*. The *C. pasteurianum nifB* and *nifN* genes are fused.

precursor does indeed bind to the *nifQ* product. Such a location is consistent with the idea that the *nifQ* product has a function early in FeMo-cofactor biosynthesis and that this function could involve transfer of a Mo-containing precursor to another polypeptide, perhaps *nifB* product.

XI.   *nifB*

Homologues to the *K. pneumoniae nifB* gene[4,28] have been identified in *A. vinelandii*,[118] *A. chroococcum*,[66] *R. capsulatus*,[133,153] *R. leguminosarum*,[81,207] *R. meliloti*,[28] *B. japonicum*,[178] and *Anabaena* 7120.[170] Alignment of the available *nifB* product sequences shows a remarkable clustering of Cys residues located near the N-terminus. A high concentration of polar and charged residues is also present in this region (Figure 20-9). Buikema et al.[28] proposed that this Cys-rich region within the *nifB* product provides a near-surface, hydrophilic pocket critical for binding of an FeMo-cofactor precursor. Nevertheless, the *nifB* product has not been isolated, nor is its specific function in FeMo-cofactor biosynthesis known. A unique situation appears to exist for the *nifB* gene from *C. pasteurianum*. Based on DNA sequence data, the *nifB* and *nifN* genes are fused (Wang, Chen, and Johnson, unpublished). That is, in *C. pasteurianum, nifN* and *nifB* are expressed as a fused protein. It is known that, in *A. vinelandii, nifE* and *nifN* form a tetrameric complex that is required for FeMo-cofactor biosynthesis (see later). Fusion of *nifB* and *nifN* in *C. pasteurianum* hints that in other organisms the *nifB* product physically interacts with the *nifEN* product complex during FeMo-cofactor biosynthesis. The Cys cluster conserved in other *nifB* product sequences is also present in the *C. pasteurianum nifB* product sequence (see Figure 20-9).

A. vinelandii[186,218] and K. pneumoniae[93,200] *nifB* mutants have been characterized. Such mutants do not reduce C_2H_2, nor do they exhibit the $S = {}^3/_2$

EPR signal characteristic of the native MoFe protein. "Apo-MoFe" protein present in crude extracts of *nifB* mutants are fully reconstituted by addition of FeMo-cofactor extracted from purified wild-type MoFe protein. MoFe protein reconstituted in this way also regains a fully developed $S = \frac{3}{2}$ signal identical to the native enzyme. This association of both MoFe protein activity and the EPR signal with FeMo-cofactor was the first strong evidence that FeMo-cofactor provides or is part of the substrate-binding and reduction site.[218] Apo-MoFe protein has been purified to near homogeneity from *K. pneumoniae*[93] and *A. vinelandii nifB* mutants.[186] Such preparations from both sources are contaminated with an approximately 20-kDa protein of unknown origin. The possible role of the 20-kDa protein in MoFe protein maturation is not known. Apo-MoFe protein purified from *K. pneumoniae* contains some Mo, although this might represent adventitious binding. Isolation of apo-MoFe protein that can be reconstituted by addition of FeMo-cofactor clearly demonstrates that the P clusters and FeMo-cofactor are separate entities.

XII. *nifE* and *nifN*

Homologues to the *K. pneumoniae nifE* and/or *nifN*[4,217] genes have been identified in *A. vinelandii*,[23,25,40,131] *A. chroococcum*,[66] *R. meliloti*,[2] *2 R. capsulatus*,[164,165] *A. brasilense* Sp7,[74] *Rhizobium* ORS571,[46] *C. pasteurianum*,[259] and *B. japonicum*.[58] The *nifE* and *nifN* gene products have a significant degree of primary sequence identity when they are compared to the MoFe protein α- and β-subunits, respectively[2,4,23,25,40,165,217,259] (see Figure 20-5). This sequence identity, cotranscription of the *nifEN* gene products in several organisms, and the location of the *nifEN* transcription unit downstream from the nitrogenase structural gene cluster in several organisms indicates that the *nifEN* gene cluster might have originated from a duplication of the *nifDK* genes. Biochemical reconstitution studies involving *nifE* and *nifN* mutants from both *K. pneumoniae* and *A. vinelandii* demonstrated that the products of both genes are required for FeMo-cofactor biosynthesis.[25,200] The preceding observations and the finding that MoFe protein subunits are not required for FeMo-cofactor biosynthesis (see later) led to the suggestion that the *nifEN* products form a complex analogous to the MoFe protein and that this complex provides a scaffold for FeMo-cofactor biosynthesis[25]; that is, it is possible that FeMo-cofactor is assembled onto a nifEN products complex through the action of certain other FeMo-cofactor biosynthetic gene products (e.g., the *nifQ* and *nifB* products). Although this hypothesis remains unproven and the sequence of events during FeMo-cofactor biosynthesis is unknown, some support for it comes from identification of a tetrameric *nifEN* product complex isolated from an *A. vinelandii nifB* mutant.[185]

XIII. *nifV*

Homologues to the *K. pneumoniae nifV* gene[4,15] have been identified in *A. vinelandii*,[15] *A. chroococcum*,[66] and *R. capsulatus* (Masepohl and Klipp, unpublished). Alignment of the primary sequences of *nifV* products reveals a high level of conserved identity. In *C. pasteurianum* the *nifV* gene is split into two separate genes (see Figure 20-7), each coding for a product having sequence identity with the N- or C-coding regions of the other *nifV* gene products (Wang, Chen, and Johnson, unpublished). The *nifV* homologous coding fragments from *C. pasteurianum* were expressed, both separately and in combination, in an *A. vinelandii nifV* deletion strain (Wang, Dean, Johnson, and Chen, unpublished). In these experiments, the *nifV* phenotype of the *A. vinelandii* mutant was only rescued when both *C. pasteurianum nifV*-like products were expressed together.

nifV mutants from *K. pneumoniae*[149,200] and *A. vinelandii*[114,115] remain capable of slow diazotrophic growth. The specific biochemical defect in nitrogenase from *K. pneumoniae nifV* mutants was studied in detail by the Sussex group.[92,157] When compared to the wild-type enzyme, nitrogenase from a *nifV* mutant is capable of reducing N_2 only at a very low rate but remains able to reduce C_2H_2 and hydrogen at moderate rates.[157] Unlike the situation with wild-type nitrogenase, hydrogen evolution catalyzed by the *nifV* mutant enzyme is substantially inhibited in the presence of CO. These altered properties of nitrogenase from *nifV* mutants were specifically ascribed to the MoFe protein. Alteration in the substrate reduction and inhibition patterns of *nifV* mutants, which was also extended to other artificial substrates,[75,140] implied an alteration in the nitrogenase substrate binding or reduction site of these mutants. To test whether *nifV* mutants were altered specifically in FeMo-cofactor, Hawkes et al.[92] extracted FeMo-cofactor from a *nifV* mutant and used this preparation to reconstitute MoFe protein in extracts of a nitrogenase derepressed *nifB* mutant. (Recall that *nifB* mutants accumulate an inactive MoFe protein having no FeMo-cofactor.) MoFe protein reconstituted in this way exhibits substrate reduction and inhibition patterns characteristic of MoFe protein from *nifV* mutants, indicating that *nifV* mutants do have an altered form of FeMo-cofactor. These results provide direct biochemical-genetic evidence that FeMo-cofactor includes or is part of the substrate reduction site. An *in vitro* FeMo-cofactor biosynthetic system[220] was later used to confirm and substantially extend these initial findings (see later). The *in vitro* biosynthetic system was also used to demonstrate that the *nifV* product is required for homocitrate synthesis and that homocitrate is an organic component of FeMo-cofactor. Collett and Orme-Johnson (unpublished) recognized some sequence conservation when they compared the *K. pneumoniae nifV* product primary sequence to the LEU4 gene product (α-isopropyl malate synthase) from *Saccharomyces cerevisiae*.[9] We have compared the analogous *leuA* gene product from *Salmonella typhimurium*[196] to the *nifV* gene product from *A. vinelandii* (Figure 20-10). This comparison reveals a great deal of primary sequence conservation, indicating that homocitrate synthesis and α-isopropylmalate synthesis are likely to be catalyzed by similar mechanisms.

XIV. In Vitro FeMo-cofactor Biosynthesis

Neither *K. pneumoniae* nor *A. vinelandii* require MoFe protein subunits for the biosynthesis of FeMo-cofactor.[69,201,247] These results imply that FeMo-cofactor is synthesized and inserted into an immature form of the MoFe protein. This information provided the rationale for formulation of an *in vitro* assay for FeMo-cofactor biosynthesis. Shah and his colleagues[220] have developed such an assay as an approach to elucidate the specific functions and the sequential involvement of certain *nif*-specific gene products in FeMo-cofactor biosynthesis. In development of this assay, extracts prepared from mutant strains having complementary defects in FeMo-cofactor biosynthesis were mixed in the presence of ATP and Mo in an attempt to reconstitute MoFe protein activity. In this way, minimum requirements for *in vitro* FeMo-cofactor biosynthesis were shown to include the *nifB* product, the *nifEN* products, a factor (homocitrate) whose accumulation requires *nifV* product, Fe protein, ATP, and Mo.

The value of *in vitro* FeMo-cofactor biosynthesis is that it permits functional assay for certain *nif*-specific gene products during their attempted purification. For example, Paustian et al.[185] used *in vitro* FeMo-cofactor biosynthesis to monitor purification of the *nifEN* products. They were able to demonstrate that *nifE* and *nifN* gene products form an $\alpha_2\beta_2$ complex analogous to the MoFe protein. It was also found, as predicted from primary sequences of the *nifEN* products, that the complex contains an Fe–S cluster. The purified *nifEN* product complex, as isolated from a *nifB* mutant, does not contain Mo. This result is consistent with the idea that the *nifEN* complex provides a scaffold for FeMo-cofactor biosynthesis and that, once completed, FeMo-cofactor escapes from the complex intact.

Both biochemical–genetic experiments and the FeMo-cofactor biosynthetic assay have been used to demonstrate that Fe protein is required for FeMo-cofactor biosynthesis.[69,202,219] Although the specific function of Fe protein in FeMo-cofactor biosynthesis is not understood, a number of observations bear on this point. Fe protein is known to bind MgATP, and it is also known that MgATP is required for FeMo-cofactor biosynthesis. Also, sequence similarities between the MoFe protein and the *nifEN* product complex suggest that Fe protein might be capable of physically interacting with the *nifEN* products complex in a fashion analogous to Fe protein–MoFe protein interaction. Such formation of an Fe protein-*nifEN* product complex,

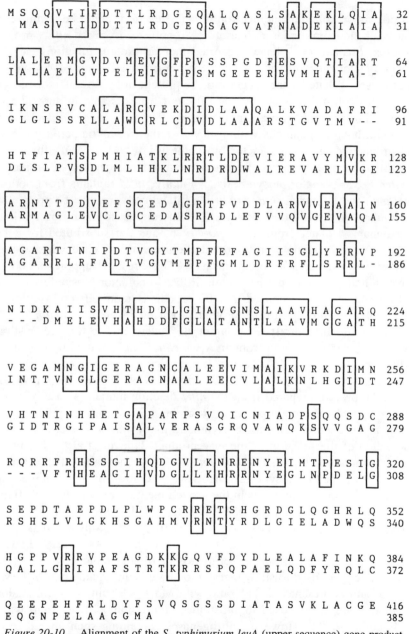

```
M S Q Q V I I F D T T L R D G E Q A L Q A S L S A K E K L Q I A    32
  M A S V I I D D T T L R D G E Q S A G V A F N A D E K I A I A    31

L A L E R M G V D V M E V G F P V S S P G D F E S V Q T I A R T    64
I A L A E L G V P E L E I G I P S M G E E E R E V M H A I A - -    61

I K N S R V C A L A R C V E K D I D L A A Q A L K V A D A F R I    96
G L G L S S R L L A W C R L C D V D L A A A R S T G V T M V - -    91

H T F I A T S P M H I A T K L R R T L D E V I E R A V Y M V K R   128
D L S L P V S D L M L H H K L N R D R D W A L R E V A R L V G E   123

A R N Y T D D V E F S C E D A G R T P V D D L A R V V E A A I N   160
A R M A G L E V C L G C E D A S R A D L E F V V Q V G E V A Q A   155

A G A R T I N I P D T V G Y T M P F E F A G I I S G L Y E R V P   192
A G A R R L R F A D T V G V M E P F G M L D R F R F L S R R L -   186

N I D K A I I S V H T H D D L G I A V G N S L A A V H A G A R Q   224
- - - D M E L E V H A H D D F G L A T A N T L A A V M G G A T H   215

V E G A M N G I G E R A G N C A L E E V I M A I K V R K D I M N   256
I N T T V N G L G E R A G N A A L E E C V L A L K N L H G I D T   247

V H T N I N H H E T G A P A R P S V Q I C N I A D P S Q Q S D C   288
G I D T R G I P A I S A L V E R A S G R Q V A W Q K S V V G A G   279

R Q R R F R H S S G I H Q D G V L K N R E N Y E I M T P E S I G   320
- - - V F T H E A G I H V D G L L K H R R N Y E G L N P D E L G   308

S E P D T A E P D L P L W P C R R E T S H G R D G L Q G H R L Q   352
R S H S L V L G K H S G A H M V R N T Y R D L G I E L A D W Q S   340

H G P P V R R V P E A G D K K G Q V F D Y D L E A L A F I N K Q   384
Q A L L G R I R A F S T R T K R R S P Q P A E L Q D F Y R Q L C   372

Q E E P E H F R L D Y F S V Q S G S S D I A T A S V K L A C G E   416
E Q G N P E L A A G G M A                                         385
```

Figure 20-10. Alignment of the *S. typhimurium leuA* (upper sequence) gene product and the *A. vinelandii nifV* gene product (lower sequence). Conserved residues are boxed. The entire *leuA* gene sequence is not shown.

and possibly a concomitant electron transfer reaction, could be required for a biosynthetic event or for escape of FeMo-cofactor from the *nifEN* product complex. Shah et al.[219] have suggested that Fe protein functions early in FeMo-cofactor biosynthesis and that it is not involved as an Fe–S donor in that process. Whatever the function of Fe protein in FeMo-cofactor biosynthesis, this role must be somewhat different from its role in substrate reduction. For example, certain mutant strains from *K. pneumoniae* and *A. vinelandii* having mutations within *nifH* lose their catalytic activity but are still productive in FeMo-cofactor biosynthesis.[69,202] Also, ADP-ribosylated Fe protein becomes catalytically inactive but remains capable of promoting *in vitro* FeMo-cofactor biosynthesis.[219] Finally, *nifM* mutants from both *A. vinelandii* and *K. pneumoniae* produce a catalytically inactive Fe protein but accumulate an active MoFe protein.[115,200] In contrast to these observations, *nifH* mutants isolated from *A. vinelandii,* having Ser substituted for either of the Cys residues proposed to provide Fe protein Fe-S cluster ligands, are deficient in FeMo-cofactor biosynthesis.[105] These results support, but do not prove, a redox role for the Fe protein in FeMo-cofactor biosynthesis.

Robinson et al.[203] proposed that Fe protein is required for insertion of FeMo-cofactor into apo-MoFe protein produced in an *A. vinelandii nifH* deletion strain. In contrast, Paustian et al.[186] were able to fully reconstitute apo-MoFe protein prepared from an *A. vinelandii nifB* mutant by the simple addition of isolated FeMo-cofactor. It is difficult to assess the significance of these apparently contradictory conclusions, because it is not known whether apo-MoFe protein prepared from the *nifH* deletion mutant is strictly equivalent to apo-MoFe protein prepared from the *nifB* mutant.

The *in vitro* FeMo-cofactor biosynthetic assay was also instrumental in elucidation of the function of the *nifV* product. Hoover et al.[103] discovered that *nifV* mutant extracts of *K. pneumoniae* were incapable of directing *in vitro* FeMo-cofactor biosynthesis. However, addition of a low-molecular-weight factor that accumulates in the growth medium of wild-type N_2-fixing *K. pneumoniae* cells, but not *nifV* mutants, could promote FeMo-cofactor biosynthesis in *nifV* mutant extracts.[99,103] NMR spectroscopy and mass spectrometry were used to identify this "V factor" as homocitrate.[102] *In vitro* FeMo-cofactor biosynthesis experiments involving radioactively labeled homocitrate were used to show that homocitrate is an organic part of FeMo-cofactor and is present in a 1:1 ratio with Mo.[100] Certain chemical analogs of homocitrate can be used to substitute for the homocitrate group within FeMo-cofactor.[101,110] However, nitrogenases containing such altered forms of FeMo-cofactor exhibit altered substrate reduction and inhibition patterns when compared to wild-type nitrogenase. On the basis of these experiments, Hoover et al.[101] suggested that FeMo-cofactor obtained from *nifV* mutants have citrate substituted for homocitrate. The further use of homocitrate analogs in FeMo-cofactor biosynthesis should prove valuable as a method to analyze the substrate binding and reduction site of nitrogenase.

XV. *nifT*, *nifY*, and *nifX*

The *nifT* gene has only been identified by DNA sequence analysis and only in *A. vinelandii*[114] and *K. pneumoniae*.[4,17] It is the smallest of the *nif* genes encoding a peptide of only 72 amino acids in *A. vinelandii*. *A. vinelandii nifT* insertion and deletion mutants exhibit normal diazotrophic growth.[114]

A *nifX* gene has been identified in *K. pneumoniae*,[4,17,190] *A. vinelandii*,[114] *R. capsulatus*,[164,165] and *B. japonicum* (Hennecke, unpublished). The *nifX* product is not required for normal diazotrophic growth for *K. pneumoniae*, *A. vinelandii*, or *R. capsulatus*. A *nifY* gene has been identified only in *K. pneumoniae*[4,17,190] and *A. vinelandii*.[114] Mutant strains having *nifY* deleted have no obvious phenotype. Interestingly, alignment of *nifX* and *nifY* product sequences shows some significant sequence conservation.[4,114,165] Such sequence conservation is in line with the suggestion that the *nifHDKTY* transcription unit bears an evolutionary relationship with the *nifENX* transcription unit[24,40] (see Figure 20-7). Whether there is a functional significance attached to *nifX*–*nifY* product sequence conservations is not yet known. However, simultaneous inactivation of both *nifX* and *nifY* products in *A. vinelandii* (our unpublished results) or in *K. pneumoniae* (Schramm, Hennecke, and Klipp, unpublished) does not have a severe effect upon diazotrophic growth in the respective double-mutant strains. Moreno-Vivian et al.[165] recognized sequence identity when they compared a region conserved between *nifX* and *nifY* products to the *nifB* product primary sequence. They interpreted this sequence conservation to indicate that an evolutionary relationship exists among *nifB*, *nifX*, and *nifY* and proposed that *nifX* and *nifY* products could have dispensable functions related to FeMo-cofactor biosynthesis. Gosink et al.[79] have used gene fusion techniques to accomplish the controlled overexpression of the *nifX* product in *K. pneumoniae*. In this way they showed that elevated accumulation of *nifX* product results in a substantial reduction in nitrogenase structural gene mRNA accumulation and a consequent reduction in *nif*-specific protein synthesis. These results provide strong evidence that the *nifX* product is a negative regulatory element, possibly involved in balancing *nif* gene expression.

XVI. *nif*-Associated Genes

Comparison of the physical and genetics maps of *nif*-specific genes from *K. pneumoniae*, *A. vinelandii*, *R. capsulatus*, *B. japonicum*, and *C. pasteurianum* (see Figure 20-7) reveals the following five important differences: (1) Spatial arrangements of individual *nif* genes are not always conserved at the interspecies level. For example, the *nifW* and *nifZ* genes are located adjacent to one another in *K. pneumoniae* and *A. vinelandii* but not in *R. capsulatus*. (2) There are a number of potential genes (*nif*-associated genes) interspersed among *nif* gene clusters in certain organisms that are not present within the *K. pneumoniae nif* gene cluster. (3) The tight clustering of all *nif* genes in *K. pneumoniae* is not conserved in all other organisms. For example, in *A. vinelandii* the *nifABQ* gene cluster is not genetically linked to the major *nif* cluster. (4) In *C. pasteurianum* certain genes corresponding to *nif* genes from *K. pneumoniae* are either split into two genes (i.e., *nifV*) or exist as a pair of fused genes (i.e., *nifN-B*).(5) Certain *nif*-specific genes are duplicated in some organisms. A duplication of the *nifA–nifB* gene region from *R. capsulatus* is one example.

The significance and functions of genes whose products appear to be associated with N_2 fixation in certain organisms, but not in *K. pneumoniae*, are not known. However, strong interspecies conservation in deduced primary sequences of the products of some of these genes (see, for example, Figure 20-11) implies that certain of them have important functions related to N_2 fixation. Mutational analysis has shown that none of the *nif*-associated genes identified in *R. capsulatus*[164] (Masepohl and Klipp, unpublished) or *A. vinelandii*[114,115,118] so far encodes a product essential for N_2 fixation under standard laboratory conditions. Nevertheless, examination of the primary sequences of *nif*-associated gene products has revealed some interesting features that could aid in elucidation of their respective functions. The products corresponding to orf2 from *A. vinelandii* and *R. capsulatus* (see Figure 20-7) are periodic proteins having a 24–amino acid motif that is repeated in tandem seven times (Jacobson, Dean, Klipp, and Masepohl, unpublished observation). At least one periodic protein has been shown to have a metal-binding function.[147] The product of orf10 from *A. vinelandii* exhibits a high degree of primary sequence identity when compared to a family of hydrophilic inner-membrane proteins that have ATP-binding functions and are involved in active transport (e.g., *chlD*[120] and *malK*[77] products). The *A. vinelandii* orf9 product probably also has a nucleotide-binding function as deduced from examination of the primary sequence (our unpublished results). The *A. vinelandii* orf14 product has some sequence identity when compared to the product of a gene involved in arsenic resistance, *arsC*.[37,223] The orf14 gene product also has a high level of sequence identity when compared to the product of a potential gene located immediately downstream from the *draG* gene from *R. rubrum*[71] (Wang, unpublished observation). A number of *nif*-associated genes encode ferredoxinlike proteins (see Figure 20-1). Potential roles for such ferredoxins include electron transport to nitrogenase, electron transfer reactions involved in protein maturation, or as Fe–S donors during metallocluster formation.

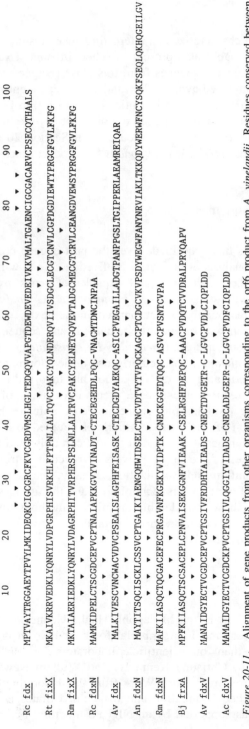

| | 10 | 20 | 30 | 40 | 50 | 60 | 70 | 80 | 90 | 100 |

Rc fdx   MPTVAYTRGGAEYTPVYLMKIDEQKCIGGCGRCFKVCGRDVMSLHGLTEDGQVVAPGTDEMDEVEDEIVKKVMALTGAENCIGCGACARVCPSECQTHAALS

Rt fixX   MKAIVKRRVEDKLYQNRYLVDPGRPHISVRKHLFPTNLIALTQVCPAKCYQLNDRRQVIIVSDGCLEGCTCNVLCGPDGDIEWTYPRGGFGVLFKFG

Rm fixX   MKTAIAERIEDKLYQNRYLVDAGRPHITVRPEHRSPSLNLLALTRVCPAKCYELNETGQVEVTADGCMEGCTCRVLCEANGDVEWSYPRGGFGVLFKFG

Rc fdxN   MAMKIDPELCTSCGDCEPVCPTNAIAPKKGVYVINADT-CTECEGEHDLPQC-VNACMTDNCINPAA

Av fdx   MALKIVESCVNCHACVDVCPSEAISLAGPHFEISASK-CTECDGDYAEKQC-ASICPVEGAILLADGTPANPPGSLTGIPPERLAEAMREIQAR

An fdxN   MAYTITSQCISCKLCSSVCPTGAIKIAENGQHWIDSELCTNCVDTVYTVPQCKAGCPTCDGCVKVPSDYWEGWFANYNRVIAKLTKKQDYWERWFNCYSQKFSEQLQKHQGEILGV

Rm fdxN   MAFKIIASQCTQCGACEFECPRGAVNFKGEKYVIDPTK-CNECKGGFDTQQC-ASVGPVSNTCVPA

Bj frxA   MPFKIIASQCTSCSACEPLCPNVAISEKGGNFVIEAAK-CSELRGHFDEPQC-AAACPVDQTCVVDRALPRYQAPV

Av fdxV   MANAIDGYECTVCGDCEPVCPTGSIVFRDDHYAIEADS-CNECTDVGETR-C-LGVCPVDLCIQPLDD

Ac fdxV   MAMAIDGYECTVCGDCKPVCPTGSIVLQGGIYVIDADS-CNECADLGEPR-C-LGVCPVDFCIQPLDD

Figure 20-11. Alignment of gene products from other organisms corresponding to the orf6 product from *A. vinelandii*. Residues conserved between adjacent sequences are indicated by an *. Arrowheads indicate residues conserved in all species. Organisms are indicated as follows: Av, *A. vinelandii*; Bj, *B. japonicum*; Rc, *R. capsulatus*; An, *Anabaena* 7120. The *Anabaena* 7120[22] and *A. vinelandii*[114] sequences are available. The *R. capsulatus* and *B. japonicum* sequences are from the unpublished work of Masepohl and Klipp, and Ebeling and Hennecke, respectively.

Kahn et al.[125] have characterized an interesting gene cluster, *fixGHI*, from *R. meliloti* that is probably also present in other rhizobia. Examination of the deduced gene product sequences from this cluster suggests they form a membrane-bound complex having redox and ATP-dependent cation pump functions. Intact *fixGHI* genes are essential for N_2 fixation in *R. meliloti*. Other examples of *nif*-associated genes required for N_2 fixation in certain organisms are *fixABC* (see previous discussion in Section 20.3), *fixW* (*R. leguminosarum*[98]), and "*nifO*" (*A. caulinodans*[127]). Product functions for these genes are unknown.

XVII. Summary Statement

Considerable progress has been made over the past decade concerning the biochemical–genetic analysis of biological N_2 fixation. We now have a nearly complete picture of the organization and sequence of genes involved in N_2 fixation in the free-living diazotrophs *K. pneumoniae*, *A. vinelandii*, and *R. capsulatus*. We can reasonably expect that within the next 5 years all genes involved in N_2 fixation will be identified, cloned, and sequenced, at least for the following organisms: *A. vinelandii*, *R. capsulatus*, *Anabaena*, *R. meliloti*, and *B. japonicum*. A particularly useful development concerning the biochemical–genetic analysis of N_2 fixation has involved the successful application of *in vitro* biosynthesis as a method for assay of certain *nif*-specific genes whose products participate in FeMo-cofactor biosynthesis. The further application of this approach, perhaps coupled to *in vitro* DNA-directed *nif* product synthesis,[5,210,272] will almost certainly lead to a detailed description of the roles of the major players involved in nitrogenase component protein maturation and stabilization. Finally, the successful application of site-directed mutagenesis technologies and the use of synthetic analogs of homocitrate show considerable promise as methods for elucidation of the individual functions of the various metal centers contained within the nitrogenase component proteins.

Acknowledgments

We thank Werner Klipp, Bernd Masepohl, Hauke Hennecke, Robert Haselkorn, Jacques Meyer, and Juan Imperial for their help in preparation of this chapter. We also thank our scientific collaborators Barbara Burgess, Jim Howard, Jim Beynon, Christina Kennedy, and William Newton for their contributions to our research effort. Work from this laboratory is supported by the National Science Foundation and the United States Department of Agriculture. MRJ was supported by an NIH Postdoctoral Fellowship.

References

1. Adams, T.H., McClung, C.R., and Chelm, B.K., "Physical organization of the *Bradyrhizobium japonicum* nitrogenase gene region," *J. Bacteriol. 159*, 857–862 (1984).

2. Aguilar, O.M., Reilander, H., Arnold, W., and Puhler, A., "*Rhizobium meliloti nifN* (*fixF*) gene is part of an operon regulated by a *nifA*-dependent promoter and codes for a polypeptide homologous to the *nifK* gene product," *J. Bacteriol. 169*, 5393–5400 (1987).

3. Ahombo, G., Willison, J.C., and Vignais, P.M., "The *nifHDK* genes are contiguous with a *nifA*-like regulatory gene in *Rhodobacter capsulatus*," *Molec. Gen. Genet. 205*, 442–445 (1986).

4. Arnold, W., Rump, a, Klipp, w., Priefer, U.B., and Puhler, A., "Nucleotide sequence of a 24,206-base-pair DNA fragment carrying the entire nitrogen fixation gene cluster of *Klebsiella pneumoniae*," *J. Molec. Biol. 203*, 715–738 (1988).

5. Austin, S., Henderson, N., and Dixon, R., "Requirements for transcriptional activation *in vitro* of the nitrogen-regulated *glnA* and *nifLA* promoters from *Klebsiella pneumoniae*: Dependence on activator concentration," *Molec. Microbiol. 1*, 92–100 (1987).

6. Avtges, P., Kranz, R.G., and Haselkorn, R., "Isolation and organization of genes for nitrogen fixation in *Rhodopseudomonas capsulata*," *Molec. Gen. Genet. 201*, 363–369 (1985).

7. Avtges, P., Scolnick, P.A., and Haselkorn, R., "Genetic and physical map of the structural genes (*nifII, D, K*) coding for the nitrogenase complex of *Rhodopseudomonas capsulata*," *J. Bacteriol. 156*, 251–256 (1983).

8. Badenoch-Jones, J., Holton, T.A., Morrison, C.M., Scott, K.F., and Shine, J., "Structural and functional analysis of nitrogenase genes from the broad-host-range *Rhizobium* strain ANU240," *Gene 77*, 141–153 (1989).

9. Beltzer, J.P., Chang, L.-F.L., Hinkkanen, A.E., and Kohlaw, G.P., "Structure of Yeast LEU4," *J. Biol. Chem. 261*, 5160–5167 (1986).

10. Benemann, J.R., Yoch, D.C., Valentine, R.C., and Arnon, D.I., "The electron transport system in nitrogen-fixation by *Azotobacter*, I. Azotoflavin as an electron carrier," *Proc. Natl. Acad. Sci. USA 64*, 1079–1086 (1969).

11. Benemann, J.R., Yoch, D.C., Valentine, R.C., and Arnon, D.I., "The electron transport system in nitrogen fixation by *Azotobacter*. IV. Requirements for NADPH-supported nitrogenase activity," *Biochim. Biophys. Acta 226*, 206–212 (1971).

12. Bennett, L.T., Cannon, F., and Dean, D.R., "Nucleotide sequence and mutagenesis of the *nifA* gene from *Azotobacter vinelandii*," *Molec. Microbiol. 2*, 315–321 (1988).

13. Bennett, L.T., Jacobson, M.R., and Dean, D.R., "Isolation, sequencing, and mutagenesis of the *nifF* gene encoding flavodoxin from *Azotobacter vinelandii*," *J. Biol. Chem. 263*, 1364–1369 (1988).

14. Berman, J., Gershoni, J.M., and Zamir, A., "Expression of nitrogen fixation genes in foreign hosts," *J. Biol. Chem. 260*, 5240–5243 (1985).

15. Beynon, J., Ally, A., Cannon, M., Cannon, F., Jacobson, M.R., Cash, V.L., and Dean, D., "Comparative organization of nitrogen fixation-specific genes from *Azotobacter vinelandii* and *Klebsiella pneumoniae*: DNA sequence of the *nifUSV* genes. *J. Bacteriol. 169*, 4024–4029 (1987).

16. Beynon, J., Cannon, M., Buchanan-Wollaston, V., and Cannon, F., "The *nif* promoters of *Klebsiella pneumoniae* have a characteristic primary structure," *Cell 34*, 665–671.

17. Beynon, J., Cannon, M., Buchanan-Wollaston, V., Ally, A., Setterquist, R., Dean, D., and Cannon, F., "The nucleotide sequence of the *nifT*, *nifY*, *nifX* and *nifW* genes of *K. pneumoniae*," *Nucleic Acids Res. 16*, 9860 (1988).

18. Biggins, D.R., Kelly, M., and Postgate, J.R., "Resolution of nitrogenase of *Mycobacterium flavum* 103 into two components and cross reaction with nitrogenase components from other bacteria," *Eur. J. Biochem. 20*, 140–143 (1971).

19. Bogusz, D, Houmard, J., and Aubert, J.P., "Electron transport to nitrogenase in *Klebsiella pneumoniae*. Purification and properties of the *nifJ* protein," *Eur. J. Biochem. 120*, 421–426 (1981).

20. Bohme, H., and Haselkorn, R., "Molecular cloning and nucleotide sequence analysis of the gene coding for heterocyst ferrodoxin from the cyanobacterium *Anabaena* sp. strain PCC 7120," *Molec. Gen. Genet. 214*, 278–285 (1988).

21. Bohme, H., and Schrautemeier, B., "Electron donation to nitrogenase in a cell free system from heterocysts of *Anabaena variabilis*," *Biochim. Biophys. Acta 891*, 115–120 (1987).

22. Borthakur, D., Basche, M., Buikema, W.J., Borthakur, P.B., and Haselkorn, R., "Expression, nucleotide sequence and mutational analysis of two open reading frames in the *nif* gene region of *Anabaena* sp. strain PCC 7120," *Molec Gen. Genet., 221*, 227–234 (1990).

23. Brigle, K.E., and Dean, D.R., "Revised nucleotide sequence of the *Azotobacter vinelandii nifE* gene," *Nucleic Acids Res. 16*, 5214 (1988).

24. Brigle, K.E., Newton, W.E., and Dean, D.R., "Complete nucleotide sequence of the *Azotobacter vinelandii* nitrogenase structural gene cluster," *Gene 37*, 37–44 (1985).

25. Brigle, K.E., Weiss, M.C, Newton, W.E., and Dean, D.R., "Products of the iron–molybdenum cofactor-specific biosynthetic genes, *nifE* and *nifN*, are structurally homologous to the products of the nitrogenase molybdenum–iron protein genes, *nifD* and *nifK*," *J. Bacteriol. 169*, 1547–1553 (1987).

26. Brigle, K.E., Setterquist, R.A., Dean, D.R., Cantwell, J.S., Weiss, M.C., and Newton, W.E., "Site-directed mutagenesis of the nitrogenase MoFe protein of *Azotobacter vinelandii*," *Proc. Natl. Acad. Sci. USA 84*, 7066–7069 (1987).

27. Bui, P.T., and Mortenson, L.E., "Mechanism of the enzymic reduction of N_2: The binding of adenosine 5'-trophosphate and cyanide to the N_2-reducing system," *Proc. Natl. Acad. Sci. USA 61*, 1021–1027 (1968).

28. Buikema, W.J., Klingensmith, J.A., Gibbons, S.L., and Ausubel, F.M, "Conservation of structure and location of *Rhixobium meliloti* and *Klebsiella pneumoniae nifB* genes," *J. Bacteriol. 169*, 1120–1126 (1987).

29. Buikema, W.J., Szeto, W.W., Lemley, P.V., Orme-Johnson, W.H., and Ausubel, F.M., "Nitrogen fixation specific regulatory genes of *Klebsiella pneumoniae* and *Rhizobium meliloti* share homology with the general regulatory gene *ntrC* of *K. pneumoniae*," *Nucleic Acids Res. 13*, 4539–4555 (1985).

30. Bulen, W.A., Burns, R.C., and LeComte, J.R., "Nitrogen fixation: Hydrosulfite as electron donor with cell-free preparations of *Azotobacter vinelandii* and *Rhodospirillum rubrum*," *Proc. Natl. Acad. Sci. USA 53*, 532–539 (1965).

31. Bulen, W.A., and LeComte, J.R., "The nitrogenase system from *Azotobacter*: Two enzyme requirements for N₂ reduction, ATP dependent H₂ evolution and ATP hydrolysis," *Proc. Natl. Acad. Sci USA, 56,* 979–986 (1966).

32. Burgess, B.K., Stiefel, E.I., and Newton, W.E., "Oxidation-reduction properties and complexation reactions of the iron–molybdenum cofactor of nitrogenase," *J. Biol. Chem. 255* 353–356 (1980).

33. Cannon, M., Cannon, F., Buchanan-Wollaston, V., Ally, D., Ally, A., and Beynon, J., "The nucleotide sequence of the *nifJ* gene of *Klebsiella pneumoniae," Nucleic Acids Res. 16,* 11379 (1988).

34. Cannon, F.C., Riedel, G.E., and Ausubel, F.M., "Overlapping sequences of *Klebsiella pneumonia nif* DNA cloned and characterized," *Molec. Gen. Genet. 174,* 59–66 (1979).

35. Chang, C.L., Davis, L.C., Rider, M., and Takemoto, D.J., "Characterization of *nifH* mutations of *Klebsiella pneumoniae," J. Bacteriol. 170,* 4015–4022 (1988).

36. Chen, K.C.-K., Chen, J.-S., and Johnson, J.L., "Structural features of multiple *nifH*-like sequences and very biased codon usage in nitrogenase genes of *Clostridium pasteurianum," J. Bacteriol. 166* 162–172 81986).

37. Chen, C.-M., Misra, T.K., Silver, S., and Rosen, B.P., "Nucleotide sequence of the structural genes for an anion pump," *J. Biol. Chem. 261,* 15030–15038 (1986).

38. Conradson, S.D., Burgess, B.K., and Holm, R.H., "Fluorine-19 chemical shifts as probes of the structure and reactivity of the iron–molybdenum cofactor of nitrogenase," *J. Biol. Chem. 263,* 13743–13749 (1988).

39. Dahlen, J.V., Parejko, R.A., and Wilson, P.W., "Complementary functioning of two components from nitrogen-fixing bacteria," *J. Bacteriol. 98,* 325–326 (1969).

40. Dean, D.R., and Brigle, K.E., "*Azotobacter vinelandii* nifD- and *nifE*-encoded polypeptides share structural homology," *Proc. Natl. Acad. Sci. USA 82,* 5720–5723 (1985).

41. Dean, D.R., Brigle, K.E., May, H.D. and Newton, W.E., "Site-directed mutagenesis of the nitrogenases MoFe protein." In Bothe, H., de Bruijn, F.J., and Newton, W.E. (eds.), *Nitrogen Fixation: Hundred Years After,"* Stuttgart: Gustav Fischer, 1988, pp. 107–113.

42. de Zamaroczy, M., Delorme, F., Elmerich, E., "Regulation of transcription and promoter mapping of the structural genes for nitrogenase (*nifHDK*) of *Azospirillum brasilense* Sp7," *Molec. Gen. Genet. 220,* 88–94 (1989).

43. Defrancesco, N., and Potts, M., "Cloning of *nifHD* from *Nostoc commune* UTEX 584 and of a flanking region homologous to a part of *Azotobacter vinelandii nifU* Gene," *J. Bacteriol. 170,* 3297–3300 (1988).

44. Deistung, J., Cannon, F.C., Cannon, M.C., Hill, S., and Thorneley, R.N.F., "Electron transfer for nitrogenase in Klebsiella pneumoniae," *Biochem. J. 231,* 743–753 (1985).

45. Deistung, J., and Thorneley, R.N.F., "Electron transfer to nitrogenase," *Biochem. J. 239,* 69–75 (1986).

46. Denefle, P., Kush, A., Norel, F., Paquelin, A., and Elmerich, C., "Biochemical and genetic analysis of the *nifHDKE* region of *Rhizobium* ORS571," *Molec. Gen. Genet. 207,* 280–287 (1987).

47. Detroy, R.W., Witz, D.F., Parejko, R.A., and Wilson, P.W., "Reduction of N₂ by complementary functioning of two components from nitrogen-fixing bacteria," *Proc. Natl. Acad. Sci. USA 61,* 537–541 (1968).

48. Dilworth, M.J., "Acetylene reduction by nitrogen-fixing preparations from *Clostridium pasteurianum*," *Biochim. Biophys. Acta 127*, 285–294 (1966).

49. Dixon, R., Eady, R.R., Espin, G., Hill, S., Iaccarino, M., Kahn, D., and Merrick, M., "Analysis of regulation of *Klebsiella pneumoniae* nitrogen fixation (*nif*) gene cluster with gene fusions," *Nature 286*, 128–132 (1980).

50. Drummond, M.H., "The base sequence of the *nifF* gene of *Klebsiella pneumoniae* and homology of the predicted amino acid sequence of its protein product to other flavodoxins," *Biochem. J. 232*, 891–896 (1985).

51. Drummond, M.H., "Structure predictions and surface charge of nitrogenase flavodoxins from *Klebsiella pneumoniae* and *Axotobacter vinelandii*," *Eur. J. Biochem. 159*, 549–553 (1986).

52. Drummond, M., Whitty, P., and Wooton, J., "Sequence and domain relationships of *ntrC* and *nifA* from *Klebsiella pneumoniae*: Homologies to other regulatory proteins," *EMBO J. 5*, 441–447 (1986).

53. Drummond, M.H., and Wootton, J.C., "Sequence of *nifL* from *Klebsiella pneumoniae*: Mode of action and relationship to two families of regulatory proteins," *Molec. Microbiol. 1*, 37–44 (1987).

54. Dunham, W.R., Hagen, W.R., Braaksma, A., Grande, H.J., and Haaker, H., "The importance of quantitative Mossbauer spectroscopy of MoFe-protein from *Azotobacter vinelandii*," *Eur. J. Bioche,. 146*, 497–501 (1985).

55. Dusha, I., Kovalenko, S., Banfalvi, Z., and Kondorosi, A., "*Rhizobium meliloti* insertion element ISRm2 and its use for identification of the *fixX* gene," *J. Bacteriol. 169*, 1403–1409 (1987).

56. Eady, R.R., Smith, B.E., Cook, K.A., and Postgate, J.R., "Nitrogenase of *Klebsiella pneumoniae*: Purification and properties of the component proteins," *Biochem. J. 128*, 655–675 (1972).

57. Earl, C.D., Ronson, C.W., and Ausubel, F.M., "Genetic and structural analysis of the *Rhizobium meliloti fixA, fixB, fixC,* and *fixX* genes," *J. Bacteriol. 169*, 1127–1136 (1987).

58. Ebeling, S., Hahn, M., Fischer, H.-M., and Hennecke, H., "Identification of *nifE*-, *nifN*- and *nifS*-like genes in *Bradyrhizobium japonicum*," *Molec. Gen. Genet. 207* 503–508 (1987).

59. Ebeling, S., Noti, J.D., and Hennecke, H., "Identification of a new *Bradyrhizobium japonicum* gene (*frxA*) encoding a ferredoxin-like protein," *J. Bacteriol. 170*, 1999–2001 (1988).

60. Elmerich, C., Houmard, J., Sibold, L., Manheimer, I., and Charpin, N., "Genetic and biochemical analysis of mutants induced by bacteriophage Mu DNA integration into *Klebsiella pneumoniae* nitrogen fixation genes," *Biochimie 165*, 181–189 (1978).

61. Emerich, D.W., and Burris, R.H., "Interactions of heterologous nitrogenase components that generate catalytically inactive complexes," *Proc. Natl. Acad. Sci. USA 73*, 4369–4373 (1976).

62. Emerich, D.W., and Burris, R.H., "Nitrogenase from *Bacillus polymyxa*: Purification and properties of the component proteins," *Biochim. Biophys. Acta 536*, 172–183 (1978).

63. Emerich, D.W., and Burris, R.H., "Complementary functioning of the component proteins of nitrogenase from several bacteria," *J. Bacteriol. 134*, 936–943 (1978).

64. Emerich, D.W., Ljones, T., and Burris, R.H., "Nitrogenase: Properties of the catalytically inactive complex between *Azotobacter vinelandii* MoFe protein and the *Clostridium pasteurianum* Fe protein," *Biochim,. Biophys. Acta 527*, 359–369 (1978).

65. Euler, W.B., Martinsen, J., McDonald, J.W., Watt, G.D., and Wang, Z.-C., "Double integration and titration of the electron paramagnetic resonance signal in the molybdenum iron protein of *Azotobacter vinelandii*," *Biochemistry 23*, 3021–3024 (1984).

66. Evans, D., Jones, R., Woodley, P., and Robson, R., "Further analysis of nitrogen fixation (*nif*) genes in *Azotobacter chroococcum*: Identification and expression in *Klebsiella pneumoniae* of *nifS*, *nifV*, *nifM* and *nifB* genes and localization of *nifE/N-*, *nifU*, *nifA* and *fixABC*-like genes," *J. Gen. Microbiol. 134*, 931–942 (1988).

67. Evans, D.J., and Leigh, G.J., "Serinato-, tyrosanoto-, and prolinato-derivatives of Fe₄S₄ clusters," *J. Chem. Soc. Chem. Commun. 1988*, 395–396 (1988).

68. Fani, R., Allotta, G., Bazzicalupo, M., Ricci, F., Schipani, C., and Polsinelli, M., "Nucleotide sequence of the gene encoding the nitrogenase iron protein (*nifH*) of *Azospirillum brasilense*," *Molec. Gen. Genet. 220*, 81–87 (1989).

69. Filler, W.A., Kemp, R.M., Ng, J.C., Hawkes, T.R., Dixon, R.A., and Smith, B.E., "The *nifH* gene product is required for the synthesis or stability of the iron–molybdenum cofactor of nitrogenase from *Klebsiella pneumoniae*," *Eur. J. Biochem. 160*, 371–377 (1986).

70. Fischer, H.-M., Bruderer, T., and Hennecke, H., "Essential and nonessential domains in the *Bradyrhizobium japonicum* NifA protein: Identification of indispensible cysteine residues potentially involved in redox activity and/or metal binding," *Nucleic Acids Res. 16*, 2207–2224 (1988).

71. Fitzmaurice, W.P., Saari, L.L., Lowery, R.G., Ludden, P.W., and Roberts, G.P., "Genes coding for the reversible ADP-ribosylation system of dinitrogenase reductase from *Rhodospirillum rubrum*," *Molec. Gen. Genet. 218*, 340–347 (1989).

72. Fuhrmann, M., and Hennecke, H., "*Rhizobium japonicum* nitrogenase Fe protein gene (*nifH*)," *J. Bacteriol. 158*, 1005–1011 (1984).

73. Fujita, Y., Takahashi, Y., Kohchi, T., Ozeki, H., Ohyama, K., and Matsubara, H., "Identification of a novel *nifH*-like (*frxC*) protein in chloroplasts of the liverwort *Marchantia polymorpha*," *Plant Molec. Biol. 13*, 551–562 (1989).

74. Galimand, M., Perroud, B., Delorme, F., Paquelin, A., Vieille, C., Bozouklian, H., and Elmerich, C., "Identification of DNA regions homologous to nitrogen fixation genes *nifE*, *nifUS* and *fixABC* in *Azospirillum brasilense* Sp7," *J. Gen. Microbiol. 135*, 1047–1059 (1989).

75. Gemoets, J.P., Bravo, M., McKenna, C.E., Leigh, G.J., and Smith, B.E., "Reduction of cyclopropene by NifV-and wild-type nitrogenases from *Klebsiella pneumoniae*," *Biochem. J. 258*, 487–491 (1989).

76. George, S.J., Armstrong, F.A., Hatchikian, E.C., and Thomson, A.J., "Electrochemical and spectroscopic characterization of the conversion of the 7Fe into the 8Fe form of ferredoxin III from *Desulfovibrio africanus*. Identification of a [4Fe–4S] cluster with one non-cysteine ligand," *Biochem. J. 264*, 275–284 (1989).

77. Gilson, E., Nikaido, H., and Hofnung, M., "Sequence of the *malK* gene in *E. coli* K12," *Nucleic Acids Res. 10*, 7449–7458 (1982).

78. Golden, J.W., Robinson, S.J., and Haselkorn, R., "Rearrangement of nitrogen fixation genes during heterocyst differentiation in the cyanobacterium *Anabaena*," *Nature 314*, 419–423 (1985).

79. Gosink, M.M., Franklin, N.M., and Roberts, G.P., "The product of the *Klebsiella pneumoniae nifX* gene is a negative regulator of the nitrogen fixation (*nif*) regulon," *J. Bacteriol. 172*, 1441–1447 (1990).

80. Govensky, D., and Zamir, A., "Structure–function relationships in the α subunit of *Klebsiella pneumoniae* nitrogenase MoFe protein from analysis of *nifD* mutants," *J. Bacteriol. 171*, 5729–5735 (1989).

81. Gronger, P., Manian, S., Reilander, H., O'Connell, M., Priefer, U.B., and Puhler, A., "Organization and partial sequence of a DNA region of the *Rhizobium leguminosarum* symbiotic plasmid pRL6JI containing the genes *fixABC, nifA, nifB* and a novel open reading frame," *Nucleic Acids. Res. 15*, 31–49 (1987).

82. Gubler, M., and Hennecke, H., "*fixA, B,* and *C* genes are essential for symbiotic and free-living, microaerobic nitrogen fixation," *FEBS Lett. 220*, 186–192 (1986).

83. Haaker, H., and Klugkist, J., "The bioenergetics of electron transport to nitrogenase," *FEMS Microbiol. Rev. 46*, 57–71 (1987).

84. Haaker, H., and Veeger, C., "Involvement of the cytoplasmic membrane in nitrogen fixation by *Azotobacter vinelandii*," Eur. J. Biochem. *77*, 1–10 (1977).

85. Hageman, R.V., and Burris, R.H., "Nitrogenase and nitrogenase reductase associate and dissociate with each catalytic cycle," *Proc. Natl. Acad. Sci. USA 75*, 2699–2702 (1978).

86. Hageman, R.V., and Burris, R.H., "Changes in the EPR signal of dinitrogenase from *Azotobacter vinelandii* during the lag period before hydrogen evolution begins," *J. Biol. Chem. 254*, 11189–11192 (1979).

87. Hagen, W.R., Dunham, W.R., Braaksma, A., and Haaker, H., "On the prosthetic group(s) of component II from nitrogenase: EPR of the Fe protein from *Azotobacter vinelandii*," *FEBS Lett. 187*, 146–150 (1985).

88. Hagen, W.R., Eady, R.R., Dunham, W.R., and Haaker, H., "A novel $S = 3/2$ EPR signal associated with native Fe-proteins of nitrogenase," *FEBS Lett. 189*, 250–254 (1985).

89. Hagen, W.R., Wassink, H., Eady, R.R., Smith, B.E., and Haaker, H., "Quantitative EPR of an $S = 7/2$ system in thionine-oxidized MoFe proteins of nitrogenase: A redefinition of the P-cluster concept," *Eur. J. Biochem. 169*, 457–465 (1987).

90. Hausinger, R.P., and Howard, J.B., "The amino acid sequence of the nitrogenase iron protein from *Azotobacter vinelandii*," *J. Biol. Chem. 257*, 2483–2490 (1982).

91. Hausinger, R.P., and Howard, J.B., "Thiol reactivity of the nitrogenase Fe-protein from *Azotobacter vinelandii*," *J. Biol. Chem. 258*, 13486–13492 (1983).

92. Hawkes, T.R., McLean, P.A., and Smith, B.E., "Nitrogenase from *nifV* mutants of *Klebsiella pneumoniae* contains an altered form of the iron–molybdenum cofactor," *Biochem. J. 217*, 317–321 (1984).

93. Hawkes, T.R., and Smith, B.E., "Purification and characterization of the inactive MoFe protein (NifB⁻Kpl) of the nitrogenase from *nifB* mutants of *Klebsiella pneumoniae*," *Biochem. J. 209*, 43–50 (1983).

94. Hearst, J.E., Alberti, M., and Doolittle, R.F., "A putative nitrogenase reductase gene found in the nucleotide sequences from the photosynthetic gene cluster of *R. capsulata*," *Cell 40*, 219–220 (1985).

95. Hill, S., and Kavanagh, E.P., "Roles of *nifF* and *nifJ* gene products in electron transport to nitrogenase in *Klebsiella pneumoniae*," *J. Bacteriol. 141*, 470–475 (1980).

96. Hiratsuka, K., and Roy, K.L., "Sequence of a 1.4 kb *Eco*RI fragment from *Azotobacter vinelandii nif* DNA," *Nucleic Acids Res. 16*, 1207 (1988).

97. Holland, D., Zilberstein, A., Zamir, A., and Sussman, J., "A quantitative approach to sequence comparisons of nitrogenase MoFe protein α-and β-subunits including the newly sequenced *nifK* gene from *Klebsiells pneumoniae*," *Biochem. J. 247*, 277–285 (1987).

98. Hontelez, J.G.J., Lankhorst, R.K., Katinakis, P., van den Bos, R.C., and van Kammen, A., "Characterization and nucleotide sequence of a novel gene *fixW* upstream of the *fixABC* operon in *Rhizobium leguminosarum*," *Molec. Gen. Genet. 218*, 536–544 (1989).

99. Hoover, T.R., Imperial, J., Ludden, P.W., and Shah, V.K., "Homocitrate cures the NifV⁻phenotype in *Klebsiella pneumoniae*," *J. Bacteriol. 170*, 1978–1979 (1988).

100. Hoover, T.R., Imperial, J., Ludden, P.W., and Shah, V.K., "Homocitrate is a component of the iron–molybdenum cofactor of nitrogenase," *Biochemistry 28*, 2768–2771 (1989).

101. Hoover, T.R., Imperial, J., Liang, J., Ludden, P.W., and Shah, V.K., "Dinitrogenase with altered substrate specificity results from the use of homocitrate analogues for *in vitro* synthesis of the iron–molybdenum cofactor," *Biochemistry 27*, 3647–3652 (1988).

102. Hoover, T.R., Robertson, A.D., Cerny, R.L., Hayes, R.N., Imperial, J., Shah, V.K., and Ludden, P.W., "Identification of the V factor needed for synthesis of the iron–molybdenum cofactor of nitrogenase as homocitrate," *Nature 329*, 855–857 (1987).

103. Hoover, T.R., Shah, V.K., Roberts, G.P., and Ludden, P.W., "*nifV*-dependent, low-molecular-weight factor required for *in vitro* synthesis of iron–molybdenum cofactor of nitrogenase," *J. Bacteriol. 167*, 999–1003 (1986).

104. Houmard, J., Bogusz, D., Bigault, R., and Elmerich, C., "Characterization and kinetics of the biosynthesis of some nitrogen fixation (*nif*) gene products in *Klebsiella pneumoniae*," *Biochimie 62*, 267–275 (1980).

105. Howard, J.B., Davis, R., Moldenhauer, B., Cash, V.L., and Dean, D., "Fe:S cluster ligands are the only cysteines required for nitrogenase Fe protein activities," *J. Biol. Chem. 264*, 11270–11274 (1989).

106. Howard, K.S., McLean, P.A., Hansen, F.B., Lemley, P.V., Koblan, K.S., and Orme-Johnson W.H., "*Klebsiella pneumoniae nifM* gene product is required for stabilization and activation of nitrogenase iron protein in *Escherichia coli*," *J. Biol. Chem. 261*, 772–778 (1986).

107. Huynh, B.H., Henzl, M.T., Christner, J.A., Zimmerman, R., Orme-Johnson, W.H., and Munck, E., "Nitrogenase XII. Mossbauer studies of the MoFe protein from *Clostridium pasteurianum* W5," *Biochim. Biophys. Acta 623*, 124–138 (1980).

108. Iismaa, S.E., and Watson, J.M., "A gene upstream of the *Rhizobium trifolii nifA* gene encodes a ferredoxin-like protein," *Nucleic Acids Res. 15*, 3180 (1987).

109. Imam, S., and Eady, R.R., "Nitrogenase of *Klebsiella pneumoniae*: Reductant-independent ATP hydrolysis and the effect of pH on the efficiency of coupling of ATP hydrolysis to substrate reduction," *FEBS Lett. 110*, 35–38 (1980).

110. Imperial, J., Hoover, T.R., Madden, M.S., Ludden, P.W., and Shah, V.K., "Substrate reduction properties of dinitrogenase activated *in vitro* are dependent upon the presence of homocitrate or its analogues during iron–molybdenum cofactor synthesis," *Biochemistry 28*, 7796–7799 (1989).

111. Imperial, J., Ugalde, R.A., Shah, V.K., and Brill, W.J., "Role of the *nifQ* gene product in the incorporation of molybdenum into nitrogenase in *Klebsiella pneumoniae*," *J. Bacteriol. 158*, 187–194 (1984).

112. Imperial, J., Ugalde, R., Shah, V.K., and Brill, W.J., "Mol⁻ mutants of *Klebsiella pneumoniae* requiring high levels of molybdate for nitrogenase activity," *J. Bacteriol. 163*, 1285–1287 (1985).

113. Ioannidis, I., and Buck, M., "Nucleotide sequence of the *Klebsiella pneumoniae nifD* gene and predicted amino acid sequence of the α-subunit of the nitrogenase MoFe protein," *Biochem. J. 247*, 287–291 (1987).

114. Jacobson, M.R., Brigle, K.E., Bennett, L.T., Setterquist, R.A., Wilson, M.S., Cash, V.L., Beynon, J., Newton, W.E., and Dean, D.R., "Physical and genetic map of the major *nif* gene cluster from *Azotobacter vinelandii*," *J. Bacteriol. 171*, 1017–1027 (1989).

115. Jacobson, M.R., Cash, V.L., Weiss, M.C., Laird, N.F., Newton, W.E., and Dean, D.R., "Biochemical and genetic analysis of the *nifUSVWZM* cluster from *Azotobacter vinelandii*," *Molec. Gen. Genet. 219*, 49–57 (1989).

116. Jeng, D.Y., Morris, J.A., and Mortenson, L.E., "The effect of reductant in inorganic phosphate release from adenosine 5' triphosphate by purified nitrogenase of *Clostridium pasteurianum*," *J. Biol. Chem. 245*, 2809–2813 (1970).

117. Jensen, B.B., and Burris, R.H., "N₂O as a substrate and as a competitive inhibitor of nitrogenase," *Biochemistry 25*, 1083–1088 (1986).

118. Joerger, R.D., and Bishop, P.E., "Nucleotide sequence and genetic analysis of the *ninfB-nifQ* region from *Azotobacter vinelandii*," *J. Bacteriol. 170*, 1475–1487 (1988).

119. Joeger, R.D., Jacobson, M.R., Premakumar, R., Wolfinger, E.D., and Bishop, P.E., "Nucleotide sequence and mutational analysis of the structural genes (*anfHDGK*) for the second alternative nitrogenase from *Azotobacter vinelandii*," *J. Bacteriol. 171*, 1075–1086 (1989).

120. Johann, S., and Hinton, S.M., "Cloning and nucleotide sequence of the *chlD* locus," *J. Bacteriol. 169*, 1911–1916 (1987).

121. Jones, R., and Haselkorn, R., "The DNA sequence of the *Rhodobacter capsulatus nifH* gene," Nucleic Acids Res. *16*, 8735 (1988).

122. Jones, R., and Haselkorn, R., "The DNA sequence of the *Rhodobacter capsulatus ntrA*, *ntrB* and *ntrC* gene analogues required for nitrogen fixation," *Molec. Gen. Genet. 215*, 507–516 (1989).

123. Jones, R., Taylor, W., and Robson, R., "A model for the MgATP binding site of nitrogenase iron protein," In *Progess*, Evans, H.J., Bottomley, P.J., Newton W.E. (eds.), *Nitrogen Fixation Research*. Dordrecht: Martinus Nijhoff, 1985, p. 634.

124. Jones, R., Woodley, P., and Robson, R., "Cloning and organization of some genes for nitrogen fixation from *Azotobacter chroococcum* and their expression in *Klebsiella pneumoniae*," *Molec. Gen. Genet. 197*, 318–327 (1984).

125. Kahn, D., David, M., Domergue, O., Daveran, M.-L., Ghai, J., Hirsch, P.R., and Batut, J., "*Rhizobium meliloti fixGHI* sequence predicts involvement of a specific cation pump in symbiotic nitrogen fixation," *J. Bacteriol. 171*, 929–939 (1989).

126. Kaluza, K., and Hennecke, H., "Fine structure analysis of the *nifHDK* operon encoding the α and β subunits of dinitrogenase from *Rhizobium japonicum*," *Molec. Gen. Genet. 196*, 35–42 (1984).

127. Kaminski, P.A., Norel, F., Desnoues, N., Kush, A., Salzano, G., and Elmerich, C., "Characterization of the *fixABC* region of *Azorhizobium caulinodans* ORS571 and identification of a new nitrogen fixation gene," *Molec. Gen. Genet. 214*, 496–502 (1988).

128. Kanatzidis, M.G., Hagen, W.R., Dunham, W.R., Lester, R.K., and Coucouvanis, D., "Metastable Fe/S cluster," *J. Am. Chem Soc. 107*, 953–961 (1985).

129. Kelly, M., "Comparisons and cross reactions of nitrogenase from *Klebsiella pneumoniae, Azotobacter chroococcum* and *Bacillus polymyxa,*" *Biochim. Biophys. Acta 191*, 527–540 (1969).

130. Kennedy, C., Eady, R.R., Kondorosi, E., and Rekosh, D.K., "The molybdenum–iron protein of *Klebsiella pneumoniae,*" *Biochem. J. 155*, 383–389 (1976).

131. Kennedy, C., Gamal, R., Humphrey, R., Ramos, J., Brigle, K., and Dean, D., "The *nifH, nifM* and *nifN* genes of *Azotobacter vinelandii,* Characterisation by Tn5 mutagenesis and isolation from pLAFR1 gene banks," *Molec. Gen. Genet. 205*, 318–325 (1986).

132. Kent, H.M., Ioannidis, I., Gormal, C., Smith, B.E., and Buck, M., "Site-directed mutagenesis of the *Klebsiella pneumoniae* nitrogenase," *Biochem. J. 264*, 257–264 (1989).

133. Klipp, W., Masepohl, B., and Puhler, A., "Identification and mapping of nitrogen fixation genes of *Rhodobacter capsulatus,* duplication of the *nifA/nifB* region," *J. Bacteriol. 170*, 693–699 (1988).

134. Klipp, W., Reilander, H., Schluter, H., Schluter, A., Krey, R., and Puhler, A., "The *Rhizobium meliloti fdxN* gene encoding a ferredoxinlike protein is necessary for nitrogen fixation and is cotranscribed with *nifA* and *nifB,*" *Molec. Gen. Genet. 216*, 293–303 (1989).

135. Klugkist, J., Haaker, H., and Veeger, C., "Studies on the mechanism of electron transport to nitrogenase in *Azotobacter vinelandii,*" *Eur. J. Biochem. 155*, 41–46 (1986).

136. Klugkist, J., Voorberg, J., Haaker, H., and Veeger, C., "Characterization of three different flavodoxins from *Azotobacter vinelandii,*" *Eur. J. Biochem. 155*, 33–40 (1986).

137. Kurtz, D.M. Jr., McMillan, R.S., Burgess, B.K., Mortenson, L.E., and Holm, R.M., "Identification of iron–sulfur centers in the iron–molybdenum proteins of nitrogenase," *Proc. Natl. Acad. Sci. USA 76*, 4986–4989 (1979).

138. Lammers, P.J., and Haselkorn, R., "Sequence of the *nifD* gene coding for the α-subunit of dinitrogenase from the cyanobacterium *Anabaena,*" *Proc. Natl. Acad. Sci. USA 80*, 4723–4727 (1983).

139. Li, J.-G., Burgess, B.K., and Corbin, J.L., "Nitrogenase reactivity: Cyanide as a substrate and inhibitor," *Biochemistry 21*, 4393–4402 (1982).

140. Liang, J., and Burris, R.H., "N_2O reduction and HD formation by nitrogenase from a *nifV* mutant of *Klebsiella pneumoniae,*" *J. Bacteriol. 171*, 3176–3180 (1989).

141. Lindahl, P.A., Day, E.P., Kent, T.A., Orme-Johnson, W.H., and Munck, E., "Mossbauer, EPR, and magnetization studies of the *Azotobacter vinelandii* Fe protein," *J. Biol. Chem. 260*, 11160–11173 (1985).

142. Ljones, T., and Burris, R.H., "ATP hydrolysis and electron transfer in the nitrogenase reaction with different combinations of iron protein and molybdenum–iron protein," *Biochim. Biophys. Acta 275*, 93–101 (1972).

143. Ljones, T., and Burris, R.H.; "Nitrogenase: The reaction between the Fe protein and bathophenanthrolinedisulfonate as a probe for interactions with MgATP," *Biochemistry 17*, 1866–1872 (1978).

144. Lowe, D.J., "Simulation of the electron-paramagnetic-resonance spectrum of the iron-protein of nitrogenase," *Biochem. J. 175*, 955–957 (1978).

145. Lowery, R.G., Chang, C.L., Davis, L.C., McKenna, M.-C., Stephens, P.J., and Ludden, P.W., "Substitution of histidine for arginine-101 of dinitrogenase reductase disrupts electron transfer to dinitrogenase," Biochemistry *28*, 1206–1212 (1989).

146. Lowery, R.G., and Ludden, P.W., "Purification and properties of dinitrogenase reductase ADP-ribosyltransferase from the photosynthetic bacterium *Rhodospirillum rubrum*," *J. Biol. Chem. 263*, 16714–16719 (1988).

147. Ludwig, A., Jarchau, T., Benz, R., and Goebel, W., "The repeat domain of *Escherichia coli* haemolysis (HylA) is responsible for its Ca^{2+}- dependent binding to erythrocytes," *Molec. Gen. Genet. 214*, 553–561 (1988).

148. Lundell, D.J., and Howard, J.B., "Isolation and partial characterization of two different subunits of the molybdenum–iron protein of *Azotobacter vinelandii* nitrogenase," *J. Biol. Chem. 253*, 3422–3426 (1978).

149. MacNeil, T., MacNeil, D., Roberts, G.P., Supiano, M.A., and Brill, W.J., "Fine-structure mapping and complementation analysis of *nif* (nitrogen fixation) genes in *Klebsiella pneumoniae*," *J. Bacteriol. 136*, 252–266 (1978).

150. Makaroff, C.A., Paluh, J.L., and Zalkin, H., "Mutagenesis of ligands to the [4Fe–4S] center of *Bacillus subtilis* glutamine phosphoribosylphrophosphate amidotransferase," *J. Biol. Chem. 261*, 11416–11423 (1986).

151. Martin, A.E., Burgess, B.K., Iismaa, S.E., Smartt, C.T., Jacobson, M.R., and Dean, D.R., "Construction and characterization of an *Azotobacter vinelandii* strain with mutations in the genes encoding flavodoxin and ferredoxin I," *J. Bacteriol. 171*, 3162–3167 (1989).

152. Martin, A., Burgess, B.K., Stout, C.D., Cash, V., Dean, D.R., Jensen, G.M., and Stephens, P.J., "Site directed mutagenesis of *Azotobacter vinelandii* ferredoxin I: [Fe : S] cluster driven rearrangement," *Proc. Natl. Acad. Sci. USA 87*, 598–602 (1990).

153. Masepohl, B., Klipp, W., and Puhler, A., "Genetic characterization and sequence analysis of the duplicated *nifA/nifB* gene region of *Rhodobacter* capsulatus," *Molec. Gen. Genet. 212*, 27–37 (1988).

154. Mazur, B.J., and Chui, C.-F., "Sequence of the gene coding for the β-subunit of dinitrogenase from the blue-green alga *Anabaena*," *Proc. Natl. Acad. Sci. USA 79*, 6782–6786 (1982).

155. Mazur, B.J., Rice, D., and Haselkorn, R., "Identification of blue-green algal nitrogen fixation genes by using heterologous DNA hybridization probes," *Proc. Natl. Acad. Sci. USA 77*, 186–190 (1980).

156. McKenna, C.E., and Huang, C.W., "*In vivo* reduction of cyclopropene by *Azotobacter vinelandii* nitrogenase," *Nature 280*, 609–611 (1979).

157. McLean, P.A., and Dixon, R.A., "Requirement of *nifV* gene for production of wild-type nitrogenase enzyme in *Klebsiella pneumoniae*," *Nature 292*, 655–656 (1981).

158. McLean, P.A., Papaefthymiou, V., Orme-Johnson, W.H., and Munck, E., "Isotopic hybrids of nitrogenase: Mossbauer study of MoFe protein with selective ^{57}Fe enrichment of the P-cluster," *J. Biol. Chem. 262*, 12900–12903 (1987).

159. McLean, P.A., Wink, D.A., Chapman, S.K., Hickman, A.B., McKillop, D.M., and Orme-Johnson, W.H., "A new method for extraction of iron–molybdenum cofactor (FeMoco) from nitrogenase adsorbed to DEAE-cellulose. 1. Effects of anions, cations, and preextraction treatments," *Biochemistry 28*, 9402–9406 (1989).

160. Merrick, M., Filser, M., Dixon, R., Elmerich, C., Sibold, L., and Houmard, J., "The use of translocatable genetic elements to construct a fine-structure map of the *Klebsiella pneumoniae* nitrogen fixation (*nif*) gene cluster," *J. Gen. Microbiol. 117*, 509–520 (1980).

161. Mevarech, M., Rice, D., and Haselkorn, R., "Nucleotide sequence of a cyanobacterial *nifH* gene coding for nitrogenase reductase," *Proc. Natl. Acad. Sci. USA 77*, 6476–6480 (1980).

162. Meyer, J., Gaillard, J., and Moulis, J.-M., "Hydrogen-1 nuclear magnetic resonance of the nitrogenase iron protein (Cp2) from *Clostridium pasteurianum*," *Biochemistry 27*, 6150–6156 (1988).

163. Miller, R.W., and Eady, R.R., "Cyanamid: A new substrate for nitrogenase," *Biochim. Biophys. Acta 952*, 290–296 (1988).

164. Moreno-Vivian, C., Hennecke, S., Puhler, A., and Klipp, W., "Open reading frame 5 (ORF5), encoding a ferredoxinlike protein, and *nifQ* are cotranscribed with *nifE*, *nifN*, *nifX*, and ORF4 in *Rhodobacter capsulatus*," *J. Bacteriol. 171*, 2591–2598 (1989).

165. Moreno-Vivian, C., Schmehl, M., Masepohl, B., Arnold, W., and Klipp, W., "DNA sequence and genetic analysis of the *Rhodobacter capsulatus nifENX* gene region: Homology between *nifX* and *nifB* suggests involvement of *nifX* in processing of the iron–molybdenum cofactor," *molec. Gen. Genet. 216*, 353–363 (1989).

166. Morgan, T.V., Lundell, D.J., and Burgess, B.K., "*Azotobacter vinelandii* ferredoxin I: Cloning, sequencing and mutant analysis," *J. Biol. Chem. 263*, 1370–1375 (1988).

167. Morgan, T.V., Prince, R.C., and Mortenson, L.E., "Electrochemical titration of the $S = 3/2$ and $S\ 1/2$ states of the iron protein of nitrogenase," *FEBS Lett. 206*, 4–8 (1985).

168. Mozen, M.M., and Burris, R.H., "The incorporation of ^{15}N-labelled nitrous oxide by nitrogen fixing agents," *Biochim. Biophys. Acta 14*, 577–578 (1954).

169. Mulligan, M.E., Buikema, W.J., and Haselkorn, R., "Bacterial-type ferredoxin genes in the nitrogen fixation regions of the Cyanobacterium *Anabaena* sp. Strain PCC 7120 and *Rhizobium meliloti*," *J. Bacteriol. 170*, 4406–4410 (1988).

170. Mulligan, M., and Haselkorn, R., "Nitrogen fixation (*nif*) genes of the cyanobacterium *Anabaena* species PCC 7120," *J. Biol. Chem. 264*, 19200–19207 (1989).

171. Munck, E., Rhodes, H., Orme-Johnson, W.H., Davis, L.C., Brill, W.J., and Shah, V.K., "Nitrogenase. VIII. Mossbauer and EPR spectroscopy," *Biochim. Biophys. Acta 400*, 32–53 (1975).

172. Murphy, P.M., and Koch, B.L., "Compatibility of the components of nitrogenase from soybean bacteriods and free-living nitrogen-fixing bacteria," *Biochim. Biophys. Acta 253*, 295–297 (1971).

173. Murrell, S.A., Lowery, R.G., and Ludden, P.W., "ADP-ribosylation of dinitrogenase reductase from *Clostridium pasteurianum* prevents its inhibition of nitrogenase from *Azotobacter vinelandii*," *Biochem. J. 251*, 609–612 (1988).

174. Nieva-Gomez, D., Roberts, G.P., Klevickis, S., and Brill, W.J., "Electron transport to nitrogenase in *Klebsiella pneumoniae*," *Prorc. Natl. Acad. Sci. USA 77*, 2555–2558 (1980).

175. Norel, F., and Elmerich, C., "Nucleotide sequence and functional analysis of the two *nifH* copies of *Rhizobium ORS571*," *J. Gen. Microbiol. 133*, 1563–1567 (1987).

176. Normand, P., and Bousquet, J., "Phylogeny of nitrogenase sequences in *Frankia* and other nitrogen-fixing microorganisms," *J. Molec. Evol. 29*, 436–447 (1989).

177. Normand, P., Simonet, P., and Bardin, R., "Conservation of *nif* sequences in *Frankia,*" *Molec. Gen. Genet. 213,* 238–246 (1988).

178. Noti, J.D., Folkerts, O., Turken, A., and Szalay, A., "Organization and characterization of genes essential for symbiotic nitrogen fixation from *Bradyrhizobium japonicum* I110," *J. Bacteriol. 167,* 74–783 (1986).

179. Ohyama, K., Fukuzawa, H., Kohchi, T., Shirai, H., Sano, T, Sano., S, Umesono, H., Shiki, Y., Takeuchi, M., Chang, Z., Aota, S.-I., Inokuchi, H., and Ozeki, H., "Chloroplast gene organization deduced from complete sequence of liverwort *Marchantia polymorpha* chloroplast DNA," *Nature 322,* 572–574 (1986).

180. Orme-Johnson, W.H., Hamilton, W.D., Jones, T.L., Tso, M.-Y.W., Burris, R.H., Shah, V.K., and Brill, W.J., "Electron paramagnetic resonance of nitrogenase and nitrogenase components from *Clostridium pasteurianum* W5 and *Azotobacter vinelandii* OP," *Proc. Natl. Acad. Sci. USA 69,* 3142–3145 (1972).

181. Orme-Johnson, W. H., and Holm, R. H., "Identification of Fe–S clusters in proteins," *Metho. Enzymol. 53,* 268–274 (1978).

182. Pagani, S., Eldridge, M., and Eady, R. R., "Nitrogenase of *Klebsiella pneumoniae.* Rhodanese-catalysed restoration of activity of the inactive 2Fe species of the Fe protein," *Biochem. J. 244,* 485–488 (1987).

183. Paul, W., and Merrick, M., "The nucleotide sequence of the *nifM* gene of *Klebsiella pneumoniae* and identification of a new *nif* gene: *nifZ,*" *Eur. J. Biochem. 170,* 259–265 (1987).

184. Paul, W., and Merrick, M., "The roles of the *nifW, nifZ* and *nifM* genes of *Klebsiella pneumoniae* in nitrogenase biosynthesis," *Eur. J. Biochem. 178,* 675–682 (1989).

185. Paustian, T. D., Shah, V. K., and Roberts, G. P., "Purification and characterization of the *nifN* and *nifE* gene products from *Azotobacter vinelandii* mutant UW45," *Proc. Natl. Acad. Sci. USA 86,* 6082–6086 (1989).

186. Paustian, T. D., Shah, V. K., and Roberts, G. P., "Apo-dinitrogenase: Purification, association with a 20 kDa protein, and activation by FeMoco in the absence of dinitrogenase reductase," *Biochemistry, 29,* 3515–3522 (1990).

187. Pope, M. R., Murrell, S. A., and Ludden, P. W., "Covalent modification of the iron protein of nitrogenase from *Rhodospirillum rubrum* by adenosine diphosphoribosylation of a specific arginine residue," *Proc. Natl. Acad. Sci. USA 82,* 3173–3177 (1985).

188. Pretorius, I.-M., Rawlings, D. E., "O'Neill, E. G., Jones, W. A., Kirby, R., and Woods, D. R., "Nucleotide sequence of the gene eroding the nitrogenase iron protein of *Thiobacillus ferrooxidans,*" *J. Bacteriol. 169,* 367–370 (1987).

189. Puhler, A., Burkardt, H. J., and Klipp, W., "Cloning of the entire region for nitrogen fixation from *Klebsiella pneumoniae* on a multicopy plasmid vehicle in *Escherichia coli,*" *Molec. Gen. Genet. 176,* 17–24 (1979).

190. Puhler, A., and Klipp, W., "Fine structure analysis of the gene region for N₂-fixation (*nif*) of *Klebsiella pneumoniae.* In Bothe and Trebst (eds.), *Biology of Inorganic Nitrogen and Sulfur.* Berlin: Springer-Verlag, 1980, pp. 276–286.

191. Quinto, C., de la Vega, H., Flores, M., Leemans, J., Cevallos, M. A., Pardo, M. A., Azpiroz, R., de Lordes Girard, M., Calva, E., and Palacios, R., "Nitrogenase reductase: a functional multigene family in *Rhizobium phaseoli,*" *Proc. Natl. Acad. Sci. USA 82,* 1170–1174 (1985).

192. Raina, R., Reddy, M. A., Ghosal, D., and Das, H. K., "Characterization of the gene for the Fe-protein of the vanadium dependent alternative nitrogenase of *Azotobacter vinelandii* and construction of a Tn5 mutant," *Molec. Gen. Genet. 214,* 121–127 (1988).

193. Rawlings, D. E., "Sequence and structural analysis of the α- and β-dinitrogenase subunits of *Thiobacillus ferrooxidans*," *Gene 69*, 337–343 (1988).

194. Rawlings, J., Shah, V. K., Chisnell, J. R., Brill, W. J., Zimmermann, R., Munck, E., and Orme-Johnson, W. H., "Novel metal cluster in the iron–molybdenum cofactor of nitrogenase," *J. Biol. Chem. 253*, 1001–1004 (1978).

195. Rees, D. C., and Howard, J. B., "Crystallization of the *Azotobacter vinelandii* nitrogenase Fe protein," *J. Biol. Chem. 258*, 12733–12734 (1983).

196. Ricca, E., and Calvo, J. M., "The nucleotide sequence of *leuA* from Salmonella typhimurium," *Nucleic Acids Res. 18*, 1290 (1990).

197. Rice, D., Mazur, B. J., and Haselkorn, R., "Isolation and physical mapping of nitrogen fixation genes from the cyanobacterium *Anabaena* 7120," *J. Biol. Chem. 257*, 13157–13163 (1982).

198. Riedel, G. E., Ausubel, F. M., and Cannon, F. M., "Physical map of chromosomal nitrogen fixation (*nif*) genes of *Klebsiella pneumoniae*," *Proc. Natl. Acad. Sci. USA 76*, 2866–2870 (1979).

199. Roberts, G. P., and Brill, W. J., "Gene-product relationships of the *nif* regulon of *Klebsiella pneumoniae*," *J. Bacteriol. 144*, 210 216 (1980).

200. Roberts, G. P., MacNeil, T., MacNeil, D., and Brill, W. J., "Regulation and characterization of protein products coded by the *nif* (nitrogen fixation) genes of *Klebsiella pneumoniae*," *J. Bacteriol. 136*, 267–279 (1978).

201. Robinson, A. C., Burgess, B. K., and Dean, D. R., "Activity, reconstitution, and accumulation of nitrogenase components in *Azotobacter vinelandii* mutant strains containing defined deletions within the nitrogenase structural gene cluster," *J. Bacteriol. 166*, 180–186 (1986).

202. Robinson, A. C., Dean, D. R., and Burgess, B. K., "Iron–molybdenum cofactor biosynthesis in *Azotobacter vinelandii* requires the iron protein of nitrogenase," *J. Biol. Chem. 262*, 14327–14332 (1987).

203. Robinson, A. C., Chun, T. W., Li, J.-G., and Burgess, B. K., "Iron–molybdenum cofactor insertion into the apo-MoFe protein of nitrogenase involves the iron protein–MgATP complex," *J. Biol. Chem. 264*, 10088–10095 (1989).

204. Robson, R. L., "Identification of possible adenine nucleotide-binding sites in nitrogenase Fe- and MoFe-proteins by amino acid sequence comparison," *FEBS Lett. 173*, 394–398 (1984).

205. Robson, R., Woodley, P., and Jones, R., "Second gene (*nifH**) coding for a nitrogenase iron protein in *Azotobacter chroococcum* is adjacent to a gene coding for a ferredoxin-like protein," *EMBO J. 5*, 1159–1163 (1986).

206. Robson, R. L., Woodley, P., Pau, R. N., and Eady, R. R., "Structural genes for the vanadium nitrogenase from Azotobacter chroococcum," *EMBO J. 8*, 1217–1224 (1989).

207. Rossen, L., Ma, Q.-S., Mudd, E. A., Johnston, A. W. B., and Downie, J. A., "Identification and DNA sequence of *fixZ*, a *nifB*-like gene from *Rhizobium leguminosarum*," *Nucleic Acids Res. 12*, 7123–7134 (1984).

208. Ruvkun, G. B., and Ausubel, F. M., "Interspecies homology of nitrogenase genes," *Proc. Natl. Acad. Sci. USA 77*, 191–195 (1980).

209. Saari, L. L., Triplett, E. W., and Ludden, P. W., "Purification and properties of the activating enzyme for iron protein of nitrogenase from the photosynthetic bacterium *Rhodospirillum rubrum*," *J. Biol. Chem. 259*, 15502–15508 (1984).

210. Santero, E., Hoover, T., Keener, J., and Kustu, S., "In vitro activity of the nitrogen fixation regulatory protein NIFA," *Proc. Natl. Acad. Sci. USA, 86*, 7346–7350.

211. Schatt, E., Jouanneau, Y., and Vignais, P. M., "Molecular cloning and sequence analysis of the structural gene of ferredoxin I from the photosynthetic bacterium *Rhodobacter capsulatus*," *J. Bacteriol. 171*, 6218–6226 (1989).

212. Schollhorn, R., and Burris, R. H., "Acetylene as a competitive inhibitor of N_2 fixation," *Proc. Natl. Acad. Sci. USA 58*, 213–216 (1967).

213. Scott, D. J., May, H. D., Newton, W. E., Brigle, K. E., and Dean, D. R., "Role for the nitrogenase MoFe protein α-subunit in FeMo-cofactor binding and catalysis," *Nature 343*, 188–190 (1990).

214. Scott, K. F., Rolfe, B. G., and Shine, J., "Biological nitrogen fixation: Primary structure of the *Klebsiella pneumoniae nifH* and *nifD* genes," *J. Molec. Appl. Genet. 1*, 71–81 (1981).

215. Scott, K. F., Rolfe, B. G., and Shine, J., "Biological nitrogen fixation: Primary structure of the *Rhizobium trifolii* iron protein gene," *DNA 2*, 149–155 (1983).

216. Scott, K. F., Rolfe, B. G., and Shine, J., "Nitrogenase structural genes are unlinked in the nonlegume symbiont *Parasponium Rhizobium*," *DNA 2*, 141–148 (1983).

217. Setterquist, R., Brigle, K. E., Beynon, J., Cannon, M., Ally, A., Cannon, F., and Dean, D. R., "Nucleotide sequence of the *nifE* and *nifN* genes from *Klebsiella pneumoniae*," *Nucleic Acids Res. 16*, 5215 (1988).

218. Shah, V. K., and Brill, W. J., "Isolation of an iron–molybdenum cofactor from nitrogenase," *Proc. Natl. Acad. Sci. USA 74*, 3249–3253 (1977).

219. Shah, V. K., Hoover, T. R., Imperial, J., Paustian, T. D., Roberts, G. P., and Ludden, P. W., "Role of *nif* gene products and homocitrate in the biosynthesis of iron–molybdenum cofactor," In *Nitrogen Fixation: Hundred Years After*. Bothe, H., de Bruijn, F. J., and Newton, W. E. (eds.), Stuttgart: Gustav Fischer, 1988, pp. 115–120.

220. Shah, V. K., Imperial, J., Ugalde, R. A., Ludden, P. W., and Brill, W. J., "*In vitro* synthesis of the iron–molybdenum cofactor of nitrogenase," *Proc. Natl. Acad. Sci. USA 83*, 1636–1640 (1986).

221. Shah, V. K., Stacey, G., and Brill, W. J., "Electron transport to nitrogenase: Purification and characterization of pyruvate:flavodoxin oxidoreductase, the *nifJ* gene product," *J. Biol. Chem. 258*, 12064–12068 (1983).

222. Sibold, L., "The polar effect on *nifM* mutations in the *nifU, -S, -V* genes of *Klebsiella pneumoniae* depends on their plasmid or chromosomal location," *Molec. Gen. Genet. 186*, 569–571 (1982).

223. Silver, S., Nucifora, G., Chu, L., and Misra, T. K., "Bacterial resistance ATPases: Primary pumps for exporting toxic cations and anions," *Trends Biochem. Sci. 14*, 76–80 (1989).

224. Silverstein, R., and Bulen, W. A., "Kinetic studies of the nitrogenase-catalyzed hydrogen evolution and nitrogen reduction reactions," *Biochemistry 9*, 3809–3815 (1970).

225. Simpson, F. B., and Burris, R. H., "A nitrogen pressure of 50 atmospheres does not prevent evolution of hydrogen by nitrogenase," *Science 224*, 1095–1097 (1984).

226. Smith, B. E., Buck, M., Eady, R. R., Lowe, D. J., Thorneley, R. N. F., Ashby, G., Deistung, J., Eldridge, M., Fisher, K., Gormal, C., Ioannidis, I., Kent, H., Arber, J., Flood, A., Garner, C. D., Hasnain, S. and Miller, R., "Recent studies on the structure and function of molybdenum nitrogenase," In Bothe, H., de Bruijn, F. J.,

and Newton, W. E. (eds.), *Nitrogen fixation: Hundred Years After*. Stuttgart: Gustav Fischer, 1988, pp. 91–100.

227. Smith, B. E., Eady, R. R., Lowe, D. J., and Gormal, C., "The vanadium–iron protein of vanadium nitrogenase from *Azotobacter chroococcum* contains an iron–vanadium cofactor," *Biochem. J. 250*, 299–302 (1988).

228. Smith, B. E., and Lang, G., "Mossbauer spectroscopy of the nitrogenase proteins from *Klebsiella pneumoniae*," *Biochem. J. 137*, 169–180 (1974).

229. Smith, B. E., Lowe, D. J., and Bray, R. C., "Studies by electron paramagnetic resonance on the catalytic mechanism of nitrogenase of *Klebsiella pneumoniae*," *Biochem. J. 135*, 331–341 (1973).

230. Smith, B. E., O'Donnell, M. J., Lang, G., and Spartalian, K., "A Mossbauer spectroscopic investigation of the redox behaviour of the molybdenum–iron protein from *Klebsiella pneumoniae* nitrogenase," *Biochem. J. 191*, 449–455 (1980).

231. Smith, B. E., Thorneley, R. N. F., Eady, R. R., and Mortenson, L. E., "Nitrogenases from *Klebsiella pneumoniae* and *Clostridium pasteurianum*. Kinetic investigations of cross reactions as a probe of the enzyme mechanism," *Biochem. J. 157*, 439–447 (1976).

232. Souillard, N., Magot, M., Possot, O., and Sibold, L., "Nucleotide sequence of regions homologous to *nifH* (nitrogenase Fe protein) from the nitrogen-fixing archaebacteria *Methanococcus thermolithotrophicus* and *Methanobacterium ivanovii*: Evolutionary implications," *J. Molec. Evol. 27*, 65–76 (1988).

233. Souillard, N., and Sibold, L., "Primary structure and expression of a gene homologous to *nifH* (Fe protein) from the archaebacterium *Methanococcus voltae*," *Molec. Gen. Genet. 203*, 21–28 (1986).

234. Steinbauer, J., Wenzel, W., and Hess, D., "Nucleotide sequence and deduced amino acid sequences of the *Klebsiella pneumoniae nifK* gene coding for the β-subunit of nitrogenase MoFe protein," *Nucleic Acids Res. 16*, 7199 (1988).

235. Stephens, P. J., McKenna, C. E., Smith, B. E., Nguyen, H. T., McKenna, M.-C., Thomson, A. J., Devlin, F., and Jones, J. B., "Circular dichroism and magnetic circular dichroism of nitrogenase proteins," *Proc. Natl. Acad. Sci. USA 76*, 2585–2589 (1979).

236. Sundaresan, V., and Ausubel, F. M., "Nucleotide sequence of the gene coding for the nitrogenase Fe protein from *Klebsiella pneumoniae*," *J. Biol. Chem. 256*, 2808–2812 (1981).

237. Swisher, R. H., Landt, M. L., and Reithel, F. J., "The molecular weight of, and evidence for two types of subunits in, the molybdenum–iron protein of *Azotobacter vinelandii* nitrogenase," *Biochem. J. 163*, 427–432 (1977).

238. Tanaka, M., Haniu, M., Yasunobu, K. T., and Mortenson, L. E., "The amino acid sequence of *Clostridium pasteurianum* iron protein, a component of nitrogenase," *J. Biol. Chem. 252*, 7093–7100 (1977).

239. Tanaka, M., Haniu, M., Yasunobu, K. T., and Yoch, D. C., "The amino acid sequence of the *Azotobacter vinelandii* flavodoxin," *Biochem. Biophys. Res. Commun. 66*, 639–644 (1975).

240. Telser, J., Hoffman, B. M., LoBrutto, R., Ohnishi, T., Tsai, A.-L., Simpkin, D., and Palmer, G., "Evidence for N coordination to Fe in the [2Fe–2S] center in yeast mitochondrial complex III," *FEBS Lett. 214*, 117–121 (1987).

241. Thomann, H., Morgan, T. V., Jin, H., Burgmayer, S. J. N., Bare, R. E., and Stiefel, E. I., "Protein nitrogen coordination to the FeMo center of nitrogenase from *Clostridium pasteurianum*," JACS *109*, 7913–7914 (1987).

242. Thöny, B., Kaluza, K., and Hennecke, H., "Structural and functional homology between the α and β subunits of the nitrogenase MoFe protein as revealed by sequencing the *Rhizobium japonicum nifK* gene," *Molec. Gen. Genet. 198*, 441–448 (1985).

243. Thorneley, R. N. F., and Eady, R. R., "Nitrogenase of *Klebsiella pneumoniae:* Evidence for an adenosine triphoshate–induced association of the iron–sulphur protein," *Biochem. J. 133*, 405–408 (1973).

244. Torok, I., and Kondorosi, A., "Nucleotide sequence of the *R. meliloti* nitrogenase reductase (*nifH*) gene," *Nucleic Acids Res. 9*, 5711–5723 (1981).

245. Tsai, L. B., and Mortenson, L. E., "Interaction of the nitrogenase components of *Anabaena cylindrica* with those of *Clostridium pasteurianum*," *Biochem. Biophys. Res. Commun. 81*, 280–287 (1978).

246. Tso, M.-Y. W., and Burris, R. H., "The binding of ATP and ADP by nitrogenase components from *Clostridium pasteurianum*," *Biochim. Biophys. Acta 309*, 263–270 (1973).

247. Ugalde, R. A., Imperial, J., Shah, V. K., and Brill, W. J., "Biosynthesis of iron–molybdenum cofactor in the absence of nitrogenase," *J. Bacteriol. 159*, 888–893 (1984).

248. Ugalde, R. A., Imperial, J., Shah, V. K., and Brill, W. J., "Biosynthesis of the iron–molybdenum cofactor and the molybdenum cofactor in *Klebsiella pneumoniae:* Effect of sulfur source," *J. Bacteriol. 164*, 1081–1087 (1985).

249. van Lin, B., and Bothe, H., "Flavodoxin from *Azotobacter vinelandii*," *Arch. Microbiol. 82*, 155–172 (1972).

250. Vaughn, S. A., and Burgess, B. K., "Nitrite, a new substrate for nitrogenases," *Biochemistry 28*, 419–424 (1989).

251. Wahl, R. C., and Orme-Johnson, W. H., "Clostridial pyruvate oxidoreductase and the pyruvate-oxidizing enzyme specific to nitrogen fixation in *Klebsiella pneumoniae* are similar enzymes," *J. Biol. Chem. 262*, 10489–10496 (1987).

252. Walker, M. N., and Mortenson, L. E., "Evidence for the existence of a fully reduced state of molybdoferredoxin during functioning of nitrogenase and the order of electron transfer from reduced ferredoxin," *J. Biol. Chem. 249*, 6356–6358 (1974).

253. Walker, G. A., and Mortenson, L. E., "Effect of magnesium adenosine 5′-triphosphate on the accessibility of the iron of clostridial azoferredoxin, a component of nitrogenase," *Biochemistry 13*, 2382–2388 (1974).

254. Walters, M. A., Chapman, S. K., and Orme-Johnson, W. H., "The nature of amide ligation to the metal sites of FeMoco," *Polyhedron 5*, 561–565 (1986).

255. Wang, Z.-C., Burns, A., and Watt, G. D., "Complex formation and O_2 sensitivity of *Azotobacter vinelandii* nitrogenase and its component proteins," *Biochemistry 24*, 214–221 (1985).

256. Wang, S.-Z., Chen, J.-S., and Johnson, J. L., "Nucleotide and deduced amino acid sequence of *nifD* encoding the α-subunit of nitrogenase MoFe protein of *Clostridium pasteurianum*," *Nucleic Acids Res. 15*, 3935 (1987).

257. Wang, S.-Z., Chen, J.-S., and Johnson, J. L., "The presence of five *nifH*-like sequences in *Clostridium pasteurianum:* Sequence divergence and transcription properties," *Nucleic Acids Res. 16*, 439–454 (1988).

258. Wang, S.-Z., Chen, J.-S., and Johnson, J. L., "Distinct structural features of the α and β subunits of nitrogenase molybdenum–iron protein of *Clostridium pasteurianum:* An analysis of amino acid sequences," *Biochemistry 27,* 2800–2810 (1988).

259. Wang, S.-Z., Chen, J.-S., and Johnson, J. L., "Nucleotide and deduced amino acid sequence of *nifE* from *Clostridium pasteurianum,"* *Nucleic Acids Res. 17,* 3299 (1989).

260. Watt, G. D., Bulen, W. A., Burns, A., and Hadfield, K. L., "Stoichiometry, ATP/2e values, and energy requirements for reactions catalyzed by nitrogenase from *Azotobacter vinelandii,"* *Biochemistry 14,* 4266–4272 (1975).

261. Watt, G. D. and Burns, A., "Kinetics of dithionite ion utilization and ATP hydrolysis for reactions catalyzed by the nitrogenase complex from *Azotobacter vinelandii,"* *Biochemistry 16,* 264–270 (1977).

262. Watt, G. D., Burns, A., Lough, S., and Tennent, D. L., "Redox and spectroscopic properties of oxidized MoFe protein from *Azotobacter vinelandii,"* *Biochemistry 19,* 4926–4932 (1980).

263. Watt, G. D., Burns, A., and Tennent, D. L., "Stoichiometry and spectral properties of the MoFe cofactor and noncofactor redox centers in the MoFe protein of nitrogenase from *Azotobacter vinelandii,"* *Biochemistry 20,* 7272–7277 (1981).

264. Watt, G. D., and McDonald, J. W., "Electron paramagnetic resonance spectrum of the iron protein of nitrogenase: Existence of a $g = 4$ spectral component and its effect on spin quantization," *Biochemistry 24,* 7226–7231 (1985).

265. Watt, G. D., and Wang, Z.-C., "Further redox reactions of metal clusters in the molybdenum–iron protein of *Azotobacter vinelandii* nitrogenase," Biochemistry *25,* 5196–5202 (1986).

266. Watt, G. D., Wang, Z.-C., and Knotts, R. R., "Redox reactions of and nucleotide binding to the iron protein of *Azotobacter vinelandii,"* *Biochemistry 25,* 8156–8162 (1986).

267. Weinman, J. J., Fellows, F. F., Gresshoff, P. M., Shine, J., and Scott, K. F., "Structural analysis of the genes encoding the molybdenum–iron protein of nitrogenase in the *Parasponium Rhizodium* strain ANU289," *Nucleic Acids Res. 12,* 8329–8344 (1984).

268. Willing, A. H., Georgiadis, M. M., Rees, D. C., and Howard, J. B., "Crosslinking of nitrogenase components: Structure and activity of the covalent complex," *J. Biol. Chem. 264,* 8499–8503 (1989).

269. Willison, J. C., Ahombo, G., Chabert, J., Magnin, J.-P., and Vignais, P. M., "Genetic mapping of the *Rhodopseudomonas capsulata* chromosome shows non-clustering of genes involved in nitrogen fixation," *J. Gen Microbiol. 131,* 3001–3015 (1985).

270. Wilson, M., Setterquist, R., Weiss, M., Newton, W., and Dean, D., "Analysis of the *nifE* and *nifN* genes from *Azotobacter vinelandii,"* In Bothe, H., de Bruijn, F. J., and Newton, W. E. (eds.), *Nitrogen Fixation: Hundred Years After.* Stuttgart: Gustav Fischer, 1988, p. 325.

271. Wink, D. A., McLean, P. A., Hickman, A. B., and Orme-Johnson, W. H., "A new method for extraction of iron–molybdenum cofactor (FeMoco) from nitrogenase adsorbed to DEAE-cellulose. 2. Solubilization of FeMoco in a wide range of organic solvents," *Biochemistry 28,* 9407–9412 (1989).

272. Wong, P.-K., Popham, D., Keener, J., and Kustu, S., "*In vitro* transcription of the nitrogen fixation regulatory operon *nifLA* of *Klebsiella pneumoniae,"* *J. Bacteriol. 169,* 2876–2880 (1987).

273. Yamane, T., Weininger, M. S., Mortenson, L. E., and Rossman, M. G., "Molecular symmetry of the MoFe protein of nitrogenase," *J. Biol. Chem. 257,* 1221–1223 (1981).

274. Yates, M. G., "Electron transport to nitrogenase in *Azotobacter chroococcum: Azotobacter* flavodoxin hydroquinone as an electron donor," *FEBS Lett. 27,* 63–67 (1972).

275. Yoch, D. C., and Arnon, D. I., "Comparison of two ferredoxins from *Rhodospirillum rubrum* as electron carriers for the native nitrogenase," *J. Bacteriol. 121,* 743–745 (1975).

276. Yoch, D. C., Benemann, J. R. Valentine, R. C., and Arnon, D. I., "The electron transport system in nitrogen fixation by *Azotobacter.* II. Isolation and function of a new type of ferredoxin," *Proc. Natl. Acad. Sci. USA 64,* 1404–1410 (1969).

277. Yun, A. C., and Szalay, A. A., "Structural genes of dinitrogenase and dinitrogenase reductase are transcribed from two separate promoters in the broad host range cowpea *Rhizobium* strain IRc78," *Proc. Natl. Acad. Sci. USA 81,* 7358–7362 (1984).

278. Zimmermann, R., Munck, E., Brill, W. J., Shah, V. K., Henzl, M. T., Rawlings, J., and Orme-Johnson, W. H., "Nitrogenase X: Mossbauer and EPR studies on reversibly oxidized MoFe protein from *Azotobacter vinelandii* OP," *Biochim. Biophys. Acta 537,* 185–207 (1978).

279. Zumft, W. G., Mortenson, L. E., and Palmer, G., "Electron-paramagnetic-resonance studies on nitrogenase," *Eur. J. Biochem. 46,* 525–535 (1974).

21

Regulation of Nitrogen Fixation Genes in Free-Living and Symbiotic Bacteria

M. J. Merrick

I. Introduction

Nitrogen fixation is an energy-demanding process, since the nitrogenase enzyme has an absolute requirement for ATP and the reduction of 1 mole of N_2 to ammonia utilizes 16 mole of ATP. It is therefore not surprising that organisms regulate tightly both the synthesis of nitrogenase and its subsequent activity, in response to a number of environmental factors. Owing to the marked oxygen sensitivity of nitrogenase, the environmental oxygen tension is a major regulatory factor for most diazotrophs. Likewise, the availability of fixed nitrogen is a significant regulatory effector in free-living diazotrophs but is of less importance for symbiotic organisms that are adapted to exporting fixed nitrogen to their host. Finally, as nitrogenase is a metalloprotein, the availability of the requisite metals can also have marked regulatory effects, particularly in those organisms that can synthesize nitrogenases with different metal centers (e.g., *Azotobacter*).

Regulation of N_2 fixation can be achieved by a number of mechanisms. In all organisms studied so far, control is exerted at the level of transcription of the N_2-fixation (*nif*) genes, and it is the mechanism of this control that forms the subject of this chapter. In certain organisms, including *Rhodospirillum,* regulation can also occur posttranslationally by the covalent modification and consequent inactivation of the nitrogenase Fe protein (see Chapter 3). In other cases (e.g., *Azotobacter*), the enzyme can be subject to reversible conformational protection from O_2 damage (see Chapter 18). In some cyanobacteria *nif* gene expression is a developmentally regulated process so that nitrogenase is only expressed in certain specialized cells (heterocysts) and is thereby spatially separated from the O_2-evolving process of photosynthesis (see Chapter 4).

Despite the considerable taxonomic diversity to be found among N_2-fixing organisms, research over the last decade has revealed a surprising degree of uniformity in the mechanisms underlying the regulation of *nif* genes in many, though not all, diazotrophs. Our present knowledge varies considerably from organism to organism but by far the most detailed description comes from studies on the facultative anaerobe *Klebsiella pneumoniae,* which has become a paradigm for genetic studies of *nif.* For this reason much of the following review is based on studies in *K. pneumoniae* with discussions of the exceptions and differences that occur in other organisms where appropriate.

The chapter is divided into two main sections, the first of which considers the archetypal *nif* gene promoter and the roles of the products of the *rpoN* and *nifA* genes that are required for transcriptional activation in the great majority of N_2-fixing organisms. In the second section I have discussed the factors that influence both the expression and activity of the *nifA* gene product (NifA), factors that vary considerably from one organism to another and that reflect the physiological and environmental influences on N_2 fixation in different species.

II. NifA-, RpoN-Dependent Activation of *nif* Gene Expression

The determination of the DNA sequences of the majority of *nif* gene promoters in *K. pneumoniae* by Beynon et al.[12] was the first indication that the structure of *nif* promoters, and hence the regulation of *nif* genes, was atypical in terms of what was known at that time about bacterial promoter sequences. The *K. pneumoniae nif* promoters have a highly conserved sequence between positions −11 and −26 with respect to the point of transcription initiation. This sequence, which has a consensus of TGG-N8-TTGCA, is quite unlike the consensus TTGACA-N17-TATACA found in the −35, −10 region of most prokaryotic promoters, and this difference suggested that these promoters may be recognized by a modified form of RNA polymerase. Expression from these promoters was known to be dependent on the products of two genes, *rpoN* (then known as *ntrA* or *glnF*) and *nifA.* The *nifA* gene product had been identified as a potential *nif*-specific transcriptional activator protein,[36,19] and it was subsequently proposed by de Bruijn and Ausubel[33] that the *rpoN* gene product might be an RNA polymerase sigma factor that could modify the recognition properties of core RNA polymerase. Studies in the laboratories of Sydney Kustu[61] and Boris Magasanik[66] have since demonstrated that this is indeed the case and that *rpoN* encodes a novel sigma factor (see later). Furthermore, RNA polymerase carrying this sigma factor is totally dependent on an activator protein such as NifA, or the related NtrC, for transcription initiation.

A. The −24, −12 Consensus Sequence

Promoters that conform to the −24, −12 consensus are often now misleadingly referred to as "nif" promoters even when located upstream of genes

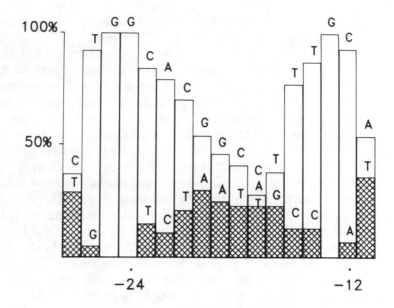

Consensus   c T G G C A C G g c c t T T G C A

Figure 21-1. Histogram summarizing the sequences of 64 σ⁵⁴-dependent promoters from 22 species (taken from Morett and Buck, 1989). The frequency of each base is plotted at each position between −27 and −11 with respect to the transcription start, and the derived consensus is given below the histogram. Guanine residues at positions −25, −24, and −13 are totally conserved.

other than *nif* genes. Such promoters should be described correctly as RpoN dependent, as not all *nif* genes are expressed from this class of promoter, and many non-*nif* genes such as *E. coli glnAp2* and *fdhF, Pseudomonas putida xylCAB* and *Caulobacter crescentus flaN* and *flbG* are *rpoN* dependent (see ref. 90 for review). Nevertheless, a large number of *nif* gene promoters have now been sequenced from many different organisms, and the great majority of these conform to the −24, −12 consensus and are known to be *rpoN* dependent. These sequences have been analyzed by Morett and Buck[105] and a revised consensus based on over 60 promoter sequences has been determined (Fig. 21-1). Within this consensus three residues remain invariant, namely, the GG pair at −25, −24 and the G at −13. The −12 position is almost totally conserved as C with the exception of a few rhizobial promoters where it is replaced by A. Furthermore, in every case where the sequence has been unambiguously demonstrated to constitute an *rpoN*-dependent promoter the spacing between the GG and GC pairs is 10 nucleotides. The −24, −12 sequence was predicted to constitute the recognition site for a holoenzyme form of RNA polymerase that contains the *rpoN* product, and this has been demonstrated by protection assays using a number of promoters both *in vitro* and *in vivo*.[91,118,103,105]

The contribution to promoter activity of the various nucleotides in the consensus sequence has been analyzed by mutagenesis of both the *K. pneumoniae nifH* and *nifL* promoters. Deletions that remove one or more base pairs between the -12 and -24 elements eliminate promoter activity,[20] demonstrating a stringent requirement for the 10-bp spacing between the two elements and suggesting that they must reside on the same face of the DNA helix. Modification of residues between -13 and -24 has relatively little effect, causing a reduction of at most 50% in promoter activity.[24,114,79] In some cases mutations in these semiconserved residues can actually increase promoter activity, and in general the change in promoter activity observed with a particular mutation depends on whether the change increases homology to the consensus or deviates from it. By contrast, mutation of any one of the conserved bases at -12, -13, -24, and -25 has a strong promoter-down phenotype. This is not invariably true for the -12 position, so that whilst a C to T transition at -12 in the *nifL* promoter causes a marked down phenotype, the same mutation in the *nifH* promoter is silent, providing that the activator protein is NifA. However, an effect of the mutation can be observed if NtrC is the activator. It is notable that the absence of an absolute requirement for a C at -12 under all conditions is reflected in the natural occurrence of an A at this position in the *nifHa, Hb,* and *Hc* promoters of *Rhizobium phaseoli,* the *nifH* promoter of *Rhizobium trifolii,* and the *nifN* promoter of *Rhizobium meliloti.*

Mutational analyses of other RpoN-dependent promoters—namely, the *C. crescentus flaN* and *flbG* promoters[111,107] and the *S. typhimurium argTr* promoter[134]—have confirmed the roles of the GG dinucleotide at -24, -25 and of the G at -13 as well as the invariant 10-bp spacing between these nucleotides. However, one exception has been reported in the *Bradyrhizobium japonicum fixRnifA* promoter where a G to T transversion at -25 did not impair expression.[147]

B.   The Upstream Activator Sequence (UAS)

It was observed by Buchanan-Wollaston et al.[19] that when the *K. pneumoniae nifH* promoter was cloned on a multicopy plasmid, it inhibited chromosomal *nif* gene expression. This effect was proposed as being due to titration of a limiting regulatory factor, possibly NifA. Mutations were subsequently isolated that relieved this "multicopy inhibition,"[18] and one of these was a G to T transversion at -136 in the *nifH* promoter. The location of this mutation focused attention on the upstream region of the promoter and led Buck et al.[25] to identify a highly conserved motif, TGT-N_{10}-ACA, which is present in almost all NifA-activated promoters between 80 and 150 bp upstream of the transcript start. This sequence conforms to a consensus sequence for protein binding sites on DNA and as such was proposed as a potential NifA-binding site.[25]

Detailed analysis of this sequence in *K. pneumoniae nif* promoters demonstrated that its deletion greatly reduced transcriptional activation by NifA but that the low level of expression conferred by the related activator protein NtrC was not affected.[25] Deletion of the upstream activator sequence (UAS) also relieved multicopy inhibition. Finally, it was demonstrated that, although the UAS was by itself transcriptionally inactive, its ability to effect activation was not dependent on orientation and it was able to increase promoter activity significantly when located up to 2 kb from the -24, -12 sequence.

Results comparable to those obtained in *K. pneumoniae* were reported by Alvarez-Morales et al.[4] for the *nifH* and *nifDK* promoters of *Bradyrhizobium japonicum,* when analyzed both in a heterologous system in *Escherichia coli* and in a homologous *B. japonicum* system. In *B. japonicum* deletion of the UAS had a marked effect on promoter activity both in microaerobic cultures and in root nodules.[4] When the two tandem *nifH* UASs were deleted from the *B. japonicum* chromosome, the mutant strain had no symbiotic nitrogenase activity.[53] However, a similar deletion analysis of the *R. meliloti nifH* and *fixABC* promoters found the UAS to be required for full expression in a heterologous system (using *K. pneumoniae* NifA to activate the promoters in *E. coli*), but not under conditions of symbiosis in alfalfa nodules.[11]

Not all NifA-dependent promoters have an obvious UAS motif. There is no UAS upstream of *K. pneumoniae nifM*[116] nor upstream of the NifA-dependent *fixA* and *fixB* promoters of *B. japonicum,* although an imperfect UAS (TGT-N_{10}-ACC) is present at -132 in the *fixB* promoter.[55] A deletion between -41 and -203 upstream of *fixA* has no effect on expression, and deletion between -83 and -261 in the *fixB* promoter only reduces *fixB* expression by 50%. However, when the *B. japonicum nifH* UASs are deleted, expression of *fixB*, which lies 3 kb downstream, is reduced by 80%, suggesting that the *nifH* UASs are a major influence on *fixB* expression.[53] In the same way, *K. pneumoniae nifM* expression may be influenced by the *nifU* UAS that lies 4 kb upstream.

The effects of mutations in the UAS motif have been analyzed in detail in the *K. pneumoniae nifH* UAS.[22] Mutations that alter the two-fold rotational symmetry of the UAS or the spacing between the TGT and ACA motifs (e.g., to 8, 9, or 11 bp) reduce promoter activity, and sequences flanking the TGT-ACA motif also have some effect on NifA-mediated activation. Furthermore, when the *nifH* UAS is replaced by a binding site for the transcriptional activator NtrC, NtrC-dependent activation of the promoter is increased 10-fold. Hence, it is apparent that the activator specificity of the promoter is markedly dependent on the presence of the appropriate up-stream sequences.

The requirement for an upstream activator binding site to achieve maximal activation of most NifA-dependent promoters is not a unique property of *nif* genes but appears to be a characteristic of *rpoN*-dependent promoters. Similar observations have been made for the *E. coli glnA* and *fdhF* promoters,[123,14] the *C. crescentus flgB* promoter,[107] and the NtrC-dependent *K. pneumoniae nifLA* promoter.[102] The function of these upstream binding sites will be discussed later in the context of a general model for the mode of action of *rpoN*-dependent promoters.

C. The Structure and Function of RpoN

The role of RpoN in *nif* gene expression was initially demonstrated by the lack of expression of *pnif-lacZ* fusions in *rpoN* mutants of *E. coli*,[98,113] and although *rpoN* mutants of *K. pneumoniae* were among mutants initially selected as being Nif⁻, they were not recognized as such until later. *rpoN* mutants have since been isolated in a variety of diazotrophs, including *K. pneumoniae*,[98,99] *A. vinelandii*,[149] *R. meliloti* strain 1021,[131] strain 5419[45] and strain 104A14,[135] *Rhizobium* sp. NGR234,[141] and *Rhodobacter capsulatus*.[85] As expected, the mutants are Nif⁻ in all cases, but also exhibit complex pleiotropic phenotypes that reflect the variety of σ⁵⁴-dependent promoters present in these organisms. In *K. pneumoniae, A. vinelandii,* and *R. meliloti, rpoN* mutants are defective in nitrate assimilation[149,131] (Merrick, unpublished). *R. meliloti rpoN* mutants are unable to transport dicarboxylic acids[131] and *A. vinelandii rpoN* mutants also fail to utilize these carbon sources (Toukdarian and Kennedy, unpublished). Although *rpoN* (*glnF*) mutants were originally isolated in *S. typhimurium* as glutamine auxotrophs,[50] comparable mutants do not cause glutamine auxotrophy in *K. pneumoniae* or *A. vinelandii,* because of differences in the precise mode of expression of the glutamine synthetase structural gene (*glnA*). In *Rhizobium* the situation is complicated by the presence of two distinct glutamine synthetases, GSI (encoded by *glnA*) and GSII (encoded by *glnII*). In *R. meliloti* and *B. japonicum, glnII* is under *rpoN* control,[135,94] but *rpoN* mutants are not glutamine auxotrophs, because of the presence of GSI.

The demonstration that the *rpoN* product is a novel RNA polymerase sigma factor was reported almost simultaneously by Hirschman et al.[61] and Hunt and Magasanik.[66] It was already apparent that RpoN was somewhat distinct from other known sigma factors, given the highly conserved −24, −12 sequence that characterizes RpoN-dependent promoters. By contrast most other sigma factors, including the major vegetative sigma factor σ⁷⁰, recognize a consensus sequence located around positions −35 and −10 with respect to the transcript start, and in no case is the consensus as highly conserved as that for RpoN. The sequence of the *K. pneumoniae rpoN* gene[100] confirmed that the primary amino acid sequence of RpoN was quite different from that of the family of proteins to which most other sigma factors appear to be related.[51] Gribskov and Burgess identified a significant degree of amino acid sequence similarity between sigma factors from *E. coli, B. subtilis,* phage T4, and phage SPO1—most notably in two regions that they designated regions 2 and 4. Genetic evidence has since implicated region 2 in establishing contacts between the sigma factor and the −10 region of the promoter, and region 4 in contacts with the −35 region (see ref. 57 for review). However, no strong sequence similarity to either region 2 or region 4 is present in the primary amino acid sequence of *K. pneumoniae* RpoN or in the predicted amino acid sequences from five other *rpoN* genes sequenced since.

K. pneumoniae RpoN has a predicted molecular mass of 54 kDa, and the sigma factor is therefore now designated σ^{54}. This molecular mass is notably different from that of some 75 kDa determined by SDS polyacrylamide gel electrophoresis (SDSPAGE),[33,101] and in this respect σ^{54} resembles other sigma factors that also have aberrant mobilities on SDSPAGE, a property that has been attributed to their acidic nature. Six *rpoN* genes have now been sequenced—from *K. pneumoniae,*[100] *R. meliloti* strain 1021[131] and strain 104A14,[135] *A. vinelandii,*[96] *R. capsulatus*[74,3] and *Pseudomonas putida*[69,83]— and the proteins encoded by these genes are extremely similar. When the amino acid sequences of the proteins are aligned, three distinct regions (or domains) can be distinguished.[96]

The N-terminal domain of RpoN is around 50 amino acids long and is highly conserved. It is rich in glutamine residues, and deletions in this domain have been found to produce mutants in which RNA polymerase containing σ^{54} ($E\sigma^{54}$) cannot progress from the initial closed complex with the promoter to a transcriptionally active open complex (Sasse-Dwight and Gralla, cited in ref. 90). Hence, this domain may be involved in an interaction between $E\sigma^{54}$ and the activator protein. The region following the N-terminal domain is very variable in length, ranging from 24 amino acids in *R. capsulatus* to 97 amino acids in *R. meliloti,* and shows essentially no conservation of sequence. The reason for this variation and the function (if any) of this region is unknown. The variation in the length of this region means that the molecular mass of the *rpoN* product varies from 40 kD in *R. capsulatus* to 58 kD in *R. meliloti.*

The C-terminal domain is approximately 380 amino acids long and shows significant similarity throughout its length, but two regions within this domain are particularly conserved. The first of these is notably similar to the helix-turn-helix (HTH) motifs that have been shown to characterize a large group of DNA-binding proteins.[115] The region is particularly homologous to the HTH motifs found in the *lac* and *gal* repressors of *E. coli*[96] and might be involved in DNA recognition by σ^{54}. It should be noted that the conserved region 4 in the major sigma factor family[57] also conforms to a HTH motif, but the primary amino acid sequence (particularly of the second "recognition" helix) is quite distinct from that in σ^{54}. The second highly conserved region in RpoN is a run of nine totally conserved amino acids. This region is rich in aromatic and basic residues and in this sense is similar to region 2 in the major sigma factor family that Helmann and Chamberlain[57] have suggested could be involved in open complex formation.

Analysis of the *rpoN* sequences from the five organisms described earlier have revealed that in all but one case—namely, *R. capsulatus*—there is also considerable similarity in the sequences that flank *rpoN*. Upstream of *rpoN* in *K. pneumoniae, R. meliloti,* and *P. putida* is an open-reading-frame (ORF) transcribed in the same direction as *rpoN* and encoding a polypeptide of unknown function. This polypeptide is homologous to a family of ATP-binding proteins with a very diverse range of functions.[2,60] The inability to isolate insertion mutations in the *R. meliloti* upstream gene led Albright et al.[2] to propose that it may code for an essential function. In *R. meliloti* the gene is transcriptionally separate from *rpoN* and appears to be in an operon with one or more genes upstream. Hence, the relationship (if any) between this gene and *rpoN* is unknown, although the marked conservation of its linkage to *rpoN* is intriguing.

In *K. pneumoniae,* sequence analysis has identified two ORFs downstream of *rpoN* that are potentially cotranscribed with *rpoN*.[99] The first of these ORFs (ORF95) is conserved in *A. vinelandii, R. meliloti,* and *P. putida;* the second (ORF162) is conserved at least in *P. putida.* The ORFs have been shown to encode polypeptides of approximately 12 kDa and 16 kDa in both *K. pneumoniae*[99] and *P. putida,*[84] and their functions have been investigated in *K. pneumoniae* by insertional mutagenesis.[99] Mutations in both genes produce an elevated level of expression (two- to sixfold) from *rpoN*-dependent promoters, suggesting that the genes may both have a role in modulating σ^{54} activity. The method by which this might be achieved and the physiological role of such an effect is unknown, although the highly conserved linkage of the genes to *rpoN* and their potential cotranscription with *rpoN* suggests a fundamental role for their products.

In both *K. pneumoniae* and *R. meliloti, rpoN* is transcribed at a low constitutive level.[101,131] By contrast, in the only organism where the flanking ORFs are not conserved—namely, *R. capsulatus*—*rpoN* expression is regulated by both the oxygen and nitrogen status of the cell.[85] This regulation, together with the location of *R. capsulatus rpoN* between the *nifK* and *nifA* genes, and the somewhat reduced homology when compared with all other σ^{54} amino acid sequences, could indicate that this *rpoN* gene is actually *nif*-specific and that a second *rpoN* gene of more general function might exist elsewhere in the *R. capsulatus* genome.

D. The Structure and Function of nifA

The *nifA* gene was initially identified in *K. pneumoniae*,[38,76] and its likely ubiquity in diazotrophs was subsequently indicated by the ability of *K. pneumoniae nifA* to activate expression of *nif* genes from both *Azotobacter*[77] and *R. meliloti*.[144] As predicted from the presence of conserved UAS sequences in *nif* promoters, *nifA* homologues have since been identified in *Azotobacter*, *Rhodobacter*, *Herbaspirillum*, and the three major groups of rhizobia—*Rhizobium*, *Azorhizobium*, and *Bradyrhizobium*. The sequence of the *Thiobacillus ferrooxidans nifH* promoter[119] also suggests the presence of a *nifA* homologue in this organism and potential *nifA* mutants have been isolated in *Azospirillum*.[117,137]

The true homology of these *nifA* genes is best assessed from DNA sequence data, and since the *K. pneumoniae nifA* sequence was reported[26,39] *nifA* sequences have been obtained from 11 other diazotrophs: *Klebsiella oxytoca*,[81] *Azotobacter vinelandii*,[10] *A. chroococcum*,[43,44] *R. capsulatus*,[95] *Herbaspirillum seropedicae* (deSouza and Yates, unpublished), *R. leguminosarum* strains 3855[52] and PRE,[129] *R. meliloti*,[26] *R. trifolii*,[67] *B. japonicum*,[147] and *Azorhizobium caulinodans* ORS751.[122,108] Alignment of the predicted amino acid sequences of the products of these genes shows them to be highly homologous and to conform to the domain structure (Figure 21-2) originally proposed by Drummond et al.[39]

In all but one case—namely, *R. trifolii* NifA—the proteins have an N-terminal domain of between 175 and 255 residues. The primary sequence of this domain is not highly conserved, and its role is not presently understood. In NtrC, a transcriptional activator with significant homology to NifA, the N-terminal domain mediates an interaction with a partner sensor protein, NtrB, which responds to the cellular N-status and in N-limiting conditions phosphorylates an aspartate residue in the N-terminus of NtrC.[154] By analogy, it has been proposed that the N-terminal domain of NifA may interact with a partner sensory protein such as NifL in *K. pneumoniae*, although there is no evidence for phosphorylation of NifA, and NifL homologues are not known in rhizobia. Deletion of the *K. pneumoniae* NifA N-terminal domain reduces activity three- to fourfold,[40] and genetic studies support the suggestion that this domain may regulate interaction between NifA and NifL (see later).

Genetic analysis of the N-terminal domain in rhizobia suggests that it is not required for NifA activity. In *R. trifolii* NifA this domain is not present,[67] and in both *B. japonicum* and *R. meliloti* the complete domain can be deleted without activation of the *nif* promoters being significantly impaired.[46,13,65] Hence, if the N-terminal domain of rhizobial NifAs has a regulatory function, the signal to which it responds is presently unknown.

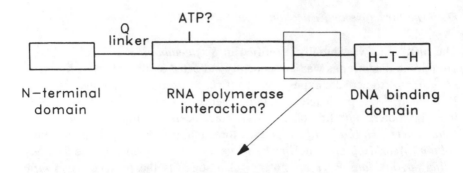

```
        *           *                      *   *
Rc    RCQFPGNERELENCVNRAAALSDGAIVLAEELACRQGACLSAEL----FRLQDGTSPIGGLAVGRVIT
Ac    RCYFPGNVRELENCIRRTATLAHDAVITPHDFACDSGQCLSAMLWKGSAPKPVMPHVPPAPTPLTPLS
Hs    NCYWPGNVRELENCVERTATHMRGDLITEVHFSCQQNKCLTKVL-------------HEPGQQQPVV
Bj    SCKFPGNVRELENCIERTATLSAGTSIVRSDFACSQGQCLSTTLWKSTSYGKTDPAAPMQPVPAKSII
Rm    KCKFPGNVRELENCVRRTATLARSKTITSSDFACQTDQCFSSRLWKGVHCSHGHIEI-DAPAGTTPLL
Rl    KCAFPGNVRELDNCVQRTATLASSNTITSSDFACQQDQCSSALLRKADGDGIGNDAM-NGLNSRDTMS
Rt    KCEFPGNIRELQNCTQRTATLARSDVIVPQDLACEQGRCYSPILKKAVAEQVGKGAIHGLARGETESM
Av    SHRWPGNVRELENCLERSAIMSEDGTITR-------------------------------DVVS
Kp    EYSWPGNVRELENCLERSAVLSESGLIDR-------------------------------DVIL
```

Figure 21-2. The domain structure of NifA proteins. The proteins comprize three major domains: an N-terminal domain of unknown function, a central domain proposed as being involved in interaction with $E\sigma^{54}$ polymerase and containing a region with homology to known ATP-binding sites, and a C-terminal domain containing a helix-turn-helix motif required for DNA binding. The primary amino acid sequences from nine NifAs for the variable region between the central and C-terminal domains are aligned below the figure, and this alignment distinguishes two classes. One class, including all the rhizobial NifAs, contains four conserved cysteine residues (*), and the activity of most members of this class has been shown to be O_2 sensitive. By contrast *K. pneumoniae* and *A. vinelandii* NifAs contain only one of the conserved cysteines and the activity of *K. pneumoniae* NifA is O_2-insensitive. The NifA sequences are from *R. capsulatus* (Rc), *A. caulinodans* (Ac), *H. seropideciae* (Hs), *B. japonicum* (Bj), *R. meliloti* (Rm), *R. leguminosarum* (Rl), *R. trifolii* (Rt), *A. vinelandii* (Av), and *K. pneumoniae* (Kp).

The N-terminal domain is linked to a large central domain by a short interdomain linker, termed a Q-linker by Wootton and Drummond.[156] Such linkers are typically 15–25 residues long and are relatively rich in glutamine, arginine, glutamate, serine and proline. In *K. pneumoniae* NifA, insertions of four or eight amino acids within the Q-linker have no effect on activity of the protein, consistent with the proposal that this domain serves simply to tether the N-terminal and central domains to each other.[156]

The central domain of NifA consists of a strongly conserved block of 238 amino acids. This central domain is characteristic not only of NifA proteins but of all other known σ^{54}-specific activator proteins whose genes have been sequenced to date—namely, NtrC from *K. pneumoniae*,[26,39] *R. meliloti*,[145] and *B. parasponia*;[112] DctD from *R. meliloti*;[130] XylR from *P. putida*;[68] and HydG from *E. coli*.[143] This homology suggests that the central domain may be involved in an interaction with Eσ,[54] and although the details of this proposed interaction are unknown, one aspect can be predicted. A common feature of the central domain (with the exception of HydG) is a conserved amino acid sequence GESGTGKE that is highly homologous to a consensus sequence found in many ATP-binding proteins. Activation of transcription by NtrC involves ATP hydrolysis,[118] and hence it has been proposed that transcription initiation by all σ[54]-dependent activators, including NifA, proceeds by a similar mechanism that involves binding of ATP by the activator and its subsequent hydrolysis.[90]

Adjacent to the central domain is a variable region of 30–75 residues followed by a conserved C-terminal domain of some 50 residues that contains a strong helix-turn-helix (HTH) motif proposed by Drummond et al.[39] to constitute the DNA-binding domain of the protein. Mutational analysis of this region in *K. pneumoniae* NifA supports this proposal.[106] Furthermore, when this domain is deleted in either *K. pneumoniae* NifA or *R. meliloti* NifA, transcriptional activation by the truncated NifA protein, although reduced, can still be detected at certain promoters. Hence, DNA binding can be separated from the positive control function of the protein.[106,65]

Interaction of NifA with the *K. pneumoniae nifH* UAS has been demonstrated *in vivo* by methylation protection experiments.[104] Such experiments were also used to show that expression of the isolated DNA-binding domain of NifA, although insufficient to activate transcription, is sufficient to protect the UAS. As predicted, mutations in the UAS or mutations in the NifA HTH motif reduce occupancy of the UAS. However, mutations in the −24, −12 sequence or in *rpoN* do not prevent NifA binding, demonstrating that NifA binding can occur in the absence of interaction of Eσ[54] with the −24, −12 sequence.

A. vinelandii is capable of synthesizing two alternative nitrogenases in addition to the conventional molybdenum nitrogenase (see Chapter 19) and each of the alternative enzymes has its own set of structural genes (*vnfHDGK* and *anfHDGK*), each of which has been shown to have its own specific activator gene (*vnfA* and *anfA*[72]). The predicted amino acid sequences of these two proteins are homologous to the NifA family. As with other NifAs, the N-termini are only slightly homologous, whereas the central domains are highly conserved. Both proteins have a potential HTH motif in the C-terminal domain, but it is notable that the amino acid sequences of these helices are different from each other and from the consensus for the NifA family. This is consistent with the absence of a typical UAS sequence in the upstream regions of *vnf* and *anf* promoters, which presumably each have their own distinctive activator binding sites.

Fischer and Hennecke[48] observed that the NifA protein of *B. japonicum*, unlike that of *K. pneumoniae*, was sensitive to aerobiosis, both in *B. japonicum* and in *E. coli*. *B. japonicum* NifA activity is drastically reduced within minutes upon a shift to aerobiosis, and this inactivation is irreversible.[89] The NifA family of proteins can be subdivided into two groups on the basis of the presence or absence of two extra cysteine residues in a Cys–X–X–X–X–Cys motif just beyond the C-terminus of the central domain (see Figure 21-2).[46] One group lacks the CXXXXC motif and includes *K. pneumoniae, A. vinelandii,* and *A. chroococcum* NifAs, together with the VnfA and AnfA proteins of *A. vinelandii.* Given the properties of *K. pneumoniae* NifA, it may be predicted that all this group of NifA proteins are likely to be oxygen insensitive. By contrast, all rhizobial NifA protein sequences and those of *R. capsulatus* and *H. seropedicae* contain the CXXXXC motif, and since NifA from three members of this group (*B. japonicum, R. meliloti,* and *R. capsulatus*) has now been shown to be oxygen sensitive,[46] (Klipp et al., unpublished) is likely that this is true for all NifA proteins with a CXXXXC motif. Two additional conserved Cys residues are located near the end of the central domain, one of these is present in all NifA proteins and one only in those having the CXXXXC motif (see Figure 21-2).

Fischer et al.[46] examined the role of all these four conserved Cys residues in *B. japonicum* NifA by mutating them individually to serine. The two cysteines in the CXXXXC motif were also changed individually to histidine.[47] In all these cases the resultant mutant proteins were inactive, as were mutants in which one residue was deleted to give a CXXXC motif.[47] Changes of residues in the four–amino acid spacer between the cysteines (e.g., Ser 473 to Glu or Gln and Gln476 to Glu or Lys) had no effect on NifA activity, but insertion of an extra residue to give CXXXXXC significantly reduced activity. In summary, these experiments suggest that the interdomain linker plays a vital role in rhizobial NifA proteins, and Fischer et al.[46] proposed that this role may involve the binding of a metal ion. Furthermore, they demonstrated that the activity of *B. japonicum* NifA is very sensitive to the presence of chelating agents in the growth medium, whereas *K. pneumoniae* NifA is not. This inhibition can be overcome by the addition of iron to the medium leading to the hypothesis that in anaerobic conditions rhizobial NifA proteins bind a metal ion such as Fe^{2+}, allowing the formation of an active protein. In aerobic conditions the metal ion could become oxidized and therefore unable to bind to the protein that would become inactive. A comparable proposal has been made for the *E. coli* regulatory protein Fnr,[140] but in neither case have the proteins yet been demonstrated to contain a metal.

More detailed analysis of NifA structure and function is presently thwarted by problems in purifying the protein. Although it can be readily overproduced, the protein synthesized in these conditions is almost completely insoluble, and extensive attempts to resolubilize it have not yet succeeded.[7,150] Nevertheless, an *in vitro* assay for NifA activity has been developed based on the expression from a *nifH–lacZ* reporter plasmid in an *E. coli* S-30 transcription–translation system. In this system the NifA protein can neither be synthesized *in vitro* from an appropriate plasmid[132] or can be derived from the small soluble NifA fraction remaining in cell-free extracts.[7]

E. The Mechanism of NifA-Mediated Activation

As already mentioned, the mechanism of transcriptional activation at all σ^{54}-dependent promoters is likely to be similar, and in the case of *nif* promoters a fairly detailed model is now available.

The identification of the UAS and the ability of NifA to activate transcription when the UAS is some distance from the downstream promoter element raised the question of how NifA is able to act from a distance to influence transcription. Ptashne[120] discussed a number of models for such activation, including (1) binding and subsequent sliding of the protein to its site of interaction and (2) looping out of the DNA to bring proteins bound at a distance together. In the latter case the introduction of one half turn (but not one full turn) of the helix between sites that interact at a distance is predicted to place these sites on opposite faces of the helix and so diminish their interaction.

Experiments in which the *nifH* UAS was displaced upstream by 5, 11, 15, and 21 bp demonstrated that activation of p*nifH* by NifA is reduced by the introduction of half-helical turns but to a far lesser extent by full-helical turns.[23] Furthermore, the *nifH* UAS does not function efficiently when placed close to the -24, -12 sequence (i.e., at positions closer than -100). These results led Buck et al.[23] to propose that NifA and the downstream promoter element interact by virtue of DNA loop formation and that the function of the UAS is to increase the effective local concentration of NifA in the vicinity of the -24, -12 sequence and to orient NifA appropriately. DNA loop formation has since also been shown to be important in activation of the *K. pneumoniae nifLA* promoter by NtrC[103] and activation of *E. coli fdhF*.[14] Hence, the upstream binding of an activator protein and the formation of a DNA loop seem to be common aspects of many σ^{54}-dependent promoters.

Promoter activation is not, however, always completely eliminated by deletion or mutation of the UAS, and in some cases activation can occur when the activator does not have an upstream binding site (e.g., the direct activation of *K. pneumoniae pnifH* by NtrC).[24] In these cases the activator can apparently interact with Eσ^{54} at the downstream promoter element in the absence of specific DNA binding upstream and loop formation, in which case neither upstream binding of activator nor DNA loop formation is an absolute prerequisite for activation.

The function of the activator protein is not face-of-the-helix dependent for all σ^{54}-dependent promoters, and both *K. pneumoniae nifF*[103] and *E. coli glnAp2*[93,125] do not show such dependence. The stringency of the requirement for the activator protein to be positioned on a particular face of the helix with respect to the −24, −12 sequence is related to affinity of that sequence for $E\sigma^{54}$. Not all σ^{54}-dependent promoters show a strong requirement for the activator protein to be bound upstream and one such example is the *R. meliloti nifH* promoter that is efficiently activated in the absence of a UAS or by a mutant form of NifA that is unable to bind to the UAS.[21,65,104] Comparison of the sequences of promoters that are not strictly dependent on a UAS suggested that the nucleotide sequence in the −17 to −11 region, in particular the presence of T residues in positions −16 to −14, might be critical in the response of a σ^{54}-dependent promoter to unbound activator.[21] This was tested by mutating the *K. pneumoniae nifH* promoter in this region and changing the −17 to −14 sequence from CCCT to TTTT. The resultant mutant promoter was far more responsive to a mutant NifA in which the DNA-binding domain had been deleted. Likewise, the promoter mutation suppressed the requirement for the activator to be located on the correct face of the helix with respect to the −24, −12 sequence, but efficient activation of the mutant promoter remained dependent on the presence of a UAS. Analysis of these promoters by *in vivo* dimethyl sulfate footprinting demonstrated that, whereas interaction of $E\sigma^{54}$ with the −24, −12 sequence cannot be detected with the wild-type *K. pneumoniae nifH* promoter, it can be detected with the mutant promoter and with the *R. meliloti nifH* promoter.[105] By using $KMnO_4$ footprinting the protection observed in the −24, −12 region can be shown to be due to the formation of a closed promoter complex, and the isomerization of this closed complex to an open complex is entirely dependent on the presence of NifA.[105]

Hence, the degree to which the −24, −12 sequence of a σ^{54}-dependent promoter differs from the consensus for that sequence determines the affinity of the promoter for $E\sigma^{54}$. When the affinity of the promoter for $E\sigma^{54}$ is low, a high local concentration of activator, as is achieved by binding to an upstream site, is critical for efficient interaction to take place and for the resultant open complex to be formed.[105] By contrast, if the affinity of the promoter for $E\sigma^{54}$ is high, then productive interactions between activator and polymerase can apparently take place without upstream binding of the activator. In intermediate situations (e.g., the *nifH* promoter mutant described by Buck and Cannon[21]), the requirement for upstream binding may be maintained but the requirement for stereospecific alignment of the activator is relaxed. Such considerations suggest that the fidelity of activation of σ^{54}-dependent promoters relies upon the formation of a relatively weak closed promoter complex that can only initiate transcription efficiently when the specific activator protein is bound at the appropriate upstream activator binding site.[21,103]

Recent studies have identified a third factor, in addition to NifA and $E\sigma^{54}$, that interacts with at least some *nif* promoters. Beynon et al.[12] reported protection from DNAaseI digestion of an AT-rich region from around −30 to

−60 in the *K. pneumoniae nifH, nifE, nifU,* and *nifB* promoters using protein extracts from a *K. pneumoniae nif* deletion strain. This factor has now been identified as integration host factor (IHF).[132] The binding of IHF to DNA is known to create bends in DNA,[126] and as the IHF binding site lies between the major NifA binding site (UAS) and the recognition site for $E\sigma^{54}$, it is possible that IHF facilitates bending of the DNA and thereby increases the likelihood of productive contacts between NifA and $E\sigma^{54}$. Whether an involvement of IHF is a characteristic of NifA-dependent *nif* promoters in other diazotrophs remains to be established.

F. Exceptions to the Rule

Although the great majority of diazotrophs appear to have *nif* promoters that are dependent on *rpoN* and *nifA* for their activation, this is not a universal rule and there are exceptions, although none of these is understood in any great detail.

i. Anabaena

In the heterocystous cyanobacterium *Anabaena* deprivation of ammonia or nitrate causes heterocyst differentiation, and within the heterocyst, nitrogenase and hydrogenase are induced together with glutamine synthetase, which is required for assimilation of fixed nitrogen. Nitrogenase synthesis is also controlled by a novel mechanism of gene rearrangement that occurs coordinately with heterocyst differentiation (see Chapter 4). Relatively little is known concerning the regulatory mechanisms controlling *nif* gene transcription. However both *nifH* and the glutamine synthetase gene (glnA) share promoter sequences that do not conform to the consensus −35, −10 or to the σ^{54}-specific −24, −12 sequence and that are utilized in conditions of N-limitation. It has therefore been suggested that these sequences may identify an *Anabaena nif–ntr* consensus sequence that is recognized by a novel form of RNA polymerase σ factor that mediates nitrogen control in this organism.[151]

ii. Clostridium

In *Clostridium pasteurianum* a total of six genes that hybridize to *K. pneumoniae nifH* and one gene homologous to *nifE* have been cloned and sequenced.[28,153] One of these genes (*nifH1*) encodes the Fe protein of the previously purified nitrogenase enzyme. Four of the other five *nifH* copies are definitely transcribed in N-limiting conditions and the functions of these copies are presently unknown, although they could encode one or more alternative nitrogenase Fe proteins. Inspection of the sequences upstream of these genes shows them all to have sequences typical of other Gram-positive organisms or *E. coli* in the -35 and -10 regions and no sequence homologies to consensus *nif* promoters to be present. The genes do share common upstream sequences with a consensus ATCAATAT-N_{6-10}-ATGGATTC in the -100 region, but the role of these sequences is not known.

iii. Archaebacteria

A number of species of archaebacteria have been shown to be capable of fixing N_2.[136] Nucleotide sequences have been determined for *nifH* homologues from three species (see Chapter 5), but no sequences corresponding to *nif* consensus promoters have been identified in these studies. Nevertheless in *Methanococcus themolithotrophicus* expression of ORF*nifH1* is regulated by the nitrogen source,[136] suggesting that a mode of nitrogen control distinct from the typical *ntr* system regulates *nif* transcription in these organisms.

iv. Desulfovibrio

The *nifH* gene of *Desulfovibrio gigas* has been sequenced and a potential *rpoN*-dependent promoter sequence is present upstream of the coding sequence.[78] An upstream activator sequence homologous to those in NifA-dependent promoters was not found and no significant expression of a *pnifH–lacZ* fusion was detected in *E. coli*. By contrast, the *D. gigas nifH* promoter was nitrogen regulated in *K. pneumoniae*, but this nitrogen control was not affected by *rpoN, ntrC,* or *nifA* mutations. It is not known whether the control observed in *K. pneumoniae* reflects a mechanism operative in *D. gigas* and what the mediators of this control are.

III. Factors Controlling NifA Expression and Activity

As discussed earlier, the majority of *nif* promoters depend for their expression on an RNA polymerase holoenzyme containing the *rpoN* encoded sigma factor σ^{54}, together with an activator protein encoded by *nifA* or its homologues (e.g., *vnfA* and *anfA* in *A. vinelandii*). Expression of *rpoN* is essentially constitutive and there is no evidence for marked posttranslational modification of σ^{54} activity. Hence, the major control of *nif* expression is mediated by regulation either of *nifA* expression or of NifA activity. The means by which this regulation is achieved varies considerably from one organism to another, although certain common features are apparent. For this reason the following discussion is divided according to the organism.

A. *K. pneumoniae*

In *K. pneumoniae nif* gene expression is subject to two major controls—namely, by the nitrogen status and by the oxygen status of the cells—and these factors have effects both on *nifA* transcription and on NifA activity. Of the two levels of control the mechanisms regulating transcription are by far the best understood. The *K. pneumoniae nifA* gene is coordinately transcribed with *nifL,* and the *nifLA* operon has a σ^{54}-dependent promoter for which the activator protein is NtrC. It is the activity of NtrC that is ultimately regulated by the N-status of the cells, thereby determining whether *nifA* is expressed.

In enteric and many other bacteria there is a general nitrogen regulation (*ntr*) system that controls the expression of many genes concerned with nitrogen metabolism. This *ntr* system comprises four gene products; a uridylyltransferase (UTase) encoded by *glnD*, a small tetrameric effector protein (P_{II}) encoded by *glnB* and a pair of regulatory proteins encoded by a single operon *ntrBC*. The system has been elegantly dissected at both the genetic and biochemical level in enteric bacteria (for review see ref. 124) and is understood in some detail.

The UTase, a 90-kDa monomer, is considered to be the primary sensor of the cellular N status and responds to the intracellular ratio of the α-ketoglutarate to glutamine pools. When cells are N-limited, UTase mediates the uridylylation of P_{II} by transfer of a uridylyl group onto a tyrosine residue on each of the four P_{II} subunits (Figure 21-3). The uridylylated form of P_{II} (P_{II}UMP) promotes the phosphorylation of NtrC (a dimer of subunit molecular mass 56 kDa) by NtrB (a dimer of subunit molecular mass 34 kDa). NtrB achieves the phosphorylation of NtrC in a two-step process involving an autophosphorylation step (in which a histidine residue in NtrB is phosphorylated) followed by transfer of the phosphate to an aspartate residue in the N-terminus of NtrC.[75,142,154] The phosphorylation of NtrC stimulates its DNA binding properties and renders it able to function as a transcriptional activator.[92,102,110]

Under conditions of N-excess this cascade of events is reversed. UTase now acts as a uridylyl-removing enzyme, converting P_{II}UMP to its deuridylylated form. P_{II} no longer stimulates the kinase activity of NtrB, and NtrB now promotes the dephosphorylation of NtrC (see Figure 21-3). As a result, the activator and DNA-binding properties of NtrC are diminished and expression from NtrC-dependent promoters is switched off. The UTase, P_{II} system also regulates posttranslational modification (by adenylylation) of glutamine synthetase (GSI) in many organisms and is probably a widespread nitrogen-sensing system in prokaryotes.

The *ntrBC* genes are present in many other bacteria, including rhizobia (see later), and their products are members of a large family of bacterial

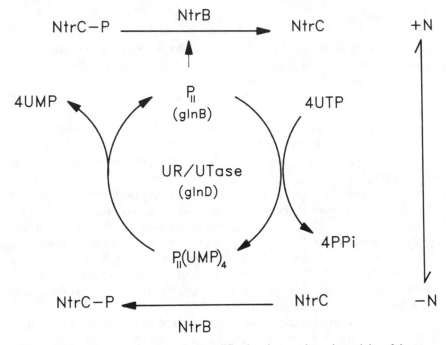

Figure 21-3. The cascade of covalent modification that regulates the activity of the transcriptional activator NtrC in response to changes in intracellular nitrogen status. UR/Utase = uridylyl-removing enzyme, uridylyltransferase. +N, nitrogen excess; −N, nitrogen limitation.

regulatory proteins, all of which probably function in a similar manner.[142] In most cases a pair of proteins acts in concert to coordinate the regulation of a group of genes in response to an environmental stimulus. One of the pair is a histidine protein kinase that acts as a sensor protein, and the partner protein or "response regulator" is usually (but not always) a transcriptional activator. The response regulator can be phosphorylated by transfer of a phosphoryl group from the kinase to an aspartate residue in the N-terminal domain of the regulator, thereby provoking some conformational change in the receptor and modifying its regulatory properties (e.g., DNA binding). Similar regulatory protein pairs are involved in such diverse functions as chemotaxis, osmotic regulation, sensing of phosphate levels and regulation of dicarboxylic acid transport. One consequence of the conservation of the mode of action of these diverse protein pairs is the possibility that a sensor protein from one pair may be able to modulate the activity of the receptor protein of another pair, and such "cross-talk" has been demonstrated experimentally.[109] Hence the cell has the ability to integrate environmental stimuli that may affect different metabolic processes to different degrees.

In the *K. pneumoniae nifLA* promoter there are two tandem NtrC binding sites 120–140 bp upstream of the −24, −12 consensus.[102,155] Neither of these sites is highly homologous to the consensus NtrC binding site (GCAC-N7-GTGC), and as such they are predicted to have a low affinity for NtrC. By comparison the *glnA* promoter (*glnA*p2) contains a number of upstream NtrC binding sites, the major one of which is homologous to the consensus sequence,[56] and *in vitro* experiments indicate that 10-fold lower concentrations of NtrC are required to activate *glnA*p2 transcription than are required for *nifLA* transcription.[6] The consequence of this difference in NtrC affinity between *pnifLA* and *glnA*p2 is that when cells become N-limited and NtrC becomes phosphorylated a significantly greater concentration of NtrC-P must build up in the cell for activation of *pnifLA* than for *glnA*p2. Hence, the *nif* regulon is only expressed under conditions of fairly severe N-limitation. The NtrC binding sites function cooperatively to activate *nifLA* expression, and this activation is face-of-the-helix dependent (Figure 21-4) such that the introduction of 5- or 15-bp spacers between the −24, −12 sequence and the NtrC binding sites depress promoter activity.[103]

The properties of the *nifLA* promoter determine its response not only to N-limitation but also to oxygen limitation. Studies with *nif–lac* fusions demonstrated that *pnifLA* was far less sensitive to the extracellular oxygen concentration than the other *nif* promoters but that *pnifLA* expression was reduced by dissolved O_2 tensions (DOTs) greater than 600 μM.[36] It has been suggested that DNA supercoiling may play a role in the anaerobic induction of some genes,[157] and Kranz and Haselkorn[86] found that *nif* expression in *K. pneumoniae* was prevented by an inhibitor of DNA gyrase. Hence, it was suggested that anaerobic–aerobic regulation of *nif* transcription might be controlled by the level of DNA supercoiling.

Dissection of this effect shows that the *K. pneumoniae nifH* promoter is not sensitive to the DNA gyrase inhibitor coumermycin A1 under anaerobic growth conditions, whereas the *nifLA* promoter is. Furthermore, *pnifLA* activity is highly sensitive to changes in DNA topology *in vitro* and transcription is completely dependent on negative supercoiling at physiological salt concentrations.[37]

Consistent with this result is the observation that *nifLA* expression is significantly decreased in a gyrase mutant of *E. coli*.[34] In summary, the O_2 sensitivity of the *nifLA* promoter appears to be a consequence of its dependence on DNA supercoiling for activity, although precisely which aspect of transcriptional activation is affected is not yet known.

Although regulation of *nifLA* transcription is clearly a major factor in nitrogen and oxygen control of *nif* expression in *K. pneumoniae*, the most immediate effect of changes in these environmental factors is on NifA activity. Studies with *nif–lac* fusions showed that *pnifLA* is far less sensitive to changes in either oxygen or nitrogen status than are the other *nif* pro-

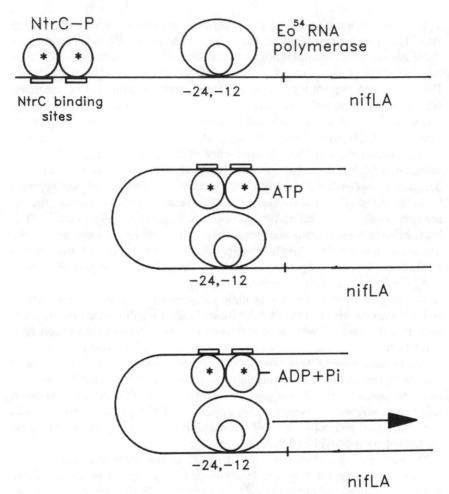

Figure 21-4. The interaction of NtrC and Eσ⁵⁴ polymerase in transcriptional activation of the *nifLA* promoter of *K. pneumoniae*.

moters. Furthermore, the sensitivity of the other promoters was abolished by mutations in *nifL*.[97] Hence, it was proposed that NifL acts as a sensor of changes in both the fixed nitrogen and oxygen status of the cell and responds to increases in either of these factors by inhibiting NifA activity in some manner.

This model has much in common with that discussed earlier for the family of protein pairs to which NtrB and NtrC belong. However, NifL and NifA do not appear to belong to this family. The primary amino acid sequence of NifL shares little homology with the sensor family of proteins that includes NtrB,[41] and the domain that contains the conserved histidine residue is not present. The most C-terminal portion of NifL is homologous to that of NtrB and related proteins, but the significance of this homology is not known. Likewise, the N-terminal domain of NifA is not homologous to that of NtrC and does not contain a sequence homologous to that which contains the conserved aspartate residue in the receptor family. Furthermore, although many of the regulatory protein pairs such as *envZ* and *ompR* or *ntrB* and *ntrC* are coordinately transcribed, the sensor protein often produced at much lower levels than the receptor protein, as might be expected if the sensor protein has a catalytic function. By contrast, NifL and NifA appear to be synthesized in more or less equimolar amounts, which is more in keeping with two proteins which interact to form a complex; indeed, immunochemical evidence suggests that such a NifLA complex can be formed.[58] Finally, as NtrC phosphorylation requires NtrB, deletions of *ntrB* significantly reduce transcriptional activation by NtrC.[92] Deletions of *nifL,* on the other hand, have no effect on NifA-mediated activation, so that NifL is clearly only required to inactivate NifA and not to activate it.[5] When the N-terminal domain of NifA is deleted, the mutant protein becomes particularly sensitive to inhibition by NifL, such that NifA activity is inhibited even in the absence of O_2 or fixed nitrogen.[40] This result suggests that the NifA N-terminus normally functions to block the action of NifL in derepressing conditions and that NifL acts directly upon the central and/or C-terminal domain of NifA by protein–protein binding.

So what are the signals to which NifL responds? With respect to nitrogen sensing one possibility is that NifL interacts with the known nitrogen sensor P_{II}, but this appears not to be so. Mutations in the P_{II} structural gene, *glnB*, which prevent the protein from being uridylylated, do not affect N-sensing by NifL.[63] Apparently, there is another (as yet unidentified) nitrogen-sensing system present in *K. pneumoniae* that operates independently of P_{II}.

The method by which NifL responds to the cellular O_2 status is also unknown at present. Drummond and Wootton[41] identified an amino acid sequence (Cys–Ala–Asp–Cys–Gly) in NifL that they suggested might be part of a haem binding site in the protein, but subsequent studies demonstrated that haem is not involved in oxygen control by NifL.[138] The Cys–Ala–Asp–Cys–Gly motif is, however, strongly homologous to the Cys–X–X–Cys–Gly motif required for the coordination of iron atoms in rubredoxins and ferredoxins,[1] and the observation (described earlier) that metal ions have a role in the response of rhizobial NifA proteins to O_2[46] raised the possibility that NifL may function in a similar manner.

The ability of NifL to antagonize NifA-mediated transcriptional activation of the *K. pneumoniae nifH* promoter is indeed inhibited by metal deprivation or the presence of iron chelators in the medium.[58] This inhibition is reversed by the addition of ferrous or manganous ions to be the medium but is unaffected by other transition metals. The activity of NifA alone is insensitive to the presence of metal ions. The effect of metal deprivation was apparent in conditions of nitrogen excess or high DOT, so the effect is not specific to the O_2-sensing properties of the protein; rather, it appears that the metal is required to maintain NifL in an active (repressive) form irrespective of the redox environment.

B. Azotobacter

Regulation of nif expression in Azotobacter is complicated by the ability of the organism to synthesize alternative nitrogenases, two in A. chroococcum and three in A. vinelandii. Expression of all three nitrogenase systems is repressed by high concentrations of fixed nitrogen, and expression of the individual systems is determined by the metal availability (see Chapter 19).

Both A. vinelandii and A. chroococcum have a major nif gene cluster containing homologues of all the K. pneumoniae nif genes except for nifA, B, and Q, which are in a separate cluster. These two clusters also contain a number of ORFs with no homologues in K. pneumoniae (see Chapter 19). Certain nif genes—namely, nifU, S, V, M, and B—are apparently required for synthesis of all the alternative enzymes and must therefore be expressed regardless of the metal content of the medium[71] (Kennedy et al., unpublished). Other genes are system-specific, so that synthesis of the vanadium enzyme by A. vinelandii or A. chroococcum requires a specific set of structural genes, vnfH, D, G, K and homologues of nifEN designated vnfEN.[15] Likewise, synthesis of the third nitrogenase by A. vinelandii requires another set of structural genes, anfH, D, G, K, but may utilize vnfEN (see Chapter 19). Finally, as described earlier, each of the alternative gene clusters also has its own nifA homologue (i.e., vnfA and anfA).

The interrelationship of these different gene systems and the coordination of their regulation have been investigated using mutations in different regulatory genes and lacZ fusions to various promoters. Sequence analysis of the nif, vnf, and anf genes revealed that they all (with the exception of nifA, where the promoter has not yet been identified) have well-conserved −24, −12 promoter sequences characteristic of σ^{54}-dependent operons,[70,71,73,121,127,128] and as predicted, all three systems are inactivated in an rpoN mutant[149] (Kennedy et al., unpublished). As already discussed, the nif operons have well-conserved UAS motifs and in the case of nifHDK the expected NifA dependence has been demonstrated.[10] No characteristic upstream notif has yet been identified in the promoters of vnf and anf genes, although given the homology between nifA, vnfA, and anfA such sequences might be expected to exist. In summary, it seems likely that the transcriptional activation of each of the three gene systems occurs in a similar fashion to that described for K. pneumoniae. Discrimination between expression of different systems must therefore occur either by selective expression of the different *nifA* homologues or by control of the activity of their products.

At present the precise pathway of regulation of the three *nifA* homologues in *A. vinelandii* is unknown. Homologues of *ntrB* and *ntrC* are present in *A. vinelandii* but, unlike *K. pneumoniae, ntrC* mutations do not affect expression of any of the three nitrogenase systems[149] (Walmsley et al., unpublished). An ORF immediately upstream of *nifA* encodes a polypeptide with significant homology to NifL and NtrB.[10] This ORF appears to encode a *nifL* homologue and mutations in this gene relieve nitrogen repression of *nif* and result in excretion of significant amounts of ammonia during diazotrophic growth on molybdenum (Bali et al., unpublished). A short ORF is located upstream of *vnfA* but this is not apparently homologous to any previously identified gene. The DNA sequence of *anfA* suggests that it is transcribed as a single cistron.[72]

Selection for Nif⁻ regulatory mutants in *A. vinelandii* has identified a further regulatory locus designated *nfrX*. Mutations in *nfrX* affect expression of the *nif* and *anf* systems but not of the *vnf* system and mutants can be restored to Nif⁺ by constitutively expressed *nifA* from *K. pneumoniae* or *A. vinelandii*.[133] The nucleotide sequence of *nfrX* has been determined and it is predicted to encode a 105kDal polypeptide, the N-terminus of which is highly homologous to that of the uridylyltransferase (*glnD* product) of *E. coli* (Contreras et al., unpublished). *In vivo* complementation experiments have shown that the *glnD* and *nfrX* products are functionally interchangeable and *nfrX* therefore appears to encode a UTase-like protein. The Nif⁻ phenotype of *nfrX* mutants is supressed by mutations in *nifL* suggesting that the *nfrX* product modifies, directly or indirectly, the activity of the *nifL* product (Contreras et al., unpublished).

C. Rhodobacter

Four potential *nif* regulatory genes were identified in *R. capsulatus* by transposon mutagenesis[85] and designated R1, R2, R3, and R4. Subsequent hybridization and sequencing experiments showed that R1 and R2 are homologous to *ntrC* and *ntrB*, respectively, and R4 is homologous to *rpoN*.[74,85] The proposed *nifR3* gene lies immediately upstream of *ntrBC* (*nifR2R1*), but Jones and Haselkorn[74] were unable to identify an open reading frame in a region of 4.7 kb upstream of *ntrB* (*nifR2*), and the reason for the Nif⁻ phenotype of *nifR3* mutations is presently unclear. Mutations in *ntrB* and *ntrC* are Nif⁻ but do not have an Ntr⁻ phenotype (i.e., they can utilize proline, glutamine, or arginine as nitrogen sources). At least two more regions homologous to *ntrBC* have been identified in *R. capsulatus*,[87] and therefore the genes identified as *nifR1* and *nifR2* may be *nif*-specific in their action. Expression of *ntrBC* increases only two-fold in derepressing conditions, whereas *rpoN* expression increases 10-fold[85] and the majority of *rpoN* transcription is NtrC dependent.[88]

Three classes of *R. capsulatus* mutants have been isolated in which *nif* gene expression is not repressed by ammonium or glutamine in the medium.[87] The gene mutated in one of these classes (*nifR5*) has been identified as *R. capsulatus glnB*.[87,88] The two other classes were not fully analyzed, but one class was selected in a *ntrBC* deletion background, suggesting that in these strains the mutation allows *nif* expression to be activated by an NtrC homologue. The ability to select ammonium constitutive mutants of this type suggests that *R. capsulatus* may not have a *nifL* homologue but that ammonium repression may be exerted directly at the level of *nifA* expression. Alternatively, if a NifL equivalent is present then its activity would be predicted to be regulated by P_{II}, unlike the situation in *K. pneumoniae*.

Genetic studies in *R. capsulatus* have shown that *nifA* and *nifB* are duplicated in this organism,[82] and DNA sequence analysis indicated that the two copies of *nifA* are identical.[95] Mutations in either copy of *nifA* do not prevent N_2 fixation, although they reduce expression of NifD and NifK polypeptides. A double *nifA* mutant is unable to fix N_2. Recent studies have identified an alternative nitrogenase system in *R. capsulatus* analogous to the third system of *A. vinelandii* (Klipp et al., unpublished), and it is possible that one of the *nifA* copies has a role in regulation of this system.

The DNA sequences extending at least 130 bp upstream of each of the *nifA* copies show very little homology to each other and do not contain sequence characteristic of -35, -10 or σ^{54}-dependent promoters. Therefore, it is not clear at present whether *nifAI* and/or *nifAII* are cotranscribed with genes located upstream or whether they are expressed from promoters immediately upstream. Expression of *nifAII* is strictly dependent upon NtrC but not on RpoN,[88] and expression studies with *nifAI* have not yet been reported.

In summary, expression of at least *nifAII* is likely to be subject to nitrogen control by a conventional *ntr* system, and *rpoN* is probably under similar control. The presence of a potentially *nif*-specific form of RpoN is apparently unique to *R. capsulatus*. Oxygen control of *nif* expression in *R. capsulatus* is postulated to be regulated both by DNA supercoiling[86] and by the oxygen sensitivity of NifA activity (see earlier).

D. Other Free-Living Diazotrophs

Very little is known about factors controlling NifA expression and activity in other free-living N_2 fixers. In *Azospirillum,* mutants with *nifA* and *ntrC*-like phenotypes have been reported,[49,117,137] but only in *Herbaspirillum seropedicae* has a *nifA* gene been cloned and sequenced (de Souza and Yates, unpublished). The sequence upstream of *H. seropedicae nifA* contains a σ^{54}-

dependent promoter sequence and potential NtrC and NifA binding sites, suggesting that the gene may be regulated by both activators under certain conditions. The *glnA* gene of *A. brasilense* has been cloned and sequenced, and an ORF immediately upstream of *glnA* encodes a polypeptide that is highly homologous to P_{II}.[17,64] This location for the P_{II} gene is also found in *Rhizobium*[30,62,64,94] and *R. capsulatus,*[88] and although P_{II} may well be involved in nitrogen control of *nif* expression in *A. brasilense,* mutations in this gene have not yet been reported.

A plasmid carrying the cloned *glnA* gene from *Thiobacillus ferrooxidans* is reported to suppress the Ntr⁻ phenotype of an *E. coli ntrBC* deletion strain, suggesting that *T. ferrooxidans* may have a functional equivalent of NtrC, but this has not been characterized further.[8]

E. Rhizobiaceae

In the three N_2-fixing genera of *Rhizobiaceae nif* genes are regulated in a NifA-, σ^{54}-dependent manner. However, the regulation of N_2 fixation in these organisms differs from that in the free-living diazotrophs in that the major factors controlling *nif* expression appear to be the oxygen status of the cells and the state of development of the symbiosis. In the symbiotic state the organism exports fixed nitrogen to the host plant, and therefore *nif* expression is not expected to be repressed by the presence of ammonium or glutamine. Consequently, although regulation is still ultimately effected by control of the activity or expression of NifA, the mechanisms involved differ from those described in free-living diazotrophs. The organization and regulation of *nifA* has been studied in each of the three genera, and in each case there are significant differences (Figure 21-5).

i. Rhizobium

Three species of *Rhizobium* have been analyzed—*R. meliloti, R. leguminosarum,* and *R. trifolii*—and *nifA* regulation in these three organisms appears to be essentially the same. In each case *nifA* is located downstream of the *fixABCX* gene cluster and upstream of *nifB*.[26,42,52,67] The precise transcriptional organization has been studied in most detail in *R. meliloti,* where it has been shown that *nifA* can be transcribed either from the *fixABCX* promoter (P2) or from a promoter between *fixABCX* and *nifA* (*pnifA*)[80] (Figure 21-5). The *fixABCX* promoter is a typical σ^{54}-dependent promoter, whereas no characteristic promoter features are apparent in the sequence upstream of *nifA*. However, a comparison of *pnifA* in all three species of *Rhizobium* does identify an AT-rich region with a consensus TAATa/tTT between −30 and −40 with respect to the transcript start.[67]

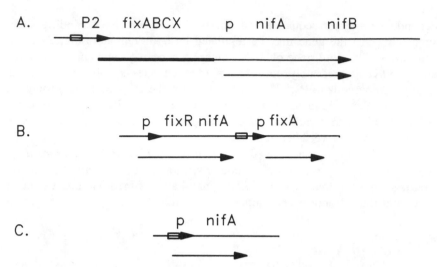

Figure 21-5. The transcriptional organization of the *nifA* gene in the three major groups of Rhizobiacea. (A). *Rhizobium* sp.; (B) *B. japonicum*; (C) *A. caulinodans.* ⟶, −24, −12 promoter sequence; □, consensus NifA binding site.

Both P2 and p*nifA* have little activity during vegetative growth but are expressed significantly during symbiosis when at least 50% of transcripts originating at P2 continue on into *nifA*.[80] The P2 promoter can be activated by NifA and therefore at the onset of symbiosis. Once a significant level of NifA has been produced from p*nifA*, activation at P2 would allow read-through transcription to boost NifA levels further. In free-living conditions the low level of expression of *fixABCX* from P2 is entirely dependent on NtrC;[145] but because N_2 fixation by *R. meliloti* in free-living, nitrogen-limiting conditions has not been observed and as *ntrC* is not required for symbiotic N_2 fixation, this NtrC-specific activation of *nifA* expression may not be significant during symbiosis.[145]

Although *nifA* expression is very low in free-living *R. meliloti,* the gene can be induced asymbiotically to levels exceeding those in nodule bacteroids when the external O_2 concentration is reduced to microaerobic levels.[35] This induction can occur in the presence of fixed nitrogen, and NifA function is independent of fixed nitrogen availability. Deletion analysis of the *nifA* promoter demonstrated that a region between −62 and −45 is essential for induction, suggesting that this region may be required for binding of a positive activator that is capable of responding to microaerobic conditions.[152]Mutants which are defective in microaerobic induction of *nifA* have been isolated[152] and localized to a cluster of *fix* genes on the pSym megaplasmid of *R. meliloti* that had previously been shown to be symbiotically activated in a *nifA*-independent manner.[32]

The *fix* cluster identified by David et al.[32] contains at least four operons and more than nine genes.[9] By studying the patterns of *nif* and *fix* transcription in Tn5 mutants of the *fix* cluster, two genes, designated *fixLJ*, were identified as having a regulatory role.[31] Expression of p*nifA* is reduced more than 10-fold in *fixL* and *fixJ* mutants, suggesting that the products of *fixLJ* are involved in activation of p*nifA* expression. *FixLJ* are also required for expression of the linked *fixN* gene cluster which, like *nifA*, is induced in microaerobic conditions.[31]

The DNA sequence of *fixLJ* showed that the two genes comprise a single operon and encode polypeptides of 51 kDa (FixL) and 23 kDa (FixJ), respectively. These two proteins are highly homologous to other members of the two-component regulatory systems described earlier. The C-terminus of FixL is homologous to the C-termini of the sensory class of proteins such as NtrB and EnvZ, and its N-terminus contains two potential transmembrane helices, which is consistent with FixL being a membrane-bound protein. FixJ does not show any homology to the central domain of NtrC, and the other σ^{54}-dependent activators. However, the N-terminus of FixJ is homologous to that of the transcriptional activator class, including NtrC and OmpR, and its C-terminus is homologous to a subclass of these activators, which includes *E. coli* UhpA, a regulator of the sugar-phosphate transporter gene *uhpT*. These homologies led David et al.[31] to propose a model in which the transmembrane protein FixL acts as a sensor of microaerobiosis and transduces this signal to FixJ by covalent modification, in a manner similar to that described for NtrB and NtrC. FixJ would then directly or indirectly stimulate transcription of *nifA* and *fixN* (Figure 21-6). The C-terminal domain of FixJ does not share strong homology with the helix-turn-helix DNA-binding motif of NtrC, and although a weak motif may be present in FixJ and the homologous protein UhpA, a direct interaction of FixJ with the *nifA* and *fixN* promoters remains to be demonstrated. When *fixJ* is cloned under control of the *lac* promoter, it is sufficient to activate expression of the *R. meliloti nifA* promoter in *E. coli*.[59] This expression is independent of the O_2 status of the cells, as expected if FixL is required for O_2 sensing. The ability of FixJ to activate in the absence of FixL is probably attributable to "crosstalk" between FixJ and other *E. coli* proteins homologous to FixL.

Surprisingly, FixJ is not sufficient to activate *fixN* expression in *E. coli*.[9] This observation led to the identification of a further regulatory gene, *fixK*, within the same *fix* cluster, which is required for activation of *fixN* expression both microaerobically and symbiotically. FixK is not required for activation of *nifA* expression, on the contrary it appears to regulate *nifA* negatively. The nucleotide sequence of *fixK* indicates that it encodes a polypeptide with significant homology to the *E. coli* cAMP receptor protein Crp and the *E. coli* oxygen regulatory protein Fnr. The most marked homology is to the helix-turn-helix DNA-binding motif of Crp and Fnr, strongly suggesting that

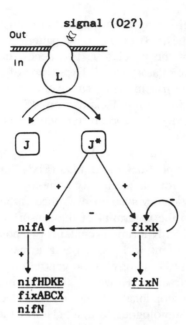

Figure 21-6. Model for the regulation of *nif* gene expression in *R. meliloti* by the products of the *fixJ, fixL,* and *fixK* genes. J*, proposed covalently modified form of FixJ. +, positive control; −, negative control. (From Batut et al., 1989.)

FixK is a DNA-binding protein. Expression of *pfixK* is strictly dependent on *fixL* and *fixJ* and is strongly induced during symbiosis and by microaerobiosis. Furthermore, *fixK* is subject to negative autoregulation, as are the homologous genes *crp* and *fnr* in *E. coli*. The precise physiologic role of FixK is unclear. The homology between FixK and Fnr suggests a possible role in oxygen regulation, but FixK clearly does not play a primary role in oxygen control, since both *pnifA* and *pfixK* respond to microaerobiosis in the absence of FixK.[9] Alternatively, FixK may respond to some other physiological signal the nature of which is still unknown.

The identification of *fixK* led to an elaboration of the model proposed by David et al.,[31] in which *nifA* expression is regulated both positively, by FixJ, and negatively by FixK[9] (see Figure 21-6). The strong conservation of amino acid sequence between the DNA-binding motifs of FixK and Fnr suggests that they may recognize similar DNA sequences, and indeed the promoter of *fixN,* which is positively controlled by FixK, contains a sequence TTGAT–N4–ATCAA that agrees with the consensus Fnr binding site proposed by Spiro and Guest.[139] Despite the fact that both *nifA* and *fixK* are activated by FixJ, no striking homology can be detected between their promoter sequences, and the mechanism of activation of these promoters requires further analysis together, perhaps with the identification of other FixJ-dependent promoters.

NifA regulation in *R. leguminosarum* appears to differ somewhat from that in *R. meliloti*. A DNA region capable of promoting induction of *R. meliloti fixN* was cloned from *R. leguminosarum* by heterologous complementation and found to contain an ORF with the potential to encode a 27-kDa polypeptide (FnrN) with significant similarity to Fnr and FixK.[29] When compared with FixK, both FnrN and *E. coli* Fnr (EcFnr) contain an additional N-terminal domain that includes a cysteine motif (Cys–X_2–Cys–X_3–His–X_3–Cys in FnrN and Cys–X_2–His–Cys–X_2–Cys–X_5–Cys in EcFnr). This domain has no counterpart in FixK, and by analogy with EcFnr it could be involved in the oxygen-dependent activity of FnrN.[29] Furthermore, *fnrN* expression appears to be independent of *fixJ*, so that in *R. leguminosarum* FnrN activity could respond directly to changes in oxygen status instead of expression being regulated by FixLJ. The *fnrN* promoter contains two potential Fnr binding sites, and the gene could therefore also be autoregulatory.

ii. Azorhizobium

The *nifA* gene of *Azorhizobium caulinodans* strain ORS571 is apparently transcribed as a monocistronic operon (see Figure 21-6) and the transcript start is consistent with the use of a σ^{54}-dependent promoter sequence.[122] In the free-living state *nifA* expression is subject to both nitrogen control and oxygen control, being repressed completely by rich nitrogen sources such as ammonium or nitrate and being reduced fivefold in the presence of air. Studies with *ntrC* mutants suggest that the nitrogen control is at least in part mediated by NtrC, but *nifA* expression can still be repressed further in an *ntrC* mutant, indicating that some of this regulation is NtrC-independent. No autoactivation by NifA was detected; in fact, NifA appears to negatively control its own expression. A potential NifA binding site is located at -82 and could account for this negative control. This situation is similar to that described by Cannon et al.[27] in the *K. pneumoniae nifU* promoter, where NifA competes with IHF for binding at a site around -80.

Both Ratet et al.[122] and Nees et al.,[108] who sequenced *A. caulinodans nifA* independently, identified a nucleotide sequence TTGAT–N4–ATCAA at position -134, which is highly conserved in the promoters of anaerobically induced *E. coli* genes, such as *nirB* and *narG*. In *E. coli* these genes are positively controlled by the Fnr protein; on the basis of this observation, Ratet et al.[122] constructed an Fnr$^-$ strain of *A. caulinodans* and found that *nifA* expression was significantly less O_2 sensitive in this background. These data suggest that an Fnr-like control system may mediate *nifA* repression in the presence of O_2, and an obvious candidate for such control would be a homologue of the *R. meliloti* FixK protein[9] described earlier.

iii. Bradyrhizobium

In *Bradyrhizobium japonicum* the *nifA* gene is part of an operon, *fixRnifA*, which is not preceded by *fixABCX* as in *Rhizobium* (see Figure 21-6). In *B. japonicum* the *fixABCX* genes are split with *fixA* being immediately downstream of *fixRnifA* and *fixBCX* being unlinked.[54,147] The predicted *fixR* product is a polypeptide of 29.6 kDa that is not homologous to any previously identified regulatory proteins in other diazotrophs, and mutations in *fixR* do not impair N_2 fixation.[147] The *fixRnifA* operon is expressed aerobically, anaerobically, and in bacteroids; and in all cases the transcript start is the same and is consistent with the use of a σ^{54}-dependent promoter sequence.[147]

Deletion analysis of the *fixRnifA* promoter indicates that expression is enhanced significantly by an upstream activator binding site between nucleotides -24 and -86. Protein extracts from *B. japonicum* wild-type, *fixR* mutant, and *nifA* mutant cells all retard a 32-bp synthetic double-stranded oligonucleotide corresponding to the *fixR* upstream sequence from -50 to -81. This retardation is observed whether cells were grown aerobically or anaerobically.[146] Finally, when the expression of chromosomal *fixRnifA* is examined in *B. japonicum*, and *fixR* promoter is subject to significant autoactivation by NifA in anaerobic conditions.[146] This activation is not dependent on sequences upstream of -50. Hence, *B. japonicum nifA* appears to be under dual control. There is a basal level of expression in aerobic conditions from a -24, -12 promoter activated by an as yet unidentified *trans*-acting protein factor, and the NifA synthesized under these conditions is inactive. Expression is elevated anaerobically when O_2 limitation allows the formation of active NifA and a fivefold increase in transcription as a result of autoactivation at the *fixRnifA* promoter.

With respect to regulation of NifA activity, as described earlier the translation products of all the rhizobial *nifA* genes sequenced so far have the Cys–X–X–X–X–Cys motif between the central and C-terminal domains, and it therefore seems likely that all rhizobial NifA proteins are O_2 sensitive. There is no evidence for a NifL homologue in any rhizobia and with the possible exception of the FixR product of *B. japonicum* the current evidence suggests that the activity of rhizobial NifAs is not modulated by other proteins.

IV. Conclusion

This review of the regulation of N_2-fixation genes, although not exhaustive, has attempted to highlight some general mechanisms that are emerging as our understanding of *nif* gene control becomes more complete and as more diazotrophs are being studied. The information we have to date allows some general conclusions to be drawn and some fairly comprehensive models to be developed.

The positive control of *nif* gene expression by NifA together with RNA polymerase and RpoN is very widespread and indeed almost universal among the major organisms in which genetic analysis of *nif* control has been carried out to date. There is presently no report of the occurrence of an RpoN homologue in Gram-positive bacteria, in cyanobacteria, or in archaebacteria. Thus, it is not surprising that these groups have different, albeit presently poorly understood, systems of control. However, with these exceptions the mechanism of transcriptional activation of *nif* genes seems likely to be highly conserved.

With respect to the environmental stimuli which ultimately control either NifA expression or activity, there are clearly two main factors—namely, the availability of fixed nitrogen sources and the levels of O_2 in the immediate environment. The degree to which either or both of these factors affect *nif* expression depends on the physiology and ecology of the organism concerned. Hence, nitrogen control is a major influence in free-living diazotrophs that are fixing N_2 for their own immediate needs but less so in symbiotic organisms that export fixed nitrogen to their hosts. However, the distinction is not absolute, as exemplified by *Azorhizobium caulinodans,* in which *nif* expression is subject to some nitrogen control. It has been proposed that this control may reflect the physiological conditions to which *A. caulinodans* is exposed while infecting the stems of the host plant *S. rostrata*.[122] The organism's proliferation in the stem infection site, in the absence of fixed nitrogen sources (e.g., from the soil), may require diazotrophic growth of the bacteria before the intracellular symbiotic state is established and N-sources are available from the plant host. Furthermore, nitrogen control of *nif* expression in *A. caulinodans,* which is primarily a symbiotic N_2 fixer, appears to use the same regulatory apparatus (i.e., the NtrBC system) as that used in free-living organisms.

Likewise, although different organisms have apparently evolved different ways of dealing with the problem of oxygen regulation, these differences are not allied uniquely to the symbiotic or nonsymbiotic lifestyle of the organism. For example, *Rhodobacter* sp. has an O_2-sensitive NifA protein similar to that found in the rhizobia, and whether this reflects some particular aspect of the organism's physiology or just an underlying evolutionary relationship remains to be determined.

One significant environmental factor that has received little attention until recently, in terms of possible regulatory effects, is the availability of such metals as molybdenum. Given the clear effects of metal availability on the expression of alternative nitrogenase genes in *Azotobacter,*[16] this is an area that will undoubtedly receive much more attention in the future.

Finally, it should be noted that the study of *nif* gene regulation has revealed a number of fundamental aspects of prokaryotic gene regulation that have implications far beyond the immediate area of N_2 fixation. The analysis of *nif* promoter structure and function was a major factor in the recognition of the novel sigma factor encoded by RpoN, and systems regulated by RpoN are now known to be widespread in bacteria (see refs. 90, 148). Similarly, the analysis of the role of *ntrB* and *ntrC* (both of which were initially sequenced in *K. pneumoniae*) in N-control of *nif* expression contributed to the recognition of the widespread family of two-component regulatory proteins.[112]

There is still undoubtedly much to be learned from further studies of these systems, particularly with respect to the detailed molecular mechanisms of activation of gene transcription and in the understanding of the processes of signal transduction that link gene expression to changes in the cell's oxygen or nitrogen status. Such studies may yet uncover further fundamental concepts as well as provide the basis for manipulation of the N_2-fixation process so as to improve its contribution to the world's agriculture.

Acknowledgments

I would like to thank my colleagues Martin Buck, Ray Dixon, and Christina Kennedy for their constructive comments during the writing of this chapter.

References

1. Adman, E., K. D. Watenpaugh, and L. H. Jensen, "NH—S hydrogen bonds in *Peptococcus aerogenes* ferredoxin, *Clostridium pasteurianum* rubredoxin and *Chromatium* high potential iron protein," *Proc. Natl. Acad. Sci. USA 72*:4854–4858 (1975).

2. Albright, L. M., C. W. Ronson, B. T. Nixon, and F. M. Ausubel, "Identification of a gene linked to *Rhizobium neliloti ntrA* whose product is homologous to a family of ATP-binding proteins," *J. Bacteriol. 171*:1932–1941 (1989).

3. Alias, A., F. J. Cejudo, J. Chabert, J. C. Willison, and P. M. Vignais, "Nucleotide sequence of wild-type and mutant *nifR4* (*ntrA*) genes of *Rhodobacter capsulatus:* Identification of an essential glycine residue," *Nucleic Acids Res. 17*:5377 (1989).

4. Alvarez-Morales, A., M. Betancourt-Alvarez, K. Kaluza, and H. Hennecke, "Activation of the *Bradyrhizobium japonicum nifH* and *nifDK* operons is dependent on promoter-upstream DNA sequences," *Nucleic Acids Res. 14*:4207–4227 (1986).

5. Arnott, M., C. Sidoti, S. Hill, and M. Merrick, "Deletion analysis of the nitrogen fixation regulatory gene *nifL* of *Klebsiella pneumoniae,*" *Arch. Microbiol. 151*:180–182 (1989).

6. Austin, S., N. Henderson, and R. Dixon, "Requirements for transcriptional activation in vitro of the nitrogen regulated *glnA* and *nifLA* promoters from *Klebsiella pneumoniae:* Dependence on activator concentration," *Mol. Microbiol. 1*:92–100 (1987).

7. Austin, S., N. Henderson, and R. Dixon, "Characterisation of the *Klebsiella pneumoniae* nitrogen-fixation regulatory proteins NIFA and NIFL *in vitro,*" *Eur. J. Biochem. 187*:353–360 (1990).

8. Barros, M. E. C., D. E. Rawlings, and D. R. Woods, "Cloning and expression of the *Thiobacillus ferrooxidans* glutamine synthetase gene in *Escherichia coli*," *J. Bacteriol. 164*:1386–1389 (1985).

9. Batut, J., M.-L. Daveran-Mingot, M. David, J. Jacobs, A. M. Garnerone, and D. Kahn, "*fixK*, a gene homologous with *fnr* and *crp* from *Escherichia coli*, regulates nitrogen fixation genes both positively and negatively in *Rhizobium meliloti*," *EMBO J. 8*:1279–1286 (1989).

10. Bennett, L. T., F. C. Cannon, and D. Dean, "Nucleotide sequence and mutagenesis of the *nifA* gene from *Azotobacter vinelandii*," *Mol. Microbiol. 2*:315–321 (1988).

11. Better, M., G. Ditta, and D. R. Helinski, "Deletion analysis of *Rhizobium meliloti* symbiotic promoters," *EMBO J. 4*:2419–2424 (1985).

12. Beynon, J., M. Cannon, V. Buchanan-Wollaston, and F. Cannon, "The *nif* promoters of *Klebsiella pneumoniae* have a characteristic primary struture," *Cell 34*:665–671 (1983).

13. Beynon, J. L., M. K. Williams, and F. C. Cannon, "Expression and functional analysis of the *Rhizobium meliloti nifA* gene," *EMBO J. 7*:7–14 (1988).

14. Birkmann, A., and A. Bock, "Characterisation of a *cis* regulatory DNA element necessary for formate induction of the formate dehydrogenase gene (*fdhF*) of *Escherichia coli*," *Mol. Microbiol. 3*:187–195 (1989).

15. Bishop, P. E., and R. D. Joerger, "Genetics and molecular biology of alternative nitrogen fixation systems," *Ann. Rev. Plant Physiol. Plant Mol. Biol., 41*:109–125 1990.

16. Bishop, P. E., R. Premakumar, R. D. Joerger, M. R. Jacobson, D. A. Dalton, J. R. Chisnell, and E. D. Wolfinger, "Alternative nitrogen fixation systems in *Azotobacter vinelandii*," pp. 71–79. In H. Bothe, F. J. de Bruijn, and W. E. Newton (eds.), *Nitrogen Fixation: Hundred Years After*. Proceedings of the 7th International Congress on Nitrogen Fixation, Cologne, FGR March 13–20, 1988. Fischer, Stuttgart, 1988.

17. Bozouklian, H., and C. Elmerich, "Nucleotide sequence of the *Azospirillum brasilense* Sp7 glutamine synthetase structural gene," *Biochimie 68*:1181–1187 (1986).

18. Brown, S. E., and F. M. Ausubel, "Mutataions affecting regulation of the *Klebsiella pneumoniae nifH* (nitrogenase reductase) promoter," *J. Bacteriol. 157*:143–147 (1984).

19. Buchanan-Wollaston, V., M. C. Cannon, J. L. Beynon, and F. C. Cannon, "Role of the *nifA* gene product in the regulation of *nif* expression in *Klebsiella pneumoniae*," *Nature 294*:776–778 (1981).

20. Buck, M., "Deletion analysis of the *Klebseilla pneumoniae* nitrogenase promoter: Importance of spacing between conserved sequences around positions -12 and -24 for activation by the *nifA* and *ntrC* (*glnG*) products," *J. Bacteriol. 166*:545–551 (1986).

21. Buck, M., and W. Cannon, "Mutations in the RNA polymerase recognition sequence of the *Klebsiella pneumoniae nifH* promoter permitting transcriptional activation in the absence of NifA binding to upstream sequences," *Nucleic Acids Res. 17*:2597–2611 (1989).

22. Buck, M., W. Cannon, and J. Woodcock, "Mutational analysis of upstream sequences required for transcriptional activation of the *Klebsiella pneumoniae nifH* promoter," *Nucleic Acids Res. 15*:9945–;9956 (1987).

23. Buck, M., W. Cannon, and J. Woodcock, "Transcriptional activation of the *Klebsiella pneumoniae* nitrogenase promoter may involve DNA loop formation," *Mol. Microbiol. 1*:243–249 (1987).

24. Buck, M., H. Khan, and R. Dixon, "Site-directed mutagenesis of the *Klebsiella pneumoniae nifL* and *nifH* promoters and in vitro analysis of promoter activity," *Nucleic Acids Res. 13*:7621–7638 (1985).

25. Buck, M., S. Miller, M. Drummond, and R. Dixon, "Upstream activator sequences are present in the promoters of nitrogen fixation genes," *Nature 320*:374–378 (1986).

26. Buikema, W. J., W. W. Szeto, P. V. Lemley, W. H. Orme-Johnson, and F. M. Ausubel, "Nitrogen fixation specific regulatory genes of *Klebsiella pneumoniae* and *Rhizobium meliloti* share homology with the general nitrogen regulatory gene *ntrC* of *K. pneumoniae*," *Nucleic Acids Res. 13*:4539–4555 (1985).

27. Cannon, W. V., R. Kreutzer, H. M. Kent, E. Morett, and M. Buck, "Activation of the *Klebsiella pneumoniae nifU* promoter: identification of multiple and overlapping upstream NifA binding sites," *Nucleic Acids Res., 18*:1693–1701 1990.

28. Chen, K. C-K., J-S. Chen, and J. L. Johnson, "Structural features of multiple *nifH*-like sequences and very biased codon usage in nitrogenase genes of *Clostridium pasteurianum*," *J. Bacteriol. 166*:162–172 (1986).

29. Colonna-Romano, S., W. Arnold, A. Schluter, P. Boistard, A. Puhler, and U. B. Priefer, "Rhizobium leguminosarum FnrN shows structural and functional homology to *Rhizobium meliloti* FixK," *Mol. Gen. Genet., 223*:138–147 (1990).

30. Colonna-Romano, S., A. Riccio, M. Guida, R. Defez, A. Lamberti, M. Iaccarino, W. Arnold, U. Priefer, and A. Pühler, "Tight linkage of *glnA* and a putative regulatory gene in *Rhizobium leguminosarum*," *Nucleic Acids Res. 15*:1951–1964 (1987).

31. David, M., M.-L. Daveran, J. Batut, A. Dedieu, O. Domergue, J. Ghai, C. Hertig, P. Boistard, and D. Kahn, "Cascade regulation of *nif* gene expression in *Rhizobium meliloti*," *Cell 54*:671–683 (1988).

32. David, M., O. Domergue, P. Pognonec, and D. Kahn, "Transcription patterns of *Rhizobium meliloti* symbiotic plasmid pSym; identification of *nifA*-independent *fix* genes," *J. Bacteriol. 169*:2239–2244 (1987).

33. de Bruijn, F. J., and F. Ausubel, "The cloning and characterization of the *glnF* (*ntrA*) gene of *Klebsiella pneumoniae*: Role of *glnF* (*ntrA*) in the regulation of nitrogen fixation (*nif*) and other nitrogen assimilation genes," *Mol. Gen. Genet. 192*:342–353 (1983).

34. Dimri, G. P., and H. K. Das, "Transcriptional regulation of nitrogen fixation genes by DNA supercoiling," *Mol. Gen. Genet. 212*:360–363 (1988).

35. Ditta, G., E. Virts, A. Palomares, and K. Choong-Hyun, "The *nifA* gene of *Rhizobium meliloti* is oxygen regulated," *J. Bacteriol. 169*:3217–3223 (1987).

36. Dixon, R., R. R. Eady, G. Espin, S. Hill, M. Iaccarino, D. Kahn, and M. Merrick, "Analysis of regulation of *Klebsiella pneumoniae* nitrogen fixation (*nif*) gene cluster with gene fusions," *Nature 286*:128–132 (1980).

37. Dixon, R., N. C. Henderson, and S. Austin, "DNA supercoiling and aerobic regulation of transcription from the *Klebsiella pneumoniae nifLA* promoter," *Nucleic Acids Res. 16*:9933–9946 (1988).

38. Dixon, R., C. Kennedy, A. Kondorosi, V. Krishnapillai, and M. Merrick, "Complementation analysis of *Klebsiella pneumoniae* mutants defective in nitrogen fixation," *Mol. Gen. Genet. 157*:189–198 (1977).

39. Drummond, M., P. Whitty, and J. Wootton, "Sequence and domain relationships of *ntrC* and *nifA* from *Klebsiella pneumoniae*: homologies to other regulatory proteins," *EMBO J. 5*:441–447 (1986).

40. Drummond, M. H., A. Contreras, and L. A. Mitchenall, "The function of isolated domains and chimaeric proteins constructed from the transcriptional activators NifA and NtrC of *Klebsiella pneumoniae*," *Mol. Microbiol. 4*:29–37 (1990).

41. Drummond, M. H., and J. C. Wootton, "Sequence of *nifL* from *Klebsiella pneumoniae*: Mode of action and relationship to two families of regulatory proteins," *Mol. Microbiol. 1*:37–44 (1987).

42. Earl, C. D., C. W. Ronson, and F. M. Ausubel, "Genetic and structural analysis of the *Rhizobium meliloti fixA, fixB, fixC,* and *fixX* genes," *J. Bacteriol. 169*:1127–1136 (1987).

43. Evans, D., R. Jones, P. Woodley, and R. Robson, "Further analysis of nitrogen fixation (*nif*) genes in *Azotobacter chroococcum*: Identification and expression in *Klebsiella pneumoniae* of *nifS, nifV, nifM* and *nifB* genes and localization of *nifE/N-*, *nifU-, nifA-* and *fixABC*-like genes," *J. Gen. Microbiol. 134*:931–942 (1988).

44. Evans, D. J., "The molecular genetics of nitrogen fixation in *Azotobacter chroococcum*," Ph.D. thesis, University of Sussex, 1987.

45. Finan, T. M., I. Oresnik, and A. Bottacin, "Mutants of *Rhizobium meliloti* defective in succinate metabolism," *J. Bacteriol. 170*:3396–3403 (1988).

46. Fischer, H-M., T. Bruderer, and H. Hennecke, "Essential and non-essential domains in the *Bradyrhizobium japonicum* NifA protein: Identification of indispensible cysteine residues potentially involved in redox activity and/or metal binding," *Nucleic Acids Res. 16*:2207–2224 (1988).

47. Fischer, H-M., S. Fritsche, B. Herzog, and H. Hennecke, "Critical spacing between two essential cysteine residues in the interdomain linker of the *Bradyrhizobium japonicum* NifA protein," *FEBS Lett. 255*:167–171 (1989).

48. Fischer, H. M., and H. Hennecke, "Direct response of *Bradyrhizobium japonicum* *nifA*-mediated *nif* gene regulation to cellular oxygen status," *Mol. Gen. Genet. 209*:621–626 (1987).

49. Fischer, M., E. Levy, and T. Geller, "Regulatory mutation that controls *nif* expression and histidine transport in *Azospirillum brasilense*," *J. Bacteriol. 167*:423–426 (1986).

50. Garcia, E., S. Bancroft, S. G. Rhee, and S. Kustu, "The product of a newly identified gene, *glnF*, is required for synthesis of glutamine synthetase in *Salmonella*," *Proc. Natl. Acad. Sci. USA 74*:1662–1666 (1977).

51. Gribskov, M., and R. R. Burgess, "Sigma factors from *Escherichia coli, Bacillus subtilis,* phage SPO1 and T4 are homologous proteins with conserved secondary structures," *Nucleic Acids Res. 14*:6745–6763 (1986).

52. Grönger, P., S. S. Manian, H. Reiländer, M. O'Connell, U. B. Priefer, and A. Pühler, "Organisation and partial sequence of a DNA region of the *Rhizobium leguminosarum* symbiotic plasmid pRL6JI containing the genes *fixABC, nifA, nifB* and a novel open reading frame," *Nucleic Acids Res. 15*:31–49 (1987).

53. Gubler, M., "Fine-tuning of *nif* and *fix* gene expression by upstream activator sequences in *Bradyrhizobium japonicum*," *Mol. Microbiol. 3*:149–159 (1989).

54. Gubler, M., and H. Hennecke, "*fixA, B* and *C* genes are essential for symbiotic and free-living, microaerobic nitrogen fixation," *FEBS Lett. 200*:186–192 (1986).

55. Gubler, M., and H. Hennecke, "Regulation of the *fixA* gene and *fixBC* operon in *Bradyrhizobium japonicum*," *J. Bacteriol. 170*:1205–1214 (1988).

56. Hawkes, T., M. Merrick, and R. Dixon, "Interaction of purified NtrC protein with nitrogen regulated promoters from *Klebsiella pneumoniae*," *Mol. Gen. Genet. 201*:492–498 (1985).

57. Helmann, J. D., and M. Chamberlin, "Structure and function of bacterial sigma factors," *Ann. Rev. Biochem. 57*:839–872 (1988).

58. Henderson, N., S. A. Austin, and R. A. Dixon, "Role of metal ions in negative regulation of nitrogen fixation by the *nifL* gene product from *Klebsiella pneumoniae*," *Mol. Gen. Genet. 216*:484–491 (1989).

59. Hertig, C., R. Y. Li, A.-M. Louarn, A.-M. Garnerone, M. David, J. Batut, D. Kahn, and P. Boistard, "*Rhizobium meliloti* regulatory gene *fixJ* activates transcription of *R. meliloti nifA* and *fixK* genes in *Escherichia coli*," *J. Bacteriol. 171*:1736–1738.

60. Higgins, C. F., I. D. Hiles, G. P. C. Salmond, D. R. Gill, J. A. Downie, I. J. Evans, I. B. Holland, L. Gray, S. D. Buckel, A. W. Bell, and M. A. Hermodson, "A family of related ATP-binding subunits coupled to many distinct processes in bacteria," *Nature 323*:448–450 (1986).

61. Hirschman, J., P. K. Wong, K. Sei, J. Keener, and S. Kustu, "Products of nitrogen regulatory genes *ntrA* and *ntrC* of enteric bacteria activate *glnA* transcription in vitro: Evidence that the *ntrA* product is a sigma factor," *Proc. Natl. Acad. Sci. USA 82*:7525–7529 (1985).

62. Holtel, A., S. Colonna-Romano, M. Guida, A. Riccio, M. J. Merrick, and M. Iaccarino, "The *glnB* gene of *Rhizobium leguminosarum* biovar *viceae*," *FEMS Microbiol. Lett. 58*:203–208 (1989).

63. Holtel, A., and M. J. Merrick, "The *Klebsiella pneumoniae* P_{II} protein (*glnB* gene product) is not absolutely required for nitrogen regulation and is not involved in NifL-mediated *nif* gene regulation," *Mol. Gen. Genet. 217*:474–480 (1989).

64. Holtel, H., and M. Merrick, "Identification of the *Klebsiella pneumoniae glnB* gene: Nucleotide sequence of wild-type and mutant alleles," *Mol. Gen. Genet. 215*:134–138 (1988).

65. Huala, E., and F. M. Ausubel, "The central domain of *Rhizobium meliloti* NifA is sufficient to activate transcription from the *R. meliloti nifH* promoter," *J. Bacteriol. 171*:3354–3365 (1989).

66. Hunt, T. P., and B. Magasanik, "Transcription of *glnA* by purified *Escherichia coli* components: Core RNA polymerase and the products of *glnF*, *glnG*, and *glnL*," *Proc. Natl. Acad. Sci. USA 82*:8453–8457 (1985).

67. Iismaa, S. E., and J. M. Watson, "The *nifA* gene product from *Rhizobium leguminosarum* biovar *trifolii* lacks the N-terminal domain found in other NifA proteins," *Mol. Microbiol. 3*:943–955 (1989).

68. Inouye, S., A. Nakazawa, and T. Nakazawa, "Nucleotide sequence of the regulatory gene *xylR* of the TOL plasmid from *Pseudomonas putida*," *Gene 66*:301–306 (1988).

69. Inouye, S., M. Yamada, A. Nakazawa, and T. Nakazawa, "Cloning and sequence analysis of the *ntrA* (*rpoN*) gene of *Pseudomonas putida*," *Gene 85*:145–152 (1989).

70. Jacobson, M. R., K. E. Brigle, L. Bennett, R. A. Setterquist, R. A. Wilson, V. L. Cash, J. Beynon, W. E. Newton, and D. R. Dean, "Physical and genetic map of the major *nif* gene cluster from *Azotobacter vinelandii*," *J. Bacteriol. 171*:1017–1027 (1989).

71. Joerger, R. D., and P. E. Bishop, "Nucleotide sequence and genetic analysis of the *nifB-nifQ* region from *Azotobacter vinelandii*," *J. Bacteriol. 170*:1475–1487 (1988).

72. Joerger, R. D., M. R. Jacobson, and P. E. Bishop, "Two *nifA*-like genes required for the expression of alternative nitrogenases in *Azotobacter vinelandii*," *J. Bacteriol.* *171*:3258–3267 (1989).

73. Joerger, R. D., M. R. Jacobson, R. Premakumar, E. D. Wolfinger, and P. E. Bishop, "Nucleotide sequence and mutational analysis of the structural genes (*anfHDK*) for the second alternative nitrogenase from *Azotobacter vinelandii*," *J. Bacteriol.* *171*:1075–1086 (1989).

74. Jones, R., and R. Haselkorn, "The DNA sequence of the *Rhodobacter capsulatus ntrA, ntrB* and *ntrC* gene analogues required for nitrogen fixation," *Mol. Gen. Genet.* *215*:507–516 (1989).

75. Keener, J., and S. Kustu, "Protein kinase and phosphoprotein phosphatase activities of nitrogen regulatory proteins NTRB and NTRC of enteric bacteria: Roles of the conserved amino-terminal domain of NTRC," *Proc. Natl. Acad. Sci. USA* *85*:4976–4980 (1988).

76. Kennedy, C., "Linkage map of the nitrogen fixation (*nif*) genes in *Klebsiella pneumoniae*," *Mol. Gen. Genet.* *157*:199–204 (1977).

77. Kennedy, C., and R. Robson, "Activation of *nif* gene expression in *Azotobacter* by the *nifA* gene product of *Klebsiella pneumoniae*," *Nature* *301*:626–628 (1983).

78. Kent, H. M., M. Buck, and D. J. Evans, "Cloning and sequencing of the *nifH* gene of *Desulfovibrio gigas*," *FEMS Microbiol. Lett.* *610*:73–78 (1989).

79. Khan, H., M. Buck, and R. Dixon, "Deletion loop mutagenesis mutagenesis of the *nifL* promoter from *Klebsiella pneumoniae*: Role of the −26 to −12 region in the promoter function," *Gene* *45*:281–288 (1986).

80. Kim, C.-H., D. R. Helinski, and G. Ditta, "Overlapping transcription of the *nifA* regulatory gene in *Rhizobium meliloti*," *Gene* *50*:141–148 (1986).

81. Kim, Y. M., K. J. Ahn, T. Beppu, and T. Uozomi, "Nucleotide sequence of the *nifLA* operon of *Klebsiella oxytoca* NG13 and characterisation of the gene products," *Mol. Gen. Genet.* *205*:253–259 (1986).

82. Klipp, W., B. Masepohl, and A. Pühler, "Identification and mapping of nitrogen fixation genes of *Rhodobacter capsulatus*: Duplication of a *nifA-nifB* region," *J. Bacteriol.* *170*:693–699 (1988).

83. Köhler, T., J. M. Cayrol, J. L. Ramos, and S. Harayama, "Nucleotide and deduced amino acid sequence of the RpoN σ-factor of *Pseudomonas putida*," *Nucleic Acids Res.* *23*:10125 (1989).

84. Köhler, T., S. Harayama, J. L. Ramos, and K. N. Timmis, "Involvement of *Pseudomonas putida* RpoN σ factor in regulation of various metabolic functions," *J. Bacteriol.* *171*:4326–4333 (1989).

85. Kranz, R. G., and R. Haselkorn, "Characterisation of the *nif* regulatory genes in *Rhodopseudomonas capsulata* using *lac* gene fusions," *Gene* *40*:203–215 (1985).

86. Kranz, R. G., and R. Haselkorn, "Anaerobic regulation of nitrogen-fixation genes in *Rhodopseudomonas capsulata*," *Proc. Natl. Acad. Sci. USA* *83*:6805–6809 (1986).

87. Kranz, R. G., and R. Haselkorn, "Ammonia-constitutive nitrogen fixation mutants of *Rhodobacter capsulatus*," *Gene* *71*:65–74 (1989).

88. Kranz, R. G., V. M. Pace, and I. M. Caldicott, "Inactivation, sequence, and *lacZ* fusion analysis of a regulatory locus required for repression of nitrogen fixation genes in *Rhodobacter capsulatus*," *J. Bacteriol.* *172*:53–62 (1990).

89. Kullik, I., H. Hennecke, and H.-M. Fischer, "Inhibition of *Bradyrhizobium japonicum nifA*-dependent *nif* gene activation by oxygen occurs at the NifA protein level and is irreversible," *Arch. Microbiol. 151*:191–197 (1989).

90. Kustu, S., E. Santero, D. Popham, and J. Keener, "Expression of σ^{54} (*ntrA*)-dependent genes is probably united by a common mechanism," *Microbiol. Rev. 53*:367–376 (1989).

91. Kustu, S., K. Sei, and J. Keener, "Nitrogen regulation in enteric bacteria," pp. 139–154. In I. R. Booth and C. F. Higgins (eds.), *Regulation of Gene Expression—25 Years On*, Cambridge University Press, Cambridge (1986).

92. MacFarlane, S., and M. J. Merrick, "Analysis of the *Klebsiella pneumoniae ntrB* gene by site-directed *in vitro* mutagenesis," *Mol. Microbiol. 1*:133–142 (1987).

93. Magasanik, B., "Reversible phosphorylation of an enhancer binding protein regulates the transcription of bacterial nitrogen utilisation genes," *Trends Biochem. Sci. 13*:475–479 (1988).

94. Martin, G. B., M. F. Thomashow, and B. K. Chelm, "*Bradyrhizobium japonicum glnB*, a putative nitrogen-regulatory gene, is regulated by NtrC at tandem promoters," *J. Bacteriol. 171*:5638–5645 (1989).

95. Masepohl, B., W. Klipp, and A. Pühler, "Genetic characterization and sequence analysis of the duplicated *nifA/nifB* gene region of *Rhodobacter capsulatus*," *Mol. Gen. Genet. 212*:27–37 (1988).

96. Merrick, M., J. Gibbins, and A. Toukdarian, "The nucleotide sequence of the sigma factor gene *ntrA* (*rpoN*) of *Azotobacter vinelandii*: Analysis of conserved sequences in NtrA proteins," *Mol. Gen. Genet. 210*:323–330 (1987).

97. Merrick, M., S. Hill, H. Hennecke, M. Hahn, R. Dixon, and C. Kennedy, "Repressor properties of the *nifL* gene product of *Klebsiella pneumoniae*," *Mol. Gen. Genet. 185*:75–81 (1982).

98. Merrick, M. J., "Nitrogen control of the *nif* regulon in *Klebsiella pneumoniae*: Involvement of the *ntrA* gene and analogies between *ntrC* and *nifA*," *EMBO J. 2*:39–44 (1983).

99. Merrick, M. J., and J. R. Coppard, "Mutations in genes downstream of the *rpoN* gene (encoding σ^{54}) of *Klebsiella pneumoniae* affect expression from σ^{54}-dependent promoters," *Mol. Microbiol. 3*:1765–1775 (1989).

100. Merrick, M. J., and J. R. Gibbins, "The nucleotide sequence of the nitrogen-regulation gene *ntrA* of *Klebsiella pneumoniae* and comparison with conserved features in bacterial RNA polymerase sigma factors," *Nucleic Acids Res. 13*:7607–7620 (1985).

101. Merrick, M. J., and W. D. P. Stewart, "Studies on the regulation and function of the *Klebsiella pneumoniae ntrA* gene," *Gene 35*:297–303 (1985).

102. Minchin, S. D., S. Austin, and R. A. Dixon, "The role of activator binding sites in transcriptional control of the divergently transcribed *nifF* and *nifLA* promoters from *Klebsiella pneumoniae*," *Mol. Microbiol. 2*:433–442 (1988).

103. Minchin, S. D., S. Austin, and R. A. Dixon, "Transcriptional activation of the *Klebsiella pneumoniae nifLA* promoter by NTRC is face-of-the-helix dependent and the activator stabilizes the interaction of sigma 54-RNA polymerase with the promoter," *EMBO J. 8*:3491–3499 (1989).

104. Morett, E., and M. Buck, "NifA-dependent *in vivo* protection demonstrates that the upstream activator sequence of *nif* promoters is a protein binding site," *Proc. Natl. Acad. Sci. USA 85*:9401–9405 (1988).

105. Morett, E., and M. Buck, "*In vivo* studies on the interaction of RNA polymerase-σ^{54} with the *Klebsiella pneumoniae* and *Rhizobium meliloti nifH* promoters: The role of NifA in the formation of an open promoter complex," *J. Mol. Biol. 210*:65–77 (1989).

106. Morett, E., W. Cannon, and M. Buck, "The DNA-binding domain of the transcriptional activator protein NifA resides in its carboxy terminus, recognises the upstream activator sequences of nif promoters and can be separated from the positive control function of NifA," *Nucleic Acids Res. 16*:11469–11488 (1988).

107. Mullin, D. A., and A. Newton, "Ntr-like promoters and upstream regulatory sequence *ftr* are required for transcription of a developmentally regulated *Caulobacter crescentus* flagellar gene," *J. Bacteriol. 171*:3218–3227 (1989).

108. Nees, D. W., P. A. Stein, and R. A. Ludwig, "The *Azorhizobium caulinodans nifA* gene: Identification of upstream-activating sequences including a new element, the "anaerobox." *Nucleic Acids Res. 16*:9839–9853 (1988).

109. Ninfa, A. J., E. Gottlin Ninfa, A. N. Lupas, A. Stock, B. Magasanik, and J. Stock, "Crosstalk between bacterial chemotaxis signal transduction proteins and regulators of transcription of the Ntr regulon: Evidence that nitrogen assimilation and chemotaxis are controlled by a common phosphotransfer mechanism," *Proc. Natl. Acad. Sci. USA 85*:5492–5496 (1988).

110. Ninfa, A. J., and B. Magasanik, "Covalent modification of the *glnG* product, NR_I, by the *glnL* product, NR_{II}, regulates the transcription of the *glnALG* operon in *Escherichia coli*," *Proc. Natl. Acad. Sci. USA 83*:5909–5913 (1986).

111. Ninfa, A. J., D. A. Mullin, G. Ramakrishnan, and A. Newton, "*Escherichia coli* σ^{54} RNA polymerase recognises *Caulobacter crescentus flbG* and *fluN* flagellar gene promoters in vitro," *J. Bacteriol. 171*:383–391 (1989).

112. Nixon, B. T., C. W. Ronson, and F. M. Ausubel, "Two-component regulatory systems responsive to environmental stimuli share strongly conserved domains with the nitrogen assimilation regulatory genes *ntrB* and *ntrC*," *Proc. Natl. Acad. Sci. USA. 83*:7850–7854 (1986).

113. Ow, D. W., and F. Ausubel, "Regulation of nitrogen metabolism genes by *nifA* gene product in *Klebsiella pneumoniae*," *Nature 301*:307–313 (1983).

114. Ow, D. W., Y. Xiong, Q. Qu, and S.-C. Shen, "Mutational analysis of the *Klebsiella pneumoniae* nitrogenase promoter: Sequences essential for positive control by *nifA* and *ntrC* (*glnG*) products," *J. Bacteriol. 161*:868–874 (1985).

115. Pabo, C. O., and R. T. Sauer, "Protein-DNA recognition," *Ann. Rev. Biochem. 53*:293–321 (1984).

116. Paul, W., and M. Merrick, "The nucleotide sequence of the *nifM* gene of *Klebsiella pneumoniae* and identification of a new *nif* gene: *nifZ*," *Eur. J. Biochem. 170*:259–265 (1987).

117. Pedrosa, F. O., and M. G. Yates, "Regulation of nitrogen fixation (*nif*) genes of *Azospirillum brasilense* by *nifA* and *ntr* (*gln*) type gene products," *FEMS Microbiol. Lett. 23*:95–101 (1984).

118. Popham, D. L., D. Szeto, J. Keener, and S. Kustu, "Function of a bacterial activator protein that binds to transcriptional enhancers," *Science 243*:629–635 (1989).

119. Pretorius, I. M., D. E. Rawlings, E. G. O'Neill, W. A. Jones, R. Kirby, and D. R. Woods, "Nucleotide sequence of the gene encoding the nitrogenase iron protein of *Thiobacillus ferrooxidans*," *J. Bacteriol. 169*:367–370 (1987).

120. Ptashne, M., "Gene regulation by proteins acting nearby and at a distance," *Nature* 322:697–701 (1986).

121. Raina, R., M. A. Reddy, D. Ghosal, and H. K. Das, "Characterization of the gene for the Fe-protein of the vanadium dependent alternative nitrogenase of *Azotobacter vinelandii* and construction of a Tn5 mutant," *Mol. Gen. Genet.* 214:121–127 (1988).

122. Ratet, P., K. Pawlowski, J. Schell, and F. J. de Bruijn, "The *Azorhizobium caulinodans* nitrogen-fixation regulatory gene, *nifA*, is controlled by the cellular nitrogen and oxygen status," *Mol. Microbiol.* 3:825–838 (1989).

123. Reitzer, L. J., and B. Magasanik, "Transcription of *glnA* in *E. coli* is stimulated by activator bound to sites far from the promoter," *Cell* 45:785–792 (1986).

124. Reitzer, L. J., and B. Magasanik, "Ammonia assimilation and the biosynthesis of glutamine, glutamate, aspartate, asparagine, L-alanine, and D-alanine," pp. 302–320. In F. C. Neidhard (ed.), *Escherichia coli* and *Salmonella typhimurium*: Cellular and Molecular Biology, Vol. 1. American Society for Microbiology, Washington DC. (1987).

125. Reitzer, L. J., B. Movsas, and B. Magasanik, "Activation of *glnA* transcription by nitrogen regulator I (NR$_I$)-phosphate in *Escherichia coli:* Evidence for a long-range physical interaction between NR$_I$-phosphate and RNA polymerase," *J. Bacteriol.* 171:5512–5522 (1989).

126. Robertson, C. A., and H. A. Nash, "Bending of the bacteriophage/attachment site by *Escherichia coli* integration host factor," *J. Biol. Chem.* 263:3554–3667 (1988).

127. Robson, R. L., P. R. Woodley, and R. Jones, "Second gene (*nifH**) coding for a nitrogenase iron-protein in *Azotobacter chroococcum* is adjacent to a gene coding for a ferrodixin-like protein," *EMBO J.* 5:1159–1163 (1986).

128. Robson, R. L., P. R. Woodley, R. N. Pau, and R. R. Eady, "Structural genes for the vanadium nitrogenase from *Azotobacter chroococcum*," *EMBO J.* 8:1217–1224 (1989).

129. Roelvink, P. W., J. G. J. Hontelez, A. van Kammen, and R. C. van den Bos, "Nucleotide sequence of the regulatory *nifA* gene of *Rhizobium leguminosarum* PRE: Transcriptional control sites and expression in *Escherichia coli*," *Mol. Microbiol.* 3:1441–1447 (1989).

130. Ronson, C. W., P. M. Astwood, B. T. Nixon, and F. M. Ausubel, "Deduced products of C4-dicarboxylate transport regulatory genes of *Rhizobium leguminosarium* are homologous to nitrogen regulatory gene products, *Nucleic Acids Res.* 15:7921–7934 (1987).

131. Ronson, C. W., B. T. Nixon, L. M. Albright, and F. M. Ausubel, "*Rhizobium meliloti ntrA (rpoN)* gene is required for diverse metabolic functions," *J. Bacteriol.* 169:2424–2431 (1987).

132. Santero, E., T. Hoover, J. Keener, and S. Kustu, "*In vitro* activity of the nitrogen fixation regulatory protein NIFA," *Proc. Natl. Acad. Sci. USA.* 86:7346–7350 (1989).

133. Santero, E., A. Toukdarian, R. Humphrey, and C. Kennedy, "Identification and characterisation of two nitrogen fixation regulatory regions *nifA* and *nfrX* in *Azotobacter vinelandii* and *Azotobacter chroococcum*," *Mol. Microbiol.* 2:303–314 (1988).

134. Schmitz, G., K. Nikaido, and G. F.-L. Ames, "Regulation of a transport operon in *Salmonella typhimurium:* Identification of sites essential for nitrogen regulation, *Mol. Gen. Genet.* 215:107–117 (1988).

135. Shatters, R. G., J. E. Somerville, and M. L. Kahn, "Regulation of glutamine synthetase II activity in *Rhizobium meliloti* 104A14, "*J. Bacteriol.* 171:5087–5094 (1989).

136. Sibold, L., and N. Souillard, "Genetic analysis of nitrogen fixation in methanogenic archebacteria," pp. 705–710. In H. Bothe, F. J. de Bruijn, and W. E. Newton (eds.), *Nitrogen Fixation: Hundred Years After*, Gustav Fischer, Stuttgart (1988).

137. Singh, M., A. K. Tripathi, and W. Klingmuller, "Identification of a regulatory *nifA*-type gene and physical mapping of cloned new *nif* regions of *Azospirillum brasilense*," *Mol. Gen. Genet. 219*:235–240 (1989).

138. Smith, A., S. Hill, and C. Anthony, "A haem-protein is not involved in the control by oxygen of enteric nitrogenase synthesis," *J. Gen. Microbiol. 134*:1499–1507 (1988).

139. Spiro, S., and J. R. Guest, "Regulation and over-expression of the *fnr* gene of *Escherichia coli*," *J. Gen. Microbiol. 133*:3279–3288 (1987).

140. Spiro, S., R. E. Roberts, and J. R. Guest, "FNR-dependent repression of the *ndh* gene of *Escherichia coli* and metal ion requirement for FNR-regulated gene expression," *Mol. Microbiol. 3*:601–608 (1989).

141. Stanley, J., J. van Slooten, D. N. Dowling, T. Finan, and W. J. Broughton, "Molecular cloning of the *ntrA* gene of the broad host-range *Rhizobium* sp. NG234 and phenotypes of site-directed mutagenesis," *Mol. Gen. Genet. 217*:528–532 (1989).

142. Stock, J. B., A. J. Ninfa, and A. M. Stock, "Protein phosphorylation and regulation of adaptive responses in bacteria," *Microbiol. Rev. 53*:450–490 (1989).

143. Stoker, K., W. N. M. Reijnders, L. F. Oltman, and A. H. Stouthamer, "Initial cloning and sequencing of *hydHG*, an operon homologous to *ntrBC* and regulating the labile hydrogenase activity in *Escherichia coli* K-12," *J. Bacteriol. 171*:4448–4456 (1989).

144. Sundaresan, V., J. G. Jones, D. W. Ow, and F. Ausubel, "*Klebsiella pneumoniae nifA* product activates the *Rhizobium meliloti* nitrogenase promoter," *Nature 301*:728–732 (1983).

145. Szeto, W. W., B. T. Nixon, C. W. Ronson, and F. M. Ausubel, "Identification and characterisation of the *Rhizobium meliloti ntrC* gene: *R. meliloti* has separate regulatory pathways for activating nitrogen fixation genes in free-living versus symbiotic cells," *J. Bacteriol. 169*:1423–1432 (1987).

146. Thöny, B., D. Anthamatten, and H. Hennecke, "Dual control of the *Bradyrhizobium japonicum* symbiotic nitrogen fixation regulatory operon *fixRnifA*: Analysis of *cis*- and *trans*-acting elements," *J. Bacteriol. 171*:4162–4169 (1989).

147. Thöny, B., H.-M. Fischer, D. Anthamatten, T. Bruderer, and H. Hennecke, "The symbiotic nitrogen fixation regulatory operon (*fixRnifA*) of *Bradyrhizobium japonicum* is expressed aerobically and is subject to a novel, *nifA*-independent type of activation," *Nucleic Acids Res. 15*:8479–8499 (1987).

148. Thöny, B., and H. Hennecke, "The −24/−12 promoter comes of age," *FEMS Microbiol. Rev., 63*:341–358 (1989).

149. Toukdarian, A., and C. Kennedy, "Regulation of nitrogen metabolism in *Azotobacter vinelandii:* Isolation of *ntr* and *glnA* genes and construction of *ntr* mutants," *EMBO J. 5*:399–407 (1986).

150. Tuli, R., and M. J. Merrick, "Over-production and characterisation of the *nifA* gene product of *Klebsiella pneumoniae*—the transcription activator of *nif* gene expression," *J. Gen. Microbiol. 134*:425–432 (1988).

151. Tumer, N. E., S. J. Robinson, and R. Haselkorn, "Different promoters for the *Anabaena* glutamine synthetase gene during growth using molecular or fixed nitrogen," *Nature 306*:337–342 (1983).

152. Virts, E. L., S. W. Stanfield, D. R. Helinski, and G. S. Ditta, "Common regulatory elements control symbiotic and microaerobic induction of *nifA* in *Rhizobium meliloti*," *Proc. Natl. Acad. Sci. USA 85*:3062–3065 (1988).

153. Wang, S.-Z., J.-S. Chen, and J. L. Johnson, "The presence of five *nifH*-like sequences in *Clostridium pasteurianum:* Sequence divergence and transcription properties," *Nucleic Acids Res. 16*:439–454 (1988).

154. Weiss, V., and B. Magasanik, "Phosphorylation of NR1 of *E. coli,*" *Proc. Natl. Acad. Sci. USA 85*:8919–8923 (1988).

155. Wong, P. K., D. Popham, J. Keener, and S. Kustu, "In vitro transcription of the nitrogen fixation regulatory operon *nifLA* of *Klebsiella pneumoniae*," *J. Bacteriol. 169*:2876–2880 (1987).

156. Wootton, J. C., and M. Drummond, "The Q-linker: A class of interdomain sequences found in bacterial multidomain regulatory proteins," *Prot. Eng. 2*:535–543 (1989).

157. Yamamoto, N., and M. L. Droffner, "Mechanisms determining aerobic or anaerobic growth in the facultative anaerobe *Salmonella typhimurium*," *Proc. Natl. Acad. Sci. USA 82*:2077–2081 (1985).

22

Isolated Iron–Molybdenum Cofactor of Nitrogenase

William E. Newton

I. Introduction

A. Nomenclature

The name iron–molybdenum cofactor[101] (commonly represented as FeMo-co or FeMo-cofactor) is a misnomer. FeMo-cofactor is really not a cofactor or a coenzyme; rather, it is a unique prosthetic group, peculiar to nitrogenase, with properties and function more in line with those of the iron–sulfur (Fe–S) clusters of ferredoxins. Except during its biosynthesis (see Section X), it remains tightly bound within the protein matrix of the molybdenum–iron (MoFe) protein, which is the larger of the two components of the nitrogenase enzyme complex.[6] In its protein-bound form, it is commonly referred to as an "M" (for magnetic) center, of which there are two per MoFe protein.[24,65] The "M" centers have been reported to be as close together[113] as 20 Å and as far apart[3] as 70 Å. A barrage of biophysical techniques has shown that the "M" centers and the isolated FeMo-cofactors are very similar in structure and properties, and although these centers are not identical, these investigations have clearly shown that the FeMo-cofactor is not significantly changed by the extraction process. The "M" center is, contains, or is part of the substrate-reducing site of nitrogenase.[33,100] Therefore, the extruded FeMo-cofactor represents the substrate-reducing site, which has been simplified for study by removal of both the MoFe–protein matrix and its companion prosthetic groups, the "P" clusters.

B. Scope

This chapter covers current knowledge of the FeMo-cofactor as isolated from the MoFe protein of nitrogenase, including its relationship (or lack of it) to the molybdenum-containing centers of other enzymes, *e.g.*, nitrate reductase and sulfite oxidase. It attempts complete coverage of the subject material

rather than merely updating two previous reviews.[9,114] The chapter starts
with a historical perspective, which describes events leading to the discovery
of extrudable Mo-containing factors and the uniqueness of the nitrogenase-
derived factor. Then follows an account of the mutant strains, which form
the basis of the assay for the nitrogenase FeMo-cofactor, and which were
vital to its discovery. The next section details the assays, isolation methods,
composition, oxidation-reduction and magnetic properties, structural insights, and reactivity of isolated FeMo-cofactor. Finally, the chapter closes
with a brief description of the nitrogen-fixation (*nif*) genes and products,
which are related to the biosynthesis of the FeMo-cofactor, and the *in vitro*
synthesis method.

II. Historical Perspective

A. Is there a Single, Common Molybdenum Cofactor?

A commonality among molybdoenzymes in terms of their Mo-containing
prosthetic groups was first indicated by the common genetic determinant for
nitrate reductase and xanthine dehydrogenase found in a series of *Aspergillus
nidulans* mutants.[84] This result led to the suggestion that the same molybdenum-containing factor might be shared among all molybdenum-containing
enzymes. Subsequently, a *Neurospora crassa* mutant,[112] which was designated *nit1* and which lacked nitrate reductase activity because of the non-
incorporation of Mo, was used to determine the requirements for reconsti-
tuting this activity[47,50,68,69] and, in turn, to give insights into the molybdenum
factor. Successful complementation of *nit*1 crude extracts, *i.e.*, reactivation
of nitrate reduction, occurred with a variety of molybdoenzymes from plant,
animal and microorganism sources, after appropriate treatment. All required
acid treatment to pH 2.5, which destroyed their native activity, followed by
complementation of the *nit*1 extract at pH 6.8. The enzymes tested successfully included xanthine oxidase, sulfite oxidase and nitrate reductase,
plus the *partially purified* nitrogenase MoFe protein. No synthetic molybdenum compound nor molybdate (MoO_4^{2-}) activated these crude extracts
alone. However, in the presence of the products of acid treatment of the
various molybdoenzymes, added molybdate enhanced the resulting activity
and, as shown by using $^{99}MoO_4^{2-}$, became incorporated into the newly activated enzyme. These data, together with other biochemical[48] and genetic[57]
results, supported the thesis that all molybdoenzymes, including the MoFe
protein of nitrogenase, contained a common acid-releasable Mo-containing
factor, which could be incorporated into, and so activate, the non-functional
nitrate reductase of the *nit*1 strain of *N. crassa*.

B.   Search for a Cofactor-activable Apo-MoFe Protein

This situation led to a search among normally N_2-fixing organisms for either mutant strains, which were unable to grow on N_2 gas, or growth conditions that would generate a functionally incompetent nitrogenase in order to provide a nitrogenase-based assay for the *supposed* common Mo-factor. In the latter area, wild-type *A. vinelandii* cells, grown in media lacking a fixed nitrogen source (which causes the nitrogenase proteins to be synthesized in a process called derepression) and without added MoO_4^{2-}, produced fully active Fe protein but no detectable MoFe protein.[66] If, however, tungstate (a competitive antagonist with respect to molybdate in N_2 fixation[5,45]) was incorporated into the medium, the cells produced MoFe protein, but in an inactive form. The inactive form was rapidly activated by cells *in vivo* on addition of MoO_4^{2-} to the medium in the absence of protein synthesis.[66] This inactive MoFe protein could not be activated *in vitro* by adding MoO_4^{2-} to extracts of these same cells.[67] Activation, however, did occur when an ATP-generating system was added along with the molybdate.[87] Further, activation could be also accomplished by adding acid-treated (to pH 2.6; which completely inactivates all inherent activity) native MoFe protein to extracts of the tungstate-derepressed cells.[67] These data indicated that the MoFe protein from tungstate-derepressed cells lacked the molybdenum factor.

Although an extract of tungstate-grown cells might then appear to be functionally equivalent to the *nit*1 extract because it can incorporate the factor from an added solution of acid-treated native MoFe protein directly into its inactive MoFe protein, a major difference is its ability to biosynthesize the absent molybdenum factor (all the neccessary bio-machinery must, therefore, be present) from added molybdate and ATP. However, the success in activating the inactive MoFe protein from tungstate-grown cells by using acid-treated wild-type cell extracts did suggest the likelihood of an assay based on activating non-functional MoFe proteins in non-N_2-fixing (Nif⁻) mutant strains.

One such mutant,[104] which played a pivotal role in the isolation and characterization of the nitrogenase factor and in distinguishing it from that of all other molybdoenzymes, was a member of a series produced by N-nitro-N-nitrosoguanidine mutagenesis of *A. vinelandii* and designated as UW45. This strain produces a fully functional Fe protein, but its MoFe protein, although immunologically detectable, is catalytically inactive.[67] Just like the extracts of tungstate-grown cells, extracts of UW45 are rapidly activated toward substrate reduction by added acid-treated native MoFe protein. In contrast, neither UW45 cells *in vivo* nor extracts *in vitro* are activated by added molybdate with or without ATP.[87] Obviously, the bio-machinery neccessary to synthesize the missing factor is inoperative in the UW45 mutant strain. The responsible lesion has since been shown[105] to be in *nifB*, one of the six currently recognized genes (*nifB*, *E*, *N*, *V*, *Q*, and *H*; see Section X) whose products are involved in biosynthesis of the nitrogenase factor.[62] Together with the *A. vinelandii* UW45 strain, related *nifB⁻* strains of *Klebsiella pneumoniae*[34,90] and *nifE⁻* and *nifN⁻* strains of *A. vinelandii*[4] form the basis of the current assay systems for the nitrogenase factor.

C. All Cofactors are not Created Equal!

With both the UW45 and *nit*1 assays now in hand, it was shown that acid-treated xanthine oxidase would activate *nit*1 extracts towards reduction of nitrate to nitrite but would *not* activate the nitrogenase in the *A. vinelandii* UW45 extracts. Moreover, in contrast to previous reports, acid-treated *purified* MoFe protein failed to activate extracts of the *nit*1 mutant strains of *N. crassa* either aerobically or anaerobically.[88] Further, acid-treated extracts of wild-type *A. vinelandii* cells, grown fully repressed on excess ammonium, activated only the *nit*1 extracts, while acid-treated extracts of fully derepressed, *i.e.*, N₂-fixing, wild-type *A. vinelandii* cells successfully complemented the activities of both *nit*1 and UW45 extracts. In addition, acid-treated extracts of nitrogenase-derepressed UW45 cells activated *nit*1 extracts, but failed in the UW45 assay. Finally, it was shown that the *A. vinelandii* UW45 strain can grow with nitrate as its sole fixed nitrogen source. These data clearly indicate that the activating factor for nitrate reduction is quite distinct from that required for N₂ fixation.[88] Only with *partially purified* preparations of the nitrogenase MoFe protein, which were shown to be contaminated with other proteins,[8,88] could *nit*1 activation be accomplished.

Both activating factors have been separated from their "host" proteins by acid denaturation and extraction into the organic solvent, *N*-methylformamide (NMF).[88,101] The nitrogenase-derived factor was named iron-molybdenum cofactor (FeMo-cofactor or FeMo-co) and that from all other molybdoenzymes was called molybdenum cofactor (or Mo-co). This terminology is based on the observations that the isolated FeMo-cofactor contains nonheme iron, while the Mo-cofactor cannot because, for sulfite oxidase and fungal nitrate reductase at least, the "host" protein contains no non-heme iron.[88] Using derepressed *A. vinelandii* cells as the source, the two cofactors were found to be distinct chromatographically on Sephedex G-100 in NMF.[88] These two cofactors also have been shown to have quite separate and distinguishable genetic determinants[82] as well as activation properties and chemical compositions.[88] The chemistry and biology of the molybdenum cofactor have been reviewed elsewhere[19] and will not be addressed further here.

III. FeMo-Cofactor Assay Systems

A. The Typical Protocol

Assays for FeMo-cofactor activity are based on its ability to reconstitute crude extracts of *nifB⁻*, *nifE⁻* or *nifN⁻* mutant strains of *A. vinelandii* or

(FeMo-cofactor)   (ATP system)

     +   ----->   (dithionite) -----> C_2H_4

(Mutant extract)   (C_2H_2)

Stage 1   Stage 2

Figure 22-1. Two-stage assay for FeMo-cofactor activity.

K. pneumoniae toward acetylene (C_2H_2) reduction. C_2H_2 is an alternative substrate of nitrogenase, whose reduction product, ethylene (C_2H_4), is very easily monitored and accurately quantified by gas chromatography.[32] This assay is a stringent test of FeMo-cofactor integrity and also presents a challenge that the ultimate synthetic model must meet.

The assay is typically conducted in two stages (see Figure 22-1). The first involves anaerobic incubation of FeMo-cofactor, either directly as an acid-treated cell extract or after extrusion into NMF, with a cell-free extract of an appropriate derepressed mutant strain at room temperature for 30–45 minutes.[101] Usually, a titration is performed to determine the specific activity of the FeMo-cofactor sample, such that increasing quantities of FeMo-cofactor (*ca.* 0.02–2.00 nmol as Mo) are added to the crude extract (*ca.* 50–200 μL containing *ca.* 1–6 mg total protein). Because NMF is inhibitory to the following activity assay, the *volume* of FeMo-cofactor solution added should be such that there is no more than 5% vol/vol NMF in the assay. After incubation, 25–100 μL of the reconstituted sample are assayed for C_2H_2 reduction after anaerobic transfer to a sealed vial (5–15 mL capacity) containing dithionite (as the electron source), an ATP-generating system (as ATP is required, but ADP is inhibitory) and an excess of the nitrogenase Fe-protein component (to ensure that it is not limiting) under an atmosphere of 10% C_2H_2 (higher concentrations are inhibitory) in argon.

FeMo-cofactor activities are most often reported on a per Mo basis because FeMo-cofactor contains only one Mo atom and its exact size and composition are as yet unknown. Activities of approximately 200–300 nmol C_2H_4 produced min^{-1} (ng atom Mo)$^{-1}$ are considered acceptable. Considerably higher values are likely to be artifactual because they represent activities greater than those so far achieveable on a per Mo basis with the pure nitrogenase MoFe protein.

The ability of FeMo-cofactor to reconstitute the activity of crude extracts of derepressed cells of *nifB⁻*, *nifE⁻*, or *nifN⁻* strains is independent of its bacterial source,[101] providing, of course, that it is derived from an extract of a properly derepressed culture containing an active MoFe protein. This identity is not necessarily expected, because the two nitrogenase component proteins, when isolated from a variety of N_2-fixing organisms, do not always "cross-react" to form an active nitrogenase.[23] The observation of but a single FeMo-cofactor, which is common to all nitrogenases, is a vital simplifying insight into understanding how nitrogenase is structured and how it works.

B. The FeMo-cofactor Insertion Process

Although preincubation of FeMo-cofactor with an extract containing the apoprotein for about 30 minutes before assaying has been routinely used, some inactivation (15–35%) of both the apoprotein and FeMo-cofactor occurs during this period.[34] It was also not clear that it actually takes this amount of time for FeMo-cofactor to be inserted into the protein or even that any insertion takes place during this period rather than after addition to the assay mixture. Because neither FeMo-cofactor nor the apoprotein have any innate activity, which means that the preincubation mixture must be added to the Stage-2 mixture to assay, it is difficult to distinguish between insertion at Stage 1 or 2. A kinetic study,[34] during which C_2H_2 reduction was initiated by adding FeMo-cofactor to an assay mixture containing an extract of a *K. pneumoniae nifB⁻* mutant (either strain Kp5058 or UNF 1718), showed that, under these conditions, an induction period of less than 1.5 minutes is required before a steady-state rate of reduction is achieved. This period, in turn, reflects the time required for full FeMo-cofactor insertion, which is best described as tight binding to a single type of site, and complete activation of the apoprotein. The steady-state rate achieved by this method was identical to that obtained by the two-stage protocol. These data indicate that the *30-minute* incubation is unnecessary.

The question still remained, however, of whether insertion occurs spontaneously or requires some additional factor(s), and of whether these factors, if needed, are provided by the crude extract used or by the assay mixture. The provision of proteins or factors by the extract was eliminated by adding extracts of other mutants, which would be expected to enrich the reconstitution mixture in all proteins but the MoFe protein to partially purified apo-MoFe protein, none of these increased the specific activity.[34,91] No consensus, however, has yet been reached with respect to whether components of the assay mixture—namely, the nitrogenase Fe protein and/or MgATP—are required for FeMo-cofactor insertion. One report states that purified apo-MoFe protein from the *A. vinelandii nifB⁻* mutant, UW45, is activated by FeMo-cofactor alone, that MgATP has no effect, and that although added Fe protein stimulates the activation, it does so only by protecting the apo-MoFe protein from degradation by the NMF added with the FeMo-cofactor.[86] In contrast, the apo-MoFe protein of the *A. vinelandii nifH⁻* mutant, DJ54, is reported not to be activated by FeMo-cofactor alone. It requires the Fe protein–MgATP complex to bind to the apo-MoFe protein and effect a conformational change that exposes the FeMo-cofactor-binding site in a process that is similar to the one that occurs during turnover.[91]

Can these two sets of results be reconciled? Only one experiment was performed similarly in both reports, and this involved incubation of either a purified apo-MoFe protein[86] or a crude extract[91] with excess FeMo-cofactor, followed by chromatography over anaerobic DEAE-cellulose. By using the appropriate elution conditions, which should retain free FeMo-cofactor on the column, any reconstituted holo-MoFe protein should be partially purified and be detected by increased specific activity. On assaying fractions for C_2H_2 reduction, an increase was obtained in one report[86] only. One explanation of these conflicting results would be if FeMo-cofactor, either free or loosely bound, was coeluted with the apo-MoFe protein. Then, on addition to the assay mixture, it would be inserted under the influence of the Fe protein-MgATP complex to give the positive result. Assuming similar chromatography, however, there is no reason to expect coelution of FeMo-cofactor and apo-MoFe protein in one experiment and not the other. A more likely possibility is that the different mutations (*nifB*⁻ vs. *nifH*⁻) have different and as yet unrecognized effects on apo-MoFe protein production or maturation.

C. Inhibitors of FeMo-cofactor Insertion

The apo-MoFe protein purified from the *K. pneunomiae nifB*⁻ mutant strains Kp5058 and UNF1718 lacks N_2-reduction ability but contains some Mo (0.45 atom/molecule).[34] This Mo might be present in the form of a FeMo-cofactor precursor, which although inactive is still able to bind to the apoprotein. Because tetrathiomolybdate (MoS_4^{2-}) and trithiomolybdate ($MoOS_3^{2-}$) are potential precursors to FeMo-cofactor[13,14] and are products of the oxidative degradation of both the "M" centers[126] and isolated FeMo-cofactor,[77] the interaction of MoS_4^{2-} with the partially purified *nifB*⁻ apo-MoFe protein was studied.[34] After a 15-minute incubation and anaerobic Sephadex G-50 gel filtration, not only had the Mo content increased (to 1.9 atom/molecule) but the reconstitution activity had decreased by 60%, indicating that species other than FeMo-cofactor can bind to the apo-MoFe protein and, in doing so, inhibit its reconstitution by FeMo-cofactor. Similar inhibition occurs when either O_2-damaged FeMo-cofactor or the "molybdenum–iron cluster," a component extracted from acid-denatured MoFe protein with methylethylketone,[102] is added to the apo-protein *before* FeMo-cofactor.[107] Neither of these biologically derived materials (possibly containing thiomolybdates) nor MoS_4^{2-} had any significant inhibitory effect if added *after* reconstitution by FeMo-cofactor. In contrast, WS_4^{2-} and $MoO_nS_{(4-n)}$ (with $n = 4$, 3, or 2) have no inhibitory effect when added at any time. These data indicate that both FeMo-cofactor and MoS_4^{2-} are bound irreversibly by the protein in a reaction that is essentially complete in less than 3 minutes.[107] Using ^{99}Mo-labeled FeMo-cofactor, the binding of FeMo-cofactor and MoS_4^{2-} were found to be mutually exclusive. Thus, MoS_4^{2-} most likely prevents FeMo-cofactor binding by direct competition at the FeMo-cofactor-binding site and vice versa, although an indirect effect that is generated by MoS_4^{2-} interacting with a site distant from the FeMo-cofactor-binding site is a possibility.

Inhibition of FeMo-cofactor reconstitution activity also occurs with dimethylaminoethanol.[41] This effect was discovered in experiments designed to determine if FeMo-cofactor could be photoreduced to its fully reduced redox state (see Section VI.B) using light, deazaflavin, and demethylaminoethanol as the electron donor. Not only did photoreduction not occur, but the resulting FeMo-cofactor solution no longer reactivated an *A. vinelandii* UW45 extract, even though it retained its fully developed EPR spectroscopic signature. This result was also observed when dimethylaminoethanol alone was added to a FeMo-cofactor solution.[73] Although no other details are available, these data suggest that FeMo-cofactor remains intact during treatment with dimethylaminoethanol but that it is inhibited somehow from binding to the apo-MoFe protein.

IV. Isolation Methods

Simultaneously, but essentially independently of the acid-denaturing protocols described earlier, procedures were being developed to extrude the various Fe–S cores from proteins. These techniques involved anaerobic protein denaturation by organic solvents [e.g., N, N'-dimethylformamide (DMF)] followed by capture of the extruded Fe–S cluster by an appropriate, discriminating "acceptor molecule" either an apoprotein or a thiol.[81] In a judicious combination of both protocol types, plus recognizing the utility of the *A. vinelandii* UW45 extract assay, a procedure was devised for isolating the activating cofactor from purified native MoFe protein.[101] This procedure, which represents a clear milestone in FeMo-cofactor research, is outlined in Figure 22-2 along with the various alternatives and options that were introduced later.

A. The Original Isolation Procedure

The original isolation procedure[101] occurs in two main steps. After dilution with water, the first step employs aqueous acid-base treatment as used in the very early investigations, but now strictly anaerobic. Then, because the FeMo-cofactor, unlike the Mo-cofactor, remains tightly bound to the precipitated denatured protein, the second step involves extraction of the FeMo-cofactor from the protein with an appropriate organic solvent, which initially had to be N-methylformamide (NMF). The acid treatment presumably un-

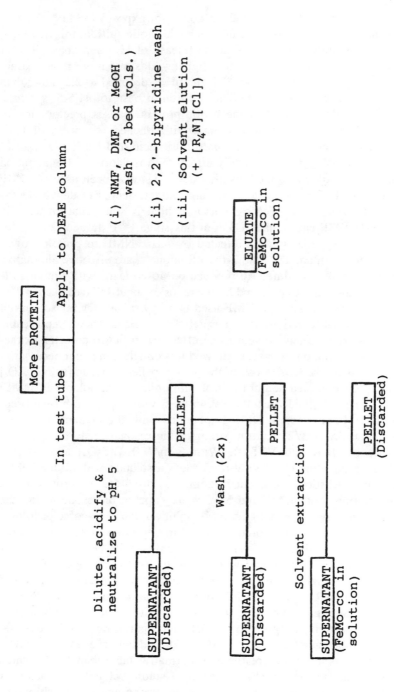

Figure 22-2. Flow diagram of a generalized FeMo-cofactor isolation scheme.

folds and denatures the protein and, in doing so, exposes the "P" clusters. Because these are Fe–S clusters containing acid-labile sulfide, this treatment causes their disruption and results in the release of H_2S and about 50% of the iron content of the MoFe protein. Higher yields and more rapid extraction of FeMo-cofactor are obtained[108] if the pH is dropped to 2.2 rather than the originally reported pH of 2.8. The addition of base to pH 5.5, which is close to the isoelectric point of the MoFe protein, causes precipitation of the denatured protein with FeMo-cofactor still attached. A variety of acids and bases can replace the originally used citric acid and $Na_2H(PO_4)$, all of which have been shown to be equally effective in isolation.[124] Denaturation with the organic solvent, dimethylsulfoxide,[108] has also been used. A DMF wash, into which FeMo-cofactor does not extract, serves to dehydrate the pellet and so protects the FeMo-cofactor, which is somewhat unstable in water. *Acidic* NMF can replace DMF at this stage.[124] Extraction of FeMo-cofactor from the pellet is then effected by *basic* NMF to give a brown-green solution. Formamide[124] and 2-pyrrolidinone[120] are effective alternative extractants for this procedure. All produce FeMo-cofactor with similar EPR signals, reconstitution activities, and Mo:Fe ratios of about 1:7 (see Section V).

The acid–base status of the NMF used is a key parameter for successful extraction. This observation was interpreted to indicate that a replaceable proton is required for any solvent to be effective in FeMo-cofactor extraction.[124] A later variation, however, showed this conclusion is not necessarily valid.[53] By making the DMF wash of the protein pellet 5 mM in $[Et_4N]_2[S_2O_4]$ (the usually encountered sodium salt of dithionite is not soluble in DMF) and 2 mM in $[Et_4N][OH]$, a fully UW45-reactivatable, green-brown preparation of FeMo-cofactor in DMF is obtained with the characteristic, but slightly altered, $S = 3/2$ electron paramagnetic resonance (EPR) spectrum (see Sections VII.A and VIII.D). With 5 mM $[Et_4][S_2O_4]$ and 50 mM $[Et_4N][OH]$, acetonitrile (MeCN) also produces a solution with all the FeMo-cofactor characteristics intact, but in doing so, dissolves the protein pellet also. In addition, both DMF and MeCN are capable of resolubilizing the residue obtained by evaporating an NMF solution of FeMo-cofactor to dryness to give fully active FeMo-cofactor.[53] Although these results raised the question of whether dithionite had to be solubilized for FeMo-cofactor extraction into DMF,[53] subsequent work showed that the presence of the organic cation is crucial.[61] In fact, not only could FeMo-cofactor be produced in DMF containing 0.5 M $[Bu_4N][Cl]$ by the "column" method (see Section IV.C),[123] but, after evaporation to dryness, it could be redissolved in MeCN (to 10 mM), acetone (Me_2CO; to 1 mM), tetrahydrofuran (THF; to <100 μM), dichloromethane (CH_2Cl_2; to >100 μm), and benzene (to <100 μM). All these solvents are more volatile and thermally stable than NMF, which should be an advantage for physicochemical studies of FeMo-cofactor. A variation of this procedure[54] uses the "nitrogenase complex," which is produced from crude extracts of derepressed *A. vinelandii* cells by centrifugation at 150,000 *g* for 18 hours,[7] and so circumvents the need for purified MoFe protein.

B. Partial Purification Protocols

Isolated FeMo-cofactor in NMF has been partially purified by several techniques. For many experimental procedures, fairly concentrated solutions of FeMo-cofactor are required. These have been obtained mostly by vacuum concentration at 30° C, which results in purification by precipitation of a certain amount of inorganic salts and proteinaceous material.[8] Chromatography over either an anaerobic anion-exchange[96] (FeMo-cofactor has been shown to be anionic[73]) or gel-filtration column[28,108] is also effective. The gel-filtration procedure has shown that considerable quantities of low-molecular-weight materials can be removed, along with about 10% of the Fe content and some Mo, both of which appear not to be FeMo-cofactor associated, and is effective in transferring FeMo-cofactor to other solvents.[27,28] This latter procedure simply involves developing columns in the appropriate solvent after applying FeMo-cofactor as a NMF solution.

C. Column-Based Isolation Procedure

An alternative "column" procedure, which eliminates a separate acid–base denaturation step, has been developed and involves the initial binding of the MoFe protein to an aqueous anaerobic DEAE-cellulose column.[61] After washing with several bed volumes of degassed NMF, DMF, or a 1:9 mixture of the two, during which the column matrix is stirred, FeMo-cofactor is eluted using solvent containing 0.2–0.5 M [R_4N][X] (for NMF, R = Et, X = Cl or Br; for DMF or the mixture, R = Bu, X = Cl or Br), with Cl^- being the more effective at displacing the FeMo-cofactor from the column and/or protein. Apparently, the strong interaction of the MoFe protein with the DEAE-cellulose is sufficient for FeMo-cofactor extraction to occur without acid treatment. Yields of FeMo-cofactor averaged about 80%, with reconstitution activities averaging about 200 nmoles C_2H_4 produced min^{-1} (ng atom Mo)$^{-1}$, both of which are comparable with other isolation procedures. The use of DMF, however, results in the extraction of all iron [Fe:Mo = (16 ± 2):1] from the protein, about 50% of which can be removed by treatment with 50 mM 2, 2'-bipyridine, whereas NMF extracts contain only FeMo-cofactor-related iron [Fe:Mo = (7 ± 1):1].[61]

FeMo-cofactor can be extracted into a wide range of organic solvents from the MoFe protein bound to a DEAE-Sepharose column, if the column is first pretreated with DMF to facilitate the use of water-immiscible solvents, followed by equilibration with the extracting solvent.[123] Reasonable yields, concentrations, and activities are obtained in MeCN, Me_2CO, and CH_2Cl_2 (all containing 0.1 M [Bu_4N][Cl]), whereas THF and benzene (containing saturating about 0.05 M [Bu_4N][Br]) are much less efficient for extraction. The Fe-to-Mo ratios in all these preparations are, however, two to three times those normally found, indicating that most of the P-cluster Fe is extracted also. However, FeMo-cofactor isolated in this way, in either DMF or CH_3CN, does not exhibit[123] the usual $S = 3/2$ EPR signal, but one centered around $g = 2$; a quite different result to that reported earlier.[53] When a methanol (MeOH) wash, which circumvents any potential carryover of DMF into the second solvent, is used instead of DMF, [Bu_4N][Cl]-containing solutions of MeCN, MeOH and CH_2Cl_2 do not effectively extract FeMo-cofactor; only Me_2CO (30% recovery) works.[123] Thus, although both DMF and MeOH completely denature the MoFe protein on the column, their effects on the protein must be different. It may be that DMF has the additional capability of percolating the protein pockets containing the FeMo-cofactor and displacing the ligands binding it to the polypeptide. A further observation is that salt-containing solutions of either MeCN or CH_2Cl_2 alone appear incapable of extracting FeMo-cofactor. MeCN, Me_2CO, THF, and CH_2Cl_2 are ineffective as prewashes because they appear not to denature column-bound MoFe protein. Thus, the ability for FeMo-cofactor solubilization, protein denaturation, and FeMo-cofactor extraction appear to be independent properties of solvents.[123].

V. Composition

A. Inorganic Components

As part of the original report[101] in 1977 of the isolation of FeMo-cofactor, the atomic ratio of its three inorganic components, Mo–Fe–S, was reported as 1:8:6. Since then, despite considerable effort and some variation in individually determined S:Mo and Fe:Mo measurements, this ratio has not changed significantly.

i. The S–Mo Ratio

Conventional biochemical analysis for S^{2-} in many FeMo-cofactor samples gave a ratio of 1:4[8,124] but this same technique underestimated the S–Mo ratio of Mo–Fe–S inorganic compounds.[8] Using ^{35}S labeling, which was introduced by growing cells on $^{35}SO_4^{2-}$, isolating ^{35}S-labeled MoFe protein, and then extracting FeMo-cofactor, widely varying values of 3.5:1[108] and 8.7($\pm$1.0):1[70] for the S–Mo ratio in FeMo-cofactor were measured. Because the samples used were prepared differently and were not purified to any great extent, varying amounts of impurities, possibly S-containing or Mo-containing, could remain associated with either the FeMo-cofactor or the contaminating protein and might affect the analyses. An 8.7:1 S–Mo ratio was arrived at after subtracting 3.2 S of the total of 11.9 determined per Mo because that amount was found to be associated with contaminating denatured MoFe protein.[70] A similar ratio of 8.8(±0.6):1 was later recorded by purifying native MoFe protein, which was assembled from ^{35}S-labeled FeMo-cofactor and apo-MoFe protein, to a constant relationship among activity and radioactivity.[109] The best value for the S–Mo ratio, therefore, appears to be in the range of 9(±1):1.

ii. The Fe–Mo Ratio

The Fe-Mo ratio has tended to drop since the original report. Conventional chemical analysis has suggested that as-isolated FeMo-cofactor contains close to eight Fe atoms per Mo atom but that on concentration this ratio drops to about 7:1,[124] and on chromatographic purification, it decreases further to 6.3(±0.5):1.[28,96] The value of 6(±1):1 for Fe:Mo also results from the "column" isolation protocol.[61] Mössbauer spectroscopy on the holo-MoFe protein has been interpreted to indicate that the "M" centers (*i.e., protein-bound* FeMo-cofactor) contain eight,[21] five-to-seven (most probably six)[43] or five-to-eight[111] Fe atoms per Mo, whereas ENDOR (electron–nuclear double resonance) estimates have been revised downward from at least six[37] to at least five[117] Fe atoms. Mössbauer studies of the *isolated* FeMo-cofactor provide a quantitation of 5(±1):1.[78] The ratio of Fe to Mo appears then to reside in the range 6.5(±1.5):1.

iii. Possible Presence of Any Other Inorganic Components

The possibility of other inorganic components (e.g., Cl^-) of FeMo-cofactor has been the subject of speculation, but it has been difficult to test because all MoFe protein solutions contain NaCl. Chromatographically purified FeMo-cofactor,[28,35] plus the recent development of a higher-sensitivity XAS (x-ray absorption spectroscopy) method at low energies for use with dilute solutions of biological samples, [36] has now made such investigations feasible. The chlorine K-edge XANES (x-ray absorption near edge structure) spectra of various G-25-purified NMF solutions of FeMo-cofactor, which have, thus, been freed of all nonligating small molecules and ions, clearly show that Cl^- plays no part in ligating intact FeMo-cofactor.[35] The sulfur K-edge XANES of the same solutions,[35] however, do show, in addition to the expected S^{2-}, the presence of an unprecedented form of oxidized sulfur, which corresponds to thiosulfate bound as $-SSO_3^{2-}$.

Thus, the inorganic composition of isolated FeMo-cofactor covers the range of 6.5 (± 1.5) Fe atoms, 9 (± 1) S^{2-}, no Cl^-, at least one $S_2O_3^{2-}$, and one Mo atom.

B. Organic Components

i. Possible Association of a Peptide with FeMo-cofactor

An early major concern was whether or not FeMo-cofactor contained a peptide. Several lines of evidence suggest that components other than Mo, Fe, and S^{2-} must be present. For example, the overall negative charge[73] on FeMo-cofactor and the detection of three neighboring, and likely donor, O (or N) atoms by XAS[13,14] are both strong indicators of the presence of some additional component. NMF solutions of as-isolated FeMo-cofactor do contain proteinaceous material, and this observation led to the suggestion that a peptide might be an integral part of FeMo-cofactor.[30,57] However, careful study of this material indicated that its composition paralleled that of the MoFe protein[73,108] and was, therefore, likely to be simply contaminating, denatured MoFe protein. Purification by either vacuum concentration and filtration[124] or chromatography[108] produced FeMo-cofactor preparations with no amino acid at a concentration above about 0.2 residues per Mo atom. Therefore, FeMo-cofactor is not associated with either individual amino acids or a peptide. These preparations also contain no common sugars.[124] Direct chemical analysis proves that neither lipoic acid nor coenzyme A (less than 0.5 nmol detectable PO_4^{3-} per ng atom Mo) are present in either the FeMo-cofactor or the parent MoFe protein[124] despite a spectroscopy-based report to the contrary.[51]

ii. Possible Presence of Exogenously Added Entities

The only obvious candidates remaining as potential ligands to FeMo-cofactor after elimination of the entities discussed earlier were the reagents added during the isolation procedure. None of these were found to be absolutely necessary, however, because a variety of acids, bases, and solvents were found to be interchangeable in successfully isolating FeMo-cofactor[61,108,124] These data, plus the interactions of the FeMo-cofactor with metal-liganding agents (e.g., citrate,[108] o-phenanthroline,[124] EDTA,[10] and thiophenol[10,89]) indicate that FeMo-cofactor has a labile coordination sphere, most of which appears available for rapid ligand exchange.

iii. NifV Factor and Homocitrate

The simplistic picture described earlier was complicated by insight into the NifV⁻ phenotype. nifV is one of six nif genes, nifBENVQH, whose products are involved in FeMo-cofactor biosynthesis (see Section X). It was found that lesions in *nifV* resulted in a defective MoFe protein and a nitrogenase that was slightly compromised in its ability to reduce C_2H_2; that was highly compromised with respect to N_2 fixation; and that, in contrast to the native enzyme, exhibited a H_2-evolution activity sensitive to carbon monoxide (CO).[59] This defect was found to reside in the FeMo-cofactor centers, because when they were extracted from the NifV⁻ MoFe protein and used to reconstitute the FeMo-cofactor-deficient apo-MoFe protein from a NifB⁻ strain, the NifV⁻ phenotype transferred with the FeMo-cofactor.[33] Very minor (or no) spectral differences among the *protein-bound* native FeMo-cofactor and the *protein-bound* NifV⁻ FeMo-cofactor have been observed from applying a variety of biophysical techniques.[22,33,60,109,116] Thus, only very minor structural or electronic changes result from the incomplete processing accorded the NifV⁻ FeMo-cofactor.

Further insight into the nature of the defect in the NifV⁻ FeMo-cofactor came when a low-molecular-weight factor, termed the V factor, which is only produced in the presence of the *nifV*-gene product,[42] was isolated from both crude extracts[42] and the growth medium[41] of nitrogenase-derepressed wild-type *K. pneumoniae* cells. The V factor was shown not to be involved in inserting FeMo-cofactor into apo-MoFe protein,[42] but it was required for the *in vitro* synthesis of FeMo-cofactor[105] (see Section X), a technique that was instrumental to its isolation. V factor was identified as (*R*)-homocitrate,[41] and the *nifV*-gene product was suggested to be homocitrate synthase. Homocitrate was shown to repair the defect in the NifV⁻ phenotype when present in the medium on which *K. pneumoniae* NifV⁻ cells were being derepressed for nitrogenase synthesis.[39] However, homocitrate neither binds nonspecifically to nor exchanges into isolated NifV⁻ FeMo-cofactor[40] (see also Section IX.B). Citrate was suggested as the likely replacement for homocitrate in NifV⁻ nitrogenase.[38]

To determine whether homocitrate was incorporated intact into the MoFe protein and in what stoichiometry with respect to Mo, ³H-labeled homocitrate was used in the *in vitro* FeMo-cofactor synthesis system and the FeMo-cofactor so produced incorporated into the apo-MoFe protein of the *nifB⁻* mutant *A. vinelandii* strain, UW45.[40] This MoFe protein, plus added native MoFe protein as a carrier, was purified and found to contain 84 (±5)% of the radioactivity expected for a 1:1 ratio of Mo to homocitrate. FeMo-cofactor was then extracted from this labeled purified MoFe protein and was found to contain >80% of the radioactivity initially associated with the protein.[40] This isolated and labeled FeMo-cofactor was then used to reconstitute a second sample of apo-MoFe protein, and >80% of the ³H present in the isolated FeMo-cofactor was found to copurify with the newly reconstituted MoFe protein. Thus, the ³H labeled entity (either homocitrate or a fragment) remains closely associated with the FeMo-cofactor. The proof that this entity was intact homocitrate came from its isolation in >87% yield from acid-denatured holo-MoFe protein and its characterization by chromatography, by the *in vitro* synthesis of FeMo-cofactor and by proton nuclear magnetic resonance spectroscopy.[40]

The easy isolation of homocitrate from the holo-MoFe protein[40] contrasts with the unsuccessful attempts to release it from isolated FeMo-cofactor following O_2 degradation.[42] This failure has been attributed[40] to inhibition by the products of the O_2-degradation reaction at the third stage of the four-stage assay used to detect homocitrate. The C_2H_2-reducing activity, which was monitored in the assay, is coupled to homocitrate release via the *in vitro* synthesis of FeMo-cofactor and its subsequent insertion into the apo-MoFe protein. The inhibition is explained by the presence in the assay of thiomolybdates, which are known to be formed when the FeMo-cofactor is exposed to O_2[77] and to be effective in preventing the binding of FeMo-cofactor to the apo-MoFe protein.[107]

C. FeMo-cofactor's Minimal Core Composition

By combining all the data discussed in this section, a likely minimal core composition for isolated FeMo-cofactor is established. This core is $[Mo–Fe_6–S_9–(S_2O_3{}^{2-})–(homocitrate)]$.

VI. Oxidation–Reduction Properties

The "M" centers within the MoFe protein have been shown to have available three (at least) interconvertible oxidation states[65,121,122,125] (see Eq. 1). The ["M"(s-r)] state is encountered when the MoFe protein is

$$
\begin{array}{ccccc}
 & (a) & & (b) & \\
[\text{``M''(ox)}] & \longleftrightarrow & [\text{``M''(s-r)}] & \longleftrightarrow & [\text{``M''(red)}] \\
(S = 0; & & (S = 3/2; & & (S = \text{integer}; \\
\text{EPR-silent}) & & \text{EPR-active}) & & \text{EPR-silent})
\end{array}
\tag{1}
$$

isolated during the normal purification protocol in the presence of excess dithionite. It is characterized spectroscopically by the biologically unique $S = 3/2$ electron paramagnetic resonance (EPR) signal. This state is designated as the semireduced state (s-r) because it is both oxidizable and reducible. Each of the two ["M"(s-r)] centers in the purified MoFe protein is oxidized at a potential of about -0.05 V versus the normal hydrogen electrode (NHE) in a one-electron process [designated as (a) in Eq. 1] to ["M"(ox)] by redox-active dyes [e.g., thionine ($E_{1/2} = +0.02$ V)[121,122,125]]. ["M"(ox)] is EPR-silent and diamagnetic.[125] This process is reversible, but it exhibits considerable hysteresis with ["M"(ox)] being rereduced to ["M"(s-r)] at about

-0.29 V versus NHE.[121] ["M"(red)] is produced when ["M"(s-r)] is reduced by one electron apparently[65] in a process [designated as (b) in Eq. 1] occurring at about -0.47 V versus NHE.[52] ["M"(red)] is the substrate-reducing state within the MoFe protein and it is achieved only under turnover conditions (i.e., in the presence of a fully operational nitrogenase system, which includes the Fe protein, MgATP, and dithionite[80,110]). ["M"(red)] is EPR-silent and has integer spin.[65]

Inasmuch as the FeMo-cofactor is the extruded "M" center and, thus, should retain as its principal function the transfer of electrons from reductant to substrate, redox processes corresponding to processes (a) and (b) in Eq. 1 for the protein-bound "M" center should be observable for isolated FeMo-cofactor. Fortunately, FeMo-cofactor retains the characteristic EPR signal, albeit slightly changed, of the protein-bound "M" centers when it is extracted into NMF in the presence of excess dithionite.[89] This state, therefore, corresponds to the semireduced state of the "M" centers and is designated as FeMo-co(s-r). Just as the EPR signal of the "M" centers facilitated redox studies of the holo-MoFe protein, so the EPR signal of the isolated FeMo-co(s-r) has been used to monitor the oxidation–reduction properties of FeMo-cofactor.

A. The FeMo-co(s-r)–FeMo-co(ox) Redox Process

This process was first demonstrated chemically,[10,73] using EPR as the monitor of redox change. The same redox-active dyes (e.g., methylene blue) that were effective in oxidizing the "M" centers[24,63,110,111,122,125] also oxidize EPR-active FeMo-co(s-r) to the EPR-silent FeMo-co(ox). In this reaction mixture, the methylene blue also oxidizes the excess dithionite present. This reaction was fully reversible with excess dithionite (see Figure 22-3). O_2 also produces the same oxidized EPR-silent state, but it usually involves some degradation of the FeMo-cofactor because rereduction with excess dithionite leads to only partial recovery of the $S = 3/2$ EPR signal.[73]

To characterize the (s-r)-to-(ox) couple more completely in terms of both the potentials and numbers of electrons involved, electrochemical methods compatible with the NMF-based and small-volume availability of the FeMo-cofactor system had to be developed, as did methods to remove the excess dithionite, which would otherwise complicate any redox investigations. The latter problem was solved serendipitously. On anaerobic storage of NMF solutions of FeMo-cofactor either at room temperature (over hours) or at $-80°$ C (over weeks), a "spontaneous oxidation" reaction occurs.[96] This reaction results in the removal of all dithionite and produces FeMo-co(ox). Presumably, dithionite autocatalytically disproportionates into thiosulfate,[35] sulfite, and protons and in doing so alters the solution conditions such that FeMo-co(s-r) now becomes oxidized to FeMo-co(ox).[76,96] The changing so-

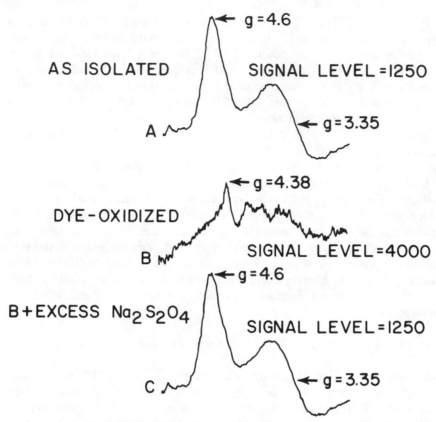

Figure 22-3. EPR spectra in the $g = 4$ region of the FeMo-cofactor at 14 K as it is cycled chemically through its (s-r)-(ox)-(s-r) oxidation states. (Reproduced with permission of the American Society for Biochemistry and Molecular Biology.)

lution conditions due to either dithionite disproportionation or NMF decomposition (or both) severely complicate any investigation of FeMo-cofactor[76,95] (see Sections VIII.D and VIII.E). *Basic* NMF, which is freshly distilled after treatment with either barium oxide or sodium bicarbonate, gives a stable, reproducible system suitable for most studies. Microelectrochemical cells for operation in anaerobic enclosures have been designed, built, and tested[25,96,99] to overcome the low-availability problem.

An electrochemical response from FeMo-cofactor was elicited only at carbon surfaces. This response did not require a redox mediator. No metal electrode (e.g., Au, Ag, Pb, Pt, and Hg) was effective. Cyclic voltammetry on volumes as small as 22 μL of a basic 2.82-mM NMF solution indicates that FeMo-cofactor undergoes the (ox)-to-(s-r) redox process at a formal redox potential (E$_0'$) of −0.34 (±0.02) V versus NHE.[25,95,96,99] This process is only quasi-reversible with the anodic and cathodic peak potentials sepa-

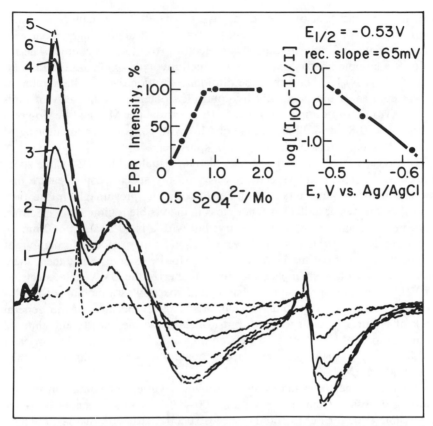

Figure 22-4. EPR spectra of FeMo-cofactor at 10 K as it is incrementally reduced with 0.25-electron equivalents of aqueous dithionite from oxidized (spectrum 1) through to the semireduced (spectrum 5) state. Insets demonstrate that complete semireduction requires one electron per FeMo-co(ox). (Reproduced with permission from the *Journal of the American Chemical Society*, vol. 107, p. 5367. Copyright 1985 American Chemical Society.)

rated by about 100 mV (or more), about twice the 59-mV separation expected for a reversible, diffusion-controlled redox process for a single species in solution. Correlation of this couple with the (ox)-to-(s-r) conversion was confirmed by monitoring concurrently (1) the development of the $S = 3/2$ EPR signal, (2) the change in the solution equilibrium rest potential, and (3) the relative sizes of the anodic and cathodic peaks as a sample of FeMo-co(ox) is titrated with aliquots of a standardized dithionite solution[96] (see Figure 22-4).

The results of this experiment show that the EPR signal is fully developed after addition of 0.8 electrons per Mo, indicating that $n = 1$ for the (ox)-to-(s-r) process. Additional aliquots of dithionite have no effect on the EPR signal. A Nernst plot of these data gave E_0' of -0.3 V versus NHE and a slope of 65 mV, which is equivalent to $n = 0.9$, all in good agreement with previous data and indicative of a one-electron (ox)-to-(s-r) couple. The results of a complementary experiment, which involved oxidizing FeMo-co(s-r) with $[Fe(CN)_6]^{3-}$, confirmed the dithionite titration data.[96] The n value of 1 was confirmed independently by controlled potential coulometry of multiple FeMo-co(ox) samples (0.89 ± 0.16 electrons per Mo) and FeMo-co(s-r) samples (0.83 ± 0.22 electrons per Mo) with concomitant monitoring of the EPR signal intensity (0.76 ± 0.07 spins per Mo).[99]

The (ox)-to-(s-r) redox change may also be visualized by UV-visible spectroscopy.[27,28] The visible spectrum was the only spectroscopic feature discussed in the original report[101] of FeMo-cofactor [presumably in the (s-r) state], and it was described as featureless in the visible region. This spectrum became of generally lowered intensity, but with a band at about 460 nm on exposure to O_2, which we now know is indicative of the formation of thiomolybdates[77] from the FeMo-cofactor's Mo–Fe–S core. Since then, there have been very few attempts either to characterize the UV-visible spectrum of FeMo-cofactor or to use UV-visible spectroscopy as a monitor of FeMo-cofactor reactions. This situation has arisen partly because of the general lack of features in the spectrum, but also because the usually encountered solutions of FeMo-cofactor, which contain NMF (cutoff at 265 nm) and dithionite ($e_{313} = 8000\ M^{-1}\ cm^{-1}$), are opaque to wavelengths of less than about 400 nm.

After G-25 gel filtration, which removes a variety of contaminants, including excess dithionite, the region 400–270 nm in NMF now becomes accessible and, for FeMo-co(s-r), shows spectral structure that is similar to known metal–sulfur charge transfer bands.[27] When this spectrum is compared with that of G-25 gel-filtered FeMo-co(ox), a generally increased intensity for FeMo-co(ox) over the range 600–270 nm is observed with the structural features around 300 nm being more pronounced.[27,28] These spectral differences produce a red-brown coloration for FeMo-co(ox) compared with the green-brown color of FeMo-co(s-r). The spectral changes were correlated with oxidation state by measuring the solution equilibrium rest potential electrochemically for each sample. Although these differing spectral features and intensities are diagnostic of oxidation state, the optical spectra appear to vary slightly from batch to batch and to show a dependence on their storage and treatment history. No electronic spectrum is available for FeMo-co(red).

The question now arises as to how the added reducing electron is accommodated on the $[Mo–Fe_6–S_9–(S_2O_3{}^{2-})–(homocitrate)]$ core of FeMo-co(ox). Early EPR studies, using ^{95}Mo-labeled[83] or ^{57}Fe-labeled[65] MoFe protein, showed that the $S = 3/2$ EPR signal was broadened only with ^{57}Fe. These results strongly indicate that Mo is not a significant factor in the EPR chromophore. And because the EPR chromophore and FeMo-cofactor are now known to be the same entity and intimately involved in this redox process, the inference is that the Fe atoms are the electron acceptor and the Mo atom is uninvolved. This view, however, has to be modified in light of the more recent EPR results on ^{95,96}Mo-labeled MoFe protein[31] and the ENDOR results on ^{95}Mo-labeled MoFe protein,[119] both of which indicate that the Mo atom is integrated into the $S = 3/2$ spin system and so could be involved in the redox process.

Direct information on the electron acceptor has been obtained through the application of Mössbauer spectroscopy[78] and sulfur K-edge and molybdenum L-edge x-ray absorption spectroscopy[35] (XAS) to FeMo-cofactor in both its (ox) and (s-r) states. The change in ^{57}Fe isomer shift from 0.32 to 0.37 mm/sec observed to accompany the reduction of FeMo-co(ox) to FeMo-co(s-r) indicates that the reducing electron resides, at least in part, on iron. However, the observed change is considerably less than the ^{57}Fe isomer shift change (about 0.08 mm/sec), which occurs on redox for the complexes $[Fe_4S_4(SCH_2Ph)_4]^{2-/3-}$, $[Fe_6S_6Cl_6]^{2-/3-}$, or $[Mo_2Fe_6S_8(SPh)_9]^{3-/5-}$. Because it is known that Fe is the electron acceptor in all these complexes, even those containing Mo^{11}, and because all entities, including FeMo-cofactor, are Fe–S clusters with similar numbers of iron atoms, this result indicates that the reducing electron on FeMo-co(s-r) must be partly accommodated elsewhere than on iron.[78]

Orbitals on FeMo-co(s-r) with a large molybdenum component would be obvious candidates for accommodating the remainder of the electron density. But when the Mo L-edge x-ray absorption near-edge structure (XANES) for both FeMo-co(ox) and FeMo-co(s-r) were recorded and analyzed,[35] they were superimposible, suggesting that the Mo atom is uninvolved in the (ox)-to-(s-r) redox process of FeMo-cofactor. However, because the unpaired spin density on the FeMo-cofactor is delocalized mainly over the iron atoms, only a relatively slight Mo involvement would be expected, leading to possible insensitivity of the Mo edge to oxidation state changes. Even so, this situation implies a sulfur-derived component as the most likely remaining acceptor, and sulfides are known to take part in redox changes on polynuclear metal–sulfur clusters.[64] This proposal could not be tested in the original experiments because FeMo-co(s-r) was produced using excess dithionite, which obscures the S K-edge completely. By modifying the XANES cell such that oxidation–reduction of FeMo-cofactor is under electrochemical control, the need for dithionite is eliminated. The sulfur K-edge and Mo L-edge XANES were recorded in this spectro-electrochemical cell for both FeMo-co(ox) and FeMo-co(s-r).[74] These data confirmed that the Mo L-edge does not shift on redox. In contrast, the S K-edge suffered a dramatic change, showing that sulfur is indeed involved in the (ox)-to-(s-r) process.

Mössbauer spectroscopy has also shown that when dithionite is used as reductant, FeMo-co(ox) appears never to be completely reduced (only about 80%) to FeMo-co(s-r) and that FeMo-co(ox), produced by prolonged storage, always contains some (about 15%) FeMo-co(s-r).[78] These observations may be related to the mechanism of the "spontaneous oxidation" reaction,[96] in which an equilibrium situation may be set up by the autocatalytic disproportionation of dithionite.

B. The FeMo-co(s-r)–FeMo-co(red) Redox Process

As the (ox)-to-(s-r) redox process for both isolated FeMo-cofactor and the protein-bound "M" centers parallel one another, after taking into account the different solvent systems,[96] a similar (s-r)-to-(red) redox process may be anticipated. Production of FeMo-co(red) would be equivalent to isolating the substrate-reducing level of the holo-MoFe protein.

The first attempts to demonstrate the existence of the FeMo-co(red) state used either chemical or photochemical means. The synthetic ion,[29] $[Fe_4S_4(SC_2H_5)_4]^{3-}$ with an E_0' of -1.09 V versus NHE in DMF, was selected as a possible analog of the "P" clusters of the MoFe protein. It failed, however, to reduce FeMo-co(s-r) to FeMo-co(red) in NMF because it did not perturb the $S = 3/2$ EPR signal.[72] It did reduce FeMo-co(ox) to FeMo-co(s-r), indicating that its anionic nature was not the reason for lack of reaction. Apparently, an E_0' of -1.09 V in DMF is insufficient for reduction of FeMo-co(s-r) in NMF. Production of FeMo-co(red) was claimed by photoreduction of FeMo-co(s-r) in NMF solution using deazaflavin with N,N'-ethylenediaminetetraacetate (EDTA) as electron donor under CO.[89] This claim was based on the loss of the $S = 3/2$ EPR signal under these conditions in a process that was reported to be reversible by the addition and removal of CO.[89] Unfortunately, other electron donors (e.g., dimethylaminoethanol) did not produce a similar result and EDTA alone caused the loss of the EPR signal in the absence of deazaflavin, CO, and light.[10] This reaction is not obviously redox in nature[66] (see Section IX.B).

A cyclic voltammogram (see Figure 22-5) of FeMo-co(ox) at a glassy carbon electrode shows, in addition to the FeMo-co(ox)-to-FeMo-co(s-r) change, a reduction process at about -1.1 V versus NHE with the corresponding oxidation at -0.9 V versus NHE. These waves are assigned to the FeMo-co(s-r)/FeMo-co(red) couple.[96,97] At the relatively fast scan rates of about 0.4 V s^{-1}, the two reductions are approximately equal in height, which suggests that the (s-r)-to-(red) couple also involves one electron.[96,97]

This conclusion has been confirmed by a comparative x-ray absorption spectro-electrochemical study of FeMo-co(s-r) and FeMo-co(red),[74] during which coulometric measurements showed that interconversions of FeMo-co(s-r) with FeMo-co(red) involved one electron. The Mo L-edge XANES

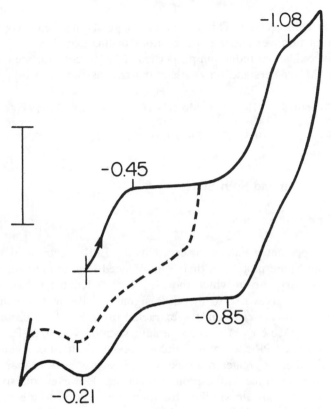

Figure 22-5. Cyclic voltammogram of FeMo-cofactor in NMF solution at a glassy carbon electrode. The dotted line represents reversal
of the scan after the first cathodic peak to illustrate the (ox)-(s-r) couple. FeMo-cofactor concentration, 0.86 mM; sweep rate, 100 mV s^{-1};
bar represents a current of 4 μA; peaks are marked in V versus NHE.
(Reproduced with permission from the *Journal of the American
Chemical Society*, vol. 107, p. 5366. Copyright 1985 American
Chemical Society.)

spectra collected on such samples indicated, surprisingly, that the Mo atom
again appears to be *uninvolved* in the (s-r)-to-(red) process, although the
possible insensitivity of the Mo edge to redox process must be taken into
account. Just as in the (ox)-to-(s-r) process, the S K-edge XANES showed
that a sulfur-based component accommodates the added reducing equivalent,
at least partially. These same data also showed that the $S_2O_3^{2-}$ entity had
been lost.[74] As neither Fe XANES nor [57]Fe Mössbauer spectra has been
collected on FeMo-co(red), the extent of any involvement of iron in the (s-
r)-to-(red) redox process is unknown. After coulometry at an applied potential sufficient to produce FeMo-co(red) and then removal of the potential,
the EPR spectrum of the resulting sample indicated that FeMo-co(red) had

"relaxed" to FeMo-co(s-r). This observation suggests that FeMo-co(red) may be capable of turnover under the conditions of this experiment.

Thus, the oxidation–reduction properties of FeMo-cofactor are quite similar to the "M" centers and may be summarized as shown in Eq. 2.

$$
\begin{array}{ccccc}
\text{FeMo-co(ox)} & 1e^- & \text{FeMo-co(s-r)} & 1e^- & \text{FeMo-co(red)} \\
(S = 0; & <\!-\!> & (S = 3/2; & <\!-\!> & (S = \text{integer;} \\
\text{EPR-silent)} & -0.34 \text{ V} & \text{EPR-active)} & ca.\ -1.0 \text{ V} & \text{EPR-silent)}
\end{array} \tag{2}
$$

VII. Magnetism and Spin State

A. The FeMo-co(s-r) State

The unique apparent g values observed in the EPR spectrum of the "M" centers of the MoFe protein in their semireduced state are characteristic of a $S = 3/2$ spin system in which the $M_S = \pm 1/2$ Kramer's doublet is the ground state that gives rise to the EPR signal.[65,83] Because this signal remains with FeMo-co(s-r) when it is extracted from the MoFe protein[89] (see Figure 22-6), FeMo-co(s-r) also has a spin state of $S = 3/2$. Each of the components of the EPR spectrum of FeMo-co(s-r) is shifted in g value, however, which indicates greater rhombicity, and each is broader than the corresponding feature in the MoFe-protein spectrum. However, the spectra are similar enough to indicate similar structures. In addition to the major features, the EPR spectra of both the MoFe protein and FeMo-co(s-r) exhibit at 13–16 K a weaker resonance at $g = 6$,[31,89] which is where one resonance is predicted for the excited state $M_S = \pm 3/2$ Kramer's doublet. In both spectra, the intensity of this resonance diminishes as the temperature is lowered to 4 K, indicating that this resonance is indeed due to an excited state. These data are strong support for the $S = 3/2$ spin-state designation.

Additional support for the assignment of the $S = 3/2$ spin state comes from Mössbauer spectroscopy, which shows that FeMo-co(s-r) is a complex iron-containing entity with half-integer electronic spin,[78,89] and low-temperature magnetic-circular-dichroism spectroscopy, which confirms that the magnetic properties of the protein-bound "M" center and isolated FeMo-co(s-r) are very similar.[92]

Using ^{19}F nuclear magnetic resonance with $C_6H_5CF_3$ as a susceptibility reference, a magnetic moment of 3.78 Bohr magnetons (BM) was determined for an as-isolated 1 M (in Mo) NMF solution of FeMo-cofactor.[55] This value is close to the spin-only value of 3.87 BM expected for a $S = 3/2$ spin system and apparently offered a strong confirmation of FeMo-cofactor's spin state. Using the known affinity of thiolate for FeMo-cofactor,[10,89] the same report[55] lists the ^{19}F chemical shift of the reporter ligand,

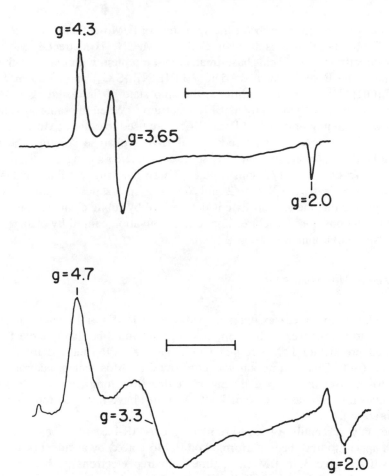

Figure 22-6.   Comparison at 13 K of the $S = {}^3/_2$ EPR signals of the "M" centers in the MoFe protein (top) and after isolation as the FeMo-cofactor (bottom).

p-$CF_3C_6H_4S^-$, bound to the FeMo-cofactor sample used for the susceptibility measurements as ~14 ppm. Unfortunately, this chemical shift is now known[12] to be characteristic of FeMo-co(ox) not FeMo-co(s-r), which has a ^{19}F chemical shift of ~31 ppm. The FeMo-cofactor sample used had most likely spontaneously oxidized[96] during storage, although this would not explain the susceptibility result unless FeMo-co(ox) has considerable paramagnetism at ambient temperature. Whatever the cause, these NMR-based susceptibility data for FeMo-co(s-r) must now be considered questionable. All other evidence is consistent with the $S = 3/2$ formulation for FeMo-co(s-r) in NMF solution.

Contrasting results in terms of the spin state of FeMo-cofactor were found when FeMo-co(s-r) was isolated in DMF or MeCN. To extract FeMo-co-factor directly from the acid–base-treated MoFe protein with either of these solvents, they have to be made 5 mM in $[EtN_4]_2[S_2O_4]$ and 2–50 mM in $[Et_4N][OH]$. Under these conditions, the spin state of the resulting FeMo-co(s-r) is $S = 3/2$ as judged by its EPR spectrum.[53] When the same solvents, but now containing only 0.1 M $[Bu_4N][Cl]$, are used to elute FeMo-cofactor from the MoFe protein denatured on a DEAE-sepharose column, the resulting FeMo-co(s-r) exhibits an EPR spectrum centered aroung $g = 2$, which is indicative of a $S = 1/2$ spin state.[123] These data suggest that, in DMF and MeCN at least, the $S = 3/2$ and $S = 1/2$ spin states are rather close in energy and that the ground state is determined by subtle changes in FeMo-cofactor composition, structure, ligation, or solvation caused by changes in the solvent environment.

B. The FeMo-co(ox) State

As similar redox processes operate for both the ["M"(s-r)] center and isolated FeMo-co(s-r) (see earlier), the assumption would be that one-electron-oxidized product of FeMo-co(s-r) would be the EPR-silent, diamagnetic FeMo-co(ox). This assumption was confirmed by Mössbauer spectroscopy of FeMo-co(ox) in a strong magnetic field at low temperature, when no magnetic splitting was observed.[78] MCD studies have not been reported for FeMo-co(ox).

The only susceptibility measurements are NMR-based.[55] These results, which appear to have been performed on FeMo-co(ox) by accident (see Section VII.A), give 3.78 BM (i.e., three unpaired electrons). Although the absolute value of the susceptibility is untenable and, in fact, was arrived at via many assumptions concerning size and composition of FeMo-cofactor, the result does indicate paramagnetism at ambient temperature in contrast to the diamagnetism observed at 4.2 K by Mössbauer spectroscopy. Assuming no effect of thiolate binding on FeMo-co(ox)'s spin state (thiolate binding does not change the spin state of FeMo-co(s-r)[10,89]), paramagnetism at ambient temperatures is also indicated by the isotropically shifted ^{19}F NMR resonances of p-$CF_3C_6H_4S^-$ bound to FeMo-co(ox).[12] The temperature dependence of this chemical shift shows Curie law behavior, which suggests that FeMo-co(ox) has a low-lying S = integer (possibly 1) excited state, which appears to be populated only at the higher temperatures used for NMR. Alternatively, the thiolate-FeMo-co(ox) complex [and possibly the thiolate-FeMo-co(s-r) complex] may undergo a conformational change at intermediate temperatures, which results in different electronic structures at high and low temperatures.[12]

C. The FeMo-co(red) State

The ["M"(red)] state is achieved in the MoFe protein only with the fully constituted, catalytically competent nitrogenase system.[80,110] Under these turnover conditions, the $S = 3/2$ EPR signal diminishes to about 10% of its initial intensity. The Mössbauer spectrum of this state shows[65] no paramagnetic hyperfine structure at low temperature, only quadrupole doublets, which have slightly higher [57]Fe isomer shifts than for ["M"(s-r)]. These observations indicate reduction of ["M"(s-r)] in the substrate-reducing system by an even number of electrons (most likely one) to give an $S =$ integer spin state. Considerably less information is available with respect to the spin state of isolated FeMo-co(red). Cyclic voltammetry[25,96,97] and coulometry[74] show that FeMo-co(s-r) accepts one electron to produce FeMo-co(red). This result, in turn, suggests that FeMo-co(red) should be formulated with $S =$ integer spin and so is predictable not to exhibit an EPR signal. This prediction remains untested so far.

All data indicate that the "M" centers and FeMo-cofactor have the same electronic structures in the three oxidation states available to them; this conclusion is consistent with the view that extraction has little effect on this prosthetic group.

VIII. Structural Insights

The structure of FeMo-cofactor is still unknown. In fact, as outlined earlier, even its exact composition remains uncertain. The major obstacle in determining the structure of FeMo-cofactor is the lack of an effective crystallization protocol, particularly with respect to generating a sufficiently pure and structurally homogeneous sample.[79,89] If suitable crystals could be grown, the solution of the FeMo-cofactor's structure through x-ray diffraction methods should be trivial. In the current situation, attempts to gain insight into its structure have taken many alternative approaches, which range from actual decomposition into recognizable fragments to the application of a variety of spectroscopic techniques.

A. Symmetry of FeMo-cofactor

Mössbauer spectroscopy applied to the *isolated* [57]Fe-labeled FeMo-cofactor suggests equivalent environments for the Fe atoms due to the narrow linewidths, which are comparable to those for [4Fe–4S] centers, of the quadrupole doublets observed.[78] A single-site fit to these data suggests that the electric field gradient tensor could be identical for all the Fe sites. Such identity places a strong constraint on site geometry. These results suggest that *isolated* FeMo-cofactor is not without some overall symmetry, which contrasts with the conclusion drawn from an analysis of the ENDOR ^{57}Fe magnetic hyperfine tensor components,[116,117] which suggests low symmetry for *protein-bound* FeMo-cofactor. However, because the ENDOR data show that three of FeMo-cofactor's iron atoms have negative and two have positive hyperfine couplings, complex spin coupling among an unknown number of components is indicated. Therefore, any attempt to determine symmetry in this situation by comparing magnitudes of tensors is on a much weaker basis than the conclusions drawn from the Mössbauer data. Magnetic circular dichroism spectra suggest that the geometry of the isolated FeMo-cofactor may be more regular than the protein-bound form,[92] and Fe K-edge x-ray absorption spectroscopy[2] indicates an ordered Fe-Mo-S framework in FeMo-cofactor.

B. Search for Identifiable Fragments

i. FeMo-cofactor: Based on Known Iron–Sulfur Clusters?

As might be expected for a molecule that contains six or seven Fe atoms, the first types of subcomponent sought were Fe–S clusters. In contrast to ferredoxins and the intact MoFe protein, neither [4Fe–4S] nor [2Fe–2S] clusters are elicited when FeMo-cofactor in NMF is treated with excess thiol and the products of the reaction monitored by either EPR[89] or ^{19}F NMR (when a fluorinated thiol has to be used).[49] Also unlike ferredoxins, FeMo-cofactor in NMF does not it give up its iron to either 2,2'-bipyridine[89] or *o*-phenenthroline.[124] These data are unambiguous and show that FeMo-cofactor does not contain [nFe–nS] (n = 2, 4) clusters and probably no other recognizable Fe–S cluster, including the 3-Fe and 6-Fe types, both of which tend to disproportionate into [4Fe–4S] clusters. This conclusion, in turn, suggests that FeMo-cofactor represents a completely new cluster type.

ii. Thiomolybdates: Building Blocks of FeMo-cofactor?

Because thiomolybdates are the starting materials for the synthesis of Mo–Fe–S compounds, they are also obvious candidates as potential fragments of FeMo-cofactor. If indeed they are released when FeMo-cofactor is degraded, it would be safe to assume that they have a role in the formation of FeMo-cofactor and in maintaining its integrity. Their presence would be a strong indicator of the nature of the Mo environment.

For the MoFe protein, it has been shown that acid treatment (for one day at pH 3), followed by neutralization, releases both $[MoS_4]^{2-}$ and $[MoOS_3]^{2-}$, which were separated by gel chromatography and identified by their characteristic electronic spectra.[126] Unfortunately, no attempt was made to correlate the quantities of thiomolybdates produced with the Mo content of the

MoFe protein used. Detection of these products, however, indicates the presence of rather robust Mo–S bonds in the intact MoFe protein and, therefore, probably in FeMo-cofactor also. The thiomolybdates must have been produced during the neutralization step, because they are innately unstable toward loss of S^{2-} in an acidic medium. The conditions under which thiomolybdates are released from the MoFe protein differ from those for the extraction of FeMo-cofactor. Both the pH used and the time of exposure to that pH are different, as is the need for an organic solvent. Apparently, these conditions are sufficient to disrupt the Fe–S portion of the Mo–S–Fe bonds in FeMo-cofactor (see later) while leaving the more robust Mo–S bonds intact.

To answer the question of why two thiomolybdates were formed, it was hypothesized that $[MoOS_3]^{2-}$ (and other products?) was likely formed from $[MoS_4]^{2-}$ during the acid–base treatment that released them.[126] If such chemistry is occurring, then the direct relationship between these products and their biological precursors is obscured. For example, $[MoS_4]^{2-}$ could be produced from $[MoOS_3]^{2-}$ by reaction with S^{2-} produced on decomposition of the Fe-S-containing P clusters of the MoFe protein.

These questions have been addressed[77] through the degradation of isolated FeMo-cofactor in NMF by O_2. Three protocols were used: (1) a titration using aliquots of air; (2) dilution with air-saturated NMF or MeOH; and (3) exposure to O_2 in the presence of $[(PPh_3)_2Cu(NO_3)]$ (PPh_3 = triphenylphosphine) to trap $[MoOS_3]^{2-}$. The first two experiments monitored the formation of thiomolybdates by visible spectroscopy and the third by detection of the product, $[OMoS_3(CuPPh_3)_3X]$ (X = halide). With the exception of the dilution experiment using air-saturated MeOH, all results were consistent with the formation of $[MoOS_3]^{2-}$ as the major product (~75% yield) of FeMo-cofactor degradation and thus indicated a MoS_3 core.[77] The MeOH-dilution experiment, however, produced $[MoS_4]^{2-}$ in about 72% yield. By using Mo-Fe-S model compounds and through known thiomolybdate chemistry, interconversions among the two thiomolybdates in either NMF or MeOH solution were ruled out as a major factor in their formation. The formation of $[MoS_4]^{2-}$ from FeMo-cofactor with air-saturated MeOH, but not with air-saturated NMF, was rationalized[77] on the basis of the insolubility of dithionite in MeOH. Thus, the rapid oxidative break-up of FeMo-cofactor by O_2, which occurs when dithionite is absent, produces local high concentrations of S^{2-} formed from the other five or six sulfides in intact FeMo-cofactor. These S^{2-} ions then react with the proposed MoS_3 core to give $[MoS_4]^{2-}$.

iii. The Molybdenum–Iron Cluster

If MEK is used to extract HCl-treated MoFe protein (rather than NMF extraction of acid–base-treated MoFe protein), the molybdenum-iron cluster

(MoFe-cluster) is isolated[102] as a dark brown solution, which contains about 70% of the Mo and a Fe–Mo ratio of 6($\pm$0.5):1. The same entity is produced if acid–base (citrate–phosphate)-precipitated MoFe protein is extracted with acidified MEK, but in only 40% yield. This latter experiment is important, however, because it demonstrates that the Fe in the MoFe-cluster has the same origin as the Fe in the FeMo-cofactor. MoFe-cluster is incompetent with respect to the FeMo-cofactor assay,[102] but it is a potent inhibitor of FeMo-cofactor binding to the apo-MoFe protein.[107] Thus, these entities appear to share major similarities but also to have significant differences.

The relationship of the MoFe-cluster to FeMo-cofactor is highlighted further by inspection of its EPR spectral properties. In MEK, the MoFe-cluster exhibits an axial $S = 1/2$ spectrum of the Hipip-type with g values at about 2.05 and 2.01 as the major component,[102] but a broad, shallow resonance between g values of about 3–6, representing up to approximately 10% of the total intensity, is also visible. When the MoFe-cluster is transferred to NMF in the presence of mercaptoethanol by vacuum evaporation of the MEK, the Hipip-type $S = 1/2$ spectrum disappears and is replaced by a FeMo-cofactor-type $S = 3/2$ EPR spectrum with g values at 4.6, 3.3, and 2.0. Although the mercaptoethanol is obviously lost during the vacuum evaporation (thiolated-FeMo-cofactor[10,89] has g values at 4.5, 3.6, and 2.0), it is apparently required to stabilize MoFe-cluster in MEK.[102] The resulting $S = 3/2$ EPR signal accounts for only about 50% of the spectral intensity; the remainder is due to a second signal at $g = 2$. These data suggest that the MoFe-cluster, possibly like FeMo-cofactor in DMF, may exist in spin equilibrium with the distribution among the $S = 1/2$ and $S = 3/2$ states being solvent sensitive.

Initially, it was suggested that the MoFe-cluster was the extruded "M" center of the MoFe protein[102] and that the "M" center was a subcomponent of FeMo-cofactor.[89] This hypothesis was based on MoFe-cluster's similar spectral properties, but lower Fe content (6 versus 8 per Mo) relative to FeMo-cofactor. The lower Fe content correlates with the number of Fe atoms determined for the "M" center within the MoFe protein by Mössbauer[43] and ENDOR[37,117] spectroscopies. Further, MoFe-cluster's lower Fe content was suggested to be the likely source of its different properties compared to FeMo-cofactor.[89] But because the Fe content of FeMo-cofactor has steadily decreased toward six atoms as purification has proceeded, this explanation now appears unlikely. Because the MoFe-cluster is not a source of homocitrate for the *in vitro* synthesis of FeMo-cofactor, a more likely explanation for the differences is that MoFe-cluster lacks homocitrate (or some other, as yet unrecognized, component of FeMo-cofactor). With no organic component, not even citrate, the MoFe-cluster would be unable to emulate even the NifV⁻ phenotype, but it could still bind to the apo-MoFe protein and inhibit activation by FeMo-cofactor. Unfortunately, this close relationship of MoFe-cluster to FeMo-cofactor helps very little with structural insights because the structure of MoFe-cluster is also unknown and the prospects for its crystallization are unclear.

C. Structural Information Through Mo, Fe, and S X-ray Absorption Spectra

X-ray absorption spectroscopy (XAS) has proved to be a very powerful probe of the local environment of inorganic elements, particularly metal ions, in many enzymes and proteins. Analysis of XAS data provides structural and electronic information about the x-ray-absorbing atom under study. Two independent, but complementary, aspects of XAS exist. The first is XANES (x-ray absorption near edge structure), which is concerned with the absorption edge itself and regions of the spectrum just prior to and after the edge. Analysis of these features provides details of the electronic structure, site symmetry, and first coordination sphere of the absorbing atom. The second aspect is EXAFS (extended x-ray absorption fine structure), which begins after the edge and provides details of the number and types of neighboring atoms around the absorber and gives their spherically averaged distance from the absorber.[18]

XAS, both XANES and EXAFS, at the Mo K-edge has been applied to both the "M" centers and the isolated FeMo-cofactor,[13,14,17] whereas XANES and EXAFS at the Fe K-edge[1,2] have been applied to FeMo-cofactor only. Mo L-edge[35] and S K-edge[35] XANES have also been applied to FeMo-cofactor, and the results of these investigations were described under Sections V and VI.

i. Studies at the Mo K-edge

The earliest XAS studies,[17] which compared the Mo environments of the "M" centers in the *C. pasteurianum* and *A. vinelandii* MoFe proteins, showed that there were striking similarities in the Mo K-edge and EXAFS regions. The basic features of these spectra were found to be preserved in FeMo-cofactor showing that the intact MoFe protein and the cofactor share quite similar Mo sites. Although the FeMo-cofactor data were not of sufficient quality for the determination of metrical information from the EXAFS, the position (at 20,011 eV) and shape of the Mo K-edge, which were identical to those of the two MoFe proteins,[17] indicated primarily sulfur ligation and the *absence* of Mo=O bonds.[18] This latter structural feature is found in Mo-cofactor, again highlighting its difference to FeMo-cofactor. Analysis of the MoFe-protein data obtained at that time[17] suggested a Mo site composed of three or four bound S atoms at a distance of 2.36 Å and two or three Fe atoms at 2.72 Å, plus possibly one or two additional S atoms at about 2.49 Å.

This first investigation was then followed up by a comparative study of the XANES[13] of a variety of synthetic Mo–S and Mo–Fe–S compounds, the semireduced, EPR-active MoFe protein and more concentrated, purified FeMo-co(s-r) solutions, including samples treated with 10 equivalents of either thiophenol (C_6H_5SH) or selenophenol (C_6H_5SeH). The conclusions drawn were that (1) treatment of FeMo-cofactor with thiol or selenol produced similar effects and resulted in Mo environments more similar to that in the MoFe protein, although the changes were relatively small; and (2) the Mo site structure is most closely approached by synthetic compounds with MoS_3O_3 coordination spheres. Thus, instead of a second S shell composed of one or two atoms at 2.49 Å from Mo, these data indicated a shell of about three singly-bonded light (most probably O) atoms.[13]

These conclusions based on the XANES data are completely consistent with the results obtained on the same entities through complementary Mo EXAFS studies.[14] The EXAFS data for the semireduced MoFe protein, FeMo-co(s-r) and thiolated-FeMo-co(s-r) are compared in Figure 22-7. The extended data range of these spectra allows for more accurate analysis than was possible earlier. For all the samples studied, the EXAFS results reveal the Mo atom to be in an environment of oxygen (or nitrogen), sulfur, and iron atoms. The number of each type of neighboring atom changes on removal of FeMo-cofactor from the protein and on its subsequent treatment with thiol–selenol.[14] These data are listed in Table 22-1. The oxygen or nitrogen shell (EXAFS cannot distinguish between the two) is listed as O because no N ligation is detectable by electron spin echo envelope modulation (ESEEM) spectroscopy with the isolated FeMo-cofactor[115] and these EXAFS-detectable interactions persist in the MoFe protein. This interpretation, in turn, suggests that the histidinyl N-coordination observed by ESEEM[115] for only the MoFe protein occurs at Fe rather than Mo of the "M" centers.

The Mo–O distances are consistent with those expected for anionic ligands, but are too short for neutral solvent molecules. In the protein, the two ligating oxygens could be donated by amino acids—such as aspartate, glutamate, or tyrosine—but as they remain in the isolated FeMo-cofactor, which contains no amino acids, they are more likely donated by homocitrate (see Section V.B). The additional oxygen that appears after extraction of FeMo-cofactor likely arises exogenously from deprotonated NMF,[120,124] OH^-, or other entity[14] added during isolation. The Mo–S distances are typical of sulfides bridging to iron and bound to hexacoordinate Mo, as are the Mo–Fe distances. Because the additional sulfur and iron neighbors, which appear when either thiol or selenol is added, are at the same distances as the other sulfur and iron atoms, a similar structural function is implied (i.e., S bridg-

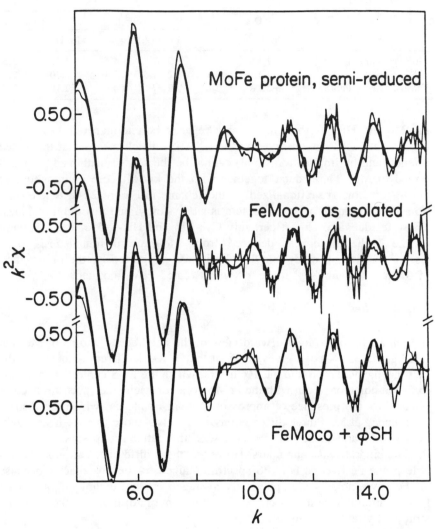

Figure 22-7. The k^2-weighted EXAFS data of semireduced MoFe protein (top), FeMo-co(s-r) (middle), and PhS-FeMo-co(s-r) (bottom). Note the similarities and the obvious difference in the "beat" region ($k = 9$) between the protein and cofactor samples. The light lines are the actual data and the heavy lines are derived by Fourier filtering to facilitate visual comparison; ϕSH represents thiophenol. (Reproduced with permission from the *Journal of the American Chemical Society*, vol. 109, p. 7510. Copyright 1987 American Chemical Society.)

Table 22-1. Types, Numbers (#) and Distances of Neighboring Atoms from Mo as Deduced by EXAFS Analysis

Sample	O shell		S shell		Fe shell	
	Mo–O (Å)	#	Mo–S (Å)	#	Mo–Fe (Å)	#
MoFe protein (s-r)	2.12	1.7	2.37	4.5	2.68	3.5
FeMo-co(s-r)	2.10	3.1	2.37	3.1	2.70	2.6
FeMo-co(s-r) + PhS(Se)H	2.11	2.6	2.37	4.0	2.70	3.3

ing Mo and Fe). Further, this S and Fe must be endogeneous because no Mo–Se interaction is observed in the EXAFS, which means that the added selenol (and by implication, thiol) cannot be the source of this additional S bound to Mo. These data, together with the known core composition of FeMo-cofactor, are rationalized by the scheme in Figure 22-8, where ligation of FeMo-cofactor to the protein is *via* cysteinyl and histidinyl residues. These residues are almost certainly Cys-275 and His-195 of the α-subunit of the MoFe protein as deduced from site-directed mutagenesis studies.[20,46,100] Binding of cysteinyl to Fe is also supported by the observation of thiolate binding[10,12,89] to Fe[55,79] with isolated FeMo-cofactor.

ii. Studies at the Fe K-edge

In contrast to the straightforward view of the neighboring atoms of Mo that XAS at the Mo K-edge gives, XAS at the Fe K-edge is complicated by the fact that the MoFe protein has P clusters, which are Fe-S centers, in addition to FeMo-cofactor. Even with the FeMo-cofactor itself, the spectrum is complicated by the presence of approximately six Fe atoms, which are distributed among at least three different environments—those that are near neighbors of Mo and those that are not, with this latter group subdivided into thiolate binders and nonbinders. However, these differences are not observable in the Fe K-edge EXAFS spectra obtained for FeMo-cofactor because all Fe environments are spherically averaged by this technique. Although this situation makes the spectra easier to deconvolute, it also makes the data structurally less meaningful.

The first analysis reported[1] Fe–Mo, Fe–Fe, and Fe–S interactions with 3.5 S atoms at 2.25 Å, 2.3 Fe atoms at 2.66 Å, and 0.4 Mo atoms at 2.76 Å. The averaged number of nearest-neighbor Mo atoms to each Fe is consistent with the Mo-derived data,[14] but the Mo–Fe distance is not. In addition, a Fe–O(N) interaction at 1.81 Å involving 1.2 O (or N) atoms is observed,[1] indicating the presence of additional ligands. Unfortunately, this potentially insightful result could not be confirmed by a second investigation,[2] which suggested that extraneous (non-cofactor) Fe in the sample might be responsible for the observed Fe–O interaction. The second data set also

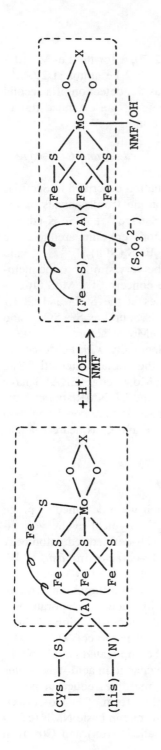
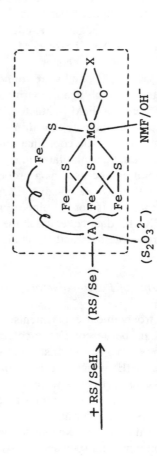

Figure 22-8. Likely structures and interconversions for the extrusion of FeMo-co(s-r) from the MoFe protein and on subsequent treatment with thiol (selenol) as visualized from the Mo K-edge XAS. --- signifies homocitrate; A represents homocitrate; X_E^0 signifies endogenous components; X_E^0 signifies the remaining XAS-undetectable Fe_2S_5 plus any as yet undiscovered component(s); ( )represents Mo K-edge XAS invisibility; −N−(his)- and −S−(cys)- represent his-195 and cys-275 of the α subunit of the MoFe protein. (Reproduced with permission from the *Journal of the American Chemical Society*, vol. 109, p. 7515. Copyright 1987 American Chemical Society.)

3886

reported a shorter, more acceptable distance of 2.70 Å for the Fe–Mo (0.8 atoms) distance, plus 3.0 S atoms at 2.20 Å, and 2.2 Fe atoms at 2.64 Å. A second major difference to the first data set was the detection of a second shell of 1.3 Fe atoms at 3.68 Å. These results suggest an extended framework for FeMo-cofactor consisting of Mo–S_2–Fe units.

iii. Studies at the Mo L-edges, S K-edge, Se K-edge, and Cl K-edge

In addition to the compositional and redox information described in Sections V and VI, structural insights have been forthcoming from analysis of the data obtained at the Mo L-edges, S K-edge, Cl K-edge,[35] and Se K-edge.[79] The Mo L_2- and L_3-edge XANES of FeMo-cofactor were compared to those of the dicubane clusters, $[Mo_2Fe_6S_8(SPh)_9]^{3-/5-}$ and $[Mo_2Fe_6S_8(OCH_3)_3(SPh)_6]^{3-}$, and were found to be very similar to the methoxy-bridged cluster,[35] which again confirms the concept of a MoS_3O_3 distorted octahedral core for FeMo-cofactor. Studies at the S K- and Cl K-edges of FeMo-cofactor indicate that thiosulfate is bound as $S—SO_3^{2-}$ and that Cl^- is not involved in the coordination of FeMo-cofactor.[35]

Because selenophenol and thiolates interact identically with FeMo-cofactor,[12,14] selenophenol can be used as a probe for the thiolate interaction site on FeMo-cofactor through Se K-edge EXAFS. Mo K-edge EXAFS have clearly indicated that Mo is not the binding site[14] and ^{19}F NMR studies have suggested Fe.[55] EXAFS data collected at the Se K-edge[79] are consistent with these other spectroscopic data and show clearly Fe and C as nearest neighbors.

D. Multiple Forms of FeMo-cofactor

In follow-up electrochemical experiments,[76] which were designed to probe further the redox properties of FeMo-cofactor,[96,99] atypical cyclic voltammograms were sometimes found. Instead of one well-defined reduction peak at −0.4 V versus NHE, which is typical of FeMo-cofactor in *basic* NMF, two (or more) less distinct peaks were observed between −0.3 V and −0.7 V. This behavior was typical of samples that were kept at room temperature (or above) for hours (e.g., during concentration by vacuum evaporation of NMF), a situation that is now known to cause disproportionation of dithionite, release of protons, and spontaneous oxidation of FeMo-co(s-r) to FeMo-co(ox).[95,96] This electrochemical behavior could be duplicated simply by diluting FeMo-cofactor into *acidic* NMF. Thus, the change in acid–base status of the solvent was responsible for eliciting the multiple reduction peaks, each of which is indicative of a separate and distinct form of FeMo-co(ox).

Each of the three forms of FeMo-co(ox) detected, one in basic NMF (called B(ox)) and a 2:1 mixture of two other forms [called A(ox) and C(ox)] in

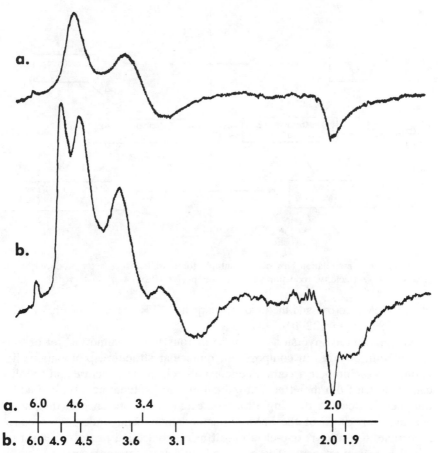

a.

b.

| a. | 6.0 | 4.6 | 3.4 | | 2.0 | |
| b. | 6.0 4.9 | 4.5 | 3.6 | 3.1 | 2.0 | 1.9 |

Figure 22-9. EPR spectra at 12 K of the dominant forms of FeMo-co(s-r) in alkaline (a; form R(s-r)) and in acidic (b; forms N(s-r) and W(s-r)) NMF solution. Apparent *g* values are indicated on the abscissa. (Reproduced with permission of the American Society for Biochemistry and Molecular Biology.)

acidic NMF, is converted to a separate FeMo-co(s-r) form on reduction as shown by their distinct EPR spectra[76] (see Figure 22-9). Form B(ox) is converted to R(s-r), where R indicates the *regular* form of the EPR signal; A(ox) is converted to N(s-r), where N indicates a *narrowing* of the spacing between the two *g* values around 4; and C(ox) is reduced to W(s-r), where W indicates a *widening* of the spacing around *g* = 4. Subspecies of each of these three major forms have also been detected.[95] Addition of a nine-fold excess of thiophenol (PhSH) converts all populations of FeMo-cofactor into a single PhS–FeMo-co(ox) or a single PhS–FeMo-co(s-r) form under all conditions[95] as shown by a single reducible form in the cyclic voltammogram

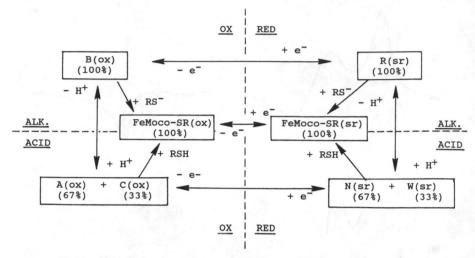

Figure 22-10. Interrelationships of the multiple forms of FeMo-cofactor produced by the effects of solvent acidity, oxidation state, and thiophenol (RSH).

of PhS–FeMo-co(ox) and the unique, sharpened EPR spectrum of PhS–FeMo-co(s-r) (See Figure 22-10).

Although interconversions of the numerous forms cannot as yet be correlated with changes in composition, molecular structure, protonation, ligation, or stoichiometry, certain conclusions related to structure and stability can be drawn from the effects of oxidation state, solvent acidity, and added thiophenol. Because the single form detected in the presence of thiophenol accounts for 100% of the FeMo-cofactor content and undergoes a quasireversible (ox)-to-(s-r) one-electron diffusion-controlled process, it must retain the greatest integrity. The increased number of forms that exist in acidic versus basic NMF and in the (ox) versus (s-r) oxidation state suggests that FeMo-cofactor assumes a greater heterogeneity of forms under the former versus the latter conditions. Increased structural or compositional integrity for FeMo-co(s-r) is also indicated by the single oxidizing wave observed in the cyclic voltammogram under all conditions.

i. Biological Significance of the Multiple Forms of FeMo-cofactor

If the multiple forms of FeMo-cofactor are simply solution artifacts that are manifested only when FeMo-cofactor is extracted from its protein matrix, their observation would still be significant if only to forewarn researchers involved in either attempts to crystallize FeMo-cofactor or in comparisons of data from different experimental sessions. It is more likely, however, that many of these forms correlate with biologically significant events. The B(ox)–R(s-r) and the A(ox)–N(s-r) redox pairs could correspond to the "M" centers either free from or constrained by protein ligation, respectively, whereas C(ox)/W(s-r) may approach the arrangement required to effect substrate reduction.[76] This last suggestion is based on the observation that, after R(s-r) is electrolyzed at potentials sufficient to produce FeMo-co(red) and effect turnover (see Section IX.D) and then the potential is switched off, W(s-r) is produced. In a related series of experiments, it was found that this same treatment caused the loss of the $S_2O_3^{2-}$ ligand.[74] Thus, it appears that R(s-r) must lose $S_2O_3^{2-}$ to become catalytically active and in doing so becomes W(s-r).

E. Concerns About Oxidation State

The observation of the spontaneous oxidation phenomenon for NMF solutions of FeMo-cofactor containing dithionite casts some concern over much of the spectral data described earlier. In many cases, although the samples clearly started out as FeMo-co(s-r), the oxidation state of the FeMo-cofactor sample may have changed during storage and/or travel. This situation is likely to have been compounded by the accompanying acidification of the medium and the further distribution of FeMo-cofactor into its multiple forms. Thus, although most of the data are presented in the literature as being collected on the (s-r) state, they may well have been collected on FeMo-co(ox) or on a mixture of the two oxidation states.

No structural data are available for FeMo-co(red).

IX. Chemical Reactivity

A. With O_2 and Water

In the original report of FeMo-cofactor isolation, the acute sensitivity of FeMo-cofactor toward O_2 is noted.[101] Destruction of its reconstitution activity is complete in less than one minute at 0° C. This activity loss is accompanied by the loss of visible spectral intensity and the production of thiomolybdates.[77,101] Controlled additions of O_2 have been used to study the oxidation–reduction behavior of FeMo-cofactor. This technique was found to be of limited value, as even less than stoichiometric amounts of O_2 caused irreversible losses of EPR signal intensity and activity.[73]

FeMo-cofactor is also susceptible to anaerobic hydrolysis.[101] Activity is lost irreversibly on addition of water or aqueous buffers to an NMF solution of FeMo-cofactor. Reports range from a 50% loss of activity in two hours at 0° C[101] to a 35% loss in 35 minutes at 30° C.[34]

How, and at what site(s), either O_2-promoted or hydrolytic degradation occurs is not known.

B. With Ligands Other Than Nitrogenase Substrates

As the putative, extruded, catalytic center, FeMo-cofactor would be expected to bind a variety of small molecules and ions because ligation is a required component of any catalytic system, along with redox capability. The redox capability of FeMo-cofactor has been amply demonstrated (see Section VI). FeMo-cofactor's ability to bind a variety of molecules usually has been demonstrated by changes in its $S = 3/2$ EPR signal. As the $S = 3/2$ EPR spectrum of FeMo-cofactor is only visible at liquid helium temperatures (below about 20 K), care has to be taken in relating these observations to the ambient-temperature situation. EPR spectroscopy has been used to demonstrate the binding of several thiols and selenols, EDTA, o-phenanthroline (o-phen), and 2,2'-bipyridine (bipy).

EDTA binds to FeMo-cofactor in a reaction that completely eliminates the EPR signal without affecting the reconstitution activity.[10] This reaction is nonstoichiometric, requiring approximately a 40-fold excess of EDTA, and is completely reversible on addition of an equivalent quantity of Zn^{2+}. o-Phen behaves identically, except that Fe^{2+} is used to reverse its ligation.[124] Bipy, which normally mimics o-phen, has no effect on either the EPR signal or the reconstitution activity.[124] It is not clear how these reactions cause loss of the EPR signal, although di-(or oligo)-merization, redox or increased relaxation have been suggested.[124] Whatever the mechanism, these results show that the EPR chromophore is accessible to exogenous ligands.

In contrast, thiophenol binds to FeMo-cofactor but does not eliminate the EPR signal.[10,89] Its effect on the EPR signal is a significant sharpening of all three lines and a shift in the g values from 4.6, 3.35, and 2.0 to 4.5, 3.6, and 2.0 (see Figure 22-11), such that the resulting signal more closely resembles that of the MoFe protein. The shift has been used to quantify the stoichiometry of thiolate binding,[10] which was found to be 1 thiolate per Mo at 14 K. As discussed in Section VIII.C, the binding site is not Mo,[14] but is rather a specific Fe atom.[55,79] Selenophenol behaves similarly.[14,79] Neither affects the reconstitution activity. The nature of the thiolate–FeMo-cofactor interaction has also been probed at ambient temperature by ^{19}F NMR using p-CF$_3$C$_6$H$_4$S$^-$ as a reporter ligand.[12] These NMR data indicate a dynamic equilibrium between free and bound thiolate at ambient temperature rather than irreversible complex formation as observed at 14 K.

Citrate also binds to the FeMo-cofactor without altering the EPR signal because FeMo-cofactor prepared in the presence of citrate exhibits the same EPR signal but has a different elution profile on Sephadex G-100 in NMF from that of FeMo-cofactor prepared in its absence.[108] Thus, citrate binds to the native, homocitrate-containing FeMo-cofactor in contrast to homocitrate which does not bind to the isolated NiFV$^-$, citrate-containing FeMo-cofactor[40] (see Section V.B.iii). Neither organic acid is capable of displacing

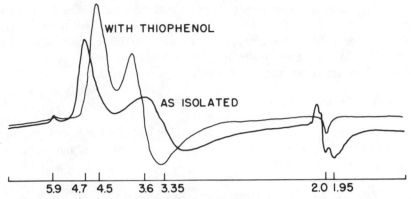

Figure 22-11. The effect of added thiophenol (light line) on the EPR spectrum of FeMo-cofactor (heavy line) in NMF at 12 K. Apparent *g* values are indicated on the abscissa. (Reproduced from *Nitrogen Fixation: The Chemical-Biochemical-Genetic Interface*, with the permission of Plenum Publishing Corp.)

the other from preformed FeMo-cofactor.[40,101,108] Mo K-edge XAS[14] also cannot distinguish between FeMo-cofactor isolated by citric acid treatment rather than by hydrochloric acid treatment. Apparently, there are two different sites on FeMo-cofactor capable of binding these organic acids. The first site (possibly on Mo; see Section VIII.C.i) is occupied (by homocitrate usually) during biosynthesis and is then no longer reactive. The second site is part of the outer labile coordination sphere, which surprisingly appears able to distinguish among citrate ($n = 1$) and homocitrate ($n = 2$) in $HO_2C-CH_2-C(OH)(CO_2H)-(CH_2)_n-CO_2H$.

C. With Nitrogenase Substrates

Cyanide (CN^-) has been shown[109] to bind directly to FeMo-cofactor through its effect on the $S = 3/2$ EPR signal. Binding is complete at ratios of <2.5 CN^- per Mo and results in a change in *g* values from 4.6, 3.3, and 2.0 to 4.3, 3.8, and 2.0. The titration data indicate that each FeMo-cofactor binds two CN^- ions with different binding constants. An investigation of cyanide and methylisocyanide (CH_3NC) binding to FeMo-cofactor has also been performed via ^{19}F NMR using $p-CF_3C_6H_4S^-$ bound to FeMo-cofactor as a reporter ligand.[15] Again, these ambient-temperature data do not agree fully with those obtained at liquid helium temperatures. Although they confirm interaction between CN^- (or CH_3NC) and FeMo-cofactor, these NMR data suggest a single binding site and a dynamic equilibrium between bound and free CN^- (or CH_3NC). As suggested for the differences found with the thiolates alone, the differences for the two data sets are likely due to an increase in the formation constant with decreasing temperature.[15] The binding of CN^-

(or CH_3NC) does not involve Mo. Neither addendum displaces the thiolate, which indicates the formation of a ternary complex, nor does their presence affect the reconstitution activity of the treated FeMo-cofactor.[15] No effect on the ^{19}F NMR spectrum of the thiolated FeMo-cofactor was observed with either CO or azide ion, indicating that they do not bind.[15]

D. Catalytic Activity

Attempts to demonstrate catalytic activity for isolated FeMo-cofactor have been of two types. The first type uses chemical reductants and the second type uses electrochemistry.

i. Chemically Driven Systems

The first chemical experiments involved diluting FeMo-cofactor into aqueous borate pH 9.6 buffer containing sodium borohydride under a C_2H_2 atmosphere.[103] These conditions were based on a system used to develop chemical models for FeMo-cofactor with C_2H_2 reduction as a monitor.[93] Under these conditions, FeMo-cofactor produces only C_2H_4, but at a rate of <10% of that of nitrogenase in a reaction that is strongly inhibited by CO. This selectivity shown by FeMo-cofactor contrasts with that of various Mo-based chemical models, which often produce ethane (C_2H_6) and butadiene (C_4H_6) as well,[16,93] and it was taken as support for the concept of FeMo-cofactor as the active site of nitrogenase.[103] These activity data alone do not make a convincing case because many Mo[16,93] and/or Fe[75] compounds will also catalyze the reduction of C_2H_2 to C_2H_4 using borohydride as reductant. In fact, FeMo-cofactor, which had been O_2 degraded for 18 hours at 4° C and was completely without reconstitution activity, was 84% as active as native FeMo-cofactor in this system.[103]

The more stringent criterion of cyclopropene reduction has also been applied to isolated FeMo-cofactor in a similar borate buffer-borohydride system.[56] In contrast to the complete nitrogenase system, which produces a 2:1 mixture of propene and cyclopropane as products,[58] the FeMo-cofactor–borohydride system produces only cyclopropane[56] just like many chemical models.[71,94]

Because the nature of the species responsible for the catalysis is unknown and because these systems cannot reproduce the specificity shown by nitrogenase, it is very difficult to assess the relevance of these systems to nitrogenase and to use them to gain mechanistic insight.

ii. Electrochemically Driven Systems

In a series of cyclic voltammetric experiments, FeMo-cofactor's potential for catalysis was indicated by the behavior of the second reduction wave as the sweep rate was varied. If the sweep rate was relatively fast (400 mV s^{-1}), the currents for the (ox)-to-(s-r) and (s-r)-to-(red) processes were approximately equal and of a magnitude appropriate for one-electron couples. If, however, the sweep rate was decreased (to 40 mV s^{-1}), the magnitude of the current for the second wave was dramatically increased.[97,98] This behavior suggests that FeMo-co(red), which is produced by the second process, undergoes a catalytic reaction with some component in the solution. Further support for this interpretation comes from cyclic voltammetry. The observed loss of the second oxidation wave at slow sweep rates, changes in both the shape and peak potential of the second reduction wave with changes in sweep rate, and the current for the second reduction wave becoming independent of sweep rate at slow sweep rates are all indicative of an electrocatalytic process.

These indications of electrocatalysis were followed up by controlled potential coulometric experiments in which FeMo-cofactor (285 nmoles) in NMF solution was electrolyzed at -1.2 V versus NHE for one hour.[97,98] At this applied potential, which is more negative than the reduction potential of the second wave, charge (380 mC) was accumulated continuously. This amount of charge is equivalent to 14 electrons per FeMo-cofactor (i.e., a turnover number of 14 per hour). Neither the substrate nor the product of this reaction is known, but protons and H_2 are the suggested possibilities. Although this rate is slow (<1%) compared to the nitrogenase rate, it clearly demonstrates catalytic capability for FeMo-cofactor when it attains the FeMo-co(red) state, which is the equivalent of the substrate-reducing state of the "M" centers in the MoFe protein.

X. Synthesis of FeMo-cofactor

A. Biosynthesis

Of the 20 known *nif*-specific genes in *K. pneumoniae*, six are involved with FeMo-cofactor biosynthesis. These are *nifBENVQH*. FeMo-cofactor biosynthesis occurs in the absence of the MoFe-protein subunits,[118] which indicates that the apo-MoFe protein and FeMo-cofactor are synthesized separately and then combined at a late stage in MoFe protein biosynthesis. Many unanswered questions remain concerning FeMo-cofactor biosynthesis and its subsequent insertion into the apo-MoFe protein.

The function of the product of *nifB* in FeMo-cofactor biosynthesis remains unknown. The *nifV* gene product is involved in the synthesis of homocitrate[41] for incorporation into FeMo-cofactor as discussed in Section V.B. *nifH* encodes the polypeptide for the Fe-protein component of nitrogenase, but a catalytically competent Fe protein is not required for FeMo-cofactor biosynthesis.[26] It is not clear how the *nifH* product is involved. The *nifQ* product appears to be concerned with the processing (not transport) of Mo specially for FeMo-cofactor because the NifQ⁻ phenotype can be overcome by elevating the levels of Mo in the growth medium and it has no effect on either Mo uptake or Mo-cofactor biosynthesis.[44] The products of *nifEN* share primary sequence identity with the products of *nifDK*, which encode the two types of subunits of the MoFe protein. This similarity is has led to the suggestion that the NifE and NifN proteins form a tetrameric complex, like the MoFe protein, on which FeMo-cofactor is built through the combined action of other *nif* gene products before being transferred to the apo-MoFe protein.[4]

B. In Vitro Synthesis

Using this scant information, an *in vitro* system was developed for FeMo-cofactor synthesis.[105] This system has great utility in determining the involvement of various *nif* gene products in FeMo-cofactor biosynthesis. Without this system, the identification of the *nifV* factor as homocitrate and its involvement with FeMo-cofactor, for example, may never have been realized. The *in vitro* synthesis system requires, at least, molybdate, ATP, homocitrate, and the Fe protein, plus the *nifBEN* gene products. The *nifB* gene product was originally supplied by an extract of *K. pneumoniae* strain UN1100, while the *nifEN* gene products *had* to be supplied by an *A. vinelandii* NifB⁻ source, which originally was an extract of a strain, such as UW45. When these components are mixed anaerobically, the FeMo-cofactor formed activates the apo-MoFe protein in both extracts.[105] In addition to its vital role in identifying and quantifying the homocitrate content of FeMo-cofactor, this system has now been used in the purification of the NifEN protein complex,[85] in incorporating analogs of homocitrate into FeMo-cofactor[106] and may assist in defining the steps in the biosynthesis of FeMo-cofactor and its insertion into the apo-MoFe protein.

XI. Summary and Conclusions

Since the discovery and isolation of FeMo-cofactor[101] as a Mo–Fe–S-containing green–brown NMF solution in 1977, much effort has been expended in attempts to learn more about its size, composition, structure, and reactivity. Despite these efforts, FeMo-cofactor remains an elusive entity. Although we now have knowledge of its minimal core, we do not know its exact size or total composition. FeMo-cofactor has also resisted all attempts at crystallization for structure determination by x-ray diffraction techniques. These problems have their roots in the likely existence of a heterogeneous population in any preparation, the possibility of different compositions depending on the isolation procedure used and on the spontaneous oxidation phenomenon leading to a mixture of structural forms and oxidation states. This lability remains unpredictable, uncontrollable, and inexplicable in structural terms.

In spite of these severe complications, impressive progress has been made. Structural insights have been gained through applying the XAS technique to the Mo, Fe, and S (and Se) components and from Mössbauer and ENDOR spectroscopy. These data have been complemented by redox and reactivity studies, which have led to insights into the dynamic operation of FeMo-cofactor. This information, together with the emerging structural data on the MoFe protein,[3,113] will be invaluable in reaching the objective of all research in this area, understanding the mechanism of biological N_2 fixation, and suggesting how it might be manipulated for agricultural benefit. It is hoped that the information summarized here and the interpretations offered will assist in the next steps in these endeavors.

References

1. Antonio, M. R., Teo, B.-K., Orme-Johnson, W.H., Nelson, M.J., Groh, S.E., Lindahl, P.A., Kauzlarich, S.M., and Averill, B.A., "Iron EXAFS of the iron–molybdenum cofactor of nitrogenase," *J. Am. Chem. Soc. 104*, 4703–4705 (1982).

2. Arber, J.M., Flood, A.C., Garner, C.D., Gormall, C.A., Hasnain, S.S., and Smith, B.E., "Iron K-edge x-ray absorption spectroscopy of the iron–molybdenum cofactor of nitrogenase from *Klebsiella pneumoniae*," *Biochem. J. 252*, 421–425 (1988).

3. Bolin, J.T., Ronco, A.E., Mortenson, L.E., Morgan, T.V., Williamson, M. and Xuong, N.-H., "The structure of the nitrogenase MoFe protein: Spatial distribution of the intrinsic metal atoms determined by x-ray anomalous scattering," *Nitrogen Fixation: Achievements and Objectives*, Gresshoff, P.M., Stacey, G., Roth, L.E., and Newton, W.E., eds. New York: Chapman and Hall, 1990, pp. 117–124.

4. Brigle, K.E., Weiss, M.C., Newton, W.E., and Dean, D.R., "Products of the iron–molybdenum cofactor-specific biosynthetic genes, *nifE* and *nifN*, are structurally homologous to the products of the nitrogenase molybdenum–iron protein genes, *nifD* and *nifK*," *J. Bacteriol. 169*, 1547–1553 (1987).

5. Bulen, W.A., "Effect of tungstate on the uptake and function of molybdate in *Azotobacter agilis*," *J. Bacteriol. 82*, 130–134 (1965).

6. Bulen, W.A., and LeComte, J.R., "The nitrogenase system from *Azotobacter vinelandii:* Two enzyme requirements for N_2 reduction, ATP-dependent hydrogen evolution and ATP hydrolysis," *Proc. Natl. Acad. Sci. USA 56*, 979–986 (1966).

7. Bulen, W.A., and LeComte, J.R., "Nitrogenase complex and its components," *Methods in Enzymology*, Vol. XXIVB, San Pietro, A., ed. New York: Academic Press, 1972, pp. 456–470.

8. Burgess, B.K., Jacobs, D.J., and Stiefel, E.I., "Large-scale purification of high activity *Azotobacter vinelandii* nitrogenase," *Biochim. Biophys. Acta 614*, 196–208 (1980).

9. Burgess, B.K., and Newton, W.E., "Iron–molybdenum cofactor and its complementary protein from mutant organisms," *Nitrogen Fixation: The Chemical-Biochemical-*

Genetic Interface, Müller, A., and Newton, W.E., eds. New York: Plenum Press, 1983, pp. 83–110.

10. Burgess, B.K., Stiefel, E.I., and Newton, W.E., "Oxidation–reduction properties and complexation reactions of the iron–molybdenum cofactor of nitrogenase," *J. Biol. Chem. 255,* 353–356 (1980).

11. Christou, G., Mascharak, P.K., Armstrong, W.H., Papaefthymiou, G.C., Frankel, R.B., and Holm, R.H., "Electron transfer series of MoFe₃S₄ double cubane clusters: Electronic properties of components and the structure of [(C₂H₅)₄N]₅[Mo₂Fe₆S₈(Sc₆H₅)₉]," *J. Am. Chem. Soc. 104,* 2820–2831 (1982).

12. Conradson, S.D., Burgess, B.K., and Holm, R.H., "Fluorine-19 chemical shifts as probes of the structure and reactivity of the iron–molybdenum cofactor of nitrogenase," *J. Biol. Chem. 263,* 13743–13749 (1988).

13. Conradson, S.D., Burgess, B.K., Newton, W.E., Hodgson, K.O., McDonald, J.W., Rubinson, J.F., Gheller, S.F., Mortenson, L.E., Adams, M.W.W., Mascharak, P.K., Armstrong, W.A., and Holm, R.H., "Structural insights from the Mo K-edge absorption near edge structure of the molybdenum–iron protein of nitrogenase and its iron–molybdenum cofactor by comparison with synthetic Fe–Mo–S clusters," *J. Am. Chem. Soc. 107,* 7935–7940 (1985).

14. Conradson, S.D., Burgess, B.K., Newton, W.E., Mortenson, L.E., and Hodgson, K.O., "Structural studies of the molybdenum site in the MoFe protein and its FeMo cofactor by EXAFS," *J. Am. Chem. Soc. 109,* 7507–7515 (1987).

15. Conradson, S.D., Burgess, B.K., Vaughn, S.A., Roe, A.L., Hedman, B., Hodgson, K.O., and Holm, R.H., "Cyanide and methylisocyanide binding to the isolated iron–molybdenum cofactor of nitrogenase," *J. Biol. Chem. 264,* 15967–15974 (1989).

16. Corbin, J.L., Pariyadath, K., and Stiefel, E.I., "Ligand effects and product distributions in molybdothiol catalyst systems," *J. Am. Chem. Soc. 98,* 7862–7864 (1976).

17. Cramer, S.P., Gillum, W.O., Hodgson, K.O., Mortenson, L.E., Stiefel, E.I., Chisnell, J.R., Brill, W.J. and Shah, V.K., "The molybdenum site of nitrogenase. 2. A comparative study of Mo-Fe proteins and the iron–molybdenum cofactor by x-ray absorption spectroscopy," *J. Am. Chem. Soc. 100,* 3814–3819 (1978).

18. Cramer, S.P., Hodgson, K.O., Stiefel, E.I., and Newton, W.E., "A systematic x-ray absorption study of molybdenum complexes. The accuracy of structural information from x-ray absorption fine structure," *J. Am. Chem. Soc. 100,* 2748–2761 (1978).

19. Cramer, S.P., and Stiefel, E.I., "Chemistry and biology of the molybdenum cofactor," *Molybdenum Enzymes,* Spiro, T., ed. New York: Wiley, 1985, pp. 411–441.

20. Dean, D.R., Scott, D.J., and Newton, W.E., "Identification of FeMoco domains within the nitrogenase MoFe protein," *Nitrogen Fixation: Achievements and Objectives,* Gresshoff, P.M., Stacey, G., Roth, L.E., and Newton, W.E., eds. New York: Chapman and Hall, 1990, pp. 95–102.

21. Dunham, W.R., Hagen, W.R., Braaksma, A., Grande, H.J., and Haaker, H., "The importance of quantitative Mössbauer spectroscopy of MoFe protein from *Azotobacter vinelandii,*" *Eur. J. Biochem. 146,* 497–501 (1985).

22. Eidsness, M.K., Flank, A.M., Smith, B.E., Flood, A.C., Garner, C.D., and Cramer, S.P., "EXAFS of *Klebsiella pneumoniae* nitrogenase MoFe protein from wild-type and *nifV* mutant strains," *J. Am. Chem. Soc. 108,* 2746–2747 (1986).

23. Emerich, D.W., and Burris, R.H., "Complementary functioning of the component proteins of nitrogenase from several bacteria," *J. Bacteriol. 134,* 936–943 (1978).

24. Euler, W.B., Martinsen, J., McDonald, J.W., Watt, G.D., and Wang, Z.-C., "Double integration and titration of the electron paramagnetic resonance signal in the molybdenum iron protein of *Azotobacter vinelandii*," *Biochemistry 23*, 3021–3024 (1984).

25. Feldman, B.J., Gheller, S.F., Bailey, G.F., Newton, W.E., and Schultz, F.A. "Electrochemical cells for voltammetry, coulometry and protein activity assays of small-volume biological samples," *Anal. Biochem. 185*, 170–175 (1990).

26. Filler, W.A., Kemp, R.M., Ng, J.C., Hawkes, T.R., Dixon, R.A., and Smith, B.E., "The *nifH* gene product is required for the synthesis or stability of the iron–molybdenum cofactor of nitrogenase from *Klebsiella pneumoniae*," *Eur. J. Biochem. 160*, 371–377 (1986).

27. Frank, P., Gheller, S.F., Hedman, B., Newton, W.E., and Hodgson, K.O., "Spectroscopic studies of the purified FeMo cofactor from *Azotobacter vinelandii*," *Nitrogen Fixation: Hundred Years After*, Bothe, H., de Bruijn, F.J., and Newton, W.E., eds. Stuttgart: Gustav Fischer, 1988, p. 64.

28. Frank, P., Gheller, S.F., Newton, W.E., and Hodgson, K.O., "Purification and spectroscopic characteristics in *N*-methylformamide of the *Azotobacter vinelandii* Fe-Mo cofactor," *Biochem. Biophys. Res. Commun. 163*, 746–754 (1989).

29. Frankel, R.B., Herskovitz, T., Averill, B.A., Holm, R.H., Krusic, P.J., and Phillips, W.D., "Synthetic analogs of the active sites of iron–sulfur proteins. VII. Some electronic properties of $[Fe_4S_4(SR)_4]^{3-}$, analogs of reduced bacterial ferredoxins," *Biochem. Biophys. Res. Commun. 58*, 974–982 (1974).

30. Ganelin, V.L., L'vov, N.P., Sergeev, N.S., Shaposhnikov, G.L., and Kretovitch, V.L., "Isolation and properties of a molybdenum-containing peptide from component 1 of the nitrogen-fixing complex of *Azotobacter vinelandii*," *Dokl. Akad. Nauk SSSR 206*, 1236–1238 (1972).

31. George, G.N., Bare, R.E., Jin, H., Stiefel, E.I., and Prince, R.C., "E.p.r.-spectroscopic studies on the molybdenum–iron protein of nitrogenase from *Clostridium pasteurianum*," *Biochem. J. 262*, 349–352 (1989).

32. Hardy, R.W.F., Burns, R.C., and Holsten, R.D., "Applications of the acetylene–ethylene assay for measurements of nitrogen fixation," *Soil Biol. Biochem. 5*, 47–81 (1973).

33. Hawkes, T.R., McLean, P.A., and Smith, B.E., "Nitrogenase from *nifV* mutants of *Klebsiella pneumoniae* contains an altered form of the iron–molybdenum cofactor," *Biochem. J. 217*, 317–321 (1984).

34. Hawkes, T.R., and Smith, B.E., "The inactive MoFe protein (NifB⁻Kp1) of the nitrogenase from *nifB* mutants of *Klebsiella pneumoniae*," *Biochem. J. 223*, 783–792 (1984).

35. Hedman, B., Frank, P., Gheller, S.F., Roe, A.L., Newton, W.E., and Hodgson, K.O., "New structural insights into the iron–molybdenum cofactor from *Azotobacter vinelandii* nitrogenase through sulfur K and molybdenum L x-ray absorption edge studies," *J. Am. Chem. Soc. 110*, 3798–3805 (1988).

36. Hedman, B., Frank, P., Penner-Hahn, J.E., Roe, A.L., Hodgson, K.O., Carlson, R.M.K., Brown, G., Cerino, J., Hettel, R., Troxel, T., Winick, H., and Yang, J., "Sulfur K edge x-ray absorption studies using the 54-pole wiggler at SSRL in undulator mode," *J. Nucl. Instrum. Meth. Phys. Res., Sect. A 246*, 797–800 (1986).

37. Hoffman, B.M., Venters, R.A., Roberts, J.E., Nelson, M.J., and Orme-Johnson, W.H., "⁵⁷Fe ENDOR of the nitrogenase MoFe protein," *J. Am. Chem. Soc. 104*, 4711–4712 (1982).

38. Hoover, T.R., Imperial, J., Liang, J., Ludden, P.W., and Shah, V.K., "Dinitrogenase with altered substrate specificity results from the use of homocitrate analogues for *in vitro* synthesis of the iron–molybdenum cofactor," *Biochemistry 27*, 3647–3652 (1988).

39. Hoover, T.R., Imperial, J., Ludden, P.W., and Shah, V.K., "Homocitrate cures the NifV⁻ phenotype in *Klebsiella pneumoniae*," *J. Bacteriol. 170*, 1978–1979 (1988).

40. Hoover, T.R., Imperial, J., Ludden, P.W., and Shah, V.K., "Homocitrate is a component of the iron–molybdenum cofactor of nitrogenase," *Biochemistry 28*, 2768–2771 (1989).

41. Hoover, T.R., Robertson, A.D., Cerny, R.L., Hayes, R.N., Imperial, J., Shah, V.K., and Ludden, P.W., "Identification of the V factor needed for synthesis of the iron–molybdenum cofactor of nitrogenase as homocitrate," *Nature 329*, 855–857 (1987).

42. Hoover, T.R., Shah, V.K., Roberts, G.P., and Ludden, P.W., "*nifV*-dependent, low-molecule-weight factor required for *in vitro* synthesis of iron–molybdenum cofactor of nitrogenase," *J. Bacteriol. 167*, 999–1003 (1986).

43. Huynh, B.H., Münck, E., and Orme-Johnson, W.H., "Nitrogenase XI. Mössbauer studies on the cofactor centers of the MoFe protein from *Azotobacter vinelandii*," *Biochem. Biophys. Acta 527*, 192–203 (1979).

44. Imperial, J., Ugalde, R.A., Shah, V.K., and Brill, W.J., "Role of the *nifQ* gene product in the incorporation of molybdenum into nitrogenase of *Klebsiella pneumoniae*," *J. Bacteriol. 158*, 187–194 (1984).

45. Keeler, R.F., and Varner, J.E., "Tungstate as an antagonist of molybdate in *Azotobacter vinelandii*," *Arch. Biochem. Biophys. 70*, 585–590 (1957).

46. Kent, H.M., Ionnidis, I., Gormal, C., Smith, B.E., and Buck, M., "Site-directed mutagenesis of the *Klebsiella pneumoniae* nitrogenase," *Biochem. J. 264*, 257–264 (1989).

47. Ketchum, P.A., Cambier, H.Y., Frazier, III, W.A., Mandansky, C.H., and Nason, A., "*In vitro* assembly of *Neurospora* assimilatory nitrate reductase from protein subunits of a *Neurospora* mutant and the xanthine oxidizing or aldehyde oxidase systems of higher animals," *Proc. Natl. Acad. Sci. USA 66*, 1016–1023 (1970).

48. Kondorosi, A., Barabas, I., Sva, Z., Orosz, L., Sik, T., and Hotchkiss, R.D., "Common genetic determinants of nitrogenase and nitrate reductase in *Rhizobium*," *Nature (New Biol.) 246*, 153–154 (1973).

49. Kurtz, Jr., D.M., McMillan, R.S., Burgess, B.K., Mortenson, L.E., and Holm, R.H., "Identification of iron–sulfur centers in the iron–molybdenum proteins of nitrogenase," *Proc. Natl. Acad. Sci. USA 76*, 4986–4989 (1979).

50. Lee, K.-Y., Pan, S.-S., Erikson, R.H., and Nason, A., "Involvement of molybdenum and iron in the *in vitro* assembly of assimilatory nitrate reductase utilizing *Neurospora* mutant *nit1*," *J. Biol. Chem. 249*, 3941–3952 (1974).

51. Levchenko, L.A., Roschupkina, O.S., Sadkov, A.P., Marakushev, S.A., Mikhailov, G.M., and Borod'ko, Yu. G., "Spectroscopic investigation of the FeMo-cofactor. Coenzyme A as one of the probable components of an active site of nitrogenase," *Biochem. Biophys. Res. Commun. 96*, 1384–1392 (1980).

52. Lough, S., Burns, A., and Watt, G.D., "Redox reactions of the nitrogenase complex from *Azotobacter vinelandii*," *Biochemistry 22*, 4062–4066 (1983).

53. Lough, S.M., Jacobs, D.J., Lyons, D.M., Watt, G.D., and McDonald, J.W., "Solubilization of the iron–molybdenum cofactor of *Azotobacter vinelandii* nitrogenase into

dimethylformamide and acetonitrile," *Biochem. Biophys. Res. Commun. 139*, 740–746 (1986).

54. Lough, S.M., Jacobs, D.J., Parker, S.F., McDonald, J.W., and Watt, G.D., "A facile method of isolation for the iron molybdenum cofactor of nitrogenase and bacterial ferritin from extracts of *Azotobacter vinelandii*," *Inorg. Chim. Acta 151*, 227–232 (1988).

55. Mascharak, P.K., Smith, M.C., Armstrong, W.H., Burgess, B.K., and Holm, R.H., "Fluorine-19 chemical shifts as structural probes of metal–sulfur clusters and the cofactor of nitrogenase," *Proc. Natl. Acad. Sci. USA 79*, 7056–7060 (1982).

56. McKenna, C.E., L'vov, N.P., Ganelin, V.L., Sergeev, N.S., and Kretovich, V.L., "Existence of a low-molecular-weight factor common to various molybdenum-containing enzymes," *Dokl. Akad. Nauk SSSR 217*, 228–231 (1974).

57. McKenna, C.E., Jones, J.B., Eran, H., and Huang, C.W., "Reduction of cyclopropene as a criterion of active-site homology between nitrogenase and its Fe-Mo cofactor," *Nature 280*, 611–612 (1979).

58. McKenna, C.E., McKenna, M.-C. and Higa, M.T., "Chemical probes of nitrogenase. 1. Cyclopropene. Nitrogenase-catalyzed reduction to propene and cyclopropane," *J. Am. Chem. Soc. 98*, 4657–4659 (1976).

59. McLean, P.A., and Dixon, R.A., "Requirement of *nifV* gene for production of wild-type nitrogenase enzyme in *Klebsiella pneumoniae*," *Nature 292*, 655–656 (1981).

60. McLean, P.A., True, A.E., Chapman, S., Nelson, M., Teo, B.-K., Münck, E., Hoffman, B.M., and Orme-Johnson, W.H., "Mo-EXAFS, Mössbauer spectroscopy and iron, proton and Mo ENDOR of the MoFe protein from *Klebsiella pneumoniae nifV* mutants," *Nitrogen Fixation Research Progress*, Evans, H.J., Bottomley, P.J., and Newton, W.E., eds. Dordrecht: Martinus Nijhoff, 1985, p. 620.

61. McLean, P.A., Wink, D.A., Chapman, S.K., Hickman, A.B., McKillop, D.M., and Orme-Johnson, W.H., "A new method for extraction of iron–molybdenum cofactor (FeMoco) from nitrogenase adsorbed to DEAE-cellulose. 1. Effects of anions, cations and pre-extraction treatments," *Biochemistry 28*, 9402–9406 (1989).

62. Merrick, M.J., "Organisation and regulation of the nitrogen fixation genes in *Klebsiella* and *Azotobacter*," *Nitrogen Fixation: Hundred Years After*, Bothe, H., de Bruijn, F.J., and Newton, W.E., eds. Stuttgart: Gustav Fischer, 1988, pp. 293–302.

63. Mortenson, L.E., Zumft, W.G., and Palmer, G., "Electron paramagnetic resonance studies on nitrogenase," *Biochim. Biophys. Acta 292*, 422–435 (1973).

64. Müller, A., Hellman, W., and Newton, W.E., "Zum verstandnis der zentralen rolle des schwefels in multi-metall-aggregaten mit verschiedener elektronenpopulation: $[S_2WS_2CoS_2WS_2]^{n-}$ ($n = 2,3$)," *Z. Naturforsch. 38b*, 528–529 (1983).

65. Münck, E., Rhodes, H., Orme-Johnson, W.H., Davis, L.C., Brill, W.J., and Shah, V.K., "Nitrogenase VIII. Mössbauer and EPR spectroscopy. The MoFe protein component from *Azotobacter vinelandii* OP," *Biochim. Biophys. Acta 400*, 32–53 (1975).

66. Nagatani, H.H., and Brill, W.J., "Nitrogenase V. The effects of Mo, W, and V on the synthesis of nitrogenase components in *Azotobacter vinelandii*," *Biochim. Biophys. Acta 362*, 160–166 (1974).

67. Nagatani, H.H., Shah, V.K., and Brill. V.K., "Activation of inactive nitrogenase by acid-treated component I," *J. Bacteriol. 120*, 697–701 (1974).

68. Nason, A., Antoine, A.D., Ketchum, P.A., Frazier, III, W.A., and Lee, D.K., "Formation of assimilatory nitrate reductase by *in vitro* inter-cistronic complementation in *Neurospora crassa*," *Proc. Natl. Acad. Sci. USA 65* 137–144 (1970).

69. Nason, A., Lee, K.-Y., Pan, S.-S., Ketchum, P.A., Lamberti, A., and DeVries, J., "*In vitro* formation of assimilatory reduced nicotinamide adenine dinucleotide phosphate: nitrate reductase from a *Neuroapora* mutant and a component of molybdenum enzymes," *Proc. Natl. Acad. Sci. USA 68*, 3242–3246 (1971).

70. Nelson, M.J., Levy, M.A., and Orme-Johnson, W.H., "Metal and sulfur composition of iron–molybdenum cofactor of nitrogenase," *Proc. Natl. Acad. Sci. USA 80*, 147–150 (1983).

71. Newton, W.E., "Chemical approaches to nitrogen fixation," *Biomimetic Chemistry*, Dolphin, D., McKenna, C., Murakami, Y., and Tabushi, I., eds. Washington, D.C.: American Chemical Society, 1980, pp. 351–377.

72. Newton, W.E., Burgess, B.K., Cummings, S.C., Lough, S., McDonald, J.W., Rubinson, J.F., Conradson, S.D., and Hodgson, K.O., "Structural aspects and reactivity of the iron–molybdenum cofactor from nitrogenase," *Advances in Nitrogen Fixation Research*, Veeger, C., and Newton, W.E., eds. The Hague: Martinus Nijhoff, 1984, p. 160.

73. Newton, W.E., Burgess, B.K., and Stiefel, E.I., "Chemical properties of the FeMo cofactor from nitrogenase," *Molybdenum Chemistry of Biological Significance*, Newton, W.E., and Otsuka, S., eds. New York: Plenum Press, 1980, pp. 191–202.

74. Newton, W.E., Cantwell, J.S., Feldman, B.J., Gheller, S.F., Schultz, F.A., Frank, P., Hedman, B., and Hodgson, K.O., "X-ray absorption spectroelectrochemistry of the nitrogenase iron–molybdenum cofactor," *Nitrogen Fixation: Achievements and Objectives*, Gresshoff, P.M., Stacey, G., Roth, L.E., and Newton, W.E., eds. New York: Chapman and Hall, 1990, p. 165.

75. Newton, W.E., Corbin, J.L., Schneider, P.W., and Bulen, W.A., "On potential model systems for the nitrogenase enzyme," *J. Am. Chem. Soc. 93*, 268–269 (1971).

76. Newton, W.E., Gheller, S.F., Feldman. B.J., Dunham, W.R., and Schultz, F.A., "Isolated iron–molybdenum cofactor of nitrogenase exists in multiple forms in its oxidized and semi-reduced states," *J. Biol. Chem. 264*, 1924–1927 (1989).

77. Newton, W.E., Gheller, S.F., Hedman, B., Hodgson, K.O., Lough, S.M., and McDonald, J.W., "Elicitation of thiomolybdates from the iron–molybdenum cofactor of nitrogenase," *Eur. J. Biochem. 159*, 111–115 (1986).

78. Newton, W.E., Gheller, S.F., Sands, R.H., and Dunham, W.R., "Mössbauer spectroscopy applied to the oxidized and semi-reduced states of the iron-molybdenum cofactor of nitrogenase," *Biochem. Biophys. Res. Commun. 162*, 882–891 (1989).

79. Newton, W.E., Gheller, S., Schultz, F.A., Burgess, B.K., Conradson, S.D., McDonald, J.W., Hedman, B., and Hodgson, K.O., "Redox and compositional insights into the iron–molybdenum cofactor of *Azotobacter vinelandii* nitrogenase as a guide to synthesis of new Mo–Fe–S clusters," *Nitrogen Fixation Research Progress*, Evans, H.J., Bottomley, P.J., and Newton, W.E., eds. Dordrecht: Martinus Nijhoff, 1985, pp. 605–610.

80. Orme-Johnson, W.H., Hamilton, W.D., Jones, T.L., Tso, M.-Y., Burris, R.H., Shah, V.K., and Brill, W.J., "Electron paramagnetic resonance of nitrogenase and nitrogenase components from *Clostridium pasteurianum W5* and *Azotobacter vinelandii OP*," *Proc. Natl. Acad. Sci, USA 69*, 3142–3145 (1972).

81. Orme-Johnson, W.H., and Holm, R.H., "Identification of iron–sulfur clusters in proteins," *Methods in Enzymology*, Vol. LIII, Fleischer, S., and Packer, L., eds. New York: Academic Press, 1978, Part D, pp. 268–274.

82. Pagan, J.D., Scowcroft, W.R., Dudman, W.F., and Gibson, A.H., "Nitrogen fixation in nitrate reductase-deficient mutants of cultered rhizobia," *J. Bacteriol. 129*, 718–723 (1977).

83. Palmer, G., Multani, J.S., Cretney, W.C., Zumft, W.G., and Mortenson, L.E., "Electron paramagnetic resonance studies on nitrogenase," *Arch. Biochem. Biophys. 153*, 325–332 (1972).

84. Pateman, A.L., Cove, D.J., Rever, B.M., and Roberts, D.B., "A common cofactor for nitrate reductase and xanthine dehydrogenase which also regulates the synthesis of nitrate reductase," *Nature 301*, 58–60 (1964).

85. Paustian, T.D., Shah, V.K., and Roberts, G.P., "Purification and characterization of the *nifE* and *nifN* gene products from *Azotobacter vinelandii* mutant UW45," *Proc. Natl. Acad. Sci, USA 86*, 6082–6086 (1989).

86. Paustian, T.D., Shah, V.K., and Roberts, G.P., "Apo-dinitrogenase: Purification, association with a 20-kilodalton protein, and activation by the iron–molybdenum cofactor in the absence of dinitrogenase reductase," *Biochemistry 29*, 3515–3522 (1990).

87. Pienkos, P.T., Klevickis, S., and Brill, W.J., "*In vitro* activation of inactive nitrogenase component I with molybdate," *J. Bacteriol. 145*, 248–256 (1981).

88. Pienkos, P.T., Shah, V.K., and Brill, W.J., "Molybdenum cofactors from molybdoenzymes and *in vitro* reconstitution of nitrogenase and nitrate reductase," *Proc. Natl. Acad. Sci. USA 74*, 5468–5471 (1977).

89. Rawlings, J., Shah, V.K., Chisnell, J.R., Brill, W.J., Zimmermann, R., Münck, E., and Orme-Johnson, W.H., "Novel metal cluster in the iron–molybdenum cofactor of nitrogenase," *J. Biol. Chem. 253*, 1001–1004 (1978).

90. Roberts, G.P., MacNeil, T., MacNeil, D., and Brill, W.J., "Regulation and characterization of protein products coded by the *nif* (nitrogen fixation) genes of *Klebsiella pneumoniae*," *J. Bacteriol. 136*, 267–279 (1978).

91. Robinson, A.C., Chun, T.W., Li, J.-G., and Burgess, B.K., "Iron–molybdenum cofactor insertion into the apo-protein of nitrogenase involves the iron protein–MgATP complex," *J. Biol. Chem. 264*, 10088–10095 (1989).

92. Robinson, A.E., Richards, A.J.M., Thomson, A.J., Hawkes, T.R., and Smith, B.E., "Low-temperature magnetic-circular-dichroism spectroscopy of the iron-molybdenum cofactor and the complementary cofactor-less MoFe protein of *Klebsiella pneumoniae* nitrogenase," *Biochem. J. 219*, 495–503 (1984).

93. Schrauzer, G. N., and Doemeny, P.A., "Chemical evolution of a nitrogenase model. II. Molybdate–cysteine and related catalysts in the reduction of acetylene to olefins and alkanes," *J. Am. Chem. Soc. 93*, 1608–1618 (1971).

94. Schrauzer, G.N., Hughes, L.A., Palmer, M.R., Strampach, N., and Grate, J.W., "The chemical evolution of a nitrogenase model. XIX. Simulation of the enzymatic reduction of cyclopropene," *Z. Naturforsch. 35b*, 1439–1443 (1980).

95. Schultz, F.A., Feldman, B.J., Gheller, S.F., and Newton, W.E., "Effects of oxidation state, solvent acidity and thiophenol on the electrochemical properties of the iron–molybdenum cofactor from nitrogenase," *Inorg. Chim. Acta 170*, 115–122 (1990).

96. Schultz, F.A., Gheller, S.F., Burgess, B.K., Lough, S., and Newton, W.E., "Electrochemical characterization of the iron–molybdenum cofactor from *Azotobacter vinelandii* nitrogenase," *J. Am. Chem. Soc. 107*, 5364–5368 (1985).

97. Schultz, F.A., Gheller, S.F., Feldman, B.J., and Newton, W.E., "Electrochemical, spectroscopic and structural aspects of the iron–molybdenum cofactor of nitrogenase,"

Nitrogen Fixation: Hundred Years After, Bothe, H., de Bruijn, F.J., and Newton, W.E., eds. Stuttgart: Gustav Fischer, 1988, pp. 121–126.

98. Schultz, F.A., Gheller, S.F., and Newton, W.E., "Iron–molybdenum cofactor of nitrogenase: Electrochemical determination of the electron stoichiometry of the oxidized/semi-reduced couple," *Biochem. Biophys. Res. Commun. 152*, 629–635 (1988).

99. Schultz, F.A., Gheller, S.F., and Newton, W.E., "Electrochemistry of the iron–molybdenum cofactor of nitrogenase: Evidence for multiple speciation and electrocatalytic behavior," *Redox Chemistry and Interfacial Behavior of Biological Molecules*, Dryhurst, G., and Niki, K., eds. New York: Plenum Press, 1988, pp. 203–216.

100. Scott, D.J., May, H.D., Newton, W.E., Brigle, K.E., and Dean, D.R., "Role for the nitrogenase MoFe protein α-subunit in FeMo-cofactor binding and catalysis," *Nature 343*, 188–190 (1990).

101. Shah, V.K., and Brill, W.J., "Isolation of an iron–molybdenum cofactor from nitrogenase," *Proc. Natl. Acad. Sci. USA 74*, 3249–3253 (1977).

102. Shah, V.K., and Brill, W.J., "Isolation of a molybdenum–iron cluster from nitrogenase," *Proc. Natl. Acad. Sci. USA 78*, 3438–3440 (1981).

103. Shah, V.K., Chisnell, J.R., and Brill, W.J., "Acetylene reduction by the iron–molybdenum cofactor from nitrogenase," *Biochem. Biophys. Res. Commun. 81*, 232–236 (1978).

104. Shah, V.K., Davis, L.C., Gordon, J.K., Orme-Johnson, W.H., and Brill, W.J., "Nitrogenase III. Nitrogenaseless mutants of *Azotobacter vinelandii*: Activities, cross-reactions and EPR spectra," *Biochim. Biophys. Acta 292*, 246–255 (1973).

105. Shah, V.K., Imperial, J., Ugalde, R.A., Ludden, P.W., and Brill, W.J., "*In vitro* synthesis of the iron–molybdenum cofactor of nitrogenase," *Proc. Natl. Acad. Sci. USA 83*, 1636–1640 (1986).

106. Shah, V.K., Madden, M.S., and Ludden, P.W., "*In vitro* synthesis of the iron–molybdenum cofactor and its analogs: Requirement for a non-*nif* gene product for the synthesis, and altered properties of dinitrogenase," *Nitrogen Fixation: Achievements and Objectives*, Gresshoff, P.M., Stacey, G., Roth, L.E., and Newton, W.E., eds. New York: Chapman and Hall, 1990, pp. 87–93.

107. Shah, V.K., Ugalde, R.A., Imperial, J., and Brill, W.J., "Inhibition of iron–molybdenum cofactor binding to component I of nitrogenase," *J. Biol. Chem. 260*, 3891–3894 (1985).

108. Smith, B.E., "Studies on the iron–molybdenum cofactor from the nitrogenase MoFe protein of *Klebsiella pneumoniae*," *Molybdenum Chemistry of Biological Significance*, Newton, W.E., and Otsuka, S., eds. New York: Plenum Press, 1980, pp. 179–190.

109. Smith, B.E., Bishop, P.E., Dixon, R.A., Eady, R.R., Filler, W.A., Lowe, D.J., Richards, A.J.M., Thomson, A.J., Thorneley, R.N.F., and Postgate, J.R., "The iron–molybdenum cofactor of nitrogenase," *Nitrogen Fixation Research Progress*, Evans, H.J., Bottomley, P.J., and Newton, W.E., eds. Dordrecht: Martinus Nijhoff, 1985, pp, 597–603.

110. Smith, B.E., Lowe, D.J., and Bray, R.C., "Nitrogenase of *Klebsiella pneumoniae*: Electron-paramagnetic-resonance studies on the catalytic mechanism," *Biochem. J. 130*, 641–643 (1972).

111. Smithe, B.E., O'Donnell, M.J., Lang, G., and Spartalian, K., "A Mössbauer spectroscopic investigation of the redox behaviour of the molybdenum–iron protein from *Klebsiella pneumoniae* nitrogenase," *Biochem. J. 191*, 449–455 (1980).

112. Sorger, G.L., "Nitrate reductase electron transport systems in mutant and wild-type strains of *Neurospora,*" *Biochim. Biophys. Acta 118,* 484–494 (1966).

113. Sosfenov, N.I., Andrianov, V.I., Vagin, A.A., Strokopytov, B.V., Vainshtein, B.K., Shilov, A.E., Gvozdev, R.I., Likhtenstein, G.I., Mitsova, I.Z., and Blazhchuk, I.S., "X-ray diffraction study of the MoFe protein nitrogenase from *Azotobacter vinelandii,*" *Sov. Phys. Dokl. 31,* 933–935 (1986) (Engl.); *Doklady Akad. Nauk. SSSR 291,* 1123–1127 (1986).

114. Stiefel, E.I., and Cramer, S.P., "Chemistry and biology of the iron–molybdenum cofactor of nitrogenase," *Molybdenum Enzymes,* Spiro, T.G., ed. New York: Wiley-Interscience, 1985, pp. 89–116.

115. Thomann, H., Morgan, T.V., Jin, H., Burgmayer, S.J.N., Bare, R.E., and Stiefel, E.I., "Protein nitrogen coordination to the FeMo center of nitrogenase from *Clostridium pasteurianum,*" *J. Am. Chem. Soc. 109,* 7913–7914 (1987).

116. True, A.E., McLean, P.A., Nelson, M.J., Orme-Johnson, W.H., and Hoffman, B.M., "Comparison of wild-type and *nifV* mutant molybdenum–iron proteins of nitrogenase from *Klebsiella pneumoniae* by ENDOR spectroscopy," *J. Am. Chem. Soc. 112,* 651–657 (1990).

117. True, A.E., Nelson, M.J., Venters, R.A., Orme-Johnson, W.H., and Hoffman, B.M., "^{57}Fe hyperfine coupling tensors of the FeMo cluster in *Azotobacter vinelandii* MoFe protein: Determination by polycrystalline ENDOR spectroscopy," *J. Am. Chem. Soc. 110,* 1935–1943 (1988).

118. Ugalde, R.A., Imperial, J., Shah, V.K., and Brill, W.J., "Biosynthesis of iron–molybdenum cofactor in the absence of nitrogenase," *J. Bacteriol. 159,* 888–893 (1984).

119. Venters, R.A., Nelson, M.J., McLean, P.A., True, A.E., Levy, M.A., Hoffman, B.M., and Orme-Johnson, W.H., "ENDOR of the resting state of nitrogenase molybdenum–iron proteins from *Azotobacter vinelandii, Klebsiella pneumoniae* and *Clostridium pasteurianum:* ^{1}H, ^{57}Fe, ^{95}Mo and ^{33}S studies," *J. Am. Chem. Soc. 108,* 3487–3498 (1986).

120. Walters, M.A., Chapman, S.K., and Orme-Johnson, W.H., "The nature of amide ligation to the metal sites of FeMoco," *Polyhedron 5,* 561–565 (1986).

121. Watt, G.D., Burns, A., Lough, S., and Tennent, D.L., "Redox and spectroscopic properties of oxidized MoFe protein from *Azotobacter vinelandii,*" *Biochemistry 19,* 4926–4932 (1980).

122. Watt, G.D., Burns, A., and Tennent, D.L., "Stoichiometry and spectral properties of the MoFe cofactor and noncofactor redox centers in the MoFe protein from *Azotobacter vinelandii,*" *Biochemistry 20,* 7272–7277 (1981).

123. Wink, D.A., McLean, P.A., Hickman, A.B., and Orme-Johnson, W.H., "A new method for extraction of the iron-molybdenum cofactor (FeMoco) from nitrogenase adsorbed to DEAE-cellulose. 2. Solubilization of FeMoco in a wide range of organic solvents," *Biochemistry 28,* 9407–9412 (1989).

124. Yang, S.-S., Pan, W.-H., Friesen, G.D., Burgess, B.K., Corbin, J.L., Stiefel, E.I., and Newton, W.E., "Iron–molybdenum cofactor from nitrogenase," *J. Biol Chem. 257,* 8042–8048 (1982).

125. Zimmermann, R., Münck, E., Brill, W.J., Shah, V.K., Henzl, M.T., Rawlings, J., and Orme-Johnson, W.H., "Nitrogenase X: Mössbauer and EPR studies on reversibly oxidized MoFe protein from *Azotobacter vinelandii,*" *Biochim. Biophys. Acta 537,* 185–207 (1978).

126. Zumft, W.G., "Isolation of thiomolybdate compounds from the molybdenum–iron protein of clostridial nitrogenase," *Eur. J. Biochem. 91,* 345–350 (1978).

Index